Diversity Amid Globalization

Diversity Amid Globalization

World Regions, Environment, Development

Les Rowntree
San Jose State University

Martin Lewis
Duke University

Marie Price
George Washington University

William Wyckoff
Montana State University

PRENTICE HALL
Upper Saddle River, NJ 07458

Library of Congress Cataloging-in-Publication Data

Diversity amid globalization: world regions, environment, development / Les Rowntree
...[et al.].
p. cm.
Includes bibliographical references.
ISBN 0-13-376427-3
1. Geography. I. Rowntree, Lester

G128 .D58 2000
910--dc21

99-052981

Executive Editor: Daniel Kaveney
Editor in Chief: Paul F. Corey
Assistant Vice President of Production & Manufacturing: David W. Riccardi
Executive Managing Editor: Kathleen Schiaparelli
Manufacturing Manager: Trudy Pisciotti
Buyer: Michael Bell
Editors in Chief of Development: Carol Trueheart and Ray Mullaney
Development Editor: Barbara Muller
Art Director: Joseph Sengotta
Assistant to Art Director: John Christiana
Art Manager: Gus Vibal
Senior Marketing Manager: Christine Henry
Assistant Editor: Amanda Griffith
Cover Designer: Bruce Kenselaar
Art Editor: Grace Hazeldine
Photo Research: Linda Sykes
Photo Research Administrator: Melinda Reo
Editorial Assistant: Margaret Ziegler
Cover Photographs: clockwise from top left: Holden Clay, Pacific Stock;
RAGA, The Stock Market; Victor Englebert, Englebert Photography, Inc.;
James Montgomery, Bruce Coleman, Inc.
Production Editor: Tim Flem/PublishWare
Page Layout: PublishWare

Printed in the United States of America

10 9 8 7 6 5 4 3 2

ISBN 0-13-376427-3

Prentice-Hall International (UK) Limited, *London*
Prentice-Hall of Australia Pty. Limited, *Sydney*
Prentice-Hall Canada Inc., *Toronto*
Prentice-Hall Hispanoamericana, S.A., *Mexico*
Prentice-Hall of India Private Limited, *New Delhi*
Prentice-Hall of Japan, Inc., *Tokyo*
Pearson Education Asia Pte., Ltd.
Editora Prentice-Hall do Brasil, Ltda., *Rio de Janeiro*

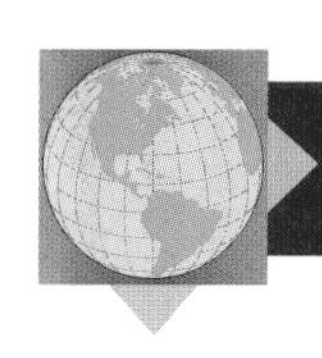

BRIEF CONTENTS

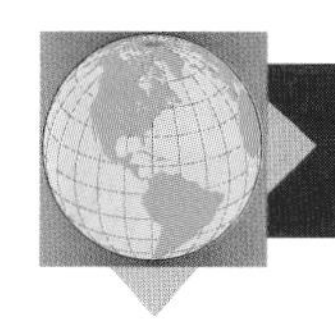

CONTENTS

5 The Caribbean 170

6 Sub-Saharan Africa 210

7 Southwest Asia and North Africa 264

14 Australia and Oceania 584

PREFACE

Places fascinate geographers. From an early age, we accumulate maps and dream about faraway corners of the world. For those of us lucky enough to make a living in geography, we also want to understand why the world works the way it does, how its unique regions have taken shape, and how those regions are increasingly interconnected. Those fundamental curiosities brought us together to create something new and different: an interpretation of world regional geography that was both deeply appreciative of global diversity and one that looked with a fresh and penetrating eye at those aspects of modern life that tie us all together.

The result has been our own odyssey of exploration, a pooling of expertise and enthusiasm that hopefully can offer students a perspective that can change their lives long after they leave the classroom. As an introduction to the field of geography, we believe our book offers students a view of the world that will serve them well in the early twenty-first century. It is a view that weds environment, people, and place; ponders how those relationships play out in particular regions; and assesses how powerful processes of globalization are reshaping those relationships in new and often unanticipated ways. *Diversity Amid Globalization* is the product of that shared vision, a vision we believe demonstrates the essential and invaluable role that geography can play in all our lives.

Objective and Approach

Diversity Amid Globalization is an issues-oriented textbook for college and university world regional geography classes that explicitly recognizes the geographic changes accompanying today's rapid rate of globalization. With this focus, we join the many who argue that globalization is the most fundamental reorganization of the planet's socioeconomic, cultural, and geopolitical structure since the Industrial Revolution. The explicit recognition of this premise provides the point of departure for this book. As geographers, we think it is essential that students understand two interactive tensions. First, they need to appreciate and critically ponder the consequences of converging environmental, cultural, political, and economic systems through forces of globalization. Second, they need to deepen their understanding of the creation and persistence of geographic diversity and difference. The interaction and tension between these opposing forces of homogenization and diversification forms a current running throughout the book and is reflected in our title, *Diversity Amid Globalization.*

Chapter Organization

As with all other world regional geography textbooks, *Diversity Amid Globalization* is structured to explain and describe the major world regions of Asia, Africa, the Americas, and so on. These 12 regional chapters, however, depart somewhat from traditional world regional textbooks. Instead of filling the regional chapters with descriptions of individual countries, we place that important material in readily accessible ancillaries, specifically, on the textbook Web site and in the student manual. This leaves us free to develop five important thematic sections as the organizational basis for each regional chapter. We begin with "Environmental Geography," which discusses the physical geography of each region, as well as current environmental issues. Next, we assess "Population and Settlement" geography, in which demography, land use, and settlement (including cities) are discussed. We also provide a section on "Cultural Coherence and Diversity," which examines the geography of language and religion, yet also explores current cultural tensions resulting from the interplay of globalization and diversity. The section on each region's "Geopolitical Framework" then treats the dynamic political geography of the region, including microregionalism, separatism, ethnic conflicts, and supranational organizations. Finally, we conclude each regional treatment with a section titled "Economic and Social Development," in which we analyze each region's economic framework as well as its social geography (including gender issues).

This regional treatment follows two substantive introductory chapters that provide the conceptual and theoretical framework of human and physical geography necessary to understanding our dynamic world. In the first chapter, students are introduced to the notion of globalization; this section also asks students to ponder both the costs and benefits of the globalization process, a critical perspective that is becoming increasingly common and important to understand. Following this, the geographical foundation for each of the five thematic sections is examined. This discussion draws heavily upon the major concepts fundamental to an introductory university geography course. The second chapter, "The Changing Global Environment," presents the themes and concepts of global physical geography, including landforms and geology, climate, hydrology, and biogeography. This chapter concludes with a discussion of world agricultural and food issues as a means of drawing together physical and human geography.

Chapter Features

Within each regional chapter, several unique features complement the thematic pedagogy of our approach:

- *Comparable Maps.* Of the fourteen or fifteen maps found in each regional chapter, eight of these maps are constructed with the same theme, similar data, and on the same base map so that readers can easily draw comparisons between different regions. More specifically, these maps are an introductory regional placename and feature map; a map of the physical geography of the region; a climate map; an environmental issues map; a population dot map; a language geography map; a map showing the geopolitical issues of each region; and, last, an economic trade map showing how the region interacts economically with other world regions. In addition to these eight maps, each regional chapter also has seven or eight other maps illustrating major themes, such as ethnic tensions, social development, or recent geopolitical changes.
- *Comparable Regional Data Sets.* Again, to facilitate comparison between regions, as well as to provide important insight on the characteristics of each region, each chapter contains three tables: the first provides population data, including natural increase and total fertility rate for each country within the region. The second is an economic table that presents data (again for each country within the region) on GNP, GNP per capita, purchasing power parity (PPP), and real economic growth. The last standardized table provides indicators of social development and the status of women within the region.
- *Sidebars.* Within each chapter, sidebars complement text material. These sidebars focus on several goals. Each chapter begins with a sidebar titled "Setting the Boundaries," which discusses the region under study in general terms. This sidebar gives a good overview of the region and discusses its role in the world today, often raising questions about its geographic coherence by accentuating the diversity within the region, and, finally, it looks at the vexatious issue of regional boundaries. After this, sidebars take on different themes; while some probe deeper into environmental issues with a case study, others elaborate on some form of popular culture, such as food or music, common to the region. Commonly, chapters include sidebars that discuss ethnic and political tensions within the region. Each regional chapter also contains a sidebar titled "Local Voices," in which the authors step aside to let local peoples—through song, literature, or the Internet—talk about their region.
- *Review and Research Questions.* There are two sets of review and research questions at the end of each regional chapter. The first set helps readers review basic terminology and concepts. The second set, "Thinking Geographically," asks readers to draw together more abstract themes and problems by doing further research on an issue. Resources for answering these critical thinking exercises are found in the chapter bibliography, in the instructor's manual, and on the textbook Web site.
- *Regional Films and Novels.* We provide a short list of films and novels that we believe captures the sense of place and flavor of the different world regions. Our experience is that these two art forms can be very important in conveying to students a more comprehensive feel for a foreign region by putting into prose or visual form the landscape, the people, and the issues that bring meaning to land and life in that part of the world.

Supplements: The Teaching and Learning Package

We have been pleased to work with Prentice Hall to produce a cutting-edge supplements package that will enhance teaching and learning. Not only does this package contain the traditional supplements that students and professors have come to expect from authors and publishers, but this supplements package also folds together new digital technologies that complement and enhance the learning experience.

For the Instructor

- **Presentation Manager 3.0 to accompany *Diversity Amid Globalization*** (0-13-020259-2). This sophisticated presentation management software makes it easy to create dynamic, customized presentations for use in your classroom. The CD-ROM includes hundreds of images (photos, illustrations, and maps) from the text as well as additional graphics. With Presentation Manager 3.0, you can:
  - Organize items in any order you choose
  - Preview resources by chapter
  - Search the digital library by keyword
  - Integrate media from your hard drive, a network, or the Internet
  - Edit lecture notes
  - Control the pace and flow of your lecture by using on-screen mouse or keyboard commands
- **Transparencies** (0-13-020262-2). 210 full-color transparencies provide the instructor with all of the maps from the text, all enlarged for excellent classroom visibility.
- **Test Item File** (0-13-020254-1).
- **Prentice Hall Custom Test**: Windows (0-13-020257-6); Macintosh (0-13-020255-X).
- **Instructor's Manual** (0-13-020253-3). Intended as a resource for both new and experienced instructors, the Instructor's Manual includes a variety of lecture outlines, additional source materials, teaching tips,

advice about how to integrate visual supplements (including Web-based resources), and various other ideas for the classroom.

For the Student

- **Student Study Guide** (0-13-0230261-4). The Study Guide includes additional learning objectives, a complete chapter outline, critical thinking exercises, problems and short essay work using actual figures from the text, and a self-test with an answer key in the back.
- ***Diversity Amid Globalization* WWW Site** (www.prenhall.com/rowntree). The Web site, tied chapter-by-chapter to the text, provides students with three main resources for further study:
  - **On-line Study Guide:** The on-line study guide provides immediate feedback to study questions, access to current geographical issues, and links to interesting and relevant sites on the Web. Additionally, instructors can create a customized syllabus that links directly to the Web site.
  - **Virtual Field Trips:** Integrated with the book's Web site, these field trips give students the on-the-ground feeling of visiting a place. The field trips include information about the history, places, people, and current events of places in each world region.
  - **Country-by-Country Data:** Also integrated with the Web site, this material summarizes the vital statistics for each country within a region.
- **Science on the Internet: A Student's Guide** by Andrew T. Stull (0-13-021308-X). The perfect tool to help students harness the power of the *Diversity Amid Globalization* Web site and the WWW, this resource gives clear step-by-step instructions to access regularly updated resources, as well as navigation strategies and an overview of the WWW.
- ***GeoTutor* CD-ROM** by Charles A. Stansfield and Jerry Westby. Included FREE with every copy of *Diversity Amid Globalization*, ***GeoTutor*** helps students develop basic geographic skills and allows them to explore the political, cultural, economic, and physical geography of the world through a series of interactive mapping exercises. Additionally, ***GeoTutor*** contains a full digital reference atlas of the world.
- **Rand McNally Atlas of World Geography.** This atlas includes 126 pages of up-to-date, accurate regional maps and 20 pages of illustrated world information tables. Available free when packaged with *Diversity Amid Globalization*. Please contact your local Prentice Hall representative for details.
- **Prentice Hall/*The New York Times* Themes of the Times: Geography.** This unique newspaper-format supplement features recent articles about geography from the pages of *The New York Times*. Available FREE from your local Prentice Hall representative, this supplement encourages students to make connections between the classroom and the world around them.

Acknowledgments

We have many people to thank for their help in the conceptualization, writing, rewriting, and production of *Diversity Amid Globalization*. First, we'd like to thank the hundreds of students in our world regional geography classes who have inspired us with their energy, engagement, and curiosity; challenged us with their critical insights; and demanded from us a textbook that better meets their needs to understand the diverse people and places of our dynamic world.

Next, we are deeply indebted to many professional geographers and educators for their assistance, advice, inspiration, encouragement, and constructive criticism as we labored through the different stages of this book. Among the many who provided invaluable comments on various drafts of *Diversity Amid Globalization* are:

Dan Arreola, Arizona State University
Bernard BakamaNume, Texas A&M University
William H. Berentsen, University of Connecticut
Kevin Blake, University of Wyoming
Craig Campbell, Youngstown State University
David B. Cole, University of Northern Colorado
Malcolm Comeaux, Arizona State University
Jeremy Crampton, George Mason University
James Curtis, California State University, Long Beach
Dydia DeLyser, Louisiana State University
Caroline Doherty, Northern Arizona University
Vernon Domingo, Bridgewater State College
Jane Ehemann, Shippensburg University
Gary Gaile, University of Colorado
Steven Hoelscher, Louisiana State University
Richard H. Kesel, Louisiana State University
Robert C. Larson, Indiana State University
Alan A. Lew, Northern Arizona University
Max Lu, Kansas State University
James Miller, Clemson University
Bob Mings, Arizona State University
Anne E. Mosher, Syracuse University
Tim Oakes, University of Colorado
Bimal K. Paul, Kansas State University
Michael P. Peterson, University of Nebraska–Omaha
Richard Pillsbury, Georgia State University
Scott M. Robeson, Indiana State University
Susan C. Slowey, Blinn College
Philip W. Suckling, University of Northern Iowa
Curtis Thomson, University of Idaho
Nina Veregge, University of Colorado
Gerald R. Webster, University of Alabama
Emily Young, University of Arizona
Bin Zhon, Southern Illinois University at Edwardsville
Dr. Henry J. Zintambila, Illinois State University

In addition, we wish to thank the many publishing professionals who have been involved with this project. It has been a privilege to work with you. We thank Editor-in-Chief Paul F. Corey, for his early—and continued—support for this book project; Geosciences Editor and good friend Dan Kaveney, for his daily engagement, professional guidance, enduring patience, unyielding discipline, high standards, and warm companionship; Developmental Editor Barbara Muller, for her clear articulation of what this project needed to be successful, her guidance on how that could be attained, and her firm insistence that we meet those standards; production editor Tim Flem, for his daily miracles and steady hand on the tiller so that somehow thousands of pages of manuscript were turned into a finished book; copyeditor Roberta Dempsey, for her eagle eye and graceful hand; photo researcher Linda Sykes, for her creative solutions to indulging four geographers in their demanding quest for outstanding pictures from strange parts of the world; editorial administrative assistant Margaret Ziegler, for gracefully meeting the incessant needs of four demanding authors who rarely remembered that she was multaneously doing the same for many other geoscienc thors; and Sam Chernawsky for his excellent research. To of you, your professionalism is truly appreciated

Finally, the authors want to thank that special group of friends and family who were there when we needed you most: early in the morning, late at night; in foreign countries and familiar places; when we were on the verge of crying, yet needed to laugh; for your love, patience, companionship, inspiration, solace, understanding, and enthusiasm: Rob Crandall, Margaret Conkey, Maureen Hays, Rowan Rowntree, Paul Starrs, Kären Wigen, Linda, Tom, and Katie Wyckoff. Words cannot thank you enough.

Les Rowntree

Martin Lewis

Marie Price

William Wyckoff

Focus on Globalization

As the title *Diversity Amid Globalization* implies, this book focuses on the tensions inherent as forces of cultural, political, and economic globalization interact with the persisting reality of geographical diversity in today's world. Throughout the book, these conflicting forces of homogenization and diversification are juxtaposed, suggesting how they shape different regional settings in particular ways.

Thematic Structure

To facilitate students' abilities to draw comparisons and contrasts between regions, each of the twelve regional chapters employs a thematic structure. This structure organizes each chapter into five thematic sections:

- **Environmental Geography**
- **Population and Settlement**
- **Cultural Coherence and Diversity**
- **Geopolitical Framework**
- **Economic and Social Development**

This structure allows a more penetrating treatment of themes and concepts than the traditional country-by-country approach, and facilitates comparisons of specific topics between regions.

Local Voices

Each regional chapter provides a sidebar entitled **Local Voices,** in which the authors step aside to let local people talk or write about their concerns. These sidebars highlight regional issues and topics such as food, music, environmental problems, cultural tensions, and work conditions that will interest students and provide further insight into the region.

For example, the chapter on Latin America contains viewpoints from indigenous people, discussion of the cocaine trade, and depictions of how urban squatters build their communities.

LOCAL VOICES The Environmentalist Message of the Kogi

The Kogi live in the Sierra de Santa Marta in Colombia. A small indigenous group that has struggled to survive for five centuries, they continue to perceive themselves as the "Elder Brothers" and the rest of the world as "Younger Brothers." One of the fundamental beliefs of the Kogi is that they are the guardians of the environment and that Younger Brother is recklessly destroying the Earth. Kogi spiritual leaders, the Mamas, expressed their environmentalist agenda before a BBC film crew in the 1980s.

We are the Elder Brothers
We have not forgotten the old ways
How could I say that I do not know how to dance?
We still know how to dance.
We have forgotten nothing.
We know how to call the rain.
If it rains too hard we know how to stop it.
We call the summer.
We know how to bless the world and make it flourish.

But now they are killing the Mother
The Younger Brother, all he thinks about is plunder.

The Mother looks after him too, but he does not think.

He is cutting into her flesh.
He is cutting into her arms.
He is cutting off her breasts.
He takes out her heart.
He is killing the heart of the world.

When the final darkness falls everything will stop.
The fires, the benches, the stones, everything.
All the world will suffer.

If that happened and all we Mamas died,
and there was no one doing our work,
well, the rain wouldn't fall from the sky.

▲ **Figure 1 The Kogi of Colombia** A Kogi man walks with his poporo (a gourd filled with lime powder) and a satchel of coca leaves. The chewing of coca leaf with lime, which reduces hunger and fatigue, has been practiced by indigenous peoples in the Andes for centuries.

It would get hotter and hotter from the sky,
and the trees wouldn't grow,
and the crops wouldn't grow.

Or am I wrong, and they would grow anyway?

Source: Alan Ereira, *The Elder Brothers.* (New York: Alfred Knopf, 1992), pp. 113–14.

Setting the Boundaries

Setting the Boundaries sidebars describe the extent and major characteristics of each region, helping students grasp the region's features in a global context.

Setting the Boundaries

The United States and Canada are commonly referred to as "North America," but that regional terminology can sometimes be confusing. In many geography textbooks, the realm is called "Anglo America" because of its close and abiding connections with Britain and its Anglo-Saxon cultural traditions. The increasingly visible cultural diversity of the realm, however, has discouraged the widespread use of the term in more recent years. While more culturally neutral, the term "North America" also has its problems. As a physical feature, the North American continent commonly includes Mexico, Central America, and often the Caribbean. Culturally, however, the United States–Mexico border seems a better dividing line, although the growing Hispanic presence in the Southwest, as well as ever-closer economic links across the border, make problematic even that regional division. North America's boundaries often extend northeastward to include Greenland, a colony of Denmark that is home to more than 55,000 indigenous people. To the west, North America reaches to Alaska's Aleutian Islands and to the state of Hawaii in the mid-Pacific Ocean. While the future may find Mexico even more intimately tied to its northern neighbors, for now the "North American" realm concentrates on Canada and the United States, two of the world's largest and most affluent nation-states.

Thematic Structure

Comparable, **standardized maps and data tables** in each chapter aid students' understanding of a specific world region, and help them make comparisons between regions in both cartographic and tabular form.

Of the fourteen or fifteen **maps** in each regional chapter, no less than eight are *organized around the same theme and designed to convey comparable material.* These are:

- a **chapter-opening map** with countries and place names
- a **physical map**, showing landforms, hydrology, and tectonic boundaries
- a **climate map,** with climograph call-outs giving temperature and precipitation data for specific cities
- a **"transformation of the Earth"** map with call-outs to environmental issues and solutions within the region
- a **population map** for the region
- a **map of regional languages**
- a **geopolitical map** with call-outs to current issues and tensions
- a **map of global economic connections** illustrating the region's world connections in trade and investment.

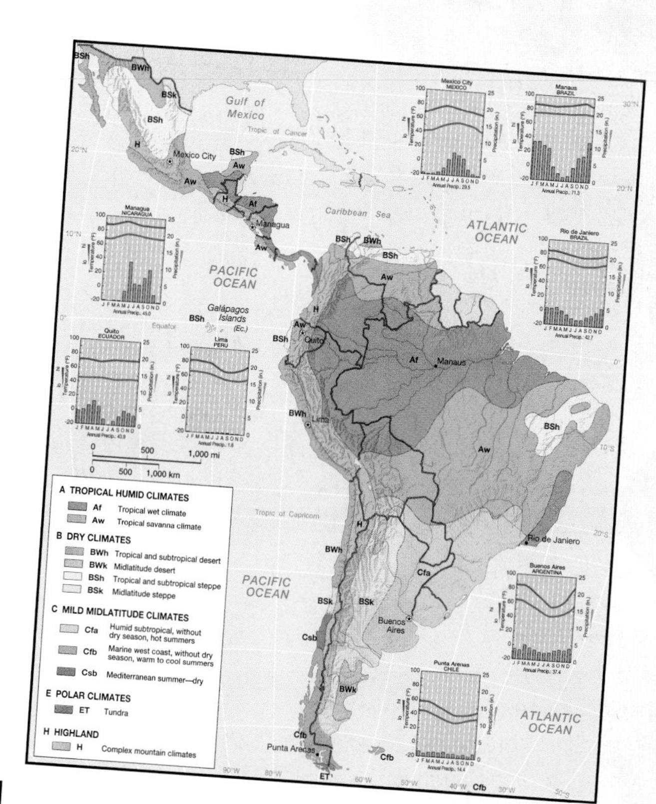

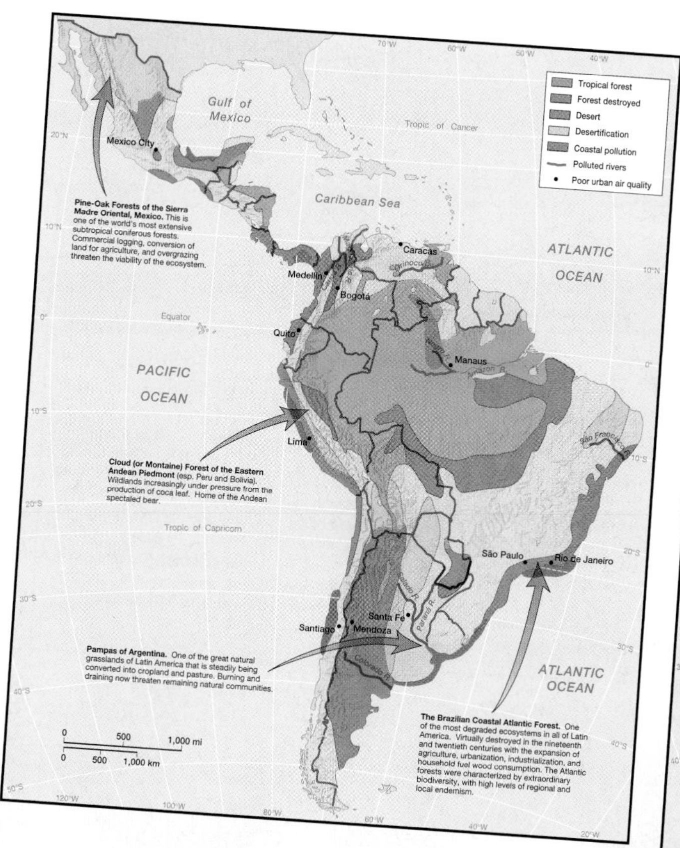

Standardized Regional Data and Tables

Each regional chapter has three tables:

- **population indicators**
- **economic indicators**
- **social development indicators**

Table 4.2 Economic Indicators

Country	GNP Per Capita ($U.S., 1996)	Total GNP (Millions of $U.S., 1996)	PPP* ($Intl, 1996)	Real Annual Growth % Per Capita, 1990–1996
Costa Rica	2,640	9,081	6,470	2.4
El Salvador	1,700	9,868	2,790	3.5
Guatemala	1,470	16,018	3,820	0.5
Honduras	660	4,012	2,130	1.2
Mexico	3,670	341,718	7,660	–0.3
Nicaragua	380	1,705	1,760	–0.2
Panama	3,080	8,249	7,060	3.6
Argentina	8,380	295,131	9,530	3.9
Bolivia	830	6,302	2,860	1.8
Brazil	4,400	709,591	6,340	2.0
Chile	4,860	70,060	11,700	6.4
Colombia	2,140	80,174	6,720	3.0
Ecuador	1,500	17,531	4,730	0.8
Paraguay	1,850	9,179	3,480	–1.5
Peru	2,420	58,671	4,410	4.8
Uruguay	5,760	18,464	7,760	3.8
Venezuela	3,020	67,333	8,130	–0.3

*Purchasing Power Parity.
Source: World Bank Atlas, 1998.

Table 7.2 Economic Indicators

Country	GNP Per Capita ($U.S., 1996)	Total GNP (Millions of $U.S., 1996)	PPP* ($Intl, 1996)	Real Annual Growth % Per Capita, 1990–1996
Algeria	1,520	43,726	4,620	–1.9
Bahrain	—	—	—	—
Cyprus	—	—	13,970	3.8
Egypt	1,080	64,275	2,860	2.2
Gaza	—	—	—	2.6
Iraq	—	—	—	—
Iran	—	—	5,360	1.0
Israel	15,870	90,310	18,100	3.2
Jordan	1,650	7,088	3,570	4.0
Kuwait	—	—	—	15.7
Lebanon	2,970	12,118	6,060	5.4
Libya	—	—	—	—
Morocco	1,290	34,936	3,320	.2
Oman	—	—	8,680	–.3
Qatar	—	—	16,330	–5.1
Saudi Arabia	—	—	9,700	–3.1
Sudan	—	—	—	—
Syria	1,160	16,808	3,020	4.3
Tunisia	1,930	17,581	4,550	1.3
Turkey	2,830	177,530	6,060	1.7
United Arab Emirates	—	—	17,000	–4.8
West Bank	—	—	—	—
Western Sahara	—	—	—	—
Yemen	380	6,016	790	–2.2

*Purchasing power parity.
Source: *The World Bank Atlas*, 1998.

Thinking Geographically

Two sets of review questions close each chapter. The first set helps review major concepts and terms covered. The second set of questions, called **Thinking Geographically**, requires students to look beyond the text to analyze and draw conclusions about key global issues.

Thinking Geographically

1. Discuss the processes driving tropical deforestation in Latin America and how they compare with deforestation in other areas of the world.
2. Agrarian reform has been advocated throughout Latin America as a means to reduce social and economic inequalities. Has agrarian reform worked in the region? How have land distribution programs faired? Where else in the world have agrarian reforms been attempted?
3. How is neoliberalism influencing the way Latin America interacts with the rest of the world? What are the social and environmental costs of neoliberalism? Is this a model for understanding the impact of globalization in the developing world?
4. After examining the language map, what conclusions can be drawn about the patterns of Indian survival in Latin America?
5. Given the dominance of cities in this region, what particular urban environmental problems face cities in the developing world? How might Latin America's megacities use their size and density to reduce the environmental problems associated with urbanization?
6. Indigenous homelands or reservations can be a source of cultural repression or cultural survival. Compare the creation of comarcas in Panama or communal land-use zones in Honduras with Indian reservations in the United States. What different political processes are behind the creation of these indigenous territories?
7. Discuss the social, environmental, and economic consequences behind the modernization of agriculture. How does Latin America's experience with modern agricultural systems compare with North America?

Presentation Manager 3.0 to accompany *Diversity Amid Globalization*

Diversity Amid Globalization is to be accompanied by Prentice Hall's **Presentation Manager 3.0.** This sophisticated presentation management software makes it easy to create dynamic, customized presentations for use in your classroom. The CD-ROM includes hundreds of images (photos, illustrations, and maps) from the text as well as additional graphics and animations. With Presentation Manager 3.0 you can:

- Organize items in whichever order you choose
- Preview resources by chapter
- Search the digital library by keyword
- Integrate media from your hard drive, a network, or the Internet
- Edit lecture notes
- Control the pace and flow of your lecture by using on-screen mouse or keyboard commands.

Companion Website: http://www.prenhall.com/rowntree

The *Diversity Amid Globalization* Website, tied chapter by chapter to the text, provides students with three main resources for further study:

On-line Study Guide: The on-line study guide provides immediate feedback to study questions, access to current geographical issues, and links to interesting and relevant sites on the Web. Additionally, instructors can create a customized syllabus that links directly to the Website.

Virtual Field Trips: Integrated with the book's Website, these field trips give students the "on-the-ground" feeling of visiting a country or interacting with the people who live there. The field trips include information about the history, places, people, and current events of places in each world region.

Country-by-Country Data: Also integrated with the Website, this material summarizes the vital statistics for each country within a region.

Instructor's Manual (0-13-020253-3)

Intended as a resource for both new and experienced instructors, the Instructor's Manual includes a variety of lecture outlines, additional source materials, teaching tips, advice about how to integrate visual supplements (including Web-based resources), and various other ideas for the classroom.

Transparencies (0-13-020262-2): 210 full-color transparencies provide the instructor with all of the maps from the text, enlarged for excellent classroom visibility.

Test Item File (0-13-020254-1)

Prentice Hall Custom Test: Windows (0-13-020257-6); Macintosh (0-13-020255-X)

Science on the Internet—A Student's Guide, 1999

by Andrew T. Stull (0-13-021308-X)

The perfect tool to help students harness the power of the Rowntree Website and the World Wide Web, this resource gives clear step-by-step instructions to access regularly updated geoscience resources, as well as navigation strategies and an overview of the World Wide Web. *Available FREE when packaged with the text.*

Prentice Hall/*The New York Times* Themes of the Times—Geography

This unique newspaper-format supplement features recent articles about geography from the pages of *The New York Times.* Available *FREE* from your local Prentice Hall representative, this supplement encourages students to make connections between the classroom and the world around them.

Study Guide (0-13-020261-4)

The Study Guide includes additional learning objectives, a complete chapter outline, critical thinking exercises, problems and short essay work using actual figures from the text, and a self-test with an answer key in the back.

Develop students' basic geographic skills

Included FREE with every copy of *Diversity Amid Globalization*, **GeoTutor** helps students develop basic geographic skills and allows them to explore the political, cultural, economic, and physical geography of the world. Students can use this highly interactive CD-ROM to draw their own thematic maps, use check-up questions to assess their understanding, and use the glossary to hear difficult-to-pronounce words.

In addition, **GeoTutor** contains a full digital reference atlas of the world, and the **GeoTutor** Website (http://www.prenhall.com/stansfield) helps students keep abreast of the continually changing fabric of world geography by providing timely updates of the material on the CD-ROM.

GeoTutor's full reference atlas helps students complete the activities on the CD-ROM.

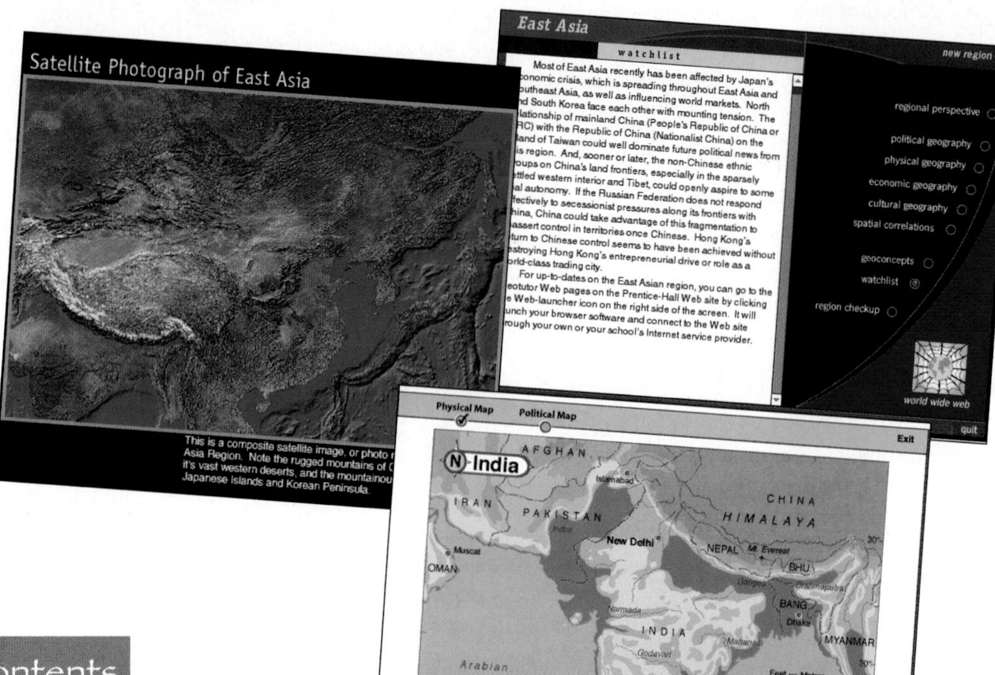

GeoTutor Contents

- Introduction to Geographic Literacy
- United States and Canada
- Europe
- Russia and the Newly Independent States
- Australia, New Zealand, and Pacific Islands
- East Asia
- South Asia
- Southeast Asia
- Africa South of the Sahara
- The Middle East and North Africa
- Latin America
- Appendix
- Glossary and Gazetteer
- Reference Atlas

Minimum System Requirements	
Windows	**Macintosh**
Pentium Processor	PowerPC
Windows® 95/98/NT 4.0X	System 7.0 or above
640x480 pixel screen resolution	640x480 pixel screen resolution
16 MB RAM	16 MB RAM
4x CD-ROM drive	4x CD-ROM drive
Sound Card/Speakers	Internet Browser 3.0 or above
Internet Browser 3.0 or above	Color monitor set to Thousands of Colors (16-bit)
Color monitor set to High Color (16-bit)	

ABOUT THE AUTHORS

Les Rowntree teaches both geography and environmental studies at San Jose State University in California, and serves as Chair for the interdisciplinary Department of Environmental Studies. As an environmental geographer, Rowntree's teaching and research interests focus on international environmental issues, the human dimensions of global change, biodiversity and conservation, and human-caused landscape transformation. He sees world regional geography as a way to engage and inform students by giving them the conceptual tools needed to assess global issues critically in their larger context. Recently Rowntree has done research in Mexico, Australia, and Europe, as well as in his native California. He is currently a member of an international research team examining environmental change in southwestern France, focusing on forest and woodland changes within the context of different European environmental policies.

Martin Lewis is Associate Research Professor of Geography at Duke University, and is Co-Director of Duke's program in Comparative Area Studies. He has conducted extensive research on environmental geography in the Philippines and on the intellectual history of global geography. His publications include *Wagering the Land: Ritual, Capital, and Environmental Degradation in the Cordillera of Northern Luzon, 1900-1986* (1992), and with Kären Wigen, *The Myth of Continents: A Critique of Metageography* (1997). He lives in Durham, North Carolina, with his wife and two children.

Marie Price is an Associate Professor of Geography and International Affairs at George Washington University, where she is Director of Latin American Studies. A Latin American specialist, Dr. Price has conducted research in Belize, Mexico, Venezuela, Cuba, and Bolivia. She has also done research in Sub-Saharan Africa. Her studies have explored human migration, natural resource use, environmental conservation, and regional development. Price brings to ***Diversity Amid Globalization*** a special interest in regions as dynamic spatial constructs that are shaped over time through both global and local forces. Her publications include articles in the *Annals of the Association of American Geographers, Geographical Review, Journal of Historical Geography, CLAG Yearbook, Studies in Comparative International Development,* and *Focus.*

William Wyckoff is a geographer in the Department of Earth Sciences at Montana State University specializing in the cultural and historical geography of North America. He has written and co-edited several books on North American settlement geography, including *The Developer's Frontier: The Making of the Western New York Landscape* (1988), and *The Mountainous West: Explorations in Historical Geography* (1995). In 1990, he received the Burlington Northern Corporation's Award for Outstanding Teaching. A World Regional Geography instructor for 18 years, Wyckoff hopes that the fresh approach taken in ***Diversity Amid Globalization*** will more effectively highlight the tensions evident in the world today as global change impacts particular places and people in dramatic and often unpredictable ways.

C H A P T E R 1

Diversity Amid Globalization

The most important challenges facing the world in the twenty-first century are associated with **globalization**, the increasing interconnectedness of people and places through converging processes of economic, political, and cultural change. Once-distant regions are now increasingly linked together through commerce, communications, and travel. Many observers argue that globalization is the most fundamental reorganization of the planet's socioeconomic structure since the Industrial Revolution. While few dispute the widespread changes brought about by globalization, not everyone agrees on the implications, or whether the benefits outweigh the costs.

Although economic activities may be the prime mover behind globalization, the consequences are far ranging and affect all aspects of land and life in the new millennium. Cultural patterns, political arrangements, and social development are all undergoing profound change. Additionally, because natural resources are now global commodities, the planet's physical environment is also implicated. Local ecosystems are impacted and altered by financial decisions made in places thousands of miles away, and the cumulative effect of these far-ranging activities has profound and possibly detrimental implications for the world's climates, oceans, waterways, and forests.

These immense, widespread global changes make understanding our world both a challenging and a thoroughly necessary task (Figure 1.1). Our future depends on comprehending globalization in its varied manifestations since our lives are now deeply intertwined with this worldwide phenomenon. Although understanding globalization cuts across many academic disciplines, world regional geography is an effective starting point because of its focus on regions, environment, geopolitics, culture, economic, and social development. For this reason the themes and structure of this book are organized around the tensions and issues resulting from globalization (Figure 1.2).

Diversity Amid Globalization: A Geography for the Twenty-first Century

This chapter introduces the framework for studying world regional geography by first examining the varied aspects of globalization in contemporary life. While the economic implications of globalization dominate most discussions, the cultural, environmental, and

◀ **Figure 1.1 World geography** Viewed from space, both the diversity and the similarities of Earth's physical fabric are apparent. The challenge of world regional geography is to move closer so that human activities, as well as the physical environment, are examined. *(European Space Agency/Science Photo Library/Photo Researchers, Inc.)*

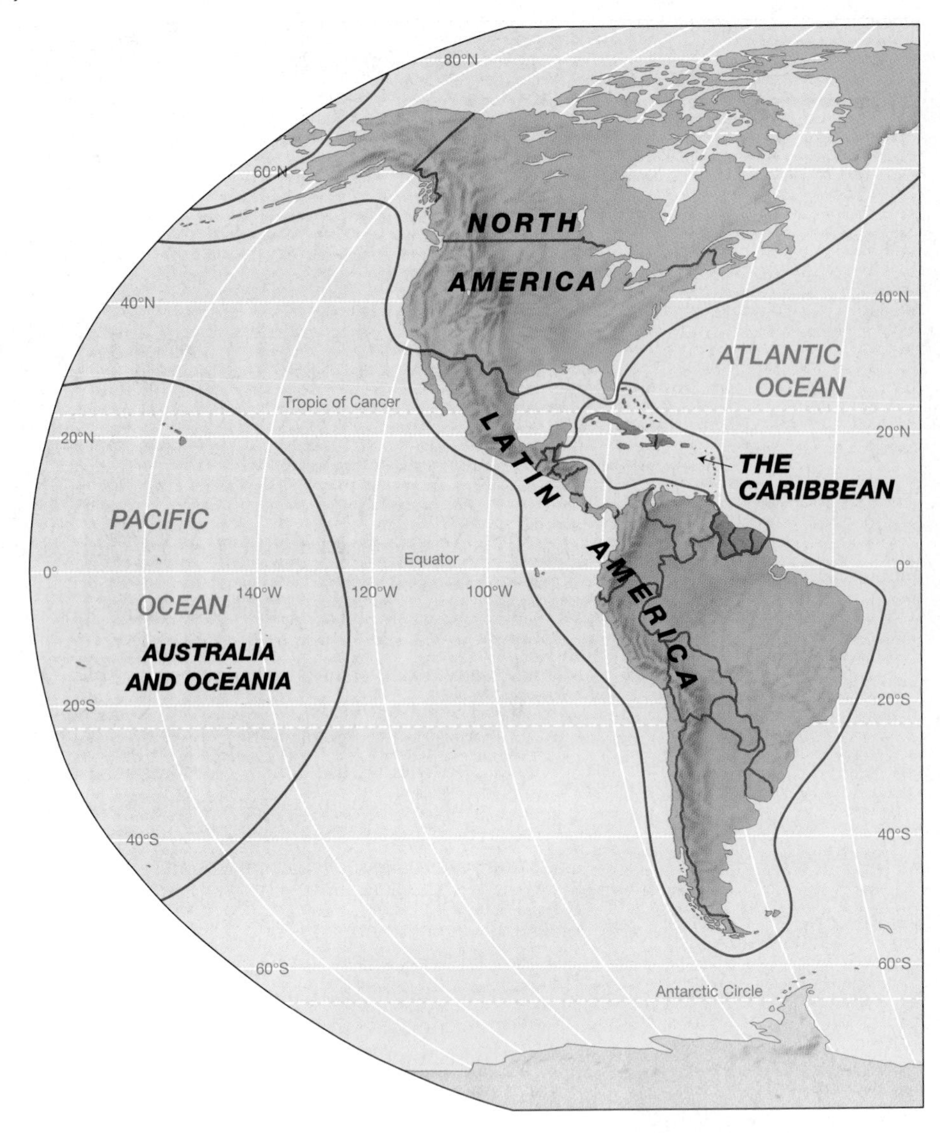

▶ **Figure 1.2 World regions** These regions form the basis for the 12 regional chapters in this book. Several areas of the world are treated in more than one chapter, and they are designated on the map with a striped pattern. For example, western China is discussed in two chapters: Chapter 11, East Asia, and Chapter 10, Central Asia. Note also that three countries on the South American continent are discussed as part of the Caribbean region because of their close cultural similarities with the island region.

geopolitical expressions must also be considered and critiqued. Following this is an overview of the major concepts of global geography: human-environment interaction; areal differentiation and integration; regions; landscapes; and global-to-local scales. The third and last section introduces the five organizational themes found in each regional chapter of this textbook.

Converging Currents of Globalization

Few argue that the major component of globalization is the economic reorganization of the world. Although different forms of a world economy have been in existence for centuries, such as the European colonization of the New World, the most recent round of economic integration is primarily a product of the last several decades. The attributes of this reorganization are now familiar, yet bear repeating—global communication systems that link all regions on the planet instantaneously; global transportation systems capable of moving goods quickly by air, sea, and land; transnational conglomerate corporate strategies that have created global corporations more economically powerful than many sovereign nations; new and more flexible forms of capital accumulation; international financial institutions that facilitate 24-hour trading; global agreements that promote free trade; market economies that have replaced state-controlled economies; privatized firms and services formerly operated by governments; a plethora of planetary goods and services that have arisen to fulfill consumer demand (real or imaginary); and, of course, an army of international workers, managers, and executives who have given this economic juggernaut a unique human geography.

As a result of this global reorganization, economic growth in some areas of the world has been unprecedented over recent decades. Additionally, international corporations, along with their managers and executives, have amassed vast amounts of wealth from this economic restructuring. However, not everyone has profited from economic globalization, nor have all world regions shared equally in the benefits. In Latin America, for example, the percentage of the population classified as poor increased from 33 percent in 1980 to 50 percent in 1990, largely as a result of economic restructuring. As some regions move ahead and prosper, others fall even further behind.

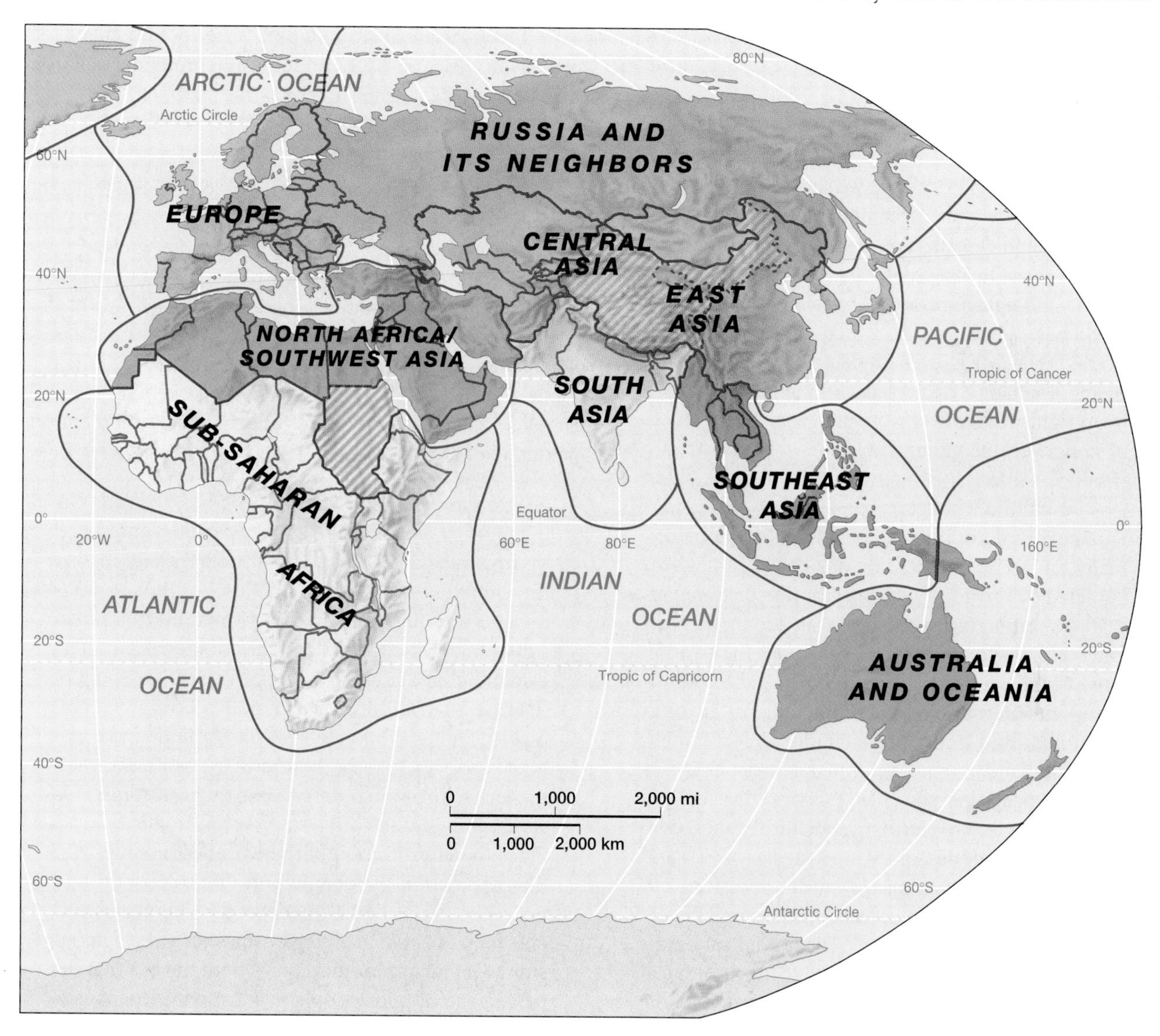

Furthermore, these economic changes also trigger fundamental cultural change. Often accompanying globalization is the spread of a homogenous global consumer culture that threatens to displace and mask local diversity. This frequently sets up deep and serious social tensions between traditional cultures and new, external globalizing currents. Global TV, movies, and videos promote images of Western style and culture that are imitated by millions throughout the world. NBA T-shirts, sneakers, and caps are now found in small villages and large world cities alike (Figure 1.3). Fast-food franchises are changing—some would say corrupting—traditional diets with the explosive growth of McDonald's, Burger King, and Kentucky Fried Chicken outlets in most of the world's cities, including Beijing, Moscow, Singapore, and Nairobi. While

◀ **Figure 1.3 Globalization** Busy Hennessy Road in Hong Kong gives clues to many different aspects of globalization, from expressions of Western culture (food, clothes, signboards) in a major Asian city, to the global reach and influence of the city's economic sector. *(MacDuff Everton/The Image Bank)*

these stylistic changes may seem innocuous and innocent since they resonate with familiarity to North Americans, they illustrate the more profound structural cultural change the world is experiencing through globalization.

Although much media attention is given to the rapid spread of Western consumer culture, we should not overlook that nonmaterial culture is also becoming more dispersed and homogenized through globalization. Speech, clearly, is an example. Many a tourist in Russia or Asia has been startled by locals speaking an English made up largely of Hollywood or pop song clichés. There is also much more than these clichés—social values, ideas, even fundamental organizational structures are also being dispersed globally. Expectations about human rights, the role of women in society, the intervention of nongovernmental organizations—these, too, are expressions of globalization, bringing with them far-reaching implications for cultural change.

There are also profound geopolitical facets to globalization. To many, an essential dimension of globalization is that it is not hindered by territorial, national, or jurisdictional restrictions. Globalization is just that—global, and its processes transcend traditional boundaries. For example, the creation of the United Nations following World War II was a step toward creating an international governmental structure in which all nations could find representation. Unfortunately, however, the simultaneous emergence of the Soviet Union as a military and political superpower led to a rigid division into Cold War blocs that inhibited further geopolitical integration. With the peaceful end of the Cold War in the late 1980s and early 1990s, the former communist countries of eastern Europe and the Soviet Union were opened almost immediately to global trade and cultural exchange. These political developments coincided with the economic and technological changes we now see as the recent wave of globalization. Some observers argue that economic imperialism has replaced the ideological divisions of the Cold War, and hence economic activity and politics are now more intertwined than ever. They offer as an example governmental activity in the World Trade Organization (WTO) and other trading blocs in which national interests are secondary to expanded economic activity. Although the Cold War may be over, regional military alliances and pacts are still very much a reality, illustrating the close linkage between geopolitics and globalization. Perhaps the best example is NATO expansion into former communist eastern Europe, along with the implications of this process for diplomatic relationships with Russia.

Beyond geopolitics, the expansion of a globalized economy is also creating environmental problems throughout the world as international conglomerates disrupt local ecosystems in their incessant search for natural resources and manufacturing sites. Landscapes and resources that were previously used only by small groups of local peoples are now thought of as global commodities to be exploited and traded on the world marketplace. As a result, indigenous peoples are often deprived of their traditional resource base and displaced into marginal environments. At a larger scale, economic globalization is also aggravating worldwide environmental problems, such as climate change, regional air pollution, water pollution, and deforestation. Because of these pervasive problems, environmental issues are a major theme in this textbook.

Critics of Globalization

Advocates of globalization argue it is a logical and inevitable expression of contemporary international capitalism that will benefit all nations and all peoples by creating greater global commerce and wealth. These new riches, they say, will eventually trickle down figurative geographic and social ladders to enrich even the poorest of peoples in all the world's different regions. But not everyone believes this, and the critics of economic globalization are increasingly outspoken in their challenges. Some of the main points of this globalization critique include the following:

1. Globalization is not a "natural" process. Instead, it is the product of an explicit economic policy promoted by free trade advocates, capitalist countries (mainly the United States, Japan, and Europe), financial interests, international investors, and multinational firms. In addition, the processes of globalization so evident today are much more pervasive than those of the past, such as during the period of European colonialism. Thus, while global-scale economic and political linkages have been around for centuries, their modern expression differs strikingly from conditions earlier in world history.
2. Since the globalization of the world economy is evidently creating greater inequity between rich and poor, the "trickle-down" model of developmental benefits for all people in all regions has yet to be demonstrated. Evidence for this comes from the fact that the percentage of poor people in most world regions is actually increasing, not decreasing. A recent United Nations report notes that, in terms of social conditions, 60 countries are worse off now than 30 years ago. At a global scale, the richest 20 percent of the world's population consumes 86 percent of the world's resources, while the poorest 80 percent uses only 14 percent of global resources.
3. Globalization promotes free market, export-oriented economies at the expense of localized, sustainable activities. World forests, for example, are increasingly cut for export timber rather than serving local needs. As part of the economic structural adjustment package, the World Bank and the International Monetary Fund encourage developing countries to expand their resource exports so they will have more hard currency to make payments on their foreign debts. This strategy, however, often leads to overexploitation of local resources.
4. The "free market" economic model commonly promoted for developing countries is not the one that Western industrial countries used for their own economic development. More to the point, the world's wealthiest countries, such as Germany, Japan, and

even to some extent the United States, used a model in which the government played a strong role in directing investment, managing trade, and subsidizing chosen sectors of the economy.

While this book does not pretend to resolve the controversy over globalization, it does encourage readers to reflect on these critical points as they might apply to different world regions and locales.

Diversity in a Globalizing World

The ever-increasing globalization of the nations and peoples of the contemporary world has led many observers to foresee a future world far more uniform and homogeneous than that of today. Optimists imagine a universal global culture uniting all humankind into a single community untroubled by war, ethnic strife, or resource shortage—a global utopia of sorts. More common, however, is the view that the world is becoming blandly homogeneous as different places, different peoples, and different environments lose their distinctive character and become indistinguishable from their neighbors. While diversity may be the hardest thing for a society to live with, it may also be the most dangerous thing to live without. Nationality, ethnicity, race, gender—all of these are the legitimate legacy of humanity. If they are blurred, denied, and perhaps even repressed through global homogenization, humanity loses one of its coveted traits.

Our concern with geographic diversity takes many forms and transcends celebration of traditional cultures and unique places. People have many ways of making a living throughout the world. Different kinds of agriculture produce different food resources, and different uses of natural resources produce different results. As a globalized economy becomes increasingly fixated on mass-produced goods, it is important to recognize this broader spectrum of economic diversity. Furthermore, a stark reality of today's economic landscape is unevenness: while some places and people prosper, others suffer from unrelenting poverty. Unfortunately, this also is diversity amid globalization. As the gap between rich and poor becomes ever greater throughout the world, this critical dimension of human geography cannot be overlooked.

Even if there is a strong tendency toward conflation of the world's diverse environments into a universal, homogenized landscape of globalization, there is still much that is different and distinctive about the different regions of Earth and its various cultures. The politics of diversity demands increasing attention these days as we find worldwide tensions over ethnic separateness, regional autonomy, and political independence. Small groups of people throughout the world seek self-rule of territory they can call their own. Today, most wars are fought within countries, not between them. Not that one should highlight the unique over the common by celebrating ethnic factionalism. Instead, we seek a balance between describing the commonalities of the world's regions while at the same time explaining the differences and distinctiveness of places, environments, and landscapes (Figure 1.4).

▲ **Figure 1.4 Local cultures** This family in Ghana reminds us that although few places are beyond the reach of globalization, there are, nevertheless, strong local landscapes, economies, and cultures. For example, most of the material culture in this photograph is local in origin and comes from local cultural tradition. *(Caroline Penn/Panos Pictures)*

In summary, globalization can be defined as the increasing interconnectedness of people and places through converging processes of economic, political, and cultural change. Further, it is a dominant feature of the contemporary world because of its pervasive influences and uneven benefits. Although economic restructuring is a prime mover, globalization is much more than simply the growth and expansion of international trade. Within this context of globalization and interdependence is the tension and interplay of geographic diversity against these converging and homogenizing forces. Since much of this environmental, cultural, political, economic, and social difference predates contemporary globalization, it could be thought of as the extant and legitimate fabric of a highly diverse world. An equally important theme, though, is how this geographic diversity comes in conflict with globalization, of how globalization is suppressed, renegotiated, protected, preserved, or extinguished. Because of the importance of this theme, diversity and globalization should be examined as an inseparable and synergistic pair—often in conflict, yet also often complementary.

Geography Matters: Environments, Regions, Landscapes

Geography is one of the most fundamental sciences, a discipline awakened and informed by a long-standing human curiosity about our surroundings and the world environment. The term *geography* has its roots in the Greek words for "describing the Earth," and this discipline has been carried forward since classical times by all cultures and civilizations. With the inherent satisfaction of knowing about different environments comes pragmatic needs that complement exploration, resource exploitation, world commerce, and travel. In some ways geography can be compared to history: while that field describes and explains what has happened over time, geography describes and explains Earth's spatial dimension, of how the world differs from place to place.

Given the broad scope of the geographical charge, it is no surprise that geographers have many different approaches to studying the world based upon the conceptual emphasis guiding their study. At the most basic level, geography can be broken into two complementary pursuits, physical and human geography. As the term suggests, physical geography examines climate, landforms, soils, vegetation, and hydrology, while human geography concentrates on the spatial analysis of economic, social, and cultural systems. For example, a physical geographer in the Amazon Basin of Brazil might be interested primarily in the ecological diversity of the tropical rain forest or how the destruction of that dense vegetation changes the local climate and hydrology. The human geographer, on the other hand, might focus on the social and economic factors explaining migration of settlers into the rain forest or the tensions and conflicts over land and resources between these migrants and indigenous peoples.

Another basic division is that between focusing on a specific topic or theme (such as climatology or cultural geography) and synthesizing various topics applied to a specific region or area of the world, such as Latin America or Europe. The former is referred to as *thematic* or *systematic geography,* while the latter is called *regional geography.* These two perspectives are complementary and by no means mutually exclusive. This textbook, for example, draws upon a world regional scheme for its overall architecture, yet each chapter is organized thematically.

Human-Environment Interaction

A fundamental theme in all geographic inquiry is that which probes into the basic yet highly complex relationship between humans and their environment. Though this analysis may take many different forms, three overarching questions form the foundation:

1. Where and under what conditions does nature control, constrain, or disrupt human activities, and what kinds of societies are most vulnerable to these environmental influences (Figure 1.5)?
2. How have human activities affected natural systems such as the world's climate, vegetation, rivers and oceans, and what are the consequences of these actions?

▲ **Figure 1.5 Natural hazards** One pressing question in geographic inquiry is how people and societies are affected by natural hazards, such as this flood in Bangladesh. As population numbers increase in a given area, often more and more people become susceptible to natural hazards, such as hurricanes, floods, and drought. *(James Blair/National Geographic Society)*

3. Is there any evidence that humans can live in balance with nature, in a way that our activities do not adversely affect environmental systems so that human use of natural resources might be sustainable over a longer period of time? Or is environmental damage an inevitable consequence of human settlement?

While there are no simple answers to these questions, the study of world regional geography offers valuable insights from different cultural and regional perspectives. The first question, for example, can frame discussions of how human vulnerability differs among cultures to flooding, earthquakes, and famine. Further, the second question gets at the ever-important issue of what kind of human activities cause the most (or the least) environmental damage. This comparative perspective can be helpful for generating solutions to existing problems, as well as constraining those actions known to be environmentally harmful. Finally, the third question drives discussions about resource and environmental planning for the near and long-term future, an extraordinarily important issue concerned with the concept of appropriate carrying capacities for localities, regions, and, ultimately, for Earth.

Areal Differentiation and Integration

At a fundamental level geography can be considered the spatial science, charged with the study of Earth's space or surface area, the uniqueness of places, and the similarities between them. One component of that responsibility is explaining the differences that mark off one piece of the world from another. The geographical term for this is **areal differentiation** (*areal* means simply "pertaining to area"). Why is one part of Earth humid and verdant, while another just a few hundred kilometers away is an arid desert (Figure 1.6)? This is a question of spatial or areal differentiation, which is defined as the systematic study of the differences between parts of Earth's surface.

Geographers are also interested in the connections between areas and in how areas are linked to each other and interact. This concern is one of **areal integration**, or the study of how areas interact with each other. How and why are the economies of Singapore and the United States so closely intertwined although these two countries are situated in entirely different physical, cultural, and political environments? Or how are two different wet and dry areas within a hundred miles of each other ecologically linked? These are questions of areal integration. Increasingly, different areas and environments of the world are linked together through the forces of globalization.

Regions

The human intellect seems driven to make sense out of the universe by lumping together phenomena into categories of similarity. Biology has its taxa of living objects, history its eras and periods of time, geology its epochs of Earth history. Geography, too, makes sense of the world by compressing and synthesizing vast amounts of information into spatial categories of similar traits. The resulting areal units are referred to as *regions* (see "Defining the Region: The Metageography of World Regions").

Sometimes the unifying threads of a region are physical, such as climate and vegetation, resulting in regional designation such as the Sahara Desert or the tropical rain forest. Other times the threads are economic and cultural, as in the use of the popular term *Midwest* for the central United States. All human beings need to compress large amounts of information into some stereotype; often the geographic region is just that, a spatial stereotype for a section of Earth that has some special signature or characteristic that sets it apart from other places (Figure 1.7).

Some caution is advised. First, few regions are homogeneous throughout. While there may be a unifying theme that unites an area, there will also be diversity and differences within it. This sort of common sense is readily applied to those areas we know well, yet it often becomes an unwitting casualty when reading about or examining distant parts of the world. A second caution is urged regarding borders and about those artificial lines drawn around a region that delimit it as a unique area. Except with political borders between states, rarely are those boundaries quite as abrupt as they appear on a generalized map. Mountain ranges usually rise gradually from lowland areas, and deserts transition only gradually into woodlands; manufacturing districts lessen in intensity toward the periphery; and a linguistic or cultural region gradually blends into another. Usually those traits used to define a geographic region are most explicit and clear in the core area. Moving away from this

◀ **Figure 1.6 Areal differentiation** The California–Mexico border, which shows different agricultural landscapes on opposite sides of the border, reminds us of the many variables that shape the landscape into different patterns. Explaining these differences (and the similarities) is a central focus of geography. *(Earth Satellite Corporation/SPL/Photo Researchers, Inc.)*

DEFINING THE REGION The Metageography of World Regions

One of the perplexing issues of world regional geography is the actual definition and demarcation of the regions themselves. For example, historically, Europe has been separated from Asia along the Straits of Bosporus, the narrow waterway that divides Turkey into "European Turkey" to the west, with "Asiatic Turkey" lying to the east; although the country is unified by a common language, religion, and culture, any traveler to Istanbul can see the fallacy of this traditional division of major geographic realms.

The problem of dividing the world into a small number of regional units has been apparent since the earliest days of geography associated with Greek civilization. The Greeks conceptualized the world as divided into a few continents based on their own experiences as seafaring explorers. To them it made sense to divide the great landmasses of Europe, Asia, and Africa based upon seas and waterways; thus, the Bosporus and the Red Sea became the landmarks separating these continent-based world regions. European explorers didn't argue with this threefold division even though once again a large country—Russia—had to be separated into "European Russia" west of the Ural Mountains and "Asiatic Russia" to the east.

In spite of this awkward problem, European explorers continued to divide the world into what they called "continents": North and South America were referred to as separate continents, although they were clearly joined by the narrow isthmus of Panama (or *pan-America,* as some call it); Australia was added as a continent, even though many argued it was simply a large island. But if Australia was a continent, why not Madagascar; shouldn't it also be a continent? Not according to European explorers who were quite comfortable with considering it a part of the African continent. Finally, Antarctica was given continental status even though it was totally uninhabited.

This seven-part "myth of the continents" seemed primarily useful so Europeans could differentiate themselves from other cultures, specifically Asians and Africans. It seemed much easier to draw cultural and ethnic boundaries if they were placed on distinct continents, separated from Europe by mountains, seas, and waterways. However, this scheme makes little sense in today's world. There is, for example, no such thing as "Asian culture." No common traits link the Japanese people of East Asia to those of Southwest Asia in, say, Saudi Arabia.

It was only World War II and the aid programs that followed that focused greater emphasis on cultural and geopolitical regions instead of continents. Asia was subdivided into a handful of regions roughly comparable to historically defined civilizations; thus East Asia, centered on China and Japan, was differentiated from South Asia, anchored by India. While this postwar scheme represented a major advance over that of the continental scheme, it was not flawless, nor was any one regional division accepted by all. Though progress was made, there is still a good deal of quibbling about where to draw the boundaries for different world regions. For those readers interested in that discussion, each of our regional chapters contains a sidebar titled "Setting the Boundaries" that examines in more detail this problematic issue of regional definition and boundary-drawing.

core toward the outlying boundaries shows these defining traits to be less and less apparent, weakening in intensity toward the region's periphery.

Space into Place: The Cultural Landscape

Human beings shape and transform undifferentiated, abstract, worldly space into distinct places that are unique and heavily loaded with meaning and symbolism. This diverse fabric of "placefulness" is of great interest to geographers because it tells us much about the human condition as it varies across the surface of Earth. It tells us how humans interact with nature and between themselves; where there are tensions and where there is peace; where people are rich and where they are poor. Consequently, the systematic study of place is a critical element to the larger task of areal differentiation and integration.

A common tool for the analysis of place is the concept of the **cultural landscape,** which is, simply stated, the visible, material expression of human settlement, past and present. Landscape is the tangible expression of the human habitat. Landscape visually reflects the most basic needs of humans: shelter, food, work. Furthermore, the cultural landscape also acts as a glue of sorts that brings people together (or keeps them apart), since it can be a marker of cultural values, attitudes, and symbols. Because cultures vary greatly around the world, so do landscapes (Figure 1.8).

Most of the French countryside, for example, with its modest-sized agricultural fields enclosed by hedgerows, with small vineyards adjacent to stone houses clustered together in villages with its town square—this landscape looks very different from a typical farm landscape of the midwestern United States, with its large fields covering several square miles, with people separated from each other by large distances, comfortable in their isolated wood frame houses far from the nearest town. From the contrasting look of these two landscapes, we can infer much about the social and economic systems underlying these different societies. While these two landscapes may be relatively familiar and easily imagined in our minds, all landscapes, be they in India, Africa, or Asia, convey much important information about land, life, and the environment throughout the world.

Increasingly, however, we see the uniqueness of places being eroded by the homogeneous landscapes of globalization: shopping malls, fast-food outlets, business towers, theme parks, industrial complexes. Although these landscapes are familiar and often taken for granted, understanding the forces behind their spread is important because these landscapes tell us much about the expansion of global economies and

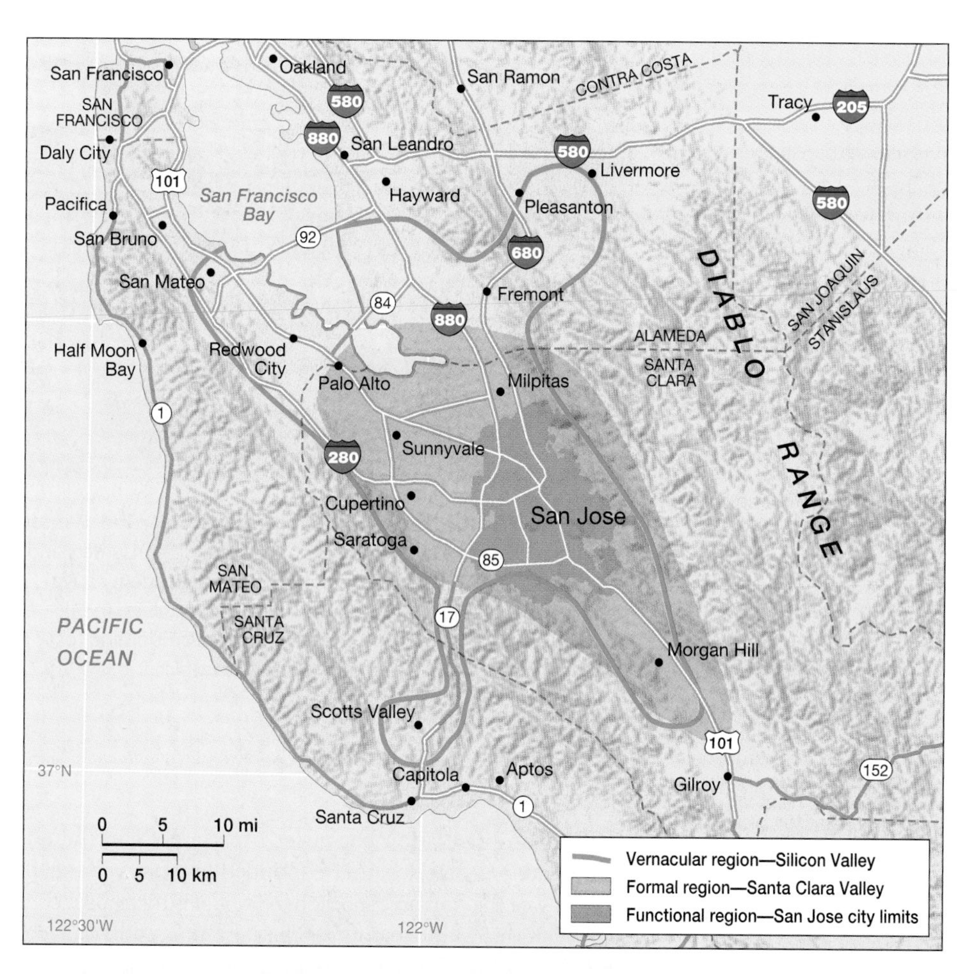

◀ **Figure 1.7 Geographic regions** This map illustrates three different kinds of regions commonly used by geographers. The vernacular region is a region with vague borders that is used by the public to characterize a general area. Other examples of vernacular regions include "the Midwest." In contrast, formal regions have distinct boundaries because they are constructed by using one specific trait. In the case of the Santa Clara Valley, it is the topographic trait of the valley. Last, a functional region is based on a certain activity or organization, such as the civic government of San Jose.

▲ **Figure 1.8 The cultural landscape** A major geographic tool is landscape analysis, which looks behind the tangible expression of human life and settlement to understand how we shape the environment into distinctive forms that give places their special—and spatial—identities. This village is in Yunnan province, China. *(M. Yamashita/Woodfin Camp & Associates)*

cultures. While seeing a modern shopping mall in Hanoi, Vietnam, may resonate with familiarity to someone from North America, this new landscape conveys a much more profound message about change and culture at the turn of the century as still another component of globalized world culture is implanted into a once remote and distinctive city.

Scales: Global to Local

There is a sense of scale to all systematic inquiry, whatever the discipline. In biology, for example, some scientists study the smaller units of cells, genes, or molecules, while others take a larger view, examining plants, animals, or ecosystems. Similarly, some historians may focus on a specific individual at a specific point in time, while others take a broader view of international events changing over several decades or throughout a century. Geographers also work at different scales. While some may concentrate on analysis of a local landscape—perhaps a single village in southern India—others will focus on the regional picture, on the spatial characteristics and interaction of a larger sphere, such as of all southern India. Others do research at a still larger global scale, examining emerging trade networks between Asian countries and North America, for example, or how India's monsoon might be connected and affected by the Pacific Ocean's El Niño. Even though geographers may be working at different scales, they never lose sight of the interactivity and connectivity between local, regional, and global scales, of how that village in southern India might be linked to world trade patterns, or how the late arrival of the monsoon (delayed, perhaps, by El Niño) could affect agriculture and food supplies in different parts of India.

This ability to move between different scales—global, regional, local—is critical for understanding contemporary world regional geography because of the way globalization links together all peoples and places. Few villages today, however remote, are without contact to the modern world. Global economies draw upon local crops and resources, and, conversely, fluctuations in world commodity prices affect the well-being of those local people. Global TV and videos introduce foreign styles, ideas, mannerisms, and expectations into small towns and remote settlements in formerly isolated parts of the world. Global tourism in its different faces brings strangers (some welcome, some not) to far-flung localities, often corrupting and compromising the very uniqueness the tourists sought. Further, very few places and people are isolated from global politics and tensions, from the influences of superpowers, supranational organizations, or, increasingly, from the ambitions and agendas of regional separatists and national splinter groups.

In this book we use the phrase "global-to-local" to capture the flexibility in the scale of analysis necessary to understanding placing regions, places, landscapes, and people into their globalized context. Since our premise is that no place in this contemporary world is completely isolated from the larger currents of the contemporary world, we need a perspective that facilitates examining the linkages connecting people and places with the larger, dynamic world (Figure 1.9).

▲ **Figure 1.9 Global-to-local connections** Today, most localities are linked to the world economy as both consumers and producers. As a result, fluctuations in the global economy have repercussions that ripple into the smallest communities. In this photo, women in a small village in China produce clothing for world markets. *(James Montgomery/Bruce Coleman Inc.)*

Themes and Issues in World Regional Geography

Geography is that academic discipline responsible for the description of Earth and explanation of the patterns on its surface. Although this task involves different and often contrasting approaches and perspectives, the themes and concepts held in common give coherence to the field and offers an effective perspective for evaluating how globalization affects different parts of the world. A general background on global environmental geography is found in Chapter 2, "The Changing Global Environment." This chapter sets the scene for the regional studies that follow by discussing the global environmental fabric fundamental to human settlement—climate, topography, vegetation, hydrology, the oceans—as well as the linkages between these environmental issues and the forces of globalization. After the two introductory chapters, this textbook is organized regionally, grouping the world's countries into a spatial framework that begins with the region most familiar to readers, North America, then moving to Latin America, the Caribbean, Africa, Europe, the Middle East, Russia, and the different regions of Asia, before concluding with Australia and Oceania. Each of the 12 regional chapters, in turn, has the same five-part thematic structure: Environmental Geography; Population and Settlement; Cultural Coherence and Diversity; Geopolitical Framework; and, last, Economic and Social Development. A guide to the issues and concepts discussed in each regional chapter, along with explanations of the data tables used, follows.

Population and Settlement: People on the Land

The human population has never been larger and there is considerable debate whether continued growth will benefit or aggravate the human condition. Not only is there a concern with aggregate numbers, but there is also interest in the geographic pattern of human settlement on the planet since we are distributed differentially. While some parts of the world are densely populated with landscapes of intertwined towns and cities, other places remain almost empty.

Global population growth is one of the most vexing issues facing the world. With six billion humans on Earth, we currently add another 86 million each year, or about 10,000 new people each hour. Most of this population growth—about 90 percent—takes place in the developing world, especially in the countries of Africa, Latin America, and South and East Asia (Figure 1.10). Because of this rapid growth in developing countries, perplexing questions dominate discussions of all global issues. Can these countries absorb this rapidly increasing population and still achieve the necessary economic and social development to ensure some level of well-being and stability for the total population? What role should developed countries such as the United States and Canada play in this debate over population, now that their populations have nearly stabilized? While population is a complicated and contentious topic, several points help focus the issue. They are, briefly:

1. We must understand and conceptualize the very different rates of population growth found in various regions of the world. While some countries are growing rapidly, other areas have essentially no natural increase; instead, any population increase comes from in-migration. India is an example of the former, Europe the latter.
2. Population planning takes many forms, from the rigid "one-child" policies of China to the pro-natal, "more

children, please" programs of western Europe. While some programs attempt to achieve their goals through coercion and force, others rely on incentives and social cooperation (Figure 1.11).

3. Not all attention should be focused on natural growth, however, since migration is increasingly a cause of population change in this globalized world. While most international migration is driven by a desire for a better life in the richer regions of the developed world, we should not lose sight of the millions of migrants who are refugees from civil strife, persecution, or environmental disasters.
4. Last, the greatest migration in human history (at least in aggregate numbers) is now going on as people move from rural to urban environments. If current rates continue, shortly after the turn of the century, more than half the world's population will live in cities. Once again it is the developing countries of Africa, Latin America, and Asia that are experiencing the most dramatic changes caused by migration to the city.

Population Growth and Change

Because of the centrality of the population topic, each regional chapter contains a table with pertinent data for every country within that region (Table 1.1). Although the statistics in these tables might seem daunting at first glance, this important information guides our understanding of population growth and of whether those rates of growth might continue in the near future. These data also convey a sense of urbanization, which is the rural-to-urban population ratio of each country.

The most common population statistic used is the rate of **natural increase** (abbreviated as RNI), which depicts the annual growth rate for a country or region in percentage increase. This statistic is produced by subtracting a population's number of deaths in a given year from its annual number of births to find the net number added through natural increase. Gains or losses through migration are not considered in this figure. Instead of using raw numbers for a large population, demographers usually divide the gross numbers of births or deaths by the total population, producing a figure per 1,000 of the population. These are referred to as either the *crude birthrate* or the *crude death rate*. For example, recent data for the world as a whole show a crude birthrate of 23 per 1,000 and a crude death rate of 9 per 1,000. Thus the natural growth is 14 births per 1,000. This is expressed as the rate of natural increase simply by converting that figure to a percentage. Therefore, the current RNI for the world is 1.4 percent per year. Because birthrates vary greatly between countries (and even regions of the world), rates of natural increase also vary greatly. In Africa, for example, many countries have crude rates of more than 40 births per 1,000 people, and some are closer to 50 per 1,000. Since the death rates are generally less than 20 per 1,000, we find RNIs approaching or even exceeding 3 percent per year; this indicates a rapidly growing population that will double its size in just 23 years.

Though crude birthrates give some insight to current conditions in a country, demographers also use the **total fertility rate,** or TFR. The TFR is actually a synthetic rate that measures the fertility of a statistically fictitious, yet average, group of women moving through their childbearing years. If women marry early and have many children over a long span of years, the TFR will be expressed as a high number; conversely, if data show women marry late and have few children, the number will be correspondingly low. From population data collected in the second half of the 1990s, the TFR for the world is 2.9 children; this, then, is the average number of children borne by a statistically average woman. In contrast to this global figure, for the same period the TFR for Africa is 5.6, and for slow/no-growth Europe, 1.4. Again, the country-by-country TFR data are found in the tables contained in each regional chapter.

One of the best indicators as to the momentum (or lack of momentum) for continued population growth is to examine the youthfulness of that population. This conveys a sense of the proportion of the population soon entering into their reproductive years. The common breakpoint for this measure is *the percentage of a population under the age of 15*. Equally important, this figure also gives some clue to the health and nutritional needs of a society providing for those most vulnerable to famine, undernutrition, and disease. The global average has 32 percent of the population younger than age 15. In fast-growing Africa, that figure is 44 percent, with several countries very close to 50 percent. In contrast, Europe averages 19 percent, which is close to the North American average of 21 percent (Figure 1.12).

The Demographic Transition

Population growth rates change through time, and this is well documented from the historical record. More specifically, there appears to be a decline in growth as a country becomes increasingly industrialized and urbanized; at least that was the case historically in Europe and North America. From this historical data demographers generated the **demographic transition** model, a four-stage conceptualization that tracks changes in birth- and death rates through time as a population urbanizes. While the demographic transition model seems valid for describing historical population change, there is considerable debate about whether it offers an accurate tool for predicting change in developing countries currently undergoing industrialization and urbanization. If this model is an accurate predictor of population growth, some population workers and politicians argue that the best form of birth control might be financial aid that furthers economic development. However, there are also a large number of skeptics who question the application of the demographic transition model as a template for all developing countries. They draw their findings from countries where low birthrates are not linked to economic development, such as in the former communist countries of eastern Europe. Because of these contrasting views, there is a spirited debate in international aid circles as to the best course of action regarding family planning.

Figure 1.10 World population This world population map shows the differing densities of population in the regions of the world. East Asia stands out as the most populated region, with high densities in Japan, Korea, and eastern China. The second most populated region is South Asia, dominated by India, which is second only to China in population. In North Africa and Southwest Asia, population clusters are often linked to the availability of water for irrigated agriculture, as is apparent with the population cluster along the Nile River. Population clusters in Europe, North America, and Latin America give clues to high rates of urbanization.

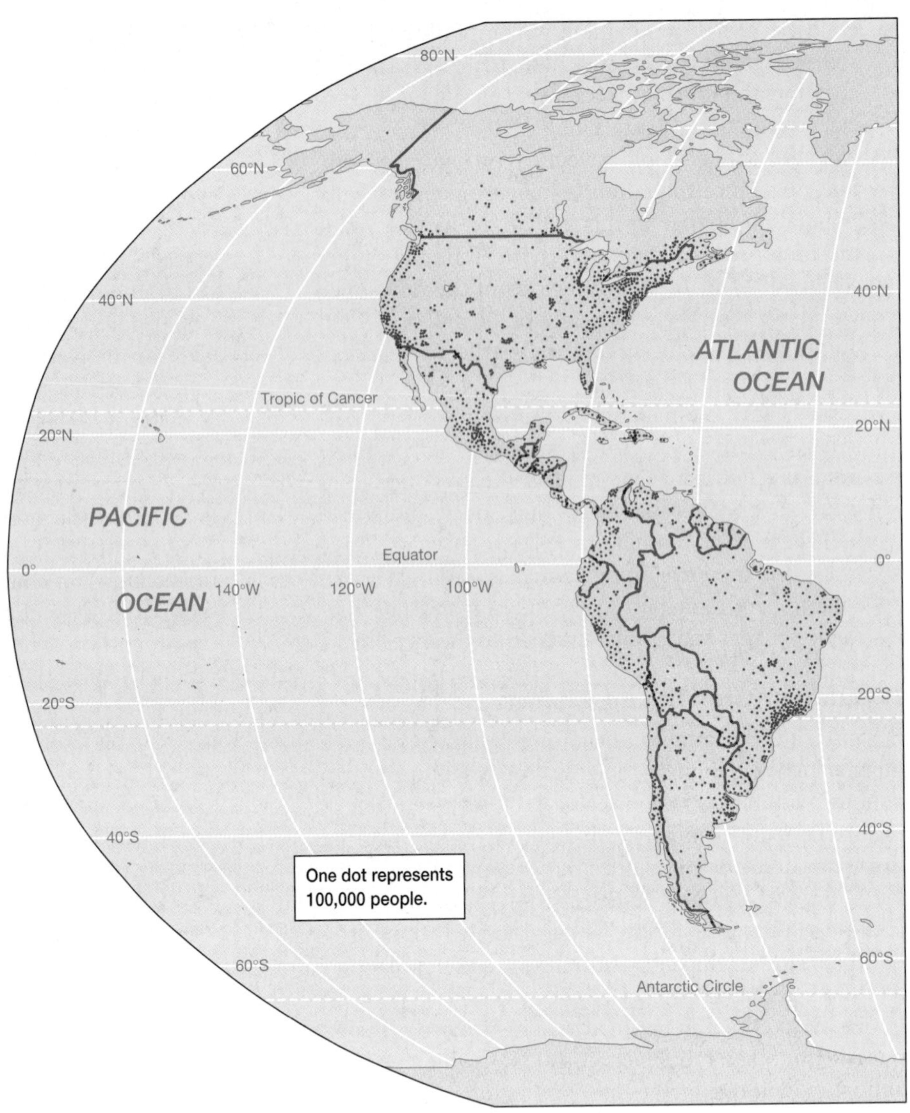

▲ **Figure 1.11 Family planning policies** Many countries in the developing world have concluded that unrestrained population growth may keep them from realizing their developmental goals, and therefore have put family planning policies in place. This poster urging smaller families is in Vietnam. *(Chris Stowers/Panos Pictures)*

The demographic transition model is at the center of this discussion (Figure 1.13).

In the demographic transition model, Phase 1 is characterized by both high birth and death rates, leading to a very slow rate of natural increase. Historically, this stage is associated with Europe's pre-industrial stage, a period that also predates common public health measures such as sewage treatment or sanitary water supplies, a scientific and medical understanding of disease transmission, and the most fundamental aspects of modern medicine. Little wonder that death rates were high and life expectancy short. Unfortunately, in some parts of the world these conditions are still common today.

In Phase 2, the death rates fall dramatically, creating a rapid rise in the RNI. Again, both historically and contemporaneously, this decrease in death rates is usually associated with the onset of public health measures and modern medicine. One of the assumptions of the demographic transition model is that these services become increasingly available after some degree of economic development takes place and, further, they are more accessible to the growing number of urban dwellers. However, even as death rates fall, it takes time for

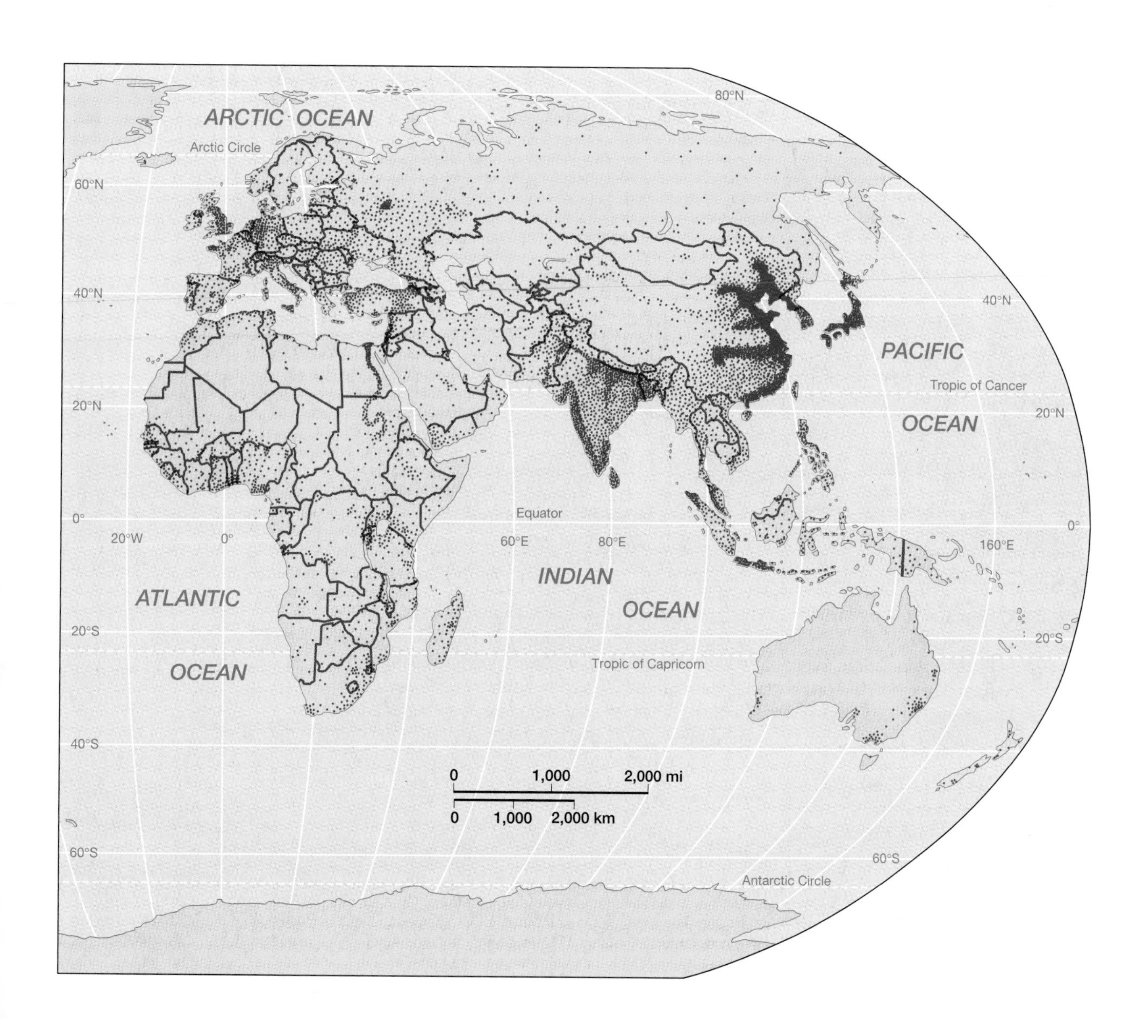

Table 1.1 Demographic Indicators (Median Values for World Regions)

Region	Population[a]	Natural Increase	TFR[b]	%<15[c]	% Urban
North America	300.8	0.55	1.8	21	76
Latin America	461.5	2.2	3.1	35	63
The Caribbean	38.4	1.2	2.0	32	50
Sub-Saharan Africa	624.3	2.5	5.8	45	29
Southwest Asia and North Africa	396.1	2.3	4.4	39	70
Europe	520.7	0.2	1.5	19	65
Russia and Its Neighbors	216.6	–0.4	1.3	21	68
Central Asia	90.1	1.6	2.9	37	38
East Asia	1440.4	0.9	1.7	23	74
South Asia	1297.7	2.8	4.6	41	22
Southeast Asia	511.9	2.1	3.2	35	37
Australia and Oceania	29.22	2.4	3.4	36	46

[a]Population in millions, 1998.
[b]Total fertility rate.
[c]Percentage of population younger than 15 years of age.
Source: *Population Reference Bureau, World Population Data Sheet,* 1998.

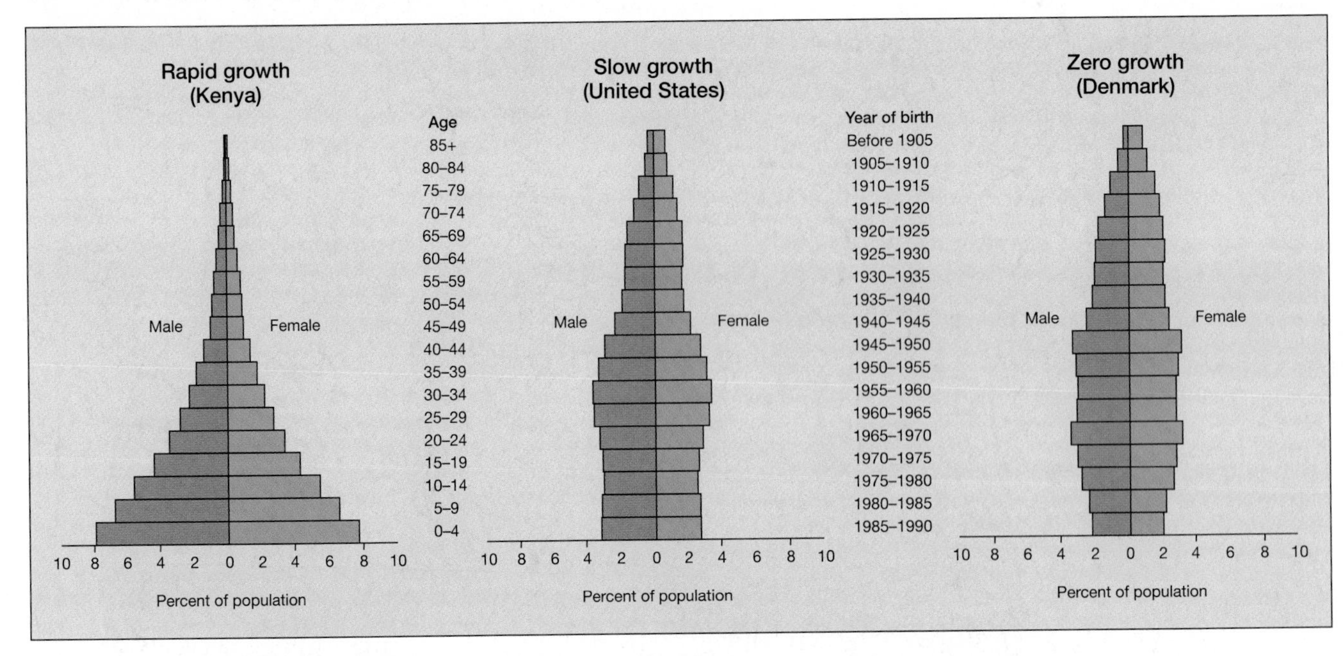

▲ **Figure 1.12 Population pyramids** The term *population pyramid* comes from the form assumed by a rapidly growing country such as Kenya when data for age and sex are plotted graphically as a percentage of its total population. Rapid natural growth will probably continue in this country because of the high percentage of the population entering fertility years. In contrast, slow and zero-growth countries have a very narrow base, which indicates relatively few people in the childbearing ages. *(From Knox and Marston, 1998,* Human Geography, *Upper Saddle River, NJ: Prentice Hall)*

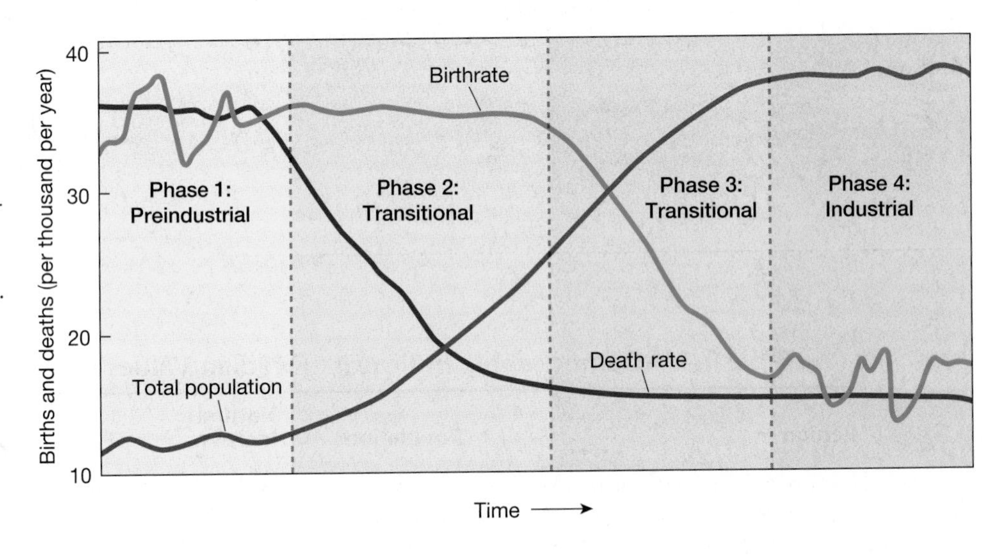

▶ **Figure 1.13 Demographic transition** As a country goes through industrialization, its population moves through the four phases in this diagram, referred to as the *demographic transition*. In the first phase, population growth is low because high birthrates are offset by high death rates. Rapid growth takes place in Phase 2 as death rates decline. The transition ends with low growth once again in Phase 4, resulting from low birthrates. *(From Knox and Marston, 1998,* Human Geography, *Upper Saddle River, NJ: Prentice Hall)*

people to respond with a lower birthrate, which begins in Phase 3. At this point the RNI also decreases. This, then, is the transitional stage as people apparently become aware of the advantages of smaller families in an urban and industrial setting. In Phase 4, a very low RNI results from a combination of low birthrates and very low death rates. As a result, there is very little natural population increase.

For many professionals, the demographic transition model is a reasonable predictor of where and when we can expect population growth rates to slow in the developing world. However, it also has its detractors, demographers who see too many complicating variables and differences between countries for universal application in the late twentieth century. Minimally, the demographic transition concept is a useful tool for organizing information about the vast difference in birthrates and death rates we find around the world (Figure 1.14).

Migration Patterns

Never before in human history have so many people been on the move and lived outside of their home country. Today, about 125 million, or 2 percent, of the total world population can be considered migrants of some sort. Estimates suggest that figure may increase by four million annually. Much of this international migration is directly linked to the new globalized economy. About half of these migrants live either in the developed world or in those developing countries with

▲ **Figure 1.14 Fertility and mortality** Birthrates and death rates vary widely around the world. Fertility rates result from an array of variables, including state family planning programs and the level of a woman's education. This family is in Bangladesh, which is working hard to reduce the birthrate through education programs. *(Trygve Bolstad/Panos Pictures)*

▲ **Figure 1.15 Refugees** About 27 million people, or 20 percent of the world's migrant population, are refugees from ethnic warfare and civil strife. Most of these refugees are in Africa and western Asia. These women wait in line for water at a refugee camp in southern Sudan. *(Liz Gilbert/Sygma Photo News)*

vibrant industrial, mining, or petroleum extraction activities. In the oil-rich countries of Kuwait and Saudi Arabia, for example, the labor force is primarily foreign migrants. In terms of total numbers, fully a third of the world's migrants live in seven industrial countries: Japan, Germany, France, Canada, the United States, Italy, and the United Kingdom. Since industrial countries usually have very low birthrates, immigration accounts for a large proportion of total population growth. For example, about one-third of the annual growth in the United States comes solely from in-migration.

But not all migrants move for economic reasons. War, persecution, famine, and environmental destruction cause people to flee to safe havens elsewhere. Accurate data on refugees are often difficult to obtain, but UN officials estimate that currently some 27 million (or about 20 percent of the migrant population) should be considered refugees. More than half of these are in Africa and western Asia (Figure 1.15). While the causes of migration are often complicated, three interactive concepts are helpful in understanding this process. First, there are those *push* forces, such as civil strife, environmental degradation or unemployment that drive people from a place. Second, there are the *pull* forces, such as better economic opportunity or health services, that attract migrants to certain locations, whether within or beyond national boundaries. Connecting the two are the informational *networks* of families, friends, and sometimes labor contractors who provide information on the mechanics of migration, transportation details or news of housing and jobs in their new destination (Figure 1.16).

One of the most pervasive push forces in the world today is the loss of jobs in agriculture. In the last half century, the share of jobs in agriculture in the world dropped from 66 percent to less than 50 percent, and this is expected to decrease even further in the early twenty-first century. One of the challenges facing the world will be finding jobs for millions of displaced farmers. One estimate is that the developing regions of Latin America, South Asia, Africa, and Southeast Asia will have to create as many new nonagricultural jobs in the next 50 years as currently exist. That is a daunting prospect.

Settlement Geography

Geography has long been concerned with the different patterns of human settlement in the landscape. For example, in some regions, such as Europe and South Asia, it is common for people to cluster together in villages and towns, which contrasts with the dispersed settlement pattern of North America, where people scatter across the countryside in individual farms and homesteads. Additionally, the density, pattern, and dynamic of settlement conveys much information about regional issues, both historically and currently. In many regions, such as Africa, Latin America, and Asia, it was common for European colonial powers to relocate indigenous settlement to suit their own purposes. More specifically, local peoples were sometimes forcibly moved from lowland areas to uplands better suited to the Europeans' proclivity for cooler climates. Plantation labor needs were also met by relocating native peoples from villages to an agglomerated settlement; and, of course, there was the massive disruption in Africa of traditional settlement patterns by the slave trade.

Today, settlement geography gives insight to the profound changes resulting from globalization: whether people are leaving the countryside or staying; whether they are moving to small towns or large cities; whether jobs from new plants and factories are decentralized into the countryside or agglomerated around large cities; whether mechanization is replacing farmworkers who then must seek other options in other places; whether civil and ethnic strife ravage the countryside, forcing people to flee as refugees; whether the rural environment is viable and productive, or degraded and sterile. This is not to suggest that settlement should be viewed only as problematic, as only landscapes of chaos and disarray. While those unfortunate aspects cannot be overlooked, there is also much to learn from how people arrange themselves on the

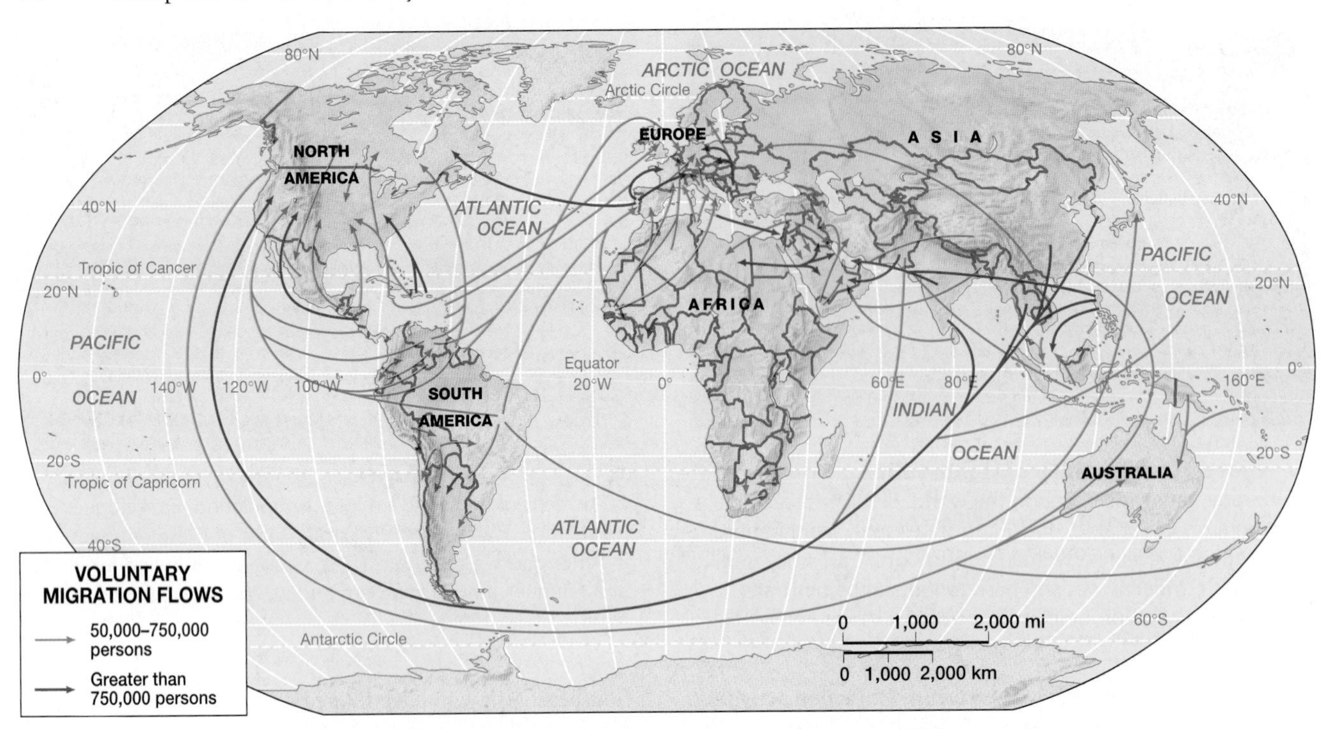

▲ **Figure 1.16 World migration** Roughly 125 million people, or 2 percent of the world's population, can be considered migrants. Most population experts predict this figure will increase by about four million annually. Globalization accounts for much of this migration, particularly for the migration from underdeveloped to developed countries. For example, the migration arrows to Europe, North America, and Southwest Asia are a function of economic globalization. *(Adapted from Knox and Marston, 1998,* Human Geography, *Upper Saddle River, NJ: Prentice Hall)*

land. Settlement also conveys information about social organizations, economic structures, cultural values, political systems, population growth and change, and about the meaning and emotion that people attach to the environment. When taken together, a settlement pattern conveys the sense of a place, the uniqueness of a certain landscape, the home territory for a specific group of human beings in this complex world (Figure 1.17).

An Urban World Cities are the focal points of the contemporary, globalizing world, the fast-paced centers of deep and widespread economic, political, and cultural change. Because of this vitality and the options cities offer to uprooted rural peoples, they are magnets for migration. The scale and rate of growth of some world cities is staggering: estimates are that both Mexico City and São Paulo (Brazil), both cities of more than 20 million, are adding nearly 10,000 new people each week. Urban planners predict these two cities will actually double in size within the next 15 years. Assuming that predictions are correct on the migration rate to cities, the world is quickly approaching the point at which more than half the planet's people will be urban dwellers. Evidence for this comes from data on the **urbanized population,** which is that percentage of a country's population living in cities. Currently, just less than half of the world's population lives in cities; however, given the high rate of migration to cities, projections are that at a global scale the world will be 60 percent urbanized by the year 2025.

Tables in each of this book's regional chapters include data on the urbanization rate for each country. For example, more than 75 percent of the populations of Europe, Japan, Australia, and the United States live in cities. Generally speaking, most countries in which three-quarters of the population lives in cities are also highly industrialized, since most manufacturing tends to cluster around urban centers (Figure 1.18). In contrast, the urbanized rate for developing countries is usually less than 50 percent; figures closer to 30 percent are not uncommon. Sub-Saharan Africa, for example, stands at just more than 30 percent. These data imply a low level of industrial activity. Urbanization figures also tell us where there is high potential for urban migration. If the statistic for urbanized population is low, such as for Zimbabwe (Africa), which is currently at 31 percent, the probability for high rates of urban migration in the next decades is clearly high. Accompanying the movement of people from countryside to cities will be profound changes in family structure, national policies, and market and manufacturing economics; these changes will be accompanied by widespread changes in the landscape and environment of the country as people move from rural to urban settlements.

Conceptualizing the City Given the important role of cities in world regional geography, it seems appropriate to offer a brief conceptual framework for thinking about urban settlements. While the regional chapters contain further details on specific cities, the following points offer a start:

(a)

(b)

▲ **Figure 1.17 Settlement landscapes** City and village landscapes differ widely because of the interplay between contemporary and historical forces, socioeconomic and cultural. (a) A small Peruvian village in the Andes that is populated by indigenous people and expresses their local traditions. (b) A town in Venezuela, where the Spanish colonial presence is still expressed in the plaza, street pattern, and building architecture. *(Rob Crandall/Rob Crandall Photographer)*

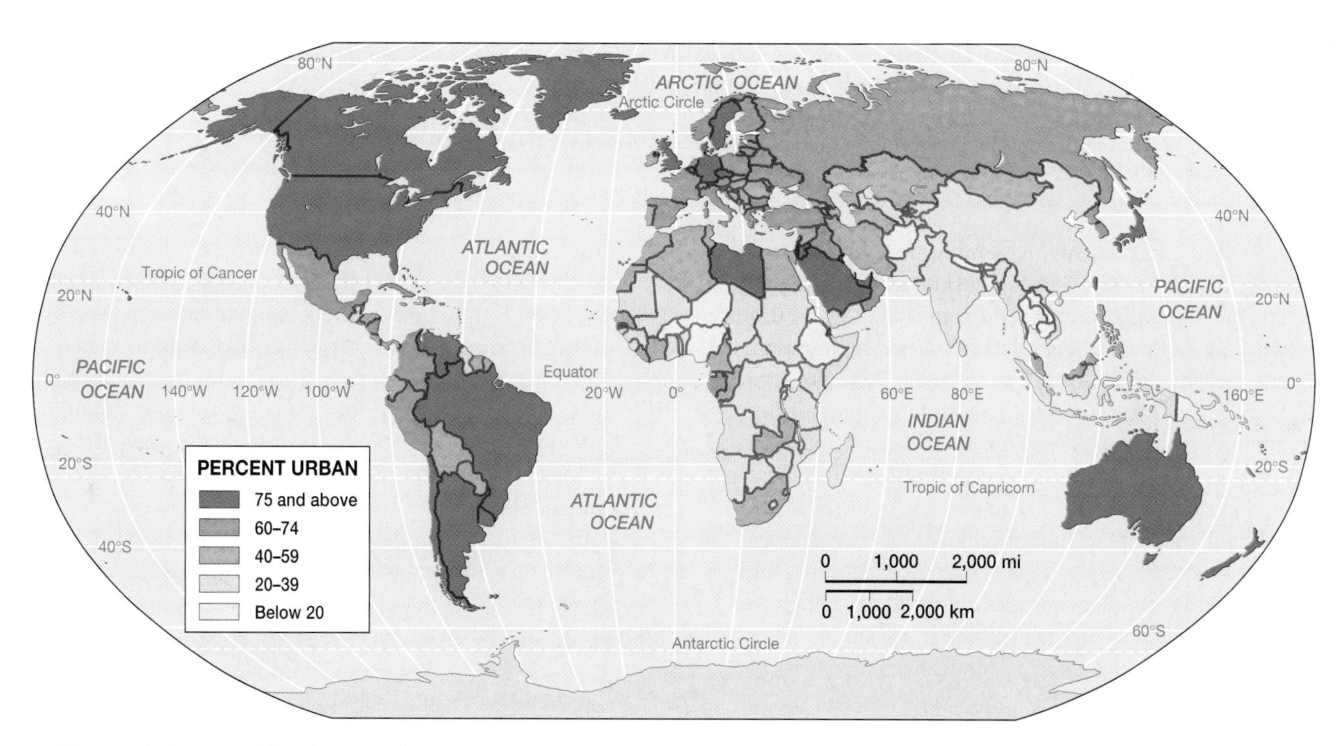

▲ **Figure 1.18 World urbanization** In general, developed countries have a higher percentage of their population living in cities as a result of the clustering of industry and business in urban places. This can be seen by contrasting Europe and North America with Africa. However, China is one example of a government that attempts to restrict the migration of people to cities to prevent overcrowding, thus the urbanization is lower than expected. *(Adapted from Rubenstein, 1999,* The Cultural Landscape: An Introduction to Human Geography, *Upper Saddle River, NJ: Prentice Hall)*

1. Cities are interdependent with other cities and towns, both regionally and globally, and are linked together in an *urban system* of interactive settlements. As some cities expand and grow, or come to specialize in an important function such as banking or manufacturing, the relationships and interactions within that urban system may change. The largest city within the city system may have *urban primacy,* a term used for a particular city that is either disproportionately large and/or dominates economic, political, and cultural activities within the country (Figure 1.19). In many countries one mega-city usually stands out in that country's urban system. Examples include Bangkok, Mexico City, Cairo, Paris, and Buenos Aires.

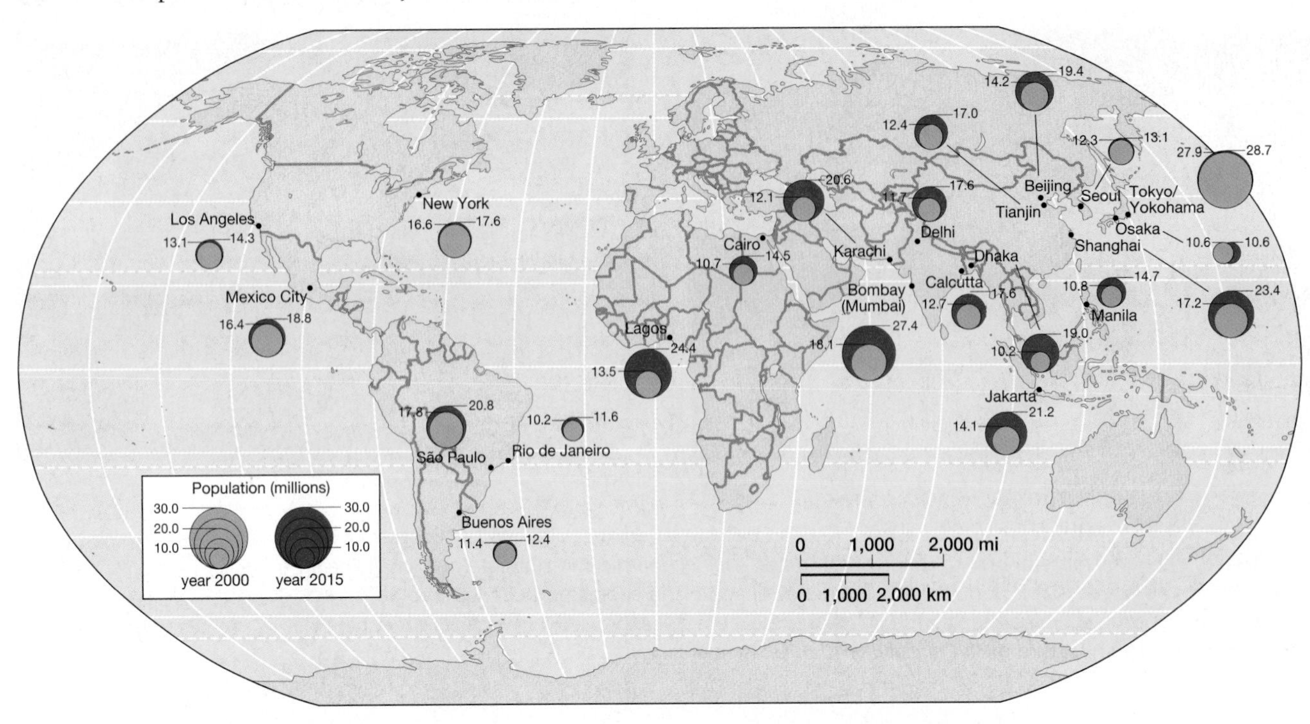

▲ **Figure 1.19 Growth of world cities** This map shows the largest cities in the world, along with predicted growth by the year 2015. Note that the greatest population gains are expected in the large cities of the developing world, such as Lagos, Nigeria, Karachi, Pakistan, and Bombay (Mumbai). In contrast, large cities in the developed world are predicted to grow slowly over the next decade. Tokyo, for example, will add fewer than a million people.

2. Although cities are highly complicated settlements, there are often shared traits of **urban structure.** This term refers to distribution and patterning of land use within the city. Land use is linked to specific functions such as the central business district (CBD), governmental activities, housing, retailing, manufacturing, and industry. Because these functions are not randomly distributed but instead follow well-understood location principles, it is possible to generate *models of urban structure* that convey the commonalties between cities within a specific region. One of the best examples is the model of Latin American cities discussed in Chapter 4.
3. Closely related to urban structure is the landscape of the city, the **urban form** or physical arrangement of buildings, streets, parks, and architecture that gives each city its unique sense of place. For example, Cairo clearly looks and feels very different from New Delhi or Caracas (Figure 1.20). The reasons for these differences in urban landscapes can be many, ranging from different histories, economies, and planning policies to ideologies and cultural values. In the regional chapters, we explain why certain cities look and feel as they do.
4. Because urban migration has taken place so rapidly in many countries, resulting in explosive growth, many cities have reached a state of **overurbanization,** in which the urban population grows more quickly than support services such as jobs, housing, transportation, sewers, and water supply. Unfortunately, this a fairly common situation in the developing world. The thousands of "street people" of Calcutta are legend, though this phenomena of homelessness is by no means restricted to South Asia. **Squatter settlements,** illegal developments of makeshift housing on land neither owned nor rented by the new urban migrants are widespread in the rapidly growing cities of the developing world. Often they are built on steep hillside slopes or on river floodplains that expose occupants to the dangers of landslide and floods. At other times, squatter settlements are found in the open space of public parks where they are regularly destroyed by government authorities, only to be quickly rebuilt by the migrants who have no other alternatives (Figure 1.21).

Even to those who have found more permanent housing, water, and sewer connections are often rudimentary at best. Water service can be sporadic and restricted to a few hours a day, and open sewers are a disturbingly common landscape feature of many cities. The World Bank estimated recently that only 40 percent of the people in less-developed cities lived in homes that were connected to sewers. Although difficult to measure accurately in developing countries, urban unemployment (and underemployment) are of epidemic proportion, estimated to be on the order of 50 percent in many cities of the developing world. Without regularly paid work, urban migrants draw upon their own resources to survive by begging, run-

▲ **Figure 1.20 Urban form** Caracas, Venezuela, shows how city landscapes express both global and local influences. The high-rise buildings of the financial district, largely a product of globalization, contrast with the diversity of low-rise residences of local design and architecture. *(Robert Frerck/Odyssey Productions)*

▲ **Figure 1.21 Squatter settlements** Because of the massive migration of people to world cities, adequate housing for the rapidly growing population becomes a daunting problem. Often migrant housing needs are filled with illegal squatter settlements, such as this one in New Delhi. *(Porterfield/Chickering/Photo Researchers, Inc.)*

ning errands, being paid to stand in lines, street vending, or scavenging public dumps. Because cities are experiencing unprecedented growth and there appears to be no end in sight to this massive migration, the multitude of problems accompanying this migration give good reason for our concern about the quality of life of the world's recent urban migrants.

Cultural Coherence and Diversity: The Geography of Tradition and Change

Culture, many argue, is the adhesive binding together of the world's diverse social fabric. A cursory read of the daily newspaper, however, raises questions as to whether the world is literally coming unglued since the frequency and intensity of cultural conflict seems pervasive and ever-increasing. With the recent rise of global communication systems (satellite TV, movies, video, etc.), stereotypic Western culture is spreading at a rapid pace. While this is willingly accepted by many throughout the world, other groups and countries resist this new form of cultural imperialism through protests, censorship, and restrictions on film, TV, and music.

In response to the spread of a common, homogenized world culture, many people throughout the globe are revisiting their traditional and historical identities as ethnic or folk groups. This is not simply a matter of quaint customs and costumes; instead, it often leads to overt conflict with neighboring groups with different group identities and self-interests. Moreover, we are seeing a new interest in small group autonomy and self-determination, not just in the former Soviet Union lands, where ethnic identities were suppressed under communism for most of this century, but also in other large nation-states—including those in western Europe—where new voices are heard demanding regional autonomy, separate territories, and control over natural resources. Religious wars, too, seem to be on the increase. Some of this is coincidental with the return of religious choice to former Soviet lands, yet other tensions result from intense missionary and proselytizing activity by Christian and Islamic sects. Religious fundamentalism is on the rise as a reaction to globalizing culture, and this, too, breeds cultural tensions.

The geography of cultural cohesion and tension, then, entails an examination of tradition and change, of language and religion, of group belonging and identity, and of those complex and varied currents that underlie and pervade twenty-first century factionalism and separatism (Figure 1.22).

Culture in a Globalizing World

Given the diversity of cultures around the world coupled with the dynamic changes associated with globalization, traditional definitions of "culture" must be stretched somewhat to provide a conceptual framework applicable to the turn of the

▲ **Figure 1.22 Ethnic tensions** Contemporary cultural change is characterized by convergence and homogeneity as people are pulled together by globalized culture, or factionalism and separatism, often fueled by ethnic tensions, such as the religious strife shown here in India between Hindu and Muslim peoples. *(Robert Nickelsberg/Liaison Agency, Inc.)*

century. Though social scientists offer varied definitions of culture, there is enough overlap and agreement between these different definitions to provide a starting point. **Culture** is learned, not innate, and is shared, not individual, behavior; it is held in common by a group of people, empowering them with what, for lack of a better term, could be called a "way of life." Clearly, the people of southern India have a different way of life than, say, those in central Africa; the concept of culture is useful for explaining those differences.

Additionally, culture has both abstract and material dimensions: speech, religion, ideology, livelihood, and value systems, but also technology, housing, foods, and music. These artifacts and expressions of culture are germane to the study of world regional geography because they tell us much about the way people interact with the environment and, further, use the landscape to shape and reinforce their culture. Finally, given the widespread influences of globalization, it is best to think of culture as dynamic rather than static. That is, culture is a process, not a state, something constantly adapting and adjusting to new conditions. As a result, there are always tensions between the conservative, traditional elements of a culture and the newer, innovative forces promoting change and modernity (Figure 1.23).

Given that group culture is expressed in many different ways and each group has different internal characteristics, there is value to creating a hypothetical spectrum of cultures ranging from the most tradition-bound to those more contemporary and fluid. The term *folk culture* is used to describe a rural, largely self-sufficient homogeneous group of people with a conservative, traditional way of life. Families or clans keep the social fabric together, and individualism is generally weak and unrewarded. Although there is little division of labor typically in rural cultures, gender roles are usually strong. The cultural landscape of folk cultures—house architecture, field types, crops, public monuments, and symbols—offers clues to how these people interact with the environment. Additionally, folk cultures usually have a strong handicraft tradition that produces many kinds of material culture, such as baskets, toys, or unique kitchen items. Folk cultures were far more prevalent historically than today, since industrial society has eroded and changed what may have been historically a common social fabric. North American examples of folk cultures would include religious groups such as the Mennonites and the Amish or the remnants of Appalachian Mountain culture. In Europe, folk culture can still be found in parts of rural Ireland, eastern Europe, or in the remote mountain valleys of the Alpine chain.

More fluid than folk cultures with defining traits based upon common ancestry, race, religion, or language is an *ethnic culture* in which there is usually a strong sense of tradition controlled by clear lines of authority through family, clan, tribe, or church. Ethnic group identity is often constructed from the shared feeling that its members are a minority within a larger cultural context. Many parts of Latin America contain good examples of ethnic cultures, groups made up of indigenous peoples such as the various Maya groups of southern Mexico and Guatemala or Amazon tribal groups.

In contrast to folk and ethnic cultures, *popular culture* is primarily urban-based, encompasses great heterogeneity, and is constantly changing and fluid. Relationships between people are often ephemeral, perhaps even shallow, with a much weaker family structure than in traditional society. Labor division is high, with distinct specialization of jobs; gender roles may be less rigidly defined; material goods are mass produced in international factories. The quintessential landscape of popular culture includes the mall, the franchise strip, the parking lot, the suburban housing tract. Although many disparage popular culture because of its faddishness and superficiality, it probably now has more adherents around the globe than folk or ethnic cultures.

As a subset or variant of popular culture, today one finds an emerging *world culture* that appears to be empowered by globalization. Transnational economies, electronic communication, and international politics produce citizens of indeterminate nationality and home base who travel widely, work in different countries, and, as a result, generate tastes and cultural values that are an amalgamation of international stimuli. Its members are equally at home in Hong Kong, New York, Singapore, or any of the world's mega-cities, all of which have interchangeable landscapes of offices, hotels, restaurants, and discos. Increasingly, it seems, the one real home base for this world culture is the Internet, where instantaneous electronic communication has replaced traditional forms of social interaction.

These four categories offer a scheme to help comprehend the wide spectrum of cultural expressions found in the contemporary world. For many human beings, membership in one culture or another is beyond choice. Until recently, social scientists argued that a defining characteristic of folk and ethnic cultures was that individuals were born into them; people could only choose to leave or disassociate through time, but their early membership was decided at birth. However, this seems a questionable position today, given the fluidity of cultural opportunities and the possibilities for multiple cultural memberships. It seems reasonable that a person might choose to exercise a narrow cultural identity at one point in time or, in a particular place, while participating in popular or world cultures at other times and places. As an example, one could point to the proclivity of some Amish youths to interact with popular culture while still taking their primary identity from a folk culture. This sort of cultural dual membership is becoming an increasingly common characteristic of contemporary globalized culture.

When Cultures Collide

Cultural change often takes place within the context of international tensions. Sometimes one cultural system will replace another. At other times, resistance may stave off change, though more commonly, a third, newer form of culture results that is an amalgamation of two different cultures. Historically, colonialism was the major perpetuator of these conflicts; today, globalization in its varied forms can be thought of as the major vehicle of cultural collision and change.

Cultural imperialism is the active promotion of one cultural system at the expense of another. Though there are still

(a)

(b)

(c)

(d)

▲ **Figure 1.23 Folk, ethnic, popular, global culture** A wide spectrum of cultural groups characterize the contemporary world's cultural geography. At one end of the spectrum are folk and ethnic cultures; popular and global cultural identities are at the other end. While the man with his laptop and cell phone (a) illustrates the homogeneity and placelessness of global culture, it is not the case with either the Peruvian Chonguinada group (b) or the Garifuna women in Honduras (c). These ethnic and folk groups represent local diversity bound to a specific place. In between these two extremes lies popular culture (d). *([a–c] Rob Crandall/Rob Crandall Photographer; [d] Jim Sugar Photography/Corbis)*

many varied expressions of cultural imperialism today, perhaps the most explicit, even severe examples come from the colonial period when European cultures spread worldwide, overwhelming, eroding, and often replacing indigenous cultures—Spanish culture into Latin America, French culture into Africa, British culture into India. New languages were mandated, new education systems implanted, new administrative institutions took the place of the old. New dress styles, dialects, gestures, and organizations were added to existing cultural systems and many of these persist today as cultural vestiges of an earlier colonial era. In India the makeover was so complete among certain social groups that wags are fond of saying with only slight exaggeration "the last true Englishman will be an Indian" (see "Local Voices: Who Are They, and Why Should We Listen?").

Today, cultural imperialism is probably less linked to an explicit colonizing force, but more often as a fellow traveler with globalization. Though many expressions of cultural

LOCAL VOICES Who Are They, and Why Should We Listen?

One of the best ways to learn about a foreign place is by listening to the people who live there: by listening to them talk about their country, their village, their neighborhood, their interests and concerns; about the problematic issues they face and the solutions they propose. Hearing these "local voices" is particularly important these days because the ubiquitous and loud voices of globalization threaten to mask the diversity of local expression with satellite TV, global music videos, Hollywood movies, and 1970s TV reruns.

While the messages of globalization are important and should not be dismissed, we wish to temper them somewhat by offering an opportunity to listen to the locals. Thus, each chapter has a sidebar titled "Local Voices" that offers brief exposure to diverse individuals and groups speaking about their homeland. Sometimes these voices come from the rich description of regional fiction, written by a single author; other sidebars draw upon a group of local voices, perhaps students from a foreign university finding their voice through an Internet Web site. Still other examples are issue-oriented, as we listen to an environmental group present its views on the problems of pollution and nature protection in the homeland, or listen to a separatist faction argue its case for territorial autonomy. Some local voices represent mainstream ideas, while others are less often heard; voices from the margins of globalization, so to speak.

The common theme of these local voices is that they add to our understanding of global geography by taking us down to ground level and providing us with local perspectives on the diverse fabric of places, people, environments, and issues that constitute world regional geography. From this view we gain insight into how local peoples are responding to the challenges of globalization.

imperialism carry a Western (even U.S.) tone—such as McDonald's, MTV, Marlboro cigarettes, or even the use of English as the dominant language of the Internet—these result more from a search for new consumer markets than from an explicit conspiracy to spread modern American culture throughout the world.

The reaction to cultural imperialism is called **cultural nationalism**, which is the process of protecting and defending a certain cultural system against diluting or offensive cultural expressions while at the same time actively promoting national cultural values and behaviors. Often, cultural nationalism takes the form of explicit legislation or official censorship that simply outlaws the offending cultural traits. Examples of legislated cultural nationalism are common. France has long fought the Anglicization of its language by banning "Franglais," the use of English words such as "Le weekend" in official French. More recently, France has also sought to protect the national music and film industries by legislating that radio DJs play a certain percentage (40 percent at this writing) of French songs and artists each broadcast day. France is also erecting tax and tariff barriers to Hollywood's products. Many Muslim countries censor Western cultural influences by restricting and censoring international TV, which they consider the cause and source for many undesirable cultural influences. Most Asian countries, as well, are increasingly protective of their cultural values and are demanding changes to tone down MTV and other international TV networks.

Even though one cultural system may be promoted at the expense of another, or a culture may resist foreign influences, a more common product than outright exclusion is the blending of forces to form a third, new synergistic form of culture. This is called **cultural syncretism** (Figure 1.24). To characterize India's culture as "British" is an oversimplified and exaggerated view of England's colonial influence. Instead, Indians have adopted many British traits by adapting them to their own circumstances, infusing them with local color and twists. India's use of English, for example, has produced a unique form of "Indlish" that often befuddles visitors to South Asia. Nor should we forget how India has added words now common to our own vocabulary: khakis, pajamas, veranda, bungalow. In sum, both cultures, Anglo and Indian, have been changed because of the British colonial presence in South Asia. While this particular illustration of cultural syncretism is more overt than many, the concept is nevertheless an important one for understanding the dynamic cultural geography of the contemporary world.

Language and Culture in Global Context

Language and culture are so intertwined that in the minds of many it is the defining characteristic that differentiates and defines cultural groups. Furthermore, since language is the

▲ **Figure 1.24 Cultural syncretism** Often, when two cultures are thrown together, a new, third culture is created, such as the case in India, resulting from the fusion of British colonial culture and local South Asian influences. This photograph shows British Victorian architecture and double-decker buses in Mumbai (formerly Bombay), India. *(Martin Jones/Corbis)*

agreed-upon means for communication, it also folds together many other aspects of cultural identity, such as politics, religion, and commerce, but also folkways and customs. Language is fundamental to the glue of cultural cohesiveness (Figure 1.25). It not only brings people together but it also sets them apart; it can be an important component of national or ethnic identity, a way of creating and maintaining boundaries necessary to the reinforcement of regional identity and character. Already noted is how cultural nationalism often focuses on protecting a national language against undesired change from "foreign" words.

Because there are common historical (and even prehistoric) roots to a number of languages, linguists have grouped together the thousands of different languages found throughout the world into a handful of *language families*. These are simply a first-order grouping of languages into large units based upon common ancestral speech. For example, about half of the world's population speak languages of the Indo-European family, which includes European languages such as English and Spanish, but also Hindi and Bengali, which are widespread languages of South Asia. Again, this grouping demonstrates that these diverse languages evolved from common linguistic roots. Within language families are smaller units that also give clues to the common geography of people and cultures. *Language branches and groups* (also called *subfamilies*) are closely related subsets within a family in which there are usually similar sounds, cognates, and grammar. Well known are the similarities between German and English, or between French and Spanish. Because of those similarities, these languages are placed into the same linguistic grouping (Figure 1.26).

Even within a given language, there is often a distinctiveness associated with specific regions or places; this, of course, is a *dialect*. Though these regional forms may have their own unique pronunciation and grammar (think of the distinctive differences, for example, between British, North American, and Australian English), they are—sometimes with considerable effort—mutually intelligible. Additionally, when people from different cultural groups cannot communicate directly in their native languages, they often agree upon a third language to enable them to communicate on specific topics. Swahili has long served that purpose in polyglot Sub-Saharan Africa, and, historically, French was the **lingua franca** of international politics and diplomacy. Today, English is increasingly the common language of international business (Figure 1.27).

A Geography of World Religions

Along with language, another extremely important and defining trait of cultural groups is religion. Indeed, with the erosion of atheistic communism in the former Soviet Union and the search for differentiating cultural identities in this era of a totalizing global culture, some argue that religion is becoming increasingly important in defining cultural identity. Recent ethnic violence and unrest in far-flung places such as the Balkans and Indonesia illustrate that point. Additionally, widespread missionary activity with its resulting conversion has both spread certain religions and aggravated religious tensions between certain groups (Figure 1.28).

Universalizing religions attempt to appeal to all peoples regardless of location or culture; these religions usually have a proselytizing or missionary program that actively seeks new converts. In contrast, there are also **ethnic religions** identified closely with a specific ethnic or tribal group. Sometimes their religion is the defining characteristic of the group, as is the case with Judaism and Hinduism, faiths that normally do not actively seek new converts. Christianity, a universalizing religion, is the world's largest, both in areal extent and number of adherents. Though fragmented into separate branches and churches, Christianity as a whole claims almost two billion people, or about a third of the world's population. The largest number of Christians can be found in Europe, Africa, Latin America, and North America. Islam, which has spread from its desert origins on the Arabian peninsula as far east as Indonesia and the Philippines, has about 1.2 billion members. While not as severely fragmented as Christianity, Islam should not be thought of as a uniform religion because it also has split into separate groups. The major sects are *Shiite* Islam, which comprises about 11 percent of the total Islamic population and represents a majority in Iraq and Iran, and the mainstream *Sunni* Islam, with a distribution running from the Arab-speaking lands of North Africa to non-Arabic language countries such as Afghanistan and Indonesia. Both of these segments are experiencing a fundamentalist revival, proponents of which are interested in maintaining a purity of faith distanced from Western influences (Figure 1.29).

Hinduism, which is closely linked to India, has about 750 million adherents and, unlike Christianity and Islam, is polytheistic, worshipping many deities. Historically, Hinduism is linked to the caste system with its segregation of peoples based upon ancestry and occupation. More recently though, because the caste system is now anathema to India's democratic ideology, these connections between religion and caste are now much less explicit than in the past. Buddhism, which derived from Hinduism 2,500 years ago as a reform movement, is widespread in Asia, extending from Sri Lanka to Japan and Mongolia to Vietnam. In its spread Buddhism fused and coexisted with many other faiths, making it difficult to accurately estimate the number of adherents. Estimates of the total Buddhist population range from 350 million to 900 million people.

Judaism, the parent religion of Christianity, is also closely related to Islam. Though tensions are often high between Jews and Muslims because of the Palestine issue, these two religions, along with Christianity, actually share historical and mythological roots in the Hebrew prophets and leaders. Judaism now numbers about 18 million, having lost perhaps a third of its total population during the systematic extermination of Jews by the Nazis during World War II.

Last, it should be noted that in some parts of the world, religious practice has declined for different reasons, giving way to *secularization,* in which people consider themselves either nonreligious or outright atheistic. Though difficult to measure, social scientists estimate that about one billion people fit into this category worldwide. Perhaps the best example of secularization comes from the former communist lands of Russia and eastern Europe, where there was overt hostility

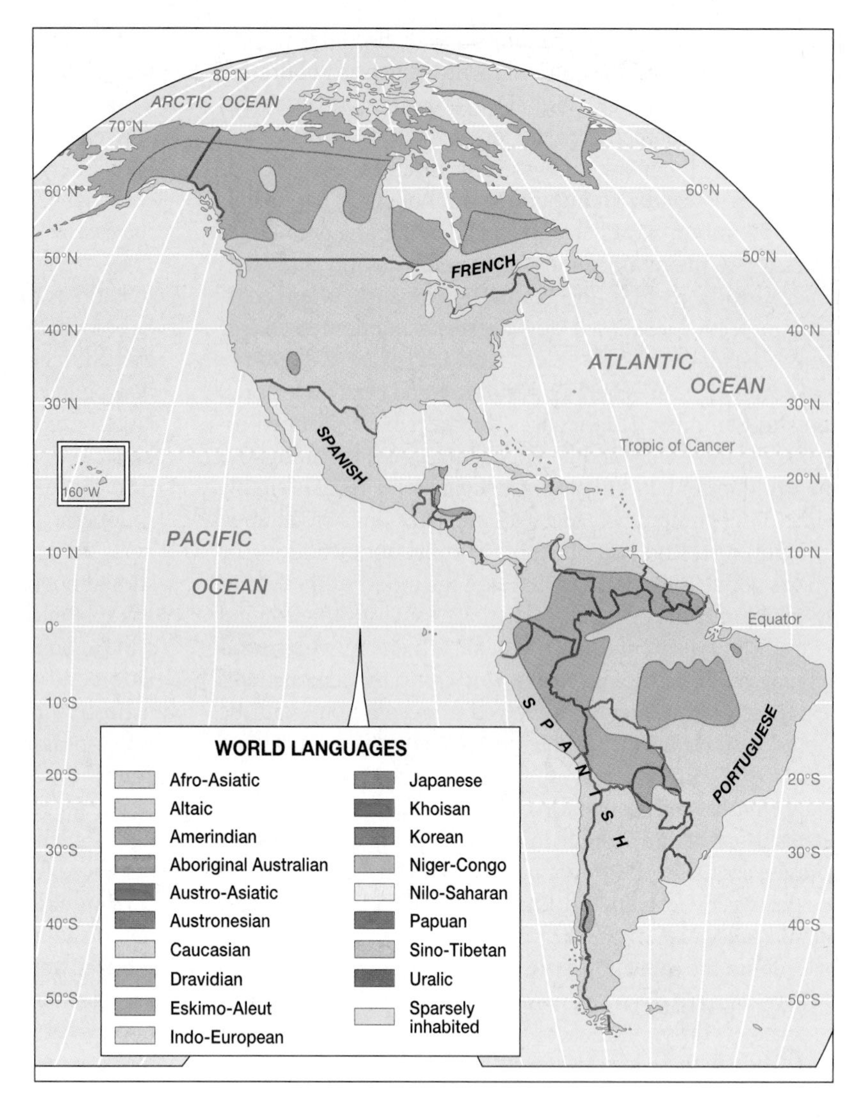

▶ **Figure 1.25 World languages** Most languages of the world belong to a handful of major language families. About 50 percent of the world's population speaks a language belonging to the Indo-European language family, which includes languages common to Europe, but also major languages in South Asia, such as Hindi. They are in the same family because of their linguistic similarities. The next largest family is the Sino-Tibetan family that includes languages spoken in China, the world's most populous country. *(Adapted from Rubenstein, 1999,* The Cultural Landscape: An Introduction to Human Geography, *Upper Saddle River, NJ: Prentice Hall)*

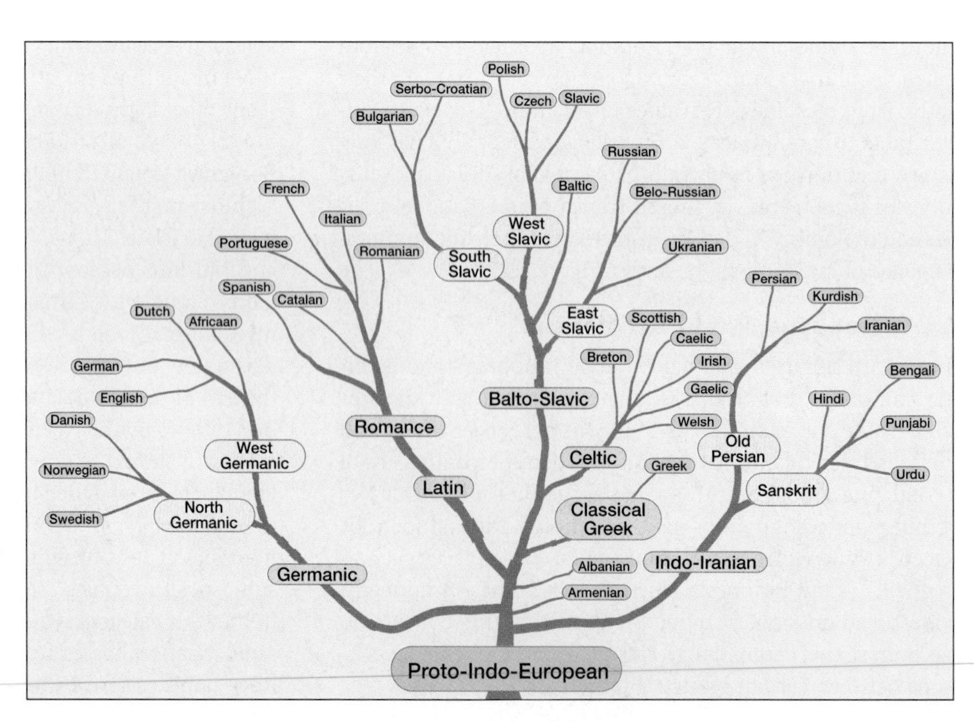

▶ **Figure 1.26 Language groups and branches** The notion of a language family, such as the one shown here for Indo-European, is that similar historical roots have imparted linguistic similarities to each group and language branch. Although the exact time and place of the origin of proto-Indo-European languages has yet to be agreed upon and remains a fascinating geographical and linguistic issue, there is little question that the Indo-Iranian languages of South Asia and Germanic languages of Europe and North America share a common root. The relationships between the later groups and branches, however, is much more clear and can be usually historically documented.

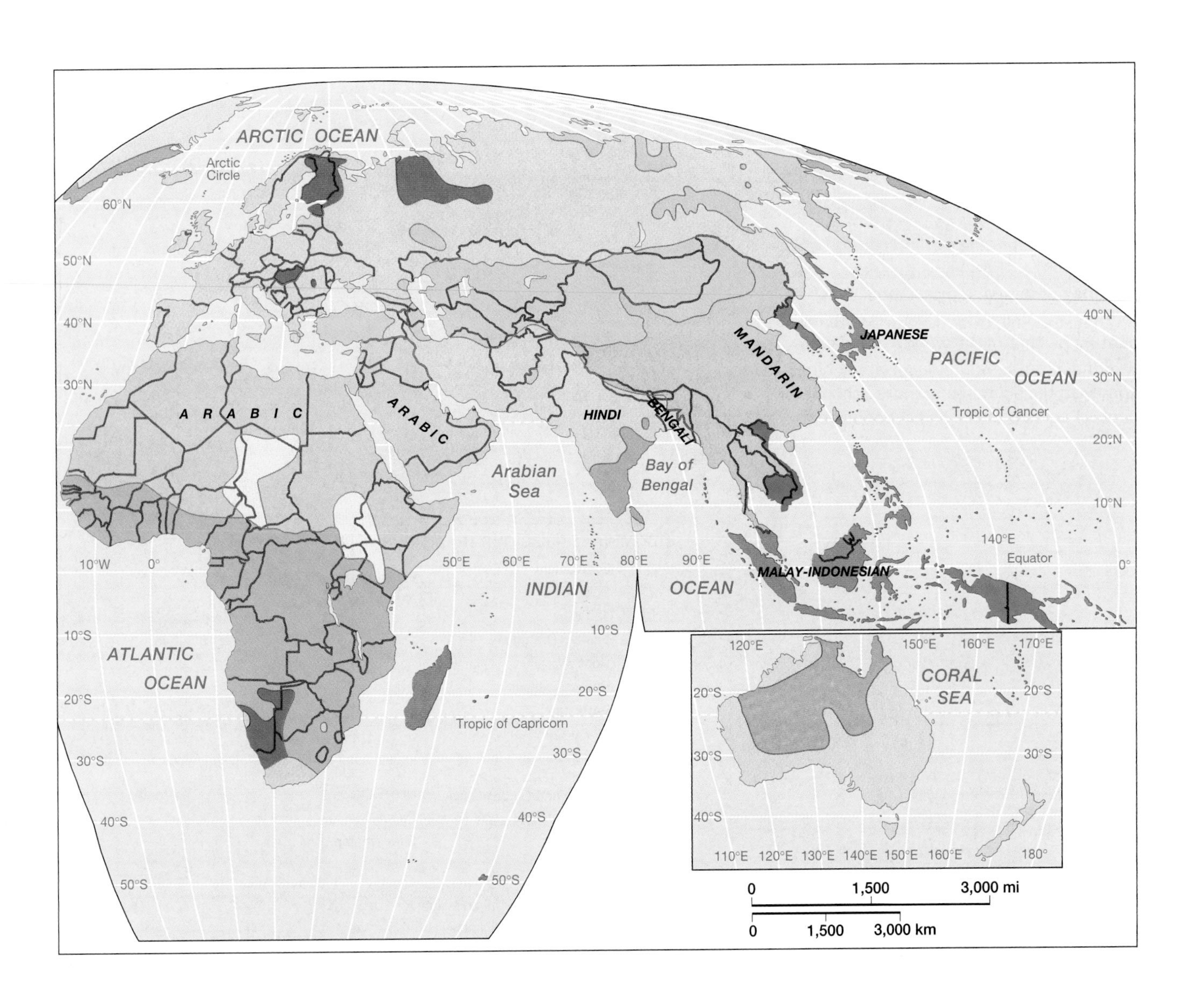

◀ **Figure 1.27 English as global language** A traffic sign in Saudi Arabia reminds us of how some familiar English phrases have become almost universal. In many ways, English has become a global lingua franca for many diverse activities, including aviation traffic communications, popular music, and road signs. *(Ray Ellis/Photo Researchers, Inc.)*

▶ **Figure 1.28 Major religious traditions** This map shows the geographical mosaic of major religious traditions found throughout the world. For most people, religious tradition is a major component of cultural and ethnic identity. While Christians of different sorts account for about 34 percent of the world's population, this tradition is highly fragmented. Within Christianity, there are about twice the number of Roman Catholics as Protestants. Adherents to Islam account for about 20 percent of the world's population. The Sunni branch within Islam is three times larger than the Shiite sect. Hindus account for about 14 percent of the global population.

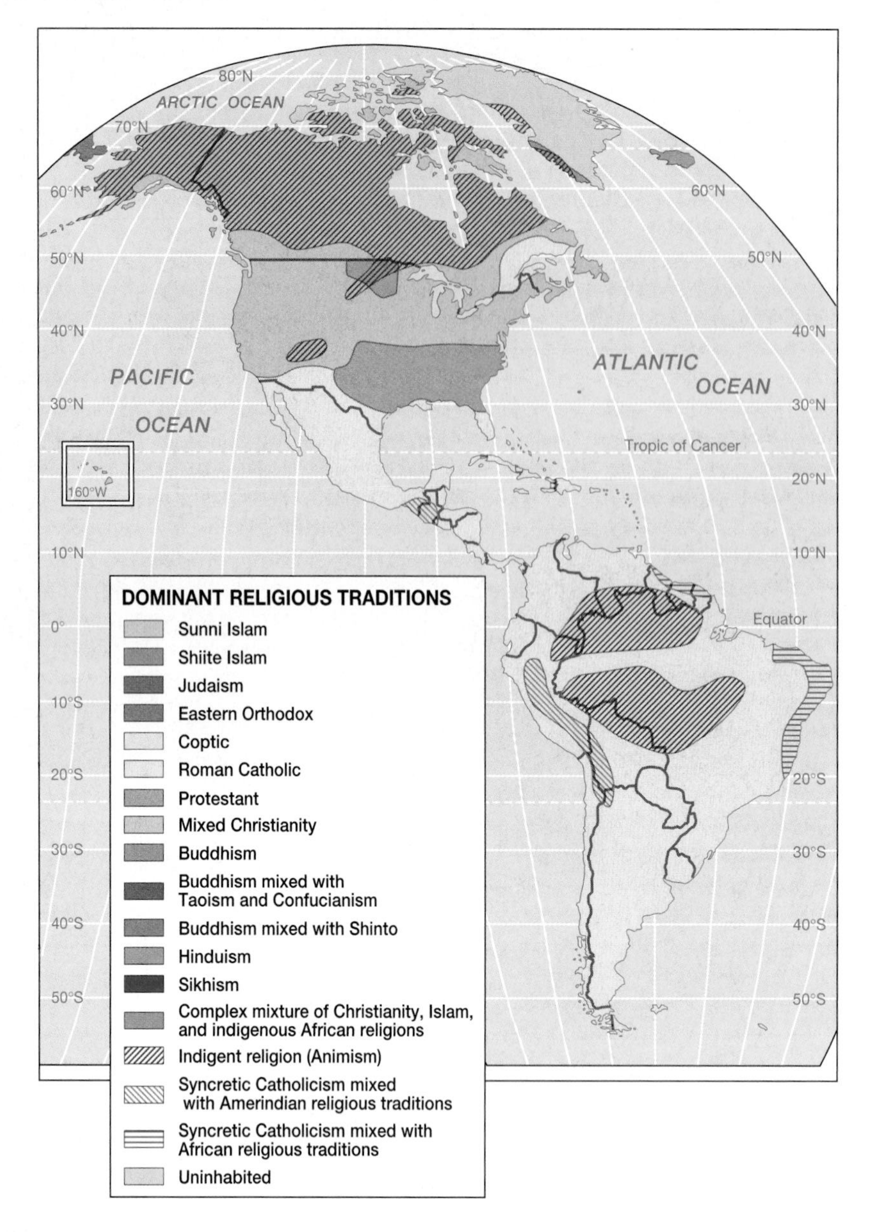

▲ **Figure 1.29 Religious landscapes** Minarets, which serve to call the faithful to prayer, surround this mosque in Istanbul, Turkey, creating an instantly recognizable landscape linked to Islam. This would be true also of the towering steeple of a Roman Catholic cathedral, regardless of world location. *(Rob Crandall/Rob Crandall Photographer)*

between government and church. Since the fall of Soviet communism in 1989, however, many of these countries have experienced a modest religious revival.

Geopolitical Framework: Fragmentation and Unity

The term *geopolitics* is used to describe and explain the close link between geography and political activity. More specifically, geopolitics focuses on the interactivity between power, territory, and space, at all scales from the local to the global, through time. One of the dominant characteristics of the late twentieth century has been the rapidity, scope, and character of political change in various regions of the world.

With the dissolution of the Soviet Union in 1991 came opportunities for self-determination and independence in eastern Europe and central Asia. As a result, there have been fundamental changes in economic, political, and even cultural alignments and alliances. While religious freedom helps

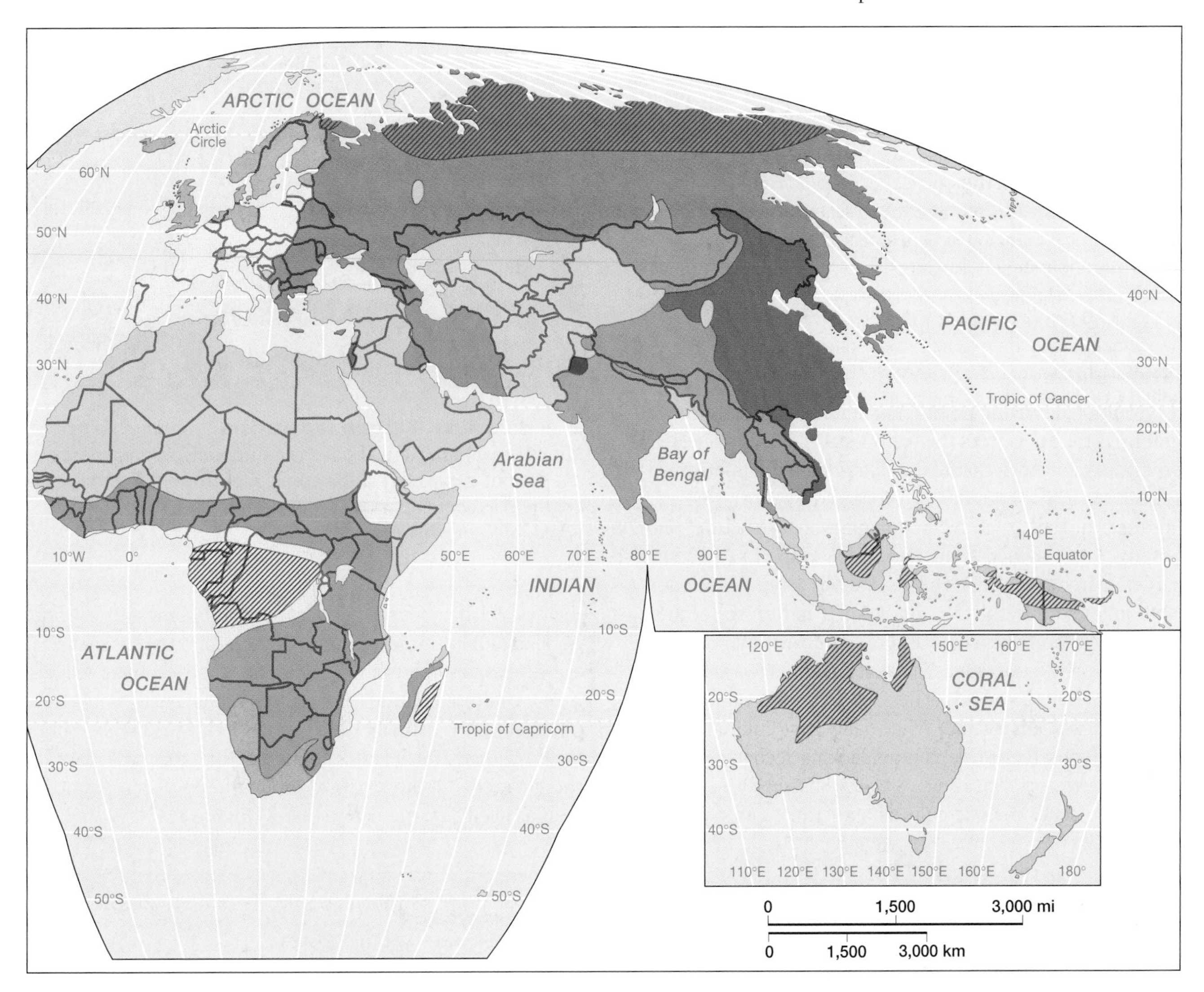

drive national identities in some new Central Asia republics, eastern Europe is primarily concerned with economic and political links to western Europe. Russia itself wavers perilously between different geopolitical pathways. Phrases and terms common to an earlier generation may disappear, and the "Cold War" between the two superpowers, the United States and Soviet Union, may become an artifact of the past (Figure 1.30). Although optimism at the end of the Cold War was captured briefly with the term "The New World Order," hopes of a new global stability were also short-lived as it became quickly apparent that contraction of the superpowers' protective concerns over their respective domains left those countries susceptible to internal tensions. Ethnic and cultural factions, giddy with the prospect of self-rule, promoted diverse and contradictory agendas that brought many countries to the brink of anarchy. Continued strife in former Yugoslavia and the Caucasus serve as examples. Factionalism, ethnic tensions, even regional separatism, then, have become increasingly widespread in the late twentieth century.

▲ **Figure 1.30 End of the Cold War** With the fall of Soviet communism, many nations rushed to remove the symbols of their former governments to forget the past and make room for a new future. In this photo, a monument to Lenin is toppled in Bucharest, Romania. *(Bi/Liaison Agency, Inc.)*

While wide-ranging international conflicts remain a nagging concern of diplomats and governments (particularly in South Asia, where the proliferation of nuclear weapons has inflamed historical animosities), perhaps more common is strife and tension within—rather than between—nation-states. Civil unrest, ethnic tensions, terrorism, the factionalism and separatism mentioned above—all of these have created a new fabric and scale of political tension. So widespread and so problematic is this tension that some scholars talk about the nation-state as an archaic form that will soon be replaced by tribal and ethnic micro-states or medieval fiefdoms.

▲ **Figure 1.31 Basque separatism** Although often outspoken and militant about their autonomy, Spanish Basques are just one expression of the ethnic separatism that is still common in contemporary Europe and, for that matter, in the world. *(Alberto Arzoz/Panos Pictures)*

Nation-states

A conventional starting point for examination of political geography is the concept of the nation-state. The hyphen links two diverse concepts, that of the *state,* which is a political entity with territorial boundaries recognized by other countries and internally governed by an organizational structure, with the term *nation,* which refers to a large group of people who share numerous cultural elements, such as language, religion, tradition, or simple cultural identity. The **nation-state,** then, is a relatively homogeneous cultural group with its own political territory. Nation-states, however, do not always fit neatly into the boundaries of states. In fact such nation/state congruence is relatively rare. While many large cultural groups consider themselves to be nations lacking recognized, self-governed territory (such as native Hawaiians in Hawaii), many, if not most, states include cultural and ethnic groups within their established boundaries who seek autonomy and self-rule. In Spain, for example, both the Catalans and the Basques seek political autonomy from the centralized government (Figure 1.31). Uniform cultural (or national) homogeneity within any political unit is actually rather rare. On a world scale, out of the more than 200 different political entities that now make up the global geopolitical fabric, only several dozen countries would qualify as true nation-states (Figure 1.32).

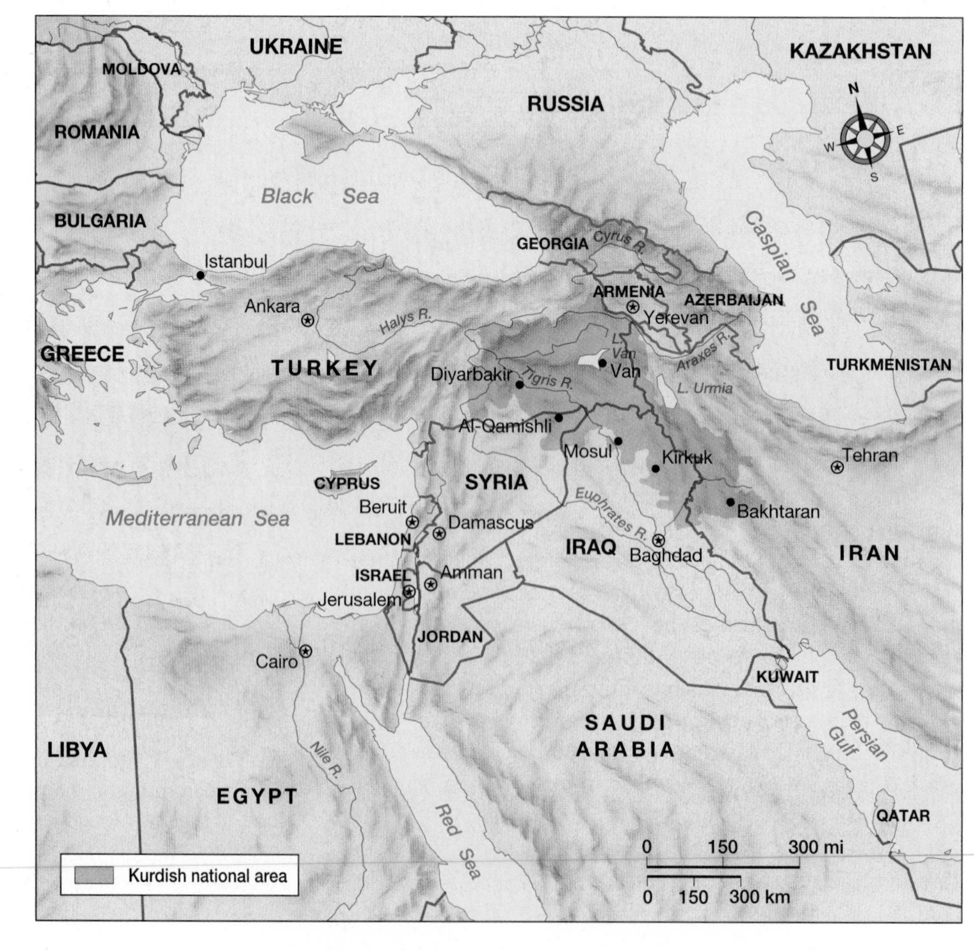

▶ **Figure 1.32 A nation without a state** Not all nations or large cultural groups control their own political territory, commonly known as a state. For example, the Kurdish people of Southwest Asia traditionally occupy a large cultural territory that is currently in four different political states: Turkey, Iraq, Syria, and Iran. As a result of this political fragmentation, the Kurds are considered minorities, and each of the these four political units attempts in its own way to limit (and even eliminate) Kurdish unity and power so as to weaken claims for an independent Kurdish state.

Though the nation-state may be the ideal political model generated in Europe centuries ago, *multinational* states containing different cultural and ethnic groups are much more numerous than homogeneous ones. Little question, though, that *micronationalism,* or a group identity with an agenda of self-rule within existing nation-states, has been on the rise for the last half century and remains today a considerable—and problematic—source of geopolitical tension throughout the world.

Centrifugal and Centripetal Forces

Those cultural and political forces acting to weaken or divide an existing state are called **centrifugal forces,** since they pull outward or away from the center. Many of these have already been mentioned: linguistic minorities, ethnic separatism, territorial autonomy, disparities in income and well-being, and so on. Separatist tendencies in French-speaking Quebec (Canada) are a good example. Counteracting these dissipating forces are those that promote political unity and reinforce the state structure. These **centripetal forces** could be a shared sense of history, a need for military security, an overarching economic structure, or simply the advantages that come from a larger political apparatus that builds and maintains the infrastructure of highways, airports, and schools.

The overriding question, however, is whether these currently modest and controlled centrifugal forces will increase in strength and furor so that they dominate the centripetal, leading, then, to new, smaller independent units. Also important is whether this process takes place through violent struggle or peaceful means. Given the current widespread nature of these internal tensions, there is much evidence to suggest these separatist struggles will continue to dominate regional geopolitics for decades to come.

Boundaries and Frontiers

A key component of a state's definition is that its territory is demarcated by borders recognized by a large number of other states. Although these boundaries may be one of the most recognizable features of a world map, the issue of agreed-upon borders is often a complicated and contentious one. Border wars have been frequent in the twentieth century; and if the current trend toward an increasing number of independent states continues, boundary disputes will become an even more important part of the geopolitical landscape (Figure 1.33).

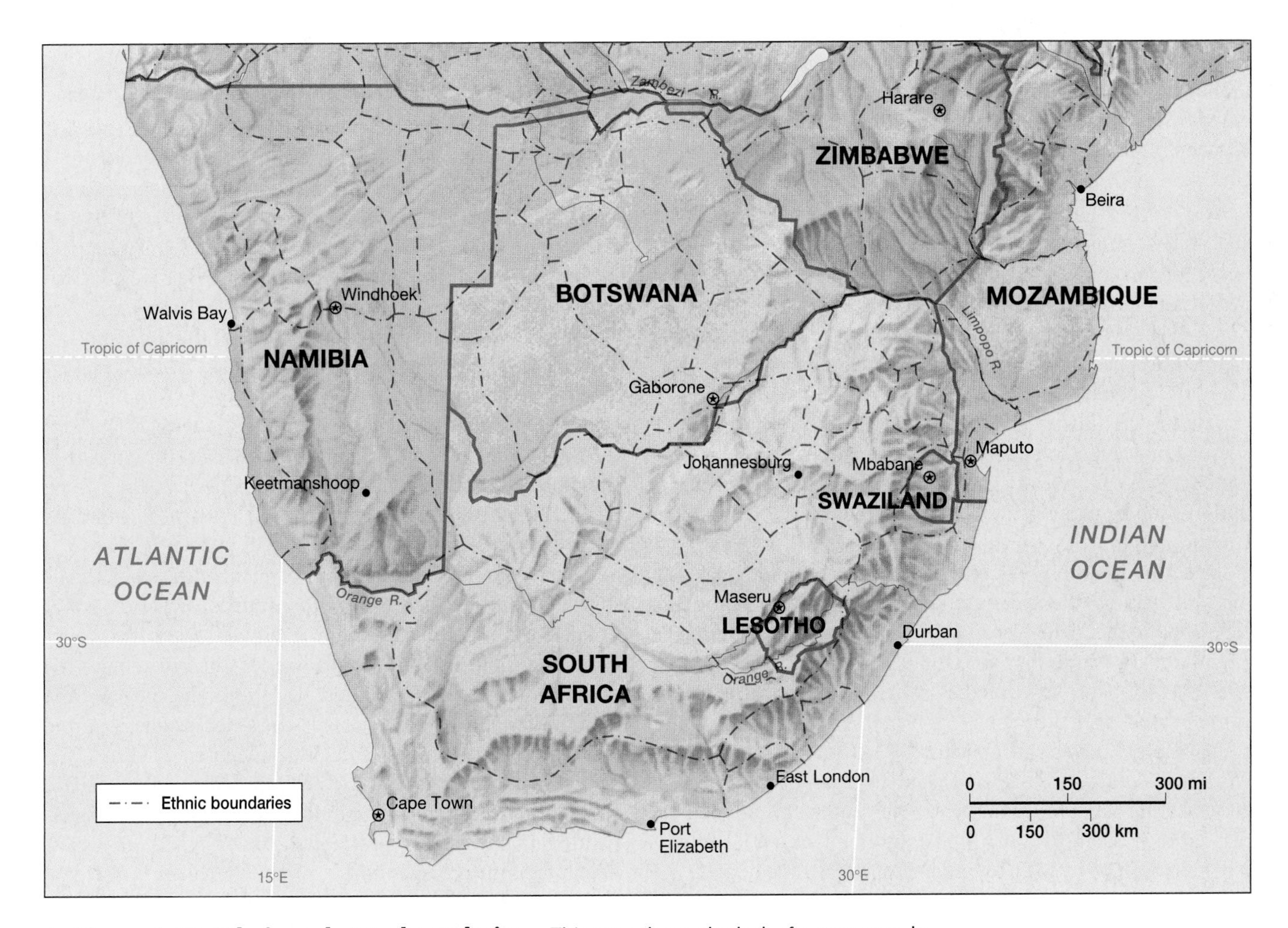

▲ **Figure 1.33 Ethnic and state boundaries** This map shows the lack of congruence between ethnic group territory and modern political borders in southern Africa. Instead, political boundaries are drawn geometrically (such as the Nambia–Botswana border) or along physical features such as rivers and mountains (South Africa–Botswana border). In very few places are there ethnographic borders that are sensitive to ethnic territories.

A cursory glance at the world political map reminds us that boundaries come in many forms. Some are drawn along physical features such as mountain ridges, rivers, or coastlines. Though historically it was thought that using clearly marked landscape features would lessen border tensions, this is not always the case. The Rio Grande river between Texas and Mexico is notoriously porous to illegal boundary crossings (Figure 1.34). Similarly, the sinuous coastline of southern Spain offers North Africans ample opportunity for illegal entry into Europe. **Ethnographic boundaries,** which follow agreed-upon cultural traits such as language or religion, have long been a model for international politics, since they complement the ideal model of the nation-state. Many boundaries in Europe were adjusted ethnographically after World War I. Similar attempts were made to define the new states of Bosnia and Serbia with ethnographic boundaries after the dissolution of the former Yugoslavia in the early 1990s.

In contrast, **geometric boundaries** are perfectly straight lines drawn without regard for physical or cultural features and usually follow a parallel of latitude or a meridian of longitude. Historically, geometric boundaries were often drawn by colonial powers as a convenient way of quickly demarcating spheres of influence; the use of the 49th parallel to draw the western boundary between Canada and the United States is an excellent example. Other illustrations can be found in the arid lands of North Africa. However, when space is divided by geometric borders without consideration of the diverse cultural fabric on the landscape, border wars and even civil wars often result.

▲ **Figure 1.34 U.S.–Mexico border** Contrasted with the international border farther to the east that is relatively easy to cross illegally, this portion of the border between the United States and Mexico near Tijuana has become a notorious symbol of the unresolved migration issues between the two countries. *(A. Ramey/Woodfin Camp & Associates)*

Colonialism and Decolonialization

One of the overarching themes in world regional geography is the waxing and waning of European colonial power over much of the world. **Colonialism** refers to the formal establishment of governmental rule over a foreign population. As a result, the colony has no independent standing in the world community but, instead, is seen only as an appendage of the colonial power. Generally speaking, the main period of colonialization by European states was from 1500 through the mid-1900s, though even today there remain a few remnant colonies (Figure 1.35).

Decolonialization refers to the process of a colony gaining (or regaining) control over its territory and establishing a separate, independent government. As was the case with the Revolutionary War in the United States, this process often begins as a violent struggle. As these wars of independence became increasingly prevalent in the mid-twentieth century, some colonial powers recognized the inevitable and began working toward peaceful disengagement. Recently, India celebrated (albeit rather modestly) a half-century of independence from Britain, and in 1997 Hong Kong was peacefully restored to China by the United Kingdom. However, decades, even centuries, of colonial rule are not easily erased in former colonies where the impress is still found in government, education, agriculture, and the economy. While some countries may enjoy special status and receive continued aid from their former colonial master, others remain disabled and disadvantaged because of a much-reduced resource base. Because the consequences of colonialism differ greatly from place to place, the final accounting is far from complete (Figure 1.36).

International and Supranational Organizations

Increasingly, international organizations that link together two or more states for the express purpose of military, trade, political, or environmental cooperation have emerged on the global scene. At the top of this organizational chart, of course, is the United Nations, which has lately become a major peacekeeping force on the international scene. Other well-known organizations are the Organization of Petroleum Exporting Countries (OPEC), which includes oil-producing countries around the world, and smaller, yet equally powerful regional groups, such as the North Atlantic Treaty Organization (NATO) and the Association of South-East Asian Nations (ASEAN). Regional trade alliances are also a common part of the global landscape, with the North American Free Trade Association (NAFTA) linking three North American countries into a free trade area.

Different from an international organization is one that is *supranational,* defined as an organization of nation-states linked together with a common goal, yet within the group, individual state power may become lessened to achieve the organization's goals. Put differently, the higher goals of the supranational organization may often necessitate a partial loss of state sovereignty by individual members (Figure 1.37). The best example is that of the European Union, or EU, which began as an economic union, yet over the years has expanded its powers so that it now holds elections, has a parliament, promotes a common monetary policy, and works toward reducing border formalities between member states. To participate fully in the EU, member states must abdicate traditional sovereignty over coinage and border control. These EU goals remain controversial, however, as discussed in Chapter 8, Europe.

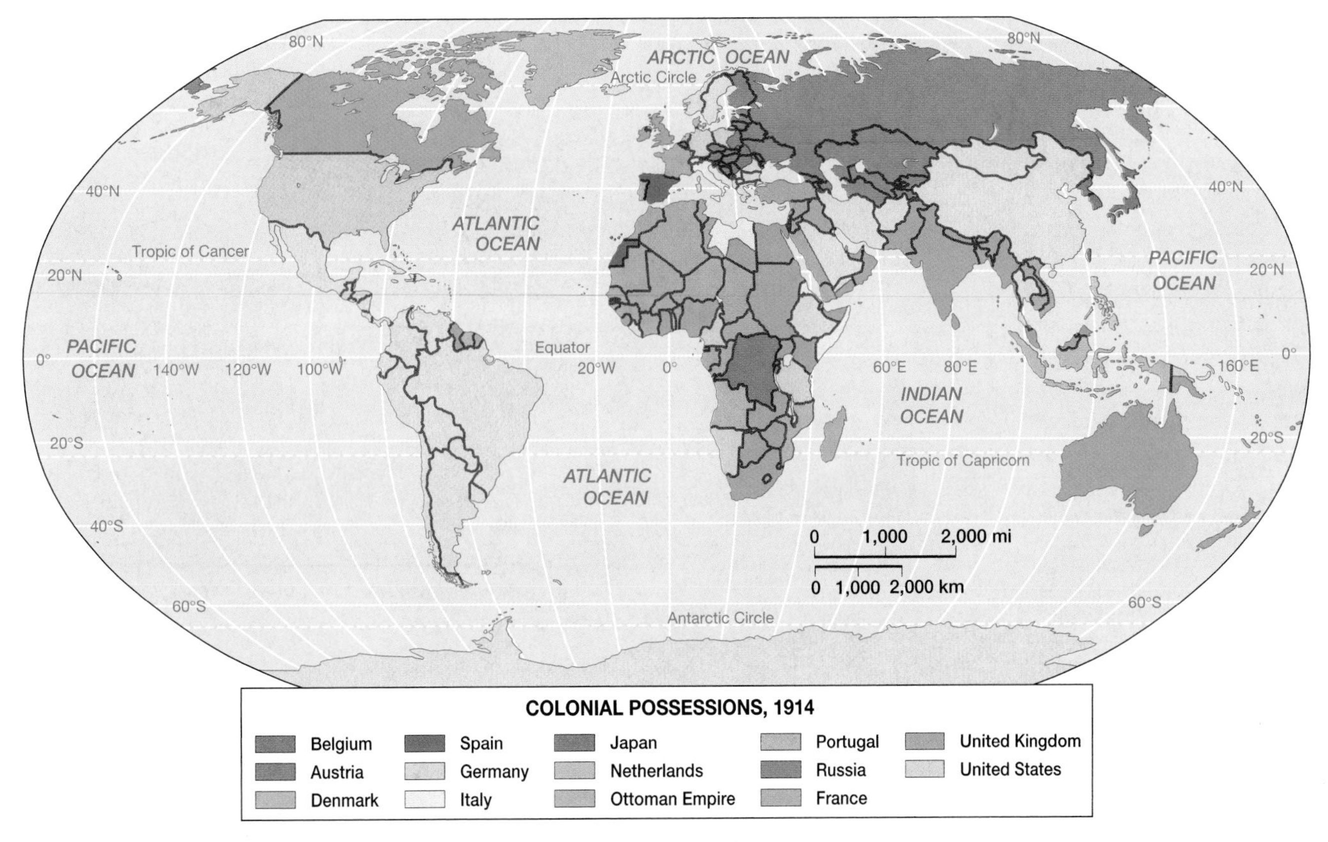

▲ **Figure 1.35 The colonial world, 1914** This world map shows the extent of colonial power and territory just prior to World War I. At that time most of Africa was under colonial control, as was Southwest Asia, South Asia, and Southeast Asia. Australia and Canada were very closely aligned with England. Also note that in Asia, Japan controlled colonial territory in Korea and northeastern China, which was known as Manchuria at that time. *(Adapted from Rubenstein, 1999,* The Cultural Landscape: An Introduction to Human Geography, *Upper Saddle River, NJ: Prentice Hall)*

◀ **Figure 1.36 Colonial remnant in Vietnam** The red star flag of communist Vietnam flies in front of the Hotel de Ville in Ho Chi Min City (formerly Saigon). This juxtaposition of the contemporary government's symbol flying from an artifact of the French colonial period captures the process of decolonialization and independence. *(Catherine Karnow/Woodfin Camp & Associates)*

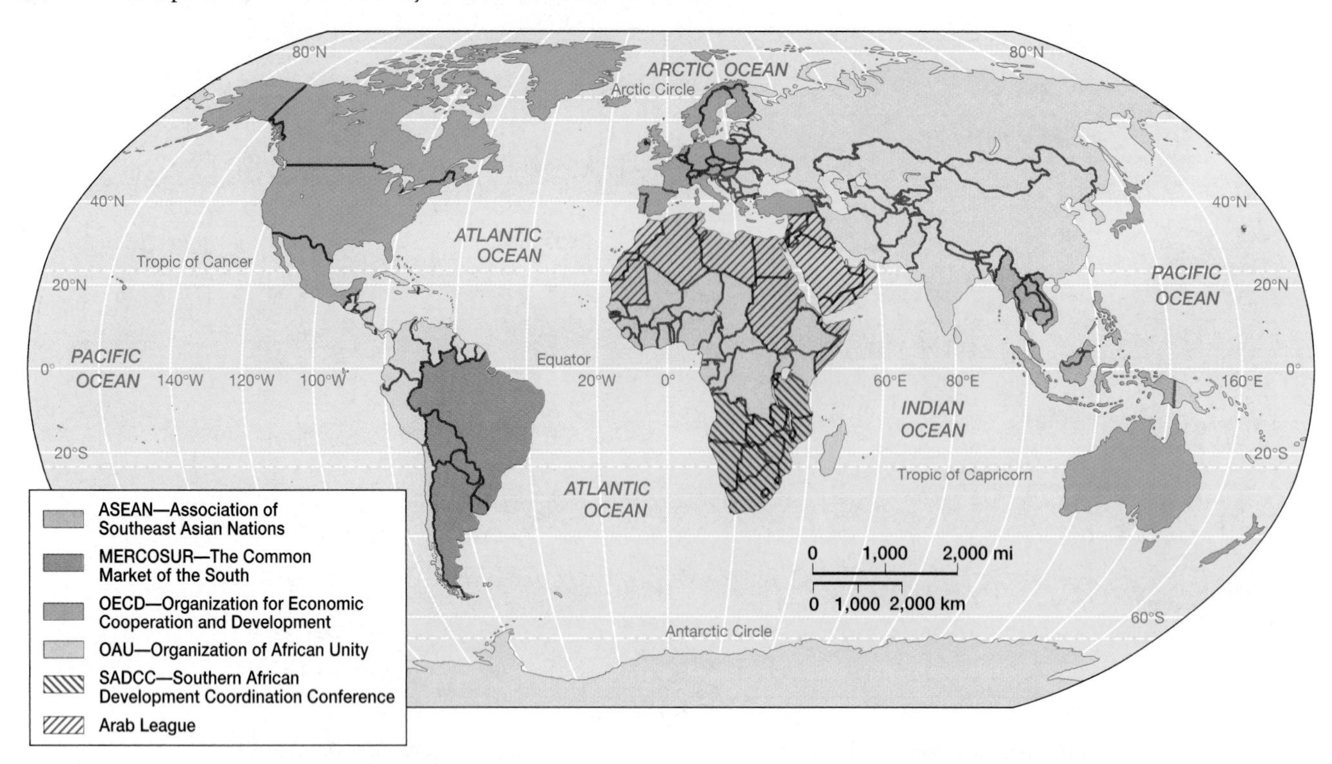

▲ **Figure 1.37 Supranational organizations** Increasingly, countries are joining international or supranational organizations that have specific goals and agendas. Sometimes this is for mutual economic and trade advantages, which is the purpose of MERCOSUR in Latin America. In other cases, it is to work on political problems, which is the case for ASEAN in Southeast Asia. Further, there are supranational organizations, such as the OAU and the Arab League, that find a common purpose in a number of cultural, economic, and political issues.

Economic and Social Development: The Geography of Wealth and Poverty

The pace and magnitude of global economic change and development has changed dramatically in the last decade so that it now moves faster and also is more widespread than before. As a result, we talk of global assembly lines, transnationals, commodity chains, and electronic offices (Figure 1.38). Few regions of the world are untouched by these changes. The overarching question, however, is whether the positive changes of economic globalization outweigh the negative. While answers are varied, sometimes elusive, and often inconclusive, an important first step in world regional geography is to link economic change to social development.

Economic development is commonly accepted as desirable change because it brings increased prosperity to individuals, regions, and nations, which, by conventional thinking at least, usually translates into social improvements such as better health care, education systems, and more enlightened labor practices. The operative assumption is that, as the economy of an area improves or develops, so will the social infrastructure in that region. One of the more troublesome expressions of the recent economic boom, though, has been the geographic unevenness of prosperity and infrastructural improvement. While some regions prosper, others languish and, in fact, fall behind so that the gap between rich and poor regions becomes ever greater. This geographic unevenness in development, prosperity, and social infrastructure has become a characteristic signature of the late twentieth century. According to the World Bank, more than 1.3 billion people live on less than a dollar a day; about 60 percent of them are in Sub-Saharan Africa and South Asia.

Additionally, these regional inequities are problematic because of their inseparable interaction with political, environmental, and social concerns. Political instability and civil strife within a nation, for example, are often driven by economic disparity between a poor periphery and an affluent, industrial core. In those backwaters, poverty, social tensions, and environmental degradation often drive civil unrest that ripples through the rest of the country. The World Bank's *World Development Reports* document these inequities. In China's Gansu province, for example, per capital income is almost 50 percent lower than the national average, and because of poor environmental conditions, most of the population lives in poverty while other regions of China prosper. Other examples of this intranational geographic unevenness are common: Italy, Poland, Mexico, Kenya, and India all have striking disparities in wealth. This theme of economic and social unevenness is a major storyline in the regional chapters that follow.

More- and Less-Developed Countries

Until recently, economic development has been centered in North America, Japan, and Europe, which has led to a conceptualization that these countries constitute the core of the world economy while all other areas make up a less-developed

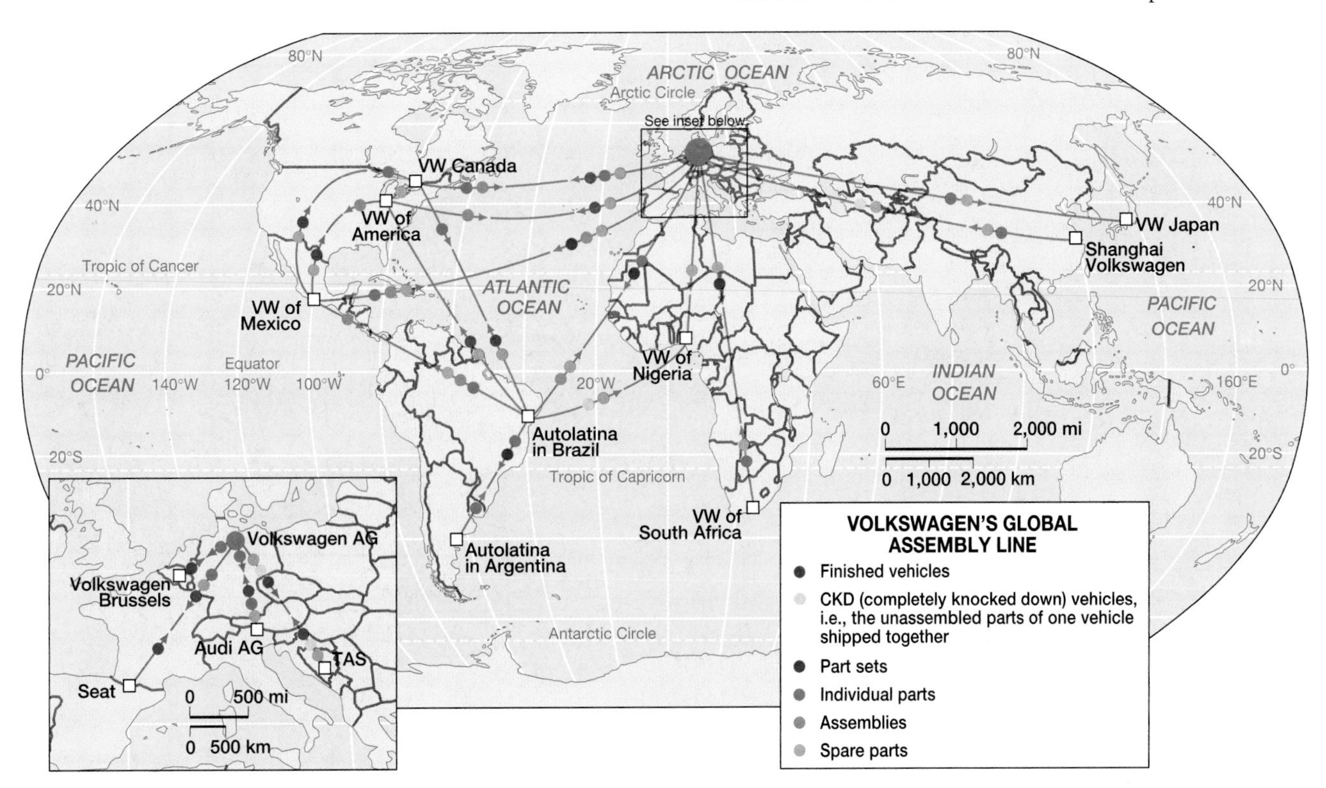

▲ **Figure 1.38 The global assembly line** One important facet of economic globalization is the increasingly common global assembly-line production. This map shows the global network of Volkswagen car processing plants and the flow of parts between them and the main assembly plant in Wolfsburg, Germany. This global network results from a number of factors, ranging from processing near-raw materials to finding the best match between labor needs and workers' wages. *(Adapted from Knox and Marston, 1998,* Human Geography, *Upper Saddle River, NJ: Prentice Hall)*

periphery within this world system. Though an oversimplification of sorts, this core-periphery dichotomy does contain some truth. All the G-7 countries—the exclusive club of the world's richest nations made up of the United States, Canada, France, England, Germany, Italy, and Japan—are located in this Northern Hemisphere core. Additionally, another assumption of this core-periphery model is that the developed core achieved its wealth by exploiting the periphery, either through historical colonial relationships or through more recent economic imperialism (Figure 1.39). Today, for example, much is made of "North–South tensions," a phrase implying that the rich and powerful countries of the Northern Hemisphere are at odds with the poor and less powerful of the south. This simplistic dichotomy is based upon the historical arrangement of northern powers—mainly European—dominating and exploiting southern colonies located in Latin America, Sub-Saharan Africa, and Southeast Asia. However, because of its geographic inaccuracies and erroneous stereotypes, many conclude the term "North–South" is useless, and therefore should be avoided.

The term *Third World* is often used to refer to the developing world, a diffuse and varied mosaic of countries that are less developed than North America, Japan, or Europe. This phrase carries connotations of a low level of economic development, unstable political organizations, and a rudimentary social infrastructure. Originally, the term Third World was a product of Cold War jargon and was used to describe those countries that were independent and not allied with either the democratic, mainly capitalist (First World) or communist (Second World) superpowers. Today, however, this term is commonly used to describe an economic level. Perhaps the more applicable terms are those that capture the complex spectrum of economic and social development with relational terms such as "More-Developed Country" (MDC), "Developing," or "Less-Developed Country" (LDC) (Figure 1.40).

Indicators of Economic Development

The terms *development* and *growth* are often used interchangeably when referring to international economic activities. There is, however, also value in keeping them separate. *Development* has both qualitative and quantitative dimensions. A dictionary definition uses phrases such as "expanding or realizing potential; bringing gradually to a fuller or better state." When we talk about economic development, then, we usually imply structural changes, such as a shift from agricultural to manufacturing activity, with accompanying changes in the uses of labor, capital, and technology. Along with these changes are assumed improvements in standard of living, education, even political organization. The structural changes undergone by Southeast Asian countries in the last several decades are good examples of development in this sense.

Growth, in contrast, is simply the increase in size of a system. The agricultural or industrial output of a country may grow, as it has for India in the last decade, and this growth

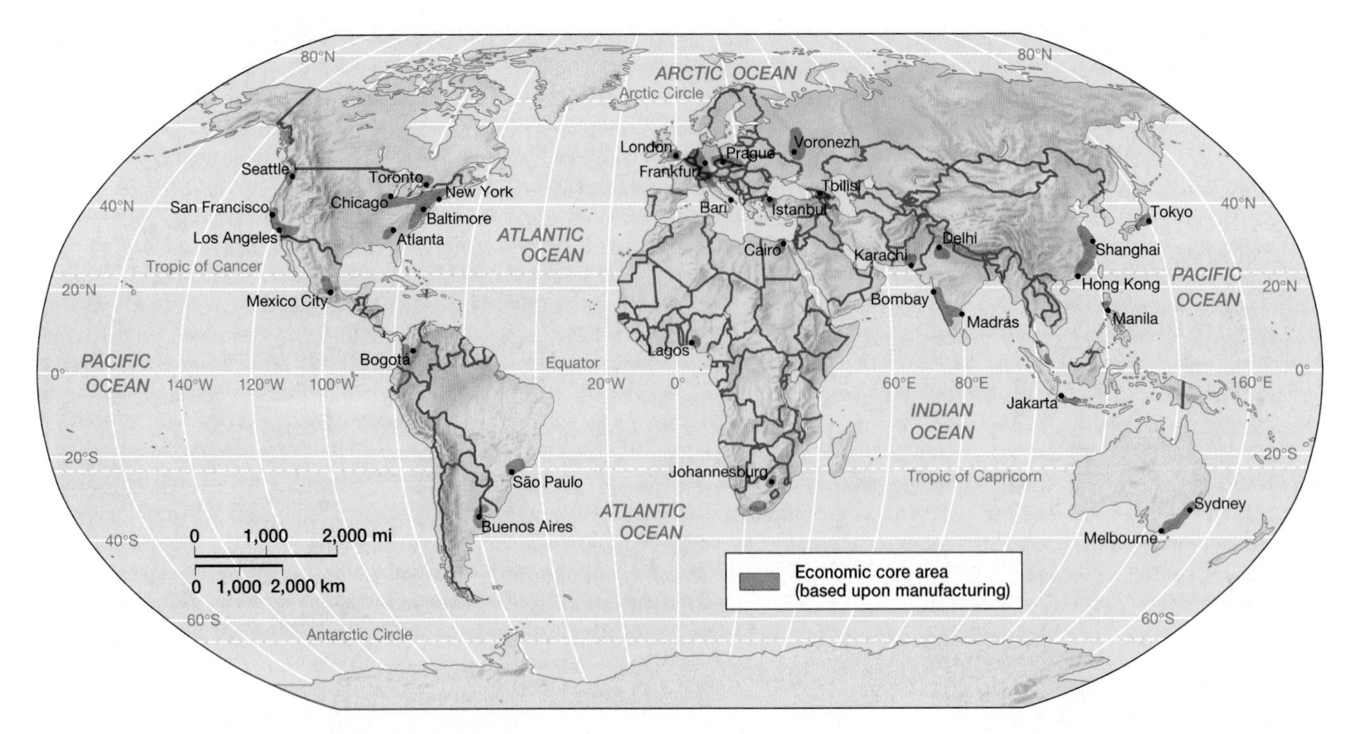

▲ **Figure 1.39 World economic core areas** Within most world regions there are centers of economic activity where manufacturing and business are clustered. Often these are the more prosperous and most urbanized areas within the region. In contrast, outlying areas within that region may lie in the shadow of these robust economic core areas, suffering from the consequences of underdevelopment. This economic diversity, both at a global scale and within regions (and even within states), has given rise to the concept of "core-periphery" interactions. One of the assumptions of this model is that economic cores prosper only by exploiting poorer periphery territories.

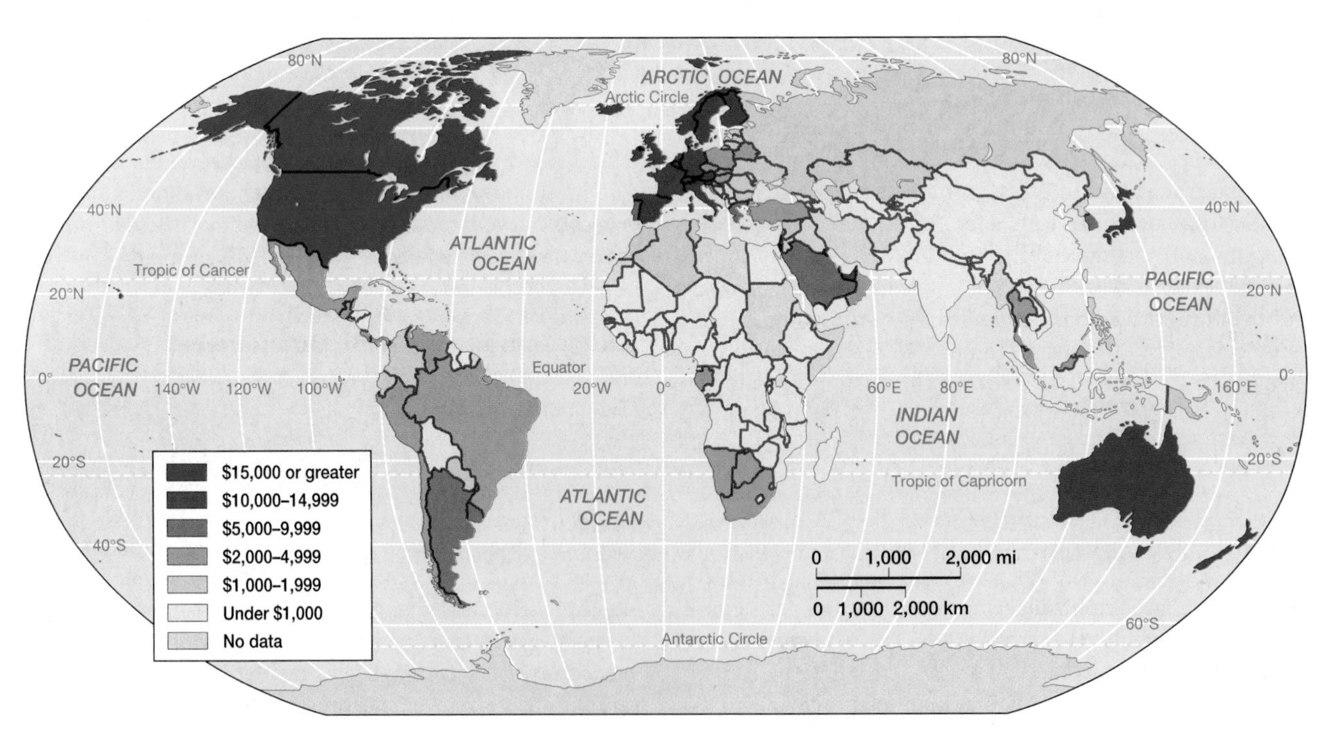

▲ **Figure 1.40 World per capita GNP** Per capita gross national product (GNP) is still used as one of the best measures for levels of economic development between countries. As can be seen on this map, the highest category of per capita GNP is associated with Europe, North America, Japan, and several Southwest Asia oil-producing countries. By economic measures, these are the "developed countries." At the other end of the spectrum, the lowest category characterizes those countries commonly referred to as "developing countries." *(Adapted from Clawson and Fisher, eds., 1998,* World Regional Geography: A Development Approach, *Upper Saddle River, NJ: Prentice Hall)*

may—or may not—have positive implications for development in that particular region. Many expanding economies have also experienced an increased amount of poverty. When something grows, it gets bigger; when it develops (sticking with the dictionary meaning, at least), the system gets better. Critics of the world economy are often heard to say we need less growth and more development. This critique should not be dismissed as unrealistic or naïve.

A table of economic development indicators is included in each of the regional chapters, but a few introductory comments are necessary to put these data in proper perspective. The traditional measure for the size of a country's economy is the value of all final goods and services produced within its borders (gross domestic product) plus the net income from abroad, which totals the *gross national product* (or GNP). Though widely used, GNP is also recognized as incomplete and even misleading in many cases because it ignores non-market economic activity such as bartering or household work, nor does it take into account degradation or depletion of natural resources that may constrain future economic growth. For example, if a country clear-cuts its forest, thereby limiting future economic growth, this would actually increase its GNP for that particular year (Table 1.2). Nor does GNP say anything about the distribution of wealth within a country, the cost of living, or the social organizations or infrastructure supporting the economy. Shutting schools and diverting that money to buying weapons would actually increase its GNP, though the economy might suffer in the future because of these acts. For this reason it is important to couple GNP data with an array of other economic and social indicators.

Until recently, GNP data were usually calculated upon the market exchange rate for that country's national currency, even though that GNP data might be inflated or undervalued depending on the strength or weakness of that currency. Today, however, most GNP figures have been adjusted by **purchasing power parity (PPP)**, which gives us a comparable figure for a standard "market basket" of goods and services purchased with the local currency (Figure 1.41). Not only does this provide a more accurate sense of the local cost of living, but these data can also be used to adjust GNP for currency inflation. For example, because conventional per capita GNP data for Japan was higher than for the United States, conclusions were drawn that the standard of living was also higher. Recent calculations based upon PPP data provide a different conclusion, suggesting the standard of living in the United States is significantly higher than in Japan.

The *economic growth rate*, which is simply the annual rate of expansion for GNP, can also be distorted by inflated currencies and, consequently, should also be standardized by PPP. China's economic growth rate, for example, averaged nearly 9 percent per year during the 1980s and early 1990s based upon conventional GNP data. When adjusted for PPP-based currency, this annual growth rate is closer to 5 percent. Increasingly, world organizations are using PPP-based measures of economic activity as data for making international comparisons about how costs of living vary from country to country. In this text, we include PPP-based data in our tables.

Indicators of Social Development

Though economic growth is a major component of development, there is equal interest in the conditions and quality of human life in this rapidly changing world. While the standard assumption is that economic development and growth will spill over into the social infrastructure so that there is also a concomitant improvement of public health, working conditions, gender roles, and education, this assumption must be supported with objective data. For that reason we include several measures of social development in the regional chapters (Table 1.3).

Table 1.2 Economic Indicators (Median Values for World Regions)

Region	GNP per Capita ($U.S., 1996)	Total GNP (Millions of $U.S., 1996)	PPP* ($Intl, 1996)	Real Annual Growth % per Capita, 1990–1996
North America	23,520	8,003,416	24,700	1.05
Latin America	2,420	16,018	6,340	1.8
The Caribbean	2,370	553	6,000	0.7
Sub-Saharan Africa	350	1,879	1,030	–0.5
Southwest Asia and North Africa	1,520	34,936	6,060	1.7
Europe	14,350	100,934	15,290	0.9
Russia and Its Neighbors	1,200	22,452	2,230	–13.5
Central Asia	550	3,642	1,970	–12.7
East Asia	8,220	366,484	10,540	5.3
South Asia	390	13,475	1,600	2.7
Southeast Asia	1,160	83,298	3,550	5.9
Australia and Oceania	2,470	5,049	4,070	1.3

*Purchasing power parity.

Source: *The World Bank Atlas,* 1998.

▶ **Figure 1.41 Village food market** The concept of purchasing power parity (PPP), which is a common market basket of goods, is used to generate a sense of the true cost of living for different countries. This outdoor produce market is in Chichicastenango, Guatemala. *(Rob Crandall/Rob Crandall Photographer)*

Table 1.3 Social Indicators and Status of Women (Median Values for World Regions)

Region	Life Expectancy at Birth		Under Age 5 Mortality, per 1,000 Live Births		Secondary School Enrollment %		Female Labor Force Participation (% of total)
	Male	Female	1960	1995	Male	Female	
North America	75	77	31	9	101	99	45
Latin America	66	72	141	34	40	49	32
The Caribbean	71	77	—	—	—	—	42
Sub-Saharan Africa	46	49	287	175	19	11	43
Southwest Asia and North Africa	68	71	233	54	66	60	28
Europe	72	78	36	10	91	90	44
Russia and Its Neighbors	62	74	—	28	80	95	49
Central Asia	63	71	—	66	89	91	45
East Asia	72	78	124	32	94	96	45
South Asia	50	59	236	128	46	23	40
Southeast Asia	63	69	216	111	56	39	43
Australia and Oceania	68	73	—	—	68	64	43

Sources: *Population Reference Bureau Data Sheet, 1998,* Life Expectancy (M/F); *World Resources Institute, 1998–99,* Under-5 Mortality Rate; *Population Reference Bureau Data Sheet, 1996,* Secondary School Enrollment (M/F); *The World Bank Atlas, 1998,* Female Participation in Labor Force.

One indicator of social development is *life expectancy,* which, as the term suggests, is the average length of life expected at birth for a hypothetical male or female based upon national death statistics. A large number of social factors influence life expectancy, such as availability of health services, nutrition, prevalence (or absence) of disease, accidents, sanitation, and homicide. Additionally, life expectancy is affected by biological factors, which may explain why females often live longer than males. However, when social and biological factors are folded together, life expectancy data does provide important insights into local conditions.

Life expectancy figures vary widely between countries in different world regions, and, in general, they have been improving over time. For the world as a whole, life expectancy in 1975 was 58 years, whereas today it is 66. Over the same period, Africa's has gone from 46 to 55, although in some African countries this figure has dropped recently because of the AIDS epidemic. In Russia, also, life expectancy has fallen lately with the deterioration of economic conditions. This compares to western Europe, which has been almost static at 73 years. Nonetheless, national life expectancy data often mask differences within a country, such as the wide disparity between rich and poor, the differences between ethnic groups, or even the differences between specific regions and places. Within the United States, for example, the gap in life expectancy between African Americans and whites is seven years, although this is an improvement over the 15-year gap recorded in 1900. Local-scale data also tell important stories. For example, because of the high homicide rates for African American males in major urban places such as New York City, Detroit, and Chicago, their calculated life expectancy is shorter than that of a male in Bangladesh, one of the world's poorest countries.

The *mortality rate under 5 years,* which is the measure of the number of children who die per 1,000 of the population, is another important indicator of social conditions since it gives insight to conditions such as food availability, health services, and public sanitation. In the first five years of life, a child moves from the personal protection and nurturing provided by the mother into a larger social environment; these first five years are a time of high vulnerability to nutritional deprivation, infection and disease, accidents, and other human tragedies. These child mortality data are given for two points in time, 1960 and 1993, to indicate whether there is a clear trend or improvement over the last 30 years.

Secondary school enrollments, which is given as a percentage of males and females in a country, provide us with information on, first, the availability of schools, and second, whether this educational system is open to both sexes. Where percentages for both sexes are low, we can assume that social capital or political will is lacking to develop the educational sector. When the percentage of females attending secondary school is considerably lower than males, we can assume that cultural constraints come into play that limit female education (Figure 1.42).

In many countries, women play a prominent role in economic development; thus, data on the *percentage of females in the labor force* are important indicators. However, three points must qualify these data. First, women's work has traditionally been undervalued and hidden from the national census because it is considered part of the unpaid sector of household and subsistence. Therefore, the long hours women spend gathering firewood, drawing water, cooking, tending gardens and fields, and raising children are not tabulated as "work" or "employment" and do not appear in the labor force statistics. Second, many women work in the "informal" economic sector, earning subsistence wages selling vegetables or handicrafts in village markets, cleaning houses, or doing other forms of temporary work. National employment data rarely include this important work. Third, even when employed in the formal economic sector, such as on a factory assembly line, women are usually paid significantly lower wages than men and are also disproportionately concentrated in jobs with little security or chance of advancement (see "Women in the Workforce: The First to Go, as Asian Economies Shrink"). Moreover, this work is usually in addition to the traditional household and family responsibilities, domestic tasks, and multitude of responsibilities captured by the phrase "the second shift."

Even though women may comprise the majority of workers in developing countries, the quality and conditions of their work often mimic the early and unpleasant days of European industrialization, when human beings worked long hours under dangerous conditions for low wages with little job security. While it took Europe almost a century to enact legislation to improve these working conditions, it seems reasonable to expect that the economic, political, and social currents of the early twenty-first century might move faster to provide humane working conditions (Figure 1.43).

▲ **Figure 1.42 Women in secondary school** One important indicator of social development and status of women in a given country is the percentage of females enrolled in secondary school. When considerably lower than the same figure for males, one can infer there are cultural and political factors that constrain a woman's life. *(Ron Giling/Panos Pictures)*

The Vision of Sustainable Development

As the environmental and social costs of globalized economic development become increasingly apparent, there is widespread discussion of the notion of **sustainable development** on the part of international aid workers, development planners, environmentalists, and community groups. This can be defined simply as an agenda of economic change and growth

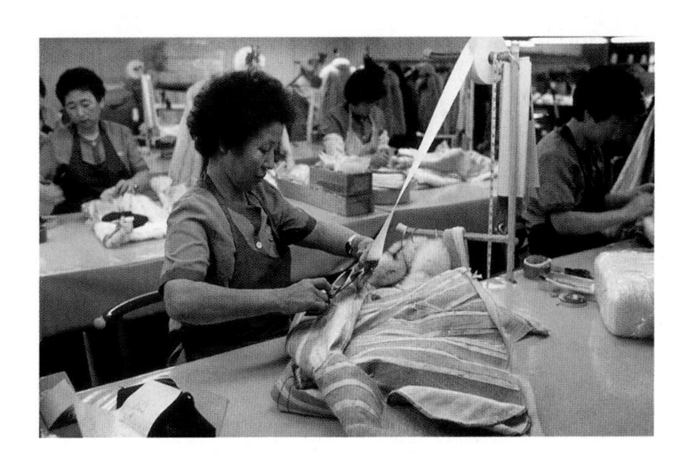

▲ **Figure 1.43 Women in the workforce** Women workers have become an important component of the modern globalized economy. Women are often paid less than men, and usually don't have the job security to prevent them from being the first be fired or laid off as the global economy fluctuates. *(Nathan Benn/Woodfin Camp & Associates)*

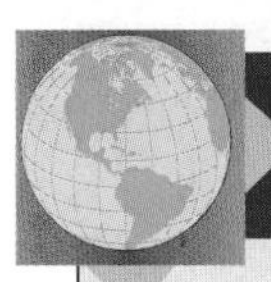

WOMEN IN THE WORKFORCE The First to Go, as Asian Economies Shrink

As globalization led to booming Asian economies in the 1990s, more and more women were brought into the workforce; while some women found jobs as assembly workers in new offshore factories, other women were given managerial responsibilities in jobs previously reserved for males; still others were hired as office ornaments—"flowers of the workplace," as they're called in Japanese.

In the late 1990s, though, with fiscal crises and shrinking economies badgering many Asian countries, these women were the first to be fired. As a result, the Asian financial crisis was particularly disastrous for women. "The impact on women and girls is just catastrophic," said the American envoy to the Asian Development Bank in the Philippines. From the high-rise office suites of Japan to the peasant villages of Thailand, the bottom line is that resources have also been allocated disproportionately to men and boys. During the good times of the Asian boom, there were plenty of leftovers for women; but in lean times, it is the women who pay the most.

Much of this discrimination comes from the revered status of the male in Asian cultures, with companies and firms attempting to minimize the pain of layoffs in a male-dominated family structure. "In a crisis, first of all we would have to fire the women," said the owner of a Japanese construction company. "We would retain men because they are the pillar of household earnings. . . . I would never want to dismiss the main income-earner because that might destroy a whole family." When the Japanese economy first slowed in the 1990s, the job market for women was referred to as *hyogaki* or "ice age"; now Japanese women call it the *cho-hyogaki* or "super ice age."

Even those females not yet in the workforce are affected. Young girls are pulled out of school in rural Indonesia because families can't afford the $2 charge for uniforms. "People say it's better for girls to stay at home, so that they can save money for the boys," says a 17-year-old girl on the island of Sumba. In Indonesia girls are six times more likely than boys to drop out of school before completing the fourth grade. Once they drop out of school, rarely do they go back, and many of these dropout girls are then pressured or sold into prostitution.

The problem is not limited to rural villages. Even in the cities of South Korea and Japan, families are pulling their daughters out of private cram schools in order to allocate this money to the boys who will carry on the family name. Is that fair? Many Asian women appear resigned to the situation. "It would be best if everyone could get opportunities," said one South Korean woman, "but I think it's right that a son gets the most attention."

Source: Adapted from "With Asia's Economies Shrinking, Women Are Being Squeezed Out," by Nicholas D. Kristof. *New York Times*, June 11, 1998.

that seeks a balance with issues of environmental protection and social equity so that the short-term needs of contemporary society do not compromise those of the future. Some reduce the definition to its essence by referring to sustainable development simply as "intergenerational equity." For example, instead of wholesale harvesting of Indonesian rain forests as a way of amortizing current foreign debt, a sustainable development plan would emphasize the paced, longer-term exploitation of this valuable resource so that the economic benefits from cutting the forest might be spread out in a more equitable manner over time so that it is self-sustaining.

An explicit assumption of sustainable development is the notion of carrying capacity, the biological concept that defines the maximum number of users (humans, insects, animals) that can be supported by an ecosystem without compromising or radically depleting a shared resource base. Though some skeptics of sustainable development maintain this vision is elusive and unattainable, that is not the case when sustainability is applied to a particular resource system over a specific period of time. While forests provide the best example of sustainability for a renewable resource system (at least when the harvesting rate is the same as the regrowth rate), even the exploitation of nonrenewable resources, such as mining or oil extraction, can also be paced over a longer period of time so that economic and social benefits are shared by a greater number (Figure 1.44).

▲ **Figure 1.44 Sustainable development in Ireland** Partially in reaction to the economic, social, and environmental costs of globalization, there is increasing interest in sustainable development—resource use in which the future is not sacrificed to the present. *(Porterfield/Chickering/Photo Researchers, Inc.)*

Another key concept underlying sustainability, albeit a more radical one, is that of "localization." This refers to the orienting of economic development away from the international markets and transnational corporate profits accompanying globalization and, instead, placing more emphasis on local economic needs. Additionally, "localization" empowers regions with genuinely participatory means for groups to take part in shaping their own economic, social, and environmental futures. The European Union (EU) has been particularly active in working through the details of sustainable development through their program, "Landscape and Life: Appropriate Scales for Sustainable Development," which currently conducts pilot studies of sustainable agricultural methods in Ireland, Germany, the Netherlands, and Sweden.

Conclusion

As regions and countries become increasingly interdependent, globalization is driving a fundamental reorganization of the world's economies, and cultures, through such vehicles as trade agreements, supranational organizations, military alliances, and cultural exchange. At one level the world seems to be converging and becoming more homogeneous because of globalization. As Southeast Asian economies rise and fall, for example, these financial fluctuations ripple through other regions, countries, and cities. A complicated fabric of global-to-local connections characterize the turn of the century's world geography. Nonetheless, even with the convergence accompanying globalization, there is still great diversity in this world. While some areas prosper, others stagnate; as global TV promotes a common world culture, small group identity becomes increasingly important; as nation-states become world players through supranational agreements, separatist groups seek autonomy and independence. This tension and interplay between globalization and diversity—between the global, the regional, and the local—gives world regional geography its focus.

Our description and analysis of each world region is organized around five issue-oriented themes: Environmental Geography; Population and Settlement; Cultural Coherence and Diversity; Geopolitical Framework; and Economic and Social Development. Though the specific issues differ somewhat from region to region, there are certain general themes that link the chapters together. Environmental issues, be they at the global or local scale, pervade all world regions. Examples are global climate change, water pollution, air pollution, forest destruction, and the protection of biodiversity. Because of their importance, the next chapter, "The Changing Global Environment," discusses these issues in more detail.

In most regions of the developing world, population and settlement issues revolve around four issues: rapid population growth; family planning; migration to new centers of economic activity (both within and outside the region), and the rapid pace of urbanization, along with the concern about whether cities can keep up with the ever-increasing demand for jobs, housing, transportation, and public facilities. Another major theme in our treatment of global cultural geography is the tension between the forces of cultural homogenization resulting from globalization and the countercurrents of small-scale cultural and ethnic identity. Throughout the world small groups are setting themselves apart from larger national cultures with renewed interest in ethnic traits, languages, religion, territory, and shared histories. This is not always a simple a matter of colorful folklore and revitalized local customs; in many parts of the world, this cultural diversity is translated into politics with outspoken calls for regional autonomy or separatism. These expressions are particularly pronounced in those regions of the world recently freed from the cultural constraints of Soviet communism or European colonialism.

Because of cultural factionalism, the geopolitical issues of many of the world's regions are dominated by matters of ethnic strife and territorial disputes within the nation-state, border tensions with neighbors of different cultural traditions (Pakistan and India come to mind as examples), and fabricating new kinds of military alliances and agreements to deal with the ever-changing nature of national security. The ideological differences between former Cold War global superpowers now seems eclipsed by more localized internal and regional tensions, including the reaction, response, alignment, and alliances of external countries and organizations. The tensions between NATO, the United States, and Russia over the Kosovo war in 1999 serve as an example.

Last, the theme of economic and social development is dominated by one issue: the increasing disparity between the rich and the poor, between countries and regions that already have wealth—and are getting even richer through globalization—and those countries and regions that do not. Economic inequities are everywhere and are found at every scale from global to local, including pockets of poverty within the world's richest countries. At a global scale there is the core of developed countries surrounded by a periphery of less-developed nations aspiring to the same success; at the smaller scale of regions and individual nations, one finds vibrant cores and centers of economic development linked to the globalized world, while more remote backwater areas languish and stagnate. Often the same blatant inequities in social development, schools, health care, and working conditions accompany these disparities in wealth.

Understanding this complex world is a challenging, yet necessary task. Think of this book as a beginning rather than an end, as a way to gain skills in using the conceptual tools of geography to engender a critical thinking about the complicated issues and themes of world regional geography related to diversity amid globalization.

Key Terms

areal differentiation (page 7)
areal integration (page 7)
centrifugal forces (page 29)
centripetal forces (page 29)
colonialism (page 30)
cultural imperialism (page 20)
cultural landscape (page 8)
cultural nationalism (page 22)
cultural syncretism (page 22)
culture (page 20)
decolonialization (page 30)
demographic transition (page 11)
ethnic religion (page 23)
ethnographic boundaries (page 30)
geometric boundaries (page 30)
globalization (page 1)
lingua franca (page 23)
nation-state (page 28)
natural increase (page 11)
overurbanization (page 18)
purchasing power parity (PPP) (page 35)
squatter settlements (page 18)
sustainable development (page 37)
total fertility rate (TFR) (page 11)
universalizing religion (page 23)
urban form (page 18)
urban structure (page 18)
urbanized population (page 16)

Questions for Summary and Review

1. Why is globalization a signature of twenty-first century, and what has facilitated this?
2. What role did the end of the Cold War play in furthering globalization?
3. What geographic concept(s) best captures the appearance of places and the "look of the land"?
4. Give examples of how changing scale from "global-to-local" might change an analysis of areal differentiation and integration for a specific area.
5. If the rate of natural increase (RNI) for the world remains consistent at 1.4 percent per year, what will be the planet's population in 10 years?
6. Explain the four stages of the demographic transition. How do birthrates and death rates differ in each stage? At what point is RNI highest?
7. Why is there an association between urbanized population rates and economic development in different countries of the world? Under what conditions would this association not hold?
8. What are the expressions and implications of overurbanization for cities in the developing world?
9. Why is it difficult to define "culture" in this globalized world? What conceptual and geographic solutions are helpful to solving this problem?
10. What are the differences between language families, branches, and groups? Give examples by noting their geographic distribution in the world.
11. Using the concepts presented in the section on "Geopolitical Framework," explain what happened to the "New World Order."
12. Using a world map or atlas, come up with five examples each of ethnographic and geometric boundaries. Is there any clue as to the problems associated with these different borders?
13. What are the shortcomings of GNP? How are these rectified?
14. Why are secondary school enrollments for men and women a good indicator of social development?
15. What are the strengths and shortcomings of sustainable development in a globalized world?

Thinking Geographically

1. Select an economic, political, or cultural activity in your city and discuss how it has been influenced by globalization.
2. What natural hazards create the most problems in your local area? What groups are most (and least) vulnerable to these problems? How have these hazards and vulnerabilities changed over the last several decades?
3. Using the concepts of areal integration and city systems, discuss the global links between your city and other urban areas, both nationally and internationally.
4. How would you characterize your location in terms of regions? That is, in what physical, economic, and cultural regions are you located?
5. How important is migration, both nationally and internationally, to your locality? Using the concepts of push and pull forces, along with that of informational networks, discuss why people leave, and arrive, how they get jobs, and how many find housing in your area.
6. Compare the issues and problems with urban growth in your city to that of a similarly sized city in a foreign country. Think, too, about how urban structure and form might be similar and different between these two cities.
7. Discuss the cultural geography of your area, noting the distribution, landscapes, and interaction associated with different cultural groups.
8. Drawing upon information in current newspapers, magazines, TV, and the Internet, apply the concepts of cultural imperialism, nationalism, and syncretism to a region or place experiencing cultural tensions.
9. Choose a large nation-state of interest. Then elucidate the different centrifugal and centripetal forces within that country. Based upon your findings, discuss the future of that country in, say, 10 years.
10. Using the concept of ethnographic borders as a focal point, critique the way boundaries have been redrawn recently in

the former Yugoslavia. Note those areas where the new boundaries seem to work well contrasted with areas where there are still ethnic tensions.

11. Apply the core-periphery concept to a country of your choice by delimiting what you think to be cores of economic development and contrasting them with peripheries of much lower development. If possible, collect and analyze data that test your findings.

12. Using the data table of social indicators found in the regional chapters of this book, try to determine what traits are shared by those countries where there is a low percentage of women enrolled in secondary schools. What general statements result from your inquiry?

13. Apply—and critique—the definition of sustainable development to these projects: (a) a large dam and hydroelectric project in the Amazon; (b) clear-cut timber harvesting in British Columbia, with logs sold to Japanese lumber firms; (c) intensive usage of chemical fertilizer and irrigation water on alfalfa agriculture in the Colorado high plains; (d) agroecology in the Amazon rain forest.

Bibliography

Anderson, Benedict. 1983. *Imagined Communities: Reflections on the Origin and Spread of Nationalism.* London: Verso.

Barber, Benjamin R. 1995. *Jihad vs. McWorld.* New York: Times Books.

Buttimer, Anne. 1998. "Close to Home: Making Sustainability Work at the Local Level." *Environment* 40(3), 12–40.

Cohen, Saul. 1991. "Global Geopolitical Change in the Post-Cold-War Era." *Annals of the Association of American Geographers* 81, 551–580.

Cosgrove, Denis. 1994. "Contested Global Visions: One-World, Whole Earth, and the Apollo Space Photographs." *Annals of the Association of American Geographers* 84, 270–294.

Dickinson, Robert E. 1969. *The Makers of Modern Geography.* New York: Frederick A. Praeger.

Fukuyama, Francis. 1992. *The End of History and the Last Man.* New York: Avon.

Gallopin, Gilberto C., and Raskin, Paul. 1998. "Windows on the Future: Global Scenarios and Sustainability." *Environment* 40(3), 6–31.

Gilbert, A. 1988. "The New Regional Geography in English and French-Speaking Countries." *Progress in Human Geography* 12, 208–228.

Gupta, A., and Ferguson, J. 1992. "Beyond 'Culture': Space, Identity, and the Politics of Difference." *Cultural Anthropology* 7, 6–23.

Huntington, Samuel P. 1993. "The Clash of Civilizations?" *Foreign Affairs* 72, 23–49.

Huntington, Samuel P. 1996. *The Clash of Civilizations and the Remaking of World Order.* New York: Simon & Schuster.

Kane, Hal. 1995. *The Hour of Departure: Forces That Create Refugees and Migrants.* Washington, DC: Worldwatch Institute.

Kaplan, Robert D. 1994. "The Coming Anarchy." *The Atlantic Monthly,* 273(2), 44–76.

Katzner, Kenneth. 1995. *The Languages of the World.* London: Routledge.

Knox, Paul, and Marston, Sallie. 1997. *Human Geography: Places and Regions in Global Context.* Upper Saddle River, NJ: Prentice Hall.

Lewis, Martin W. 1991. "Elusive Societies: A Regional-Cartographical Approach to the Study of Human Relatedness." *Annals of the Association of American Geographers* 81, 605–626.

Lewis, Martin W., and Wigen, Karen. 1997. *The Myth of Continents: A Critique of Metageography.* Berkeley: University of California Press.

Mander, Jerry, and Goldsmith, Edward. 1996. *The Case Against the Global Economy and for a Turn Toward the Local.* San Francisco: Sierra Club Books.

Martin, Philip, and Widgren, Jonas. 1996. *International Migration: A Global Challenge.* Washington, DC: Population Reference Bureau.

McFalls, Joseph, Jr. 1995. *Population: A Lively Introduction.* Washington, DC: Population Reference Bureau.

Mikesell, Marvin. 1983. "The Myth of the Nation-State." *Journal of Geography* 82, 257–260.

Murphy, Alexander B. 1991. "Regions as Social Constructs: The Gap Between Theory and Practice." *Progress in Human Geography* 15, 22–35.

Riley, Nancy. 1997. *Gender, Power, and Population Change.* Washington, DC: Population Reference Bureau.

Robertson, Roland, and Khondker, H. H. 1998. "Discourses of Globalization: Preliminary Considerations." *International Sociology* 13, 25–40.

Rodrik, D. 1997. *Has Globalization Gone Too Far?* Washington, DC: Institute for International Economics.

Rosenau, James N. 1997. "The Complexities and Contradictions of Globalization." *Current History,* 360–364.

Schaeffer, Robert K. 1997. *Understanding Globalization: The Social Consequences of Political, Economic, and Environmental Change.* Lanham, MD: Rowman & Littlefield.

Taylor, Peter J. 1994. "The State as Container: Territoriality in the Modern World System." *Progress in Human Geography* 18, 151–162.

Thrift, Nigel. 1993. "For a New Regional Geography 3." *Progress in Human Geography* 17, 92–100.

The World Bank. 1997. *The World Bank Atlas.* Washington, DC: International Bank for Reconstruction and Development/The World Bank.

World Resources Institute. 1998. *World Resources: A Guide to the Global Environment. 1998–99.* New York: Oxford University Press.

C H A P T E R 2

The Changing Global Environment

The human imprint is everywhere on Earth, from the highest mountains to the deepest ocean depths; from dry deserts to lush tropical forests; from frozen Arctic ice caps to the cloudless atmosphere (Figure 2.1). Hundreds of spent oxygen canisters clutter the heights of Mount Everest; radioactive debris accumulates in deep offshore waters of the North Atlantic; deserts in North America bloom with irrigated cotton, while tropical forests in Brazil are laid waste by logging; farmland pesticides appear in Arctic ice caps; and carbon dioxide from burnt fossil fuels collects in the skies. While many changes to the global environment are intentional and have improved the human habitat, other environmental changes are inadvertent, accidental, and often harmful to human welfare (Figure 2.2). Because of the importance of environmental issues, a study of the changing global environment is central to the study of world regional geography.

Environmental issues are also intertwined with globalization and diversity (see "Globalization, Trade Agreements, and the Environment"). The destruction of tropical rain forests, for example, is a response to international demand for wood products and beef. Similarly, global warming through human-caused climate change is also closely linked to patterns of world industry and commerce, as is the diversion of streams and rivers for irrigated agriculture. With six billion people on Earth, there is a long list of the ways humans interact with—and change—the natural environment.

The purpose of this chapter is to provide an overview of Earth's environmental systems—geology, climate, hydrology, and vegetation—to set the scene for better understanding the environmental geography of the 12 world regions taken up in the following chapters. This chapter concludes with a discussion on world food supply issues, since that topic folds together many different aspects of the environment.

Geology and Human Settlement: A Restless Earth

Geology offers a common point for any discussion of the global environment since it shapes the fundamental form of Earth's surface by giving distinctive character to world landscapes through the physical fabric of mountains, hills, valleys, and plains. The geological environment is also critical to a wide spectrum of human activities and concerns, such as the relationship between soil fertility and agriculture or the distribution of mineral resources such as iron and coal. Finally, the geologic environment presents humans with challenges

◀ **Figure 2.1 Earth's landscapes** Jewel of the Hawaiian Islands, Kauai features an abundant and diverse natural landscape. However, Kauai's mosaic of volcanic landforms, tropical vegetation, and spectacular river valleys has been much modified by humans, which is true of most of the world's environments. *(Kevin O. Mooney/Odyssey Productions)*

▲ **Figure 2.2 Trash on Mt. Everest** Few places on Earth have escaped human impacts in the early twenty-first century. Even isolated Mt. Everest is now littered with base camp garbage and discarded oxygen canisters at 17,000 feet. *(Mountain Light Photography, Inc.)*

and hazards in the form of devastating earthquakes, landscapes, or explosive volcanoes. Clearly, a basic understanding of the physical processes shaping Earth's landscapes is crucial to comprehending human settlement in different parts of the world.

Plate Tectonics

The starting point for understanding the dynamic geology of Earth is with **plate tectonics,** a geophysical theory that postulates that the surface of Earth is made up of a large number of geological plates that move slowly across the surface. This theory explains and describes both the inner workings of our planet as well as many surface landscape features. Additionally, plate tectonics explains the world distribution of earthquakes and volcanoes.

Earth's interior is apparently separated into three major zones with very different physical characteristics. These three parts are the core, the mantle, and the outer crust. Tectonic plate theory is built upon the assumption that there is a significant heat exchange taking place within Earth's interior. More specifically, this heat exchange results from cooling of the inner core. As a result, there is a exchange of plastic-like molten material through the mantle area (Figure 2.3).

This heat exchange takes the form of numerous **convection cells,** which are large areas of very slow-moving molten rock within Earth. Much like a stove provides heat for boiling water within a kettle, radioactive decay deep within Earth's core drives these convection cells. As the molten material reaches Earth's surface, it cools and becomes more dense. This causes the material to sink back into the mantle. Although geophysicists

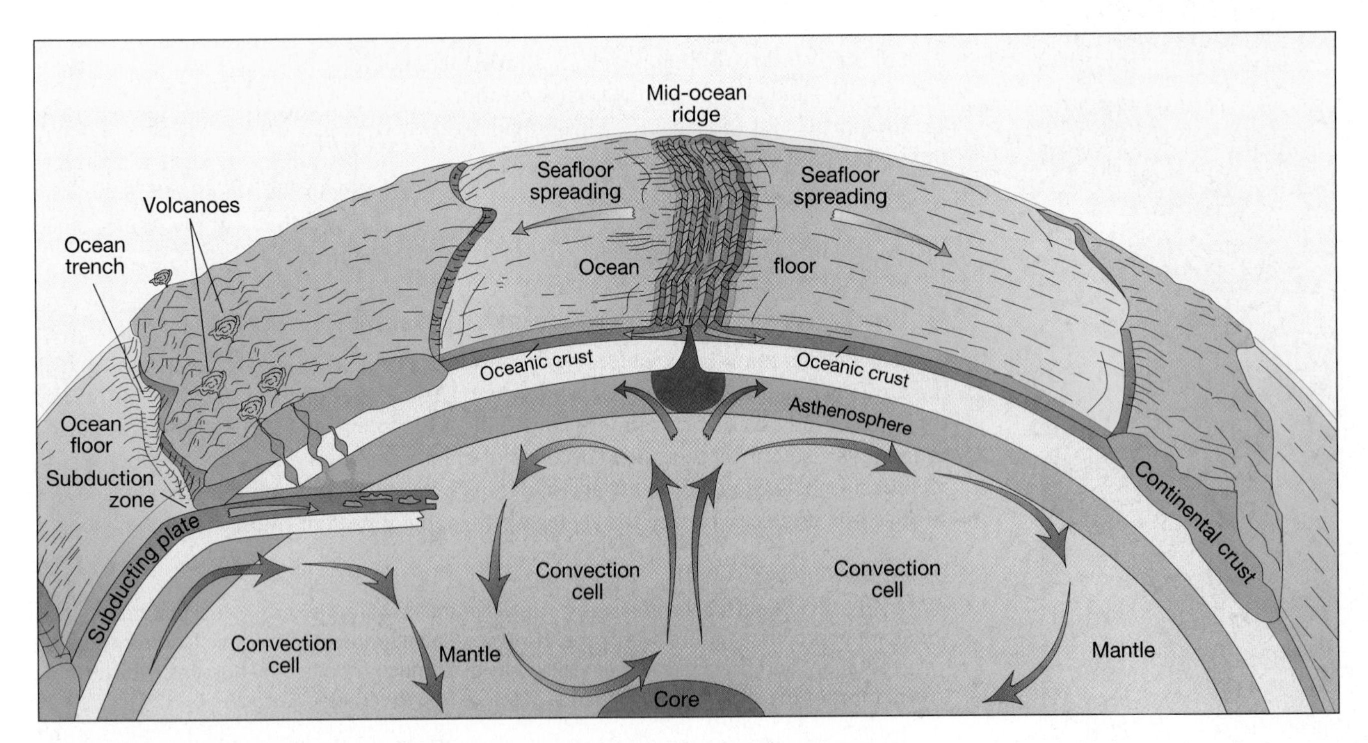

▲ **Figure 2.3 Plate tectonics** Plate tectonic theory suggests that vast convection cells in Earth's mantle drive molten rock toward the surface along mid-ocean ridges, creating new crust. Elsewhere, continental plates collide or are subducted beneath the surface in deep ocean trenches. *(Adapted from McKnight, 1996,* Physical Geography: A Landscape Appreciation, *Fifth Ed., Upper Saddle River, NJ: Prentice Hall)*

Globalization, Trade Agreements, and the Environment

Economic globalization depends heavily on unfettered world trade, and to facilitate this international commerce many countries have entered into free agreements and pacts under the World Trade Organization (WTO). An example is the North American Free Trade Agreement, or NAFTA, that binds together Canada, Mexico, and the United States. Although these trade agreements may be an economic boom, their effect on the environment can be troublesome. More specifically, national and local control for environmental protection can be sacrificed to global trade. Several illustrations follow.

- In 1997 the U.S. Environmental Protection Agency (EPA) weakened its Clean Air Act regulations to comply with a World Trade Organization ruling barring U.S. limits on contaminants in foreign gasoline. Venezuela claimed that these EPA limits acted as an unfair trade barrier against their petroleum.
- In October 1998 the World Trade Organization ruled against the U.S. ban on shrimp imports from nations whose fishing fleets do not use devices to keep endangered sea turtles out of the shrimp nets. As a result of this ruling, the Clinton administration had to revise the U.S. Endangered Species Act to comply with the WTO ruling.
- In response to a recent WTO ruling, U.S. consumers will no longer find "dolphin-safe" tuna on supermarket shelves. This is because the Congress had to weaken the U.S. Marine Mammal Protection Act to comply with the WTO decision that the U.S. was restricting free trade by discriminating against those countries catching tuna with mile-long nets that also snare and kill thousands of dolphins each year.
- In June 1999 the Canadian-based Methanex Corporation filed a $970 million lawsuit against the state of California under NAFTA. This lawsuit results from California's recent decision to ban MTBE, the gasoline additive that is blamed for polluting the state's groundwater and lakes. This is a similar lawsuit to a 1998 lawsuit by a U.S. company that forced Canada to overturn its ban on a similar gasoline additive. In both cases, the companies pleaded loss of business from the bans.

The MTBE case is "just what we predicted would happen under NAFTA and what we predict will happen under the WTO," said Congressman George Miller of California. "It's happened with dolphin-safe tuna, and it could happen with lots of other laws. This is the New World Order's assault on democracy. . . . Local [environmental] legislation can be nullified because a secret traded tribunal says so."

Source: Adapted from Robert Collier and Glen Martin, "U.S. Laws Diluted by Trade Pacts," *San Francisco Chronicle*, July 24, 1999.

postulate some convection cells move material at different rates, as a general rule the rate of movement is measured in inches or centimeters per year, which roughly approximates a rate similar to the growth of human fingernails.

On the Earth's surface, these convection cells drag an array of **tectonic plates** in different directions, sometimes forcing them into collision, sometimes spreading them apart in different directions. On top of these slowly moving tectonic plates sit parts of the major continents and the major ocean basins. The map of world tectonic plates (Figure 2.4) shows these plates vary significantly in size. It is important to note that these continents are not synonymous with tectonic plates. In fact, rarely is there a close match between the two. More often, continents straddle several tectonic plates, a fact that explains much of the world's mountainous landscapes. For example, western North America sits atop two tectonic plates, the Pacific Plate and the North American Plate, on what is called a **convergent plate boundary.** This means that the two plates are converging, or being forced together by convection cells deep within the Earth. The infamous San Andreas Fault that runs north–south through coastal California is the actual plate boundary. Major mountain ranges have resulted from the tectonic convergence, or collision of these two plates. The volcanic Cascade Range of Oregon and Washington and the massive Sierra Nevada of California are two examples. Similarly, in Latin America, the Andes Mountains are also products of two colliding plates, the eastward-moving Nazca Plate and the westward-moving South American plate.

Often in these collision zones, one tectonic plate will dive below another, creating a **subduction zone,** characterized by deep trenches where the ocean floor has been pulled downward by tectonic movement. This is true off the western coast of South America and, as well, near the Philippines, where the Mariana Trench forms the deepest ocean depths in the world at 35,000 feet (10,700 meters).

In other parts of the world, tectonic plates move away from each other in opposite directions, forming a **divergent plate boundary.** As these plates diverge, magma often flows from the Earth's interior, creating mountain ranges with active volcanoes. Iceland, in the North Atlantic Ocean, is a landscape formed on a divergent plate boundary that bisects the Atlantic Ocean. In other places, divergent boundaries form deep depressions, or **rift valleys,** such as that occupied by the Red Sea between northern Africa and Saudi Arabia. To the west, in Africa, a splinter of this boundary has created the extensive African rift, a part of which is the famous Olduvai Gorge, where tectonic activity has preserved the earliest traces of our human ancestors in volcanic soils (Figure 2.5).

Geologic evidence suggests that there was a time some 250 million years ago when all of the world's plates were tightly consolidated into a supercontinent centered on present-day Africa. Through time, this large area called *Pangaea* was broken up as convection cells moved the tectonic plates

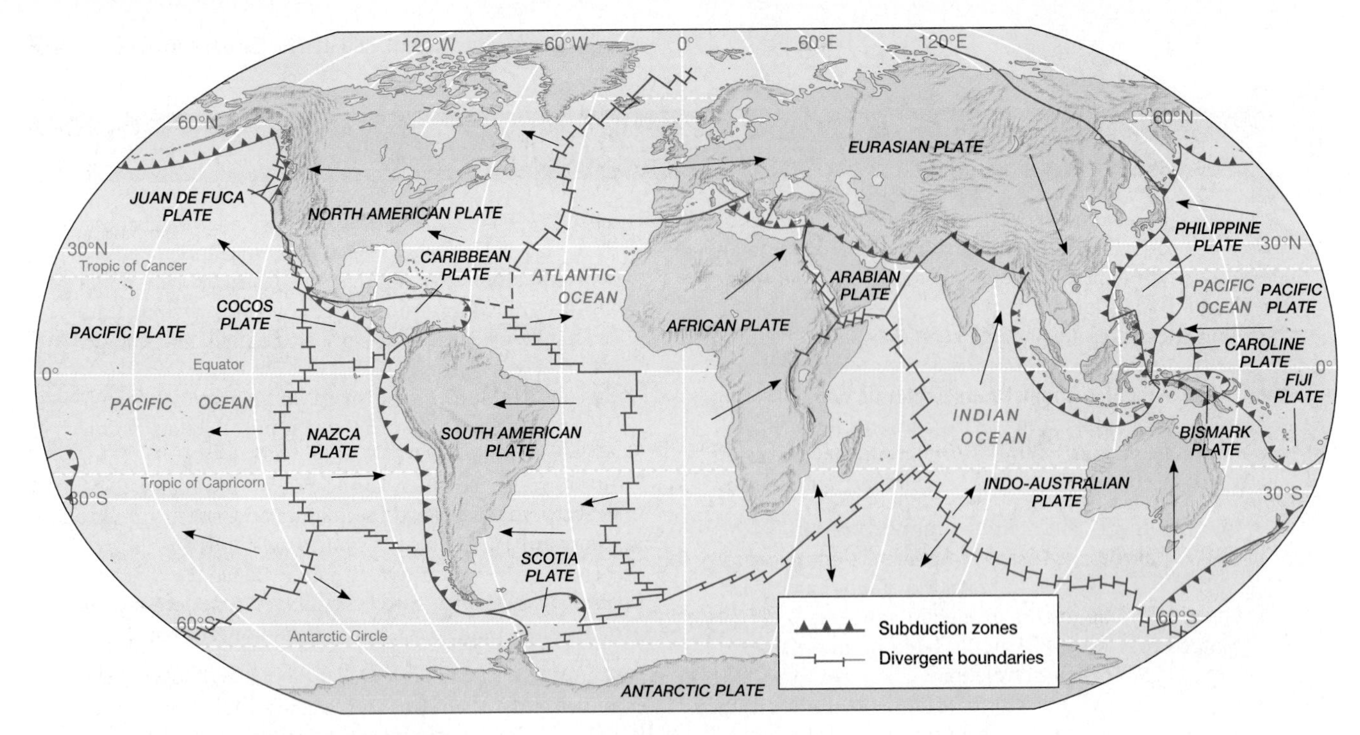

▲ **Figure 2.4 Global tectonic plates** Akin to a fractured jigsaw puzzle, Earth's tectonic plates vary greatly in size and shape. Where plates converge and collide, active volcano and earthquake zones frequently appear, creating new mountains and significant environmental hazards. In other settings, one plate may dive beneath another, creating ocean trenches more than 30,000 feet deep. *(Adapted from McKnight, 1996,* Physical Geography: A Landscape Appreciation, *Fifth Ed., Upper Saddle River, NJ: Prentice Hall)*

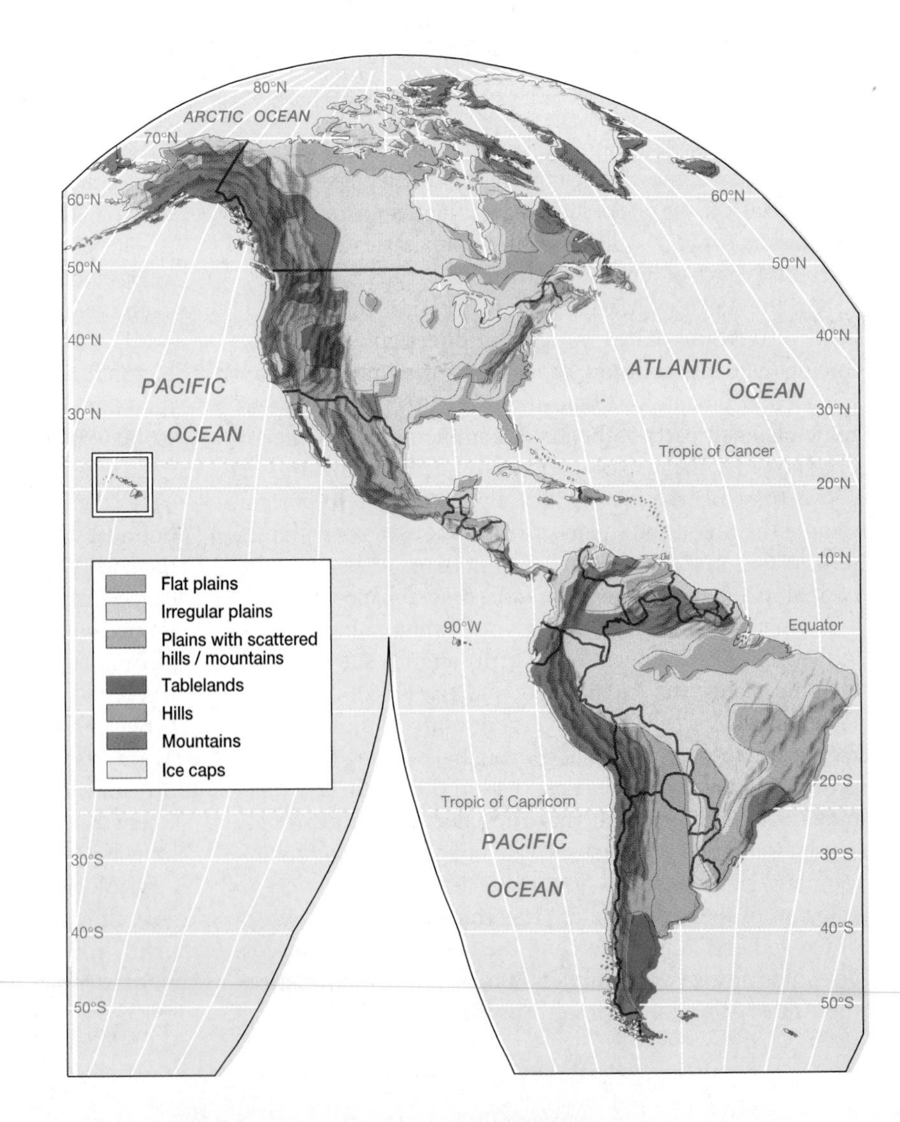

▶ **Figure 2.6 Global landforms** A vast mosaic of uplifts and plains, the world's major landform regions are the product of many dynamic processes acting upon the planet's complex surface for millions of years. While many mountain ranges follow major tectonic boundaries, other uplifts are the product of more local and regional forces. *(Adapted from McKnight, 1996,* Physical Geography: A Landscape Appreciation, *Fifth Ed., Upper Saddle River, NJ: Prentice Hall)*

▲ **Figure 2.5 African rift valley** Where tectonic plates pull apart or diverge from one another, large structurally defined rift valleys often create dramatic landforms. Here in East Africa, these settings were occupied by our first human ancestors. *[Altitude (Y. Arthus-B.)/Peter Arnold, Inc.]*

apart. A clue to this former continent can be imagined from the jigsaw puzzle fit of South America to Africa, or North America to Europe.

While tectonic plate theory explains many of the world's large mountain ranges, it does not account for all highlands. For example, both the Himalaya and Alpine mountain ranges have been created by the colliding forces of tectonic plates, as have the western mountains of North America. However, a glance at the world map of landforms (Figure 2.6) shows that there are numerous mountain ranges far removed from plate boundaries. In North America, the Rocky Mountains serve as an illustration, as do the Ural Mountains of Russia. This reminds us that there are other, more local geologic forces than tectonic plates that also play an important role in shaping the landscape.

Geologic Hazards: Earthquakes and Volcanoes

Although the human toll from geologic hazards is not nearly as great as deaths from other natural disasters, such as floods or severe storms, earthquakes and volcanoes nevertheless can have a major impact on human settlement and activities. In August 1999, more than 14,000 people died when a strong earthquake struck western Turkey. In January 1995, Kobe, Japan, suffered $100 million in damage from a serious earthquake that killed more than 5,000 people (Figure 2.7). In

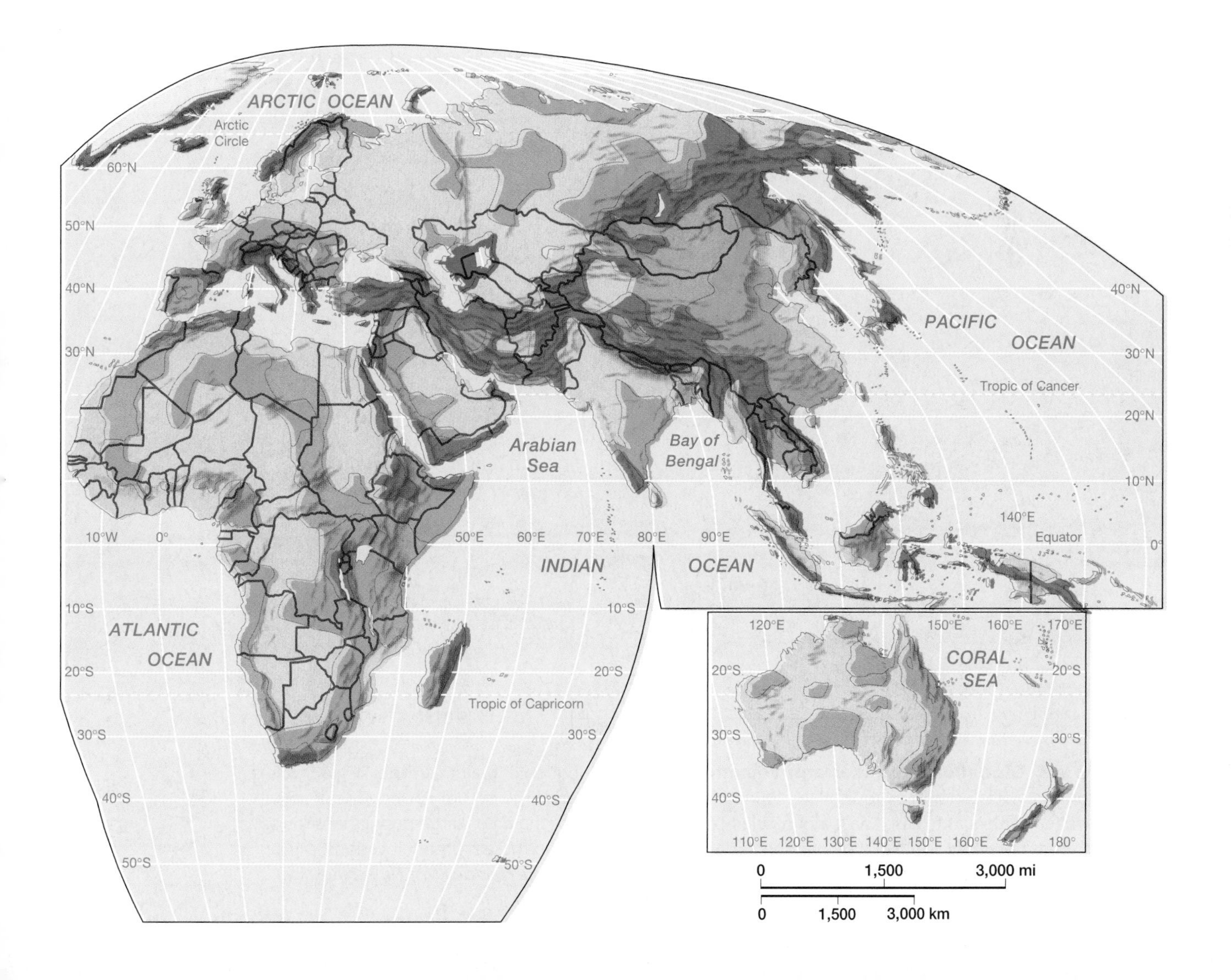

▲ Figure 2.7 Kobe earthquake Widespread urban damage resulted from the Kobe, Japan, earthquake of January 1995. Increasingly large populations in quake-prone areas suggest that future human generations will face even larger environmental hazards in such settings. *(Mike Yamashita/Woodfin Camp & Associates)*

1976 in rural China, a quarter-million people were killed when a strong earthquake hit Tangshan. Generally, each year we can expect more than 100 earthquakes to cause considerable damage to human settlement somewhere on Earth.

Predicting earthquakes remains much more difficult than warning people about hurricanes, tidal waves (tsunamis), or even volcanic explosions. Further, no scientific breakthrough is expected in the next decade that will make earthquake prediction more accurate. Instead, many cities in earthquake-prone regions emphasize building codes for stronger structures, land-use planning that prevents buildings in hazardous areas, drills for taking cover during an earthquake itself, and post-quake search and rescue. Unfortunately, these measures are expensive; hence, it is usually only wealthier countries that have been able to reduce the toll from seismic disasters. Conversely, damage from earthquakes is usually higher in those countries least able to cope with such a disaster. This disparity is illustrated by the higher loss of life and more widespread damage from a similar-magnitude earthquake in, say, Iran or Mexico, than would be found in Japan or California.

However, this does not mean that developed countries can dismiss the threat of major earthquakes. On the contrary, professionals estimate that when the inevitable "Big One" strikes California, the loss of life could be between 30,000 and 50,000 people, depending on the location of the earthquake in terms of proximity to the major urban areas of Los Angeles and San Francisco. Another important variable is the time of day of the disaster. If a major earthquake struck at night, for example, casualties would be far fewer than if it came during the day when people are at work, at school, and on the streets. Despite planning and emergency preparedness, the threat of major destruction and a disastrous loss of life from a catastrophic earthquake is a reality along North America's West Coast, from San Diego to Seattle, and even inland to Salt Lake City, because of the geologic activity associated with the collision of the North American and Pacific plates. The same scenario is true for many other regions of the world as well because of the ceaseless movement of convection cells and tectonic plates.

Volcanic eruptions are also found along most tectonic plate boundaries, and these also can cause major destruction (Figure 2.8). In 1985, for example, 23,000 deaths re-

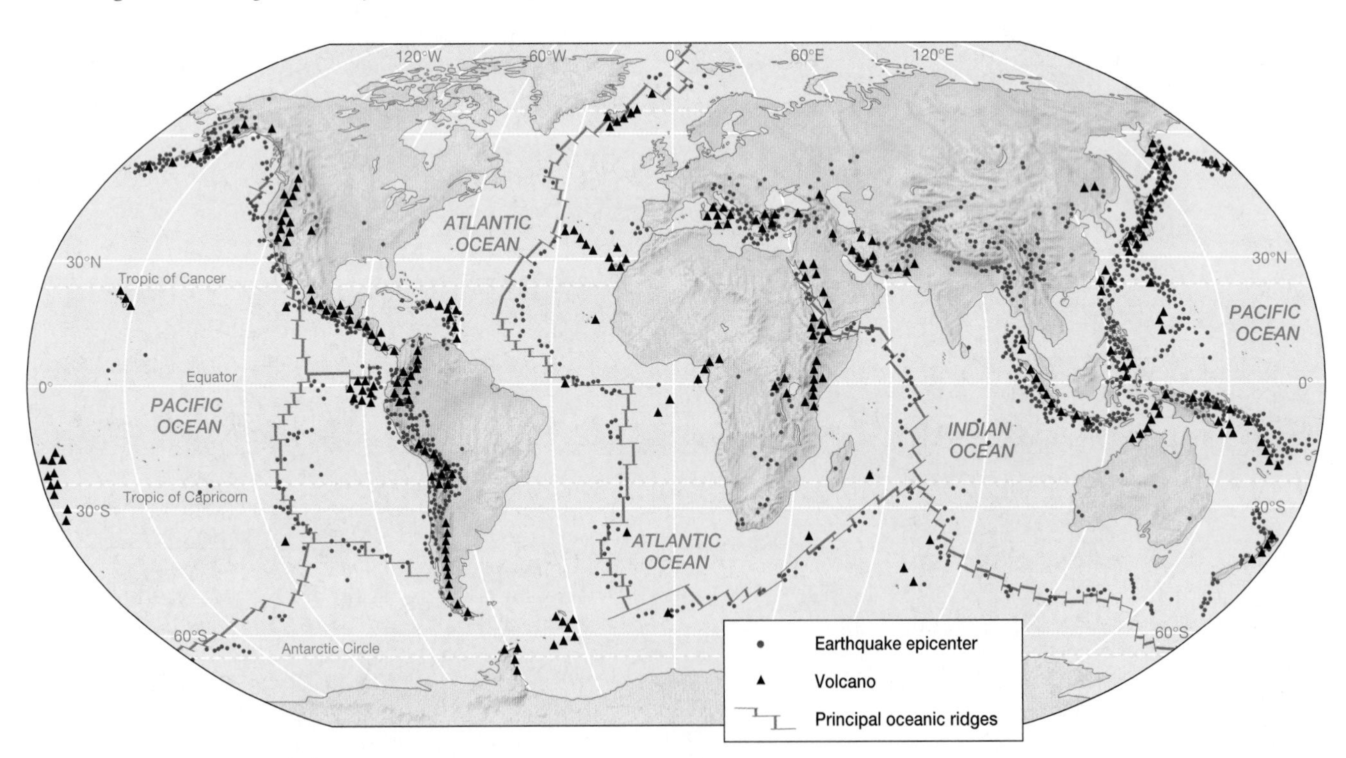

▲ Figure 2.8 Global earthquakes and volcanoes The distribution of major earthquakes and volcanoes is strongly associated with tectonic plate boundaries around the world. The circum-Pacific zone of activity from the western Americas to East Asia is a particularly dramatic feature on the map. Large populations found near these zones of tectonic activity will continue to pay the price of living in these high-risk locations. *(Adapted from McKnight, 1996,* Physical Geography: A Landscape Appreciation, *Fifth Ed., Upper Saddle River, NJ: Prentice Hall)*

sulted in Colombia, South America, from a volcanic eruption. In most cases these eruptions can be predicted days in advance, which usually provides enough time for evacuation. During the 1991 eruption of Mount Pinatubo in the Philippines, 60,000 people were evacuated, although 800 did die in the disaster. Because of this predictability, the loss of life from volcanoes is generally a fraction of that from earthquakes. In the twentieth century, it is estimated that 75,000 people have been killed during volcanic eruptions, compared with the approximate 1.5 million who have died in earthquakes (Figure 2.9).

Unlike earthquakes, volcanoes do provide some benefits to people in certain areas. In Iceland, New Zealand, and Italy, geothermal activity produces energy to heat houses and power factories. In other parts of the world, such as the islands of Indonesia, volcanic ash has enriched soil fertility for food crops; there is also the considerable gain to local economies from tourism attracted by scenic volcanoes in places such as Hawaii, Japan, and the Pacific Northwest.

▲ **Figure 2.9 Plymouth, Montserrat** Recent volcanic eruptions on the Caribbean island of Montserrat have ravaged the island's small population and reminded residents of the region of the costly and unpredictable consequences of living amid a tectonically active natural landscape. *(John McConnico/AP/Wide World Photos)*

Global Climates: An Uncertain Forecast

Human settlement and food production throughout the world are closely linked to patterns of local weather and climate. Where it is dry, such as in the arid parts of Southwest Asia, life and landscape differ considerably from the wet tropical areas of Southeast Asia. Additionally, people in different parts of the world adapt to weather and climate in widely varying ways, depending on their culture, economy, and technology. Whereas some desert areas of California are covered with high-value irrigated agriculture that produces vegetables for the global marketplace, most of the world's arid regions support very little agriculture; thus, they barely participate in global commerce. Moreover, when drought hits one portion of the world (such as Russia's grain belt or Africa's Sahel), the socioeconomic repercussions are felt throughout this world. In many ways, climate links us together in our globalized economy, providing opportunities to some, hardships to other, and challenges to all in the struggle to supply the world with food (see "Hurricane Mitch: An Unnatural Disaster?").

HURRICANE MITCH An Unnatural Disaster?

Hurricane Mitch swept through Central America in December 1998, causing widespread damage and destruction. In Honduras alone, more than 6,000 people were killed, with thousands more listed as missing; damage was estimated at more than $5 billion. Throughout Central America, towns and cities were flooded, roads and bridges destroyed, and agricultural fields wiped out. Was this a natural disaster, or one of human making?

Although Mitch was unusually violent, with days of rain and high winds, those areas that bore the brunt of the hurricane were not necessarily the areas to suffer the most damage. That is, the killer floods and mudslides that did major damage did not follow the path of the storm itself. "The areas affected the worst were those that are heavily deforested," said George Pilz, head of the natural resources department at the Panamerican Agricultural College at Zamorano, Honduras. In the deforested areas, floods surged unobstructed, soil became waterlogged, and denuded hills collapsed with mudslides. "In areas with forests, there was less damage," Pilz said.

Until recently, tropical forests still blanketed much of the country. However, in the 1990s, huge melon farms and cattle ranches covered the fertile lowlands, along with tropical fruit plantations owned by U.S.-owned companies. With the best land devoted to export markets and global trade, hundreds of thousands of Honduran subsistence farmers were pushed off the lowlands into the steep hillsides covered by tropical forests. Clearing land for subsistence crops meant clearing the forest. With the forest gone, soil nutrients were lost after a few years, so the farmers had to move deeper into the forested mountains. After a decade of tropical forest clearance by these peasant farmers, much of the upland forest was gone. With its demise, the landscape was poised for destruction by tropical storms. Not only did these farmers suffer most from Hurricane Mitch with loss of life and subsistence, but the cleared forest lands in the mountains also contributed to the widespread flooding that devastated the lowlands.

Disasters such as Hurricane Mitch are not "natural" disasters and, unfortunately, they are not uncommon. Furthermore, they will continue—and probably increase—throughout the developing world until governments address the short-term management of their natural resources and the deep social inequities that usually determine the way land is used. As government officials plan the rebuilding of Central America, they should remember that local people need secure access to good land. Otherwise, the region will remain vulnerable to the next "unnatural" disaster.

Source: Adapted from Barbara Goldoflas, "Unnatural Disasters," *San Francisco Chronicle*, January 11, 1999.

Understanding the complex meteorological processes that influence our different global climates presents quite an intellectual challenge. Nevertheless, it is highly rewarding because understanding provides insight into some of the reasons the human condition varies so widely around the world.

Climatic Controls

Although weather and climate differ tremendously around the world, there is an accepted set of atmospheric processes that control and influence these meteorological conditions. More specifically, there are five main factors that must be understood—solar energy, latitude, interaction between land and water, world pressure systems, and global wind patterns.

Solar Energy Both the surface of Earth and the atmosphere immediately above it are heated by solar energy from the sun as our globe revolves around that large planet. This fact is one of the most important variables that produces the different world climates, for it explains the great differences between the tropical climates near the equator and the cold climates closer to the poles. Most of the incoming solar energy, or **insolation,** is absorbed by Earth's land and water surfaces. These, in turn, heat the lower atmosphere through the process of re-radiation. This re-radiated energy is trapped by clouds and water moisture in the air adjacent to Earth, providing a warm envelope that makes life possible on our planet (Figure 2.10). Because there is some similarity between this process and the way a garden greenhouse traps sunlight to make the structure's interior warmer than the outside, this natural process of atmospheric heating is known as the **greenhouse effect.** Were it not for this process, Earth would be far too cold for human habitation; some scientists suggest the climate would be much like that on the planet Mars.

Latitude Because of the curvature of Earth, the highest amounts of insolation are received in the equatorial region, which is the area 23.5° latitude north and south of the equator. Poleward of this equatorial region, Earth receives far less insolation. As a result, not only are the Tropics much warmer than the middle or high latitudes, but there is also a buildup of heat energy that must be redistributed through other processes, namely global wind systems, ocean currents, even massive tropical storms such as typhoons and hurricanes (Figure 2.11).

Interaction Between Land and Water Because land and water differ in their abilities to absorb and re-radiate insolation, the global arrangement of oceans and continents

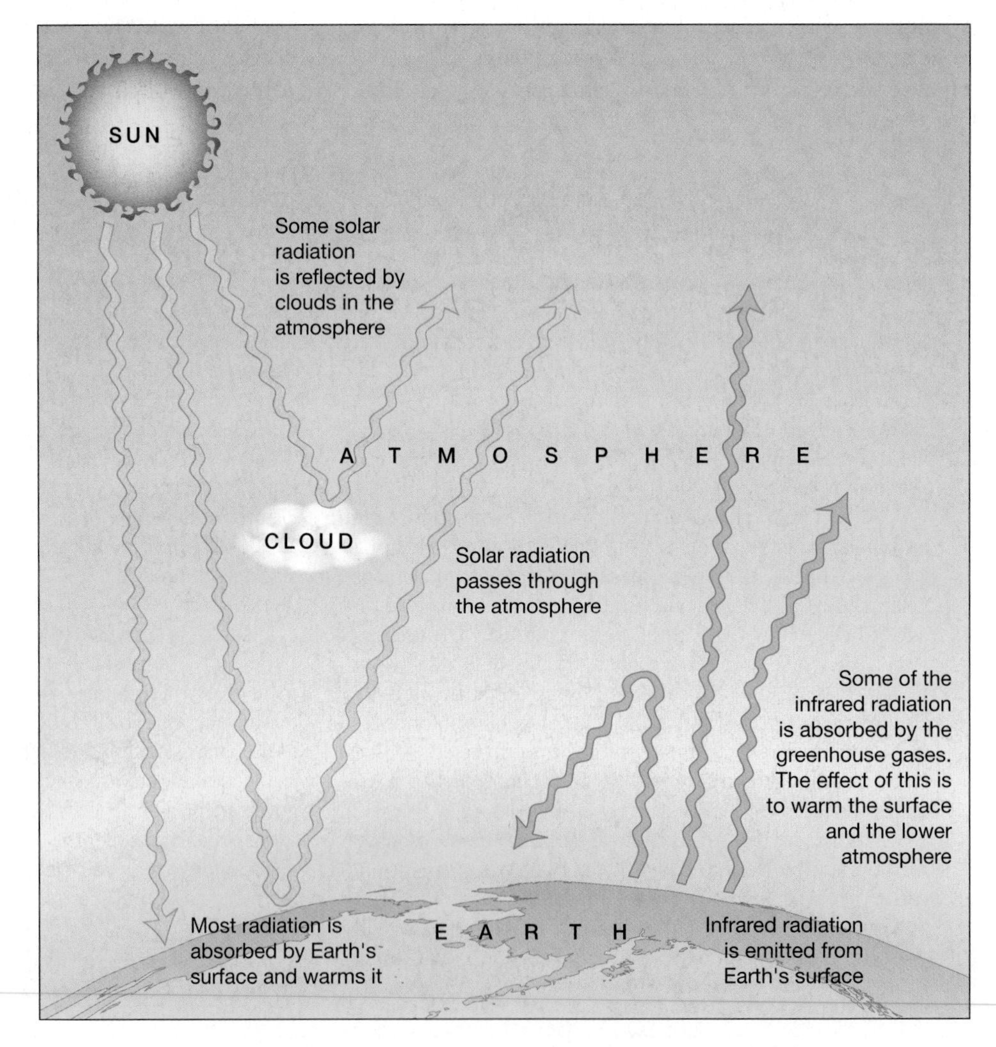

▶ **Figure 2.10 Solar radiation** Essential for Earth's survival, incoming solar radiation, or insolation, warms the planet as heat is absorbed by the surface. This heat is then radiated back to the lower atmosphere with infrared radiation, which is retained by the lower atmosphere.

▲ **Figure 2.11 Cumulus clouds** A vivid signature of Earth's atmosphere, these massive cumulus clouds display the transfer of heat and moisture in the tropics. As heat energy builds up in low latitudes, ocean currents and wind and weather systems redistribute it globally. *(Jules Bucher/Photo Researchers, Inc.)*

is a major influence on world climates. More specifically, land areas heat and cool faster than do water bodies, which explains why temperature extremes such as hot summers and cold winters are always found inland away from the coast. Conversely, since water bodies retain solar heat longer (although they also take longer to warm), oceanic or maritime climates always have moderate temperatures without the same seasonal extremes as found inland. The climatic characteristics of land and water differ so much that geographers use the term **continentality** to describe inland climates with hot summers and cold, snowy winters such as those found in interior North America or Russia. In contrast, **maritime climates** are those close to the ocean. They generally have cool, cloudy summers (Oregon, Washington, and British Columbia are good examples), with winters that are cold, yet lack the subzero temperatures of interior locations.

Global Pressure Systems The uneven heating of Earth due to latitudinal differences and the arrangement of oceans and continents produces a regular pattern of high- and low-pressure cells that, in turn, drives the world's wind and storm systems (Figure 2.12). For example, the interaction between high- and low-pressure systems in the North Pacific Ocean produces the storms that are driven onto the North American continent in both winter and summer. The same processes in the North Atlantic Ocean produce winter and summer weather for Europe. Farther south, however, closer to the equatorial zone, large oceanic cells of high pressure cause different conditions. These high-pressure cells expand during

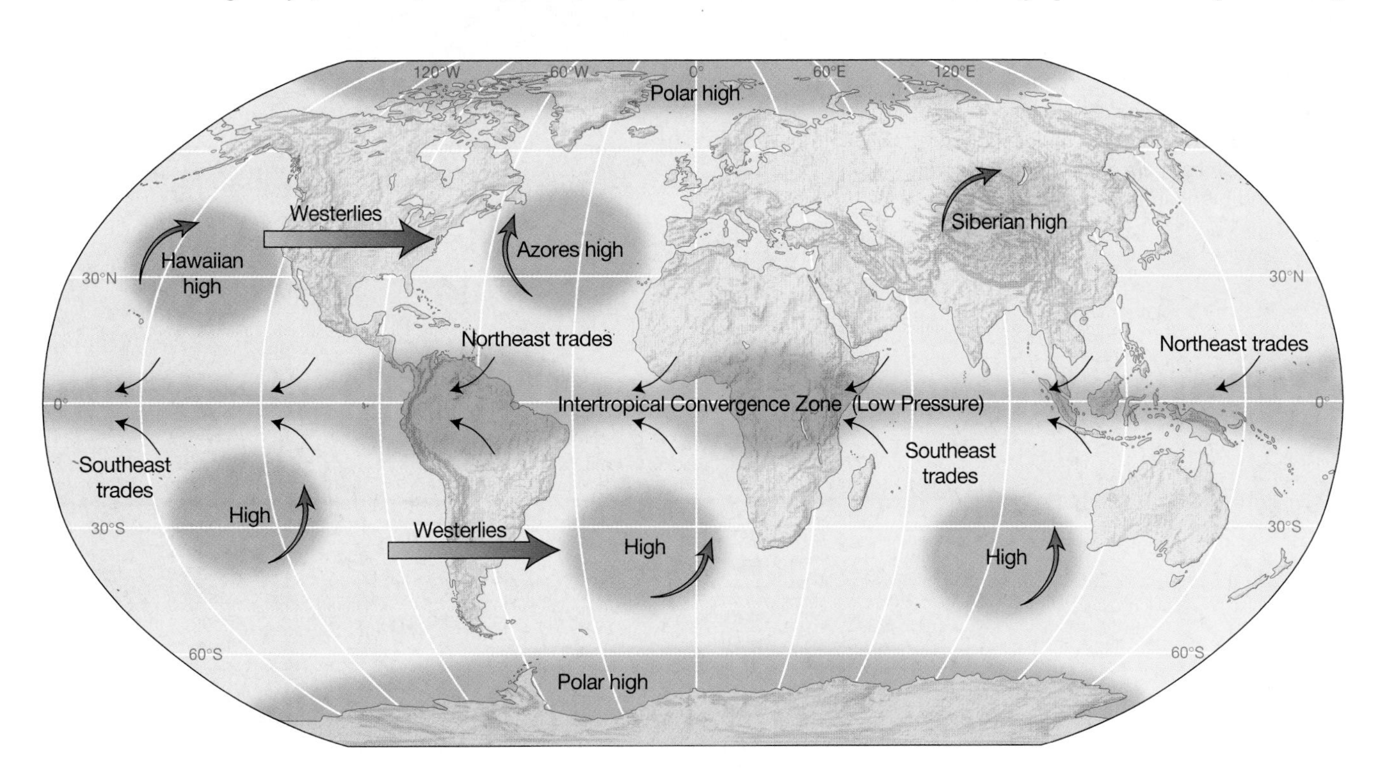

▲ **Figure 2.12 Global pressure systems** Predictable patterns of low- and high-pressure result from the unequal heating of Earth's surface and the positioning of the continents. The Intertropical Convergence Zone often provides unsettled weather near the equator while fast-flowing westerlies move storm systems across the middle latitudes.

the warm summer months because of the inflow of equatorial air. As they enlarge, these cells produce the warm, rainless summers of the Mediterranean area of Europe and California. Farther south in the equatorial zone, summer weather spawns the strong tropical storms known as typhoons in Asia and hurricanes in North America and the Caribbean. This same pattern of different pressure systems is found in the Southern Hemisphere.

Global Wind Patterns These same pressure systems also produce global wind systems. It is important to remember that air flows from high pressure to low (just as water flows from high elevations to lower ones); thus, winds will flow away from high-pressure and into low-pressure cells. This important fact explains the monsoon in India, for example, which arrives in June as moisture-laden air masses flow from the warm Indian Ocean over land into the low-pressure area over northern India and Tibet. In the winter the opposite is true: as high pressure builds over these same areas, winds flow outward, from cold Tibet and the snowy Himalayas toward the low pressure over the warm Indian Ocean. (More detail on the monsoon is found in Chapter 12, South Asia.)

World Climate Regions

The interaction of the meteorological processes discussed previously produces the world's weather and climate. Before going further, it is important to note the difference between the two terms. *Weather* is the short-term day-to-day (or even hourly) expression of atmospheric processes; our weather can be rainy, cloudy, sunny, hot, windy, calm, or stormy all within a short period. Measures of this ever-changing weather are taken at regular intervals each day, often hourly. Data are compiled on temperature, pressure, precipitation, humidity, and so on. Over a period of time, statistical averages from these daily observations are analyzed that provide a quantitative picture of common or usual conditions. From this long-term view, a sense of the regional *climate* is generated. Usually, at least 30 years of daily weather data are required before climatologists and geographers construct a picture of an area's climate. In summary, weather is the short-term expression of highly variable meteorological processes, whereas climate is the long-term, average conditions.

Since weather observations have been taken for far longer than 30 years in most parts of the world, it is possible to use these data to generate a picture of the climate for thousands of

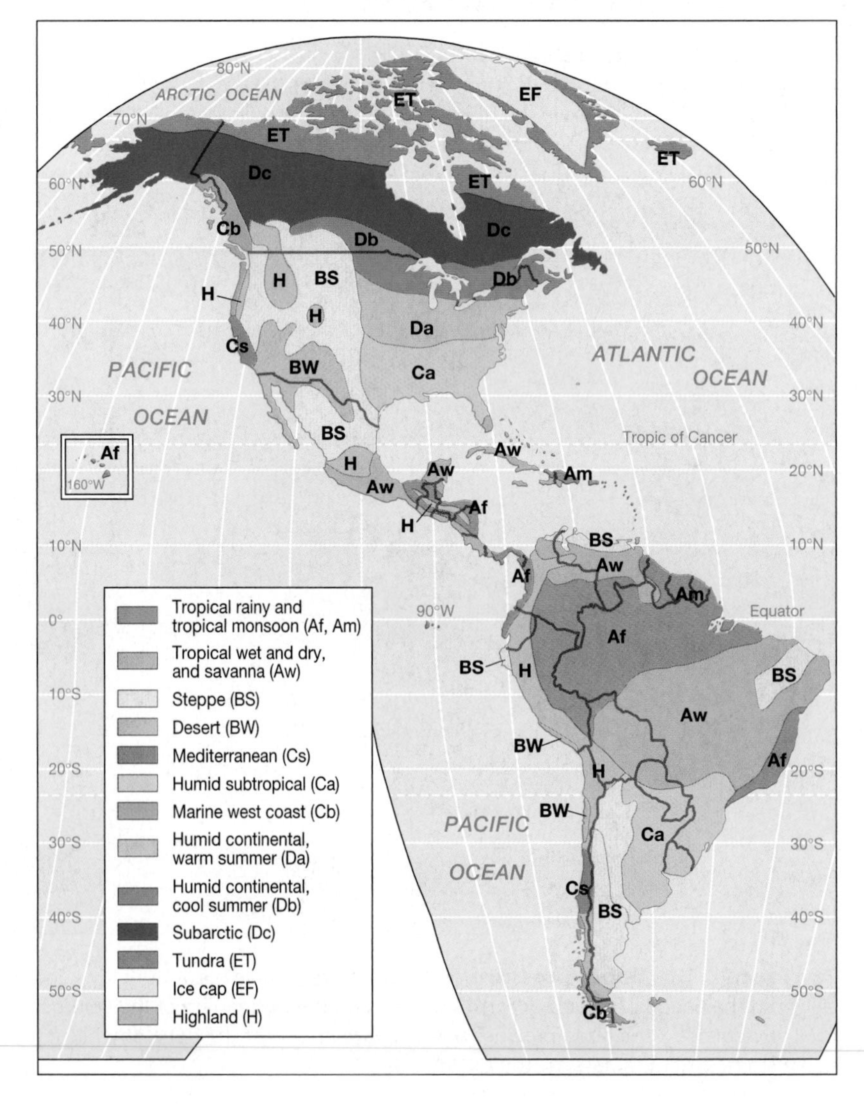

▶ **Figure 2.13 World climate regions** A standard scheme, called the *Köppen system* after the Austrian geographer who devised the system in the early twentieth century, is used to describe the world's diverse climates. This system uses a combination of letters to describe the general climate type, along with its precipitation and temperature characteristics. More specifically, the *A* climates are tropical; the *B* climates are dry; the *C* climates are generally moderate and found in the middle latitudes, and the *D* climates are associated with continental and high-latitude locations.

places all over the globe. Where similar conditions prevail over a larger area, boundaries are drawn cartographically around that area—called a **climate region**—so it can be folded into a larger classification of world climates. Knowing the climate type for a given part of the world not only conveys a clear sense of average rainfall and temperatures, but it also translates to larger inferences about human activities and settlement. Two examples suffice. If an area is categorized as desert, then we infer that rainfall is so limited that any agricultural activities must have irrigation. Without this supplemental watering, no crops can be grown, so people must turn to other strategies to obtain food. In contrast, if we see an area characterized by the climatic designation for tropical monsoon, we can infer that there is quite enough rainfall for agriculture.

A standard scheme of climate types is used throughout this book, and each regional chapter contains a map showing the different climate regions in some detail. Figure 2.13 shows these climatic regions generalized to a world scale. These regional climate maps also contain **climographs**, which are graphs of average high and low temperatures and precipitation for the 12 months of a year. Each climograph is for a specific city or locale, and the temperature and precipitation data illustrate the different climate regions and categories. Two lines for temperature data are given on each climograph: the upper one plots average high temperatures for each month, below that are the average low temperatures. These monthly averages convey a good sense of what a typical day might be like during the different seasons. Besides temperatures, the climographs also contain bar graphs depicting average monthly precipitation. Not only is the total amount of rain and snowfall important, but the seasonality of this precipitation provides valuable information for making inferences about agriculture and food production. In many midlatitude regions of the world, moisture falls as rainfall during the warm summer months, which is good for agriculture. However, in other parts of the world, precipitation falls primarily as snowfall during the cold winter months when it cannot be used by crops (Figure 2.14). In a few areas, such as in California and the Mediterranean region of Europe and northern Africa, the summers are dry, with rain falling during the winters. Only irrigated agriculture can thrive during the summer season.

Global Warming

Human activities connected with economic development and industrialization appear to be changing the world's climate in ways that may have significant and problematic consequences.

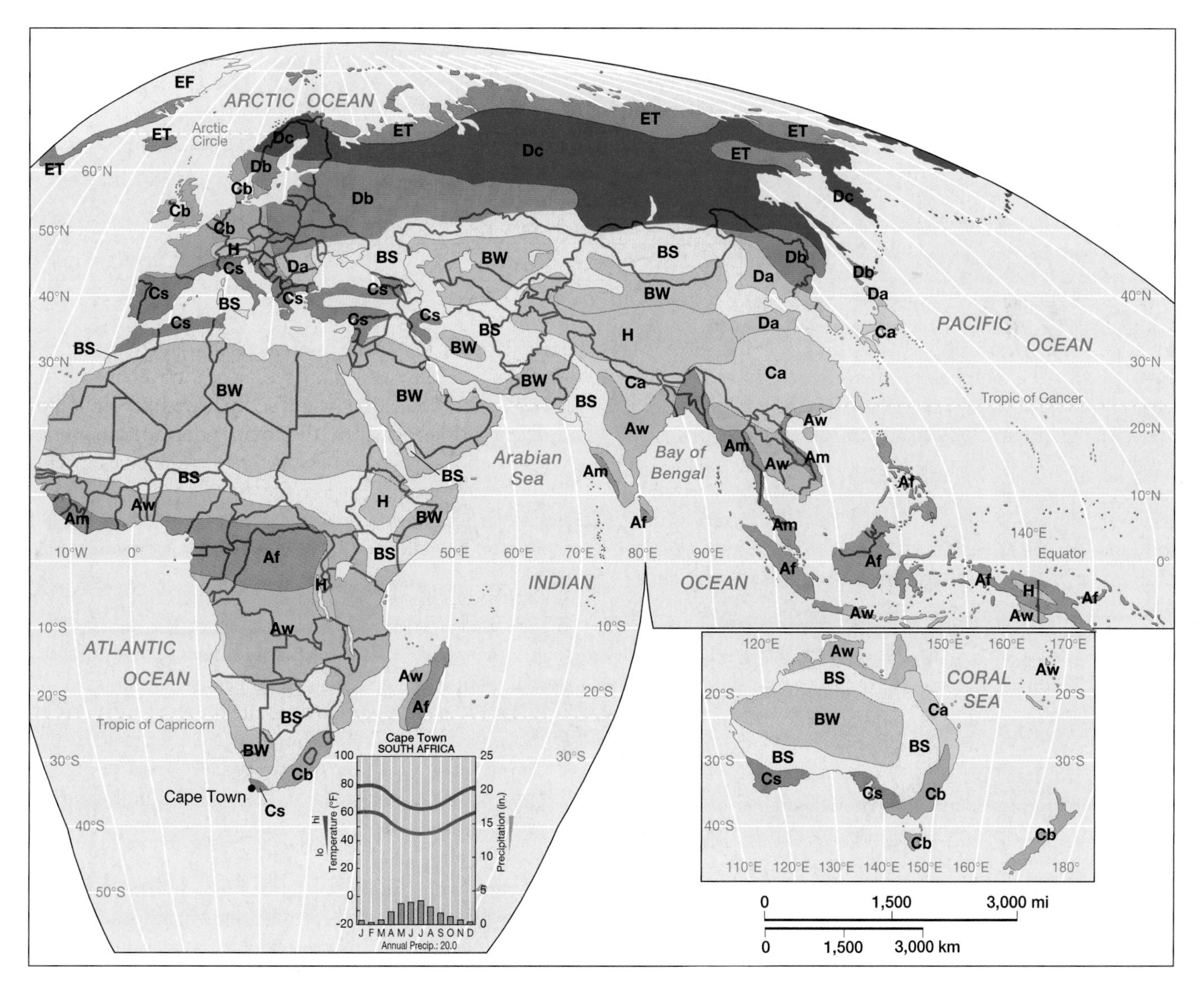

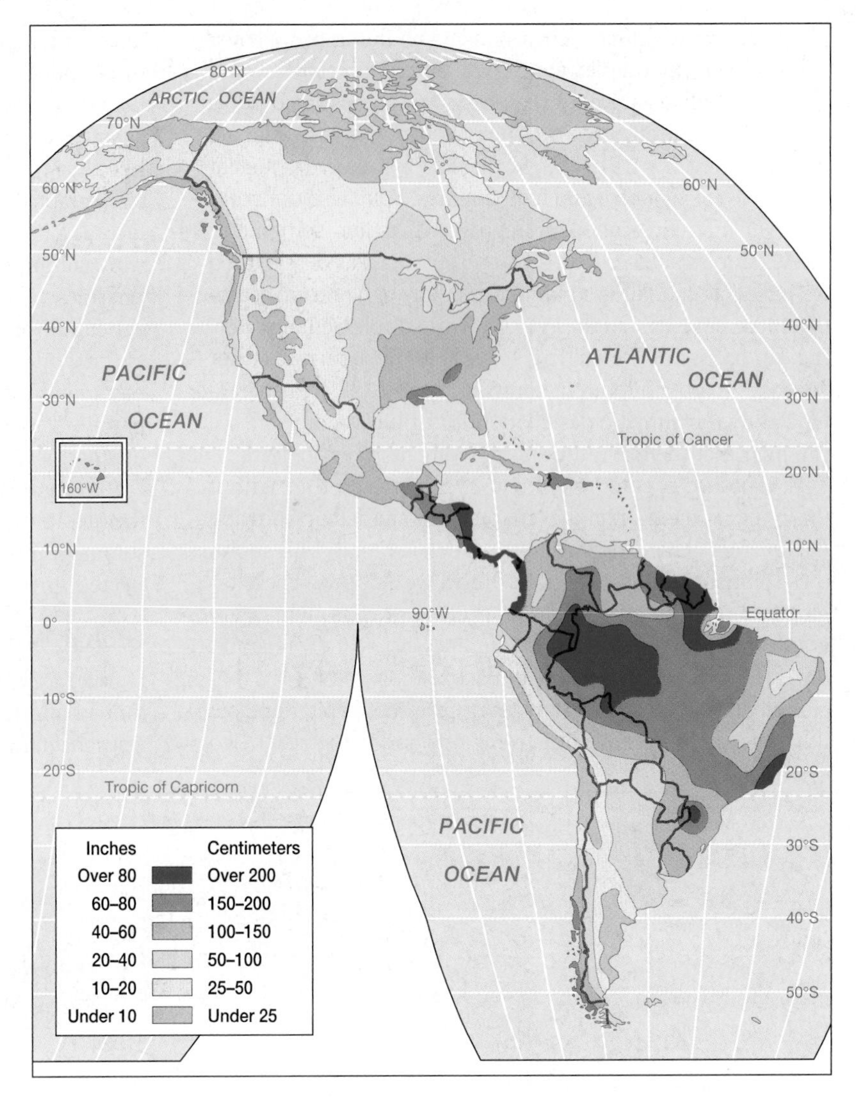

▶ **Figure 2.14 Global precipitation** This map shows the range of annual precipitation totals for the world. As a general rule, the highest amounts are found in the Tropics and, further, particularly in those places where mountains wring moisture from the tropical clouds. To illustrate, on the island of Kauai, Hawaii, annual totals of more than 400 inches are found in the uplands. Directly north and south of these wet Tropics lie expansive deserts and steppes with low rainfall. This is due to the subtropical high-pressure cells that dominate weather in these regions. Poleward of the arid lands lie the midlatitudes, where precipitation falls as both rain and snow and varies highly depending on storm tracks, coastal location, and topography. *(Adapted from Clawson and Fisher, 1998,* World Regional Geography, *Upper Saddle River, NJ: Prentice Hall)*

More specifically, **anthropogenic**, or human-caused, emissions to the lower atmosphere are increasing the natural greenhouse effect so that worldwide global warming is taking place. This warming, in turn, may change rainfall patterns; increase aridity; melt polar ice caps, causing an increase in sea levels; and possibly lead to a greater intensity in tropical storms. Moreover, because of climate change from global warming, the world may experience a dramatic change in food production. While some prime agricultural areas, such as the lower Midwest of North America, may lose out because of increased dryness, other areas, such as the Russian steppes, may actually gain a more favorable climate for agriculture. Though it is too early to tell if and when this will happen, the implications for world food supplies could be profound.

Causes of Global Warming As noted earlier, Earth has a natural greenhouse effect that provides us with a warm atmospheric envelope that supports human life. This comes from the trapping of incoming and outgoing (or re-radiated) solar radiation by water moisture and an array of natural greenhouse gases in the atmospheric layer close to Earth. Although these natural greenhouse gases have varied somewhat over long periods of geologic time, they seem to have been relatively stable until recently. However, within the last 130 years, coinciding with the Industrial Revolution in Europe and North America, the composition and amount of these greenhouse gases has changed rather dramatically. This has primarily resulted from the burning of fossil fuels associated with industrialization. Four major greenhouse gases account for the bulk of this atmospheric change (Figure 2.15).

1. *Carbon dioxide (CO_2)* accounts for more than half of the human-generated greenhouse gases. The increase in atmospheric CO_2 is primarily a result of burning coal, petroleum, and gasoline; the latter is a function of the automotive age. To illustrate this increase, note that in 1860 CO_2 was measured at 280 parts per million (ppm) in the atmosphere. Today, CO_2 is at 350 ppm, and this is expected to rise to 450 ppm by the year 2050.
2. *Chlorofluorocarbons (CFCs),* which make up nearly 25 percent of the human-generated greenhouse gases, come mainly from widespread usage of aerosol sprays and refrigeration, including air conditioning. Although

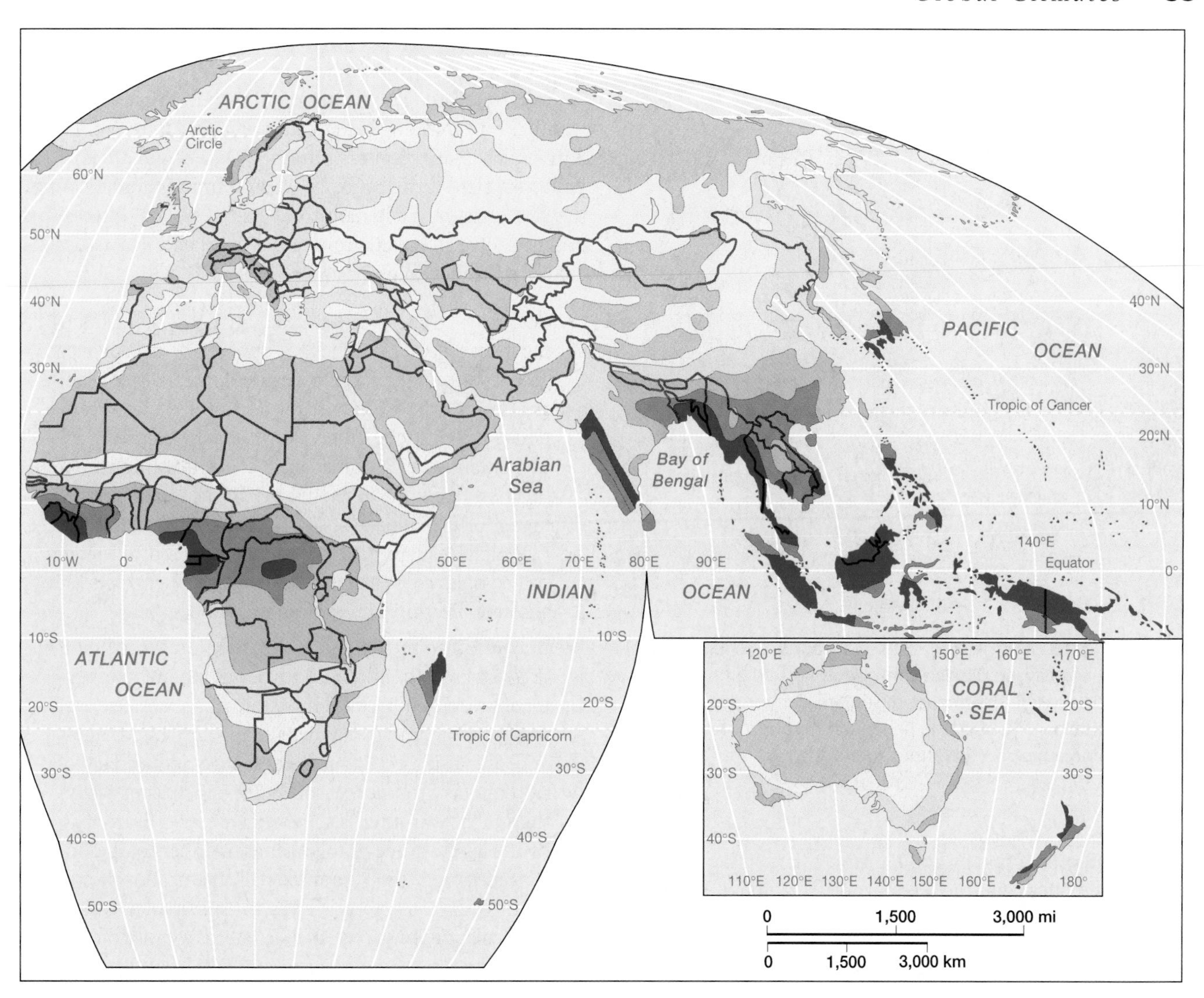

CFCs have been banned in North America and most of Europe, they are still increasing in the atmosphere at the rate of 4 percent each year. These gases are highly stable, with a very long resident time in the atmosphere, lasting perhaps as long as 100 years. As a result their role in global warming is highly significant. Recent research has shown that a molecule of CFC absorbs a thousand times more infrared radiation from Earth than a molecule of CO_2.

3. *Methane (CH_4)* comes primarily from vegetation burning associated with rainforest clearing, with additional emissions resulting from the anaerobic activity in flooded rice fields, as a by-product of cattle and sheep digestion, and from leakage of pipelines and refineries associated with natural gas production. Currently, CH_4 accounts for about 15 percent of anthropogenic greenhouse gases.

4. *Nitrous oxide (N_2O)* is responsible for just over 5 percent of the human-caused greenhouse gases. It results primarily from the widespread usage of chemical fertilizers associated with modern agriculture.

Effects of Global Warming The complexity of the global climate system leaves some uncertainty about how the world's climate may change as a result of human-caused greenhouse cases. Increasingly, though, the high-powered computer models used by science to refine our understanding of climate change are reaching consensus on the possible effects of global warming.

Unless countries of the world drastically reduce their emission of greenhouse gases in the next few years, the average global temperature will increase 2 to 4 °F (1 to 2 °C) by the year 2030. While this small amount may not seem dramatic at first glance, this change in average global temperature is about the same magnitude of change as the cooling that caused the Ice Age glaciers to cover much of Europe and North America 20,000 years ago.

This change in climate could cause a shift in major agricultural areas. For example, the Wheat Belt in the United States may receive less rainfall and become warmer, endangering grain production as we know it today. While more northern countries, such as Canada and Russia, may experience a longer growing season because of global warming, the soils in these two areas are not nearly as fertile as in the United States. As a result, scientists are predicting a decrease in the

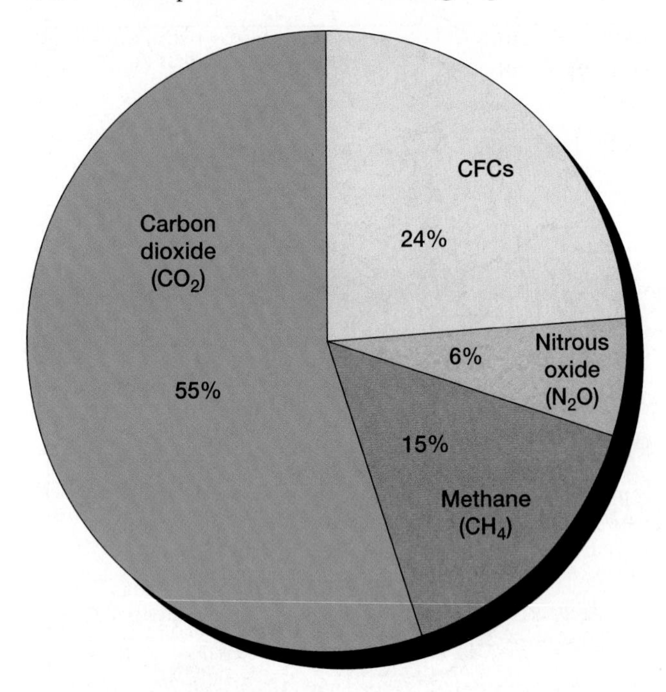

▲ **Figure 2.15 Major greenhouse gases** A buildup of human-induced carbon dioxide contributes more than half of the world's increased greenhouse gases. Smaller amounts of chlorofluorocarbons (CFCs), methane, and nitrous oxide add more warming agents to the atmosphere.

world's grain production by 2030. Further, the southern areas of the United States and the Mediterranean region of Europe can expect a warmer and drier climate that will demand even more irrigation for crops.

Warmer global temperatures will cause an increase in sea level from the thermal expansion of the oceans and, additionally, a warmer global climate may also cause polar ice sheets (particularly those in Antarctica) to melt. Although projections on this predicted sea level rise vary widely—from several inches to several feet in the next century—even the smallest rise in ocean level will endanger low-lying island nations and coastal areas in Europe, Asia, and North America. Island nations in the Pacific and the Indian Ocean are particularly outspoken on this matter since they may well be flooded; these nations formed a political lobby to advocate a reduction in international controls on atmospheric emissions.

The International Debate on Limiting Greenhouse Gases
By the early 1990s, the United Nations recognized that climate change was "a common concern of humankind" and that all nations have a "common but differentiated responsibility" for combating global warming. At the Rio de Janiero Earth summit in 1992, the first legal instrument addressing global warming was formulated. Signatories were bound by international law to reduce their greenhouse gas emissions by agreed-upon target dates. Within a year, 167 countries had signed. Unfortunately, those countries emitting the most greenhouse gases, such as the United States, Japan, India, and China, did not come anywhere close to meeting the emission reductions agreed upon in Rio.

Five years later, another attempt at an international treaty was made at a meeting held during late 1997 in Kyoto, Japan. Several key issues resulted from the Kyoto negotiations; these points speak to the tensions between countries and regions of the world that are reluctant to sacrifice national interests in order to solve global environmental problems. The issues included the following:

- Within the United States, the country that contributes the most emissions to the atmosphere, the debate over global warming remains contentious, highly partisan, and unresolved. While many see the need for greenhouse gas controls, others emphatically resist any action because they are afraid such controls will constrain business, slow the economy, and increase the cost of living for U.S. citizens. An increase in the cost of living might result if the costs of emission controls were passed on to consumers in the form of higher prices for gasoline, heating, air conditioning, and so on.
- In contrast to the United States, the European Union (EU) is one of the strongest advocates for emission regulations and cutbacks. This is primarily because Europe is more energy efficient than the United States and, as a result, could more easily meet reduced emission quotas. Another reason Europe is so assertive on the global warming issue is that western European countries are grouped together as members of the European Union, the supranational organization representing 15 states, which gives them some advantage in calculating emissions. To illustrate, as a group the EU can meet requirements for a cutback to 1990 emissions because since that year the British coal industry has effectively closed down. As a result, EU emissions are much lower now than earlier. This trade-off could compensate for higher emissions since 1990 in another EU country, such as Italy or France.
- One of the most vexatious tensions is between the developed nations and less-developed countries. Historically, the industrial world (North America, Europe, Japan) created the global warming problem with unrestricted emissions. Furthermore, today these countries contribute more than half the problematic greenhouse gases to the atmosphere. Many argue these developed countries should not only be required to take stringent steps to curb their own emissions, but they should also subsidize and underwrite emission controls in the developing countries (Figure 2.16). Understandably, the less-developed countries are reluctant to sign any kind of emission control agreement that will constrain their economic future when—as they argue—they have contributed very little to the global warming problem up to now. China, for example, which has the world's largest supply of soft coal, emphatically resists mortgaging its economic future, by agreeing not to use that coal to fuel its industrialization.

▲ **Figure 2.16 Bangladesh brick factory** Increasing industrialization in the developing world poses new threats to global environmental health. Often free of stringent pollution controls, Third World manufacturing operations offer new economic opportunities but at a high environmental price. *(S. Noorani/ Woodfin Camp & Associates)*

Because of issues such as these, a timely solution to global warming seems elusive. While the technology exists for controlling most greenhouse gas emissions, the political will and a strong commitment from the major global polluters appears to be lacking.

Water on Earth: A Scarce and Polluted Resource

Water is one of the most critical—and scarcest—resources on Earth. Not only is it needed to sustain human life itself, but also to support agricultural systems, industry, transportation, and even recreation. However, water is unevenly distributed about the world; some regions have abundant water resources, others have serious shortages. For example, about 40 percent of the world's population lives in arid or semiarid regions, such as Southwest Asia, North Africa, or Central Asia.

The Global Water Budget

A world map shows that most of Earth is covered by water. Specifically, more than 70 percent of the surface area of the world is covered by oceans. This means that of the total global water budget, 97 percent is salt water, with only 3 percent fresh water. Of that miniscule amount of fresh water, almost 99 percent is locked up in polar ice caps and mountain glaciers. Of the remaining 1 percent, about half is in groundwater and the other half is accessible in rivers and lakes.

Another way of conceptualizing this limited amount of fresh water is to think of the total global water supply as 100 liters, or 26 gallons. Of that amount, only 3 liters (0.8 gallon) would be fresh water; of that, only a mere 0.003 liters, or about half a teaspoon, would be available to humans. Given this limited amount of water, coupled with growing demands for its use, political and economic tensions over effective water usage are bound to increase. International planners use the concept of **water stress** to describe and predict where water resource problems will be greatest (Figure 2.17). Water stress data are generated from the amount of fresh water available on a per capita basis in different parts of the world. By dividing known population growth rates into the amount of water available, a picture of future problem areas appears.

Africa stands out as the location of future difficulties because it is predicted that three-quarters of the population will experience water shortages or water stress problems by the year 2025. Other problem areas will be northern China, India, much of Southwest Asia, Mexico, and even parts of Russia. Aggravating water shortages throughout the world is the great amount that is polluted. Most commonly this pollution results from urban sewage, which is usually dumped untreated into streams and rivers flowing through cities. In India, for example, less than a third of all sewage receives any kind of treatment; in many developing countries, the figure is closer to 25 percent. If the world is to solve its water shortages, the solution will have to come from combined efforts to increase water supplies, while at the same time cleaning up and protecting those rivers and lakes that are now unsafe and unusable because of pollution.

Flooding

Ironically, within the context of global water shortages, floods cause the most deaths of all the different natural disasters, far more than earthquakes, hurricanes, or volcanic eruptions, More specifically, floods are responsible for about 40 percent of deaths from natural disasters. Further, as the world's population increases, this figure seems to be increasing (Figure 2.18). This can be attributed to several causes. As the population of a densely settled area increases, either from natural growth or from in-migration, people are often forced to settle in hazardous areas, such as river floodplains and deltas. Bangladesh is a good example. Such a move increases their risk due to flooding. Another factor is that the incidence and magnitude of flooding is increasing because of deforestation. Without a forest cover to absorb and slow rainfall runoff, flooding increases downstream. Last, urbanization can also increase flooding because cities are made up of hard and impervious surfaces such as pavement and parking lots that do not absorb rainfall. With urban expansion, it is common that river flooding increases downstream unless specific measures are taken to control this altered drainage. When these three factors are put together—population increases, deforestation, and urbanization—we see why the toll from flooding is increasing.

The Oceans

Not only are the oceans a major influence on global climate, but they are also an important source of food for much of the world's population. More specifically, fish from the oceans is the primary protein source for about 20 percent of the world's people, most of whom live in East Asia, Africa, and Latin America. However, because of a combination of pollution

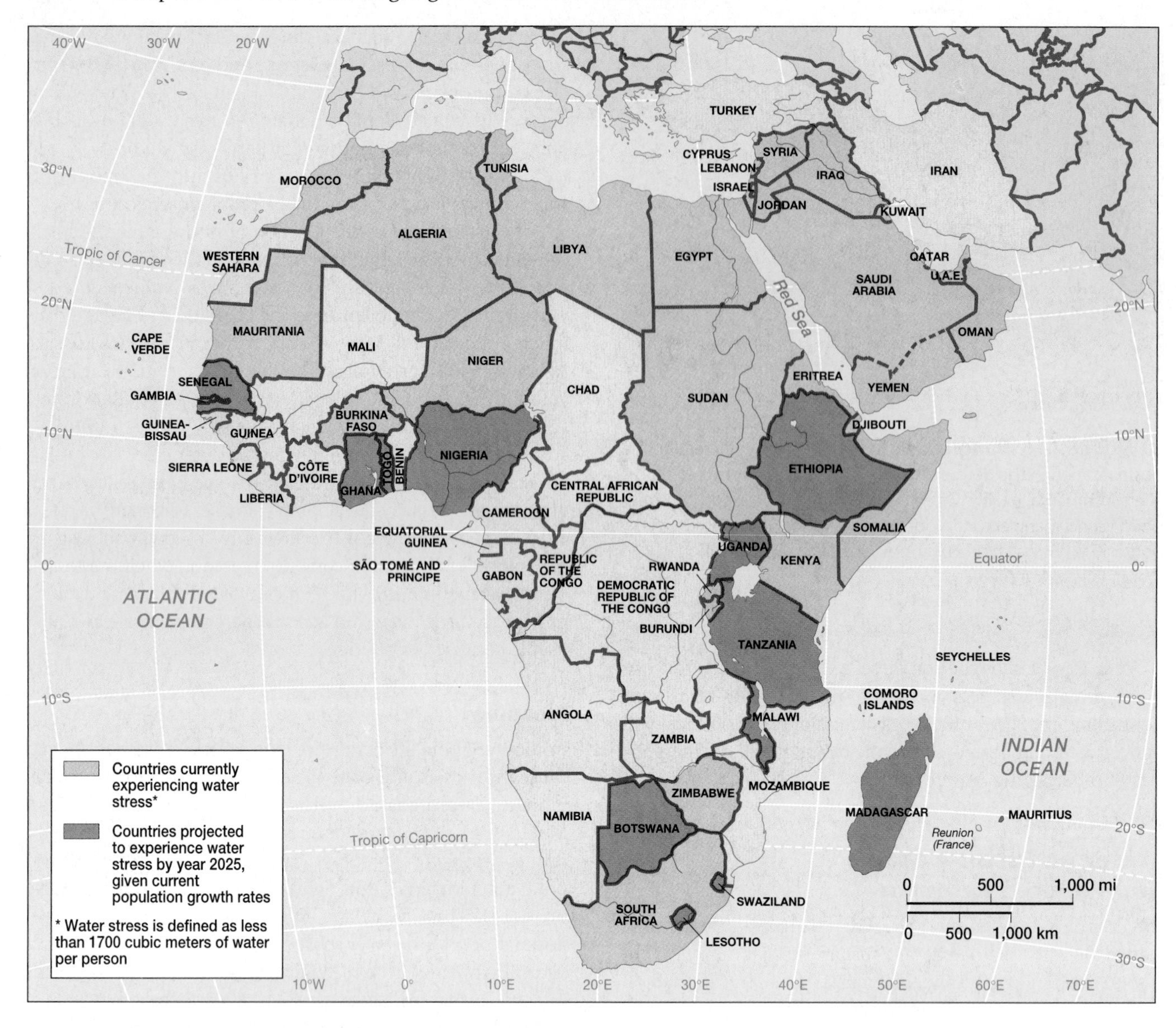

▲ **Figure 2.17 Water stress in Africa and Southwest Asia** International resource planners use the concept of water stress to target where populations currently suffer from shortages of water, as well as where shortages might occur in the future. Projections to the year 2025 are based upon current water availability, combined with present-day population growth rates. As can be seen on this map, many countries in Africa and Southwest Asia either do—or will—experience severe water stress.

and overfishing, this important food source has approached a state of crisis. Several years ago the United Nations' Food and Agricultural Organization (FAO) estimated that two-thirds of the world's fish stocks were endangered from overfishing. This is particularly the case in the northern portions of both the Atlantic and Pacific oceans. The cause for this depletion appears to be the rapid expansion of industrial fishing during the 1970s and 1980s. Most experts agree that the number of commercial fishing boats should be reduced by almost a third to enable a natural balance to be restored.

The problem, however, is not only that of overfishing by globalized industrial fishing fleets in deep oceanic waters, there is also a serious crisis in subsistence coastal fishing done by local peoples. Here the problem is primarily the destruction of fisheries by pollution and urban development into the mangrove swamps that serve as nursery areas for coastal fish species. In the Philippines, for example, about 20 percent of the fish catch is from coastal reef fish. However, at this time almost three-quarters of the country's coastal reefs have been damaged by a combination of urban sewage and sediment accumulations due to erosion from commercial logging. As a result, the traditional supply for many Philippine coastal peoples is now in jeopardy (see "Case Study: Fishing with Poison in the Pacific")(Figure 2.19).

▲ **Figure 2.18 Flooding in the Philippines** Growing world populations create more opportunities for natural disasters. Increasingly, human settlements such as Makati City (near Manila) in the Philippines are located in flood-prone areas and the trend is likely to continue in the future. *(Bullit Marquez/AP/Wide World Photos)*

▲ **Figure 2.19 Cyanide fishing** Many coral reef ecosystems in the tropical Pacific have been subjected to harmful cyanide. Fishermen use the poison to stun and harvest local fish populations. While lucrative, the practice often results in severe overfishing of fragile oceanic environments. *(Lynn Funkhouser/Peter Arnold, Inc.)*

CASE STUDY Fishing with Poison in the Pacific

Fishing with cyanide, the most lethal poison known to science, is common through the Pacific islands, as local people and international firms alike capture tons of reef fish alive by stunning them with poison. These fish are bound for two international markets—upscale restaurants throughout Asia where diners select live fish from tanks for their dinner, and, second, the worldwide demand for ornamental aquarium fish. Whether the fish ends up on a dinner plate in Hong Kong or in a tank in a dentist's office in California, it was probably captured using cyanide.

In the last few decades, tons of this poison have been squirted onto coral reefs throughout the Pacific, including the Philippines, Indonesia, Malaysia, New Guinea, and Micronesia. Cyanide fishing is fairly simple. Fishermen first crush cyanide pellets into bottles filled with seawater. Then they dive down to coral formations and squirt the cyanide solution into crevices where fish hide. The cyanide stuns the fish, which makes them easy to capture, although in some cases the fishermen have to pry coral heads apart with metal crowbars to retrieve the stunned fish.

Cyanide fishing can be very rewarding and lucrative for the exporters and importers. Additionally, military, police, and other officials are well paid for looking the other way. It is the divers, however, who are exploited and pay a heavy toll, as does the environment. For the divers, the health threat is not just from handling cyanide on a regular basis, but also from the amount of time they spend underwater breathing through tubes attached to air compressors on surface boats. Not only do they commonly suffer from decompression sickness ("the bends"), but they also suffer from breathing bad air from faulty air compressors.

This sort of fishing is also very destructive to coral reef ecosystems. Since large percentages of the fish that are captured live actually die in transit due to their poisoned state, many more fish must be taken to ensure suppliers of their quota. As a result, overfishing of these reef environments is widespread. Additionally, cyanide kills reef invertebrates, as well as the coral itself. Recent reports from the outer islands of the Philippines and Indonesia report widespread areas of empty and devastated coral reefs. While deadly in any marine environment, cyanide fishing is particularly tragic in the Pacific, which contains 70 percent of the world's coral reefs. This kind of fishing for the export trade also threatens the subsistence of island and coastal people who depend on fish protein as a crucial component of their diet.

Source: Adapted from Charles Barber and Vaughan Pratt, "Poison and Profits: Cyanide Fishing in the Indo Pacific." *Environment* 40(8), October 1998.

Human Impacts on Plants and Animals: The Globalization of Nature

One aspect of Earth's uniqueness compared to other planets is the cloak of vegetation covering the continents, containing a rich diversity of plants and animals. Geographers and biologists think of this vegetation as the "green glue" that binds together life, land, and atmosphere because this cloak is both a product of—and also affects—climate, geology, and hydrology. Humans are also very much a part of this interaction. Not only are we evolutionary products of a specific **bioregion**, or assemblage of local plants and animals (the tropical savanna of Africa, in this case), but our subsequent human prehistory was deeply intertwined with the domestication of certain plants and animals. From this process of domestication has come agriculture; thus, we are also closely bound to this "green glue." Over time, humans changed the natural pattern of plants and animals dramatically by plowing grasslands, burning woodlands, cutting forests, and hunting animals. The pace of such change has accelerated in recent decades, and now these actions have led to a crisis in the biological world—forests are devastated, plants and animals exterminated, and watersheds denuded. As a result, the very vitality of our life-giving biosphere is threatened in many parts of the world.

Many of these problems can be explained by the globalization of nature and of local ecologies. Until the last half-century, for example, tropical forests were primarily homes for small populations of indigenous peoples who made modest demands on local environments for their sustenance and subsistence. Today, however, these same tropical forests are capital for multinational corporations that clear-cut forests for international trade in wood products or search out plants and animals to meet needs of far-removed populations. As a result, Japanese lumber companies cut South American rain forests; German pharmaceutical corporations harvest medicinal plants in Africa; North American bears are killed by poachers who then sell bear gall bladders on the Asian black market as elixirs and sexual stimulants. The list of problems and issues regarding human impacts on vegetation is long; therefore, many of these topics are discussed in the regional chapters of this book.

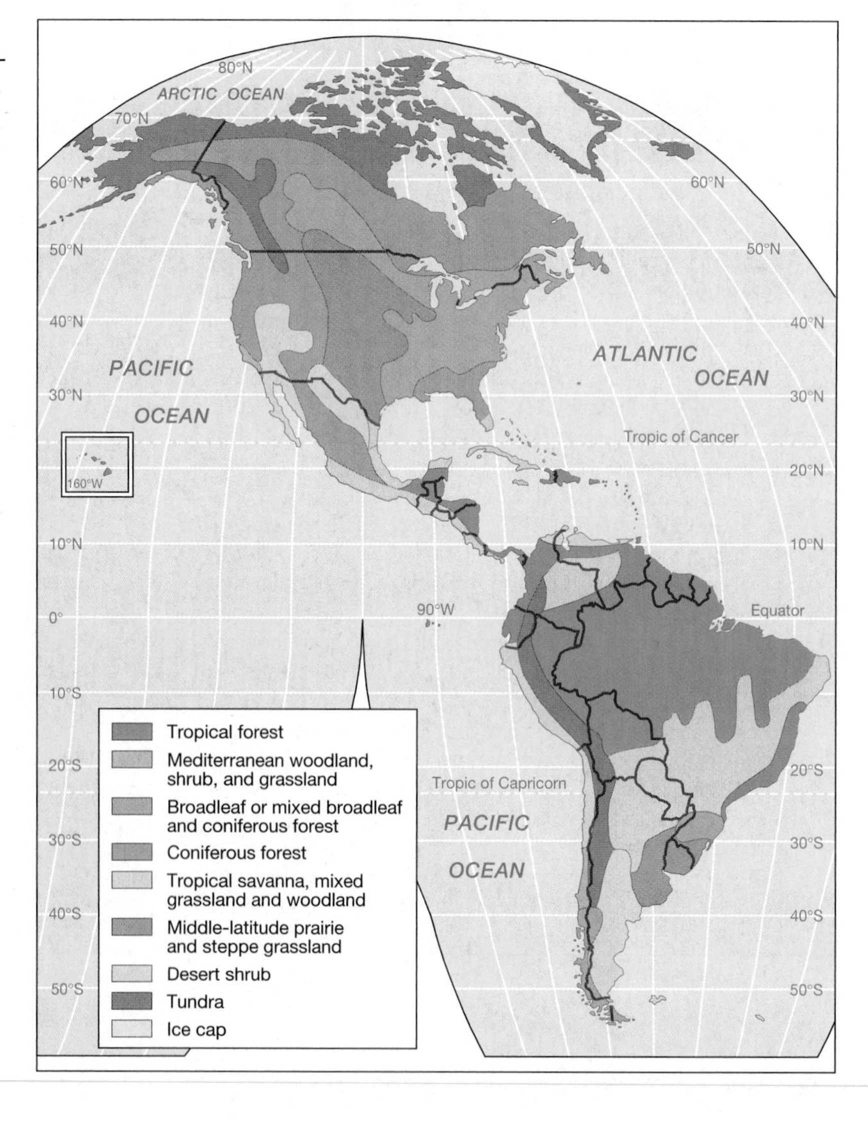

▶ **Figure 2.20 World bioregions** Although global vegetation has been greatly modified by humans by clearing land for agriculture and settlements, and cutting forests for lumber and paper pulp, there is still a recognizable pattern to the world's bioregions, ranging from tropical forests to arctic tundra. Each bioregion has its own array of ecosystems containing plants, animals, and insects. These species constitute the biodiversity so necessary for robust gene pools. Put differently, biodiversity can be thought of as the genetic library that keeps life going on Earth. *(Adapted from Clawson and Fisher, 1998,* World Regional Geography, *Upper Saddle River, NJ: Prentice Hall)*

Biomes and Bioregions

Biome is the biogeographical term used to describe a grouping of the world's flora and fauna into a large ecological province or region. In this book we use the synonym *bioregion*. Although some biologists would argue, we use the two terms interchangeably. Biomes and bioregions are closely connected with climate regions, since the major characteristics of a climate region—temperature, precipitation, and seasonality—are also the major factors influencing the distribution of natural vegetation and animals (Figure 2.20).

In the preindustrial world, there was a very close correlation between climate and vegetation. That linkage is less clear now, however, because of the ways humans have altered the environment. Desert areas now bloom with agricultural crops because of irrigation; once-extensive forests are gone, replaced in many cases with pastures or farm fields; plants and animals earlier restricted to particular environments are now domesticated and thrive worldwide. Nevertheless, even with these human-induced changes, large tracts of nature's original pattern can still be found in the landscape. A brief overview of the more important bioregions follows.

Tropical Forests and Savannas

The tropical forests are largely coincidental with the equatorial climate zones of high average annual temperatures, long days of sunlight throughout the year, and heavy amounts of rainfall. This bioregon covers about 7 percent of the world's land area, which is roughly the size of the contiguous United States, and is found in Central and South America, Sub-Saharan Africa, Southeast Asia, Australia, and, in small tracts, on many tropical Pacific islands. More than half of the known plant and animal species live in the tropical forest bioregion, making it the most diverse of all biomes.

The dense tropical forest vegetation is usually arrayed in three distinct levels that are adapted to decreasing amounts of sunlight closer to the forest floor. The tallest trees, which are close to 200 feet (61 meters) receive open sunlight; the middle level (around 100 feet, or 31 meters) gets filtered sunlight; the third level is the forest floor, where plants can survive with very little direct sunlight (Figure 2.21). Although a high amount of organic material accumulates on the forest floor because of falling leaves, tropical forest soils tend to be very low in stored nutrients. These nutrients are instead stored in

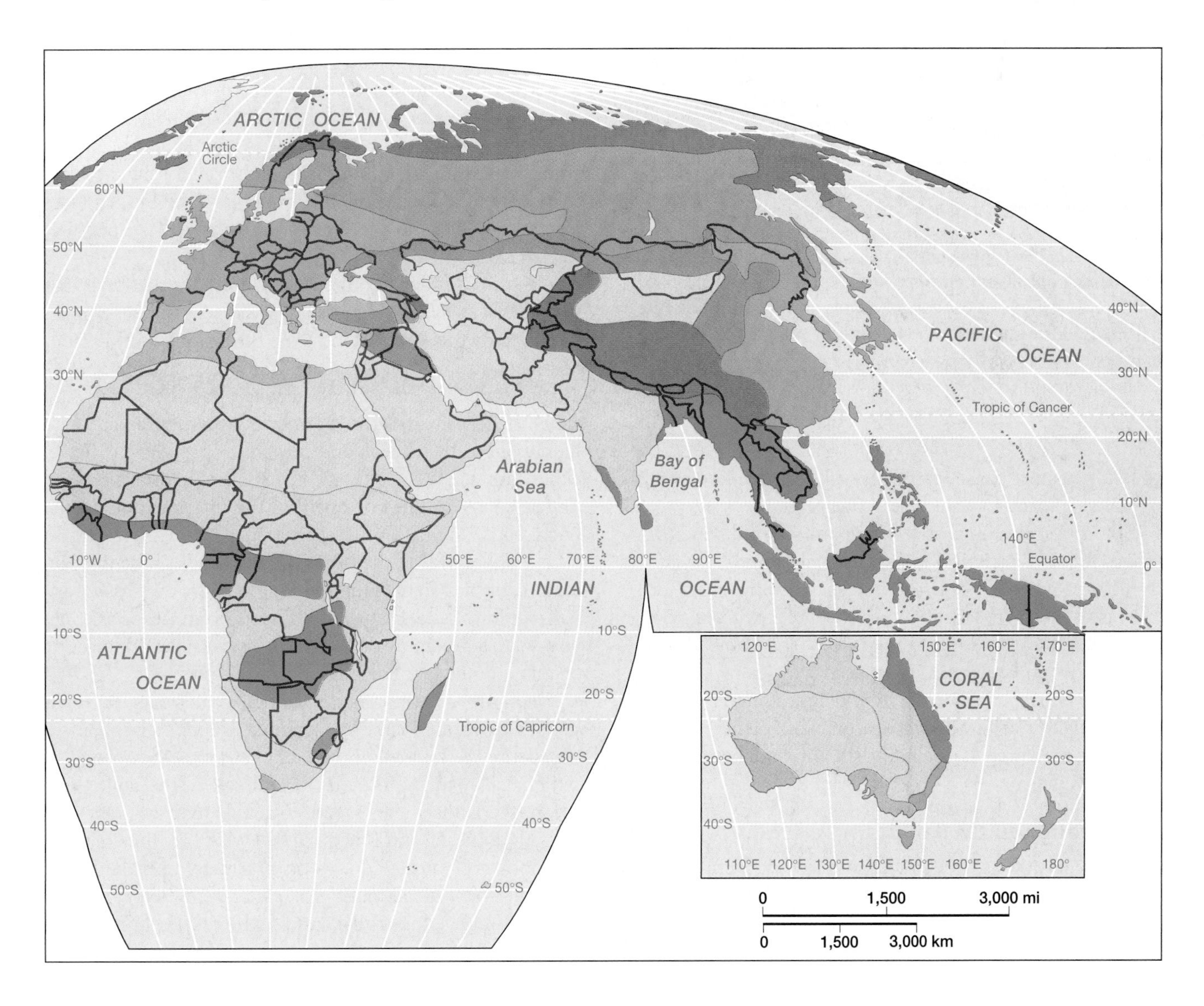

▲ **Figure 2.21 Tropical rain forest** As fragile as they are diverse, tropical rain forest environments feature a complex, multilayered canopy of vegetation. Plants on the forest floor are well-adapted to receiving very little direct sunlight. *(Gary Braasch/Woodfin Camp & Associates)*

the living plants themselves. As a result, tropical forest soils are not well suited for intensive agriculture.

If the tropical rainfall is seasonal and has a distinct dry season, the dense tropical rain forest is replaced by a more open forest where trees drop their leaves during the hot, dry season. At the poleward margins of the equatorial zone, where there are even longer dry seasons, the tropical forest is missing completely and is replaced with tropical grassland or savanna, characterized by widely scattered trees and lush grassland during the wet season. This is the environment of large mammals, including the ancestral home of the earliest humans.

Deforestation in the Tropics

For different reasons, tropical forests are now being devastated at an unprecedented rate, creating a crisis that tests our political, economic, and ethical systems. Although deforestation rates differ from region to region, each year an area of tropical forest about the size of Wisconsin or Pennsylvania is denuded. Spatially, almost half of this activity is in the Amazon Basin of South America. However, actual deforestation rates appear to be faster in Southeast Asia, where some estimates suggest logging rates three times faster there than in the Amazon (see "Case Study: Illegal Logging in Cambodia"). If those estimates are true, Southeast Asia could be completely stripped of forests within 15 years. Behind this widespread cutting of tropical forests lies the recent globalization of international wood products commerce. By all accounts, Japan was the first to globalize its timber business by reaching far beyond its national boundaries to purchase timber in North and South America and Southeast Asia. This was a two-pronged strategy for meeting the increased demand for wood products in Japan while at the same time protecting its own rather limited forests for recreational use. Currently, about one half of all tropical forest timber is destined for Japan. Unfortunately, much of it is used for throwaway items such as chopsticks and newspapers.

Another factor contributing to the rapid destruction of tropical rain forests is the world's seemingly insatiable appetite for beef. Cattle species originally bred to survive the hot weather of India are now found throughout the tropical world, raised on grassland pastures created by cutting tropical forests. Unfortunately, because tropical forest soils are poor in nutrients, cattle ranching is not a sustainable activity in these new grassland areas. After a few years, soil nutrients become exhausted, making it necessary to move ranching activities into newly cleared forest land. More forest is cut; more pasture created; and the process goes on at the expense of forest lands.

A third factor explaining tropical forest destruction is that these forest areas are often the last settlement frontiers for the rapidly growing population of the developing world. To illustrate, Brazil has used much of its interior Amazon rain forest for settlement in order to alleviate population pressure along its densely settled eastern coast where rural lands are monopolized by the rich and powerful. Brazil had a choice: either address the troublesome issue of land reform and break up the coastal estates or open up the interior rain forests. Brazil chose the path of least resistance, allowing settlers to clear and homestead the tropical forest lands of the interior. Many countries do the same, looking at the vast tracts of tropical rain forest as a safety valve of sorts, a landscape that can be used to temporarily deflect the pressures of land hunger by opening up the land for migration and settlement.

Often these three processes work together. First, international logging companies are granted timber concessions to log the forest. Following this, cutover lands are opened to settlement by migrants from other areas. At the same time, deforested lands are also made available for cattle ranching by both domestic and foreign firms. When all three of these demands are combined—wood products, beef, and land hunger—the tropical rain forest, along with the plants, animals, and indigenous tribal groups who dwell within are the losers.

CASE STUDY Illegal Logging in Cambodia

On the road south from Cambodia's capital city of Phnom Penh, Route 4 crosses over a spectacular mountain pass into a wide valley flanked by mountains covered with tropical forest. Until recently, the road was dotted with sawmills processing logs brought down from the mountains east and west. To the west are legal timber concessions; to the east, Bokor National Park, which was supposedly off-limits to logging. Yet these forests are now laid bare.

However, for now these sawmills along Route 4 are quiet as the Cambodian government responds to international pressure to crack down on illegal logging. But the damage has been done. Half of Cambodia's forests have been cut in the last several decades, most during an unbroken period of civil unrest. Government and military officials collaborated to strip the forests in an atmosphere of unbridled profit-taking. Much of the illegal logging was done under the eye of military and police officials who plundered national parks and nature preserves, immune to penalty by high-level political connections and powerful weaponry.

The World Bank estimated that if the government had not stepped in with strong reform measures, Cambodia's forests would be completely worthless by the year 2003. "Cambodia is a textbook example of bad forestry management. You name it, they've done it," said one World Bank official. "The political environment for the last five years has been very difficult, and the worst victim was the forestry sector."

Cambodia is one of the world's poorest countries and is highly dependent on foreign aid. Earlier, international donors made it clear to the Cambodian government that their aid would stop unless real progress was made in forestry reform. In response, Prime Minister Hun Sen implemented a plan to curb deforestation that included cancellation of major timber concessions, increased revenues from foreign logging firms, and strong prohibitions against logging national parks and nature preserves. These new regulations will be enforced by Cambodian army troops. Government officials claim that 95 percent of illegal logging has now been stopped as more than 800 small unlicensed sawmills have been shut down.

This case study of illegal tropical forest logging is not unique to Cambodia. Unfortunately, illegal timber cutting is widespread throughout the Tropics, from Africa to Latin America, and certainly throughout Southeast Asia. All too common is illegal collaboration among government, military, and police officials that allows timber cutting in national parks and nature reserves. Either this is done overtly by international logging companies or, more commonly, by small local timber cutters who then sell the illegal logs to giant international firms.

Source: Adapted from Mark Lioi, "Cambodia Slows Its Destruction of Jungles," *San Francisco Chronicle*, July 28, 1999.

Deserts and Grasslands

Large areas of arid and semiarid climate lie poleward of the Tropics, and here are found the world's extensive deserts and grasslands. Fully a third of Earth's land area qualifies as true desert, with less than 10 inches (25 centimeters) of rainfall per year. In those areas receiving more than 10 inches of rainfall, grassy plants appear, often forming a verdant cover during the wet season; in the higher elevations where evapotranspiration is lower because of cooler temperatures, extensive grasslands may cover the landscape. In North America, the midsection of both Canada and the United States is covered by grassland known as **prairie.** In other parts of the world, such as Central Asia, Russia, and Southwest Asia, shorter, less-dense grasslands form the **steppe.**

The boundary between desert and grassland has always fluctuated naturally because of changes in the climate; during wet periods, grasslands might expand, only to contract once again during drier decades. This transition zone between the two is a rather precarious environment for humans, as the United States found out during the 1930s when the semiarid grasslands of the western prairie lands turned into the notorious "Dust Bowl." At that time, hundreds of farmers saw their fields devastated by wind erosion and drought—a calamity that led to the exodus of thousands from these once-productive farmlands. Farming these marginal lands may actually worsen the situation, leading to **desertification,** or a spread of desert-like conditions (Figure 2.22). This has happened on a large scale throughout the world—in Africa, Australia, South Asia, to name just a few regions. In fact, current estimates are that an area about the size of Brazil has become desertified through poor cropping practices, overgrazing, and the buildup of salt in soils from irrigation that ruins the land. In northern China an area the size of Denmark became desertified in the 30 years between 1950 and 1980 with the expansion of farming into marginal lands. Earlier, this region averaged three sandstorms a year; today, 25 such storms are common each year. Although some scientists say the case is somewhat overstated, the United Nations recently estimated that about 60 percent of the world's rangelands are threatened by desertification. If such an amount of rangeland were to become desert, according to the UN, the agricultural livelihood of some 1.2 billion people will be threatened.

Temperate Forests

The large tracts of forests found in middle and high latitudes are called *temperate forests.* This vegetation is quite different from the low-latitude forests found in the equatorial regions. In the temperate forests, two major tree types dominate. One type

▲ **Figure 2.22 Desertification** Climatic fluctuations and human misuse combine to produce desertification. In such localities, marginal lands are overcropped or grazed heavily, resulting in an expansion of nearby deserts. Globally, many rangelands remain threatened by desertification. *(Mark Edwards/Still Pictures/Peter Arnold, Inc.)*

▲ **Figure 2.23 Clear-cut forest** Commercial logging in Washington's Olympic Peninsula has dramatically reshaped the landscape. Elsewhere in the Pacific Northwest, environmental lobbies have successfully halted such economic activities, often sparking local controversy. *(Calvin Larsen/Photo Researchers, Inc.)*

are softwood coniferous or evergreen trees, such as pine, spruce, or firs that are found in higher elevations and higher latitudes. The second category comprises the deciduous trees that drop their leaves during the winter season. Examples are elm, maple, or beech. Because these trees are hardwood, they are generally less favored by the timber industry than softwood species.

In North America, conifers dominate the mountainous West, Alaska, and Canada's western mountains, while deciduous trees are found on the eastern seaboard of the United States north to New England. There, the two tree types intermix before giving way to the softwood forests of Maine and the Maritime provinces of eastern Canada. In the coniferous forest of western North America, the struggle between timber harvesting and environmental concerns remains controversial. While timber interests argue they must meet high demand for lumber and other wood products through increased cutting, environmental concerns such as the protection of habitat for endangered species (for example, the spotted owl) have caused the government to place large tracts of forest off limits to commercial logging (Figure 2.23). Further complicating the future of western forests are global market forces. More specifically, many Japanese timber firms will pay premium prices for logs cut from U.S. and Canadian forests, thus outbidding domestic firms for these scarce resources. Since these trees are often cut from public lands, this facet of globalization raises an interesting question over the appropriate usage of public forests that are maintained by tax money.

In Europe, hardwoods were the natural tree cover in western countries such as France and Great Britain before these once-extensive forests were cleared away for agriculture. In the higher latitudes of Norway and Sweden, coniferous species prevail in the remaining forests. These conifers also form extensive forests across Germany, eastern Europe, and through Russia into Siberia, creating an almost unbroken landscape of dense forests. The Siberian forest remains a resource that could become a major source of income for Russia. Some argue that if these Siberian forests are put on the block of global trade, it will reduce logging pressure on North America's western forests. This in turn could make it easier to enact and enforce comprehensive environmental protection in the United States. This illustrates once again how the forces of globalization are intertwined with the fate of local ecosystems.

Food Resources: Environment, Diversity, Globalization

If the human population continues to grow at expected rates, food production must be doubled between now and the year 2025 just to provide each person in the world with a basic subsistence diet. Every minute of each day, there are 170 people born who need food; during the same minute, about 10 acres of existing cropland are lost because of environmental problems. Soil erosion and desertification are the main culprits in this loss.

Because of the close ties between food production and environmental geography, this issue of food availability draws together those topics discussed earlier, such as landforms, climate, water, and vegetation. Food scarcity also involves the many facets of geography discussed in Chapter 1—economics, geopolitics, and culture. Furthermore, the topic of food and hunger is framed by the larger issues of globalization and diversity, of the tensions between local and world forces. Many experts argue that food scarcity will be the defining issue of the next several decades, just as the ideological tensions between superpowers defined decades in the recent past. Frequent discussion of environment, population, agriculture, and food supplies appear in the world regional chapters of this book. In the next several pages, we set the scene for these discussions by introducing some of the important concepts underlying world food resource issues.

Industrial and Traditional Agriculture

Food is produced in a great variety of ways around the world, and these different production strategies and techniques yield widely varying results. A useful, though necessarily simplified starting point for assessing these strategies is the differentiation between *industrial* and *traditional* food production systems. Industrialized agriculture is practiced on about 25 percent of the world's croplands. It is characterized by the usage of large amounts of fossil fuel, both to power farm machinery and to provide the industrial basis for chemical fertilizers and pesticides (Figure 2.24). Additionally, irrigation water is generally used to a high degree. Because of the high level of mechanization, labor is low. In the United States, for example, less than 2 percent of the total workforce is employed in agriculture.

In contrast, traditional agriculture is practiced by almost half the people on Earth. Besides being labor-intensive, traditional agriculture is also characterized by less mechanization, less fertilizer, infrequent applications of pesticides, and, as a result, generally lower yields. Until recently, traditional agriculture was synonymous with **subsistence agriculture,** which is defined as farming that produces only enough crops or livestock for a farm family's survival. However, because of the increasing influence of market economies through globalization, many traditional farmers now also grow cash crops to supplement their subsistence needs (Figure 2.25).

The amount of capital (which includes labor) necessary for an agricultural system provides two further evaluative reference points. They are *intensive* and *extensive* farming. Intensive farming can be found in industrial and traditional systems. Intensive industrial farming requires high inputs of capital to support fossil fuels, machinery, and fertilizers and generally produces high yields. Intensive traditional farming primarily requires high inputs of labor. Padi rice agriculture in Southeast Asia is an example. Extensive farming is used where labor and capital inputs are considerably lower, such as in open-range livestock ranching in North America or nomadic herding in North Africa. In both cases, large amounts of grazing land are needed to support herd animals. Another example of extensive agriculture would be traditional wheat farming in places such as Turkey, Iran, or Poland, where animals rather than machinery are used for plowing and harvesting; thus, the capital inputs for fossil fuels or tractors is nonexistent.

▲ **Figure 2.24 Industrial farming in Iowa** Harvesting corn in Iowa reveals the imprint of technology upon modern agriculture. In such settings, large capital investments in machinery, fuel, and agrochemicals produce high yields while keeping labor costs low. *(Andy Levin/Photo Researchers, Inc.)*

▲ **Figure 2.25 Subsistence farming in Burkina Faso** These West African women are harvesting sorghum. Such traditional agricultural operations remain pivotal in feeding people in the less-developed world. They require low inputs of technology, but are often quite labor-intensive. *(Mark Edwards/Still Pictures/Peter Arnold, Inc.)*

The Green Revolution

During the last 40 years of the twentieth century, the world's population has doubled. Even more remarkably, during the same time, global food production also doubled to keep pace with this population explosion. This increase in food production came primarily from the expansion of intensive, industrial agriculture into areas that previously produced subsistence crops through extensive and traditional means.

More specifically, since 1950 the increases in global food production have come from three interconnected processes that are known as the first stage of the **Green Revolution.** These three processes are, first, the change from traditional mixed crops to monocrops, or single fields, of genetically altered, high-yield rice, wheat, and corn seeds; second, intensive applications of water, fertilizers, and pesticides; and, third, further increases in the intensity of agriculture by reducing the fallow or field resting time between seasonal crops.

Since the 1970s, a second stage of this Green Revolution has evolved. This emphasized new strains of fast-growing wheat and rice specifically bred for tropical and subtropical climates. When combined with irrigation, fertilizers, and pesticides, these new varieties allow farmers to grow two or even three crops a year on a parcel that previously supported just one. Using these methods, India actually doubled its food production between 1970 and 1992.

However, many argue these agricultural revolutions also carry high environmental and social costs. Since these crops draw heavily on fossil fuels, there has been a 400 percent increase in the agricultural use of fossil fuels during the last several decades. As a result, Green Revolution agriculture now consumes almost 10 percent of the world's annual oil output. Cheap oil prices have facilitated much of this increased agricultural usage of fossil fuel, yet if oil prices once again increase as they have in the recent past, one must wonder how this will affect food prices in the less-developed world. The environmental costs of the Green Revolution countries include damage to habitat and wildlife from diversion of natural rivers and streams to agriculture; pollution of rivers and water sources by pesticides and chemical fertilizers applied in heavy amounts to fields; and the increase in regional air pollution from factories and chemical plants that produce these agricultural chemicals.

In addition, there is evidence that the Green Revolution can also bring high social costs and disruption to some areas. Because the fiscal costs to the farmer participating in the Green Revolution are higher than with traditional farming, either capital or the access to capital through bank loans is a necessary ingredient. Money is needed for hybrid seeds, fertilizers, pesticides, and even new machinery. For those with high social standing, a family support system, or good credit, the rewards can be high, but for those without access to loans or family support, the toll can be heavy. Usually traditional farmers cannot compete against Green Revolution crops in the regional marketplace. As a result, they often struggle at a poverty level, while Green Revolution farmers prosper. In some wheat-growing areas of India, the social and economic distance between the well-off Green Revolution farmers and poor, traditional agriculturalists has become a major source of economic, social, and political tension. What was once an area where all farmers shared a common plight has become a highly stratified society of rich and poor. These social costs can be condoned only because of the pressing need for food in this rapidly growing country.

Problems and Projections

Even though agriculture has been able to keep up with population growth in the last decades, few are completely confident that this can be done once again in the twenty-first century (Figure 2.26). While fuller discussion of food and agriculture issues is found in the regional chapters of this book, four key points offer a starting point for understanding the issues.

1. While overall food production remains an important global issue, it is in fact local and regional problems that often keep people from obtaining food. To many experts, the issue is less one of global food production and more the daunting problems of widespread poverty and civil unrest at local levels that keep people from growing, buying, or receiving adequate food supplies. If the total global agricultural output were somehow shared equally among all people in the world, each person would have approximately four times his or her basic daily need. Given this view, some say there is already enough food to feed the world's population. The real issue, they argue, is distribution and purchasing power.
2. Further, political problems are usually more responsible for food shortages and famines than are natural events such as drought and flooding. Distribution of world food supplies is often highly politicized, starting at the global level and continuing right on down to the local. Food aid goes to political friends and allies, while enemies go without. During the Cold War, both the Soviet Union and the United States provided food aid to their ideological allies, but never to those of the other camp. Put differently, the abundance of food available in the world was selectively distributed depending on political outlook and allegiance. Whether or not this politicization of world food supplies will continue in the post-Cold-War era remains to be seen.
3. Globalization is causing dietary preferences to change worldwide, and the implications of this change could be profound. Currently, two-thirds of the world population is primarily vegetarian, eating only small portions of meat because it is treated as a luxury item. Yet many would increase their meat diet if it were affordable. Because of recent economic booms in some developing countries, an increasing number of people who were previously vegetarian

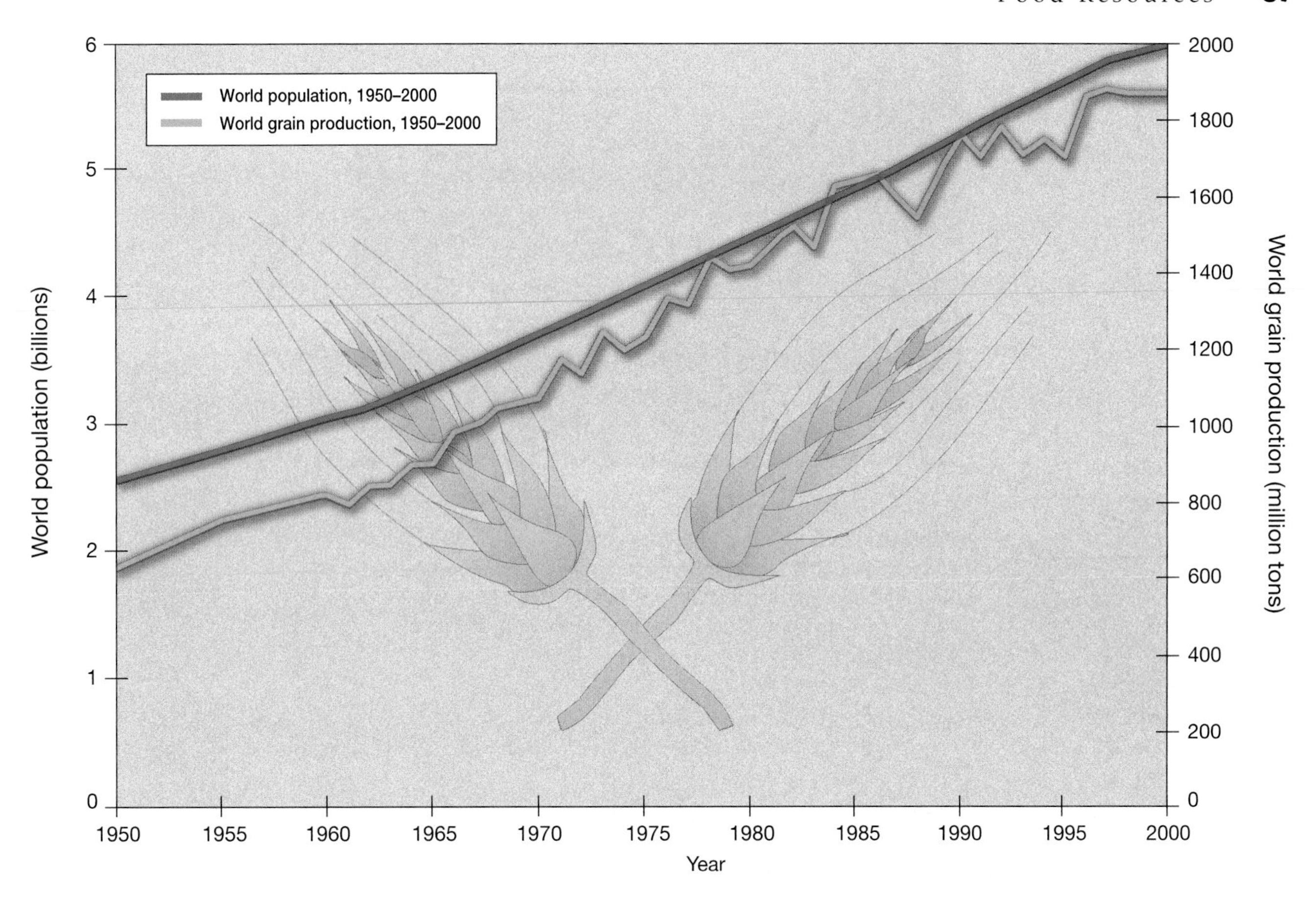

▲ **Figure 2.26 World population and grain production, 1950–2000** Grain production has been able to keep pace with population growth since the mid-twentieth century as the Green Revolution has contributed to higher crop yields. However, the future remains uncertain. Costs associated with modernized agriculture, increasingly stressed environments, and political instability may outweigh any future gains in food production.

are now eating meat, having moved up the food chain from a subsistence diet based upon local resources to a global diet containing a greater proportion of meat. In many cases this comes not just from a new economic prowess but also from changing cultural tastes and values as people are exposed to new products through economic and cultural globalization. There are, however, constraints on the world's ability to supply meat (Figure 2.27). Some food experts say that the global food production system could sustain less than half the current world population if everyone ate the standard meat-rich diet of North America, Europe, and Japan. Given this constraint, at some point the world will have to face the reality of a limited meat supply.

4. Given these different issues, most food supply experts agree that the two world regions of greatest concern are Africa and South Asia. Until 1970, Africa was self-sufficient in food, but since that time there have been serious disruptions and even breakdowns in the food supply system. The causes for these problems in Africa are twofold—rapid population increase and civil disruption from tribal warfare. As a result, one out every four people face food shortages in Sub-Saharan Africa. More detail on these issues is found in Chapter 6 on Sub-Saharan Africa. South Asia's future is also problematic. The United Nations predicts that by the year 2010, almost 200 million people in South Africa will suffer from chronic undernourishment. Further discussion of these problems is found in Chapter 12, South Asia.

There is, however, some good news in this otherwise bleak picture. Because population growth rates are generally declining in the industrializing areas of East Asia, and because food production there is still increasing, the UN predicts the percentage of undernourished people in that region will actually drop by almost 5 percent in the next 10 years. Similarly, gains are being made in Latin America because of lower population growth and higher agricultural production. Predictions are that the proportion of chronically hungry in Latin America will fall to about 6 percent in 2010, which is half the rate of the 1990 figure.

▲ **Figure 2.27 Cattle ranching in Brazil** Growing global demands for meat are reshaping the world's agricultural landscapes. Cattle ranching in western Brazil's rain forest reflect these changing market conditions, but the practice is also introducing new environmental problems. *(Martin Wendler/Peter Arnold, Inc.)*

Conclusion

Environmental geography is fundamental to the study of world regional geography because of its close links to issues of globalization and diversity. For example, understanding the natural diversity of the world and the uniqueness of its local landscapes necessitates understanding those processes shaping a region's geology, climate, vegetation, and hydrology. The interaction of those natural forces forges the character of a place. Additionally, these natural processes provide humans with the fabric of our habitat in terms of environmental resources.

Global environmental change has always been a part of the human backdrop. Sometimes this environmental change results from purely natural processes, such as the unpredictable climatic fluctuations that produce a drought year or an unusually wet season. Increasingly, however, global environmental change is being driven by human activities. These changes take place at all scales, from the global to the local. While some environmental change, such as the logging of an Alaskan rain forest, are an expected by-product of world and national economic activity, other environmental changes are often unanticipated, inadvertent, and often accidental. For example, decades ago only a few scientists saw the large-scale consequences of global climate change resulting from the atmospheric by-products of fossil fuel consumption. Today, however, the problems of global warming are accepted by most.

One lesson learned from these experiences is that global environmental science is a relatively new endeavor, and, further, is constantly evolving as we strive to understand the world's complex environmental interactions. Because of the newness and evolving nature of global environmental geography, we cannot always expect absolutely accurate and certain scientific predictions as to how our global environment will change in the next few decades. Negotiating solutions to any environmental problem is difficult under the best of circumstances. Unfortunately, this becomes even more difficult when there is scientific uncertainty about the consequences. The current controversy over the effects and implications of global climate change illustrate this problem, and this scientific uncertainty complicates finding an appropriate international solution.

Globalization is both a help and a hindrance to world environmental problems. From a positive perspective, some would argue that the world's nation-states are increasingly willing to sign international agreements to solve environmental problems. Examples would be treaties on whaling, ocean pollution, fisheries, and the protection of wildlife species. Because of these supranational agreements, much progress has been made in some areas of environmental protection.

A conflicting view argues globalization is the villain that has aggravated global environmental problems. This is the case, say the critics, because of environmental damage resulting from superheated economic activity that exploits the widest possible range of international resources, regardless of the consequences. Additionally, since unrestricted world free trade is a handmaiden of globalization, these trade agreements often conflict with national and local environmental protection, as was discussed in "Globalization, Trade Agreements, and the Environment."

The conflict and controversy between globalization and environment is an important theme discussed further in the following regional chapters. Each regional section begins with a section on the environmental geography of the region. Those discussions draw upon the concepts and discussion presented in this introduction and may, in fact, refer readers back to this chapter for background information. Additionally, each regional chapter contains a map of environmental issues. On these maps call-outs highlight the most pressing environmental controversies and, if a noteworthy solution has taken place, this also is mentioned. Besides the environmental issue maps, each regional chapter provides cartographic displays of the physical and environmental geography of each region, along with a separate map of climates. Taken together, these resources should provide readers with a good sense of the environmental geography for each of the 12 world regions.

Key Terms

anthropogenic (page 54)
biome (page 61)
bioregion (page 60)
climate region (page 53)
climograph (page 53)
continentality (page 51)
convection cells (page 44)
convergent plate boundary (page 45)
desertification (page 63)
divergent plate boundary (page 45)
greenhouse effect (page 50)
Green Revolution (page 66)
insolation (page 50)
maritime climate (page 51)
plate tectonics (page 44)
prairie (page 63)
rift valley (page 45)
steppe (page 63)
subduction zone (page 45)
subsistence agriculture (page 65)
tectonic plates (page 45)
water stress (page 57)

Questions for Summary and Review

1. Explain how convection cells are linked to the tectonic plate theory.
2. Describe several different kinds of tectonic plate boundaries, along with the landforms and landscapes often associated with these boundaries.
3. Explain why there are usually more human casualties from earthquakes than from volcanic eruptions.
4. Describe the natural greenhouse effect.
5. How do continental and maritime climates differ? What causes this difference?
6. What are the major atmospheric pollutants causing global climate change and warming?
7. Describe the major unresolved issues in achieving an international agreement to limit global warming.
8. Which natural disaster takes the highest human toll—earthquakes, volcanic eruptions, or flooding? Why?
9. Describe the ecological characteristics of a tropical rain forest.
10. What are the different causes of tropical forest deforestation? How are they linked to globalization?
11. What is *desertification?* Why, and where, is it a problem?
12. How does industrial agriculture differ from traditional agriculture?
13. Describe the characteristics of the Green Revolution.
14. Explain the different variables important to providing the world's population with adequate food. How might these change in the next 10 years?

Thinking Geographically

1. What are the most threatening natural hazards in your region? Earthquakes, tornados, hurricanes, floods, drought? Contact local agencies or visit the public library to find copies of disaster preparedness plans for your community. After reading these, discuss them with your class to determine if they are adequate for the danger faced in your area. If not, suggest how they might be changed and updated.
2. What climate region do you live in? What are the major weather problems faced by people in your area? How do they adjust or adapt to these problems? Has your climate changed over the last decades? You can gain some insight into this by plotting annual temperature and precipitation data over the last 30 years to see if there is a trend toward warmer or drier conditions.

3. Visit the many different Web sites on the Internet about global warming. Once you get an overview of the different positions, concentrate on one or two of the most contentious issues, such as the debate within the United States between environmentalists and business interests, or the difference in opinion between developed and developing countries. Make sure you also look at the Web sites for the small island nations threatened by flooding. If possible, set up a debate in your class that will bring out the different issues, perspectives, and agendas held by vested interests and different countries.
4. How has the vegetation in your area been changed by human activities in the last 100 years? Has this change led to the extinction of any plants or animals or placed them on the endangered list? If so, what is being done to protect them or restore their habitat?
5. Study the globalization of wood products by acquainting yourself with the source areas for different items such as building lumber, paper, furniture, or other items found in a local import store. More specifically, does the lumber used in construction in your area come from Canada or the United States? What items in local stores come from tropical forests? In what part of the world—Asia, Africa, or Latin America—are these forests located?
6. Select a foreign country of your choice for a more detailed analysis of its food issues. In your study, consider the following points. In general, has food supply kept up with population growth? Has this country or region suffered recently from food shortages or famine? If so, were these from natural causes, such as drought or floods, or from distribution problems resulting from civil disruption? What segments of the population have food security and what peoples have recurring problems obtaining adequate food? How is this inequity best explained? Finally, have their food preferences or diets changed recently? If so, how and why?

Bibliography

Alexandatos, Nikos, ed. 1995. *World Agriculture: Toward 2010. An FAO Study.* Chichester, UK: John Wiley.

Allen, J. C., and Barnes, D. F. 1985. "The Causes of Deforestation in Developing Countries." *Annals, Association of American Geographers* 75, 163–184.

Barber, Charles, and Pratt, Vaughan. "Poison and Profits: Cyanide Fishing in the Indo Pacific." *Environment* 40(8), 4ff.

Blaikie, Piers, and Brookfield, Harold. 1995. *Land Degradation and Society.* London: Methuen.

Blaikie, Piers, et al. 1994. *At Risk: Natural Hazards, People's Vulnerability, and Disasters.* London: Routledge.

Botkin, Daniel, and Keller, Edward. 1995. *Environmental Science.* New York: John Wiley.

Brown, Lester. 1996. *Tough Choices: Facing the Challenge of Food Scarcity.* New York: W. W. Norton.

Brown, Lester; Flavin, Christopher; and French, Hilary. 1999. *State of the World.* New York: W. W. Norton.

Brown, Lester; Renner, Michael; and Halweil, Brian. *Vital Signs 1999: The Environmental Trends That Are Shaping Our Future.* New York: Worldwatch Institute, W. W. Norton.

Christopherson, Robert. 1997. *Geosystems: An Introduction to Physical Geography,* 3rd ed. Upper Saddle River, NJ: Prentice Hall.

Gelbspan, Ross. 1997. *The Heat Is On: The High Stakes Battle Over Earth's Threatened Climate.* Reading, MA: Addison Wesley.

Gleick, Peter, ed. 1993. *Water in Crisis: A Guide to the World's Fresh Water Resources.* New York: Oxford University Press.

Global Environment Facility. 1998. *Valuing the Global Environment: Actions and Investments for a 21st Century.* Washington, DC: Global Environment Facility.

Goudie, Andrew. 1994. *The Human Impact on the Natural Environment,* 4th ed. Cambridge, MA: The MIT Press.

Hecht, Susanna, and Cockburn, Alexander. 1990. *The Fate of the Forest: Developers, Destroyers, and Defenders of the Amazon.* New York: HarperCollins.

Hertsgaard, Mark. 1998. *Earth Odyssey: Around the World in Search of Our Environmental Future.* New York: Broadway Books.

Hidore, John. 1996. *Global Environmental Change: Its Nature and Impact.* Upper Saddle River, NJ: Prentice Hall.

Livernash, Robert, and Rodenburg, Eric. 1998. "Population Change, Resources, and the Environment." *Population Bulletin* 53(1).

Mackenzie, Fred. 1998. *Our Changing Planet,* 2nd ed. Upper Saddle River, NJ: Prentice Hall.

Marchak, Patricia. 1995. *Logging the Globe.* Montreal: McGill-Queens University Press.

Miller, G. Tyler. 1993. *Living in the Environment,* 9th ed. Belmont, CA: Wadsworth.

Moser, Susanne. 1996. *Human Driving Forces and Their Impacts on Land Use/Land Cover.* Washington, DC: Association of American Geographers.

Ojima, D. S., et al. 1995. "The Global Impact of Land Use Changes." *Bioscience* 44(5), 300–303.

Ott, Hermann. 1998. "The Kyoto Protocol: Unfinished Business." *Environment* 40(6), 16ff.

Postel, Sandra. 1996. *Dividing the Waters: Food Security, Ecosystem Health, and the New Politics of Scarcity.* Worldwatch Paper 132. Washington, DC: Worldwatch Institute.

Schaeffer, Robert. 1997. *Understanding Globalization: The Social Consequences of Political, Economic, and Environmental Change*. Lanham, MD: Rowman & Littlefield.

Sponsel, Leslie; Headland, Thomas; and Bailey, Robert. 1996. *Tropical Deforestation: The Human Dimension*. New York: Columbia University Press.

Soroos, Marvin. 1998. "The Thin Blue Line: Preserving the Atmosphere as a Global Common."*Environment* 40(2), 6ff.

Turner, B. L., et al. 1990. *The Earth as Transformed by Human Action*. Cambridge: Cambridge University Press.

World Resources Institute. 1999. *World Resources, 1999–2000*. New York: Oxford University Press.

RUSSIA
ARCTIC OCEAN
0 250 500 mi
0 250 500 km
80°N
GREENLAND
(DENMARK)
Baffin Bay
Prudhoe Bay
Yukon R.
ALASKA
(U.S.)
Fairbanks
Anchorage
Mackenzie R.
YUKON
Yukon R.
NUNAVUT
Iqaluit
NORTHWEST TERRITORIES
Whitehorse
Juneau
Yellowknife
50°N
NORTH AMERICA
Political Map
Over 1,000,000
500,000–1,000,000
(selected cities)
Selected smaller cities
(National capitals shown in red)
Slave R.
Peace R.
Churchill
Hudson Bay
BRITISH COLUMBIA
ALBERTA
SASKATCHEWAN
MANITOBA
Athabasca R.
CANADA
QUEBEC
Fraser R.
Edmonton
Vancouver
Calgary
ONTARIO
St. Lawrence R.
40°N
Regina
Seattle
Winnipeg
Quebec
WASHINGTON
Columbia R.
Montreal
Portland
Ottawa
MONTANA
NORTH DAKOTA
PACIFIC OCEAN
OREGON
MINN.
VT.
N.H.
Toronto
Albany
Minneapolis-St. Paul
WIS.
MICH.
Rochester
N.Y.
MA.
IDAHO
SOUTH DAKOTA
Mississippi R.
Buffalo
CT.
Milwaukee
Detroit
Hartford
WYOMING
Cleveland
PENN.
N.J.
UNITED STATES
IOWA
Chicago
IND.
OHIO
Pittsburgh
Philadelphia
Reno
Sacramento
NEVADA
Salt Lake City
Cheyenne
NEBRASKA
Missouri R.
Indianapolis
Dayton
Columbus
Baltimore
DEL.
San Francisco
UTAH
Denver
ILL.
Cincinnati
W.VA.
Washington, D.C.
30°N
CALIFORNIA
COLORADO
Kansas City
St. Louis
Richmond
KANSAS
MO.
Ohio R.
Louisville
VA.
Norfolk
Las Vegas
KY.
N.C.
Nashville
Charlotte
Los Angeles
ARIZONA
Oklahoma City
Memphis
TENN.
S.C.
San Diego
Phoenix
Albuquerque
OKLAHOMA
ARK.
Atlanta
NEW MEXICO
Birmingham
Tucson
Dallas-Ft. Worth
MISS.
ALA.
GEORGIA
TEXAS
Jacksonville
LA.
New Orleans
Orlando
Rio Grande
Houston
Tampa-St. Petersburg
FLORIDA
San Antonio
Gulf of Mexico
Miami
MEXICO
22°N
Honolulu
PACIFIC OCEAN
HAWAII
20°N
0 75 150 mi
0 75 150 km
160°W
158°W
156°W
130°W
120°W
90°W

C H A P T E R 3

North America

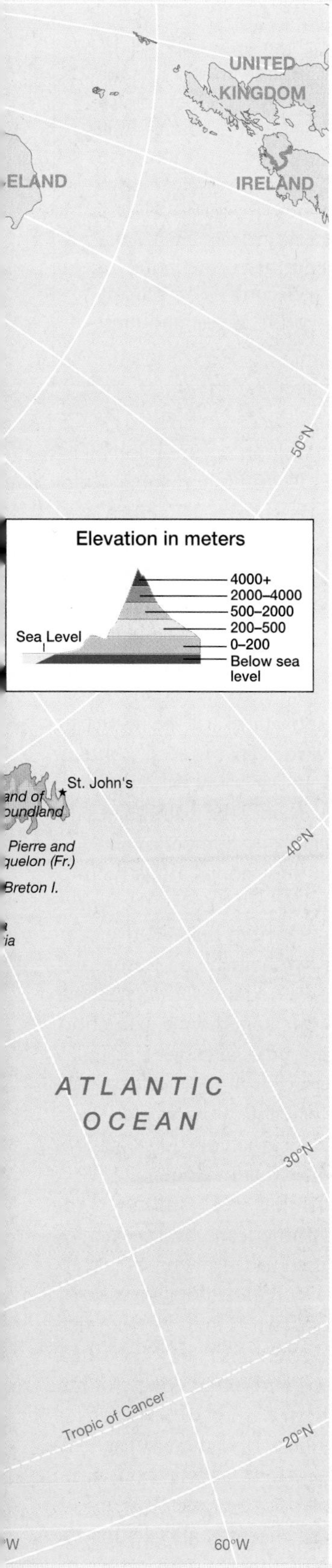

North America encompasses the United States and Canada, a culturally diverse and resource-rich region that has seen unparalleled human modification and economic development over the past two centuries (Figure 3.1; see also "Setting the Boundaries"). The result is one of the world's most affluent regions where two highly urbanized and mobile societies are oriented around the processes of globalization and the highest rates of resource consumption on Earth. Indeed, the realm superbly exemplifies a **postindustrial economy** in which human geographies are shaped by modern technology, by innovative financial and information services, and by a popular culture that dominates both North America and the world beyond.

Globalization has fundamentally reshaped North America. Stroll down any busy street in Toronto, Tucson, or Toledo and ponder the landscape (Figure 3.2). Count the ways in which international products, foods, culture, and economic connections shape the everyday scene. The simple visual lesson is revealing: today the North American region stands at the center of the globalization process and the results are transforming the cultural and economic geographies of the region. From farmers to computer workers, most North Americans are employed in occupations that are either directly or indirectly linked to the global economy. Sizable foreign-born populations in each nation also provide direct links to every part of the world. Tourism brings in millions of additional foreign visitors and billions of dollars that are spent everywhere from Las Vegas to Disney World. In more subtle ways, North Americans embrace globalization in their everyday lives. They consume ethnic foods, tune in to international sporting events on television, enjoy the sounds of salsa and Senegalese music, surf the Internet from one continent to the next, and invest their pensions in global mutual funds.

Globalization is also a two-way street and the outward reach of North American capital, culture, and power is ubiquitous. By any measure of multinational corporate investment and global trade, the region plays a dominant role that far outweighs its population of 300 million residents. North American automobiles, consumer goods, information technology, and investment capital circle the globe. In addition, North America remains a major exporter of such raw materials as wheat, coal, metals, and wood products. North

◀ **Figure 3.1 North America** The North American region includes both Canada and the United States. It contains one of the world's most highly urbanized and culturally varied populations. The region also includes some of the world's largest metropolitan areas as well as its most thinly settled frontiers. Globally, North America plays a pivotal role in industrial output and world trade. With a population of more than 300 million and extensive economic development, the region is also one of the largest consumers of natural resources on the planet.

Setting the Boundaries

The United States and Canada are commonly referred to as "North America," but that regional terminology can sometimes be confusing. In many geography textbooks, the realm is called "Anglo America" because of its close and abiding connections with Britain and its Anglo-Saxon cultural traditions. The increasingly visible cultural diversity of the realm, however, has discouraged the widespread use of the term in more recent years. While more culturally neutral, the term "North America" also has its problems. As a physical feature, the North American continent commonly includes Mexico, Central America, and often the Caribbean. Culturally, however, the United States–Mexico border seems a better dividing line, although the growing Hispanic presence in the Southwest, as well as ever-closer economic links across the border, make problematic even that regional division. North America's boundaries often extend northeastward to include Greenland, a colony of Denmark that is home to more than 55,000 indigenous people. To the west, North America reaches to Alaska's Aleutian Islands and to the state of Hawaii in the mid-Pacific Ocean. While the future may find Mexico even more intimately tied to its northern neighbors, for now the "North American" realm concentrates on Canada and the United States, two of the world's largest and most affluent nation-states.

American foods and popular culture are diffusing globally at a rapid pace. A new McDonald's restaurant opens every four hours somewhere on the planet! North American music, cinema, and fashion have also spread rapidly around the world. It is no accident that more than two-thirds of the world's Internet hosts are based in North America.

Politically, North America is home to the United States, the last remaining global superpower. Such status brings the country onto center stage in times of global tensions, whether they are in Latin America, East Africa, or South Asia. In addition, North America's largest metropolitan area of New York City (20 million people) is home to the United Nations and other global political and financial institutions. North of the United States, Canada is the other political unit within the region. While slightly larger in area than the United States [3.83 million square miles (9.97 million square kilometers) versus 3.68 million square miles (9.36 million square kilometers)], Canada's population is only about 10 percent that of the United States' (30 million vs. 270 million).

▲ **Figure 3.2 Toronto's cultural landscape** Toronto's varied ethnic population is celebrated during the Caribana Parade through the city. Powerful forces of globalization have reshaped the cultural and economic geographies of dozens of North American cities. *(Canadian Tourism Commission)*

Widespread abundance and affluence characterize North America. The region is extraordinarily rich in natural resources such as navigable waterways, good farmland, fossil fuels, and industrial metals. Combine that continental good fortune with the acquisitive nature of its European colonizers and an accelerating pace of technological innovation, and the results are reflected everywhere on the modern scene. Indeed, contemporary North America displays both the bounty and the price of the development process. On one hand, the realm shares the benefits of modern agriculture, globally competitive industries, excellent transport and communications infrastructure, and two of the most highly urbanized societies in the world. The cost of abundance, however, has been high: Native populations were all but eliminated by encroaching Europeans, forests were logged, grasslands converted into farms, precious soils eroded, numerous species threatened with extinction, great rivers diverted, and natural resources often wasted. Today, although home to only about 6 percent of the world's population, the region consumes 30 percent of the world's commercial energy budget and produces carbon dioxide emissions at a per person rate almost ten times that of Asia.

Nevertheless, economic growth has vastly improved the standard of living for many North Americans, and they enjoy high rates of consumption and varied urban amenities that are the envy of the less-developed world. Satellite dishes, sushi, and shopping malls are within easy reach of most North American residents. Amid this material abundance, however, there are persisting disparities in income and in the quality of life. Poor rural and inner-city populations still struggle to match the affluence of their wealthier neighbors, and these social geographies of poverty have been slow to disappear, even given the unprecedented economic growth of the past 50 years.

North America's unique cultural character also defines the realm. The cultural glue that holds this region together involves a common process of colonization, a heritage of Anglo dominance, and a shared set of civic beliefs in representative democracy and individual freedom. But the history of the region has also juxtaposed Native Americans, Europeans, Africans, and Asians in fresh ways, and the results are two societies unlike any other. Bicultural Canada is shaped by both its French and English antecedents, and major political tensions are associated with the geographic concentration of its French-speaking citizens within the province of Quebec. Added to Canada's cultural variety is a diverse mix of other European, African, and Asian immigrants who arrived in the country after 1900. South of the Canadian border, cultural pluralism characterizes an ethnically complex United States. The country witnessed early and enduring English and African contributions that were later supplemented by other European, Latin American, and Asian influences. Adding to the mix is a popular culture that today exerts a powerful homogenizing influence on North American society.

Environmental Geography: A Land of Plenty

North America's physical geography is incredibly diverse; its varied environmental settings include the rich lowlands of the humid Midwest, the deserts of the American Southwest, and the cool rain forests of the British Columbia coast (Figure 3.3). The sheer scale of the physical setting is notable: it is a region that varies in latitude from Hawaii (20° north) to northernmost Canada and Greenland (80° north). Furthermore, it includes landforms that range from the sprawling lowlands of the continental interior to the rugged, tectonically active mountain chains that stretch down the western coast from Alaska to southern California. Understanding the basic characteristics of that physical setting is essential to appreciating the scope and speed of the human transformations that have since taken place across the entire realm.

Fertile Plains and Rugged Mountains

The North American landscape is dominated by vast interior lowlands bordered by more mountainous topography (Figure 3.4) in the western portion of the region. Erosional and depositional signatures of coastlines, river basins, and glaciers are reminders that dynamic physical processes continue to transform the modern face of North America.

Each landform region within the region is associated with well-defined topographic characteristics that provide diverse settings for human settlement. The Gulf-Atlantic Coastal Plain stretches from southern New York to Texas and includes a sizable portion of the lower Mississippi Valley. These plains are flat, sometimes poorly drained, and are prone to coastal and river flooding. Much of the region was submerged in recent geologic time and is covered by depositional layers of sand, gravel, and clay. The present coastline is an intricate landscape of drowned river valleys, bays, swamps, and low barrier islands (Figure 3.5). Decent farmland, good accessibility, and early European settlement contributed to the area's extensive development and later urbanization. Today, large North American cities such as New York, Philadelphia, Miami, New Orleans, and Houston can be found within its bounds.

The nearby Piedmont, Appalachian Mountains, and Interior Highlands provinces provide more varied terrain than the coastal plains. The Piedmont consists of rolling hills and low mountains that are much older and less easily eroded than the nearby lowlands. West and north of the Piedmont are the Appalachian Mountains, an internally complex zone of higher and rougher country that typically crests between 3,000 and 6,000 feet (915 and 1,829 meters, respectively). From Pennsylvania south, folded layers of sedimentary rock, often rich in coal, create a diverse setting of ridges, valleys, and plateaus. North of Pennsylvania, the underlying rock is crystalline and surface forms are more irregular. Here the highlands reach the Atlantic and include New England and the Canadian Maritimes. The result is the famed and highly indented coastline that extends from southern Maine to the rugged, fjordlike landscapes of western Newfoundland. Also notable are signatures of glacial erosion north of central New York, revealed in the region's numerous lakes, broad and scoured U-shaped valleys, and thin soils. Far to the southwest, the smaller Interior Highlands are similar in appearance to southern Appalachia. They include southern Missouri's Ozark Mountains and the Ouachita Plateau of northern Arkansas. Overall, the Appalachian Mountains and Interior Highlands support lower population densities than much of eastern North America.

The Interior Lowlands are North America's largest landform region. These mainly depositional plains have accumulated thick sediments, and they extend from west central Canada to the coastal lowlands near the Gulf of Mexico. Eastward, they include the southern Great Lakes and the lower Ohio River Valley. Within this vast domain, gently sloping beds of sedimentary rock offer few barriers to extensive agricultural development. In addition, the Great Lakes (along with the St. Lawrence River) and the Mississippi Valley provide easy access into much of the continental interior. To the west, the Great Plains are formed from younger sediments that have been eroded from the Rocky Mountains. Glacial forces, particularly north of the Ohio and Missouri rivers, also have actively carved and reshaped the Interior Lowlands landscape.

North and east of the Interior Lowlands, the Canadian Shield and the Hudson Bay Lowlands comprise the remainder of eastern North America. Although sparsely settled and poorly endowed with agricultural potential, portions of the Shield are rich in minerals, a function of the ancient, crystalline rock that dominates the province. More than any other landform region, the lake-studded, soil-poor Shield bears the mark of glacial erosion because the entire area was once covered by enormous sheets of continental ice. The smaller lowlands along Hudson Bay resemble the Shield but are underlaid with more recent sedimentary deposits.

The western third of North America features a staggering variety of landforms that resist easy generalization. Geographers organize the area into the Rocky Mountains, the Intermontane Basins and Plateaus, and the complex Pacific Mountains and Valleys provinces. Active mountain-building (including large earthquakes and volcanic eruptions), alpine glaciation, and often spectacular processes of erosion produce

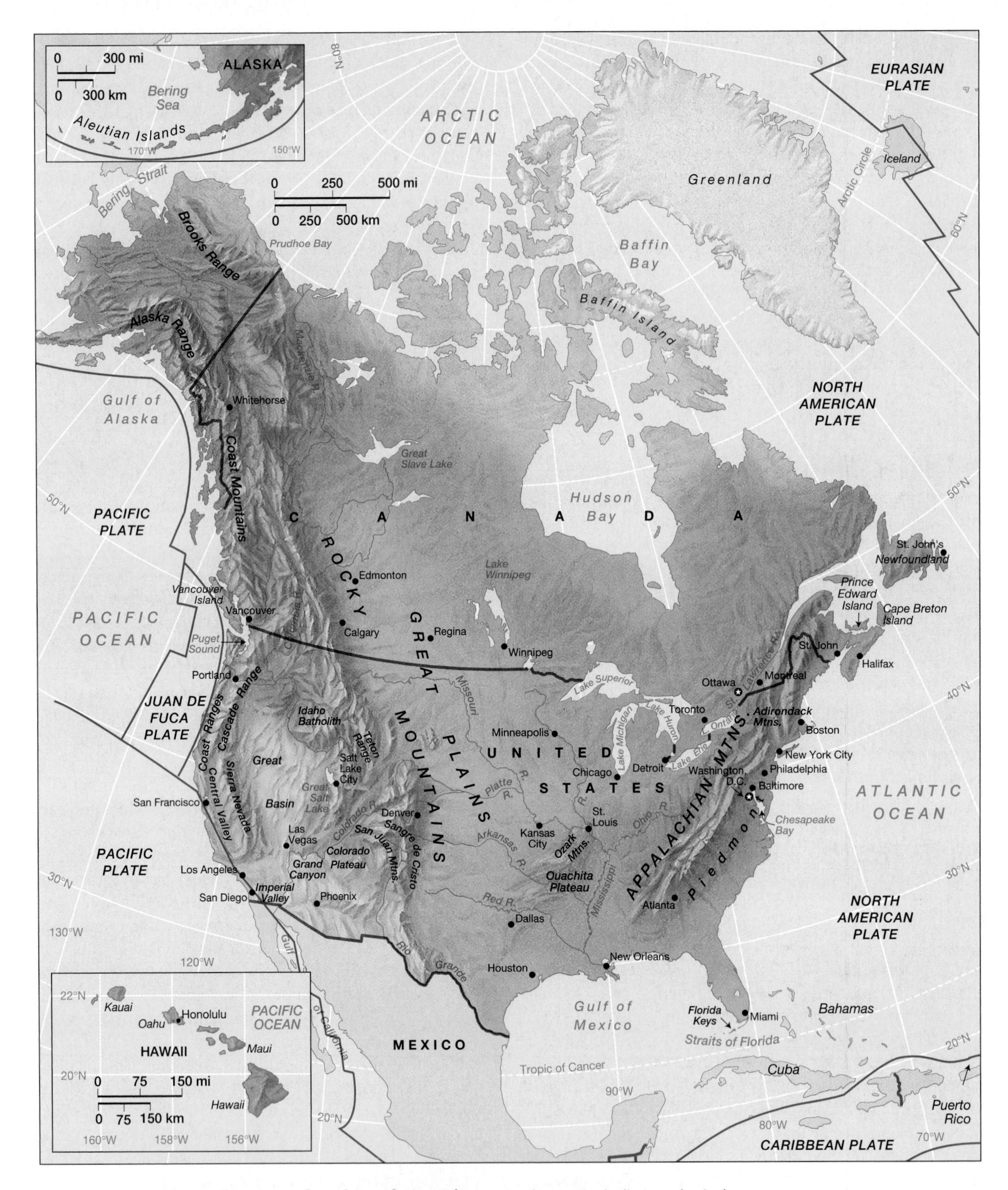

▲ **Figure 3.3 Physical geography of North America** North America's diverse physical setting includes both Arctic tundra and tropical forests. Stretching from northern Canada to the Hawaiian Islands, the region reveals varied climates, vegetation, and landforms. While old mountain systems such as the Appalachian and Adirondack ranges dominate in the east, western North America is home to active mountain-building, partly generated by the converging tectonic plates along the Pacific Coast.

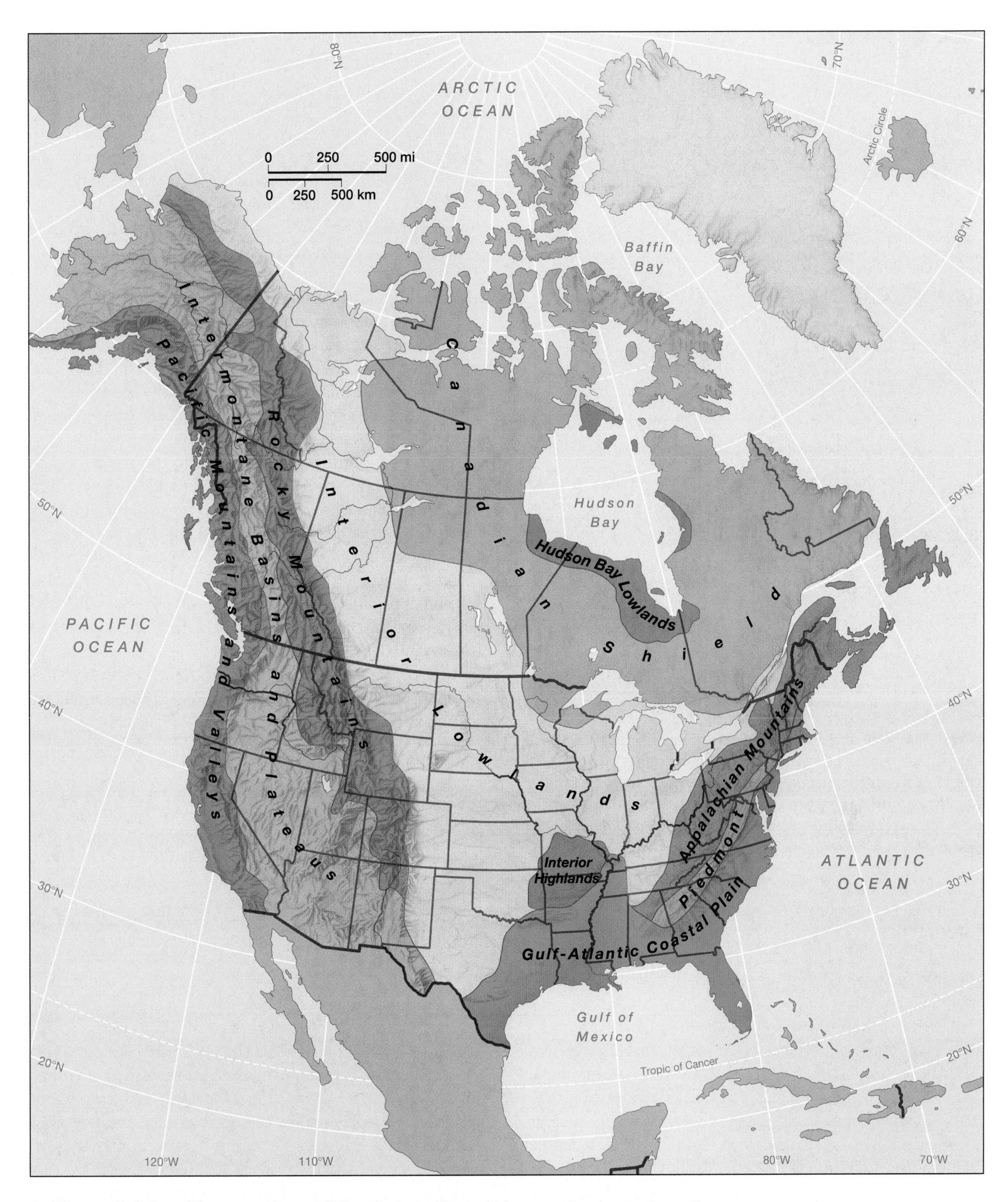

▲ **Figure 3.4 Landform regions of North America** This generalized map shows the major landform regions of North America. The Canadian Shield dominates eastern Canada. To the south, vast lowlands offer little topographic relief from Alberta to the Gulf Coast. The most dramatic landforms are found in the west, where rugged mountain ranges and plateaus alternate with structural valleys and basins. *(Modified from McKnight, 1992,* Regional Geography of the United States and Canada, *Upper Saddle River, NJ: Prentice Hall)*

▲ **Figure 3.5 Satellite image of the Chesapeake Bay** This view of the Middle Atlantic Coast reveals the intricate shoreline of the Chesapeake Bay (lower center). The coastal sector is characterized by drowned river valleys, barrier islands, and sandy beaches. The Piedmont zone and Appalachian system appear to the northwest. *(Earth Satellite Corporation/Science Photo Library/Photo Researchers, Inc.)*

a regional topography quite unlike that of eastern North America. Indeed, millions of westbound travelers, from the time of Lewis and Clark to the present, have eagerly scanned the approaching horizon for their first peek at this varied mountain country.

The Rocky Mountains are a series of uplifts, many more than 10,000 feet (3,048 meters) in height, that stretch from Alaska's Brooks Range to northern New Mexico's Sangre de Cristo Mountains. Spectacular fault-blocks such as the Grand Teton Range dominate, but broader igneous uplifts such as those found in Colorado's San Juan Mountains and in central Idaho add geological variety. The Rockies form a dramatic backdrop to cities such as Calgary and Denver, and their twin riches of metals resources and spectacular scenery have drawn European and American settlers for the last 150 years. Perhaps their greatest resource is the moisture they capture as accumulated snowfall. Thirsty residents nearby depend on that seasonal surplus. The human geographies of the Platte, Rio Grande, Columbia, and Colorado river basins are increasingly shaped by the challenges of getting water where it is needed for everything from hay fields and vegetable gardens to hydroelectric facilities and swimming pools.

West of the Rocky Mountains, the intermontane region includes the Colorado Plateau, where uplifted blocks of highly colorful sedimentary rock are eroded into spectacular buttes and mesas (Figure 3.6). Further west, Nevada's vast and sparsely settled basin and range country features numerous north/south-trending mountain ranges interrupted by structural basins with no outlet to the sea. To the north, lava plateaus and low mountains are interspersed with river valleys to create a complex tangle of landforms found from eastern Oregon to the Yukon.

The Pacific Mountains and Valleys province includes the mountainous and rain-drenched coasts of southeast Alaska and British Columbia; the Coast Ranges of Washington, Oregon, and California; the lowlands of the Puget Sound, Willamette Valley, and Central Valley; and the complex uplifts of the Cascade Range and Sierra Nevada. Much of the geological variety of the region is related to the crustal contact zone between the North American and Pacific tectonic plates. Mountain-building is so dynamic and frequent that it is often chronicled in the daily newspaper headlines. Major California earthquakes and smoking Cascade volcanoes offer an ongoing display of landscape changes that have direct and sometimes painful consequences for regional residents. Nature has also bestowed its riches upon the province: spectacular harbors include developments at Vancouver, Seattle, and San Francisco. Where water is available, some of North America's most productive farmland lies in California's Central Valley and in the Willamette Valley south of Portland.

Patterns of Climate and Vegetation

North American climates and vegetation are as varied as its landforms. Not surprisingly, the region's size, latitudinal range, and varied terrain have contributed to a diversity of temperature and precipitation patterns that have, in turn, shaped geographies of vegetation. Several factors explain North America's climatic variations (Figure 3.7). Latitudinal position is most crucial because it determines how a locality is situated with respect to global precipitation belts and seasonal temperature variations. Most populated parts of the realm are in the middle latitudes where westerly winds and shifting jet streams control the movement of storms from west to east.

Oceans and landforms also shape climate patterns within the region. Coastal zones, particularly in western North America, have maritime climates in which seasonal temperature extremes are moderated by the nearby ocean and prevailing onshore winds. Much of the North American interior, however, has a continental climate that exhibits greater seasonal variations in temperature, including hotter summers and colder winters. Landforms also shape climate, especially in western North America where mountains alter patterns of temperature and precipitation. Temperatures generally decrease with increasing elevation, while precipitation varies in more complex ways. The windward side of the uplift, typically to the west in the middle latitudes, receives more abundant precipitation as air masses are forced upward and release their

◀ **Figure 3.6 Landforms of the Colorado Plateau** Colorful sedimentary rocks are a part of the unique regional character of the Colorado Plateau. Such eroded uplifts are common across northern Arizona and southern Utah. The region's aridity and lack of vegetation help to highlight the dramatic colors of these buttes and mesas. *(© Carr Clifton)*

moisture. The leeward, or eastern slopes of the uplift, often experience the **rain shadow effect** as the air mass warms and dries, thus creating a distinctly more arid climate. Local variations can be striking, particularly when crossing the high Cascades or Sierra Nevada. Wet conditions to the west contrast sharply with the arid Great Basin east of the mountains.

North American climates strongly influence patterns of natural vegetation. Much of humid North America south of the Great Lakes is characterized by a relatively long growing season, 30 to 60 inches (76.2 to 152.4 centimeters) of precipitation annually (Figure 3.7, Columbus and Philadelphia), and a deciduous broadleaf forest that has been extensively modified by commercial agriculture. From the Great Lakes north, the coniferous evergreen or **boreal forest** dominates the cooler continental interior and these conditions extend to sparsely settled portions of eastern Quebec. Near Hudson Bay (Figure 3.7, Churchill) and across harsher northern tracts, trees give way to **tundra,** a mixture of low shrubs, grasses and flowering herbs that grow briefly in the short growing seasons of the high latitudes.

The drier continental climates found from west Texas to Alberta (Figure 3.7, Cheyenne) feature tremendous seasonal ranges in temperature, frequent winds (including tornadoes), and variable rates of precipitation that average between 10 and 30 inches (25.4 and 76.2 centimeters) annually. Indeed, the region straddles the transition zone between humid and arid portions of the North American interior, a boundary broadly defined as the 20-inch line of annual precipitation that runs near the 100th meridian. Much of this subhumid region is blessed with fertile soils and was originally clothed in **prairie** vegetation dominated by tall grasslands in the East and by short grasses and scrub vegetation in the West. As with favored portions of the humid East, however, the interior grasslands have seen their original vegetative cover replaced by exotic agricultural adaptations, particularly where dryland wheat can grow or where irrigation mitigates the threat of drought.

Western North American climates and vegetation are greatly complicated by the Pacific Ocean and the region's complex landforms. Marine west coast climates dominate north of San Francisco (Figure 3.7, Vancouver), while a dry summer Mediterranean climate holds sway across central and southern California (Figure 3.7, Los Angeles). Vegetation varies from coastal British Columbia's coniferous forests to the mixed oak woodlands and chaparral scrub found south of San Francisco. Variations related to elevation complicate these broad patterns and cities also modify local climates and patterns of native vegetation. Further east, the intermontane interior and the Rocky Mountains experience the typical seasonal variations of the middle latitudes, but the patterns are modified by the effects of topography. Vegetation is strongly influenced by altitude. A single mountain front in the Rockies, for example, might include sagebrush and pinyon pine on its lower slopes; a denser cover of ponderosa pine, western fir, and quaking aspen above; and an alpine zone of meadows, tundra, and snowfields on the ridge crest.

Natural Processes

Today's North American climates and vegetation are merely a snapshot in time, reflecting dynamic natural processes. Climate cycles have brought glacial advances and retreats, the most recent dating to about 15,000 years ago when ice sheets expanded across Canada and much of the eastern United States north of the Missouri and Ohio rivers. Although these glaciers melted, the current period could be followed by another era of extraordinary change in which

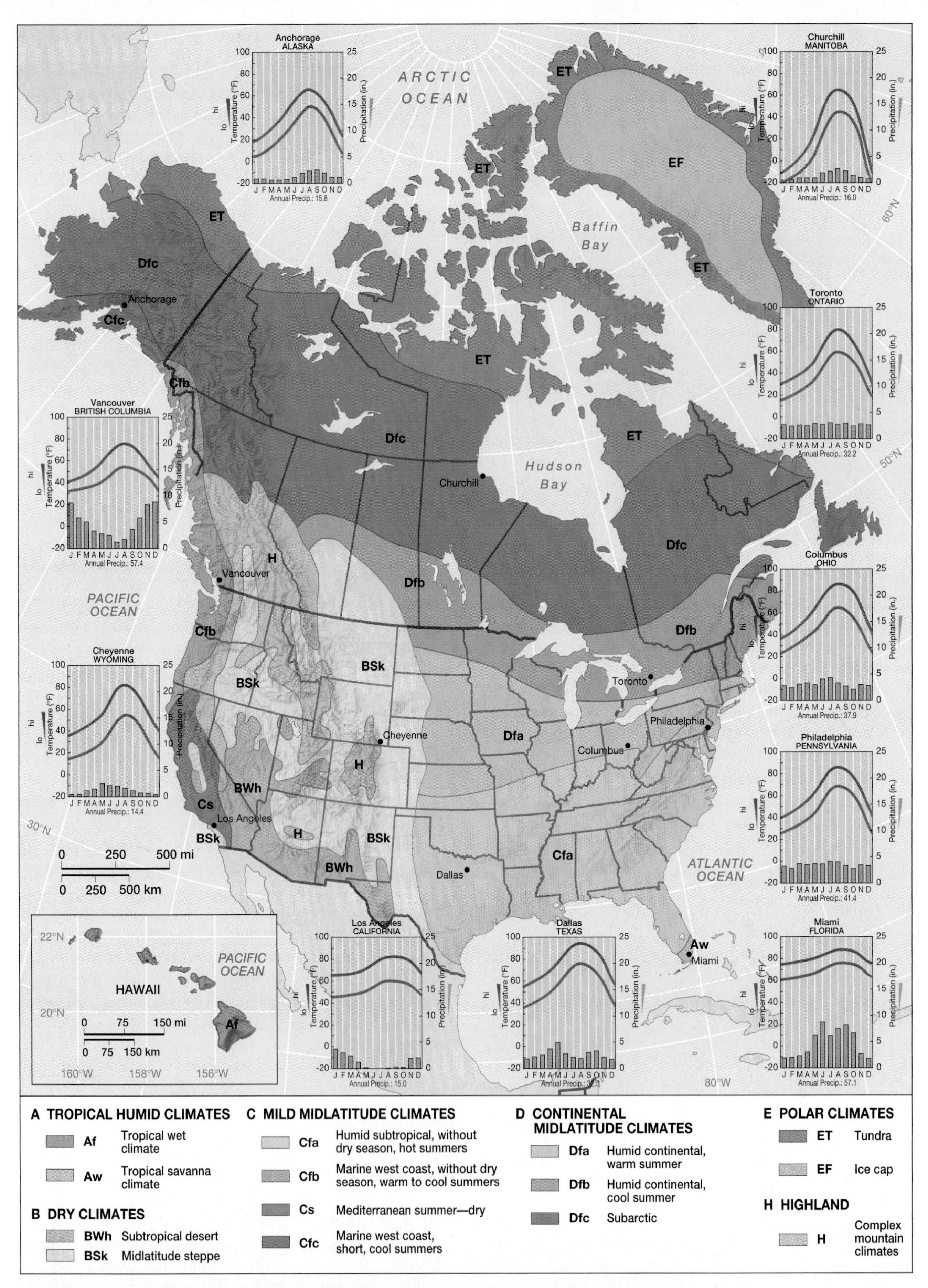

▲ **Figure 3.7 Climate map of North America** North American climates include everything from tropical savanna (Aw) to tundra (ET) environments. Most of the region's best farmland and densest settlements lie in the mild (C) or continental (D) midlatitude climate zones.

climate and vegetation belts are rapidly redistributed. Within such a scenario, North America's humid subtropics might find themselves an outpost of the subarctic! On the other hand, if global warming continues, perhaps fueled by natural as well as human causes, many of North America's cities, including New York, Miami, and Houston, would be under water, amid rising world sea levels.

Other dynamic natural processes exhibit shorter cycles, but still have profound consequences for many North American localities. Witness the endless oscillation of ocean currents and temperatures that drive the El Niño phenomenon and its impacts on many North American localities. Similarly, North Americans must adjust to periodic drought cycles on the Great Plains and prairies. Although the Dust Bowl era of the 1930s was the most famous of these dry spells, there is no reason to suspect that the cyclical pattern will not continue in the future.

The Costs of Human Modification

The story of North America's dynamic environment becomes even more complex when we consider how extensive human modifications have transformed the region's soils and vegetative cover, water resources, and atmosphere (Figure 3.8). The speed and magnitude of that transformation are remarkable, largely taking place over the past 200 years and leaving few areas of North America free from human imprint. Some changes are beneficial, including agricultural adaptations that easily feed the region's population. Other impacts have been less welcome, the unintended consequences of activities with lasting and destructive effects on the natural environment and its human occupants. Reacting to these issues, environmental movements in Canada and the United States have increasingly shaped national policies over the past 30 years. The most pressing environmental issues in North America today center on questions of public health, resource conservation, and the long-term costs of economic productivity.

Transforming Soils and Vegetation North American agriculture and urbanization have profoundly altered the environment. Particularly after Europeans arrived on the scene, countless new species of plants and animals invaded the continent, fundamentally reshaping its ecology to include exotics such as wheat, cattle, and horses. Forest cover was removed from millions of acres. Similarly, grasslands were plowed under and replaced with grain and forage crops not native to the region. With twentieth-century urbanization, many earlier agricultural transformations were swept aside to make way for suburbs and shopping centers. Indeed, more than two million acres of agricultural land are lost annually to growing North American cities. From the non-native lawns of southern California to the wheat fields of Montana and Alberta, North Americans today are living in rural and urban environments that bear little resemblance to settings across the region a mere two centuries ago.

Soil erosion is one of the most damaging consequences of this human transformation. Within the United States, almost two billion tons of soil is lost annually to erosion and nearly 40 percent of the country's intensively farmed croplands are losing topsoil faster than it is being replaced. Certain regions have been especially hard hit. Many of North America's semiarid grain-farming districts, such as eastern Washington, southern Saskatchewan, and the central Great Plains, are particularly prone to wind and water erosion. Indeed, the Dust Bowl conditions of the 1930s were caused by a combination of periodically occurring drought and by farmers plowing up too much marginal land in ways that accelerated erosion. Hilly portions of the Southeast have also been vulnerable, particularly agricultural regions in northern Missouri, western Mississippi and Tennessee, and the Piedmont zones from Georgia to North Carolina.

Managing Water North Americans consume huge amounts of water. While conservation efforts and better technology have slightly reduced per capita rates of water use over the past 20 years, city dwellers still average more than 170 gallons daily. When society's overall demands are figured to include total agricultural and industrial needs, every North American consumes more than 1,500 gallons of water per day! Approximately 45 percent of the water used in the United States is employed in manufacturing and energy production, 40 percent in agriculture, and the remainder for home and business use. While water is essential to North America's survival and level of affluence, managing it effectively and efficiently is a growing environmental challenge (see "Environment: Sink or Swim? Life on the Modern Mississippi"). It is a fundamental problem of geography: water resources are not equally spread across North America nor are demands for water evenly distributed across time and space. The result is a mammoth problem of moving water from one place to another, while at the same time maintaining or enhancing its quality so that it can be safely used where it is needed.

Many North American localities are threatened by water shortages. Metropolitan areas in eastern Canada and the United States struggle with outdated municipal water supply systems built when urban populations were a fraction of their current levels. The combined challenges of aging infrastructure and continually expanding suburban populations translate into the increasing probability of future water shortages in these settings. Beneath the Great Plains, the waters of the Ogallala Aquifer are also being depleted. The largest in North America, this aquifer is a huge reserve of precious groundwater created during the last Ice Age, and today it irrigates about 20 percent of all U.S. cropland. Huge center-pivot irrigation systems are steadily tapping into the supply at a rate of 21 million acre-feet of water annually, water tables across much of the region have fallen more than 100 feet (30 meters) in the past 50 years, and now the costs of pumping are rising steadily. Farther west, California's elaborate system of water management is a reminder of that state's ever-growing demands (Figure 3.9). Aqueducts move the precious resource from an already overutilized Colorado River system into southern California's agricultural and metropolitan regions. Both the eastern (Los Angeles Aqueduct) and western (Central Valley Project) slopes of the Sierra Nevada are also tapped to satisfy the thirsts of Golden State homeowners, farmers, and industrialists.

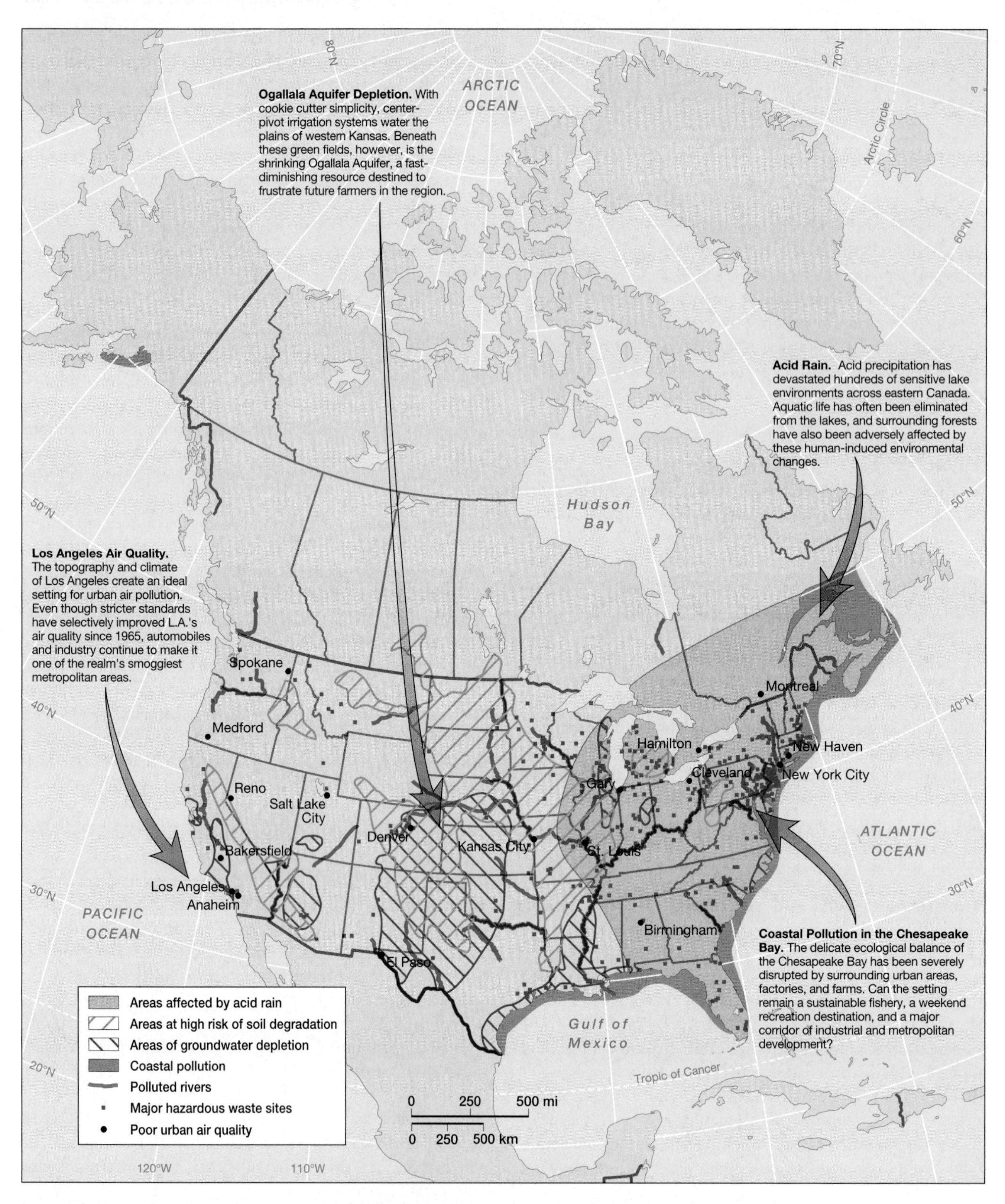

▲ **Figure 3.8 Environmental issues in North America** Many environmental issues threaten North America. Acid rain damage is widespread in regions downwind from industrial source areas. Elsewhere, widespread water pollution, cities with high air pollution, and zones of accelerating groundwater depletion pose health dangers and economic costs to residents of the region. Since 1970, however, both Americans and Canadians have become increasingly responsive to the dangers posed by these environmental challenges.

▲ **Figure 3.9 California aqueduct** Thirsty Californians have extensively modified the geography of water in their state. Large aqueducts have dramatically reconfigured the distribution of this precious resource, promoting tremendous agricultural and metropolitan expansion within western North America. *(Alexander Lowry/Photo Researchers, Inc.)*

Issues of water quality are equally problematic. North Americans are exposed to water pollution in varied ways, and even mandates, such as the U.S. Clean Water Act or Canada's Green Plan, cannot stop ongoing abuses. Mining operations and industrial users such as chemical, paper, and steel plants generate toxic wastewater or metals that enter surface and groundwater supplies. America's most toxic localities include the petroleum-rich Texas and Louisiana Gulf coasts and the older industrial centers of the Northeast and Midwest. Toxic threats in the West are found near industrial sites, in former mining regions from Arizona to Montana, and in nuclear fuel and chemical warfare storage areas such as Hanford, Washington; Rocky Flats (near Denver), Colorado; and Tooele (near Salt Lake City), Utah. In addition, the widespread use of fertilizers and pesticides in modern North American agriculture raises land productivity, but many of these chemicals are carried into natural runoff. In the U.S. Southwest, another critical water problem is salinization, a process in which irrigation water leaves behind a toxic buildup of salts in the soil. Municipal sewage also is costly to treat, and cities are faced with growing metropolitan populations and aging technology. Some cities such as San Diego, California, are pondering a "flush now, drink later" plan that would repurify raw urban sewage with multiple-barrier filtration methods and then sell the water back to municipal users. Clearly, however, daunting challenges loom in utilizing such an approach.

Altering the Atmosphere North Americans humanize the very air they breathe and, in doing so, they change local and regional climates as well as the chemical composition of the atmosphere. Urban heat islands, particularly on clear, calm nights, generate nighttime temperatures some 9 to 14 °F (5 to 8 °C) warmer than nearby rural areas. There are also many North American cases in which industrial emissions alter local climates by enhancing levels of downwind precipitation. Most significantly, air pollution affects large numbers of plants, animals, and people in a myriad of localities. At the local level, industries, utilities, and automobiles contribute carbon monoxide, sulfur, nitrogen oxides, hydrocarbons, and particulates to the urban atmosphere. Although both Canadian and American governments have acted to control emissions, widespread public health problems remain.

At a broader scale, North America is plagued by **acid rain,** industrially produced sulfur dioxide and nitrogen oxides in the atmosphere that damage forests, poison lakes, and kill fish. Geographically, many acid rain producers are located in the Midwest and southern Ontario, where industrial plants, power-generating facilities, and motor vehicles contribute emissions. Prevailing winds transport the pollutants and deposit damaging acidic precipitation across the Ohio Valley, Appalachia, the northeastern United States, and eastern Canada. Although tougher controls have reduced pollution levels, much damage has already been done to sensitive forests and lakes in New York's Adirondack Mountains, southern Quebec, and the Maritimes.

The Price of Affluence

Globalization has brought many benefits to North America, but with the accompanying urbanization, industrialization, and heightened consumption, the realm is also paying an environmental price for its affluence. For all its economic growth and material wealth, the twentieth century in North America will also be remembered for its toxic waste dumps, frequently unbreathable air, and wildlands lost to development. Still, many environmental initiatives in the United States and Canada have addressed local and regional-scale problems. For example, the improved water quality of the Great Lakes over the past 30 years is an achievement to which both nations contributed and which benefit both. Tougher air quality standards have selectively reduced emissions in many North American cities. Though forces still oppose them, preservation groups in both countries have secured new tracts of endangered forest, alpine, and wetland environments, including acreage in the Rocky Mountains and in the Alaskan and Canadian North. Whatever the outcome of future policy debates, North America's humanized environment will continue to reflect both the costs and rewards of living in one of the world's most developed regions.

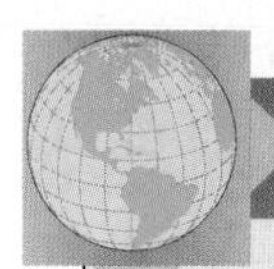

ENVIRONMENT Sink or Swim? Life on the Modern Mississippi

Back in Mark Twain's time, life on the Mississippi River had challenges that ranged from shifting channels to unsavory waterfront characters. Today, the Mississippi poses a new set of problems for residents along America's greatest river. Record spring and early summer rains in 1993 produced one of the continent's greatest floods of the past 200 years. More than 14 million acres (6 million hectares) were inundated in the deluge. Almost $20 billion in crop and property damage resulted as swirling waters of the Mississippi and Missouri rivers overwhelmed levees and forced more than 50,000 people from their homes (Figure 1). Was this a random 500-year flood or does the Mississippi's behavior contain more important environmental lessons?

What the flood dramatically demonstrates are the unanticipated consequences incurred as people interact with their environment. Since Mark Twain's time, billions of federal dollars have been spent controlling the Mississippi River: its channel was narrowed and deepened to handle commercial barge traffic, and thousands of miles of levees were constructed to hold back the river's flow. At the same time, floodplain acreage filled with farms, suburban housing tracts, and shopping malls. Federally subsidized crop and flood insurance further emboldened residents to challenge the odds in such settings.

Even more daunting are the prospects of how global climate change may impact flood probabilities in the valley or along North America's thousands of miles of coastline. Indeed, global warming might increase the chances for more cataclysmic floods on the Mississippi and rising sea levels may offer the same troubling scenario for increasingly developed coastal settings. Should billions of dollars be spent on development in such places, only to see it periodically washed away? Who should pay for insuring such activities? What risks *are* worth taking in these localities? Although Huck Finn never worried about such things, modern life along the Mississippi demands that residents confront such issues and that they ponder solutions that can help them stay above water in the future.

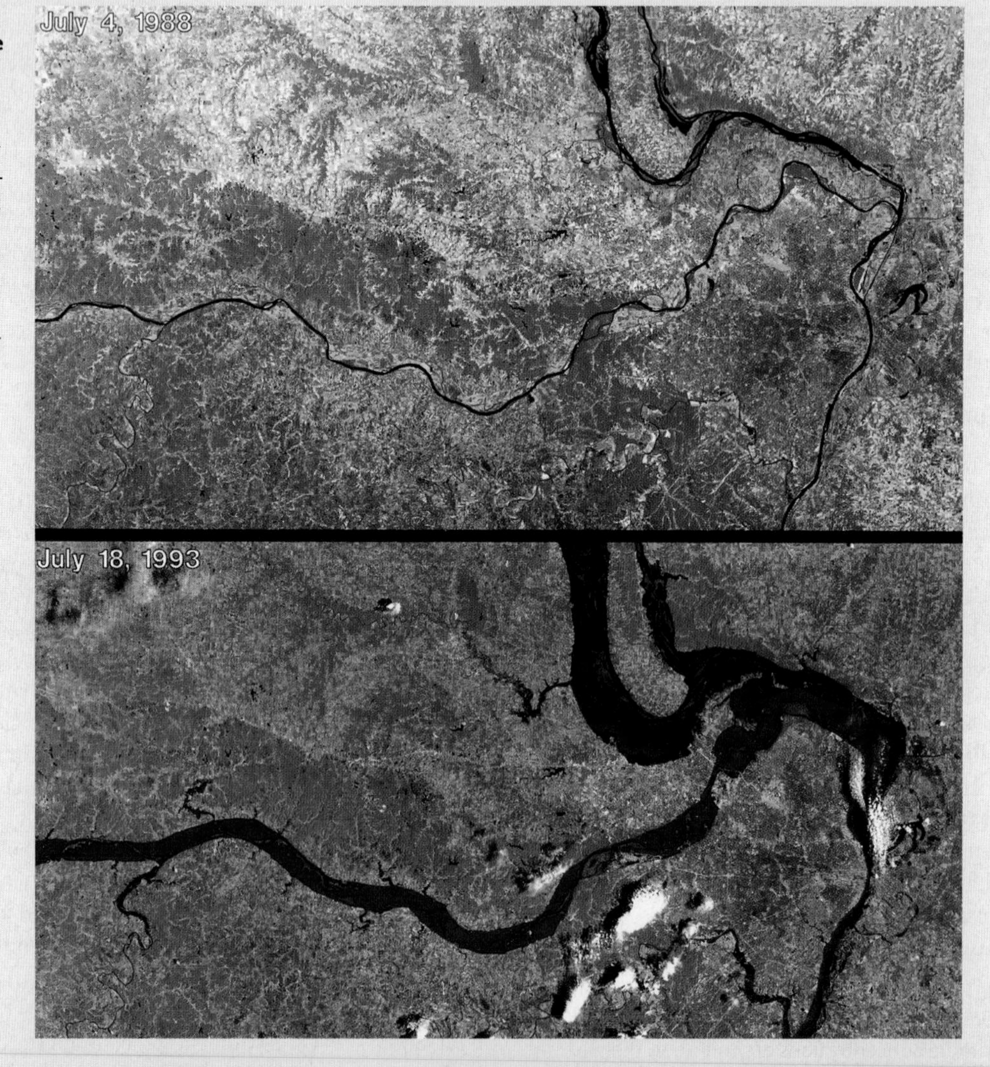

▶ **Figure 1 Satellite views of the confluence of the Missouri and Mississippi rivers** The upper image, showing the setting in the drought year of 1988, contrasts dramatically with the lower image, taken during the peak of the 1993 floods as millions of acres in the region were inundated. St. Louis is just downriver from the confluence. *(Space Imaging EOSAT)*

Population and Settlement: Refashioning a Continental Landscape

The North American landscape is the product of four centuries of extraordinary human change. During that period, Europeans, Africans, and Asians converged upon the realm, displaced a continental expanse of Native American peoples, and created a new geography of human settlement. Today, more than 300 million people live within the region, and they are some of the world's most affluent and highly mobile populations. The contemporary North American scene dramatically displays how its population has refashioned the settlement landscape to meet the needs of a modern, postindustrial society. Above all, urbanization shapes contemporary North American settlement, and more than 75 percent of the region's population now lives in cities, a striking contrast to 1850 when only about one in five (20 percent) people lived in an urban area.

Modern Spatial and Demographic Patterns

Metropolitan clusters dominate North America's population geography, producing strikingly uneven patterns of settlement across the region (Figure 3.10). The largest number of cities and the densest collection of rural settlement are found south of the Canadian Shield and east of the 100th meridian. Within this broad region, Canada's "Main Street" corridor contains most of that nation's urban population, led by the two cities of Toronto (4.2 million) and Montreal (3.3 million). The federal capital of Ottawa (1 million) and the industrial center of Hamilton (620,000) are also within the Canadian urban corridor. Indeed, more than 60 percent of Canada's population lives in southern Ontario and along Quebec's St. Lawrence River Valley. **Megalopolis**, the largest settlement agglomeration in the United States includes Washington, D.C. (4.6 million), Baltimore (2.5 million), Philadelphia (6 million), New York City (20 million), and Boston (5.6 million). Beyond these two national core areas, other sprawling urban centers in eastern North America include Chicago (8.6 million) in the Great Lakes region and the cities of Dallas–Ft. Worth (4.6 million), Houston (4.3 million), and Atlanta (3.5 million) in the South. Another smaller, but dynamic string of cities extends from Vancouver (1.8 million) to Los Angeles (15.5 million) on the western periphery of the realm. Beyond these clusters, much of the West and North are sparsely settled lands interrupted by sizable urban centers such as Denver (2.3 million) and Edmonton (860,000).

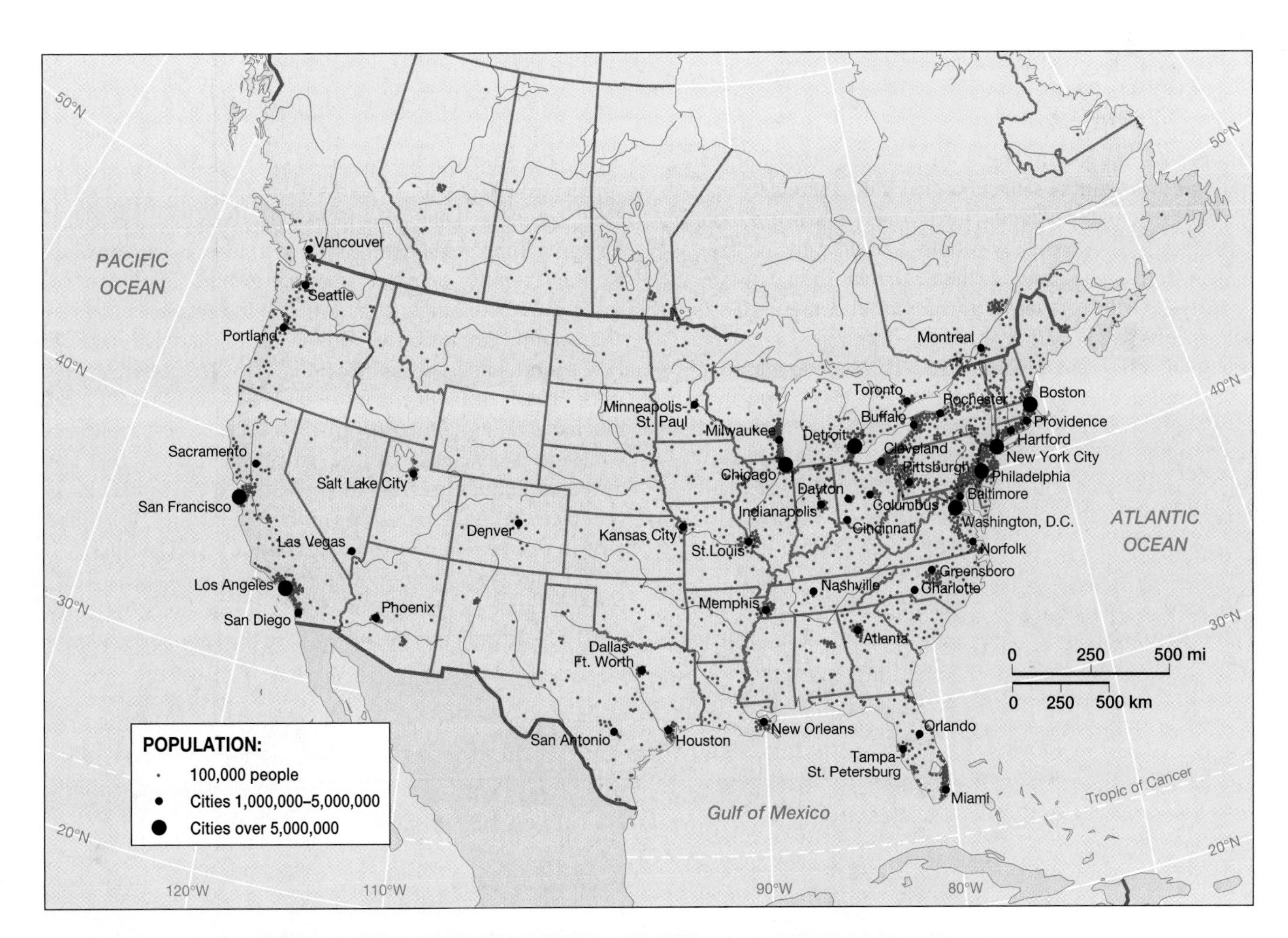

▲ **Figure 3.10 Population map of North America** North America's geography of population reveals a strikingly clustered pattern of large cities interspersed with more sparsely settled zones. Notable concentrations are found on the eastern seaboard between Boston and Washington, D.C., along the shores of the Great Lakes, and across the Sunbelt from Florida to California.

North America's population has increased greatly since the beginning of European colonization. Before 1900, high rates of natural increase produced large families. In addition, waves of foreign immigration swelled settlement, a pattern that continues today. In Canada, a population of less than 300,000 Indians and Europeans in the 1760s grew to an impressive 3.2 million a century later. For the United States, a late colonial (1770) total of around 2.5 million increased more than tenfold to more than 30 million by 1860. Thereafter, both countries saw even higher rates of immigration in the late nineteenth and twentieth centuries, although birthrates gradually fell after 1900. After World War II, birthrates rose once again in both countries, resulting in the "baby boom" generation born between 1946 and 1965. Today, however, as with much of the developed world, rates of natural increase in North America are below 1 percent annually (Table 3.1), and the overall population is growing older. Still, the region continues to attract immigrants: more than 24 million foreign-born migrants now live in North America. These growing numbers, along with higher birthrates among immigrant populations, have led demographic experts to raise long-term population projections for the twenty-first century. Indeed, predictions that the region's population will reach 370 million (335 million in the United States; 37 million in Canada) by 2025 may prove conservative.

Occupying the Land

Europeans began occupying North America about 400 years ago. It is important to remember that Europeans were hardly peopling an empty land. North America was populated for at least 12,000 years by peoples as culturally diverse as those who came to conquer them. Native Americans were broadly distributed across the realm and made diverse adaptations to its many natural environments. Their pre-contact numbers are impossible to reconstruct precisely, although cultural geographers estimate A.D. 1500 populations at 3.2 million for the continental United States and another 1.2 million for Canada, Alaska, Hawaii, and Greenland. European diseases and disruption decimated these Native American populations as contacts increased. Northern Inuits, coastal fishing peoples, southwestern pueblo-dwellers, and varied eastern woodlands societies hardly could have anticipated the magnitude of coming changes after 1600 as the European world expanded its reach. The continental sweep of settlement that followed took shape in three stages, and the twentieth-century outcome fundamentally reordered North America's human geography.

The first stage of this dramatic new settlement geography began with a series of European colonial footholds, mostly within the coastal regions of eastern North America (Figure 3.11). Established between 1600 and 1750, these regionally distinct societies were anchored on the north by the French settlement of the St. Lawrence Valley and nearby areas of the Canadian Maritimes. English Puritans dominated nearby southern New England, imposing their own brand of cultural orthodoxy by the 1640s. Farther south, both the Dutch colony of New Netherlands (later English-controlled New York) and the English Quaker colony of Pennsylvania attracted a varied collection of farmers, merchants, and tradesmen. Bicultural European and African settlements concentrated in the plantation South, with the largest colonies in English-controlled Virginia and South Carolina. Additional French settlements concentrated along the Gulf of Mexico (New Orleans was founded in 1718) and there was an early Spanish presence in the Southwest (Santa Fe was founded in 1610).

The second stage in the Europeanization of the North American landscape took place between 1750 and 1850, and it was highlighted by the infilling of much of the better agricultural land within the eastern half of the continent (Figure 3.11). Restrictive English colonial policies failed to deter frontier settlement in the upper Ohio and Tennessee valleys. Following the American Revolution (1776) and a series of Indian conflicts, pioneers surged across the Appalachian Mountains. They found much of the Interior Lowlands region almost ideal for agricultural settlement. As a result, most of the the Midwest and interior South were occupied by 1850, and these subregions increasingly became tied to the national and global economy. Expansion in early Canada was more modest. Few major changes in settlement came as the region shifted from French to English control in 1763. More importantly, much of southern Ontario, or Upper Canada, was opened to widespread development in 1791. In the decades that followed, thousands from Britain, Ireland, and the United States occupied the fertile farmlands of the region and sparked the growth of cities such as Toronto, London, and Hamilton.

The third stage in North America's settlement expansion accelerated after 1850 and continued until just after 1910 (Figure 3.11). During this period, most of the region's remaining

Table 3.1 Demographic Indicators

Country	Population[a]	Natural Increase	TFR[b]	%<15[c]	% Urban
Canada	30.6	0.5	1.6	20	77
United States	270.2	0.6	2.0	22	75

[a]Population in millions, 1998.
[b]Total fertility rate.
[c]Percent of population younger than 15 years of age.
Source: *Population Reference Bureau. World Population Data Sheet,* 1998.

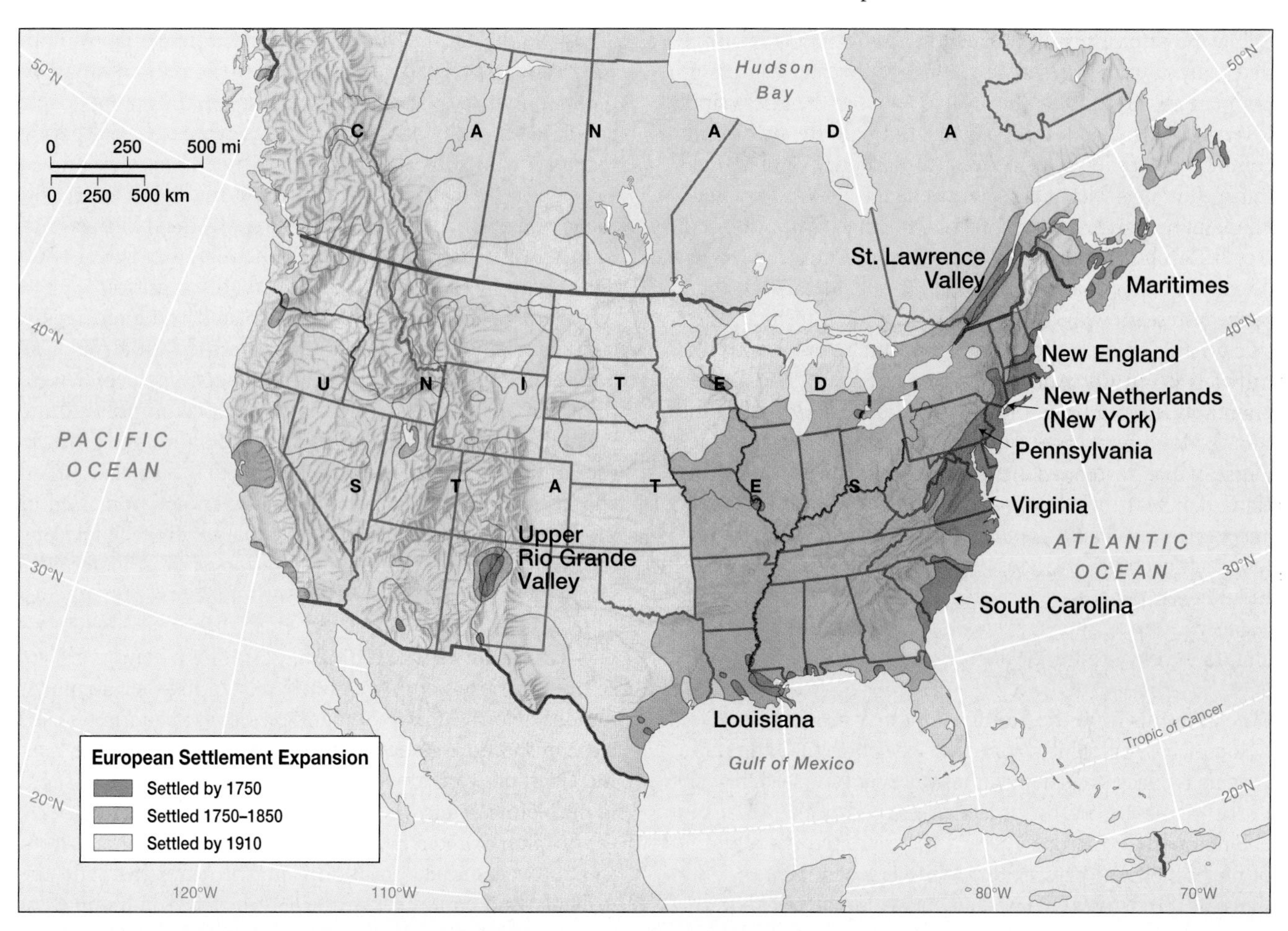

▲ **Figure 3.11 European settlement expansion** Sizable portions of North America's East Coast and the St. Lawrence Valley were occupied by Europeans before 1750. The most remarkable surge of settlement, however, occurred during the next century as vast areas of land were opened to European dominance and Native American populations were dramatically disrupted.

agricultural lands were settled by a mix of native-born and immigrant farmers. Often, pioneers tested their luck and ingenuity in quite unfamiliar environmental surroundings. Farmers were challenged and sometimes defeated by aridity, mountainous terrain, and short, northern growing seasons. In the American West, settlers were attracted by opportunities in California, the Oregon country, Mormon Utah, and the Great Plains. In Canada, few flocked to the northern Shield, but thousands did occupy southern portions of Manitoba, Saskatchewan, and Alberta. Mineral rushes led to initial development in areas such as Colorado, Montana, and British Columbia's Fraser Valley. After 1910, an increasingly urban and industrial West beckoned millions of new migrants, but the overall spatial extent of settlement expansion had run its course, leaving few large areas of the realm to be newly occupied thereafter.

Incredibly, in a mere 160 years, much of the North American landscape was domesticated as expanding populations sought new land to occupy and as a globe-encircling capitalist economy demanded resources to fuel its growth. It was one of the largest and most rapid human transformations of the landscape in the history of the human population. This European-led advance forever reshaped North America in its own image and in the process also changed the larger globe in lasting ways by creating a "New World" destined to reshape the Old.

North Americans on the Move

From the mythic days of Davy Crockett and Daniel Boone to the twentieth-century sojourns of John Steinbeck and Jack Kerouac, North Americans have been on the move. Indeed, almost one in every five Americans moves annually, suggesting that residents of the region are readily willing to change addresses in order to improve their income or their quality of life. Although interregional population flows are complex in both the United States and Canada, several twentieth-century trends dominate the picture.

Westward-moving Populations The most persistent regional migration trend in North America has been the tendency for people to move west. Indeed, the dominant thrust in the past two centuries (Figure 3.11) has been to follow the setting sun, and many North Americans continue that pattern to the present. By 1990, more than half of the population of the United States lived west of the Mississippi River, a dramatic shift from colonial times. In the nineteenth century, migrants moved westward along well-defined transport routes. New Englanders settled southern Michigan, Pennsylvanians preferred central Indiana or Illinois, and North Carolinians favored Tennessee and Kentucky. Today, migration patterns are more complex, but since 1960, some of the fastest-growing states have been in the American West (including California, Texas, Arizona, and Nevada) as well as in the western Canadian provinces of Alberta and British Columbia. In fact, all of the states and provinces west of the Rocky Mountains have grown by at least 50 percent since 1960, while many eastern North American settings (the Canadian Maritimes, Appalachia, and the Midwest) have experienced sustained population losses due to outmigration. Job creation consistently has been higher in western localities. In addition, people are attracted to the perceived amenities of the West, including its recreational opportunities, climate, and its informal, outdoor-oriented lifestyles.

Black Exodus from the South African Americans displayed a somewhat different pattern of interregional migration. Originally, black slave populations were concentrated strongly in the plantation South. In fact, at the end of the American colonial period, African Americans constituted the majority of population (55 percent) in southern states such as South Carolina. Even after the legal emancipation of blacks in the 1860s, most remained economically bound to the rural South where they worked as sharecroppers, often for their former owners. Conditions changed, however, in the twentieth century. Many blacks migrated because of declining demands for labor in the agricultural South and growing industrial opportunities in the North and West. Two waves of migration, one between 1910 and 1920 and the other from 1940 to 1960, propelled blacks into many new American settings. Overwhelmingly, migrants ended up in cities where job opportunities beckoned. Boston, New York, and Philadelphia became key destinations for blacks in Georgia and the Carolinas, while Midwest cities such as Detroit and Chicago attracted many blacks from Alabama, Mississippi, and Louisiana. Particularly in the later migration wave, Los Angeles and San Francisco/Oakland drew many blacks to the West. Since 1970, however, more blacks have moved from north to south. Indeed, Sun Belt jobs and federal civil rights guarantees now attract many northern urban blacks to growing southern cities. The net result is still a profound change from 1900: at the beginning of the century, more than 90 percent of African Americans lived in the South, while today only about half of the nation's 33 million blacks reside within the region.

Rural to Urban Migration Another persistent trend in North American migration has taken people from the country to the city. Two centuries ago, only 5 percent of North Americans lived in an urban area (of more than 2,500 people), whereas today more than 75 percent of the North American population resides in a city. Shifting economic opportunities account for much of the transformation: as mechanization on the farm reduced the demand for labor, many young people left for new employment opportunities in the city. Well-paying manufacturing jobs and a growing service economy proved to be powerful magnets for many rural residents. Historically, rates of urbanization jumped first in the Northeast as industrialization spread through the region between 1800 and 1850. The scene of rapid urban expansion shifted westward to the Great Lakes region later in the nineteenth century as industrial growth transformed cities such as Toronto and Chicago. In the twentieth century, people were drawn by many urban employment opportunities in more peripheral zones, resulting in large migrations to cities in the South and West. The consequences of this rural-to-urban shift transcend mere job relocation. Larger processes of modernization and globalization were greatly facilitated by the overwhelmingly urban orientation of North America's population.

Growth of the Sun Belt South Twentieth-century moves to the American South are clearly related to other dominant trends in North American migration, yet the pattern deserves closer inspection. Particularly after 1970, southern states from the Carolinas to Texas grew much more rapidly than states in the Northeast and Midwest. For example, during the 1980s, the South received a net inflow of 1.7 million migrants from the Midwest and 1.6 million from the Northeast, and the general trend continued in the 1990s. The South's buoyant economy, modest living costs, attractive recreational opportunities and its appeal to snow-weary retirees all contributed to its growth. Movements have been selective, however; many rural agricultural and mountain counties within the South have seen few new residents, while amenity-rich coastal settings and job-generating metropolitan areas witness spectacular growth. Texas and Florida have seen the largest increases in population, but gains have also transformed many settings in the Carolina Piedmont, Georgia, Kentucky, Tennessee, and along the southern Gulf Coast (Figure 3.12).

The Counterurbanization Trend During the 1970s, certain nonmetropolitan areas in North America began to see significant population gains, including many rural settings that had traditionally lost population. Selectively, that pattern of **counterurbanization**, in which people leave large cities and move to smaller towns and rural areas, continues today. Some participants in counterurbanization are part of the growing retiree population in both Canada and the United States, but a substantial number are younger, so-called *lifestyle migrants*. They find or create employment in affordable smaller cities and rural settings that are rich in amenities and often removed from the perceived problems of urban America. One recent example of the phenomenon was the 1990s exodus of native-born whites from urban California to less-populated counties in the interior West. Communities such as Bend, Oregon; Coeur d'Alene, Idaho; and Kalispell, Montana,

▲ **Figure 3.12 Downtown Atlanta, Georgia** Sunbelt cities such as Atlanta have been transformed by rapid job creation. Healthy growth in office space, specialty retailing, and entertainment districts has fueled downtown Atlanta's expansion and reshaped the look of the central city skyline. *(Bill Bachmann/Photo Researchers, Inc.)*

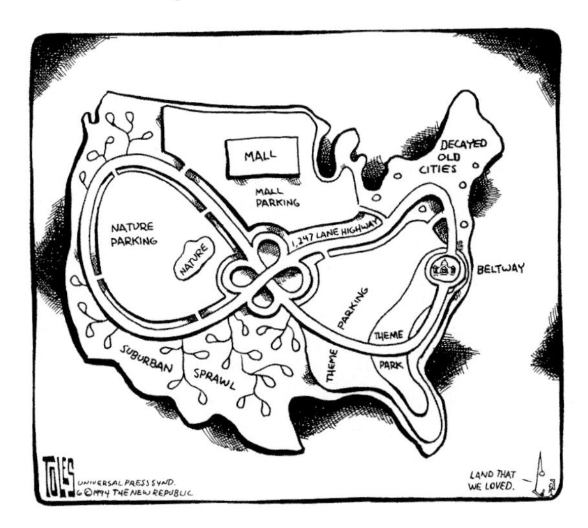

▲ **Figure 3.13 Contemporary North American geography** This light-hearted North American map by Toles is a reminder of how rapidly the continental scene has been transformed in the past century. *(Universal Press Syndicate)*

witnessed a flood of new migrants, many of them former residents of the Golden State. In fact, during the 1990s, nonmetropolitan population growth exceeded metropolitan growth in the western states. Other smaller communities outside the West such as Mason City, Iowa; Mankato, Minnesota; and Traverse City, Michigan, also are seen as desirable destinations for migrants interested in downsizing from their metropolitan roots.

Settlement Geographies: The Decentralized Metropolis

North America's settlement landscape bears witness to the population movements, shifting regional economic fortunes, and technological innovations that shaped twentieth-century geographical change across the realm. The ways in which settlements are organized on the land, the actual appearance of cities, suburbs, and farms, as well as the very ways in which North Americans socially construct their communities, have changed greatly in the past century. Today's everyday experiences of cloverleaf interchanges, sprawling suburbs, outlet malls, and theme parks would have struck most 1900-era residents as utterly extraordinary (Figure 3.13).

Settlement landscapes of North American cities boldly display the consequences of **urban decentralization** in which metropolitan areas sprawl in all directions and suburbs take on many of the characteristics of traditional downtowns. Although both Canadian and U.S. cities have experienced decentralization, the impact has been particularly profound in the United States, where inner-city problems, poorer public transportation facilities, widespread automobile ownership, and fewer regional-scale planning initiatives have encouraged many middle-class urban residents to move beyond the central city. Even beyond North America, observers note there is a globalization of urban sprawl: many Asian, European, and Latin American cities are taking on attributes of their North American counterparts as they experience similar technological and economic shifts. Indeed, much as they have in Seattle or Albuquerque, suburban Wal-Marts, semiconductor industrial parks, and shopping malls may become increasingly familiar sites on the peripheries of Kuala Lumpur or Mexico City.

Historical Evolution of the City in the United States

Changing transportation technologies decisively shaped the evolution of the city in the United States. Initially, the Pedestrian/Horsecar City (pre-1888) of the mid-nineteenth century was compact, essentially limiting urban growth to a three- or four-mile ring around downtown conveniently accessible by foot or horse-powered trolley cars. The invention of the electric trolley in 1888, however, propelled the urbanized landscape farther into new "streetcar suburbs," often five or 10 miles from the city center. Indeed, the Electric Streetcar City (1888–1920) offered commuters affordable mass transit at the dizzying speed of 15 or 20 miles per hour. A star-shaped urban pattern resulted, with growth extending outward along and near the streetcar line.

The biggest technological revolution came after 1920 with the widespread adoption of the automobile. The Recreational Automobile City (1920–1945) promoted the infilling of areas left beyond the reach of the streetcar, and added even more distant suburban rings in the surrounding countryside. Essentially, it allowed many middle-income residents, particularly whites, to leave the central city in favor of lower density and less ethnically complex suburban settings. Following World War II, the Freeway City (1945 to the present) promoted even more dramatic decentralization along commuter routes as built-up areas appeared 40 to 60 miles from downtown. The perception that inner-city crime, congestion, and racial tensions were worsening after 1960 accelerated the movement of whites toward the presumed peace of the suburbs.

Urban decentralization also reconfigured land-use patterns in the city, producing metropolitan areas today that are strikingly different from their counterparts of the early twentieth

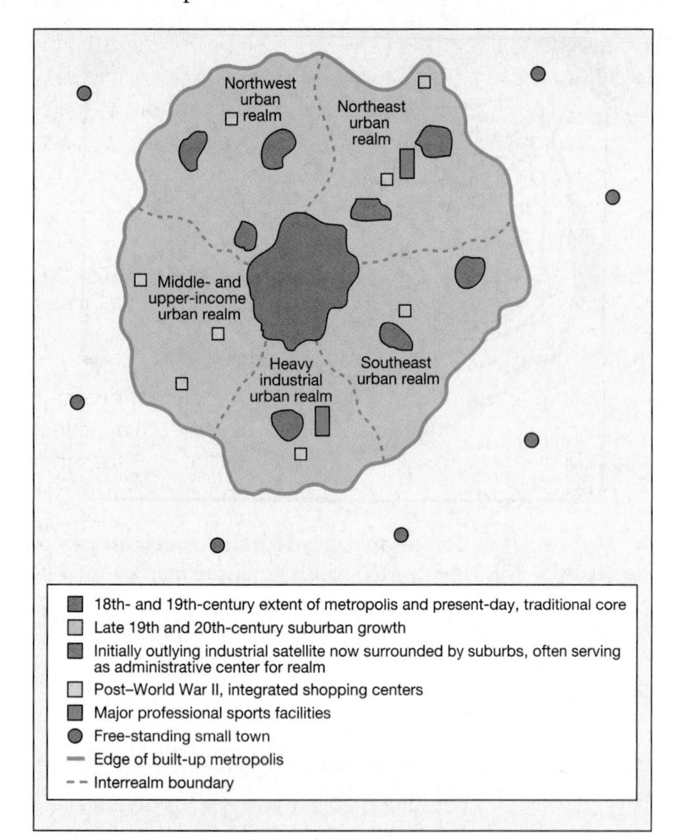

▲ **Figure 3.14 North American urban realms model** The modern decentralized North American metropolis offers a complex mix of urban and suburban land uses. Increasingly, much of urban North America's most dynamic economic activity takes place beyond the traditional urban core. *(Reprinted from Knox and Marston, 1998,* Human Geography, *Upper Saddle River, NJ: Prentice Hall)*

▲ **Figure 3.15 Tysons Corner, Virginia** North America's edge-city landscape is superbly illustrated by Tysons Corner, Virginia. Far from a traditional metropolitan downtown, this sprawling complex of suburban offices and commercial activities reveals how and where many North Americans will live their lives in the twenty-first century. *(Robert Llewellyn)*

century. In the city of the early twentieth century, idealized in the **concentric zone model**, urban land uses are neatly organized in rings around a highly focused central business district (CBD) that contains much of the city's retailing and office functions. Residential districts beyond the CBD are added as the city expands, with higher-income groups seeking more desirable, peripheral locations. The model nicely summarized the urban geography of many cities in the United States during the 1920s and elements of the model are still relevant today, particularly in examining how eras of economic expansion add rings of residential construction around the periphery of a city.

The **urban realms model**, however, highlights new suburban growth characterized by a mix of peripheral retailing, industrial parks, office complexes, and entertainment facilities (Figure 3.14). These areas of activity, often called "edge cities," have fewer functional connections with the central city than they have with other suburban centers. For most suburban residents of the edge city, jobs, friends, and entertainment are located in other peripheral realms versus the old downtown. Tysons Corner, Virginia, is a superb example of the edge-city landscape on the dynamic periphery of the North American metropolis (Figure 3.15).

The Consequences of Sprawl The rapid evolution of the North American city continues to transform the urban landscape and those who live within it. As suburbanization accelerated in the 1960s and 1970s, many inner cities, especially in the Northeast and Midwest, suffered absolute losses in population, increased levels of crime and social disruption, and a shrinking tax base that often brought them to the brink of bankruptcy. The inner city, particularly in the United States, still faces daunting educational challenges, homeless populations, and immense planning problems. Poverty rates average almost three times those of nearby suburbs. Unemployment, while declining somewhat in the 1990s, continues above the national average. Inner-city populations often lack the training, family stability, and social networks that can aid in obtaining employment. Above all, central cities within the United States remain places of racial tension, the product of decades of discrimination, segregation, and poverty. While there are growing numbers of middle-class blacks and Hispanics, many exit the central city for the suburbs, further isolating the urban underclass that remains behind.

Amid these challenges, inner-city landscapes are also enjoying a selective renaissance. Referred to as **gentrification,** the process involves the displacement of lower-income residents of central city neighborhoods with higher-income residents, the rehabilitation of deteriorated inner-city landscapes, and the construction of new shopping complexes, entertainment attractions, or convention centers in selected downtown locations. The older and more architecturally diverse housing stock of the central city is also a draw, serving as specialty shops and restaurants for a cosmopolitan urban clientele and offering residential opportunities for upscale singles who wish to live near downtown. Seattle's Pioneer Square, Toronto's Yorkville district, and Baltimore's Harborplace exemplify how such new public and private investments shape the central city (Figure 3.16).

▲ **Figure 3.16 Baltimore's Harborplace** Inner-city revitalization is exemplified by the dramatic transformation of Baltimore Harbor. The elaborate development offers visitors food, entertainment, and specialty shopping in an upscale setting that caters to a cosmopolitan urban clientele. *(Medford Taylor/National Geographic Society)*

The suburbs are also changing. Construction of new corporate office centers, fashion malls, and industrial facilities has created true "suburban downtowns" that are no longer simply bedroom communities for central city workers. The edge-city lifestyle has also transformed America into a continent of suburban commuters in which people live in one suburb and work in another. The average daily travel mileage now exceeds 30 miles per person in cities such as Atlanta, Georgia, and Birmingham, Alabama. As urban settlements of the twenty-first century grow in area but decline in density, it will become more difficult to build roads, schools, and other public infrastructure in cost-effective ways. Another outcome of sprawl is growing congestion: time wasted on the road cost Americans an estimated $50 billion annually during the 1990s. Commuters face the greatest daily delays in Los Angeles; San Francisco; Washington, D.C.; and Houston, where they spend more than 1.5 work-weeks per year stranded on grid-locked freeways! Another consequence of edge-city sprawl is the rapid loss of surrounding rural land. Since 1945, U.S. metropolitan areas have consumed about one million acres (405,000 hectares) of rural land annually, including some of the nation's most productive and intensely utilized farmland. Indeed, particularly on the outer suburban fringe, the boundary between city and country is blurring: as urbanites reach toward the amenities and privacy of the rural landscape, modern technologies and metropolitan growth bring the countryside inexorably closer to the city.

Settlement Geographies: Rural North America

Rural North American landscapes trace their origins to initial European settlement. Although many New World communities replicated elements of the Old World, North Americans favored more dispersed rural settlement patterns than their European antecedents. Traditionally, many European farmers lived in nucleated settlements in which people grouped themselves together in a village and farmed surrounding fields. Things changed, however, on the other side of the Atlantic Ocean. An abundance of available acreage, liberal land disposal laws, and perhaps a cultural predilection for independence and privacy all contributed to a heritage of dispersed settlement in North America in which settlers lived on their own farmsteads, often at some distance from their nearest neighbors.

In portions of the United States settled after 1785, the federal government surveyed and sold much of the rural landscape. Surveys were organized around the simple, rectangular pattern of the federal government's township-and-range survey system, which offered a convenient method of dividing and disposing of the public domain in six-mile-square townships. In a repeating pattern across the U.S. interior, grid towns were laid out amid a seemingly endless geometry of straight lines that neatly squared off the surrounding rural landscape (Figure 3.17). Canada developed a similar system of regular surveys that stamped much of southern Ontario and the western provinces with a strikingly rectilinear character. Exceptions to the predictable pattern were found in former American colonies such as Virginia, where irregular surveys dominated, and in Canada's St. Lawrence Valley, in which distinctive long lot lines lay perpendicular to the river to give more farmers access to the water.

Commercialized agriculture and technological changes further reorganized the settlement landscape. Railroads opened new corridors of development wherever tracks were laid, provided access to markets for commercial crops, and facilitated the establishment of towns that served surrounding rural populations. By 1900, several transcontinental lines spanned North America, radically transforming the farm economy and the pace of rural life. After 1920, however, even more profound changes accompanied the arrival of the automobile, farm mechanization, and better rural road networks. The need for farm labor declined with mechanization, and many smaller market centers became unnecessary as farmers equipped with automobiles and trucks could travel farther and faster to larger, more diverse towns. Other technological innovations, including center-pivot irrigation systems, better yielding seeds, and new fertilizers and insecticides, also made their mark on the rural settlement landscape.

Today, many rural North American settings face population declines as they adjust to the changing conditions of modern agriculture. Both U.S. and Canadian farm populations fell by more than two-thirds during the last half of the twentieth century. Typically, larger but fewer farms dot the modern rural scene and many young people leave the land to obtain employment elsewhere. In both countries, large specialized farms dominate agricultural production, even though more than 75 percent of farms remain family-operated. The population drain in settings such as rural Iowa, southern Saskatchewan, or eastern Montana also impacts towns within these regions. Fewer shoppers visit local stores, doctors and dentists retire or move, and even schools and churches are forced to close or consolidate. Indeed, the very fabric of rural and small-town communities is threatened. The visual legacies of abandonment provide poignant reminders of painful adjustments:

▶ **Figure 3.17 Iowa settlement patterns** The regular rectangular look of this Iowa town and the nearby rural setting reveals the North American penchant for simplicity and efficiency. In the United States, the township-and-range survey system stamped such predictable patterns across vast portions of the North American interior. *(Craig Aurness/Corbis)*

weed-choked driveways, empty farmhouses, roofless barns, and the empty marquees of small-town movie houses tell the story more powerfully than any census or government report.

Other rural settings show signs of economic health and population growth. No single explanation suffices for these burgeoning hinterlands. Some localities are merely feeling the fringe effects of encroaching edge cities. Suddenly, almost overnight, such places are overrun with suburbanites who snap up farmhouses once beyond the reach of the city. Other growing rural settings lie beyond direct metropolitan influence but are seeing new populations who seek amenity-rich environments removed from city pressures. These impulses toward counterurbanization are shaping the settlement landscape from British Columbia's Vancouver Island to Michigan's Upper Peninsula. Newly subdivided land parcels, a plethora of real estate offices, the appearance of resort and golf complexes, and the telltale in-migration of espresso bars all signal the changes afoot in such surroundings. These rural landscapes mirror broader changes unfolding across North America that promise to continue shaping its population geography in profound and unexpected ways.

Cultural Coherence and Diversity: A Geographic Mosaic

North America's cultural geography is both globally dominant and internally pluralistic. On one hand, history and technology have produced a contemporary North American cultural force that is second to none in the world. Many people outside the United States speak of cultural imperialism to describe the increasing global dominance of American popular culture and often see it as threatening the vitality of other cultural values. Yet North America is also a mosaic of different peoples that retain part of their traditional cultural identities. In fact, North Americans celebrate their pluralistic folk roots and acknowledge the realm's multicultural character.

Cultural Coherence and Pluralism

Powerful forces forged a common dominant culture within the realm. Historically, both Canada and the United States were strongly tied to Great Britain, and these links transcended colonial politics. With some exceptions, most early Europeans who played key cultural, economic, and political roles within the region were from the British Isles. Key Anglo legal institutions and social mores solidified the common set of core values that many North Americans shared with Britain and, eventually, with one another. Traditional beliefs, defined within the limits of the Anglo world view, emphasized representative government, separation of church and state, liberal individualism, privacy, pragmatism, and social mobility. The American Revolution (1776) and the Confederation of Canada (1867), while distancing the realm from direct British authority, were nevertheless assertions of cultural continuity that combined the Anglo world view with converging national cultural identities. As settlement spread and economic linkages forged truly continental connections in the late nineteenth century, these core values offered a set of experiences shared by many North Americans, both native- and foreign-born. From those traditional roots, particularly within the United States, consumer culture blossomed after 1920, producing a common, if more diffuse, set of shared experiences that included mass advertising, movies, national radio and television programming, and an increasingly secular society oriented toward convenience and consumption. Since World War II, these modern technologies and values, with the help of growing political and multinational corporate imperatives, have cumulatively shaped the world in North America's cultural image.

But North America's cultural coherence coexists with pluralism, the persistence and assertion of distinctive cultural identities. Closely related is the concept of **ethnicity,** in which a group of people with a common background and history identify with one another, often as a minority group within a larger society. Indeed, the roots of these impulses are deep and sometimes divisive, and they have created ongoing tensions within the two countries that often have a powerful geographical expression. For Canada, the early and enduring French colonization of Quebec complicates its contemporary cultural geography. Canadians face the challenge of creating a truly bicultural society where issues of language and political representation are central concerns. Within the United States, given its unique immigration history, a greater diversity of ethnic groups exists, and key geographical divisions of social space are often local as well as regional. The result can be a mosaic of Anglos, African Americans, Hispanics, and Asians often coexisting within a single city yet retaining their cultural identities and claims to distinctive areas of local territorial dominance. Numerous subcultures further complicate postindustrial cultural geographies in both countries. Expressed geographically, these are places where individuals with similar lifestyles and cultural preferences congregate, and they range from retirement communities in Florida to white supremacist compounds in northern Idaho.

Peopling North America

North America is a region of immigrants. Quite literally, global-scale migrations made North America possible. Decisively displacing Native Americans in most portions of the realm, immigrant populations created a new cultural geography of ethnic groups, languages, and religions. Early migrants often had considerable cultural influence, even though their numbers were minuscule compared to the flood of later arrivals. Over time, varied immigrant groups and their changing destinations produced a culturally heterogeneous landscape that continues to evolve today. Also varying between groups was the pace and degree of **cultural assimilation**, the process in which immigrants were absorbed by the larger, host society.

Migration to the United States In the United States, variations in the number and source regions of migrants produced five distinctive chapters in the country's history (Figure 3.18). In Phase 1 (prior to 1820), English and African influences dominated. Other Europeans, particularly Irish, Dutch, French, and Germans, were also important, but it was the English who played the pivotal role. Slaves, mostly from West Africa, contributed additional cultural influences in the South. Northwest Europe served as the main source region for immigrants between 1820 and 1870 (Phase 2). The emphasis, however, shifted away from English dominance to include greater numbers of Irish and Germans. The pace of immigration also increased greatly: only 130,000 migrants arrived in the 1820s versus more than 2.8 million in the 1850s.

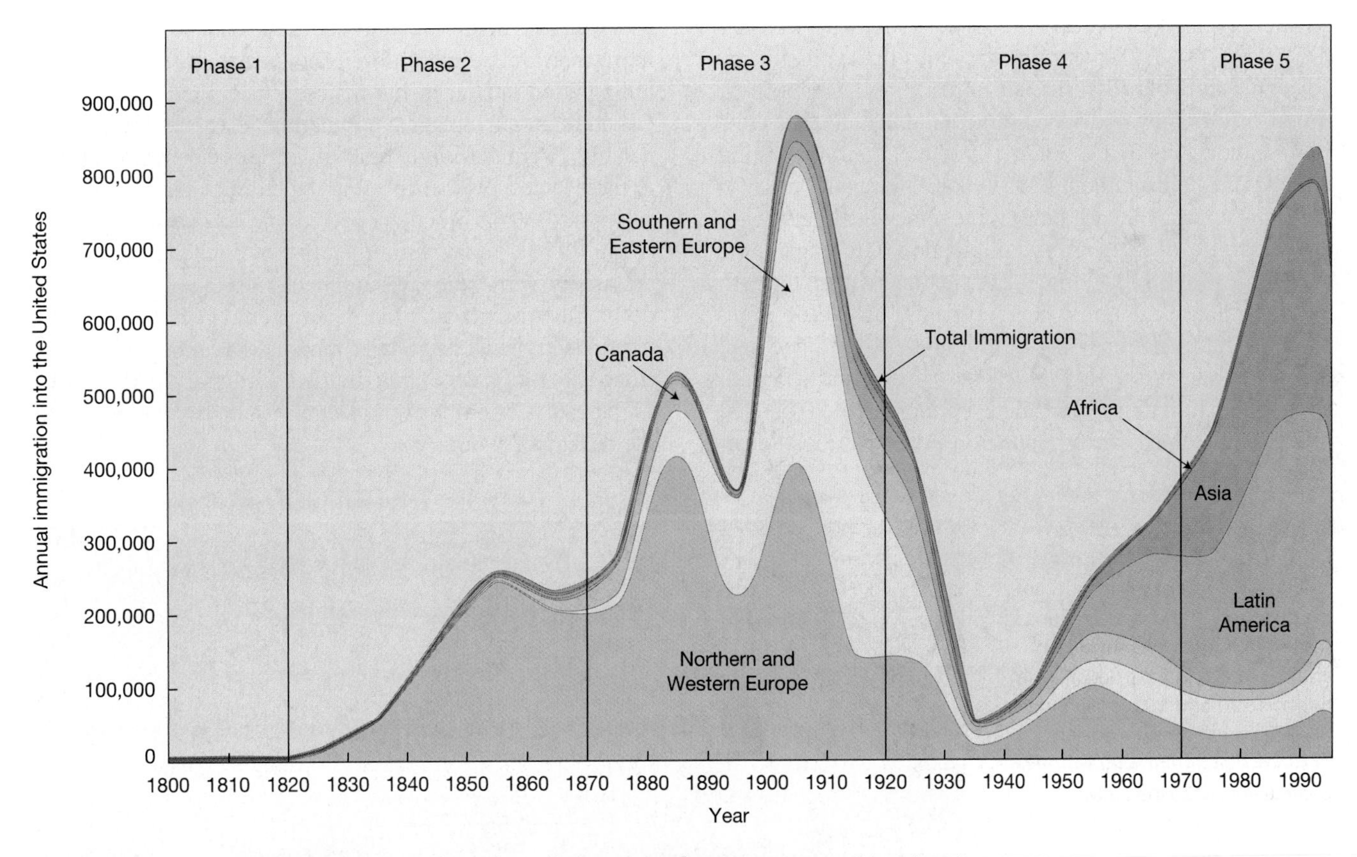

▲ **Figure 3.18 United States immigration, by year and group** Annual immigration rates peaked around 1900, declined in the early twentieth century, and then surged again, particularly since 1970. The source areas of these migrants have also shifted. Note the decreased role Europeans currently play versus the growing importance of Asians and Latin Americans. *(Modified from Rubenstein, 1999,* An Introduction to Human Geography, *Upper Saddle River, NJ: Prentice Hall)*

As Figure 3.18 shows, immigration reached a much higher peak around 1900 when almost one million foreigners entered the United States *annually*. During Phase 3 (1870–1920), the source regions of this great wave of migrants also shifted. Northwest Europeans declined in importance while southern (particularly Italians) and eastern (Poles, Russians, Austro-Hungarians) Europeans dominated. Political strife and poor economies in Europe sharply contrasted with news of available land, expanding industrialization, and well-paying jobs in America. Many later migrants went to cities in the Northeast and Midwest, gaining employment in manufacturing and urban services. Between 1880 and 1920, Scandinavians also swept through the northern interior, where they settled on still-available farmland. By 1910, almost 14 percent of the nation was foreign-born. Very few of these immigrants, however, targeted the job-poor American South, creating a cultural divergence that persists to the present.

Twentieth-century migrations also profoundly shaped the nation's cultural geography. Between 1920 and 1970 (Phase 4), more immigrants came from neighboring Canada and Latin America, but overall totals plunged, a function of more restrictive federal immigration policies (the Quota Act of 1921 and the National Origins Act of 1924), the Great Depression, and the disruption caused by World War II. Since 1970 (Phase 5) the realm witnessed a sharp reversal in numbers, however, and now arrivals match those of the early twentieth century (Figure 3.18). The current surge was made possible by economic and political instability abroad, a growing postwar American economy, and a loosening of immigration laws (the Immigration Acts of 1965 and 1990, and the Immigration Reform and Control Act of 1986). Illegal immigration is also on the rise, probably accounting for somewhere between 1 and 3 percent of the country's population. Once again, the geography of source regions and destinations shifted for both legal and illegal migrants. Most migrants since 1960 originated in Latin America or Asia. By 1990, Mexico had sent more legal migrants to the United States than any other country, and Mexican immigrants made up more than 20 percent of America's foreign-born population. Although most Mexican migrants live in California or Texas, many from the Caribbean move to Florida or New York.

In percentage terms, Asian newcomers are the fastest-growing immigrant group, and they account for almost 4 percent of the population. Asian migrants often move to large cities. In fact, more than 40 percent of the nation's Asian population lives in the Los Angeles (1.7 million), San Francisco (1.2 million), or New York City (1.2 million) metropolitan areas. Beyond these key gateway cities, Asians are also moving to growing communities in Washington, D.C.; Chicago; Seattle; and Houston. Filipinos, Vietnamese, Chinese, and Indians dominate the flow, but Korean and Japanese communities are also well established, particularly in the West Coast states and Hawaii.

The Canadian Pattern The peopling of Canada broadly parallels the U.S. story, but there are some important differences. Early French arrivals concentrated in the St. Lawrence Valley and constituted a well-defined cultural nucleus of 60,000 residents by the mid-eighteenth century. Between 1750 and 1810, however, many new migrants came from Britain, Ireland, and the United States. The War of 1812 and restrictive British policies discouraged U.S. immigration to Canada after 1810, but the numbers of Irish, Scots, and English rose through the middle of the century. Canada then experienced the same surge and reorientation in migration flows seen in America around 1900. Between 1900 and 1920, more than three million foreigners ventured to Canada, an immigration rate far higher than for the United States, given Canada's much smaller population. Eastern Europeans, Italians, Ukrainians, and Russians were prominent participants in these later movements. Some settled in Ontario cities such as Toronto, while others pioneered on the Canadian prairies (Figure 3.19). Today, more than half of Canada's immigrants are Asians, with Vancouver serving as a principal West Coast destination, particularly for large incoming Chinese populations. As in the United States, more liberal immigration laws since the 1960s encouraged movement to Canada, and its 17 percent foreign-born population is among the highest in the developed world.

▶ **Figure 3.19 Ukrainian imprints on the Canadian prairie** In Canada, immigrants from southern and eastern Europe settled in many different urban and rural settings, particularly between 1900 and 1920. Ukrainians often favored farming opportunities in the prairie provinces. *(Brian K. Miller/Bruce Coleman, Inc.)*

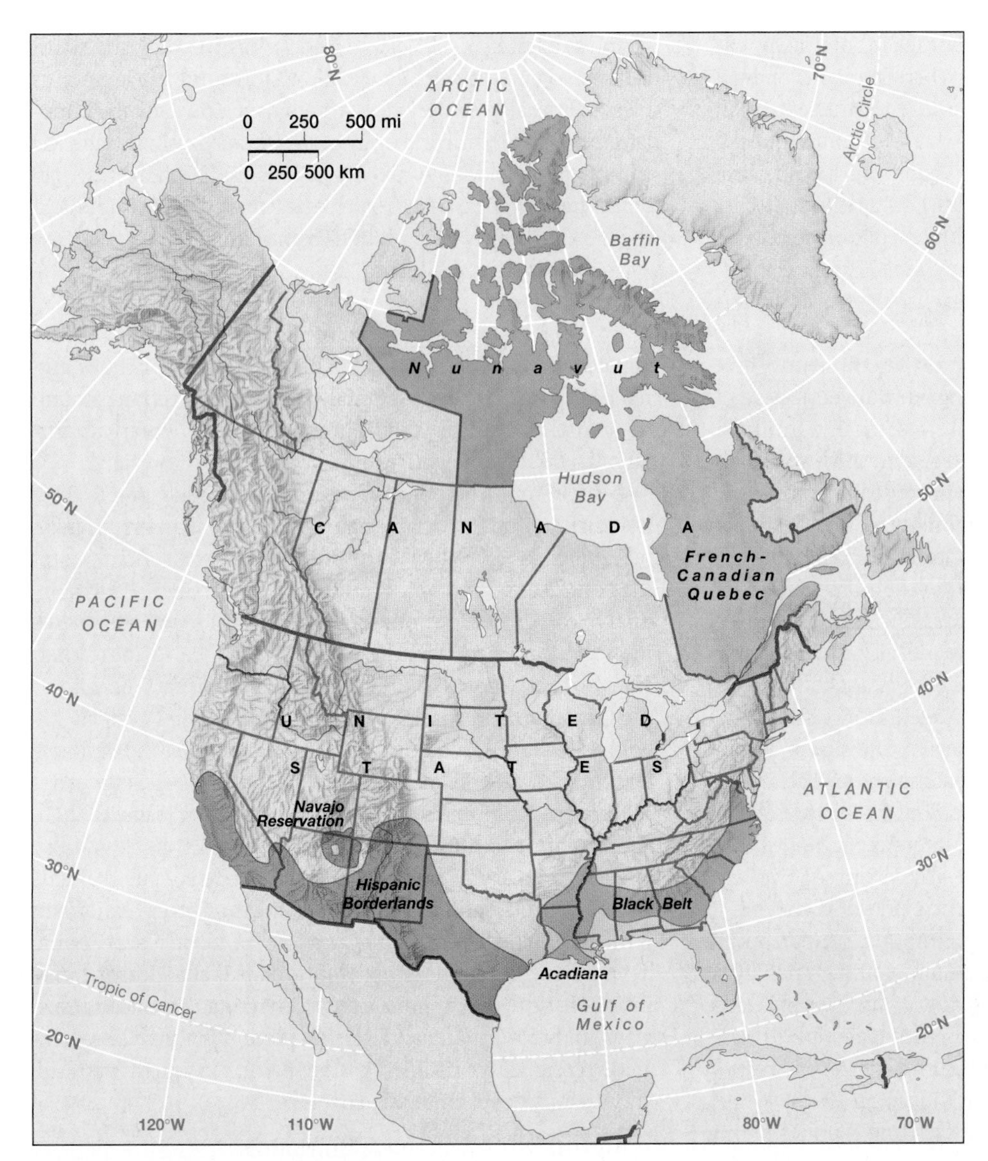

◀ **Figure 3.20 Selected cultural regions of North America** From northern Canada's Nunavut to the Southwest's Hispanic Borderlands, different North American cultural groups strongly identify with traditional local and regional homelands. Such patterns stamp the modern landscape with an enduring personality. *(Modified from Jordan, Domosh, and Rowntree, 1998,* The Human Mosaic, *Upper Saddle River, NJ: Prentice Hall)*

Culture and Place in North America

Cultural and ethnic identity are often strongly tied to place. North America's cultural diversity is expressed geographically in two ways. First, similar people congregate near one another and derive meaning from the territories they commonly occupy. Second, culture marks the visible scene: the everyday landscape is filled with the artifacts, habits, language, and values of different groups. Boston's Italian North End simply looks and smells different than nearby Chinatown, and rural French Quebec is a world away from a Hopi village in Arizona. In exploring the relationship between culture and place in North America, both large, regional-scale cultural homelands as well as smaller, more local-scale ethnic neighborhoods illustrate the connection.

Persisting Cultural Homelands French-Canadian Quebec superbly exemplifies the cultural homeland: it is a culturally distinctive nucleus of settlement in a well-defined geographical area, and its ethnicity has survived over time, stamping the cultural landscape with an enduring personality (Figure 3.20). More than 80 percent of the population of Quebec speaks French, and language remains the cultural glue that holds the homeland together. Indeed, policies adopted after 1976 strengthened the French language within the province by requiring French instruction in the schools and by mandating national bilingual programming by the Canadian Broadcasting Corporation (CBC). Also fostering unity is the group's minority status within a larger Anglo-dominated Canada (only about 25 percent of Canadians are French) and the feeling that the French-Canadians are second-class citizens within their own country. Ironically, many Quebeçois feel that their greatest cultural threat may not come from Anglo-Canadians at all, but rather from immigrants arriving in the province. Southern Europeans or Asians in Montreal, for example, show little desire to learn French, preferring instead to put their children in English-speaking private schools.

Another well-defined cultural homeland is the Hispanic Borderlands (Figure 3.20). It is similar in geographical magnitude to French-Canadian Quebec, significantly larger in total population, but more diffuse in its cultural and political articulation. Historical roots of the homeland are deep, extending back to the sixteenth century when Spaniards opened the region to the European world. The homeland's core is in northern New Mexico, including Santa Fe and much of the surrounding rural hinterlands. A rich legacy of Spanish place names, earth-toned Catholic churches, and traditional Hispanic settlements dot the rolling highlands of northern New Mexico and southern Colorado (Figure 3.21). From California to Texas, other historical sites, place names, missions, and presidios also mark the Hispanic legacy.

Unlike Quebec, however, massive twentieth-century migrations from Latin America brought an entirely new wave of Hispanic settlement to the Southwest. More than 25 million Hispanics now live in the United States, with more than half in California, Texas, New Mexico, and Arizona combined. While almost two-thirds have cultural roots in Mexico, many Hispanics residing in the United States originated elsewhere in Latin America or the Caribbean. Within the homeland, Hispanics have created a distinctive Borderlands culture that mixes many elements of Latin and North America. These newer migrants augment the rural Hispanic presence in agricultural settings such as the lower Rio Grande Valley in Texas and the Imperial and Central valleys in California. Cities such as San Antonio and Los Angeles also play leading roles in articulating the Hispanic presence within the Southwest. Regionally distinctive Latin foods and music add internal cultural variety to the region. New York City, Chicago, and Miami serve as key points of Hispanic cultural influence beyond the homeland.

African Americans also retain a cultural homeland, but it has diminished in intensity due to outmigration (Figure 3.20). Reaching from coastal Virginia and North Carolina to east Texas, a zone of enduring African American population (the Black Belt) remains a legacy of the Cotton South when a vast majority of American blacks resided within the region. Today, while many blacks have left for cities, dozens of rural counties in the region still have large black majorities. More broadly, the South is home to many black folk traditions, including music such as black spirituals and the blues, which have now become popular far beyond their rural origins. Regrettably, even though the rural neighborhoods of the black homeland differ greatly in density and appearance from black urban neighborhoods in the North, poverty plagues both communities.

A second rural homeland in the South is Acadiana, a zone of persisting Cajun culture in southwestern Louisiana (Figure 3.20). This homeland was created in the eighteenth century when French settlers were expelled from the Maritimes and relocated to Louisiana. Nationally popularized today through their food and music, the Cajuns retain an enduring attachment to the bayous and swamps of southern Louisiana.

Native American populations are also strongly tied to their homelands. Indeed, many native peoples maintain intimate relationships with their surroundings, weaving elements of the natural environment together with their material and spiritual lives. Almost four million Indians, Inuits, and Aleuts live in North America, and they claim allegiance to more than 1,100 tribal bands. Particularly in the American West and in the Canadian and Alaskan North, native peoples control sizable reservations, including the 16-million-acre (6.5-million-hectare) Navajo Reservation in the Southwest, as well as self-governing Nunavut in the Canadian North (Figure 3.20). Although these homelands preserve traditional ties to the land, they have also been settings for increasing cultural tensions. Within the United States, many Native American groups, given the special legal status of their reservations, have built gambling casinos and tourist facilities that bring in much-needed capital but also challenge traditional life ways. In both Canada and Alaska, friction has also been created among native peoples, private natural resource interests, and various governmental agencies.

▲ **Figure 3.21 The enduring Hispanic legacy** Spanish Catholic missionaries reconfigured the cultural geography of New Mexico many centuries ago. That legacy persists to the present, preserved in the earth-toned churches that remain, as well as in the important role still played by the region's Hispanic Catholic population. *(Robert Frerck/Odyssey Productions)*

A Mosaic of Ethnic Neighborhoods North America's cultural mosaic is also enlivened by smaller-scale ethnic signatures that shape both rural and urban landscapes. For example, distinctive rural communities that range from Amish settlements in Pennsylvania to Ukrainian neighborhoods in southern Saskatchewan add cultural variety. When much of the agricultural interior was settled, immigrants often established close-knit communities. Among others, German, Scandinavian, Slavic, Dutch, and Finnish neighborhoods took shape, held together by their common origins, languages, and religions. Although many of these ties weakened over time, rural landscapes of Wisconsin, Minnesota, the Dakotas, and the Canadian prairies still display many of these cultural imprints. Folk architecture, distinctive settlement patterns, ethnic place names, and the simple elegance of rural churches selectively survive as signatures of cultural diversity upon the visible scene of rural North America.

Within North American cities, ethnic neighborhoods also enrich the urban landscape and reflect both global-scale and internal North American migration patterns. Complex social and economic processes are clearly at work. Employment opportunities historically fueled population growth in North American cities, but the cultural makeup of the incoming labor force varied, depending on the timing of the economic expansion and the relative accessibility of an urban area to different cultural groups. For instance, Chicago's economy boomed in the late nineteenth century and attracted varied southern and eastern Europeans as well as sizable German and Irish communities (Figure 3.22). Continued twentieth-century growth in the manufacturing and service sectors brought southern blacks, Hispanics, and Asians to the city. By contrast, the ethnic geography of Los Angeles, typical of many western North American cities, displays fewer of the older immigrant influences while emphasizing the cultural impacts of more recent migrants (Figure 3.22).

The degree and persistence of ethnic clustering in North American cities also varies. Some cultural groups, particularly Europeans, initially congregated in well-defined ethnic neighborhoods but later dispersed into the suburbs. Thus, originally tight-knit neighborhoods of Germans, Irish, and Italians often saw their communities decline in importance in the twentieth century. On the other hand, surviving cultural prejudices and the practical advantages of clustering encouraged, indeed compelled, many urban black, Hispanic, and Asian communities to remain strongly clustered upon the urban landscape. In the Los Angeles example, marked ethnic territories remain a local fixture; much of East Los Angeles is heavily Hispanic, other Los Angeles communities such as Compton and Inglewood are black districts, and the San Gabriel Valley suburbs of Alhambra and Monterey Park are increasingly dominated by Chinese (Figure 3.22). Particularly in the United States, ethnic concentrations of nonwhite populations increased in many cities during the twentieth century as whites exited for the perceived cultural safety of the suburbs. In terms of central-city population, blacks make up 75 percent of Detroit and more than 60 percent of Atlanta, while Los Angeles is now more than 40 percent Hispanic. Often these ethnic neighborhoods are termed **ghettos** (black communities) or **barrios** (Hispanic communities), especially when high levels of poverty and unemployment further isolate them from the urban mainstream. Canadian central cities, while often multicultural in character, have more successfully retained sizable middle-class white populations than their counterparts in the United States.

Patterns of North American Religion

Many religious traditions also shape North America's human geography. Reflecting its colonial roots, Protestantism is dominant within the United States, accounting for about 60 percent of the population. Still, many local and regional churches add variety to the Protestant mix. For example, Baptists represent religious majorities in states such as Mississippi, Alabama, and Georgia, but black and white denominations are often separated within the region. In addition, electronic televangelism has enhanced the visibility of diverse fundamentalist Protestant denominations, particularly in the politically conservative Bible Belt, which stretches from Virginia to Texas and Missouri. In the upper Midwest, Scandinavian migrations produced enduring Lutheran concentrations. Elsewhere, hybrid American religions sprang from broadly Protestant roots. New England and upstate New York have rich utopian traditions exemplified by groups such as the Oneida Community, Harmony Society, and the Amana Colony. By far the most successful modern articulation of this impulse are the Latter-day Saints (Mormons), regionally concentrated in Utah (77 percent of the population) and Idaho (27 percent of the population), and claiming almost 5 million North American members. Within Canada, almost 40 percent of the population are Protestant, with the United Church of Canada and the Anglican Church claiming the largest numbers of adherents.

Roman Catholicism remains important in regions that received large numbers of Catholic immigrants. French-Canadian Quebec is a bastion of Catholic tradition and makes Canada's population (44 percent) distinctly more Catholic than that of the United States (25 percent). Still, major regional concentrations of American Catholics persist. Cities such as New York, Boston, and Chicago are home to large Catholic communities, reflecting enduring impacts of Irish, German, Italian, Polish, and Hispanic residents. Catholic populations are on the rise in southern Florida and the Southwest as Hispanics increasingly shape the cultural geographies of those regions.

Millions of other North Americans practice religions outside Protestant and Catholic traditions or they are nonbelievers. Orthodox Christians congregate in the urban Northeast, where many Greek, Russian, and Serbian Orthodox communities were established between 1890 and 1920. The telltale domes of Ukrainian Orthodox churches still dot the Canadian prairies of Alberta, Saskatchewan, and Manitoba. More than seven million Jews live in North America, concentrated in East and West Coast cities. In the United States, the rapidly growing Nation of Islam (Black Muslims) also has a strong urban orientation, reflecting its appeal to many economically dispossessed African Americans. There has been additional recent growth in other North American Islamic (six million), Buddhist (one million), and Hindu (one million) populations where sizable numbers of Asians have settled. While only about 8 percent of the U.S. population classify themselves as nonbelievers, a recent survey showed that 30 percent of the population claimed to have a largely secular lifestyle in which religious traditions were rarely practiced.

The Globalization of American Culture

Simply put, North America's cultural geography is becoming more global at the same time that global cultural geographies are becoming more North American (influenced particularly by the United States). As a dominant player in globalization, the United States is inevitably changing in the process, but to what extent? Similarly, we can ask more precisely how the globalization of American culture will transform the larger world. Are basic values shifting with globalization, is a new "global culture" being forged, or are most of the technological innovations and elements of popular culture incorporated into existing traditions? There are no simple answers to these questions, but they should be central concerns to human geographers and to any citizen of the twenty-first century.

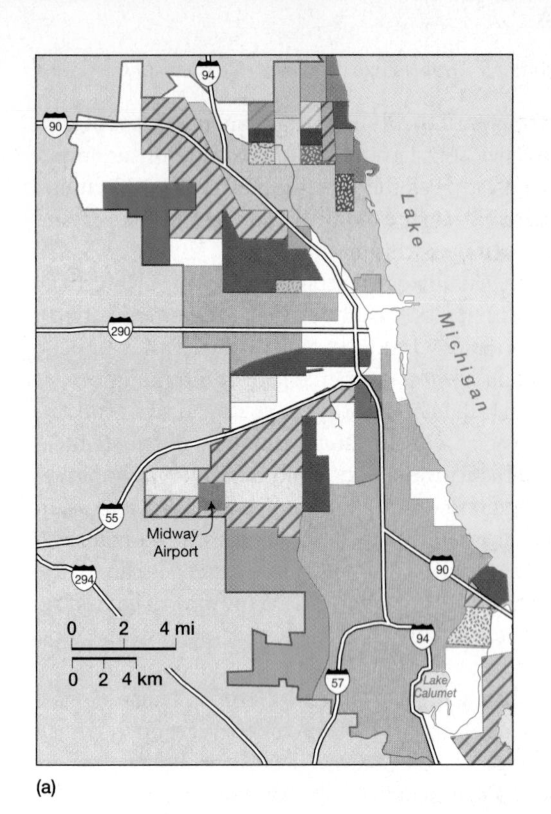

(a)

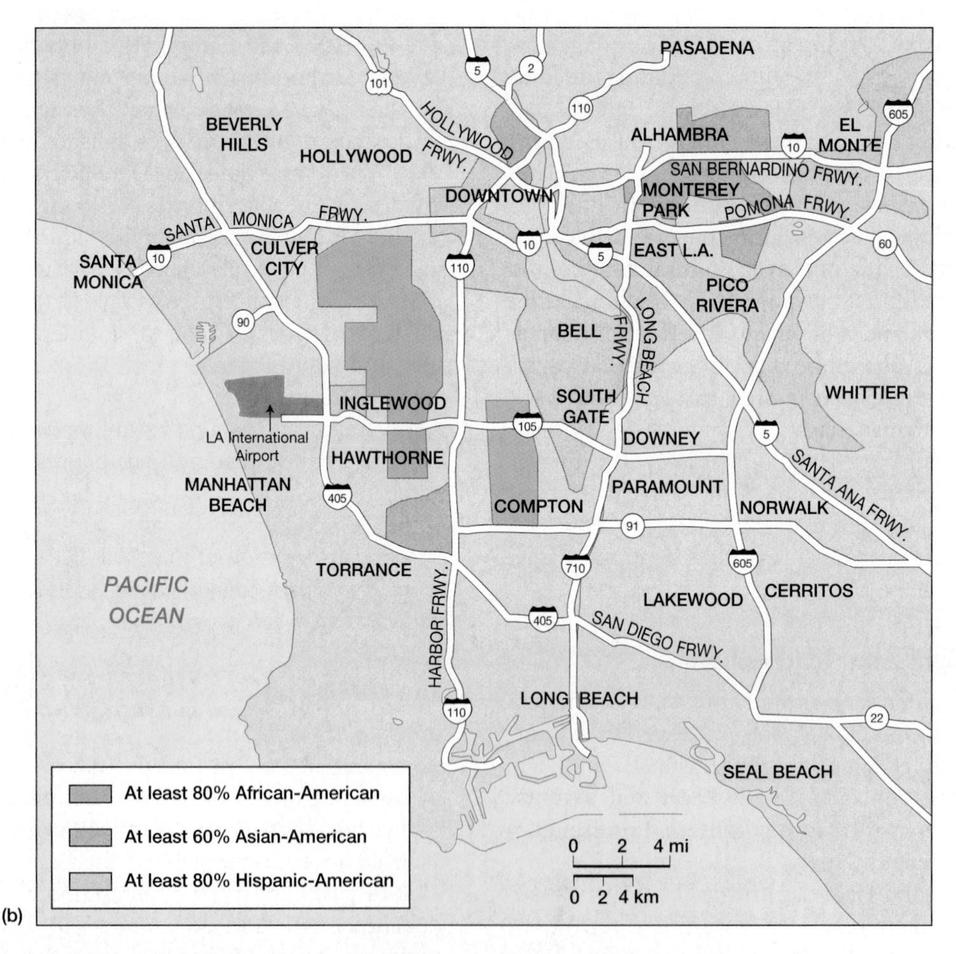

(b)

▲ **Figure 3.22 Ethnicity in Chicago and Los Angeles** Varied economic opportunities attracted a wide variety of migrants to North American cities, producing an intricate ethnic mosaic of distinctive neighborhoods and communities. In Chicago (a) and Los Angeles (b), several cycles of economic expansion have drawn a diverse collection of residents from around the globe. *(Reprinted from Rubenstein, 1999,* An Introduction to Human Geography, *Upper Saddle River, NJ: Prentice Hall)*

North Americans: Living Globally In their everyday lives North Americans both celebrate the benefits and confront the challenges posed by globalization (see "'All Right Already!' Globalization and Noo Yawk City's Languages"). For Americans, one key issue revolves around the English language, which some have described as the key "social glue" that holds the nation together. Since 1980, the continuing flood of non-English-speaking immigrants into the country has sharpened the debate. Many in the United States argue that English should be the country's only officially recognized language in order to compel immigrants to learn it and thus speed their cultural and economic assimilation into the host culture. On the other hand, immigrant groups suggest they need to maintain their traditional languages, both to function within their ethnic communities and to preserve their cultural heritage. In reaction, a growing number of states, especially in the South and West, have recognized English as their "official language," and several programs in bilingual education, including those in Florida and California, have been challenged so that immigrants are forced to learn English in school. At the same time, foreign-language media outlets flourish in many multicultural markets around the country. New York City's WSKQ, known as "Mega," is a leading Spanish-language radio station, and Los Angeles is home to Korean, Cantonese, and Japanese television broadcasts. Satellite dishes and cable programming have also given birth to multilingual offerings such as the Filipino Channel, Native American Nations, and TV Asia.

North Americans are going global in other ways. Most obviously, they travel much more widely than ever before. Residents of the United States logged 52.1 million international flights in 1997, almost four times the rate 20 years earlier. In diet, the popularity of ethnic restaurants has peppered the realm with a bewildering variety of Cuban, Ethiopian, Basque, and Pakistani eateries. Chinese and Italian foods dominate the international taste buds of all North Americans, with Mexican food a rapidly growing choice in the United States. The growing affinity for foreign beverages mirrors the pattern; imported beer sales in the United States are growing 14 percent annually with the top producers selling 159 million cases during 1997 (Figure 3.23). Americans also have increased their consumption of foreign red wines and they have rapidly Europeanized their coffee-drinking habits. In fashion, Gucci, Armani, and Benetton are household words for millions who keep their eyes on European styles. While British pop music has been an accepted part of North American culture for four decades, the beat of German techno bands, Gaelic instrumentals, and Latin rhythms also have become an increasingly seamless part of daily life within the realm. Indeed, from acupuncture and homeopathic cures to soccer and New Age religions, North Americans are tirelessly borrowing, adapting, and absorbing the larger world around them.

The Global Diffusion of U.S. Culture In a parallel fashion, the lives of billions of people beyond the region are forever changed by U.S. culture. Although the economic and military rise of the United States was notable by 1900, it was not until after World War II that the country's popular culture reshaped global human geographies in fundamental ways. The Marshall Plan and Peace Corps initiatives exemplified the growing presence of the United States on the world stage even as European colonialism waned. Rapid improvements in global transportation and information technologies, much of it engineered in the United States, also

"ALL RIGHT, ALREADY!" Globalization and Noo Yawk City's Languages

What do Ed Koch, Archie Bunker, Fran Drescher, and Joe Pesci have in common? Their Noo Yawk accent, naturally! Cultural geographers and linguists have long marveled at the distinctive cadence, pronunciation, and idiomatic expressions of New York City residents. Recently, however, experts fear that the city's classic vernacular may be the victim of a more homogenized America. This is only the latest chapter in how "talking the tawk" of New York City represents the intersection of globalization and the evolution of English in North America.

Historically, New York City was the meeting place for millions of immigrants and thus its streets were filled with richly varied accents from the far corners of Earth. Even as immigrants learned English, they often retained distinctive elements that contributed to the city's rich linguistic traditions. Arguably, their varied contributions produced one of the world's first truly globalized dialects. Even today, Jewish New Yorkers remain more apt to have the nasal "a" sound in their speech, while Hispanics sometimes mix "z" and "s" sounds and drop word endings. New York speech also owes its origins to the Irish, who may be responsible for toilet sounding like "terlet" or girl being turned into "goil." Such classic New Yorkese is also strongly linked to the city's lower socioeconomic groups; it is not surprising that there are more than two dozen listings for voice and diction coaches in the Manhattan Yellow Pages who hold out hope for clients intent on losing "da woist" of their accents!

While some attributes of the city's dialect have worked into national speech patterns, other elements are fading, victims of the latest chapter of globalization that spreads a sanitized mass media version of English not only throughout New York City but also to the world beyond. What better proof of global cultural convergence than bland New Yorkers! As New York's own Jimmy Breslin put it, "If you're going to homogenize it, you make it no good." Whether that's true or not, New York City's changing accents vividly tell the story of larger shifts in American society and how it has been shaped by varied cultural influences, both from home and abroad.

Source: Adapted from "Talking the Tawk," *New York Times*, September 21, 1998.

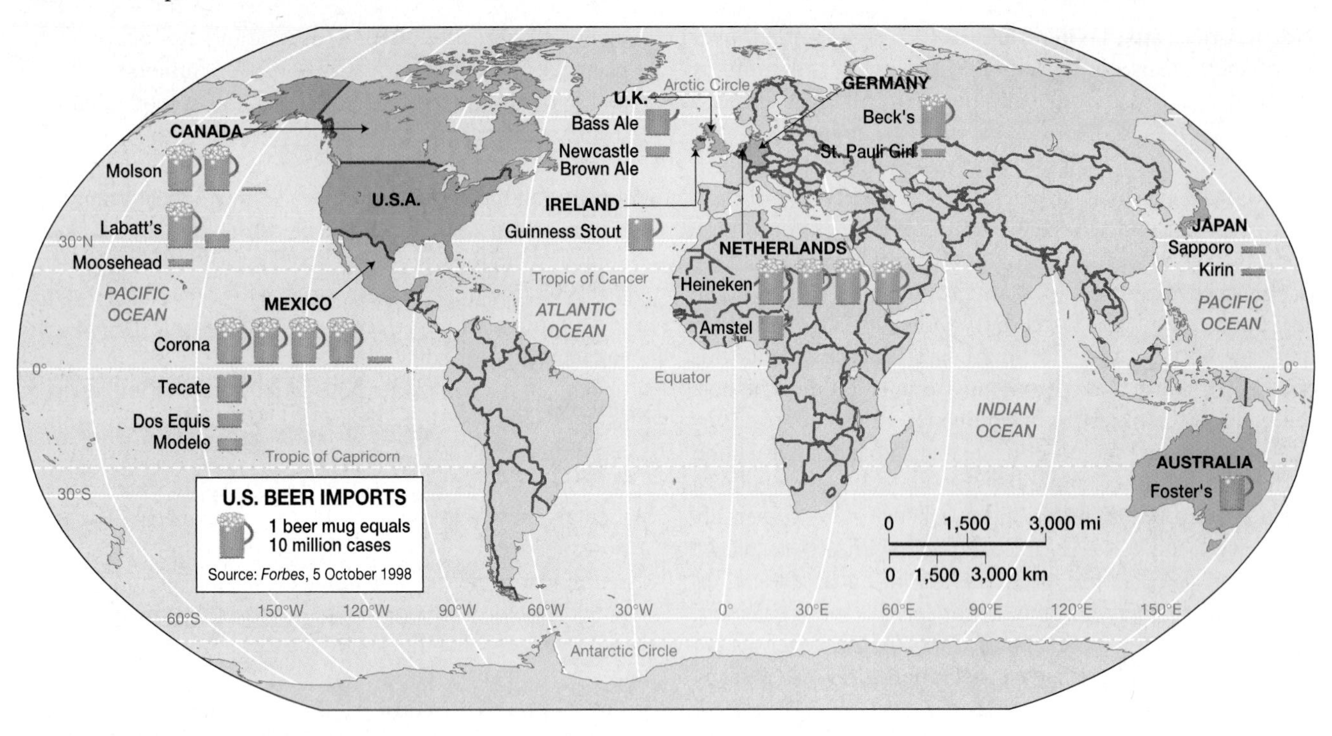

▲ **Figure 3.23 Annual beer imports to the United States** Whether they are aware of it or not, North Americans are increasingly eating and drinking globally. Rising beer imports, including many upscale foreign labels, exemplify the pattern. The nation's beer drinkers know no bounds to their thirsts, drawing diversely from Asian, Australian, European, and Latin American producers. *(Data from* Forbes Magazine, *October 5, 1998)*

brought the world more surely under the region's spell. Perhaps most critical was the marriage between growing global demands for consumer goods and the rise of the multinational corporation that was superbly structured to meet and cultivate those needs. Today, many U.S. companies dominate such transactions, and with the new industries, products, and services come subtle but fundamental shifts in attitudes and values that challenge older ways of life. The results, however, are not a simple Americanization of traditional cultures or a single, synthesized global culture shaped in the U.S. image. Clearly, however, millions of people, particularly the young, are strongly attracted by images of individualism, wealth, freedom, youth, and mobility.

Global information flows illustrate the country's influence in promoting its own cultural image of the world. Companies such as Time Warner, Inc., and Walt Disney Co. increasingly dominate multiple entertainment media. The emergence of English as the *de facto* global language has obviously been an advantage to a variety of U.S. media outlets. U.S. book and magazine publishing is thriving, with an expanding international appetite for everything from technical journals to romance novels and science fiction. *Cosmopolitan* magazine produces 36 different foreign editions and its international circulation tops four million copies, almost twice its readership in the United States. The United States also dominates television; as television sets, cable networks, and satellite dishes spread, so do U.S. sitcoms, CNN, and MTV. Even the Asia-wide STAR TV network, which broadcasts from Hong Kong, carries a preponderance of North American and English-language programming. Movie screens and VCRs offer a similar story. In western Europe, the United States claims 70 percent of the film market, and more than half of Japan's movie entertainment is dubbed fresh from Hollywood. In the video market, Blockbuster Entertainment now has 2,000 outlets in more than two dozen foreign countries. The country also dominates cyberspace, both in terms of the number of host sites on the Internet as well as leading the way in technological innovation.

The United States shapes the popular cultural landscape of every corner of the globe. The scope of global corporate advertising, distribution networks, and mass consumption brings Cokes® and Big Macs™ to Moscow and Beijing, golf courses to Thai jungles, Mickey and Minnie Mouse to Tokyo and Paris, and Avon cosmetics to millions of beauty-conscious Chinese (Figure 3.24). Western-style business suits become the professional costume of choice, while T-shirts and jeans offer standardized global comfort on days away from work. In the built landscape, central city skylines become indistinguishable from one another, suburban apartment blocks take on a global sameness, and one airport hotel looks the same as the next one eight time zones away.

The country's cultural control has not gone unchallenged. For example, Canadian government agencies routinely chastise their radio, television, and film industries for letting in too much U.S. cultural influence. As an antidote to their overbearing southern neighbor, the government requires certain

▲ **Figure 3.24 Tokyo Disneyland** While some traditional cultures resist American influences, many people around the world have embraced globalization, especially when it is brought by franchised ambassadors such as Mickey and Minnie Mouse. Tokyo Disneyland is only one of a growing number of Magic Kingdoms around the planet. *(AP/Wide World Photos)*

levels of Canadian content in much of their media programming. Indeed, Canada hosted a 1998 conference on U.S. cultural imperialism in which 19 countries (excluding the United States) explored ways to derail the global juggernaut of Holiday Inns and Kentucky Fried Chicken franchises. The French have also been critical of U.S. dominance in such media as the Internet. Public subsidies to France's Centre National de la Cinematographie are designed to foster film-making for a national audience deluged with English-language productions. Elsewhere, Afghanistan and Iran have banned satellite dishes and many U.S. films, although illegal copies of top box-office hits often find their way through national borders. In one way or another, U.S. cultural influences manage to reach beyond that nation's borders, a reminder that the region's cultural values will continue to have an increasingly greater impact on the larger world beyond.

Geopolitical Framework: Patterns of Dominance and Division

In disarmingly simple fashion, North America's political geography brings together two of the world's largest states. The creation of those states, however, was neither simple nor preordained but rather the complex outcome of historical processes that might have created quite a different North American map than we presently have. Once established, those states have coexisted in a close relationship of mutual economic and political interdependence. President John F. Kennedy summarized the links in a speech to the Canadian parliament in 1962 when he said, "Geography has made us neighbors, history has made us friends, economics has made us partners and necessity has made us allies." That cozy continental relationship, however, has not been without its tensions, and some continue today. In addition, both nations have had to deal with fundamental internal political complexities that have not only tested the limits of their federal structures, but even challenged their very existence as states.

Creating Political Space

The United States and Canada sprang from very different political roots. The United States broke cleanly and violently from Great Britain. The American Revolution created a powerful sense of nationalism that sped the process of spatial expansion and produced a political rhetoric oriented around the country's preordained role as a continental, indeed global, power. Canada, by contrast, was a country of convenience, born from a peaceful, incremental separation from Britain and then assembled as a collection of distinctive regional societies that only gradually acknowledged their common political destiny.

Uniting the States The creation of the United States replaced one continental division of political space with another, and that process of annihilation and invention proceeded in ways that no one could have anticipated. Turning back the clock to the early eighteenth century reveals a political geography very much in the making. Beyond scattered frontiers of diverse European settlement lay vast domains of Native-American-controlled political space. Although boundaries were not formally surveyed or mapped, native peoples carved up the continent in an elaborate geography of homelands, allied territories, and enemy terrain. In the next two centuries, that dynamic and intricate pattern was swept away, suddenly and irrevocably disrupted beyond recognition.

With amazing rapidity, Europe and then the United States imposed their own political boundaries across the realm. The 13 English colonies, sensing their common destiny after 1750, finally united two decades later and clashed violently with their colonial parent. By the 1790s, young America's political claims reached the Mississippi River, the new republic was busily coercing land cessions from native peoples, and the Ordinance of 1787 provided a template for western territory and state formation that served as the model of expansion for the next century. Soon the Louisiana Purchase (1803) nearly doubled the national domain, creating in a pen stroke a new political geography that was as vast as it was unexplored. Still, the precise evolution of the young United States remained undecided. Under different historical circumstances, it might well have included independent Texan (in existence between 1836 and 1845), Mormon, and Californian nations within the West, along with an expanded Mexico and a more intrusive British North America. But larger political imperatives intervened. By mid-century, Texas was annexed, treaties with Britain secured the Pacific Northwest, and an aggressive war with Mexico captured much of the Southwest. Alaska (1867) and Hawaii (1898) eventually rounded out the present political domain of 50 states.

Assembling the Provinces Canada was created under quite different circumstances. The modern pattern of provinces assembled in a slow and uncertain fashion. After the American Revolution, England's remaining territorial claims in the region came under the control of British North America. The Quebec Act of 1774 allowed for continued French settlement in the St. Lawrence Valley and provided the initial template for governing the region. Soon, however, Anglo settlers near Lake Ontario and Lake Erie pressed for more local colonial representation. The result was the Constitutional Act of 1791, which divided the colony into Upper Canada (Ontario) and Lower Canada (Quebec). Frustration with that system led to the Act of Union in 1840, thereby reuniting the two Canadas. Meanwhile, Maritime Canada passed through several administrative eras in which Nova Scotia and New Brunswick incrementally moved toward their modern boundaries. Various questions of colonial governance resurfaced, and in 1867 the British North America Act united the provinces of Ontario, Quebec, Nova Scotia, and New Brunswick in an independent Canadian Confederation. The peaceful separation from the mother country also guaranteed to Quebec special legal and cultural privileges that presaged some of the modern power struggles within the country.

Once created, the Canadian Confederation grew in piecemeal fashion, more out of geographical convenience than from any compelling nationalism at work to unite the northern portion of the continent. Within a decade, the Northwest Territories (1870), Manitoba (1870), British Columbia (1871), and Prince Edward Island (1873) joined Canada, and the continental dimensions of the country took shape. The 1886 completion of the transcontinental Canadian Pacific Railway cemented the link to British Columbia and accelerated development within the interior. Soon, Yukon Territory (1898) separated from Northwest Territories; Alberta and Saskatchewan gained provincial status (1905); and Manitoba, Ontario, and Quebec were enlarged (1912) north to Hudson Bay. Newfoundland finally joined in 1949. The recent addition of Nunavut Territory (1999), carved from the Northwest Territories, represents the latest change in Canada's political geography.

Continental Neighbors

Geopolitical relationships between Canada and the United States have always been intimate: their common 5,525-mile (8,900-kilometer) boundary compels both nations to pay close attention to one another. More than that, their status as continental neighbors has generated tremendous interaction, trade, and mutual cooperation, while at the same time offering potential for conflict and controversy.

Canadians, for good reason, have worried about being in the political shadow of the United States. Historically, the War of 1812 included U.S. invasions of British North American territory, a military and geopolitical offense not soon forgotten. Tensions between the two powers rose again in the 1830s and 1840s as boundary issues in Maine and the Oregon country threatened to send both sides to arms. Indeed, Canada's independence in 1867 was probably linked in no small way to fears over its southern neighbor. U.S. rhetoric in the post-Civil-War era bristled with the language of Manifest Destiny, and open talk of annexing Canadian lands may have hastened British actions to unite the region.

During the twentieth century, however, political cooperation outweighed lingering suspicions, and the two countries saw both their political and economic destinies increasingly intertwined. There was also the recognition that the inevitable conflicts that did occur would need to be swiftly adjudicated if the two countries were to benefit from their proximity and shared interests. By 1909, the Boundary Waters Treaty created the International Joint Commission, an early step in the common regulation of cross-boundary issues involving water resources, transportation, and environmental quality. That tradition of cooperation was reinforced with the joint administration of the St. Lawrence Seaway project in 1959, opening the Great Lakes region to better global trade connections. The two states joined in cleaning up Great Lakes pollution and in plans to reduce acid rain in eastern North America. Internationally, both cooperated in creating a common North American defense and in serving in the North Atlantic Treaty Organization (NATO).

It has been in the realm of trade relations that the close political ties between these neighbors have mattered most. The United States receives 80 percent of Canada's exports and supplies more than 75 percent of its imports. Conversely, Canada is the United States' most important trading partner, accounting for roughly 20 percent of its exports and imports. These historically close ties, fed by advantages of accessibility and nourished through shared economic growth, have been further strengthened by an increasing number of trade agreements wedding the countries together. One landmark reached in 1989 was the signing of the bilateral Free Trade Agreement (FTA). Five years later, the larger **North American Free Trade Agreement (NAFTA)** extended the alliance to Mexico and laid out a 15-year plan for drastically reducing all barriers to trade or capital investment among the three nations. Paralleling the success of the European Union (EU), NAFTA has forged the world's largest trading bloc, including 400 million consumers and a huge free trade zone that stretches from beyond the Arctic Circle to Latin America. While NAFTA has resulted in job dislocations in both Canada and the United States, it has also stimulated total economic activity and trade within the realm as investment and employment migrate to more optimal locations. If Mexican wages improve, it may also ease the problem of illegal immigration along the porous 2,000-mile border between Mexico and the United States (Figure 3.25).

Political conflicts still divide the two countries. Environmental issues produce cross-border tensions. Many U.S. waterways begin in Canada, while elsewhere Canada receives much of its water from upstream settings in the United States. Inevitable conflicts result when environmental degradation from one nation affects the other. For example, Montana's North Flathead River flows out of British Columbia where Canadian logging and mining operations periodically threaten fisheries and recreational lands south of the border. Similarly, industrial pollution in the United States often becomes Canada's problem when fouled waters flow through the Great Lakes system. Agricultural and natural resource competition also engenders periodic fits of controversy between the two neighbors. For example, recent problems developed when Canadian wheat and potato growers were accused of dumping their products into United

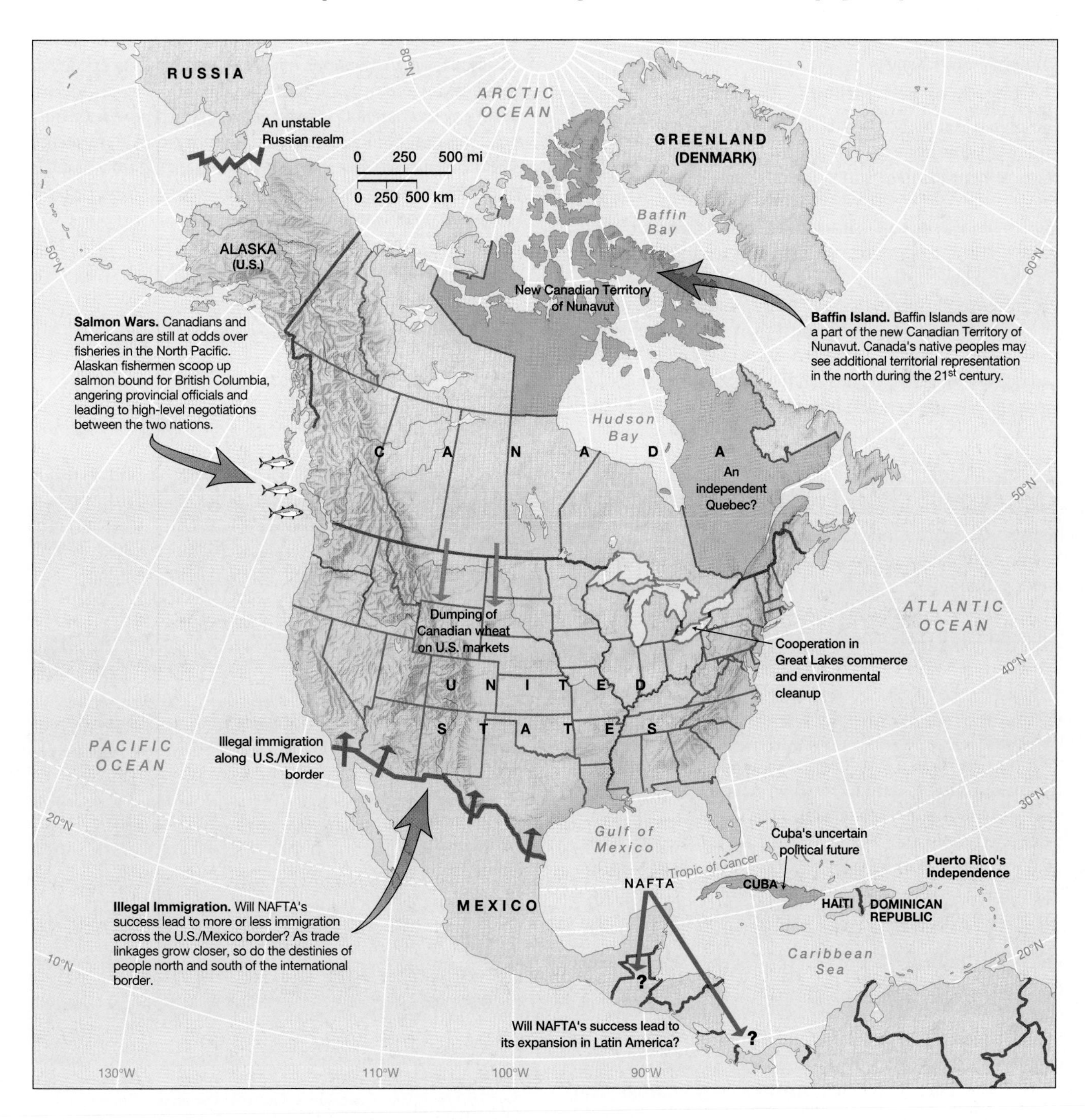

▲ **Figure 3.25 Geopolitical issues in North America** Although Canada and the United States share a long and peaceful border, many political issues still divide the two countries. In addition, internal political conflicts, particularly in bicultural Canada, cause tensions. Shifting relations with Latin America and the Caribbean remain important issues for their neighbors to the north.

States markets, thus depressing prices and profits for U.S. farmers. Salmon wars have also flared along the Pacific Coast, where U.S. and Canadian fishermen compete keenly for fish claimed by either side. In 1997 national leaders had to mediate the controversy when Canadian boats blockaded an Alaskan ferry docked at a harbor in British Columbia. Occasional foreign policy disputes also divide the pair, including long-standing political differences over Cuba. Still, these neighbors coexist in a remarkably harmonious marriage, a political conjoining born both from the heart as well as the pocketbook, and one that is likely to be sustained in the twenty-first century.

The Legacy of Federalism

The United States and Canada are **federal states** in that both nations allocate considerable political power to units of government beneath the national level. Other nations, such as France, have traditionally been **unitary states** in which power is centralized at the national level. Federalism, however, leaves many political decisions to local and regional governments and often allows distinctive cultural and political groups to be recognized as distinct entities within a country. Although both nations have federal constitutions, their origins and evolution are quite distinctive. The U.S. Constitution (1787), created out of a violent struggle with the powerful British nation, specifically limited centralized authority, giving all unspecified powers to the states or people. In contrast, the Canadian Constitution (1867) that created a federal parliamentary state was an act of British Parliament. Originally, it reserved most powers to central authorities and maintained many political links between Canada and the Crown.

Ironically, the evolution of the United States as a federal republic produced an increasingly powerful central government while Canada's geopolitical balance of power shifted toward more provincial autonomy and a relatively weak national government. For example, the federal government largely controls U.S. public lands, but in Canada provincial authorities retain power over public Crown lands. In addition, Canadian central authorities traditionally faced powerful regional political identities, particularly in Quebec. The close connection to Britain did little to foster Canadian nationalism. Indeed, a distinctive Canadian citizenship apart from that of Britain came about only after World War II, and the modern Canadian flag is less than 50 years old. It was only with the Constitution Act of 1982 that Canada formally transferred all legal authority from the British to Canadian Parliament.

Quebec's Challenge The greatest challenge facing Canada is determining what role Quebec will play in the country's future (Figure 3.25). Will it secede from Canada, remain a province, or seek another form of political autonomy from the rest of the country? Quebec's distinctive French society has deep historical roots. Over time, economic disparities between Anglo and French populations have reinforced differences between the two groups, with the French-Canadians often suffering when compared with their wealthier neighbors in Ontario. Beginning in the 1960s, a separatist political party in Quebec increasingly voiced French-Canadian concerns. When the party won provincial elections in 1976, it quickly moved to declare French the official language of Quebec and to schedule a provincial referendum on remaining within Canada (Figure 3.26). Although only 41 percent of the electorate voted for separation in 1980, the referendum precipitated the drafting of a new federal constitution two years later. Many French-Canadians were not impressed, however, fearing that their distinctive cultural identity might be threatened. They refused to sign the document, leading federal lawmakers in Ottawa to propose another solution, the Meech Lake Accord of 1990. The Accord included stronger guarantees of Quebec's special status as a "distinct society." Manitoba and Newfoundland proceeded to derail the agreement, arguing it gave Quebec too much autonomy that was not being extended to other provinces. After a national compromise failed (the Charlottetown Agreement), Quebec again held a referendum on separating from Canada in 1995. This time, only a razor-thin majority (50.6 percent) voted to remain within the larger country.

Canada's future remains clouded by the Quebec issue. Federalism has given the larger nation the flexibility to accom-

▲ **Figure 3.26 Canada's French legacy** Four centuries of French settlement in Canada, particularly within Quebec, leave many cultural and political issues unanswered. Most French-speaking Canadians want to preserve their language and culture. Can they do it within the present political framework? *(R. Sidney/The Image Works)*

modate a "distinct society," but that same federal structure has proven cumbersome in bringing the country together in a lasting agreement that all parties can adopt. In late 1998, Canada's Supreme Court ruled that even if Quebec elects to separate, it would need to negotiate with the rest of the country; provincial elections in Quebec increased uncertainties over the timing of a future referendum on separation. Many questions remain. How would a sovereign Quebec function? How would Quebec's sizable non-French population react? Would the rest of Canada accede to its wishes? Would other Canadian provinces ponder independence? Final answers to these questions or a solution to Canada's political challenges are not likely soon.

Native Peoples and National Politics Another challenge to federal political power has come from North American Indian and Inuit populations, both in Canada and the United States. Within the United States, the renewed assertion of Native American political power began in the 1960s and marked a decisive turn away from earlier policies of assimilation. Since the Indian Self Determination and Education Assistance Act of 1975, the trend has been toward increased Native American autonomy. The Indian Gaming Regulatory Act (1988) offered potential economic independence for many tribes. In the western American interior, where Indians control roughly 20 percent of the land, tribes are solidifying their hold on resources, reacquiring former reservation acreage, and participating in political interest groups such as the Native American Fish and Wildlife Society and the Council of Energy Resource Tribes. In Alaska, native peoples acquired title to 44 million acres (18 million hectares) of land in 1971 under the Alaska Native Claims Settlement Act.

In Canada even more ambitious challenges to a weaker centralized government have yielded dramatic results. As natives pressed their claims for land and political power in the 1970s, Canada established the Native Claims Office (1975) and began negotiating settlements with various groups, particularly within the country's vast northern interior. Agreements with native peoples in Quebec, Yukon, and British Columbia turned over millions of acres of land to aboriginal control and increased native participation in managing remaining public lands. By far, the most ambitious agreement has been to create the new Canadian territory of Nunavut out of the eastern portion of the Northwest Territories in 1999 (Figure 3.25). Nunavut is home to 22,000 people (80 percent Inuit) and is the largest territorial/provincial unit within Canada. Its creation represents a new level of native self-government in North America (see "Local Voices: Nunavut Meets the Internet").

A Global Reach

The geopolitical reach of the United States, in particular, has taken its influence far beyond the bounds of the realm. The Monroe Doctrine (1824) asserted that U.S. interests were hemispheric and that they transcended national boundaries, but it was not until after 1895 that the United States accelerated its global-scale expansion. Two principal settings served as early laboratories for political imperatives in the United States. In the Pacific, the United States claimed the Philippines as a prize of the Spanish-American War (1898), and further annexations of Guam (1898) and the Hawaiian Islands (1898) presaged the country's twentieth-century dominance of the region. In Central America and the Caribbean, the growing role of the American military between 1898 and

LOCAL VOICES Nunavut Meets the Internet

Baffin Island resident Adamee Itorcheak is wired to the world and he is helping fellow citizens of Canada's Nunavut Territory surf the Web as well. Itorcheak lives in Iqaluit on the southern portion of Baffin Island and he runs Nunanet Communications, the only Internet service provider within a thousand miles of the village. Globalization first arrived in Iqaluit during World War II when a refueling airstrip was built on the edge of town. Gradually thereafter, regular mail and passenger air service, telephone connections, and satellite television have linked the villagers with the larger world. Although some native Inuits have greeted cyberspace with suspicion, there has also been a cultural heritage of borrowing that portends well for the future of Itorcheak's high-technology gamble. The native word for e-mail roughly translates to "letters through the artificial brain" and Baffin Islanders increasingly see the virtues of such quick communications in a land where travel is often difficult.

Itorcheak's Nunanet service has 900 customers and serves a varied clientele, including businessmen, seal hunting guides, and educators. Itorcheak had to teach his first few hundred customers how to use the Internet, but now he is busily supplementing his service provider business as a consultant to local entrepreneurs and government agencies. His customers monitor wholesale fish prices in Japan, set up dogsled excursions for incoming tourists, and disseminate information on preserving native languages and customs to other Arctic communities.

Whether Nunanet survives or not, the Internet has a firm foothold in Nunavut, another reminder of the role this technology is playing in the diffusion of information in the twenty-first century. Its diverse uses also illustrate how the complex process of globalization transforms particular places in varied ways. Thanks to Itorcheak, Baffin Island will never be quite the same and, conversely, the new connection brings the newly wired Arctic a bit closer to the outer world.

Source: Adapted from "Across Tundra and Cultures, Entrepreneur Wires Arctic," *Wall Street Journal*, October 19, 1998.

1916 shaped politics in Cuba, Puerto Rico, Panama, Nicaragua, Haiti, Mexico, and elsewhere. Further, the country's role in World War I raised its stakes in European affairs.

While the 1920s and 1930s briefly returned the United States to isolationist policies, World War II and its aftermath forever redefined the country's role in world affairs. Victorious in both Atlantic and Pacific theaters, postwar America emerged from the conflict as the world's dominant political power. Quickly, however, a resurgent Soviet Union challenged the United States, and the Cold War began in the late 1940s. In response, the Truman Doctrine promised aid to struggling postwar economies and actively challenged communist expansion in Europe and elsewhere. The United States also fashioned multinational political and military agreements such as the North Atlantic Treaty Organization (NATO) and the Organization of American States (OAS), which were designed to cast a broad umbrella of U.S. protection across much of the noncommunist world. Violent conflicts in Korea (1950–1953) and Vietnam (1961–1973) pitted U.S. political interests against communist attempts to extend their Asian dominance beyond the Soviet Union and China (Figure 3.27). Tensions also ran high in Europe as the Berlin Wall Crisis (1961) and nuclear weapons deployments by NATO- and Soviet-backed forces brought the world closer to another global war. The Cuban missile crisis (1962) reminded Americans that traditional political boundaries provided little defense in a world uneasily brought closer together by technologies of potential mass destruction.

Even as the Cold War gradually receded during the 1980s, the global political reach of the United States continued to expand. A few examples suggest the pattern. Interventionist policies in Central America favored regimes friendly to the United States. President Carter's successful Middle East Peace Treaty between Israel and Egypt (1979) guaranteed a continuing diplomatic and military presence in the eastern Mediterranean. When Iraq's Saddam Hussein threatened Persian Gulf oil supplies in 1990, the United States led the United Nations coalition to contain the aggression. Indeed, since the breakup of the Soviet Union (1991), the global political presence of the United States has made it the world's only surviving superpower and a force to be reckoned with in every corner of the world. It remains to be seen precisely how the United States accepts that role and what political counterforces may arise to resist it.

▲ **Figure 3.27 The Vietnam War** U.S. involvement in the Vietnam War during the 1960s and 1970s reflected fear of global communism and determination to resist communist expansion in Asia. More than 50,000 Americans lost their lives in the conflict that ended with a communist victory in Southeast Asia and a generation of Americans politically divided. *(Halstead/Liaison Agency, Inc.)*

Economic and Social Development: Geographies of Abundance and Affluence

North America possesses the world's most powerful economy and its most affluent population. Its 300 million people consume huge quantities of global resources, but also produce some of the world's most sought-after manufactured goods and services. North America's size, geographic diversity, and resource abundance have all contributed to the realm's global dominance in economic affairs. More than that, however, it has been the region's human capital—the skills and diversity of its population—that have enabled North Americans to achieve high levels of economic development (Table 3.2). Even amid this continental affluence, however, many North Americans still struggle to escape from poverty. Indeed, since 1950, the social and economic gap between rich and poor has widened across the region, and these disparities will continue to challenge both Canada and the United States in the twenty-first century.

An Abundant Resource Base

Good fortune blessed North America with a varied storehouse of natural resources. The realm's climatic and biotic variety, its soils and terrain, and its abundant energy, metals, and forest

Table 3.2 Economic Indicators

Country	GNP per Capita ($U.S., 1996)	Total GNP (Millions of $U.S., 1996)	PPP* ($Intl, 1996)	Real Annual Growth % per Capita, 1990–1996
Canada	19,020	569,899	21,380	0.6
United States	28,020	7,433,517	28,020	1.5

*Purchasing power parity.

Source: *The World Bank Atlas*, 1998.

resources provided diverse raw materials for development. Indeed, the direct extraction of natural resources still makes up almost 4 percent of the U.S. economy and 7 percent of the Canadian economy. Some of these North American resources are then exported to global markets, while other required raw materials are imported to the region.

Opportunities for Agriculture North Americans have created one of the most productive food-producing systems in the world (Figure 3.28). Farmers within the realm practice highly commercialized, mechanized, and specialized agriculture that emphasizes the importance of efficient transportation systems, global markets, and large capital investments in farm machinery. Agricultural productivity climbed rapidly during the twentieth century, with crop yields rising at the same time that demands for labor declined. Today, agriculture employs only a small percentage of the labor force, both in the United States (2.5 percent) and Canada (3 percent). At the same time, farm consolidations have sharply reduced the number of operating units while average farm sizes have steadily risen. For example, in the United States, the two million farms currently in operation represent a 20 percent percent decline just since 1975. On the other hand, average farm size has increased from 420 acres in 1975 to more than 470 acres in 1998. As a result, farm populations have plummeted in many communities from Illinois to Alberta. Still, agriculture remains a critically important part of the North American economy. In 1997 farms directly produced more than $220 billion in sales. In addition, agriculture remains a dominant land use across much of the realm. Although suburbanization has taken its toll, almost half of the United States is classified as cropland, pasture, or potentially productive range lands. Given Canada's northern latitude, only about 7 percent of its lands is potentially agricultural, but its extensive forests have made it the world's largest exporter of timber, pulp, and newsprint.

The geography of North American farming represents the marriage between diverse environments, varied continental and global markets for food, and historical patterns of settlement and agricultural evolution. In the Northeast, dairy operations and truck farms specializing in fresh produce capitalize on the proximity of the region to major cities in Megalopolis

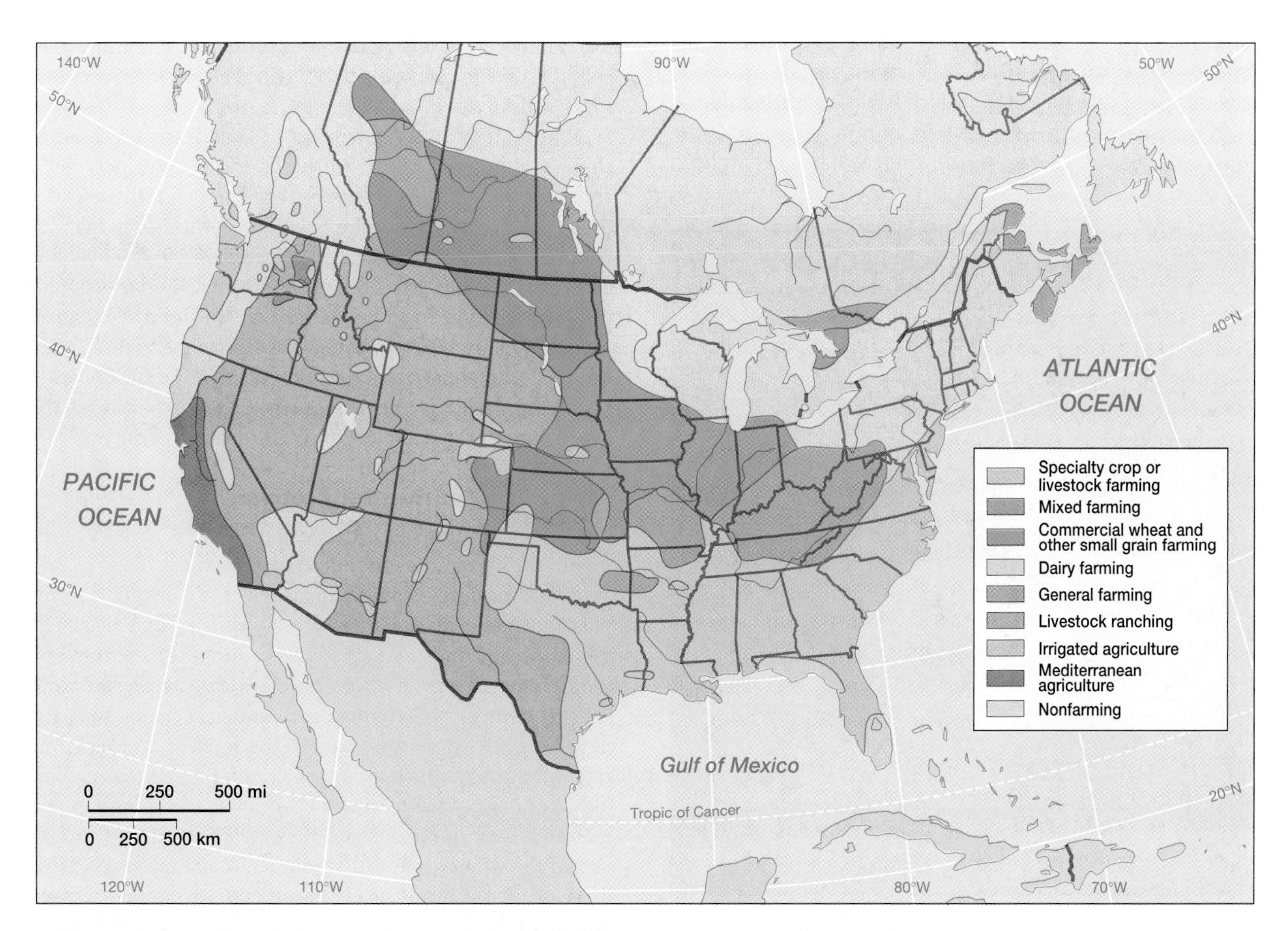

▲ **Figure 3.28 Agricultural regions of North America** Varied environmental settings, settlement histories, and economic conditions have produced the modern map of North American agricultural regions. Endowed with an extraordinarily favorable natural setting, North Americans have created one of the world's most productive food-producing systems. *(Modified from Clawson and Fisher, 1998,* World Regional Geography, *Upper Saddle River, NJ: Prentice Hall)*

and southern Canada. Corn and soybeans dominate croplands across the U.S. Midwest and western Ontario, where a tradition of mixed farming combines feed grains with commercially raised livestock, particularly cattle and hogs. Indeed, more than 15 percent of the agricultural wealth of the United States sprouts from the three heartland states of Illinois, Iowa, and Nebraska. North of the Corn Belt, cooler, shorter growing seasons support more dairy operations across Wisconsin and central Michigan (Figure 3.29). To the south, only remnants of the old Cotton Belt remain, largely replaced with a diverse geography of subtropical specialty crops, poultry, catfish, and livestock production and commercial forestry. West of the 100th meridian, extensive, highly mechanized cash grain operations stretch from Kansas to Saskatchewan and Alberta. Depending on surface and groundwater resources, irrigated agriculture across western North America also offers opportunities for more intensive operations. Indeed, California's agricultural output, nourished by large agribusiness operations in the irrigated Central Valley, accounts for more than 10 percent of the overall farm economy of the United States.

Industrial Raw Materials North Americans produce and consume huge quantities of other natural resources. While well endowed with a variety of energy and metals resources, the scale and diversity of the North American economy has also translated into the need to import additional raw materials. Petroleum use in the United States exemplifies the pattern. The country produces about 12 percent of the world's oil, but consumes more than 25 percent. As a result, the United States imports half of its oil, much of it coming from the Americas (Venezuela, Mexico, and Canada), the Middle East (Saudi Arabia), and Africa (Nigeria). The country also produces growing quantities of natural gas (about 25 percent of the world's total), but it still must import supplies from neighboring Canada. Within the realm, the major areas of oil and gas production are (1) the Gulf Coast of Texas and Louisiana; (2) the Central Interior, including West Texas, Oklahoma, and eastern Kansas; (3) Alaska's North Slope; and (4) Central Canada (especially Alberta).

▲ **Figure 3.29 Wisconsin dairy farm** Many family-owned dairy operations dominate the agricultural landscape of the upper Midwest. Cooler summers and shorter growing seasons in such areas often preclude mixed farming, but dairy operations benefit from their proximity to major metropolitan markets. *(Francois Gohier/Photo Researchers, Inc.)*

The most abundant fossil fuel in the United States is coal, but its relative importance in the overall energy economy declined in the twentieth century as industrial technologies changed and as environmental concerns grew. Still, the country's 400-year supply of coal reserves (23 percent of the world's total) will be an important energy resource, both for domestic consumption as well as for export. Today, U.S. coal is largely used in the production of electricity (the nation's number 1 source) and in a variety of industrial processes such as steelmaking. The nation's leading coal-producing region is Appalachia, but demand has been increasing for the environmentally cleaner low-sulfur coals of the western interior (Great Plains, Intermountain West, and Alberta).

North America remains a major producer of metals resources, although global competition, rising extraction costs, and environmental concerns pose challenges for this sector of the economy. Still, the realm is endowed with more than 20 percent of the world's copper, lead, and zinc reserves, and it accounts for more than 20 percent of global gold, silver, and nickel production. In addition, iron ore remains in abundance, although many U.S. industries now import it from Canada (Quebec–Labrador) rather than depend on declining domestic supplies in the upper Great Lakes. The Canadian Shield and the western interior remain key regions of metals production.

Creating a Continental Economy

The timing of European settlement in North America proved pivotal in its rapid economic transformation. The region's abundant resources came under the control of varied European powers armed with new technologies that reshaped the landscape and reorganized its economic geography. By the nineteenth century, North Americans themselves were actively contributing to those technological changes, and as new resources were unlocked in the interior and new immigrant populations flooded the continent, the region took full advantage of the marriage between its storehouse of raw materials and the Industrial Revolution. In the twentieth century, although natural resources remained important, new industrial innovations and greatly expanded service employment diversified the economic base and extended its global reach.

Connectivity and Economic Growth Dramatic improvements in North America's transportation and communication systems laid the foundation for urbanization, industrialization, and the commercialization of agriculture. Indeed, the region's economic miracle was a function of its **connectivity** or how well its different locations became linked with one another through an improved transportation and communication infrastructure. Those links greatly facilitated the potential for interaction between locations and dramatically reduced the cost of distance. Before 1830, North American connectivity gradually improved with the help of better roads, but it still took a week to travel from New York City to Ohio. For commercial traffic, canal construction also facilitated the movement of bulky goods. More than 1,000 miles (1,600 kilometers) of canals crisscrossed the eastern United States, including the Erie Canal (completed in 1825), which linked New York City with the North American interior. Other canals connected the Great Lakes and Ohio River systems, further integrating the Midwest to the Northeast economic core.

Tremendous technological breakthroughs revolutionized North America's economic geography between 1830 and 1920. Railroads cemented together whole new economic relationships. By 1860, there were more than 30,000 miles (48,387 kilometers) of track laid in the United States and the network grew to more than 250,000 miles (403,226 kilometers) by 1910. Farmers in the Midwest and Plains found ready markets for their products in cities hundreds of miles away. Industrialists collected raw materials from vast distances, processed them, and shipped manufactured goods to their final destinations. After 1869, travelers could board a train in Chicago and conveniently cross the continent in a few days. The telegraph brought similar changes to information: long-distance messages flowed across eastern North America by the late 1840s and 20 years later undersea cables firmly linked the realm to Europe, another milestone in the larger process of globalization.

The realm's integrated transportation and communications systems were redefined after 1920 as automobiles, mechanized farm equipment, paved highways, commercial air links, national radio broadcasts, and dependable transcontinental telephone service further reconfigured the cost of distance across the region. Cheap, mass-produced automobiles reached many middle-class North Americans by the 1920s, and the subsequent construction of major North American highways by 1970 produced the world's largest integrated road system. After World War II, continental connectivity also benefited from the St. Lawrence Seaway between the Atlantic Ocean and Great Lakes (1959), vast improvements in jet airline connections, and the increasing prevalence of television. Perhaps most profoundly, the region has taken the lead in the global information age, integrating computer, satellite, and Internet technologies in a web of connections that facilitates the flow of knowledge both within the realm and beyond.

The Sectoral Transformation Changes in employment structure signaled North America's economic modernization just as surely as its increasingly interconnected society. The **sectoral transformation** refers to the evolution of a nation's labor force from one highly dependent on the *primary* sector (natural resource extraction) to an economy with more employment in the *secondary* (manufacturing or industrial), *tertiary* (services), and *quaternary* (information processing) sectors. The sectoral transformation thus reflects how innovation changes employment opportunities in a society. For example, with agricultural mechanization, lower demands for primary sector workers are replaced by new opportunities in the expanding industrial sector. In the twentieth century, new services (trade, retailing) and information-based activities (education, data processing, research) created employment opportunities in these postindustrial sectors of the economy.

Both the United States and Canada have witnessed dramatic changes in employment that reveal the formative impacts of technological innovation and economic restructuring. In the late nineteenth century, primary sector employment in agriculture and mining dominated the occupational structure of the two countries. Rapid industrialization after 1870 contributed to steady growth in the secondary sector, and a century later it accounted for about 30 percent of the workforce. Modern technology has had the opposite effect on the primary sector, however, as machines replaced labor and employment in the sector has fallen to less than 4 percent of the workforce. In the later twentieth century, both nations also experienced relative declines in manufacturing employment when compared with the rapid growth of the tertiary and quaternary sectors. Today these two latter sectors employ more than 70 percent of the labor force, both in Canada and the United States. These recent trends also reveal the tangible imprint of globalization upon the North American labor force. Much of North America's sustained growth in the tertiary and quaternary sectors is directly tied to the ability of those industries to export innovations in services, financial management, and information processing to a worldwide clientele.

Regional Economic Patterns North America's industries reveal important regional patterns of concentration. **Location factors** are the varied influences that explain *why* an economic activity is located where it is, and North America's economic geography suggests that these factors change over time and that many influences, both within and beyond the realm, shape patterns of economic activity. Patterns of industrial location illustrate the concept (Figure 3.30). The historical Manufacturing Core includes Megalopolis (Boston, New York, Philadelphia, and Baltimore), southern Ontario (Toronto and Hamilton), and much of the industrial Midwest. Several location factors account for the region's nineteenth-century development and the ongoing importance of its diverse industrial economy. Its proximity to *natural resources* (farmland, coal, and iron ore); increasing *connectivity* (canals and railroad networks, highways, air traffic hubs, and telecommunications centers); a ready supply of *productive labor;* and a growing national, then global, *market demand* for its industrial goods spurred sustained *capital investment* within the region. Traditionally, the Core has dominated in the production of steel, automobiles, machine tools, and agricultural equipment as well as playing a pivotal role in producer services such as banking and insurance.

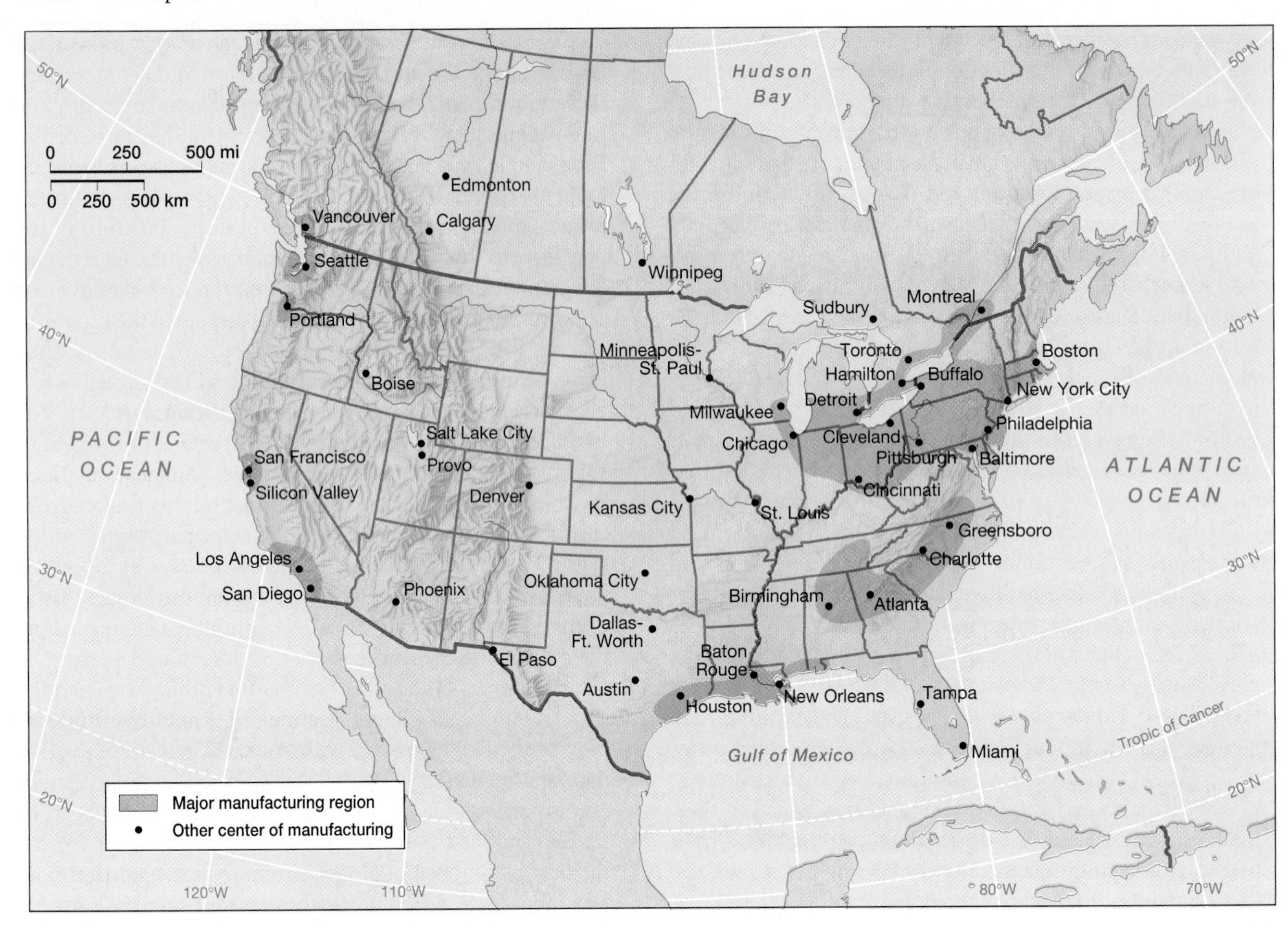

▲ **Figure 3.30 Major manufacturing regions** North America's traditional manufacturing core includes southern New England, the Mid-Atlantic region, the U.S. Midwest, and southeastern Canada. During the twentieth century, however, major industrial expansions transformed the economic landscape from the Sunbelt South to the Pacific Coast. *(Modified from Clawson and Fisher, 1998,* World Regional Geography, *Upper Saddle River, NJ: Prentice Hall)*

In the last half of the twentieth century, however, industrial and service-sector growth has gravitated to peripheral centers in the South and West. Indeed, many older industrial centers in the Core experienced drastic job losses in manufacturing, thus reducing its overall dominance within the realm. Conversely, cities of the Piedmont Manufacturing Belt have seen industrial expansion after 1960, partly because lower labor costs and Sun Belt amenities attracted new investment. The Gulf Coast Industrial Region is strongly tied to nearby fossil fuels that provide raw materials for its many energy refining and petrochemical industries (Figure 3.31). The varied West Coast Industrial Region stretches discontinuously from Vancouver, British Columbia, to San Diego, California, and it reveals the increasing importance of Pacific Basin trade. Large aerospace operations in the West also suggest the role of *government spending* as a location factor. Silicon Valley is now North America's leading region of manufacturing exports. Its proximity to Stanford, Berkeley, and other universities demonstrates the importance of *access to innovation and research* for many fast-changing high-technology industries and the advantages of *agglomeration economies* where many companies with similar and often integrated manufacturing operations locate near one another (Figure 3.32). Many smaller industrial centers also thrive throughout North America, suggesting that some industries can be relatively footloose and succeed in many localities. Places such as Provo, Utah, and Austin, Texas, specializing in high-technology industries, demonstrate the growing role of *lifestyle amenities* in shaping industrial location decisions, both for entrepreneurs and the skilled workers that need to be attracted to such opportunities.

North America and the Global Economy

Together with Europe and Japan, North America plays a pivotal role in the global economy. In prosperous times, the realm benefits from global economic growth, but in periods of international instability, globalization means that the region is more vulnerable to economic downturns. These interconnections transcend the abstract statistics of trade flows and

▲ **Figure 3.31 Gulf Coast petroleum refining** Petroleum-related manufacturing has transformed many Gulf Coast settings. Much of Houston's twentieth-century growth has been fueled by the dramatic expansion of oil-related industries. The port of Houston remains a major center of North America's refining and petrochemical operations. *(Walter Frerck/ Odyssey Productions)*

foreign investment. Increasingly, North American workers and localities find their futures directly tied to export markets in Latin America, the rise and fall of Asian imports, or the pattern of global investments in U.S. stock and bond markets. No community remains untouched by these links and the consequences will shape the lives of every North American in the twenty-first century (see "Latinized Miami: Emergent Global City").

The United States, with Canada's firm support, played a formative role in creating the modern global economy and in shaping many of its key institutions. In 1944 allied nations met at Bretton Woods, New Hampshire, to discuss economic affairs. Under U.S. leadership, the group set up the International Monetary Fund (IMF) and the World Bank and gave these global organizations the responsibility of defending the world's monetary system and making key postwar investments in infrastructure. The United States also spurred the creation (1948) of the General Agreement on Tariffs and Trade (GATT). Renamed the **World Trade Organization (WTO)** in 1995, its 130 member states are dedicated to reducing global barriers to trade. In addition, the United States and Canada participate in the **Group of Seven (G-7)**, a collection of powerful countries (including Japan, Germany, Great Britain, France, Italy, and sometimes Russia) that confers regularly on key global economic and political issues.

Patterns of Trade Global geographies of trade reveal North America's prominence in both the sale and purchase of goods and services within the international economy (Figure 3.33). Both countries import diverse products from many global sources. Dominated overwhelmingly by the United States, Canada's imports ($186 billion in 1997) include large quantities of manufactured parts, vehicles, computers, and foodstuffs. Imports to the United States ($1.05 trillion in 1997) continue to grow, creating a persistent global trade deficit for the country. Canada, Japan, Mexico, China, and world oil exporters supply the United States with a diversity of raw materials, low-cost consumer goods, and high-quality vehicles and electronics products.

Outgoing trade flows suggest what North Americans produce most cheaply and efficiently. Canada's exports ($201 billion in 1997) include large quantities of raw materials (grain, energy, metals, and wood products), but manufactured goods are becoming increasingly important, particularly in its pivotal trade with the United States. Since 1994, trade initiatives with the Pacific Rim offer Canadians new opportunities for export growth. The United States also enjoys many lucrative

◀ **Figure 3.32 Silicon Valley** The high-technology industrial landscape of California's Silicon Valley contrasts sharply with the look of traditional manufacturing centers. Here, similar industries form complex links, benefiting from their proximity to one another and to nearby universities such as Stanford and Berkeley. *(George Hall/Woodfin Camp & Associates)*

LATINIZED MIAMI: Emergent Global City

Walk through bustling Miami International Airport and you are immediately aware of the unique role this city plays in linking the economic and cultural worlds of Latin America with those of North America. Spanish-speaking travelers and international flight announcements dominate airport conversation as planes arrive and depart for destinations throughout Central and South America. Indeed, this bustling city of 3.5 million has seen much of its post-1960 economic growth fueled by its increasingly close ties with the Latin American world.

Much of the recent Latin connection came with Castro's takeover of Cuba. Between 1960 and 1980, hundreds of thousands of refugees left Cuba and went to Miami's Cuban community. Many brought capital and entrepreneurial skills, establishing hundreds of Cuban businesses in the city. By the late 1970s, Latin America's economic growth demanded closer ties with the United States. What better place to serve as a gateway than Miami? In 1977 Miami's Chamber of Commerce launched the Greater Miami Foreign-Trade Zone, Inc., an organization to foster economic links with Latin America. Two years later, the Florida International Bankers Association (FIBA) was formed to facilitate international banking in the city. Since then, other business groups, such as the Brazilian-American Chamber of Commerce of Florida, and the Ecuadorian American Chamber of Commerce of Greater Miami, have tied their home countries to the southern Florida economy.

As a result, modern Miami has become the leading North American trade center for Latin America. Airplane connections, money flows, and even the illegal drug trade demonstrate Miami's centrality. Newspapers such as the *Colombian Post, Venezuela Al Dia,* or the *Diario Las Americas* keep local residents and Latin visitors in touch with news from the home country. Agencies such as Elite International Realty (with offices in Caracas, Lima, Rio de Janeiro, and São Paulo) cater to the needs of an exclusive Latin American clientele who plan to spend time in the city. As journalist Joel Garreau noted in the 1980s, Miami has become a new sort of financial and cultural capital for Latin America, and in the process has attained the status of a global city.

▲ **Figure 1 A Latin American bank office in Miami** Many major financial institutions in Miami have key connections to Latin America, making the city one of the region's primary contact points with the global economy. Language, landscapes, and money flows confirm the Latin link, and Miami's economy and society are now forever wed to their hemispheric neighbors. *(Robert Frerck/Odyssey Productions)*

economic ties to the world beyond. By 1996, international sales of U.S. software and entertainment products totaled more than $60 billion annually, more than any other industry. Sales of automobiles, aircraft, computer and telecommunications equipment, financial and tourism services, and food products also contributed to the nation's flow of exports ($937 billion in 1997). The destinations for these products and services spanned the globe, with Canada, Japan, Mexico, and the European Union serving as major purchasers.

Patterns of Investment in North America Patterns of capital investment and corporate power place the North American realm at the center of global money flows and economic influence. Given its relative stability, the region attracts huge inflows of foreign capital, both as investments in North American stocks and bonds as well as foreign direct investment (FDI) by international companies. For Canada, the wealth and proximity of the United States has meant that 80 percent of foreign-owned corporations in the country are based in the United States. These investments include a large U.S. presence in Canada's automobile, chemical, electronics, and natural resource industries. In the United States, sustained economic growth and supportive government policies have encouraged large foreign investments, particularly since the late 1970s. Today, the United States is the largest destination of foreign investment in the world. For example, Japanese carmakers build U.S. assembly plants, foreign banks set up U.S. operations, and many parcels of the country's most prestigious real estate are purchased by eager Japanese and European buyers. Predictably, Canadian companies also aggressively participate in making investments south of the border. In addition, billions of dollars flow into U.S. stock and bond markets, especially when global economic turbulence increases the appeal of relatively stable U.S. investments.

Doing Business Globally The growing impact of U.S. investments in foreign stock markets and the dominance of U.S.-controlled multinational companies suggest how flows of outbound capital are transforming the way business is done throughout the world. Since 1980, aging U.S. baby boomers

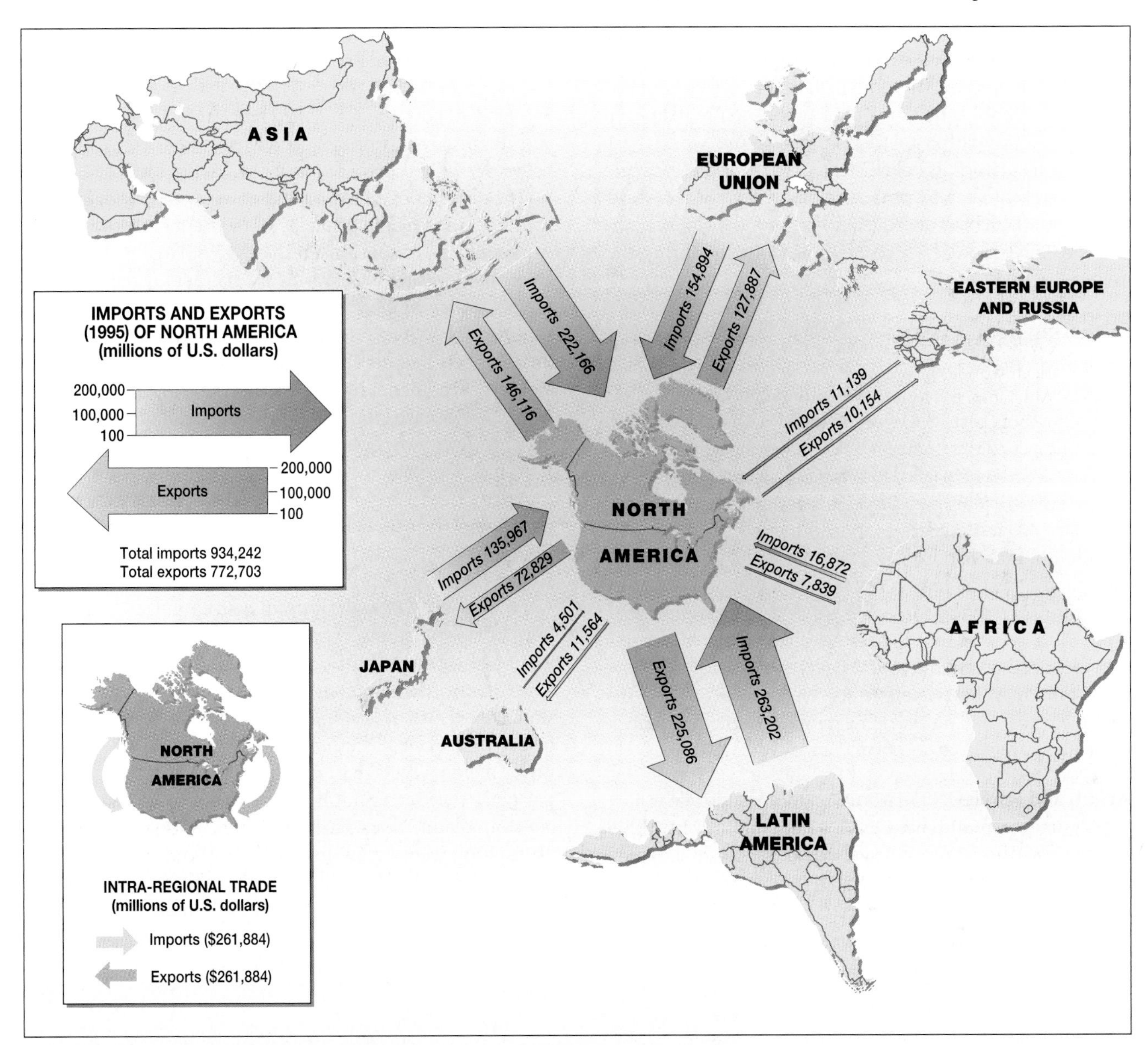

▲ **Figure 3.33 North American trade** North Americans play a pivotal role in world trade. A large flow of goods links the region with nearby Latin America, particularly Mexico. Asian trade has also grown greatly in the past 30 years, and the European Union's recent economic consolidation may create new twenty-first-century trade opportunities with that region of the world. *(Data from Euromonitor,* International Marketing Data and Statistics, *1997)*

have poured billions of pension fund and investment dollars into Japanese, European, and "emerging" stock markets. U.S. investments in foreign countries also flow through direct investments made by multinational corporations based in the United States. In 1998, 15 of the world's 20 largest public companies were based in the United States. Coca-Cola, Intel, and Procter & Gamble are household words from Bangladesh to Bolivia. The economic influence of these corporations exceeds that of many nation-states. General Electric is the world's largest public company. Valued at $300 billion, it is "bringing good things to life" across the world at an annual sales rate of more than $90 billion. Indeed, GE now generates more than 40 percent of its sales from foreign customers. The success of these complex multinational entities is increasingly related to their penetration of foreign markets. Their global orientation is explicit: the Internet version of IBM's annual report, for example, is available to readers in Chinese, English, German, Italian, Japanese, and Spanish!

The geographical organization of multinational corporations parallels their global financial reach. U.S. companies such as AT&T have equipment assembly plants in France, China, and Thailand; service contracts in Mexico and Saudi Arabia; and joint ventures with dozens of other firms in places such as Japan and the Dominican Republic. The processes of globalization in the automobile and electronics industries are also creating cars and computers literally constructed from parts and by workers from every corner of the world. Put simply, economic opportunities, more than ever before, are gravitating to those parts of the world best equipped to perform them efficiently. The fact that U.S. investments dominate that process in many parts of the world will have incalculable consequences for both the global economy and the region's own economic geography during the twenty-first century.

Persisting Social Issues

Profound economic and social problems still shape the human geography of North America. Even with its continental wealth, great differences persist between rich and poor. In 1996 high median household incomes in the United States ($36,700) and Canada ($30,150) failed to reveal the differences in wealth within the two countries. Broader measures of social well-being also suggest disparities in health care and education. In addition, both nations face enduring issues of gender inequity and aging that remain key social and political issues for the future. One of the consequences of globalization is that many of these economic and social challenges are increasingly defined beyond the realm. Poverty in the rural American South may be related to low Asian wage rates, for example, and a viral outbreak in Hong Kong might be only a plane flight away from suburban Vancouver.

Wealth and Poverty The North American landscape vividly displays contrasting signatures of wealth and poverty. Elite suburbs, gated and guarded neighborhoods, upscale shopping malls, and posh resorts are all geographical expressions of the spatial exclusivity that characterizes many wealthier North American communities, particularly within the United States (Figure 3.34). On the other hand, signatures of poverty are displayed in a great range of local and regional settings. Substandard housing, abandoned property, aging infrastructure, and unemployed workers are visual reminders of the gap between rich and poor within the realm (Figure 3.35).

The distribution of wealth and poverty varies widely across the United States and Canada. Many of America's wealthiest communities are suburbs on the edge of large metropolitan areas. Just outside New York City, Fairfield, Connecticut is the nation's wealthiest county; similar settings are found in suburban New Jersey, Maryland, and Virginia, as well as on the peripheries of dozens of southern and western cities. Resort and retirement communities are havens for the rich, as well: Palm Beach, Florida, and Aspen, Colorado, have some of the nation's most desirable real estate and costliest housing. In terms of average income, the Northeast and West remain America's richest regions, although the South is now within 8 percent of the national norm. In Canada, Ontario and British Columbia are the country's wealthiest provinces, with Vancouver's high house prices vying with those of San Francisco.

Poverty levels have declined in both countries since 1980, but poor populations are still clustered in a variety of geographical settings. In the late 1990s, poverty rates in the United States fell to about 13 percent of the population; a similar measure of low-income Canadians declined to just below 20 percent. The problems of the rural poor remain major regional social issues in the Canadian Maritimes, Appalachia, the Deep South, and the Southwest. Incomes in Newfoundland, Nova Scotia, and New Brunswick continue 10 to 20 percent below the national norm. Similarly, West Virginia, New Mexico, and Mississippi rank as America's three poorest states in median income (1997). In the United States,

▶ **Figure 3.34 Gated America** Wealthier North Americans have had a tremendous impact on the region's cultural landscapes, and their influence far outweighs their relatively modest numbers. Many gated communities in North American suburbs and resorts display a desire for privacy, safety, and social exclusivity. *(James Patelli)*

◀ **Figure 3.35 Inner-city neighborhood** Many poor inner-city neighborhoods such as this community in New York City suffer from substandard housing and aging infrastructure. Ethnic bonds often supply the social support necessary to ease the challenges of high unemployment and troubled family life. *(Erica Lananer/Black Star)*

ethnicity is often linked with rural poverty, particularly in the case of southern blacks, Hispanics in the borderlands, and Native Americans on reservation lands in the Southwest. Most poor people in the United States, however, live in central city locations, and the links between ethnicity and poverty are even stronger in these communities. Nationally, 27 percent of the country's black and Hispanic populations live below the poverty line, and the great majority resides in central city ghettos or barrios.

Twenty-first Century Challenges Measures of social well-being in North America compare favorably with most world regions (Table 3.3). Still, many economic and social challenges confront the realm. As workers, North Americans compete globally, and the declining percentage of stable, high-paying jobs with long-term security and benefits suggests that companies and employees are scrambling to adjust to the uncertainties of the world economy.

Education is also a major public policy issue in Canada and the United States. Although political parties differ in their approach, most public officials agree that more investment in education can only improve North America's chances for competing successfully in the global marketplace. In the United States, dropout rates average about 13 percent for 18- to 24-year-olds, but rates are much higher than those in many high-poverty neighborhoods, both in cities and in rural areas. Another challenge, particularly in the United States, is debating an effective national education policy in the face of a very strong tradition of local control of educational issues.

Since World War II, both nations have seen great changes in the role that women play in society, but the "gender gap" is yet to be closed when it comes to salary issues, working conditions, or political power. Women widely participate in the workforce in both countries (Table 3.3). In the United States, although women are as educated as men (55 percent of American college students are female), they still earn only

Table 3.3 Social Indicators and Status of Women

Country	Life Expectancy at Birth		Under Age 5 Mortality, per 1,000 Live Births		Secondary School Enrollment %*		Female Labor Force Participation (% of total)
	Male	Female	1960	1993	Male	Female	
Canada	75	81	33	8	104	103	45
United States	74	79	30	10	98	97	46

*These data cover the percentage of people in a certain age group enrolled in secondary school. Older or younger people attending high school can boost the percentage above 100%.

Sources: *Population Reference Bureau, Data Sheet, 1998,* Life Expectancy (M/F); *World Resources Institute, 1998–99,* Under-5 Mortality Rate; *Population Reference Bureau, Data Sheet, 1996,* Secondary School Enrollment (M/F); *The World Bank Atlas, 1998,* Female Participation in Labor Force.

▶ **Figure 3.36 Tomorrow's baby boom landscape?** This Arizona retirement community, boldly stamped on the desert landscape, suggests the twenty-first-century destination for many aging "baby boomers." As North Americans grow older, entire subregions may be devoted to age-dependent demands for housing, recreation, and health services. *(Cont/Frank Fournier/Woodfin Camp & Associates)*

about 75 cents for every dollar that men earn. Although the gap is shrinking, corporate America's "glass ceiling" still makes it extraordinarily difficult for women to advance to top managerial positions and pay scales. Even more daunting is the fact that women head the vast majority of poorer single-parent families in the United States and more than 40 percent of these women are unwed mothers. Canadian women, particularly full-time working single mothers, are also greatly disadvantaged, averaging only about 65 percent of the salaries of Canadian men. In addition, political power remains largely in male hands. Even though Canadian women have voted since 1918 and U.S. women since 1920, females are still in the minority in the Canadian Parliament and the U.S. Congress in the late 1990s.

Health care and aging are also key concerns within a world realm of graying baby boomers. A recent report on aging in the United States predicted that 20 percent of the nation's population will be elderly (older than 65) by 2050 and that the most elderly senior citizenry (age 85+) are the fastest-growing parts of the population. With fewer young people to support their parents and grandparents, officials debate the merits of reforming social security programs. Whatever the outcome of such debates, the geographical consequences of aging are already abundantly clear. Whole sections of the United States—from Florida to southern Arizona—become increasingly oriented around retirement (Figure 3.36). Communities cater to seniors with special assisted-living arrangements, health-care facilities, and recreational opportunities. Florida and California have the largest number of senior citizens; farm belt states struggling to hold on to their younger residents have higher percentages of the aged in their populations.

Although Canada and the United States have distinctive health-care delivery systems, they share many of the health-related benefits and costs of living within the realm. As long average life spans and low childhood mortality suggest (Table 3.3), the two countries reap many rewards from modern health-care systems that offer the latest in high technology. Still, North Americans voice concern over escalating health-care costs and uneven access to premium care.

The rising incidence of chronic diseases associated with aging (heart disease, cancer, and stroke are the three leading causes of death) will continue to pressure both health-care systems. Another added cost has been the care and treatment of the realm's 850,000 AIDS victims. Although the pace of new infections finally began falling in the late 1990s, the price of the disease will still be broadly borne in the twenty-first century, particularly among poorer black and Hispanic populations in which rates of new infection are still on the rise.

The realm also reveals striking variations in the geographies of disease. Certain populations are therefore much more likely to bear the risks and costs of particular health-care problems. Gay men in larger North American cities have borne much of the AIDS burden since the early 1980s. Elsewhere, a combination of diet, poor medical care, and genetic factors may contribute to the existence of the "stroke belt" in the eastern Carolinas and Georgia, where death rates from the disease are double the national norm. Although cancer is a leading cause of death in both countries, certain regions and populations appear prone to variations of the disease. For example, soaring rates of bladder and lung cancer in southern Louisiana appear strongly linked to the region's petrochemical industry and to problems of toxic waste and pollution. High rates of stomach cancer in the northern Midwest, however, may be more related to that region's ethnic stock and traditional diet. Hopefully, as North Americans age in the twenty-first century, medical geographers can help unlock these locational mysteries and better explain the origins and progression of diseases through the population.

Conclusion

North Americans have reaped the natural abundance of their realm and in the process they have transformed the environment, created a highly affluent society, and extended their global economic, cultural, and political reach. In a remarkably short period, a unique mix of varied cultural groups contributed to the settlement of a huge and resource-rich continent that is now the world's most urbanized realm. Along the way, North Americans produced two societies that are closely intertwined, yet they face distinctive national political and cultural issues. For both, the twenty-first century brings uncertainties. In Canada, the nation's very existence remains problematic as it works through the persisting realities of its bicultural character. For the United States, social problems linked to its ethnic diversity and enduring poverty remain central concerns, along with the environmental price tag associated with its high rates of resource consumption. Precisely how the United States fits into the global economy and the global balance of political power also remain uncertain. One thing is clear, however; North Americans will play central, if not dominant, roles in the processes of globalization, and the results of their participation will enduringly shape not only the realm itself, but the larger world as well.

Key Terms

acid rain (page 83)
barrio (page 97)
boreal forest (page 79)
concentric zone model (page 90)
connectivity (page 109)
counterurbanization (page 88)
cultural assimilation (page 93)
ethnicity (page 92)
federal state (page 104)
gentrification (page 90)
ghetto (page 97)
Group of 7 (G-7) (page 111)
location factors (page 109)
Megalopolis (page 85)
North America Free Trade Agreement (NAFTA) (page 102)
postindustrial economy (page 73)
prairie (page 79)
rain shadow effect (page 79)
sectoral transformation (page 109)
tundra (page 79)
unitary state (page 104)
urban decentralization (page 89)
urban realms model (page 90)
World Trade Organization (WTO) (page 111)

Questions for Summary and Review

1. Describe North America's major landform regions and suggest ways in which their physical setting has shaped patterns of human settlement.
2. How are patterns of commercial agriculture related to underlying climatic patterns within North America? What key economic and technological trends have shaped agriculture?
3. What were some of the dominant North American migration flows during the twentieth century?
4. Describe the principal patterns of land use within the modern U.S. metropolis. Include a discussion of (a) the central city and (b) the suburbs/edge city.
5. How have (a) railroads, (b) the township-and-range survey system, and (c) freeways shaped North America's settlement geography?
6. What are the distinctive eras of immigration in U.S. history and how do they compare with those of Canada?
7. How do the political origins of the United States and Canada differ?
8. Cite examples of globalization that illustrate the impact of North American cultural, political, and economic influence elsewhere in the world.
9. Briefly summarize North America's endowment of natural resources and describe where they can be found within the realm.
10. What is the sectoral transformation and how does it help us understand economic change in North America?
11. Cite five types of location factors and illustrate each with examples from your local economy.

Thinking Geographically

1. Compare and contrast the black experience in the Black Belt and in urban America. Assess the unique challenges faced by each group.
2. Summarize the ethnic background of (a) your own family and (b) your home community. Discuss how these patterns parallel or depart from larger North American trends.
3. Discuss the strengths and weaknesses of federalism and cite examples from both Canada and the United States.
4. "An independent Quebec is a far more viable and coherent nation-state than a divided Canada." Do you agree or disagree? Why?

5. The environmental price for North American development has often been steep. Suggest why it may or may not be worth the price and defend your answer.
6. Who will America's leading trading partner be in 2050? Suggest why this may be the case.
7. Assume you are a leader of a small African nation. Describe what the chief benefits and costs of North American globalization have been on your part of the world.

Regional Novels and Films

Novels

Willa Cather, *My Antonia* (1918, Houghton Mifflin)

Ivan Doig, *This House of Sky* (1978, Harcourt Brace Jovanovich)

William Faulkner, *The Bear* (1964, Random House)

F. Scott Fitzgerald, *The Great Gatsby* (1925, Scribner)

Frederick Philip Grove, *Settlers of the Marsh* (1925, Ryerson Press)

William Kennedy, *Ironweed* (1983, Viking Press)

Jack Kerouac, *On the Road* (1957, New American Library)

E. Annie Proulx, *The Shipping News* (1993, Scribner)

Ole Rolvaag, *Giants in the Earth* (1927)

John Steinbeck, *The Grapes of Wrath* (1939, Viking Press)

Wallace Stegner, *Angle of Repose* (1971, Doubleday)

George Stewart, *Storm* (1941, Random House)

Mark Twain, *The Adventures of Huckleberry Finn* (1885)

Nathaniel West, *The Day of the Locust* (1939)

Tom Wolfe, *Bonfire of the Vanities* (1987, Farrar, Straus, Giroux)

Films

Avalon (1990, U.S.)

Chinatown (1974, U.S.)

Dances with Wolves (1990, U.S.)

The Deer Hunter (1978, U.S.)

Fargo (1996, U.S.)

The Godfather (1972, U.S.)

Gone with the Wind (1939, U.S.)

Little Big Man (1970, U.S.)

Pulp Fiction (1994, U.S.)

To Kill a Mockingbird (1962, U.S.)

West Side Story (1961, U.S.)

Bibliography

Adams, John S. 1970. "Residential Structure of Midwestern Cities." *Annals of the Association of American Geographers* 60, 37–62.

Allen, James P., and Turner, Eugene J. 1987. *We the People: An Atlas of America's Ethnic Diversity*. New York: Macmillan.

"American Pop Penetrates Worldwide." 1998. *The Washington Post*, October 25.

Birdsall, Stephen S., and Florin, John W. 1992. *Regional Landscapes of the United States and Canada*. New York: John Wiley & Sons.

Borchert, John R. 1967. "American Metropolitan Evolution." *Geographical Review* 57, 301–32.

Braus, Patricia. 1998. "Strokes and the South." *American Demographics* 20(5), 26–29.

Butzer, Karl W., ed. 1992. "The Americas Before and After 1492: Current Geographical Research." *Annals of the Association of American Geographers* 82, 345–368.

"Canada and the United States. Border Wars." 1997. *The Economist*, August 16.

Castells, Manuel. 1996. *The Information Age. Economy, Society and Culture. Volume 1: The Rise of the Network Society*. Cambridge: Blackwell.

Conzen, Michael P., ed. 1990. *The Making of the American Landscape*. Boston: Unwin Hyman.

"Culture Wars." 1998. *The Economist*, September 12.

Ford, Larry R. 1994. *Cities and Buildings. Skyscrapers, Skid Rows, and Suburbs*. Baltimore: Johns Hopkins University Press.

Garreau, Joel. 1981. *The Nine Nations of North America*. Boston: Houghton Mifflin.

Garreau, Joel. 1991. *Edge City: Life on the New Frontier*. New York: Doubleday.

Gentilcore, R. Louis, ed. 1990. *Historical Atlas of Canada. Volume 2: The Land Transformed*. Toronto: University of Toronto Press.

Getis, Arthur, and Getis, Judith, eds. 1995. *The United States and Canada: The Land and the People*. Dubuque, IA: William C. Brown.

Harris, R. Cole, ed. 1987. *Historical Atlas of Canada. Volume 1: From the Beginning to 1800.* Toronto: University of Toronto Press.

Harris, R. Cole, and Warkentin, John. 1974. *Canada Before Confederation.* New York: Oxford University Press.

Heubusch, Kevin. 1998. "Small Is Beautiful." *American Demographics* 20(1), 43–49.

Hugill, Peter J. 1993. *World Trade Since 1431: Geography, Technology, and Capitalism.* Baltimore: Johns Hopkins University Press.

Kacapyr, Elia. 1998. "The Well-Being of American Women." *American Demographics,* 20(8), 30–32.

Kerr, Donald, and Holdsworth, Deryck, eds. 1993. *Historical Atlas of Canada. Volume 3: Addressing the Twentieth Century.* Toronto: University of Toronto Press.

Knox, Paul L., and Marston, Sallie A. 1998. *Human Geography. Places and Regions in Global Context.* Upper Saddle River, NJ: Prentice Hall.

McKnight, Tom L. 1997. *Regional Geography of the United States and Canada,* 2nd ed. Upper Saddle River, NJ: Prentice Hall.

Meinig, Donald W. 1986. *The Shaping of America: A Geographical Perspective on 500 Years of History. Volume 1, Atlantic America, 1492–1800.* New Haven: Yale University Press.

Meinig, Donald W. 1993. *The Shaping of America: A Geographical Perspective on 500 Years of History. Volume 2, Continental America, 1800–1867.* New Haven: Yale University Press.

Mitchell, Robert D., and Groves, Paul A., eds. 1987. *North America: The Historical Geography of a Changing Continent.* Totowa, NJ: Rowman and Littlefield.

National Geographic Society. 1985. *Atlas of North America: Space Age Portrait of a Continent.* Washington, DC: National Geographic Society.

Riebsame, William E., ed. 1997. *Atlas of the New West: Portrait of a Changing Region.* New York: W. W. Norton.

Shortridge, James R. 1991. "The Concept of the Place-Defining Novel in American Popular Culture." *Professional Geographer* 43, 280–91.

Statistics Canada. 1998.

United States Bureau of the Census. 1995. *Sixty-Five Plus in the United States.*

Vance, James E., Jr. 1995. *The North American Railroad. Its Origin, Evolution, and Geography.* Baltimore: Johns Hopkins University Press.

Warf, Barney. 1998. "Reach Out and Touch Someone: AT&T's Global Operations in the 1990s." *Professional Geographer* 50, 255–67.

"The World's 100 Largest Public Companies." 1998. *Wall Street Journal,* September 28.

Yeates, Maurice H. 1990. *The North American City,* 4th ed. New York: Harper & Row.

Zelinsky, Wilbur. 1992. *The Cultural Geography of the United States: A Revised Edition.* Englewood Cliffs, NJ: Prentice Hall.

LATIN AMERICA
Political Map
Over 1,000,000
500,000–1,000,000 (selected cities)
Selected smaller cities
0 500 1,000 mi
0 500 1,000 km
UNITED STATES
MEXICO
GUATEMALA
HONDURAS
EL SALVADOR
NICARAGUA
COSTA RICA
PANAMA
VENEZUELA
COLOMBIA
ECUADOR
PERU
BOLIVIA
BRAZIL
PARAGUAY
CHILE
ARGENTINA
URUGUAY
PACIFIC OCEAN
ATLANTIC OCEAN
Caribbean Sea
Galápagos Is. (ECUADOR)
Falkland Is. (U.K.)
Tropic of Cancer
Tropic of Capricorn
Equator
Tijuana
Ciudad Juárez
Monterrey
Guadalajara
Mexico City
Veracruz
Puebla
Guatemala City
Tegucigalpa
San Salvador
Managua
San José
Panama
Barranquilla
Cartagena
Maracaibo
Valencia
Caracas
Ciudad Bolívar
Medellín
Bogotá
Cali
Quito
Guayaquil
Iquitos
Manaus
Belém
São Luis
Trujillo
Rio Branco
Callao
Lima
Cuzco
Arequipa
La Paz
Santa Cruz
Sucre
Potosí
Arica
Brasília
Goiânia
Salvador
Belo Horizonte
Rio de Janeiro
São Paulo
Santos
Curitiba
Asunción
Pôrto Alegre
Córdoba
Rosario
Valparaíso
Santiago
Buenos Aires
La Plata
Montevideo
Concepción
Bahía Blanca
Puntas Arenas
Orinoco R.
Negro R.
Amazon R.
Tapajós R.
Xingu R.
Araguaia R.
São Francisco R.
Paraná R.
Uruguay R.
Río de la Plata
30°N
20°N
10°N
0°
10°S
20°S
30°S
40°S
50°S
120°W
100°W
80°W
70°W
60°W
50°W
40°W
20°W

Latin America

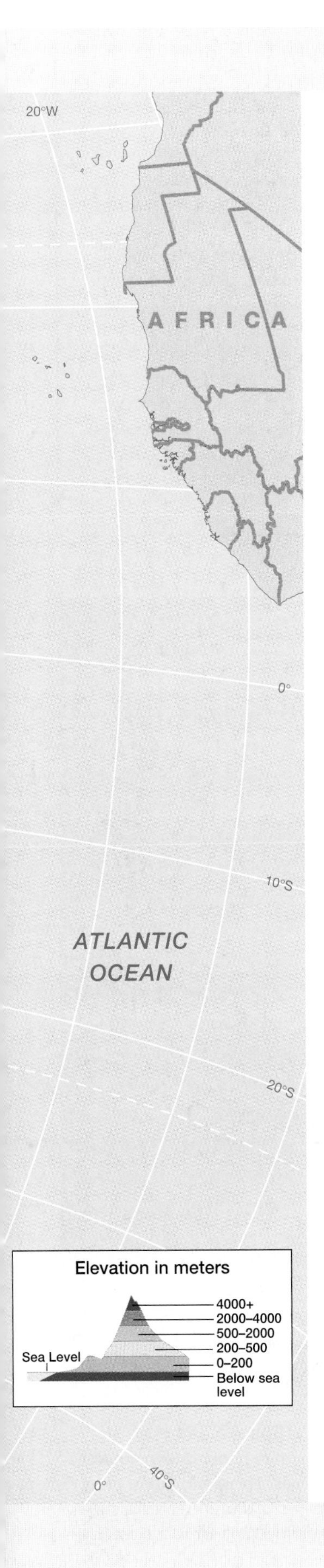

Latin America has been a popularly accepted world region for nearly a century. Beginning with Mexico and extending to the tip of South America, the region's identity stems largely from its shared colonial history rather than from the disparate levels of development seen today. More than 500 years ago, the Iberian countries of Spain and Portugal began their conquest of the Americas. Iberia's mark is still visible throughout the area: officially two-thirds of the population speak Spanish and the rest speak Portuguese. Iberian architecture and town design add homogeneity to the colonial landscape. The vast majority of the population nominally practices Catholicism, even though Protestantism has recently made inroads. These European traits blended with those of different Amerindian peoples. The Indian presence remains especially strong in Bolivia, Peru, Ecuador, Guatemala, and southern Mexico, where large and diverse indigenous populations maintain their native languages, dress, and traditions. Later, other culture groups were added to this mix of indigenous and Iberian peoples. The legacy of slavery imparted a strong African presence, primarily on the coasts of Colombia and Venezuela and throughout Brazil. In the nineteenth and twentieth centuries new waves of settlers came from Spain, Italy, Germany, Japan, and Lebanon. The end result is one of the world's most racially mixed regions. The modern states of Latin America are multiethnic, with distinct indigenous and immigrant profiles, and very different rates of social and economic development (Figure 4.1; see also "Setting the Boundaries").

For five centuries, Latin America has provided the world with many valuable commodities. The early Spanish empire concentrated on extracting precious metals, sending their galleons laden with silver and gold across the Atlantic. The Portuguese became prominent producers of dyewoods, sugar products, gold, and later coffee. In the late nineteenth and early twentieth centuries, exports to North America and Europe fueled the region's economy. Most countries specialized in one or two products: bananas and coffee, meats and wool, wheat and corn, petroleum and copper. Such a primary export tradition, according to the region's economists, led to an unhealthy economic dependence. They argued that Latin American economies were too specialized and faced unequal terms of trade that inhibited overall development. Since the 1950s the countries of the region have worked to

◀ **Figure 4.1 Latin America** Roughly equal in size to North America, Latin America supports a larger population and far greater ecological diversity. The 17 countries included in this region share a history of Iberian colonization. Seventy percent of the region's 460 million people live in cities, making it the most urbanized region of the developing world. It is noted for its production of primary exports; however, rates of economic development vary greatly between states.

Setting the Boundaries

The boundaries of this region are relatively unproblematic, beginning at the Rio Grande (called the Rio Bravo in Mexico) and ending at Tierra del Fuego. French geographers are credited with inventing the term "Latin America" in the nineteenth century as a means to distinguish the Spanish- and Portuguese-speaking republics of the Americas, plus Haiti, from the English-speaking territories. There is nothing particularly "Latin" about the area, other than the predominance of romance languages. The term stuck because it was vague enough to be inclusive of different colonial histories while also offering a clear cultural boundary from Anglo-America, the region referred to as North America in this text.

Since the region is so large, geographers often divide Latin America. The continent of South America is typically distinguished from Middle America (which includes Central America, Mexico, and the Caribbean). The term "Middle America" was created to distinguish an area culturally distinct form North America but physically part of it since most of Mexico and all of Cuba rest on the North American plate. Such a division has its merits, but it also separates countries with very similar histories (such as Mexico and Peru) while joining countries that have very little common (such as El Salvador and Jamaica). In this text the Americas are divided slightly differently. Latin America will consist of the Spanish- and Portuguese-speaking countries of Central and South America, including Mexico. Chapter 5 will examine the Caribbean, consisting of the islands of the Antilles, the Guianas, and Belize. So divided, the important Indian and Iberian influences of mainland Latin America will be emphasized. Similarly, the Caribbean's unique colonial and demographic history will be discussed in Chapter 5.

industrialize and diversify their production, but they continue to be major producers of primary goods for North America, Europe, and Japan.

Extractive industries prevail, in part, because of the area's impressive natural resources. Latin America is home to Earth's largest rain forest, the greatest river by volume, and massive reserves of natural gas, oil, and copper. With its vast territory, its tropical location, and its relatively low population density (Latin America has half the population of India and nearly seven times the area), the region is also recognized as one of the world's great reserves of biological diversity. How this diversity will be managed in the face of global demand for natural resources is an increasingly important question for the countries of this region (Figure 4.2).

Roughly equal in area to North America, Latin America has a much larger and faster-growing population of 460 million people. Its most populous state, Brazil, has 162 million people, making it the fifth largest country in the world. Throughout the region, population growth rates have slowed dramatically in the last 30 years, so the typical Latin American woman now has three children, compared with six in the 1960s. Unlike most areas of the developing world today, Latin America is decidedly urban. Prior to World War II, most people lived in rural settings and worked as farmers. Today three-quarters of Latin Americans are city dwellers. Even more startling is the number of **megacities.** São Paulo, Mexico City, Buenos Aires, and Rio de Janeiro all have more than 10 million inhabitants. In addition, there are more than 30 cities of at least one million residents.

The urban nature of life in Latin America provides a landscape of contrasts. Modern automotive and high-tech industries, high-rise apartments and stately residential suburbs are juxtaposed with shantytowns, street vendors, and artisans who fill the ranks of the urban poor. Collectively, many Latin American states fall into the middle-income category and support a significant middle class. But national debt, political

▲ **Figure 4.2 Chilean wood chips** A mountain of wood chips awaits shipment to Japanese paper mills from the southern Chilean port of Punta Arenas. The exploitation of wood products, from both native and plantation forests, supports Chile's booming export economy. *(Rob Crandall/Rob Crandall Photography)*

scandal, currency devaluation, and triple-digit inflation have triggered grave economic hardships throughout the region, especially during the 1980s. Today, neoliberal policies that encourage foreign investment, export production, and privatization have been adopted by many states. These policies exemplify the forces of economic globalization on Latin America. The results are mixed, with some states experiencing impressive economic growth but increased disparity between rich and poor.

Environmental Geography: Neotropical Diversity

Much of the region is characterized by its tropicality. Travel posters of Latin America showcase verdant forests and brightly colored parrots. The diversity and uniqueness of the **neotropics** (tropical ecosystems of the Western Hemisphere) have long been attractive to naturalists eager to understand their unique flora and fauna. It is no accident that Charles Darwin's insights on evolution were inspired by his two-year journey in tropical America. Even today, scientists throughout the region work to understand complex ecosystems, discern new species, conserve genetic resources, and interpret the impact of human settlement, especially in neotropical forests. Not all of the region is tropical. Important population centers extend below the Tropic of Capricorn, most notably Buenos Aires and Santiago. Much of northern Mexico, including the city of Monterrey, is north of the Tropic of Cancer. Yet it is Latin America's tropical climate and vegetation that prevail in popular images of the region.

The movement of tectonic plates explains much of the region's basic topography. As the South and North American plates slowly drift westward, the Nazca and Cocos plates in the Pacific are subducted below them (Figure 4.3). In this contact zone, deep oceanic trenches exist along the Pacific coast. The submerged plates have folded and uplifted the mainland's surface, creating the geologically young western mountains such as the Andes and the Central American highlands. In contrast, the Atlantic side is characterized by humid lowlands interspersed with upland plateaus called **shields.** The Brazilian shield is the largest one, followed by the Patagonia and Guiana shields. Across these lowlands meander some of the great rivers of the world, including the Amazon. An exception to the pattern of Atlantic lowlands is the Central Plateau of Mexico, whose eastern escarpment forms the Sierra Madre Oriental, which averages 7,000 to 8,000 feet (2,100 to 2,400 meters). On the North American plate, the Sierra Madre Oriental is geologically similar to the Appalachian Mountains in the eastern United States.

The Caribbean plate, which contains most of Central America and part of Colombia, is moving slowly to the east. This plate's movement corresponds to volcanic activity in both the isthmian zone (Guatemala, El Salvador, Nicaragua, and Costa Rica) and the Lesser Antilles (to be discussed in Chapter 5). The volcanic axis of Central America is also vulnerable to frequent and strong earthquakes. The devastating effects of the 1970 earthquake that leveled Managua, Nicaragua, for example, are still evident today because much of that inner city was never rebuilt.

River Basins and Lowlands

Three great river basins drain the Atlantic lowlands of South America: the Amazon, Plata, and Orinoco. Within these basins are vast interior lowlands, less than 600 feet (200 meters) in elevation, that lie over young sedimentary rock. From north to south they are the Llanos, the Amazon lowlands, the Pantanal, the Chaco, and the Pampas. With the exception of the Pampas, most of these lowlands are sparsely settled and offer limited agricultural potential except for grazing land for livestock. Yet the pressure to open new areas for settlement and to exploit natural resources have created pockets of intense economic activity in the lowlands. Areas such as the Amazon and the Chaco that were once thought of as **static frontiers** (open lands unable to support permanent settlement), have witnessed marked increases in resource extraction and settlement since the 1970s.

Amazon Basin The Amazon drains an area of roughly 2.4 million square miles (6.1 million square kilometers), making it the largest river system in the world by volume and area, and the second longest by length. Everywhere in the basin rainfall is more than 60 inches (150 centimeters) a year, and in many places more than 80 inches (200 centimeters). The basin's largest city, Belém, averages close to 100 inches (250 centimeters) a year. Although there is no real dry season, there are definitely drier and wetter times of year, with August and September being the driest months. In the basin, rainfall is likely most days, but showers often pass quickly, leaving bright blue skies. The enormity of this watershed and its hydrologic cycle is underscored by the fact that 20 percent of all freshwater discharged into the oceans comes from the Amazon.

The river is navigable to Iquitos, Peru, some 2,100 miles (3,600 kilometers) upstream. More than 200 tributaries drain the eastern slope of the Andes, as well as portions of the Brazilian and Guianan shields. Since the Amazon Basin draws from nine different countries, it would seem that this watershed would be an ideal network to integrate the northern half of South America. Ironically, compared with the other great rivers of the world, settlement in the basin continues to be sparse, in large part due to the poor quality of the forest soils. The best soils are found in the floodplain, where natural levees reach heights of 20 feet (6 meters). Thousands of years of alluvium deposited on these levees make them extremely fertile as well as safe from normal flooding. Consequently, most of the older settlements, such as Manaus, are found on these levees (Figure 4.4).

Plata Basin The region's second largest watershed begins in the tropics and discharges into the Atlantic in the midlatitudes. Three major rivers make up this system: the Paraná, the Paraguay, and the Uruguay. The Paraguay River and its tributaries drain the eastern Andes of Bolivia, the Brazilian shield, and the Chaco. The Paraná primarily drains the

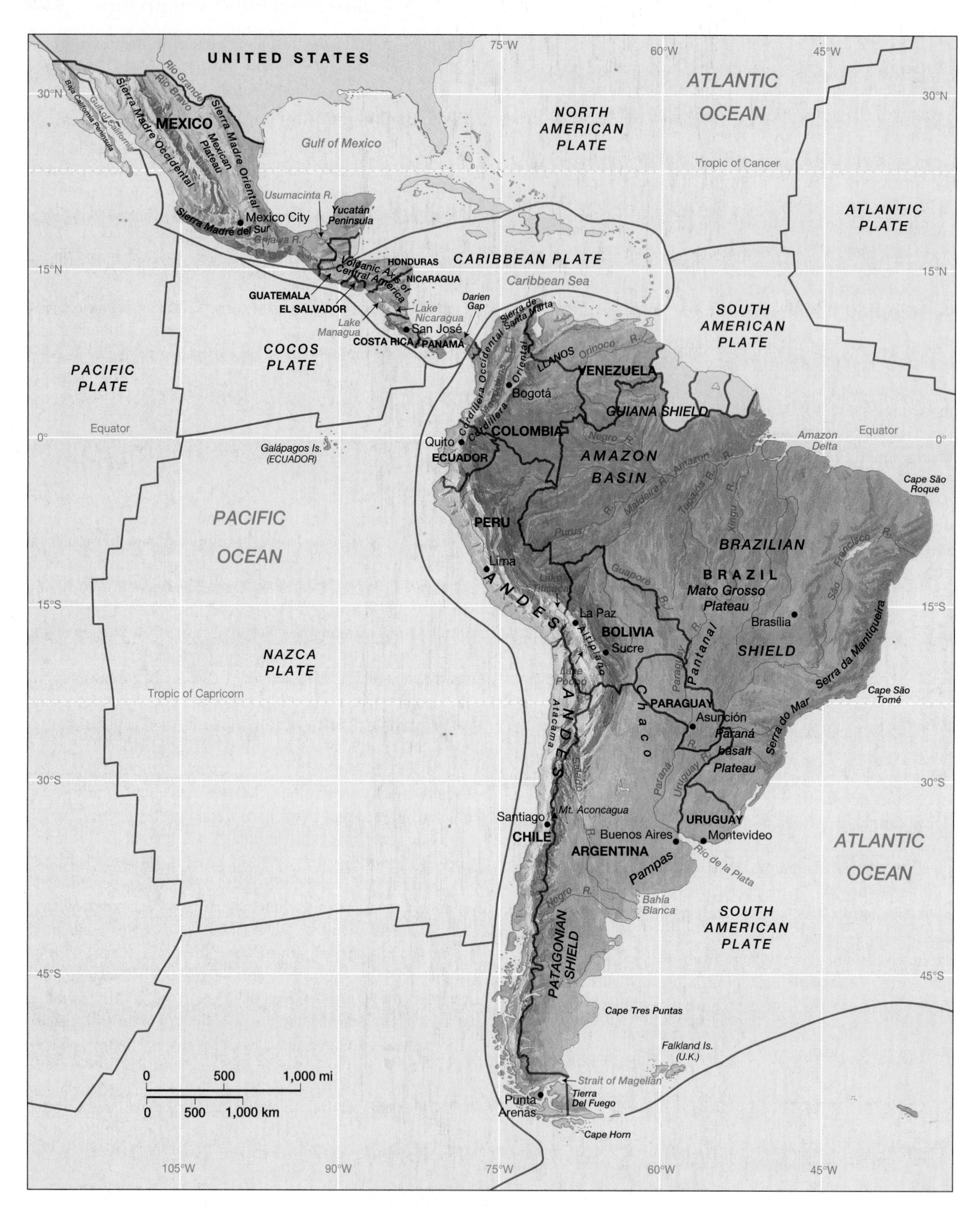

▲ **Figure 4.3 Physical geography of Latin America** Centered in the tropics but extending into the midlatitudes, Latin American landforms include mountains, shields, highland plateaus, vast river basins, and grassy plains. Due to the movement of continental plates, earthquakes and volcanic eruptions are always threatening, especially on the Pacific coast where the region's highest mountains are found. The lower landforms on the Atlantic side include the Amazon, Plata, and Orinoco river basins, as well as the highly eroded surfaces of the Brazilian, Guiana, and Patagonian shields.

◀ **Figure 4.4 Amazonian floodplains** More water flows through the Amazon Basin than any other river. Historically, settlement has been along the river's natural levees and floodplains (called *varzeas*). Agricultural interest in the fertile varzeas is growing as farmers seek to develop sustainable agricultural practices within this tropical ecosystem. *(Victor Englebert/Englebert Photography, Inc.)*

Brazilian uplands before the Paraguay River joins it in northern Argentina. The Paraná and the considerably smaller Uruguay empty into the Rio de la Plata estuary that begins north of Buenos Aires.

Unlike the Amazon, much of the Plata Basin is now economically productive through large-scale mechanized agriculture, especially soybean production. Arid areas such as the Chaco and inundated lowlands such as the Pantanal support only livestock. The Plata Basin is also home to the region's largest hydroelectric plant, the Itaipú on the Paraná, which generates all of Paraguay's electricity and much of Southern Brazil's. As agricultural output in this watershed grows, sections of the Paraná River are being canalized and dredged to enhance the region's fluvial transportation network. In the late 1990s plans also existed to canalize the Paraguay River so that eastern Bolivia could have a water route to the Atlantic. The costs of such a project are enormous, and if they are carried out, much of the Pantanal (a unique Brazilian wetland) will be drained.

Orinoco Basin The third largest river basin by area is the Orinoco in the northern part of South America. The Orinoco River meanders through much of southern Venezuela and part of western Colombia, giving character to a tropical grassland called the *Llanos*. Although it is just one-sixth the size of the Amazon watershed, its discharge roughly equals that of the Mississippi River. Like the Amazon, this basin is home to very few individuals; 90 percent of Venezuela's population lives north of the river. With the exception of the industrial developments between Ciudad Guayana and Ciudad Bolívar, cities are rare. Much of the Orinoco drains the Llanos, which are seasonally inundated by several feet of water during the rainy season. Since the colonial era these grasslands have supported large cattle ranches. Cattle is still important but the Llanos have become a dynamic area of petroleum production for both Colombia and Venezuela.

Grijalva-Usumacinta In Mexico and Central America the largest watershed by volume, the Grijalva-Usumacinta Basin, also flows through a sparsely populated tropical forest zone in Mexico and Guatemala. In the Mexican state of Tabasco, the Usumicinta joins the Grijalva and flows into the Bay of Campeche, accounting for nearly half of Mexico's freshwater river flow. Hydroelectric dams have been built or are planned, and an active agricultural frontier exists. This frontier is contested, however, by indigenous and environmental groups that are fighting to slow the incorporation of this area into the modern economy. Political interest in the basin has intensified over the years because the watershed may be critical for satisfying the water and energy demands of Mexico City (Figure 4.5).

Rio Grande (Rio Bravo) Delimiting the boundary between Mexico and the state of Texas, what is called the Rio Grande in the United States is the Rio Bravo for Mexicans. This watershed has been extensively managed to enable agricultural development in northern Mexico as well as provide water for the states of Texas and New Mexico. With headwaters in the Sierra Madre Occidental, the Rio Bravo and its tributaries carry the snowmelt from the mountains through parts of arid northern Mexico. Dams have been built on some of its major tributaries (such as the Rio Conchos) to produce electricity and supply cities and towns with water. The floodplain of the lower Rio Grande has emerged as one of Mexico's major agricultural areas, with nearly 700,000 acres (280,000 hectares) of irrigated land planted mostly in sorghum. The rise of border cities and industrialization have combined to degrade this watershed. Surface water is scarce, and what does exist is badly polluted.

▶ **Figure 4.5 Satellite image of Mexican deforestation** This view from space shows the clearing of Mexican forests adjacent to the Guatemalan border. Frontier settlement programs brought a wave of Mexican farmers and ranchers into this area in the 1970s and 1980s. Environmentalists and Indian-rights groups fear that Guatemala's forests may soon meet a similar fate. *(Earth Satellite Corporation/SPL/Photo Researchers, Inc.)*

Western Mountains and Eastern Shields

Historically, the most important areas of settlement in tropical Latin America were not along the region's major rivers but across its shields, plateaus, and fertile intermontane basins. In these localities the combination of fertile soils, benign climate, and sufficient rainfall produced the region's most productive agricultural areas, and its densest settlement. The Mexican Plateau, for example, is a massive upland area ringed by the Sierra Madre Mountains. The southern end of the plateau, known as the Mesa Central, is rimmed by volcanic mountains that give way to fertile basins. Similarly, the elevated and well-watered basins of Brazil's southern mountains provide an ideal setting for agriculture. These especially fertile areas are able to support much higher population densities, so it is not surprising that the two largest cities, Mexico City and São Paulo, emerged from these settings. The Latin American highlands also lend a special character to the region. Luxuriant tropical valleys nestled below snow-covered mountains hint at the diversity of ecosystems found in close proximity to each other. The most dramatic of these highland areas, the Andes, runs like a spine up the length of the South American continent.

The Andes Beginning in northwestern Venezuela and ending at Tierra del Fuego, the Andes are relatively young mountains that extend nearly 5,000 miles (8,000 kilometers). Created by the collision of oceanic and continental plates, the mountains are a series of folded and faulted sedimentary rocks with intrusions of crystalline and volcanic rock. The Andes are still forming, so active volcanism and regular earthquakes are common in this zone. The end result is an ecologically and geologically diverse mountain chain with some 30 peaks higher than 20,000 feet (6,000 meters). Due to the violent and complex origins of this mountain chain, many rich veins of precious metals and minerals are found here. For many Andean countries, their initial economic wealth came from the mining of silver, gold, tin, copper, or iron.

Given the length of the Andes, it is typically divided into its northern, central, and southern components. In Colombia the northern Andes actually split into three distinct mountain ranges before merging near the border of Ecuador. High-altitude plateaus and snow-covered peaks distinguish the Central Andes of Ecuador, Peru, and Bolivia. The Andes reach their greatest width here. Of special interest is the treeless high plain of Peru and Bolivia called the **Altiplano.** The floor of this elevated plateau ranges from 11,800 feet (3,600 meters) to 13,000 feet (4,000 meters), and it has limited use for grazing. Two high-altitude lakes, Titicaca on the Peruvian and Bolivian border and the smaller Poopó in Bolivia, are located in the Altiplano as well as many mining sites. The southern Andes are shared by Chile and Argentina. The highest peaks of the Andes are found in the southern Andes, including the highest peak in the Western Hemisphere, Aconcagua, at almost 23,000 feet (6,958 meters). South of Santiago, Chile,

▲ **Figure 4.6 Bolivian Altiplano** The Altiplano is an elevated plateau straddling the Bolivian and Peruvian Andes. This stark, windswept land is mostly inhabited by Amerindians. Considered one of the poorer areas of the Andes, mining and pastoral activities dominate. *(Loren McIntyre/Woodfin Camp & Associates)*

however, the mountains are lower and the chain less compact. The impact of glaciation is most evident in the southernmost extension of the Andes (Figure 4.6).

The Uplands of Mexico and Central America The Mexican Plateau and the Volcanic Axis of Central America are the most important elevated lands in terms of settlement. Most of major cities of Mexico and Central America are found here. The Mexican Plateau is a large, tilted block that has its highest elevations in the south, about 8,000 feet (2,500 meters) around Mexico City, to just 4,000 feet (1,200 meters) at Ciudad Juárez. The southern end of the plateau is called the Mesa Central, which contains a number of flat-bottomed basins interspersed with volcanic peaks. Many of the basin floors were once lake beds that filled with fertile alluvium and later proved ideal for agriculture. The Mesa Central supports Mexico's highest population densities, including the cities of Mexico City, Guadalajara, and Puebla. Yet this productive region is also under severe stress. The once ample ground and surface waters of the Mesa Central have been overtapped and badly contaminated. Water shortages have threatened the agricultural productivity of the Mesa Central, which was long thought of as Mexico's breadbasket. Throughout the Mexican Plateau are rich seams of silver, copper, and zinc. The quest for silver drove much of the economic activity of colonial Mexico. Today, the Mexican economy is driven by petroleum and gas production along the Gulf Coast and not by the metals of the plateau.

Along the Pacific coast of Central America lies a chain of volcanoes that stretches from Guatemala to Costa Rica. The Volcanic Axis of Central America is a handsome landscape of rolling green hills, elevated basins with sparkling lakes, and conical volcanic peaks. More than 40 volcanoes are found here, many of which are still active. Their legacy is a rich volcanic soil that yields a wide variety of domestic and export crops. Most of Central America's population is also concentrated in this zone, either in the capital cities or the surrounding rural villages. The bulk of the agricultural land is tied up in large holdings that export beef, cotton, and coffee. Yet most of the farms are small subsistence properties that produce corn, beans, squash, and assorted fruits. A major rift (a large crustal fracture) east of the Volcanic Axis exists in Nicaragua. In this low-lying valley are Lakes Managua and Nicaragua, the largest in Central America. Surrounded by mountains, this rift zone presently yields sorghum and cotton for export (Figure 4.7).

The Shields South America has three major shields—large upland areas of exposed crystalline rock that are similar to upland plateaus found in Africa and Australia. (The Guiana shield will be discussed in Chapter 5.) The Brazilian and Patagonian shields vary in elevation between 600 and 5,000 feet (200 and 1,500 meters) and are remnants of the ancient

▲ **Figure 4.7 Guatemalan farmer** A peasant farmer works his small plot of subsistence and cash crops. Behind him looms a volcano. The fertile volcanic soils, ample rainfall, and temperate climate of the Guatemala highlands have supported dense populations for centuries. *(Sean Sprague/Panos Pictures)*

▶ **Figure 4.8 Brazilian oranges** Most estate-grown oranges in Brazil are processed into frozen concentrate and exported. São Paulo and Paraná have some of the finest soils in Brazil. In addition to oranges, coffee and soybeans are widely cultivated. *(Stephanie Maze/National Geographic Society)*

landmass of Gondwanaland. The Brazilian shield is the largest and most important one in terms of natural resources and settlement. Far from a uniform land surface, the Brazilian shield covers much of Brazil, from the Amazon Basin in the north to the Plata Basin in the south. It is studded with isolated low ranges and flat-topped plateaus in the north. In southeastern Brazil a series of mountains (Serra da Mantiqueira and Serra do Mar) reach elevations of 9,000 feet (2,700 meters). In between these ranges are elevated basins that offer a mild climate and fertile soils. In one of these basins is the city of São Paulo, the largest urban conglomeration in Latin America. The other major population centers are on the coastal fringe of the plateau, where large protected bays made the sites of Rio de Janeiro and Salvador attractive to Portuguese colonists. Finally, the Paraná basalt plateau located on the southern end of the Brazilian shield is celebrated for its fertile red soils *(terra roxa)* that yield coffee, oranges, and soybeans. The basalt plateau, much like the Deccan plateau in India, is an ancient lava flow that resulted from the breakup of Gondwanaland. So fertile is this area that the economic rise of São Paulo is attributed to the expansion of commercial agriculture, especially coffee, into this area (Figure 4.8).

The vast low-lying shield of Patagonia lies south of the Plata Basin. Often described as the end of the world, much of Patagonia is open steppe country with few settlements. Located in the midlatitudes, Patagonia captured the imagination of many nineteenth-century travelers. Because most transoceanic traffic prior to the building of the Panama Canal used the Strait of Magellan, many voyagers had the opportunity to observe Patagonia and write about it. Beginning south of Bahia Blanca and extending to Tierra del Fuego, the region to this day remains sparsely settled and hauntingly beautiful. It is treeless and covered by scrubby steppe vegetation (Figure 4.9). Sheep were introduced to Patagonia in the late nineteenth century, spurring a wool boom. More recently, offshore oil production has renewed the economic importance of Patagonia.

Climate

In tropical Latin America average monthly temperatures in localities such as Managua, Quito, or Manaus show little variation (Figure 4.10). Precipitation patterns, however, do vary and create distinct wet and dry seasons. In Managua, for example, January is typically a dry month and October is a wet one (see the Managua graph in Figure 4.10). The tropical lowlands of

▲ **Figure 4.9 Patagonian wildlife** Guanaco thrive on the thin steppe vegetation found throughout Patagonia. Native to South America, the numbers of guanaco fell dramatically due to hunting and competition with introduced livestock. *(Rob Crandall/Rob Crandall Photography)*

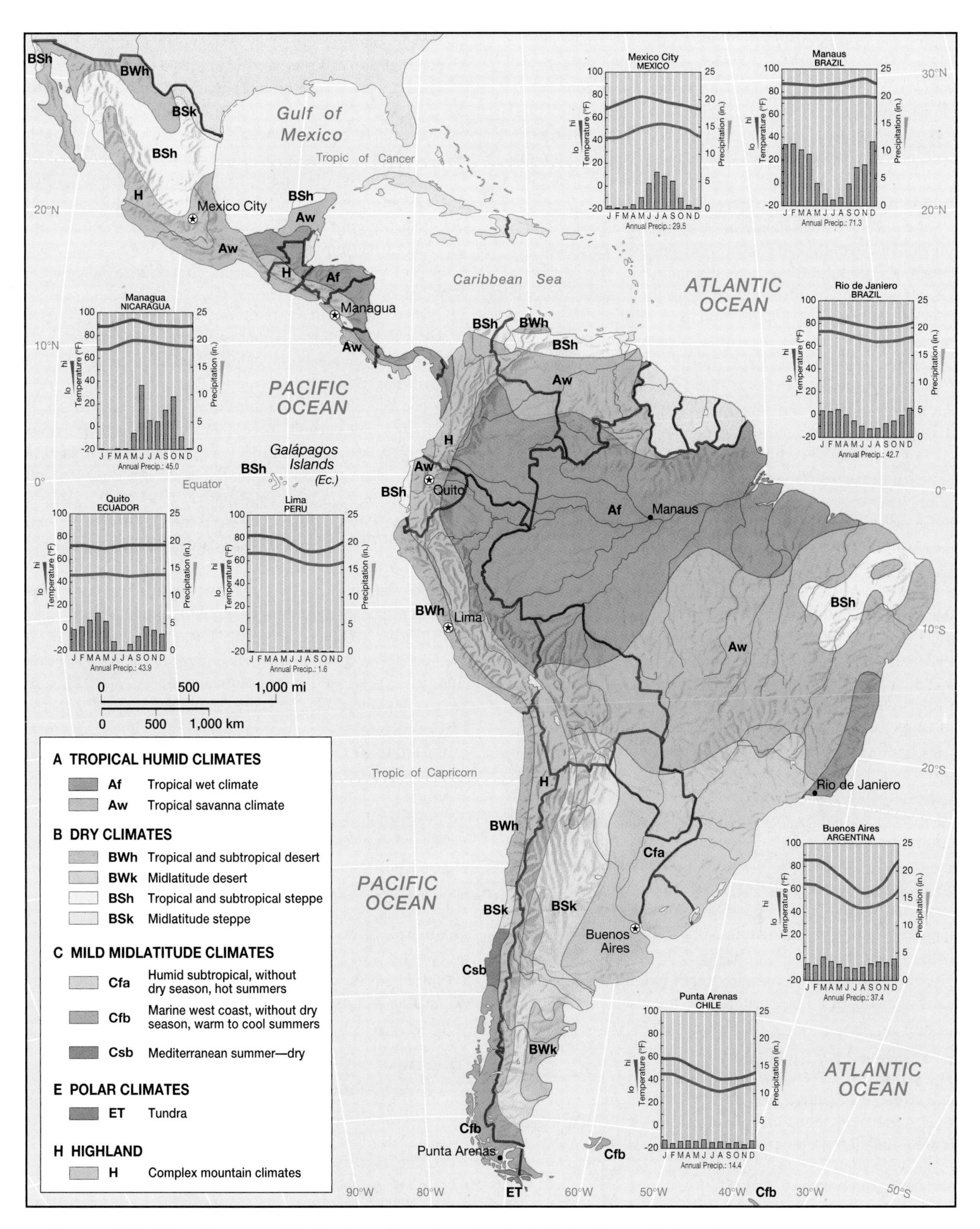

▲ **Figure 4.10 Climate map of Latin America** Latin America includes the world's largest rain forest (Af) and driest desert (BWh), as well as nearly every other climate classification. Latitude, elevation, and rainfall play an important role in determining the region's climates. Note the contrast in rainfall patterns between humid Quito and arid Lima. *(Temperature and precipitation data from Pearce and Smith, 1984.* The World Weather Guide. *London: Hutchinson)*

Latin America, especially east of the Andes, are usually classified as tropical humid climates that are covered in forest or savanna, depending on the amount of rainfall. The region's desert climates are found along the Pacific coast of Peru and Chile, Patagonia, northern Mexico, and the Bahia of Brazil. Thus, a city such as Lima, Peru, which is clearly in the tropics, averages only 1.5 inches (4 centimeters) of rainfall a year due to the hyperaridity of the Peruvian coast.

Midlatitude climates, with hot summers and cold winters, prevail in Argentina, Uruguay, and parts of Paraguay. Of course the midlatitude temperature shifts in the Southern Hemisphere are the inverse of those in the Northern Hemisphere (cold Julys and warm Januarys). Chile's climate is a mirror image of the west coast of Mexico and the United States, with the Atacama Desert in the north (like Baja California), a Mediterranean dry summer around Santiago (similar to Los Angeles), and a marine west coast climate with no dry season south of Concepción (such as the coasts of Oregon and Washington). In the mountain ranges, complex climate patterns result from changes in elevation. To appreciate how humans adapt to tropical mountain ecosystems, one must understand the concept of **altitudinal zonation,** the relationship between cooler temperatures at higher elevations and changes in vegetation.

Altitudinal Zonation First described in the scientific literature by Prussian naturalist Alexander von Humboldt in the early 1800s, the practical applications of altitudinal zonation have been intimately understood by all the region's native inhabitants. Humboldt systematically recorded declines in temperature as he ascended higher elevations, a phenomenon referred to today as the **environmental lapse rate** (averaging a temperature decline of 3.5 °F for every 1,000 feet in elevation or 6.5 °C for every 1,000 meters). He also noted changes in vegetation by elevation, demonstrating that plant communities common to the midlatitudes could thrive in the tropics at higher elevations. These different altitudinal zones are commonly referred to as the **tierra caliente** (hot land) from sea level to 3,000 feet (900 meters); the **tierra templada** (temperate land) from 3,000–6,000 feet (900–1,800 meters); the **tierra fría** (cold land) at 6,000–12,000 feet (1,800–3,600 meters); and the **tierra helada** (frozen land) above 12,000 feet (3,600 meters). Exploitation of these zones allows agriculturists, especially in the uplands, access to a great diversity of domesticated and wild plants (Figure 4.11).

The concept of altitudinal zonation is most relevant for the Andes, the highlands of Central America, and the Mexican Plateau. Traditional Andean farmers, for example, might use the high pastures of the Altiplano for the grazing of llamas and alpacas, the tierra fría for potato and quinoa production, and the lower temperate zone to produce corn. All the great pre-contact civilizations, especially the Incas and the Aztecs, systematically extracted resources from these zones, thus ensuring a diverse and abundant resource base.

El Niño El Niño (a reference to the Christ child) is probably the most talked-about weather phenomenon in Latin America and the world. Its name comes from a warm Pacific current that usually arrives along coastal Ecuador and Peru in December, around Christmastime. Every decade or so, an abnormally large current arrives that produces torrential rains, signaling the arrival of an El Niño year. The 1997–98 El Niño was especially bad. Devastating floods occurred in Peru and Ecuador. Heavy May rains in Paraguay and Argentina caused the Paraná River to rise 26 feet (8 meters) above normal. Flooding drove some 350,000 people from their homes in Peru. In Argentina 150,000 fled. At least 900 people were killed by floods and storms in Latin America, the single most devastating being a hurricane that hit Acapulco, Mexico, and took at least 200 lives. The other less talked about aspect of El Niño is drought. While the Pacific coast of South and North America experienced record rainfall in the 1997–98 El Niño, Colombia, Venezuela, northern Brazil, and Central America battled drought. In addition to crop and livestock losses estimated to be in the billions of dollars, hundreds of brush and forest fires left their mark. The amount of smoke produced by forest fires in northern Mexico in the spring of 1998 was so great that it caused haze in the southeastern United States. The indirect costs of drought, such as fire-charred hillsides vulnerable to landslides, are impossible to calculate.

Drought Drought conditions seem to regularly threaten certain areas of Latin America. In northeastern Brazil, from Fortaleza to Salvador, lies an area of relatively high rural population density with a notoriously precarious environment.

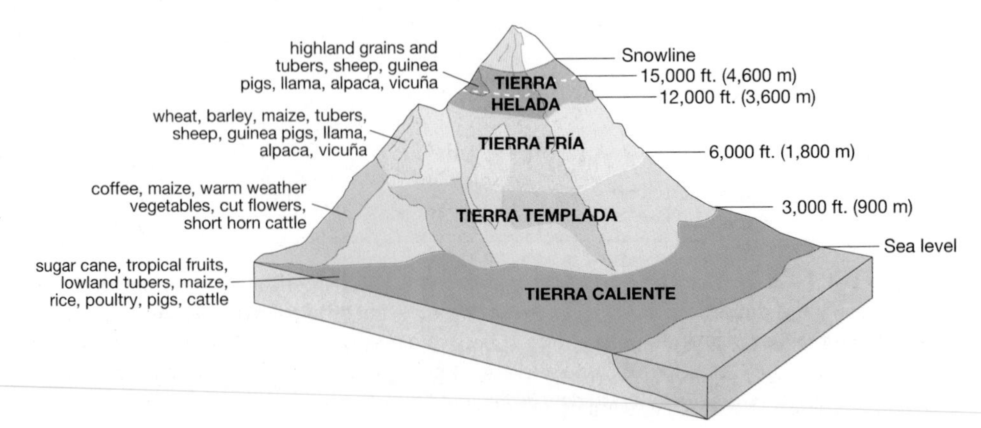

▶ **Figure 4.11 Altitudinal zonation** Tropical highland areas support a complex array of ecosystems. In the tierra fría zone (6,000–12,000 feet), for example, midlatitude crops such as wheat and barley can be grown. This diagram depicts the range of crops and wildlife found in different elevations in the Andes.

Here, when the intertropical convergence zone (ITCZ) fails to oscillate, devastating droughts occur. The worst was in the 1870s when a half million people perished due to drought. Recent droughts in the 1970s, 1983, and 1991 and 1992, while not as deadly, destroyed crops and triggered outflows of migrants to Brazil's southern cities and the Amazon.

Today, few Latin American droughts cause the fatalities seen in Sub-Saharan Africa. Yet too little or late rainfall can produce serious economic hardships. Due to the 1996 drought in northern Mexico, basic grain production declined by three million tons and half a million cattle were sent prematurely to slaughterhouses. Similarly, droughts in the densely settled highlands of Central America and the Andes often result in agricultural losses and in disruptions in electric power. In 1992 several cities in Colombia instituted the practice of rolling blackouts (often for several hours a day) as a means to ration electricity use. A combination of below-average rainfall and reservoir sedimentation had drastically reduced the country's production of hydroelectricity.

Environmental Concerns for Latin America

Given the territory's immense size and relatively low population density, Latin America has not experienced the same levels of environmental degradation witnessed in East Asia and Europe. The worst environmental problems are found in cities, their surrounding rivers and coasts, and intensely farmed zones. Yet vast areas remain relatively untouched, supporting an incredible diversity of plant and animal life. Throughout the region national parks offer some protection to unique communities of plants and animals. A growing environmental movement in countries such as Costa Rica and Brazil has yielded both popular and political support for green initiatives. In short, Latin Americans enter the twenty-first century with a real opportunity to avoid many of the environmental missteps seen in other regions of the world. At the same time, economic pressures brought on by global market forces are driving governments to aggressively exploit their natural resources (minerals, fossil fuels, forests, soils). The challenge lies in managing the region's immense natural resources and balancing the economic benefits of extraction with the ecological soundness of conservation (see "Local Voices: the Environmentalist Message of the Kogi").

Deforestation The environmental issue most associated with this region is deforestation. The Amazon Basin and portions of the eastern lowlands of Central America and Mexico still maintain unique and impressive stands of tropical forest (Figure 4.12). Other areas, such as the Atlantic coastal forests of Brazil and the Pacific forests of Central America, have nearly disappeared due to agriculture, settlement, and ranching. In the midlatitudes, the ecologically unique evergreen rain forest of southern Chile (the Valdivian forest) is being cleared to export wood chips to Asian markets. The coniferous forests of northern Mexico are also falling, in part because of a bonanza for commercial logging stimulated by the North American Free Trade Association (NAFTA) agreement.

In terms of biological diversity, however, the loss of tropical rain forest is the most critical. Tropical rain forests account for only 6 percent of Earth's landmass, but at least 50 percent of the world's species are found in this biome. Moreover, only the Amazon contains the largest undisturbed stretches of rain forest in the world. Unlike Southeast Asian forests, where hardwood extraction motivates forest clearance, Latin American forests are usually seen as an agricultural frontier that state governments divide in an attempt to appease landless peasants or reward political cronies. Thus forests usually fall to the ax and fire, with colonists and politicians carving up the forest to create permanent settlements, slash-and-burn plots, or large cattle ranches. In addition, some tropical forest clearance has been motivated by the search for gold (Brazil, Venezuela, and Costa Rica) and the production of coca leaf for cocaine (Peru, Bolivia, and Colombia).

Brazil has incurred more criticism than other countries for its Amazon forest policies. Through these policies some 15–20 percent of the Brazilian Amazon was deforested. In states such as Rondônia, where colonists streamed in along a popular road known as BR364, close to 60 percent of the state is deforested. What most alarmed environmentalists and forest dwellers (Indians and rubber tappers) was the dramatic increase in the rate of clearing, especially in the 1980s. Should those rates continue, it is feared that only small unsustainable remnants of forest would remain by the mid-twenty-first century. Yet there are signs that forest clearing has slowed, due in part from pressure from environmental, indigenous, and development agencies, and due to Brazil's own economic crisis that limited resources directed toward the region (Figure 4.13).

Grassification The conversion of tropical forest into pasture, called **grassification,** is a serious problem affiliated with modern ranching. Particularly in Chiapas, Mexico, Central America, and the Brazilian Amazon a hodgepodge of development policies from the 1960s through the 1980s encouraged deforestation to make room for cattle. The preference for ranching as a status-conferring occupation seems to be a transfer from Iberia. The image of the *vaquero* (cowboy) looms large in the region's history. Even poor farmers appreciate the value of having livestock. Like a savings account, cattle can be quickly sold for cash. Although there are many natural grasslands such as the Llanos, the Chaco, and the Pampas suitable for grazing, it was the rush to convert forest into pasture that made ranching a scourge on the land. Ultimately, a cattle bust occurred in the 1980s when world prices collapsed. Even in cases where domestic demand for beef increased, ranching in remote tropical frontiers was seldom economically self-sustaining. All told, nearly 300 million acres of tropical forest were destroyed in the 1980s, and much of this went to pasture.

Degradation of Farmlands The pressure to modernize agriculture has produced a series of unintended environmental problems. As peasants were encouraged to adopt new hybrid varieties of corn, beans, and potatoes, an erosion of genetic

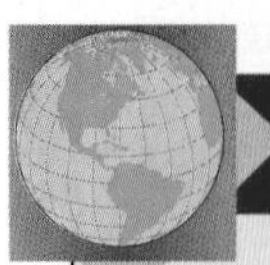

LOCAL VOICES The Environmentalist Message of the Kogi

The Kogi live in the Sierra de Santa Marta in Colombia. A small indigenous group that has struggled to survive for five centuries, they continue to perceive themselves as the "Elder Brothers" and the rest of the world as "Younger Brothers." One of the fundamental beliefs of the Kogi is that they are the guardians of the environment and that Younger Brother is recklessly destroying Earth. Kogi spiritual leaders, the Mamas, expressed their environmentalist agenda in front of a BBC film crew in the 1980s.

We are the Elder Brothers
We have not forgotten the old ways
How could I say that I do not know how to dance?
We still know how to dance.
We have forgotten nothing.
We know how to call the rain.
If it rains too hard we know how to stop it.
We call the summer.
We know how to bless the world and make it flourish.

But now they are killing the Mother.
The Younger Brother, all he thinks about is plunder.

The Mother looks after him too, but he does not think.

He is cutting into her flesh.
He is cutting into her arms.
He is cutting off her breasts.
He takes out her heart.
He is killing the heart of the world.

When the final darkness falls everything will stop.
The fires, the benches, the stones, everything.
All the world will suffer.

If that happened and all we Mamas died,
and there was no one doing our work,
well, the rain wouldn't fall from the sky.

▲ **Figure 1 The Kogi of Colombia** A Kogi man walks with his poporo (a gourd filled with lime powder) and a satchel of coca leaves. The chewing of coca leaf with lime, which reduces hunger and fatigue, has been practiced by indigenous peoples in the Andes for centuries. *(Victor Englebert/Englebert Photography, Inc.)*

It would get hotter and hotter from the sky,
and the trees wouldn't grow,
and the crops wouldn't grow.

Or am I wrong, and they would grow anyway?

Source: Alan Ereira, *The Elder Brothers.* (New York: Alfred Knopf, 1992), pp. 113–14.

diversity occurred. Efforts to preserve dozens of native domesticates are under way at agricultural research centers in the central Andes and Mexico. It is feared, nonetheless, that many useful native plants are lost. Modern agriculture also depends on chemical fertilizers and pesticides that eventually run off into surface and groundwater supplies. Consequently, many rural areas suffer from contamination of local water supplies. More troublesome still is the direct exposure of farm workers to toxic agricultural chemicals. Mild exposure to pesticides and fertilizers typically occur due to mishandling, resulting in rashes and burns. Yet there are areas such as Sinaloa, Mexico, where the widespread application of chemicals parallels a rise in serious birth defects.

Soil erosion and fertility decline occur in all agricultural areas. Certain soil types in Latin America are particularly vulnerable to erosion, most notably the volcanic soils and the reddish oxisols found in the humid lowlands. The productivity of the Paraná basalt plateau, for example, has declined over the decades due to the ease with which these volcanic soils erode and the failure to apply soil conservation methods. The oxisols of the tropical lowlands, by contrast, can quickly degrade into a baked claypan surface when the natural cover is removed, making permanent agriculture nearly impossible. Ironically, the consolidation of the large-scale modern farms in the basins and valleys of the highlands tends to push peasant subsistence farmers into marginal areas. On these hillside farms, gullies and slides plague the productivity of small farmers. Lastly, the sprawl of Latin American cities consumes both arable land and water, resulting in land-use decisions that are unsustainable (Figure 4.14).

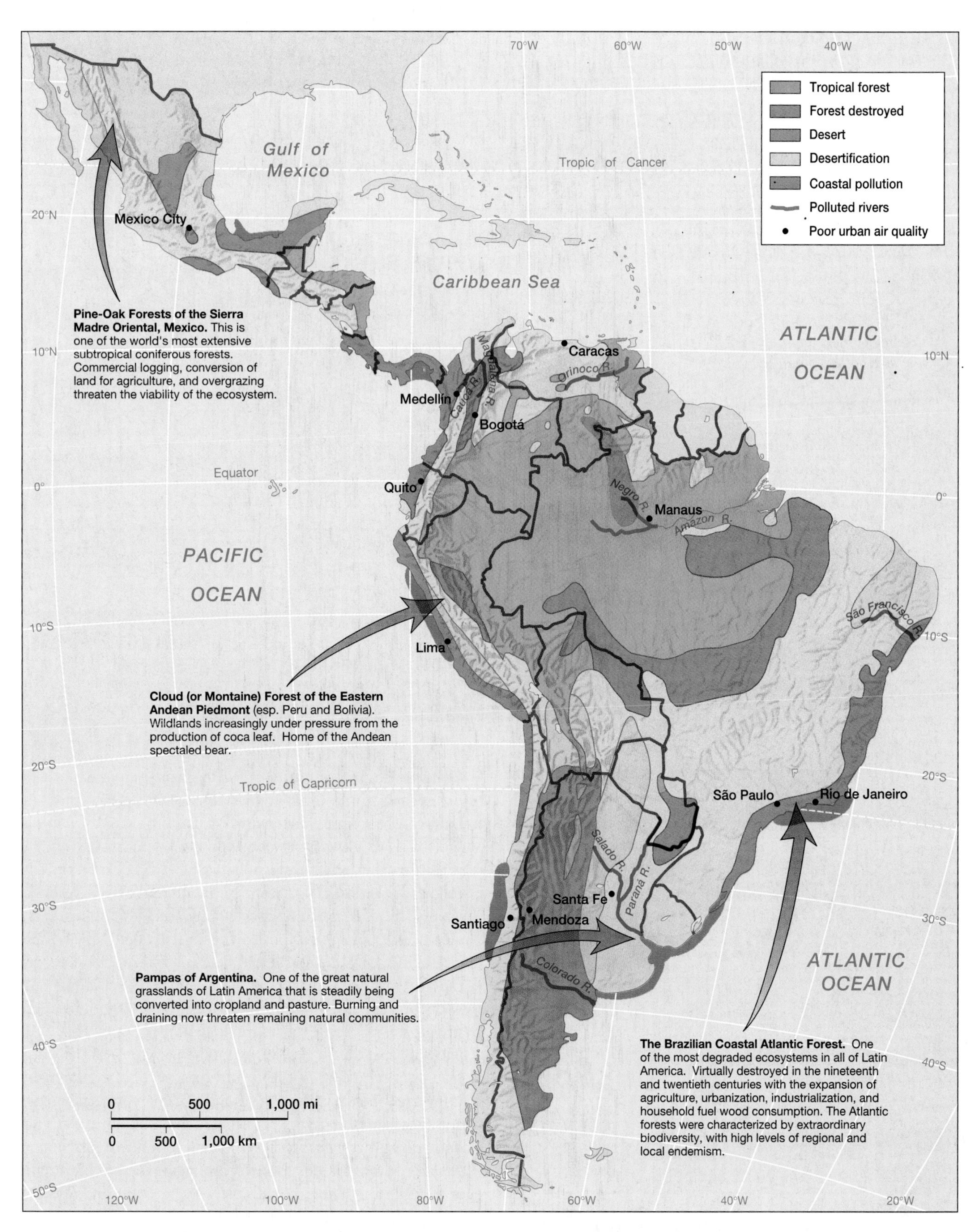

▲ **Figure 4.12 Environmental issues in Latin America** Tropical forest destruction, desertification, water pollution, and poor urban air quality are some of the pressing environmental problems facing Latin America. Yet, vast areas of tropical forest are still present, supporting a wealth of genetic and biological diversity. *(Adapted from* DK World Atlas, *1997, pp. 7, 55. London: DK Publishing)*

▶ **Figure 4.13 Frontier settlement in Rondônia, Brazil** Colonization along Amazonian highways, such as BR364, sparked a wave of forest clearing in the 1980s. Paired satellite images from Rondônia, Brazil, in 1975 (left) and 1992 (right) reveal the extent of forest clearing that occurred along this major highway and its side roads. *(EROS Data Center, U.S. Geological Survey)*

▲ **Figure 4.14 Degraded Bolivian farmlands** The farmlands of Cochabamba have historically been Bolivia's bread basket. Agricultural productivity has declined in recent decades due to increased aridity and severe soil erosion. In this photo, soil erosion has produced deep gullies that divide fields and remove valuable topsoil. *(Rob Crandall/Rob Crandall Photography)*

Urban Environmental Challenges For most Latin Americans air pollution, inadequate water, and garbage removal are the pressing environmental problems of everyday life. Consequently, many environmental activists from the region focus their efforts on making urban environments cleaner by introducing green legislation and rousing popular action. In this most urbanized region of the developing world, city dwellers do have better access to water, sewers, and electricity than their counterparts in Asia and Africa. Moreover, the density of urban settlement seems to encourage the widespread use of mass transportation—both public and private bus and van routes make getting around most cities fairly easy. Yet the inevitable environmental problems that come from dense urban settings ultimately require expensive infrastructural remedies. The money for such projects is never enough, thanks to currency devaluation, inflation, and foreign debt. Since many urban dwellers tend to reside in unplanned squatter settlements, retroactively servicing these communities with utilities is difficult and costly. In addition, many urban squatters continue certain rural practices such as keeping animals. It is estimated that the chronic landslide problem in Rio de Janeiro is exacerbated by hundreds of goats stripping the hillsides of vegetation.

Industrialization is also a major cause of pollution. Factories, electricity generation, and transportation all contribute to urban pollution. A lax attitude toward enforcing environmental laws tends to be the norm. The consequences are, in the worst cases, a serious threat to people and the environment. In the 1980s the Brazilian industrial center of Cubatão, near São Paulo, became synonymous with environmental catastrophe. For years people complained of headaches and nausea from the belching factory smokestacks, but their complaints were not taken seriously. In 1984 a leak occurred in a gasoline pipeline that ran through one of the poorest squatter settlements. The smell of leaking gas went unnoticed because of the omnipresent stench of industrial pollutants in the valley. When the gas was finally ignited, as many as 200 people were incinerated in the resulting explosion and fire. A year later, a break in an ammonia pipeline forced the evacuation of 6,000 people and the hospitalization of 65. While industry downplayed these disasters as the risk of doing business, traumatized residents mobilized to address the worst abuses. The events in Cubatão, more than the destruction of rain forest, are credited with invigorating the environmental movement in Brazil.

The Case of Mexico City Even if there were no shortages of money, this capital city of 16–18 million people, located at an 8,000 foot elevation and spread across 4,000 square kilometers, would be a challenge to manage. Much like Denver, Colorado, the city's high setting surrounded by mountains supports its legendary and perpetual smog. Most modern-day visitors to Mexico City have no idea that mountains surround them because they are rarely seen. Air quality has been a major issue for Mexico City since the 1960s, driven in part by the city's phenomenal rate of growth. (Between 1950 and 1980 the city's annual rate of growth was 4.8 percent.) Steps were finally taken in the 1980s to reduce emissions from factories and cars. Unleaded gas is now widely available and some of the worst polluting factories in the Valley of Mexico have closed. Still, the health costs of breathing such contaminated air are real, as elevated death rates due to heart disease, influenza, and pneumonia suggest. By the 1990s, Mexico City's growth rate had slowed to less than 1 percent a year. The greater availability of jobs in northern Mexico coupled with thousands of people displaced by the 1985 earthquake lowered rates of in-migration.

One of Mexico City's most relentless environmental problems is water. The city gets three-quarters of its water from an underlying aquifer that has been steadily overdrawn so that water levels show a 20–30 feet (6–10 meters) loss since the mid-1980s. Besides the long-term threat of the aquifer drying up, pumping it has resulted in subsidence throughout the basin, which destroys building foundations and water and sewer lines. Moreover, there is troubling evidence that the aquifer is at risk of contamination, especially in areas where unlined drainage canals can leak pollutants into the surrounding soil, which then leach into the aquifer. Remedies to many of these water problems are costly, but efforts to charge industrial consumers higher rates, implement modern conservation efforts, and recharge the most depleted areas of the aquifer are being tested (Figure 4.15).

▲ **Figure 4.15 Air pollution in Mexico City** Notorious for its poor air quality, Mexico City's high elevation and immense size make management of air quality difficult. While lead levels have declined with the introduction of unleaded gasoline, respiratory ailments are on the rise. *(Tom Owen Edmunds/The Image Bank)*

Population and Settlement: The Dominance of Cities

The great river basin civilizations found in Asia do not exist in Latin America. In fact, the great rivers of the region are surprisingly underutilized as areas of settlement or corridors for transportation. Whereas the major population clusters of Central America and Mexico are in the interior plateaus and valleys, the interior lowlands of South America are relatively empty. Historically, the highlands supported most of the region's population during the pre-Hispanic and colonial eras. In the twentieth century population growth and immigration to the Atlantic lowlands of Argentina and Brazil, along with continued growth of Andean coastal cities such as Guayaquil, Barranquilla, and Maracaibo, have reduced the demographic importance of the highlands. Major highland cities such as Mexico City, Guatemala City, Bogotá, and La Paz still dominate their national economies, but the majority of large cities are on or near the coasts (Figure 4.16).

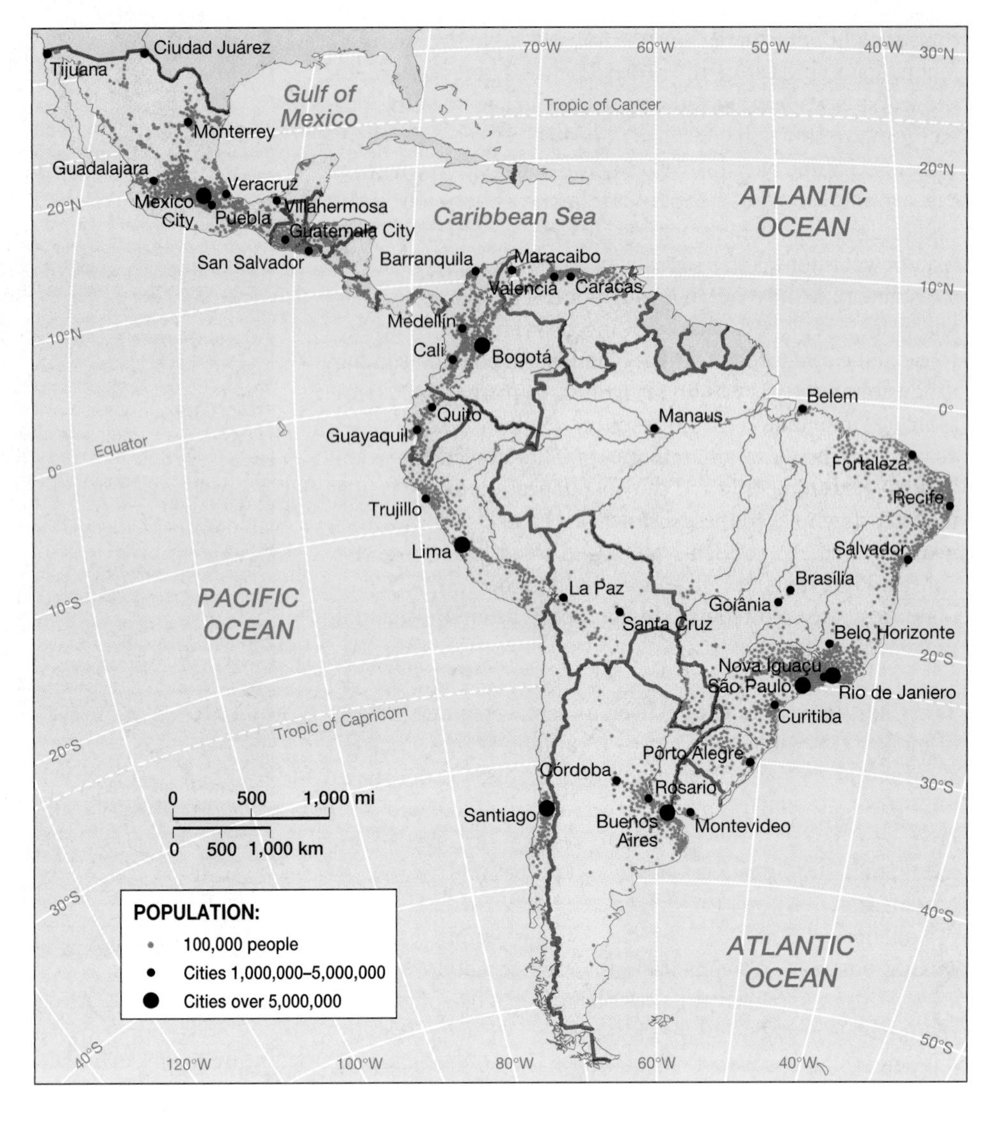

▶ **Figure 4.16 Population map of Latin America** The concentration of population in urban and coastal settlements is evident in this map. Population density in central and southern Mexico, as well as Central America, is quite high. In South America, the majority of people live on or near the coasts, leaving the interior of the continent lightly populated.

Like the rest of the developing world, Latin America experienced dramatic population growth in the 1960s and 1970s. In 1950 Latin America totaled 150 million people, which equaled the population of the United States at that time. By 1995 the population had tripled to 450 million; in comparison, the United States will not reach 300 million until 2010. Latin America outpaced the United States because infant mortality rates declined and life expectancy soared. In 1950 Brazilian life expectancy was only 43 years; by the 1980s it was 63. In fact, most countries in the region experienced a 15- to 20-year improvement in life expectancy between 1950 and 1980, which pushed up growth rates. Four countries account for 75 percent of the region's population: Brazil with more than 160 million, Mexico with nearly 100 million, and Colombia and Argentina with roughly 40 million each (Table 4.1).

During the 1980s, growth rates in Latin America suddenly began to slow, and during the 1990s most countries have reported rates of less than 2 percent. This relatively sudden shift surprised demographers, who predicted in 1985 that the region's population would reach 750 million in 2025. Today's projection is for only 650 million (100 million fewer). One of the reasons for this fertility decline is the shift to urban living, which tends to encourage smaller family size.

The Latin American City

A quick glance at the population map of Latin America shows a concentration of people in cities. One of the most significant demographic shifts was the movement out of rural areas to cities, which began in earnest in the 1950s. In 1950 just one-quarter of the region's population was urban; the rest lived in small villages and in the countryside. Today the pattern is nearly reversed, with almost 70 percent of the population living in cities. In the most urbanized countries, such as Argentina, Chile, and Uruguay, more than 85 percent of the population live in cities (Table 4.1). This preference for urban life is attributed to cultural factors as well as economic ones. Under Iberian rule, residing in a city conferred status and offered opportunity. Initially only Europeans were allowed to live in the colonial cities, but this exclusivity was not enforced. Over the centuries these colonial cities became the hubs for transportation and communication, underscoring the primary role of cities in structuring regional economies.

Table 4.1 Demographic Indicators

Country	Population[a]	Natural Increase	TFR[b]	%<15[c]	% Urban
Costa Rica	3.5	1.9	2.8	33	44
El Salvador	5.8	2.5	3.9	39	50
Guatemala	11.6	3.1	5.1	44	38
Honduras	5.9	2.8	4.4	42	44
Mexico	97.5	2.2	3.1	36	74
Nicaragua	4.8	3.2	4.6	44	63
Panama	2.8	1.8	2.7	33	55
Argentina	36.1	1.1	2.5	29	89
Bolivia	8.0	2.6	4.8	41	58
Brazil	162.1	1.4	2.5	32	76
Chile	14.8	1.4	2.4	29	85
Colombia	38.6	2.1	3.0	33	71
Ecuador	12.2	2.2	3.6	36	61
Paraguay	5.2	2.7	4.4	41	52
Peru	26.1	2.2	3.5	35	71
Uruguay	3.2	0.8	2.4	25	90
Venezuela	23.3	2.1	3.1	38	86

[a]Population in millions, 1998.
[b]Total fertility rate.
[c]Percentage of population younger than 15 years of age.
Source: *Population Reference Bureau, World Population Data Sheet,* 1998.

Latin American cities are noted for high levels of **urban primacy**, a condition in which a country has a primate city that is three to four times larger than any other city in the country. Examples of primate cities are Lima, Caracas, Guatemala City, Santiago, Buenos Aires, and Mexico City. In the case of Brazil, São Paulo and Rio de Janeiro are an example of **dual primacy** (a pair of cities that dominate all others in the country in terms of size and economic importance). Primacy is often viewed as a liability since too many national resources are concentrated into one urban center. In an effort to decentralize, governments have intentionally built cities far from existing primate cities (Ciudad Guayana in Venezuela or Brasília in Brazil). Despite these efforts, the tendency toward primacy remains.

Urban Form Latin American cities have a distinct urban morphology that reflects both their colonial origins and contemporary growth (Figure 4.17). Usually a clear central business district (CBD) exists in the old colonial core. Radiating out from the central business district is older middle and lower-class housing found in the zones of maturity and *in situ* accretion. In this model residential status diminishes in concentric circles from the center to the periphery. The exception to this is the elite spine, a newer commercial and business strip that extends from the colonial core to the newer parts of the city. Along the spine, one finds superior services, roads, and transportation. The city's best residential zones, as well as shopping malls, are usually on either side of the spine. Close to the elite residential sector, a limited area of middle-class tract housing is typically found. Most major urban centers also have a periférico (a ring road or beltway highway) that circumscribes the city. Industry is located in isolated areas of the inner city and in larger industrial parks beyond the ring road.

Beyond the periférico is a zone of peripheral squatter settlements where many of the urban poor live in the worst housing. Services and infrastructure are extremely limited; the roads are unpaved, water is often trucked in, and sewer systems are nonexistent. The dense ring of squatter settlements (variously called *ranchos, favelas, barrios jovenes,* or *pueblos nuevos*) that encircle Latin American cities are an expression of the speed and intensity in which these zones were created. In some cities more than half the population live in these self-built homes of marginal to poor quality. These kinds of dwellings are recognizable throughout the developing world, yet the practice of building one's own home on the "urban frontier" has a longer history in Latin America than in most Asian and African cities. The combination of a rapid inflow of migrants (at times reaching 1,000 people per day), the inability of governments to meet pressing housing needs, and the eventual official recognition of many of these neighborhoods with land title and utilities meant that this housing strategy was never discouraged. Each successful colonization encouraged more.

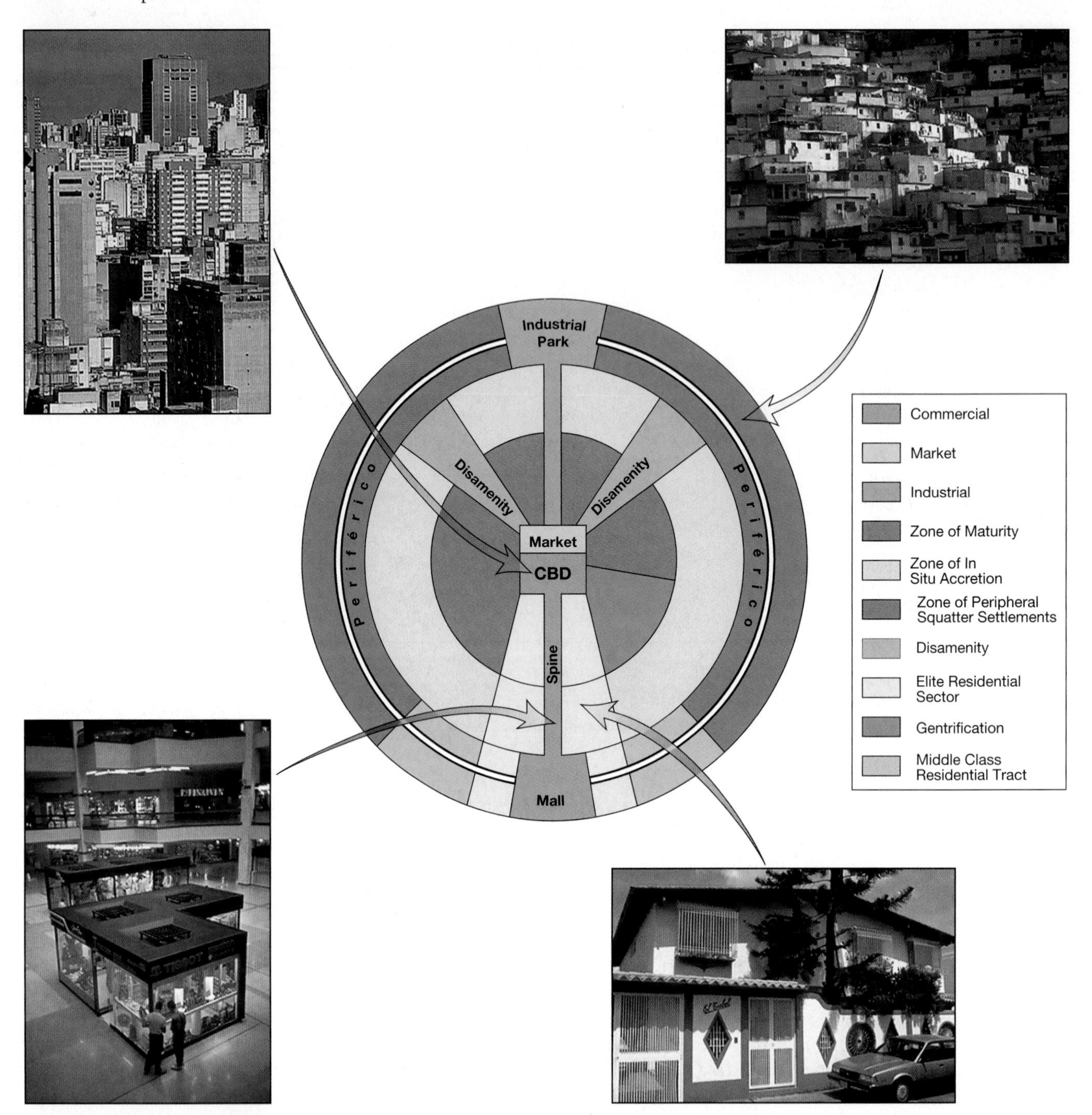

▲ **Figure 4.17 Latin American city model** This urban model highlights the growth of Latin American cities and the class divisions within them. While the central business district (CDB), elite spine, and residential sector may have excellent access to services and utilities, life in the Zone of Peripheral Squatter Settlements is much more difficult. In many Latin American cities, half the population resides in squatter settlements. *[Model reprinted from Ford, 1996. "New and Improved Model of Latin American City Structure."* Geographical Review, *86(3), 437–40. Photos by Rob Crandall/Rob Crandall Photography.]*

Among the inhabitants of these neighborhoods, the **informal sector** is a fundamental force that houses, services, and employs them. Definitions of the informal sector are much debated. It usually includes self-employed, low-waged jobs (such as street vending, shoe shining, and artisan manufacturing) that are unregulated and untaxed. Other scholars include illegal activities such as drug smuggling, sale of contraband items such as illegally copied videos and tapes, and prostitution as part of the informal sector. One of the most interesting expressions of informality is the actual housing found in the squatter settlements. The creation of these landscapes reflects a conscious and organized effort on the part of the urban poor to make a place for themselves in Latin American cities (see "Land Invasion: Lima, Peru").

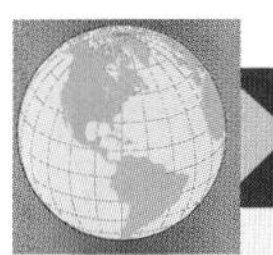

LAND INVASION Lima, Peru

The following description by William Mangin, an urban anthropologist, details the formation and development of a *barriada* (squatter settlement) in Lima, Peru. These illegal invasions of open spaces are carefully timed and planned to reduce the risk of being evicted by state authorities. The one-room houses are quickly built using straw mats. Later, if the settlements go unchallenged, squatters rebuild their houses with brick.

Many barriada invasions had been arranged for the eve of a religious or national holiday. The next holiday was the Independence Day vacation, July 28th, 29th, 30th; so they picked the night of the 27th. It would give them a holiday to provide a patriotic aura as well as three days off from work to consolidate their position. They thought of naming their settlement after the dictator's popular wife, but, after taking into account the vicissitudes of current politics, they decided to write to her about their pitiful plight, but to name the place after a former general-dictator, long dead, who freed the slaves . . . .

Each family bought its own straw mats and poles for the house, small groups made arrangements for trucks and taxis. Each household was asked to get a Peruvian flag or make one out of paper. No two remember the details of the invasion the same way, but about thirty of the expected forty-five families did invade during the night. A newspaper photographer was notified by the invaders and he arrived about the time the houses were being finished. The members had discussed previously what lots they would take, and how the streets would be laid out and there was very little squabbling during the first day. By early morning when the police arrived there were at least thirty one-room straw houses flying Peruvian flags and the principal streets were outlined with stones . . . .

▲ **Figure 1 Lima Squatter Settlement** In arid Lima, squatters initially build their homes using straw mats. As settlements become established, residents will invest in adobe and cinder block to improve their homes. Life on the urban frontier is harsh. Water is trucked in, electricity is irregular, and travel towards the city center is costly and slow. *(Rob Crandall/Rob Crandall Photography)*

Blas and some friends quickly expanded the simple invasion one-room houses to a three-room straw mat house, and they outlined the lot with stones. He worked hard on Sundays and some nights, sometimes alone, sometimes with friends from the barriada or outside. He soon managed to get a brick wall six-and-a-half feet high around his property.

Source: William Mangin, *Peasants in Cities* (Boston: Houghton Mifflin, 1970), pp. 50–52.

Rural-to-Urban Migration Since the 1950s peasants have poured into the cities of Latin America in a process referred to as **rural-to-urban migration.** Conditions in rural areas deteriorated with consolidation of lands, mechanization of agriculture, and increased population pressure. As life in the countryside worsened, many rural households started to send family members to the cities for employment as domestics, construction workers, artisans, and vendors. Once in the cities, rural migrants generally found conditions better, especially access to education, health care, electricity, and clean water. It was not poverty alone that drove people out of rural areas but individual choice and an urban preference. Migrants believed in, and often realized, greater opportunities in cities, especially the capital cities. Those who came were usually young (in their twenties) and better educated than those who stayed behind. Women slightly outnumbered men in this migrant stream. The move itself was made easier by extended kin networks formed by earlier migrants who settled in discrete areas of the city and aided new arrivals. Moreover, these migrants maintained their links to their rural communities by periodically sending remittances and making return visits.

As better roads and transport evolved, the connections between rural and urban areas were reinforced. One indication is a pattern of seasonal returns to the countryside. In Caracas, for example, during the Christmas and Easter holidays the city empties out as thousands of city dwellers jam into buses and cars to visit relatives in the countryside. Latin America is the most urbanized in the developing world, with some states, such as Argentina, Uruguay, and Venezuela, having a proportionally larger percentage of their population in cities than does the United States.

Patterns of Rural Settlement

Throughout Latin America a distinct rural lifestyle exists, especially among peasant subsistence farmers. While the majority of people live in cities, approximately 140 million people do not. In Brazil alone at least 40 million people live in rural areas. Interestingly, the absolute number of people living in rural areas in the 1990s is roughly equal to the numbers in the 1960s. Yet rural life has definitely changed. The links between rural and urban areas are much improved, making rural folks less isolated. In addition to village-based subsistence production, in most rural areas highly mechanized capital-intensive farming occurs. Much like the region's cities, the rural landscape is divided by extremes of poverty and wealth. The root of social and economic tension in the countryside is the uneven distribution of arable land.

Rural Landholdings The control of land in Latin America was the basis for political and economic power. Historically, colonial authorities granted large tracts of land to the colonists, who were also promised the service of Indian laborers as part of the **encomienda** system. These large estates typically took up the best lands along the valley bottoms and coastal plains. The owners were often absentee landlords, spending most of their time in the city and relying on a mixture of hired, tributary, and slave labor to run their operations. Passed down from one generation to the next, many estates can trace their ownership back several centuries. The allocation of large blocks of land also denied peasants access to land, so they were then forced to labor for the estates. This entrenched pattern of large estates is called **latifundia.**

Although the pattern of estate ownership is well documented, peasants have always farmed small plots for their subsistence. The practice of **minifundia** can lead to permanent or shifting cultivation. Small farmers typically plant a mixture of crops oriented to subsistence as well as distant markets. Peasant farmers in Colombia or Costa Rica, for example, produce corn, fruits, and various vegetables alongside coffee bushes that produce beans for export. Strains on the minifundia system occur when demographic pressures create land scarcity or political elites reallocate open land for their needs. Much of the turmoil in twentieth-century Latin America surrounded the question of land, with peasants demanding its redistribution through the process of agrarian reform. Governments have addressed these concerns in different ways. The Mexican revolution in 1910 yielded a system of communally held lands called **ejidos.** In the 1950s Bolivia crafted agrarian reform policies that led to the expropriation of estate lands and their reallocation to small farmers. Both of these programs suffered resistance and proved to be politically costly. Eventually, the path chosen by most governments was to make frontier lands available to land-hungry peasants. The opening of tropical frontiers, especially in South America, was a widely practiced strategy that changed national settlement patterns and began waves of rural-to-rural migration and even urban-to-rural migration.

Agricultural Frontiers The creation of agricultural frontiers served several purposes: providing peasants with land, tapping unused resources, and shoring up political boundaries. Frontier colonization efforts in South America are some of the more noteworthy cases. In addition to settlement along Brazil's Trans-Amazon highway, Peru developed its Carretera Marginal (Marginal Highway) in the 1960s in an effort to lure colonists into the cloud and rain forests of eastern Peru. In Bolivia, Colombia, and Venezuela, agricultural frontier schemes in the lowland tropical plains attracted peasant farmers and large-scale investors. Mexico sent colonists, some displaced by dam construction, into the forests of Tehuantepec. Guatemala developed its northern Petén region. El Salvador had no frontier left, but many desperately poor Salvadorans poured into the neighboring states of Honduras and Belize in search of land. In short, although the dominant demographic trend has been a rural-to-urban movement, an important rural-to-rural flow has changed previously virgin areas into agricultural communities.

In the 1960s Brazil began a period of rapid frontier expansion by constructing several new Amazonian highways, a new capital (Brasília), and state-sponsored mining operations. It was the Brazilian military who directed the opening of the Amazon to provide an outlet for landless peasants and to extract the region's many resources. Yet the generals' plans did not deliver as intended. Throughout the basin thin forest soils were incapable of supporting permanent agricultural colonies and in the worst cases degraded into baked claylike surfaces devoid of vegetation. Government-promised land titles, agricultural subsidies, and credit were slow to reach small farmers, even if they were fortunate enough to be given land on *terra roxa,* a nutrient-rich purple clay soil. Instead, a disproportionate amount of money went to subsidizing large cattle ranches through tax breaks and improvement deals where "improved" land meant cleared land. Despite a concerted effort by Brazilian officials to settle the entire region, most commercial activities and residents are concentrated in a few sites.

Nearly four times more people lived in the Amazon in the 1990s than in the 1960s. The 10 million people living in the Brazilian Amazon in the mid-1990s is projected to reach 15 million by 2010; thus, increased human modification of the Brazilian Amazon is inevitable. As Figure 4.18 shows, however, much of the population growth is concentrated in the states of Pará and Rondônia, the port towns of Belém and Manaus, and the mining centers of Marabá and Boa Vista. Since the Brazilian Amazon is larger than India and 60 percent of its inhabitants live in cities, it seems that portions of it can be conserved or managed without limiting the potential of the region's inhabitants to make a living. Certainly the creation of extractive reserves for rubber tappers in the Acre area is a creative strategy to support sustainable economic uses of the forest.

Population Growth and Movements

The high growth rates in Latin America throughout the twentieth century are attributed to natural increase as well as immigration. The 1960s and 1970s were decades of tremendous growth due to high fertility rates and increasing life expectancy. In the 1960s, for example, a typical Latin American woman had six to seven children. By the 1980s the TFR (total fertility rate) was half this in some countries. As Table 4.1 shows, the 1998 TFR for Brazil was 2.5, Colombia 3.0, and 3.1 for Mexico. A number of factors explain this: more urban families, which tend to be smaller than rural ones; increased participation of women in the workforce; higher education levels of women; state support of family planning, and better access to birth control. The exceptions to this trend are the poor and more rural countries, such as Guatemala and Bolivia,

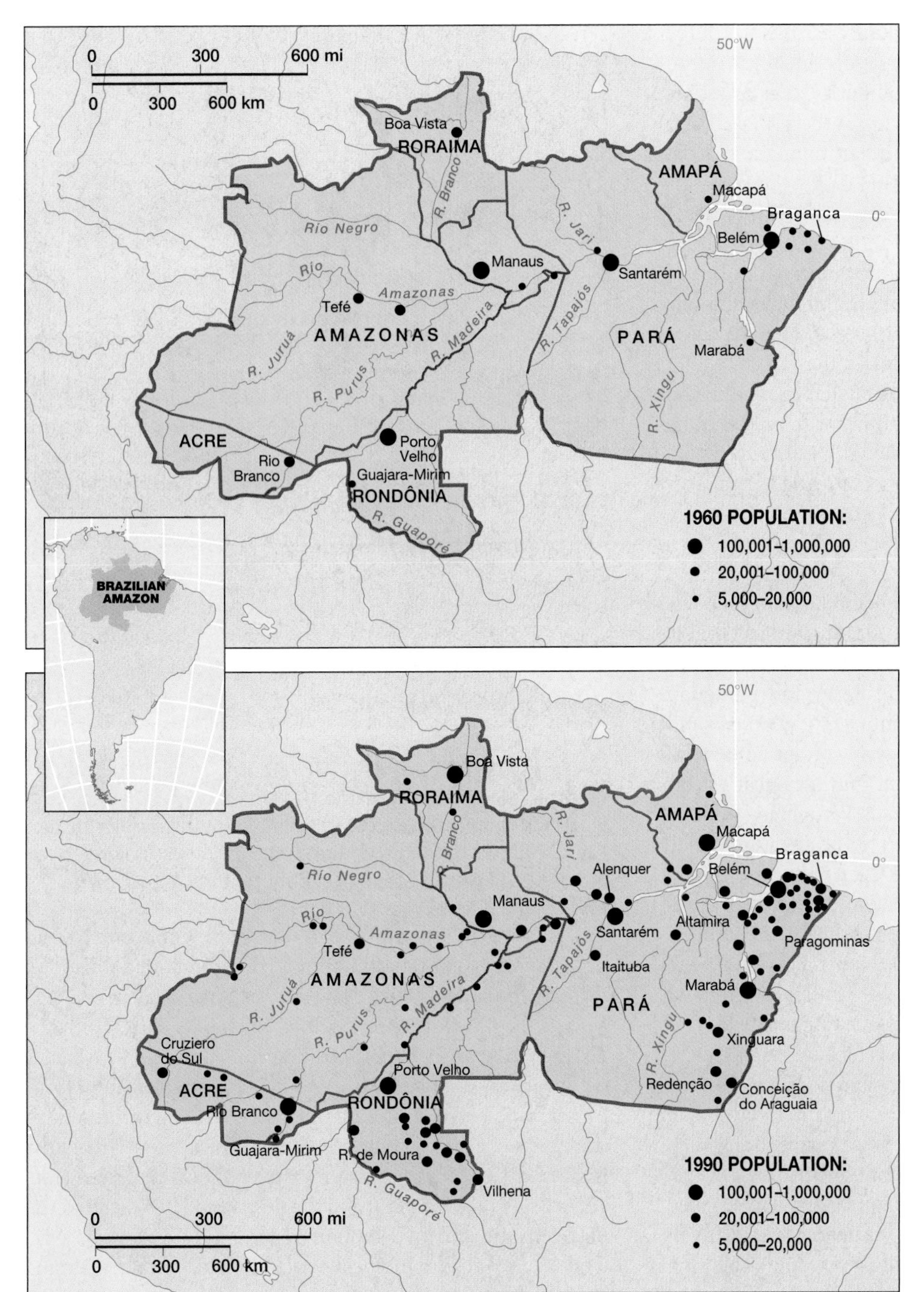

◀ Figure 4.18 Urbanization in the Brazilian Amazon These two maps illustrate the pace of urbanization in the Brazilian Amazon. In 1960 only four cities of 100,000 or more people existed. By 1990 there were eight such cities and dozens of new urban settlements with more than 5,000 people. Much of the urban growth occurred in the states of Pará and Rondônia. Instead of becoming the agricultural frontier that planners dreamed of in the 1960s, the majority of people in the Brazilian Amazon live in cities. *(Reprinted from Knox and Marston,* Human Geography, *©1998, Prentice Hall, Inc., Upper Saddle River, NJ)*

where TFR hovers around five children per woman. Cultural factors may also be at work, as Amerindian peoples in the region tend to have more children.

Even with family sizes shrinking, nearing replacement rates in Uruguay and Chile, there is built-in potential for continued growth because of the relative demographic youth of these countries. The average percentage of the population below the age of 15 is 35 percent. In North America the similar cohort is 22 percent of the population and in Western Europe it is just 18 percent. This means that a proportionally larger segment of the population has yet to enter into its child-bearing years.

Waves of immigrants into Latin America and migrant streams within Latin America have influenced population size and patterns of settlement. Beginning in the late nineteenth century, new immigrants from Europe and Asia added to the region's size and ethnic complexity. Important population shifts within countries have also occurred in recent decades, as witnessed by the growth of Mexican border towns or the demographic expansion of the Bolivian plains. In an increasingly globalized economy, even more Latin Americans live and work outside the region, especially in the United States.

European Migration After gaining their independence from Iberia, Latin America's new leaders sought to develop their territories through immigration. Firmly believing in the dictum "to govern is to populate," many countries set up immigration offices in Europe to attract hard-working peasants to till the soils and "whiten" the **mestizo** (people of mixed European and Indian ancestry) population. Southern Cone countries such as Argentina, Chile, Uruguay, and southern Brazil were the most successful in attracting European immigrants from the 1870s until the depression of the 1930s. During this period, some eight million Europeans arrived (more than came during the entire colonial period), with Italians, Portuguese, Spaniards, and Germans being the most numerous. Some of this immigration was state-sponsored, such as the nearly one million laborers (including entire families) brought to the coffee estates surrounding São Paulo at the turn of the century. Other migrants came seasonally, especially the Italian peasants who left Europe in the winter for agricultural work in Argentina and were thus nicknamed "the swallows." Still others paid their own passage, intending to permanently settle and prosper in the growing commercial centers of Buenos Aires, São Paulo, Montevideo, and Santiago.

Asian Migration Less well known are the Asian immigrants who also arrived during this period. Although considerably fewer in number, over time they established an important presence in the large cities of Brazil, Peru, and Paraguay. Beginning in the mid-nineteenth century, the Chinese and Japanese who settled in Latin America were contracted to work on the coffee estates in southern Brazil, and the sugar estates and guano mines of Peru. In the 1990s Peruvians elected a Japanese-Peruvian, Alberto Fujimori, to be president. The Japanese in Brazil are the most studied Asian immigrant group. Between 1908 and 1978, a quarter-million Japanese immigrated to Brazil; today the country is home to 1.3 million people of Japanese descent. Initially, most Japanese were landless laborers; yet by the 1940s they had accumulated enough capital so that three-quarters of the migrants had their own land in the peripheral areas of São Paulo and Paraná states. As a group, the Japanese have been closely associated with the expansion of soybean and orange production. Today Brazil leads the world in exports of orange juice concentrate, with most of the oranges grown on Japanese-Brazilian farms. Increasingly, second- and third-generation Japanese have taken professional and commercial jobs in Brazilian cities; many of them have married outside their ethnic group and are losing their fluency in Japanese. Even with intermarriage, they continue to maintain a strong sense of their ethnic identity with more than two-thirds of Japanese-Brazilians speaking both Portuguese and some Japanese (Figure 4.19).

The latest Asian immigrants are from South Korea. Unlike their predecessors, most of the Korean immigrants came with enough capital to invest in small business and settled in cities rather than in the countryside. According to official South Korean statistics, 120,000 Koreans went to Paraguay between 1975 and 1990. While many have stayed in Paraguay, there seems to be a pattern of immigration to Brazil. Recent Korean immigrants in São Paulo have created more than 2,500 small businesses. Unofficial estimates of the number of Koreans living in Brazil range from 40,000 to 120,000. As a group they are decidedly commercial in orientation and urban in residence; their cities of choice are Asunción and Ciudad del Este in Paraguay and São Paulo in Brazil.

▲ **Figure 4.19 Japanese-Brazilians** Retired Japanese-Brazilians play the board game "Go" in a city plaza. Most Brazilians of Japanese ancestry live in the southern states of São Paulo, Paraná, and Santa Catarina. The majority immigrated to Brazil in the first half of the twentieth century. *(Gary Payne/ Liaison Agency, Inc.)*

Latino Migration and Hemispheric Change Movement within Latin America and between Latin America and North America has had a significant impact on sending and receiving communities alike. Within Latin America international migration is shaped by shifting economic and political realities. Thus Venezuela's oil wealth during the 1960s and 1970s attracted between one and two million Colombian immigrants who tended to work as domestics or agricultural laborers. Argentina has long been a destination for Bolivian and Paraguayan laborers. And of course, farmers in the United States have depended on Mexican laborers for most of the twentieth century.

Political turmoil also sparked waves of international migrants: Chilean intellectuals fled to neighboring countries in the 1970s when General Pinochet wrested power from the socialist government led by Salvador Allende. Nicaraguans likewise fled when the socialist Sandanistas came to power in the 1980s. The bloody civil wars in El Salvador and Guatemala sent waves of refugees into neighboring countries such as Mexico and the United States. With democratization on the rise in the region, many of today's immigrants are classified as economic migrants, not political asylum seekers.

Presently, Mexico is the largest country of origin of legal immigrants to the United States, followed by the Philippines,

▲ **Figure 4.20 Mexican–U.S. border crossing** Mexican day workers cross the border into El Paso, Texas, from Ciudad Juárez. Mexicans have long used these busy border crossings to enter the United States. Other Latino immigrants, especially from Central America, now join them. *(Rob Crandall/ Rob Crandall Photography)*

China, Korea, and Vietnam. Three million Mexicans immigrated to the United States legally during the 1980s and early 1990s, accounting for one-quarter of all legal immigration. In addition, Mexicans were estimated to make up 30 percent of the 3.4 million illegal immigrants residing in the United States in 1992. Mexican labor migration to the United States dates back to the late 1800s when relatively unskilled labor was recruited to work in agriculture, mining, and railroads. This practice was formalized in the 1940s through the 1960s with the "Bracero Program," which authorized temporary employment residence to five million Mexican laborers (much like the "guest workers" recruited from Southern Europe and Turkey by West Germany). Today roughly 60 percent of the Hispanic population (both foreign-born and native-born) in the United States claims Mexican ancestry. Mexican immigrants are concentrated in two states, California and Texas. Although Mexicans continue to have the greatest presence among Latinos in the United States, the number of immigrants from El Salvador, Guatemala, Colombia, and Brazil has grown since the 1980s (Figure 4.20).

Population shifts from one country to another change both demographic and cultural patterns. The cultural complexity of Latin America is attributable, in part, to immigration; today's emigrants from Latin America are weaving their culture into the fabric of North American and European societies. Interestingly, many migrants maintain close contact with their home countries, a phenomenon that some scholars have labeled **transnationalism.** A cultural and economic outcome of globalization, transnationalism highlights the social and economic links that form between home and host countries. Technological advances that make communication both faster and cheaper, as well as improved banking and courier services, allow immigrants to maintain contacts with their home countries in ways that earlier generations could not. Among these transnational migrants, a dual or hybrid identity is maintained, which also is seen as a cultural expression of globalization. Thus Ecuadorians working in New York, for example, maintain regular contact with their rural villages in Ecuador while also developing vital Ecuadorian social networks in New York City (Figure 4.21).

Patterns of Cultural Coherence and Diversity: Repopulating a Continent

The Iberian colonial experience imposed a political and cultural coherence on Latin America that makes it distinguishable today as a world region. Yet this was not a simple transplanting of Iberia across the Atlantic. Often a syncretic process unfolded in which European and Indian traditions blended as indigenous groups were subsumed into either the Spanish or the Portuguese empires. In some areas such as southern Mexico, Guatemala, Bolivia, and Peru, Indian cultures have showed remarkable resilience, as evidenced by the survival of Amerindian languages. Yet the prevailing pattern is one of forced assimilation in which European religion, language, and political organization was imposed on the surviving fragments of native society. Later other cultures, arriving as both forced and voluntary migrants, added to the region's cultural mix. Perhaps the single most important factor in the dominance of European culture in Latin America was the demographic collapse of native populations.

Demographic Collapse

It is hard the grasp the enormity of cultural change and human loss due to this cataclysmic encounter between two worlds. Throughout the region archaeological sites are poignant reminders of the complexity of pre-contact civilizations. Dozens of stone temples found throughout Mexico and Central America, where the Mayan and Aztec civilizations flourished, attest to the ability of these societies to thrive in the area's tropical forests and upland plateaus. In the Andes,

▲ **Figure 4.21 Ecuadorians in New York City** A band of Ecuadorian musicians from Otavalo entertain New Yorkers with traditional Andean songs. Ecuadorians, along with Colombians, Brazilians, and Dominicans, are one of many Latino immigrant groups adding to New York City's ethnic diversity. *(Martha Cooper)*

stone terraces built by the Incas are still being used by Andean farmers; earthen platforms for village sites and raised fields for agriculture are still being discovered and mapped. Ceremonial centers such as Cuzco—the core of the great Incan empire that was nearly leveled by the Spanish—and the Incan site of Machu Picchu—not discovered by Europeans until the early 1900s—are evidence of the complexity of pre-contact civilizations (Figure 4.22). The Spanish, too, were impressed by the sophistication and wealth they saw around them, especially in the incomparable Tenochtitlán, where Mexico City sits today. Tenochtitlán was the political and ceremonial center of the Aztecs, supporting a complex metropolitan area with some 300,000 residents. By comparison, the largest city in Spain at the time was considerably smaller.

The most telling figures of the impact of European expansion are demographic. It is widely believed that pre-contact America had 54 million inhabitants; by comparison, Western Europe in 1500 had approximately 42 million. Of the 54 million, about 47 million were in Latin America and the rest were in North America and the Caribbean. There were two major population centers: one in Central Mexico with 14 million people and the other in the Central Andes (highland Peru and Bolivia) with nearly 12 million. By 1650, after a century and a half of colonization, the indigenous population was one-tenth its pre-contact size. The human tragedy of this population loss is difficult to comprehend. The relentless elimination of 90 percent of the indigenous population was largely caused by epidemics of influenza and smallpox, but warfare, forced labor, and starvation due to a collapse of food production systems also contributed to the death rate.

▲ **Figure 4.22 Machu Picchu** This complex, ancient city near Cusco was not discovered by Europeans until the early 1900s. Located above the humid Urubamba River valley, Machu Picchu is one of Peru's major tourist destinations. *(Rob Crandall/Rob Crandall Photography)*

The tragedy of conquest did not end in 1650, the population low point for Amerindians, but continued throughout the colonial period and to a much lesser extent continues even today. Even after the indigenous population began its slow recovery in the Central Andes and Central Mexico, there were still tribal bands in southern Chile (the Mapuche) and Patagonia (Araucania) that experienced the ravages of disease three centuries after Columbus landed. Even now, the isolation of some Amazonian tribes has made them vulnerable to disease. Conflicts with outsiders who invade their territories in search of land or gold still occur. In an all too familiar story, a common cold can prove deadly to forest dwellers who lack the needed immunities to defend themselves.

The Columbian Exchange Historian Alfred Crosby likens the contact period between Europe (Old World) and the Americas (New World) as an immense biological swap, which he terms the **Columbian exchange.** According to Crosby, Europeans benefited greatly from this exchange, and Amerindian peoples suffered the most from it. The human ecology of both sides of the Atlantic, however, was forever changed through the introduction of new diseases, peoples, plants, and animals. Take, for example, the introduction of Old World crops. The Spanish, naturally, brought their staples of wheat, olives, and grapes to plant in the Americas. Wheat did surprisingly well in the highland tropics and became a widely consumed grain over time. Grapes and olive trees did not fare as well; eventually grapes were produced commercially in the temperate zones of South America. The Spanish grew to appreciate the domestication skills of Indian agriculturalists who had developed valuable starch crops such as corn, potatoes, and bitter manioc, as well as exotic condiments such as hot peppers, tomatoes, pineapple, cacao, and avocados. Corn never became a popular food for Europeans (with the exception of Romania), but over time it became a feed crop of some importance. After initial reluctance, northern and eastern Europeans widely consumed the potato as a basic food. Domesticated in the highlands of Peru and Bolivia, the humble potato has an impressive ability to produce a tremendous volume of food in a very small area, even when climatic conditions are not ideal. This root crop is credited for driving Europe's rapid population increase in the eighteenth century when peasant farmers from Ireland to Poland became increasingly dependent on this one basic food. This potato dependence also made them vulnerable to potato blight, a fungal disease that emerged in the nineteenth century and came close to unraveling Irish society.

Tropical crops transferred from Asia and Africa reconfigured the economic potential of the region. Sugar cane became the dominant cash crop of the Caribbean and the Atlantic tropical lowlands of South America. With sugar production came the importation of millions of African slaves. Coffee, a later transfer from East Africa, emerged as one of the leading export crops throughout Central America, Colombia, Venezuela, and Brazil in the nineteenth century. Introduced African pasture grasses enhanced the forage available to livestock.

The movement of Old World animals across the Atlantic had a profound impact on the Americas. Initially these animals hastened Indian decline by introducing animal-borne disease and by producing feral offspring that consumed everything in their path. For the most part, native agriculturalists did not have to contend with grazing animals, and their vast gardens of corn, beans, and squash proved to be attractive fodder for the rapidly multiplying swine, cows, and horses. The utility of these animals was eventually appreciated by native survivors. Draft animals were adopted, as was the plow, which facilitated the preparation of soil for planting. Wool became a very important fiber for indigenous communities in the uplands. And, slowly, pork, chicken, and eggs added protein and diversity to the staple diets of corn, potatoes, and cassava. Ironically, the horse, which was a feared and formidable weapon of the Europeans, became a tool of resistance in the hands of skilled riders who inhabited the plains of the Chaco and Patagonia. Much like native peoples of North America, these tribal groups challenged European conquest by using their horsemanship in combat or for flight. With the major exception of disease, many transfers of plants and animals ultimately benefited both sides of the Atlantic. Still, it is clear that the ecological and material basis for life in Latin America was completely reworked through this exchange process initiated by Columbus.

Indian Survival Presently, Mexico and Guatemala in the north and Ecuador, Peru, and Bolivia in the south have the largest indigenous populations. Not surprisingly, these are the areas that had the densest native population at contact. Indigenous survival also occurs in isolated settings, where the workings of national and global economies are slow to penetrate. The Sierra de Santa Marta in Colombia, home of the Kogi, or the Gran Sabana in Venezuela where the Pemon live are examples of relatively small groups that have managed to maintain a distinct Indian way of life despite pressures to assimilate.

In many cases Indian survival comes down to one key resource—land. Indigenous peoples who are able to maintain a territorial home, formally through land title or informally through long-term occupance, are more likely to preserve a distinct ethnic identity. Because of this close association between identity and territory, native peoples are increasingly insisting on a recognized space within their respective countries. Yet these efforts to define indigenous territory are seldom welcomed by the state.

Today the state of Panama recognizes four **comarcas** that encompass six different native groups; the most successful is Comarca San Blas on the Caribbean coast where 40,000 Kuna live. A comarca is a loosely defined territory similar to a province or a homeland. What distinguishes the comarca from an Indian reservation is that the native people have defined the territory and assert political and resource control within its boundaries. From Amazonia to the highlands of Chiapas, many native groups are demanding formal political and territorial recognition as a means to redress centuries of injustice. Whether these efforts will actually reshape political and cultural space for Latin America's Indians is still uncertain (see "Mapping for Indigenous Survival").

Patterns of Ethnicity and Culture

The Indian demographic collapse enabled Spain and Portugal to refashion Latin America into a European likeness. Yet instead of creating a neo-Europe in the tropics, a complex ethnic blend evolved. Beginning with the first years of contact, unions between European sailors and Indian women began the process of racial mixing that over time became a defining feature of the region. The courts of Spain and Portugal officially discouraged racial mixing, but not much could be done about it. Spain, which had a far larger native population with which to deal than the Portuguese in Brazil, became obsessed with the matter of race and maintaining racial purity among its colonists. An elaborate classification system was constructed to distinguish emerging racial castes. Thus in Mexico in the eighteenth century, a Spaniard and an Indian union resulted in a mestizo. An offspring of a mestizo and a Spanish woman was a castizo. However the children from a castizo woman and a Spaniard were considered a Spaniard in Mexico but a quarter mestizo in Peru. Likewise mulattoes were the progeny of European and African unions and zambos were the offspring of African and Indian.

After generations of intermarriage, such a classification system collapsed under the weight of its complexity and four broad categories resulted: blanco (European ancestry) mestizo (mixed ancestry), indio (Indian ancestry), and negro (African ancestry). The blancos (or Europeans) continue to be well represented among the elites, yet the vast majority of people are of mixed racial ancestry. In Venezuela, for example, the phrase "cafe con leche" (coffee with milk) is used to describe the racial makeup of the majority of the population who share European, African, and Indian characteristics. Throughout Latin America, more than other regions of the world, miscegenation (or racial mixing) is the norm, which makes the process of mapping racial or ethnic groups especially difficult.

Languages Roughly two-thirds of Latin Americans are Spanish speakers and one-third speak Portuguese. These colonial languages were so prevalent by the nineteenth century that they were the unquestioned languages of government and instruction for the newly independent Latin American republics. In fact, until recently many countries actively discouraged, and even repressed, Indian tongues. It took a constitutional amendment in Bolivia in the 1990s to legalize native-language instruction in primary schools and to recognize the country's multiethnic heritage (more than half the population is Indian and the languages Quechua, Aymara, and Guaraní are widely spoken)(Figure 4.23).

Because Spanish and Portuguese dominate, there is a tendency to neglect the influence of indigenous languages in the region. Mapping the use of indigenous languages, however, yields important pockets of Indian resistance and survival. Despite the insistence on European languages by the colonial authorities, the use of Indian languages persisted. In the Central Andes of Peru, Bolivia, and southern Ecuador, more than 10 million people still speak Quechua and Aymara, along with their knowledge of Spanish. In Paraguay and lowland Bolivia there are four million Guarani speakers, and in southern

Mapping for Indigenous Survival

Within and around the Rio Plátano Biosphere Reserve in northeastern Honduras are the communities of Miskito, Pech, Garífuna, and ladinos. This is the largest area of road-free rain forest in Central America. For centuries indigenous people have inhabited this area, maintaining its biological diversity while extracting resources through traditional economic activities of hunting, gathering, slash-and-burn agriculture, and fishing.

Through participatory mapping techniques using global positioning systems and local knowledge of resource zones, indigenous people are outlining their communal land-use zones (the resource area needed for subsistence). With technical assistance by geographer Peter Herlihy, the "resourcesheds" that native groups rely on were mapped (Figure 1). By mapping communal land-use zones, indigenous claims to this area are being articulated in a powerful but nonconfrontational way. Potential problem areas between groups are readily apparent. For example, the several ethnic groups rely on the Tinto-Ibans at the park's northern border. More importantly, mapping indigenous land use acknowledges the Amerindian presence in and around the park.

The biggest challenge to maintaining the biosphere's biological and cultural diversity is dealing with some 6,000 *ladino* settlers (the Honduran term for mestizos) on the southwestern perimeter who are clearing land to create permanent agricultural colonies. One strategy is to legally secure the territorial claims of indigenous people within the biosphere and also recognize the rights of existing ladino settlements outside of the biosphere zone while discouraging further expansion into the park itself. A new participatory mapping initiative forms one component of a Honduran-German project to protect and manage the Rio Plátano Biosphere Reserve. Through this process, indigenous and ladino communities have come together to define their own land-use zoning system and management guidelines.

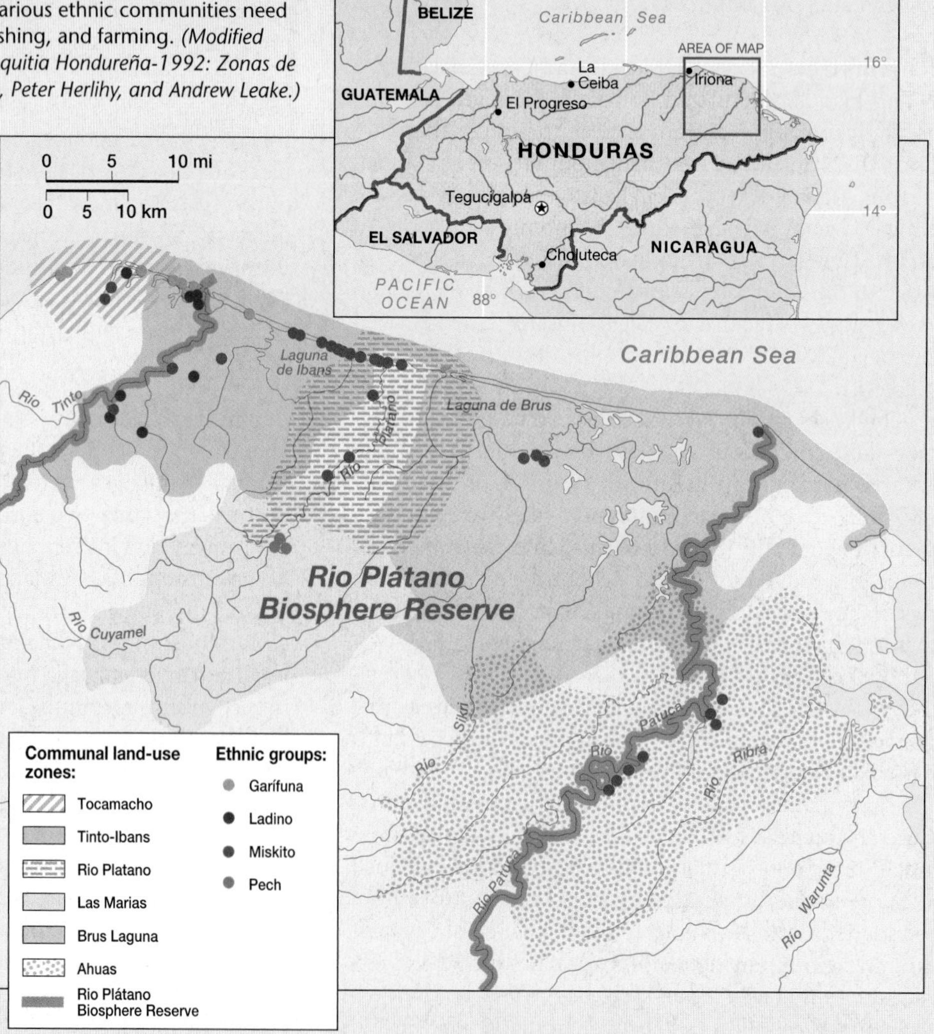

▶ **Figure 1 Communal Land-use Zones and the Rio Plátano Biosphere Reserve** Various ethnic communities need land for subsistence hunting, fishing, and farming. *(Modified from "Tierra Indigenas de la Mosquitia Hondureña-1992: Zonas de Subsistencia" by Mopawi, Masta, Peter Herlihy, and Andrew Leake.)*

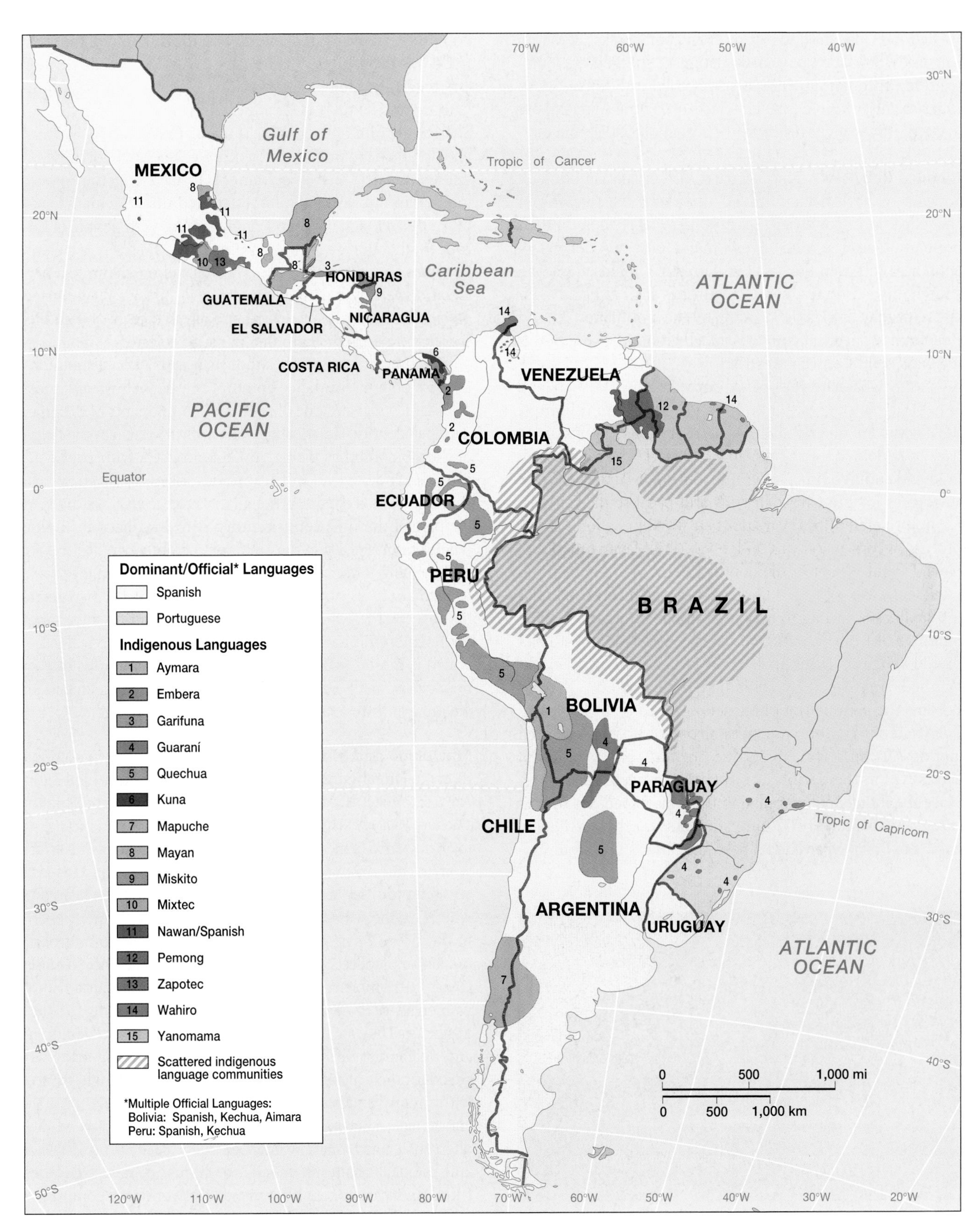

▲ **Figure 4.23 Language map of Latin America** The dominant languages of Latin America are Spanish and Portuguese. Nevertheless, there are significant areas where indigenous languages persist and, in some cases, are recognized as official languages. Smaller language groups exist in Central America, the Amazon Basin, and southern Chile. *(Adapted from the* Atlas of the World's Languages, *1994, New York: Routledge)*

Mexico and Guatemala at least six to eight million speak Mayan languages. Small groups of native language speakers are found scattered throughout the sparsely settled interior of South America and the more isolated forests of Central America, but many of these languages have fewer than 10,000 speakers.

Blended Religions Like language, the Roman Catholic faith appears to have been imposed upon the region without challenge. Most countries report 90 percent or more of their population as Catholic. Every major city has dozens of churches, and even the smallest hamlet maintains a graceful church on its central square. In countries such as El Salvador and Uruguay, a sizable portion of the population attend Protestant evangelical churches, but the Catholic core of this region is still intact (Figure 4.24).

Exactly what native peoples absorbed of the Christian faith is unclear. Throughout Latin America, **syncretic religions,** the blending of different belief systems, enabled animist practices to be folded into Christian worship. These blends took hold and endured, in part because Christian saints were easy surrogates for pre-Christian gods and because the Catholic Church tolerated local variations in worship as long as the process of conversion was under way. The Mayan practice of paying tribute to spirits of the underworld seems to be replicated today in Mexico and Guatemala via the practice of building small cave shrines to favorite Catholic saints and leaving offerings of fresh flowers and fruits. One of the most celebrated religious icons in Mexico is Guadeloupe, a dark-skinned virgin seen by an Indian shepherd boy; she has become the patron saint of Mexico.

Syncretic religious practices also evolved and endured among African slaves. By far the greatest concentration of slaves was in the Caribbean, where slaves were used to replace the indigenous population that was wiped out by disease (see Chapter 5). Within Latin America the Portuguese colony of Brazil received the most Africans—at least four million. In Brazil, where the volume and the duration of the slave trade was the greatest, the transfer of African-based religious and medical systems is most evident. West-African-based religious systems such as Batuque, Umbanda, Candomble, and Shango are often mixed with or ancillary to Catholicism and widely practiced in Brazil. So accurate were some of these religious transfers that it is common to have Nigerian priests journey to Brazil to learn forgotten traditions. In many parts of southern Brazil, Umbanda is as popular with people of European ancestry as with Afro-Brazilians. Typically, a person becomes familiar with Umbanda after falling victim to a magician's spell by having some object of black magic buried outside his or her home. In order to regain control of their life, the victim needs the help of a priest or priestess.

▲ **Figure 4.24 The Catholic Church** Churches, such as the Dolores Church in Tegucigalpa, are important religious and social centers. The vast majority of people in Latin America define themselves as Catholic. Many churches built in the colonial era are valued as architectural treasures and are beautifully preserved. *(Rob Crandall/Rob Crandall Photography)*

The syncretic blend of Catholicism with African traditions is most obvious in the celebration of carnival, Brazil's most popular festival and one of the major components of Brazilian national identity. The three days of carnival known as the Reign of Momo combines Christian Lenten beliefs with pagan influences and African musical traditions epitomized by the rhythmic samba bands. Although the street festival was banned for part of the nineteenth century, Afro-Brazilians in Rio de Janeiro resurrected it in the 1880s with nightly parades, music, and dancing. Fifty years later the street festival had given rise to formalized samba schools and helped break down racial barriers. By the 1960s, carnival became an important symbol for Brazil's multiracial national identity. Today, the festival—which is most associated with Rio—draws thousands of participants from all over world, although increased incidences of violent crime and robbery have tarnished its image.

Machismo and Marianismo Cultural traits often ascribed to men **(machismo)** and women **(marianismo)** in Latin America reflect the influence of Iberia and the Catholic faith. The term *macho* (which means "male") is widely used in English to characterize men who have great faith in their ability to attract women, but machismo in the Latin American context is also about honor, risk-taking, and self-confidence. Thus someone who is macho would not allow his authority to be questioned, in the home or in public. The female cultural counterpart, marianismo, reflects the ideal woman and is a reference to María (Mary), the mother of Jesus Christ. Following marianismo, women strive to be patient, loving, gentle, and willing to suffer in silence. They are keepers of the home, nurturers of children, and deferential to their husbands. Women are also acknowledged for their higher moral authority and, accordingly, treated with dignity and respect. The traditional standards for male and female behavior that are a result of these stereotypes are often challenged. Many women, for example, have professional and political positions outside the home. In recent decades Latin American women have entered the workforce, comprising over one-third of all workers.

The Global Reach of Latino Culture

Latin American culture, vivid and diverse as it is, is widely recognized throughout the world. Whether it is sultry pulse of the tango or the fanaticism in which Latinos embrace soccer as an art form, aspects of Latin American culture have been absorbed into a globalizing world culture. For example, Latin American writers such as Gabriel García Marquez or Isabel Allende have obtained worldwide recognition. As for popular culture, various musical styles as well as telenovelas (soap operas) are heard and seen around the world.

Telenovelas Popular nightly soap operas are a mainstay of Latin American television. These tightly plotted series are filled with intrigue and double deals. Unlike their counterparts in the United States, they end, usually after 100 episodes. Once standard fare for the working class, many telenovelas take hold and absorb an entire nation. During particularly popular episodes, the streets are noticeably calm as millions of people tune in to catch up on the lives of their favorite heroines. Brazil, Venezuela, and Mexico each produce scores of telenovelas, but the Mexican ones are international mega-hits.

Televisa, a Mexican production agency, has aggressively marketed its inventory of soap operas to an eager global public. Mexican telenovelas are avidly watched in countries as diverse as Croatia, Russia, China, South Korea, Iran, the United States, France, and throughout Latin America. Predictably scripted as Mexican Cinderella stories, these sagas of poor underclass women (often domestics) falling in love with members of the elite, battling jealous rivals, and ultimately emerging triumphant seem to resonate with fans around the world. When one Mexican telenovela star, Veronica Castro, visited Russia, her plane had trouble arriving at the gate due to the crowds that came to greet her. In addition to their broad appeal, telenovelas are big business, perhaps Mexico's largest international export. While Hollywood and Bombay grind out movies, much of Mexico's entertainment industry is geared toward producing this vernacular and loved art form.

National Identities Viewed from the outside, there is considerable homogeneity to this region, yet distinct national identities and cultures flourish in Latin America. Since the early days of the republics, countries celebrated particular elements from their pasts when creating their national histories. In the case of Brazil, the country's interracial characteristics were highlighted to proclaim a new society in which the color lines between Europeans and Africans ceased to matter. Mexico turned to its Aztec past, celebrating the architectural and cultural achievements of its predecessors while at the same time forging an assimilationist strategy that discouraged surviving indigenous culture and language.

Musical and dance traditions evolved and became emblematic of these new societies: the tango in Argentina and the vallenato and cumbia in Colombia, the mariachi in Mexico, the huaynos in Peru, or the samba in Brazil are easily distinguished styles that often become popular anthems for the nations in the region (Figure 4.25). Literature also reflects the perceived cleavages in identity found in Latin America. Writers such as Isabel Allende, Gabriel García Marquez, Mario Vargas Llosa, Carlos Fuentes, or Jorge Amado situate their stories in their native countries and, in so doing, celebrate the unique characteristics of Chileans, Colombians, Peruvians, Mexicans, or Brazilians. Distinct political cultures evolved, which at times led to expansionist policies that brought neighbors into conflict. The geopolitical dimensions of these intraregional disputes will be discussed later.

Geopolitical Framework: Redrawing the Map

It is Latin America's colonial history, more than its present conditions, that holds this region together. For the first 300 years after Columbus's arrival, Latin America was a territorial prize sought by various European countries but effectively settled by Spain and Portugal. By the nineteenth century, the independent states of Latin America had formed but they continued to experience foreign influence and, at times, overt political pressure, especially from the United States. At other times a more neutral Pan-American vision

(a)

(b)

▲ **Figure 4.25 Musical traditions** (a) Mariachi musicians pose in their performance costumes. Mariachi blends European horn and string instruments with lyrics that celebrate Mexican life. (b) A samba band marches in the streets of Rio de Janeiro. Samba is the quintessential music of carnival and draws inspiration from African rhythmic traditions. *[(a) Werner Bertsch/Bruce Coleman, Inc.; (b) Murilo Dutra/The Stock Market]*

of American relations and hemispheric cooperation has held sway, represented by the formation of the **Organization of American States (OAS).** The present organization was chartered in 1948 but its origins date back to 1889. Yet there is no doubt that the U.S. policies toward trade, economic assistance, political development, and at times military intervention are often seen as compromising to the sovereignty of these states.

Within Latin America there have been cycles of intraregional cooperation and antagonism. Certainly neighboring countries have fought over territory, closed borders, imposed high tariffs, and cutoff diplomatic relations. Even today there are a dozen long-standing border disputes in Latin America that occasionally erupt into armed conflict. The 1990s witnessed a revival in the trade block concept with the formation of **Mercosur** (the Southern Cone Common Market that includes Brazil, Uruguay, Argentina, and Paraguay) and NAFTA (Mexico, United States, and Canada). It is possible that as these economic ties strengthen, these trade blocks could form the basis for a new alignment of political and economic interests in the region.

Iberian Conquest and Territorial Division

Because it was Christopher Columbus who claimed the Americas for Spain, the Spanish were the first active colonial agents in the Western Hemisphere. In contrast, the Portuguese presence in the Americas was an unintentional benefit of the **Treaty of Tordesillas** brokered by the Catholic Pope in 1493–94. At that time Portuguese navigators had charted much of the coast of Africa in an attempt to find a water route to the Spice Islands (Moluccas). With the help of Christopher Columbus, Spain sought a western route to the Far East. With Columbus's discovery of the Americas, Spain and Portugal turned to the Pope for deliberation as to how these new territories would be divided. Without consulting other European powers, the Pope divided the Atlantic world in half—the eastern half containing the African continent was awarded to Portugal, the western half with most of the Americas was given to Spain. This treaty was never recognized by the French, English, or Dutch who also asserted territorial claims in the Americas, but it did provide the legal apparatus for the creation of a Portuguese territory in America—Brazil—which would later become the largest and most populous state in Latin America (Figure 4.26).

The Treaty of Tordesillas presented a number of interesting geographical problems. It was decreed that a line would be drawn 370 leagues west of the Portuguese-controlled Cape Verde islands, yet exactly how long a league was and which island was the starting point was not specified. Today it is estimated that the line falls somewhere between the 48th and 49th west line of longitude, which would award the eastern third of what is now modern-day Brazil to the Portuguese. To complicate matters further, Columbus was convinced that the West Indian islands he discovered were in Asia off the coast of Cathay (China). The existence of the Pacific Ocean, let alone the continents of North and South America, were unknown when the treaty was signed. Thus the treaty established a far larger Spanish colonial territory than any of the participants realized; Spain ultimately extended its claims across the Pacific to the islands of the Philippines.

Six years after the treaty was signed, Portuguese navigator Alvares Cabral inadvertently reached the coast of Brazil on a voyage to southern Africa. The Portuguese soon realized that this territory was on their side of the Tordesillas line. Initially they were unimpressed by what Brazil had to offer; there were no spices or major indigenous settlements. Quickly, however, they came to appreciate the utility of the coast as a provisioning site, as well as a source for Brazil wood used to produce a valuable dye. Portuguese interest in the territory intensified in the late sixteenth century with the development of sugar estates and the expansion of the slave trade and in the seventeenth century with the discovery of gold in the Brazilian interior.

Spain, in contrast, aggressively pursued the conquest and settlement of its new American territories from the very start. After discovering little gold in the Caribbean, by the mid-sixteenth century Spain's energy was directed toward developing the silver resources of Central Mexico and the Central Andes (most notably Potosí in Bolivia). Gradually the economy diversified to include some agricultural exports, cacao and sugar, as well as a variety of livestock. In terms of foodstuffs, the colonies were virtually self-sufficient. Some basic manufactured items such as crude woolen cloth and agricultural tools were also produced, but, in general, manufacturing was forbidden in the Spanish American colonies as a means to kept them dependent on Spain.

Revolution and Independence It was not until the 1800s, with a rise of revolutionary movements between 1810 and 1826, that Spanish authority on the mainland was challenged. Ultimately, elites born in the Americas gained control, displacing the representatives of the crown. In Brazil the evolution from Portuguese colony to independent republic was a slower and less violent process that spanned eight decades (1808–89). In the nineteenth century Brazil was declared a separate kingdom from Portugal with its own monarch but later became a republic.

It was the territorial division of Spanish- and Portugese-America into administrative units that provided the legal basis for the modern states of Latin America (Figure 4.26). The Spanish colonies were first divided into two viceroyalties (New Spain and Peru) and within these were various subdivisions that later became the basis for the modern states. (In the eighteenth century the Viceroyalty of Peru, which included all of Spanish South America, was divided to form three viceroyalties: La Plata, Peru, and New Granada.) Unlike the Portuguese territory of Brazil, which evolved from a colony into a single Republic, the former Spanish colonies experienced fragmentation in the nineteenth century. Prominent among revolutionary leaders was Venezuelan-born Simon Bolívar (Figure 4.27), who advocated his vision of a single state, **Gran Colombia,** which for a short time (1822–30) combined Colombia, Venezuela, Ecuador, and Panama into one political unit.

▲ **Figure 4.26 Shifting political boundaries** The evolution of political boundaries in Latin America begins with the 1494 Treaty of Tordesillas that gave much of the Americas to Spain and a slice of South American to Portugal. The larger Spanish territory was gradually divided into viceroyalties and audiencias that formed the basis for many modern national boundaries. As the 1830 map shows, the borders of these newly independent states were far from fixed. Bolivia would lose its access to the coast, Peru would gain much of Ecuador's Amazon, and Mexico would be stripped of its northern territory by the United States. *(From Lombardi, Cathryn L. and John V. Lombardi.* Latin American History: A Teaching Atlas. *© 1993. Reprinted by permission of the University of Wisconsin Press)*

Similarly, in 1823 the **United Provinces of Central America** was formed to avoid annexation by Mexico. By the 1830s, this union also broke apart into the states of Guatemala, Honduras, El Salvador, Nicaragua, and Costa Rica. Today the former Spanish mainland colonies include 16 states (plus three Caribbean islands) with a total population of 300 million. If the Spanish colonial territory had remained a unified political unit, it would now have the third largest population in the world, following China and India.

Persistent Border Conflicts As the colonial administrative units turned into states, it became clear that the territories were not clearly delimited, especially the borders that stretched into the sparsely populated interior of South America. This would later become a source of conflict as new states struggled to demarcate their territorial boundaries. Numerous border wars erupted in the nineteenth and twentieth centuries, and the map of Latin American has been redrawn many times. Some of the more noted conflicts were: the War of the Pacific (1879–82), in which Chile expanded to the north and Bolivia lost its access to the Pacific; warfare between Mexico and the United States in the 1840s, which resulted in the present border under the Treaty of Hidalgo (1848) and Mexico's loss of its hold over the southwestern United States; and the War of the Triple Alliance (1864–70), the bloodiest war of the post-colonial period, which occurred when Argentina, Brazil, and Uruguay allied themselves to defeat Paraguay in its claim to control the upper Paraná River Basin. It is estimated that the adult male population in Paraguay was reduced by nine-tenths as a result of this latter conflict. Sixty years later, the Chaco War (1932–35) resulted in a territorial loss for Bolivia in its eastern lowlands and a gain for Paraguay. In the 1980s Argentina lost a war with Great Britain over control of the Falkland, or Malvinas, Islands in the South Atlantic. And as recently as 1998, Peru and Ecuador skirmished over a disputed boundary in the Amazon Basin (Figure 4.28).

Outright war in the region is less common than ongoing and nagging disputes over international boundaries. Any of a dozen dormant claims erupt from time to time given the political climate between neighbors. These include Guatemala's claim to Belize, Venezuela and Guyana's border dispute, Ecuador's claims to the Peruvian eastern lowlands, and Chile and Argentina's dispute over the international boundary within the Beagle Channel. Many of these disputes are based on historical territorial claims dating back to the colonial period.

One of the more interesting geopolitical conflicts in the Southern Cone is based on territorial claims to Antarctica. Seven claimant states, including Argentina and Chile as well as Norway, the United Kingdom, France, New Zealand, and Australia, plus five nonclaimant states (including the United States) seek resource rights to Antarctica. Even though an Antarctic Treaty has existed since 1959 stating that the landmass should be used for peaceful purposes, Chile and Argentina have incorporated their Antarctic territorial claims on national maps and even postage stamps. A Chilean president, Pinochet, spent a week touring the Chilean claim in the 1970s. Argentina held a national cabinet meeting on its portion of Antarctica in the 1970s and then sent a pregnant Argentine woman to give birth to the continent's first child. For the most part, these nationalist claims are symbolic. Recent treaties show an international willingness toward cooperation and the banning of commercial exploitation of the continent.

▲ **Figure 4.27 Simon Bolívar** This statue of Simon Bolívar is in the Andean city of Mérida, Venezuela. Such heroic images of Bolívar (The Liberator) are found throughout South America, especially in his native state of Venezuela. *(Rob Crandall/Rob Crandall Photography)*

The Trend Toward Democracy Early in the twenty-first century, most of the 17 countries in this region will celebrate their bicentennial. Compared with most of the developing world, these countries have been independent for a long time. Yet political stability is not a hallmark of the region, among the countries in the region some 250 constitutions have been written since independence, and military coups have been alarmingly frequent. Since the 1980s, however, the trend is toward democratically elected governments, the opening of markets, and broader popular participation in the political process. Where once dictators outnumbered elected leaders, by the 1990s each country in the region had a democratically elected president. (Cuba, the one exception, will be discussed in Chapter 5.)

Democracy may not be enough for the millions frustrated by the slow pace of political and economic reform. In survey after survey, Latin Americans register their dissatisfaction with politicians and the government. Most of the newly elected democratic leaders are also free-market reformers who are quick to eliminate state-backed social safety nets such as food subsidies, government jobs, and pensions. By the 1990s, many of the poor and middle class had grown skeptical as to whether this brand of democracy could make their lives better. The political left, however, has yet to produce an alternative to privatization and market-driven policies. For now the status quo continues, although popular frustration with falling incomes, rising violence, and chronic underemployment are a recipe for political instability (Figure 4.29).

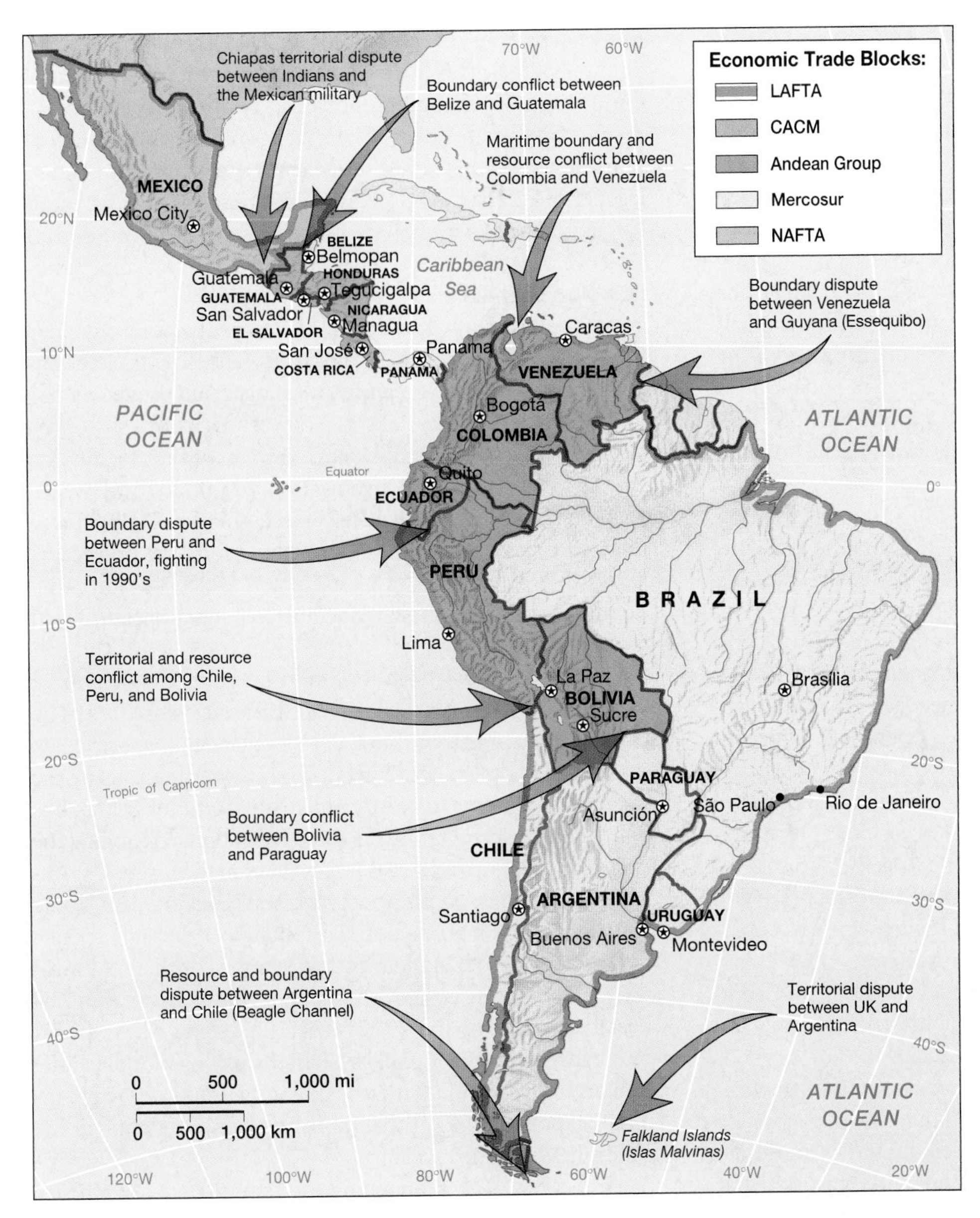

◀ **Figure 4.28 Border disputes and trade blocks in Latin America** Of the five economic trade blocks depicted, Mercosur and NAFTA are the most dynamic. In fact, several members of the Andean Group express interest in joining Mercosur while members of the Central American Common Market would like to be included in NAFTA. Most of Latin America's disputed boundaries are not being actively challenged. Yet these longstanding but dormant border disputes often contribute to tense relations between neighbors. *(Data from Child, 1985.* Geopolitics and Conflict in South America. *New York: Praeger Press; and Allcock, 1992.* Border and Territorial Disputes, *Third Edition. Harlow, Essex, UK: Longman Group)*

Supranational and Subnational Organizations

At the same time that democratically elected leaders struggle to address the pressing needs of their countries, political developments at the supranational and subnational scales pose new challenges to their authority. The most discussed **supranational organizations** (governing bodies that include several states) are the trade blocks. **Subnational organizations** (groups that represent areas or people within the state) often form along ethnic or ideological lines and can provoke serious internal divisions. Indigenous groups seeking territorial recognition (like the Kuna) and insurgent groups espousing Marxist ideology have challenged the authority of the state. Finally, the financial and political force of drug cartels, especially for smaller countries, transcends state boundaries and undermines judicial systems.

Trade Blocks Beginning in the 1960s, regional trade alliances were attempted in an effort to foster internal markets and reduce trade barriers. The Latin American Free Trade

▲ **Figure 4.29 Protesting political violence** Women protestors demand justice for the alleged killing of "disappeared" labor organizers in Honduras. Political violence at the hands of state, paramilitary, and guerrilla organizations is a tragic undercurrent affecting the lives of many Latin Americans. *(Rob Crandall/Rob Crandall Photography)*

Association (LAFTA), the Central American Common Market (CACM), and the Andean Group have existed for decades, but their ability to influence economic trade and growth is limited at best (Figure 4.28). In the 1990s Mercosur and NAFTA emerged as supranational structures that could influence development. For Latin America, the lessons from Mercosur are causing politicians to rethink the value of regional trade.

Mercosur was formed in 1991 with Brazil and Argentina, the two largest economies in South America, with the smaller states of Uruguay and Paraguay as members. Since its formation, trade among these countries has grown tremendously, so that expansion of this trading group appears likely. In 1996 Chile, which previously had declined joining in hopes of being asked to join NAFTA, signed a free trade agreement with Mercosur, making it an associate member. Bolivia, too, is eager to join because it sees Brazil and Argentina as its best potential market for natural gas and grains. Even Venezuela has expressed some interest in joining the trade group. This is significant in two ways: it reflects the growth of these economies and the willingness to put aside old rivalries (especially long-standing antagonisms between Argentina and Brazil) for the economic benefits of cooperation. Mercosur represents a market of 220 million people with a combined GNP of $900 billion and total trade of $175 billion in 1996. The size and productivity of this market has not gone unnoticed. Mercosur's leaders are negotiating directly with the European Union to develop separate trade agreements. With Chile on board, expansion across the Pacific with Asian markets is also likely.

If enthusiasm for Mercosur lasts, transportation and communication between Southern Cone countries will certainly improve. Plans abound to rework fundamentally the flow of goods and communication in this area. As privatization of the telephone companies has benefited telecommunication, major joint engineering projects are being considered. Studies are under way to improve navigation along the Paraná and Paraguay rivers, which are already important arteries for the transport of grains. Other schemes include building a bridge across the Plata River estuary to form a more direct link between the capital cities of Uruguay and Argentina. A tunnel through the Andes between Chile and Argentina is another dream, as well as a regional network of natural gas pipelines that would link Bolivia, Argentina, and southern Brazil. In short, the elaboration of Mercosur could change the way individuals in the member countries relate to one another and think about themselves. Much like the European Union, which fostered a sense of European identity, Mercosur may over time shape a Southern Cone identity.

Insurgencies, Drug Traffickers, and Violence Guerrilla groups such as the Shining Path *(Sendero Luminoso)* in Peru or the FARC (Revolutionary Armed Forces of Colombia) in Colombia have controlled large territories of their countries through the use of patronage, extortion, kidnapping, and violence. In the 1980s the Shining Path terrorized the rural highlands of Peru and caused citywide blackouts in Lima. Thousands of peasants fled the highlands, especially Ayacucho, to live as virtual refugees in Lima. With the capture of the Shining Path's leader in 1992, its destructive force was greatly diminished. In Colombia, however, two major well-armed paramilitary groups exist. Their tactics of kidnapping, extortion, blowing up pipelines, and closing down public roads are legendary. Yet their exact political ambitions are unclear. Rural Colombians are especially vulnerable in this era of political violence. In 1992 the Colombian murder rate was nine times that of the United States.

More powerful than the guerrilla groups, and a greater threat to the state, are the drug cartels. The drug trade began in earnest in the 1970s with Colombia at its center. By the 1980s the Medellín Cartel was a powerful and wealthy crime syndicate that used narco-dollars to bribe or murder anyone who got in its way. The organization was profoundly weakened by the death of its leader Pablo Escobar in 1993, but by then the Cali Cartel was poised to take control. Its power was reduced by key arrests of its leaders in 1995. No one cartel dominates the drug trade today; instead, it has decentralized into dozens of smaller syndicates that are productive and creative. Despite costly efforts to eradicate coca production, more coca was grown in the late 1990s than ever before (see "The Coca Trade").

Billions of dollars are generated for Latin Americans by the sale of illegal drugs to North Americans and Europeans. Initially, most Latin governments cared little about controlling the drug trade as it brought in much-needed hard currency. Within the region drug consumption was scarcely a problem. Some drug lords even became popular folk heroes, lavishly spending money on housing, parks, and schools for their communities. The social cost of the drug trade to Latin America became evident by the 1980s when the region was crippled by a badly damaged judicial system. By paying off police, the military, judges, and politicians, the drug syndicates wield incredible political power that threatens the social fabric of the states in which they work.

Crime, especially kidnapping and homicides, add to the region's social instability. With regard to violent crime, no one country has suffered more than Colombia. Colombia is the world's kidnapping capital, accounting for half of all such crimes in the world and costing Colombians millions in ransom. Colombia's high murder rate also helped to boost the murder rate for all of Latin America to six times the world average in 1996. Drug trafficking and income inequality, especially in the cities, are blamed for the rise in crime. Others blame democratic governments for being too tolerant, arguing for the hard hand of a dictator to eradicate crime. Polls show that concern for personal security—resulting in a robust trade in bulletproof attire—is what worries most voters.

Economic and Social Development: Dependent Economic Growth

Most Latin American economies fit into the broad middle-income category set by the World Bank. Clearly part of the developing world, its people are much better off than those in Sub-Saharan Africa, South Asia, and China. Still the economic

The Coca Trade

Exact numbers are unknown, but for Bolivia, Peru, and Colombia nobody doubts that the coca leaf (the source of cocaine) is the most lucrative agricultural activity. Moreover, this illegal activity absorbs a large segment of the rural population. It is such an important component of the national economy there is little incentive to get rid of it. Coca is a perennial shrub native to the eastern slopes of the Andes. Today several hundred thousand acres of coca are cultivated in Peru's Huallaga valley, the Yungas and Chapare of Bolivia, and, increasingly, in the Amazonian forests of Colombia and Brazil, much of it in biologically diverse rain forest and pre-montane forest. Geographer Ken Young estimates that in Peru alone as many as 2.5 million acres (1 million hectares) of tropical forest were degraded in the 1980s due to coca cultivation.

Coca leaf was domesticated by Andean agriculturalists because of its ability to suppress hunger and fatigue. Chewing coca leaf also reduces nausea and headache brought on by exposure to high altitudes. Mate de coca, a common tea sold in the Andes, is often served to people suffering from altitude sickness. Coca is even an important ingredient for Coca-Cola, which requires decocainized leaves. Yet it was the Colombian cartels in Medellín and Cali that emerged as the important middlemen to connect production zones in the Central Andes with consumers in North America and Europe. The Colombians organized the production of cocaine in clandestine laboratories and developed a labyrinth of shipping techniques and routes, which include everything from stuffing sacks of cocaine into carnation boxes bound for New York to driving carloads of cocaine across the Mexican border.

Other drugs also contribute to local economies. Marijuana is an easily grown weedy plant that is found throughout Mexico, Central America, and Colombia, as well as in other parts of the world. A recent surge in heroin production (from poppies) in Mexico and Ecuador suggests possible competition with poppy growers in Southeast and South Asia. The production of coca, however, is still exclusive to Latin America.

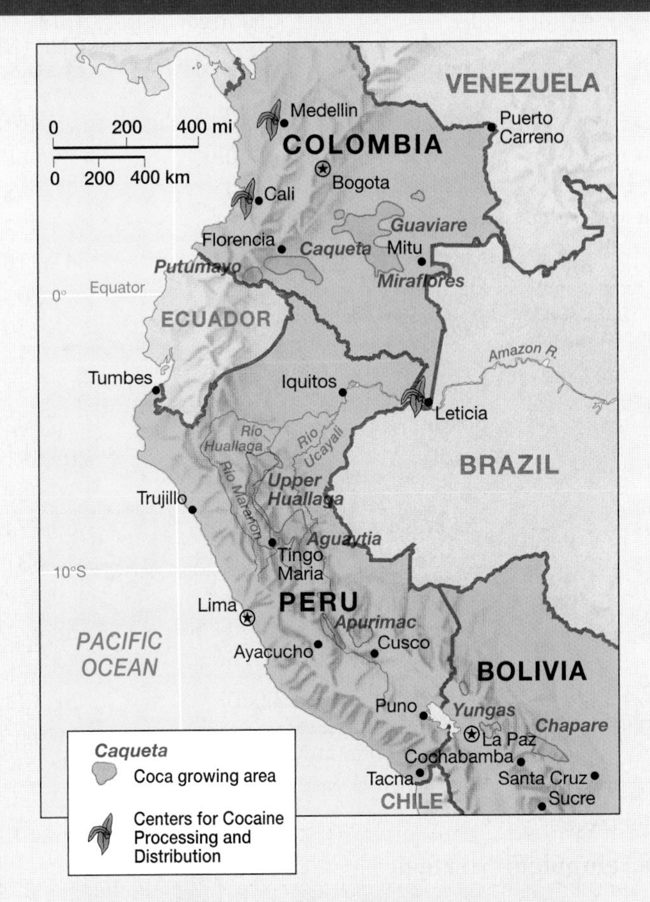

▲ **Figure 1 Coca-growing areas in South America** Although the oldest coca-growing regions are in Bolivia and Peru, Colombian traffickers turned the processing and distribution of cocaine into an international narcotics trade. In the mid-1990s, Peru led the world in coca production, followed by Colombia and Bolivia.

contrasts are sharp, both between states and within them (Table 4.2). Generally, the Southern Cone states (including southern Brazil and excluding Paraguay) are the richest. Argentina leads the way with a per capita GNP of more than $8,000 in 1998. Venezuela and Mexico, Latin America's leading oil producers, also rank well in per capita economic measures, although both slipped in the 1990s. The poorest countries in the region lie in Central America and the Andes. In 1998 Nicaragua, Honduras, and Bolivia had per capita income figures lower than $1,000. Even with its middle income status, extreme poverty is evident throughout the region; some 10–15 percent of Latin Americans live on less than one dollar a day.

The programs of the 1960s and 1970s accelerated industrialization and infrastructural development, but they also fostered debt and rural displacement. All sectors of the economy were radically transformed. Agricultural production increased with the application of green revolution technology and mechanization. State-run industries reduced the need for imported goods, and the service sector ballooned due to government and private jobs. Yet this rush to modernize produced its victims. In rural areas poverty, landlessness, and inadequate systems of credit stymied the productivity of small farmers. As rural people were pushed off the land, there were not enough jobs in industry to absorb them, so they created their own niche in the urban informal economy. In the end, most countries made the transition from predominantly rural and agrarian economies dependent upon one or two commodities to more economically diversified and urbanized countries with mixed levels of industrialization.

Table 4.2 Economic Indicators

Country	GNP per Capita ($U.S., 1996)	Total GNP (Millions of $U.S., 1996)	PPP[a] ($Intl, 1996)	Real Annual Growth % per Capita, 1990–1996
Costa Rica	2,640	9,081	6,470	2.4
El Salvador	1,700	9,868	2,790	3.5
Guatemala	1,470	16,018	3,820	0.5
Honduras	660	4,012	2,130	1.2
Mexico	3,670	341,718	7,660	–0.3
Nicaragua	380	1,705	1,760	–0.2
Panama	3,080	8,249	7,060	3.6
Argentina	8,380	295,131	9,530	3.9
Bolivia	830	6,302	2,860	1.8
Brazil	4,400	709,591	6,340	2.0
Chile	4,860	70,060	11,700	6.4
Colombia	2,140	80,174	6,720	3.0
Ecuador	1,500	17,531	4,730	0.8
Paraguay	1,850	9,179	3,480	–1.5
Peru	2,420	58,671	4,410	4.8
Uruguay	5,760	18,464	7,760	3.8
Venezuela	3,020	67,333	8,130	–0.3

[a]Purchasing power parity.

Source: *The World Bank Atlas,* 1998.

Development Strategies

Since the 1950s Latin America experimented with various development strategies, from closed economies reliant on **import substitution** (policies that foster domestic industry by imposing inflated tariffs on all imports) to state-run nationalized industries and various attempts at agrarian reform. By the 1960s Brazil, Mexico, and Argentina all seemed poised to enter the ranks of the industrialized world. Multilateral agencies such as the World Bank and the Inter-American Development Bank loaned money for big development projects: continental highways, dams, mechanized agriculture, and power plants. Yet the dream of the 1960s became nightmarish by the 1980s. Argentina, which was richer than Japan in 1960, ranked thirty-fourth in the world in 1995. Oil wealth helped to foster a large middle class in Mexico that was badly shaken by the debt crisis in the early 1980s. And Brazil, the tenth largest economy in the world, is the developing world's largest debtor. Brazil also maintains the region's worst income disparity; in 1994 just 1 percent of the population controlled 15 percent of the wealth, whereas 25 percent controlled only 12 percent.

Industrialization Since the 1960s most government development policies emphasized manufacturing. Various strategies have been employed from growth poles, nationalized industries, and import substitution. The results have been mixed. Today at least 15–20 percent of the labor force in Mexico, Argentina, Brazil, Chile, Colombia, Peru, Uruguay, and Venezuela is employed in manufacturing. Yet this is far short of the hoped-for levels of industrialization, especially when one considers the size of the urban population. What is more, the most industrialized areas tend to be around the capitals or in planned **growth poles** such as Ciudad Guayana in Venezuela and the Mexican border cities of Ciudad Juárez and Tijuana (Figure 4.30).

There are cases in which industry flourished in noncapital cities and without direct state support. The cities of Monterrey, Mexico; Medellín, Colombia; and São Paulo, Brazil, all developed important industrial sectors initially from local investment. Long before the city of Medellín (two million people) was associated with cocaine, it was a major center of textile production, indeed more industrialized than the larger capital of Bogotá. A popular image of Medellín's inhabitants—the South American Yankee—surfaced in the 1950s. Celebrated as hard-working and entrepreneurial to the core, a strong sense of regional pride also emerged so that being from the department of Antioquia (where Medellín is capital) meant more than being Colombian. Similar assessments exist for Monterrey (three million people), a city not well known outside Mexico, but within Mexico is perceived as innovative, resourceful, and solidly middle class.

The industrial giant of Latin America is metropolitan São Paulo in Brazil. Rio de Janeiro has greater name recognition and was the capital before Brasília was built, but it does not have the economic muscle of São Paulo. This city of 19 mil-

▲ **Figure 4.30 Mexican autoworkers** Plants like this one in Puebla, Mexico, have assembled Volkswagen Beetles for more than two decades. Automakers from the United States, Europe, and Japan have built plants throughout Mexico. In a global economy, Mexico offers lower labor costs and ready access to markets in Mexico and North America. *(Ron Giling/Panos Pictures)*

lion, Latin America's largest, began to industrialize in the early 1900s when the city's coffee merchants started to diversify their investments. Since then a combination of private and state-owned industries have agglomerated around São Paulo. Within a 60-mile radius of the city center, automobiles, aircraft, chemicals, processed foods, and construction materials are produced. There is also heavy industry and industrial parks. With the port of Santos nearby and the city of Rio de Janeiro a few hours away, São Paulo is the uncontested financial center of Brazil. What stuns most first-time visitors to São Paulo is the forest of high-rises that greets them, a tropical version of Manhattan, only larger.

Maquiladoras The Mexican assembly plants that line the border with the United States are characteristic of manufacturing systems in an increasingly globalized economy. More than 2,000 of these plants existed in the mid-1990s, employing more than half a million people and assembling automobiles, consumer electronics, and apparel. Part of a border development program that began in the 1960s, the Mexican government allowed the duty-free import of machinery, components, and supplies from the United States to be used for manufacturing goods for export back to the United States. Initially, all products had to be exported, but changes in the law in 1994 allow up to half of the goods to be sold in Mexico. The program was slow to develop but took off in the 1980s as foreign companies realized tremendous profits from the inexpensive cost of Mexican labor. An autoworker in a General Motors plant in Mexico averages $10 a day while the same worker in the United States earns more than $200 a day in wages and benefits.

Considerable controversy surrounds this form of industrialization. Critics exist on both sides of the border. Organized labor in the United States complains that well-paying manufacturing jobs are being lost to low-cost competitors, while environmentalists decry serious industrial pollution resulting from lax government regulation. Mexicans worry that these plants are poorly integrated with the rest of the economy and that many of the workers are young unmarried women who are easily exploited. With NAFTA, some of the advantages offered by early border-development programs have eroded. Increasingly, foreign-owned maquiladora plants are being constructed near the population centers of Monterrey and central Mexico. Mexican workers continue to come to the border because maquiladoras offer steady work above the minimum wage (see "A Profile of Ciudad Juárez").

Primary Export Dependency

Historically, Latin America's abundant natural resources were its wealth. In the colonial period silver, gold, and sugar generated tremendous wealth for the colonists. With independence in the nineteenth century a series of export booms introduced commodities such as bananas, coffee, cacao, grains, tin, rubber, copper, wood, and petroleum to an expanding world market. One of the legacies of this export-led development was a tendency to specialize in one or two major commodities, a pattern that continued into the 1950s. During that decade Costa Rica earned 90 percent of its export earnings from bananas and coffee; Nicaragua earned 70 percent from coffee and cotton; 85 percent of Chilean export income came from copper; half of Uruguay's export income came from wood. Even Brazil generated 60 percent of its export earnings from coffee in 1955. By the 1990s that figure was less than 5 percent (although it remains the world's leader in coffee production) and soy products were the biggest foreign exchange earner for Brazil, accounting for 10 percent of its total exports in 1995.

Agricultural Production Since the 1960s the trend in Latin America has been to diversify and to mechanize agriculture. Nowhere is this more evident than in the Plata Basin, which includes southern Brazil, Uruguay, northern Argentina, Paraguay, and eastern Bolivia. Soybeans, used for oil and animal feed, transformed these lowlands in the 1980s and early 1990s. Added to this are acres of rice, cotton, and orange groves, as well as the more traditional plantings of wheat and sugar. The speed in which the plains were converted into fields alarmed environmentalists. Throughout the 1990s, efforts to conserve some of the ecosystem have increased. The swath of red shown in Figure 4.31 is a national park straddling the border of Brazil and Argentina along the Iguazú River. The checkerboard pattern surrounding the park is cultivated land.

Similar large-scale agricultural frontiers exist along in the piedmont zone of the Venezuelan Llanos (mostly grains) and the Pacific slope of Central America (cotton and some tropical fruits). In northern Mexico water supplied from dams along the Sierra Madre Occidental has turned the valleys in Sinaloa into intensive producers of fruits and vegetables for consumers in the United States. The relatively mild winters in northern Mexico also allow growers to produce strawberries and tomatoes during the winter months.

Profile of Ciudad Juárez

Ciudad Juárez is big, probably one million people, and much larger than El Paso, Texas, its sister city across the border. Maquiladora-based manufacturing profoundly changed the look, size, and rhythm of the city. The factories are east of the old town, as are most of the middle- and upper-class houses. To the west squatters have settled in shantytowns that take up three-fifths of the city's land area. Every morning buses grind their way through the west side of town, slowing picking up factory workers. The trip can take up to an hour each way. The workers, young men and women, are neatly dressed. Their day of labor will earn them anywhere from $5–10.

Founded in 1659 as Paso del Norte on the floodplain of the Rio Grande, it was renamed in 1888 in honor of Mexican President Benito Juárez. Yet it is the sprawling, shabby newness of the place, and not its history, that impresses the visitor. Figures tell only part of the story. It is a city of young workers with a high birthrate, but no one really knows how many people live there. Projections for the year 2000 range from 1.3 to 2.4 million people. Since the 1980s several hundred acres of housing were added to the city each year. In the west the roads are mostly dirt and the houses are built of cinder blocks, adobe, and plywood. Access to electricity is spotty and water is usually delivered by truck. The east side of town is another world. In between the assembly plants are the nicest suburbs with sidewalks, garages, and neatly landscaped yards. Drive down the commercial strip and stop at McDonald's, Burger King, or the Seven Eleven.

Most geographers find border zones intriguing cultural landscapes that invite the question of how a line in the sand can create two distinct faces. At this particular bend of the Rio Grande, there are three faces: El Paso to the north and the dual faces of Ciudad Juárez's west and east sides.

▲ **Figure 1 Ciudad Juárez** The core of the city sits along the Rio Grande/Bravo while the periphery radiates out for several miles. The industrial zone, filled with maquilas, is east of downtown. *[Modified from* The Mexican Border Cities *(1993), p. 39, by Daniel Arreola and James Curtis. Tucson, AZ: University of Arizona Press.]*

(a)

(b)

▲ **Figure 2 Elite and shanty housing in Ciudad Juárez** (a) The majority of the city's newcomers reside in self-built homes like these on the urban periphery. (b) In contrast, small elite suburbs are interspersed with assembly plants on the east side of the city. *(Photos by Rob Crandall/Rob Crandall Photography)*

In each of these cases the agricultural sector is capital-intensive and dynamic. By using machinery, high-yielding hybrids, chemical fertilizers, and pesticides, many corporate farms are extremely productive and profitable. What these operations fail to do is employ many rural people, which is especially problematic in countries where a third or more of the population depend upon agriculture for their livelihood. As industrialized agriculture becomes the norm

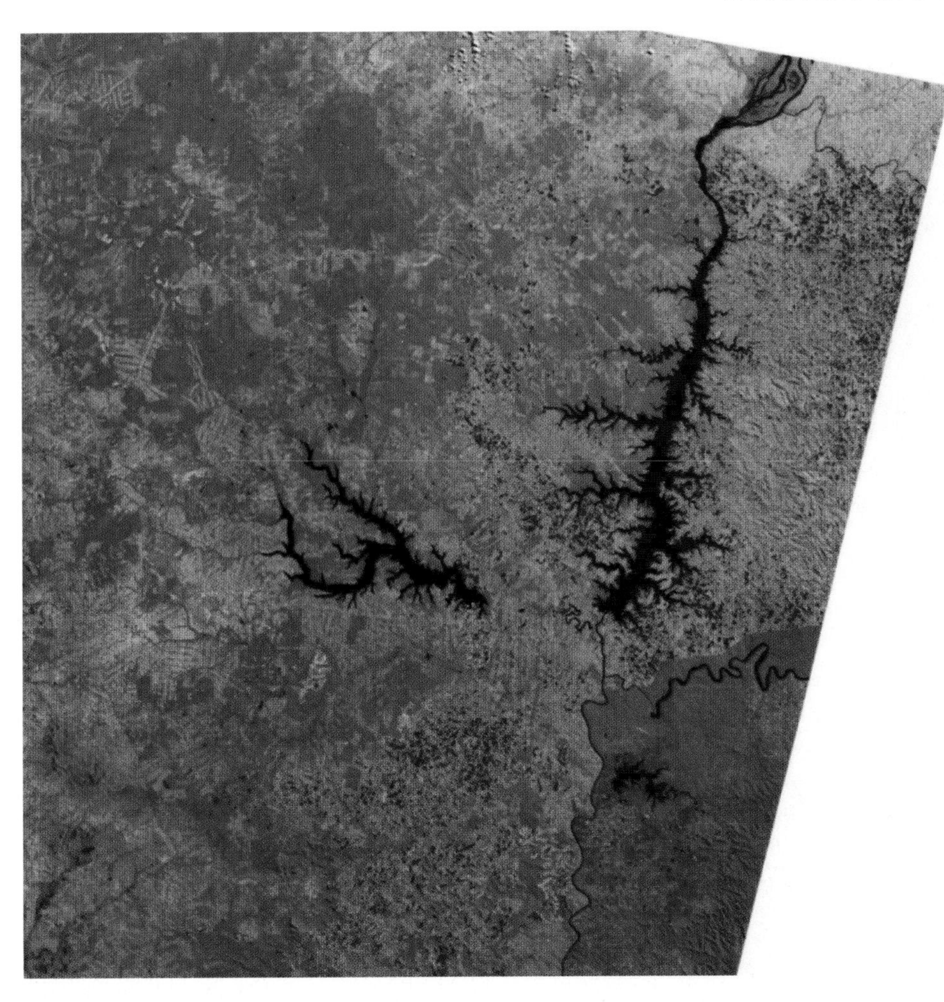

◀ **Figure 4.31 Soybean farms and national parks along the Paraná River** This satellite image shows some of the most productive farmland in South America, where Brazil, Paraguay, and Argentina share borders. The large reservoir is the Itaipú Dam on the Paraná River, the region's largest hydroelectric project. The red area below is Iguaçu National Park, where the river and famous falls of the same name are jointly protected on the Argentina/Brazil border. *(Earth Satellite Corporation/SPL/Photo Researchers, Inc.)*

in Latin America, subsistence peasant producers are further marginalized. The overall trend is that agricultural production is increasing while proportionally fewer people are employed by it. In absolute terms, however, the number of people living in rural areas is about the same as it was in 1960 (roughly 100 million). The major difference over the last 40 years is that many of these people are worse off as traditional rural support networks break down and small farmers are forced onto marginal lands that are vulnerable to drought and erosion. Peasant farmers who are able to produce a surplus of corn or wheat have its value undercut by cheaper imported grains.

The poorest countries of the central Andes and Central America still show 30–40 percent of their economically active populations in agriculture. Much of this agriculture is subsistence farming—small plots of corn, potatoes, or wheat grown by families. These countries also tend to be the least diversified economically, still dependent on one or two agricultural exports for foreign exchange. On the plantations of bananas, sugar, coffee, cotton, or coca, the wage labor tends to be seasonal, so workers migrate between their subsistence plots and wage work, barely eking out an existence. Even when a country attempts to diversify its economy—Honduras began producing nontraditional exports such as shrimp and melon—the rural poor are not the beneficiaries of these programs. Instead, they typically lose their land and move to cities or to new agricultural frontiers.

Mining and Forestry The exploitation of silver, zinc, copper, iron ore, bauxite, gold, oil, and gas is the economic mainstay for many countries in the region. The oil-rich nations of Venezuela, Mexico, and Ecuador are able to meet their own fuel needs and to earn vital state revenues from oil exports. Venezuela is most dependent on revenues from oil, earning up to 90 percent of its foreign exchange from crude petroleum and petroleum products. In 1998 Venezuela also exported more oil to the United States than any other country, including oil-rich Saudi Arabia. Vast oil reserves also exist in the Llanos of Colombia, yet a costly and vulnerable pipeline that connects the oil fields to the coast is a regular target of guerrilla groups. By the late 1990s, Colombian oil production had improved but it is far from reliable (Figure 4.32).

Like agriculture, mining has become more mechanized and less labor intensive. Even Bolivia, a country dependent upon tin production, cut 70 percent of its miners from state payrolls in the 1990s. The measure was part of a broad-based austerity program, yet it suggests that the majority of the miners were not needed. Similarly the vast copper mines of northern Chile are producing record amounts of copper with fewer miners. Gold mining, in contrast, continues to be labor intensive,

▲ **Figure 4.32 Oil production** A portable drilling platform on Lake Maracaibo, Venezuela. Oil production accelerated the pace of development for countries like Venezuela, Mexico, and Ecuador. But these economies struggled in the 1990s as oil prices declined. *(Rob Crandall/Rob Crandall Photography)*

offering employment for thousands of prospectors. Gold rushes are occurring in remote tropical regions of Venezuela, Brazil, Colombia, and Costa Rica. The famous Serra Pelada mine in Brazil, an anthill of workers with ladders, shovels, and burlap sacks filled with gold-bearing soil, is the most dramatic example of labor-intensive mining. Generally, gold miners tend to work in small groups. Moreover, many gold strikes are made illegally on Indian lands or within the borders of national parks (as is the case in Costa Rica). Invariably, if a large enough strike is made, the state steps in to regulate it and to get its share of the profits (Figure 4.33).

▲ **Figure 4.33 Amazonian gold** Serra Pelada, in the Brazilian state of Pará, is one of the most productive gold mines in Latin America. *(Peter Frey/The Image Bank)*

Logging is another important, and controversial, extractive activity. Ironically, many of the forest areas cleared for cattle were not systematically harvested. More often than not, all but the most valuable trees were burned. Logging concessions are commonly awarded to domestic and foreign timber companies who export boards and wood pulp. These one-time arrangements are seen as a quick means for foreign exchange, particularly if prized hardwoods like mahogany are found. Logging can mean a short-term infusion of cash into a local economy. In 1995, $31 million in forest products was exported from the Bolivian state of Santa Cruz (a relative newcomer to the tropical hardwood market). Yet rarely do long-term conservation strategies exist, making this system of extraction unsustainable. Interest in certification programs, ones that designate when a wood product has been produced sustainably, is beginning to grow. This is due to consumer demand for certified wood, mostly in Europe. Unfortunately, such programs are small and the lure of profit usually overwhelms the impulse to conserve for future generations.

Several countries rely on plantation forests of exotic pines, teak, and eucalyptus to supply domestic fuelwood, pulp, and board lumber. These plantation forests are single species and fall far short of the complex ecosystems occurring in natural forests. Still, growing trees for paper or fuel reduces the pressure on other forested areas. Leaders in plantation forestry are Brazil, Venezuela, Chile, and Argentina. Considered Latin America's economic star in the 1990s, Chile relied on timber and wood chips to boost its export earnings. Thousands of hectares of exotics (eucalyptus and pine) have been planted, systematically harvested, cut into boards, or chipped for wood pulp (see Figure 4.2). Japanese capital is heavily involved in this sector of the Chilean economy. The recent expansion of the wood chip business, however, led to a dramatic increase in the logging of native forests. By 1992, more than half of all wood chips were from native species. As forests fell, an entrepreneur from the United States, Douglas Tompkins, amassed 670,000 acres (270,000 hectares) of native forest in an effort to conserve it as a private park. Tompkins's property nearly bisected the narrow country and produced a firestorm of controversy about national security and foreign domination. As of 1998, the proposed park was still being debated but the forest within it remained intact.

Latin America in the Global Economy

In order to conceptualize Latin America's place in the world economy, **Dependency Theory** was advanced in the 1960s by scholars from the region. The premise of the theory is that expansion of European capitalism created the region's condition of underdevelopment. For the developed "cores" of the world to prosper, the "peripheries" became dependent and impoverished. Dependent economies, as those in Latin America, were export-oriented and vulnerable to fluctuations in the global market. Even when they experienced economic growth, it was subordinate to the economic demands of the core (Europe and North America).

Economists who embraced this interpretation of Latin America's history were convinced that economic development could only occur through self-sufficiency, growth of internal markets, agrarian reform, and greater income equality. In short, they argued for vigorous state intervention and an uncoupling from the economic cores. Policies such as import substitution industrialization and nationalization of key industries were partially influenced by this view. Dependency theory has its detractors. In its simplest form it becomes a means to blame forces external to Latin America for the region's problems. Implicit in dependency theory is also the notion that the path to development taken by Europe and North America cannot be easily replicated. This was a radical idea for its time.

As Figure 4.34 shows, there is considerable intraregional trade in Latin America that did not exist in the 1960s. The vigor of Mercosur has boosted trade between Argentina and Brazil to new heights. The legacy of a dependent relationship with the core is also evident. Latin America's biggest trading partner is the United States. The second largest is the European Union, followed by Japan. Trade with all other regions in the world is considerably smaller (Figure 4.34).

Neoliberalism and Globalization By the 1990s governments and the World Bank became champions of neoliberalism as a sure path to economic development.

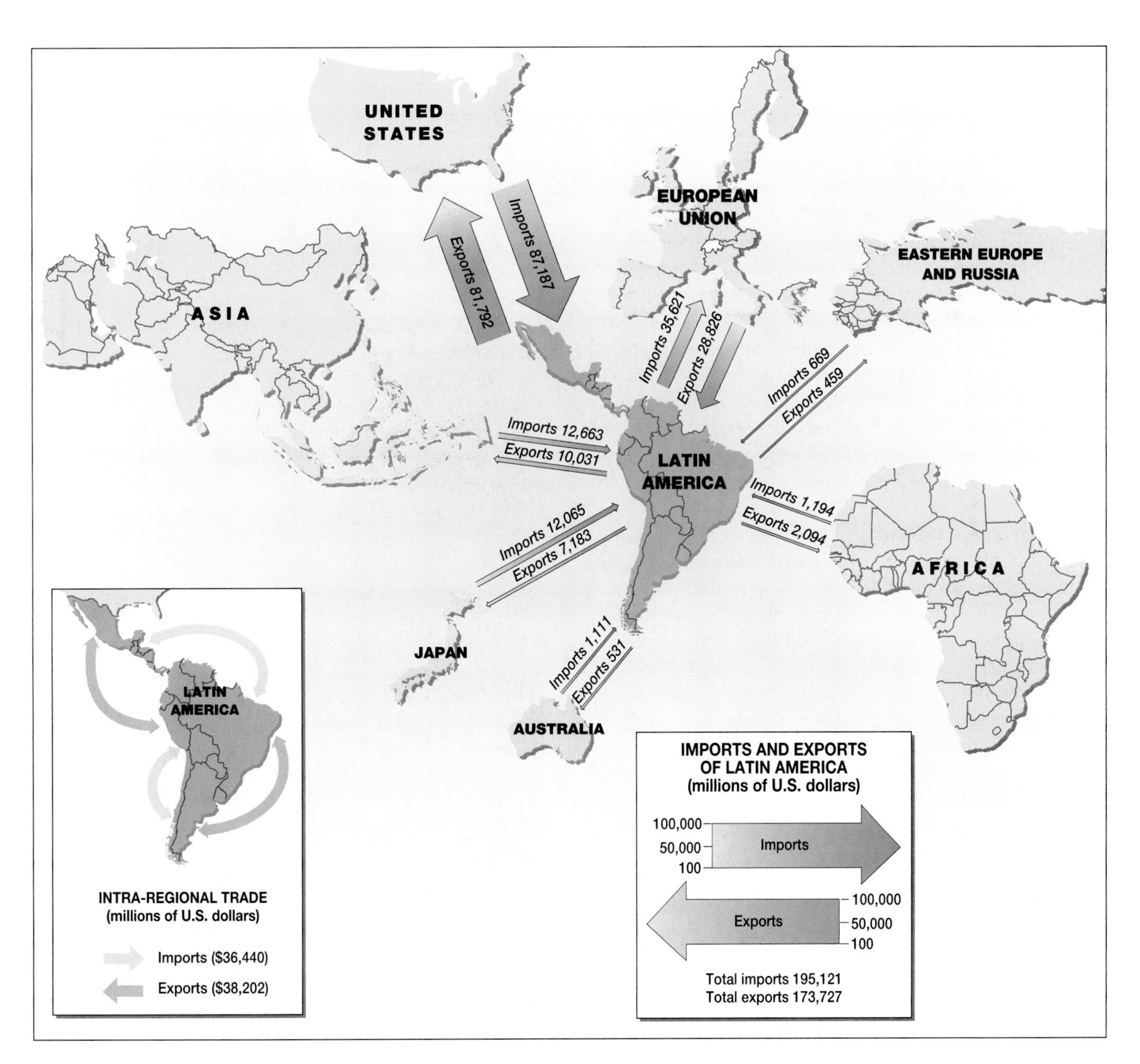

▲ **Figure 4.34 Latin American trade** Latin American trade has well established links with North America and Europe. Yet in the past two decades intra-regional trade has surged. Many credit the growth in intra-regional trade to the success of Mercosur. The bulk of trade between Latin America and North America is between Mexico and the United States. *(Data from Euromonitor,* International Marketing Data and Statistics, *1997)*

Neoliberal policies stress privatization, export production, and few restrictions on imports. They epitomize the forces of globalization by turning away from policies that stressed state intervention and self-sufficiency. Chile is an outspoken proponent of neoliberalism. Its growth rate between 1985 and 1994 averaged 6.2 percent, the region's healthiest. In 1995 the Chilean economy grew 10.4 percent, placing it in the same league as the Asian tigers. Consequently, it is the most studied and watched country in Latin America. By the numbers Chile's 14.5 million people are doing well. With unemployment at 4 percent in 1996 (lower than that of the United States), anyone who wants a job has one. When comparing the PPP index (Table 4.2), Chile tops Argentina with purchasing power of more than $11,000 per capita. So optimistic was Chile about its future that it deferred membership in Mercosur in hopes of joining NAFTA.

Chile's accomplishments should be admired but they are not readily transferable. To begin with, the move to privatize state-owned business and open the economy occurred under an oppressive military dictatorship under which government opposition was not tolerated. Thankfully, such contexts are rare in Latin America today. Much of this export-led growth is based on primary products: fruits, seafood, copper, and wood. Although many of these are renewable, Chile will need to develop more value-added goods if it hopes to be labeled "developed." This relatively small and homogeneous population in a resource-rich land does not have the same ethnic divisions that negatively affect so many states in Latin America. This experiment in neoliberalism is still new, which means the social and environmental costs associated with this economic free-for-all are still not known.

The Entrenched Informal Sector Even in prosperous Santiago, Chile, a short drive to the west shows large neighborhoods of self-built housing filled with street traders and family-run workshops. Such activities make up the informal sector, the provision of goods and services without the benefit of government regulation, registration, or taxation. Most people in the informal economy are self-employed and receive no wages or benefits except the profits they clear. The most common informal activities are housing (in many cities half of all residents live in self-help housing), small workshops, street vendors, transportation services (messengers, bicycle delivery, and collective taxis), garbage pickers, street performers, and even line-waiters. These activities are legal. Illegal informal activities also exist: drug trafficking, prostitution, and money laundering, for example. The vast majority of people who are reliant on informal livelihoods produce legal goods and services.

No one is sure how big this economy is, in part because it is difficult to untangle and quantify formal activities from informal ones. Yet visit Lima, Belém, Guatemala City, or Guayaquil and it is easy to get the impression that the informal economy *is* the economy. From the self-help housing that dominates the landscape to the hundreds of street vendors that crowd the sidewalks making it difficult to pass, it is impossible to avoid. No doubt there are advantages in the informal sector, hours are flexible, children can work with their parents, and there are no bosses. Peruvian economist Hernando de Soto even argues that this most dynamic sector of the economy should be encouraged, rather than discouraged, and offered formal lines of credit. As important as this sector may be, such widespread dependence on it signals Latin America's poverty, not its wealth. It reflects the inability for the economies of the region to absorb labor in the formal economy, especially in industry. For millions of urban dwellers, the thought of finding formal employment that offers benefits, safety, and a living wage is still a dream. With nowhere else to go, the ranks of the informally employed continue to grow (Figure 4.35).

Social Development

Marked improvements in life expectancy, child survival, and educational attainment have occurred since 1960. One telling indicator is the mortality rate for children younger than 5 years old, which dropped by two-thirds or more throughout the region between 1960 and 1993 (Table 4.3). Chile in 1960 had 138 deaths per 1,000 children under age 5. In 1995 the number was 15. Nicaragua dropped from 209 deaths per 1,000 in 1960 to 60. This indicator is important because for children younger than 5 years old to survive suggests that basic nutritional and health care needs are being met. One can also infer that resources are being used to sustain women and their children. By comparison, the United States has a death rate of 10 per 1,000 children under age 5, Japan's rate is 6. Other developing countries have much higher under-5 death rates than Latin America; Indonesia's is 111 per 1,000 and Zambia's is 203. Despite economic downturns, the region's social networks have been able to mitigate the effects on children.

Grassroots and nongovernmental organizations (NGOs) play a fundamental role in contributing to social well-being. Initiated by international humanitarian organizations, church

▲ **Figure 4.35 Peruvian street vendors** A street vendor selling produce in Huancayo, Peru. Street vending plays a critical role in the distribution of goods and the generation of income. Contrary to popular opinion, it is often regulated by city governments and by the vendors themselves. *(Rob Crandall/ Rob Crandall Photography)*

Table 4.3 Social Indicators and Status of Women

Country	Life Expectancy at Birth		Under Age 5 Mortality, per 1,000 Live Births		Secondary School Enrollment %		Female Labor Force Participation (% of total)
	Male	Female	1960	1995	Male	Female	
Costa Rica	73	78	112	16	45	49	30
El Salvador	65	72	210	40	27	30	35
Guatemala	63	68	205	67	25	23	27
Honduras	66	71	203	38	29	37	30
Mexico	69	75	141	32	57	58	31
Nicaragua	63	68	209	60	39	44	36
Panama	71	77	104	20	60	65	34
Argentina	69	76	68	27	70	75	31
Bolivia	57	63	252	105	40	34	37
Brazil	64	71	181	60	—	—	35
Chile	72	78	138	15	65	70	32
Colombia	65	73	132	36	57	68	38
Ecuador	66	71	180	40	54	56	27
Paraguay	66	71	90	34	36	38	29
Peru	67	71	236	55	66	60	29
Uruguay	72	78	47	21	—	—	41
Venezuela	70	75	70	24	29	41	33

Sources: *Population Reference Bureau, Data Sheet, 1998,* Life Expectancy (M/F); *World Resources Institute, 1998–99,* Under-5 Mortality Rate; *Population Reference Bureau, Data Sheet, 1996,* Secondary School Enrollment (M/F); *The World Bank Atlas, 1998,* Female Participation in Labor Force.

organizations, and community activists, these popular groups provide many services that state and local governments can not. Catholic Relief Services and Caritas, for example, work with the rural poor throughout the region to improve their water supplies, sanitation, and education. Other groups are secular, lobbying local governments to build schools or recognize squatters' claims. Grassroots organizations also develop cooperatives that market anything from sweaters to cheeses. Cooperative organizations are able to realize some economies of scale, as well as access to credit. In the 1980s Lima, Peru, witnessed the mobilization of 2,000 popular kitchens, run by poor women to feed poor families with local and foreign donations. At their peak in the late 1980s, it was estimated that one million meals a day were served. Had this not occurred, the human cost of Peru's economic collapse would have been far worse.

Other important gauges for social development are life expectancy, literacy, school enrollment, and access to safe water. In aggregate, 80 percent of the people in the region have access to safe drinking water, 87 percent enroll in primary school and 55 percent in secondary school, literacy rates are 85 percent, and life expectancy (men and women) is 68 years. Masked by this aggregate data are extreme variations between rural and urban areas, between regions, and along racial and gender lines. No matter how difficult city life may be, the poorest people in Latin America are in rural areas. The move to cities, for many people, is a development strategy.

A running joke in Latin America is that the economy is doing well but the people are not. Even today when Argentina or Costa Rica are heralded for making impressive economic gains, fear persists that a large segment of the population is excluded. The 1980s were dubbed the "lost decade" for Latin America due to the debt crisis, inflation, and slow-to-negative growth rates. During that period, several countries witnessed setbacks in both their economic and social indicators—most notably Peru, Bolivia, Nicaragua, Venezuela, Mexico, and Brazil. Many economic indicators improved in the 1990s. Still, the two largest countries, Mexico and Brazil, with more than half the region's population, struggle with inflation, currency devaluation, and a per capita GDP that is lower than it was in 1980 constant dollars.

Within Mexico and Brazil tremendous internal differences exist when mapping socio-economic indicators (Figure 4.36). The northeast part of Brazil lags behind the rest of the country in every social indicator. The country has a literacy rate of 80 percent but in the northeast it is only 60 percent. Moreover, when northeastern rural and urban areas are contrasted, 70 percent of city residents are literate but only 40 percent of rural ones are. In Mexico the levels of deprivation are highest in the Indian states of Chiapas, Oaxaca, and Guerrero. In

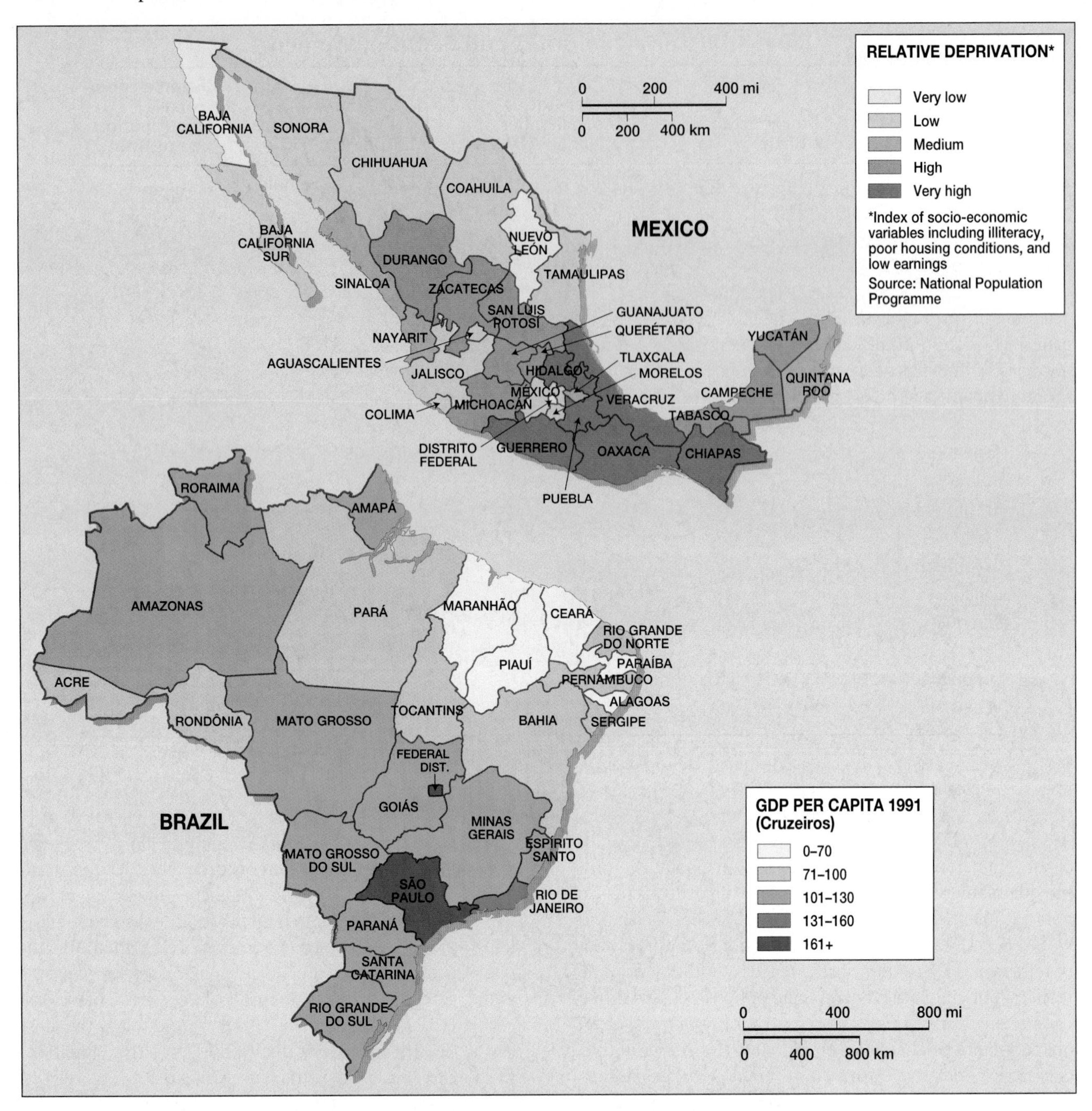

▲ **Figure 4.36 Mapping poverty and prosperity** Mexico and Brazil are the largest states in the region and yet the levels of poverty vary greatly within each country. In Mexico the southern states of Chiapas, Oaxaca, and Guerrero have the highest rates of deprivation. In Brazil, the northeastern states of Paraíba, Alagoas, Ceará, and Piauí have the country's lowest per capita GDP. *(© 1995 The Economist Newspaper Group, Inc. Reprinted with permission. Further reproduction prohibited.)*

contrast, Mexico City, the state of Nuevo Leon (Monterrey is the capital), and Baja California Norte (Mexicali is the capital) do best in terms of literacy, housing quality, and income. All countries have spatial inequities regarding income and availability of services, even though the contrasts tend to be sharper in the developing world. In the cases of Mexico and Brazil, it is hard to ignore ethnicity and race when trying to explain these patterns.

Race and Inequality There is much to admire about race relations in the Americas. The complex racial and ethnic mix that was created in Latin America fostered tolerance for diversity. That said, Indians and blacks are disproportionately represented among the poor of the region. More than ever, racial discrimination is a major political issue in Brazil. Reports of organized killings of street children, most of them Afro-Brazilian, make headlines. For decades, Brazil espoused its vision of a color-blind racial democracy. True, residential segregation by race is less evident in Brazil and interracial marriage is common, but certain patterns of social and economic inequity seem best explained by race.

There are major problems in trying to assess racial inequities in Brazil. The Brazilian census asks few racial questions and all are based on self-classification. In 1987 only 11 percent of the population called itself black and 53 percent white. Some Brazilian sociologists, however, claim that more than half the population is of African ancestry, making Brazil the second largest "African state" after Nigeria. Racial classification is always highly subjective and relative, but there are patterns that support the existence of racism. Evidence from northeastern Brazil, where Afro-Brazilians are the majority, shows death rates approaching those of some of the world's poorest countries. Throughout Brazil, blacks suffer higher rates of homelessness, landlessness, illiteracy, and unemployment. There is even a movement in the northeastern state of Bahia to extend the same legal protection given Indians to hundreds of Afro-Brazilian villages that were once settled by runaway slaves.

Similarly, in areas of Latin America where Indian cultures are strong, one also finds low socioeconomic indicators. In most countries mapping areas where native languages are widely spoken invariably corresponds with areas of persistent poverty. In Mexico the Indian south lags behind the booming north and Mexico City. Prejudice is embedded in the language, to call someone an *indio* (Indian) is an insult in Mexico. In Bolivia, women who dress in the Indian style of full, pleated skirts and bowler hats are called *cholas*. This descriptive term referring to the rural mestizo population has negative connotations of backwardness and even cowardice. No one of high social standing, regardless of skin color, would ever be called a *chola* or *cholo*.

It is difficult to separate status divisions based on class from those based on race. From the days of conquest, being European meant an immediate elevation in status over the Indian, African, and mestizo populations. Class awareness is very strong. Race does not necessarily determine one's economic standing but it certainly influences it. Most people recognize the power of the elite and envy their lifestyle. An emerging middle class also exists that is formally employed, aspires to own a home and car, and strives to give their children a university education. The vast majority, however, are the working poor who struggle to meet basic food, shelter, clothing, and transport needs. These class differences express themselves in the landscape. Go to any large city and find handsome suburbs, country clubs, and trendy shopping centers. High-rise luxury apartment buildings with beautiful terraces offer all the modern amenities, including maid's quarters. The elite and the middle class even show a preference for decentralized suburban living and dependence on automobiles, as do North Americans. Yet near these same residences are shantytowns where urban squatters build their own homes, create their own economy, and eke out a living.

The Status of Women Many contradictions exist with regard to the status of women in Latin America. The gender stereotypes of machismo and marianismo are breaking down among younger generations. Many Latina women work outside the home. In most countries the formal figures hover between 30 and 40 percent of the workforce, which is not far off from many European countries but lower than that of the United States (Table 4.3). Legally speaking, women can vote, own property, and sign for loans, although they are less likely to do so than men, which reflects the patriarchal tendencies in the society. Even though Latin America is predominantly Catholic, divorce is legal and family planning is promoted. Yet in most countries abortion remains illegal.

In almost every country the percentage of women who go to high school is higher than that of men (Table 4.3). This suggests that, among those who can send their children to high school, educating women is a priority. Only in three countries (Bolivia, Guatemala, and Peru) are men more represented in this category, which may reflect conservative attitudes about the role of women in societies where nearly half the population is of Indian heritage. Even in higher education, male and female students are equally represented today. Consequently, in the fields of education, medicine, and law, women are regularly employed.

The biggest changes for women are the trends toward smaller families, urban living, and educational parity with men. These factors have greatly improved the earning capacity of women. In the countryside, however, serious inequities remain. Rural women are less likely to be educated and tend to have larger families. In addition, they are often left to care for their families alone as husbands leave seasonally in search of employment. In general, but not in all cases, the conditions and prejudices facing rural woman have been slow to improve.

Women are increasingly taking an active role in politics. In 1990 Nicaragua elected the first woman president in Latin America, Violeta Chamorro, an owner of an opposition newspaper. A Mayan woman from Guatemala, Rigoberta Manchú, won the Nobel Peace Prize in 1992 for denouncing human rights abuses in her native country. Women are also active organizers and participants in cooperatives, microenterprises, and unions. In a relatively short period, urban women have established a formal place in the economy and a partial political voice. Moreover, evidence suggests that this trend will continue, reflecting an improvement for the status of women in the region (Figure 4.37).

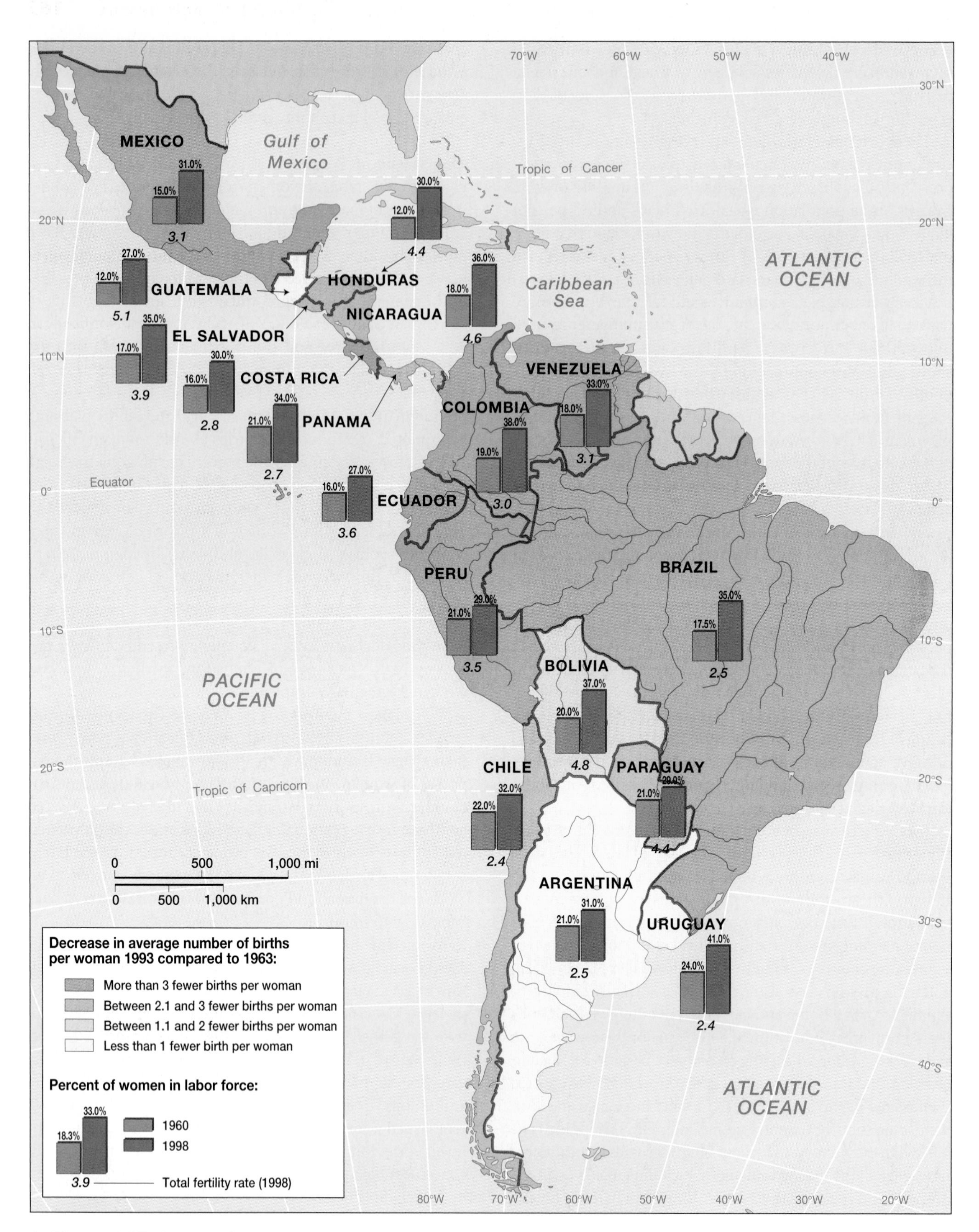

▲ **Figure 4.37 Status of Women** This map shows the large number of states that have seen total fertility rates decline by 3 or more from 1963 to 1993. Over a similar period, the number of women in the labor force grew dramatically (see bar graphs). *("Motherhood," from* The State of Women in the World Atlas *by Joni Seager. Copyright © 1997 by Joni Seager, text. © 1997 by Myriad Editions Ltd, maps & graphics. Used by permission of Viking Penguin, a division of Penguin Putnam Inc.)*

Conclusion

Latin America and the Caribbean were the first world regions to be fully colonized by Europe. In the process, perhaps 90 percent of the native population died due to disease, cruelty, and forced resettlement. The slow demographic recovery of native peoples and the continual arrival of Europeans and Africans resulted in an unprecedented level of racial and cultural mixing. It took nearly 400 years for the population of Latin America and the Caribbean to reach 54 million again, its pre-contact level. During this long period, European culture, technology, and political systems were transplanted and modified. Indigenous peoples integrated livestock and wheat into their agricultural practices, yet they held true to their preference for native corn, potatoes, and cassava. In short, a syncretic process unfolded, in which many indigenous customs were preserved beneath the veneer of Iberian ones. Over time a blending of indigenous and Iberian societies, with some African influences, gave distinction to this part of the world. The music, literature, and artistry of Latin America are widely acknowledged.

Compared with Asia, this is still a region rich in natural resources and is relatively lightly populated. Yet as population continues to grow along with economic expectations, there is considerable concern that much of this natural endowment could be squandered for short-term gains. In the midst of a boom in natural resource extraction, popular concern for the state of the environment is growing. Not only was Brazil the site of the 1992 United Nations Earth Summit, but hundreds of locally based environmental groups have formed to try to protect forests, grasslands, indigenous peoples, and freshwater supplies. This brand of environmentalism is pragmatic; it recognizes the need for economic development but hopes to improve urban environmental quality and sustainable resource use.

In Latin America the trend toward modernization began in the 1950s, and the pace of change has been rapid. Unlike other developing areas, most Latin Americans live in cities. This shift started early and reflects a cultural bias toward urban living with roots in the colonial past. Not everyone who came to the city found employment, however; thus, the dynamics of the informal sector were set in place. Even though population growth rates have declined, the overall makeup of the population is young. Serious challenges lie ahead in educating and finding employment for the cohort younger than age 15. In this matter some countries are faring better than others. In El Salvador only 27 percent of boys go to high school, whereas in Panama 60 percent go and in Argentina 70 percent (see Table 4.3). Industrial employment in the Southern Cone is up, yet in many parts of the Central Andes subsistence farming is widely practiced.

Latin America is one of the world regions that North Americans are most likely to visit in their lifetime. The trend, of course, is to visit the northern fringe of this region. Tourism is robust along Mexico's border and coastal resorts. (Cancun is now the single most popular destination for foreigners coming to Mexico). Unfortunately, there is a tendency to visit one area in the region and generalize for all of it. Although it is historically sound to think of Latin America as a major world region, extreme variations in the physical environment, levels of social and economic development, and the influence of indigenous society exist.

Since regions are fluid constructs, it is worth mentioning two important transition zones. One is along the U.S.–Mexican border. As Mexico becomes more intricately tied to the U.S. economy, might there be significant changes in the functional and formal linkages between these two countries? Most people see the cultural divide as so great between them that cultural coherence seems unlikely. At the same time, these so-called "distant neighbors" are creating a border area with strong influences from both sides. The other transition zone is the Southern Cone (including southern Brazil). Here the economic and social indicators of well-being are so much better than the Andean countries it is easy to imagine the Mercosur countries increasing their influence within the region and even globally. It is this area that has the best chance of joining the ranks of the developed world, and in the process perhaps distinguishing itself from the rest of Latin America.

Key Terms

Altiplano (page 126)
altitudinal zonation (page 130)
Columbian Exchange (page 144)
comarca (page 145)
dependency theory (page 160)
dual primacy (page 137)
ejido (page 140)
El Niño (page 130)
encomienda (page 140)
environmental lapse rate (page 130)
Gran Colombia (page 150)
grassification (page 131)
growth poles (page 156)
import substitution (page 156)
informal sector (page 138)
latifundia (page 140)
machismo (page 148)
maquiladora (page 157)
marianismo (page 148)
megacity (page 122)
Mercosur (page 150)
mestizo (page 142)
minifundia (page 140)
neoliberal policies (page 162)
neotropics (page 123)
Organization of American States (OAS) (page 150)
rural-to-urban migration (page 139)
shields (page 123)
static frontiers (page 123)
subnational organizations (page 153)
supranational organizations (page 153)
syncretic religions (page 148)
tierra caliente (page 130)
tierra fria (page 130)
tierra helada (page 130)
tierra templada (page 130)
transnationalism (page 143)
Treaty of Tordesillas (page 150)
United Provinces of Central America (page 152)
urban primacy (page 137)

Questions for Summary and Review

1. Why is this region called *Latin America?*
2. What is El Niño and how does it impact Latin America and other parts of the world?
3. What is altitudinal zonation and how is it relevant to agricultural practices?
4. What is the Columbian Exchange? Discuss ways in which it still occurs today.
5. What factors contributed to the rapid and early rate of urbanization in Latin America?
6. What are the origins of latifundia? Discuss the relationship between latifundia and minifundia.
7. Besides Iberia, what are the source countries for Latin America immigrants? What impact have these different groups had on the region?
8. Give two examples of syncretic religions. What processes create them?
9. Discuss the origins of the current political boundaries in Latin America. Where are boundary disputes occurring and why?
10. How are trade blocks reshaping international relations within Latin America?
11. Is the dependency theory useful in understanding Latin America's position in the world economy? Can the theory be applied to other regions in the developing world?
12. What social and demographic shifts account for improvements in the lives of women in Latin America?

Thinking Geographically

1. Discuss the processes driving tropical deforestation in Latin America and how they compare with deforestation in other areas of the world.
2. Agrarian reform has been advocated throughout Latin America as a means to reduce social and economic inequalities. Has agrarian reform worked in the region? How have land distribution programs fared? Where else in the world have agrarian reforms been attempted?
3. How is neoliberalism influencing the way Latin America interacts with the rest of the world? What are the social and environmental costs of neoliberalism? Is this a model for understanding the impact of globalization in the developing world?
4. After examining the language map, what conclusions can be drawn about the patterns of Indian survival in Latin America?
5. Given the dominance of cities in this region, what particular urban environmental problems face cities in the developing world? How might Latin America's megacities use their size and density to reduce the environmental problems associated with urbanization?
6. Indigenous homelands or reservations can be a source of cultural repression or cultural survival. Compare the creation of comarcas in Panama or communal land-use zones in Honduras with Indian reservations in the United States. What different political processes are behind the creation of these indigenous territories?
7. Discuss the social, environmental, and economic consequences behind the modernization of agriculture. How does Latin America's experience with modern agricultural systems compare with North America?

Regional Novels and Films

Novels

Isabel Allende, *House of the Spirits* (1982, Barecelona Plaza & Janes)

Jorge Amado, *Dona Flor and Her Two Husbands* (1967, Losada)

Jorge Amado, *Gabriela, Clove and Cinnamon* (1958, Biblioteca Ayacucha)

Jorge Luis Borges, *Labyrinths* (1962, New Directions)

Carlos Fuentes, *The Old Gringo* (1985, Fondo de Cultura Economica)

Gabriel García Marquez, *One Hundred Years of Solitude* (1967, Biblioteca Ayacucha)

Gabriel García Marquez, *Love in the Time of Cholera* (1985)

Mario Vargas Llosa, *The Real Life of Alejandro Mayta* (1984)

Films

Amazon Journal

Black Orpheus (1958, Brazil)

Children of Rio (1990, Brazil)

Coffee: A Sack Full of Power (1997, France)

El Norte (1983, U.S.)

Flowers for Guadalupe

Like Water for Chocolate (1993, U.S.)

Lines of Blood: The Drug War of Colombia

Transnational Fiesta (1992, U.S.)

Trinkets & Beads (1996, U.S.)

Bibliography

Arreola, Daniel D., and Curtis, James R. 1993. *The Mexican Border Cities, Landscape Anatomy and Place Personality*. Tucson, AZ: University of Arizona Press.

Brown, Ed. 1996. "Articulating Opposition in Latin America: the Consolidation of Neoliberalism and the Search for Radical Alternatives." *Political Geography* 15(2), 169–92.

Butzer, Karl W., guest ed. 1992. "The Americas before and after 1492: Current Geographical Research." *Annals of the Association of American Geographers* 82(3), 343–568.

Caviedes, César, and Knapp, Gregory. 1995. *South America*. Englewood Cliffs, NJ: Prentice Hall.

Clapp, Roger Alex. 1998. "Waiting for the Forest Law: Resource-Led Development and Environmental Politics in Chile." *Latin American Research Review* 33(2), 3–36.

Clawson, David L. 1997. *Latin America and the Caribbean, Lands and Peoples*. Dubuque, IA: William C. Brown Publishers.

Crosby, Alfred. 1972. *The Columbian Exchange, Biological and Cultural Consequences of 1492*. Westport, CT: Greenwood Press.

Dean, Warren. 1995. *With Broadax and Firebrand: The Destruction of the Brazilian Atlantic Forest*. Berkeley, CA: University of California Press.

Denevan, William M. 1992. *The Native Population of the Americas in 1492*, 2nd ed. Madison, WI: University of Wisconsin Press.

de Soto, Hernando. 1989. *The Other Path: The Invisible Revolution in the Third World*. New York: Harper & Row.

Dodds, Klaus-John. 1993. "Geopolitics, Cartography and the State in South America." *Political Geography* 12, 361–81.

Ezcurra, Exequiel, and Mazari-Hiriart, Marisa. 1996. "Are Megacities Viable? A Cautionary Tale from Mexico City." *Environment* 38(1), 6–15, 26–35.

Gilbert, Alan. 1994. *The Latin American City*. New York: Monthly Review Press.

Godfrey, Brian. 1992. "Modernizing the Brazilian City." *Geographical Review* 81(1), 18–34.

Hays-Mitchell, Maureen. 1994. "Streetvending in Peruvian Cities: The Spatio-Temporal Behavior of Ambulantes." *The Professional Geographer* 46(4), 425–38.

Hecht, Susanna, and Cockburn, Alexander. 1989. *The Fate of the Forest: Developers, Destroyers and Defenders of the Amazon*. London: Verson Press.

Herlihy, Peter H. 1989. "Panama's Quiet Revolution: Comarca Homelands and Indian Rights." *Cultural Survival Quarterly* 13(3), 17–24.

Lovell, W. George. "Heavy Shadows and Black Night: Disease and Depopulation in Colonial Spanish America." *Annals of the Association of American Geographers* 82(3), 426–43.

Mangin, William. 1970. *Peasants in Cities: Readings in the Anthropology of Urbanization*. Boston: Houghton Mifflin.

Mörner, Magnus. 1985. *The Andean Past, Land, Societies and Conflicts*. New York: Columbia University Press.

Place, Susan E., ed. 1993. *Tropical Rainforests, Latin American Nature and Society in Transition*. Wilmington, DE: Scholarly Resources, Inc.

Preston, David. 1996. *Latin American Development, Geographical Perspectives*, 2nd ed. Essex, England: Longman Scientific & Technical.

Price, Marie. 1994. "Ecopolitics and Environmental Nongovernmental Organizations in Latin America." *Geographical Review* 84(1), 42–58.

Sánchez-Albornoz, Nicolás. 1974. *The Population of Latin America, A History*. Berkeley, CA: University of California Press.

Slater, David. 1993. "The Geopolitical Imagination and the Enframing of Development Theory." *Transactions* 18, 419–37.

Smith, Nigel J. H. 1996. *The Enchanted Amazon Rain Forest, Stories from a Vanishing World*. Gainesville, FL: University Press of Florida.

"A Survey of Brazil." *The Economist*, April 29, 1995.

"A Survey of Mexico." *The Economist*, October 28, 1995.

"A Survey of Mercosur." *The Economist*, October 12, 1996.

Tenenbaum, Barbara, ed. 1996. *Encyclopedia of Latin American History and Culture*. New York: Charles Scribner's Sons.

Thrupp, Lori Ann. 1995. *Bittersweet Harvests for Global Supermarkets: Challenges in Latin America's Agricultural Export Boom*. Washington, DC: World Resources Institute.

Voeks, Robert. 1993. "African Medicine and Magic in the Americas." *Geographical Review* 83(1), 66–78.

West, Robert, and Augelli, John P. 1989. *Middle America, Its Lands and Peoples*, 3rd ed. Englewood Cliffs, NJ: Prentice Hall.

Williams, Robert G. 1994. *States and Social Evolution, Coffee and the Rise of National Governments in Central America*. Chapel Hill, NC: University of North Carolina Press.

Winn, Peter. 1992. *Americas, The Changing Face of Latin America and the Caribbean*. Berkeley, CA: University of California Press.

Young, Kenneth R. 1996. "Threats to Biological Diversity Caused by Coca/Cocaine Deforestation in Peru." *Environmental Conservation* 23(1), 7–15.

Zimmerer, Karl. 1996. *Changing Fortunes: Biodiversity and Peasant Livelihood in the Peruvian Andes*. Berkeley, CA: University of California Press.

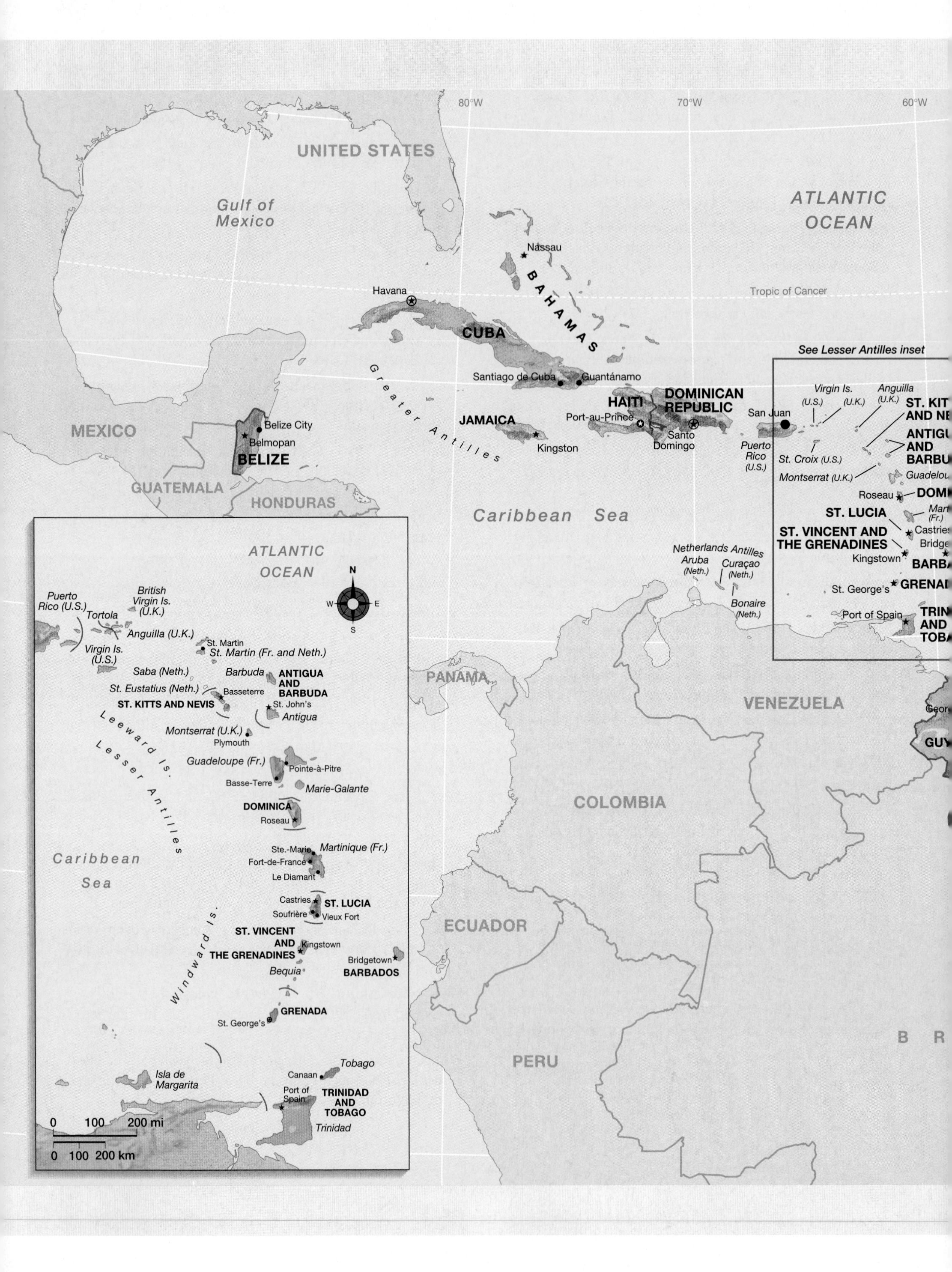

80°W
70°W
60°W
UNITED STATES
Gulf of Mexico
ATLANTIC OCEAN
Nassau
BAHAMAS
Tropic of Cancer
Havana
CUBA
See Lesser Antilles inset
Greater Antilles
Santiago de Cuba
Guantánamo
DOMINICAN REPUBLIC
HAITI
Virgin Is. (U.S.) (U.K.)
Anguilla (U.K.)
ST. KIT AND NE
JAMAICA
Port-au-Prince
Santo Domingo
San Juan
MEXICO
Belize City
Belmopan
BELIZE
Kingston
Puerto Rico (U.S.)
St. Croix (U.S.)
ANTIGU AND BARBU
Montserrat (U.K.)
Guadelou
GUATEMALA
HONDURAS
Roseau
DOMI
Caribbean Sea
ST. LUCIA
Mart (Fr.)
Castries
ST. VINCENT AND THE GRENADINES
Bridge
Netherlands Antilles
Aruba (Neth.)
Curaçao (Neth.)
Kingstown
BARB
Bonaire (Neth.)
St. George's
GRENA
Port of Spain
TRIN AND TOBA
PANAMA
VENEZUELA
Geor
GUY
COLOMBIA
ECUADOR
PERU
B R
ATLANTIC OCEAN
N
W
E
S
Puerto Rico (U.S.)
British Virgin Is. (U.K.)
Tortola
Anguilla (U.K.)
Virgin Is. (U.S.)
St. Martin
St. Martin (Fr. and Neth.)
Saba (Neth.)
Barbuda
ANTIGUA AND BARBUDA
St. Eustatius (Neth.)
Basseterre
ST. KITTS AND NEVIS
St. John's
Antigua
Leeward Is.
Montserrat (U.K.)
Plymouth
Lesser Antilles
Guadeloupe (Fr.)
Pointe-à-Pitre
Basse-Terre
Marie-Galante
DOMINICA
Roseau
Caribbean Sea
Ste.-Marie
Martinique (Fr.)
Fort-de-France
Le Diamant
Castries
ST. LUCIA
Soufrière
Vieux Fort
Windward Is.
ST. VINCENT AND THE GRENADINES
Kingstown
Bridgetown
BARBADOS
Bequia
GRENADA
St. George's
Tobago
Isla de Margarita
Canaan
Port of Spain
TRINIDAD AND TOBAGO
Trinidad
0 100 200 mi
0 100 200 km

The Caribbean

50°W

CARIBBEAN
Political Map
Over 1,000,000
500,000–1,000,000 (selected cities)
Selected smaller cities

20°N

Elevation in meters
4000+
2000–4000
500–2000
200–500
0–200
Below sea level
Sea Level

ATLANTIC
OCEAN
10°N

Paramaribo
Cayenne
FRENCH GUIANA (Fr.)
INAME
0°

0 200 400 mi
0 200 400 km

The Caribbean was the first region of the Americas to be extensively explored and colonized by Europeans. Yet its modern regional identity remains ambiguous. Today the region is home to 38 million inhabitants scattered across 25 countries and dependent territories. They range from tiny St. Kitts and Nevis with 40,000 people, to the island of Hispaniola with 15 million. In addition to the Caribbean Islands, Belize of Central America and the three Guianas—Guyana, Suriname, and French Guiana—of South America are included as part of the Caribbean. For historical and cultural reasons, the peoples of these mainland states identify with the island nations and are thus included in this chapter (Figure 5.1; see also "Setting the Boundaries").

Historically, the Caribbean became a battleground between rival European powers competing for territorial control of these tropical colonies. In the early 1900s, the United States took over as the dominant geopolitical force in the region, regularly sending troops and maintaining what some call a neocolonial presence. Like many developing areas, external control of the Caribbean produced highly dependent and inequitable economies. The plantation was the dominant production system and sugar the leading commodity. Later bananas, tobacco, and citrus were added. The combination of high population densities and the dominance of agriculture have caused serious land degradation in the Caribbean, especially deforestation and soil erosion, challenging the sustainability of even subsistence agriculture. Governments have sought to diversify their economies by developing tourism, nontraditional exports (such as flowers), offshore banking, and manufacturing to reduce the region's dependence upon primary exports.

Generally when one thinks of the Caribbean, images of white sandy beaches and turquoise tropical waters come to mind. Since the 1960s, it has earned much of its international reputation as a playground for northern vacationers. There is, of course, another Caribbean that is far poorer and economically dependent than the one portrayed on travel posters. Haiti, the poorest country in the Western Hemisphere, has the third largest population in the Caribbean with more than 7 million people. The two largest countries—Cuba (11 million) and the Dominican Republic (8 million)—also suffer from serious economic problems.

◀ **Figure 5.1 The Caribbean** Named for one of its former native inhabitants—the Carib Indians—the modern Caribbean is a culturally complex and economically dependent world region. Settled by various colonial powers, most of the region's inhabitants are a mix of African, European, and South Asian peoples. Twenty-five countries and dependent territories make up the Caribbean, yet most of the population resides on the four islands of the Greater Antilles: Cuba, Hispaniola, Jamaica, and Puerto Rico. Home to 38 million people, the Caribbean is demographically larger than Oceania.

Setting the Boundaries

This culturally diverse area of the world does not fit neatly into current world regional schemes. In fact, in most textbooks it is scarcely mentioned, being folded into Latin American or appearing as part of Middle America (along with Mexico and Central America). By recognizing the Caribbean as a distinct world region, however, one can see clearly the long-term processes of colonization and globalization that have created it.

For much of the colonial period, the islands themselves were referred to as "the Indies" or "the Spanish Main," and the sea surrounding them was called *Mar del Norte* (North Sea). It was not until the late eighteenth century that an English cartographer first applied the name Caribbean to the sea (a reference to the Carib Indians who once inhabited the islands). It took another 200 years for the term *Caribbean* to be generally applied to this region of islands and rimland.

The rationale for treating the Caribbean as a distinct area lies within its particular cultural and economic history. Culturally, the islands and portions of the littoral can be distinguished from the largely Iberian-influenced mainland of Latin America because of its more diverse European colonial history and a much stronger African imprint. Demographically, the native population of the Caribbean was virtually eliminated within the first 50 years following Columbus's arrival. Thus, unlike Latin America and North America, the Amerindians of the insular Caribbean have virtually no legacy, save a few place names. In terms of economic production, the dominance of export-oriented plantation agriculture explains many of the social, economic, and environmental patterns in the region. African slaves were thrust upon the scene as replacements for lost native labor. As Caribbean forests and savannas became sugarcane fields, a pattern was set whereby exotic species of plants replaced native ones. The legacy of plantation life is still visible in the grossly uneven distribution of land and resources.

Today organizations such as the Caribbean Common Market exist to foster the common interests of island and mainland states. Yet intraregional alliances often split along colonial lines, so that former Spanish, British, French, and Dutch colonies at times have more affinity for each other than for the region as a whole.

The majority of Caribbean people are poor, living in the shadow of North America's vast wealth. It is this contrast that inspires the concept of **isolated proximity** to explain the region's unusual and contradictory position in the world. The *isolation* of the Caribbean sustains the area's cultural diversity but also explains its limited economic opportunities. Caribbean writers note that this isolation fosters a strong sense of place and an inward orientation by the people. Yet it is the relative *proximity* of the Caribbean to North America (and, to a lesser extent, Europe) that ensures its transnational connections and economic dependence. History has shown the inhabitants of the Caribbean that their fate is not their own. Through the years the Caribbean has evolved as a distinct but economically peripheral world region. This status expresses itself today as workers flee the region in search of employment, while foreign companies are attracted to the Caribbean for its cheap labor. The economic well-being of most countries is precarious. Despite such uncertainty, an enduring cultural richness and attachment to place is witnessed here that may explain a growing countercurrent of returnees back to their homelands (Figure 5.2).

▲ **Figure 5.2 Carnival dancer** A woman in elaborate costume celebrates Carnival in Port of Spain, Trinidad. Linking a Christian pre-Lenten celebration with African musical rhythms, Carnival has become an important symbol of Caribbean identity. *(Rob Crandall/Rob Crandall Photographer)*

Environmental Geography: Paradise Lost?

Tucked between the tropic of Cancer and the equator, with year-round temperatures averaging in the high 70s, the hundreds of islands and picturesque waters of the Caribbean have often inspired comparisons to paradise. Columbus began the tradition by describing the islands of the New World as the most marvelous, beautiful, and fertile lands he had ever known, filled with flocks of parrots, exotic plants, and friendly natives. Writers today are still lured by the sea, sands, and swaying palms of the Caribbean. It was the region's tropical-

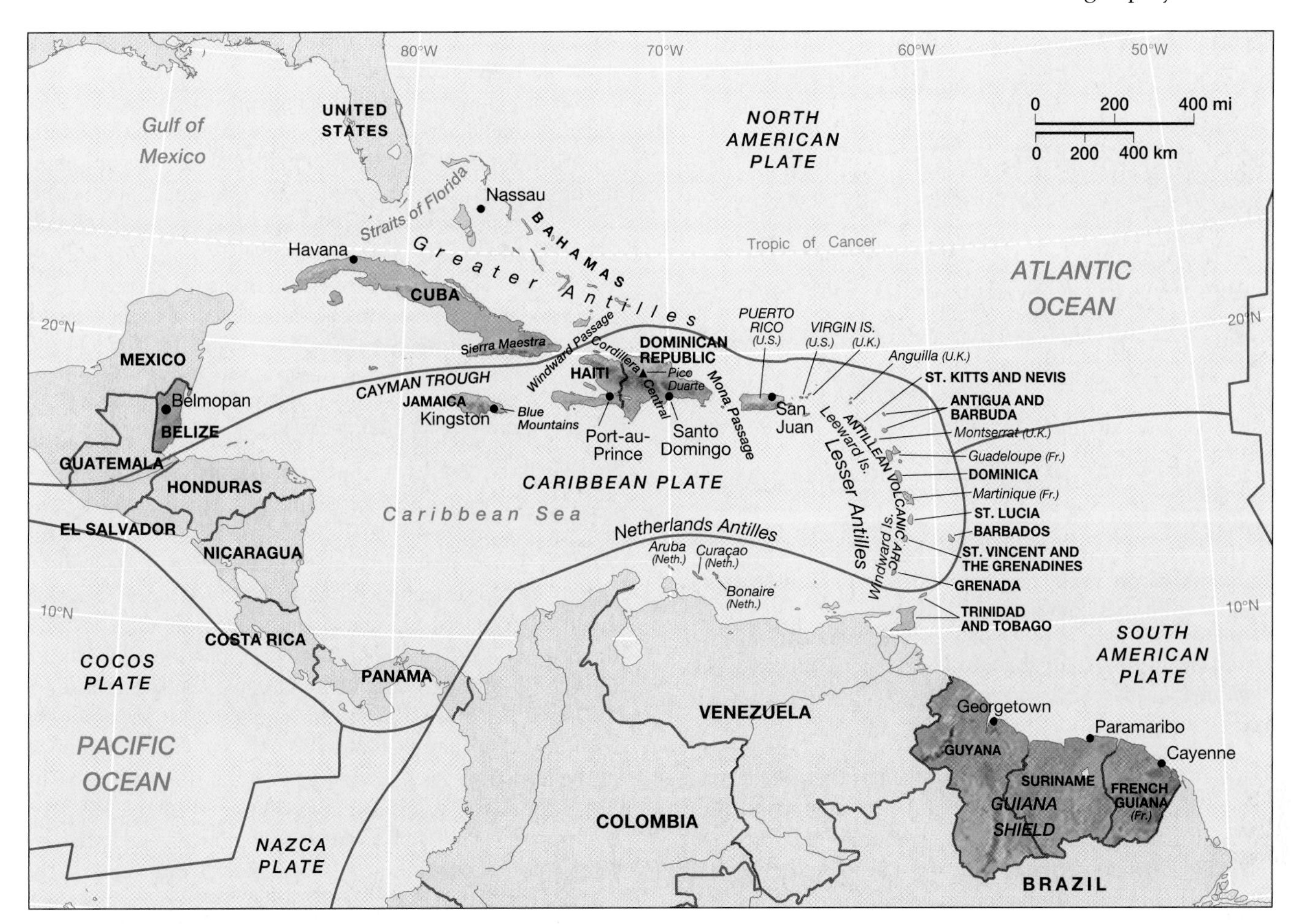

▲ **Figure 5.3 Physical geography of the Caribbean** Centered on the Caribbean Sea, the region falls neatly within the neotropical belt north of the equator. In addition to a warm year-round climate and beautiful waters, the Greater Antilles have several important mountain ranges, whereas some of the Lesser Antilles have active volcanoes. The Guianas, on the South American plate, contain some of the oldest rock surfaces found on Earth. Cuba and Belize are on the North American plate; these countries have flat limestone formations similar to Florida.

ity and accessibility to Europe via the trade winds that stimulated its colonization. Even the smallest island, if properly developed, could yield handsome profits from sugarcane, spices, and other tropical crops. Yet the colonized environment created by Europeans and Africans over the centuries was radically different from the physical setting first described by the Spanish chroniclers. It is this transformation that invokes the metaphor of a lost paradise.

The Sea, Islands, and Rimland

It is the Caribbean Sea itself, that body of water enclosed between the *Antillean* islands (the arc of islands that begins with Cuba and ends with Trinidad) and the mainland of Central and South America, that links the states of the region (Figure 5.3). Historically the sea connected people through its trade routes and sustained people with its marine resources of fish, green turtle, manatee, lobster, and crab. Noted for its clarity and biological diversity, the quantities of any one species in the Caribbean are not great, so the sea has never supported large commercial fishing. The surface temperature of the sea ranges from 73° to 84° F (23° to 29° C), over which forms a warm tropical marine air mass that influences daily weather patterns. It is this warm water and tropical setting that continues to be a key resource for the region as millions of tourists visit the Caribbean each year (Figure 5.4).

The arc of islands that stretches across the sea is its most distinguishing feature. The Antillean Islands are divided into two groups: the Greater and Lesser Antilles. Three-quarters of the region's population are found on the Greater Antilles. The rimland (the Caribbean coastal zone of the mainland) includes Belize and the Guianas, as well as the Caribbean littoral of Central America. In contrast to the islands, the rimland has low population densities and tremendous biological diversity.

Greater Antilles The four large islands of Cuba, Jamaica, Hispaniola (shared by Haiti and the Dominican Republic), and Puerto Rico make up the **Greater Antilles.** On these islands are found the bulk of the region's population, arable lands, and large mountain ranges. Given the popular interest in Caribbean coasts, it surprises many people that Pico Duarte in the Cordillera Central of the Dominican Republic is more than 10,000 feet (3,000 meters), Jamaica's Blue Mountains

▲ **Figure 5.4 Caribbean Sea** Noted for its calm turquoise waters, steady breezes, and treacherous shallows, the Caribbean Sea has both sheltered and challenged sailors for centuries. This aerial photograph shows the southern Caribbean islands of Los Roques, off the Venezuelan coast. *(Rob Crandall/Rob Crandall Photographer)*

top 7,000 feet (2,100 meters), and Cuba's Sierra Maestra is more than 6,000 feet (1,800 meters). The mountains of the Greater Antilles were of little economic interest since plantation owners preferred the coastal plains and valleys. Yet the mountains were an important refuge for runaway slaves and subsistence farmers, and thus figure prominently in the cultural history of the region.

The best farmlands are found in the central and western valleys of Cuba where a limestone base contributes to the formation of a fertile red clay soil (locally called *Matanzas*) and a gray or black soil type called *rendzinas* (also found in Antigua, Barbados, and lowland Jamaica). The rendzinas soils consist of a gravelly loam with a high organic content ideal for sugar production and were actively exploited wherever found. Surprisingly, given the area's agricultural orientation, many of the soils are nutrient poor, heavily leached, and acidic. These *ferralitic* soils are found in the wetter areas where crystalline base rock exists (as in parts of Hispaniola, the Guianas, and Belize). They are characterized by heavy accumulations of red and yellow clays and offer little potential for permanent intensive agriculture.

Lesser Antilles The **Lesser Antilles** form a double arc of small islands stretching from the Virgin Islands to Trinidad. Smaller in size and with far fewer people than the Greater Antilles, early on they were important footholds for rival European colonial powers. The islands from St. Kitts to Grenada form the inner arc of the Lesser Antilles. These mountainous islands, with peaks ranging from 4,000 to 5,000 feet (1,200 to 1,500 meters), have volcanic origins. In this subduction zone, the heavier North and South American plates go underneath the Caribbean plate, producing volcanic activity. Erosion of the island peaks and the accumulation of ash from eruptions have created small pockets of arable soils, although the steepness of the terrain places limits on agricultural development. The latest round of volcanic activity began in July 1995 on Montserrat. A series of volcanic eruptions of ash and rock have forced most people off the island. Residents of Plymouth, the capital, were forced to evacuate in 1996. The volcano has taken several lives, and evacuees from the uninhabitable southern end of the island have relocated to the northern villages, other nearby islands, and even London (Figure 5.5).

Just east of this volcanic arc are the low-lying islands of Barbados, Antigua, Barbuda, and the eastern half of Guade-

▶ **Figure 5.5 Volcanic destruction** The eruptions of the Soufrière Hills volcano on the island of Montserrat sent mudflows and volcanic ash into the capital city of Plymouth, partially submerging a clock tower and a telephone booth in the town center. Damage was widespread on the island, and many of the 11,000 residents of this British colony had to be evacuated. Eruptions began on July 18, 1995. *(Rob Huibers/Panos Pictures)*

loupe. Covered in limestone that overlays volcanic rock, these lands were much more inviting for agriculture. Trinidad and Tobago are on the South American plate and consist of sedimentary rather than volcanic rock. These islands include alluvial soils, and more importantly, sedimentary basins that contain oil reserves.

Other island groups in the Caribbean are the Dutch islands of Aruba, Bonaire, and Curaçao (known as the ABC islands) off the coast of Venezuela; the British dependencies of the Cayman Islands and the Turks and Caicos on either side of Cuba; and the Bahamian archipelago northeast of Cuba. The Bahamas encompass hundreds of barren islands and islets as well as 20 inhabited ones. While technically the Bahamas and the Turks and Caicos are in the Atlantic Ocean and not the Caribbean Sea, convention places them as part of the Caribbean region.

Rimland States This Caribbean **rimland** is the coastal zone of the mainland, beginning with Belize and extending along the coast of Central America to northern South America. This chapter focuses on the rimland states of Belize and the Guianas (Central America's rimland is discussed in Chapter 4). Much of low-lying Belize is limestone. Sugarcane dominates in the drier north, while citrus is produced in the wetter central portion of the state. The Guianas, however, are characterized by the rolling hills of the Guyana Shield. The shield's crystalline rock explains the area's overall poor soil quality. Most agriculture in the Guianas occurs on the narrow coastal plain, where sugar and rice are produced. Unlike the rest of the Caribbean, these rimland territories still contain significant amounts of forest cover. In Suriname alone, 96 percent of the territory is in forest or woodland. Timber continues to be an important export for the rimland states. Metal extraction (bauxite and gold) is also vital to the economies of Guyana and Suriname. French Guiana, which is an overseas territory of France, relies mostly upon French subsidies but exports shrimp and timber. It is also home to the French space center at Kourou.

Climate and Vegetation

Royal palms framed by a blue sky evoke the postcard image of the Caribbean. In this tropical region, it is warm year-round and rainfall is abundant. Much of the Antillean islands and rimland receive more than 80 inches (200 centimeters) of rainfall annually and can support tropical forests. Amid the forests are pockets of naturally occurring grasslands in parts of Cuba, Hispaniola, and southern Guyana. Distinctly dry areas exist, such as the rain shadow basin in western Hispaniola. Given the economic history of the region, much of the natural vegetation has been removed to accommodate agriculture and fuel needs. Today only small forest fragments remain on the islands. Tropical ecosystems in the rimland are largely intact.

Like many tropical lowlands, seasonality is defined more by changes in rainfall than temperature. Although some rain falls throughout the year, the rainy season is from July to October. In Belize City, Havana, Port-au-Prince, and Bridgetown, October is the wettest month (Figure 5.6). This is the time when the Atlantic high-pressure cell is farthest north, and easterly winds generate moisture-laden and unstable atmospheric conditions that sometimes yield hurricanes. During the slightly cooler months of December through March, rainfall declines. This drier time of year corresponds with the peak tourist season.

In the Guianas a different rainfall cycle is evident. These territories, on average, receive more rain than the Antillean islands. In Cayenne, French Guiana, an average of 126 inches (320 centimeters) fall each year. Unlike the Antilles, the Guianans experience a brief dry period in late summer (September to October). Also, January tends to be a wet period for the mainland, while it is a dry time for the islands. This difference is due to the influence of the inter-tropical convergence zone (ITCZ) on equatorial South America. The ITCZ is a circulation of equatorial air masses and winds. Within the zone, converging trade winds are forced upward by equatorial warming, creating active cloud bands and precipitation. The ITCZ shifts to the north in winter and the south in summer, causing variations in rainfall. Climatically, the Guianas are also distinguishable from the rest of the region because they are not affected by hurricanes.

Hurricanes Each year several **hurricanes** form, pounding the Caribbean as well as Central and North America with heavy rains and fierce winds. Beginning in July, westward-moving low-pressure disturbances form off the coast of West Africa, picking up moisture and speed as they move across the Atlantic. The air masses are usually no more than 100 miles across, but to achieve hurricane status they must reach velocities of more than 75 miles per hour. Hurricanes may take several paths through the region, but they typically enter through the Lesser Antilles. They then arc north or northwest and collide with the Greater Antilles, Central America, Mexico, or southern North America before moving to the northeast and dissipating in the Atlantic Ocean. The hurricane zone (Figure 5.7) lies just north of the equator on both the Pacific and Atlantic sides of the Americas. Typically, a half dozen to a dozen hurricanes form each season and move through the region, causing limited damage.

There are, of course, noted exceptions, and most long-time residents of the Caribbean have felt the full force of at least one major storm in their lifetime. The destruction caused by these storms is not just from the high winds but from the heavy downpours that can cause severe flooding and deadly coastal tidal surges. In 1998 the torrential rains of Hurricane Mitch, the most deadly tropical storm in a century, resulted in the death of at least 8,000 people in Honduras, Nicaragua, and El Salvador. Mud slides and flooding ravaged structures and roads, leaving upward of one-quarter of Honduras's population without shelter. While Hurricane Mitch largely bypassed the Caribbean, in 1988 Hurricane Gilbert took 260 lives and pounded Jamaica and the Yucatán before slamming into Texas. Eighty percent of the houses in Jamaica lost their roofs. Plantations of coconut palms on the Yucatán were leveled like matchsticks. Hugo came the following year, leaving nearly everyone in Montserrat homeless, wiping out the

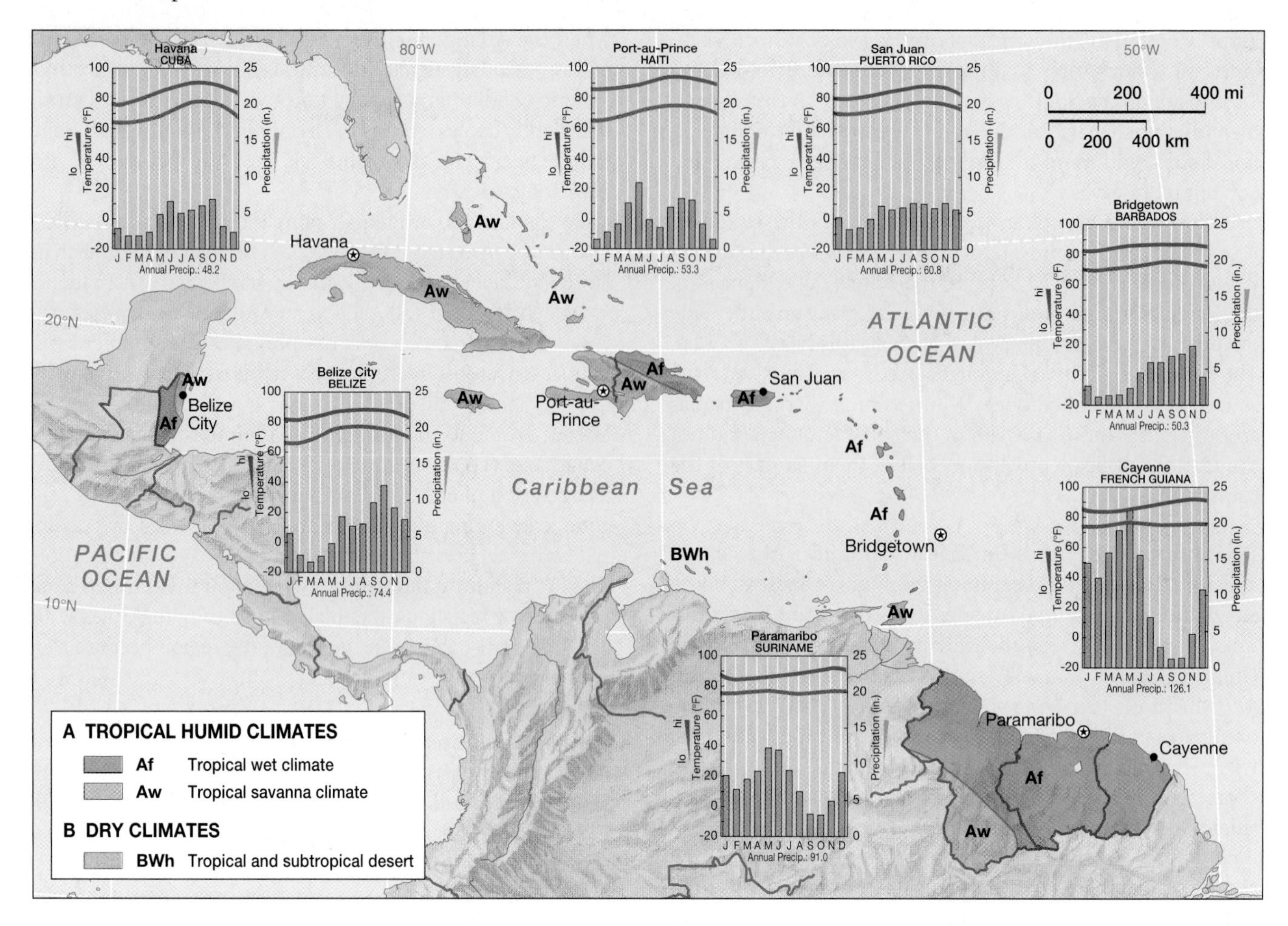

▲ **Figure 5.6 Climate map of the Caribbean** Much of the region is classified as having either a tropical wet (Af) or tropical savanna (Aw) climate. Temperature varies little across the region, with highs slightly above 80 degrees and lows around 70 degrees. Important differences in total rainfall and the timing of the dry season distinguish different places. In the Guianas, for example, the dry season is August through October, whereas the drier months for the islands are December through March. *(Temperature and precipitation data from* The World Weather Guide *(1984) by E. A. Pearce and C. G. Smith. London: Hutchinson).*

infrastructure and tourist economy of St. Croix, and damaging Puerto Rico to such an extent that troops from the mainland were sent in to help restore order.

Modern tracking equipment has improved hurricane forecasting and reduced the number of fatalities, primarily by evacuation of threatened areas. Forecasting cannot reduce the damage to crops, forests, or infrastructure. A badly timed storm can destroy a banana harvest or shut down resorts for a season or more. The sheer force of the storm can radically transform the landscape as well, turning a palm-strewn sandy beach into a treeless rocky shore covered with debris.

Forests Much of this region was covered in tropical rain forest and deciduous forest prior to the arrival of Europeans. The great clearing of the Caribbean's forests began in earnest in the Lesser Antilles in the seventeenth century and spread westward (Figure 5.8). The island forests fell not only to plant sugarcane but to provide the fuel necessary to turn the cane syrup into sugar, to construct housing and fences, and to build ships. Mostly tropical forests were removed because they were seen as unproductive; the European colonists valued cleared land. Yet unlike the midlatitudes, the newly exposed tropical soils easily eroded and ceased to be productive after several harvests, a situation that led to two distinct land-use strategies. On the larger islands and the mainland, new lands were constantly cleared and older ones abandoned or fallowed (left untilled for several seasons) in an effort to keep up sugar production. On the smaller islands such as Barbados and Antigua, where land was limited, labor-intensive efforts to conserve soil and maintain fertility were employed. In either case, the island forests were replaced by a landscape devoted to crops for world markets.

In Belize, mahogany and dye wood extraction triggered selective deforestation in addition to clearing for agriculture on the coastal plain. Much of the wetter southern half of the country is still covered in forest. In contrast to the rest of the region, the rain forests of the Guianas are mostly undisturbed because of their small populations and economies based on mining. Nevertheless, pressures to extract the most valuable hardwoods, especially mahogany, are mounting.

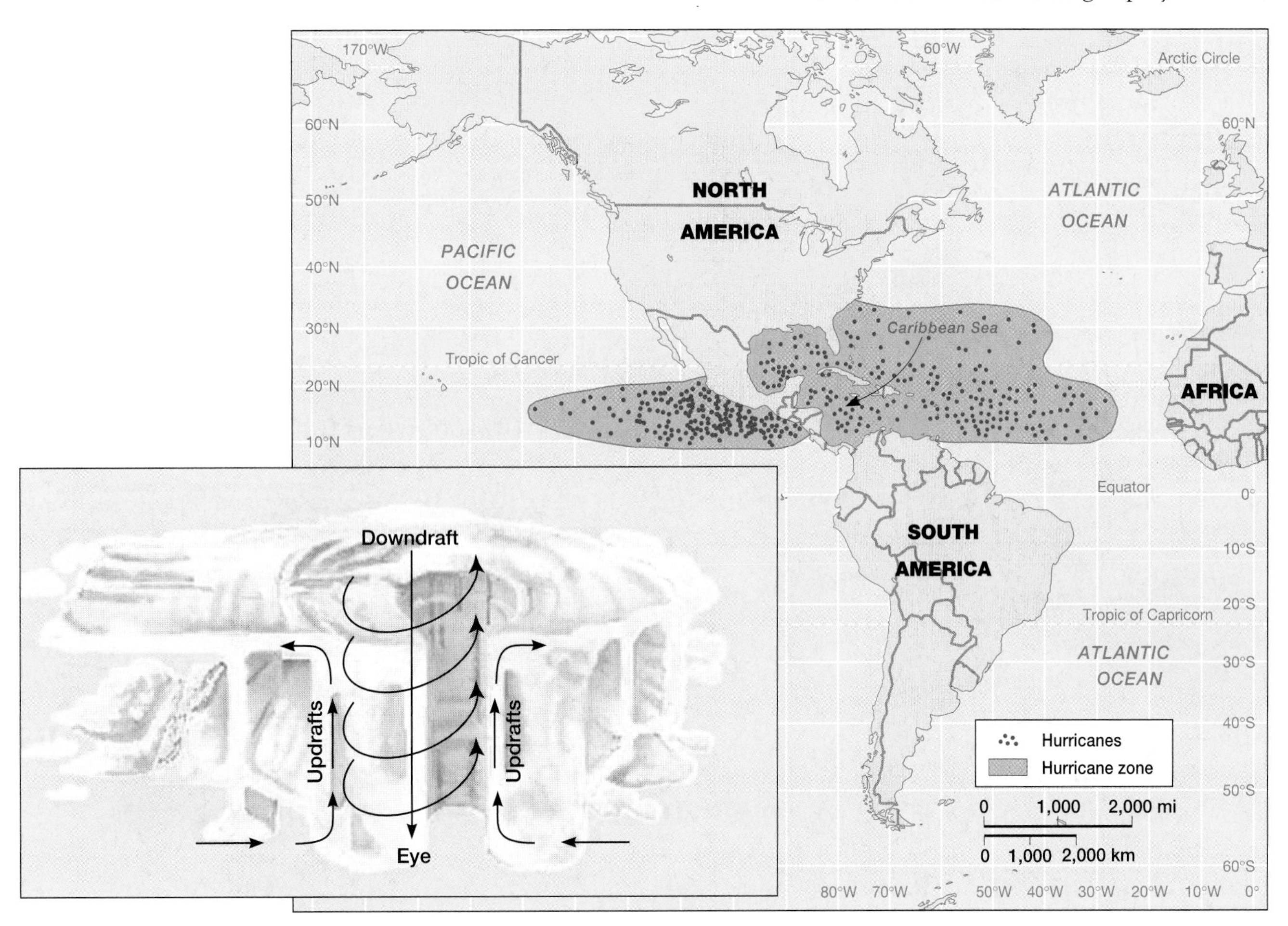

▲ **Figure 5.7 Hurricanes** From July until November each year, people of the Caribbean know several hurricanes will hit, some with deadly consequences. These tropical disturbances bring heavy rains and winds (often more than 100 miles per hour). They rip roofs off houses, destroy crops, and flood settlements. As Hurricane Mitch showed in 1998, the worst storms kill thousands of people and devastate a country's infrastructure. *(Modified from McKnight, 1996,* Physical Geography, *Upper Saddle River, NJ: Prentice Hall)*

Savannas and Mangroves In addition to tropical forests, the palm savannas and mangrove swamps are important biomes in the region. The palm savannas of Hispaniola and Cuba are quite fertile and easily adapted for agriculture. In central Cuba these natural grasslands studded with palms have the best soils in the entire country and are mostly planted in sugarcane or citrus. The savanna soils of southern Guyana, however, are acidic with little agricultural potential.

For centuries the coastal mangrove swamps were largely left alone. This wet environment is poorly suited for human settlement, although it is a vital nursery for young crustaceans and fish. Found throughout the Caribbean, especially on the calmer leeward shores, the tangled stands of woody mangrove are regularly cleared to create open beaches. Mangroves also decline when disturbances upstream increase the silt load in the water. The removal of mangrove, besides eliminating a vital marine habitant, exposes coasts to increased erosion.

Arid Zones Semiarid vegetation of thorn–scrub brush or even cactus exist in the rain shadow of the Antillean mountains, or in the generally drier leeward side of the islands where comparatively less rain falls (20 to 40 inches annually). These scrublands have limited agricultural potential unless irrigation is introduced or a drought-resistant crop such as sisal or agave is planted. Often, subsistence grazing of goats occurs in this ecological zone (Figure 5.9). Aruba, Bonaire, and Curaçao, as well as Anguilla, typify the shallow soils and sparse vegetation found in these arid areas. Too dry to support agriculture, the colonial economy hobbled along by producing salt and raising goats. It wasn't until the 1960s that the dry climate, white sands, and tropical reefs (in the case of Bonaire) facilitated the transformation of these islands into world-class resorts.

Environmental Degradation

Ecologically speaking, it is difficult to image an area that has been so completely reworked by colonization as the Caribbean. The Spanish, English, Dutch, and French reconstituted a tropical environment that best suited their needs. Native vegetation was replaced with plantation crops, and even exotic ornamentals like the Flamboyant tree, the coconut palm, and

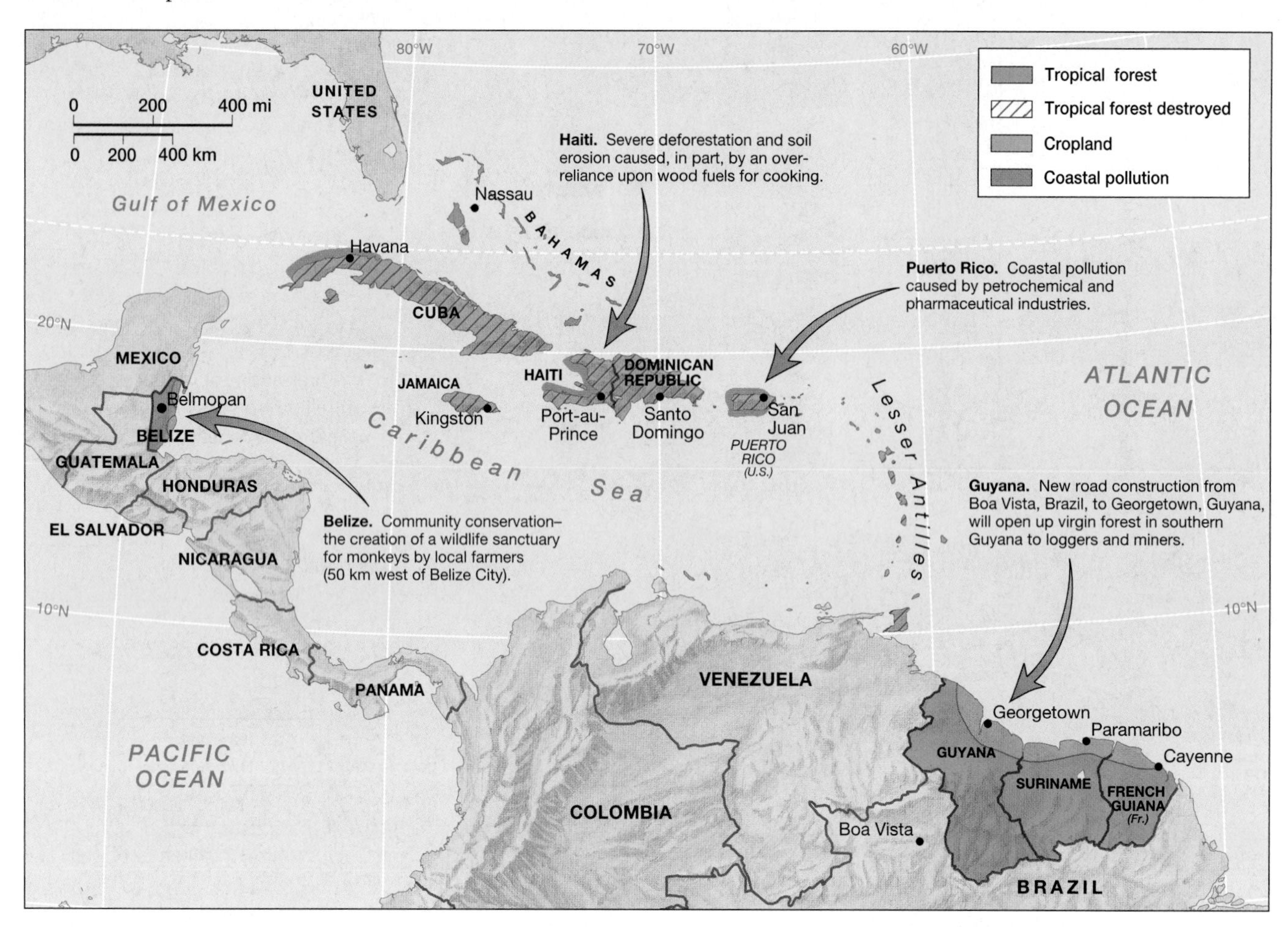

▲ **Figure 5.8 Environmental issues in the Caribbean** It is hard to imagine a region in which the environment has been so completely transformed. Most of the island forests were removed long ago for agriculture or fuel wood, and soil erosion is a chronic problem. Coastal pollution is serious around the largest cities and industrial zones. The forest cover of the rimland states, however, is largely intact and attracting the interest of environmentalists. As tourism becomes increasingly important for the Caribbean, efforts to protect the beaches and reefs along with the fauna and flora are growing. *(Adapted from* DK World Atlas, *1997, pp. 7, 55. London: DK Publishing)*

▲ **Figure 5.9 Goats and cacti** Goats are an important source of food and income for people in the semiarid zones where thorn-scrub vegetation is common. Goats can quickly overgraze areas, which can lead to increased aridity and reduced biodiversity. *(James Amos/National Geographic Society)*

bougainvillea were added to fulfill a tropical ideal. Most of the environmental problems in the region are associated with agriculture, reliance upon biofuels, and water and air pollution associated with sprawling and impoverished cities.

Agriculture's Legacy of Deforestation Plantation agriculture initiated cycles of deforestation that permanently degraded the vegetative cover, especially on the islands. Where once Amerindians practiced shifting cultivation, Europeans introduced continuous cultivation practices that depleted soils and, in the worst cases, produced desertification. All the islands have experienced soil erosion to some degree but it is most noted in Hispaniola and Cuba because of the sheer size of the islands and the number of people they support (26 million).

The connection between economic poverty and environmental degradation is clearest in Haiti. The inhabitants of this densely settled country are largely dependent upon commercial and subsistence agriculture, which has resulted in serious problems with soil erosion and declining yields. In

◀ **Figure 5.10 Deforestation on Hispaniola** The border between Haiti and the Dominican Republic is easily seen in this aerial photo. Haitians, the poorest people in the Western Hemisphere, rely almost exclusively on firewood and charcoal for their energy needs. With so many trees removed, soils are more vulnerable to erosion and agricultural yields tend to decline. Successful reforestation efforts have yet to be implemented. *(James Blair/National Geographic Society)*

addition, the majority of Haitians rely upon charcoal (made from trees) for their cooking fuel, which places additional strain on the island's vegetation. The deterioration of the resource base is evident from the air; aerial photos reveal a sharp boundary between a denuded Haiti and a forested Dominican Republic (Figure 5.10). The difference between the two countries is explained by the lack of affordable fuel alternatives. Whereas many Dominicans can afford to buy liquid or gas cooking fuel, many Haitians cannot.

Cuba has experienced a surge in charcoal production brought on by the country's economic and energy crises that began in 1990. Since much of the country's electricity is generated by imported fuel, locals have turned to Cuba's forests to make charcoal for domestic energy needs. A recent governmental report acknowledged that up to 60 percent of the national territory is being desertified due to mechanized agriculture and a surge in charcoal production.

For nearly five centuries the destruction of forest and the unrelenting cultivation of soils resulted in the extinction of many endemic Caribbean plants and animals, including various shrubs and trees, songbirds, large mammals, and monkeys. As beautiful as the Caribbean is, severe depletion of biological resources has occurred, which helps explain some of the present economic and social instability of the region.

Settling the Frontier In the rimland states biological diversity and stability are less threatened. Thus, current conservation efforts could produce important results. Even though much of Belize was selectively logged for mahogany in the nineteenth and twentieth centuries, healthy forest cover still exists that supports a diversity of mammals, birds, reptiles, and plants. Public awareness of the negative consequences of deforestation is also greater now. In the 1980s the Coca Cola corporation purchased thousands of acres around Dangriga, Belize, with the intention of turning forest and savanna into orange groves for juice concentrate. International protests, especially from U.S. environmentalists, forced this Atlanta-based multinational corporation to cancel its plans. Limited and locally owned orange production has occurred instead.

The relatively pristine interior forests of Guyana are becoming a battleground between conservationists and developers. A new, dry-season highway traverses the length of the country connecting Boa Vista, Brazil, with Georgetown, Guyana. Not only does this road improve trade between Guyana and Brazil, giving Brazil access to the North Atlantic, it also opens the forest and mineral resources (chiefly gold) in southern Guyana. While governments in both states are encouraged by the economic possibilities of this road, a coalition of conservationists (local and foreign) and indigenous peoples is attempting to delineate a vast national park in part of this region where no commercial extraction could occur.

Failures in Urban Infrastructure The growth of Caribbean cities produces localized environmental strains, especially water contamination and waste disposal. The urban poor are the most vulnerable to health problems associated with overly strained or nonexistent water and sewage services. It was estimated in 1991 that only 28 percent of urban households in the Dominican Republic had piped water. Urban residents get water from shared faucets or take their chances with surface water or collected rainwater. The expense of improving basic urban infrastructure far exceeds the capabilities of most island economies. Several dams already exist on the larger islands to supply water, while some of the smaller islands rely upon expensive desalination plants.

Still, existing freshwater supplies fall far short of domestic needs. As tourism, offshore manufacturing, and a growing urban population demand more water and produce more waste, Caribbean countries will be forced to make hard decisions about their infrastructure.

In addition to being a public health concern, water contamination poses serious economic problems for states dependent on tourism. Governments are caught between a desire to fix the problem before tourists notice and a tendency not to discuss it at all. Doing nothing, however, may not be an option. In Grenada, for example, ocean-dumped sewage from the capital, St. George, is killing the sea fans that, among other things, protect the island's most important tourist beach from erosion. In the more industrialized Puerto Rico, it is estimated that half the country's coastline is unfit for swimming, mostly due to contamination from sewage.

Throughout the Caribbean it is increasingly recognized that protecting the environment is not a luxury but a question of economic livelihood. There are various environmental groups that lobby their governments to create new laws or comply with existing ones. Although many see increased environmental awareness as a hopeful sign, both urban and rural ecosystems suffer from serious degradation.

Population and Settlement: Densely Settled Islands and Rimland Frontiers

In the Caribbean, population density is generally quite high, and like that of neighboring Latin America, increasingly urban. Eighty-five percent of the region's population is concentrated on the four islands of the Greater Antilles (Figure 5.11). Add to this Trinidad's 1.3 million and Guyana's 700,000, and most of the population of the Caribbean is accounted for by six countries and one U.S. territory (Puerto Rico). Of these territories Puerto Rico has the highest population density, with 1,100 people per square mile (424 people per square kilometer), whereas Hispaniola and Jamaica have slightly more than 500 people per square mile (200 people per square kilometer).

In absolute terms, few people inhabit the Lesser Antilles; nevertheless, some of these microstates are densely settled.

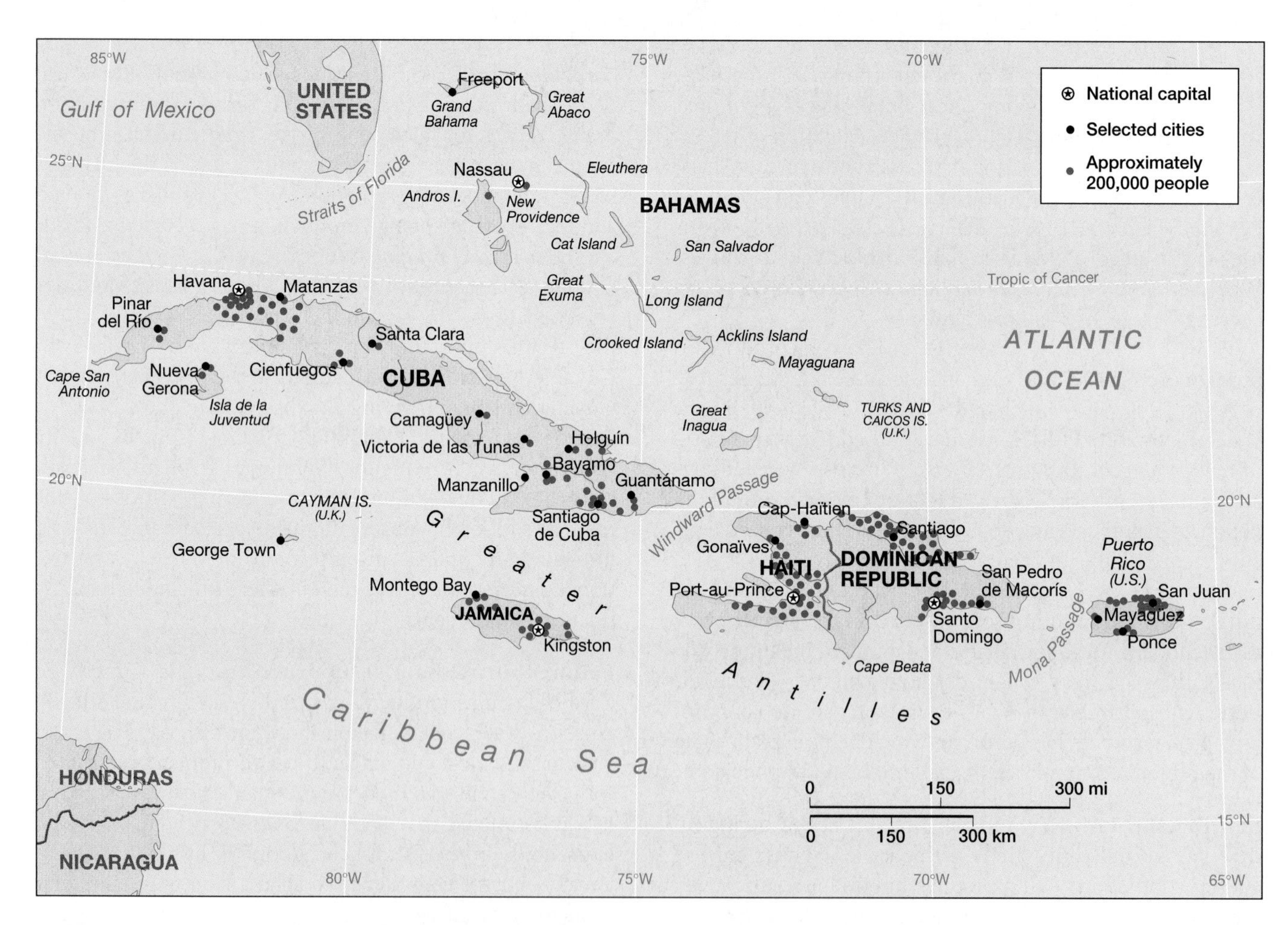

▲ **Figure 5.11 Population of the Greater Antilles** The major population centers are on the islands of the Greater Antilles. The pattern here, like the rest of Latin America, is a tendency toward greater urbanism. The largest city of the region is Santo Domingo, followed by Havana. In comparison, the rimland states are very lightly settled. All of Suriname has just 400,000 people, and Belize has just 200,000.

The small island of Barbados is an extreme example. With only 161 square miles (417 square kilometers) of territory, it has 1,800 people per square mile (695 people per square kilometer). Population densities on St. Vincent, Martinique, and Grenada, while not as high, are still more than 700 people per square mile (270 people per square kilometer). If one takes into consideration the scarcity of arable land on some of these islands, it is clear that access to land is a basic resource problem for many inhabitants of the Caribbean. The growth in the region's population coupled with its scarcity of land has forced many people into the cities or abroad. It also has forced many Caribbean states to be net importers of food.

In contrast to the islands, the mainland territories of Belize and the Guianas are lightly populated; Guyana, averages 8.5 people per square mile (3.3 people per square kilometer), Suriname only 6 people, and Belize 23 (respectively, 2.3 and 8.9 people per square kilometer). These areas are sparsely settled, in part, because the relatively poor quality and accessibility of arable land made them less attractive to colonial enterprises. Presently, Belize is the fastest-growing state in the region due to immigration from the surrounding Central American countries and high fertility rates (2.7 percent). By 2025 Belize is projected to have 400,000 people.

Demographic Trends

During the years of slave-based sugar production, mortality rates were extremely high due to disease, inhumane treatment, and malnutrition. Consequently, the only way population levels could be maintained was through the continual importation of African slaves. With the end of slavery in the mid- to late nineteenth century, and the gradual improvement of health and sanitary conditions on the islands, natural population increase began to occur. In the 1950s and 1960s many states achieved peak growth rates of 3.0 or higher, causing population totals and densities to soar. Over the past 20 years, however, growth rates have come down or stabilized. The Caribbean is still growing at a rate of 1.4 percent; its projected population in 2025 is 49 million (Table 5.1).

Table 5.1 Demographic Indicators

Country	Population[a]	Natural Increase	TFR[b]	%<15[c]	% Urban
Anguilla	0.01	3.4	2.0	28	—
Antigua and Barbuda	0.1	1.2	1.7	30	36
Bahamas	0.3	1.7	2.0	32	86
Barbados	0.3	0.5	1.7	24	38
Belize	0.2	2.7	4.1	42	51
Cayman	0.04	4.3	1.4	—	—
Cuba	11.1	0.6	1.4	22	74
Dominica	0.1	1.1	2.0	—	—
Dominican Republic	8.3	2.1	3.2	36	62
French Guiana	0.2	3.6	3.4	32	—
Grenada	0.1	2.3	3.8	38	32
Guadeloupe	0.4	1.2	2.0	26	99
Guyana	0.7	1.7	2.7	35	36
Haiti	7.5	2.1	4.8	40	33
Jamaica	2.6	1.8	3.0	32	50
Martinique	0.4	0.9	1.7	24	81
Montserrat	0.01	0.2	1.9	—	—
Netherlands Antilles	0.2	1.2	2.2	27	90
Puerto Rico	3.9	1.0	2.1	27	71
St. Kitts–Nevis	0.04	1.0	2.6	34	43
St. Lucia	0.1	1.9	2.7	37	48
St. Vincent and Grenadines	0.1	1.6	2.4	37	25
Suriname	0.4	1.8	2.6	34	70
Trinidad and Tobago	1.3	0.8	1.9	29	65
Turks and Caicos	0.01	1.9	1.8	—	—

[a]Population in millions, 1997.
[b]Total fertility rate.
[c]Percentage of population younger than 15 years of age.
Source: *PRB World Population Data Sheet*, 1998. Data for dependent territories from CIA, *World Factbook*, 1997.

▲ **Figure 5.12 Fewer children** A mother from the Bahamas with her two children. The average Caribbean woman has far fewer children now than 30 years ago. Higher levels of education, improved availability of contraception, a rise in urban living, and a large percentage of women in the labor force contribute to slower population growth rates in many countries. *(Tony Arruza/Tony Arruza Photography)*

Haiti and the Dominican Republic are still growing quickly and may double their size in the next 30 years. Given the severe strain already placed on the island's resources, it is hard to image Hispaniola supporting 30 million people unless the economy is radically restructured away from its subsistence and export agricultural base.

Fertility Decline The most significant demographic trend in the Caribbean is that of fertility decline (Figure 5.12). Cuba and Barbados have the region's lowest rates of natural increase (Table 5.1). In socialist Cuba the education of women combined with the availability of birth control and abortion means that the average woman has 1.4 children (compared to 2.0 in the United States). In Barbados similar results were achieved, albeit in a capitalist context. Barbados's population pressures were also eased by a steady out-migration of young Barbadians overseas, especially to Great Britain. The decision to emigrate, along with a preference for smaller families, have contributed to slower growth rates throughout the region.

Emigration A pattern of emigration to other Caribbean islands, North America, or Europe began in the 1950s. Barbadians generally choose England, most settling in the London suburb of Brixton with other Caribbean immigrants. In contrast, one out of every three Surinamese have moved to the Netherlands, with most residing in Amsterdam. As for Puerto Ricans, slightly more live on the island than those residing on the U.S. mainland. In the 1980s roughly 10 percent of Jamaica's population legally immigrated to North America (some 200,000 to the United States and 35,000 to Canada). Cubans have made the city of Miami their destination of choice since the 1960s. Today, they are the majority of that city's population.

Intraregional movements are also important. While perhaps one-fifth of all Haitians do not live on their island of birth, the most common destination is the neighboring Dominican Republic, followed by the United States and Canada. Dominicans also leave; the vast majority come to the United States, settling in New York. Others, however, simply cross the Mona Passage and settle in Puerto Rico. The economic implications of this labor-related migration are significant and will be discussed later.

The Rural–Urban Continuum

Initially, plantation agriculture and subsistence farming shaped Caribbean settlement patterns. Low-lying arable lands were dedicated to export agriculture and controlled by the colonial elite. Only small amounts of land were set aside for subsistence production. Over time, villages of freed or runaway slaves existed, especially in remote areas of the interior. Still the vast majority of people lived on estates, as owners, managers, or slaves. The cities that formed existed to serve the administrative and social needs of the colonizers but most were small, containing a small fraction of a colony's population. The colonists that linked the Caribbean to the world economy saw no need to develop major urban centers.

Even today the structure of Caribbean communities reflects the plantation legacy. Many of the region's subsistence farmers are descendants of former slaves, who continue to work their small plots and seek seasonal wage-labor on estates. The social and economic patterns generated by slavery still mark the landscape. Rural communities tend to be loosely organized, labor is transient, and small farms are scattered wherever pockets of available land are found. Since men tend to leave home for seasonal labor, matriarchal family structures and female-headed households are common.

The construction of **houseyards** in the Lesser Antilles typifies the blending of rural subsistence, economic survival and a matriarchal social structure. These small enclosed properties of a half acre or less include a number of dwellings, small livestock, fruit trees, herb gardens, and a protected play and work space. Typically the houseyard is owned by a woman and often her extended family of married children live there. It is not unusual for a woman to be associated with one yard for most of her life, while a male's connection to this domestic space is more tenuous (Figure 5.13).

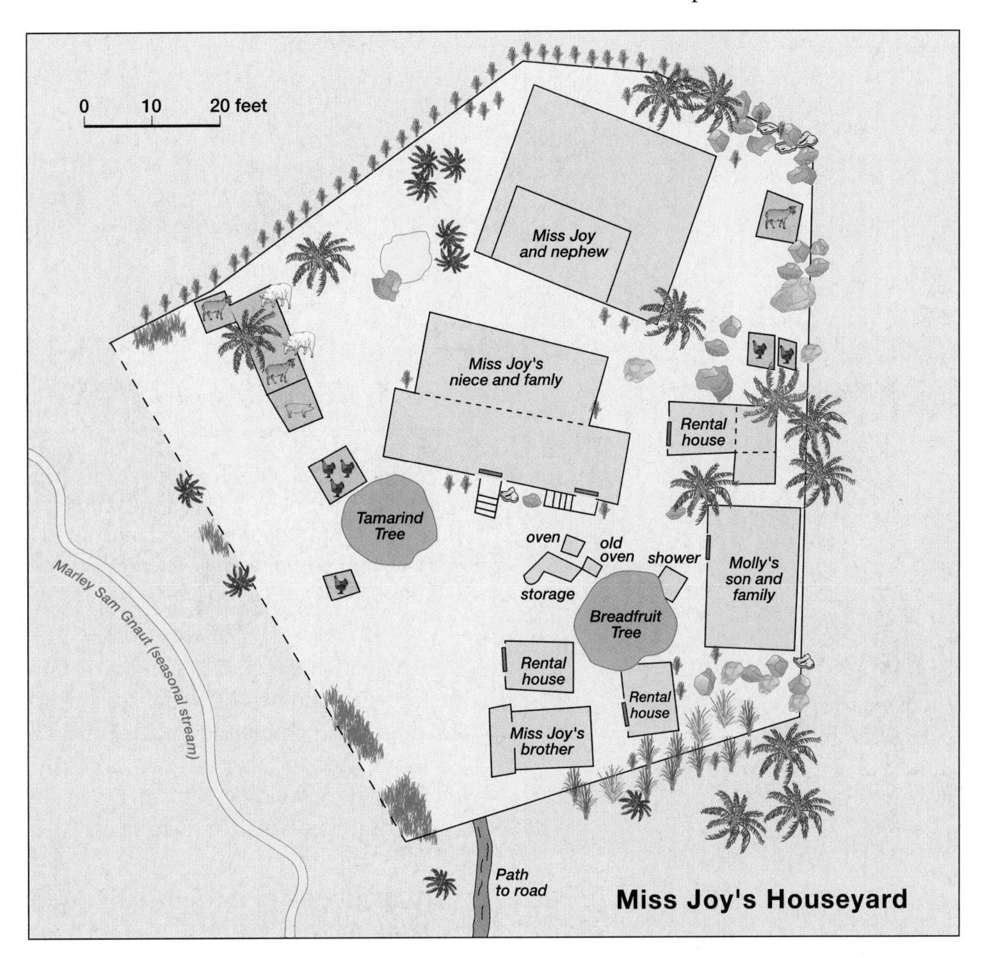

▲ **Figure 5.13 Miss Joy's Houseyard** Caribbean houseyards are an adaption to conditions of land scarcity, poverty, out-migration and matriarchal family structures. Compound life offers a social network, some subsistence, and a relatively safe and inexpensive way to live. *(Source: Lydia Pulsipher. 1993. "Changing Roles in the Life Cycles of Women in Traditional West Indian Houseyards," pp. 50–64, in* Women and Change in the Caribbean, *Janet Momsen, editor)*

Caribbean Cities Since the 1960s, the mechanization of agriculture, offshore industrialization, and rapid population growth caused a surge in rural-to-urban migration. Cities have grown accordingly, and today 60 percent of the region is classified as urban. Cuba is the most urban (74 percent) and Haiti the least (32 percent). Caribbean cities are not large by world standards as only four have one million or more residents: Santo Domingo (2.2 million), Havana (2.1 million), Port-au-Prince (1.5 million), and San Juan (1.0 million). All but Port-au-Prince were laid out by the Spanish.

Like their counterparts in Latin America, the Spanish Caribbean cities were laid out on a grid with a central plaza. Vulnerable to raids by rival European powers and pirates, these cities were usually walled and extensively fortified. The oldest continually occupied European city in the Americas is Santo Domingo in the Dominican Republic, settled in 1496. Yet it was Havana that emerged as the most important city in the region, serving as a port for all incoming and outgoing Spanish galleons. Strategically situated on Cuba's north coast at a narrow opening to a natural deep-water harbor, Havana became an essential city for the Spanish empire. Consequently, Havana possesses a handsome collection of colonial architecture, especially from the eighteenth and nineteenth centuries (Figure 5.14). Only recently did Santo Domingo edge out Havana as the Caribbean's largest city.

Other colonial powers left their mark on the region's cities. For example, Paramaribo, the capital of Suriname, has been described as a tropical, tulipless extension of Holland. In the British colonies a preference for wooden white-washed cottages with shutters was evident. Yet the British and French colonial cities tended to be unplanned afterthoughts; these port cities were built to serve the rural estates, not the other way around. Most of these cities have grown dramatically in the last 40 years. No longer small ports for agricultural exports, increasingly these cities are oriented to welcoming cruise ships and sun-seeking tourists.

Caribbean cities and towns do have their charms and reflect their melange of cultural influences. Throughout the region, houses are often simple structures (made of wood, brick, or stucco), raised off the ground a few feet to avoid flooding, and painted in soft pastels. Most people still get around by foot, bicycle, or public transportation; neighborhoods are

▲ **Figure 5.14 Old Havana** A Cuban family strolls down a narrow cobble-stoned street near the Cathedral in Old Havana, Cuba. The best examples of eighteenth and nineteenth colonial architecture in the Caribbean are found in Havana. *(Rob Crandall/Rob Crandall Photographer)*

filled with small shops and services that are within easy walking distance. Streets are narrow and the pace of life is markedly slower. Even when space is tight in town, most settlements are close to the sea and its cooling breezes. An afternoon or evening stroll along the waterfront is a common activity. Flowering shrubs and swaying palms add to the tropical ambiance.

Housing The sudden surge in urbanization in the Caribbean is best explained by an erosion of rural jobs rather than a corresponding rise in urban opportunities. Thousands poured into the cities as economic refugees, erecting shantytowns and filling the ranks of the informal sector. Squatter settlements in Port-au-Prince and Santo Domingo are especially bad, with residents living in wretched housing without the benefits of sewers and running water. Electricity exists because it is usually pirated from nearby power lines.

The one place that dramatically breaks from this pattern is Cuba. Forged in a socialist mode, Cubans are housed in uniform government-built apartment blocks like those seen throughout Russia and Eastern Europe (Figure 5.15). These

▲ **Figure 5.15 Government housing** Cubans in a state-built apartment block on the outskirts of greater Havana. Under socialism, thousands of standardized apartment blocks were built to ensure that all Cubans had access to basic modern housing. *(Rob Crandall/Rob Crandall Photographer)*

unimaginative complexes contained hundreds of identical units of one- and two-bedroom apartments where basic modern amenities (plumbing, electricity, and sewage) are provided. Compared to other large cities of the developing world, the paucity of squatter settlements makes Havana unusual (see "Local Voices: Life in Belize City").

Cultural Coherence and Diversity: A Neo-Africa in the Americas

Linguistic, religious, and ethnic differences abound in the Caribbean. A score of European colonies, millions of ethnically distinct Africans, indentured labor from India and China, and isolated Amerindian communities on the mainland challenge any notion of cultural coherence.

Within such diversity, common historical and cultural processes provide the rudimentary glue for the region. European colonization with their plantation-based economies reproduced similar social structures throughout the region. The imprint of more than seven million African slaves, creating a Neo-Africa in the Americas, is the focus of vigorous scholarly research to show the important linkages between the Caribbean with the wider Atlantic world. Last, in a process called **creolization,** African and European cultures were blended in the Caribbean. Through this mixing, European languages were transformed into vibrant local dialects and, at times, entirely new languages were created (Papiamento and French Creole). This melding also produced the rich and diverse musical traditions heard throughout the world: reggae, salsa, merengue, and calypso.

The Cultural Imprint of Colonialism

The formation of European colonies in the Caribbean destroyed indigenous society and imposed completely different social systems and cultures. Plantation-based agriculture, dependent upon forced and indentured labor, is the clearest expression of how the colonists ordered space and soci-

LOCAL VOICES Life in Belize City

The following passage describes Belize City through the eyes of a young girl, Beka Lamb. The novel *Beka Lamb* is a Belizean coming-of-age story set in the capital city in the 1960s when it was still a British colony.

At times like these, Beka was glad that her home was one of those built high enough so that she could look for some distance over the rusty zinc rooftops of the town. Many of the weathered wooden houses, built fairly close together, tilted slightly as often as not, on top of pinewood posts of varying heights. In the streets, by night and by day, vendors sold, according to the season, peanuts, peppered oranges, craboos, roasted pumpkin seeds and coconut sweets under lamp-posts that also seemed at times to lean. For a while, after heavy rainstorms, water flooded streets and yards at least to the ankles.

A severe hurricane early in the twentieth century, and several smaller storms since that time had helped to give parts of the town the appearance of a temporary camp. But this was misleading, for Belizeans loved their town which lay below the level of the sea and only through force of circumstances, moved to other parts of the country. It was a town, not unlike small towns everywhere perhaps, where each person, within his neighborhood, was an individual with well known characteristics. Anonymity, though not unheard of, was rare. Indeed, a Belizean without a known legend was the most talked about character of all.

It was a relatively tolerant town where at least six races with their roots in other districts of the country, in Africa, the West Indies, Central America, Europe, North America, Asia, and other places, lived in a kind of harmony. In three centuries miscegenation, like logwood, had produced all shades of black and brown, not grey or purple or violet, but certainly there were a few people in town known as red ibos. Creole regarded as a language to be proud of by most people in the country, served as a means of communication amongst the races. Still, in the town and in the country, as people will do everywhere, each race held varying degrees of prejudice concerning the others.

The town didn't demand too much of its citizens, except that in good fortune they be not boastful, not proud, and above all, not critical in any unsympathetic way of the town and country.

Source: Zee Edgell, *Beka Lamb*. London: Heinemann, 1982, pp. 11–12.

▲ **Figure 1 Belize City cottage** Residents of Belize City build their wooden cottages on stilts as protection against flooding. Shuttered cottages are typical throughout the British Caribbean. *(Rob Crandall/Rob Crandall Photographer)*

ety. Unlike other colonies that relied on indigenous labor, Caribbean plantations depended upon millions of foreign laborers.

The arrival of Columbus in 1492 triggered a devastating chain of events that depopulated the region within 50 years. A combination of Spanish brutality, enslavement, warfare, and disease reduced the densely settled islands supporting up to three million Caribs and Arawaks into an uninhabited territory ready for the colonizer's hand. The demographic collapse of Amerindian populations occurred throughout the Americas (see Chapter 4). The death rates were highest in the Caribbean. Only fragments of Amerindian communities survived, mostly on the rimland.

By the mid-sixteenth century, as rival European states vied for Caribbean territory, the lands they fought for were virtually uninhabited. In many ways this simplified their task, as they did not have to acknowledge indigenous land claims or work amid Amerindian societies. Instead, the Caribbean territories were reorganized to serve a plantation-based production system. The critical missing element was labor. Once slave labor from Africa, and later indentured labor from Asia, were secured, the small Caribbean colonies became surprisingly profitable. Much of Caribbean culture and society today can be traced to processes that created plantation America.

Plantation America The term **plantation America** was coined by anthropologist Charles Wagley to designate a cultural region that extends from midway up the coast of Brazil, through the Guianas and the Caribbean, and into the southeastern United States. Ruled by a European elite dependent upon an African labor force, this society was primarily coastal and produced agricultural exports. Other characteristics included a reliance upon **mono-crop production** (a single commodity such as sugar) under a plantation system that concentrated land in the hands of elite families. Such a system engendered rigid class lines as well as the formation of a multiracial society that privileged people with lighter skin. The term *plantation America* is not meant to describe a race-based division of the Americas but rather a production system that engendered specific ecological, social, and economic relations (Figure 5.16).

▲ **Figure 5.16 Sugar plantation** A historical illustration (1823) of slaves harvesting sugarcane on a plantation in Antigua. Sugar production was profitable but arduous work. Several million Africans were enslaved and forcibly relocated into the region. *(Michael Holford/Michael Holford Photographs)*

Asian Immigration Before detailing the pervasive influence of Africans on the Caribbean, the lesser-known Asian presence deserves mention. By the mid-nineteenth century, most colonial governments in the Caribbean began to free their slaves. Fearful of labor shortages, they sought **indentured labor** (workers contracted to labor on estates for a set period of time, often several years) from South and Southeast Asia.

The legacy of these indentured arrangements is clearest in Guyana, Trinidad, and Suriname. Guyana and Trinidad were British colonies and most of the contract labor came from India. Today, half of Guyana's population and 40 percent of Trinidad's claim South Asian ancestry. Hindu temples are found in the cities and villages, and many families speak Hindi in the home (Figure 5.17). In Suriname, a former Dutch colony, more than one-third of the population is South Asian and 16 percent are Javanese (from Indonesia).

China was another source of indentured labor, although fewer Chinese came than South Asians. Most of the former English colonies have small Chinese populations of not more than 2 percent. Once these East Asian immigrants fulfilled their agricultural contracts, they often became merchants and small-business owners, positions they still hold in Caribbean society.

Creating a Neo-Africa

African slaves were first introduced to the Americas in the sixteenth century, partly in response to the demographic collapse of the Amerindians. The flow of slaves continued into the nineteenth century. This forced migration of Africans to the Americas was only part of a much more complex **African diaspora**—the forced removal of people from their native area. A slave trade also crossed the Sahara to include North Africa and linked East Africa with a slave trade in the Middle East (see Chapter 6). The best documented slave route is the transatlantic one; at least 10 million Africans landed in the Americas (it is estimated that another two million died en route). More than half of these slaves went to the Caribbean (Figure 5.18).

This input of slaves, in combination with the extermination of the nearly all native inhabitants, recast the Caribbean as the area with the greatest concentration of African transfers in the Americas. The African source areas extended from Senegal to Angola, and slave purchasers intentionally mixed tribal groups in order to dilute ethnic identities. Consequently, intact transfer of religion and languages into the Caribbean did not occur; instead words, customs, and beliefs were blended.

Maroon Societies Communities of runaway slaves—termed **maroons**—offer the most compelling examples of African cultural diffusion across the Atlantic. Clandestine settlements of escaped slaves existed wherever slavery was practiced. Called maroons in English, *palenques* in Spanish, and *quilombos* in Portuguese, many of these settlements were short-lived but others have endured and allowed for the survival of African traditions, especially farming practices,

▲ **Figure 5.17 Hindu immigrants** A Hindu woman wearing a traditional sari worships in front of a temple in Trinidad. South Asian immigrants were brought to the English Caribbean colonies as indentured laborers in the nineteenth and early twentieth centuries. *(Michael Ventura Photography)*

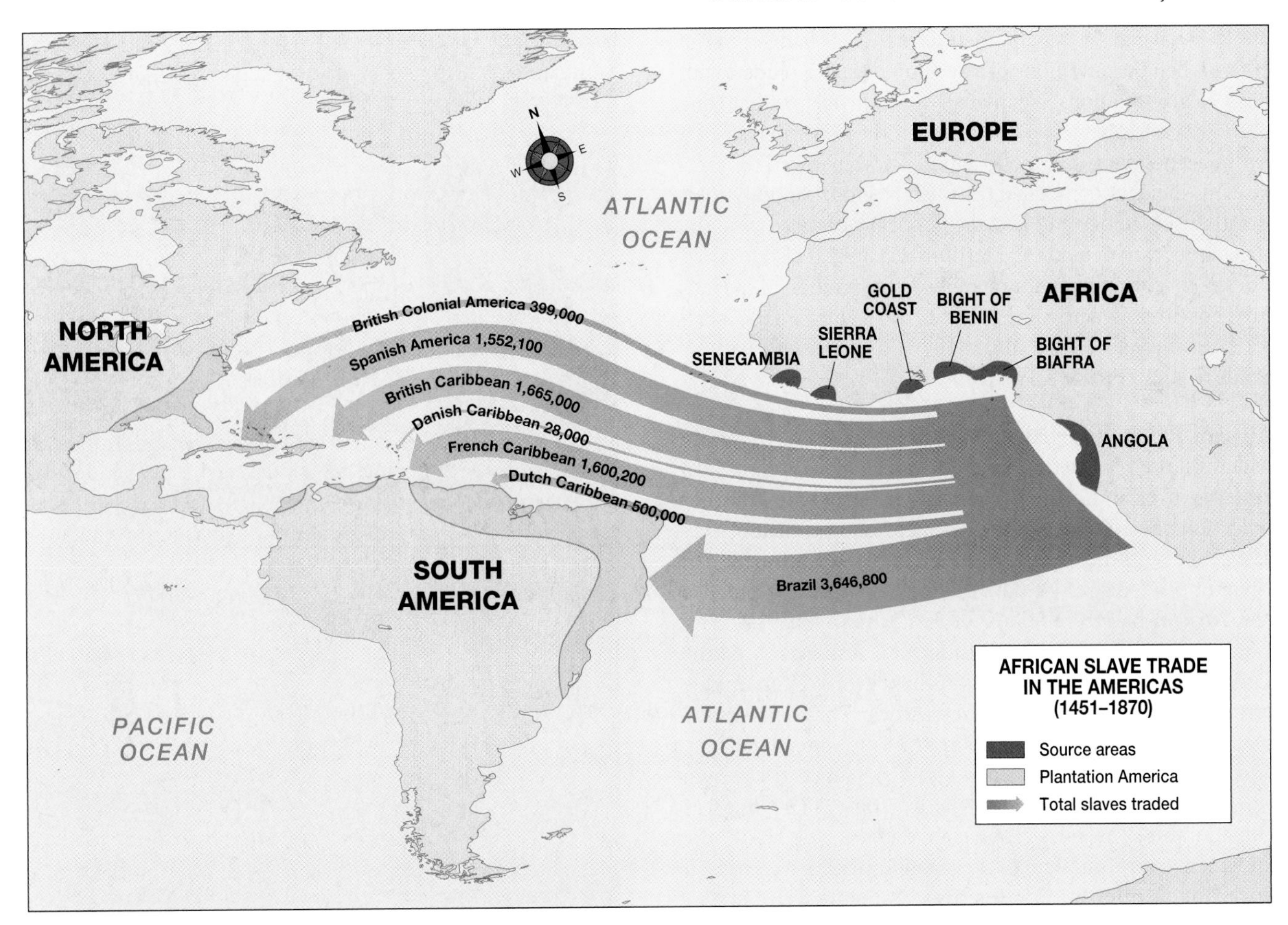

▲ **Figure 5.18 Transatlantic Slave Trade** At least ten million Africans landed in the Americas over the four centuries in which the Atlantic slave trade operated. Most of the slaves came from West Africa, especially the Gold Coast (now Ghana) and the Bight of Biafra (now Nigeria). Angola, in southern Africa, was also an important source area. *(Data based on Philip D. Curtin. 1969.* The Atlantic Slave Trade, A Census. *Madison: University of Wisconsin Press. p. 268).*

house designs, community organization, and language maintenance. The most famous maroon society was Palmares in Pernambuco, Brazil, which existed during almost the entire seventeenth century. At its height the community included 15,000 to 20,000 mostly Bantu-speaking runaway slaves, living self-sufficiently under an African political structure that included a king. Fearful of the challenge that Palmares posed to the Portuguese authority in Brazil, government troops repeatedly attacked this interior African community and eventually destroyed it.

The maroons of Jamaica formed dozens of isolated and independent communities in the forested mountains of the island's interior. As in Brazil, the maroons were a constant source of aggravation for the white plantation owners, yet these smaller communities managed to endure and are well documented. The most striking characteristic of these settlements (with names like Me No Sen, and You No Come) was their ability to maintain some African language and religious practices, especially the belief in Obi (an African god) and Obeah (black magic). With the abolition of slavery in Jamaica, the political necessity for maroon villages ceased to exist, and maroons were gradually assimilated. Today a village that began as a maroon settlement might look like any other rural community in Jamaica, except that there is a local appreciation for the origin of these communities.

The so-called Bush Negros of Suriname still manifest clear links to West Africa. Whereas other maroon societies gradually assimilated into their local populations, to this day the Bush Negros maintain a distinct identity. These runaways fled the Dutch coastal plantations in the seventeenth and eighteenth centuries, forming riverine settlements amid the interior rain forest. Six distinct Bush Negro tribes formed, ranging in size from a few hundred to 20,000. Clear manifestations of West African cultural traditions persist, including religious practices, crafts, patterns of social organization, agricultural systems, and even dress. Living relatively undisturbed for 200 years, these rainforest inhabitants fashioned a rich ritual life for themselves, involving oracles, spirit possession, and witch doctors.

Bush Negros were familiar with Surinamese society through trade with the coast and male out-migration for seasonal work, but the engagement was on their terms. This changed in the 1960s when the ambitions of a developing state collided with Bush Negro traditions. Nearly half of Saramaka territory (one of the largest tribes) was flooded with the construction of a hydroelectric dam, and 6,000 Saramaka were forced to resettle. This incident compelled the Bush Negros to modify their isolationist ideology and become politically engaged with the state in order to defend their territorial claims. It also fostered a desire to affirm their cultural roots. In recognition of the cultural continuities across the Atlantic, tribal elders made a historic trip to West Africa in the 1970s to meet with various heads of state (Figure 5.19).

African Religions Linked to maroon societies, but more widely diffused, is the transfer of African religious and magical systems to the Caribbean. These patterns are another reflection of Neo-Africa in the Americas but are most closely associated with northeastern Brazil and the Caribbean. In Chapter 4 we discussed how millions of Brazilians practice the African-based religions of Umbanda, Macuba, and Condomble along with Catholicism. Likewise in the Caribbean, Afro-religious traditions have evolved into unique forms that have clear ties to West Africa. The most widely practiced are Voodoo (also Vodoun) in Haiti, Santería in Cuba, and Obeah in Jamaica. These religions have their own priesthood and unique patterns of worship. Their impact is considerable; the father and son dictators of Haiti, the Duvaliers, were known to hire voodoo priests to scare off government opposition. Moreover, as Figure 5.20 shows, many of these religions have diffused from their areas of origin. Santería is practiced in Florida and New York by some Cuban immigrants. Likewise, belief in Obeah diffused when Jamaicans migrated to Los Angeles and Toronto.

▲ **Figure 5.19 Bush Negros and Maroons** A man weaves a basket in the Bush Negro settlement of Bolopasi, Suriname. The so-called Bush Negros fled Dutch plantations in the eighteenth and nineteenth centuries and settled in the interior rain forests of Suriname where they continue many West African cultural practices. *(Martha Cooper/The Viesti Collection, Inc.)*

Creolization and Caribbean Identity

Creolization refers to the blending of African, European, and even some Amerindian cultural elements into the unique sociocultural systems found in the Caribbean. Early anthropological work tended to shun the Caribbean as "cultureless," since an indigenous society was replaced with partial fragments of European and African traditions. Yet the Creole identities that have formed over time are complex; they illustrate the dynamics of identity formation and social change that challenge static interpretations of culture. Today, Caribbean writers (V. S. Naipaul, Derek Walcott, Jamaica Kinkaid), musicians (Bob Marley, Celia Cruz, Juan Luís Guerra), and artists (Trinidadian costume designer Peter Minshall) are internationally regarded. Collectively these artists are representative of their individual islands and of Caribbean culture as a whole.

The story of the Garifuna people illustrates creolization at work. Settled along the Caribbean rimland, from the southern coast of Belize to the northern coast of Honduras, the Garifuna (formerly called *the Black Carib*) are descendants of African slaves who speak an Amerindian language. Unions between Africans and Carib Indians on the island of St. Vincent produced an ethnic group that was predominantly African but spoke an Indian language. In the late eighteenth century Britain forcibly resettled some 5,000 Garifuna from St. Vincent to the Bay Islands in the Gulf of Honduras. Over time the Garifuna settled along the Caribbean coast of Central America, living in isolated fishing communities from Honduras to Belize (Figure 5.21). In addition to maintaining an Indian language, the Garifuna are the only group in Central America who regularly eat bitter manioc—a root crop common in lowland tropical South America. It is assumed they acquired their taste for manioc from their exposure to Carib culture. The Afro-Indian blend that the Garifuna manifest is unique, but the process of creolization is recognizable throughout the Caribbean, especially in language and music.

Language The dominant languages in the region are European: Spanish (23 million speakers), French (8 million), English (6 million), and about half a million Dutch speakers (Figure 5.22). Yet these figures only tell part of the story. In Cuba, the Dominican Republic, and Puerto Rico, Spanish is the official language, and it is universally spoken. As for the other countries, colloquial variants of the official language

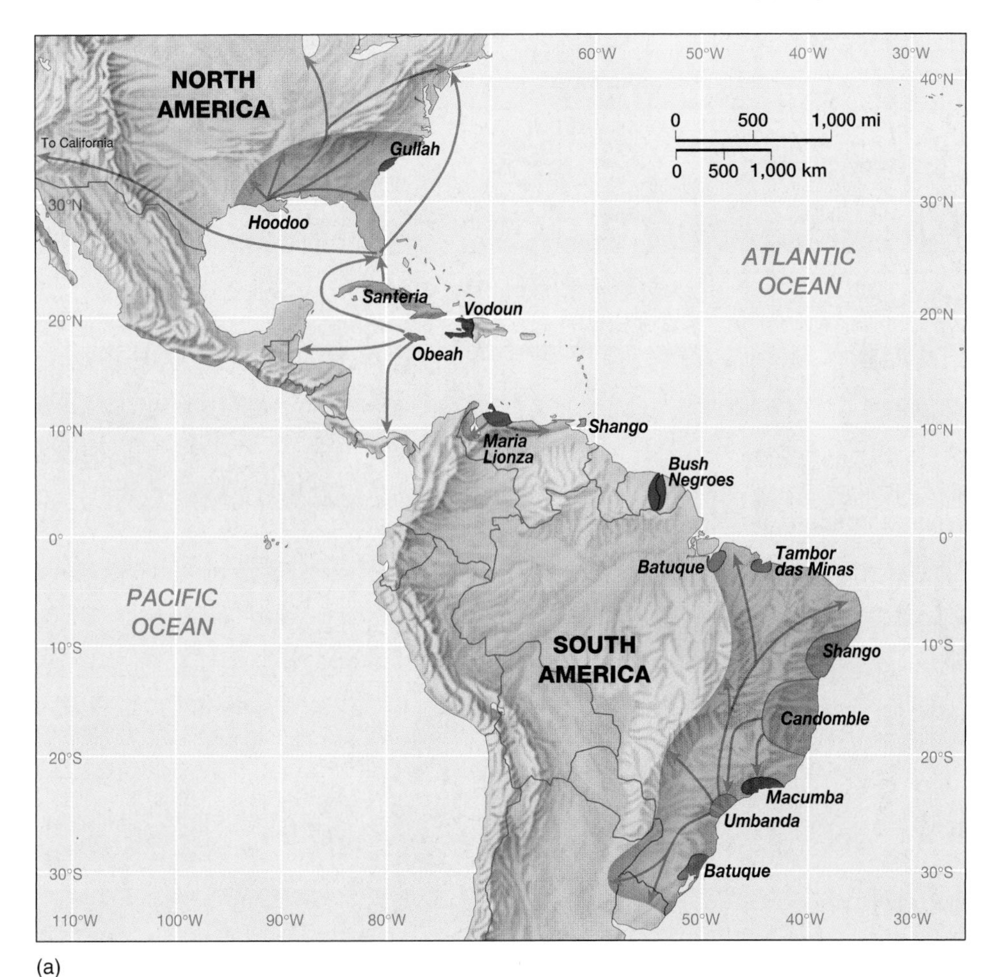

(a)

(b)

◀ **Figure 5.20 African religious influences** (a) African religious practices are found in the Americas where large concentrations of slaves existed. Practitioners of such religions as Voodoo (Vodoun), Santería, Obeah, and Shango often mix their beliefs with Christianity. (b) A worshipper during a Voodoo ceremony in Haiti. *[Source: Robert Voeks. 1993. "African Medicine and Magic in the Americas"* Geographical Review *83(1): 66–78; (b) Tony Savino/The Image Works].*

▲ **Figure 5.21 Garifuna girl** A young Garifuna girl sits on fishing nets in Batalla Village, on the north coast of Honduras. The Garifuna are mostly of African ancestry, but they speak an Amerindian language. *(Rob Crandall/Rob Crandall Photographer)*

exist, especially in spoken form, that can be difficult for a non-native speaker to understand. In some cases completely new languages emerge; in the ABC islands, Papiamento (a trading language that blends Dutch, Spanish, Portuguese, English, and African languages) is the *lingua franca* with usage of Dutch declining. Similarly French Creole or *patois* in Haiti has constitutional status as a distinct language. In practice French is used in higher education, government, and the courts, but *patois* (with clear African influences) is the language of the street, the home, and oral tradition. Most Haitians speak *patois* but only the formally educated know French (see "Language and Identity: Caribbean English").

With independence in the 1960s, Creole languages became politically and culturally charged with national meaning. During the colonial years Creole was negatively viewed as a corruption of standard European forms. Those who spoke Creole where considered uneducated or backward. As linguists began to study these languages, they found that while the vocabulary came from Europe the syntax or semantic structure had other origins, notably from the African language families. While most formal education is taught using standard language forms, the richness of vernacular expression and its ability to instill a sense of identity is appreciated. Locals rely on their ability to switch from standard to vernacular forms of speech. Thus a Jamaican can converse with a tourist in standard English and then switch to a Creole variant when a friend walks by, effectively excluding the outsider from the conversation. While this ability to switch is evident in many cultures, it is widely used in the Caribbean.

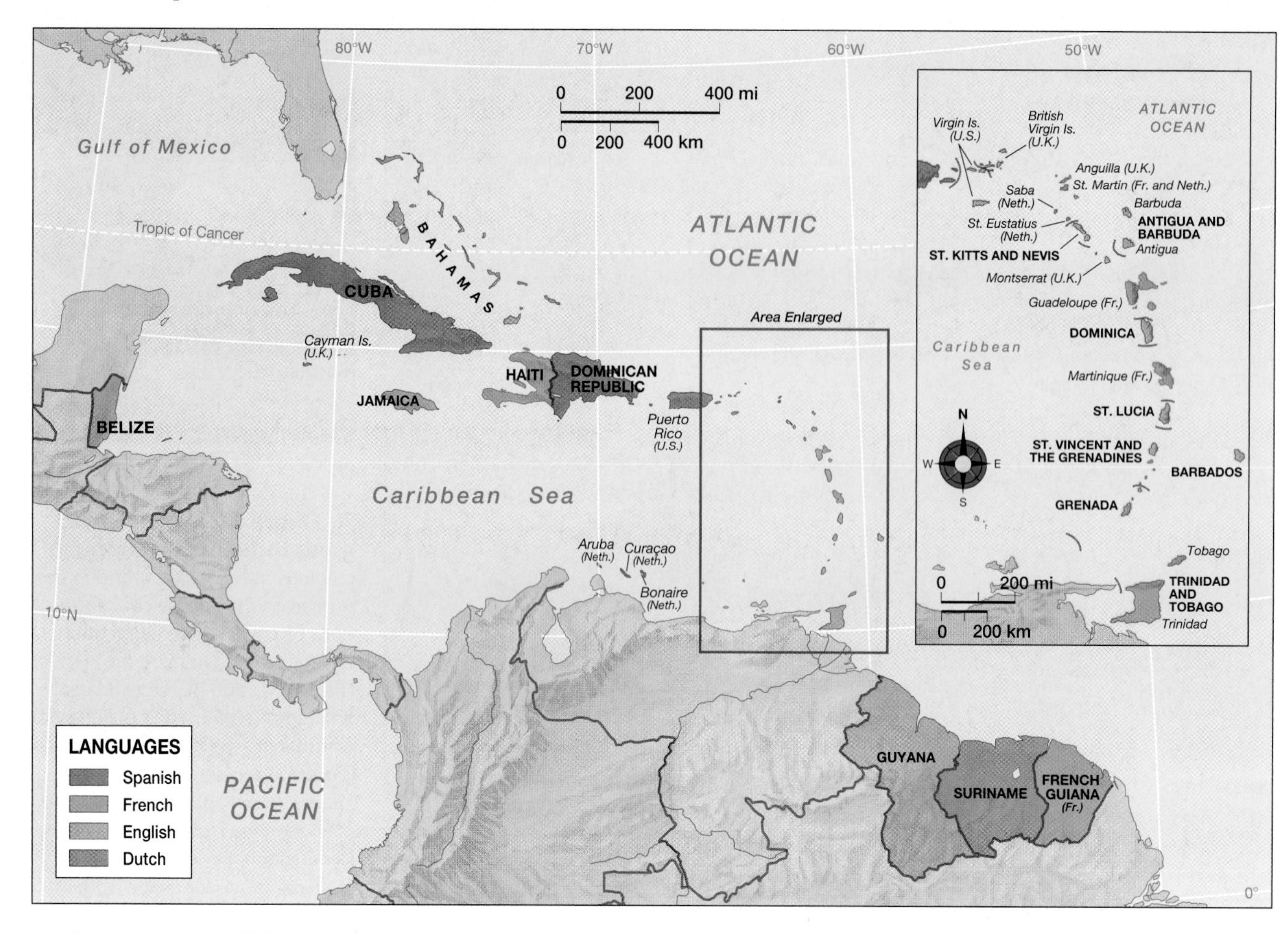

▲ **Figure 5.22 Caribbean language map** Since this region has no significant Amerindian population (except on the mainland), the dominant languages are European: Spanish (23 million), French (8 million), English (6 million), and Dutch (0.5 million). However, many of these languages have been creolized, making it difficult for outsiders to understand them.

Music The rhythmic beats of the Caribbean might be the region's best-known product. This small area is the hearth of reggae, calypso, merengue, rumba, zouk, and scores of other musical forms. The roots of modern Caribbean music reflect a combination of African rhythms with European forms of melody and verse. These diverse influences, coupled with a long period of relative isolation, sparked distinct local sounds. As circulation among Caribbean inhabitants increased, especially during the twentieth century, musical traditions were grafted onto each other but characteristic sounds remained.

The famed steel pan drums of Trinidad were created from oil drums discarded from a U.S. military base there in the 1940s. The bottoms of the cans are pounded with a sledge hammer to create a concave surface the produces different tones. During Carnival, racks of steel pans are pushed through the streets by dancers while the panmen sound off. So skilled are these musicians that they even perform classical music, and government agencies encourage troubled teens to learn steel pan (Figure 5.23).

It is the eclectic sound and the ingenious rhythms that make Caribbean music so popular. Yet it is more than good dancing music; the music is closely tied with Afro-Caribbean religions and is a popular form of political protest. In Haiti, *ra-ra* music mixes percussion instruments, saxophones, and bamboo trumpets, while weaving in funk and reggae baselines. The songs are always performed in French Creole and typically celebrate Haiti's African ancestry and the use of Voodoo. The lyrics address difficult issues such as political oppression or poverty. Consequently, ra-ra groups have been banned from performing and musicians forced into exile—most notably, singer Manno Charlemange. The music of the late Bob Marley also takes a political stand. Jamaican-born Marley sang of his life in the Kingston ghetto of Trenchtown. He was a devout Rastafarian who believed in Jah as the living force, that New World Africans should look to Africa for a prince to emerge (determined to be Haile Selassie of Ethiopia), and that *ganja* (marijuana) should be consumed regularly. It was Marley's political voice as peacemaker, however, that touched so many lives. His first hit, "Simmer Down," was to

LANGUAGE AND IDENTITY Caribbean English

Caribbean English reflects a diverse blend of influences. Lexicographers study the various loan words from African, Spanish, French, Hindi, and Amerindian languages to try and explain the origins of certain terms. Expressions unique to the region are also based on a shared history and often are politically charged. In Belize and Jamaica a *charley-price* is a very large rat. It is credited to Sir Charles Price, an eighteenth-century politician and planter who is believed to have introduced the rat to Jamaica.

Besides unique terms, patterns of speech distinguish spoken Caribbean English. For example, by repeating words, the intensity of expression is evident: *big big big, stupid stupid, fraidy-fraidy*. Or *he talk-talk till I get weary*. Typically these English Caribbean phrasings are part of a spoken vernacular that readily separates locals from outsiders; it is seldom written.

The following terms and gestures reflect the blending of African and European words that make up Caribbean Creole.

backra: A white person of authority, usually a man. A loan word traced to the Efik language in eastern Nigeria. The Efik were established middlemen in the slave trade and probably originated the term. *Backra land* refers to England. *Example:* Le[t] me go and do de backra wo[r]k.

jook: To poke, stab, or wound; to prick or pierce the skin; or to give an infection. This word has phonic correlates in various African languages, including Hausa, Fulani, Mende, and Tsonga. However, European sailors might have reinforced the spread of this word. *Example:* He start jooking me under me arm with his hands and asking if I know him.

joukoutoo: An insignificant, unschooled, or of-no-account person. Used in Grenada, this is probably from French Creole. *Example:* But you dam[ned] rude! Joukoutoo want to wash yo[ur] mout[h] [u]pon me? (But you're damned rude! Even you want to say rude things about me?)

mamaguy: To fool; to trick or deceive, especially by flattery. Used in Trinidad, Tobago, and Grenada. It probably is borrowed from a Venezuelan cock-fighting term *(mamar gallo)* that refers to a fighting cock that only poses but does not fight. *Example:* After you mamaguy me and take my money.

suck-teeth: The action or sound of sucking your teeth; to make an insulting sound with saliva. A salivary sound with similar significance found throughout Sub-Saharan Africa, especially West Africa. *Example:* Give a suck-teeth; *or* He looked up in surprise and answered with a long-drawn suck-teeth.

Source: Richard Allsopp, ed., *Dictionary of Caribbean English Usage.* Oxford: Oxford University Press, 1996.

◀ **Figure 5.23 Carnival drummer** A steel pan drummer performs while his drum cart is pushed through the streets during Carnival. Steel drums originated from discarded oil cans that local peoples fashioned into drums by hammering the tops into a concave surface. Many steel drum bands perform internationally, playing everything from calypso to classical music. *(Rob Crandall/Rob Crandall Photographer)*

▲ **Figure 5.24 Bob Marley** Jamaican reggae musician Bob Marley performs on stage. His music made him an international superstar in the 1970s. *(Kate Simon/Sygma Photo News)*

quell street violence that had erupted in Kingstown in 1964. Other songs such as "Stand-up" and "No Woman No Cry" had a message of social unity and freedom from oppression that resonated in the 1970s. Commercial success never dulled Marley's political edge. Wildly popular in Africa, one of his last concerts before his death in 1981 was in Zimbabwe to mark its independence (Figure 5.24).

Geopolitical Framework: Colonialism, Neo-Colonialism, and Independence

Caribbean colonial history is a patchwork of rival powers dueling over profitable tropical territories. By the seventeenth century, the Caribbean had become an important proving ground for European colonial ambitions. Spain's grip on the region was tentative, and rivals felt confident that they could win territory by gradually moving from the eastern edge of the sea to the west (Figure 5.25). Many territories, especially islands in the Lesser Antilles, changed hands several times. Trinidad, for example, went from Spanish, to limited French control, and finally became an English colony in the eighteenth century. In a few instances, contested colonial holdings produced contemporary border disputes. Only recently did the Guatemalan government give up its claim to Belize, arguing that the British ignored Spanish claims to the area and illegally acquired Belize. To this day, most Guatemalan-made maps show Belize as part of Guatemala. Also there are several long-standing border disputes among the Guianas. The most notable is Venezuela's claim to the western half of Guyana.

Europeans viewed the Caribbean as a strategic and profitable region to produce sugar, rum, and spices. Geopolitically, rival European powers also felt that their presence in the Caribbean checked Spanish hegemony there. Yet Europe's geopolitical dominance in the Caribbean began to wane by the mid-nineteenth century just as the U.S. presence increased. Inspired by the **Monroe Doctrine**, which claimed that the United States would not tolerate European military involvement in the Western Hemisphere, the U.S. government made it clear that it considered the Caribbean to be within its sphere of influence. This view was underscored during the Spanish-American War in 1898. Even through several English, Dutch, and French colonies persisted after this date, the United States indirectly (and sometimes directly) asserted its control over the region, ushering in a period of **neocolonialism.**

Life in the "American Backyard"

To this day, the United States maintains a proprietary attitude toward the Caribbean, referring to it as "the American Backyard." Initial foreign policy objectives were to free it from European tyranny and foster democratic governance. Yet time and again, American political and economic ambitions betrayed these goals. President Theodore Roosevelt made his priorities clear with imperialistic policies that emphasized the construction of the Panama Canal and the maintenance of open sea lanes. The United States later offered benign-sounding development packages such as the Good Neighbor Policy (1930s), the Alliance for Progress (1960s), and the Caribbean Basin Initiative (1980s). The Caribbean view of these initiatives has been guarded at best. Rather than feeling liberated, many residents believe that one kind of political dependence was traded for another: colonialism for neocolonialism.

In the early 1900s the role of the United States in the Caribbean was overtly a military and political one. The Spanish-American War (1898) secured Cuba's freedom from Spain and also resulted in Spain ceding the Philippines, Puerto Rico, and Guam to the United States; the latter two are still U.S. territories. The U.S. government also purchased the Danish Virgin islands in 1917, renaming them the U.S. Virgin Islands and developing the harbor of St. Thomas. French, English, and Dutch colonies were tolerated as long as these allies recognized the supremacy of the United States in the region. Avowedly against colonialism, the United States had become much like an imperial force.

One of the requirements of empire is the ability to impose one's will, by force if necessary. When a Caribbean state refused to abide by U.S. trade rules, navy vessels would embargo its ports. Marines landed and U.S.-backed governments were installed throughout the Caribbean basin. These were not short-term engagements; U.S. troops occupied the Dominican Republic from 1916 to 1924, Haiti from 1913 to 1934, and Cuba from 1906 to 1909 and 1917 to 1922 (Figure 5.26). Even today several important military bases are in the region, including Guantánamo in eastern Cuba. There is greater reluctance to commit troops in the area now, but as recently as 1994 President Bill Clinton sent troops to Haiti to suppress political violence and prevent a mass exodus of Florida-bound refugees.

Many critics of U.S. policy in the Caribbean complain that business interests overshadow democratic principles when determining foreign policy. U.S. banana companies settled the coastal plain of the Caribbean rimland and operated as if they were independent states. Sugar and rum manufacturers from the United States bought the best lands in Cuba and Puerto Rico. Meanwhile, truly democratic institutions remained weak, and there was little improvement in social development. True, exports increased, railroads were built, and

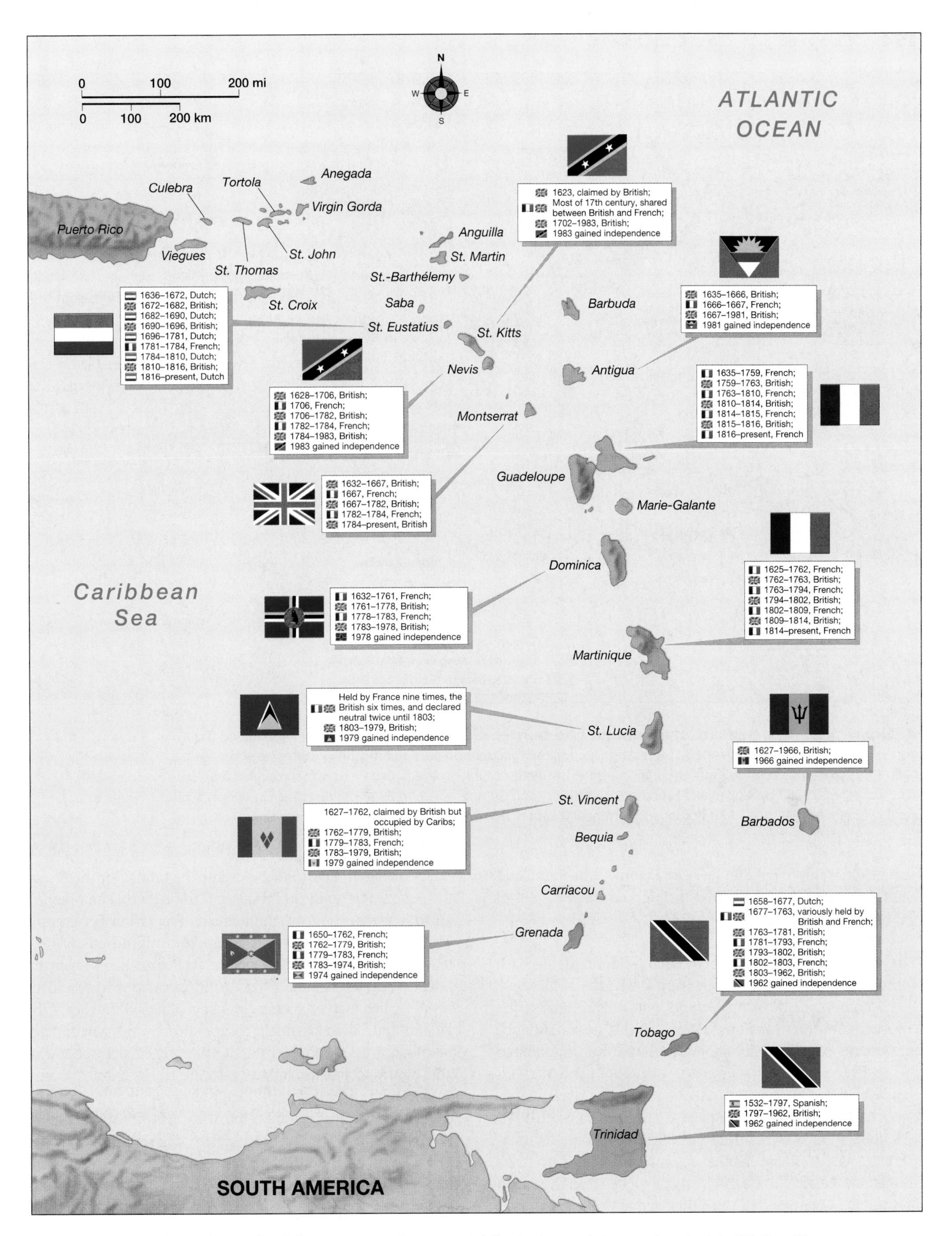

▲ **Figure 5.25 Changing colonial masters** The Lesser Antilles had several changes in colonial affiliation. The French and British traded islands such as Tobago, Grenada, Dominica, and Guadeloupe several times. Many of these territories gained their independence in the 1960s through the 1980s. *(Data source: Bonham Richardson, 1992.* The Caribbean in the Wider World, 1492–1992. *Cambridge University Press. p. 56, reprinted with the permission of Cambridge University Press; and D.P. Henige, 1970.* Colonial Governors from the Fifteenth Century to the Present. *University of Wisconsin Press)*

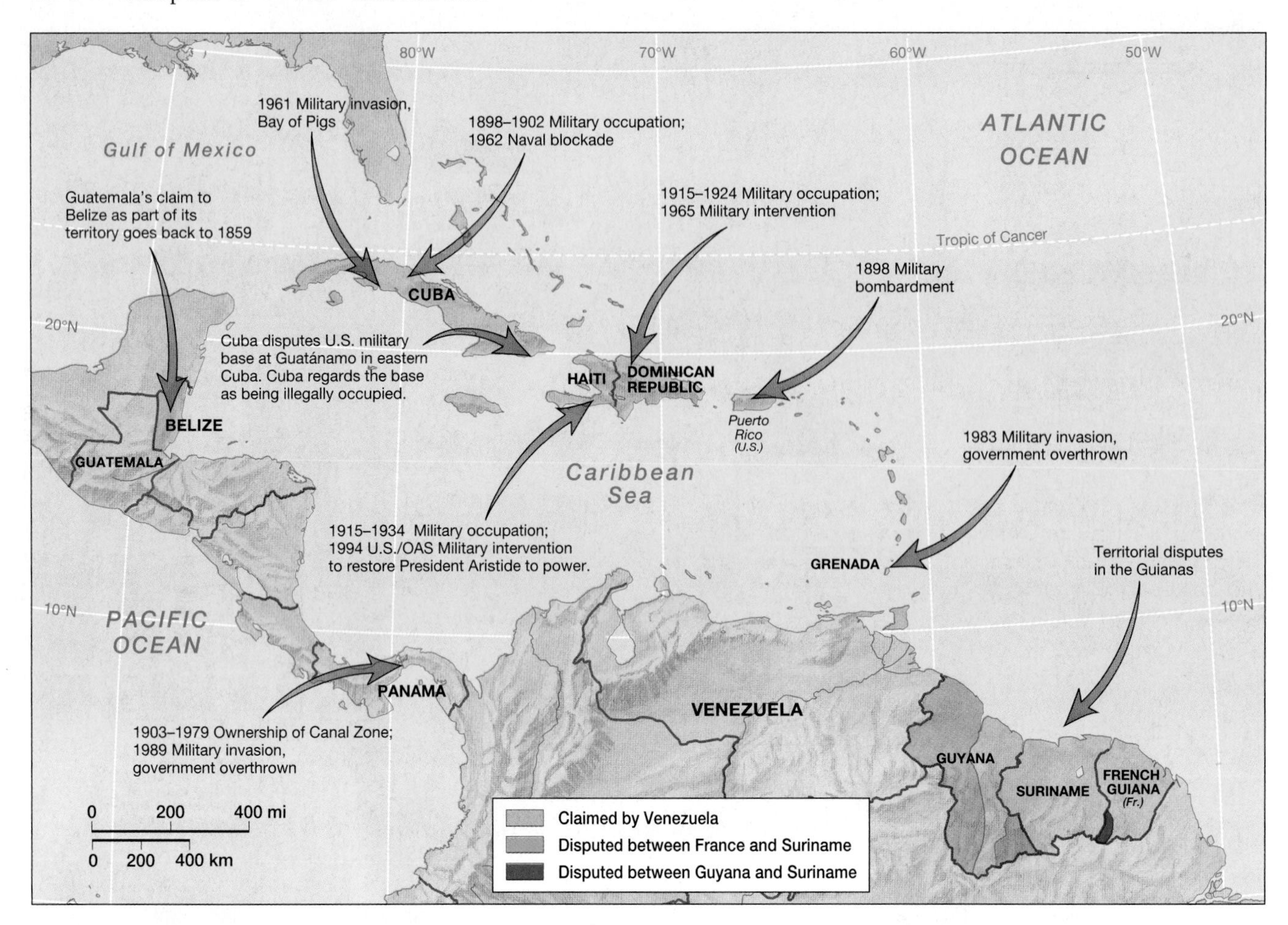

▲ **Figure 5.26 U.S. military involvement and border disputes** The Caribbean was labeled the geopolitical backyard of the United States, and U.S. military occupation was a common occurrence in the first half of the twentieth century. Border conflicts also exist, most notably in the Guianas. *(Data Sources: Barbara Tenenbaum, ed.* 1996 Encyclopedia of Latin American History and Culture, *vol. 5, p. 296, with permission of Charles Scribner's Sons; and John Allcock, 1992.* Border and Territorial Disputes, *Third Ed. Harlow, Essex, UK: Longman Group).*

port facilities improved, but levels of income, education, and health remained abysmally low throughout the first half of the twentieth century.

The Commonwealth of Puerto Rico Puerto Rico is both within the Caribbean and apart from it because of its status as a commonwealth of the United States. Throughout the twentieth century various Puerto Rican independence movements sought to uncouple the island from the United States. Even today, residents of the island are divided about their island's political future. At the same time, Puerto Rico depends upon U.S. investment and welfare programs. In 1974 U.S. food stamps were a leading source of income for two-thirds of Puerto Rican families. Commonwealth status also means that Puerto Ricans can freely move between the island and the U.S. mainland, a right they actively assert. In other ways Puerto Ricans symbolically manifest their independence; for example, they support their own "national" sports teams and send a Miss Puerto Rico to international beauty pageants.

In the 1950s Puerto Rico led the Caribbean in the transition from an agrarian economy to an industrial one. For some U.S. statesmen, Puerto Rico became the model for the rest of the region. Puerto Rican President Muñoz Marín championed an industrialization program called "Operation Bootstrap." Through tax incentives and cheap labor, hundreds of U.S. textile and apparel firms relocated to Puerto Rico. Over the next two decades, some 140,000 industrial jobs were added, resulting in a marked increase in per capita GNP. By the 1970s when Puerto Rico faced stiff competition from Asian apparel manufacturers, the government encouraged petrochemical and pharmaceutical plants to relocate to the island (Figure 5.27). By the 1990s, Puerto Rico was one of the most industrialized localities in the region with a significantly higher GNP than its neighbors (see Table 5.2). Yet it still showed many signs of underdevelopment (rampant out-migration, low rates of educational attainment, and widespread poverty and crime).

Cuba and Regional Politics The most profound challenge to U.S. authority in the region came from Cuba and its superpower ally, the former Soviet Union. Historically, Havana was the dominant city of the region. After Spain lost its

▲ **Figure 5.27 Puerto Rican factory** A pharmaceutical plant near San Juan, Puerto Rico. Operation Bootstrap helped to launch the rapid industrialization of Puerto Rico. The island is the Caribbean's most industrialized location. *(Robert Frerck/Odyssey Productions)*

mainland territories in the 1820s, it invested heavily in the sugar and tobacco industries in Cuba and Puerto Rico. Cubans formally challenged Spain's colonial rule in 1895. Over a three-year period, 400,000 Cubans died in their fight for independence. Even though the United States is credited with winning what is known as the Spanish-American War in 1898, U.S. forces were relative latecomers to this struggle. In the end, many Cubans resented the United States's presence on their island because it undermined their hard-fought independence.

In the 1950s another revolutionary effort began in Cuba led by Fidel Castro against the pro-American Batista government. Cuba's economic productivity had soared but its people were still poor, uneducated, and increasingly angry. The contrast between an average cane worker and a foreign elite was stark and inescapable. In the 1950s Havana emerged as the top Caribbean tourist destination for Americans, further demonstrating the gap between the classes. Beach resorts, casinos, and several hundred brothels cemented Havana's reputation as "Sin City." Yet the vast majority of rural workers

Table 5.2 Economic Indicators

Country	GNP per Capita ($U.S., 1996)	Total GNP (Millions of $U.S., 1996)	PPP* ($Intl, 1996)	Real Annual Growth % per Capita, 1990–1996
Anguilla	—	52	7,400	—
Antigua and Barbuda	7,330	482	8,660	2.0
Bahamas	—	—	10,180	-2.0
Barbados	—	—	10,510	-0.8
Belize	2,700	600	4,170	0.7
Cayman	23,800	860	—	—
Cuba	—	—	—	—
Dominica	3,090	228	4,390	2.3
Dominican Republic	1,600	12,765	4,390	3.1
French Guiana	—	800	6,000	—
Guadeloupe	—	3,700	9,200	—
Grenada	2,880	285	4,340	0.6
Guyana	690	582	2,280	10.4
Haiti	310	2,282	1,130	-6.9
Jamaica	1,600	4,066	3,450	0.9
Martinique	—	3,950	10,000	—
Montserrat	—	55	4,360	—
Netherlands Antilles	—	2,000	9,800	—
Puerto Rico	—	31,600	8,200	—
St. Kitts–Nevis	5,870	240	7,310	3.5
St. Lucia	3,500	553	4,920	2.8
St. Vincent and Grenadines	2,370	264	4,160	2.4
Suriname	1,000	433	2,630	-0.3
Trinidad and Tobago	3,870	5,017	6,100	0.1
Turks and Caicos	—	84	6,400	—

*Purchasing power parity.

Source: *World Bank Atlas,* 1998. Data for dependent territories from CIA, *World Factbook,* 1997.

felt betrayed by six decades of American neocolonialism. In defiance, they lent their support in 1959 to a charismatic, cigar-smoking, former law student named Fidel.

Castro tapped a deep vein of Cuban resentment against U.S. policy. After Castro's government nationalized American industries and took ownership of all foreign-owned properties, the United States responded by refusing to buy Cuban sugar and ultimately ending diplomatic relations with the state. What sealed Cuba's fate as a geopolitical enemy was its establishment of diplomatic relations with the U.S.S.R. in 1960 during the height of the Cold War. With the Soviet Union financially and militarily backing Castro, a direct U.S. invasion of Cuba was too risky. The fall of 1962 produced one of the worst episodes of the Cold War, when Soviet missiles were discovered on Cuban soil. Ultimately, the Soviet Union removed its weapons; in return, the United States promised not to invade Cuba.

The Caribbean as Shatterbelt The Cuban missile crisis signaled that American dominance in the Caribbean was not universally accepted. Cuba by itself was not a threat to the United States, but the geopolitical implications of its turn toward the Soviet Union could not be ignored. Suddenly the American backyard had turned into a geopolitical **shatterbelt,** a locality where superpowers vie for political influence. Moreover, Castro wanted to extend his socialist vision abroad, supporting at different times revolutionaries in Colombia, Venezuela, Guyana, Suriname, Bolivia, and across the Atlantic in Angola. Other Caribbean states watched the Cuban socialist experiment with interest, especially its remarkable strides in literacy and public health. By the mid-1970s Cuba had the lowest infant mortality rate in all of the Caribbean and Latin America (see "Cuba—Coping Without a Superpower").

Cuban-style socialism did not readily transfer to other countries, however, because the United States was determined to prevent it, and Soviet support of Cuba proved expensive. Still, socialism appealed to the economically marginalized masses. In 1974 Jamaican President Michael Manley declared his country a democratic socialist state and imposed hefty tariffs on North American mining companies there. The island of Grenada, under the leadership of Maurice Bishop, took a decidedly socialist turn in the late 1970s, allying itself with Cuba and Nicaragua. In the autumn of 1983, after a Revolutionary Military Council assassinated Bishop and assumed power, U.S. paratroops invaded Grenada. Ostensibly they came to liberate American medical-school students, but they also claimed that an airstrip being built with Soviet support threatened the region's stability. After the invasion, millions of U.S. dollars were pumped into the Grenadan economy, and the same airstrip that was viewed as a security threat became the island's largest commercial runway for tourism.

Even with the end of the Cold War, the Caribbean is still within the U.S. sphere of influence. U.S. policy in the region continues to emphasize democratic government and economic growth through market economies. By the 1990s, every independent country but Cuba had a democratically elected leader. The region is no longer described as a shatterbelt. While gains have been made in democratic governance, the economic and political infrastructure is still weak.

Independence and Integration

Given the repressive colonial history of the Caribbean, it is no wonder that the struggle for political independence began more than 200 years ago. Haiti was the second colony in the Americas to gain independence (in 1804); the United States was the first in 1776. The achievement of political independence, enjoyed by many states in the region, does not guarantee economic independence. Many Caribbean states struggle to meet the basic needs of their people. Surprisingly, today some Caribbean territories maintain their colonial status as an economic asset. For example, the French islands of Martinique and Guadeloupe are overseas departments of France; island residents have full French citizenship and social welfare benefits.

Perhaps the hardest task for the Caribbean is fermenting greater regional integration. Scattered islands, a divided rimland, different languages, and limited economic resources inhibit the formation of a meaningful regional trade block. More common is cooperation between groups of islands with a common colonial background.

Independence Movements Haiti's revolutionary war began in 1791 and ended in 1804. Spanish, French, and British forces were involved, as well as factions within Haiti that had formed along racial lines. In the end, a full-blown race war ensued. The island's population was halved, through casualties and emigration; ultimately, the former slaves became the rulers. Independence, however, did not allow this crown of the French Caribbean to prosper. "Plantation America" watched in horror as Haitian slaves used guerrilla tactics to gain their freedom. Fearing that other colonies might take Haiti's lead, plantation owners were on guard for the slightest hint of revolt. For its part, Haiti did not become a leader in liberation. Mired in economic and political problems, it was shunned by the European powers and never embraced by the states of the Spanish mainland when they became independent in the 1820s.

Several revolutionary pulses followed in the nineteenth century. In the Greater Antilles, the Dominican Republic finally gained its independence in 1844, after wresting control of the territory from Spain and Haiti. Cuba and Puerto Rico were freed from Spanish colonialism in 1898, but their independence was compromised by greater U.S. involvement. The British colonies also faced revolts, especially in the 1930s, yet it was not until the 1960s that independent states emerged from the English Caribbean. First the larger colonies of Jamaica, Trinidad and Tobago, Guyana, and Barbados gained their independence. Other British colonies followed throughout the 1970s and early 1980s: the Bahamas in 1973, Grenada in 1974, Dominica in 1978, St. Vincent and the Grenadines in 1979, St. Lucia in 1979, Antigua and Barbuda in 1981, Belize in 1981, and St. Kitts and Nevis in 1983. Suriname, the only Dutch colony on the rimland, became an autonomous territory in 1954 but remained part of the Kingdom of the Netherlands until 1975, when it declared itself an independent republic.

CUBA Coping Without a Superpower

In 1991 Cuba faced the reality of the "zero option": no subsidized food or fuel from the former Soviet Union. The economy was radically restructured and Cubans were asked to ride bikes, pick fruit, endure blackouts, and wait . . . wait for everything.

The first crisis was one of fuel. Dependent upon subsidized Soviet imports of petroleum, suddenly Cuba had a chronic fuel shortage that affected transportation and electricity production. The most striking impact was the emptiness of Cuban streets. In the early 1990s auto traffic in Havana nearly ceased; the few buses that ran were crammed beyond capacity. The solution became bicycles. Nearly two million of them were imported from or built with the help of China. Fuel and parts to run agricultural equipment were also scarce; thus, oxen, carts, and manual harvesting were resurrected. Many factories were reduced to shorter workweeks as they lacked the energy or parts to run. Brownouts and scheduled blackouts also became normal in the early 1990s because of fuel shortages. These were less frequent by the end of the decade, but energy costs escalated.

The second crisis was food. Cuba's overly specialized agricultural sector produced sugar and citrus, relying upon generous imports of staples from the Soviet bloc. To increase and diversify agricultural production, private farms were permitted, local farmer's markets were created, urban residents were encouraged to work in the countryside (especially during harvest), and subsistence crops replaced export ones. In the worst years of the crisis, food was rationed on a monthly basis according to family size. Cuba's sugar harvest was also cut in half due to lack of fuel and working farm equipment. By 1995, however, the sugar harvest began to recover, although the government now earns significantly less from its principal export.

Without subsidized imports of fuel and food, Cuba became desperate for hard currency. Reluctantly, the country turned to tourism to earn foreign exchange. Initially the government tightly regulated tourism, trying to isolate foreign tourists from the daily lives of Cubans and in some cases forbidding Cubans to enter tourist hotels and restaurants. This policy was deeply resented by Cubans, who prefer to interact with tourists. In fact, a brisk informal sector has emerged, catering to tourists, which includes cigar sales, family-run restaurants, and prostitution.

The crisis forced Castro to loosen government controls of the economy. Whereas having U.S. dollars was once a punishable offense, today Cubans freely carry and exchange them. The streets of Havana buzz with small private enterprises, including flea markets, street vendors, and privately owned restaurants all of which are licensed and sanctioned by the government. It is this private sector of the economy that earns hard currency. Spending a couple of hours as a tour guide for wealthy foreigners can earn one more than a month's wages.

Despite the odds against it, Cuba has managed to cope without its superpower patron. Just how Cuba's brand of tropical socialism will evolve is anyone's guess.

▲ **Figure 1 Bicycle power** Cubans riding bicycles participate in a patriotic parade at Playa Larga, Cuba. The scarcity of fuel has led to a dramatic increase in bicycle transport. *(Rob Crandall/Rob Crandall Photographer)*

▲ **Figure 2 Plowing with oxen** A Cuban farmer plows his privately owned plot in preparation for planting corn. Oxen returned to the fields once Soviet subsidies ceased to exist. *(Rob Crandall/Rob Crandall Photographer)*

Present-day Colonies Britain still maintains several Crown Colonies in the region: the Cayman Islands, the Turks and Caicos, Anguilla, and Montserrat. The combined population of these islands is just 71,000 people, yet their standard of living is high due to, in part, their specialization in a recently developed industry of offshore banking. French Guiana, Martinique, and Guadeloupe are each departments of France, and, thus, technically speaking, not colonies. Together they total 900,000 people. The Dutch islands in the Caribbean are considered autonomous countries that are part of the

Kingdom of the Netherlands. Curaçao, Bonaire, St. Martin, Saba, and St. Eustatius comprise the federation of the Netherlands Antilles. Aruba left the federation in 1986 and governs without its influence. Together the population of the Dutch islands is a quarter million people.

Regional Integration Experimentation with regional trade associations as a means to improve the economic competitiveness of the Caribbean began in the 1960s. The goal of regional cooperation was to improve employment rates, increase intraregional trade, and, ultimately, to reduce external dependence. The countries of the English Caribbean took the lead in this development strategy. In 1963 Guyana proposed an economic integration plan with Barbados and Antigua. In 1972 the integration process intensified with the formation of the **Caribbean Community and Common Market (CARICOM).** Representing the former English colonies, CARICOM proposed an ambitious regional industrialization plan and the creation of the Caribbean Development Bank to assist the poorer states. CARICOM also oversees the University of the West Indies, with campuses in Trinidad, Jamaica, and Barbados. As important as this trade group is as an institutional symbol of collective identity, it has produced limited improvements in intraregional trade. The fact that it excluded non-English speaking states underscores the deep linguistic fractures in the Caribbean.

The dream of regional integration as a means to produce a more stable and self-sufficient Caribbean has never been realized. One scholar of the region argues that a limiting factor is a small-islandist ideology. Islanders tend to keep their backs to the sea, oblivious to the needs of one's neighbors. At other times such isolationism results in suspicion, distrust, and even hostility toward nearby states. There is a desire to remain inward-looking, but economic necessity dictates engagement with partners outside the region. And so this peculiar status of isolated proximity unfolds in the Caribbean, expressing itself in uneven social and economic development trends.

Economic and Social Development: From Cane Fields to Cruise Ships

Collectively, the population of the Caribbean lies within the lower-middle income category (per capita GNP is $786 to $3,115 in U.S. dollars) set by the World Bank. This means that the population, albeit poor by U.S. standards, is economically better off than most of Sub-Saharan Africa, South Asia, and China. Given periods of economic stagnation in the Caribbean, social gains in education, health, and life expectancy are significant. Historically the Caribbean's links to the world economy was through tropical agricultural exports, yet several specialized industries such as tourism, offshore banking, and assembly plants have challenged the dominance of agriculture. These industries grew because of the region's proximity to North America and Europe, the availability of cheap labor, and the implementation of policies that created a nearly tax-free environment for foreign-owned companies. Unfortunately, growth in these sectors does not employ all the region's displaced rural laborers, so the lure of jobs in North America and Europe is still strong.

From Fields to Factories and Resorts

Agriculture used to dominate the economic life of the Caribbean. Decades of turbulent commodity prices and the decline in preferential trade agreements with former colonies have produced more hardship than prosperity. Ecologically, the soils are overworked and there are no frontier areas to expand production, save for areas of the rimland. Moreover, agricultural prices have not kept pace with rising production costs, so that wages and profits remain low. With the exception of a few mineral-rich territories such as Trinidad, Guyana, Suriname, and Jamaica, most other countries have systematically tried to diversify their economies, relying less on their soils and more on manufacturing and services.

Comparing export figures over time demonstrates the shift away from mono-crop dependence. In 1955 Haiti earned more than 70 percent of its foreign exchange through the export of coffee; by 1990 coffee accounted for only 11 percent of its export earnings. Similarly, in 1955 the Dominican Republic earned close to 60 percent of its foreign exchange through sugar, but 35 years later sugar earned less than 20 percent of the country's foreign exchange and pig iron had become the leading export. The one exception to this trend is Cuba, which earned approximately 80 percent of its foreign exchange through sugar production from the 1950s to 1990. Cuba is now trying to diversify since Russia no longer guarantees the price supports that kept sugar a lucrative commodity.

Sugar, Coffee, Bananas The evolution of the Caribbean cannot be separated from the production of sugarcane. Even relatively small territories such as Antigua and Barbados yielded fabulous profits because the demand for sugar showed no limits. Once considered a luxury crop, it became a popular necessity for European and North American laborers by the 1750s. It sweetened tea and coffee and made jams a popular spread for stale bread. In short, it made the meager and bland diets of ordinary people tolerable, and it also boosted caloric intake. Distilled into rum, sugar produced a popular intoxicant. Though it is hard to imagine today, consumption of a pint of rum a day was not uncommon in the 1800s.

Sugarcane is still grown throughout the region for domestic consumption and export. Its economic importance has declined, however, mostly due to increased competition from sugar beets grown in the midlatitudes. The Caribbean and Brazil are the world's major sugar exporters. Up until 1990, Cuba alone accounted for more than 60 percent of the value of world sugar exports. Cuba's dominance in sugar exports had more to do with its subsidized and guaranteed markets in Eastern Europe and Russia than with unprecedented productivity. Since 1990 the sugar harvest has been reduced by half.

Coffee is planted in the mountains of the Greater Antilles. Haiti has been the most dependent on coffee, relying upon peasant sharecroppers to tend the plants and harvest the beans. For other countries coffee is a valued specialty commodity. Beans harvested in the Blue Mountains of Jamaica, for example, fetch two to three times the going price for similar highland coffee grown in Colombia. Puerto Rico and Cuba are also trying to develop a niche in the gourmet coffee market. An important production distinction with coffee, as compared to sugar, is that it is mostly grown on small farms and then sold to buyers or delivered to cooperatives. Typically, farmers intermix other crops between the coffee bushes so that they can meet their subsistence needs as well as produce a cash crop. Even with this self-provisioning system, peasants often seasonally abandon their farms for work elsewhere as laborers.

The major banana exporters are in Latin America, not the Caribbean. In fact, the success of banana plantations is mixed in this region, as banana plants are especially vulnerable to hurricanes. Still, a collection of English Caribbean countries (Belize, Jamaica, Dominica, St. Vincent, St. Lucia, and Grenada) has a special status with the European Union that guarantees a market and higher prices for their bananas. Of these countries, the Lesser Antilles islands of Dominica, St. Vincent, and St. Lucia have become the most dependent upon bananas. With pressure on Great Britain by other European states to drop its preferential treatment of Caribbean banana growers, the long-term economic viability of this crop does not look good (Figure 5.28).

Even with the relative decline in the economic importance of agriculture, there are still fields of swaying sugarcane, groves of shaded coffee, and plantations of bananas and tobacco. What has changed is the amount of labor and land needed to produce similar yields. Beginning in the early 1900s and intensifying in the 1950s, sugar production was mechanized, making much of rural labor redundant. Coffee and bananas are still harvested by hand, but they grow in limited areas. In addition, the green revolution meant that the use of chemical fertilizers, pesticides, and high-yielding varieties of crops enabled yields to increase even when the area under cultivation remained the same or declined. The combination of mechanization, green revolution technology, and a shift away from mono-crop production meant that many rural laborers needed to find employment elsewhere.

Assembly-plant Industrialization One regional strategy to deal with the unemployed was to invite foreign investors to set up assembly plants and thus create jobs. This was first tried successfully in Puerto Rico in the 1950s and copied throughout the region. Dubbed "Operation Bootstrap" in Puerto Rico, leaders of the island encouraged U.S. investment by offering cheap labor, local tax breaks, and, most importantly, federal tax exemptions (something only Puerto Rico can do because of its special status as a commonwealth of the United States). Initially the program was a tremendous success, so that by 1970 nearly 40 percent of the island's gross domestic product (GDP) came from manufacturing. Higher oil prices and the imposition of new tariffs rocked the economic foundations of this policy in the 1970s. Today about 14 percent of the Puerto Rican labor force is employed in manufacturing and this sector accounts for nearly half of the island's GDP. Yet competition from other states with even lower wages, and the U.S. Congress's decision in 1996 to phase out many of the tax exemptions, may threaten Puerto Rico's ability to maintain its specialized industrial base.

▲ **Figure 5.28 Banana farm** Small banana farms struggle to be economically viable in a globalizing marketplace. At one time, small island producers such as St. Lucia, shown here, received preferential access to the European market. As the European Union debates dropping this preferential status, many banana-dependent Caribbean nations fear for their livelihood. *(Bob Krist/Corbis)*

Through the creation of **free trade zones (FTZs)**—duty free and tax-exempt industrial parks for foreign corporations—legalization of foreign ownership, pursuit of direct foreign investment, and cheap labor, the Caribbean is increasingly an attractive location to assemble goods for North American consumers. Jamaica began developing its light-manufacturing sector in the 1980s and has since seen manufacturing rise to account for nearly one-fifth of its GDP. Most production takes place in export-oriented free trade zones that receive some government support, offer tax breaks, and have no import restrictions. These zones exist in the island's major cities: Kingston, Spanish Town, and Montego Bay. Beginning with food, drink, and tobacco products, they have expanded to include textiles and garments. As part of the Caribbean Basin Initiative, the United States agreed to purchase most of the garments manufactured in Jamaica. Investment in the free trade zones is not exclusively from the United States; capital from Taiwan and Hong Kong has found its way there too.

The Dominican Republic offers another example of the globalizing trends in manufacturing. Like Jamaica, it took advantage of tax incentives, and guaranteed access to the U.S. market offered through the Caribbean Basin Initiative. The number of free trade zones in the Dominican Republic reached 33 in 1995, including an FTZ on the island's north shore, near the Haitian border, optimistically named the

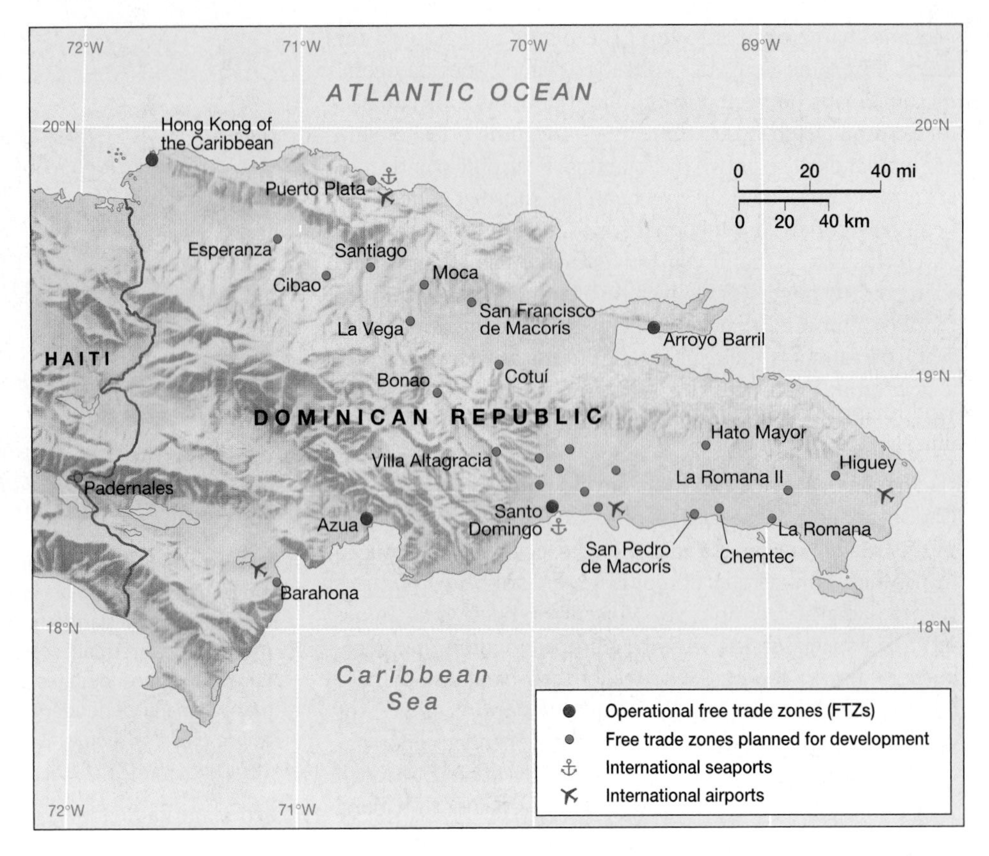

▶ **Figure 5.29 Free Trade Zones in the Dominican Republic** Another sign of globalization is the proliferation of duty-free and tax-exempt industrial parks in the Caribbean. In the Dominican Republic more than 30 of these parks were planned for development in the mid-1990s. *(Source: Barney Warf. 1995.* "Information Services in the Dominican Republic" Yearbook. *Conference of Latin Americanist Geographers. p. 15).*

"Hong Kong of the Caribbean." FTZs are found throughout the country. Firms from the United States and Canada are the most frequent investors in these zones, followed by Dominicans, South Koreans, and Taiwanese. Traditional manufacturing on the island was tied to sugar refining, whereas up to 60 percent of production in the FTZs is in garments and textiles (Figure 5.29).

In each of these cases, the growth in manufacturing depended upon national and international development policies that supported export-led growth through direct foreign investment. Certainly new jobs are being created and national economies are diversifying in the process, yet critics believe that foreign investors gain more than their host countries. Since most goods are assembled from imported materials, there is little integration with national suppliers. Often higher than local averages, wages are still miserable when compared to the developed world—often just two to three dollars a day. Moreover, as other developing countries compete with the Caribbean for the establishment of FTZs, this strategy may become less significant over time.

Offshore Banking The rise of offshore banking in the Caribbean is most closely associated with the Bahamas, which began this industry back in the 1920s. **Offshore banking centers** appeal to foreign banks and corporations by offering specialized services that are confidential and tax-exempt. Localities that provide offshore banking make money through registration fees, not taxes. The Bahamas were so successful in developing this sector that by 1983 it was the third largest banking center in the world. Its dominance began to decline due to other Caribbean competitors and concerns about corruption and laundering of drug money. By the 1990s, the Cayman Islands emerged as the region's leader in financial services. With a population of 30,000, this crown colony of Britain has 25,000 registered companies and the highest per capita GNP of the region at $23,800.

Each of the offshore banking centers in the Caribbean (Figure 5.30) try to develop special financial niches to attract clients such as banking, functional operations, insurance, or trusts. The Caribbean is an attractive locality for such services because of its proximity to the United States (home of many of the registered firms), client demand for these services in different localities, and the steady improvement in telecommunications that make such an industry possible. On the supply side, the resource-poor islands of the region see financial services as a means to bring hard currency to state coffers. Envious of the economic success of the Bahamas and the Cayman islands, countries such as Antigua, Barbados, and Belize hope that greater prosperity will come by establishing close ties to international finance.

The growth of offshore financial centers and the manufacturing zones discussed earlier are indicative of the forces of globalization (Figure 5.31). Ironically, banking does not employ that many people and can quickly unravel with the slightest signs of political instability. Offshore banking also attracts money tied to the drug trade. A major transshipment area for cocaine bound for North America and Europe, the Caribbean is also a major money laundering center. This

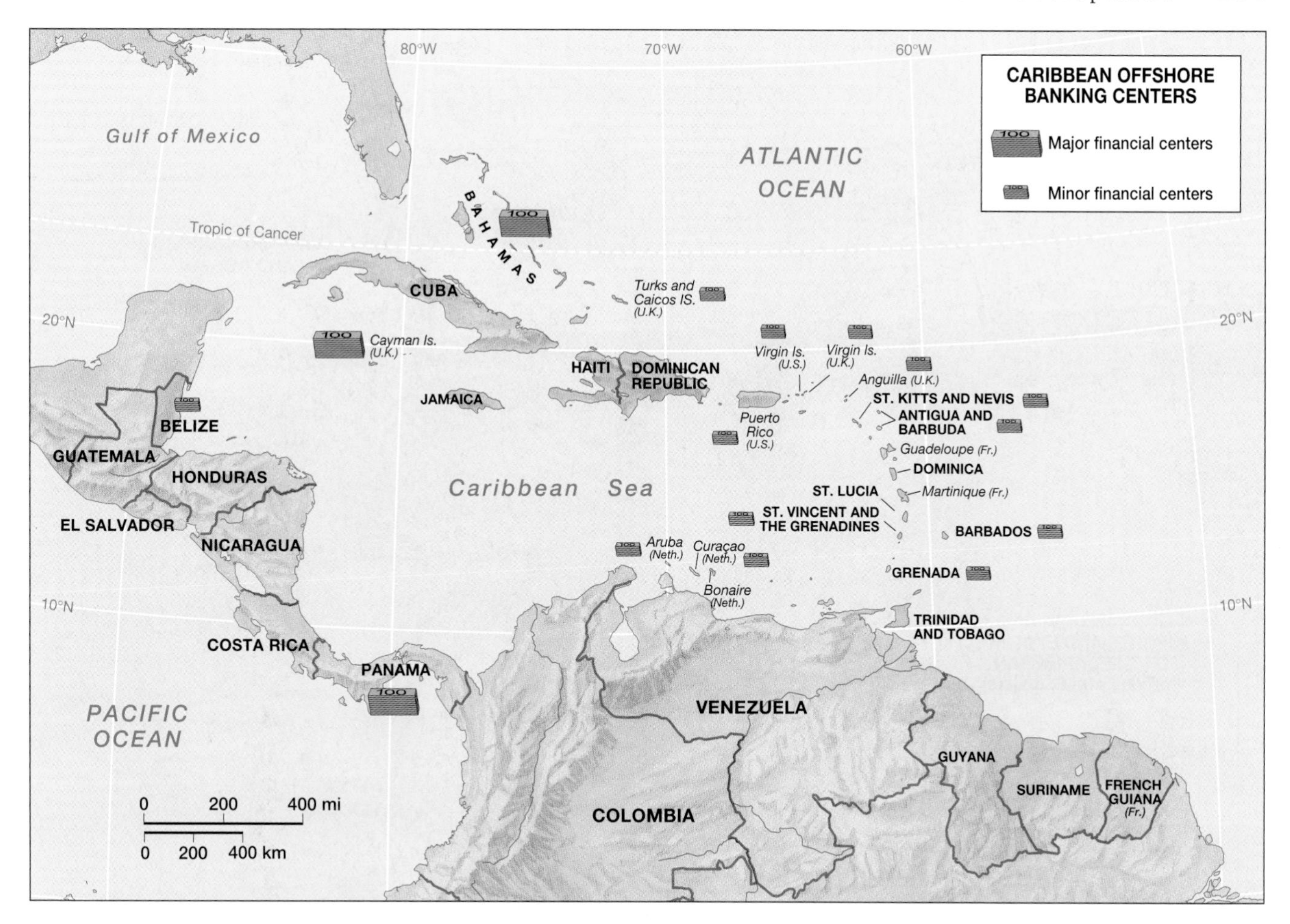

▲ **Figure 5.30 Caribbean offshore banking centers** Another development strategy used by the smaller islands is to offer specialized banking services to international corporations that are both confidential and tax-exempt. The Bahamas, Cayman Islands, and Panama dominate offshore banking in the region. Many other countries, however, are beginning to offer similar services. *[Source: Susan Roberts. 1995. "Small Place, Big Money: The Cayman Islands and the International Financial System."* Economic Geography *71(3): 237–256].*

brings its own problems to the region, such as a rise in drug consumption, corruption of local officials, and even an increase in drug-related murders. Local economies are being transformed by this flood of new investment. What is less certain is whether such development trends will greatly improve local earnings and standards of living.

Tourism Environmental, locational, and economic factors converge to buttress tourism in this part of the world. The earliest visitors to this tropical sea admired its clarity and sparkling turquoise waters. By the nineteenth century, elite North Americans fled winter to enjoy the restorative warmth of the Caribbean during its dry season. Entrepreneurs later realized that this complement of the Northern Hemisphere's winter, with the Caribbean dry season, was ideal for beach-oriented resorts. By the twentieth century, tourism was well established, with both destination resorts and cruise lines. By the 1950s the leader in tourism was Cuba, and the Bahamas was a distant second. Castro's rise to power, however, virtually eliminated this sector of the island's economy, yet it opened the door for other islands to step in.

Puerto Rico saw its tourist sector begin to grow with conferment of commonwealth status in 1952. San Juan is now the largest home port for cruise lines and the second largest cruise-ship port in the world in terms of total visitors (some 1.2 million). The cruise business, in combination with stay-over visitors, totaled 2.5 million guests in 1995. The Dominican Republic also sees more than two million visitors, many of them Dominican nationals who live overseas. Since 1980, tourist receipts have increased tenfold, making tourism the leading foreign-exchange earner. Jamaica has become similarly dependent upon tourism for hard currency and accommodates 1.5 million guests each year (Figure 5.32).

After years of neglect, Cuba is reviving tourism in an attempt to earn badly needed hard currency. From representing less than 1 percent of the national economy in the early 1980s, by the mid-1990s tourism earnings had increased tenfold. In 1995 more than 700,000 tourists, mostly Canadians and Europeans, yielded gross receipts of $1 billion. Conspicuous in their absence are travelers from the United States, forbidden to vacation in Cuba because of the U.S.-imposed travel ban. The Helms-Burton Law of 1996 even tried to reduce the flow

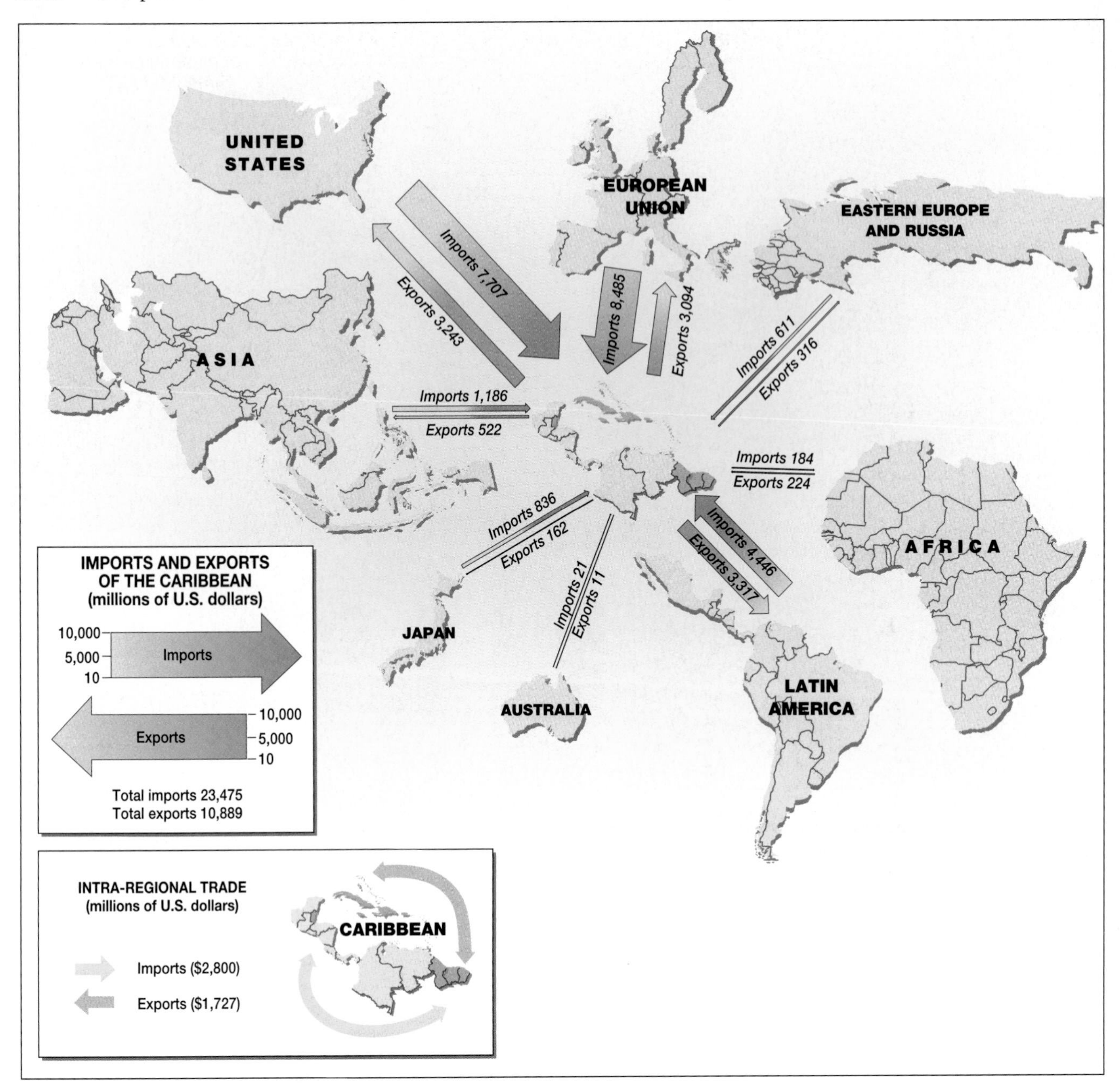

▲ **Figure 5.31 Caribbean trade** The Caribbean is a net importer of goods, most notably from Europe and the United States and increasingly from Latin America. Its trade with other world regions is limited, especially since the collapse of the Soviet Union in 1991. *(Source:* International Marketing Data and Statistics 1998. *London: Euromonitor)*

of foreign investment by threatening to sue major foreign investors in Cuba. While some investors delayed their plans because of this law, others, such as a leading Spanish hotel group, are busy building the island's tourist capacity in the hope that some day U.S. sanctions will be lifted.

As important as tourism is for the larger islands, it is often the principal source of income for smaller ones. With more than 3.5 million visitors a year, the Bahamas attributes most of its economic development and high per capita income to tourism. Nearly half the population is employed in tourism, and it represents 55 percent of the country's GDP. The Virgin Islands, Barbados, Turks and Caicos, and, recently, Belize all greatly depend upon international tourists. To show how quickly this sector can grow, consider this example: Belize began promoting tourism in the early 1980s when just 30,000 arrivals came a year. Close to North America and English-speaking, it specialized in a more rustic eco-tourism that showcased its interior tropical forests and coastal barrier reef. By 1994 the number of tourists topped 300,000 and tourism was credited for employing one-fifth of the workforce. Its contribution to the GNP equals that of agriculture and may soon surpass it (Figure 5.33).

▲ **Figure 5.32 Caribbean cruise ships** Cruise ships in port at Nassau, the Bahamas. Tourism is vital to many of the smaller islands of the Caribbean, but most ships, owned by companies outside the region, offer relatively little employment for Caribbean workers. *(Galen Rowell/Corbis)*

Tourism-led growth has its detractors. It is subject to the overall health of the world economy, so if North America experiences a recession, tourist dollars in the Caribbean dry up. It is also vulnerable to natural disasters. Hurricane Hugo virtually shut down St. Croix to visitors for an entire season, but islanders managed to rebuild most of their infrastructure by the following year. Local resentment may build as residents confront the disparity between their own lives and that of the tourists. There is also a serious problem of **capital leakage**, which is the huge gap between gross receipts and the total tourist dollars that remain in the Caribbean. Since many guests stay in hotel chains with corporate headquarters outside the region, leakage of profits is inevitable. On the plus side, tourism tends to promote stronger environmental laws and regulation. Countries quickly learn that their physical environment is the foundation for success. And while tourism does have its costs (higher energy and water consumption and demand for more imports), it is environmentally less destructive than traditional export agriculture and currently more profitable.

Social Development

While the record for economic growth in the region is inconsistent, measures of social development are stronger, in part due to Cuba's accomplishments in health care and education. With the exception of Haiti's abysmally low life expectancy of 50 years, most other Caribbean countries have an average life expectancy of 70 years or older (Table 5.3). Levels of secondary school enrollment are also very good, especially in Cuba, the Bahamas, Jamaica, Trinidad, and Barbados. Higher educational attainment and out-migration have contributed to a marked decline in natural increase rates over the last 20 years so that the average Caribbean woman has two or three children. Despite real social gains, many inhabitants are chronically underemployed, poorly housed, and dependent upon foreign remittances or subsistence agriculture. For rich and poor alike, the temptation to leave the region in search of better opportunities is always present.

▲ **Figure 5.33 Eco-tourism** A scuba diver explores the Belizean reef. Many Caribbean and Latin American countries are betting on eco-tourism as a means to generate income and better manage their environments. In Belize, for example, part of the reef was declared a marine park in the hopes of protecting species diversity and attracting more tourists. *(Doug Perrine/Innerspace Visions)*

Education Both Cuba and the English Caribbean have excelled in educating its citizens. In the Bahamas 95 percent of children receive secondary education, a rate higher than many developed countries (Table 5.3). More impressive still is Cuba's educational accomplishments, given the size of the country and its high illiteracy rates in the 1960s. Today three-quarters of the population receive secondary education. Cuba also excels in training medical personnel and advancing biomedical technology. Hispaniola is the obvious contrast to Cuba's success. Barely 20 percent of Haitians and about one-third of the Dominican Republic's citizens (30 percent of males and 43 percent of females) acquire secondary education. These troubling figures mirror other troubling social indicators

Table 5.3 Social Indicators and Status of Women

Country	Life Expectancy at Birth		Under Age 5 Mortality, per 1,000 Live Births		Secondary School Enrollment %		Female Labor Force Participation (% of total)
	Male	Female	1960	1995	Male	Female	
Anguilla	74	80	—	—	—	—	—
Antigua and Barbuda	71	76	—	—	—	—	—
Bahamas	70	77	—	—	95	95	46
Barbados	73	77	—	—	90	80	46
Belize	70	74	—	—	46	48	22
Cayman	75	79	—	—	—	—	—
Cuba	74	77	50	10	73	81	38
Dominica	74	80	—	—	—	—	—
Dominican Republic	68	72	152	44	30	43	29
French Guiana	—	—	—	—	—	—	—
Grenada	68	73	—	—	—	—	—
Guadeloupe	73	80	—	—	—	—	45
Guyana	63	69	126	—	56	59	33
Haiti	49	53	270	124	22	21	43
Jamaica	70	73	76	13	62	70	46
Martinique	75	82	—	—	—	—	47
Montserrat	74	77	—	—	—	—	—
Netherlands Antilles	72	78	—	—	—	—	42
Puerto Rico	70	79	—	—	—	—	36
St. Kitts–Nevis	64	70	—	—	—	—	—
St. Lucia	68	75	—	—	—	—	—
St. Vincent and Grenadines	71	74	—	—	—	—	—
Suriname	68	73	96	—	—	—	32
Trinidad and Tobago	68	73	73	18	74	78	37
Turks and Caicos	73	77	—	—	—	—	—

Sources: *Population Reference Bureau Data Sheet,* 1996, Secondary School Enrollment (M/F); *Population Reference Bureau Data Sheet,* 1998, Life Expectancy (M/F); World Resources Institute, *World Resources 1998–99,* Under-5 Mortality Rate (per 1,000) 1960 and 1995; *The World Bank Atlas, 1998,* Female Participation in Labor Force. Data for dependent territories from CIA, *World Factbook,* 1996.

(high illiteracy and child mortality). Political stability and economic growth have helped the Dominican Republic better its social conditions over the past decade. In fact, many Haitians have fled to the Dominican Republic because conditions, although far from ideal, are much better than in their homeland.

Education is expensive for these nations, but it is considered essential for development. Ironically, many states express frustration about training professionals for the benefit of developed countries in a phenomenon called *brain drain.* In the early 1980s the Prime Minister of Jamaica complained that 60 percent of his country's newly trained workers left for the United States, Canada, and Britain, representing a subsidy to these economies far greater than the foreign aid Jamaica received from them. During the economically depressed 1980s, Barbados, Guyana, and even the Dominican Republic and Haiti lost up to 20 percent of its most educated citizens. Brain drain occurs throughout the developing world, especially between former colonies and the mother countries. Yet given the small population of many Caribbean territories, each professional person lost to emigration can negatively impact local health care, education, and enterprise. For the time being, stronger economic performance in the Caribbean has slowed the outflow of professionals.

Status of Women The matriarchal basis of Caribbean households is often heralded as a distinguishing characteristic of the region. The rural custom of men leaving home for seasonal employment tends to nurture strong and self-sufficient female networks. Women typically run the local street markets. With men absent for long periods of time, women tend to make the household and community decisions. While giving women local power, this position does not always confer status. In rural areas female status is often undermined by the relative exclusion of women from the cash economy; men earn wages while women provide subsistence.

As Caribbean society urbanizes, women are being employed in assembly plants (the garment industry, in particular, prefers to hire women) and in tourism. With new employment opportunities, female labor force participation has surged; in many countries more than 40 percent of the workforce is female (Table 5.3). Increasingly women are the principal earners of cash, which challenges established gender roles. One impact of greater female participation in the formal economy seems to be smaller families.

Long before offshore assembly plants and resorts began hiring women, Cuba's educational and labor policies yielded the most educated and professional women in the Caribbean. Here, female doctors outnumber their male counterparts and women have achieved near parity in the sciences, which is not true for many developed countries. Yet economic hardship may undermine these achievements as professionals see their wages and purchasing power erode. Cuba has not experienced the same level of brain drain as its neighbors, mostly because of restrictive immigration policies. Yet when the United States held a lottery for 6,000 visas in 1986, more than 400,000 Cubans applied.

Labor-Related Migration Inequitable economies and limited resources conspire to force many residents to migrate. Besides the obvious economic rationale to migrate, this strategy socially impacts community and household structures as well. Historically, labor circulated within the Caribbean in pursuit of the sugar harvest—Haitians went to the Dominican Republic, residents of the Lesser Antilles journeyed to Trinidad. During the construction of the Panama Canal in the early 1900s, thousands of Jamaicans and Barbadians journeyed to the canal zone and many remained (Figure 5.34).

After World War II, better transportation and political developments in the Caribbean produced a surge of migrants to North America. This trend began with Puerto Ricans going to New York in the early 1950s and intensified in the 1960s, with the arrival of nearly half a million Cubans. Since then, large groups of Dominicans, Haitians, and Jamaicans have settled in Miami, New York, Los Angeles, and Toronto.

Often what begins as a temporary move can stretch into a longer residence. The first Cuban arrivals in 1959 thought that they were coming for a few months, but four decades later they are the best-established and most-successful immigrant group from the Caribbean in the United States, totaling one million with a strong political base in Miami, Florida. Other Caribbean migrants, though less economically successful, have created thriving enclave communities. Over the past decade Dominicans have been the largest group to arrive in New York City, where they cluster in North Manhattan and the West Bronx. Jamaicans (the fourth largest group of recent arrivals) and Guyanese tend to cluster in Flatbush, Brooklyn. In general, two-thirds of New York's new immigrants reside in the boroughs of Brooklyn and Queens.

Most migrants, with the exception of the Cubans, are part of a **circular migration** flow. In this type of migration, a man or woman typically will leave children behind with relatives in order to work hard, save money, and return home. Other times, a **chain migration** begins, in which one family member at a time is brought over. In some cases large numbers of residents from a Caribbean town or district send migrants to a particular locality in North America. Thus, chain migration can account for the formation of immigrant enclaves.

Crucial in this exchange of labor from north to south is the flow of cash **remittances** (monies sent back home). Immigrants are expected to send something back, especially when

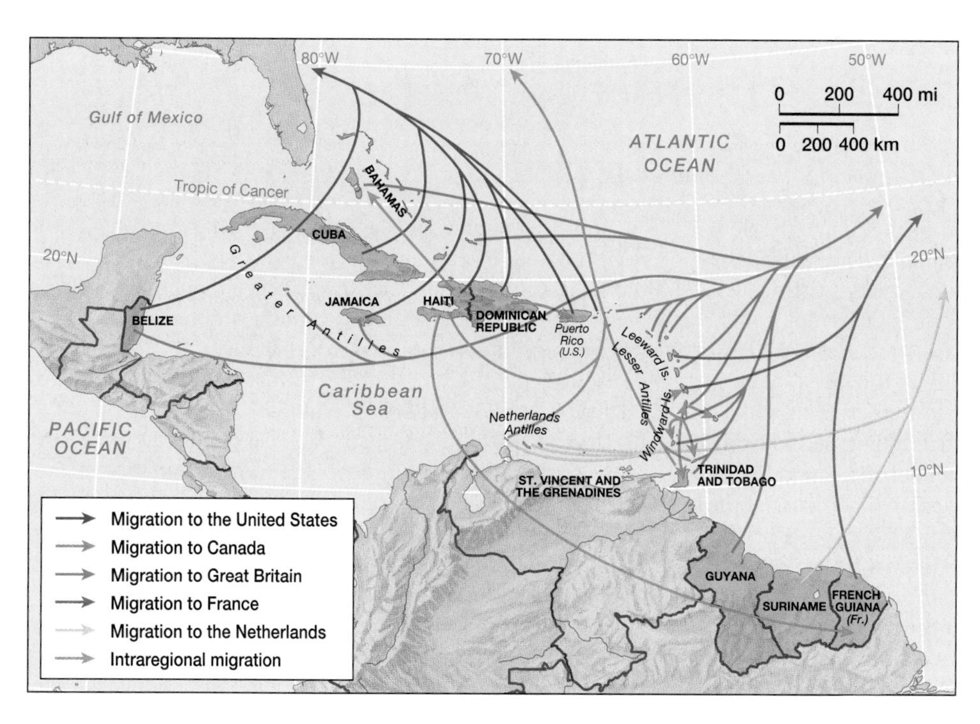

Figure 5.34 Caribbean diaspora Labor-related emigration has long been a way of life for Caribbean peoples. With relatively high education levels but limited professional opportunities, immigrants from the region head to North America, Great Britain, France, and the Netherlands. Intraregional migrations between Haiti and the Dominican Republic or the Dominican Republic and Puerto Rico also occur. *(Data from Barry Levin. 1987.* Caribbean Exodus. *Praeger Publishers)*

LOCAL TO GLOBAL The Caribbean Diaspora

"Ten years ago, it was much more fun here," said Marrion Udenhout despondently. "In 1981 all the young men left. Now it is very boring here." These youths had earned their living in Paramaribo as street peddlers. They sold clothes, soap, whatever there was to buy and sell in Suriname. In Holland, what they found to buy and sell was cocaine—South American-grown, Colombian-made, shipped to Holland for the European market by Surinamese and sold on the streets of Amsterdam by these young men. Just that week some of the young men had come back, and they were sick.

The young men were being taken to the edge of the jungle to a loekoeman—literally a seeing man—who would pass herbs over their bodies, bathe them, and try to cure them of the craving that was making them sick. "Bad things," she repeated, were happening to their people in Holland. . . .

The diaspora of the second half of the twentieth century has made Caribbeans an international people. In New York, Boston, Toronto, Montreal, Miami, London, Amsterdam, and Paris, communities have become established of people who regard their home as a distant island that they may have never seen, or barely remember, or remember in a way that has not existed for decades. Even the governments of those distant places recognize these unfamiliar exiles as a part of the greater nationality. Although the migratory experience has been painful for many Caribbeans, it has become a recognized part of the Caribbean way of life and a fixed element in the Caribbean economic system.

A Caribbean nation in which no one left would be facing a crisis. The populations have been expanding far faster than the economies. Many of the people who have left would have been jobless at home or would have forced someone else to be jobless. Families depend on migration, carefully selecting the member most likely to succeed to go first and establish the base abroad.

The governments have come to depend on these families. Remittances, money sent from abroad, have become a major source of hard currency, an important part of the economy in much of the Caribbean.

Source: Mark Kurlansky, *A Continent of Islands, Searching for the Caribbean Destiny*. Reading, MA: Addison Wesley Publishing Company, pp. 219–220.

immediate family members are left behind. Collectively, remittances add up; it is estimated that $300 million to $600 million a year is sent to the Dominican Republic by migrants in New York City alone, making remittance income the country's second leading industry. Another strategy is to work abroad and retire in the Caribbean with a pension and savings that ensure a comfortable old age. Returnees often bring cash, new skills, and increased expectations. There is anecdotal evidence that this cohort of migrants can introduce positive economic and political changes in particular localities. Still, it is unlikely that migrants and their remittances can challenge the larger structural forces that make migration attractive in the first place. Their impact is too fragmented and experienced primarily at the household level to represent a national development force (see Local to Global: The Caribbean Diaspora").

Conclusion

As a world region, the Caribbean is more integrated into the global economy than most areas of the developing world, albeit as a dependent economic periphery. Even though this is the smallest world region discussed, it offers some of the best examples of the long-term effects of globalization. This region was fashioned five centuries ago by Europeans who eliminated the native population and imported millions of Africans to support a plantation-based economy that supplied raw materials to Europe and North America. Consequently, most of the human and economic resources of the region went toward enriching other parts of the world, especially Europe and later North America. In a blatant example of colonial exploitation, little was invested in the social and economic development of these colonial localities.

Today the region includes 20 independent countries, although each maintains strong international links. Rather than harvesting sugarcane, Caribbean laborers are more likely to sew garments, assemble telephones, manage offshore banks, or cater to the needs of the 10 million tourists who visit each year. The economies of the region have diversified, although they still depend upon foreign investment, exports, and tourism. Perhaps more significant is the region's strides in social development, especially in education, health, and the status of women that distinguish it from other developing areas.

There is a definite transnational quality to life in this region as social and economic influences from Latin America, Africa, North America, and Europe are woven into the lifestyle. Rap music blares from boom boxes in the streets of Havana and Kingston, while African-based religions are intertwined with Catholicism in the countryside. Similarly, as Caribbean residents emigrate in search of opportunities, they introduce their own traditions to host countries. In Toronto, for example, Caribbean immigrants hold Canada's largest Carnival. The input of so many cultures has triggered an eclectic identity for the Caribbean, most notably through its literature, music, and arts.

The Caribbean also maintains important transatlantic ties with Europe and Africa. European colonies still exist in the region and maintain special trade relations with their former colonizers. And even independent states look toward their former rulers for help; for example, since the fall of the Soviet Union, Spain is one of Cuba's largest foreign investors. In contrast, Caribbean links with Africa are more cultural and political than economic. The cultural transfers of music, religion, and dance from West Africa to the Caribbean are well documented. Moreover, Afro-Caribbean scholars have supported the development of a Pan-African consciousness in their writings.

With the end of the Cold War, many of the microstates in the region fear that their lack of strategic significance may result in neglect by the United States and Europe and prejudice their ability to participate in world trade. It seems that the intervention of the United States in the past few years is driven more by a fear of thousands of refugees seeking sanctuary in Florida than by the ideologies and geopolitics of the past. Certainly it is difficult to untangle the Caribbean from the larger regions that surround it. Perhaps it is best to view it as the crossroads of the Americas. Whether through offshore banking, money laundering, the drug trade, or commercial airline traffic, the Caribbean is a vibrant intersection for the Western Hemisphere.

Key Terms

African diaspora (page 186)
capital leakage (page 203)
Caribbean Community and Common Market (CARICOM) (page 198)
chain migration (page 205)
circular migration (page 205)
creolization (page 184)
free trade zones (page 199)
Greater Antilles (page 173)
houseyards (page 182)
hurricanes (page 175)
indentured labor (page 186)
isolated proximity (page 172)
Lesser Antilles (page 174)
maroons (page 186)
mono-crop production (page 185)
Monroe Doctrine (page 192)
neocolonialism (page 192)
offshore banking centers (page 200)
plantation America (page 185)
remittances (page 205)
rimland (page 175)
shatterbelt (page 196)

Questions for Summary and Review

1. What characteristics define the Caribbean as a world region?
2. What environmental, economic, and locational factors contributed to the growth of tourism in the Caribbean?
3. What is the typical path of a hurricane in the Caribbean, and when are these meteorological disturbances likely to occur?
4. What significance does sugarcane play in the region's history?
5. What factors contributed to the deforestation of the Caribbean islands? Why are the rimland forests, in comparison, still intact?
6. What happened to the Amerindian population in the Caribbean?
7. How does the experience of the Garifuna people illustrate the process of creolization?
8. What cultural traditions transferred from Africa to the Caribbean via the slave trade?
9. How is the cultural history of the Caribbean reflected in its music? What musical styles are associated with the region?
10. What geopolitical role did Cuba play during the Cold War? What is Cuba's current geopolitical significance?
11. Which countries rely upon mining to bring in revenue? What is mined in this region?
12. Where are the Asian populations in the Caribbean? What explains their presence?
13. Which territories in the region are still colonies? Why have they not sought independence?
14. Where is "plantation America" and what characteristics define it?

Thinking Geographically

1. After looking at the map of offshore banking in the Caribbean, what conclusions can be drawn about where this occurs?
2. Contrast the historical African diaspora to the contemporary Caribbean one. What patterns are formed by these two distinct population movements? What social and economic forces are behind them? Are they comparable?
3. What advantages and disadvantages exist when pursuing tourism as a development strategy?
4. What advantages might Caribbean free trade zones retain over their competitors in Southeast and East Asia? What disadvantages might they face?

5. Why have U.S. actions in the region been considered neocolonial? Are there other regions in the world where the United States exerts neocolonial tendencies?
6. Compare and contrast the industrial development of Cuba with Puerto Rico. How were these two islands integrated into the world economy during the twentieth century?
7. Why does chain migration occur between the Caribbean and North America or Europe? How does a migrant benefit from being part of a chain migration? How does chain migration affect sending communities?
8. In the twenty-first century, will agricultural exports continue to be important for Caribbean economies?

Regional Novels and Films

Novels

Aimé Césaire, *Notes on a Return to the Native Land* (1968)

Patrick Chamoiseau, *Texaco* (1988, Vintage International)

Zee Edgell, *Beka Lamb* (1982, Heinemann)

Cristina Garcia, *Dreaming in Cuba* (1992, Random House)

Jamaica Kincaid, *A Small Place* (1988, Penguin)

Jamaica Kincaid, *At the Bottom of the River* (1983, Farrar, Straus & Giroux)

Shiva Naipaul, *Fireflies* (1971) and *The Chip-Chip Gatherers* (1973)

V. S. Naipaul, *The Middle Passage* (1963, Macmillan)

V. S. Naipaul, *Miguel Street* (1959, Vanguard Press)

Derek Walcott. *Sea Grapes* (1976, Farrar, Straus & Giroux) *The Star-Apple Kingdom* (1979, Farrar, Straus & Giroux) and *Omeros* (1990, Farrar, Straus & Giroux)

Alec Waugh, *Island in the Sun* (1955, Farrar, Straus, and Cudahy)

Herman Wouk, *Don't Stop the Carnival* (1965, Doubleday)

Films

Bitter Cane (1986)

Buena Vista Social Club (1999)

Cuba—In the Shadow of Doubt (1987)

The Emperor's Birthday: The Rastafarians Celebrate

Inside Castro's Cuba (1995)

The King Does Not Lie: The Initiation of a Shango Priest (1993)

Bibliography

Allsopp, Richard, ed. 1996. *Dictionary of Caribbean English Usage*. Oxford: Oxford University Press.

Domínques, Jorge. 1997. "The Powers, the Pirates and International Norms and Institutions in the American Mediterranean," in *Cuba and the Caribbean* (pp. 3–19), edited by Joseph Tulchin, Andrés Serbín, and Rafael Hernández. Wilmington, DE: Scholarly Resources.

Europa. 1997. *South America Central America and the Caribbean,* 6th ed. London: Europa Publications Limited.

Gordon, Joy. 1997. "Cuba's Entrepreneurial Socialism." *Atlantic Monthly,* January, 18–30.

Klak, Thomas, ed. 1998. *Globalization and Neoliberalism: The Caribbean Context*. Lanham, MD: Rowman and Littlefield.

Kurlansky, Mark. 1992. *A Continent of Islands: Searching for the Caribbean Destiny.* Reading, MA: Addison-Wesley.

Levinson, David. 1998. *Ethnic Groups Worldwide: A Ready Reference Handbook*. Phoenix, AZ: Oryx Press.

Mintz, Sidney. 1985. *Sweetness and Power: The Place of Sugar in Modern History.* New York: Viking Penguin.

Mintz, Sidney. 1989. *Caribbean Transformations*. New York: Columbia University Press.

Mintz, Sidney, and Price, Sally. 1985. *Caribbean Contours*. Baltimore: Johns Hopkins University Press.

Momsen, Janet H., ed. 1993. *Women and Change in the Caribbean*. London: James Currey Limited.

Pattullo, Polly. 1996. *Last Resorts: The Cost of Tourism in the Caribbean*. New York: Monthly Review Press.

Payenne, André. 1985. "Plugging the Brain Drain." *World Press Review,* August, 33–34.

Price, Richard. 1979. *Maroon Societies: Rebel Slave Communities in the Americas*. Baltimore: Johns Hopkins University Press.

Pulsipher, Lydia. 1993. "Changing Roles in the Life Cycles of Women in Traditional West Indian Houseyards," in *Women and Change in the Caribbean* (pp. 50–64), edited by Janet Momsen. Bloomington, IN: Indiana University Press.

Richardson, Bonham C. 1992. *The Caribbean in the Wider World, 1492–1992*. Cambridge, England: Cambridge University Press.

Roberts, Susan. 1995. "Small Place, Big Money: The Cayman Islands and the International Financial System." *Economic Geography,* 71(3), 237–256.

Sale, Kirkpatrick. 1990. *The Conquest of Paradise: Christopher Columbus and the Columbian Legacy*. New York: Knopf.

Sauer, Carl Ortwin. 1969. *The Early Spanish Main*. Berkeley, CA: University of California Press.

Segre, Roberto; Coyula, Mario; and Scarpaci, Joseph L. 1997. *Havana: Two Faces of the Antillean Metropolis*. New York: John Wiley and Sons.

Tenenbaum, Barbara, ed. 1996. *Encyclopedia of Latin American History and Culture*. New York: Charles Scribner's Sons.

Thornton, John. 1992. *Africa and Africans in the Making of the Atlantic World, 1400–1680*. Cambridge, England: Cambridge University Press.

U.S. Committee for Refugees. 1997. *World Refugee Survey.*

Voeks, Robert. 1993. "African Medicine and Magic in the Americas." *Geographical Review,* 83(1), 66–78.

Wagley, Charles. 1970. "Plantation-America: A Culture Sphere" in *Caribbean Studies: A Symposium* (pp. 3–13), edited by Vera Rubin. Seattle: University of Washington Press.

Warf, Barney. 1995. "Information Services in the Dominican Republic." *Yearbook, Conference of Latin Americanist Geographers,* 21, 13–23.

Watts, David. 1987. *The West Indies: Patterns of Development, Culture and Environmental Change Since 1492*. Cambridge, England: Cambridge University Press.

Wiley, James. 1996. "The European Union's Single Market and Latin America's Banana Exporting Countries." *Yearbook, Conference of Latin Americanist Geographers,* 22, 31–40.

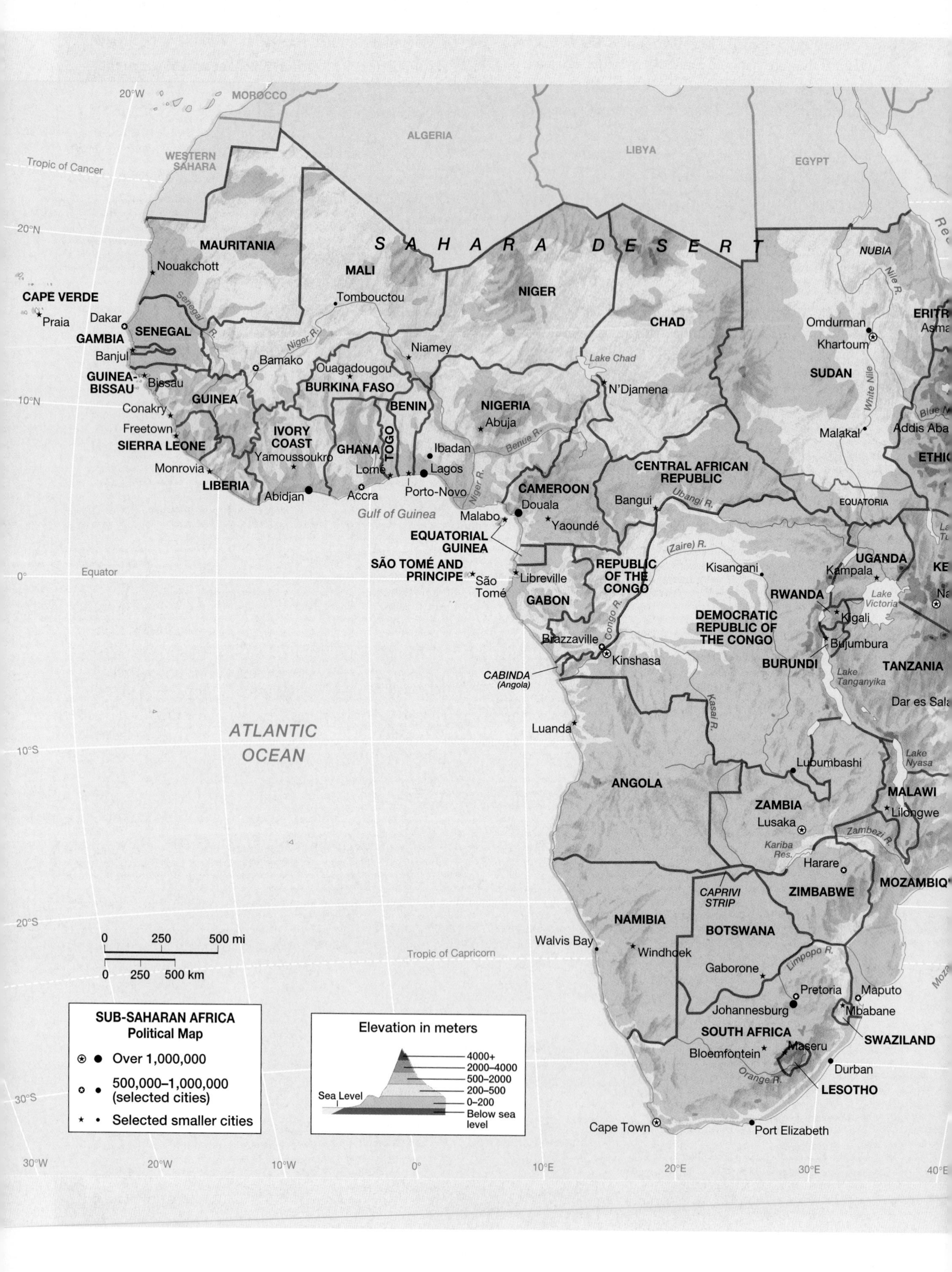

SUB-SAHARAN AFRICA
Political Map
Over 1,000,000
500,000–1,000,000 (selected cities)
Selected smaller cities
Elevation in meters
4000+
2000–4000
500–2000
200–500
0–200
Below sea level
Sea Level
0 250 500 mi
0 250 500 km
ATLANTIC OCEAN
Gulf of Guinea
S A H A R A D E S E R T
Tropic of Cancer
Tropic of Capricorn
Equator
MOROCCO
WESTERN SAHARA
ALGERIA
LIBYA
EGYPT
MAURITANIA
Nouakchott
MALI
Tombouctou
NIGER
CHAD
NUBIA
CAPE VERDE
Praia
Dakar
SENEGAL
GAMBIA
Banjul
Bamako
Niamey
Ouagadougou
BURKINA FASO
GUINEA-BISSAU
Bissau
GUINEA
Conakry
Freetown
SIERRA LEONE
Monrovia
LIBERIA
IVORY COAST
Yamoussoukro
Abidjan
GHANA
Accra
TOGO
Lomé
BENIN
Porto-Novo
NIGERIA
Abuja
Ibadan
Lagos
Lake Chad
N'Djamena
SUDAN
Omdurman
Khartoum
Malakal
White Nile
Nile R.
Niger R.
Senegal R.
Benue R.
CAMEROON
Douala
Yaoundé
Malabo
EQUATORIAL GUINEA
SÃO TOMÉ AND PRINCIPE
São Tomé
Libreville
GABON
CENTRAL AFRICAN REPUBLIC
Bangui
Ubangi R.
EQUATORIA
REPUBLIC OF THE CONGO
Brazzaville
Kinshasa
Congo R.
(Zaire) R.
CABINDA (Angola)
DEMOCRATIC REPUBLIC OF THE CONGO
Kisangani
Kasai R.
UGANDA
Kampala
Lake Victoria
RWANDA
Kigali
BURUNDI
Bujumbura
Lake Tanganyika
TANZANIA
Dar es Sala
Luanda
ANGOLA
Lubumbashi
ZAMBIA
Lusaka
Lake Nyasa
MALAWI
Lilongwe
Zambezi R.
Kariba Res.
Harare
ZIMBABWE
MOZAMBIQ
CAPRIVI STRIP
NAMIBIA
Walvis Bay
Windhoek
BOTSWANA
Gaborone
Limpopo R.
Pretoria
Johannesburg
Maputo
Mbabane
SWAZILAND
SOUTH AFRICA
Bloemfontein
Maseru
LESOTHO
Durban
Orange R.
Cape Town
Port Elizabeth
ERITR
Asma
Addis Aba
ETHIO
Blue N
KE
Na
30°W 20°W 10°W 0° 10°E 20°E 30°E 40°E
20°N 10°N 0° 10°S 20°S 30°S

Sub-Saharan Africa

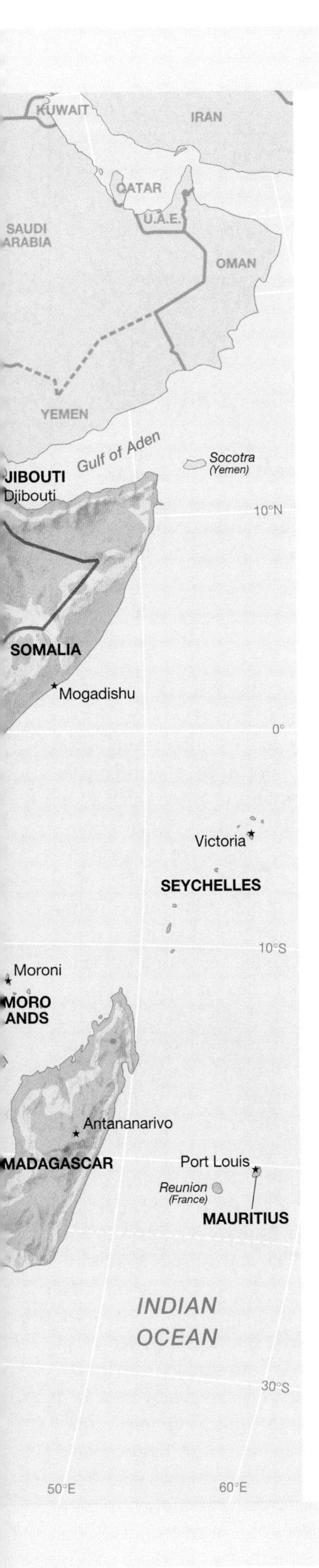

Compared with Latin America and the Caribbean, Africa south of the Sahara is poorer and more rural, and its population is very young. Some 625 million people reside in this region, which includes 48 states and one territory. Demographically, this is the world's fastest-growing region (2.6 percent rate of natural increase); in some countries, nearly half the population is younger than 15 years old. Income levels are low (a per capita GNP of $500 in 1996). Life expectancy is only 48 years. For many people, this part of the world has become synonymous with poverty, disease, violence, and refugees. Overlooked in the all-too-frequent negative headlines are initiatives by private, local, and state groups to improve the region's quality of life. Many states have reduced infant mortality, expanded basic education, and increased food production in the past two decades despite the profound economic and political crises that beleaguer this region.

Sub-Saharan Africa—that portion of the African continent lying south of the Sahara Desert—is a commonly accepted world region (Figure 6.1). Historically referred to as "Black Africa" because of the skin color of its indigenous inhabitants, such a race-based division falls apart under closer inspection. The coherence of this region has less to do with skin color and more to do with similar livelihood systems and a shared colonial experience. No common religion, language, philosophy, or political system ever united the area. Instead, diffuse cultural bonds developed from a variety of lifestyles and idea systems that evolved here. The impact of outsiders also helped to determine the region's identity. Slave traders from Europe, North Africa, and Southwest Asia treated Africans as chattel; up until the mid-1800s, millions of Africans were taken from the region and sold into slavery. In the late 1800s, the entire African continent was divided by European colonial powers, imposing political boundaries that remain to this day. In the post-colonial period, which began in the 1960s, Sub-Saharan African countries faced many of the same economic and political challenges, creating still more commonality.

The cultural complexity of the region is underappreciated. It is not uncommon for 20 or more languages to be spoken in one large state. Consequently, most Africans understand and speak several languages. Ethnic identities do not conform to the political divisions of Africa, sometimes resulting in bloody ethnic warfare, as witnessed in Rwanda in the mid-1990s. Nevertheless, throughout the region peaceful coexistence between distinct ethnic

◀ **Figure 6.1 Sub-Saharan Africa** Africa south of the Sahara includes 48 states and one territory. This vast region of rain forest, tropical savanna, and desert is home to 625 million people. Much of the region consists of broad elevated plateaus, ranging from 500 to 2,000 meters in elevation. Although the population is growing rapidly, the overall population density of Sub-Saharan Africa is low. Considered one of the least-developed regions of the world, it remains an area rich in natural resources.

Setting the Boundaries

Sub-Saharan Africa includes 43 mainland states plus the island nations of Madagascar, Cape Verde, São Tomé and Principe, Seychelles, Mauritius, and the French territory of Reunion. When setting this particular regional boundary, the major question is North Africa.

Many scholars argue for Africa's continental integrity as a culture region. The Sahara has never formed a complete barrier between the Mediterranean north and the remainder of the African landmass. Moreover, the Nile River forms a corridor several thousand miles long of continuous settlement linking North Africa directly to the center of the continent. There is no obvious place to divide the watershed between northern and Sub-Saharan Africa. Political units, such as the Organization of African Unity, are modern examples of the continent's indivisibility.

The lack of a clear divide across Africa does not invalidate its division into two world regions. North Africa is generally considered more closely linked, both culturally and physically, to Southwest Asia. Arabic is the dominant language and Islam the dominant religion of North Africa. Consequently, North Africans feel more closely connected to the Arab hearth in Southwest Asia than to the Sub-Saharan world.

The decision to view Sub-Saharan Africa as a world region still presents the problem of where to divide it from Africa north of the Sahara. Political boundaries were preserved when drawing the line so that the Mediterranean states of North Africa will be discussed with Southwest Asia. Sudan will be discussed in Chapters 6 and 7 because it shares characteristics common to both regions. Sudan is Africa's largest state in terms of area; it is one-fourth the size of the United States. In the more populous and powerful north, Muslim leaders have forged an Islamic state that is culturally and politically oriented toward North Africa and Southwest Asia. Southern Sudan, however, has more in common with the animist and Christian groups of the Sub-Saharan region. Moreover, peoples in the south continue to fight for their independence from Khartoum.

groups is the norm. The cultural impress of European colonizers cannot be ignored: European languages, religions, educational systems and political ideologies were adopted and modified. Yet the daily rhythms of life are far removed from the industrial world. Most Africans still engage in subsistence and cash crop agriculture (Figure 6.2). Women in particular are charged with tending crops and procuring household necessities. Male roles revolve around tending livestock and the public life of the village and market. As is happening elsewhere in the developing world, cities are growing rapidly, but less than 30 percent of the population is urban.

▲ **Figure 6.2 Women in the fields** Women from the Merina tribe in Madagascar plant paddy rice. The majority of Sub-Saharan people engage in subsistence agriculture. Women are typically responsible for producing food crops, while men are more likely to tend livestock. *(Carl D. Walsh/Aurora & Quanta Productions)*

The influence of African culture outside the region is striking. One legacy of the slave trade is that African-based religious systems are widely practiced in the Caribbean and Latin America, especially in Brazil. The popularity of "world music" has meant that contemporary African performers, such as the Democratic Republic of the Congo's Papa Wemba or South Africa's Ladysmith Black Mambazo, have large international audiences in Europe, Asia, and the Americas. Likewise, many of the rhythms and melodies that make up the core of popular Western music are traceable to West African musical traditions. Sub-Saharan Africa has exerted significant influence on global culture, especially when one considers prehistory and that human origins are traceable to this part of the world.

The African economy, however, is marginal when compared to the rest of the world. According to the World Bank, Sub-Saharan Africa's economic output in 1996 amounted to only 1 percent of global output, even though the region contains 10 percent of the world's population. Moreover, the gross national product of just one country, South Africa, nearly equals that of all the other countries in the region combined. Even the region's share of world primary exports (such as copper, cocoa, and coffee) has fallen since 1980. Most troubling of all, Sub-Saharan Africa's per capita food production, which made gains in the 1960s and early 1970s, has been below the region's 1961 output level for the last 25 years. No one doubts there is undercounting of subsistence activities in the region, but rapid population growth has greatly increased food demands, and it is hard to ignore this troubling decline.

Many scholars feel that Sub-Saharan Africa has benefited little from its integration (both forced and voluntary) into the global economy. Slavery, colonialism, and export-oriented mining and agriculture served the needs of consumers outside the region but undermined domestic production for those within.

Foreign assistance in the post-Independence years initially improved agricultural and industrial output but also led to mounting foreign debt and corruption, which over time undercut the region's economic gains. One consequence of mounting debt was the implementation of **structural adjustment programs** by the World Bank and the International Monetary Fund in the 1980s. These controversial economic measures are designed to reduce government spending and encourage private sector initiatives. Yet they typically trigger drastic cutbacks in government-supported services and food subsidies, which disproportionately affect the poor. Given the region's history, most scholars argue that foreign involvement in Sub-Saharan Africa has led to greater underdevelopment than development. Ironically, many of these same scholars worry that global pessimism about the region has produced a pattern of neglect. Private capital investment in Sub-Saharan Africa lags far behind investment rates in Latin America, the Caribbean, and East and Southeast Asia. It is hard to imagine how the region can economically recover without new private investment.

Small political and social movements within the region offer valuable lessons on the importance of meeting Africa's basic needs first. Thousands of nongovernmental organizations have formed to support women's groups, cooperatives, schools, and health centers. By addressing basic needs and empowering community members to solve their own problems, organizations such as the Nanas-Benz cloth traders of Togo (so named for their ownership of Mercedes-Benz automobiles) or Hadasko Women's Farmers Association of Ghana, have become progressive agents for local change. Although there are serious human rights issues throughout the region, the near bloodless transition from white rule to black rule in South Africa in 1994 underscores the politics of promise for Sub-Saharan Africa. To recognize the scale of what is possible for this region, one must first appreciate the vast yet capricious resource base that sustains the region.

Environmental Geography: The Plateau Continent

The largest landmass straddling the equator, the physical environment of Sub-Saharan Africa is vast in scale and remarkably beautiful. Called the plateau continent, the African interior is dominated by extensive uplifted areas that resulted from the breakup of **Gondwanaland.** The ancient megacontinent of Gondwanaland included Africa, South America, Antarctica, Australia, Madagascar, and Saudi Arabia. Some 250 million years ago, it began to split apart through the forces of continental drift. As this process unfolded, the African landmass experienced a series of continental uplifts that left much of the area with vast elevated plateaus. The highest areas are found on the eastern edge of the continent, where the Great Rift Valley forms a complex upland area of lakes, volcanoes, and deep valleys. In contrast, lowlands prevail in West Africa, although smaller elevated areas exist, such as the Guinea and the Andamawa highlands (Figure 6.3).

The landscape of Sub-Saharan Africa offers a palette of intense colors: deep red soils studded with plantings of subsistence crops; the blue of the tropical sky; golden savannas that ripple with the movement of animal herds; dark rivers meandering through towering rain forests; and sun-drenched deserts. Amid this beauty, however, one finds relatively poor soils, endemic disease, and vulnerability to drought. Large areas for potential agricultural development still exist, especially in southern Africa; throughout the continent vast water resources and immense mineral wealth abound.

Landforms

A series of plateaus and elevated basins dominate the African interior and explain much of the region's uniqueness. Generally, elevations increase toward the south and east of the continent. Most of southern and eastern Africa lies well above 2,000 feet (600 meters), and sizable areas sit above 5,000 feet (1,500 meters). This is typically referred to as High Africa; Low Africa includes West Africa and much of Central Africa (Figure 6.3). The higher plateaus of Kenya, Zimbabwe, and Angola are noted for their cool springlike climates and relatively abundant moisture. Steep escarpments form where plateaus abruptly end, as illustrated by the majestic Victoria Falls on the Zambezi River (Figure 6.4). Much of southern Africa is rimmed by a landform called the **Great Escarpment,** which begins in southwestern Angola and ends in northeastern South Africa, creating a formidable impediment to coastal settlement. South Africa's Drakensberg Mountains (with elevations reaching 10,000 feet, or 3,100 meters) rise up from the Great Escarpment. Due to this landform, coastal plains tend to be narrow, few natural harbors exist, and river navigation is impeded by a series of falls. Fairly wide coastal lowlands are found in parts of western Africa, Mozambique, and Somalia, but even these lack extensive areas of fertile alluvial soil.

Even though Sub-Saharan Africa is an elevated landmass, it has few significant mountain ranges. The one extensive area of mountainous topography is in Ethiopia, which lies in the northern portion of the Rift Valley zone. Yet even there the dominant features are high plateaus intercut with deep valleys rather than actual mountain ranges. Receiving heavy rains in the wet season, the Ethiopian Plateau forms the headwaters of several important rivers, most notably the Blue Nile, which joins the White Nile at Khartoum, Sudan.

A discontinuous series of volcanic mountains, some of them quite high, are associated with the southern half of the Rift Valley. Kilimanjaro at 19,000 feet (5,900 meters) is the continent's largest mountain, and nearby Mount Kenya (17,000 feet, or 5,200 meters) is the second largest. Both are snow-covered peaks that rise from the grassy upland plains. The physiographically complex Rift Valley reveals the slow but inexorable progress of geological forces. Eastern Africa is slowly being torn away from the rest of the continent, and within some tens of millions of years it will form a separate subcontinent. Such motion has already produced a great gash across the uplands of eastern Africa, much of which is occupied by elongated and extremely deep lakes (most notably Nyasa, Malawi, and Tanganyika). In central eastern Africa this rift zone splits into two separate valleys, each of which is flanked by volcanic uplands. Between the

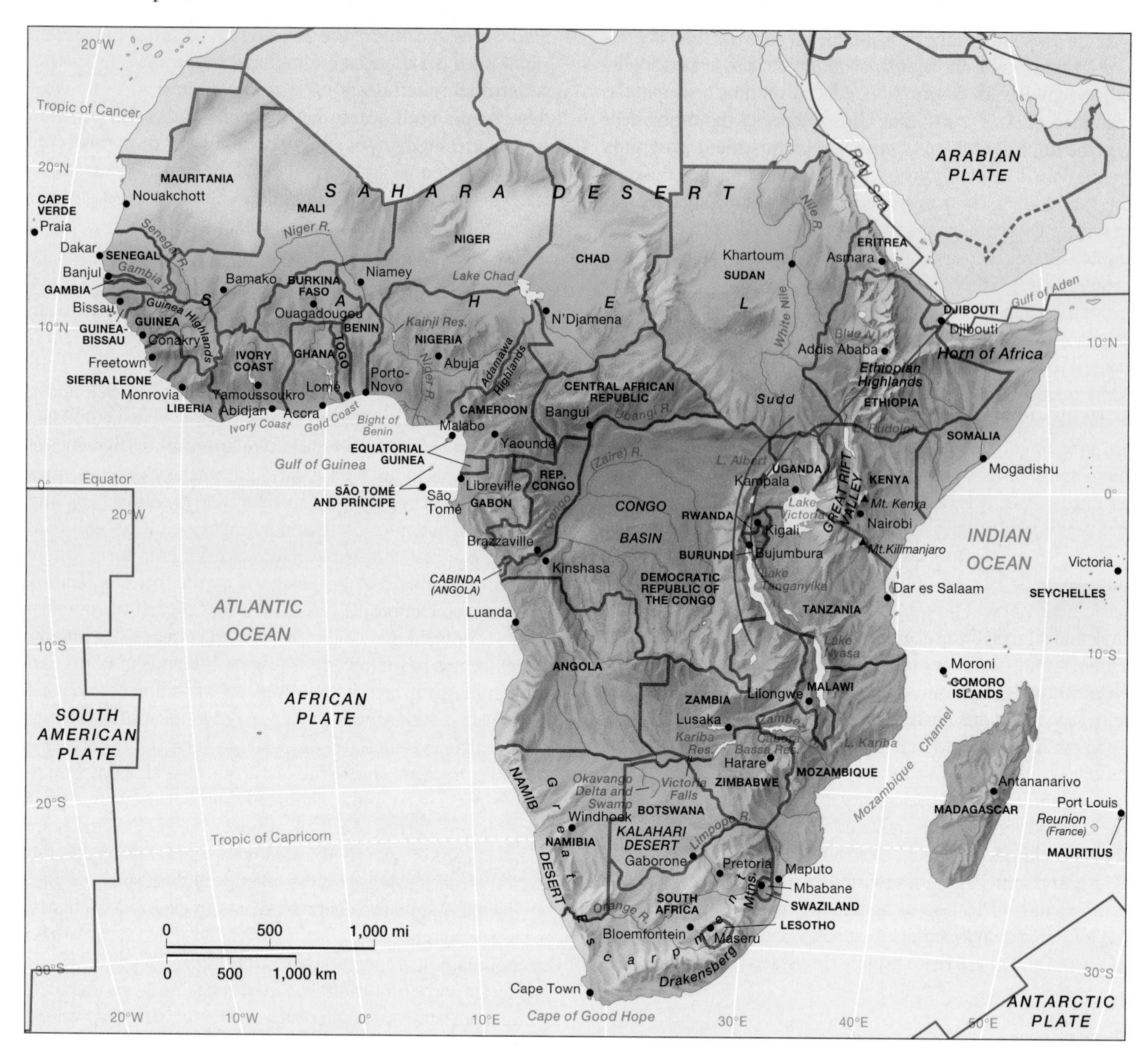

▲ **Figure 6.3 Physical geography of Sub-Saharan Africa** The so-called plateau continent was gradually uplifted with the breakup of Gondwanaland 160 million years ago. While much of the landmass is elevated, there are relatively few mountains except for those associated with the Rift Valley in eastern Africa where a complex chain of mountains, lakes, and valleys exist. Four major river basins are associated with the region: the Congo, Nile, Niger, and Zambezi. The Congo flows through the second largest rain forest in the world after the Amazon. The Nile, the lifeline for Sudan and Egypt, is the world's longest river.

eastern and western rifts lies a bowl-shaped depression, the center of which is filled by Lake Victoria—Africa's largest body of water. Not surprisingly, some of the densest areas of settlement are found amid the fertile and well-watered soils that border the Rift Valley (Figure 6.5).

Watersheds Africa south of the Sahara conspicuously lacks the broad, alluvial lowlands that influence patterns of settlement throughout Asia and other landmasses. The four major river systems are the Congo, Niger, Nile and Zambezi. Smaller rivers, such as the Orange in South Africa; the Senegal, which divides Mauritania and Senegal; and the Limpopo in Mozambique are locally important but drain much smaller areas. Ironically, most people think of Africa south of the Sahara as suffering from water scarcity and tend to discount the size and importance of these watersheds (or catchment areas) that these river systems drain.

▲ **Figure 6.4 Victoria Falls** The Zambezi River descends over Victoria Falls. A fault zone in the African plateau explains the existence of a 360 foot (110 meter) drop. The Zambezi has never been important for navigation, but it is a vital supply of hydroelectricity for Zimbabwe, Zambia, and Mozambique. *(Rob Crandall/Rob Crandall Photographer)*

▲ **Figure 6.5 Africa's Rift Valley** An aerial view of a portion of the Rift Valley in Kenya. This is one of the continent's most dramatic landforms, extending nearly 5,000 miles from Lake Nyasa to the Red Sea. The Rift Valley zone is a series of faults that have created volcanoes, escarpments, elongated lakes, and valleys. *(Altitude/Y. Arthus-B./Peter Arnold, Inc.)*

The Congo River (or Zaire) is the largest watershed in terms of drainage and discharge in the region. It is second only to the South America's Amazon River in terms of annual discharge. The Congo flows across a relatively flat basin that lies more than 1,000 feet (300 meters) above sea level, meandering through Africa's largest tropical forest, the Ituri. Entry from the Atlantic into the Congo Basin is prevented by a series of rapids and falls, making the Congo River only partially navigable. Despite these limitations, the Congo River has been the major corridor for travel within the Congo and the Democratic Republic of the Congo (formerly Zaire); the capitals of both countries, Brazzaville and Kinshasa, rest on opposite sides of the river (Figure 6.6). Political turmoil in the late 1990s, however, greatly reduced commerce on this vital waterway.

The Nile River, the world's longest, is discussed in Chapter 7 as it is the lifeblood for Egypt and Sudan. Yet this river originates in the highlands of the Rift Valley zone and it is an important physical expression of the linkages between North and Sub-Saharan Africa. The Nile begins from the lakes of the rift zone (Victoria and Edward) before descending into a vast wetland in southern Sudan known as the Sudd. In the Sudd the river divides into two main channels, before reuniting in Malaka, Sudan. Agricultural development projects in the 1970s greatly increased the agricultural potential of the Sudd, especially its peanut crop. Unfortunately, the past two decades of civil war in Sudan (between northern Muslims and southern Christians) ravaged this area, turning farmers and herders into refugees and undermining the productive capacity of this important ecosystem.

Like the Nile, the Niger River is the critical source of water for the arid countries of Mali and Niger. Beginning in the humid Guinea highlands, the Niger flows first to the northeast and then spreads out to form a huge inland delta in Mali before making a great bend southward at the margins of the Sahara near Gao. On the banks of the Niger River are the capitals of Mali (Bamako) and Niger (Niamey) as well as the historic city of Tombouctou (Timbuktu). Tombouctou flourished as a center of trans-Saharan trade in the sixteenth century, with a population of nearly 100,000. (Its population is now one-fifth that, reflecting the decline of long-distance caravan

▲ **Figure 6.6 Boat traffic on the Congo River** Heavily laden barges ferry cargo and people between the cities of Kinshasa and Kisangani in the Democratic Republic of the Congo. The Congo is Sub-Saharan Africa's largest river by volume. Meandering through a relatively flat basin, the river is an important transportation corridor. *(Georg Gerster/NGS Image Collection)*

trade.) After flowing through the desert north, the Niger River returns to the humid lowlands of Nigeria, where the Kainji Reservoir temporarily blocks its flow to produce electricity for Africa's most populace state.

The considerably smaller Zambezi River begins in Angola and flows east, spilling over an escarpment at the incomparable Victoria Falls, and finally reaching the Indian Ocean. More than other rivers in the region, the Zambezi is a major supplier of commercial energy. Sub-Saharan Africa's two largest hydroelectric installations, the Kariba on the border of Zambia and Zimbabwe and the Cabora Bassa in Mozambique are on this river. Even though the navigational offerings of the Zambezi are limited, this did not stop German colonialists from demanding access to the river as they formed the territory of Southwest Africa in the 1880s, which eventually became Namibia. The resulting Caprivi Strip is a narrow belt of Namibian territory that separates Botswana from Zambia. It is one of oddest geopolitical boundaries of the continent, but it does give arid Namibia access to the Zambezi.

Soils Scholars concerned about Africa's growing population often pay close attention to the poverty of its soils. With a few major exceptions, Sub-Saharan Africa's soils are relatively infertile and thus cannot easily support the intensive agriculture needed to feed large populations. Generally speaking, fertile soils are young soils, those deposited in recent geological time by rivers, volcanoes, glaciers, or windstorms. In older soils—especially those located in moist tropical environments—natural processes tend to wash out most plant nutrients over time. Over most of Sub-Saharan Africa, the agents of soil renewal have largely been absent; the region has few alluvial lowlands where rivers periodically deposit fertile silt, and it did not experience significant glaciation in the last ice age, as did North America.

Portions of Sub-Saharan Africa are, however, noted for their natural soil fertility and, not surprisingly, these areas support denser settlement. Some of the most fertile soils are in the Rift Valley, enhanced by the volcanic activity associated with the area. The population densities of rural Rwanda and Burundi, for example, are partially explained by the highly productive soils. The same can be said for highland Ethiopia, which supports the region's second largest population of nearly 60 million people. The Lake Victoria lowlands and central highlands of Kenya are also noted for their sizable populations and productive agricultural bases.

In the drier grasslands and semi-desert areas one finds a soil type called *alfisols*. High in aluminum and iron, these red soils are heavily leached but have more organic matter, and thus greater fertility, than comparable soils found in wetter zones. This helps to explain the tendency of farmers to plant in drier areas, such as the **Sahel** (a semi-desert transition zone between the Sahara and wetter savannas), even though they risk exposure to drought. With irrigation, many agronomists feel that the southern African countries of Zambia and Zimbabwe could greatly increase their commercial grain production.

Knowing areas of soil fertility can help explain patterns of settlement and can even suggest potential areas for development, but soil fertility alone does not determine where Africans live. The Nigerian case is instructive in this regard. Although some of its soils are fertile, many are mediocre or poor. Nigerian peoples long ago developed methods of enhancing soil fertility, allowing them to move from shifting cultivation to permanent-field agriculture and thus sustain impressive population density. In recent decades Nigeria has been able to feed its people in large part by importing food from abroad, paying for it by exporting oil. This strategy was not possible several generations ago, however, when Nigeria was even then the most populous country of Sub-Saharan Africa.

Climate and Vegetation

Sub-Saharan Africa lies in the tropical latitudes. Beginning just north of the tropic of Cancer, crossing the equator, and extending past the tropic of Capricorn in the south, it is the largest tropical landmass on the planet. Only the far south of the continent extends into the subtropical and temperate belts. Much of the region averages high temperatures from 70 to 80 °F year-round. The seasonality and amount of rainfall, more than temperature, determine the different vegetation belts that characterize the region. As Figure 6.7 shows, Addis Ababa, Ethiopia, and Walvis Bay, Namibia, have similar average temperatures, but the former is in the moist highlands and receives nearly 50 inches (127 centimeters) of rainfall annually, while Walvis Bay rests on the Namibian Desert and receives less than 1 inch (2.5 centimeters).

To understand the relationship between climate and vegetation in Sub-Saharan Africa, refer again to Figure 6.7. Imagine a series of concentric vegetation belts that begin in the western equatorial zone as forest (Am and Af), followed by woodlands and grasslands (Aw), semi-desert (BSh), and finally desert (BWh). The montane zones of East Africa, Cameroon, Guinea, and South Africa (especially along the Drakensberg Range) exhibit the forces of vertical zonation discussed in Chapter 4. Capturing more rainfall than the surrounding lowlands, these mountains often support unique forests and woodlands with high rates of endemic species. The only midlatitude climates of the region are found in South Africa. Western South Africa contains a small zone of Mediterranean climate comparable to that of southern California and noted for its production of fine wine (Csb). The eastern half of the country has a moist subtropical climate, not unlike that of Florida (Cfa).

Tropical Forests The core of Sub-Saharan Africa is remarkably moist. The world's second-largest expanse of humid equatorial rain forest lies in the Congo Basin, extending from the Atlantic Coast of Gabon two-thirds of the way across the continent, including the northern portions of the Congo and the Democratic Republic of the Congo (Zaire). The conditions here are constantly warm to hot and precipitation falls year-round (see graph for Kisangani, Figure 6.7).

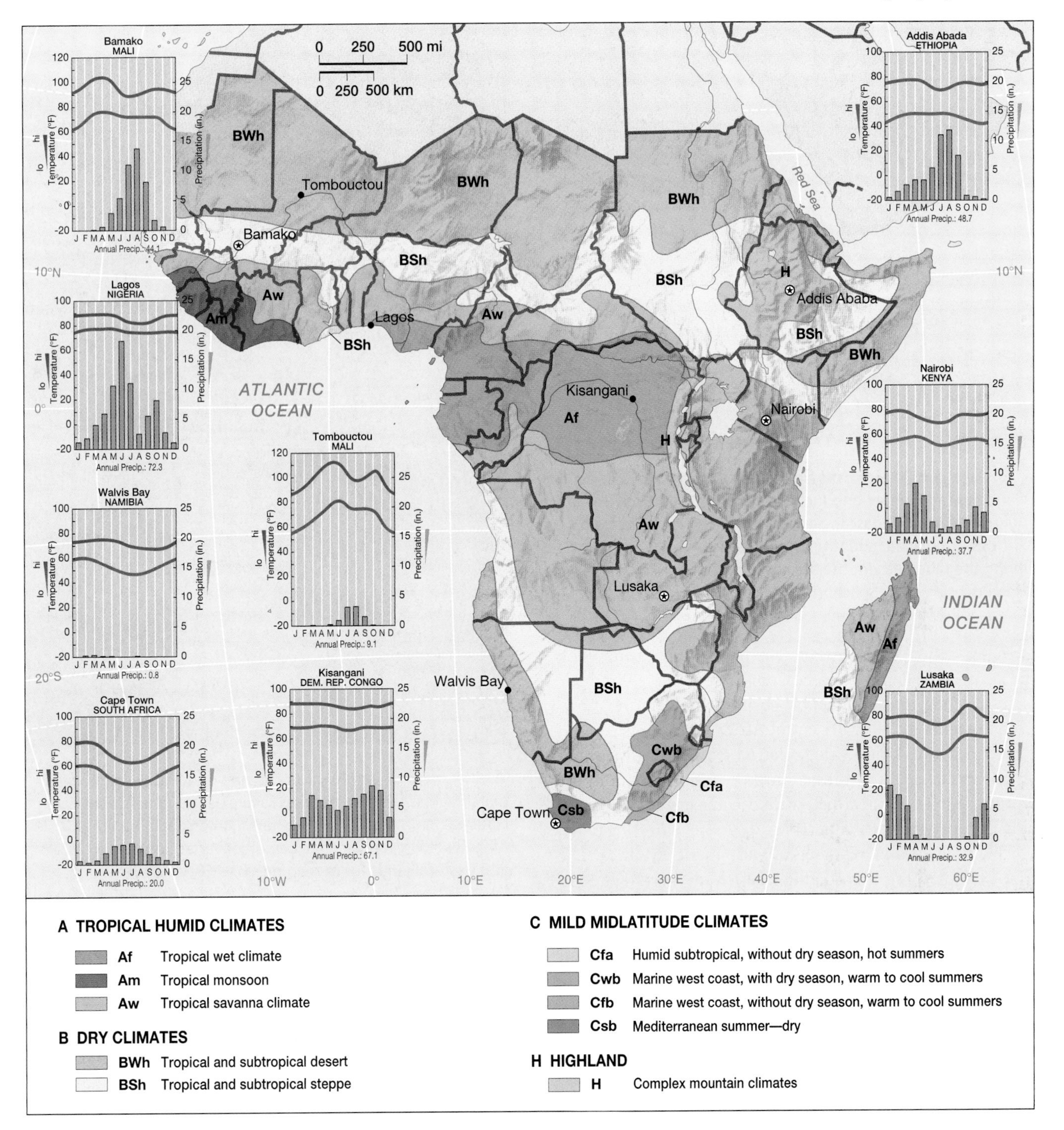

▲ **Figure 6.7 Climate map of Sub-Saharan Africa** Much of the region lies within the tropical humid and tropical dry climatic zones; thus, the seasonal temperature changes are not great. Precipitation, however, varies significantly from month to month. Compare the distinct rainy seasons in Lusaka and Lagos; Lagos is wettest in June and Lusaka receives most of its rain in January. Although there are important tropical forests in West and Central Africa, much the territory is semiarid steppe or tropical savanna. *[Temperature and precipitation data from* The World Weather Guide *(1984) by E. A. Pearce and C. G. Smith. London: Hutchinson]*

Commercial logging and agricultural clearing have degraded the western and southern fringes of this vast forest, but much of it is still intact. Considering the high rates of tropical deforestation in Southeast Asia and Latin America, the Central African case is a pleasant exception. Certainly the area's low population base of subsistence farmers does not strain the resource base. Moreover, Gabon and the Congo derive substantial foreign exchange from oil exports, making

▲ **Figure 6.8 Deforestation in Madagascar** Forest clearing for agriculture and fuel has denuded much of the countryside in Madagascar. Serious problems with soil erosion exist throughout the country, limiting agricultural productivity. During the rainy season, the discharge of topsoil into the Indian Ocean is visible from space. *(Bios/M. Gunther/Peter Arnold, Inc.)*

forest extraction less necessary. The political chaos engulfing the Democratic Republic of the Congo in the late 1990s made large-scale logging a difficult proposition. In the future, however, it seems likely that Central Africa's rain forests could suffer the same kind of degradation experienced in other equatorial areas. As deforestation proceeds elsewhere in the world, the trees of equatorial Africa become increasingly valuable, and hence more vulnerable.

Two smaller rainforest belts are also found in Africa. The first extends along the Atlantic Coast from Sierra Leone to western Ghana, and the second is located along the eastern coast of the island of Madagascar. Unlike the Congo Basin, these rainforest belts have been severely degraded by commercial logging and agricultural clearance. At present rates, several countries in West Africa will be completely deforested early in the twenty-first century. Madagascar's eastern rain forests, as well as its western dry forests, have suffered serious degradation in the past three decades (Figure 6.8). Deforestation in Madagascar is especially worrisome because the island forms a unique environment, with a large number of endemic species. Many species of fauna (including birds, reptiles, and lemurs) are now endangered and several face imminent extinction.

Savannas Wrapped around the Central African rainforest belt in a great arc lie Africa's vast tropical wet and dry savannas. Savannas are dominated by a mixture of trees and tall grasses in the wetter zones immediately adjacent to the forest belt and shorter grasses with fewer trees in the drier zones. North of the equatorial belt, rain generally falls only from May to October. The farther north one travels, the total amount of rainfall diminishes and the length of the dry season increases

▲ **Figure 6.9 African savannas** A line of wildebeests march across the tree studded savannas found in Masai Mara National Park, Kenya. The savannas, an essential habitat for Africa's wildlife, are steadily being converted into agricultural and pastoral lands for human needs. *(Nik Wheeler)*

(Figure 6.9). Climatic conditions south of the equator are similar, only reversed, with the wet season occurring between October and May, and with precipitation generally decreasing toward the south (see Lusaka in Figure 6.7). A larger area of wet savanna exists south of the equator, with substantial woodlands in southern portions of the Democratic Republic of the Congo, Zambia, and eastern Angola. These savannas are also a critical habitat for the region's large fauna. Elephant, zebra, rhinoceros, and lion are found in the wooded grasslands, although their numbers have declined in most areas since the 1980s.

A semiarid transition zone called the *Sahel,* is found north of the equator between the tropical dry savanna and the Sahara Desert (see Tombouctou/Timbuktu in Figure 6.7). Stretching across the length of the continent, from Senegal to Sudan, this precarious vegetative zone is mostly used for grazing livestock and limited agriculture. Life depends on a delicate balance of limited rain, drought-resistant plants, and a pattern of animal **transhumance** (the movement of animals between wet-season and dry-season grazing lands) that allows areas to recover. In the 1970s a series of droughts led to famine and crisis in the Sahel. It also led to serious discussions about human-induced environmental change in this region, which will be discussed later.

Deserts The vast extent of tropical Africa is bracketed by several deserts. The Sahara, the world's largest desert and one of its driest, spans the landmass from the Atlantic coast of Mauritania all the way to the Red Sea coast of Sudan. A narrow belt of desert extends to the south and east of the Sahara, wrapping around the **Horn of Africa** (the northeastern corner that includes Somalia, Ethiopia, Djibouti, and Eritrea) and pushing as far as eastern and northern Kenya. An even drier zone is found in southwestern Africa. In the Namib Desert of coastal Namibia, rainfall is a rare event, although temperatures are usually mild (see Figure 6.7 for Walvis Bay). Inland from the Namib lies the Kalahari Desert. Most of the Kalahari is not dry enough to be classified as a true desert, since it receives slightly more than 10 inches of rain a year (Figure 6.10). Its rainy season, however, is brief. Most of the precipitation is immediately absorbed by the underlying sands. Surface water is thus scarce, giving the Kalahari a desertlike aspect for most of the year.

Environmental Issues

Because much of Sub-Saharan Africa's population is rural and poor, earning its livelihood directly from the land, environmental shortages are keenly felt. Deforestation, especially in the woodlands of the savannas and Sahel, aggravate problems of soil erosion, moisture loss, and shortages of **biofuels** (wood and charcoal used for household energy needs, especially cooking). Similarly, agricultural practices in semiarid zones seem to be encouraging **desertification,** the expansion of desertlike conditions due to human-induced land degradation. In the most extreme cases, desertification exposes more people to the threat of drought and famine—crises all too familiar to the peoples of the region (Figure 6.11). Other environmental issues include the impact of mining operations (especially massive open-pit sites) on vegetation, water quality, and human health. And as Sub-Saharan cities grow in size and importance, urban environments increasingly face problems of air and water pollution as well as sewage and waste disposal. Lastly, Sub-Saharan Africa is home to an impressive array of wildlife, especially large mammals. Just how humans' and animals' competing demands for land will be worked out remains to be seen.

▲ **Figure 6.10 Kalahari Desert** Sand dunes are one of the attractions found in Kalahari Gemsbok Park in South Africa. Centered in Botswana, the Kalahari extends into parts of Namibia and South Africa. Sub-Saharan Africa has many semiarid areas but few true deserts. *(Clem Haagner/Photo Researchers, Inc.)*

Deforestation Although Sub-Saharan Africa still contains extensive forests, much of the region has relatively little tree cover. Unlike tropical America, forest clearance in the wet and dry savanna is of greater local concern than the limited commercial logging of the rain forest. North of the equator, for example, only a few wooded areas remain in a landscape dominated by grasslands, savanna (grassland with scattered trees and shrubs), and, occasionally, cropland. Highland Ethiopia was once covered with lush forests, but these have long since been reduced to a few remnant patches. Loss of woody vegetation has resulted in extensive hardship, especially for women and children who must spend many hours a day scrounging for the fuelwood needed for cooking.

In the southern tropical savanna, where human population has historically been lighter, extensive tracts of dry woodland remain intact. Zambia and sections of Mozambique, Angola, and Tanzania are still extensively wooded, although the actual tree coverage is not dense. The trees of the dry forest have little commercial value compared to some of the species found in the rain forest, but they do provide vital subsistence resources for local people as well as wildlife habitat.

Even in these more abundant areas, fuelwood scarcity is common around the larger towns and villages (Figure 6.12). In some countries, village women have organized into community-based nongovernmental organizations (NGOs) to plant trees and create greenbelts to ensure future fuel needs. One of the most successful efforts exists in Kenya under the leadership of Wangari Maathai. Maathai's Green Belt Movement has more than 50,000 members, mostly women, organized into 2,000 local community groups. Since the group's inception in 1977, millions of trees have been successfully planted. In those areas, village women now spend less time collecting fuel and local environments have improved. Kenya's success has drawn interest from other African countries, spurring a Pan-African Green Belt Movement largely organized through nongovernmental organizations interested in biofuel generation, the environment, and empowering women.

Desertification Desertification is evident along the margins of many of the world's arid areas. Throughout the Horn of Africa, the margins of the Kalahari, and even southwestern Madagascar, the process of desertification is under way.

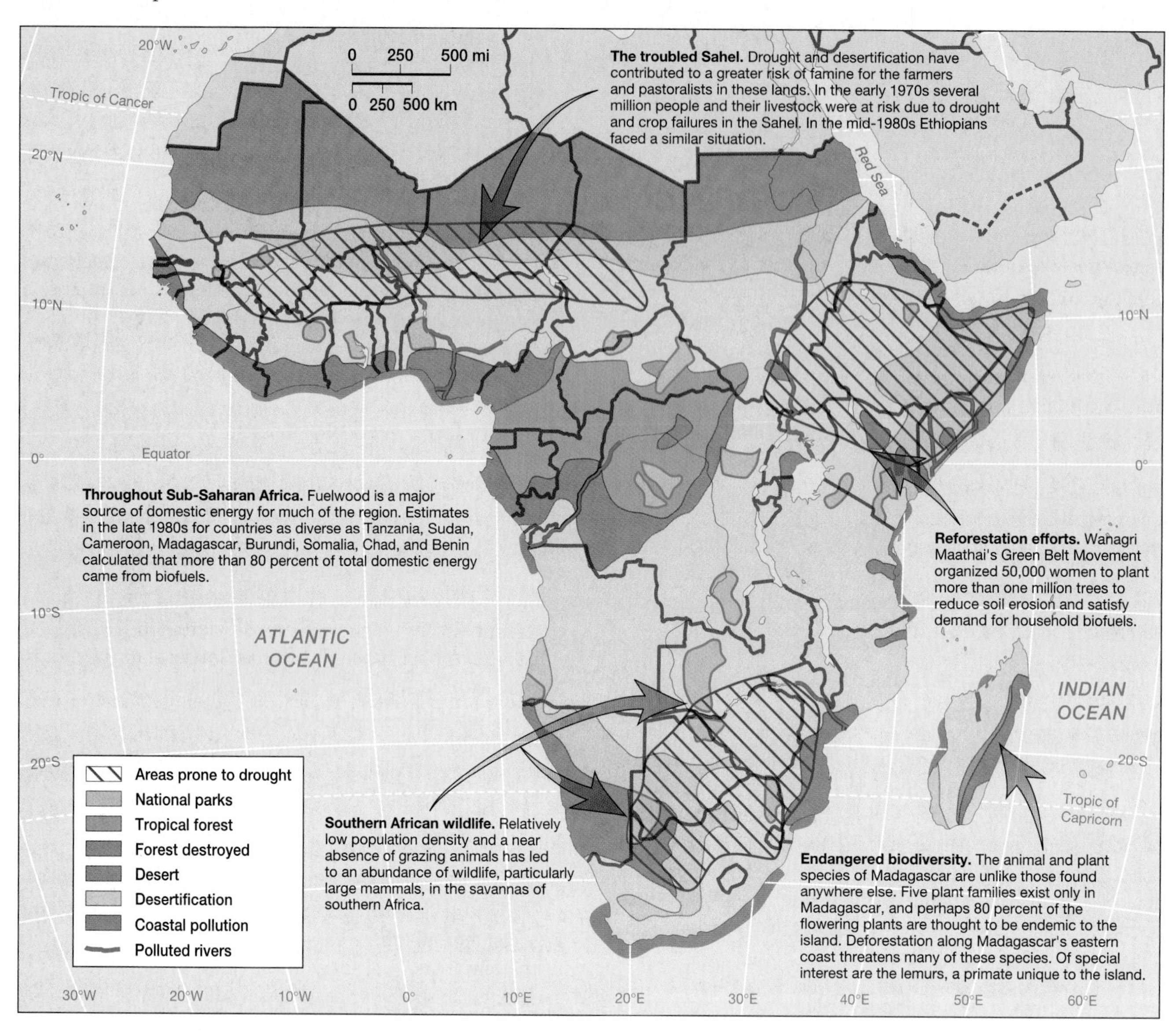

▲ **Figure 6.11 Environmental issues in Sub-Saharan Africa** Given the immense size of the Sub-Saharan Africa, it difficult to generalize about environmental problems. Dependence upon trees for fuel places strains on forests and wooded savannas throughout the region. In semiarid regions, such as the Sahel, population pressures and land-use practices seem to have exacerbated desertification. Yet Sub-Saharan Africa also supports the most impressive array of wildlife, especially large mammals, on Earth. The rapid growth of cities in the past three decades has yielded new concerns about water quality and pollution. *(Adapted from* DK World Atlas, *1997, p. 75. London: DK Publishing)*

Nowhere is it more severe than in the Sahel. The problem seems to have been compounded by recent climatic fluctuations. Over the centuries, the southern boundary of the Sahara has swung sometimes to the north and sometimes to the south in response to poorly understood rhythms in the global atmospheric system. A relatively wet period seems to have come to an abrupt end in the 1970s. Rivers in the area then began to diminish, and desert conditions began to move south. Unfortunately, tens of millions of people inhabit this desiccated area—peoples whose livelihoods had come to depend on the more abundant precipitation of the relatively wet period just past.

Human activities have exacerbated the process of desertification. The main culprits are improper cultivation and overgrazing, leading to the loss of soil. Through most of the Sahel, French colonial authorities forced villagers to grow peanuts as an export crop, a policy continued by the newly independent states of the region. Unfortunately, peanuts tend to deplete several key soil nutrients, which means that peanut farms are often abandoned after a few years as cultivators move on to fresh sites. A number of years are required for natural vegetation to become reestablished and begin holding the soil in place and absorbing rainfall. Since peanuts grow underground, moreover, the soil must be overturned at har-

▲ **Figure 6.12 Gathering fuel** A wagon loaded with firewood in rural Mali. Villagers and city dwellers alike rely upon wood for household energy needs, placing an enormous strain on Africa's woodlands. In areas with the greatest scarcity, women and children may spend many hours each day gathering fuel. *(Betty Press/Woodfin Camp & Associates)*

▲ **Figure 6.13 Desertification in the Sahel** In southern Niger, pastoralists rely upon wells to water their herds of goats and cattle. The relationship between livestock, people, and the Sahelian ecosystem is complex. Many researchers believe that increased grazing pressure brought on by more livestock and wells may have made the Sahel more vulnerable to desertification. *(Victor Englebert/Englebert Photography, Inc.)*

vest time. This typically occurs at the onset of the dry season, when dry winds from the Sahara can carry away the fine dirt of the newly harvested field. With the loss of topsoil, the ground no longer absorbs the precipitation that does fall—often in drenching torrents—during the brief rainy season. The end result of this grim process is the spread of desert-like conditions regardless of actual changes in precipitation.

Overgrazing has also been implicated in Sahelian desertification, as is true in other regions of the world. Livestock is a traditional export product of the region, and animal production dramatically expanded after World War II as the population swelled and the need for export earnings grew. In many areas, natural pasture was greatly reduced by intensified grazing pressure, leading to increased wind erosion. National and international developmental agencies, hoping to increase production further, then began to dig deep wells in areas that had previously been unutilized by herders through most of the year. The new supplies of water, in turn, allowed year-round grazing in places that could not withstand it. Large barren circles around each new well began to appear even on satellite images (Figure 6.13). Some scholars questioned whether recovery of such degraded areas was possible even if "normal" precipitation patterns were to return.

Other scholars, however, have come to a different conclusion, and the desertification phenomenon generated an unresolved controversy in the technical literature. Pessimists view the process as virtually irreversible and argue that the Sahel can no longer support the millions of people who presently reside there; a few foresee little else but a future of mass starvation. Optimists, to the contrary, contend that the land can quickly recover if appropriate practices are followed, and a few maintain that, given proper adjustments, larger populations could subsist in the area even if dry conditions continue. Debates also surround the issue of climatic fluctuation itself. Some climatologists believe that the Sahel now faces a prolonged dry period, while others argue that wetter conditions could soon return. A few suggest that in the warmer global climate of the future, precipitation should increase markedly in the area. Regardless of the causes, many of the areas experiencing desertification are the same ones most vulerable to drought and famine (see "Local to Global: African Wildlife").

Population and Settlement: Young and Restless

Sub-Saharan's population is growing quickly, with an estimated doubling time of less than 27 years. It is also very young, with 45 percent of the population younger than age 15, compared to just 19 percent for more-developed countries. Families tend to be large, with an average woman having six children. Yet high child and maternal mortality rates also exist, reflecting disturbingly low access to basic health services. The most troubling indicator for the region is its declining life expectancy, which dropped from 52 years in 1995 to 48 years in 1998, in part due to the AIDS epidemic.

Behind these demographic facts lie complex differences in settlement patterns, livelihoods, belief systems, and access to health care. Contrary to popular stereotypes, Sub-Saharan Africa is not densely populated. The entire region holds some 625 million persons—half the population that is crowded into the much smaller land area of South Asia. In fact, the overall population density of the region (74 people per square mile) is similar to that of the United States (76 people per square mile). Just six states account for half of the region's population: Nigeria, Ethiopia, the Democratic Republic of Congo, South Africa, Tanzania, and Kenya (Table 6.1). Many of the remaining Sub-Saharan countries are quite sparsely inhabited. Chad has just 15 people per square mile (six people per square

LOCAL TO GLOBAL African Wildlife

Sub-Saharan Africa is justly famous for its wildlife. In no other region of the world can one find such abundance and diversity, especially of large mammals. The survival of wildlife here reflects, to some extent, the historically low human population density and the fact that sleeping sickness (transferred by the tsetse fly) and other diseases have kept people and their livestock out of many areas. It is also true, however, that many African peoples have developed various means of successfully coexisting with wildlife.

But as is true elsewhere in the world, wildlife is quickly declining in much of Sub-Saharan Africa. West Africa now has few areas of prime habitat, and what remains is rapidly shrinking. The most noted wildlife reserves are in East Africa; in Kenya and Tanzania these reserves are major tourist attractions and are thus economically important. Even there, however, population pressure, political instability, and poverty make the maintenance of large wildlife reserves difficult. Poaching is a major problem, particularly for rhinoceroses and elephants; the price of a single horn or tusk to distant markets represents several years' wages for most Africans. Ivory is avidly sought in East Asia, especially in Japan. In China, powdered rhino horn is used as a traditional medicine, whereas in Yemen rhino horn is prized for dagger handles. During the 1980s, the region's elephant population fell by more than half, to 600,000, in part due to ivory trade. Rhino populations are far smaller, and the black rhino is considered an endangered species.

The most secure wildlife reserves now seem to be located in southern Africa (see Figure 6.11). In fact, elephant populations are considered to be too high in countries such as Zimbabwe. South Africa's wildlife reserves are so well managed that even white rhino populations are said to be too large. Some wildlife experts here contend that herds should be culled to prevent overgrazing, and that the ivory and rhino horn should to be legally sold in the international market in order to generate revenue for further conservation. Only such a market-oriented approach, they argue, will give African countries a long-term incentive to preserve habitat. Many environmentalists, not surprisingly, disagree strongly.

In 1989 a worldwide ban on the legal ivory trade was imposed, as part of the Convention on Trade in Endangered Species (CITES). While several African states such as Kenya lobbied hard for the ban, others, such as Zimbabwe, Namibia, and Botswana, complained that their herds were growing and the sale of ivory helped to pay for conservation efforts. Conservationists feared that a lifting of the ban would bring on a new wave of poaching and illegal trade. In the late 1990s the ban was lifted in Zimbabwe and ivory was exported. Whether this will have long-term repercussions on elephant survival is not yet known.

▲ **Figure 1 Southern African wildlife** An African elephant traversing the savannas of Kruger National Park in South Africa, one of the region's oldest wildlife parks. South Africa, Botswana, Zimbabwe, and Zambia have the largest elephant populations. Other countries have seen a troubling decline in elephant numbers. *(Rob Crandall/Rob Crandall Photographer)*

kilometer) and Botswana has only six people per square mile (two people per square kilometer). Such states do not consider population growth a serious issue. Many of the governments of the more densely settled territories, however, began seriously promoting family planning policies in the 1980s.

The crude population density is an imperfect indicator of whether or not a country is overpopulated. Geographers are often more interested in the **physiological density**, which is the number of people per unit of arable land. The physiologic density in Chad, where only 3 percent of the land is arable, is much higher than its crude population density. Perhaps a more telling indicator of population pressure and potential food shortages is agricultural density. **Agricultural density** is the ratio between the number of farmers per unit of arable land. Since the majority of people in Sub-Saharan Africa earn their living or subsistence from agriculture, agricultural density indicates the number of people who directly depend upon each arable square mile. The agricultural density of many Sub-Saharan countries is 10 times greater than their crude population density.

Population Trends and Demographic Debates

It is the combination of growth in particular areas of Sub-Saharan Africa (high agricultural densities and rates of natural increase) and reversals in some economic and social indicators that make demographers concerned about the region's overall well-being. The population density of the Sahel, for example, is not crowded by European or Asian standards, but it may already contain too many people for the land to support, given the sparsity and unpredictability of its rainfall.

This assertion remains controversial, however—as does the entire field of African demography. Some believe that the region could support many more people than it presently does; these people even argue that Sub-Saharan Africa as a

Table 6.1 Demographic Indicators

Country	Population[a]	Natural Increase	TFR[b]	%<15[c]	% Urban
Angola	12.0	3.2	7.2	48	42
Benin	6.0	3.2	6.3	49	36
Botswana	1.4	1.2	4.3	43	48
Burkina Faso	11.3	2.9	6.9	49	15
Burundi	5.5	2.5	6.6	47	5
Cameroon	14.3	2.8	5.9	44	44
Cape Verde	0.4	2.9	5.3	45	44
Central African Republic	3.4	2.1	5.1	42	39
Chad	7.4	3.3	6.6	44	22
Comoros	0.5	2.7	5.1	47	29
Congo	2.7	2.3	5.1	46	58
Dem. Rep. of Congo	49.0	3.2	6.6	47	29
Djibouti	0.7	2.3	5.8	41	81
Equatorial Guinea	0.4	2.6	5.9	43	37
Eritrea	3.8	3.0	6.1	44	16
Ethiopia	58.4	2.5	7.0	46	16
Gabon	1.2	2.0	5.0	38	73
Gambia	1.2	2.4	5.9	44	37
Ghana	18.9	2.9	5.5	45	35
Guinea	7.5	2.4	5.7	45	29
Guinea-Bissau	1.1	2.1	5.8	42	22
Ivory Coast	15.6	2.6	5.7	45	46
Kenya	28.3	2.0	4.5	46	27
Lesotho	2.1	2.1	4.3	42	16
Liberia	2.8	3.1	6.2	42	45
Madagascar	14.0	3.0	6.0	47	22
Malawi	9.8	1.7	5.9	48	20
Mali	10.1	3.1	6.7	47	26
Mauritania	2.5	2.5	5.4	43	54
Mauritius	1.2	1.0	2.0	27	43
Mozambique	18.6	2.2	5.6	46	28
Namibia	1.6	1.7	5.1	42	27
Niger	10.1	3.4	7.4	48	15
Nigeria	121.8	3.0	6.5	46	16
Reunion	0.7	1.6	2.3	30	73
Rwanda	8.0	2.1	6.0	47	5
São Tomé and Principe	0.2	3.4	6.2	47	46
Senegal	9.0	2.7	5.7	45	42
Seychelles	0.1	1.4	2.1	31	59
Sierra Leone	4.6	1.9	6.5	43	36
Somalia	10.7	3.2	7.0	48	24
South Africa	38.9	1.6	3.3	35	57
Sudan	28.5	2.1	5.0	43	27
Swaziland	1.0	3.3	5.6	49	22
Tanzania	30.6	2.5	5.7	46	21
Togo	4.9	3.6	6.8	46	31
Uganda	21.0	2.7	6.9	47	14
Zambia	9.5	1.9	6.1	45	39
Zimbabwe	11.0	1.5	4.4	44	31

[a]Population in millions.

[b]Total fertility rate.

[c]Percentage of population younger than 15 years of age.

Source: *Population Reference Bureau World Population Data Sheet,* 1998.

whole is still underpopulated and therefore will benefit from continued population growth. Pessimists, however, argue that the region is a demographic time bomb, and that unless fertility is quickly reduced, Sub-Saharan Africa will face massive famines in the near future. Despite these disparate views, the majority of African states officially support lowering rates of natural increase and are slowly promoting modern contraception practices—both to reduce family size and to protect people from sexually transmitted diseases.

Family Size A preference for large families is the basis for the region's demographic growth. In the 1960s, many areas in the developing world had comparable total fertility rates (TFR) of 6.0 or higher, but by the mid-1990s, only people in Sub-Saharan Africa and Southwest Asia continued to have such large families. For Southwest Asia the dominance of Islam is used to explain high fertility rates. In Sub-Saharan Africa a combination of cultural practices, rural lifestyles, and economic realities encourage large families.

Throughout the region large families are considered prestigious and guarantee a family's lineage and status. Even now most women marry young, typically when they are teenagers, which increases their opportunity to have children. Demographers often point to the limited formal education available to women as another factor contributing to high fertility (Figure 6.14). Ethnic rivalries may also encourage pro-natal practices. In ethnically divided states, such as Kenya and Nigeria, resources are often divided according to ethnic affiliation. An ethnic group's decision to reduce its numbers could, over time, weaken its political influence and thus be seen as unsound. Surprisingly, religious affiliation has little bearing on the region's fertility rates; Muslim, Christian, and animist communities all have similarly high birthrates.

▲ **Figure 6.14 Large families** A Gambian father poses with his 10 children. Large families are still common in Sub-Saharan Africa, where in many countries the average woman will have five or six children. Growth rates soared in the 1970s and 1980s, but they began to slow slightly in the 1990s. In southern Africa total fertility rates have dropped to three and four children per woman. The overall average for the region is still six children per woman. *(Mark Boulton/Photo Researchers, Inc.)*

Attitudes about family size, however, are shaped by high child mortality rates. In the 1960s, for example, it was not uncommon for a woman to lose several children before they reached the age of five. As child health improved in the 1970s and 1980s, women still continued to have the same number of children but more lived, which pushed fertility rates even higher—in some countries reaching averages of eight children per woman.

The everyday realities of rural life make large families an asset rather than a burden. Children are an important source of labor; from tending crops and livestock to gathering fuelwood, they add more to the household economy than they take. Also, for the poorest places in the developing world, such as Sub-Saharan Africa, children are likened to social security. Should the health of parents falter, there is the expectation that grown children will step in and care for them.

Government policies toward family size have shifted dramatically in the past three decades. During the 1970s, population growth was not perceived as a problem by many African governments; in fact, many equated limiting population size with a neocolonial attempt to slow regional development. By the 1980s, a shift in national policies occurred. For the first time, government officials argued that smaller family sizes and lower growth rates were needed for social and economic development. Following the United Nations International Conference on Population and Development in Cairo, Egypt, in 1994, the following ambitious goals were announced: to bring the natural increase rate down to 2.0 by the year 2020 and to increase the rate of contraceptive use to 40 percent.

Other factors are converging to slow the growth rate. As African states slowly become more urban, there is a corresponding decline in family size—a pattern seen throughout the world. Tragically, declines in natural increase are also occurring as a result of AIDS.

The Impact of AIDS on Africa In April 1999 the president of Zimbabwe, Robert Mugabe, announced that 1,200 Zimbabweans were dying each week from AIDS. This was the first public acknowledgment of the gravity of the epidemic that is reversing the gains Zimbabwe has made since independence. If it were not for AIDS (acquired immunodeficiency syndrome), life expectancy in Zimbabwe would be in the high 60s. Instead, it plummeted to 40 years in 1998 and perhaps one-quarter of the population aged 15–49 is infected with HIV (the human immunodeficiency virus that causes AIDS) or AIDS.

Zimbabwe and Botswana are ground zero for the AIDS epidemic that is just beginning to ravage the region (Figure 6.15). Two-thirds of the HIV/AIDS cases in the world are found in Africa. The virus is thought to have originated in the forests of the Congo, possibly crossing over from chimpanzees to humans sometime in the 1950s. Yet it was not until the late 1980s that the impact of the disease was widely felt in some of the more-populated parts of the region.

Infection rates are such that demographers anticipate a slower population growth rate for the entire region due to AIDS. Sub-Saharan Africa will still grow, but population es-

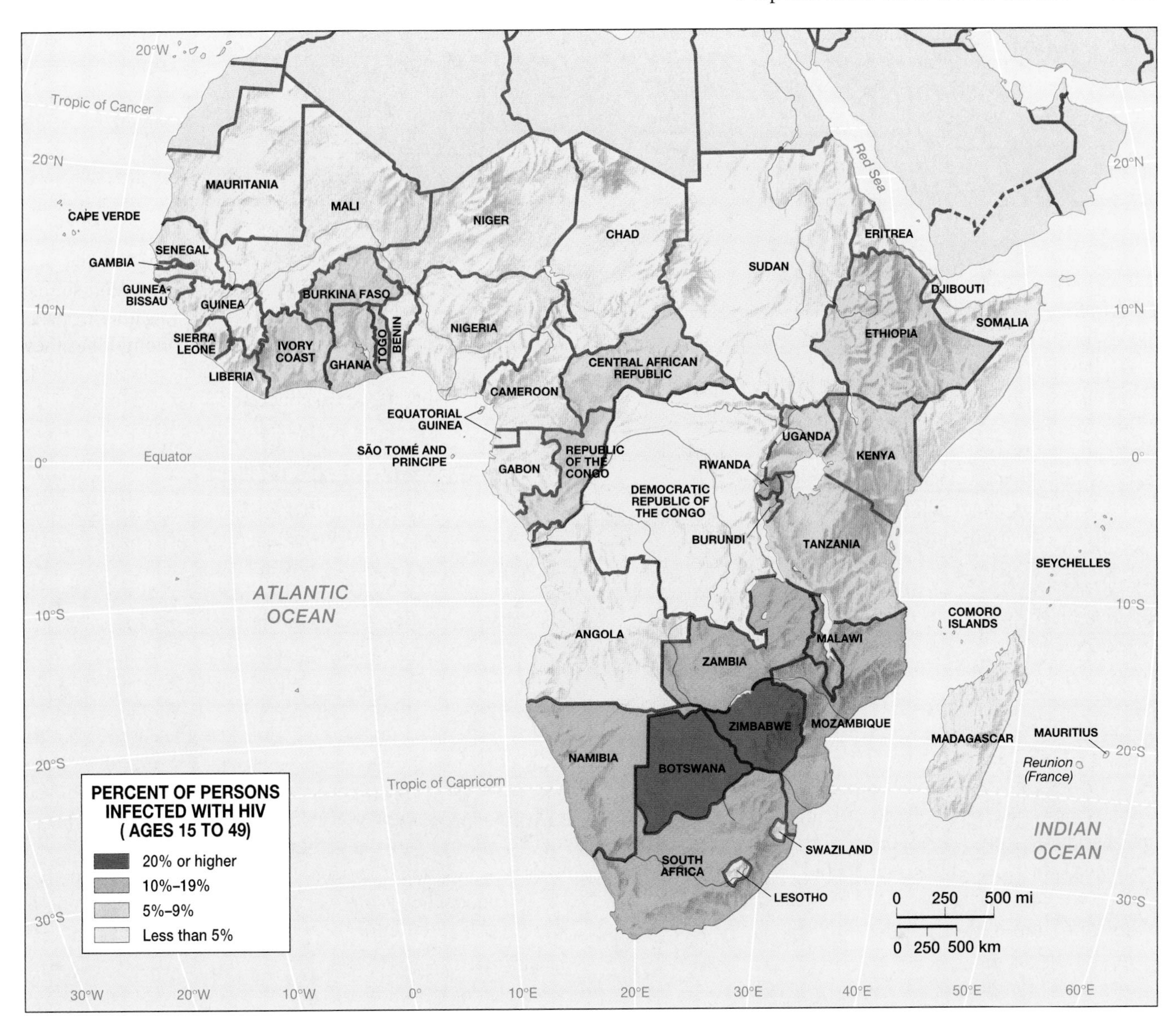

▲ **Figure 6.15 HIV prevalence** In Sub-Saharan Africa more than 20 million people are infected with HIV, two-thirds of the world total for 1997. Infection rates are highest in southern Africa, especially Botswana and Zimbabwe. *[Source: Thomas J. Goliber, 1997, "Population and Reproductive Health in Sub-Saharan Africa,"* Population Bulletin *52(4), p. 31. Washington, DC: Population Reference Bureau.]*

timates for 2025 have been trimmed by as much as 200 million. Sadly, the social and economic implications of this epidemic are hard to measure. AIDS typically hits the portion of the population that is most active economically. Time lost to care for sick family members and the outlay of workers' compensation benefits could reduce economic productivity and overburden public services in hard-hit areas. Infection rates among newborns are high, and many areas struggle to care for children orphaned by AIDS.

Unlike the developed world, where expensive and potent drug therapies have prolonged the lives of people with HIV/AIDS, Sub-Saharan Africa does not have the money for this option. The only way governments can reduce the impact of the epidemic is through prevention, mostly through educating people about how the virus is spread and convincing them to change their sexual behavior. In Uganda, state agencies, along with NGOs, began a national no-nonsense campaign for AIDS awareness in the 1980s that focused on the schools. Through explicit materials, role-playing games, and frank discussion, the prevalence of HIV among women in prenatal clinics had declined by the late 1990s. Senegal, which has a very low infection rate, also mounted a vigorous campaign to educate its citizens. One indicator of its success is a tenfold increase in condom purchases from 1988 to 1997. Yet many of the areas hardest hit, such as Botswana and Zimbabwe, were reluctant to launch serious education efforts in the late 1980s. The consequences of this inaction were deadly.

Patterns of Settlement and Land Use

Due to the dominance of rural settlement in Sub-Saharan Africa, people are widely scattered throughout the region (Figure 6.16). Population concentrations are the highest in West

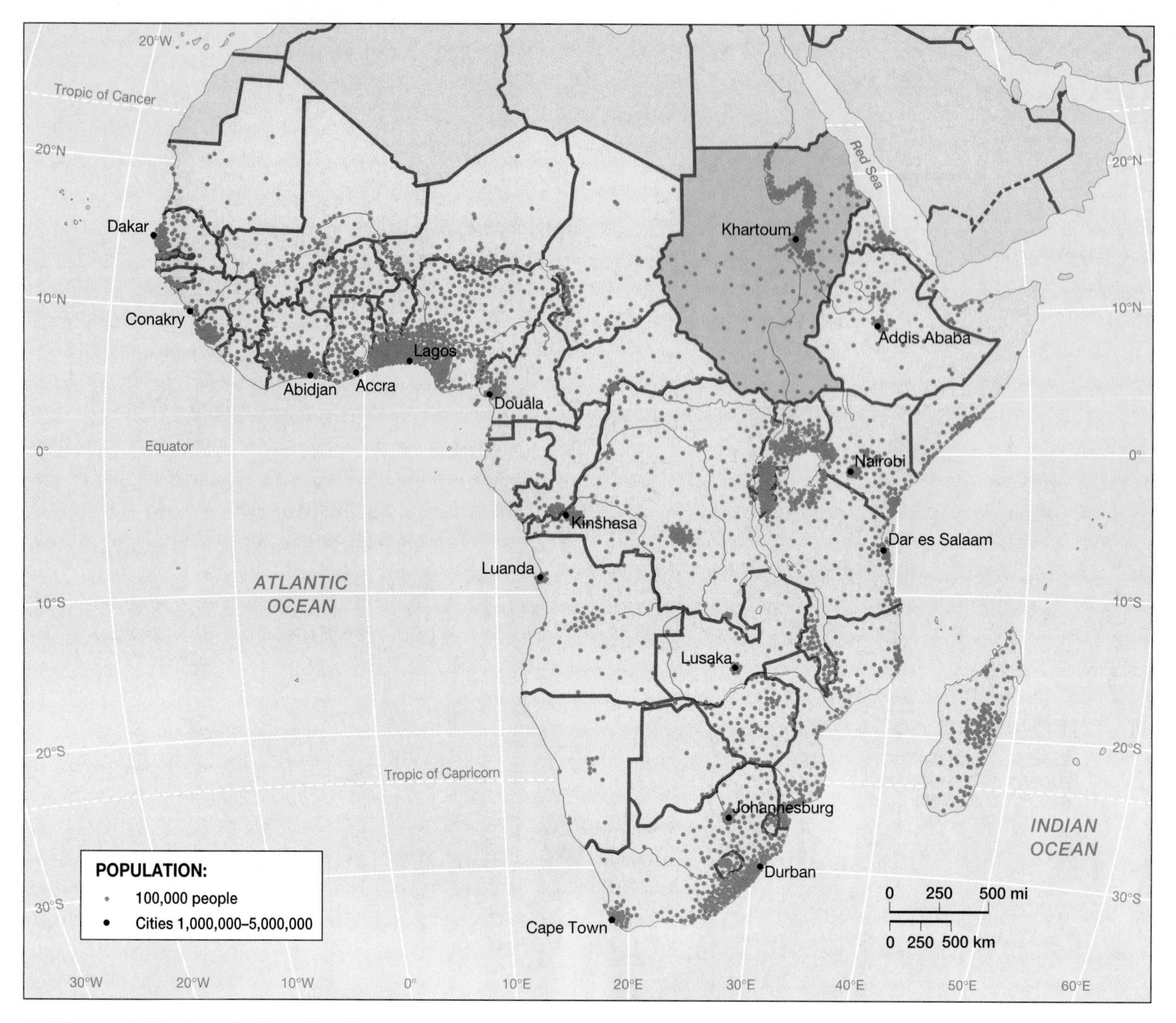

▲ **Figure 6.16 Population distribution** The majority of people in Sub-Saharan Africa live in rural areas. Some of these rural zones, however, are densely settled, such as West Africa and the East African highlands. Major urban centers, especially in South Africa and Nigeria, support millions. Overall, the region has few large cities with more than one million people.

Africa, highland East Africa, and the eastern half of South Africa. The first two areas have some of region's best soils, and indigenous systems of permanent agriculture developed there. In the latter, an urbanized economy based on mining, as well as the forced concentration of black South Africans into eastern homelands, contributed to the region's overall density.

West Africa is more heavily populated than most of Sub-Saharan Africa, although the actual distribution pattern is patchy. Density in the far west, from Senegal to Liberia, is moderate; this area is characterized by broad lowlands with decent soils, and in many areas the cultivation of wet rice has enhanced agricultural productivity. Greater concentrations of people are found along the Gulf of Guinea, from southern Ghana through southern Nigeria, and again in northern Nigeria along the southern fringe of the Sahel. Nigeria is moderately to densely settled through most of its extensive territory; with 122 million inhabitants, it stands as the demographic core of Sub-Saharan Africa. The next largest country, Ethiopia, has only half of Nigeria's population.

The population centers in East Africa are highland Ethiopia centered around Addis Ababa, the Lake Victoria basin, and the Kenyan portion of the Rift Valley. On the high plateaus of Ethiopia, agriculture is based on temperate crops such as wheat and barley, as well as varieties of lentils, peas, and potatoes. In addition to subsistence production, coffee is an important export that was first domesticated by farmers in this area hundreds of years ago. In the volcanic highland zones of Kenya, Tanzania, Uganda, Rwanda, and Burundi, subsistence agriculture relies more on maize and root crops than wheat. Export crops such as tea, coffee, and cotton are also produced.

LOCAL VOICES Ibo Agricultural Practices

Nigerian Chinua Achebe wrote *Things Fall Apart* in the 1950s prior to Nigerian independence. It is the story of an Ibo man, Okonkwo, coming to terms with how colonialism has transformed the logic and security of traditional lifeways.

The following excerpt describes the planting cycle of yams, the basis for subsistence in much of humid West Africa. The mastery of yam production allowed the Ibo to procure more food and live in denser settlements. The art of planting and tending the yams, the rainy-season wait for them to mature, and finally the celebrated harvest illustrates the intimate linkages between rural societies and their environments.

▲ **Figure 1 Nigeria's Ibo** A young Ibo boy sits in front of his village home. The Ibo (also Igbo) are Nigeria's third largest ethnic group, representing 17 percent of the population. They live in the southeastern portion of the state. *(Department of Social Anthropology, Cambridge University, Cambridge, England)*

Some days later, when the land had been moistened by two or three heavy rains, Okonkwo and his family went to the farm with baskets of seed-yams, their hoes and machetes, and the planting began. They made single mounds of earth in straight lines all over the field and sowed the yams in them.

Yam, the king of crops, was a very exacting king. For three or four moons it demanded hard work and constant attention from cock-crow till the chickens went back to roost. The young tendrils were protected from earth-heat with rings of sisal leaves. As the rains became heavier, the women planted maize, melons, and beans between the yam mounds. The yams were then staked, first with little sticks and later with tall and big tree branches. The women weeded the farm three times at definite periods in the life of the yams, neither early nor late.

And now the rains had really come, so heavy and persistent that even the village rain-maker no longer claimed to be able to intervene. He could not stop the rain now, just as he would not attempt to start it in the heart of the dry season, without serious danger to his own health. The personal dynamism required to counter the forces of these extremes of weather would be far too great for the human frame.

And so nature was not interfered with in the middle of the rainy season. Sometimes it poured down in such thick sheets of water that earth and sky seemed merged in one gray wetness. It was then uncertain whether the low rumbling of Amadiora's thunder came from above or below. At such times, in each of the countless thatched huts of Umuofia, children sat around their mother's cooking fire telling stories, or with their father in his obi warming themselves from a log fire, roasting and eating maize. It was a brief resting period between the exacting and arduous planting season and the equally exacting but light-hearted month of harvests.

Source: From *Things Fall Apart* by Chinua Achebe. New York: Fawcett Crest, 1959, pp. 35–36.

As more Africans move to cities, patterns of settlement are evolving into clusters of higher concentration. Localities that were once small administrative centers for colonial elites mushroomed into major cities. The region even has its own mega-city; Lagos topped 10 million in the 1990s. Throughout the continent, African cities are growing faster than rural areas. But before examining the Sub-Saharan urban scene, a more detailed discussion of rural subsistence is needed.

Agricultural Subsistence The main staple crops over most of Sub-Saharan Africa are millet, sorghum, and corn (maize), as well as a variety of tubers and root crops such as yams. Irrigated rice is widely grown in West Africa and Madagascar. Wheat and barley are grown in parts of South Africa and Ethiopia. Intermixed with subsistence foods are a variety of export crops—coffee, tea, rubber, bananas, cocoa, cotton, and peanuts—that are grown in distinct ecological zones and often in some of the best soils (see "Local Voices: Ibo Agricultural Practices").

Over much of the continent, African agriculture remains relatively unproductive, and population densities tend to be low. Amid these poorer tropical soils, cropping usually entails shifting cultivation (or **swidden**). This process involves burning the natural vegetation to release fertilizing ash and planting crops such as maize, beans, sweet potatoes, banana, papaya, manioc, yams, melon, and squash. Eventually, each plot is temporarily abandoned once its source of nutrients has been exhausted. Swidden cultivation is often a very finely tuned adaptation to local environmental conditions, but it is unable to support high population densities.

While South Africa has a good agricultural base, until recently the best farmlands were reserved for the small white population and hence were not densely settled. The rural black population, in contrast, was forcibly resettled onto "homelands" with poor agricultural potential in the eastern half of the country. Due in part to overcrowding, these areas have suffered from severe forms of environmental degradation, especially soil erosion and overgrazing.

Madagascar's agricultural patterns are unique and thus deserve special mention. Some 1,500 years ago, settlers from Indonesia landed on the island. These seafarers had been blown off course, or perhaps they were merely sailing into the unknown, hoping to find new land suitable for settlement. At any rate, once they landed on Madagascar, they lost contact with Southeast Asia, and within a few hundred years they had settled throughout the island. Their main focus of occupation was the central highlands, where they built irrigated rice fields and created a cultural landscape reminiscent of those in their ancestral home. Somewhat later, another wave of immigrants began to arrive, this time from the African mainland, introducing the shifting cultivation techniques commonly used there. The ancestry of Madagascar's 14 million residents seems to be evenly divided between African and Southeast Asian stock. The bulk of the population continues to reside in the eastern highlands, but deforestation and serious erosion problems are undermining the productivity on traditional subsistence practices.

Plantation Agriculture Plantation agriculture, designed to produce crops for export, is critical to the economies of many states. If African countries are to import the modern goods and energy resources they require, then they must sell their own products on the world market. Since the region has few competitive industries, the bulk of its exports are primary products derived from farming, mining, and forestry.

A number of African countries rely heavily on just one or two export crops. Coffee, for example, is vital for Ethiopia, Kenya, Rwanda, Burundi, and Tanzania. Peanuts have long been the primary foreign-exchange earner in the Sahel belt, while cotton is tremendously important for Sudan and the Central African Republic. Ghana and the Ivory Coast have long been the world's main suppliers of cocoa (the source of chocolate), Liberia produces plantation rubber, and many farmers in Nigeria specialize in palm oil (Figure 6.17). The export of such products can bring good money when commodity prices are high, but when prices collapse, as they periodically do, economic devastation may follow.

▲ **Figure 6.17 Harvesting palm oil** A man harvests palm oil near Ibadan, Nigeria. Palm oil is a major export for western Nigeria. Other important Sub-Saharan plantation crops include cocoa, rubber, tea, coffee, cotton, and peanuts. *(M. & E. Bernheim/Woodfin Camp & Associates)*

Herding and Livestock Animal husbandry is extremely important in Sub-Saharan Africa, particularly in the semiarid zones. Camels and goats are the principal animals in the Sahara and its southern fringe, but farther south, cattle are primary. Many African peoples have traditionally specialized in cattle raising and are often tied into symbiotic relationships with neighboring farmers. Such **pastoralists** typically graze their stock on the stubble of harvested fields during the dry season and then move them to drier uncultivated areas during the wet season when the pastures turn green. Farmers thus have their fields fertilized by the manure of the pastoralists' stock, while the pastoralists find good dry-season grazing. At the same time, the nomads can trade their animal products for grain and other goods of the sedentary world. Several pastoral peoples of East Africa, however, are noted for their extreme reliance on cattle and general (but never complete) independence from agriculture. The Masai of the Tanzanian/Kenyan borderlands traditionally derive a large percentage of their nutrition from drinking a mixture of milk and blood (Figure 6.18). The blood is obtained by periodically tapping the animal's jugular veins, a procedure that evidently causes them little harm.

Large expanses of Sub-Saharan Africa have been off-limits to cattle due to infestations of **tsetse flies,** which spread sleeping sickness to cattle, humans, and some wildlife species. Where wild animals, which harbor the disease but are immune to it, were present in large numbers, especially in environments containing brush or woodland (which are necessary for tsetse fly survival), cattle simply could not be raised. Some evidence suggests that tsetse fly infestations dramatically increased in the late 1800s, greatly harming African societies dependent on livestock, but benefiting wildlife populations. In colonial Uganda, for example, where the burning of brush was outlawed, cases of sleeping sickness surged. At present, tsetse fly eradication programs are reducing the threat, and cattle raising is spreading into areas that were previously forbidden. This process is beneficial for African peoples, but it obviously bodes ill for the continued survival of large numbers of wild animals. When people and their stock move into new areas in sizable numbers, wildlife almost inevitably declines.

As the tsetse fly story shows, Sub-Saharan Africa presents a difficult environment for raising livestock due to the virulence of its animal diseases. In the tropical rainforest zone of Central Africa, cattle have never survived well and the only domestic animal that thrives is the goat. Raising horses, moreover, has historically been feasible only in the Sahel and in

▲ **Figure 6.18 Masai pastoralists** A Masai man holds a cow steady while a woman collects blood being drained from the animal's neck. Pastoral groups such as the Masai live in the drier areas of Sub-Sahara Africa. The Masai live in Kenya and Tanzania. Other pastoral groups are found in the Sahel, the Horn of Africa, and East Africa. *(CORBIS/© Kennan Ward)*

South Africa. As we shall see later, the disease environment of tropical Africa has presented a variety of problems for its human populations as well.

Urban Life

Considered the least-urbanized region in the developing world, most Sub-Saharan cities are growing at twice their national growth rates. If present trends continue, half of the region's population may well be living in cities by 2025. One of the consequences of this surge in city living is urban sprawl. Rural-to-urban migration, industrialization, and refugee flows are forcing the cities of the region to absorb more people and require more resources. As in Latin America, the tendency is toward urban primacy, the condition in which one major city is dominant and at least three times larger than the next largest city. Kinshasa, the capital of the Democratic Republic of the Congo, is an example. In the 1960s less than half a million people resided there; by the late 1990s it was a city of more than four million, dwarfing the country's other cities.

▲ **Figure 6.19 Downtown Lagos, Nigeria** Lagos is home to 12 million people, making it the region's largest city. A classic primate city, the streets team with buses, collective taxis, thousands of pedestrians, and street vendors. Infrastructure has not kept up with the city's rapid growth. The demand for roads and utilities is far greater than the city's government is able to provide. *(Daniel Lainé/Corbis)*

It is estimated that Sub-Saharan Africa's largest city, Lagos, will have 12 million inhabitants in 2000 (Figure 6.19). In 1960 it was a city of only one million. Unable to keep up with the surge of rural migrants, Lagos's streets are clogged with traffic; for those living on the city's periphery, three- and four-hour commutes (one way) are common. City officials struggle to build enough roads, provide electricity, water and sewage, and employment for all of these people. In many cases the informal sector (unregulated services and trade) provides urban employment and services. Crime is another major problem for Lagos. The chances of being attacked on the streets or robbed in one's home are quite high, even though most windows are barred and houses are fenced. To deal with this problem, Lagos does have the highest density of police in Nigeria, but the weak social bonds between urban migrants, widespread poverty, and the gap between rich and poor seem to encourage lawlessness.

European colonialism greatly influenced urban form and development in the region, especially in southern Africa, where more than half the population in the late 1990s lived in cities. Africans, however, had an urban tradition prior to the colonial era, although a very small percentage of the population lived in cities. Ancient cities, such as Axum in Ethiopia, thrived 2,000 years ago. Similarly, in the Sahel, prominent trans-Saharan trade centers, such as Timbuktu (Tombouctou) and Gao, have existed for more than a millennium. In East Africa, an urban mercantile culture emerged that was rooted in Islam and the Swahili language. The prominent cities of Zanzibar, Tanzania, and Mombasa, Kenya, flourished by supporting a trade network that linked the East African highlands with the Persian Gulf. The stone ruins of Great Zimbabwe in southern Africa are a testimony to the achievements of stone working, metallurgy, and religion achieved by Bantu groups in the fourteenth century. West

Africa, however, had the most-developed precolonial urban network, reflecting both indigenous and Islamic traditions. It also supports some of the region's largest cities today.

West African Urban Traditions The West African coastline is dotted with cities, from Dakar, Senegal, in the far north, to Lagos, Nigeria, in the east. Nigeria is only 16 percent urban, yet it has half a dozen metropolitan areas with a population of more than one million. Historically, the Yoruba cities in southwestern Nigeria are the best documented. Developed in the twelfth century, cities such as Ibadan were walled and gated, with a palace encircled by large rectangular courtyards at the city center. An important center of trade for an extensive hinterland, Ibadan was also a religious and political center. Lagos was also a Yoruba settlement. Founded on a coastal island on the Bight of Benin, most of the modern city has spread onto the nearby mainland. Its coastal setting and natural harbor made this relatively small indigenous city attractive to colonial powers. When the British took control in the mid-nineteenth century, the city's size and importance grew.

Most West African cities are hybrids, combining Islamic, European, and national elements such as mosques, Victorian architecture, and streets named after independence leaders. Accra is the capital city of Ghana and home to more than one million people. Originally settled by the Ga people in the sixteenth century, by the late 1800s it had become a colonial administrative center. As Figure 6.20 suggests, the modern city is largely divided along income lines. The wealthier inhabitants live to the north and east of the central business district in an area first constructed for the European elite. The lowest-income zone is to the west, behind the major industrial area. Like many West African cities, interspersed amid the low-income neighborhoods are *zongos* (so-called stranger communities filled with ethnic groups). In Accra, Hausa traders from the north created zongos so they could assist each other and practice their Islamic faith. More recently, Accra has experienced sprawl on its perimeter, largely for the development of upper-income suburbs.

Urban Industrial South Africa The major cities of southern Africa, unlike those of West Africa, are colonial in origin. Most of these cities grew as administrative or mining centers such as Lusaka, Zambia, or Harare, Zimbabwe. The nation of South Africa is one of the most urbanized states in the entire region, and it is certainly the most industrialized. The foundations of South Africa's urban economy lie largely on its incredibly rich mineral resources (diamonds, gold, chromium, platinum, tin, uranium, coal, iron ore, and manganese). Eight metropolitan areas have more than one million people, the largest of which are Johannesburg, Durban, and Cape Town.

The form of South African cities continues to be imprinted by the legacy of **apartheid** (an official policy of racial segregation that shaped social relations in South Africa for nearly 50 years). Even though apartheid was abolished in 1994, it is still evident in the landscape. Under apartheid rules, cities such as Cape Town were divided into residential areas according to racial categories: white, coloured (mixed race), Indian (South Asian), and African (black). Whites occupied the largest and most desirable portions of the city, especially the scenic areas below Table Mountain (Figure 6.21). Blacks were crowded into the least desired areas, forming squatter settlements called *townships* in places such as Gugulethu (Figure 6.22). Today, blacks, coloureds, and Indians are legally allowed to live anywhere they want. Yet the economic disparity between racial groups, as well as entrenched animosity, hinder residential integration. Post-apartheid settlements, such as Cape Town's Delpht South, promote residential integration of black and coloured households, something that was forbidden during the apartheid years.

Cultural Coherence and Diversity: Unity Through Adversity

No world region is culturally homogeneous, but most have been partially unified in the past by widespread systems of belief and communication. Traditional African religions, how-

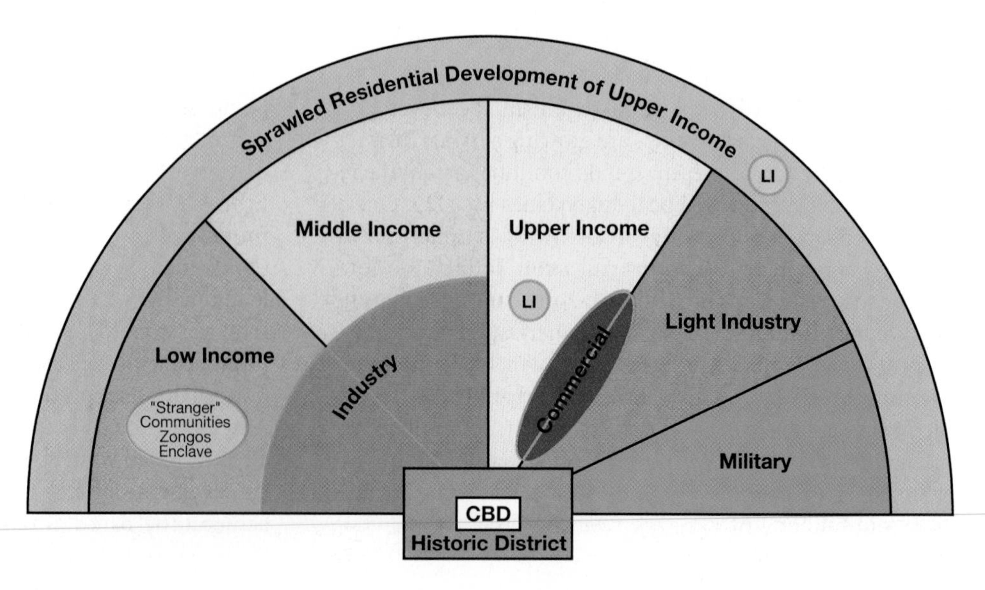

▶ **Figure 6.20 Urban model of Accra, Ghana** In the late nineteenth century Accra was a small colonial administrative center, with most of its European population living in the historic district. Like all African cities, Accra has grown tremendously since independence. The city exhibits a sectoral pattern of land use with upper-, middle-, and lower-income groups living in distinct areas. *(Adapted from Aryeetey-Attoh, 1997,* Geography of Sub-Saharan Africa, *Upper Saddle River, NJ: Prentice Hall)*

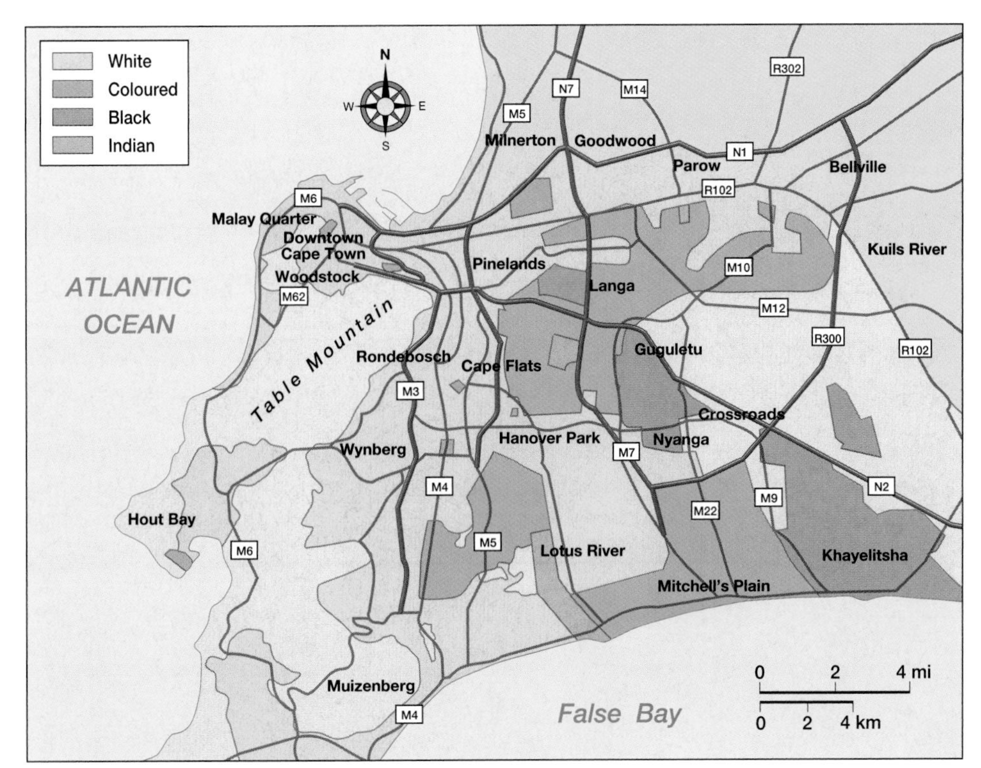

◀ **Figure 6.21 Racial segregation in Cape Town** Under apartheid, the majority of land in Cape Town was designated for white use. Coloureds, blacks, and Indians (South Asians) were crammed into far smaller areas on the less desirable flat and sandy soils east of Table Mountain. *(Source: A. J. Christopher, 1994.,* The Atlas of Apartheid, *p. 156, London: Routledge)*

▲ **Figure 6.22 Cape Town township** Gugulethu is a densely settled black township in the sandy flats east of downtown Cape Town. One of the older black townships, it was considered an undesirable area by whites because of fierce winter winds that whip through the area. Residents stack heavy rocks onto their metal roofs to keep them from blowing away. *(Rob Crandall/Rob Crandall Photographer)*

ever, were largely limited to local areas, and the religions that did become widespread, Islam and Christianity, are primarily associated with other world regions. Certainly a few indigenous religious ideas and practices were shared over large expanses of Sub-Saharan Africa, but no institutionalized form of religion ever came close to unifying the region. A handful of African trade languages similarly have long been understood over vast territories (Swahili in East Africa, Mandingo and Hausa in West Africa), but none came close to spanning the entire Sub-Saharan region. Sub-Saharan Africa also lacks a history of widespread political union or even that of an indigenous system of political relations. Powerful African kingdoms and empires existed in past centuries, but all were limited to distinct subregions of the landmass.

The lack of traditional cultural and political coherence across Sub-Saharan Africa is not surprising if one considers the region's vast scale. Sub-Saharan Africa is more than four times larger than Europe or South Asia. Had foreign imperialism not impinged on the region, it is quite possible that West Africa and southern Africa would have developed into distinct world regions of their own.

An African identity south of the Sahara was forged through a common history of slavery and colonialism, as well as struggles for independence and development. More telling, the people of the region often define themselves as African, especially to the outside world. That Sub-Saharan Africa is poor, no one will argue. And yet the cultural expressions of its people—its music, dance, and art—are joyous. Africans share an extraordinary resilience and optimism that visitors to the region often comment upon. The cultural diversity of the region is obvious, yet there seems to be a unity drawn from surviving adversity.

Language Patterns

In most Sub-Saharan countries, as in other former colonies, the persistence of multiple languages reflects the layers of ethnic, colonial, and national identity. Indigenous languages, many from the Bantu sub-family, are often localized to relatively small rural areas. More widely spoken African trade languages, such as Swahili or Hausa, serve as a lingua franca over broader areas. Overlaying indigenous languages are Indo-European and Afro-Asiatic ones (French and English;

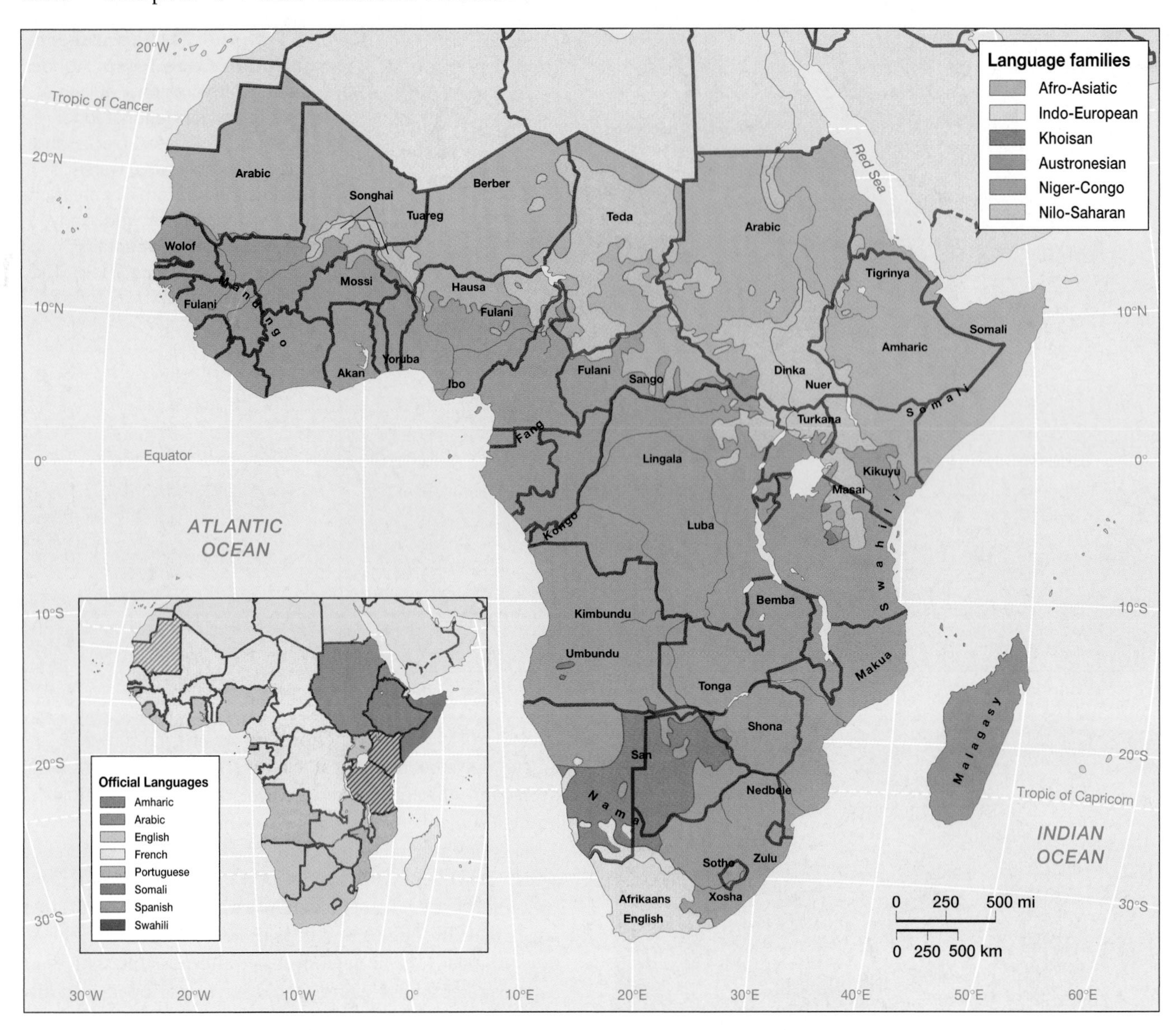

▲ **Figure 6.23 African language groups and official languages** Mapping language is a complex task for Sub-Saharan Africa. There are languages with millions of speakers, such as Swahili, and there are languages spoken by a few hundred people living in isolated areas. Six language families are represented in the region. Among these families are scores of individual languages (see labels on the map). Since most modern states have many indigenous languages, the colonial language often became the "official" language because it was less controversial than picking from one of several indigenous languages. English and French are the most common official languages in the region (see inset).

Arabic and Somali). Figure 6.23 illustrates the complex pattern of language families and major languages found in Africa today. Contrast the larger map with the smaller inset that shows current "official" languages. A comparison of the two maps shows that most African countries are multilingual, which can be a source of tension within states. In Nigeria, for example, the official language is English, yet there are 25 million Hausa speakers, 24 million Yoruba speakers, 20 million users of Igbo (or Ibo), 11 million Ful (or Fulani) speakers, and six million who speak in Efik, as well as dozens of other languages.

African Language Groups Three of the six language groups mapped in Figure 6.23 are unique to the region (Niger-Congo, Nilo-Saharan, and Khoisan), while the other three (Afro-Asiatic, Austronesian, and Indo-European) are more closely associated with other parts of the world. Afro-Asiatic languages, especially Arabic, dominate North Africa and are understood in Islamic areas of Sub-Saharan Africa as well. Amharic in Ethiopia and Somali of Somalia are also Afro-Asiatic languages. The Malayo-Polynesian language family is limited to the island of Madagascar, which many believe was first settled by seafarers from Indonesia some

1,500 years ago. Indo-European languages, especially French, English, Portuguese, and Afrikaans, are a legacy of colonialism and widely used today.

Of the three language groups found exclusively in the region, Khoisan includes only a few languages spoken in a limited area. Khoisan speakers, who probably inhabited all of southeastern Africa several thousand years ago, are now confined to the arid lands of the Kalahari (Figure 6.23). Speakers of the Nilo-Saharan languages seem to have originated in what is now southern Sudan and then spread first westward along the Sahel corridor and then to the south, into the heart of East Africa. Dinka and Nuer in Sudan, Songhai in Mali, and Turkana in Kenya are part of this group.

The Niger-Congo language group is by far the most important one in the region. This linguistic group originated in West Africa and includes Mandingo, Yoruba, Ful(ani), and Igbo, among others. Around 3,000 years ago, a people of the Niger-Congo stock began to expand out of western Africa into the equatorial zone. This group, called the Bantu, commenced one of the most far-ranging migrations in human history that introduced agriculture into large areas of central and southern Africa. One Bantu group migrated east across the fringes of the rain forest to settle in the Lake Victoria Basin in East Africa, where they formed an eastern Bantu core that later pushed south all the way to South Africa (Figure 6.24). Another group moved south, into the rain forest proper. The equatorial rainforest belt immediately adjacent to the original Bantu homeland had been very sparsely settled by the ancestors of the modern pygmies (a distinct people noted for their short stature and hunting skills). Pygmy groups, having entered into close trading relations with the Bantu newcomers, eventually came to speak Bantu languages as well. While several pygmy populations have persisted to the present, their original languages disappeared long ago.

Once the Bantu migrants had advanced beyond the rain forest into the savannas and woodlands, their agricultural techniques proved highly successful and their influence expanded (Figure 6.24). Sometime around A.D. 650, Bantu-speaking peoples reached South Africa. Over the centuries the various languages and dialects of the many Bantu-speaking groups, which were often separated from each other by considerable distances, gradually diverged from each other. Today there are several hundred distinct languages in the Bantu subfamily of the great Niger-Congo group. All Bantu languages, however, remain closely related to each other and a speaker of any one can generally learn any other without undue difficulty.

Most individual Sub-Saharan languages are limited to relatively small areas and are significant only at the local scale. One language in the Bantu sub-family, Swahili, eventually

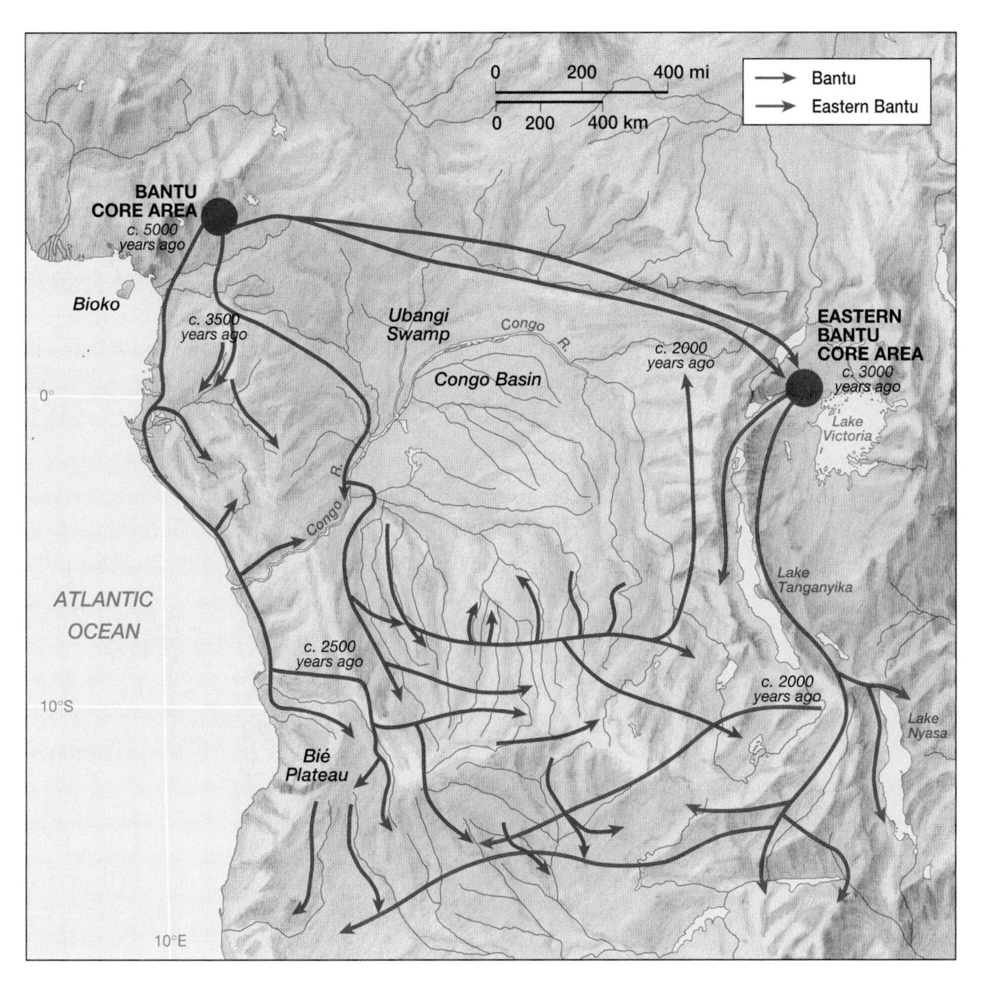

◀ Figure 6.24 Bantu migrations Bantu languages, a subfamily of the Niger-Congo language family, are widely spoken throughout Sub-Saharan Africa. The out-migration of Bantu tribes from an original core in West Africa and a secondary core in East Africa helps to explain the diffusion of Bantu languages, which include Zulu, Swahili, Bemba, Shona, Lingala, and Kikuyu, among others. *(Source: James Newman, 1995,* The Peopling of Africa, *p. 141, New Haven: Yale University Press)*

became the most widely spoken Sub-Saharan language. Swahili originated as a trade language on the East African coast, where a number of merchant colonies from Arabia were established around A.D. 1100. A hybrid society grew up in a narrow coastal band of modern Kenya and Tanzania, one speaking a language of Bantu structure, yet enriched with many Arabic words. While Swahili became the primary language only in the narrow coastal belt, it spread far into the interior as the language of trade. After independence was achieved, both Kenya and Tanzania adopted Swahili as an official language. Swahili, with some 47 million speakers, is the lingua franca of East Africa. It has generated a fairly extensive literature, and is often studied in other regions of the world.

Language and Identity Ethnic identity as well as linguistic affiliation have historically been highly unstable over much of the region. The tendency was for new groups to arise when people threatened by war fled to less-settled areas, where they often mixed with refugees from other places. In such circumstances, new languages arise quickly and divisions between linguistically distinct groups are blurred. Nevertheless, distinct **tribes** formed that consisted of a group of families or clans with a common kinship, language, and definable territory. The impetus to formalize tribal boundaries came from European colonial administrators, who were eager to establish a fixed indigenous social order to better control native peoples. In this process a cultural map of Sub-Saharan Africa evolved, albeit flawed. Some tribes were artificially divided, meaningless names were applied, and territorial boundaries were often misinterpreted.

Social boundaries between different ethnic and linguistic groups have become more stable in recent years, and a number of individual languages have emerged as particularly important vehicles of communication at the national scale. Wolof in Senegal; Mandingo and other closely related Mande languages in Mali; Mossi in Bourkina Fasso; Yoruba, Hausa, and Igbo in Nigeria; Kikuyu in central Kenya; and Zulu, Xosha, and Sotho in South Africa are all nationally significant languages spoken by millions of persons (Figure 6.25). None, however, has the status of being the official language of any country. Indeed, there are only a handful of Sub-Saharan countries in which any single language has a clear majority status, which is partially explained by the arbitrary territorial boundaries superimposed by Europeans. The more linguistically homogeneous countries of the region include Somalia (where virtually everyone speaks Somali), and the very small states of Rwanda, Burundi, Swaziland, and Lesotho.

European Languages In the colonial period, European countries used their own languages for administrative purposes in their African empires. Education in the colonial period also stressed literacy in the language of the imperial power. In the post-independence period, most Sub-Saharan African countries have continued to use the languages of their former colonizers for government and higher education. Few

▲ **Figure 6.25 Multilingual South Africa** A South African sign warns pedestrians not to cross a highway in three languages: English, Afrikaans, and Zulu. Throughout Sub-Saharan Africa, many people are multilingual because of the diversity of languages that exists in each state. *(Rob Crandall/Rob Crandall Photographer)*

of these new states had a clear majority language that they could employ, and picking any minority tongue would have aroused the opposition of other peoples. The one exception is Ethiopia, which maintained its independence during the colonial era. The official language is Amharic, although other indigenous languages are used, especially in the southwestern corner of the country.

Two vast blocks of European languages exist in Africa today: Francophone Africa, encompassing the former colonies of France and Belgium, where French serves the main language of administration; and Anglophone Africa, where the use of English prevails (see inset, Figure 6.23). Early Dutch settlement in South Africa resulted in the use of Afrikaans (a Dutch-based language) by several million South Africans. In Mauritania and Sudan, Arabic serves as the main language, although in the case of Sudan, this has resulted in severe internal tensions.

Religion

Indigenous African religions are generally classified as animist. This is a somewhat misleading catch-all term used to classify all local faiths that do not fit into one of the handful of "world religions." Most animist religions are centered on the worship of nature and ancestral spirits, but the internal diversity within the animist tradition is vast. Classifying a religion as animist says more about what it is not than what it actually is.

With such warnings in mind, one can still say that much of Sub-Saharan Africa has an animist religious tradition. If one goes back far enough, the entire region was animist—but that is, of course, true for the rest of the world as well. Both Christianity and Islam actually entered the region early in their histories, but they advanced slowly for many centuries. Since the beginning of the twentieth century, both religions have spread rapidly, more rapidly, in fact, than in any other part of the world. But tens of millions of Africans still follow animist beliefs, and many others combine animist practices and ideas with their observances of Christianity and Islam.

The Introduction and Spread of Christianity Christianity came first to northeast Africa. Kingdoms in both Ethiopia and central Sudan were converted by A.D. 300—the earliest conversions outside of the Roman Empire. The peoples of northern and central Ethiopia adopted the Coptic form of Christianity and have thus historically looked to Egypt's Christian minority for their religious leadership (Figure 6.26). At present, roughly half of the population of both Ethiopia and Eritrea profess Coptic Christianity; most of the rest are Muslim, but there are still some animist communities, especially in Ethiopia's western lowlands.

European settlers and missionaries introduced Christianity to other parts of Sub-Saharan Africa beginning in the 1600s. The Dutch, who began to colonize South Africa at this time, brought their Calvinist Protestant faith. Today, most of their descendants, the Afrikaners as well as the **coloureds** (a South African term to describe people of mixed African and European ancestry), still practice this rather puritanical faith (other coloureds are Muslims). Later European immigrants to South Africa brought Anglicanism and other Protestant creeds, as well as Catholicism. A substantial Jewish community also emerged, concentrated in the Johannesburg area. Most black South Africans eventually converted to one or another form of Christianity as well. In fact, in South Africa churches were instrumental in the long fight against white racial supremacy. Religious leaders, such as Bishop Desmond Tutu, were outspoken critics of the injustices of apartheid and worked to bring down the system.

▲ **Figure 6.26 Ethiopian Christians at prayer**
A cathedral for Coptic Christians located in the city of Addis Ababa, Ethiopia. Christianity spread into the Horn of Africa around A.D. 300. Today, the Coptic faith is practiced by half the population of Ethiopia and Eritrea, as well as a small minority in Egypt. *(Aurora & Quanta Productions)*

Elsewhere in Africa, Christianity came with European missionaries, most of whom arrived after the mid-1800s. As was true in the rest of the world, missionaries had little success where Islam had proceeded them, but they eventually made numerous conversions in animist areas. As a general rule, Protestant Christianity prevails in areas of former British colonization, while Catholicism is more important where France, Belgium, and Portugal had staked their empires. In the post-colonial era, African Christianity has diversified, at times taking on a life of its own independent from foreign missionary efforts. Still active in the region are various Pentecostal, Evangelical, and Mormon missionary groups, mostly from the United States. Yet many areas have also seen the emergence of syncretic faiths, where Christianity is complexly intertwined with traditional belief systems. It is difficult to map the distribution of Christianity in Africa, however, since it has spread irregularly across the entire non-Islamic portion of the region.

The Introduction and Spread of Islam Islam began to advance into Sub-Saharan Africa 1,000 years ago. Berber traders from North Africa and the Sahara introduced the religion to the Sahel, and by 1050 the Kingdom of Tokolor in modern Senegal emerged as the first Sub-Saharan Muslim state. Somewhat later, the ruling class of the powerful Mande-speaking mercantile empires of Ghana and Mali converted as well. In the fourteenth century the emperor of Mali astounded the Muslim world when he and his huge entourage made the pilgrimage to Mecca, bringing with them so much gold that they set off an inflationary spiral throughout Southwest Asia.

Mande-speaking traders, whose networks spanned the area from the Sahel to the Gulf of Guinea, gradually introduced the religion to other areas of West Africa. There, however, many peoples remained committed to animism, and Islam made slow and fitful progress (Figure 6.27). Even in the Sahel, syncretic forms of Islam prevailed through the 1700s. In the early 1800s, however, the pastoral Fulani people launched a series of successful holy wars designed to shear away animist practices and to establish pure Islam. Today, orthodox Islam prevails through most of the Sahel. Farther south Muslims are mixed with Christians and animists, but their numbers continue to grow and their practices seem to becoming gradually more orthodox as well (Figure 6.28).

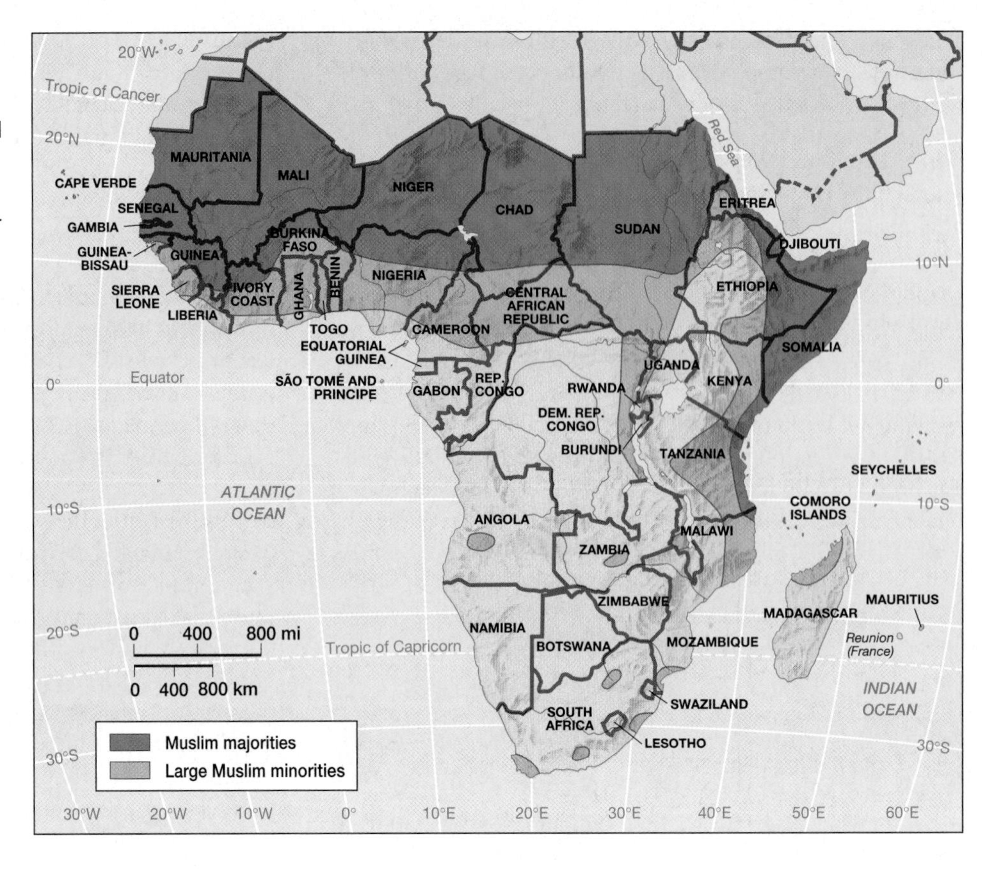

▶ **Figure 6.27 Extent of Islam** Muslim majorities prevail in the Sahelian states that border North Africa as well as Somalia and Djibouti. Throughout West and East Africa there are also large Muslim minorities. With the exception of Sudan, religion has not been a source of political tension in the region. *(Source: Claude S. Phillips, 1984,* The African Political Dictionary, *p. 196, Santa Barbara, CA: Clio Press Ltd.)*

▲ **Figure 6.28 West African mosque** A mosque rises above the houses in the town of Mankono, Ivory Coast. Islam entered this part of West Africa more than 600 years ago, but conversions were limited, with many Ivorians retaining their animist religious practices. Today, Christian, Muslim, and animist faiths are practiced in the Ivory Coast, which reflects a pattern of the religious tolerance seen throughout much of Sub-Saharan Africa. *(Victor Englebert/Englebert Photography, Inc.)*

Interaction Between Religious Traditions The southward spread of Islam from the Sahel, coupled with the northward dissemination of Christianity from the port cities, has generated a complex religious frontier across much of West Africa. In Nigeria, the Hausa are firmly Muslim, while the southeastern Ibo are largely Christians. The Yoruba of the southwest are divided between Christians and Muslims. In the more remote parts of Nigeria, moreover, animist traditions remain vital. But despite this religious diversity, there has been little overt religious animosity in Nigeria—or elsewhere in West Africa, for that matter. Certainly there have been many regional conflicts, some of them very violent, but these have usually been framed in ethnic and linguistic terms, rather than in those of religion.

Religious conflict has historically been far more acute in northeastern Africa, where Muslims and Christians have struggled against each other for centuries. Islam came early to the coastal areas of the Horn, and soon it had virtually isolated the Ethiopian highlands from the rest of the Christian world. In due time large areas on the plateau itself were brought into the Muslim sphere as well, and in the early 1500s it seemed that Islam might prevail throughout the Ethiopian highlands. Animist invasions from the lowlands in the 1600s then put both groups on the defensive, but by the 1800s the Christians had acquired modern weapons and had gained the upper hand. In more recent years, religious conflict in the Horn has remained muted and has been largely replaced by ethnic and ideological struggles. The new country of Eritrea, which is roughly half Christian and half Muslim, has in the past few years emerged as something of a model of peaceful coexistence between members of opposing faiths.

Sudan, on the other hand, is currently the scene of an intense conflict that is both religious and ethnic in origin. Here Islam was introduced in the 1300s by an invasion of Arabic-speaking pastoralists who extinguished the indigenous Coptic Christian kingdoms of the area. Within a few hundred years, central and northern Sudan were thoroughly islamized. The southern Equatoria province of Sudan, however, where tropical diseases and extensive wetlands prevented Arab incursions, remained animist. During the British colonial era, many of the Dinka, Nuer, and other peoples of this area converted to Christianity. In the post-colonial period, the Arabic-speaking Muslims of the north and center quickly emerged as the country's dominant group, and in the 1970s they began to build an Islamic state. Experiencing both religious discrimination and economic exploitation, the peoples of the south subsequently launched a massive rebellion. Human rights groups report that some southerners are being enslaved by people living in northern and central Sudan. Fighting throughout the 1980s and 1990s has been intense, with the government generally controlling the main towns and roads and the rebels maintaining power in the countryside. The war has periodically prevented the distribution of food, resulting in horrific famines and the massive flight of refugees.

The final zone of widespread islamization in Sub-Saharan Africa, the eastern coast, exhibits yet another historical-geographical pattern. Islam was introduced by merchants from southern Arabia around the year A.D. 1000. The Swahili-speaking community of the eastern coastal plain has continued to observe a relatively orthodox form of Islam ever since. Unlike West Africa, however, in East Africa the religion did not spread to any great extent. A few peoples of central Tanzania eventually accepted Islam, but the hinterland has remained largely animist, with some areas later experiencing numerous conversions to Christianity (see "Cultural Diffusion: South Asians and Hinduism in Africa").

Sub-Saharan Africa is a land of profound religious vitality. Both Christianity and Islam are spreading rapidly, but animism continues to hold widespread appeal. Many new and syncretic forms of religious expression are also emerging. With such religious turmoil, it is fortunate that religion has seldom been the cause of overt conflict. Religious vitality and tolerance are distinguishing features of Sub-Saharan Africa that deserve greater study. So too does the question of the region's influence on global culture, especially its musical traditions.

Globalization and African Culture

The slave triangle that linked Africa to the Americas and Europe set in process patterns of cultural diffusion that transferred African peoples and practices across the Atlantic. Tragically, slavery undermined the demographic and political strength of African societies, especially in West Africa from where the most slaves were taken. It is estimated that some 13 million Africans were sold into slavery from the 1500s until 1870. As Figure 6.29 shows, slavery impacted the entire region, sending Africans to the Americas, Europe, North

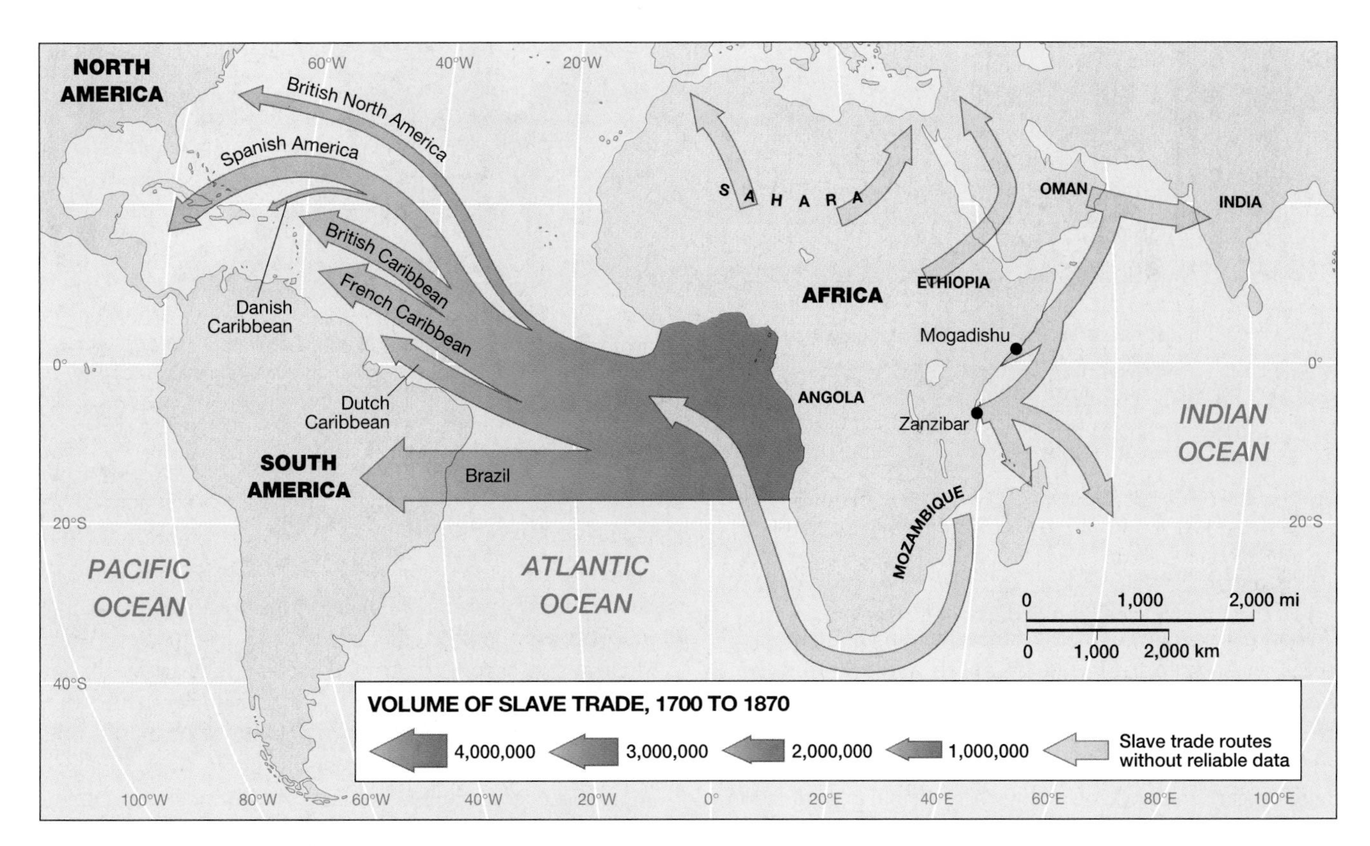

▲ **Figure 6.29 African slave trade** The horrors of the slave trade had a devastating impact on Sub-Saharan societies. From ship logs, it is estimated that 12 million Africans were shipped to the Americas to work as slaves in sugar, cotton, and rice plantations; the majority went to Brazil and the Caribbean. Yet other slave routes existed, although the data is less reliable. Africans from south of the Sahara were used as slaves in North Africa. Others were traded across the Indian Ocean into Southwest Asia and South Asia.

CULTURAL DIFFUSION South Asians and Hinduism in Africa

East Africa and South Asia lie on opposite sides of the Arabian Gulf, and it is not surprising that they should have a long history of interaction. In earlier centuries, the primary connections were between Ethiopia and India, with Ethiopia supplying substantial numbers of mercenaries and military slaves for the traditional kingdoms of South Asia. Few Indians, however, seem to have sailed to Africa. This was to change radically with British conquest in both areas. The British rulers in East and South Africa often found that they had inadequate labor at their disposal. In the east their main concern was railway building; in the south it was sugar growing in Natal, South Africa. Both propositions required large amounts of labor, and since the British territories in South Asia had a labor surplus, the practice of recruiting Indian labor to Africa added another cultural layer to the region's ethnic complexity. In west India the region of Gujarat soon became the main recruiting ground. Both Hindu and Muslim Gujaratis migrated to Africa in the hundreds of thousands.

Many Indians returned to their homeland once their labor contracts were fulfilled. Others, however, elected to stay, usually turning their efforts to business and trade. Within a few decades the Indian communities of countries such as Kenya and Uganda had emerged as business leaders. Many Indians also prospered in South Africa, but this community equally suffered under the apartheid system. It is notable that the founder of modern India, Mohandas Gandhi, began his career as a lawyer representing the Indian community in South Africa against a repressive, overly racist state.

After independence in the 1960s, the Indian merchants sometimes found themselves as economic scapegoats. When Uganda's economy began to falter in the 1970s, the country's ruthless dictator, Idi Amin, blamed the Indians, claiming that they were foreign economic predators unconcerned about the well-being of indigenous Africans. As a result, he decided to expel the entire community, many of whom fled to the United States. When the South Asians were deported, the country lost much of the business and professional expertise that had previously kept it afloat. After suffering a long interval of economic devastation and political chaos, Uganda emerged in the 1990s with a competent and reforming government that set the country on the road to economic reconstruction. One of the new government's first acts was to extend an invitation to the country's former South Asian residents. A number have returned—and Uganda now has one of the most dynamic economies in Sub-Saharan Africa.

▲ **Figure 1 South Asian merchant** A South Asian merchant sells toys to a South African couple in Johannesburg. South Asians were brought to South and East Africa as indentured laborers under British rule. Those who stayed continue to practice their religion (either Hinduism or Islam) and typically work in the service sector as merchants. *(Don L. Boroughs)*

Africa, and Southwest Asia. The vast majority, however, toiled on plantations across the Americas.

Out of this tragic diaspora came a melding of African cultures with Amerindian and European ones. Rumba, jazz, bossa nova, the blues, and even rock 'n' roll have African rhythms at their core. Brazil, the largest country in Latin America, is claimed to be the second largest "African state" (after Nigeria) because of its huge Afro-Brazilian population.

Cultural exchanges are never one way. Foreign language, religion, and dress were absorbed by Africans who remained in the region. With independence in the 1960s and 1970s, several states sought to rediscover their ancestral roots by openly rejecting European cultural elements. Ironically, the search for traditional religions found African scholars travelling to the Caribbean and Brazil to consult with Afro-religious practitioners about ceremonial elements lost to West Africa. Musical styles from the United States, the Caribbean, and Latin America were imitated and transformed into new African sounds. The complexity of cultural exchange, politics, and world markets are best illustrated through an examination of contemporary music in the Democratic Republic of the Congo and Nigeria.

Congo's Authenticity Movement The Democratic Republic of the Congo (formerly Zaire) became a major center for African music in the 1970s, with Kinshasa as its hub.

Despite the myriad of social and economic problems facing this country, its people have a reputation for leading the region in popular music and dance forms. This was not an accident but part of then-President Mobutu's "authenticity movement," which began in the 1970s. President Mobutu, who ruled the country as a president-for-life from 1965 until 1997, desired a distinct Congolese musical voice. Unlike other leaders, he invested in this vision by subsidizing musical groups and sponsoring state competitions. A leader in the authenticity movement was singer and guitarist Franco Luambo Makiadi, known throughout Africa as Franco. Franco's OK Jazz band re-Africanized the Afro-Cuban rumba sound by borrowing elements from Congolese folk music, and it became wildly popular throughout the region. Patriotic and a moralist, he even wrote national anthems that workers were required to sing once a week. Although supported by the state, Franco's relationship with the government was not trouble-free; in some instances, his records were even banned when he became too critical of the state.

The authenticity movement also supported musicians working on contemporary dance music. Soukous (both a dance step and a music style) became an international sensation in the 1980s, especially in Western Europe and Japan (Figure 6.30). Papa Wemba, a star soukou performer, made regular trips to Japan in the 1980s and 90s. Before the political turmoil of the late 1990s, Japanese tourists were frequent visitors in Kinshasa's hip music clubs. A few Japanese groups even played Soukous, performing in typical dress and singing in Lingala (the local language). Interestingly, since the state financially backed its top musicians, the music rarely acquired a political edge, despite the mounting unrest in the country. Maintaining an apolitical line and producing an irresistible beat, Soukous became one of the Congo's biggest exports, both within the region and beyond it.

▲ **Figure 6.30 Soukou performers** Tabu Ley Rochereau and members of his band perform Soukous for an appreciative audience in Lafayette, Louisiana. This form of African pop music originated in Kinshasa, Democratic Republic of the Congo, and is widely played throughout Sub-Saharan Africa. *(Philip Gould/Corbis)*

Music as Political Conscience Nigeria is the musical center of West Africa, with a well-developed and cosmopolitan recording industry. Modern Nigerian styles such as juju, highlife, and Afro-beat are influenced by jazz, rock, reggae, and gospel, but they are driven by an easily recognizable African sound. Two of the country's musical leaders are Sunny Ade and Fela Kuti. Both are from the Yoruba ethnic group, but their styles are strikingly different.

Sunny Ade became an international pop star in the 1980s and was dubbed the King of Juju Music (a combination of intricate lead guitar melodies supported by the percussive beat of talking drums). Because he sang in Yoruba, the music was not as readily accessible to Western audiences, who gradually lost interest after a few years. Within Nigeria juju is still a popular sound, filling the buses and streets with its lilting tropical rhythms.

Singer Fela Kuti, on the other hand, became a voice of political conscience for Nigerians struggling for true democracy. From an elite family and educated in England, Kuti borrowed from jazz, traditional, and popular music to produce the Afro-beat sound in the 1970s. Yet it was his searing and angry lyrics that attracted the most attention. Acutely critical of the military government, he sang of police harassment, the inequities of the international economic order, and even Lagos's infamous traffic. Singing in English and Yoruba, his message was transmitted to a larger audience and he became a target of state harassment. At times self-exiled in Ghana, in the 1980s he was jailed briefly by the Nigerian government but widespread protest eventually led to his early release. Though Fela Kuti's protest music is unpopular with the state, it has been copied by other groups, making music an important form of political expression in Nigeria, as well as in other Sub-Saharan states.

Sub-Saharan Africa has shared many musical and spiritual traditions with the world, and the creative vitality of its people continues to be recognized. In contrast to the fairly peaceful coexistence of diverse artistic and religious traditions in Sub-Saharan Africa, ethnic tensions have often boiled over into violent disputes. To understand the roots of ethnic conflict in Sub-Saharan Africa, it is necessary first to examine the region's political history, paying particular attention to the legacy of European colonization.

Geopolitical Framework: Legacies of Colonialism and Conflict

The duration of human settlement in Sub-Saharan Africa is unmatched by any other region. After all, humankind originated there, evidently evolving from a rather ape-like *Australopithicus* all the way to modern *Homo sapiens*. Over the millennia, many diverse ethnic groups formed that defy simple classification. Conflict among these groups existed, with certain groups (the Bantu, for example) overwhelming others (the Khoisan). But cooperation and coexistence among different peoples were also evident.

With the arrival of Europeans, patterns of human relatedness and ethnic relations were changed forever. As Europeans rushed to carve up the continent to serve their imperial ambitions, they instituted various policies that charged ethnic tensions and promoted hostility. Many, but not all, of the region's modern conflicts can trace their roots back to the colonial era, especially the arbitrary drawing of political boundaries. Others are attributed to struggles over national identity and political control among different ethnic groups.

Indigenous Kingdoms and European Encounters

The first significant state to emerge in Sub-Saharan Africa was Nubia, which controlled a large territory in central and northern Sudan some 3,000 years ago; 1,000 years later the Kingdom of Axum arose in northern Ethiopia and Eritrea. Both of these states were strongly influenced by political models derived from Egypt and Arabia. The first wholly indigenous African states were founded in the Sahel around A.D. 700. Kingdoms such as Ghana, Mali, Songhai, and Kanem-Bornu grew rich by exporting gold to the Mediterranean and importing salt from the Sahara, and they maintained power over lands to the south by monopolizing horse breeding and mastering cavalry warfare (Figure 6.31).

Over the next several centuries, a variety of other states emerged elsewhere in West Africa. Some were large but diffuse empires organized through elaborate hierarchies of local kings and chiefs; others were centralized states focused on small centers of power. The Yoruba of southwestern Nigeria, for example, developed a city-state form of government, and

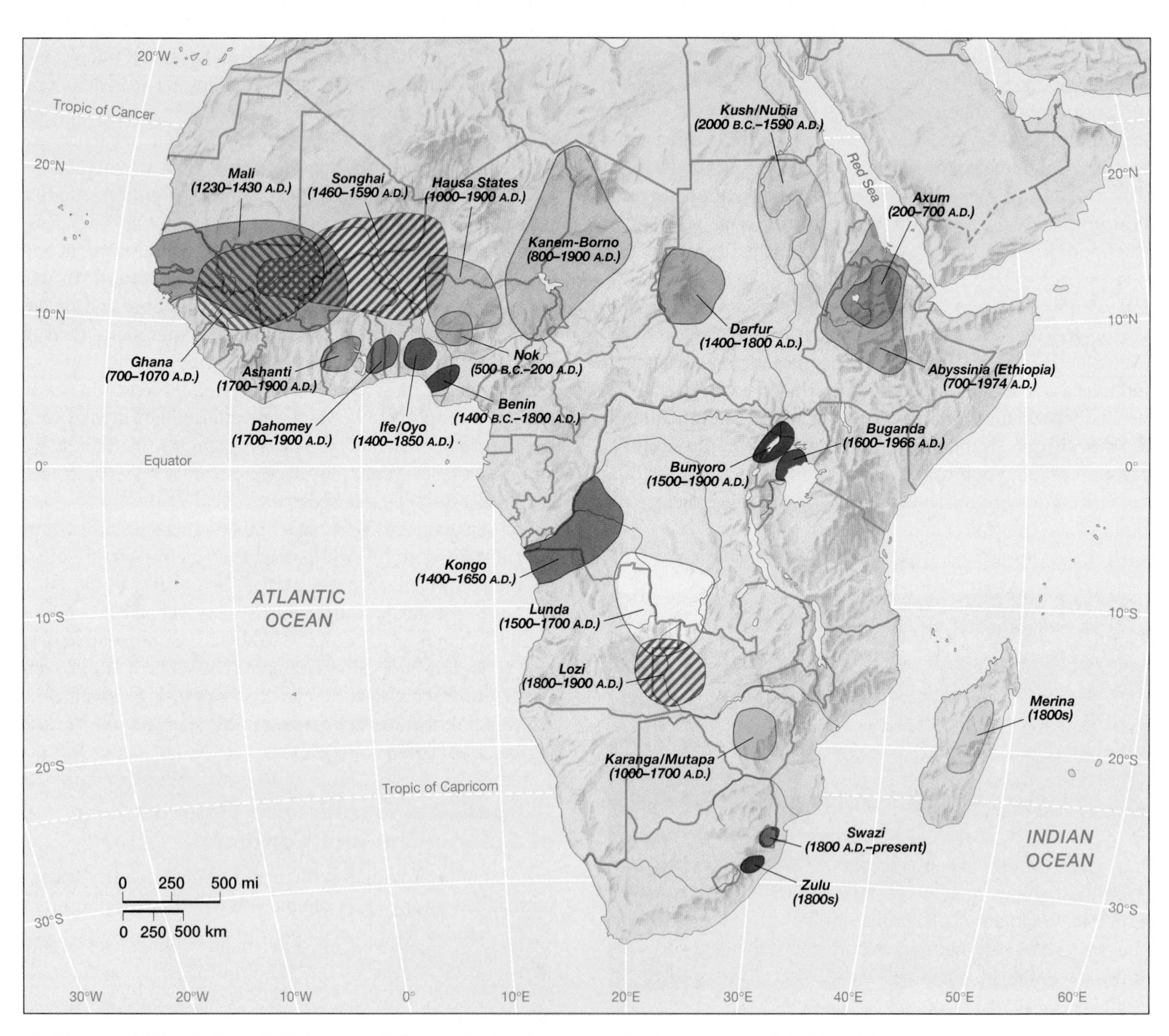

▲ **Figure 6.31 Early Sub-Saharan states and empires** Lost in the current political boundaries of Sub-Saharan Africa are the many African states and empires that existed long before Europeans advanced their territorial claims in the region. Most African kingdoms ceased to exist by 1900, but several such as Buganda (in Uganda) and Abyssinia (Ethiopia) existed well into the mid-twentieth century. *(Sources: Samuel Aryeetey-Attoh, 1997,* Geography of Sub-Saharan Africa, *p. 63, Upper Saddle River, NJ: Prentice Hall; and Robert Stock, 1995,* Africa South of the Sahara, *p. 62, New York: Guilford Press)*

it is not coincidental that their homeland is still one of the most urbanized, and densely populated, parts of Africa. The most powerful Sub-Saharan states continued to be located in the Sahel until the 1600s, when European coastal trade undercut the lucrative trans-Saharan networks. Subsequently, the focus of power moved to the Gulf of Guinea, where well-organized states took advantage of the lucrative opportunities presented by the slave trade.

Early European Encounters Prior to European colonization, Sub-Saharan Africa presented a complex mosaic of kingdoms, states, and tribal societies. With the intensification of the trans-Atlantic slave trade in the sixteenth and seventeenth centuries, various indigenous African states (such as Dahomey or Ashanti) increased their military and economic power. Certain states in West Africa were well positioned to profit from the slave trade; by selling slaves to Europeans, they could then obtain the firearms that would give them further military advantages over their enemies. Thus the slave trade seems to have accelerated the gradual process of state formation almost everywhere in the region. Tribal peoples were usually the main victims of slave raiders, since they typically lacked effective military force and were easily captured by well-armed rivals.

Unlike the relatively rapid colonization of the Americas, Europeans needed centuries to gain effective control of Sub-Saharan Africa. Portuguese traders arrived along the coast of West Africa in the 1400s, and by the 1500s they were well established in East Africa as well. The Portuguese made large profits, converted a few local rulers to Christianity, established several fortified trading posts, and acquired dominion over the Swahili trading cities of the east. They stretched themselves too thin, however, and had little ultimate success in any of their endeavors. Only where a sizable population of mixed African and Portuguese descent emerged, as along the coasts of modern Angola and Mozambique, could Portugal maintain power. Along the Swahili, or eastern, coast they were eventually expelled by Arabs from Oman, who subsequently established their own mercantile empire in the area.

The Disease Factor One of the main reasons for the Portuguese failure was the disease environment of Africa. With no resistance to malaria and other tropical diseases, roughly half of all Europeans who remained on the African mainland died within a year. Protected both by their formidable armies and by the diseases of their native lands, African states were able to maintain an upper hand over European traders and adventurers well into the 1800s. Unlike the Americas where European conquest was facilitated by the introduction of Old World diseases that devastated native populations (see Chapters 4 and 5), in Sub-Saharan Africa endemic disease limited European settlement in the region until the mid-nineteenth century.

The hazards of malaria and other tropical diseases such as sleeping sickness were compensated by the lure of profit, and soon other European traders followed the Portuguese. By the 1600s Dutch, British, and French firms dominated the lucrative export of slaves, gold, and ivory from the Gulf of Guinea. The Dutch also established a settler colony in South Africa, safely outside of the tropical disease zone, to supply their ships bound for Indonesia. For the next 200 years European traders came and went, occasionally building fortified coastal posts, but they almost never ventured inland and they seldom had any real influence on African rulers. By exporting millions of slaves, however, they had a profoundly negative impact on African society.

In the 1850s European doctors discovered that a daily dose of quinine would offer protection against malaria, radically changing the balance of power in Africa. Explorers immediately began to penetrate the interior of the continent, while merchants and expeditionary forces began to move inland from the coast. The first claims to the empire soon followed. The French quickly grabbed power along the easily navigated Senegal River, while the British established protectorates over the indigenous states of the Gold Coast (modern-day Ghana).

European Colonization

In the 1880s European colonization of the region quickly accelerated, leading to the so-called scramble for Africa. By this time, after the invention of the machine gun, no African state could long resist European force. The exact reasons for the abrupt division of Africa among the colonial powers remain controversial, but several developments seem to have been crucial. One was the British seizure in 1882 of Egypt, a territory that the French had long coveted. In compensation, the infuriated French began to seize additional lands in West Africa, in equatorial Africa, and in Madagascar.

Another precipitating factor was the desire of several new European countries to join the game of empire-building. Since Asia was either occupied by established European powers or was controlled by still-formidable indigenous empires, Africa emerged the main arena of rivalry and expansion. Even though Belgium had been a country only since 1830, its king quickly began to carve out a personal empire along the Congo River, using particularly brutal techniques. The German government—which itself dated back only to 1871—began to claim territories wherever German missionaries were active, and it had soon staked out the colonies of Togo, Cameroon, Namibia, and Tanganyika (modern Tanzania minus the island of Zanzibar). The Italians eyed the Horn of Africa, while Spain acquired a small coastal foothold in equatorial West Africa. Alarmed by such activity, the Portuguese began to push inland from their coastal possessions in Angola and Mozambique.

Also in the early 1800s, two small territories were established in West Africa so that free and runaway slaves would have a place to return to in Africa. The territory that was to become Liberia was set up by the American Colonization Society in 1822 to settle African American slaves. By 1847 it was the independent and free state of Liberia. Sierra Leone served a similar function for ex-slaves from the British Caribbean, but it remained a protectorate of Britain until the 1960s. Despite the good intentions behind the creation of these territories, they too were colonies. Liberia was imposed on existing indigenous groups who viewed their new "African" leaders with contempt.

The Berlin Conference As the scramble intensified, tensions among the participating countries mounted. Rather than risk war, 13 countries convened in Berlin at the invitation of the German chancellor Bismarck in 1884 in a gathering known as the **Berlin Conference.** During the conference, which no African leaders attended, rules were established as to what constituted "effective control" of a territory, and Sub-Saharan Africa was carved up and traded like properties in a game of Monopoly®. Exact boundaries in the interior, which was still poorly known, were not determined, and a decade of "orderly" competition remained as imperial armies marched inland. While European arms were by the 1880s far superior to anything found in Africa, several indigenous states did mount effective resistance campaigns. In central Sudan an Islamic-inspired battle against the British held out until 1900, and as late as 1914 the Darfur region of western Sudan maintained tenuous independence.

Eventually European forces prevailed everywhere, with one major exception: Ethiopia. The Italians had conquered the Red Sea coast and the far northern highlands (modern Eritrea) by 1890, and they quickly set their sights on the large Ethiopian kingdom called Abyssinia, which had itself been vigorously expanding for several decades. In 1896, however, Abyssinia vanquished the invading Italian Army, earning it the respect and recognition of the European powers. In the 1930s, fascist Italy launched a major invasion of the country, now renamed Ethiopia, to redeem its earlier defeat, and with the help of poison gas and aerial bombardment it quickly prevailed. By 1942, Ethiopia had regained its freedom.

Although Germany was a principal instigator of the scramble for Africa, it lost its own colonies after suffering defeat in World War I. Britain and France then partitioned most of Germany's African empire between themselves. Figure 6.32 shows the colonial dismemberment of the region in 1913, prior to Germany's territorial loss. The French held most of West Africa, but the British controlled populous Nigeria and several other coastal territories. The French also colonized Gabon in western equatorial Africa, Madagascar, and the small but strategic enclave of Djibouti at the southern end of the Red Sea. The British holdings were larger still, covering a continuous swath of territory in the east from Sudan to South Africa. Belgium and Portugal formed the other main colonial powers. The government of Belgium had taken direct control over the personal domain of King Leopold II and had extended it to the south, eventually reaching the mineral-rich copper belt. Portugal, the weakest European power, controlled huge territories in southwestern and southeastern Africa (Angola and Mozambique). Portugal failed in its effort to bridge the continent due to Britain's drive to the north from South Africa, organized by the imperial dreamer and diamond magnate, Cecil Rhodes.

Establishment of South Africa While the Europeans were cementing their rule over Africa after World War I, South Africa was inching toward political freedom, at least for its white population. South Africa's political history is unique in that it was one of the oldest colonies in Sub-Saharan Africa and the first to obtain its political independence from Europe in 1910. The economy of South Africa is the most productive and influential of the region, yet the legacy of apartheid made it an international pariah, especially within Africa. Territorial divisions in South Africa make for a classic study of applied political geography, in which an elite uses its control of space to ensure power. The evolution of South Africa's internal political boundaries merits extended consideration.

The original body of Dutch settlers in South Africa had grown slowly, expanding to the north and east of its original nucleus around Cape Town. As a farming and pastoral people largely isolated from the European world, these Afrikaners, or Boers, developed an extremely conservative cultural outlook marked by an intensifying belief in their racial superiority over the indigenous population. In 1806 the British, then the world's unchallenged maritime power, seized the Cape district from the Dutch. Relations between the British rulers and their new Afrikaner subjects quickly soured, in part because the British attempted to restrict the enslavement of Africans as part of a larger abolitionist movement. In the 1830s the bulk of the Afrikaner community opted to leave British territory and strike out for new lands. After being rebuffed by the powerful Zulu state in the Natal area, they opted instead to settle on the northeastern plateau known as the high veldt. By the 1850s they had established two "republics" in the area: the South African Republic (commonly called the Transvaal) and the Orange Free State (Figure 6.33).

The Boers were able to occupy the high veldt relatively easily because the area had been largely depopulated by the Zulu wars of the preceding decade. The Zulus had previously been a relatively minor group until a Zulu king introduced centralized rule and new military techniques in the 1820s. From then on Zulu armies were virtually invincible, and, rather than suffer defeat at their hands, many other people chose to flee, some migrating as far north as Central Africa. Once relocated on the high veldt, the Afrikaners found themselves threatened by Zulu armies. After a series of inconclusive wars, the British intervened in 1878 on behalf of the Dutch settlers. The first British army sent into Zululand was annihilated; the second, equipped with machine guns, prevailed. By 1900 the British had incorporated the Zulu into South Africa's Natal province.

The British had more difficulty subduing the Afrikaners. By the late 1800s it had become clear that the two Boer republics were sitting above one of the world's greatest troves of mineral wealth, inciting British imperial ambition. English-speaking people were increasingly drawn to the area, and they grew resentful of Afrikaner power. The result was the Boer War, which turned into a protracted and brutal guerilla struggle between the British Army and mobile Afrikaner bands. By 1905 the Boers relented, after which the British joined their two former republics to Cape Province and Natal to form the Union of South Africa. Five years later, South Africa was given its independence from Britain. Britons and other Europeans continued to settle there, but the Afrikaners remained the ma-

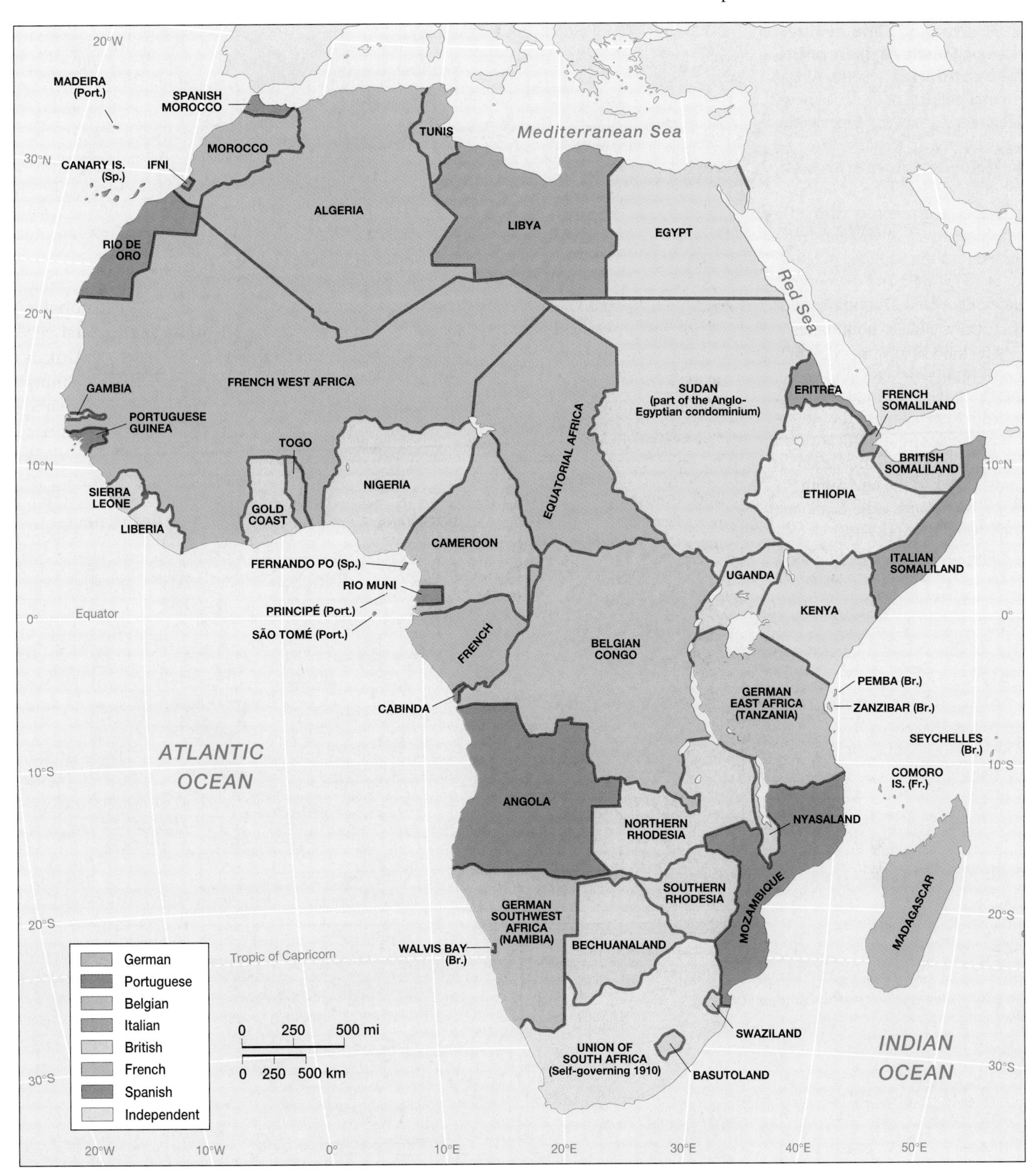

▲ **Figure 6.32 European colonization in 1913** Before 1880 the areas of Africa under direct European control were few. By 1884 when the Berlin Conference began, Africa was carved up and traded between European powers. France and Britain controlled the most territory but Germany, Portugal, Belgium, Spain, and Italy all had their claims. By 1913 the entire continent, except Ethiopia, was under foreign control. *(Source:* The Times Atlas of World History. *1989. Hammond Inc.)*

jority white group. Black and coloured South Africans, however, greatly outnumbered whites.

It wasn't until 1948 when the Afrikaner's National Party gained control of the government that they introduced their policy of "separateness" known as *apartheid.* British South Africans had enacted a series of laws that were prejudicial to nonwhite groups, but it was under Afrikaner leadership that racial separation become more formalized and systematic.

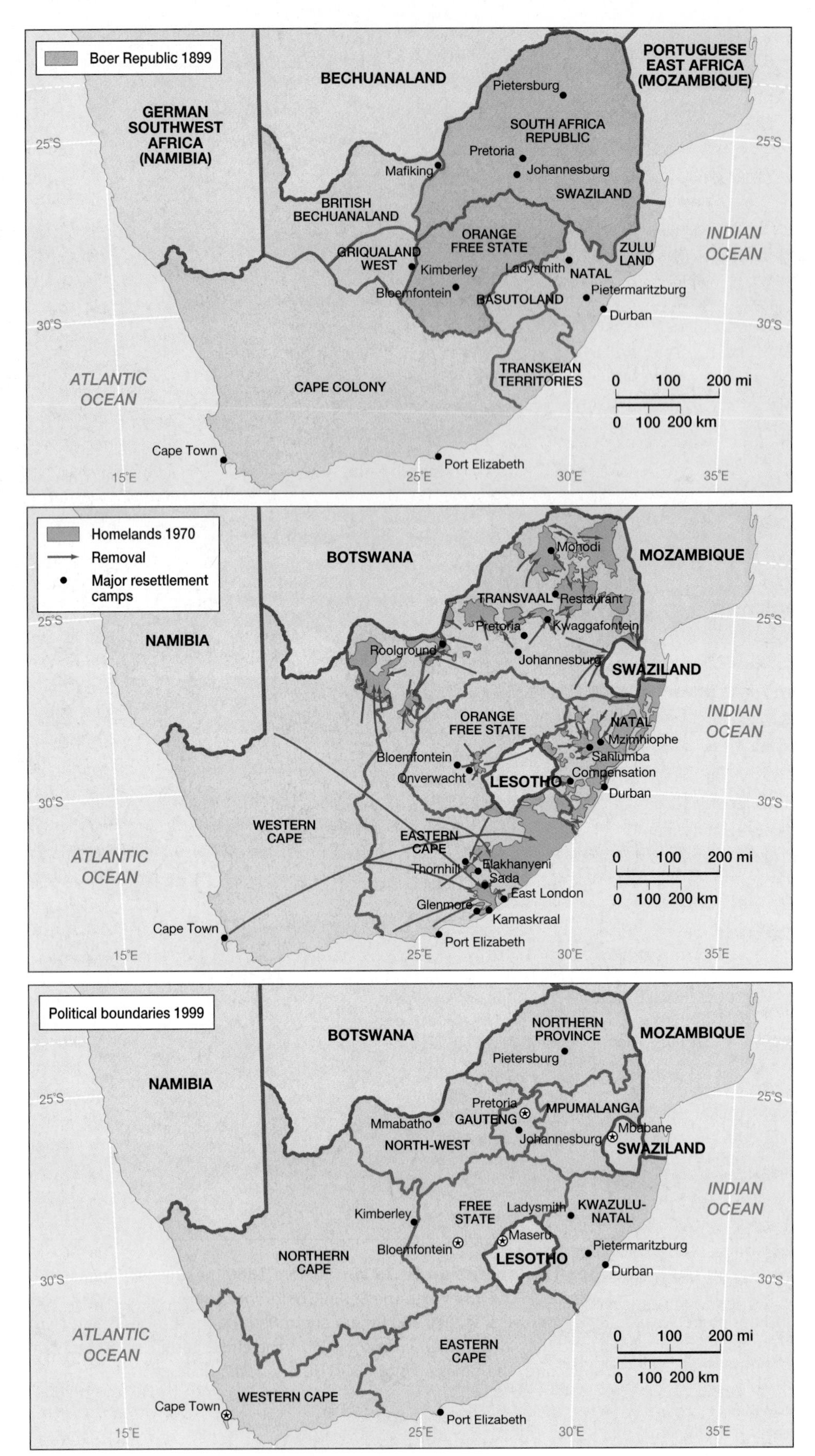

▶ **Figure 6.33 The evolution of South African political boundaries** South Africa's internal political boundaries have been redrawn several times in the past century. To appease the Afrikaans (Dutch) settlers, a Boer Republic was established in the late nineteenth century with British approval. It was later reabsorbed into the Republic of South Africa when gold and diamonds were discovered. During the apartheid years, homelands were created with the intent of having all South African blacks live in these ethnic-based rural settings. In order to accomplish this, thousands of black South Africans were forcibly relocated in the homelands. Lastly, in the new South Africa, homelands were eliminated and several new provinces were created. *(Source: A. J. Christopher, 1994,* The Atlas of Apartheid, *pp. 16, 83, London: Routledge)*

Operating at three scales—petite, meso-, and grand—apartheid managed social interaction by controlling space. Petite apartheid, like Jim Crow laws in the United States, created separate service entrances for government buildings, bus stops, and restrooms based on one's skin color. Meso-apartheid divided the city into residential sectors by race. Naturally, whites were given the best and largest urban zones. Finally, grand apartheid was the construction of black **homelands** by ethnic group. Technically blacks were to become citizens of the nominally independent homelands such as KwaZulu and Transkei (see the 1970 map in Figure 6.33). Likened to the reservations created for Native Americans in the United States, homelands were rural, overcrowded, and on marginal land. Moreover, to ensure the notion that every black had a homeland, some three million blacks were forcibly relocated into homelands during apartheid, and residence outside of the homelands was strictly regulated.

Granted its independence in 1910, South Africa was the first state in the region freed from colonial rule. Yet because of its formalized system of discrimination and racism, South Africa was hardly a symbol of liberty. Ironically, at the same time that the Afrikaners tightened their political and social control over the nonwhite population, the rest of the continent was preparing for political independence from Europe.

Decolonization and Independence

Decolonization of the region happened rather quickly and peacefully beginning in 1957. Independence movements, however, had sprung up throughout the continent, some dating back to the early 1900s. Workers' unions and independent newspapers became voices for African discontent and the hope for freedom. Black intellectuals, who typically studied abroad, where influenced by the ideas of the **Pan-African Movement** led by W. E. B. Du Bois and Marcus Garvey in the United States. Founded in 1900, the movement's slogan of "Africa for Africans" encouraged a trans-Atlantic liberation effort. Nevertheless, Europe's hold on Africa remained secure through the 1940s and early 1950s, even though other colonies in South and Southeast Asia gained their independence.

By the late 1950s, Britain, France, and Belgium decided that they could no longer maintain their African empires and thus began to withdraw. (Italy had already lost its colonies during World War II; Britain gained Somalia and Eritrea.) Once started, the decolonization process moved rapidly. By the mid-1960s, virtually the entire region had achieved independence. In most cases the transition was relatively peaceful and smooth.

Dynamic African leaders put their mark on the region during the early decades after independence. Men such as Kenya's Jomo Kenyatta, Ivory Coast's Felix Houphuët-Boigney, Julius Nyerere of Tanzania, or Ghana's Kwame Nkrumah became powerful father figures who molded their new nations (Figure 6.34). President Nkrumah's vision for Africa was the most expansive. After helping to secure independence for Ghana in 1957, his ultimate aspiration was the political unity of Africa.

▲ **Figure 6.34 A monument to Kwame Nkrumah** Charismatic independence leader Kwame Nkrumah is remembered with this monument in Accra, Ghana. Nkrumah led Ghana to an early independence in 1957; he was also a founder of the Organization of African Unity (OAU). *(Victor Englebert/Englebert Photography, Inc.)*

While his dream was never realized, it set the stage for the founding of the **Organization of African Unity (OAU)** in 1963. The OAU is a continent-wide organization. Its main role has been to mediate disputes between neighbors, although its success rate is mixed. Certainly in the 1970s and 1980s it was a constant voice of opposition against South Africa's minority rule, and it intervened in some of the more violent independence movements in southern Africa.

Southern Africa's Independence Battles Independence did not come easily to southern Africa. In southern Rhodesia (modern-day Zimbabwe) the problem was the presence of some 250,000 white residents, most of whom owned large farms. Unwilling to see power pass to the country's black majority, then some six million strong, these settlers unilaterally declared themselves the rulers of an independent, white-supremacist state in 1965. The black population continued to resist, however, and in 1978 the Rhodesian government was forced to capitulate. The renamed country of Zimbabwe was henceforth ruled by the black majority, although the remaining whites still form an economically privileged community.

In the former Portuguese colonies, independence came violently. Unlike the other imperial powers, Portugal refused to relinquish its colonies in the 1960s. As a result, the people of Angola and Mozambique turned to armed resistance. The most powerful rebel movements adopted a socialist orientation and received support from the Soviet Union and Cuba. A new Portuguese government came to power in 1974, however, and it withdrew abruptly from its African colonies. At this point Marxist regimes quickly came to power in both Angola and Mozambique. The United States, and especially South Africa, responded to this perceived threat by supplying arms to other rebel groups that opposed the new governments. As is common in much of Sub-Saharan Africa, the resulting struggles turned out to be more firmly grounded in ethnic loyalty than in Cold War ideology. Nevertheless, many of Africa's post-independence political struggles were cloaked in Cold War rhetoric that tended to increase the supply of arms to the region.

Fighting dragged on for several decades in Angola and Mozambique (Figure 6.35). The countryside in both states is now so heavily riddled with landmines that it can hardly be used. With the end of the Cold War, however, outsiders lost their interest in perpetuating these conflicts, and sustained efforts have been under way to broker a peace. At present Mozambique is at peace, but Angola's attempts at reconciliation during the 1990s have been fitful. As of 1999, UNITA, an opposition group, was at war with the government of Angola. The peace in Mozambique has held, and new investment is coming into the country. Yet the years of warfare ravaged the country's infrastructure, crippled its agricultural production, and left the country with the lowest estimated GNP in the world—$80 per capita, or less than 25 cents a day.

▲ **Figure 6.35 War in Angola** UNITA rebels continue to fight the government of Angola, controlled by the MPLA. UNITA controls much of central and southern Angola, while MPLA controls the north. Nearly three decades of war have destroyed much of the country's infrastructure. Over the years, other states became involved in the fighting, most notably South Africa and Cuba. *(Jason Lauré/Lauré Communications)*

Apartheid's Demise in South Africa While fighting continued in the former Portuguese zone, South Africa underwent a remarkable transformation. Through the 1980s, its government had remained firmly committed to white supremacy. Under apartheid, only whites enjoyed real political freedom, while blacks were denied even citizenship in their own country—technically, they were citizens of homelands. Yet since labor was needed in South African cities and mines, nonwhite groups were allowed to reside in segregated neighborhoods on the outskirts of cities called **townships.** Some 20 miles southwest of Johannesburg, for example, is the famous black township of Soweto. Home to Nelson Mandela and Desmond Tutu, it was the principal home for many blacks working in Johannesburg and a center for political resistance to apartheid. In the post-apartheid era, Soweto is a mixture of lean-tos and middle-class housing that still attracts mostly black migrants from former homelands and neighboring countries. Planners estimate that 1.5 million people lived in Soweto in 1999.

Opposition to apartheid began in the 1960s, intensifying and becoming more violent by the 1980s. Blacks led the opposition, but coloureds and Asians (who suffered severe, but less extreme, discrimination) also opposed the Afrikaner government. International pressure also mounted, and white South Africans found themselves ostracized. Many corporations refused to do business there and South African athletes (regardless of color) were banned from most international competitions, such as the Olympics and World Cup Soccer. Increasing numbers of whites also opposed the apartheid system, and many businesspeople began to believe that apartheid threatened to undermine their economic endeavors.

The first major change came in 1990, when South Africa withdrew from Namibia, which it had controlled as a protectorate since the end of World War I. South Africa now stood alone as the single white-dominated state in Africa. A few years later the leaders of the Afrikaner-dominated political party decided they could no longer resist the pressure for change. In 1994 free elections were held in which Nelson Mandela, a black leader who had been imprisoned for 27 years by the old regime, emerged as the new president. Black and white leaders pledged to put the past behind them and work together to build a new, multiracial South Africa. The homelands themselves were the first to be eliminated from the political map of the new South Africa (see the 1999 map in Figure 6.33). Residential segregation is officially illegal, but neighborhoods are still sharply divided along race lines. In 1999 peaceful elections were held and a new black president, Thabo Mbeki, replaced Nelson Mandela as the country's leader.

Unfortunately, South Africa's racial and economic problems could not be eliminated so easily. Many whites remained opposed to the new regime, and a number of Afrikaners indicated a desire to create a new, smaller, white-dominated state. Distrust also grew between blacks and coloureds and between blacks and South Asians; no longer did these groups face a common enemy and a similar set of political disadvantages. Divisions within the black community also grew more pronounced. Many Zulus felt more loyalty toward their own ethnic group than to South Africa, and members of the rather conservative Zulu Inkatha Party engaged in violent conflicts with other black political groups. Violent crime, moreover, increased in many parts of the country. Under the new system a black middle-class was rapidly emerging, but most blacks remained extremely poor (and most whites remained quite prosperous). Since the political change was not matched by a significant economic transformation, the hopes of many people were frustrated. Such a situation may not be conducive for political stability.

Enduring Political Conflict

Although most Sub-Saharan countries made a relatively peaceful transition to independence, virtually all of them immediately faced a difficult set of institutional and political problems. In several cases the old authorities had done virtually nothing to prepare their colonies for independence. Lacking an institutional framework for independent government, countries such as the Democratic Republic of the Congo confronted a chaotic situation from the beginning. Only a handful of Congolese had received higher education, let alone been trained for administrative posts. The indigenous African political framework had been essentially destroyed by colonization, and in most cases very little had been built up in its place.

Even more problematic, in the long run, was the political-geographical structure of the newly independent states. Civil servants could always be trained and administrative systems built, but little could be done to rework the region's basic political map. The fundamental problem was the fact that the European colonial powers had essentially ignored indigenous cultural and political geographies, both in dividing Africa among themselves and in creating administrative subdivisions within their own imperial territories.

The Tyranny of the Map All over Africa, different ethnic groups found themselves forced into the same state with peoples of disparate linguistic and religious backgrounds, many of whom had recently been their enemies. At the same time, a number of the larger ethnic groups of the region found their territories split between two or more countries. The Hausa people of West Africa, for example, were divided between Niger (formerly French) and Nigeria (formerly British), each of which they had to share with several former enemy groups.

Given the imposed political boundaries, it is no wonder that many African countries struggle to generate a common sense of national identity or establish stable political institutions. **Tribalism,** or loyalty to the ethnic group rather than to the state, has emerged as the bane of African political life. Especially in rural areas, tribal identities usually supercede national ones. In urban areas, however, the reverse is increasingly true. The few African countries that approach the ideal nation-state condition, one in which a single people are united under a single state, are the small, landlocked states of southern Africa (Lesotho and Swaziland) and Somalia. Even for these states the situation is far from ideal, since many members of the given nationalities reside outside of the national boundaries (Sothos and Swazis in South Africa, and Somalis in Kenya, Ethiopia, and Djibouti). Multiethnic Tanzania, however, is an example of a Sub-Saharan state that has forged a national identity, in part because of its socialistic policies in the 1970s.

Since virtually all of Africa's countries inherited an inappropriate set of colonial borders, one might assume that they would have been better off scrapping the system altogether and drawing a new political map based on indigenous identities. Such an ideal solution was impossible—as all of the leaders of the newly independent states realized. Any new territorial divisions would have created winners and losers, and thus would have proved tremendously contentious. Moreover, since ethnicity in Sub-Saharan Africa was traditionally fluid, and since many groups were territorially interspersed among their neighbors, it would have been difficult to generate a clear-cut system of division. Finally, most African ethnic groups were considered too small to form viable countries. With such complications in mind, the new African leaders, meeting together in 1963 to form the Organization of African Unity, agreed that colonial boundaries should remain. The violation of this principle, they argued, would lead to pointless wars between states and interminable civil struggles within them.

Despite the determination of Africa's leaders to build their new nations within existing boundaries, challenges to the states began soon after independence. Figure 6.36 maps the numerous post-colonial political conflicts that have disabled parts of Africa. Ethnic and secessionist conflicts, as well as military governments, have undermined African hopes for peaceful democracy.

The human cost of this turmoil is several million refugees. **Refugees** are people who flee their state because of a well-founded fear of persecution based on race, ethnicity, religion, or political orientation. In the late 1990s, some four million Africans were considered refugees. Some nine million internally displaced persons were also counted in the late 1990s. **Internally displaced persons** have fled from conflict but they still reside in their country of origin. These populations are hard to assist because they are not technically considered refugees, making it difficult for humanitarian nongovernmental organizations and the United Nations to help them (see "The Geography of Displacement: West Africa's Conflict Zone").

Ethnic Conflicts In 1994 the world was shocked by the genocide in Rwanda among the Hutus and Tutsis. Rwanda and its neighbor Burundi inherited a potentially explosive ethnic mix from the precolonial period. Both countries had been indigenous kingdoms in which a Tutsi aristocracy ruled a Hutu peasantry. During the colonial period, the practice of giving one group privileges over the other, intensified ethnic animosities. The Hutus gained political power in Rwanda after independence. In 1994 the Hutu president died in a plane crash under somewhat mysterious conditions, which Hutu leaders interpreted as the work of Tutsi rebels. In an

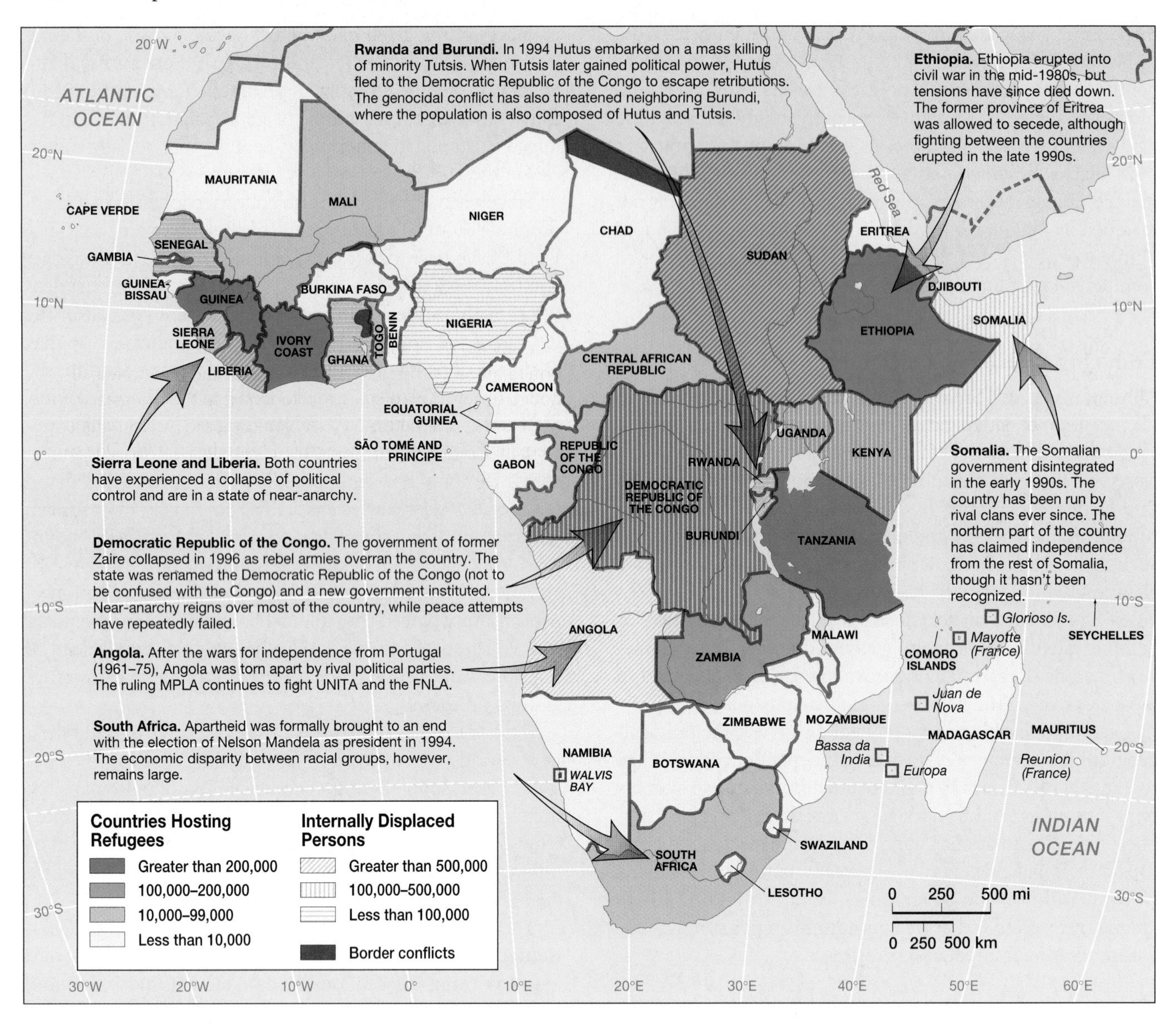

▲ **Figure 6.36 Postcolonial conflicts** The independence period has witnessed many ethnic conflicts as well as civil wars between competing political groups. In comparison to other world regions, Sub-Saharan Africa has had only one successful secessionist movement that led to the creation of Eritrea. One of the consequences of political turmoil has been several million refugees and internally displaced persons. It is estimated that there are four million internally displaced persons in Sudan. More than one million people have fled Liberia and Sierra Leone in recent years. *(Data sources: U.S. Committee for Refugees, 1998,* World Refugee Survey; *and John Allcock, et al., 1992,* Border and Territorial Disputes, *Third ed., Longman Current Affairs)*

orchestrated wave of mass slaughter, many hundreds of thousands of Tutsis were slaughtered over the next several months. Rebel Tutsi soldiers soon thereafter managed to oust the government and gain power, after which more than a million Hutus, fearing reprisals, fled the country.

The image of a million refugees living in camps on the Rwandan border filled the international media (Figure 6.37). Efforts to repatriate the refugees were stymied for two years. Ultimately, political turmoil in the Democratic Republic of the Congo (the host country) caused a massive and sudden repatriation in 1996. In four days, a half million Rwandans left the Congolese camps and walked back to their homes. The Rwandan case underscores the cruel reality of how ethnic conflict begets further turmoil. Moreover, the pattern of poor refugees being hosted by equally poor countries underscores the enormous strain that host countries, as well as international relief agencies, are under to contain and hopefully resolve these conflicts.

In 1997 a half million Somalis were living as refugees in Ethiopia, Kenya, Yemen, and Djibouti. Somalia had politically disintegrated a few years earlier. Its collapse was somewhat ironic, since Somalia is one of the very few ethnically unified states of Sub-Saharan Africa. Its people, however, have long been divided into fiercely independent clans, and it has

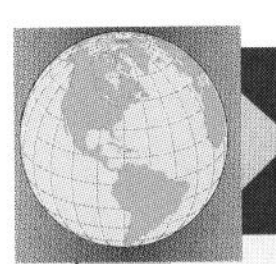

THE GEOGRAPHY OF DISPLACEMENT West Africa's Conflict Zone

Liberia and Sierra Leone both saw their states virtually collapse in the 1990s due to rebellions, ethnic conflict, and economic decay. The brutal civil war that began in Liberia in 1989 between followers of Charles Taylor against the government of Samuel Doe quickly took on ethnic dimensions and spilled over into neighboring Sierra Leone. Well-armed bands of teenagers terrorized the countryside in rural Liberia, causing massive displacement. The conflict was still under way in 1998, with regional and international efforts to try and resolve it having limited success.

In Liberia, roughly half of the pre-war population of 2.5 million was internally displaced. Many fled to the nation's capital, Monrovia, because rural areas were too dangerous. Another 750,000 fled to neighboring Ivory Coast and Guinea and settled along the border region interspersed among local populations (see Figure 1). Virtually the entire southeast corner of Liberia became depopulated and an estimated 150,000 people (6 percent of the country's population) died from 1989 to 1996. The failure of Liberia is sadly ironic, as it is the oldest independent country in the region. Ever since it was founded in the middle 1800s, however, Liberia has been divided between its political elite, the descendants of former U.S. slaves, and its much larger indigenous population.

While the levels of violence were not as great in Sierra Leone, roughly a third of its pre-war population of 4.2 million was displaced. Another 300,000 refugees fled to Guinea and even Liberia. Concentrations of internally displaced persons formed in the southern towns of Kenema and Segbwema, although in 1996 many were able to return to their homes.

In total, an estimated one million refugees and 2.4 million internally displaced persons resulted from the fighting. Due to the continuing abuse of human rights in both countries, it has been difficult for humanitarian workers to reach needy populations. Consequently, child mortality rates are disturbingly high—one out of five children die before their fifth birthday.

Source: Based on "West Africa's Conflict Zone: Current Humanitarian Situation," TSTI 97-10008. Washington, DC: U.S. State Department, 1997.

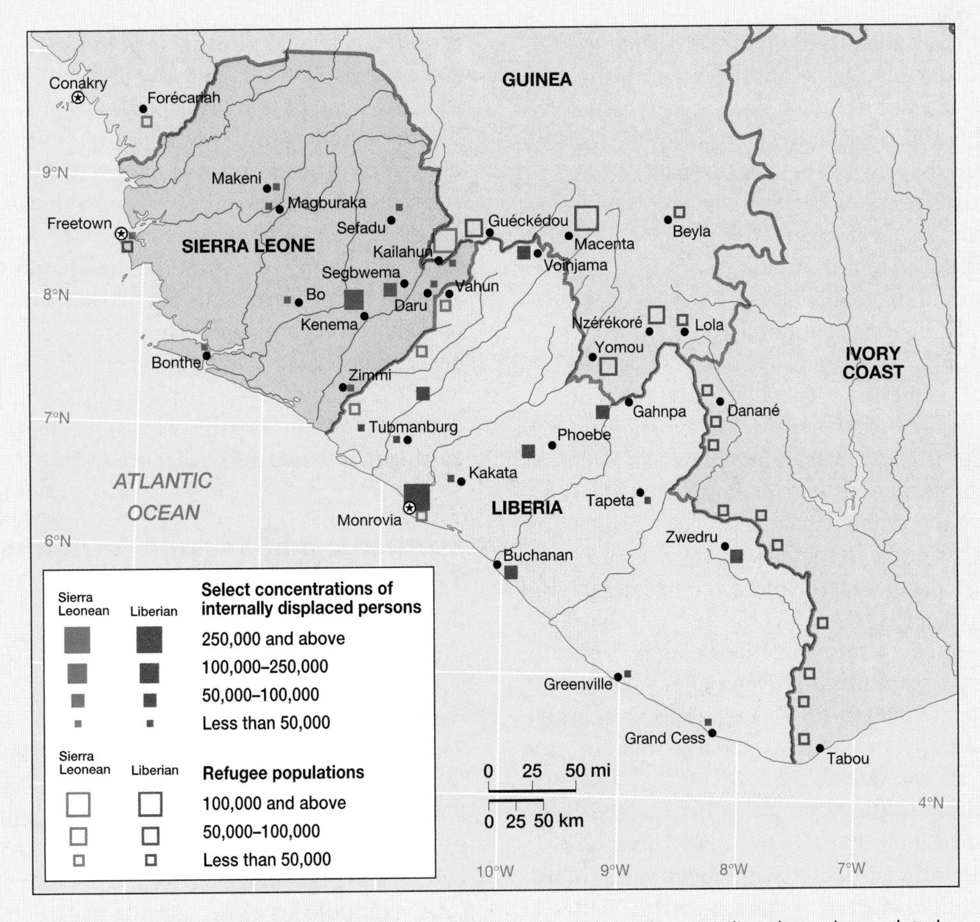

▲ **Figure 1 West African refugees** Civil war and brutal attacks on civilians have depopulated large areas of rural Liberia and Sierra Leone, forcing people into cities and neighboring countries. For these poor countries, the burden of assisting one million refugees and internally displaced people is overwhelming. *(Source: "West Africa's Conflict Zone: Current Humanitarian Situation, February 1997," U.S. Government Document TSTA-97-10008)*

▲ **Figure 6.37 Rwandan refugees** In 1994 more than one million Hutus fled Rwanda for camps in Goma, Democratic Republic of the Congo. The camps were occupied for nearly two years when political instability in the Congo (then Zaire) resulted in the forced repatriation of refugees back to Rwanda. *(Wesley Bocxe/Photo Researchers, Inc.)*

no tradition of central authority. **Clans** are social units that are a branch of a tribe or ethnic group but larger than a family. As clan warfare intensified, food could no longer be easily produced or distributed, and a massive famine followed. In 1992 United Nations forces, led by the United States, arrived to impose order and distribute food. While they were rather successful at the latter task, they were utterly ineffective at the former, and they were soon forced to withdraw. Since the 1990s Somalia has been informally divided into several clan territories, and sporadic fighting continues.

Secessionist Movements Knowing how problematic African political boundaries are, one would expect territories to secede and form new states. The Shaba (or Katanga) province in what was then the state of Zaire tried to leave its national territory soon after independence. The rebellion was crushed a couple years after it started with the help of France and Belgium. It seems that then-Zaire was unwilling to give up its copper-rich territory, and former colonialists had economic interests in the territory worth defending. Similarly, the Ibo in oil-rich southeastern Nigeria proclaimed an independent state of Biafra in 1967. After a short but brutal war, during which Biafra was essentially starved into submission, Nigeria was reunited.

Only one territory in the region actually has seceded. In 1993 Eritrea gained independence from Ethiopia after two decades of civil conflict. What is striking about this territorial secession is that Ethiopia forfeited its access to the Red Sea, making it landlocked. Yet the creation of Eritrea still did not bring about peace. After years of fighting, the transition to Eritrean independence began remarkably well. Unfortunately, border disputes between the two countries erupted in 1998, leaving thousands of troops killed or wounded. Yet if Ethiopia can accept Eritrea's independence, then perhaps it might be possible for other areas torn by ethnic warfare to do so. The bloody civil war in southern Sudan, which has produced a half million refugees and four million internally displaced persons, might possibly be resolved by granting independence to the province of Equatoria. Still, any thorough transformation of Africa's political map should not be expected

Big Man Politics Out of political turmoil, "Big Man" politics emerged. Presidents, both military and civilian, grabbed onto the reigns of power and refused to let go. Military governments, one-party states, and presidents-for-life became the norm. So too were violent changes in government, which occurred in virtually every state, and often several times. Strongmen such as General Abacha in Nigeria, President Kabila of the Congo (who ousted president-for-life Mobuto Sese-Seko in 1997), or President Daniel arap Moi of Kenya are ruthless leaders who do not welcome open political debate. Many see the fragility and corruption of Africa's political institutions as its biggest impediment to development. Furthermore, the reliance on bullets over ballots encourages disproportionate spending on the military. In countries such as Sudan, Mozambique, and Liberia, and even relatively peaceful ones such as Tanzania, military expenditures per capita far outstrip spending on health care.

Fortunately, the 1990s saw a growth in the number of states with multiple political parties and free elections. Uganda, for example, previously suffered from one of Africa's most repressive regimes and was burdened by a legacy of bitter ethnic estrangement and capricious rule. Today it has a stable government, a growing economy, and reduced ethnic tensions. In many other Sub-Saharan countries as well, authoritarian governments have recently been giving way to more liberal regimes promising increased freedom and democratic rule. Looking exclusively at debacles such as Liberia, Rwanda, or Somalia—as the U.S. press tends to do—makes it difficult to appreciate the political progress that has been occurring in much of Sub-Saharan Africa.

Economic and Social Development: The Struggle to Rebuild

By almost any measure, Sub-Saharan Africa is the poorest and least-developed world region. Only a few African countries, such as Gabon, South Africa, Seychelles, and Mauritius, are counted among the middle-income economies of the world. The rest are ranked at the bottom of most lists comparing per capita GNP or purchasing power parity. Using social indicators such as life expectancy or secondary school enrollment, the region also fares poorly. Interestingly, female labor force participation is comparable to, and even far exceeds, rates in other parts of the world. Yet the overall status of African women is mixed, and labor force participation is not necessarily an indicator of equality.

The most troubling indicators show that the region's economic base actually declined in the 1980s and early 1990s. Table 6.2 shows most countries with negative growth rates

Table 6.2 Economic Indicators

Country	GNP per Capita ($U.S., 1996)	Total GNP (Millions of $U.S., 1996)	PPP[a] ($Intl, 1996)	Real Annual Growth % per Capita, 1990–1996
Angola	270	2,972	1,030	–5.6
Benin	350	1,998	1,230	1.9
Botswana	3,020[b]	4,381[b]	7,390	1.3
Burkina Faso	230	2,410	950	–0.1
Burundi	170	1,066	590	–6.4
Cameroon	610	8,356	1,760	–3.8
Cape Verde	1,010	393	2,640	–16.7
Central African Republic	310	1,024	1,430	–1.7
Chad	160	1,035	880	–1.7
Comoros	450	228	1,770	–1.8
Congo	670	1,813	1,410	–4.3
Dem. Rep. of Congo	130	5,727	790	–10.4
Djibouti	—	—	—	—
Equatorial Guinea	530	217	2,690	15.9
Eritrea	—	—	—	—
Ethiopia	100	6,042	500	2.0
Gabon	3,950	4,444	6,300	–1.2
Gambia	—	—	1,280	–0.5
Ghana	360	6,223	1,790	1.5
Guinea	560	3,804	1,720	1.9
Guinea-Bissau	250	270	1,030	0.5
Ivory Coast	660	9,434	1,580	0.2
Kenya	320	8,661	1,130	–0.5
Lesotho	660	1,331	2,380	0.9
Liberia	—	—	—	—
Madagascar	250	3,428	900	–2.0
Malawi	180	1,832	690	–0.2
Mali	240	2,422	710	–0.2
Mauritania	470	1,089	1,810	1.7
Mauritius	3,710	4,205	9,000	3.6
Mozambique	80	1,472	500	2.6
Namibia	2,250	3,569	5,390	1.6
Niger	200	1,879	920	–2.3
Nigeria	240	27,599	870	1.2
Reunion	—	—	—	—
Rwanda	190	1,268	630	–8.2
São Tomé and Principe	330	45	—	–1.7
Senegal	570	4,856	1,650	–0.6
Seychelles	6,850	526	—	1.5
Sierra Leone	200	925	510	–3.9
Somalia	—	—	—	—
South Africa	3,520	132,455	7,450	–0.2
Sudan	—	—	—	—
Swaziland	1,210	1,122	3,320	–1.2
Tanzania	170	5,174	—	–0.2
Togo	300	1,278	1,650	–3.9
Uganda	300	5,826	1,030	4.0
Zambia	360	3,363	860	–4.8
Zimbabwe	610	6,815	2,200	–1.1

[a]Purchasing power parity.

[b]Botswana's GNP per capita and total GNP are based on 1995 figures.

Source: *The World Bank Atlas,* 1998.

between 1990 and 1996. The impact of falling economic output was made worse by rapid population growth. (The region's population growth rate of 2.6 percent in 1998 is considerably higher than Southeast Asia's 1.6 percent rate.) This means that the national economies of Africa needed to grow briskly just to maintain their living standards; instead, they declined. The economic and debt crisis that ensued prompted the introduction of structural adjustment programs. Promoted by the IMF and the World Bank, structural adjustment programs typically reduce government spending, cut food subsidies, and encourage private-sector initiatives. Yet these same policies spurred immediate hardships for the poor, especially women and children, and ignited social protest, most notably in cities.

Over the past few years there have been some signs of an impending economic turnaround. The economy of the region as a whole seems to have experienced some growth beginning in 1995, but this growth is at a slow pace. The late 1990s witnessed African economies growing at a slightly higher rate than their population growth, something that has not happened in many years. A select few African countries, moreover, have done much better over the past decade, most notably Uganda. Economic and social development will likely improve by a combination of foreign investment, debt relief, state reform, and private initiatives led by individuals and NGOs. How much Africa should push for greater integration in the world economy is hotly debated, as are the roots of the region's poverty.

Roots of African Poverty

In the past outside observers often attributed Africa's poverty to its environment. Favored explanations included the infertility of its soils, the erratic patterns of its rainfall, the paucity of its navigable rivers, and the virulence of its tropical diseases. Most contemporary scholars, however, argue that such handicaps are by no means prevalent throughout the region, and that even where they do exist they can be—and have often been—overcome by human labor and ingenuity. The favored explanations for African poverty now look much more to historical and institutional factors than to environmental circumstances.

Numerous scholars have singled out the slave trade for its debilitating effect on Sub-Saharan African economic life. Large areas of the region were effectively depopulated, and many people were forced to flee into poor, inaccessible refuges. Colonization was another blow to Africa's economy. European powers invested little in infrastructure, education, or public health and were instead interested mainly in extracting mineral and agricultural resources for their own benefit. Several plantation and mining zones did achieve a degree of prosperity under colonial regimes, but internally dynamic national economies failed to develop. In almost all cases, the rudimentary transport and communi-s systems were designed to link administration centers of extraction directly to the colonial powers, heir own hinterlands or neighboring areas. On achieving independence, Sub-Saharan African countries thus faced economic challenges that were as daunting as their political problems.

Failed Development Policies The first decade or so of independence was a time of relative prosperity and optimism for many African countries. Most of them relied heavily on the export of mineral and agricultural products, and through the 1970s commodity prices generally remained high. Some foreign capital was attracted to the region, and in many cases the European economic presence actually increased after decolonization. (This is one reason why critics refer to the period as one of "neocolonialism.")

The relatively buoyant economies of the 1960s and early 1970s were to disappear in the 1980s as most commodity prices began to decline. Sizable foreign debt began to weigh down many Sub-Saharan countries. By the end of the 1980s most of the region entered a virtual economic tailspin. By the early 1990s, the region's foreign debt (minus South Africa) was around $200 billion. Although low compared to other developing regions (such as Latin America), as a percentage of its economic output Sub-Saharan Africa's debt was the highest in the world.

Many economists contend that Sub-Saharan African governments enacted counterproductive economic policies and thus brought some of their misery down on themselves. Keen to build their own economies and reduce their dependency on the former colonial powers, most African countries followed a course of economic nationalism. Thus they set about building steel mills and other forms of heavy industry in which they were simply not competitive (Figure 6.38). In many cases inefficient and often corrupt government ministries took over large segments of the economy. Local currencies were also often maintained at artificially elevated levels, which benefited the elite who consume imported prod-

▲ **Figure 6.38 Industrialization** Heavy industry, such as this chemical plant in Kafue, Zambia, has failed to deliver Sub-Saharan Africa from poverty. In the worst cases, these industrial enterprises were unable to produce competitive products for world and domestic markets. *(Marc & Evelyne Bernheim/Woodfin Camp & Associates)*

ucts, but it undercut exports. Some former French colonies kept their currencies pegged to the franc, which almost precluded successful export strategies.

Food Policies The largest blunders made by Sub-Saharan leaders were in agricultural and food policies. The main intention was to retain a cheap supply of staple foods in the urban areas. A modern industrial economy could emerge, or so it was argued, only if manufacturing wages remained low, which in turn was feasible only if food could remain inexpensive. Another reason was simply that cheap food in potentially volatile urban areas would help maintain political stability. The main problem with this policy was that the majority of Africans were farmers—farmers who could not make money from their crops because prices were kept artificially low. Peasants thus opted to grow mainly for subsistence, rather than sell—often at a loss—to national marketing boards. At the same time, farmers were encouraged to shift into export crops, such as coffee and cocoa, peanuts and cotton. The end result was the failure of food production to expand at a time when the population was growing explosively. In the 1980s famine occurred in 22 African states, partly due to failed agricultural policies combined with drought.

It is also important to realize that although Sub-Saharan Africa is poorer than South Asia in terms of per capita GDP, its people are generally better nourished and often better housed. Starvation has followed war and drought in Africa, but on a day-to-day level food intake is adequate. Such "nutritional prosperity" can be explained in part by the region's low population density and by the fact that many people still rely largely on subsistence production, which does not figure into official economic statistics.

Many African countries could vastly increase their production of staple foods, according to prevailing economic wisdom, but only if cultivators could be offered adequate incentives. Most countries in the region began to reform their agricultural policies in the 1980s, but this has been difficult to do because, as explained above, the urban poor have grown dependent on low prices of staple goods.

Corruption Although prevalent through most of the world, corruption also seems to have been particularly rampant in several African countries. Civil servants are sometimes not paid a living wage, and are thus virtually forced to solicit bribes. According to a recent poll of international businesspeople, Nigeria ranks as the world's most corrupt country. (Skeptical observers, however, point out that several highly successful Asian economies, such as China, are also noted for high levels of corruption.)

With millions of dollars in loans and aid pouring into the region, officials at various levels were tempted to skim from the top. African states, such as the former Zaire, were dubbed kleptocracies. A **kleptocracy** is a state where corruption is so institutionalized that politicians and government bureaucrats siphon off a huge percentage of the country's wealth. Again, President Mobutu in Zaire was a legendary kleptocrat. While his country was saddled with an enormous foreign debt, he supposedly skimmed several billion dollars and deposited them in Belgian banks.

Links to the World Economy

In terms of trade, Sub-Saharan Africa's connection with the world is limited. The level of overall trade is low, both within the region and outside it. As expected, most exports are bound for the European Union, especially the former colonial powers (Figure 6.39). The United States is the second most common destination. Exports to Asia are surprisingly few—mostly oil and tropical hardwoods. The import pattern mirrors exports. Despite four decades of independence, the majority of African countries turn to Europe for imports. Increasingly, cheaper goods from Asia are making their way across the Indian Ocean to African consumers. Exchange with eastern Europe and Russia, which was vigorous for a few African states during the Cold War, has nearly disappeared.

Africa clearly lacks the infrastructure and goods to facilitate more intraregional trade. Only southern Africa has a telecommunications network of any note. Of the few roads that exist, especially in West and East Africa, the majority are not paved. Nothing comparable to MERCOSUR (the trade bloc in South America) is found in Sub-Saharan Africa. This too is a poignant legacy of colonialism, and one that keeps Africans from nurturing economic linkages with each other.

Aid versus Investment As a poor region, Sub-Saharan Africa is linked to the global economy more through the flow of financial aid and loans than through the flow of goods. Some African states such as Mali, Gabon, Mozambique, and Namibia received more than $50 per capita in assistance in 1996. By contrast, net flows of private capital are extremely low. Countries such as South Africa, Ivory Coast, Ghana, and Angola are the exceptions. Most other states in the region attracted less than $1 per capita in private foreign investment.

The reasons foreign investors shun Africa are fairly obvious. The region is generally perceived to be too poor and too unstable to merit much attention, and most foreign investors eager to take advantage of low wages put their money in Asia or Latin America. Sub-Saharan Africa, some economists contend, is thus starved for capital. Other scholars, however, see foreign investment as more of a trap—and one that the region would be wise to avoid. Recently, in a number of African countries a small boom has occurred in mineral exploration and production, attracting considerable overseas interest. Optimists interpret this as a sign of impending African economic renewal, but others counter that in the past mineral extraction failed to lead to broad-based and lasting economic gains.

Finally, the World Bank and the IMF proposed in 1996 to reduce debt levels for the poorest African states, which on average earmark 25 percent of their export earnings toward debt repayment. Most Sub-Saharan states are indebted to official creditors such as the World Bank, and not commercial banks (as is the case in Latin America and Southeast Asia). Under this

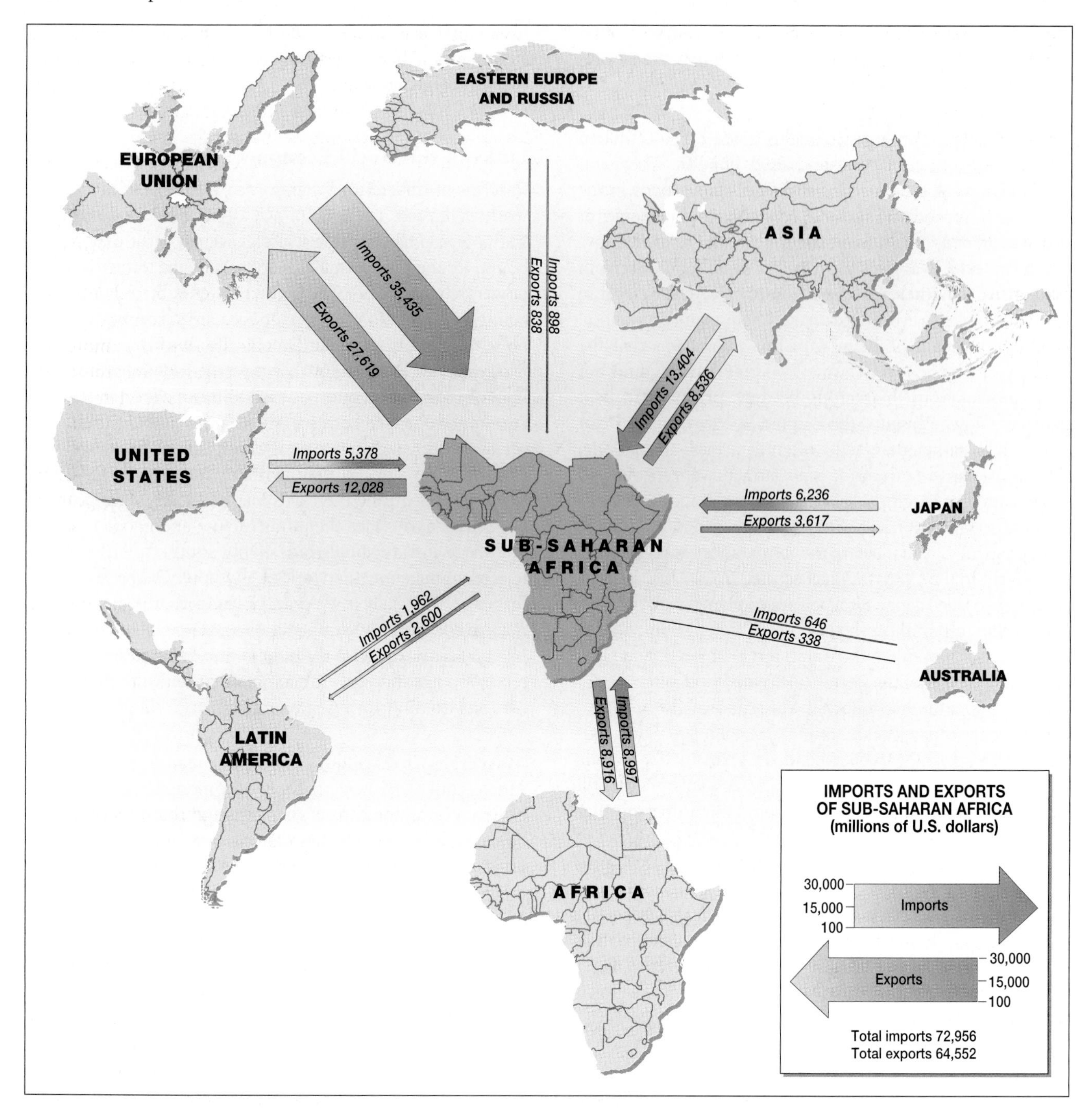

▲ **Figure 6.39 Sub-Saharan trade** Sub-Saharan Africa's major trading partners are in Europe. More goods flow between Europe and Sub-Saharan Africa than are traded among African countries. This reflects of pattern of historical development in which a few select commodities were produced for export to Europe or the United States, ignoring intraregional networks. Efforts to expand intraregional trade, especially in southern Africa, are under way. *(Data source:* International Marketing Data and Statistics 1998. *London: Euromonitor)*

new World Bank/IMF program, substantial debt reduction will be given to Sub-Saharan countries that are determined to have "unsustainable" debt burdens. If qualifying states stay the path ·f structural adjustment for six years, debts will be written off. ·· Burkina Faso, Ivory Coast, and Mozambique have ·ogram. With luck, more African countries will be ·stabilizing foreign debts.

Future Paths Sub-Saharan Africa's future economic path is highly uncertain. Will the various countries of the region opt for integration into the global system, or will they instead attempt to build a more autonomous, and less capitalistic, economic system? Certainly the latter option was tried in the several decades after independence, but the former seems to be gaining momentum. Many African countries have begun to privatize sectors of their economies, a process that often leads to faster economic growth, but which also tends to heighten income disparities and negative environmental impacts. Trade groups, such as the Southern African Development Community, have the potential to build up regional economies and infrastructure. Certainly neither option, African socialism or global capitalism, offers an easy fix for the region's problems, but both hold out promises for future improvement.

Economic Differentiation Within Africa

As in most regions, considerable disparity in levels of economic and social development persist. In many respects, the small island nations of Mauritius and the Seychelles have little in common with the mainland. With a high per capita GNP, life expectancies averaging in the low 70s, and economies built on tourism, they could more easily fit into the Caribbean were it not for their Indian Ocean location. In contrast, Chad has a per capita GNP of $160 and only 13 percent of its boys and 2 percent of its girls enrolled in high school.

Given the scale of the African continent, it is not surprising that groups of states formed trade blocs to facilitate intraregional exchange and development. The two most active regional organizations are the Southern African Development Community (SADC) and the Economic Community of West African States (ECOWAS). Both were founded in the 1970s but became more prominent in the 1990s (Figure 6.40). SADC and ECOWAS are each anchored by the region's largest economies: South Africa and Nigeria. Other prosperous states are those that exploit oil resources. The following section contrasts Africa's better-off states with its poorer ones.

South Africa South Africa is the unchallenged economic powerhouse, with a GNP almost as high as that of all the other countries in the region combined. Only South Africa has a well-developed and well-balanced industrial economy. It also boasts a healthy agricultural sector, and, more importantly, it stands as one of the world's mining superpowers. South Africa remains unchallenged in gold production and is a leader in many other minerals as well. But while South Africa is undeniably a wealthy country by African standards, and while its white minority is prosperous by any standard, it is also a country beset by severe and widespread poverty. In the townships lying on the outskirts of the major cities, and in the rural districts of the former homelands, employment opportunities remain limited and living standards marginal. According to some recent statistics, South Africa now has the highest unemployment rate in the industrialized world. It also has one of the highest rates of violent crime in the entire world.

Oil and Mineral Producers Another group of relatively well-off Sub-Saharan countries benefits from substantial oil and mineral reserves, and small populations (Figure 6.41). The prime example is Gabon, a country of noted oil wealth that is inhabited by 1.2 million people. Its neighbors, Republic of the Congo and Cameroon, also benefit from oil, but both have experienced economic declines in recent years, with that of Cameroon being particularly dramatic. Farther south, Namibia and Botswana also have the advantage of small populations and abundant mineral resources, especially diamonds. Both countries have also enjoyed sound government over the past few years and have experienced significant economic growth.

The Leaders of ECOWAS Nigeria has the largest oil reserves in the region, but its huge population has kept its per-capita GNP at a low $240. Oil money has also helped make Nigeria notoriously inefficient. A small minority of its population has grown fantastically wealthy more by manipulating the system than by engaging in productive activities. Most Nigerians, however, remain trapped in poverty. Oil money also led to the explosive growth of the former capital of Lagos, which by the 1980s became one of the most expensive—and least livable—cities in the world. As a result, the Nigerian government opted to build a new capital city in Abuja, located near the country's center, a move that has proved tremendously expensive.

Ivory Coast and Senegal, formerly the core territories of the French Sub-Saharan empire, still function as commercial centers, but they have also suffered economic downturns in the 1980s. In the mid-1990s, the Ivorian economy again began to grow. Boosters within the country call it an emerging "African elephant" (comparing it to the successful "economic tigers" of eastern Asia). Ghana, a former British colony, has also begun to mount an economic recovery.

Kenya Long the commercial and communications center of East Africa, Kenya experienced economic decline and political tension throughout the 1990s. Despite a per capita GNP of just $320, Kenya boasts good infrastructure by African standards and more than one million foreign tourists come each year to marvel at its wildlife. Agricultural exports of coffee, sisal, and tea dominate the economy. As for social indicators, Kenyan women are having three fewer babies now than in the late 1970s (down from an average of 8 to 4.5 children per woman) and those children are healthier and better educated. If Kenya can avoid political implosion, it could lead East Africa into better economic integration with the southern and western parts of the continent.

The Poorest States The poorest parts of Sub-Saharan Africa include the Sahel, the Horn, and the southeast. Sahelian countries such as Mali, Burkina Faso, Niger, and Chad have suffered from prolonged droughts and serious environmental degradation, and they have poorly developed infrastructures and commercial networks. More so than the rest of Africa, their economies remain dominated by agriculture, with very little urban or industrial development (the largest city in Chad, N'Djamena, has only 280,000 inhabitants). In the Horn of Africa, Sudan, Ethiopia, Eritrea, and Somalia have all been the victims of drought as well as war. The same is true for the world's poorest country, Mozambique, in the southeast.

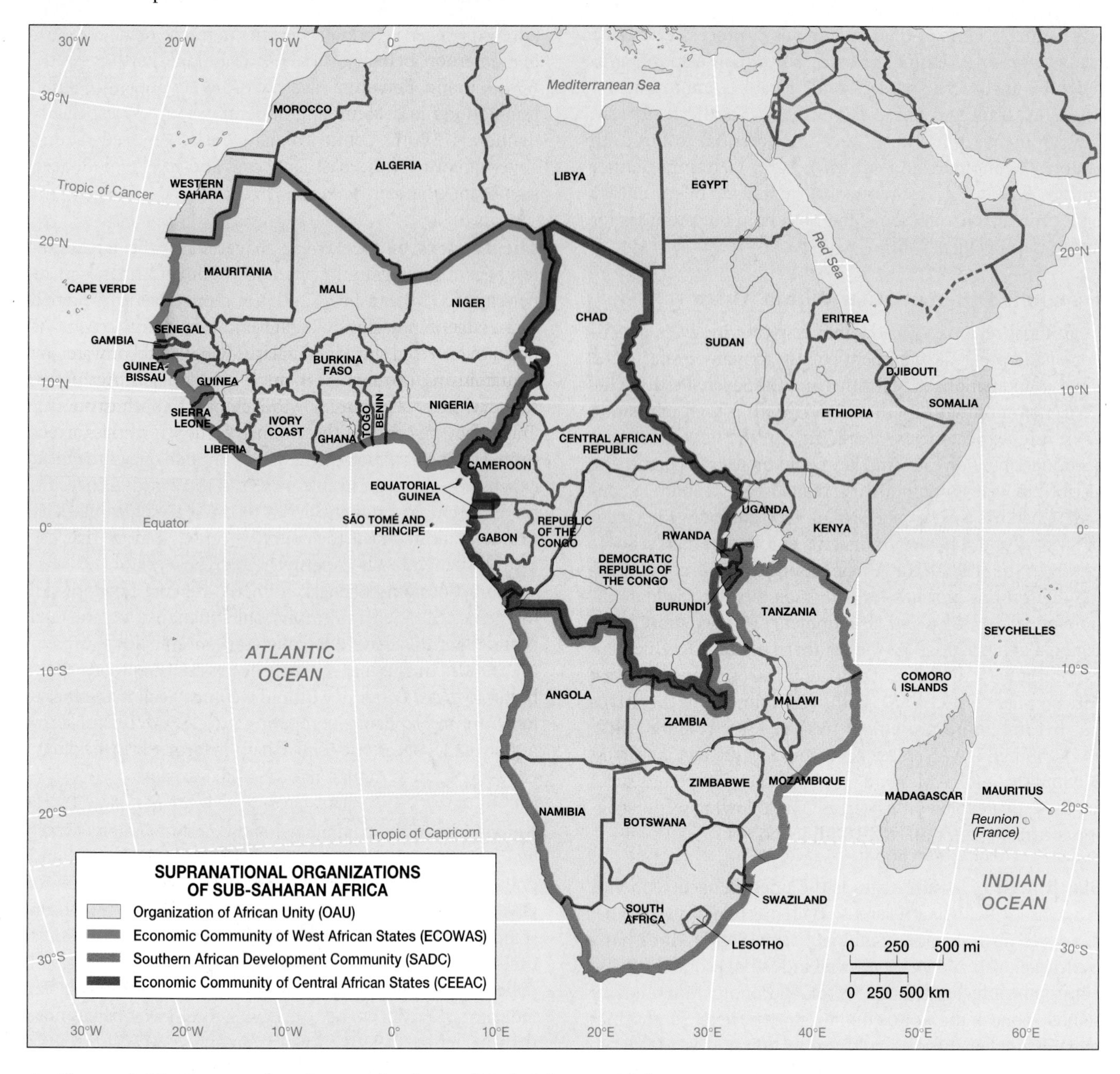

▲ **Figure 6.40 Supranational organizations of Sub-Saharan Africa** Political affiliations in Sub-Saharan Africa are both continental and regional. The OAU includes each African country. Smaller organizations such as SADC or ECOWAS represent regional affiliations. Of these, SADC shows the most economic promise.

The poor economic showing of Tanzania is less easily explained. In the 1960s and 1970s Tanzania was actually considered something of a showcase for Africa development, and it did make some advances in health and education. Its ambitious plans for building a distinctly African form of socialism, termed *Ujaama* in Swahili, largely failed. Peasants resisted attempts to form agricultural cooperatives, and large indus-·ial projects came to naught. Such failures proved bitterly ·ointing not only for Tanzania's leadership, but also for ·lopmental specialists in the United States and Eu-·eat promise in this unique and well-meaning experiment. For many years Tanzania was the world's largest per capita recipient of foreign aid, but the funds that it received seem to have had little lasting impact.

Measuring Social Development

By world standards, measures of social development in Africa are extremely low. Yet unlike Africa's economic indicators, at least the trends in social development are mostly positive. Rates of child survival have improved, especially when compared to child mortality rates in 1960 (Table 6.3). Likewise, literacy is also improving, but access to education varies enormously (Fig-

▲ **Figure 6.41 Copper miners** Miners drill for copper in Zambia. Mining is an important sector of the region's economy, especially for South Africa, Zambia, and Namibi, where it is a capital-intensive, industrial enterprise. *(Jason Lauré/Lauré Communications)*

ure 6.42). Enrollment percentages in secondary school ranges from less than 10 percent in Niger, Guinea-Bissau, and Malawi to well over 50 percent in Botswana, Mauritius, and South Africa. Gaps do exist between male and female enrollment figures. In most states, males are more likely to receive a secondary education. Although in South Africa and Namibia, girls have higher enrollment rates than boys. A high school education, however, is still a luxury for most Africans.

Life Expectancy Sub-Saharan Africa's figures on life expectancy are, however, the world's lowest. Only the poorest Asian countries, such as Afghanistan, Nepal, and Laos, stand at the average African level. Sierra Leone's particularly miserable figures of 33 years for men and 36 for women reflects, in part, recent fighting and political disorder. Despite these dismal figures, some progress has been made in enhancing life expectancy in Sub-Saharan Africa. While a life expectancy of 48 years may seem incredibly young, it must be remembered that high infant and childhood mortality figures depress these numbers; average life expectancy for adults is substantially higher. States such as Kenya, Botswana, and Zimbabwe did establish relatively high life expectancy figures in the 1980s, but the growing AIDS epidemic eroded their accomplishments.

The causes of low life expectancy are generally related to extreme poverty, environmental hazards (such as drought), and various environmental and infectious diseases (malaria, schistosomiasis, cholera, and measles). Often these factors work in combination. Malaria, for example, kills a half million African children each year. The death rate is exacerbated by poverty, undernourished children being the most vulnerable to the effects of high fevers. Cholera outbreaks occur in crowded slums and villages, where food or water are contaminated by the feces of an infected person. Again, unsanitary practices result due to lack of basic infrastructure. Tragically, diseases that are preventable, such as measles, occur when people have no access to or cannot afford vaccines.

Health Issues A paucity of doctors and health facilities, especially in rural areas, also helps to explain Sub-Saharan Africa's high mortality levels. The more successful countries in the region, however, have discovered that inexpensive rural clinics dispensing rudimentary treatments, such as oral rehydration therapy for infants with severe diarrhea, can substantially improve survival rates (see Kenya's child mortality rate, Table 6.3). Another problem is the severity of the African disease environment itself. Malaria is making a comeback in many areas, as the disease-causing organisms develop resistance to common drugs and the mosquito carriers develop resistance to insecticides. While most Africans have partial immunity to malaria, it remains a major killer, and many who survive infections remain debilitated.

Many other tropical diseases also continue to plague the region. Schistosomiasis is carried by freshwater snails and can cause chronic diarrhea and cramping. The construction of dams, reservoirs, and irrigation ditches has greatly increased the environment for these snails, yet basic eradication programs are often not implemented. Consequently, 200 million Africans are affected by schistosomiasis, which can lead to dehydration and susceptibility to other diseases. In some West African areas, up to 10 percent of villagers have lost their sight due to river-blindness, a disease spread by a small fly. Fortunately, effective pest control is reducing the incidence of river-blindness in the region. Collectively, national and international health agencies, along with local NGOs, have improved the availability of basic health care in Sub-Saharan Africa, but much work remains to be done.

Women and Development

Development gains cannot be made in Africa unless the economic contributions of African women are recognized. Officially, women are the invisible contributors to local and national economies. In agriculture, women account for 75 percent of the labor that produces more than half the food consumed in the region. Tending subsistence plots, taking in extra laundry, and selling surplus produce in local markets

Table 6.3 Social Indicators and Status of Women

Country	Life Expectancy at Birth		Under Age 5 Mortality, per 1,000 Live Births		Secondary School Enrollment %*		Female Labor Force Participation (% of total)
	Male	Female	1960	1993	Male	Female	
Angola	45	48	345	292	—	—	46
Benin	51	56	310	144	17	7	48
Botswana	40	42	170	56	49	55	46
Burkina Faso	46	47	318	175	11	6	47
Burundi	44	47	255	178	8	5	49
Cameroon	53	56	264	113	32	23	38
Cape Verde	66	73	—	—	21	20	38
Central African Republic	44	48	294	177	17	6	47
Chad	45	50	325	206	13	2	44
Comoros	57	62	—	—	21	17	42
Congo	45	49	220	109	—	—	43
Dem. Rep. of Congo	47	51	286	187	33	15	44
Djibouti	47	50	—	—	14	10	40
Equatorial Guinea	46	50	316	—	—	—	35
Eritrea	52	57	—	204	17	13	47
Ethiopia	41	42	294	202	12	11	41
Gabon	52	55	287	154	—	—	44
Gambia	43	47	375	—	25	13	45
Ghana	54	58	215	170	44	28	51
Guinea	43	47	337	226	17	6	47
Guinea-Bissau	41	44	336	235	9	4	40
Ivory Coast	51	54	300	120	33	17	33
Kenya	48	49	202	90	28	23	46
Lesotho	54	58	204	156	22	31	37
Liberia	56	61	288	217	—	—	39
Madagascar	51	53	364	164	14	14	45
Malawi	36	36	365	223	6	3	49
Mali	45	47	400	217	12	6	46
Mauritania	50	53	321	202	19	11	44
Mauritius	67	74	84	23	58	60	32
Mozambique	43	46	331	282	9	6	48
Namibia	42	42	206	79	49	61	41
Niger	45	48	320	320	9	4	44
Nigeria	49	52	204	191	32	27	36
Reunion	75	75	—	—	—	—	42
Rwanda	43	44	191	141	11	9	49
São Tomé and Principe	62	65	—	—	—	—	35
Senegal	48	50	303	120	21	11	43
Seychelles	65	76	—	—	—	—	—
Sierra Leone	33	36	384	284	22	12	36
Somalia	45	49	294	211	9	5	43
South Africa	55	60	126	69	71	84	37
Sudan	50	52	292	128	24	19	26
Swaziland	38	41	233	—	51	50	37
Tanzania	45	49	249	167	6	5	49
Togo	56	60	264	135	34	12	40
Uganda	40	41	218	185	14	8	48
Zambia	37	38	220	203	25	14	45
Zimbabwe	40	40	181	83	51	40	44

Sources: *Population Reference Bureau Data Sheet,* 1998, Life Expectancy (M/F); World Resources Institute, *World Resources 1996–97,* Under-5 Mortality Rate; *Population Reference Bureau Data Sheet, 1996,* Secondary School Enrollment (M/F); *The World Bank Atlas, 1998,* Female Participation in Labor Force.

▲ **Figure 6.42 Educating children** An all-girls school in Kenya reflects an important investment in the country's social development. Education of girls, in particular, is seen as a way to raise the status of women, lower birthrates, and improve economic opportunities. *(Betty Press/Woodfin Camp & Associates)*

all contribute to household income. Yet since many of these activities are considered informal economic activities, they are not counted. For many of Africa's poorest people, however, the informal sector is the economy. Some research suggests it can account for 30 to 50 percent of the gross domestic product in certain states. Within this sector, women dominate.

Status of Women The social position of women is difficult to gauge for Sub-Saharan Africa. Women traders in West Africa, for example, are regarded as having considerable political and economic power. By such measures as female labor force participation, many Sub-Saharan African countries show relative gender equality. And women in most Sub-Saharan societies do not suffer the kinds of traditional social liabilities encountered in much of South Asia and Southwest Asia and North Africa (see "Women and Development: The Nanas-Benz of Togo").

By other measures, however, such as the prevalence of polygamy, the practice of the "bride-price," and the denial of property inheritance, African women do suffer discrimination. Perhaps the most controversial issue regarding women's status is the practice of female circumcision, or genital mutilation. In Sudan, Ethiopia, Somalia, Eritrea, as well as parts of West Africa, almost 80 percent of girls are subjected to this practice, which is extremely painful and can have serious health consequences. Yet because these practices are considered "traditional," most African states are unwilling to legislate against them.

Regardless of their social position, most African women still live in remote villages in which educational and wage-earning opportunities remain limited, and in which bearing numerous children remains a major economic contribution to the family. As educational levels increase and as urban society expands—and as reduced infant mortality provides greater security—one can expect fertility in the region to gradually decrease. Governments can greatly quicken the process by providing birth control information and cheap contraceptives—and by investing more money in health and educational efforts aimed at women. As the economic importance of women receives greater attention, both from national and international organizations, more programs are being directed exclusively toward them.

Building from Within Major shifts in how development agencies view women, and how women view themselves, has the potential to transform the region. All across the continent, support groups and networks have formed, raising women's consciousness and harnessing their economic power. From farm-labor groups to women's market associations, investment in the organization of women has paid off. In Kenya, for example, there are hundreds of women's groups planned around tree planting to prevent soil erosion and ensure future fuel supplies.

Whether inspired by feminism, African socialism, or the free market, community organizations have made a difference in meeting basic needs in sustainable ways. No doubt the majority of the groups fall short of all of their objectives. Yet for many people, especially women, the message of creating local networks to solve community problems is an empowering one.

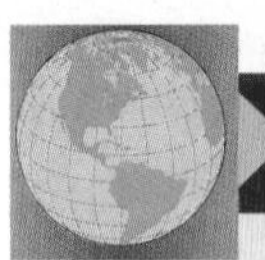

WOMEN AND DEVELOPMENT The Nanas-Benz of Togo

The *Nanas* (a mina word for chief) are women of power and prestige touring the streets of Lomé, Togo, in their Mercedes Benz automobiles. Rooted in a West African tradition of textile manufacturing and vending, the Nanas have cornered the lucrative market for dress fabric in this part of Africa. Togo's cloth traders have earned a reputation for being astute, well-connected, and successful businesswomen. Even though many of the founding traders were illiterate, it has not stopped them from acquiring loans, making investments, and conducting million-dollar transactions. Honing their skills in the markets of West Africa, and not in the business schools of the West, the largest cloth traders do thousands of dollars of trade each day with Europe and Asia. In Togo they rely upon a network of formal and informal markets, limited bookkeeping, and mostly cash payments. The story of their success and influence is legendary.

The Nanas formed the Association of Cloth Merchants, an organization for women who sell cloth, in the 1960s. Prior to that, Ghana was the undisputed center for the African cloth trade, but Togo's Nanas took advantage of political instability in Ghana during the 1960s and secured exclusive import contracts with top textile firms in Europe. By the 1980s the Association of Cloth Merchants represented some 3,000 retailers and wholesalers. The Nanas-Benz, the most successful of the group, are so named for their preference for Mercedes-Benz cars. A story is told about the president of Togo needing a fleet of cars for a state occasion in 1977. The Nanas were contacted and in short order the president had 30 Mercedes-Benz cars at his disposal.

As the Nanas prospered, so too did their business holdings, expanding to restaurants, hair salons, bakeries, and real estate, operations that employ many people. Always on the lookout for new opportunities, their latest ventures include building luxurious rental properties for Lomé-based embassies and corporations.

Active politically, the Nanas-Benz support the lowering of import tariffs and encourage women to vote. Both their political and religious activism has allowed them to enter spheres usually reserved for men. As economically and politically powerful women, they symbolize what can be possible for Africa in the modern global economy.

Source: Adapted from "Africa's Mercedes Ladies," *World Press Review* January 1985.

Conclusion

By simply looking at the numbers, it is easy to grow despondent about Africa's future. Population growth has been outstripping economic expansion for some years, and the living standards of most of the region's inhabitants have steadily declined. Many Sub-Saharan cities are now growing rapidly, but few jobs await either the rural migrants or the youngsters born and raised in the urban environment. In many respects, however, opportunities in the villages are even more limited. Frustration is often particularly acute for young people who have struggled to obtain an education, only to find that jobs are scarce.

Some pessimists think that Sub-Saharan Africa could experience an almost complete political and economic breakdown within the next few decades. Large cohorts of unemployed and disillusioned young people, they remind us, often lead to political instability. In the worst-case situation—such as currently exists in Liberia—young men may see joining a predatory armed band as their only option. As various armed forces vie for power, an entire country can be quickly devastated—as indeed occurred in the 1980s in Mozambique. Ethnic conflicts and the spread of AIDS exacerbate the region's many problems. Meanwhile, desertification and other forms of environmental degradation, which are themselves linked to runaway population growth, threaten the very survival of tens of millions of Africans, as well as the economic and political foundation of the entire region. For the confirmed pessimists, Rwanda—densely populated and torn by an ethnic conflict of genocidal proportions—foretells Africa's future.

The optimists' vision of Africa's future, however, can be equally compelling. Most of the region remains sparsely populated, and large expanses of land could be productively farmed. Countries such as Central African Republic, Democratic Republic of the Congo, and Angola could keep growing for many generations before crowding becomes a serious problem. In other words, most of the region still enjoys a crucial period of breathing space in which its problems can begin to be addressed.

Many recent developments give cause for hope. These include the initial signs of fertility decline, improvements in health and education, and the tentative movement away from authoritarianism and toward democracy. Equally important has been the seemingly successful transition to majority rule and multiracial society in South Africa. Now that peace seems to have come at long last to Mozambique, southern Africa as a whole has emerged as a promising area. As the Southern African Development Community, a group of 12 countries pledged to economic cooperation, matures, it could quickly improve infrastructure and attract more foreign investment. Positive examples include the new Trans-Kalahari highway that links Johannesburg, South Africa, with Windhoek, Namibia. Also planned is an improved highway between Johannesburg, South Africa, and Maputo, Mozambique, as well as new investment in Maputo's war-ravaged port.

In focusing on Africa's traumas, the U.S. media misses many of the region's success stories. We have heard much about Somalia, Liberia, and Rwanda in recent years, but very little about

Botswana. Yet the story of Botswana is in many respects just as instructive. It has a stable, democratic government, and it enjoys a strong and growing economic base. Botswana's levels of social development (under age 5 mortality and secondary school enrollment) are among the highest in the region. The country still faces a number of significant problems—most notably the AIDS epidemic—but thus far it has demonstrated an impressive ability to address them in a constructive manner. Let us hope Botswana, rather than Rwanda, becomes the best emblem for Sub-Saharan Africa at the beginning of the new millennium.

Key Terms

agricultural density (page 222)
apartheid (page 230)
Berlin Conference (page 242)
biofuels (page 219)
clan (page 250)
coloureds (page 235)
desertification (page 219)
Gondwanaland (page 213)
Great Escarpment (page 213)
homelands (page 245)
Horn of Africa (page 219)
internally displaced persons (page 247)
kleptocracy (page 253)
Organization of African Unity (OAU) (page 245)
Pan-African Movement (page 245)
pastoralists (page 228)
physiological density (page 222)
refugee (page 247)
Sahel (page 216)
swidden (page 227)
structural adjustment programs (page 213)
township (page 246)
transhumance (page 218)
tribalism (page 247)
tribe (page 234)
tsetse fly (page 228)

Questions for Summary and Review

1. Discuss the relationship between soil types and patterns of settlement.
2. How might water resources be exploited in the future to help in Africa's development?
3. Is desertification a natural or a human-induced process? Where does it occur?
4. What are the demographic, social, and economic consequences of AIDS in Sub-Saharan Africa?
5. Explain the factors that contribute to high population growth rates in the region.
6. What is the political significance of the 1884 Berlin Conference for Africa?
7. Why have there been relatively few boundary changes since Africa's independence?
8. How are private and grassroots organizations shaping African development?
9. Explain the history and policies the make South Africa distinct from other states in the region.
10. How might the food policies of African governments have undermined the region's ability to feed itself?
11. What are the region's key economic resources that may facilitate its integration into the global economy?

Thinking Geographically

1. What factors might explain why European conquest and settlement occurred much earlier in tropical America than in tropical Africa?
2. What are some of the cultural and political ties that unite Africa with the Americas? With Southwest Asia? With South Asia?
3. Discuss how increasing urbanization in the twenty-first century might affect the overall structure of the population of Sub-Saharan Africa.
4. More than any other region, Sub-Saharan Africa is noted for its wildlife, especially large mammals. What environmental and historical processes explain the existence of so much fauna? Why are there relatively fewer mammals in other world regions?
5. Compare and contrast the role of tribalism in Sub-Saharan Africa with that of nationalism in Europe.
6. Historically, how was Sub-Saharan Africa integrated into the global economy? Was its role similar to or different from other developing regions?
7. Sub-Saharan Africa is a particularly problematic world region because of its relationship to North Africa. Should the Sahara be considered a cultural divide between these two regions? Why or why not?

Regional Novels and Films

Novels

Chinua Achebe, *Things Fall Apart,* (1959, Fawcett Crest)

J.M. Coetzee, *Waiting for the Barbarians* (1982, Penguin)

Nadine Gordimer, *Burger's Daughter* (1979, Viking)

Nadine Gordimer, *Conservationist* (1975, Viking)

Nadine Gordimer, *Crimes of Conscience* (1991, Heinemann)

Nadine Gordimer, *Lifetimes Under Apartheid* (1986, Knopf)

Barbara Kingsolver, *The Poisonwood Bible* (1998, Harper)

Doris Lessing, *African Laughter: Four Visits to Zimbabwe* (1992, Harper)

Alan Paton, *Cry, the Beloved Country* (1948, Scribner's)

Ngugi Wa Thiong'o, *A Grain of Wheat* (1967, Heinemann)

Ngugi Wa Thiong'o, *Petals of Blood* (1977, Heinemann)

Zoë Wicomb, *You Can't Get Lost in Cape Town* (1987, Pantheon)

Films

Black Girl (1966, Senegal)

Borom Sarret (1966, Senegal)

Breaker Morant (1980, Australia)

Burden on the Land (1992, U.S.)

Ceddo (1978, Senegal)

Emitai (1978, Senegal)

General Idi Amin Dada (1975, Uganda)

The Gods Must Be Crazy (1981, Botswana/South Africa)

Haramuya (1995, Burkina Faso)

Tilai (1990, Burkina Faso)

With These Hands (1987, U.S.)

Yaaba (1990, Burkina Faso)

Bibliography

Abraham, Willie E. 1962. *The Mind of Africa.* Chicago: The University of Chicago Press.

Aryeetey-Attoh, Samuel, ed. 1997. *Geography of Sub-Saharan Africa.* Upper Saddle River, NJ: Prentice Hall.

Bell, Morag. 1986. *Contemporary Africa: Development, Culture and the State.* White Plains, NY: Longman.

Berg, Robert J., and Whitaker, Jennifer S. 1986. *Strategies for African Development.* Berkeley: University of California Press.

Best, Alan C. G., and de Blij, Harm J. 1977. *African Survey.* New York: Wiley.

Binns, Tony. 1994. *Tropical Africa.* London: Routledge.

Broughton, Simon; Ellington, Mark; Muddyman, David; and Trillo, Richard, eds. 1994. *World Music.* London: The Rough Guides.

Christopher, A. J. 1984. *Colonial Africa: A Historical Geography.* Totowa, NJ: Barnes & Noble.

Curtin, Philip D. 1969. *The Atlantic Slave Trade.* Madison: University of Wisconsin Press.

Drake, St. Clair. 1987. *Black Folk Here and There.* Los Angeles: Center for Afro-American Studies, University of California, Los Angeles.

Fage, John D. A. 1995. *A History of Africa.* London: Routledge.

Fortes, Meyer, and Evans-Pritchard, Edward E. 1940. *African Political Systems.* London: Oxford University Press.

Franke, Richard W., and Chasin, Barbara H. 1980. *Seeds of Famine: Ecological Destruction and the Development Dilemma in the West African Sahel.* Totowa, NJ: Rowman and Allanheld.

Gaile, Gary, and Ferguson, Alan. 1996. "Success in African Social Development: Some Positive Indicators." *Third World Quarterly* 17(3), 557–572.

Gleave, M. B., ed. 1992. *Tropical African Development: Geographical Perspectives.* New York: Wiley.

Goliber, Thomas. 1997. "Population and Reproductive Health in Sub-Saharan Africa." *Population Bulletin* 52(4). Washington, DC: Population Reference Bureau.

Griffiths, Ienan L. L. 1984. *An Atlas of African Affairs.* London: Methuen.

Grove, Alfred T. 1994. *The Changing Geography of Africa.* Oxford: Oxford University Press.

Kwamena-Poh, Michael; Tosh, John; Waller, Richard; and Tidy, Michael. 1982. *African History in Maps.* London: Longman.

Manning, Patrick. 1990. *Slavery and African Life: Occidental, Oriental, and African Slave.* Cambridge: University Press.

Marcus, Harold G. 1994. *A History of Ethiopia.* Berkeley: University of California Press.

Mortimore, Michael. 1989. *Adapting to Drought: Farmers, Famines and Desertification in West Africa.* Cambridge: Cambridge University Press.

Mountjoy, Alan B., and Embleton, Clifford. 1967. *Africa: A New Geographical Survey.* New York: Praeger.

Mudimbe, Vumbi Y. 1988. *The Invention of Africa: Gnosis, Philosophy, and the Order of Knowledge.* Bloomington, IN: Indiana University Press.

Murdock, George P. 1959. *Africa: Its People and Their Culture History.* New York: McGraw-Hill.

Newman, James L. 1995. *The Peopling of Africa: A Geographical Interpretation.* New Haven: Yale University Press.

Oliver, Roland, and Crowder, Michael, eds. 1981. *The Cambridge Encyclopedia of Africa.* Cambridge: Cambridge University Press.

Peil, Margaret. 1991. *Lagos: The City Is the People*. Boston: G.K. Hall & Co.

Ramsay, F. Jeffress, ed. 1999. *Global Studies: Africa.* 8th edition. Guilford, CT: Dushkin/McGraw-Hill.

Reader, John. 1997. *Africa: A Biography of the Continent*. London: Penguin Books.

Senior, M., and Okunrotifa, P. 1983. *A Regional Geography of Africa*. London: Longman.

Stock, Robert. 1995. *Africa South of the Sahara: A Geographical Interpretation.* New York: Guilford Press.

Stren, Richard, and White, Rodney, eds. 1989. *African Cities in Crisis: Managing Rapid Urban Growth.* Boulder, CO: Westview.

"A Survey of Sub-Saharan Africa." *The Economist,* September 7, 1996.

Taylor, Fraser, and Mackenzie, Fiona. 1992. *Development from Within: Survival in Rural Africa*. London: Routledge.

Thornton, John. 1992. *Africa and Africans in the Making of the Atlantic World, 1400–1680.* Cambridge: Cambridge University Press.

The UNESCO General History of Africa. 1981–present. (Various authors). Berkeley: University of California Press.

Vansina, Jan. 1990. *Paths in the Rainforest: Toward a History of Political Tradition in Equatorial Africa.* Madison: University of Wisconsin Press.

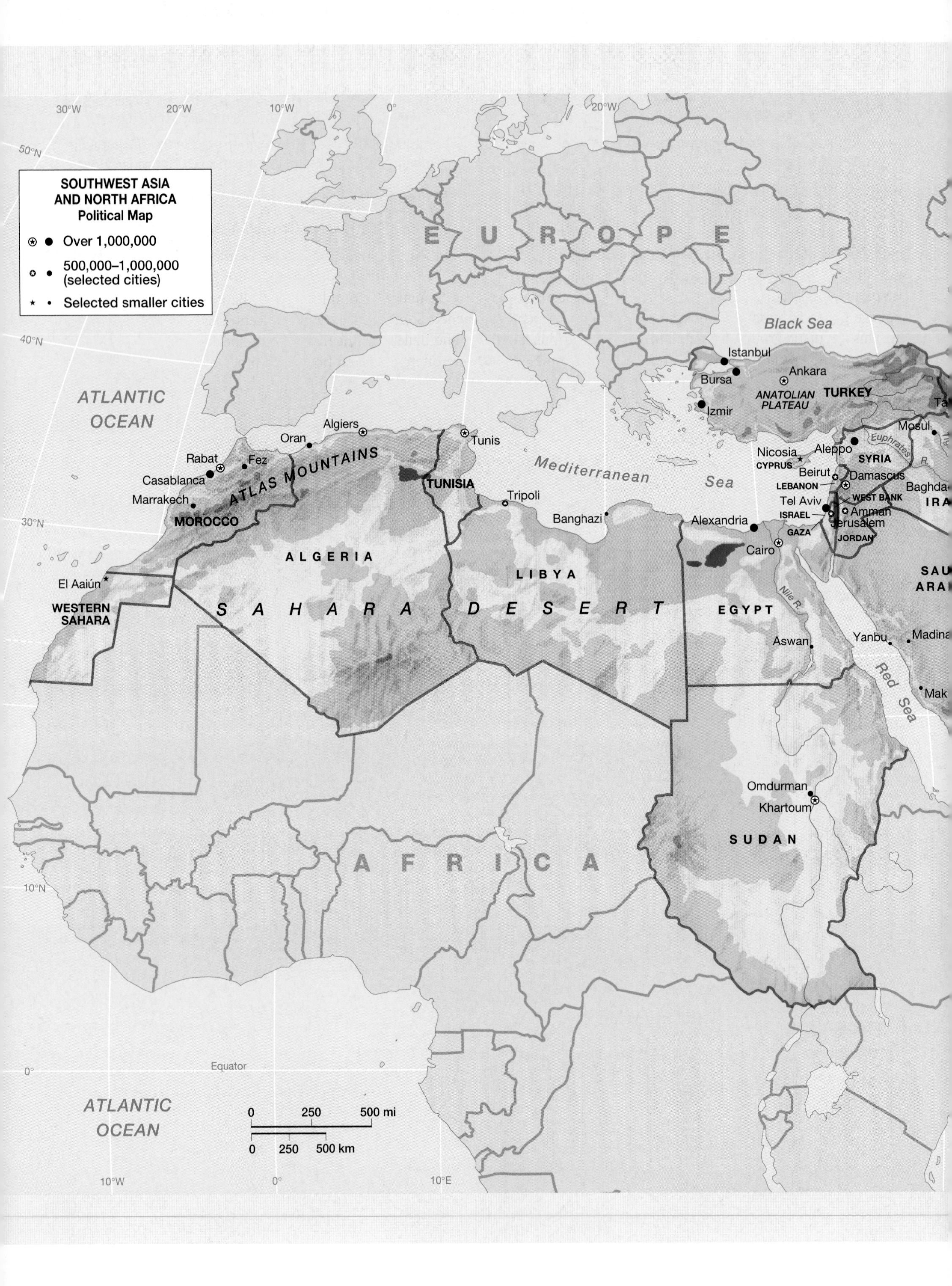

SOUTHWEST ASIA AND NORTH AFRICA
Political Map
Over 1,000,000
500,000–1,000,000 (selected cities)
Selected smaller cities
E U R O P E
Black Sea
ATLANTIC OCEAN
Mediterranean Sea
Istanbul
Bursa
Ankara
Izmir
ANATOLIAN PLATEAU
TURKEY
Mosul
Euphrates R.
Aleppo
SYRIA
Nicosia
CYPRUS
Beirut
LEBANON
Damascus
Baghdad
Tel Aviv
WEST BANK
ISRAEL
Amman
Jerusalem
GAZA
JORDAN
Algiers
Oran
Tunis
Rabat
Fez
Casablanca
ATLAS MOUNTAINS
TUNISIA
Marrakech
MOROCCO
Tripoli
Banghazi
Alexandria
Cairo
ALGERIA
LIBYA
El Aaiún
WESTERN SAHARA
S A H A R A D E S E R T
EGYPT
Nile R.
Aswan
Yanbu
Madina
Red Sea
Omdurman
Khartoum
SUDAN
A F R I C A
Equator
ATLANTIC OCEAN
0 250 500 mi
0 250 500 km
30°W
20°W
10°W
0°
20°W
50°N
40°N
30°N
10°N
0°
10°W
0°
10°E

Southwest Asia and North Africa

Climate, culture, and oil all help define the complex Southwest Asia and North Africa world region (see "Setting the Boundaries"). Straddling the historic meeting ground between Europe, Asia, and Africa, the region sprawls across thousands of miles of parched deserts, rugged plateaus, and oasis-like river valleys. It extends 4,000 miles (6,400 kilometers) between Morocco's Atlantic coastline and Iran's eastern boundary with Pakistan. More than two dozen nations are included within its borders, with the largest populations found in Egypt (65.5 million people), Turkey (64.8 million), and Iran (64.1 million) (Figure 7.1). Generally, its arid climates make it a part of the dry world, although the region's diverse physical geography causes precipitation to vary considerably. Culturally, diverse languages, religions, and ethnic identities have molded land and life within the region for centuries, strongly wedding people and place in ways that have had profound social and political implications. One traditional zone of conflict is the Middle East, where Jewish, Christian, and Islamic peoples have yet to resolve long-standing cultural tensions and political differences, particularly as they relate to the state of Israel and the pivotal Palestinian issue. In addition, Southwest Asia and North Africa's extraordinary petroleum resources place the area in the global economic spotlight. The strategic value of oil has combined with ongoing ethnic and religious conflicts to produce one of the world's least-stable political settings, one prone to geopolitical conflicts both within and between the countries of the region.

In the long sweep of history, no region of the world better exemplifies the theme of globalization than Southwest Asia and North Africa. A key global **culture hearth**, the region witnessed many cultural innovations that subsequently diffused widely to other portions of the world (Figure 7.2). As a seedbed of agriculture, several great civilizations, and three major world religions, the region has been a pivotal human crossroads for thousands of years. Great trade routes have connected North Africa with the Mediterranean and Sub-Saharan Africa. Southwest Asia has also had historic ties to Europe, the Indian subcontinent, and the vastness of Central Asia. The result has been that innovations within the region have often spread far beyond its bounds. Ponder the global importance of wheat and cattle, both domesticated in Southwest Asia. Consider, as well, the far-reaching impacts of urban-based civilizations that originated within its bounds and then formed models for city-building that spread to Europe and beyond. In addition, religions born

◀ **Figure 7.1 Southwest Asia and North Africa** This vast region extends from the shores of the Atlantic Ocean to the Caspian Sea. Within its boundaries, persisting cultural differences and strategically important petroleum reserves have contributed to recent political tensions. The region's economic geography reveals great differences in wealth that include some of the world's richest and poorest nations. In addition, the region's fragile arid environments and growing populations pose enduring twenty-first century challenges.

Setting the Boundaries

"Southwest Asia and North Africa" is both a cumbersome term and a complex region. Often the same area is simply called the "Middle East," but some experts would exclude the western parts of North Africa as well as Turkey and Iran from such a region. In addition, the Middle East carries with it a peculiarly European vantage point—Lebanon is in the "middle of the east" only from the perspective of the western Europeans who colonized the region and still shape the names we give the world today. Instead, "Southwest Asia and North Africa" offers a useful and straightforward way to describe the general limits of the region. Largely arid climates, an Islamic religious heritage, abundant fossil fuel reserves, and persistent political instability are certainly dominant in many areas, but internal variations on these themes make ready generalizations difficult.

There are also problems with simply defining the geographical limits of the region. On the northwest, a small piece of Turkey actually sits west of the Bosporus, generally considered to be the dividing line between Europe and Asia (Figure 7.1). Nearby Cyprus, located in the eastern Mediterranean, also resists easy regional definition since the population is divided between ethnic Greeks and Turks. We include it in Southwest Asia and North Africa because of its proximity to the Turkish and Syrian coasts. To the northeast, the largely Islamic peoples of Central Asia share many religious and linguistic ties with Turkey and Iran, but we have chosen to treat these groups in a separate Central Asia chapter (see Chapter 10).

African borders are also problematic. The conventional regional division of "North Africa" from "Sub-Saharan Africa" often cuts directly through the middle of modern Mauritania, Mali, Niger, Chad, and Sudan. We place all of these transitional countries in Chapter 6 on Sub-Saharan Africa, but include some discussion of Sudan within our Southwest Asia and North Africa material because of its status as an "Islamic Republic," a political and cultural designation that ties it strongly to the Muslim world.

▲ **Figure 7.2 Aleppo, Syria** For centuries, the northern Syrian city of Aleppo has served as a cultural and economic crossroads. Today, the city has more than 1.5 million residents and reveals traditional mosques and minarets next to newer apartments and office buildings. *(Jean-Leo Dugast/Panos Pictures)*

within the region (Judaism, Christianity, and Islam) have shaped many other parts of the world.

Particularly within the past century, globalization has also operated in the opposite direction: the region's strategic importance has made it increasingly vulnerable to outside influences. The twentieth-century development of the petroleum industry, largely initiated by outside U.S. and European investment, has had enormous, but selective, consequences for economic development. Global demand for oil and natural gas has powered rapid industrial change within the region, defining its pivotal role in world trade. Many key members of **OPEC (Organization of Petroleum Exporting Countries)** are found within the region, and these countries profoundly influence global prices and production targets for petroleum. Still, the regional patterns of petroleum resources are complex. Oil-rich nations such as Saudi Arabia, Kuwait, and Iran have been fundamentally transformed, while petroleum-poor neighbors such as Jordan, Sudan, and Morocco have seen less dramatic impacts. Politically, petroleum resources have also elevated the region's strategic importance, adding to many traditional cultural tensions existing within the realm. As a result, the area was a central theater of conflict in the Cold War between the United States and the Soviet Union, and international tensions remain focused there.

More recently, **Islamic fundamentalism** has advocated a return to more traditional practices within the religion and has challenged the encroachment of global popular culture as civil and religious authority have sometimes merged. The movement rejects many characteristics of modern, Western-style consumer culture. Indeed, in many ways, fundamentalism can be seen as a cultural reaction to the disruptive role that forces of globalization have played across the region. Recently, it also has provided a dynamic political element within the region that has had consequences far beyond its boundaries. Indeed, the early

▲ **Figure 7.3 Modern Istanbul** Turkey's largest city has been a major urban center for more than 2,000 years. The city was a center for Christianity before Islam arrived in the fifteenth century. Modern Western influences now add cultural diversity to this scene on Independence Avenue. *(Rob Crandall/Rob Crandall Photographer)*

twenty-first century will no doubt see the continuation of these cultural and political tensions as traditional values are juxtaposed with the omnipresence of the modern world (Figure 7.3).

The realities of the regional setting provide additional challenges for the populations of Southwest Asia and North Africa. The availability of water in this largely dry portion of the world has shaped both the physical and human geographies of the region. More fundamental to regional residents than their petroleum riches, water remains a daily and omnipresent need. Biologically, the region's plants and animals must adapt to the pervasive aridity of long dry seasons and short, often unpredictable, rainy periods. Similarly, the geography of human settlement is wedded to water. Whether it comes from precipitation, underground aquifers, or exotic streams, water has shaped patterns of human settlement and placed severe limits on agricultural development across vast portions of the region. Just as important, however, have been the human transformations that have dramatically reconfigured the region's environmental setting. With remarkable ingenuity, residents have captured water in varied ways that include underground drainage systems, elaborate deep water wells, life-giving canal networks, and huge dam projects that expand irrigable acreage. However, the region's rapidly growing population, already approaching 400 million, will stress these available resources even further in the future and act as an additional catalyst for potential economic and political instability.

Environmental Geography: Life in an Arid World

In the popular imagination, much of Southwest Asia and North Africa is a land of shifting sand dunes, searing heat, and scattered oases. Although examples of those stereotypes certainly can be found across the region, the actual physical setting, both in terms of landforms and climate, is considerably more complex (Figure 7.4). In reality, the regional terrain varies greatly, with rocky plateaus and mountain ranges more common than sandy deserts. Even the climate, although dominated by aridity, varies remarkably from the dry heart of North Africa's Sahara Desert to the well-watered highlands of northern Morocco, coastal Turkey, or western Iran. One theme is pervasive, however: a lengthy legacy of human settlement has left its mark upon the region's fragile environmental settings and the entire region will be faced with increasingly daunting ecological problems in the decades ahead.

Regional Landforms

Across North Africa, the physical setting is marked by a surprising diversity of landform features. In the far northwest, the **Maghreb** region (meaning "western island") includes the nations of Morocco, Algeria, and Tunisia and is dominated near the Mediterranean coastline by the Atlas Mountains. Indeed, the rugged flanks of the Atlas rise like a series of islands above the narrow coastal plains to the north and the vast stretches of the lower Saharan deserts to the south (Figure 7.5). The Atlas Mountains are related to Europe's nearby alpine system and are a complex series of folded highlands separated by interior plateaus. They reach heights of more than 13,000 feet (3,965 meters) in central Morocco and sweep eastward to dominate the physical settings of northern Algeria and Tunisia. South and east of the Atlas Mountains, much of the remainder of interior North Africa varies between rocky plateaus and extensive lowlands. Even here, however, the visual scene includes a diversity of desert topography: the bare and steeply sloping mountains of southern Algeria contrast markedly with the sandier reaches of the Libyan Desert or the river-carved lowlands of the Nile as it flows northward through Sudan and Egypt.

Southwest Asia is generally more mountainous than North Africa. In the **Levant**, or eastern Mediterranean region, mountains rise within 20 miles (32 kilometers) of the sea and the highlands of Lebanon reach heights of more than 10,000 feet (3,048 meters). Farther south, the Arabian Peninsula forms a massive tilted plateau, with western highlands higher than 5,000 feet (1,524 meters) gradually sloping eastward to extensive lowlands in the Persian Gulf area. The Persian Gulf separates the Arabian Peninsula from nearby Iran, reaching its narrowest point at the strategically pivotal Straits of Hormuz through which pass vast amounts of the region's petroleum exports. Both the southwest (Yemen) and southeast (Oman) corners of the peninsula are marked by particularly rugged highlands whose steep slopes and narrow canyons depart strikingly from the drifting dune fields that dominate other portions of the Arabian interior.

North and east of the Arabian Peninsula lie the two great upland areas of Southwest Asia: the Iranian and Anatolian plateaus (*Anatolia* refers to the large peninsula of Turkey, sometimes called Asia Minor) (Figures 7.4 and 7.6). Both of these plateaus, averaging between 3,000 and 5,000 feet (915 to 1,524 meters) in elevation, are ringed by highlands, several of which form formidable mountain barriers. Indeed,

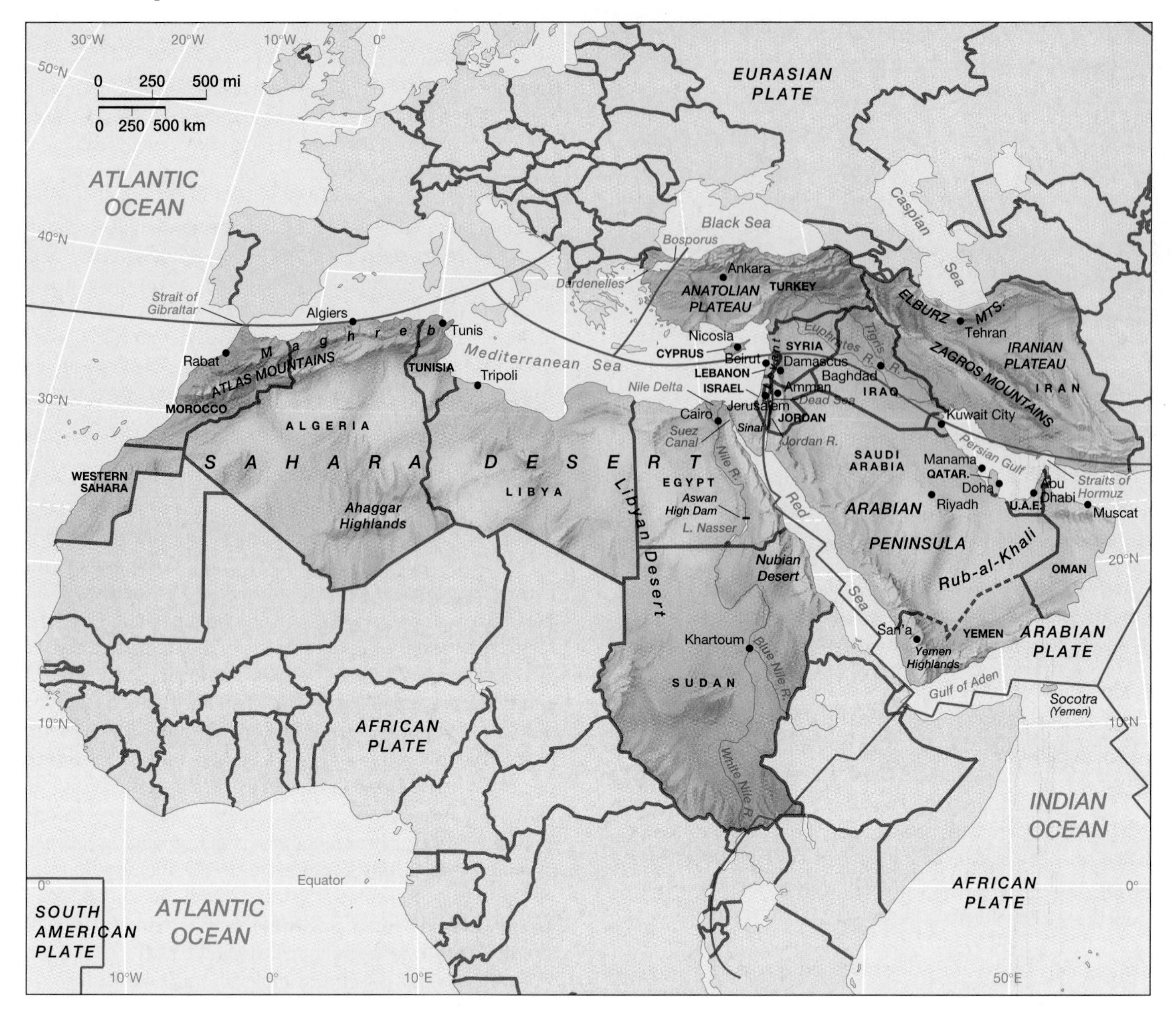

▲ **Figure 7.4 Physical geography of Southwest Asia and North Africa** Vast deserts stretch across much of the region and dominate the physical setting. Major mountains and highland zones complicate the scene by altering local climate and vegetation patterns and by offering distinctive environments for human settlement. In addition, converging tectonic plates from North Africa to Iran provide opportunities for active mountain building, and earthquake hazards are high across much of the region.

the Elburz Mountains of northern Iran reach more than 18,000 feet (5,485 meters) in height, higher than any mountains in the lower 48 United States. They offer a dramatic contrast to the lowlands next to the Caspian Sea immediately to the north. Portions of these plateaus also feature interior drainage, with streams flowing from bordering mountains into basins only to evaporate in extensive salt flats or saline lakes.

Smaller lowlands characterize other portions of Southwest Asia. Narrow coastal strips are common in the Levant, along both the southern (Mediterranean) and northern (Black Sea) Turkish coastlines, and north of the Iranian Elburz near the Caspian Sea. Iraq contains the most extensive alluvial lowlands in Southwest Asia, dominated by the Tigris and Euphrates rivers that flow southeast to empty into the Persian Gulf. Neighboring portions of Kuwait and southwestern Iran are also part of these river valley settings that once formed part of a larger sea before being filled in by river-borne sediments. Although much smaller in size, the distinctive Jordan River Valley is also a notable lowland region that straddles the strategic borderlands of Israel, Jordan, and Syria and drains southward to the Dead Sea.

The tectonic framework that created the varied landforms of Southwest Asia and North Africa is extraordinarily complex. The Arabian plate of Earth's crust has separated from the African plate, and both plates are colliding with the Eurasian plate along the Mediterranean Sea and across the mountain rim that separates the Arabian Peninsula from the Anatolian and Iranian plateaus. Some geologists identify separate small-

▲ **Figure 7.5 Atlas Mountains** Residents living within Morocco's Atlas Mountains adapt to the region's steep slopes, and their Mediterranean-style agriculture takes advantage of higher rates of annual precipitation than are found in the arid deserts below. *(Sean Sprague/Panos Pictures)*

er plates in central Turkey and Greece. The complex interactions between these various plates generate many earthquakes. Not surprisingly, Turkey and Iran are two of the most seismically active countries in the world.

Patterns of Climate

Although often termed the "Dry World," a closer look at Southwest Asia and North Africa reveals a more complex climatic pattern (Figure 7.7). Two major complicating variables come into play. First, the latitudinal breadth of the region, stretching from southern Sudan to northern Turkey, incorporates several global climate belts. Much of the region lies within a desert zone dominated by a belt of subtropical high pressure. To the north, however, North Africa's Maghreb coast, the eastern Mediterranean Levant, the interior of Turkey, and western Iran are far enough poleward to pick up seasonal winter moisture from midlatitude storms. Conversely, southern Sudan extends into the tropical wet and dry zone of Sub-Saharan Africa. In this climatic setting, the intertropical convergence zone of unsettled weather migrates northward

▲ **Figure 7.6 Anatolian Plateau** Turkey's vast interior is a mix of high mountains and elevated plateaus. The Anatolian Plateau provides a complex environment in which traditional agriculture includes midlatitude crops and livestock adapted to the region's semiarid setting. *(Adam Woolfitt/Woodfin Camp & Associates)*

during the summer and produces considerable, though unpredictable moisture. Another key climatic influence is topography: from the Atlas Mountains to the Iranian Plateau, the region's highland zones gather extra moisture from passing storms, producing much higher rates of precipitation than those found in nearby lowlands.

The Desert Lands A nearly continuous belt of arid lands stretches eastward from the Atlantic coast of southern Morocco across the continent of Africa, through the Arabian Peninsula, and into central and eastern Iran. Intervening mountains and waterways divide this vast zone into several distinct deserts. Throughout this vast dry zone, plant and animal life have adapted to the extreme conditions. Deep or extensive root systems and rapid life cycles allow desert plants to benefit from the limited moisture they receive. Similarly, animals adjust by efficiently storing water, hunting nocturnally, or migrating seasonally to avoid the worst of the dry cycle. Camels are justly famous as beasts of burden, going a week or more between water holes.

Some of the driest conditions are found across North Africa. Away from the Atlas Mountains, the Sahara Desert dominates much of the region. Much of the central Sahara receives less than 1 inch (2.5 centimeters) of rain a year, and can thus support only the most meager forms of vegetation. Large expanses of sand dune (erg) and rocky waste (reg) have virtually no plant life at all. Saharan summers feature extremely hot days and warm evenings (the world's record high, 136 °F [58 °C] was recorded in Libya), but winters are generally pleasant, and even cool, at night. Conditions are similar in northeast Africa, where the Sahara merges with the Libyan and Nubian deserts of Egypt and northern Sudan.

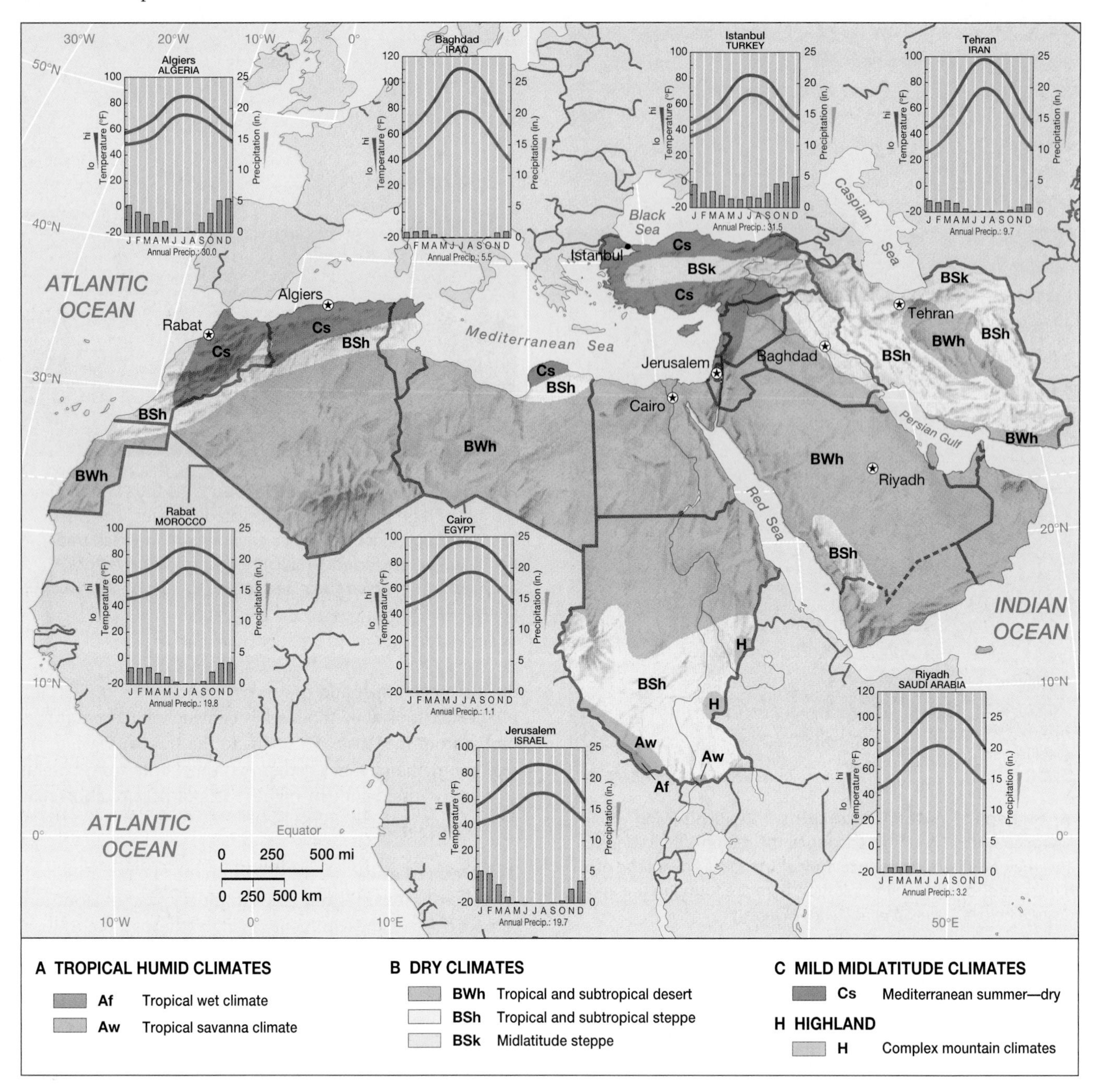

▲ **Figure 7.7 Climate map of Southwest Asia and North Africa** Dry climates dominate from western Morocco to eastern Iran. Within these zones, persistent subtropical high-pressure systems offer only limited opportunities for precipitation. Elsewhere, Mild midlatitude climates with wet winters are found near the Mediterranean Basin and Black Sea. To the south, tropical savanna climates provide summer moisture to southern Sudan.

Cairo, Egypt, averages only about 1 inch (2.5 centimeters) of rain annually. Only in southern Sudan do precipitation levels rise significantly. Here, summer rains produce between 20 and 50 inches (51 to 127 centimeters) of precipitation and tropical savannas and woodlands replace the desert vegetation to the north.

Across the Red Sea, deserts also dominate Southwest Asia. Most of the Arabian Desert is not quite as dry as the Sahara proper, although the Rub-al-Khali along Saudi Arabia's southern border is one of the world's most desolate areas. On the fringe of the summer-monsoon belt, the Yemen Highlands along the southwestern edge of the Arabian Peninsula receive much more rain than the rest of the region and thus form a more favorable site for human habitation. Highest summertime temperatures are encountered in the center of the peninsula, with Riyadh, Saudi Arabia, averaging July daytime highs of 107 °F (42 °C). While coastal locations in the United Arab Emirates experience August high temperatures

▲ **Figure 7.8 Arid Iran** Large portions of Iran are arid. Settlements are strongly oriented around zones of higher moisture or opportunities for irrigated agriculture. This view of remote desert lands to the north of the Zagros Mountains is near the city of Shiraz. *(Isabella Tree/Hutchison Picture Library)*

▲ **Figure 7.9 Algerian orange harvest** The Mediterranean moisture in northern Algeria produces an agricultural landscape similar to that of southern Spain or Italy. Winter rains create a scene that contrasts sharply with deserts found elsewhere in the region. *(Dott. Giorgio Gualco/Bruce Coleman Inc.)*

of only 103 °F (40 °C), the afternoon relative humidity often exceeds 60 percent. To the northwest, precipitation increases slightly in the central Tigris and Euphrates River valleys with Baghdad, Iraq, averaging just over 5 inches (13 centimeters) of rain annually. Another major arid zone lies across Iran, a country divided into a series of minor mountain ranges and desert basins (Figure 7.8). Hot summers are the rule, but winters are distinctly cool if not cold. For example, the Iranian capital of Tehran, positioned on the southern slopes of the Elburz Mountains, experiences wide-ranging temperatures that vary from average January maximums of only 45 °F (7 °C) to blistering July readings much higher than 100 °F (38 °C). Overall, the Iranian Plateau is not as arid as the Sahara or Arabian deserts, but it contains extensive salt flats almost devoid of vegetation.

Lands of Seasonal Rainfall Although deserts dominate regional climates, several wetter zones suggest the importance of latitude and altitude in altering the arid regime. For example, the Atlas Mountains and the nearby lowlands of northern Morocco, Algeria, and Tunisia experience a distinctly Mediterranean climate in which hot, dry summers alternate with cooler, relatively wet winters. In these areas the landscape resembles that found in nearby southern Spain or Italy (Figure 7.9). Both the Moroccan capital of Rabat and the Algerian capital of Algiers reveal the classic climatic profile of the Mediterranean world. Annual precipitation averages between 20 and 30 inches (51 to 76 centimeters), most moisture comes in winter, and summers are warm, dry, and dusty.

A second zone of Mediterranean climate extends along the Levant coastline into the nearby mountains and northward across sizable portions of northern Syria, Turkey, and northwestern Iran. The mountains of Lebanon, for example, receive heavy winter snows and even coastal Beirut gets more annual rainfall than either London or Seattle. Farther south, the Israeli capital of Jerusalem is somewhat drier, but still averages more than 20 inches (51 centimeters) of precipitation. More arid conditions are found in the rain shadow east of the mountains where Damascus, Syria, receives only 8.6 inches (22 centimeters) of precipitation annually. Elsewhere, western Turkey also benefits from seasonal Mediterranean moisture, with Istanbul's climate resembling that of the Po River valley in northeastern Italy. Eastward, precipitation decreases gradually across the Anatolian Plateau, and Mediterranean woodlands are replaced by semiarid grassland (or steppe) vegetation. Within northern and western Iran, however, the Elburz and Zagros mountains squeeze more moisture from passing air masses and many highland settlements receive more than 40 inches (102 centimeters) of moisture annually.

Legacies of a Vulnerable Landscape

The environmental history of Southwest Asia and North Africa illustrates both the shortsighted and ingenious legacies of its human occupants. Littered with examples of despoiled vegetation, blighted soils, and depleted water resources, the region reveals the hazards of lengthy human settlement in a marginal land (Figure 7.10). But the long record of its residents is also testimony to their remarkable ability to adapt to

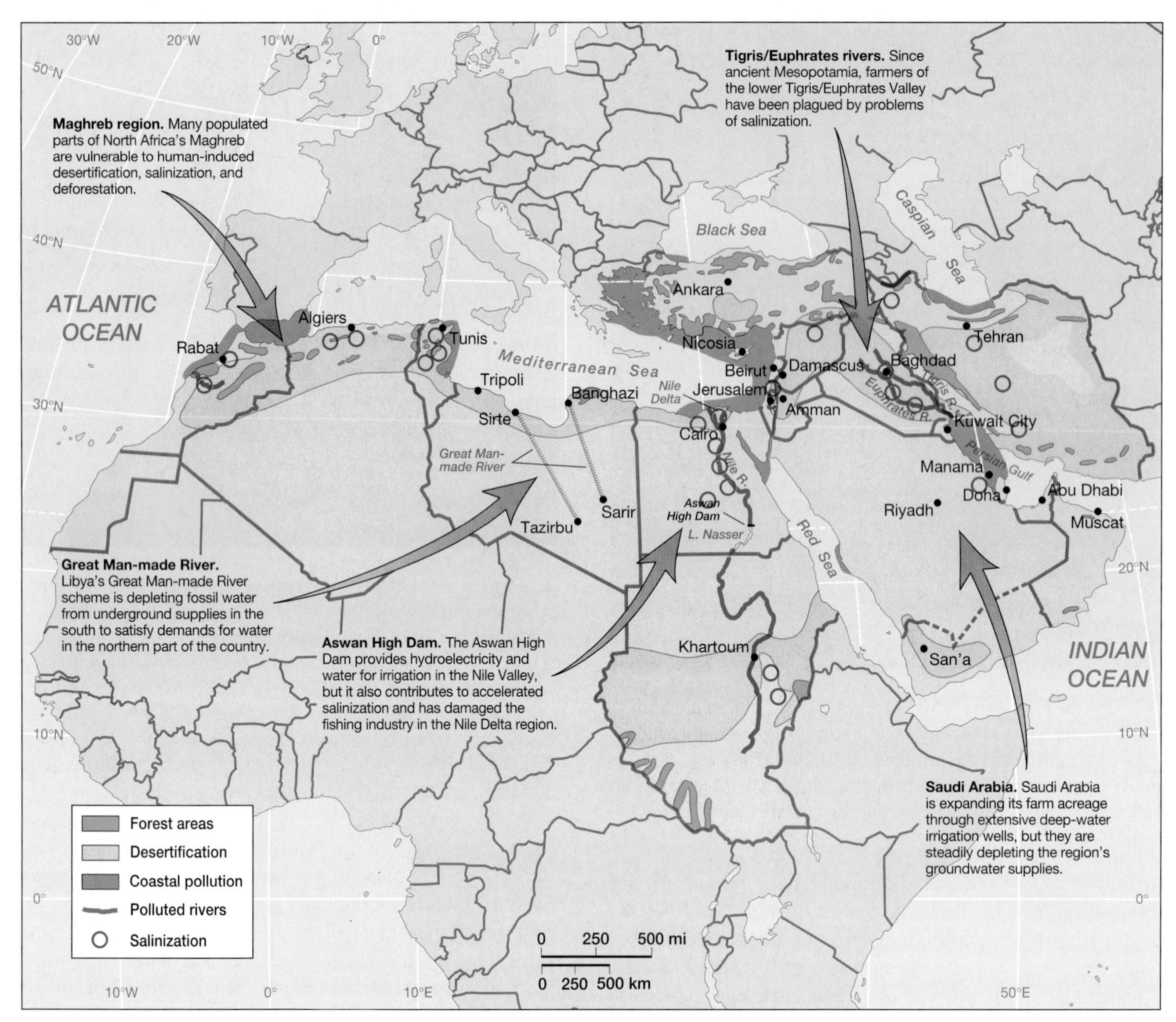

▲ **Figure 7.10 Environmental issues in Southwest Asia and North Africa** Growing populations, pressures for economic development, and pervasive aridity combine to create environmental hazards across the region. Long human occupance has contributed to deforestation, irrigation-induced salinization, and expanding desertification. Saudi Arabia's deep-water wells, Egypt's Aswan High Dam, and Libya's Great Man-made River are all recent technological attempts to expand settlement, but they may carry a high long-term environmental price tag.

and transform their regional environment and in the process to produce a truly humanized landscape that differs considerably from the natural setting. In the past 150 years, quickening technological change and rapidly expanding populations have further imperiled the regional environment. Perhaps most disturbingly, recent political conflicts have even resulted in the use of the environment as a weapon of war, an ominous sign both for regional residents and the lands they call home (see "Environmental History: War and the Middle East Environment").

Deforestation and Overgrazing Deforestation is an ancient problem in Southwest Asia and North Africa. Although much of the region is too dry for trees, the more humid and elevated lands that ring the Mediterranean once supported heavy forests. Included in these woodlands are the cedars of Lebanon, despoiled in ancient times and now reduced to a few scattered groves that survive in a largely denuded landscape (Figure 7.11). Even 4,000 years ago, monarchs in both Egypt (along the Nile River) and Mesopotamia (the area between the Tigris and Euphrates rivers) were forced to launch major expeditions to obtain lumber needed for city building and ship construction. Elsewhere, growing demands for agricultural land pressured upland forests, which were removed and replaced with grain fields, orchards, and pastures.

Human activities have conspired with natural conditions to reduce most of the region's forests to grass and scrub.

▲ **Figure 7.11 Mediterranean deforestation** Many hill slopes around the Mediterranean Basin are plagued with human-induced deforestation. The forests that once clothed lands above the Lebanese town of Kousba were removed centuries ago. *(S. Wilhelm/Bruce Coleman Inc.)*

Mediterranean forests often grow slowly, are highly vulnerable to fire, and usually fare poorly if subjected to heavy grazing. Yet intensive pressures have been applied to these lands for thousands of years. Browsing by goats in particular has often been blamed for much of the region's forest loss, but other livestock have also had an impact on the vegetative cover, particularly in steeply sloping semiarid settings that are slow to recover from grazing. Whatever the causes of forest destruction, the scarcity of tree cover in Southwest Asia and North Africa has resulted in significant costs to its human societies. Lumber, pulp, and other forest products generally have to be imported, and are thus quite expensive. Deforestation has also resulted in a millennia-long deterioration of the region's water supplies and soil resources.

Southwest Asia and North Africa are not, however, entirely without tree cover. The mountains of northern and southern Turkey and those of northern Iran retain a considerable tree cover, and scattered forests can be found in western Iran, the Levant, and the Atlas Mountains. Moreover, several governments have launched reforestation drives in recent years. For example, both Israel and Syria have expanded their coverage of wooded lands since the 1980s. Growing populations and continued stress on the region's limited woodland and grazing resources do not bode well, however, for the long term.

Salinization Salinization, or the buildup of toxic salts in the soil, is another ancient environmental issue in a region where irrigation has been practiced for centuries (Figure 7.10). The accumulation of salt in the topsoil is a common problem wherever desert lands are subjected to extensive irrigation. All fresh water contains a small amount of dissolved salt, and when water is diverted from streams into fields, salt remains in the soil after the moisture has been absorbed by the plants and evaporated by the sun. In humid climates, accumulated salts are washed away by saturating rains, but in arid climates this rarely occurs. Where irrigation is practiced, salt concentrations build up over time, leading to lower crop yields and eventually to land abandonment.

Hundreds of thousand of acres of once-fertile farmland within the region have been destroyed or degraded by salinization. The problem has been particularly acute in Iraq, where centuries of canal irrigation along the Tigris and Euphrates rivers has seriously degraded land quality in the valley. It has been experienced elsewhere in the region as well, including central Iran, Egypt, and irrigated portions of the Maghreb. While salinization dates back to ancient times, recent attempts to intensify agriculture, made necessary by the region's rapidly growing population, have only worsened the problem. Agronomists are attempting to breed salt-resistant crops and to devise inexpensive methods of soil desalinization, but their success so far has been minimal.

Managing Water Residents of the region are continually challenged by many other problems related to managing water in one of the driest portions of Earth. For thousands of years, the pervasive attitude has been that any available technology should be used to bring water to where it is needed. As technological change has accelerated, the scale and impact of water management schemes has had a growing effect on the region's environment. Sometimes the costs are justified if large new areas are brought into productive and sustainable agricultural use. In other cases, however, the long-term environmental price will far outweigh any immediate and perhaps short-lived gains in agricultural production.

Regional populations have been modifying drainage systems and water flows for thousands of years. The Iranian **qanat system** of tapping into groundwater through a series of gently sloping tunnels was widely replicated on the Arabian Peninsula and in North Africa. With simple technology, farmers directed the underground flow to fields and villages where it could be efficiently utilized. Other ingenious adaptations such as waterwheels, carefully managed flood irrigation, and increasingly elaborate canal systems also redefined the local distribution of water across the region.

In the past half century, however, the scope of environmental change has been greatly magnified. Water management schemes have become huge engineering projects, major budget expenditures, and even political issues, both within and between countries of the region. One remarkable example is Egypt's Aswan High Dam, completed in 1970 on the Nile River south of Cairo. Many benefits came with the dam's completion. Most importantly, greatly increased storage capacity in the upstream reservoir promoted more year-round cropping and an expansion of cultivated lands along the Nile, critical changes in a country that continues to see rapid population growth. The dam also generates large amounts of clean

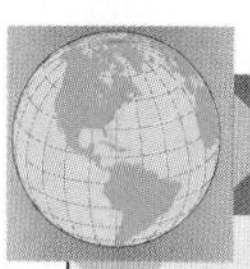

ENVIRONMENTAL HISTORY War and the Middle East Environment

One of the most disturbing dimensions of the Persian Gulf conflict in 1990 and 1991 was the decision by Iraq to use environmental devastation as a tool of warfare. Certainly there is a long legacy of "environmental warfare," including the use of fire to flush out an opponent or the more modern use of defoliants to reveal enemy movements through dense vegetation. Still, Iraq's aggressive use of such strategies in the recent conflict does not bode well for the future environmental health of a region filled with fragile ecological niches, including densely settled river valleys and vulnerable desert settings.

Most dramatically, the Iraqis used the very oil they coveted as an environmental weapon, particularly once they saw they could not win the military conflict against the UN-supported forces (Figure 1). They set fire to more than 700 wells in Kuwait, and many burned for months after the Iraqi army withdrew. In addition, millions of gallons of petroleum were released into the desert, forming vast pools of damaging oil. Iraqis also targeted the Gulf, and inflows of crude oil decimated many shoreline settings. Decades will pass before the region recovers from these few weeks of environmental mayhem.

Following the war, the weakened Hussein regime encountered additional civil resistance at home. While Kurds lobbied for independence in the north, the Marsh Arabs, a largely Shiite minority living in the far south, also resisted Iraqi rule. Their traditional homelands lie in marshes near the mouths of the Tigris and Euphrates rivers. The area provides a living through its farmlands and fisheries, as well as a denser vegetative cover that offers protection from excessive Iraqi interference. In 1991, however, Iraqi engineers constructed a causeway that drained much of the marsh country, devastating the region's ecology and forcing many Marsh Arabs from their homes. While the Iraqis claim it was a "reclamation" scheme, the real purpose was clear. The result has been to cause great environmental damage and to force many Marsh Arabs from their protective settlements.

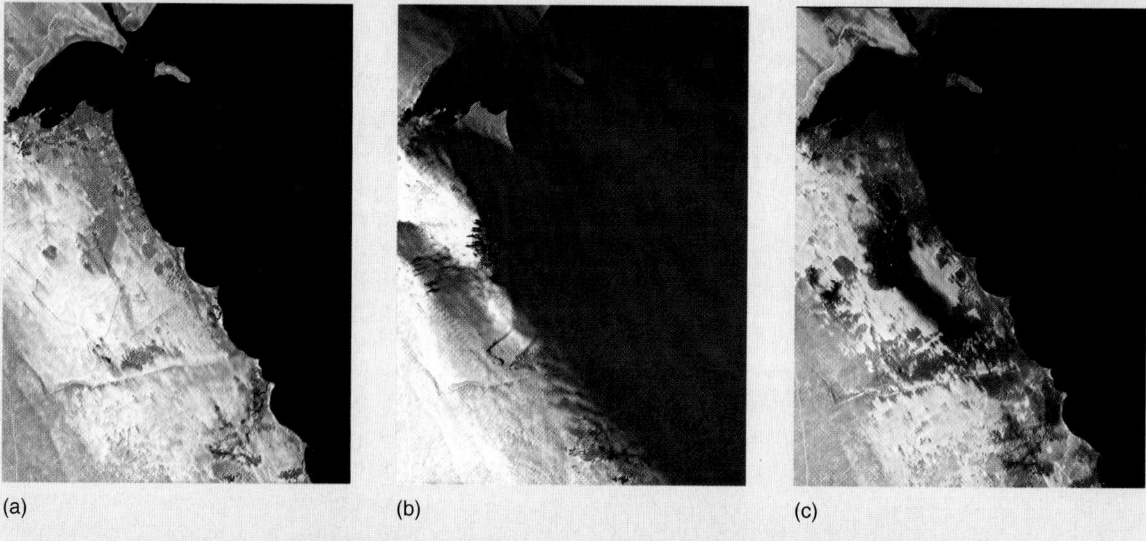

(a) (b) (c)

▲ **Figure 1 Burning oil wells in Kuwait** Landsat images tell the story of environmental warfare in Kuwait. The first image (a), taken August 1990, shows Kuwait City in the upper center of the image. Much of the city is obscured by smoke (b) as fires burned in February 1991. Petroleum and soot deposits remain as an oily legacy after the fires are extinguished in November 1991 (c). *(EROS Data Center, U.S. Geological Survey)*

electricity for the region. But the environmental costs have been high. The dam has changed methods of irrigation along the Nile, greatly increasing problems of salinization. While fresh sediments and soil nutrients once annually washed across the valley floor, more controlled irrigation has meant greatly increased inputs of costly fertilizers as well as the infilling of Lake Nasser behind the dam with accumulating sediments. Other problems with the dam include an increased incidence of *schistosomiasis* (a debilitating parasitic disease spread by waterborne snails in irrigation canals), and the collapse of the Mediterranean fishing industry near the Nile Delta, an area previously nourished by the river's silt.

Elsewhere, **fossil water,** or water supplies stored underground during earlier and wetter climatic periods, has also been put to use by modern technology. Libya's "Great Man-made River" scheme taps underground water in the southern part of the country, transports it 600 miles (965 kilometers) to northern coastal zones, and uses the precious resource to expand agricultural production in one of North Africa's driest countries. Similarly, Saudi Arabia has invested huge sums to develop deepwater wells, allowing it to greatly expand its food output (Figure 7.12). Unfortunately, these underground supplies are being depleted much more rapidly than they are recharged, thus limiting the long-term sustainability of such ventures.

▲ **Figure 7.12 Saudi Arabian irrigation** One of Saudi Arabia's largest deep-water irrigation sites is at the oasis of Al Hofuf east of Riyadh. While significantly expanding the country's food production, such efforts are rapidly depleting underground supplies of fossil water. *(Minosa/Scorpio/Sygma Photo News)*

Most dramatically, water politics have raised tensions between countries of the area that are destined to share drainage basins in a time of growing demands for the limited resource. For example, Sudan's plans to expand its irrigation networks along the upper Nile as well as Ethiopia's Blue Nile Dam project are both causes for concern in Egypt. To the north, Turkey's growing development of the upper Tigris and Euphrates rivers (the Southeast Anatolian Project) has raised tensions with Iraq and Syria, who argue that capturing "their" water might be considered a provocative political act. In addition, water politics has played into negotiations between Israel, the Palestinians, and other neighboring states, particularly in the valuable Jordan River drainage, which runs through the center of the area's most hotly disputed lands. Inevitably, the stresses of population growth, unpredictable political tensions, and costly technological solutions will continue to threaten the sustainability of water resources across the dry expanses of Southwest Asia and North Africa.

Population and Settlement: Patterns in an Arid Land

The geography of human population across Southwest Asia and North Africa demonstrates the intimate tie between water and life in this part of the world. The pattern is complex: large areas of the population map remain almost devoid of permanent settlement, while other, more moisture-favored lands suffer increasingly from problems of crowding and overpopulation (Figure 7.13). Almost everywhere across the region, its human occupants have uniquely adapted themselves to living within arid or semiarid settings, and these cultural and technological transformations have been enduringly stamped upon the visual scene. In addition, rapid increases in population, levels of urbanization, and contacts with the outside world have also left their mark and suggest that future patterns of population growth and economic development will continue reshaping the settlement landscape in dramatic and unanticipated ways.

The Geography of Population

Today, almost 400 million people live in Southwest Asia and North Africa (Table 7.1). The distribution of that population is strikingly varied: in countries such as Egypt, large zones of almost empty desert land stand in sharp contrast to crowded, well-watered locations, such as those along the Nile River. While overall population densities in such countries appear modest, **physiological densities**, a statistic that relates the number of people to the amount of arable land, are among the highest on Earth. Patterns of urban geography are also highly uneven: although less than two-thirds of the overall population is urban, many nations are overwhelmingly dominated by huge and sprawling cities that concentrate urban functions and produce the same problems of urban crowding found elsewhere in the developing world. Rates of recent urban growth have also been phenomenal: Cairo, a modest-sized city of 3.5 million people in the 1960s, has almost quadrupled in population in the past 40 years.

Across North Africa, two dominant clusters of population, both shaped by the availability of water, account for most of the region's 165 million people (Figure 7.13). In the Maghreb, the moister slopes of the Atlas Mountains and nearby better-watered coastal districts have accommodated denser populations for centuries. Today, concentrations of both rural and urban settlement extend from south of Casablanca in Morocco to Algiers and Tunis on the shores of the southern Mediterranean. Indeed, most of the populations of Morocco (27.7 million people), Algeria (30.2 million), and Tunisia (9.5 million) crowd into this crescent of more-favored country, a stark contrast to the almost empty lands south and east of the Atlas Mountains. Morocco's Casablanca and the Algerian capital of Algiers are the largest cities in the Maghreb. They have rapidly growing metropolitan populations of around 4 million residents each. Farther east, much of Libya (5.7 million) and western Egypt is very thinly settled. Nearby, however, Egypt's Nile River Valley is home to the other great North African population cluster. The vast majority of Egypt's 65 million people live within 10 miles of the river, and a similar ribbon of dense population follows the watery lifeline upriver into Sudan (28.5 million). Indeed, the region features some of the highest population densities in the world, as millions of farmers crowd along the Nile's narrow lifeline. The corridor is also home to North Africa's largest city: Cairo, Egypt (12 million), dominates that country's urban geography and sprawls for many miles along and east of the Nile.

Southwest Asian populations also reflect the realities of living in a dry world: the region's 225 million residents are clustered in favored coastal zones, moister highland settings, or in desert localities where water is available from nearby rivers or subsurface aquifers. High population densities are found in better-watered portions of the Levant, which includes clusters in Israel (6 million), Lebanon (4.1 million), and Syria (15.6 million). In Turkey (64.8 million), large rural and urban settlements concentrate along its Mediterranean and Black Sea coastlines as well as in the western portion of the Anatolian Plateau. Nearby Iran also is home to more than 60 million residents, but population densities vary considerably from thinly

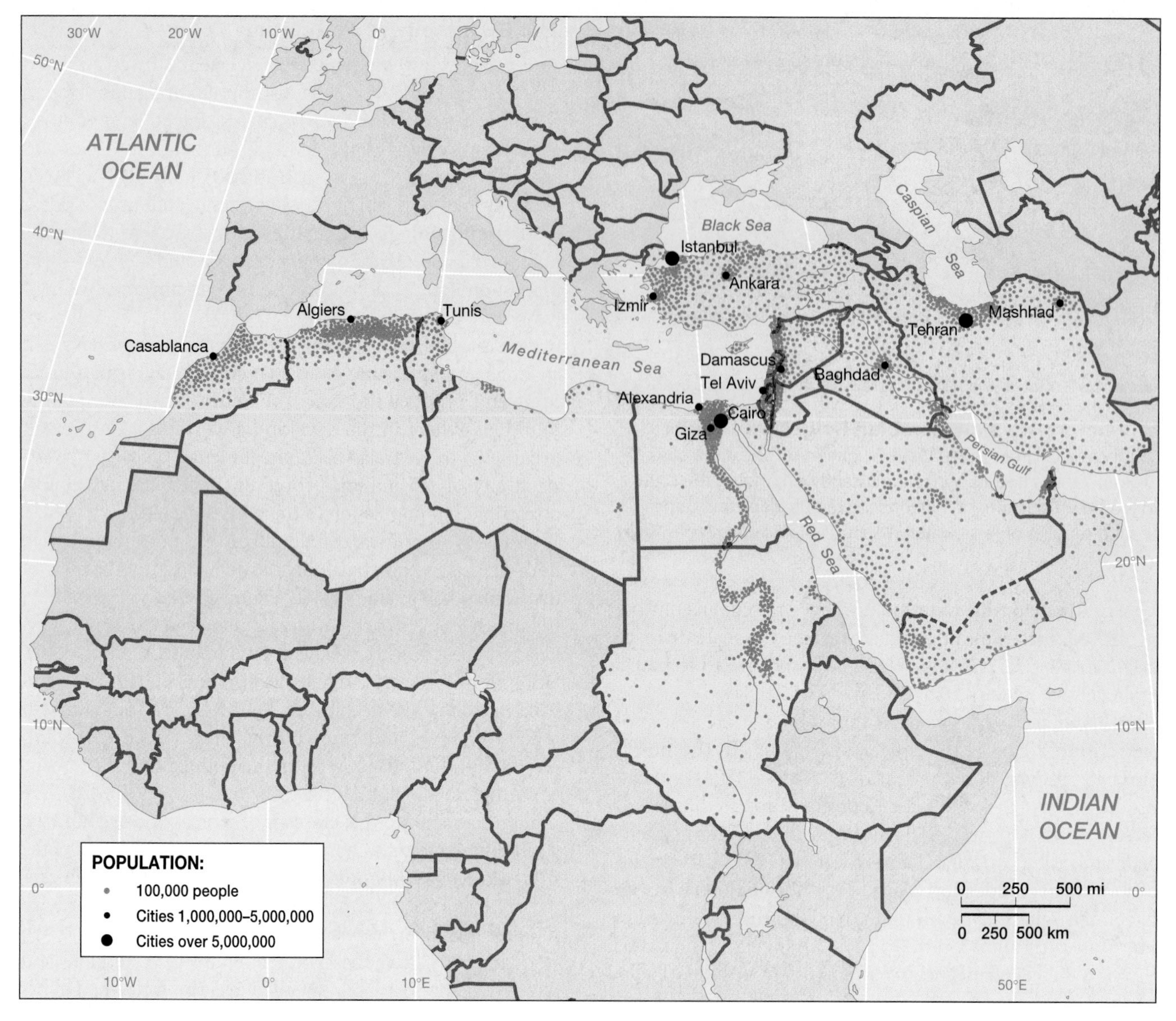

▲ **Figure 7.13 Population map of Southwest Asia and North Africa** The striking contrasts between large sparsely occupied desert zones and much more densely settled regions where water is available are clearly evident. The Nile Valley and the Maghreb region contain most of North Africa's settlement, while Southwest Asian populations cluster in the highlands and along the better-watered shores of the Mediterranean.

occupied deserts in the east to more concentrated settlements near the Caspian Sea and across the more humid highlands of the northwest. Turkey's Istanbul (7.5 million) (formerly Constantinople) and Iran's Tehran (10 million) are Southwest Asia's largest urban areas and both have grown in recent years as rural populations in those nations gravitate toward the perceived economic opportunities of these cities. Elsewhere, sizable populations are scattered through the Tigris and Euphrates Valley, the Yemen Highlands, and near oases where groundwater can be tapped to support agricultural or industrial activities.

Water and Life: Rural Settlement Patterns

Water and life are intimately linked across the rural settlement landscapes of Southwest Asia and North Africa. Indeed, the diverse environments of Southwest Asia, in particular, are home to one of the world's earliest hearths of **domestication**, where plants and animals were purposefully selected and bred for their desirable characteristics. Beginning around 10,000 years ago, increased experimentation with wild varieties of wheat and barley initiated a long and complex series of cultural innovations that subsequently included domesticated animals such as cattle, sheep, and goats. Much of the early agricultural activity focused on the **Fertile Crescent**, an ecologically diverse zone that stretches from the Levant inland through the fertile hill country of northern Syria into Iraq. Between 5,000 and 6,000 years ago, better knowledge of irrigation techniques and increasingly centralized political states promoted the diffusion of agriculture into nearby valleys such as the Tigris and Euphrates (Mesopotamia) and North Africa's Nile Valley. Since then, different peoples of the region have

Table 7.1 Demographic Indicators

Country	Population[a]	Natural Increase	TFR[b]	%<15[c]	% Urban
Algeria	30.2	2.4	4.4	39	56
Bahrain	0.6	2.0	3.2	31	88
Cyprus	0.7	0.7	2.1	25	68
Egypt	65.5	2.2	3.6	39	43
Gaza	1.1	4.6	7.4	50	—
Iran	64.1	1.8	3.0	40	61
Iraq	21.8	2.8	5.7	43	70
Israel	6.0	1.5	2.9	30	90
Jordan	4.6	2.5	4.4	41	78
Kuwait	1.9	2.3	3.2	29	100
Lebanon	4.1	1.6	2.3	34	87
Libya	5.7	3.7	6.3	50	86
Morocco	27.7	1.8	3.3	36	52
Oman	2.5	3.9	7.1	47	72
Qatar	0.5	1.7	4.1	27	91
Saudi Arabia	20.2	3.1	6.4	42	80
Sudan	28.5	2.1	5.0	43	27
Syria	15.6	2.8	4.6	45	51
Tunisia	9.5	1.9	3.2	35	61
Turkey	64.8	1.6	2.6	31	64
United Arab Emirates	2.7	2.2	4.9	30	82
West Bank	1.8	3.4	5.4	45	—
Western Sahara	0.2	2.9	6.9	—	—
Yemen	15.8	3.3	7.3	47	25

[a]Population in millions, 1998.
[b]Total fertility rate.
[c]Percentage of population younger than 15 years of age.
Source: *Population Reference Bureau World Population Data Sheet,* 1998.

adapted to its environmental diversity and limitations in distinctive ways and, in the process, they have practiced forms of agriculture appropriate to their settings and they have created their own unique imprints upon the landscape (Figure 7.14). Throughout the region, the signatures of rural settlement reflect the lasting tension between water and life, an uneasy marriage between a limited natural resource and a dynamic, unpredictable, and growing human population.

Pastoral Nomadism **Pastoral nomadism,** most common in the drier portions of the region, is a traditional form of subsistence agriculture in which practitioners depend on the seasonal movement of livestock for a large part of their livelihood. An offshoot of sedentary agriculture, pastoral nomadism is an appropriate adaptive strategy to life in arid settings where inadequate moisture and forage makes permanent settlement impossible. Arabian Bedouins, North African Berbers, and Iranian Bakhtiaris exemplify surviving examples of nomadism within the realm. Today, however, with fewer than 10 million nomads remaining, the lifestyle is in decline, victim of more constricting political borders, a declining demand for traditional beasts of burden, competing land uses, and selective overgrazing. In addition, government resettlement programs in Saudi Arabia, Syria, Egypt, and elsewhere are actively promoting a more settled lifestyle for many nomadic groups.

The settlement landscape of pastoral nomads reflects their need for mobility and flexibility as they seasonally move camels, sheep, and goats from place to place. Typically, nomads use lightweight and easily transportable tents, and tribes often split up into smaller kin-related clans to move about the landscape easily and take advantage of even limited water and pasture resources. Migratory patterns are hardly random, however: tribes have well-established territories and time-tested seasonal movement strategies that maximize the efficient use of available resources. Near highland zones such as the Atlas Mountains or the Anatolian Plateau, nomads practice **transhumance** by seasonally moving their livestock to cooler, greener high country pastures in the summer and then returning them to valley and lowland settings for fall and winter grazing. Some central Saharan groups maintain home bases in isolated mountain ranges such as

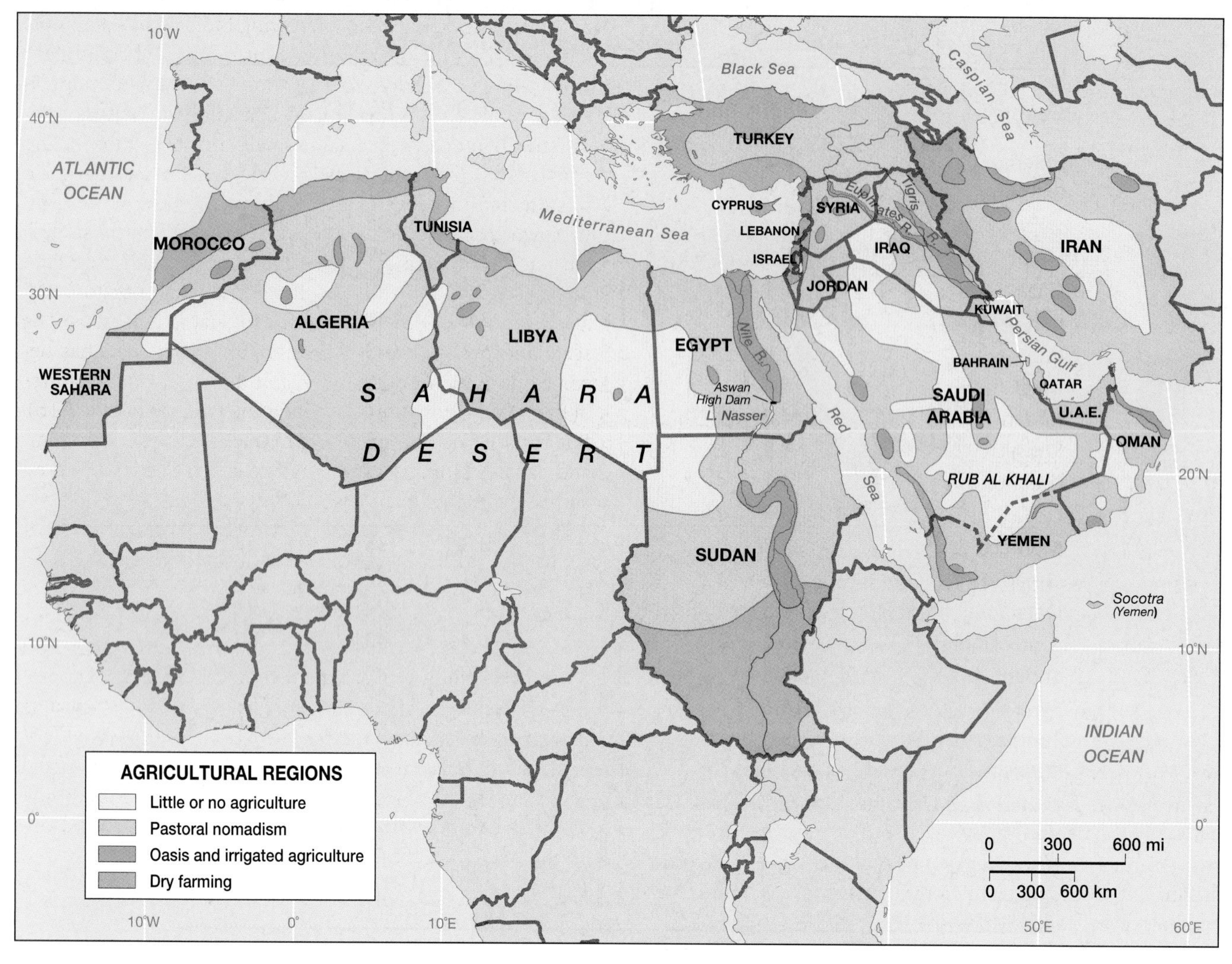

▲ **Figure 7.14 Agricultural regions of Southwest Asia and North Africa** Important agricultural zones include oases and irrigated farming where water is available. Elsewhere, dry farming is practiced in midlatitude settings where it is supplemented with irrigation. *(Modified from Clawson and Fisher, 1998,* World Regional Geography, *Upper Saddle River, NJ: Prentice Hall; and Bergman and Renwick, 1999,* Introduction to Geography, *Upper Saddle River, NJ: Prentice Hall)*

southern Algeria's Ahaggar Highlands, where the summer heat and aridity are less extreme. Elsewhere, broad horizontal movements often incorporate huge territories of desert to support small groups of a few dozen families. Where moisture conditions allow, some nomads divide further and often put women and children to work on small agricultural plots that supplement the tribe's dependence on livestock. In addition, nomads trade with sedentary agricultural populations, a symbiotic relationship in which they exchange meat, milk, hides, and wool for cereal and orchard crops available at desert oases.

Oasis Life Permanent oasis settlements dot the arid landscape where high groundwater levels or modern deep-water wells provide reliable moisture in otherwise arid locales (Figures 7.12 and 7.15). Tightly clustered, often walled villages, their sun-baked mud houses blending into the surrounding scene, sit adjacent to small, but intensely utilized fields where underground water is carefully applied to tree and cereal crops. In more recently created oases, concrete blocks and prefabricated housing add a contemporary look to the settlement scene. Surrounded by large zones of desert, these green islands of rural activity stand out in sharp contrast to the surrounding sand- and rock-strewn landscape.

Traditional oasis settlements are composed of close-knit families who sometimes work their own irrigated plots or who commonly toil for absentee landowners. Although oases are usually small, eastern Saudi Arabia's Hofuf oasis covers more than 30,000 acres (12,150 hectares). While some crops are raised for local consumption, commercial trade has always played a role in such settings. Although less pivotal today, centuries ago camel caravans and pastoral nomads frequently stopped at such settlements, trading their goods for the fruits of oasis life. In the past century, the expanding world demand for products such as figs and dates has included even these remote locations in the global economy, and many prod-

▲ **Figure 7.15 Oasis agriculture** Date palms dominate this Tunisian oasis. Often, these commercial farm products find their way to consumers in distant European and North American markets. *(Alain Le Garsmeur/Panos Pictures)*

ucts end up on the tables of hungry Europeans or North Americans thousands of miles away. From central Algeria and Libya eastward to the dry reaches of the Arabian Peninsula, oasis life continues for some residents of the region. New drilling and pumping technologies, particularly in Saudi Arabia, have even added to the size and number of oasis settlements. But oasis life across the region faces major challenges: more rapid population increases, potential future groundwater depletion, and the pressures of global cultural change and modernization threaten the economic and social integrity of these settlements. It remains to be seen whether they can continue to be viable and sustainable adaptations in the fast-changing world of the twenty-first century.

Settlement Along Exotic Rivers For centuries, the region's densest rural settlement has been tied to its great river valleys and their seasonal floods of water and enriching nutrients. In such settings, **exotic rivers** transport precious water and nutrients from distant, more humid lands into drier regions, where the resources are utilized for irrigated farming. The Nile and the Tigris and Euphrates rivers are the largest regional examples of such activity, with both systems also characterized by large, densely settled deltas. Other linear irrigated settlements can also be found near the Jordan River in Israel and Jordan, along short streams originating in North Africa's Atlas Mountains, and on the more arid peripheries of the Anatolian and Iranian plateaus. These settings, while capable of supporting sizable rural populations, are also among the most vulnerable to overuse, particularly if irrigation results in salinization. Farmlands were abandoned centuries ago in ancient Mesopotamia (modern Iraq) as toxic salts lowered crop yields. These hazards of rural settlement in dry lands persist, and the growing buildup of salts in the modern Nile Valley suggests such problems will continue.

Farming in such localities supports much higher population densities than is the case with pastoral nomadism or traditional desert oases. Fields are small, intensely utilized, and connected with closely managed irrigation systems designed to store and move water efficiently through the settlement. In Egypt, farmers of the Nile Valley live in densely settled, clustered villages near their fields; work much of their land with the same tools and technologies as their ancestors did; and grow a mix of cotton, rice, wheat, and forage crops. The scale and persistence of their agricultural imprint has become one of the world's most visible and dramatic signatures of the human transformation of Earth's surface (Figure 7.16). But rural life is also changing in such settings. New dam- and canal-building schemes in Egypt, Israel, Syria, Turkey, and elsewhere are increasing the storage capacity of river systems, which allows for more year-round agricultural activity. The use of better-yielding rice and wheat varieties and more mechanized agricultural methods have also raised food output, particularly in places such as Egypt and Israel. Some of the most efficient farms in the region are associated with the Israeli **kibbutz**, collectively worked settlements that produce grain, vegetable, and orchard crops irrigated by waters from the Jordan River and from the country's elaborate feeder canals. For example, the country's National Water Carrier system takes water from the Sea of Galilee and moves it south and west, where intensively worked agricultural operations produce key food crops for nearby Tel Aviv and Jerusalem.

▲ **Figure 7.16 Nile Valley** This Landsat image of the Nile Valley dramatically reveals the impact of water on the North African desert. Cairo lies at the southern end of the delta where it begins to widen toward the Mediterranean Sea. *(Earth Satellite Corporation/SPL/Photo Researchers, Inc.)*

The Challenge of Dryland Agriculture Mediterranean climates in portions of the region permit varied forms of dryland agriculture that depend largely on seasonal moisture to support farming. These zones include the better-watered valleys and coastal lowlands of the northern Maghreb, lands along the shore of the eastern Mediterranean, and favored uplands across the Anatolian and Iranian plateaus. A different regional variant of dry farming is also present across Yemen's terraced highlands in the moister corners of the southern Arabian Peninsula, and a smaller zone also appears in Oman just east of the Persian Gulf.

In the classic Mediterranean adaptation, the settlement landscape reflects the varied crop and livestock mix found in similar settings elsewhere in the world. Often, a multiplicity of crops and livestock surrounds the Mediterranean villages of the region. Drought-resistant tree crops diversify the scene, with olive groves, almond trees, and citrus orchards producing output, both for local consumption and for commercial sale. Elsewhere, favored locations support grape vineyards, while more marginal settings can still be used to grow wheat and barley, or to raise forage crops to feed cattle, sheep, and goats. Vulnerability to drought and the availability of more sophisticated water management strategies are leading some Mediterranean farmers to utilize more irrigated cropping, thus improving production of cotton, wheat, citrus fruits, and tobacco across portions of the region. More mechanization, crop specialization, and fertilizer use are also transforming such agricultural settings, following a pattern set earlier in nearby areas of southern Europe.

Many-Layered Landscapes: The Urban Imprint

Cities have also played a pivotal role in the region's human geography. Indeed, some of the world's oldest urban places are located in the region. Today, enduring political, religious, and economic ties wed the city and countryside. This lengthy and intimate relationship has shaped the urban landscape just as surely as cities have transformed the rural scene.

A Long Urban Legacy Cities in the region have traditionally played important functional roles as centers of political and religious authority as well as key focal points of local and long-distance trade. Urbanization in Mesopotamia (in modern Iraq) began by 3500 B.C., and cities such as Eridu and Ur reached populations of 25,000 to 35,000 residents. Similar centers appear in Egypt by 3000 B.C., with Memphis and Thebes assuming major importance amid the dense populations of the middle Nile Valley. These ancient cities were key centers of political and religious control. Temples, palaces, tombs, and public buildings dominated the urban landscapes of such settlements, and surrounding walls (particularly in Mesopotamia) offered protection from outside invasion. By 2000 B.C., however, a different kind of city was emerging, particularly along the shores of the eastern Mediterranean and at the junction points of important caravan routes. Centers such as Beirut, Tyre, and Sidon, all in modern Lebanon, as well as Damascus in nearby Syria, exemplified the growing role of trade in creating the urban landscapes of selected cities. Expanding port facilities, warehouse districts, and commercial thoroughfares suggested how trade and commerce shaped these urban settlements, and many of these early Middle Eastern trading towns have survived to the present.

The religion of Islam also left an enduring urban signature because cities traditionally served as centers of Islamic religious power and education. By the eighth century, Baghdad emerged as a focus of Islamic power, followed soon thereafter by the appearance of Cairo as a seat of religious authority and expansion. Urban settlements from North Africa to Turkey felt the influences of Islam. Indeed, the Moors carried its characteristic signature to Spain, where it enduringly shaped centers such as Córdoba and Málaga. Its impact upon the settlement landscape merged with older urban traditions across the region and established a characteristic Islamic cityscape that exists to this day (Figure 7.17). Its traditional attributes include a walled urban core, or **medina**, dominated by the central mosque and its associated religious, educational, and administrative functions. A nearby bazaar, or "suq," functions as a marketplace where products from city and

▲ **Figure 7.17 The Islamic landscape** Iranian mullahs discuss the religious questions of the day beneath the minarets of the Moussavi Mosque in Qom. Islam has left a widespread mark on the region's cultural landscapes. *(S. Franklin/Sygma Photo News)*

countryside are traded. Housing districts feature an intricate maze of narrow, twisting streets that maximize shade and accentuate the privacy of residents, particularly women. Houses have small windows, frequently are situated on dead-end streets, and typically open inward to private courtyards often shared by extended families with similar ethnic or occupational backgrounds.

In the more recent past, European colonialism added yet another layer of urban landscape features in selected cities. Particularly in North Africa, coastal and administrative centers during the late nineteenth century sprouted dozens of architectural reminders of British and French urban traditions (see "Settlement Patterns: The Changing Urban Landscape in Fez, Morocco"). Victorian building blocks, French mansard roofs, suburban housing districts, and wide, European-style commercial boulevards complicated the settlement landscapes of dozens of cities, both old and new, within the region. Centers such as Algiers (French) and Cairo (British) vividly displayed the effects of colonial control, and many of these signatures remain a legacy on the modern scene.

Signatures of Globalization Since 1950, dramatic new forces have transformed the settlement landscapes of most Southwest Asian and North African cities, and the pace of change continues to accelerate. Cities have become the key gateways to the global economy. As the region has been

SETTLEMENT PATTERNS The Changing Urban Landscape of Fez, Morocco

The settlement patterns of Fez, Morocco, dramatically display the contrasting signatures of Islamic and Western influences upon the modern scene. The northern Moroccan city, home to more than 800,000 people, is a fascinating amalgam of traditional and modern worlds, and a reminder of how the urban landscape can become a vast visual accumulation of differing cultural, economic, and technological influences.

The old city lies to the east, a classic walled medina typical of the urban Islamic world (Figure 1). The traditional city is home to a variety of central mosques, the usual assemblage of palaces and official buildings, and the kasbah, a formerly fortified quarter used by local officials for defense. The narrow, curved streets are a labyrinth of cul-de-sacs and private courtyards. Along the exterior walls, a series of city gates allow passage into the protected inner city. This intricate, densely interconnected urban landscape evolved centuries ago, in harmony with the needs of Islamic leaders who valued both its functional and symbolic characteristics.

The Treaty of Fez (1912), however, formally turned over control of Morocco to the French. Colonialism's signature quickly appeared on the urban landscape: Fez expanded to the west in a series of formal, straight boulevards and radiating avenues that echoed the baroque-style planning traditions of post-Renaissance France. In addition, the French built a rail station and hospital to serve the needs of the colonial regime. A new university was also constructed nearby. Since then, even more contemporary influences have reshaped the urban scene, with office buildings and apartment houses reflecting the needs of an ever-expanding urban population. Thus, modern Fez remains a complex accumulation of the past and present, a record of changing cultural influences visible on the contemporary map and in the everyday landscapes of its city streets.

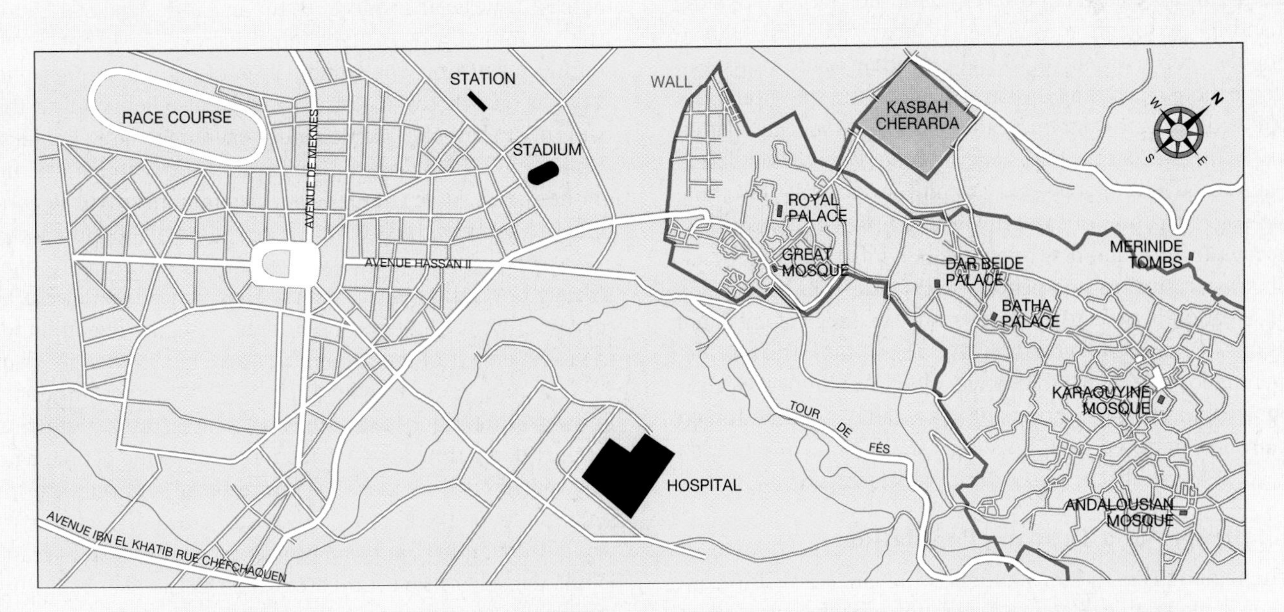

▲ **Figure 1 Fez, Morocco** The tiny neighborhoods and twisting lanes of the old walled city reveal features of the traditional Islamic urban center. To the west, however, the rectangular street patterns, open spaces, and broad avenues suggest colonial European influences. *(Modified from Rubenstein, 1999,* An Introduction to Human Geography, *Upper Saddle River, NJ: Prentice Hall)*

opened to new investment, industrialization, and tourism, the urban landscape reflects the fundamental changes taking place. Expanded airports, commercial and finance districts, industrial parks, and luxury tourist facilities all mark the impress of the global economy.

Further, as urban centers become focal points of economic growth, surrounding rural populations are drawn to the new employment opportunities, thus fueling rapid population increases. The results are both impressive and problematic. Many traditional urban centers, such as Algiers and Istanbul, have more than doubled in size in recent years. Booming demand for homes has produced scores of ugly, cramped, high-rise apartment houses in some government-planned neighborhoods, while elsewhere, sprawling squatter settlements provide little in the way of quality housing or municipal services. Crowded Cairo exemplifies the pattern. Its legendary "City of the Dead" neighborhood, now home to almost one million residents, is an urban cemetery intermingled with homes and apartment houses, a vivid acknowledgment of the premium put on living space. Many central-city neighborhoods in such settings have become so crowded and congested that wealthier urbanites have left, moving to outlying suburbs where larger homes and lower population densities offer more appealing living environments. Other suburban locales, however, have become sites for major industrial expansion, often financed through foreign investment. In some cases, entirely new industrial centers are being established. For example, Egypt's ambitious New-Town Program includes plans for more than a dozen new cities.

▲ **Figure 7.18 Abu Dhabi cityscape** The capital city of the oil-rich United Arab Emirates vividly reveals the impact of Western wealth and architecture on the modern Arab world. *(M. Attar/Sygma Photo News)*

Undoubtedly, the oil-rich states of the Persian Gulf display the most extraordinary changes on the urban landscape (Figure 7.18). Before the twentieth century, urban traditions were developed relatively weakly in the area, and even as late as 1950, only 18 percent of Saudi Arabia's population lived in cities. All that changed, however, as the global economy's demand for petroleum mushroomed. Today, the Saudi Arabian population is more urban (80 percent) than many industrialized nations, including the United States, and the capital city of Riyadh has grown to more than 2.5 million people. Particularly after 1970, other cities, such as Abu Dhabi (United Arab Emirates), Doha (Qatar), and Kuwait City (Kuwait), also bore the bold signatures of modern Western urban design, futuristic architecture, and new transportation infrastructure. In addition, investments into petrochemical industries have fueled the creation of new urban centers, such as Jubail along Saudi Arabia's Persian Gulf coastline. The result is an urban settlement landscape where traditional and global influences curiously intermingle, producing cityscapes where domed mosques, mirrored bank buildings, and oil refineries uneasily coexist beneath the dusty skies and desert sun.

Challenges of a Growing Population

The region's thirsty farms and growing cities reveal a troubling demographic dilemma. Population growth averages well over 2 percent across the region and will therefore double in less than 35 years. Areas such as the West Bank (3.4 percent), Gaza (4.6 percent), and Libya (3.7 percent) experience some of the highest rates of natural increase on the planet. In some localities, persisting patterns of poverty and the traditional ways of rural life contribute to the large rates of population increase, but even in more urban and industrialized Saudi Arabia, annual growth rates remain above 3 percent. Such settings result from the combination of high birth rates and very low death rates. Cultural variables also help explain the pattern. Traditional Islamic values encourage large families. While family planning and increased contraceptive use are on the rise in some countries such as Iran, nearby Iraq continues to promote high fertility rates among its population.

Future population pressures seem destined to continue, although overall rates of increase have slowed somewhat in the last 20 years. Still, for most countries in the region, more than 35 percent of the present population are younger than 15 years old, promising another large generation of people in need of food, jobs, and housing. The result is a growing number of frustrated and underemployed young people who may increase the chances for future social and political instability. Egypt, the Levant, and coastal North Africa face particularly daunting population problems over the next 20 years. While densely settled rural areas such as the Nile Valley increasingly experience the pressures of demographic growth, the region's large cities will bear the brunt of future population increases. Growing populations will also impose increasingly daunting demands on the region's already limited water resources. Although rapid development might mitigate the situation by providing more jobs and encouraging lower birth rates, current levels of economic expansion are anemic at best in most countries of the region, suggesting that the present problems are likely to continue.

Cultural Coherence and Diversity: Signatures of Complexity

While Southwest Asia and North Africa clearly define the heart of the Islamic and Arab worlds, a surprising degree of cultural diversity characterizes the region today. Muslims (or Moslems) practice their religion in a great variety of ways, often disagreeing profoundly on basic religious tenets as well as on how much of the modern world and its mass consumer culture should be incorporated into their daily lives. Further, while Islam certainly has its historical roots within the region, many Muslims live beyond its bounds. In addition, diverse religious minorities complicate the region's contemporary cultural geography. Most notably, both Judaism and Christianity trace their origins to the eastern Mediterranean, and modern practitioners of these faiths remain concentrated in a variety of settings from North Africa to the Levant. Linguistically, Arabic languages form an important cultural core historically centered on the region. Still, many non-Arab peoples, including Persians, Kurds, and Turks, also populate important homelands within the region. Understanding these varied patterns of cultural geography is essential to comprehending many of the region's political tensions as well as appreciating why many of its residents resist processes of globalization.

Patterns of Religion

Religion permeates the lives of most people within the region. Its centrality is hard to underestimate and it stands in sharp contrast to largely secular cultural impulses that often dominate life in many other parts of the world. Whether it is the quiet rituals of morning prayers or profound discussions regarding contemporary political and social issues, religion suffuses the daily routine of most regional residents from Casablanca to Tehran. The geographies of religion—their points of origin, paths of diffusion, and patterns of modern regional articulation—are essential elements in defining varied cultural and political zones of both conflict and coexistence. In some settings, religious differences have had the power to pit one neighbor against another, while elsewhere, ancient religious rivalries appear to be slowly healing. One thing is certain: patterns of religion will continue to matter greatly in the twenty-first century, and they will remain central in defining many of the area's future political, social, and economic challenges.

Hearth of the Judeo-Christian Tradition Both Jews and Christians trace the roots of their religions to the eastern Mediterranean, and while neither are numerically dominant across the area today, each group continues to play a pivotal cultural role across the region. The roots of Judaism lie deep in the past: Abraham, considered an early patriarch in the Jewish tradition, lived some 4,000 years ago and led his people from Mesopotamia to Canaan (modern-day Israel), near the shores of the Mediterranean. From Jewish history, recounted in the Old Testament of the Holy Bible, springs a rich religious heritage focused on a belief in one God (or **monotheism**), a strong code of ethical conduct, and a powerful ethnic identity that continues to the present. During the time of the Roman Empire (A.D. 70), most Jews were forced to leave the eastern Mediterranean after they challenged Roman authority. The resulting forced migration, or Diaspora, of the Jews took them to the far corners of Europe and North Africa. Only in the past century have the world's far-flung Jewish populations returned to the religion's hearth area, a process that accelerated greatly with the formation of the Jewish state of Israel in 1948.

Christianity also emerged in the vicinity of modern-day Israel and has left a lasting legacy across the region. An outgrowth of Judaism, Christianity was based on the teachings of Jesus and his disciples who lived and traveled in the eastern Mediterranean about 2,000 years ago. While many Christian traditions became associated with European history, some forms of early Christianity remained potent nearer to the religion's original hearth. To the south, one stream of Christian influences associated with the Coptic Church diffused into northern Africa, shaping the cultural geographies of places such as Egypt and Ethiopia. In the Levant, another group of early Christians known as the Maronites retained a separate cultural identity that survives today. Elsewhere, smaller, but enduring Christian communities also became established across other portions of northern Africa and southwestern Asia.

The Emergence of Islam Islam originated in Southwest Asia in A.D. 622, forming yet another cultural hearth of global significance. While Muslims can be found today from North American cities to islands of the southern Philippines, the Islamic world is still centered on its Southwest Asian origins. Most Southwest Asian and North African peoples still follow its religious precepts and moral doctrines. Muhammad, the founder of Islam, was born in Makkah (Mecca) in A.D. 570 and taught in nearby Medinah (Medina)(Figure 7.19). In many respects, the religion he founded represents a continuation of the Judeo-Christian tradition. Muslims believe that both Moses and Jesus were true prophets, and that both the Hebrew Bible (or Old Testament) and the Christian New Testament, while incomplete, are basically accurate. Ultimately, however, Muslims hold that the **Quran** (or Koran), a book of revelations received by Muhammad from Allah (God), represents God's highest religious and moral revelations to mankind.

The basic tenets of Islam offer an elaborate blueprint for leading an ethical and religious life. Islam literally means "submission to the will of God," and the creed rests on five essential pillars. They include: (1) repeating the basic creed ("There is no God but God, and Muhammad is his prophet"); (2) prayer, facing Makkah five times daily; (3) giving charitable contributions; (4) fasting between sunup and sundown during the month of Ramadan; and (5) making at least one religious pilgrimage, or **Hajj**, to Muhammad's birthplace of

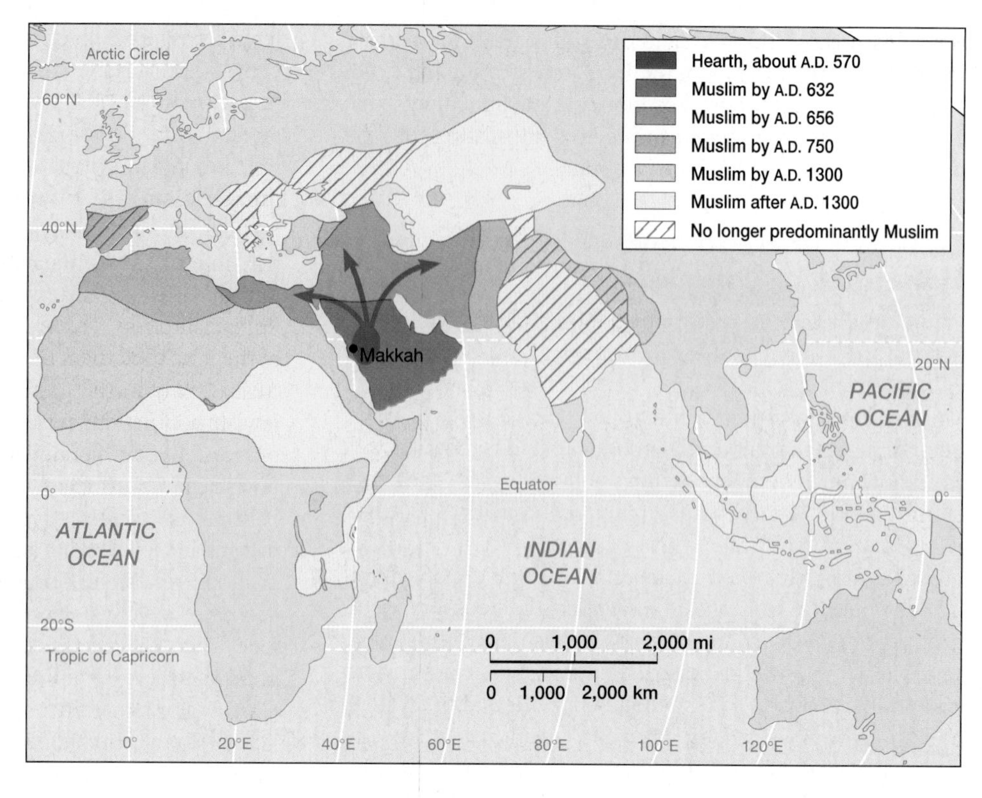

▶ **Figure 7.19 Diffusion of Islam** The rapid expansion of Islam that followed its birth is shown here. Sweeping from Spain to Southeast Asia, Islam's legacy remains strongest nearest its Southwest Asian hearth. In some settings, its influence has ebbed or come into conflict with other religions such as Christianity, Judaism, and Hinduism. *(Modified from Rubenstein, 1999,* An Introduction to Human Geography, *Upper Saddle River, NJ: Prentice Hall)*

Makkah (Figure 7.20). Islam is a more austere religion than most forms of Christianity, and its modes of worship and forms of organization are generally less ornate. Muslims regard religious images as idolatrous, and the strictest interpretations even forbid the depiction of the human form. Followers of Islam are prohibited from drinking alcohol and are instructed to lead moderate lives, avoiding excess. In traditional Islam there is no real distinction between law and religion, for the Quran itself supplies a legal framework. Nor is there clear precedence for separating political from religious authority. Many Islamic fundamentalists still argue for a **theocratic state** such as modern-day Iran, in which religious leaders (ayatollahs) guide policy.

A major religious schism divided Islam early on, and the differences endure to the present. The breakup occurred almost immediately after the death of Muhammad in A.D. 632. Key questions surrounded the succession of religious power. One group, now called the **Shiites**, favored passing power on within Muhammad's own family, specifically to Ali, his son-in-law. Most Muslims, later known as **Sunnis**, advocated passing power down through the established clergy. This group emerged victorious, Ali was martyred, and his Shiite supporters went underground. Ever since, Sunni Islam has formed the mainstream branch of the religion, to which Shiite Islam has presented a recurring, and sometimes powerful, challenge. Of the two factions, the Shiite tradition argues that a successor to Ali will someday return to reestablish the pure, original form of Islam. The Shiites are also more hierarchically organized than the Sunnis. Only in Shiism, for example, does one find individual leaders, such as Iran's ayatollahs, wielding overriding religious and even political power.

▲ **Figure 7.20 Makkah** Millions of the faithful make the pilgrimage to Makkah where they visit the holy al-Ka'ba, the black, cube-like structure in the center of the picture. In Muslim theology, Abraham and Ishmael constructed the al-Ka'ba, and the site was later honored by Muhammad. *(Elkoussy/Sygma Photo News)*

In a very short period of time, Islam diffused widely from its Arabian hearth, often following camel caravan routes and Arab military campaigns as it remarkably expanded its geographical range and converted thousands to its creed. By Muhammad's death in A.D. 632, the peoples of the Arabian Peninsula were united under its banner. Shortly thereafter, the Persian Empire fell to Muslim forces and the Eastern Roman (or Byzantine) Empire lost most of its territory to Islamic influences. By A.D. 750, Arab armies swept across North Africa, conquered most of Spain and Portugal, and established footholds in Central and South Asia. For several hundred years this vast area was united under a single Islamic caliphate (named after the caliph, the political-religious leader of the state). At first, only Arabs, who formed an elite military stratum in the conquered territories, followed Islam. In fact, for a time they discouraged conversion because Muslims did not have to pay the lucrative poll tax required of all others. Eventually, however, the diverse inhabitants of Southwest Asia and North Africa were gradually woven together into a single cultural system, albeit one with many distinct local variants. Conversion of non-Arabic peoples to Islam was slow at first, but after a few hundred years, the pace quickened. By the thirteenth century, most of the people in the region had become Muslims, while older creeds such as Christianity and Judaism became minority faiths in some areas and disappeared altogether from others.

Between 1200 and 1500, Islamic influences expanded in some areas and contracted in others. The Iberian Peninsula (Spain and Portugal) was lost to Christendom in 1492, although many Moorish (Islamic) cultural and architectural features remained behind and still shape the region today. Elsewhere, Muslims expanded their influence southward and eastward into Africa. In addition, Muslim Turks largely replaced Christian Greek influences in Southwest Asia after 1100. One group of Turks moved into the Anatolian Plateau and finally conquered the last vestiges of the Byzantine Empire in 1453. These Turks soon created the vast **Ottoman Empire** (named after one of its leaders, Osman) that included southeastern Europe (including modern-day Albania, Bosnia, and Kosovo) and most of Southwest Asia and North Africa. The legacy of the Ottoman Empire was considerable: it offered a new, distinctly Turkish interpretation of Islam, and it provided a dynamic and centralized expression of Muslim political power within the region until the Empire's disintegration in the late nineteenth and early twentieth centuries.

Modern Religious Diversity Today, Muslims form the majority population in all of the countries of Southwest Asia and North Africa except Israel, where Judaism is the dominant religion, and Cyprus, where Greek Orthodox outnumber Turkish Muslims (Figure 7.21). Still, divisions within Islam have created key cultural differences within the region. While most (73 percent) of the region is dominated by Sunni Muslims, the Shiites (23 percent) remain an important element in the contemporary cultural mix. Particularly dominant in Iran, southern Iraq, Lebanon, Sudan, and Bahrain, the Shiites also form a major religious minority in many other countries, including Algeria, Egypt, and Yemen. Strongly associated with the recent flowering of Islamic fundamentalism, the Shiites also have benefited from rapid growth rates because their brand of Islam is particularly appealing to many of the poorer, powerless, and more-rural populations within the region. While some Sunnis have been attracted to fundamentalism as well, many reject its more radical cultural and political precepts and argue for a more modern Islam that incorporates some accommodation with Western values and traditions.

While the Sunni–Shiite split is the great divide within the Muslim world, other variations of Islam can also be found in the region. One division, for example, separates the mystically inclined form of Islam—known as *Sufism*—from the more

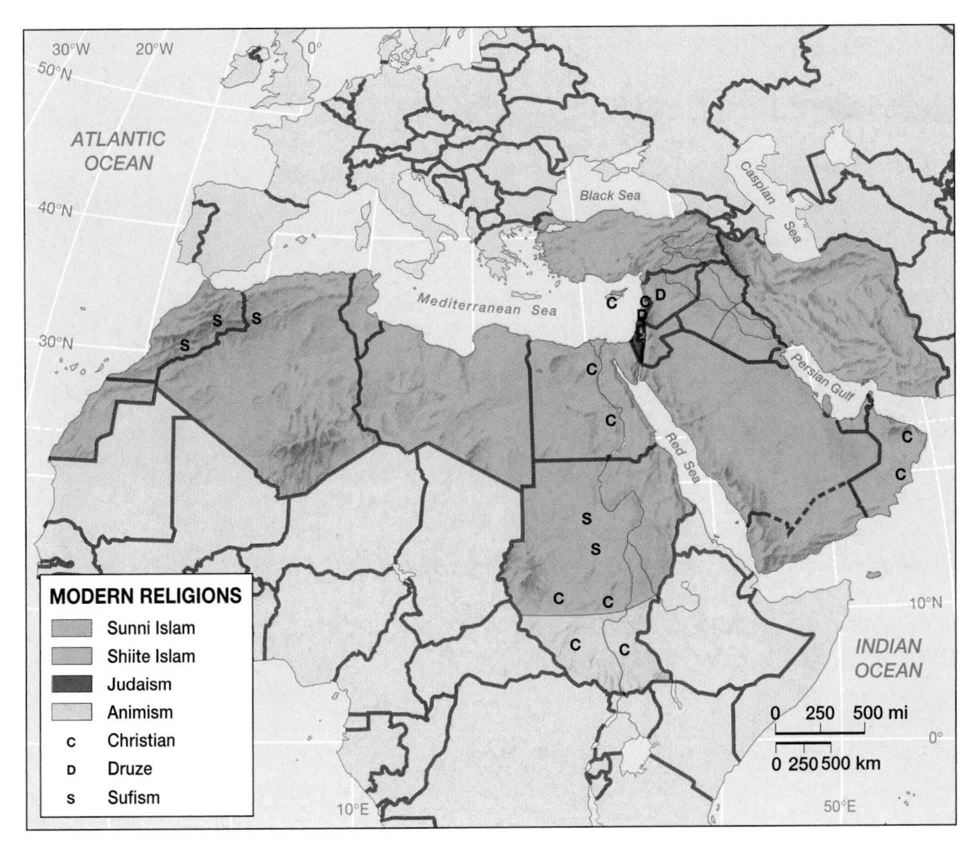

◀ **Figure 7.21 Modern religions** Islam persists as the dominant religion across the region. Most Muslims share ties to the Sunni branch, whereas Shiites are found in places such as Iran and southern Iraq. In some locales, however, Christianity and Judaism remain important; African animism is found in southern portions of Sudan. *(Modified from Rubenstein, 1999,* An Introduction to Human Geography, *Upper Saddle River, NJ: Prentice Hall)*

scripturally oriented mainstream tradition. Sufism is especially prominent in the peripheries of the Islamic world, including the Atlas Mountains of Morocco and Algeria. Elsewhere, the Druze of Lebanon practice an even more distinct variant of Islam. An offshoot of the Shiite tradition, the Druze believe that God is revealed in successive incarnations, and some religious scholars question whether they are really Muslims at all. Still, they have formed an enduring and cohesive religious minority in the Shouf Mountains east of the Lebanese capital of Beirut.

Southwest Asia is also home to many non-Islamic communities. Israel has a Muslim minority (14 percent) that is dominated by that nation's Jewish population (80 percent), while Christians comprise another 2 percent of the total. Even within Israel's Jewish community, increasing cultural differences divide Jewish fundamentalists from more reform-minded Jews. Indeed, this cultural diversity within Judaism has shaped how Israel has dealt with political and social issues involving nearby Muslims, both within and beyond its borders. In neighboring Lebanon, there was actually a slight Christian (Maronite and Orthodox) majority as recently as 1950. Christian out-migration and differential birthrates, however, have created a nation that is 60 to 70 percent Muslim. Christians also form approximately 10 percent of Syria's population, while Iraqi Christians, concentrated mostly in the rugged northern uplands, make up about 3 percent of its population.

Within the Middle East, the city of Jerusalem (now the Israeli capital) holds special religious significance for several groups and provides a microcosm of the region's religious diversity (Figure 7.22). In fact, the deeply divided sacred space of this ancient Middle Eastern city continues to create cultural and political conflict within the region as different groups argue for more control of contested neighborhoods and nearby real estate. Historically, Jews particularly revere the city's old Western Wall (the site of a Roman-era temple), Christians honor the Church of the Holy Sepulchre (the burial site of Jesus), and Muslims hold sacred religious sites in the city's eastern quarter (including where the prophet Mohammad reputedly ascended to heaven). Further complicating the traditional religious divisions of the city, Israel's victory in the 1967 war emboldened it to establish many new Jewish settlements within the city's redefined and expanded eastern suburbs. The move angered Islamic Palestinian residents of the area who saw the new settlements as rightfully part of their own homeland.

Important Jewish and Christian communities also have left a long legacy across North Africa. Roman Catholicism was once dominant in much of the Maghreb, but it disappeared several hundred years after the Muslim conquest. The Maghreb's Jewish population, on the other hand, remained prominent until the post-World War II period, when most of

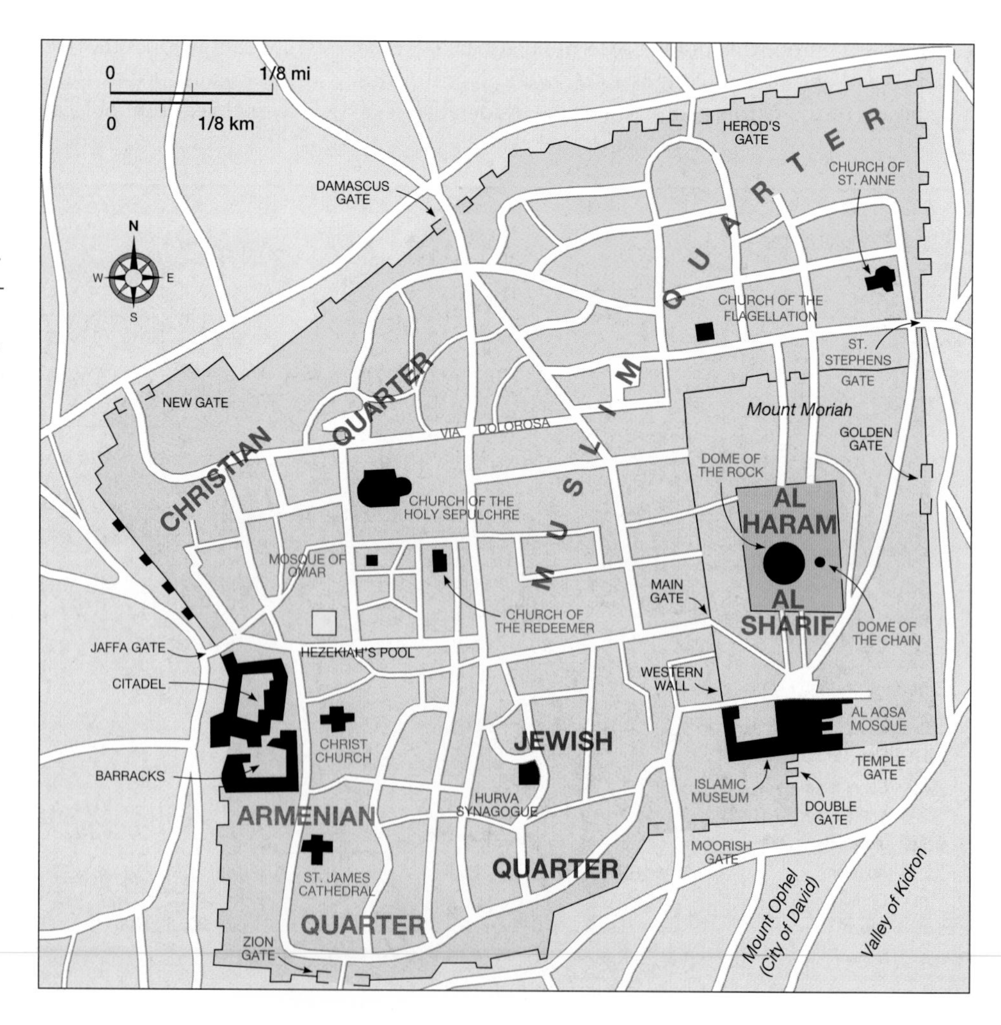

▶ **Figure 7.22 Old Jerusalem** Jerusalem's historic center reflects its varied religious legacy. Traditionally, Christians, Muslims, and Jews have occupied different quarters of the old city. Recent suburban growth has added to tensions. New Israeli settlements have displaced Arab populations and the controversy continues over the city's role as Israel's national capital. *(Modified from Rubenstein, 1999,* An Introduction to Human Geography, *Upper Saddle River, NJ: Prentice Hall)*

the region's Jews migrated to the new state of Israel. The Jewish community of Morocco, the largest in the area, now amounts to only 0.2 percent of the country's population. Egypt also saw most of its Jewish community move to Israel after World War II. In Egypt, however, Coptic Christianity has maintained a stable presence over the centuries and today includes approximately 7 percent of the country's population. In earlier years the Coptic community had a secure place in Egyptian society, and numerous Copts held high-level posts in government and business. Today, however, Egypt's Christians are being increasingly marginalized, and some of their communities have been put under pressure, even subjected to physical attack, by extremist elements within the Islamic fundamentalist movement. Other Christian communities are located in southern Sudan, but unlike those of Egypt, these are mostly recent converts from traditional African religions.

Geographies of Language

Although the region is often referred to as the "Arab World," linguistic complexity creates many important cultural divisions across Southwest Asia and North Africa (Figure 7.23). Understanding these varied geographies of language offers insights into regional patterns of ethnic identity, potential cultural conflicts that exist at linguistic borders, and the instability that often characterizes geopolitical relationships within and between states.

Semites and Berbers Afro-Asiatic languages dominate much of North Africa and Southwest Asia. Within that family, Arabic-speaking Semitic peoples can be found from Morocco to Saudi Arabia. Before the expansion of Islam, Arabic was limited to the Arabian Peninsula. Today, however, Arabic is spoken from the Persian Gulf to the Atlantic and it reaches southward into Sudan, where it borders the Nilo-Saharan speaking peoples of Sub-Saharan Africa. As the language has diffused, it has slowly diverged into local dialects. As a result, the everyday Arabic spoken on the streets of Fez, Morocco, is not easily intelligible to an Arabic speaker from the United Arab Emirates. The Arabic language also has a special religious significance for all Muslims. It was, they believe, the language in which God delivered his ultimate message to humankind, and is thus sacred. While most of the world's Muslims do not speak Arabic, the faithful often memorize certain prayers in the language, and many Arabic words have entered the other important languages of the Islamic world. Advanced Islamic learning, moreover, demands competence in Arabic. The classical Arabic of the Quran as the language of learning, and religion has remained virtually unchanged from the time of Muhammad to the present. Classical Arabic, however, gradually ceased to be used in most areas for common, everyday forms of conversation.

Hebrew is another traditional Semitic language of Southwest Asia that was recently reintroduced into the region with the creation of the Jewish state of Israel. This Semitic language originated in the Levant and was used by the ancient Israelites 3,000 years ago. Today its modern version survives as the sacred tongue of the Jewish people. When the Zionist movement for a Jewish state in the region began in the late 1800s and early 1900s, Jewish immigrants to Southwest Asia were uncertain which language they should use. Many spoke Yiddish,

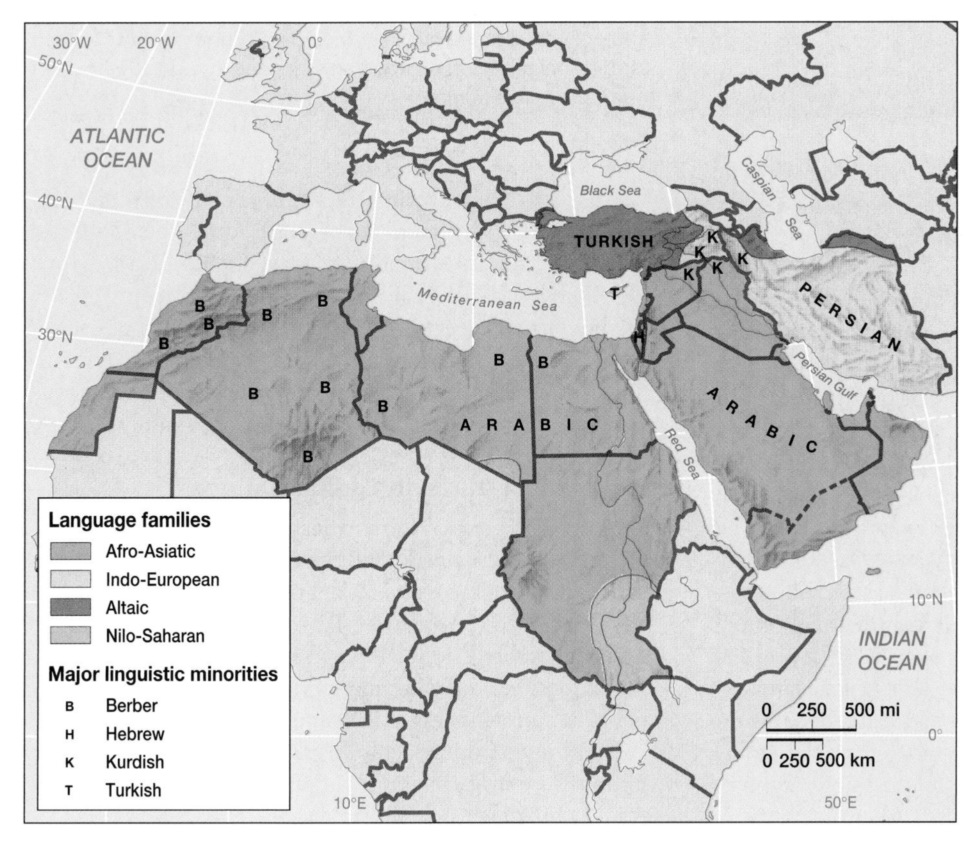

◀ Figure 7.23 Modern languages Arabic is a Semitic Afro-Asiatic language and it dominates the region's cultural geography. Turkish, Persian, and Kurdish, however, remain important exceptions, and such cultural differences within the region have often had enduring political consequences. Israel's more recent reintroduction of Hebrew further complicates the region's linguistic geography. *(Modified from Rubenstein, 1999,* An Introduction to Human Geography, *Upper Saddle River, NJ: Prentice Hall)*

a German dialect written in the Hebrew script, while others conversed in English, French, or Arabic. Ultimately they decided that Hebrew was the only form of communication common to the global Jewish community. Today Hebrew is the main and official language of Israel, although the country's non-Jewish population largely speaks Arabic. In addition, many recent Jewish immigrants speak Russian, while English is widely used as a second or third language throughout the country.

While Arabic eventually spread across North Africa, several older languages survive in the more remote areas. Older Afro-Asiatic tongues endure in the Atlas Mountains and in certain parts of the Sahara. Collectively known as Berber, these languages are related to each other but are not mutually intelligible. Most Berber languages have never been written and none has generated a significant literature. Literacy in Berber-speaking communities thus usually entails learning Arabic, the language of the dominant lowland society. The decline in pastoral nomadism and the pressures of modernization also threaten the integrity of these Berber languages. Scattered Berber-speaking communities are found as far to the east as the oases of Egypt, but Morocco is the key center of this language group.

▲ **Figure 7.24 Kurdish family** As an ethnic minority, these Kurdish-speaking settlers from eastern Turkey face cultural and political discrimination in their own country, similar to their Kurdish neighbors in nearby Iran, Iraq, and Syria. *(J. Egan/ Hutchison Picture Library)*

Persians and Kurds Although Arabic spread readily through portions of Southwest Asia, much of the Iranian Plateau and nearby mountains are dominated by older Indo-European languages. Here the principal tongue remains Persian, although, since the tenth century, the language has been enriched with Arabic words and written in the Arabic script. Persian, like other languages, developed distinct local dialects. Today Iran's official language is called *Farsi,* which denotes the form of Persian spoken in Fars, the area around the city of Shiraz. Thus, while both Iran and neighboring Iraq are largely Islamic nations, their traditional ethnic identities spring from quite different linguistic, literary, and cultural traditions and often contribute to cultural tensions between the two countries.

The Kurdish speakers of northern Iraq, northwest Iran, and eastern Turkey add further complexity to the regional pattern of languages in Southwest Asia (Figure 7.24). Kurdish, also an Indo-European language, is spoken by 10 to 15 million persons in the region. Kurdish has not historically been a written language, but the Kurds do have a strong sense of shared cultural identity. Indeed, "Kurdistan" has sometimes been called the world's largest nation without its own political state since the group remains a minority in several countries of the region. Making matters worse, the Kurds have been discriminated against and attacked by surrounding groups, particularly by antagonistic government forces in Iraq and Turkey.

The Turkish Imprint Turkish languages provide more linguistic variety across much of modern Turkey, in portions of far northern Iran, and on the northern third of Cyprus. The Turkish languages are a part of the larger Altaic language family that originated in Central Asia. Turkey remains the largest nation in Southwest Asia dominated by the family. Tens of millions of persons in other countries of Southwest and Central Asia speak related Altaic languages such as Azeri, Uzbek, or Uighur. During the era of the Ottoman Empire, Turkish speakers ruled much of North Africa and Southwest Asia, but Iran is the only other large country in the region today where the language persists. Turks also contribute to the linguistic diversity and the cultural tensions of Cyprus in the eastern Mediterranean: a Turkish minority occupies the north, while a Greek (Indo-European family) majority dominates most of the southern part of the island.

Regional Cultures in Global Context

Southwest Asian and North African peoples increasingly find themselves intertwined with cultural complexities and conflicts external to the region. Islam links the realm with a globally dispersed population. In addition, European and American cultural influences have multiplied greatly across the region since the mid-nineteenth century. Colonialism, the boom in petroleum investment, and the growing presence of Western-style popular culture have had enduring impacts upon the complex regional mosaic that stretches from the Atlas Mountains to the Indian Ocean.

Islamic Internationalism While Islam is geographically and theologically divided, all Muslims recognize the fundamental unity of their religion. This religious unity extends far beyond Southwest Asia and North Africa. Islamic communities are well established in such distant places as central China, European Russia, central Africa, and the southern Philippines. Today, Muslim congregations are also expanding rapidly in the major urban areas of western Europe and North America, largely due to migration but also through local conversions. Islam is thus emerging as a truly global religion. Even with its global reach, however, Islam remains centered on Southwest Asia and North Africa, the site of its origin and its most holy places. As Islam expands in its number of adherents and in its geographical scope, the religion's tradition of pilgrimage ensures that Makkah will become a city of increasing global significance in the twenty-first century. The global growth of Islamic fundamentalism also focuses attention on the region, where much of the contemporary movement burst upon the scene in the late twentieth century. Future prospects for fundamentalism throughout the Islamic world will no doubt hinge on its cultural and political evolution within the region. In addition, the oil wealth accumulated by many Islamic nations will be used to sustain and promote the religion. Countries such as Saudi Arabia and Libya will continue to invest in Islamic banks and economic ventures and to make donations to Islamic cultural causes, colleges, and hospitals worldwide.

Globalization and Cultural Change The region is also grappling with how its growing role in the global economy is changing traditional cultural values. Indeed, the expansion of Islamic fundamentalism is in many ways a reaction to the threat posed by external cultural influences, particularly those of western Europe and the United States. European colonialism left its own cultural legacy, not only in the architectural landscapes still found in the old colonial centers, but also in the widespread use of English and French among the Western-educated elite across the region. In oil-rich countries, huge capital investments also have had important cultural implications as the number of foreign workers has grown and as more affluent young people have embraced elements of Western-style music, literature, and clothing. Even in conservative Iran, where satellite dishes are officially banned, millions of people have access to them, beaming in multicultural programming from around the globe. While fundamentalists in Islamic republics, such as Iran and Sudan, would like to impose a complete cultural wall between themselves and the West, such actions remain problematic in today's world.

In other settings, cultural ties have created strong linkages between the realm and other world regions. Former colonial ties between Algeria and France, for example, have encouraged a sizable out-migration of North Africans to western Europe, thus increasing interaction between the former colony and her mother country. Large numbers of Turkish workers have also spent considerable time in European employment, particularly in Germany. While some have remained as guest workers, many others work periodically in Europe and then return to their Turkish homelands. Indeed, Turkey's general openness to European and U.S. investment and popular culture has forever reshaped the cultural scene, particularly in the urban areas (see "The Local and the Global: The Marlboro Man Arrives in Istanbul"). In the case of Israel, its role as the center of Judaism and its historical significance as the hearth

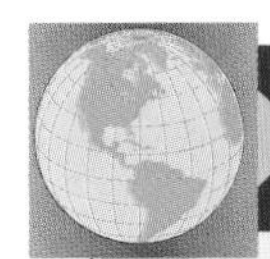

THE LOCAL AND THE GLOBAL The Marlboro Man Arrives in Istanbul

The impact of globalization has taken many forms in Turkey, but few examples are more dramatic than the arrival of the Marlboro Man in Istanbul. Even in a land where native tobacco is revered, Phillip Morris, the U.S. manufacturer of Marlboro products, has succeeded in capturing almost 25 percent of the Turkish cigarette market. Their effective use of U.S.-style marketing campaigns has especially appealed to young urban Turks, who equate the Marlboro name with a new, more progressive lifestyle. As one young man in Istanbul noted, "None of my friends smoke local cigarettes . . . that would be humiliating."

The success of Phillip Morris illustrates how many U.S. and European corporations have westernized Turkish culture. In a society where more than 40 percent of the population smokes, the company saw great potential in their Marlboro strategy. They lobbied the Turkish government to open their domestic markets to foreign competition. Then they invested more than $230 million in a new cigarette manufacturing facility, carefully engineering the local Marlboros to Turkish tastes.

Most powerfully, the company's marketing efforts have inundated the country with the Marlboro name and insignia. Salesmen dress as cowboys as they deliver their products in vans designed to look like traveling packs of cigarettes. Sidewalk diners around Istanbul are protected from the hot Turkish sun with umbrellas bejeweled in the Marlboro logo, fashion-conscious urbanites proudly swagger city streets in "Marlboro Classic Khakis," and the company's advertising campaign successfully features the same independent cowboys and sweeping western vistas as it does in the United States. Marlboros are also penetrating the Turkish countryside, a subtle but omnipresent signature of increasing globalization, both in terms of changing cultural values and in how consumers spend their money.

Source: Adapted from "How Phillip Morris Got Turkey Hooked on American Tobacco," *Wall Street Journal*, September 11, 1998.

of Christianity have made it a destination of global cultural importance. Akin to the Hajj in Islam, many of the world's Jews and Christians consider a visit to the Holy Land in Israel an essential part of their religious lives. Close economic and military connections between Israel and the United States also enhance the cultural glue that binds the region to the world beyond.

Geopolitical Framework: A Region of Persisting Tensions

Southwest Asia and North Africa continue to be at the center of a multitude of geopolitical conflicts (Figure 7.25). Some of the tensions relate directly to age-old patterns of cultural geography in which different ethnic, religious, and linguistic groups are still struggling to live with one another in a rapidly changing world of new nation-states and political relationships. The region's brief but complex ties to the era of European colonialism also contribute to present difficulties, since the boundaries of many countries grew from an imposed political geography that still shapes the scene today. Geographies of wealth and poverty also enter the geopolitical mix: some residents profit mightily from petroleum resources and industrial expansion, while others struggle just to feed their families. The result is a political climate charged with potential problems, a region in which the sounds of bomb blasts and gunfire have been an all too common characteristic of everyday life.

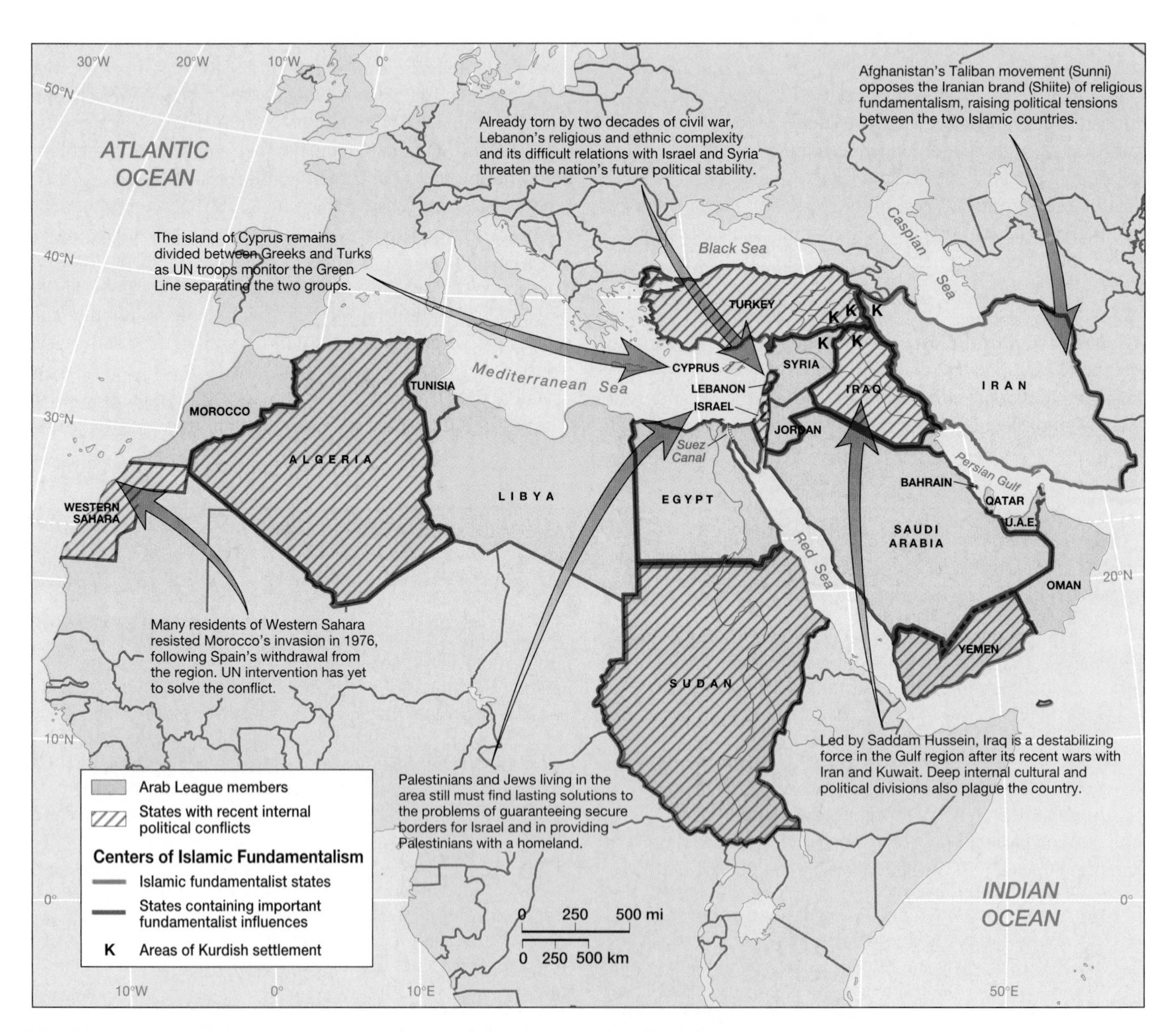

▲ **Figure 7.25 Geopolitical issues in Southwest Asia and North Africa** Political tensions continue across much of the region. While the central conflict remains oriented around Israel, its neighboring states, and the rights of resident Palestinians, other regional trouble spots periodically erupt in violence. Islamic fundamentalism challenges political stability in settings from Algeria to Sudan. Elsewhere, persisting ethnic conflicts shape daily life in Lebanon, Turkey, Cyprus, and Iraq.

The Colonial Legacy

European colonialism arrived relatively late in Southwest Asia and North Africa, but the era left an important impact upon the region's modern political geography. The Turks were one reason for Europe's tardy participation in imposing colonial rule. Between 1550 and 1850, much of the region was dominated by the Turkish Ottoman Empire, which expanded from its Anatolian hearth to engulf much of North Africa as well as nearby areas of the Levant, the western Arabian Peninsula, and modern-day Iraq. Ottoman rule imposed an outside imperial political order and a larger colonial economic framework that had lasting consequences for the region. The tide began to turn, however, in the early nineteenth century as the European presence increased and Ottoman power ebbed. Still, it took a century of shifting geopolitical fortunes for Ottoman influences to be replaced with largely European colonial dominance after World War I (1918). While much of that direct European control ebbed by the 1950s, old colonial ties persist in a variety of economic, political, and cultural contexts. Indeed, it is still common to encounter British English on the streets of Cairo, while French can still be heard in Algiers and Beirut.

▲ **Figure 7.26 Algiers** European colonial influences still abound in the old French capital of Algiers in northern Algeria. The city's modern bustle belies the increasing political and religious tensions that have torn the country apart since 1992. *(Delluc-XPN/REA/SABA Press Photos, Inc.)*

Imposing European Power French colonial ties have long been a part of the region's history. The first major European effort to build a North African empire was Napoleon's failed attempt to occupy Egypt in 1799. Thirty years later, France decided to redeem this debacle by conquering a new empire in North Africa. Although it took several decades of fierce fighting, the entire territory of modern Algeria ended up in French hands. Several million French, Italian, and other European immigrants poured into the country, taking the best lands from the Algerian people. The French government expected this territory to become an integral part of France, dominated by the growing French-speaking immigrant population (Figure 7.26). Later, France established protectorates in Tunisia (1881) and Morocco (1912), ensuring an enduring French political and cultural presence in the Maghreb. Finally, France's victory over the German/Ottoman Turk alliance in World War I produced additional territorial gains in the Levant. France controlled the northern zone encompassing the modern nations of Syria and Lebanon.

Great Britain's colonial fortunes also grew within the region before 1900. To control their sea lanes to India, Britain established a series of nineteenth-century protectorates over the small coastal states of the southern Arabian Peninsula and the Persian Gulf. In this manner such places as Kuwait, Bahrain, Qatar, the United Arab Emirates, and Aden (in southern Yemen) were loosely incorporated into the British Empire. Nearby Egypt also caught Britain's attention. Once the British-engineered **Suez Canal** linked the Mediterranean and Red seas in 1869, European banks and trading companies gained more control over the Egyptian economy. The British took more direct control in 1883. In the process, Britain also inherited a direct stake in Sudan, since Egyptian soldiers and traders had been pushing south along the Nile for decades.

Another series of British colonial gains within the region came at the close of World War I. In Southwest Asia, British and Arab forces joined to expel the Turks during the war. To obtain Arab trust, Britain promised that an independent Arab state would be created in the former Ottoman territories. At roughly the same time, however, Britain and France signed a secret agreement to partition the area. When the war ended, Britain opted to slight its Arab allies and honor its treaty with France. The desire for a large independent Arab state was thus frustrated, with one exception. The Saud family convinced the British that a smaller country (Saudi Arabia) should be established, focused on the desert wastes of the Arabian Peninsula. It became fully independent in 1932. Elsewhere, however, Britain carried out its plan to partition lands with France. Britain divided its new territories into three entities, including Palestine (now Israel) along the Mediterranean coast; Transjordan, to the east of the Jordan River (now Jordan); and a third zone that later became Iraq. Iraq, in particular, was a contrived territory that combined three dissimilar former Ottoman provinces. It included the centers of Basra in the south (an Arabic-speaking Shiite area), Baghdad in the center (an Arabic-speaking Sunni area), and Mosul in the north (a Kurd-dominated zone).

Other settings within the region felt more marginal colonial impacts. Libya, for example, was long regarded by Europeans as a desert wasteland. Italy, never a dominant colonial power, expelled Turkish forces from the coastal districts by 1911, but they did not subdue the Saharan oases until the 1930s. Spain also carved out a territorial stake within the region, gaining control over southern Morocco (now Western Sahara) in 1912. To the east, Persia and Turkey were never directly occupied by European powers, but both played into the calculus of the global geopolitics of the time. In Persia, the British and Russians agreed to establish mutual spheres of economic influence in the region (the British in the south, the Russians in the north), while respecting Persian independence. In 1935, Persia's modernizing ruler Reza Shah changed the country's name to Iran, signifying a break from tradition and de-emphasizing the country's Persian-speaking majority in favor of a more inclusive national formulation.

In nearby Turkey the old core of the Ottoman Empire was almost partitioned by European powers following World War I. After several years of fighting, however, the Turks expelled the French from southern Turkey and the Greeks from western Turkey. The key to the successful Turkish resistance was the spread of a new modern, nationalist ideology under the leadership of Kemal Ataturk. Ataturk decided to emulate the European countries and establish a culturally unified and resolutely secular state. He was quite successful, and Turkey was quickly able to stand up to European power.

Decolonization and Independence European colonial powers began their withdrawal from several Southwest Asian and North African colonies before World War II. By the 1950s, most of the countries in the region were independent, although many maintained political and economic ties with their former colonial rulers. In North Africa, Britain finally withdrew its troops from Sudan and Egypt in 1956, although its direct political control in the region began declining in the 1920s. Libya (1951), Tunisia (1956), and Morocco (1956) achieved independence peacefully during the same era, but the French colony of Algeria became a major problem. Since several million French citizens resided there, France had no intention of simply withdrawing. A bloody war for independence resulted, sparking a conflict with the mother country that began in 1954. Finally, France agreed to an independent Algeria in 1962, but the two nations continued to share a close—if not always harmonious—relationship thereafter.

Southwest Asia also lost its colonial status between 1930 and 1960, although many of the imposed colonial-era boundaries continue to shape the regional geopolitical setting to the present day. While Iraq became independent from Britain in 1932, its later instability in part resulted from its artificial borders that never recognized much of its cultural diversity. Similarly, the French division of its Levant territories into the two independent states of Syria (1946) and Lebanon (1946) greatly angered local Arab populations and set the stage for future political instability in the region. As a favor to its small Maronite Christian majority, France carved out a separate Lebanese state from largely Arab Syria, even guaranteeing the Maronites constitutional control of the national government. The action created a culturally divided Lebanon as well as a Syrian state that repeatedly has asserted its influence over its Lebanese neighbors.

Modern Geopolitical Issues

The geopolitical instability in Southwest Asia and North Africa will persist in the twenty-first century. It remains difficult to predict political boundaries that seem destined to shift, either through negotiated settlements or political conflict. Several key problems loom within the region. First, the Arab-Israeli conflict zone in the eastern Mediterranean will continue as a major trouble spot, vulnerable to periodic difficulties. Second, the catalyst of Islamic fundamentalism can be expected to spark ongoing cultural and political tensions wherever the movement achieves prominence. Third, internal political conflicts are likely where multiple cultural groups occupy different territories within a single country. Finally, future tensions between states are likely where neighboring nations develop disagreements over cultural issues, natural resources, or political boundaries.

The Arab-Israeli Conflict The 1948 creation of the Jewish state of Israel produced an enduring zone of cultural and political tensions within the eastern Mediterranean. Jewish migration to the area now known as Israel accelerated after the British took Palestine from the defeated Ottoman Empire after World War I. In 1917 Britain issued the **Balfour Declaration,** essentially a pledge to encourage the "establishment of Palestine as a home for the Jewish people." After World War II, the British withdrew from the area and the United Nations divided the region into two states, one to be predominantly Jewish, the other primarily Muslim (Figure 7.27). Most of the area's indigenous Arab Palestinian population rejected the partition, and war erupted as soon as the British departed. Jewish forces proved victorious, and by 1949, their new state of Israel gained a larger share of land than it had originally been allotted. The remainder of Palestine, including the West Bank and the Gaza Strip, passed to Jordan and Egypt, respectively. Hundreds of thousands of Palestinian refugees fled from Israel to neighboring countries, such as Egypt, Jordan, and Lebanon, where many of them remained in makeshift camps. Under these difficult conditions, the Palestinians nurtured the idea of creating their own state in the land that had become Israel. The Israelis, not surprisingly, remained adamant that the country was theirs.

Israel's relations with neighboring countries were bitter from the beginning. The supporters of Arab unity and Muslim solidarity sympathized with the Palestinians, while their antipathy toward Israel grew. In the 1950s and 1960s, leaders of Egypt, Syria, Iraq, and Jordan viewed Israel as an unwanted Western, neocolonial presence in their midst. Indeed, Israel's presence fostered pan-Arab solidarity that for a time promoted the creation of a region-wide Arab state, a political impulse that continues today in the form of the Arab League organization. The Israelis saw things very differently: many were born in the area, and many others migrated from other portions of Southwest Asia and North Africa.

Israel and its Arabic-speaking neighbors fought three major wars in 1956, 1967, and 1973. In territorial terms, the Six-Day War of 1967 was the most important conflict. In this struggle against Egypt, Syria, and Jordan, Israel gained substantial new territories in the Sinai Peninsula, the Gaza Strip, the West Bank, and the Golan Heights. Israel annexed the eastern part of the formerly divided city of Jerusalem, arousing particular bitterness among the Palestinians, since Jerusalem is a sacred city in the Muslim tradition. The center is sacred in Judaism, as well, and Israel remains adamant in its claims to the entire city. A

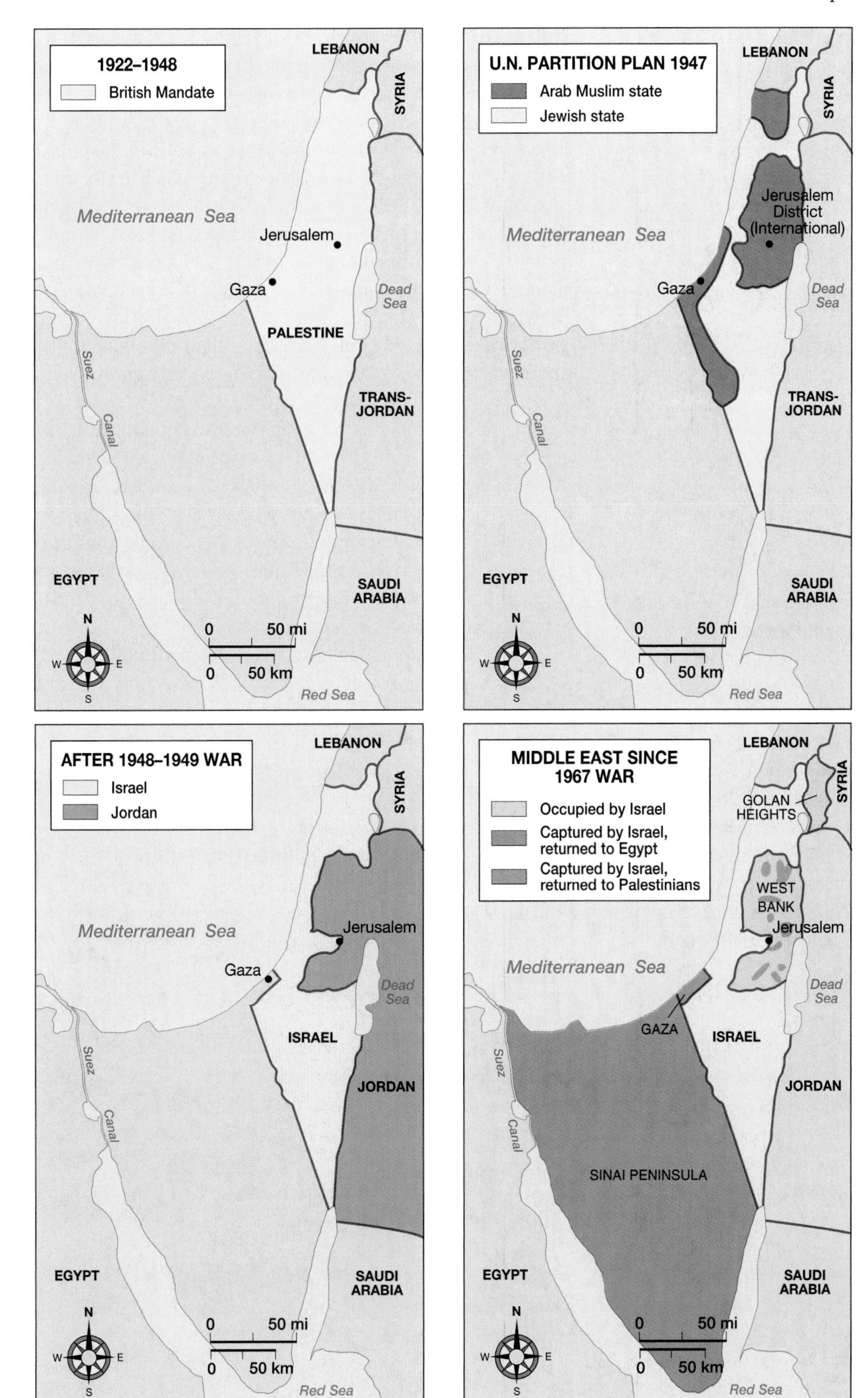

◀ **Figure 7.27 Evolution of Israel** Israel's complex evolution began with an earlier British colonial presence and a United Nations partition plan in the late 1940s. Thereafter, multiple wars with nearby Arab states produced Israeli territorial victories in settings such as Gaza, the West Bank, and the Golan Heights. Each of these regions continues to figure importantly in the country's recent relations with nearby states and with resident Palestinian populations. *(Modified from Rubenstein, 1999,* An Introduction to Human Geography, *Upper Saddle River, NJ: Prentice Hall)*

peace treaty with Egypt later resulted in the return of the Sinai Peninsula, but tensions then focused on the other occupied territories that remained under Israeli control. To strengthen its geopolitical claims, Israel also built additional Jewish settlements on the West Bank and in the Golan Heights, further angering Palestinian residents and nearby Syrians.

A cycle of violent actions and equally violent reprisals accelerated with the Palestinian uprising in 1987 that became known as the "Intifada." The conflict continued unabated until 1993, when the two sides began to negotiate a settlement. Preliminary agreements called for the construction of a quasi-independent Palestinian state in the Gaza Strip and across much

▲ **Figure 7.28 Rafah International Airport** The Palestinian-controlled Gaza Strip now features a new international airport that adds further legitimacy to the ruling power of the Palestinian Authority (PA). *(Nadav Neuhaus/Sygma Photo News)*

of the West Bank (Figure 7.28). Radical factions on both sides denounced the settlement as a sellout, and political polarization, both within the Palestinian movement and among the Israelis themselves, heightened tensions across the region. Israel also has harshly resisted ongoing Palestinian terrorism, while Palestinian leaders such as Yasser Arafat have criticized continuing construction of new Jewish settlements in the West Bank. Still, the peace process has unfolded, offering hope for the region's seven million Palestinians. A tentative agreement late in 1998 strengthened the ability of the ruling **Palestinian Authority (PA)** to control transportation and policing powers in the Gaza Strip. It also established goals for increasing PA autonomy in varied West Bank districts, such as Jericho and Hebron, and it articulated further guarantees for Israel's security.

Tensions also continue between Israel and its neighbors. The Israeli occupation of the Golan Heights remains the key point of conflict with Syria. Syrian authorities argue that peace in the region is only possible once the territory is returned to Syrian control. The government of Israel, however, counters that it is always vulnerable to attack from any potentially hostile state that might occupy this strategic highland zone. In addition, the region's relatively abundant water supplies add to Israel's determination to retain the area. To the north, Israeli forces continue their occupation of a security zone in southern Lebanon, an arrangement that angers both Lebanese and Syrian authorities. Farther east, Iraq remains a potential enemy. Indeed, Iraq's anti-Western sentiment during the Persian Gulf War quickly grew to include Israel among nearby countries subject to missile attacks. Tensions have been reduced between Israel and nearby Jordan and Egypt, but shifting political winds among the Israelis make future relations hard to predict. One thing is certain: geographical issues will remain at the center of the regional conflict. Palestinians hope for a land they can call their own, and the Israelis continue their search for more secure borders that guarantee their own political integrity in a region where they are surrounded by potentially hostile neighbors.

Politics of Fundamentalism Islamic fundamentalism is another catalyst for continuing regional geopolitical tensions. Fundamentalism dramatically appeared on the political scene in 1978 and 1979 as Shiite Muslim clerics in Iran successfully engineered the overthrow of the Shah, an authoritarian, pro-Western ruler friendly to U.S. political and economic interests. Suddenly, the Ayatollah Khomeni, a once-exiled religious leader, took power in 1979 and proclaimed an Islamic republic in which religious officials ruled both clerical and political affairs. Building on considerable domestic distrust of both the Shah and the United States, the Ayatollah fomented a revolutionary fervor that not only engulfed his own country, but that also spread to many other states in the region (Figure 7.29). Fundamentalists throughout the realm have challenged existing political regimes, advocated more traditional roles for women, and criticized the influx of too many heretical ideas from the West. Since the mid-1990s (Khomeni died in 1989), however, more-moderate leaders have come to power in Iran, and they are slowly redefining that nation's brand of fundamentalism in ways that may permit more domestic democratic reforms and reopen a political dialogue with the United States and other Western powers.

Sudan's Sunni Muslims also followed the fundamentalist path. An Islamic-led military coup in 1989 overthrew a dem-

▲ **Figure 7.29 Khartoum, Sudan** The politics of fundamentalism spread to the streets of Khartoum in 1989 as a military regime overthrew a democratic government. Today, signatures of the city's religious heritage mingle with office buildings and commercial streets. *(J. Hartley/Panos Pictures)*

ocratic government and imposed an Islamic republic. Immediately, the Sudanese fundamentalists strengthened official ties with Iran. Further, elements of the new regime became associated with revolutionary political groups in the region, provoking U.S. military reprisals in 1998. The regime also imposed Islamic law across the country and in the process antagonized both moderate Sunni Muslims as well as the nation's large non-Muslim (mostly Christian and animist) population in the south. Persisting differences between northern and southern groups within the country reflect larger global tensions on the margins of the Islamic world, and have produced intermittent but highly disruptive civil wars within the country. Two million people have died in the conflict and more than five million people have been displaced from their homes. Indeed, the present tensions may provoke the political separation of the non-Muslim south from the northern Sudanese fundamentalists.

Fundamentalism is a potent political force elsewhere in the region. Most notably, the growing role of fundamentalism in Algeria has further destabilized a country already divided by increasingly extreme political factions. Once a stable and moderate state within the region, Algeria has been plunged into an escalating cycle of internal violence and terrorism that has killed tens of thousands since 1992. Elections in late 1991 made it clear that Islamic fundamentalists were on the verge of assuming power in the country. Other Algerian nationalists and the military feared the fundamentalists. They negated the election results and suspended democracy in the country. Bombings, kidnappings, and assassinations have been a regular feature of Algerian life ever since. In 1997 and 1998, a series of murderous attacks were carried out against unarmed villagers, some of them located very close to the capital city of Algiers. Making matters worse, Islamic militants accuse the French of secretly backing forces opposed to their fundamentalist cause.

The Algerian debacle heightened concerns over fundamentalism in other countries within the region, particularly Egypt, Turkey, and Saudi Arabia. In Egypt, various fundamentalist factions have attacked government officials and tourists. Although Turkey has long been regarded as a moderate, more secular state, the rising tide of fundamentalism has challenged the political status quo and clouded the country's future potential links with the European Union. Even in Saudi Arabia, a bastion of conservative Sunni traditions and mostly pro-U.S. politics, fundamentalists threaten to destabilize the authority of the ruling Saudi family. Indeed, every corner of the region has been touched by the cultural and political fires fanned by the Islamic fundamentalists, and the trend appears likely to continue reshaping the regional geopolitical scene in the years to come.

Conflicts Within States As the Algerian and Sudanese situations suggest, the region contains many settings in which domestic cultural and political differences have led to civil war. While each country has a unique geopolitical history, broader regional and global influences also play pivotal roles in these settings. Unpredictable relations with the West, the shifting price of oil, and longer-term demographic and economic pressures suggest that these internal geopolitical conflicts will continue in the future.

Several examples illustrate the challenges faced by nations within the region. Lebanon has seen some of the most destructive fighting over the past several decades. From its birth in the 1940s, Lebanon faced potential discord arising among its many distinct religious communities. Early on, a complex power-sharing arrangement was worked out in which the president would be a Maronite Christian, the prime minister a Sunni Muslim, and the speaker of Parliament a Shiite. This arrangement worked relatively well until the 1970s. At that time, the Israeli-Palestinian struggle spilled over into neighboring Lebanon as Palestinian refugees in southern Lebanon used the area as a base for attacks in Israel. Israel responded with armed reprisals. The fighting soon spread among the Lebanese, and by 1975 civil war raged within the country. Fighting generally pitted Christians against Muslims, but the existence of numerous sects and factions on both sides made the actual struggle far more complicated (see "Local Voices: The Sorrows of Lebanon"). By the late 1970s, the capital city of Beirut was a scarred and divided war zone (Figure 7.30).

The fighting in Lebanon also involved other countries. Both Syria and Israel sent in armies, and an international peacekeeping force, backed by the United States, vainly attempted to pacify Beirut. Although the conflict flared throughout the 1980s, the nation's various armed factions reached a new agreement in 1995. While Syria continues to maintain a military force in the country and Israel keeps troops in the southern Lebanese "security zone," fighting within the country has subsided and rebuilding has begun. Most observers doubt, however, that this once-prosperous and peaceful country will ever fully recover. Indeed, its cultural geography is as diverse as ever, poorer Shiite fundamentalist populations are growing much more rapidly than other groups, and both Israeli and Syrian involvement remain persistent threats to stability.

Troubled Iraq is another nation born during the colonial era that has yet to escape the consequences of its externally

▲ **Figure 7.30 War-damaged Beirut** This Mediterranean city is slowly recovering from multiple decades of civil war. Persisting tensions among varied Muslim and Christian factions promise more challenges in the years ahead. *(C. Maurice Harvey LMPA/Hutchison Picture Library)*

LOCAL VOICES The Sorrows of Lebanon

Long a dynamic and productive part of the Middle East, Lebanon has also suffered greatly from cultural and political divisions. Lebanon's fragmented cultural geography is far older than the country itself. Indeed, ancient religious rivalries were only further complicated by long periods of colonial rule. Many of the resulting cultural tensions and political conflicts still so much a part of daily life in Lebanon were captured many years ago by Lebanese novelist and poet Kahlil Gibran (1883–1931). He describes the sorrows of Lebanon in this excerpt from *The Garden of the Prophet:*

Pity the nation that is full of beliefs and empty of religion.

Pity the nation that wears a cloth it does not weave, eats a bread it does not harvest, and drinks a wine that flows not from its own wine-press.

Pity the nation that acclaims the bully as hero, and that deems the glittering conqueror bountiful. . . .

Pity the nation that raises not its voice save when it walks in a funeral, boasts not except among its ruins, and will rebel not save when its neck is laid between the sword and the block.

Pity the nation whose statesman is a fox, whose philosopher is a juggler, and whose art is the art of patching and mimicking.

Pity the nation that welcomes its new ruler with trumpetings, and farewells him with hootings, only to welcome another with trumpetings again. . . .

Pity the nation divided into fragments, each fragment deeming itself a nation.

A vibrant and independent Lebanon, economically focused around the bustling city of Beirut, emerged in the 1940s after Gibran's poem was published. Later, however, the country fell once again into the pitfalls of civil war and foreign invasion (by Syria and Israel) during the 1970s and 1980s. Recent political improvements, though encouraging, have yet to negate the significance of Gibran's words.

Source: Kahlil Gibran, *The Garden of the Prophet* (New York: Knopf, 1933)

imposed geopolitical origins. Simmering beneath the bluster of Saddam Hussein's Iraqi regime have been enduring cultural differences within the country that periodically produce internal political conflicts. When the country was carved out of the British Empire in 1932, it contained the cultural seeds of its later troubles. One problem zone centers on the lower Tigris and Euphrates Valley south of Baghdad, where most of the country's 10 million Shiites live. Indeed, the region, focused around the city of Basra, contains some of the holiest Shiite sites in the world. Its Marsh Arab population has periodically resisted control from Baghdad, and it threatened independence just after the Persian Gulf War in the early 1990s. Saddam responded with military reprisals as well as with environmental warfare designed to drive the Shiites off their agricultural lands (see "Environmental History: War and the Middle East Environment" on page 274).

Conditions also remain grim in northern Iraq. For decades, periodic revolts by the culturally distinctive Kurds have attempted to create a Kurdish homeland politically separate from Iraq, Turkey, and other nearby nations. The Kurdish population again raised its own flags of rebellion in the 1980s, only to face the full wrath of the Iraqi military. Entire villages were bombed with poison gases. After the Persian Gulf War, the United Nations forbade Iraqi air forces from operating north of the 36th parallel, an area encompassing the main Kurdish districts. As a result, northern Iraq's political identity remains ambiguous: officially it remains part of Iraq, but local forces under UN supervision control some portions of the area. Unfortunately, the Kurds and other peoples of the region continue to quarrel with each other, reducing the chances of a peaceful settlement.

Finally, divided Cyprus in the eastern Mediterranean offers a classic example of a country torn in two by virtue of its complex historical geography. The island's ancient connections with the Greek world were disrupted in the sixteenth century when Ottoman Turks occupied the area. Although Cyprus passed into British hands three centuries later, an important Turkish minority population was well established on the island. When Cyprus gained independence in 1960, the country's national leaders hoped that the island's mixed Greek Orthodox and Islamic Turkish populations could find common ground in their newly won freedom. Unfortunately, divisiveness won out in the 1970s and civil war erupted. The conflict intensified when Turkey sent troops to the island in 1974, and war with neighboring Greece became a real possibility. One dramatic geographical consequence of the conflict was the spatial polarization of the population: Turks concentrated on the northern portion of the island, while Greeks settled to the south. As a result, UN peacekeepers established a **Green Line** that separates the two groups, a dramatic symbol of failed political compromise that runs through the heart of the country's divided capital of Nicosia (Figure 7.31). In 1983 the Turks (20 percent of the population) declared their own Turkish Republic of Northern Cyprus, an orphan state that has gone unrecognized by the rest of the world (with the exception of Turkey), and that has languished economically even as the Greek Cypriots enjoy more rapid economic development.

Conflicts Between States Political tensions within the region also involve more than one state, and these trouble spots are not limited to the Arab-Israeli conflict. Indeed, every

▲ **Figure 7.31 Nicosia, Cyprus** The United Nations-enforced Green Line separating Greeks and Turks on the island runs through the major city of Nicosia. The result is an unusual urban landscape that parallels scenes in once-divided East and West Berlin. *(Ricki Rosen/SABA Press Photos, Inc.)*

corner of the realm has been touched by geopolitical struggles that periodically erupt between nearby countries. The sources of these conflicts vary. Natural resources—including contested claims for oil, water, and farmland—often play a role. Elsewhere, persisting cultural differences still fire contemporary antagonisms, including the enduring problems introduced by boundaries imposed during the colonial era.

Varied North African settings threaten the region's political stability. One trouble spot is Western Sahara, just south of Morocco. The former Spanish colony was invaded and annexed by Morocco in the late 1970s, a move resisted by most of the local population. Although home to only about 200,000 people, the marginal desert land contains substantial mineral (especially phosphate) deposits. Algerian support for the Western Saharans has also periodically strained relations with neighboring Morocco, and UN intervention has yet to quiet the conflict. To the east, Libya's leader, Colonel Muammar al-Qaddafi, has intermittently fomented regional tensions ever since he took power in 1969. Libya has financed violent political movements directed against Israel, Western Europe, and the United States. U.S. planes bombed Libya in 1986, responding to what the United States identified as repeated acts of terrorism. In addition, Libya's relationships with nearby Egypt and Chad have been rocky at best over the past 20 years, and conflicting territorial claims still threaten the region's fragile stability. Internal conflicts within Sudan's fundamentalist Islamic state also have spilled into adjacent portions of Ethiopia, Eritrea, Chad, and Uganda, as refugee populations have fled the country. In addition, the possible involvement of Sudanese elements in the 1995 assassination attempt on Egypt's President Mubarak further strained relations between those two nations.

In Southwest Asia, regional geopolitical tensions persist, still reflecting the recent legacy of bloodshed in the Iran–Iraq (1980–88) and Persian Gulf (1990–91) wars. Iraq, led by Saddam Hussein, remains one focus of continuing instability. In 1980 Hussein invaded oil-rich but politically weakened Iran to gain a better foothold on the Persian Gulf. Hundreds of thousands of young Iraqis and Iranians perished in the conflict that included a bloody ground war as well as a costly air campaign. Eight years of bloody fighting resulted in a stalemate, and the conflict left Iraq's finances in disarray. After the war, Iraq received little economic support from wealthy Persian Gulf states even though they opposed the radicalism of the Iranian regime. Iraq was also frustrated because of its difficulty in exporting oil through its narrow opening to the Persian Gulf. It tried to convince nearby Kuwait to hand over several strategically located islands in the area, but Kuwait refused. In response, Iraq invaded and overran Kuwait in 1990, claiming it as a new province. A U.S.-led UN coalition, receiving substantial support from Saudi Arabia, expelled Iraq from Kuwait in early 1991. In many respects, however, the war in Iraq continued thereafter. The country remained under UN-imposed economic sanctions, devastating the Iraqi economy. UN inspectors also monitored the country closely in an effort to halt its nuclear and biological warfare programs. Large "no-fly" zones were imposed upon Iraqi air space north of 36° and south of 33° latitude in the northern and southern thirds of the country, limiting Iraqi territorial control over those portions of the country. In the late 1990s, tensions again increased as UN inspection teams withdrew from the country and as British and U.S. warplanes bombed strategic military and communications sites.

Elsewhere in Southwest Asia, other cultural divides and contested resource issues also threaten to ignite larger-scale political conflicts. For example, Turkey's geopolitical tensions lie both to the west and to the east. The country's relations with Greece remain troubled over the issue of Cyprus as well as conflicts in the Balkans. In addition, Turkish Kurds remain a destabilizing element in the eastern part of the country that also spills over into relations with neighboring states containing their own Kurdish minorities. Indeed, Turkey and Syria came close to war late in 1998 over questions concerning Kurdish rebels, conflicting water rights in the upper Euphrates River, and territorial disputes that date back to the late 1930s. On the eastern borders of the region, Iran faces a new political threat from Afghanistan's Taliban movement, a radical Sunni Islamic group opposed to the Shiite regime in Tehran. To the south, geopolitical issues on the Arabian Peninsula remain focused on Iraqi-Kuwaiti tensions as well as the future stability of the Saudi government.

An Uncertain Political Future

Few areas of the world pose more geopolitical question marks that Southwest Asia and North Africa. Twenty years from now, the region's political map could look quite different than it does today. The region's strategic global importance increased greatly after World War II, propelled into the international spotlight by the creation of Israel, the tremendous growth in the world's petroleum economy, the Cold War tensions between the United States and the Soviet Union, and the more recent rise of Islamic fundamentalism. Now that the Cold War between global superpowers has ended, the region is experiencing a geopolitical reorientation. Allies of the former Soviet Union, such as Libya, Syria, and Iraq, have often had to turn to other sources for military equipment and expertise. Indeed, the end of the Cold War may have actually destabilized the geopolitical situation. Before the downfall of the Soviet Union, it was unlikely that Iraq would have done anything as adventurous as invade Kuwait. Still, Russia, as successor to the Soviet state, may play an important diplomatic role as an intermediary in quelling tensions within the region.

The relationship between the United States and the various countries of the region remains complex. Israel and Turkey are close U.S. allies and recipients of large amounts of military aid, just as Iran, Iraq, Syria, and Libya remain firmly opposed to the United States. Relations with the oil-rich states of the Arabian Peninsula are more complex. Most countries in the region, especially Saudi Arabia, have been major purchasers of U.S. military hardware, and all have relied to some extent on U.S. forces for their protection. The United States has eagerly supplied such protection because of its reliance on the oil resources of the region, a mutual dependency made evident in the Gulf War of the early 1990s. In the future, however, it is uncertain whether the Arab-speaking world will readily join forces with the United States, as many people in the region view the United States as an intrusive superpower traditionally too closely allied with Israel.

The likelihood of future political conflicts in the region is high, as is the probability for drawing in outside military participants from Europe, the United States, or even South Asia. Most countries in the region spend a disproportional amount of money on defense. Traditions of peaceful democratic political regimes are few and far between, and they often run counter to fundamentalist leanings. On the other hand, many authoritarian governments within the region face opposition from groups both within and beyond their borders. The realm's petroleum riches also promise to keep it in the global spotlight. Still, there are political groups and individuals within the region that are striving to settle their differences and to establish foundations for peace. Some hope that increased economic cooperation—especially between Israel and its neighbors—will bring about more amicable relations. In the long run, a lasting settlement to the Palestinian issue as well as the question of securing stable borders for Israel seem essential before the region can move toward greater political stability.

Economic and Social Development: Lands of Wealth and Poverty

Southwest Asia and North Africa is a region of both incredible wealth and disheartening poverty. While a few of its countries enjoy great prosperity, due mainly to rich reserves of petroleum and natural gas, other nations within the region are counted among the least developed in the world (Table 7.2). Overall, recent economic growth rates have lagged those of the more-developed world. Lower oil prices in the late 1990s hampered economic growth in some areas, while persisting political instability also contributed to the regional economic malaise. These economic stumbling blocks have also had profound social consequences: investments in education, health care, and new employment opportunities have slowed considerably in many nations from the heady gains of the late 1970s and 1980s. Petroleum will no doubt figure prominently into the region's future economic relationships with the rest of the world, but many countries in the area also have focused on increasing agricultural output, investing in new industries, and promoting tourism as important ways to broaden the regional economic base.

The Geography of Fossil Fuels

The striking global geographies of oil and natural gas reveal the region's persisting importance in the world oil economy as well as the extremely uneven distribution of these resources within the region (Figure 7.32). Saudi Arabia remains one of the major producers of petroleum in the world, and Iran, the United Arab Emirates, Libya, and Algeria also make major contributions to global oil output. The region also plays an important though less dominant role in natural gas production. In the future, the distribution of fossil fuel reserves suggests that regional supplies will not be exhausted anytime soon. Overall, with only 7 percent of the world's population, the region holds a staggering 68 percent of the world's proven oil reserves. Saudi Arabia's pivotal position, both regionally and globally, is also clear: its 20 million residents live atop 25 percent of the planet's known oil supplies.

Two major geological zones supply much of the region's output of fossil fuels. The world's largest concentration of petroleum lies within the Arabian-Iranian sedimentary basin, a geological formation that extends from northern Iraq and western Iran to Oman and the lower Persian Gulf (Figure 7.33). All of the states bordering the Persian Gulf reap the benefits of oil and gas deposits within this geological basin, and it is not surprising that the world's densest concentration of OPEC members is found in the same area. A second important zone of oil and gas deposits includes eastern Algeria, northern and central Libya, and scattered developments in northern Egypt. As in the Persian Gulf region, these North African fields are tied to regional processing points and to global petroleum markets by a complex series of oil and gas pipelines and by networks of technologically sophisticated oil shipping facilities.

Even with all of these regional riches, the geography of fossil fuels is strikingly uneven across Southwest Asia and North Africa. Some nations—even those with tiny populations (Bahrain, Qatar, and Kuwait, for example)—contain incredible fossil fuel reserves, especially when considered on a per capita basis. Many other countries, however, and millions of regional inhabitants, reap relatively few benefits from the oil and gas economy. In North Africa, for example, Morocco and Sudan possess few developed petroleum reserves. Even in oil-rich Southwest Asia, the distribution of fossil fuels is amazingly fickle: Israel, Jordan, and Lebanon all lie outside favored geological zones for either petroleum or natural gas. While Turkey has some developed fields in the far southeast, it must import substantial supplies to meet the needs of its large and industrializing population.

Table 7.2 Economic Indicators

Country	GNP per Capita ($U.S., 1996)	Total GNP (Millions of $U.S., 1996)	PPP* ($Intl, 1996)	Real Annual Growth % per Capita, 1990–1996
Algeria	1,520	43,726	4,620	–1.9
Bahrain	—	—	—	—
Cyprus	—	—	13,970	3.8
Egypt	1,080	64,275	2,860	2.2
Gaza	—	—	—	2.6
Iran	—	—	5,360	1.0
Iraq	—	—	—	—
Israel	15,870	90,310	18,100	3.2
Jordan	1,650	7,088	3,570	4.0
Kuwait	—	—	—	15.7
Lebanon	2,970	12,118	6,060	5.4
Libya	—	—	—	—
Morocco	1,290	34,936	3,320	0.2
Oman	—	—	8,680	–0.3
Qatar	—	—	16,330	–5.1
Saudi Arabia	—	—	9,700	–3.1
Sudan	—	—	—	—
Syria	1,160	16,808	3,020	4.3
Tunisia	1,930	17,581	4,550	1.3
Turkey	2,830	177,530	6,060	1.7
United Arab Emirates	—	—	17,000	–4.8
West Bank	—	—	—	—
Western Sahara	—	—	—	—
Yemen	380	6,016	790	–2.2

*Purchasing power parity.

Source: *The World Bank Atlas,* 1998.

Regional Economic Patterns

As Table 7.2 suggests, remarkable economic differences characterize the region. Some oil-rich countries have prospered greatly since the early 1970s, but in many cases fluctuating oil prices, political disruptions, and rapidly growing populations have pressured prospects for economic growth. Other nations, while poor in oil and gas reserves, have seen brighter prospects through moves toward greater economic diversification. Finally, some countries in the region exemplify the ongoing reality of persisting poverty, where rapid population growth and the basic challenges of economic development combine with political instability to produce very low standards of living.

Higher-Income Oil Exporters The most prosperous countries of Southwest Asia and North Africa owe their wealth to massive oil reserves. Nations such as Saudi Arabia, Kuwait (after recovering from the Gulf War), Qatar, Bahrain, and the United Arab Emirates benefit from their fossil fuel production as well as from their relatively small populations. Since the oil-price hikes of the 1970s, these countries have reaped billions in revenues that have had a fundamental impact upon their economies. Particularly in the case of Saudi Arabia, huge investments in transportation infrastructure, in urban commercial and financial centers, as well as in other petroleum-related industries have reshaped the contemporary cultural landscape (Figure 7.34). The Saudi petroleum-processing and shipping centers of Jubail (on the Persian Gulf) and Yanbu (on the Red Sea) exemplify this commitment to expand their economic base beyond the simple extraction of crude oil. With these new industrial hubs, the Saudis participate more broadly in the benefits of the oil economy by refining and manufacturing petroleum-related products before they leave the country. While much of the oil wealth in these nations remains concentrated among the ruling elite, petrodollars also have provided real improvements for the larger population. Billions of dollars have poured into new schools, medical facilities, low-cost housing, and modernized agriculture, fundamentally raising the standard of living in these localities in the past 30 years.

Still, problems remain, even in these centers of relative wealth. Poor people do reside in Saudi Arabia and elsewhere

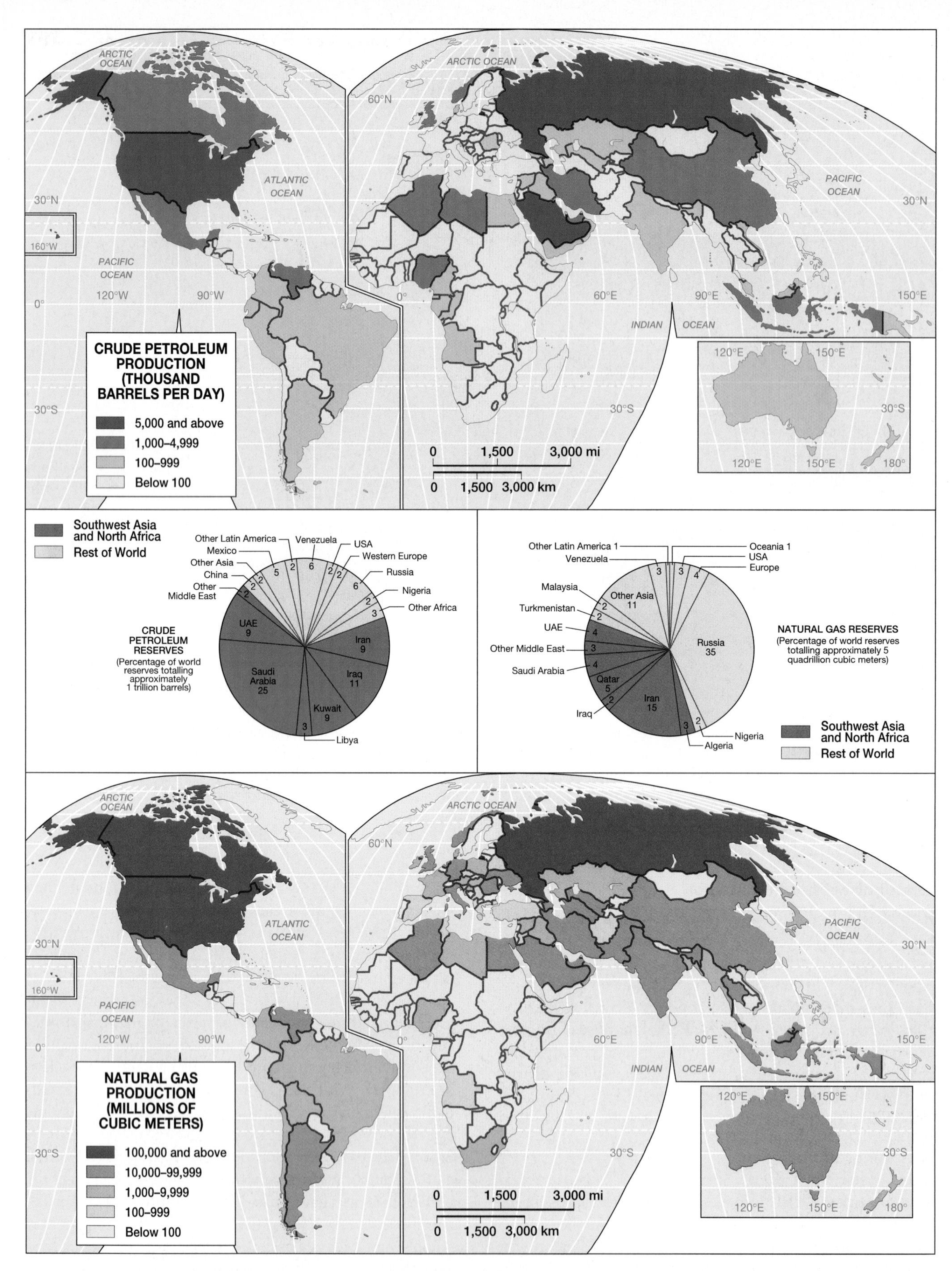

▲ **Figure 7.32 Crude petroleum and natural gas production and reserves** The regions clearly play a pivotal role in the global geography of fossil fuels. Abundant regional reserves suggest that pattern will continue. *(Modified from Rubenstein, 1999,* An Introduction to Human Geography, *Upper Saddle River, NJ: Prentice Hall)*

◀ **Figure 7.33 Persian Gulf** This satellite view of the Persian Gulf reveals one of the world's richest sources of petroleum. Sedimentary rocks, both on land and offshore, contain additional reserves that can sustain production for decades. *(Earth Satellite Corporation/Science Photo Library/Photo Researchers, Inc.)*

in the oil-rich region. For example, the Shiite minority living in eastern Saudi Arabia has a standard of living far below the urban populations elsewhere in the country. In addition, large foreign workforces are common across the oil-rich zone, and they typically are paid far less than domestic laborers. Indeed, less than 10 percent of the private workforce in Saudi Arabia is composed of Saudi citizens. Foreign men usually work in construction and in the oil industry, while foreign women mostly work as domestic servants. Labor sources include the poorer countries of the region (including Jordan, Egypt, and Yemen), as well as such distant states as Pakistan, Bangladesh, and the Philippines. While foreign laborers typically live and work under tough conditions, most are eager to receive wages that are far better than they could earn at home.

Another challenge relates to the shifting fortunes of the oil economy itself. When prices rise, as they did in the 1970s and early 1980s, these oil-rich nations gained an economic windfall. But such a dependence on oil and gas revenues clearly has a darker side: falling prices, such as those seen in the mid-1980s and again in the late 1990s, created immediate economic pain for the major Middle East producers. In fact, the UAE, Qatar, and Oman all experienced real declines in GNP during the 1990s. Since 1990, even Saudi Arabia has been forced to reduce government spending as global petroleum prices have fallen and as its external debt has ballooned. Perhaps the biggest effect of these changes has been on the psychology of Gulf Arabs who increasingly must learn to live within their more modest means. With many non-OPEC producers competing in the world marketplace (including Europe, Mexico, and Russia), OPEC's larger hold on oil markets has diminished, as have prospects for a sustained return to high petroleum prices. As a result, future economic growth in these wealthier but oil-dependent states will likely be moderate, at best. While Saudi Arabia, blessed with abundant reserves of easily accessible oil, can survive with high or low prices in the future, countries such as Bahrain and Oman are faced with the additional problem of rapidly depleting their reserves over the next 20 to 30 years.

▲ **Figure 7.34 Saudi Arabian oil refinery** Eastern Saudi Arabia's Ras Tanuna Oil Refinery links the oil-rich country to the world beyond. Huge foreign and domestic investments since 1960 have dramatically transformed many other settings in the region. *(Minosa/Scorpio/Sygma Photo News)*

Lower-Income Oil Exporters Other states in the region are important secondary players in the oil trade, but different political and economic variables have often hampered sustained economic growth. In North Africa, Algeria and Libya illustrate this scenario. Algerian oil and natural gas overwhelmingly dominate its exports, but the 1990s brought lower fossil fuel prices, political instability, and increasing shortages of consumer goods. While the country contains some excellent agricultural lands in the north, the overall amount of arable land has increased little over the past 25 years even as the country's population has grown by more than 50 percent. The pressure on the rural economy has accelerated movement to the increasingly crowded cities, and many of the nation's upwardly mobile workers have left the country for employment opportunities in western Europe. Nearby Libya also remains a major exporter of oil and natural gas, but its political hostility to most of Europe and the United States has severely limited prospects for expanded foreign investment. While Qaddafi has undertaken ambitious efforts to expand the agricultural economy and improve water availability in his desert land, the country's enduring image as an international "outlaw" nation and its heavy military buildup do not bode well for building new economic bridges with the West.

Similar economic and political challenges face Iraq in Southwest Asia. Iraq has been its own worst economic enemy. Sanctions imposed after the Persian Gulf War virtually halted Iraq's oil exports, crippling its economy. Since Iraq is not self-sufficient in most basic products, living standards plummeted to extremely low levels. In the late 1990s the United Nations allowed Iraq to export oil in limited quantities, but only in exchange for food and medicine. Even so, health care in the country has deteriorated greatly and food shortages in many areas are much more common than they were 20 years ago. Iraq certainly has the potential to be a prosperous country: its population density is moderate and its resource base is substantial. Yet as long as the current regime remains in power, Iraq will likely continue to suffer.

The situation in Iran is more complex. The country is large, populous, and has a relatively diverse economy. Iran's oil reserves are huge and have seen active commercial development since 1912. The country also has a sizable industrial base, much of it built in the 20 years prior to the fundamentalist revolution in 1979. In addition, considerable capital flowed into modernizing Iran's agricultural infrastructure during the same period, bringing with it the benefits of increased mechanization and new irrigation projects. Still, most observers agree that today Iran is relatively poor, burdened with a stagnating if not declining standard of living. Since 1980, the country's fundamentalist leaders have downplayed the role of international trade in consumer goods and services, fearing they would import unwanted cultural influences from abroad. Making matters worse were the twin challenges of a costly and bloody war with Iraq in the 1980s followed by the struggling oil economy of the 1990s. Recently, however, the country's economic prospects have brightened somewhat as new economic links with Central Asia have been completed and as more moderate political leadership has tilted toward greater economic interaction with the West.

Prospering Without Oil Some countries, while lacking in petroleum resources, have nevertheless found paths toward increasing economic prosperity. Israel, for example, supports the highest standard of living in the region, even with its political challenges. The nation's sparse natural resource base would hardly seem conducive to economic development. But the Israelis have invested large amounts of capital to create a highly productive agricultural and industrial base. The country is also emerging as a global center for high-tech computer and telecommunications products. Many U.S. and European companies maintain development and production centers there, and the country is increasingly becoming known for its fast-paced and highly entrepreneurial business culture that resembles California's Silicon Valley. Tourism is another important sector of the economy as the country attracts global visitors interested in the region's rich cultural heritage. Even so, Israel has many economic problems. Its persisting struggles with the Palestinians and with neighboring states have sapped much of its potential vitality. Defense spending commands a large share of total GNP, necessitating high tax rates. If the peace process continues, however, Israel and its neighbors may eventually trade more and engage in other forms of economic cooperation, thus boosting all of their economies.

Turkey also has a diversified economy. While its per capita GNP of $2,830 is modest even by regional standards, it has shown far greater economic dynamism in recent years than many of the major oil producers in the region. Lacking petroleum, Turkey produces varied agricultural and industrial goods for export. Almost half of the population remains employed in agriculture and the country's principal commercial products include cotton, tobacco, wheat, and fruit. The industrial economy has also grown since 1980, including exports of textiles, food, and chemicals. Turkey remains the most important tourist destination in the region as well, attracting more than six million visitors annually during the 1990s. In addition, Istanbul enjoys increasing prominence as a regional finance and investment center. Many Turkish leaders hope to bolster the economy by eventually joining the European Union. Turkey already has closer links to the West than any other country of the region except Israel (it is, for example, a member of NATO), and its trade potential with Europe is great. Still, economic integration with the West will not be easy. Some European leaders are wary of extending their economic union to what they consider to be a politically unstable non-European state. Greece also resists the closer integration of Turkey into the European economic and political block, and conservative religious leaders in Turkey oppose stronger ties to Europe.

Other oil-poor countries in the region also aspire to economic prosperity. In North Africa, Tunisia has fostered relative economic stability without the benefit of large oil reserves.

Although the country possesses a few important oilfields south of Tunis, most of its workforce is in agriculture, services (particularly tourism), or manufacturing. Recent government policies have favored economic reforms, increased private investment, and encouraged participation in global markets such as those that bring the region's citrus fruits to grocery stores in the United States. A similar situation benefits Cyprus. Although divided along ethnic lines, the Greek-dominated southern portion of the island has invested in infrastructure, education, and tourism, while favorable tax and trade policies have attracted considerable foreign capital. Even war-torn Lebanon has the potential for prosperity. At one time, Lebanon was the region's financial center and wealthiest state. Then the country was devastated by war during most of the 1970s and 1980s. A slow economic recovery, however, began in the 1990s as the political climate improved somewhat. Tourism is on the rise and the country's new telecommunications industry signals the potential for growth in that sector of the economy.

▲ **Figure 7.35 Sudanese village** Poverty and political instability remain a part of life in many Sudanese villages. The traditionally cultivated fields around the village center contrast vividly with the skyscrapers and industrial complexes found in oil-rich portions of the region. *(Liba Taylor/Hutchison Picture Library)*

Regional Patterns of Poverty The poorer countries of the region share the problems of much of the less-developed world. In North Africa, the nations of Sudan, Morocco, and Egypt each face distinctive economic challenges. For Sudan, devastating political problems have stood in the way of progress. Civil war has resulted in major food shortages, particularly in the south. In addition, the conflicts have disrupted major agricultural improvement schemes in the Gezira district between the White and Blue branches of the Nile River. The country's lack of infrastructure has seen little new investment, settlement remains overwhelmingly rural, and secondary school enrollments stand at less than 25 percent of the potential school-age population (Figure 7.35 and Table 7.3). Properly developed, Sudan's fertile soils could support a major food-producing region, but the country's sustained economic development appears forestalled by continuing political instability.

Morocco's economic picture is brighter, but the country remains poorer than either Algeria or Tunisia. Poverty is especially widespread in the Atlas Mountains, where many Berber communities have little access to modern services or infrastructure. Elsewhere in the country, economic reforms have spurred privatization, encouraged investment by foreign banks, and slowed inflation. Still, illiteracy is widespread and the country suffers from the **brain drain** phenomenon as some of its brightest young people leave for better jobs in western Europe. Population growth and a stagnating agricultural economy have also spurred internal movements from rural areas to the country's major cities, including Casablanca and Rabat, but often migrants are frustrated by ongoing problems of underemployment even in these rapidly expanding metropolitan areas.

Egypt's economic prospects are also unclear. On the one hand, the country experienced real economic growth during the 1990s as President Hosni Mubarak pushed for smaller government deficits and a multibillion-dollar privatization program to put government-controlled assets under more efficient management. Egypt also actively courts foreign investment in its expanding industrial sector, and the streets of Cairo are increasingly sprinkled with the flash of Rolls-Royces, Porsches, and Lamborghinis. Even so, most Egyptians still live in poverty. Millions till tiny plots of land along the Nile, while others scramble for living wages in low-paying service and manufacturing jobs in Cairo or Alexandria. Furthermore, the demographic clock is ticking: Egypt's 65 million people already make it the region's most populous state, and recent efforts to expand the nation's farmland have met with numerous environmental, economic, and political problems. Future prosperity hinges on the country's ability to expand its economy faster than its population.

In Southwest Asia, Yemen remains the poorest country on the Arabian Peninsula. Positioned far from most of the region's principal oilfields, Yemen's low per capita GNP puts it on par with many nations in impoverished Sub-Saharan Africa or South Asia. The largely rural country relies mostly on marginally productive subsistence agriculture and much of its mountain and desert interior lacks effective links to the outside world. The present state emerged in 1990 with the political union of the former countries of North and South Yemen, but periodic civil unrest continues between the two groups. Yemen was also a major economic victim in the Persian Gulf War. Its government supported Iraq in the conflict, prompting many nearby oil-rich states to return millions of Yemeni workers to their impoverished homeland.

Issues of Social Development

Measures of social development vary widely across the realm and are broadly associated with overall levels of economic development. Israel's population benefits from high-quality health care and education. In fact, Israel does better in many social measures than might be expected on the basis of its per

Table 7.3 Social Indicators and Status of Women

Country	Life Expectancy at Birth		Under Age 5 Mortality, per 1,000 Live Births		Secondary School Enrollment %		Female Labor Force Participation (% of total)
	Male	Female	1960	1993	Male	Female	
Algeria	66	68	243	68	66	55	25
Bahrain	68	71	—	—	98	101	19
Cyprus	75	80	—	—	94	96	38
Egypt	65	69	258	59	81	69	29
Gaza	70	74	—	—	—	—	—
Iran	67	68	233	54	74	58	25
Iraq	58	60	171	71	—	—	18
Israel	76	80	39	9	84	91	40
Jordan	66	70	149	27	52	54	22
Kuwait	72	73	128	13	60	60	29
Lebanon	68	73	91	40	73	78	28
Libya	62	67	269	100	95	95	21
Morocco	69	74	215	59	40	29	35
Oman	68	72	300	29	64	57	15
Qatar	69	74	—	—	82	84	13
Saudi Arabia	68	71	292	38	54	43	14
Sudan	50	52	292	128	24	19	29
Syria	67	68	201	39	52	42	26
Tunisia	67	69	244	36	—	—	31
Turkey	66	71	217	84	74	48	36
United Arab Emirates	73	75	240	21	84	94	14
West Bank	70	74	—	—	—	—	—
Western Sahara	46	48	—	—	—	—	—
Yemen	57	60	—	—	38	7	29

Sources: *Population Reference Bureau Data Sheet, 1998,* Life Expectancy (M/F); *World Resources Institute, 1996–97,* Under-5 Mortality Rate; *Population Reference Bureau Data Sheet, 1996,* Secondary School Enrollment (M/F); *The World Bank Atlas, 1998,* Female Participation in Labor Force.

capita GNP. Average life spans and rates of infant mortality parallel those of the United States, and the vast majority of the population is literate. Social conditions are appreciably better, however, for Israel's Jewish majority than they are for its substantial Muslim minority. Other relatively wealthy countries—particularly high-income oil exporters such as Saudi Arabia—have lower figures of social well-being than might be expected. In Saudi Arabia, for example, 24 percent of the population remains illiterate and infant mortality is four times higher than in Israel. The recent timing of Saudi Arabia's development helps explain the divergence, and future figures may improve substantially, based on recent investments in health care and education. Not surprisingly, illiteracy rates are high and average life expectancy is low in the poorer countries of the region, such as Sudan and Yemen.

The role of women in the largely Islamic region also remains a major social issue. Female labor participation rates in the workforce are the lowest in the world. In the most conservative parts of the region, few women are allowed to work outside of the home. Even in portions of Turkey, where Western influences are widespread, it is rare to see women selling goods in the marketplace or driving cars in the street. More orthodox Islamic states impose legal restrictions on the activities of women. In Saudi Arabia, for example, women are not allowed to drive. In Iran, full veiling remains mandatory in more conservative parts of the country. Yet even in a fundamentalist state such as Iran, women's roles are changing. Young working women in Tehran, for example, are much more likely to be wearing Western-style fashions than was the case 10 years ago (Figure 7.36). Educational opportunities are also increasingly open to girls across much of the region. In Sudan and Saudi Arabia, a growing number of women pursue high-level careers. Education may be segregated, but it is available. Women also have a relatively visible social position in Israel, except in fundamentalist Jewish communities.

▲ **Figure 7.36 Life in modern Tehran** More than twenty years past Iran's Islamic Revolution, young Iranian urbanites are pressing for political reforms and a more liberal society that reverses many of the conservative mandates of the nation's religious leaders. *(U Corbis/AFP)*

Global Economic Relationships

Southwest Asia and North Africa share close economic ties with the world beyond. While OPEC's fortunes have weakened with global competition, regional exports are still dominated overwhelmingly by oil, natural gas, and other petroleum-related products. Western Europe, the United States, Japan, and many less-developed countries depend on the region's fossil fuels (Figure 7.37). In the case of Saudi Arabia, crude oil shipments make up more than 70 percent of their exports, with refined oil and petrochemicals constituting another 20 percent of the total. Other countries within the region, however, are less dependent on oil-related exports. Turkey, for example, exports textiles, food products, and manufactured goods to its principal trading partners Germany, Russia, Italy, and the United States. Similarly, Israeli exports emphasize its highly skilled workforce: products such as cut diamonds, electronics, and machinery parts leave the country for destinations in the United States, western Europe, and Japan. Food and manufactured goods dominate imports in most countries. The sources for these products vary across the realm. The UAE depends on a broad assortment of suppliers, including Japan, the United States, China, and India. Conversely, the Maghreb nations of Morocco and Algeria import goods largely from western Europe, particularly France.

Flows of labor and capital also illustrate the region's interconnections with the global economy. Migrating workers, both within and beyond the region, have had major economic impacts across the region. For the Persian Gulf states, the influx of millions of workers from other parts of the Middle East and South Asia has kept wages relatively low in these less-populated countries. In addition, millions of workers have left the area, principally for jobs in western Europe. Turkish workers in Germany and Algerians and Moroccans in France exemplify the pattern and are a reminder of the increasingly close ties between the region and its European neighbors. Political instability often accentuates such movements: many wealthier Lebanese and Iranians who were unhappy with their domestic political scene in the 1980s migrated to the United States or Europe to escape the turmoil. Flows of investment capital out of the region also have had broad global impacts. As oil prices increased tenfold between 1970 and 1980, many of those petrodollars ended up in foreign bank accounts, stock markets, and real estate investments. Although the region's role in the global oil economy has declined in subsequent decades, latest estimates still put the total value of Gulf Arab foreign investments at $800 billion. These investments influence everything from the cost of office buildings in downtown Singapore to the yields on Brazilian bonds.

Future interconnections with the global economy may depend increasingly upon cooperative economic initiatives far beyond the dimming prospects of the OPEC organization. Relations with the European Union (EU) are critical. Beginning in 1996, Turkey enjoyed closer economic ties with the EU, but recent attempts at full membership in the organization have failed. Other so-called "Euro-Med" agreements have also been signed between the EU and Morocco, Tunisia, Jordan, and Israel. Supporters argue that these agreements will bring more export-oriented industries to the region. Most Arab countries, however, are wary of too much European dominance. They formed a regional political organization known as Arab League in 1945, and in 1998, 18 League members established the Arab Free-Trade Area (AFTA) designed to eliminate all intraregional trade barriers by 2008 and to spur economic cooperation within the region. Smaller organizations tie together other regional constituencies. For example, the six-member Gulf Cooperation Council provides major oil producers in the area a forum to discuss shared security and trade-related issues, and the five-member Union of the Arab Maghreb is dedicated to promoting economic integration and a free trade zone across North Africa. In addition, Saudi Arabia has played a particularly pivotal role in regional economic development through organizations such as the Islamic Development Bank and the Arab Fund for Economic and Social Development.

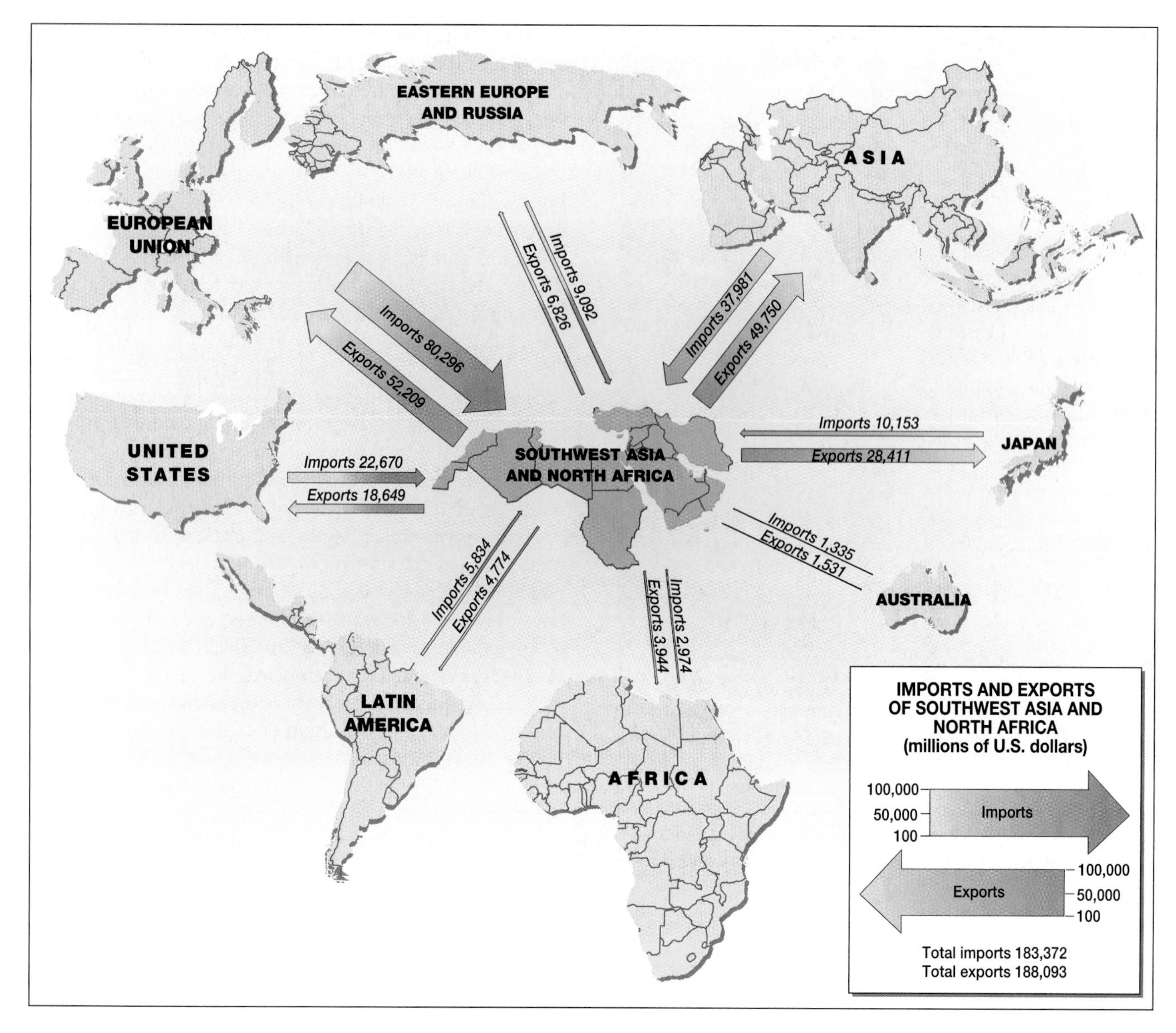

▲ **Figure 7.37 Southwest Asia and North African trade** Major trade flows connect the region with Europe, North America, and Asia, particularly Japan. Outgoing oil and petroleum-related products are counterbalanced by imports of manufactured goods and food. Increasingly, however, specialized manufacturing operations in countries such as Israel, Lebanon, and Turkey find demand for products in global markets. *(Data from Euromonitor,* International Marketing Data and Statistics, 1997*)*

Conclusion

Positioned at the meeting ground of Earth's largest land masses, Southwest Asia and North Africa has often played a pivotal role in world history and in processes of globalization that bind the planet ever more tightly together. In ancient times, the region's inhabitants were early contributors to the Agricultural Revolution, a long process of plant and animal domestication destined to reshape the cultural landscapes of virtually every corner of the globe. The realm also served as a crucible for urban civilization, offering in the process a fundamentally new and different way of human settlement that continues to reshape the distribution of global populations today. Three of the world's great religions—Judaism, Christianity, and Islam—also emerged beneath its desert skies. Ever since, the world has pondered the profundity of their messages and struggled to reconcile the cultural and political conflicts that sprang from their closely intertwined roots. More recently, from the sweeping expansion of Ottoman rule to the unpredictable machinations of OPEC oil ministers, regional forces have continued to reach beyond its boundaries, reconfiguring global political, economic, and cultural relationships.

Despite this rich legacy of global influence and power, the peoples of Southwest Asia and North Africa are struggling at the beginning of the twenty-first century. Indeed, most of the countries within the region suffer from economic stagnation. There are many causes for the present malaise. Twentieth-century population growth across the region has been dramatic. At the same time, it has been difficult and costly to expand the region's limited supplies of agricultural land and water resources. The results, apparent from the eroded soils of the Atlas Mountains to the overworked garden plots along the Nile, are a classic illustration of the environmental price paid when population growth outstrips the ability of the land to support it. Political conflicts have also played a role in disrupting processes of economic development across the region. Civil wars, conflicts between states, and regional-scale tensions have worked against initiatives for greater cooperation and trade.

Perhaps most importantly, the region must deal with the intrinsic contradictions between modernity and the more fundamentalist interpretations of Islam. Turkey, Iran, and Saudi Arabia represent three distinctive responses to those contradictions, but all three models have come under stress in recent years. Turkey has opted for the model of a modern secular society fully integrated with the global economic system, one in which religion is relegated to the private sphere. Even so, the recent rise of an explicitly Islamic Turkish political party shows that many Turks now question their more secular approach. Nearby Iran, on the other hand, created a radical Islamic theocracy, only to find recently that its citizens are questioning the wisdom of insulating themselves from Western society. Saudi Arabia has tried to play it both ways, linking its economy and security closely to the West while remaining essentially a religious state with a highly traditional and authoritarian political and social order. One thing is certain: future cultural change will still be guided by a complex response to Western influences, a mix of fascination and suspicion that will no doubt yield its own unique regional cultural amalgam.

The region still commands considerable global attention in the early years of the twenty-first century. It remains the hearth of Christendom, the spatial and spiritual core of Islam, and the political and territorial salvation of modern Judaism. Muslims worldwide are influenced by its cultural and political evolution, and the Arab-Israeli conflict will continue to be a critical element in global geopolitics. Indeed, broader regional hopes for peace seem limited until the diverse peoples of the eastern Mediterranean reach a political rapprochement that satisfies both the Palestinians and their Jewish neighbors. In addition, the accidents of geology and the thirsts of the global economy dictate that the region will maintain its prominence in world petroleum markets, although it is never likely to dominate oil prices and output to the extent it did in the late 1970s and early 1980s. What is more likely are moves toward economic diversification and integration, initiatives that gradually will draw the region closer to Europe and other participants in the global economy. Even so, Southwest Asia and North Africa will retain its distinctive regional identity, a character defined by its setting, the rich cultural legacy of its history, the selective abundance of its natural resources, and its stubbornly persistent political problems.

Key Terms

Balfour Declaration (page 292)
brain drain (page 303)
culture hearth (page 265)
domestication (page 276)
exotic rivers (page 279)
Fertile Crescent (page 276)
fossil water (page 274)
Green Line (page 296)
Hajj (page 283)
Islamic fundamentalism (page 266)
kibbutz (page 279)
Levant (page 267)
Maghreb (page 267)
medina (page 280)
monotheism (page 283)
OPEC (Organization of Petroleum Exporting Countries) (page 266)
Ottoman Empire (page 286)
Palestinian Authority (PA) (page 294)
pastoral nomadism (page 277)
physiological densities (page 275)
qanat system (page 273)
Quran (page 283)
Shiites (page 284)
Suez Canal (page 291)
Sunnis (page 284)
theocratic state (page 284)
transhumance (page 277)

Questions for Summary and Review

1. Why do Southwest Asia and North Africa form a useful world region? What are some of the problems associated with defining the region?
2. Describe the climatic changes you might experience as you travel from the eastern Mediterranean coast to the highlands of Yemen. What are some of the key climatic variables that explain these variations?
3. Discuss five important human modifications of the Southwest Asian and North African environment and assess whether these changes have benefited the region.
4. Discuss how pastoral nomadism, oasis agriculture, and dryland wheat farming represent distinctive adaptations to the regional environments of Southwest Asia and North Africa. How do each of these rural lifeways create distinctive patterns of settlement?
5. Compare the modern maps of religion and language for the region and identify three major examples where Islam dominates non-Arabic speaking areas. Explain why that is the case.
6. Describe the role played by the French and British in shaping the modern political map of Southwest Asia and North Africa. Provide specific examples of their lasting legacy.
7. Outline three regional examples where Islamic fundamentalism has redefined the domestic geopolitical setting since the late 1970s.

8. Explain why internal political tensions in both Lebanon and Cyprus are also related to neighboring states.
9. Describe the basic geography of oil reserves across the region and compare the pattern with the geography of natural gas reserves.
10. What strategies for economic development have recently been employed by nations such as Turkey, Israel, Egypt, and Morocco? How successful have they been?

Thinking Geographically

1. How might a major project for transferring water from Turkey to the Arabian Peninsula affect the development of Saudi Arabia? What would be some of the potential political and ecological ramifications of such a project?
2. How might the high birthrates of many Southwest Asian and North African countries affect their future development? Does religion play a significant role in determining fertility patterns in this part of the world?
3. What economic changes could occur if Israel and the Palestinians were to reach a lasting peace? What kinds of general connections might be found between political conflict and economic conditions throughout the region?
4. Will future relations between the region's Muslim societies and the United States improve or worsen? What factors would you cite to defend your answer?
5. What might be some of the reasons for Southwest Asia and North Africa's general failure to match the rates of economic and industrial growth found in North America?
6. Why has the idea of Arab nationalism failed to achieve any lasting geopolitical changes in the region?
7. As a ruler of a conservative Arab state, what might be the advantages and disadvantages of opening up your country to the Internet?

Regional Novels and Films

Novels

Shmuel Agnon, *Shira* (1989, Schocken)

Hanan Al-Shaykah, *Beirut Blues: A Novel* (1995, Doubleday)

Simin Daneshvar, *Savushun: A Novel About Modern Iran* (1990, Mage)

Kahlil Gibran, *The Prophet* (1923, Knopf) and *Broken Wings: A Novel* (1957, Citadel)

Barbara Hodgson, *The Tattooed Map* (1995, Chronicle Books)

Ulfat Idlibi, *Sabruya: Damascus Bitter Sweet, A Novel* (1980, Interlink)

Sahar Kalifeh, *Wild Thorns* (1988, Olive Branch Press)

Yasar Kemal, *Salman the Solitary* (1998, Harvill)

Muhsin Mahdi, ed., *The Arabian Nights* (1992, Knopf ed.)

Naguib Mahfouz, *Palace Walk* (1991, Doubleday) and *Palace of Desire* (1991, Doubleday)

Abdelrahman Munif, *Cities of Salt: A Novel* (1987, Random House)

Films

The Battle of Algiers (1965, Italy)

Casablanca (1942, U.S.)

The English Patient (1996, U.S.)

The House on Chelouche Street (1973, Israel)

Kazablan (1974, Israel)

Late Summer Blues (1987, Israel)

Lawrence of Arabia (1962, U.S.)

Lion of the Desert (1981, U.K.)

Raiders of the Lost Ark (1981, U.S.)

Sallah (1965, Israel)

The Sheltering Sky (1990, U.S.)

The Ten Commandments (1956, U.S.)

The White Balloon (1995, Iran)

Bibliography

Abun-Nasr, Jamil M. 1975. *A History of the Maghrib*. Cambridge: Cambridge University Press.

Barakat, Halim. 1993. *The Arab World: Society, Culture, and State*. Berkeley: University of California Press.

Blake, Gerald H. 1988. *The Cambridge Atlas of the Middle East and North Africa*. Cambridge: Cambridge University Press.

Canfield, Robert, ed. 1991. *Turko-Persia in Historical Perspective*. Cambridge: Cambridge University Press.

Cohen, Saul B. 1992. "Middle East Geopolitical Transformation: The Disappearance of a Shatter Belt." *Journal of Geography* 91 (January/February), 2–10.

Cole, Juan R., ed. 1992. *Comparing Muslim Societies: Knowledge and the State in a World Civilization*. Ann Arbor: University of Michigan Press.

Cressey, George B. 1960. *Crossroads: Land and Life in Southwest Asia*. Chicago: J. B. Lippincott.

Drysdale, Alasdair, and Blake, Gerald H. 1985. *The Middle East and North Africa: A Political Geography*. New York: Oxford University Press.

English, Paul Ward. 1966. *City and Village in Iran: Settlement and Economy in the Kirman Basin*. Madison: The University of Wisconsin Press.

Gellner, Ernest. 1981. *Muslim Society*. Cambridge: Cambridge University Press.

Held, Colbert C. 1994. *Middle East Patterns: Places, Peoples, and Politics*. Boulder, CO: Westview Press.

Hillel, Daniel. 1994. *Rivers of Eden: The Struggle for Water and the Quest for Peace in the Middle East*. Oxford: Oxford University Press.

Hourani, Albert. 1991. *A History of the Arab Peoples*. New York: Warner Books.

Jabbur, Jibrail. 1995. *The Bedouins and the Desert: Aspects of Nomadic Life in the Arab East*. Albany, NY: State University of New York Press.

Kliot, Nurit. 1994. *Water Resources and Conflict in the Middle East*. London: Routledge.

Lapidus, Ira M. 1988. *A History of Islamic Societies*. Cambridge: Cambridge University Press.

Lawrence, Bruce B. 1989. *Defenders of God: The Fundamentalist Revolt Against the Modern Age*. San Francisco: Harper & Row.

Lewis, Bernard. 1993. *The Arabs in History*. Oxford: Oxford University Press.

Miller, Judith. 1996. *God Has Ninety-Nine Names: Reporting from a Militant Middle East*. New York: Simon & Schuster.

Mostyn, Trevor, ed. 1988. *The Cambridge Encyclopedia of the Middle East and North Africa*. New York: Cambridge University Press.

Peters, F. E. 1994. *The Hajj: The Muslim Pilgrimage to Mecca and the Holy Places*. Princeton: Princeton University Press.

Rodenbeck, Max. 1999. *Cairo: The City Victorious*. New York: Random House.

Starr, Joyce R., and Stoll, Daniel. 1988. *The Politics of Scarcity: Water in the Middle East*. Boulder, CO: Westview Press.

Stewart, Dona. 1996. "Cities in the Desert: The Egyptian New-Town Program." *Annals of the Association of American Geographers* 86, 459–80.

Swearingen, Will D., and Bencherifa, Abdellatif, eds. 1996. *The North African Environment at Risk*. Boulder, CO: Westview Press.

Tessler, Mark. 1994. *A History of the Israeli-Palestinian Conflict*. Bloomington, IN: Indiana University Press.

GREENLAND (Denmark)
Jan Mayen (Norway)
10°W
0°
10°E
20°E
30°E
40°E
ICELAND
Reykjavik
Arctic Circle
60°N
Norwegian Sea
Oulu
FINLAND
Trondheim
Faroe Islands (Denmark)
SWEDEN
Tampere
Shetland Islands (U. K.)
NORWAY
Helsinki
Oslo
Tallinn
Orkney Islands (U. K.)
Stockholm
ESTONIA
ATLANTIC OCEAN
Skagerrak
Kattegat
Goteborg
Riga
LATVIA
Alborg
Glasgow
Edinburgh
North Sea
LITHUANIA
Vilnius
Baltic Sea
Copenhagen
DENMARK
Belfast
UNITED KINGDOM
RUSSIA
Dublin
Leeds
Gdansk
BELARUS
50°N
IRELAND
Sheffield
Cork
Hamburg
Birmingham
NETHERLANDS
Bremen
Elbe R.
Vistula R.
Warsaw
Poznan
Amsterdam
Thames R.
The Hague
Hannover
Berlin
Rotterdam
Lodz
London
Oder R.
Plymouth
Essen
Leipzig
POLAND
Brussels
English Channel
GERMANY
Wroclaw
UKRAINE
BELGIUM
Frankfurt
Prague
Krakow
Brest
LUXEMBOURG
Luxembourg
CZECH REP.
SLOVAKIA
MOLDOVA
0 200 400 mi
0 200 400 km
Paris
Seine R.
Stuttgart
Kishinev
Rhine R.
Danube
Bratislava
Vienna
Budapest
Loire R.
Munich
AUSTRIA
Bern
LIECHTENSTEIN
HUNGARY
FRANCE
SWITZERLAND
ROMANIA
Bay of Biscay
Timisoara
SLOVENIA
Zagreb
Lyon
Milan
Ljubljana
Bucharest
Bordeaux
CROATIA
Rhône R.
Torino
Po R.
Venice
Belgrade
Garonne R.
Genoa
Bologna
BOSNIA & HERZEGOVINA
Danube R.
SAN MARINO
Sarajevo
León
Florence
BULGARIA
Toulouse
Sofia
YUGOSLAVIA
40°N
MONACO
Porto
Marseille
Ebro R.
Adriatic Sea
Skopje
Bospo
Zaragoza
ANDORRA
Corsica (France)
Rome
MACEDONIA
PORTUGAL
Madrid
Tirana
Barcelona
ITALY
Thessaloniki
Tagus R.
ALBANIA
Lisbon
Dardane
SPAIN
Naples
Valencia
Aegean Sea
GREECE
Balearic Islands (Spain)
Sevilla
Sardinia (Italy)
Athens
Palermo
Cádiz
Malaga
Mediterranean Sea
Sicily (Italy)
Sparta
Strait of Gibraltar
Valletta
Crete (Greece)
MALTA
10°W
AFRICA
20°E

CHAPTER 8

Europe

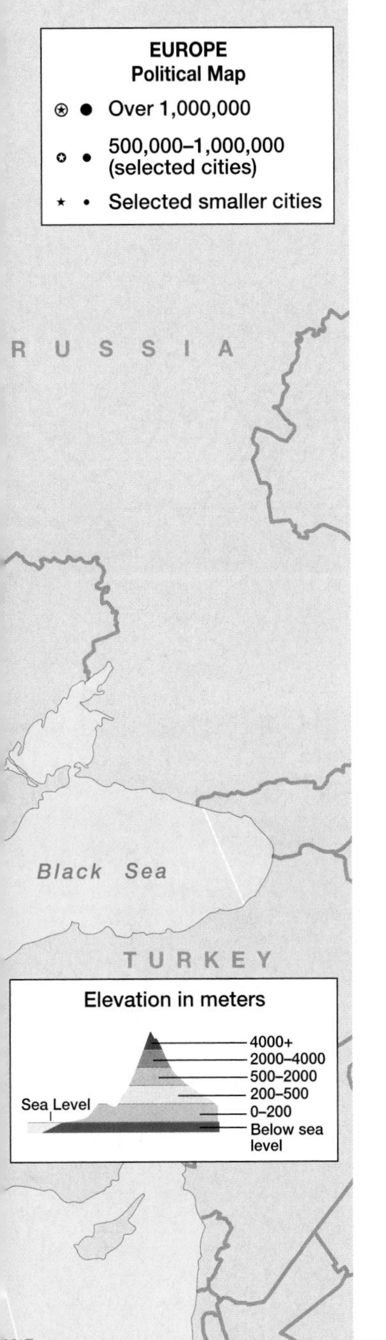

Europe is one of the most diverse regions in the world, encompassing a wide array of people and places in an area considerably smaller than North America. More than half a billion people reside in this region, living in 37 different countries that range in size from giant Germany to microstates, such as Andorra and Monaco (Figure 8.1).

The region's remarkable cultural diversity produces a geographical mosaic of different languages, religions, and landscapes. Commonly, a day's journey finds a traveler speaking two or three languages, changing money several times, and sampling distinct regional food and drink. Cultural landscapes also vary widely, with diverse house types, settlement forms, and field patterns characterizing the everyday scene. Indeed, much of Europe's regional cohesion and identity comes from a shared history that has unfolded in close geographic proximity (see "Setting the Boundaries").

Though the traveler may revel in Europe's cultural and environmental diversity, these regional differences are also entangled with Europe's troubled past, a history of neighbors warring on each other because of these very same cultural and political differences. It is often said Europe invented nationalism and its political expression, the nation-state. But nationalism has also been Europe's Achilles heel, leading it into devastating wars and destructive regional rivalries. In the twentieth century alone, Europe was the battleground of two world wars, followed by a 40-year **Cold War** that divided the continent (and the world) into two hostile, highly armed camps: Europe and the United States against the former Soviet Union.

Today, however, a spirit of cooperation prevails as Europe sets aside nationalistic agendas and works toward regional economic, political, and cultural integration through the **European Union (EU).** This supranational organization is made up of 15 countries, anchored by the western European states of Germany, France, Italy, and the United Kingdom. Because of the EU's demonstrated success, many applicants want to join the organization, including most of the former Soviet satellites in eastern Europe. Undoubtedly, the geographical reach and economic policies of the EU will continue to transform the region during the twenty-first century.

Some Europeans, however, are not convinced the EU is the answer to Europe's problems. In Norway, for example, the government has voted to join the EU on two separate occasions, only to have its population twice reject the idea in national referenda. Opponents of

◀ **Figure 8.1 Europe** Stretching from Iceland in the Atlantic to the Black Sea, Europe includes no fewer than 37 countries, ranging in size from large states, such as France and Germany, to the microstates of Liechtenstein, Andorra, San Marino, and Monaco. Currently, the population for the region is about 520 million. Europe is highly urbanized and, for the most part, affluent, particularly in the western portion. However, economic and social disparities between eastern and western Europe remain a problem.

Setting the Boundaries

The European region is small in scale compared to the United States. In fact, Europe, from Iceland to the Black Sea, would fit easily into the eastern two-thirds of North America. A more apt and specific comparison would be Canada, since Europe, too, is a northern region; more precisely, more than half of Europe lies north of the 49th parallel, the line of latitude forming the western United States–Canada border (Figure 8.3).

Europe contains no fewer than 37 countries, and that number may indeed grow in the near future. These states range in size in size from large countries, such as France and Germany, to microstates, such as Liechtenstein, Andorra, Monaco, and San Marino. Currently the population for these combined states totals more than 520 million people. Most of these states are fully independent, with their own sovereign governments and most assuredly their own unique histories. Tensions exist between and within many of these varied political units. Also at issue is regional European integration and cooperation.

The notion that Europe is a continent with clearly defined boundaries is a misconception rooted in history. The Greeks and Romans divided their worlds into the three continents of Europe, Asia, and Africa, separated by the water bodies of the Mediterranean, Red Sea, and the Bosporus Straits. A northward extension of the Black Sea was thought to separate Europe from Asia, and only in the sixteenth century was this proven false. Instead, explorers and cartographers discovered that the "continent" of Europe was firmly attached to the western portion of Asia.

Since that time, geographers have not agreed on the eastern boundary of Europe. During the existence of the Soviet Union, some took the "continent's" border all the way east to the Ural Mountains, while others drew the line at the western boundary of the Soviet Union. However, with the disintegration of the Soviet Union in 1990, the eastern boundary of Europe became even more problematic. Now, some geography textbooks extend Europe to the border with Russia, which places the two countries of Ukraine and Belarus, former Soviet republics, in eastern Europe. Though we appreciate that an argument can be made for that expanded definition of the Europe, recent events, along with a bit of crystal-gazing into the near-term future, leads us to reject that approach and, instead, to draw our eastern border at Poland, Romania, and Moldova. To the north, the three Baltic republics of Estonia, Lithuania, and Latvia are also included in Europe. Our justification is this: all of these six countries are currently clearly engaged with Europe and aspire to become even more closely linked to the West in the future. This is not the case with Ukraine and Belarus, which, in contrast, show a decidedly eastern orientation toward Russia.

the EU argue that individual countries lose control over important sovereign matters; for some people—and a few states—EU membership carries more costs than benefits. Thus, tension exists between large-scale regional cooperation and integration and smaller-scale impulses toward retaining independence and autonomy, between a new and shared sense of "Europeaness," and traditional, national, and regional identities (Figure 8.2).

▲ **Figure 8.2 European Union** Fifteen countries are bound together in the European Union (EU), with headquarters in Brussels, Belgium. Originally an organization promoting economic integration in western Europe, the EU now oversees political, cultural, and environmental matters for member states, as well as economic policy. *(Hilarie Kavanagh/Tony Stone Images)*

Europe, like most world regions, is caught up in the tension between globalization and national diversity. Given Europe's considerable impact on the rest of the world as the hearth of western civilization, as the cradle of the Industrial Revolution, and as the home of a global geography of imperialism, some would argue that Europe actually invented globalization. But while all world regions struggle with this vexing problem of trading off global convergence in the face of national interests, Europe finds itself with an added layer of complexity as it moves into the new and untested waters of economic, political, and cultural integration. Europe is truly at a juncture: either it will succeed dramatically by moving past its historical legacy of nationalistic in-fighting, or it will fail miserably by fracturing further into state-based divisions.

Environmental Geography: Human Transformation of a Diverse Landscape

Despite its small size, Europe's environmental diversity is extraordinary. Within its borders are found a startling array of landscapes from the Arctic tundra of northern Scandinavia to the barren hillslopes of the Mediterranean islands; from the explosive volcanoes of southern Italy to the glaciated seacoasts of western Norway.

Four factors explain this environmental diversity. The complex geology of this western extension of the Eurasian land-

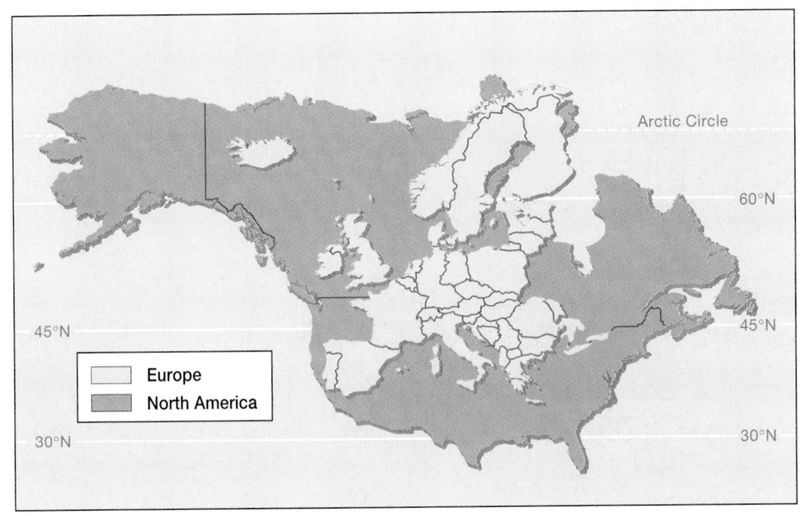

◀ **Figure 8.3 Europe: Size and northerly location** Europe is about two-thirds the size of North America, as shown in this cartographic comparison. Another important characteristic is the northerly location of the region, which affects its climate, vegetation, and agriculture. Much of Europe lies at a similar latitude as Canada, and even the Mediterranean lands are considerably further north than the U.S.–Mexico border.

mass has produced some of the newest, as well as the oldest, landscapes in the world. Europe's latitudinal position also creates opportunity for diversity (Figure 8.3). The region extends from the Arctic to the Mediterranean subtropics. These latitudinal controls are further modified by the interaction of land and sea as the Atlantic Ocean and Black, Baltic, and Mediterranean seas shape regional climates and bioregions. Last, the long history of human settlement has transformed and modified Europe's natural landscapes in fundamental ways over thousands of years. A million years ago the first fire-bearing hunter-gatherers arrived; while their imprint on the landscape was minimal, this was not the case when Neolithic agriculturists arrived in Europe about 7,000 years ago. From that time on, human hands have left a major imprint on the European environment.

Landform Regions

European landforms can be organized into four general topographic regions. The European Lowland forms an arc from the southwest of France to the northeast plains of Poland and includes southeastern England. The Alpine mountain systems extend from the Pyrenees in the west to the Balkan mountains of southeast Europe. The Central Uplands are positioned between the Alps and the European Lowland, stretching from France into eastern Europe. Finally, the Western Uplands include mountains from Spain, portions of the British Isles, and the highlands of Scandinavia (Figure 8.4).

The European Lowland This lowland (also known as the North European Plain) is the unquestionable focus of western Europe. It is an area of high population density, intensive agriculture, large cities, and major industrial regions. Though not completely flat by any means, most of this lowland lies below 500 feet (150 meters) in elevation, though it is broken in places by rolling hills, plateaus, and uplands (such as in Brittany, France), where elevations exceed 1,000 feet (300 meters). Many of Europe's major rivers, such as the Rhine, the Loire, the Thames, and the Elbe, meander across this lowland and form broad estuaries before emptying into the Atlantic. Several of Europe's great ports inhabit these strategic lowland settings, including London, Le Havre, Rotterdam, and Hamburg.

The Rhine River delta conveniently divides the unglaciated lowland to the south from the glaciated plains to the north, which were covered by a Pleistocene ice sheet until about 15,000 years ago. Because of these continental glaciers, the North European Lowlands, including Netherlands, Germany, Denmark, and Poland, are far less fertile for agriculture than the unglaciated portion in Belgium and France (Figure 8.5). Rocky, clay materials eroded by glaciers in Scandinavia were transported south. As the glacier later retreated with a warming climate, piles of glacial debris known as **moraines** were left on the plains of Germany and Poland. Elsewhere in the north, glacial meltwater created infertile outwash plains that have limited agricultural potential. Accordingly, the irregular pattern of drainage in the post-glacial period left a landscape of lakes, marshlands, and bogs that severely limits agriculture in the Baltic areas, particularly when contrasted with the unglaciated and more fertile areas of the southern European Lowland.

The Alpine Mountain System Without a doubt, the Alpine Mountain System forms the topographic spine of Europe. It consists of a series of east–west running mountains from the Atlantic to the Black Sea and the southeastern Mediterranean. Though these mountain ranges carry distinct regional names, such as the Pyrenees, Alps, Carpathians, Dinaric Alps, and Balkan mountains, they share similar geologic traits. All were created more recently (about 20 million years ago) than other upland areas of Europe, and all of these different highlands are constructed from a complex arrangement of rock types.

The *Pyrenees* form the political border between Spain and France (including the microstate of Andorra). This rugged range extends almost 300 miles (480 kilometers), stretching from the Atlantic to the Mediterranean. Within the mountains, glaciated peaks higher than 11,000 feet (3,350 meters) alternate with broad glacier-carved valleys. The Pyrenees have long been home to the Basque people in the western reaches, as well as to distinctive Catalan-speaking minorities in the

▲ **Figure 8.4 Physical geography of Europe** Much of the mountain and upland topography of Europe is a function of the gradual, northward movement of the African tectonic plate into the Eurasian plate. These tectonic forces have created the east–west trending Alpine mountain chain. Besides these tectonic forces, Pleistocene glaciation has also shaped the European region. Until about 15,000 years ago, much of the region was covered by an continental glaciers that extended south to the mouth of the Rhine river, as shown by the dotted line on the map.

east. Insurrection and separatism produce a fascinating, but often violent, human geography in these two different areas.

The centerpiece of the larger geologic system is the prototypical Alpine range itself, the *Alps,* reaching more than 500 miles (800 kilometers) from France to eastern Austria. These impressive mountains are highest in the west, reaching more than 15,000 feet (4,575 meters) in Mt. Blanc on the French-Italian border; in Austria, to the east, the mountains are less high with peaks more than 10,000 feet (3,050 meters) of note. Though easily crossed today by car or train through a system of long tunnels and valley-spanning bridges, these mountains have formed an important cultural divide between the

▲ **Figure 8.5 The European Lowland** Also known as the North European Plain, this large lowland extends from southwestern France to the plains of northern Germany and into Poland. Although this landform region has some rolling hills, most of it is less than 500 feet (150 meters) in elevation. *(P. Vauthey/Sygma Photo News)*

Mediterranean lands to the south and central and western Europe in the north. The Alps also restrict the Mediterranean's distinctive climate and bioregions from spilling over into western and central Europe.

The *Appenine* Mountains are located south of the Alps, but the two ranges are physically connected by the hilly coastline of the French and Italian Riviera. Forming the mountainous spine of Italy, the Appenines are generally lower and lack the spectacular glaciated peaks and valleys of the true Alps. Farther to the south, the Appenines take on their own distinct character with the impressive and explosive volcanoes of Mt. Vesuvius, outside of Naples, and the much higher (almost 11,000 feet, or 3,350 meters) Mt. Etna, off Italy's toe on the island of Sicily. These two active volcanoes give clues to the geophysical heritage of the Alpine mountain chain as the tectonic meeting ground of the African and Eurasian plates. The northward movement of the African Plate against the Eurasian Plate created both the Mediterranean basin and the Alpine mountain system that forms its northern rim.

To the east, the *Carpathian* Mountains define the limits of the Alpine system in eastern Europe. They are a plow-shaped upland area that extends from eastern Austria to the Iron Gate gorge, where the borders of Romania and Yugoslavia intersect, and these mountains create a formidable passage for Danube River traffic. About the same length as the main Alpine chain, the Carpathians are not nearly as high in elevation. The highest summits in Slovakia and southern Poland are less than 9,000 feet (2,780 meters).

In southeastern Europe, the complicated tail of the Alpine system is expressed in a series of uplands with several distinct regional names. The *Dinaric* Alps fringe the Adriatic Sea to the west in Croatia, Bosnia, and Yugoslavia. To the southeast are the *Balkan* Ranges that trend both north–south and east–west, reaching almost to the Black Sea in Bulgaria. Finally, the *Rhodope* Mountains occupy northern Greece and Macedonia. Elevations throughout these different mountain ranges are generally less than 10,000 feet (3,050 meters).

Central Uplands In western Europe, another much older highland region occupies an arc between the Alps and the European Lowland in France and Germany. These mountains are much lower in elevation than the Alpine system, with their highest peaks at 6,000 feet (1,830 meters). Dated to about 100 million years ago, much of this upland region is characterized by rolling landscapes of about 3,000 feet (less than 1,000 meters). Their importance to Western Europe is great because they contain the raw materials for Europe's industrial areas. In both Germany and France, for example, these uplands have provided the iron and coal necessary for each country's steel industry; in the eastern reaches, mineral resources from the Bohemian highlands have also fueled major industrial areas in Germany, Poland, and the Czech Republic.

Western Highlands The Western Highlands define the western edge of the European subcontinent, extending from Portugal in the south, through the portions of the British Isles in the west, to the highland backbone of Norway, Sweden, and Finland in the far north. These are Europe's oldest mountains, formed about 300 million years ago. Their unity as a landform region, however, comes more from similar topography and elevation than from a uniform geology.

As with other upland areas that traverse many separate countries, specific place-names for these mountains differ from country to country (Figure 8.4). A portion of the Western Highlands form the highland spine of England, Wales, and Scotland, where picturesque glaciated landscapes are found at modest elevations of 4,000 feet (1,220 meters) or less. These U-shaped glaciated valleys are also appear in Norway's uplands, where they produce a spectacular coastline of **fjords**, or flooded valley inlets similar to the coastlines of Alaska and New Zealand.

Though less elevated, the Fenno-Scandian Shield of Sweden and northern Finland is noteworthy because it is made up of some of the oldest rock formations found anywhere in the world, which are dated conservatively at 600 million years. This **shield landscape** was eroded to bedrock by Pleistocene glaciers, and because of the cold climate and sparse vegetation, still has extremely thin soils that severely limit agricultural activity (Figure 8.6). As with other heavily glaciated areas, such as the Canadian Shield area of North America, numerous small lakes dot the countryside, giving clues to the impressive erosional power of ice sheets.

Europe's Climates

Three principal climates characterize Europe. Along the Atlantic coast, a moderate and moist maritime climate dominates, modified by oceanic influences. Farther inland, continental climates

▶ Figure 8.6 Northern landscapes Northern Europe is a harsh land characterized by landscapes with little soil, glaciated rock expanses, sparse vegetation, and thousands of lakes. Pleistocene glaciers sculpted this region 20,000 years ago, and its high-latitude location maintains its Ice Age character with a very short growing season. *(Macduff Everton/ Corbis)*

prevail, with hotter summers and colder winters. Finally, dry summer Mediterranean climates are found in southern Europe, from Spain to the Greece. Here, an extensive area of warm-season high pressure inhibits summer storms and rainfall in this region, creating the seemingly endless blue skies so attractive to tourists from northern Europe.

One of the most important climate controls is that of the Atlantic Ocean. Though most of Europe is at a relatively high latitude (London, England, for example, is slightly farther north than is Vancouver, British Columbia), the mild North Atlantic current, which is a continuation of the warm Atlantic Gulf Stream, moderates coastal temperatures from Norway to Portugal and inland to the western reaches of Germany. As a result, this maritime influences gives Europe a climate that is 5 to 10 °F (2.8 to 5.7 °C) warmer than comparable latitudes without this oceanic effect (Figure 8.7).

In the **Marine West Coast climate** region, no winter months average below freezing, though cold rain, sleet, and an occasional blizzard are common winter visitors. Summers are often cloudy and overcast, with frequent drizzle and rain as moisture flows in from the ocean. Ireland, the Emerald Isle of green landscapes, offers an appropriate snapshot of this maritime climate.

With increasing distance from the ocean (or where a mountain chain limits the maritime influence, as in Scandinavia), landmass heating and cooling becomes a strong climatic control, producing hotter summers and colder winters. Indeed, all **Continental climates** average at least one month below freezing during the winter. In Europe, the transition between maritime and continental climates takes place close to the Rhine River border of France and Germany. Farther north, although Sweden and other nearby countries are close to the moderating influence of the Baltic Sea, high latitude and the blocking effect of the Norwegian mountains produces cold winter temperatures characteristic of continental climates. Precipitation in continental climates comes as rain from summer frontal systems and local thundershowers. Usually this rainfall is sufficient for nonirrigated agriculture, though supplemental watering is increasingly common where high value crops are grown.

Mediterranean climate is characterized by a distinct dry season during the summer, which results from the warm season expansion of the Atlantic (or Azores) high-pressure area. This high pressure is produced by the global circulation of air warmed in the equatorial tropics that subsides or descends between latitudes 30 and 40 degrees, thus inhibiting summer rainfall. This same phenomenon also produces the Mediterranean climates of California, western Australia, parts of South Africa, and Chile. While these rainless summers may attract tourists from northern Europe, this seasonal drought can be problematic for agriculture. It is no coincidence that traditional Mediterranean cultures such as the Arabs, Moors, Greeks, and Romans have been major innovators of irrigation technology (Figure 8.8).

Of course, within these three major climate types there are significant variations. For example, in the high-latitude Scandinavian countries, where there is a transition from Maritime to Continental climates, short (or even nonexistent) winter days give this region its unique climatological character. Even farther north in Scandinavia, these climates are replaced by Arctic tundra climates, characterized by low precipitation, long, cold winters, and the briefest of summers.

In the Alpine mountain chain, elevation and terrain shape local variation in temperature and precipitation. Further, these mountains also act as a barrier or dividing zone between Maritime and Continental climates of the north and the Mediterranean zone to the south. Crossing the Alps in March, for example, either via a high mountain road (if it is free of snow) or through an underground tunnel, quickly takes a traveler from a northern winter of snow and mud into the brilliant sunshine and blossoming vegetation of southern Europe.

Seas, Rivers, Ports, and Coastline

Europe remains a maritime region with strong ties to its surrounding seas. Even its landlocked countries, such as Austria and the Czech Republic, have access to the ocean through an interconnected network of navigable rivers and canals.

Europe's Ring of Seas Five major seas encircle Europe; these water bodies are connected to each other through narrow straits with strategic importance for controlling waterborne trade and naval movement (Figure 8.4). In the north, the Baltic Sea separates Scandinavia from north-central Europe. Denmark and Sweden have long commanded the narrow Skagerrak and Kattegat straits that connect the Baltic to the North Sea. Besides its historic role as a major fishing

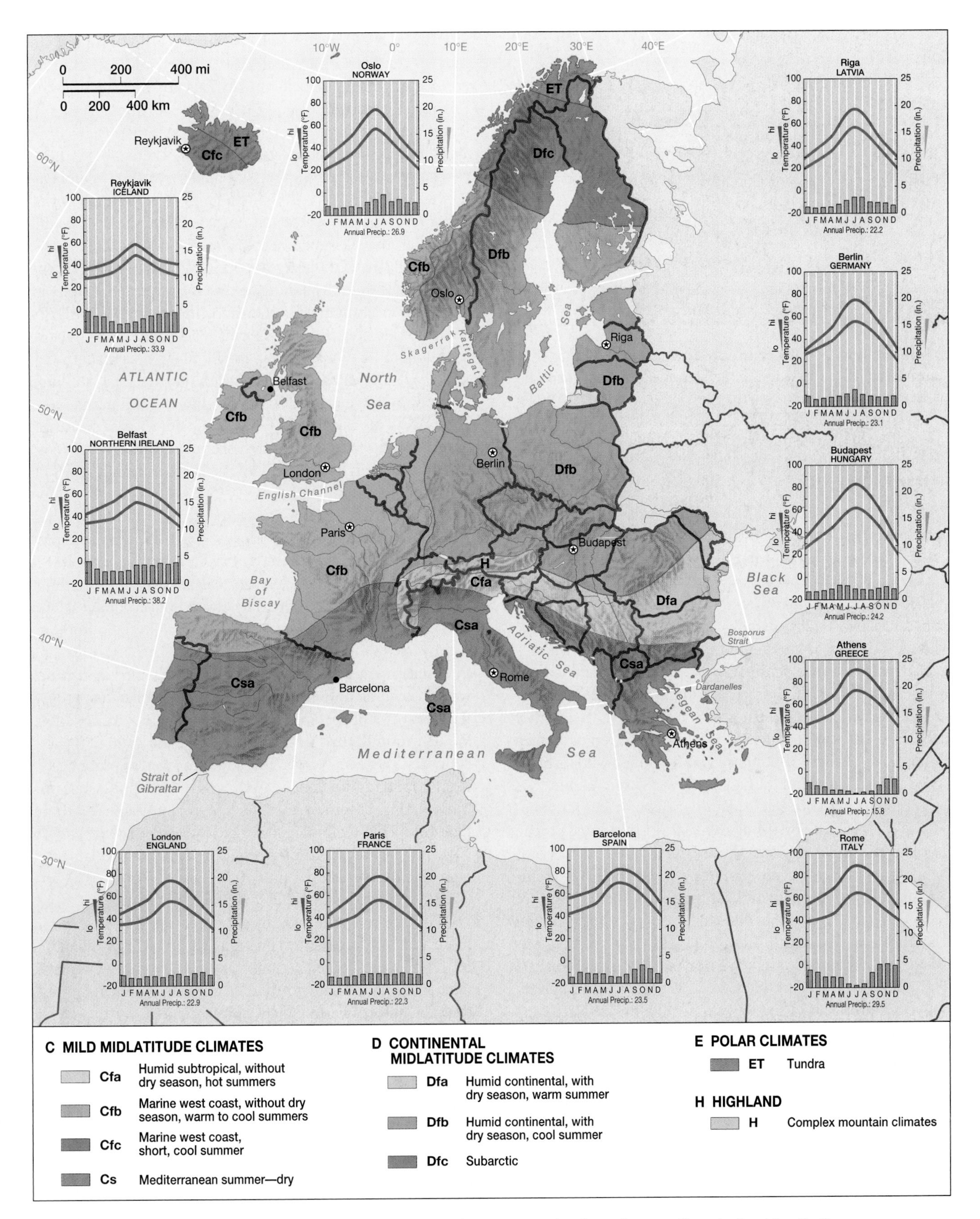

▲ **Figure 8.7 Climate map of Europe** Three major climate zones dominate Europe. Close to the Atlantic Ocean, the Marine West Coast climate is found, with cool seasons throughout the year and steady monthly rainfall. Farther inland, Continental climates are found. They have at least one month averaging below freezing, and warm to hot summers, with a precipitation maximum falling in the summer season. The dry summer Mediterranean climate is found in southern Europe. Unlike the Continental climate, in the Mediterranean region, most precipitation falls during the cool winter period.

▲ **Figure 8.8 Mediterranean agriculture** Because summer rainfall is sparse in the Mediterranean region, crop irrigation is necessary. In this photo, mechanized sprinklers irrigate a field crop in Spain. This technology draws upon a long history of irrigation in the region. Many scholars argue that irrigation was introduced into Spain by Muslim peoples sometime after the eighth century. *(David Woodfall/Tony Stone Images)*

ground, the North Sea is now well known for its rich oil and natural gas fields mined from deep-sea drilling platforms.

The English Channel (or in French, *La Manche*), separates the British Isles from continental Europe. At its narrowest point, the Dover Straits are only 20 miles (32 kilometers) wide. Though England has thought of the Channel as its own protective moat, it has been primarily a symbolic barrier, because it deterred neither the French Normans from the continent nor Viking raiders from the north. Since 1993, however, after decades of resistance, England is now connected to France through the 31-mile (50-kilometer) Eurotunnel, with its high-speed rail system carrying passengers, autos, and freight.

Gibraltar's role is legend, guarding the narrow straits between Africa and Europe at the western entrance to the Mediterranean Sea. British stewardship of this pinch point remains an enduring symbol of its once great sea-based empire. Finally, on Europe's southeastern flanks are the Straits of Bosporus and the Dardanelles, the narrows connecting the eastern Mediterranean to the Black Sea. Disputed for centuries, these pivotal waters are now controlled by Turkey. Though these straits are often used as a physical boundary between Europe and Asia, they are readily bridged in several places to facilitate truck and train transportation within Turkey and between Europe and Southwest Asia.

Rivers and Ports Europe is also a region of navigable rivers that are connected by a system of canals and locks to allow for inland barge travel from the Baltic and North seas to the Mediterranean, and between western Europe and the Black Sea. Many rivers on the European Lowland, such as the Loire, Seine, Rhine, Elbe, and Vistula, flow into Atlantic or Baltic waters. However, the Danube, Europe's longest river, flows east and south, rising in the Black Forest of Germany only a few miles from the Rhine River, and running southeastward to the Black Sea. It offers a connecting artery between central and eastern Europe. Similarly, the Rhône headwaters rise close to those of the Rhine in Switzerland, yet it flows southward into the Mediterranean. Both the Danube and the Rhône are connected through locks and canals to the rivers of the European Lowland, making it possible for barge traffic to travel between all of Europe's fringing seas and oceans.

Major ports are found at the mouths of most western European rivers, serving as transshipment points for inland waterways, as well as focal points for rail and truck networks. From south to north, these ports include Bordeaux at the mouth of the Garonne, Le Havre on the Seine, London on the Thames, Rotterdam (the world's largest port in terms of tonnage) at the mouth of the Rhine, Hamburg on the Elbe River, and, to the east in Poland, Szcezin on the Oder, and Gdansk on the Wisla. Of the major Mediterranean ports, only Marseilles, France, is close to the mouth of a major river, the Rhône. Other modern-day ports, such as Genoa, Naples, Venice, and Barcelona, are some distance from the delta harbors that served historic trade. Apparently, as Mediterranean forests were cut in past centuries, erosion on the hillslopes carried sediment down the rivers to the delta regions, effectively filling in the historic ports used by the Greeks and Romans.

Reclaiming the Dutch Coastline Much of the Netherlands landscape is a product of the people's long struggle to protect their agricultural lands against coastal and river flooding. Beginning around A.D. 900, dikes were built to protect these fertile but low-lying lands against periodic flooding from the nearby Rhine River, as well as from the stormy North Sea. By the twelfth century, these protected and reclaimed landscapes were known as **polders**, or diked agricultural settlements, a term that is still used today to describe coastal reclamation. While windmills had long been used to grind grain in the low countries, this wind power was now employed to pump water from low-lying wetlands. As a result, windmills became increasingly common on the Dutch landscape to drain marshes. This technology worked so well that the Dutch government embarked on a widespread coastal reclamation plan in the seventeenth century that converted an 18,000-acre (7,275-hectare) lake into agricultural land (Figure 8.9).

With steam- (and later, electric-) powered pumps, even more ambitious polder reclamation was possible, culminating in the massive Zuider Zee project of the twentieth century, where the large bay north of Amsterdam was dammed, drained, and converted to agricultural lands over the course of a half century. This project both alleviated a major flooding hazard and opened up new lands for farming and settlement in the heart of the Netherlands. Even new suburbs of Amsterdam are now spilling onto these reclaimed lands.

For the last several decades, major efforts have been directed toward curbing flooding where the Rhine distributaries flow into the English Channel. After a disastrous flood in 1953, the Dutch government conceived the Delta Project, which has constructed four large sea dikes that essentially keep the many different mouths of the Rhine River within their channels. By all accounts, these projects have succeeded in stabilizing the North Sea coast and providing lowland Europe with more area in its densely settled heart for both rural and urban settlement.

▲ **Figure 8.9 Polder landscape** Diked agricultural settlements, or polders, are situated along coastal Netherlands. Because these lands, such as the fields on the left side of this photo, are reclaimed from the sea, many are at or below sea level. The English Channel and North Sea are to the right. *(Adam Woolfitt/Woodfin Camp & Associates)*

Environmental Problems and Solutions, West and East

Because of its long history of agriculture, resource-extraction industrial manufacturing, and urbanization, Europe has its share of serious environmental problems. Compounding the situation is that pollution rarely respects political boundaries. Air pollution from England, for example, creates serious acid-rain problems in Sweden, and, similarly, water pollution of the upper Rhine River by factories in Switzerland creates major problems for the Netherlands, where Rhine River water is used for municipal drinking supplies. When environmental problems cross state boundaries, the solutions must come from intergovernmental cooperation (Figure 8.10).

Since the 1970s, when the European Union (EU) added environmental issues to its economic and political agenda, western Europe has been increasingly effective in addressing its varied environmental problems with regional solutions. Besides the more obvious environmental problems of air and water pollution, the EU has also taken the lead on matters of recycling, waste management, reduced energy usage, and sustainable resource use. As a result, western Europe is probably the "greenest" of the major world regions. This heightened environmental sensitivity is also found in the political arena, where the Green Party regularly gets about 10 percent of the vote and in some areas, such as southern England, close to 20 percent. In Germany this environmental party recently shared national power with the Social Democrats in what was referred to as a "Red-Green Coalition."

However, the situation is decidedly grimmer in eastern Europe. During the period of Soviet economic planning in eastern Europe (1945–90), little attention was paid to environmental issues because of the overt emphasis on short-term industrial output. Additionally, communist economics had no way to calculate environmental costs, thus they were not a concern. Since the environment was a non-issue, there were few controls on air and water pollution, environmental safety and health, or on the dumping of toxic and hazardous wastes.

Unfortunately, the costs of those decisions have been extraordinarily high, whether measured in environmental, human, or monetary terms. For example, 90 percent of Poland's rivers have no aquatic or plant life. Until recently, 95 percent of the sewage from Warsaw's 1.6 million people was untreated, and more than 50 percent of the country's forest trees show signs of damage from air pollution. Humans are also suffering. Fully one-third of the population is expected to suffer from an environmental-induced disease such as cancer or respiratory illness. Sadly, the nation's own Polish Academy of Scientists pronounced their country the most polluted in the world.

Nearby countries in eastern Europe fare no better. In the neighboring Czech Republic, almost three-quarters of the forests are dying from air pollution and acid rain, though in all fairness it should be noted that some of this airborne pollution originated in the industrial countries to the west (Figure 8.11). Nevertheless, in the Czech industrial heartland, the air pollution is so bad that human life expectancy is 11 years less than the national average. Though Bulgaria had about 100 environmental laws on its books during the communist period, they were never enforced. As a result, two-thirds of its rivers are badly polluted today, with almost half of its trees showing environmental damage. Around the uranium plant in Rokovski, Bulgaria, the death rate for children is three times the national average.

Ill-designed and poorly maintained Soviet-built nuclear power plants are also common in the east, threatening all of Europe with a repeat of the 1986 Chernobyl disaster in Ukraine. At that time, the meltdown of a nuclear reactor sent radioactive contamination into the atmosphere that not only polluted the immediate area, but was also carried into western and northern Europe by wind circulating around a high-pressure system. School children in Austria were kept indoors for a week as a safety measure. Dairy supplies were irradiated by fallout, as were vegetables and other food supplies, which were subsequently destroyed. In Scandinavia, radioactive fallout was absorbed by lichens and moss, the fodder for large reindeer herds that provide livelihood for Sami peoples of the Arctic region. For several years, reindeer meat, their major food source, was unacceptably high in radioactivity, causing widespread misery and deprivation for the Sami. Another nuclear meltdown of a Chernobyl-type reactor took place in Lithuania in 1992, this time resulting in limited damage because of a different wind pattern that did not spread the radiation.

Solutions to eastern Europe's environmental problems will not come easily because of several complicating factors. First, most eastern European states are still struggling with their post-1990 economic and political transitions, and it is still unclear whether these new governments will either enact or enforce the necessary environmental controls. Instead, there is widespread concern that environmental regulations will inhibit economic recovery by burdening industry with higher operating costs.

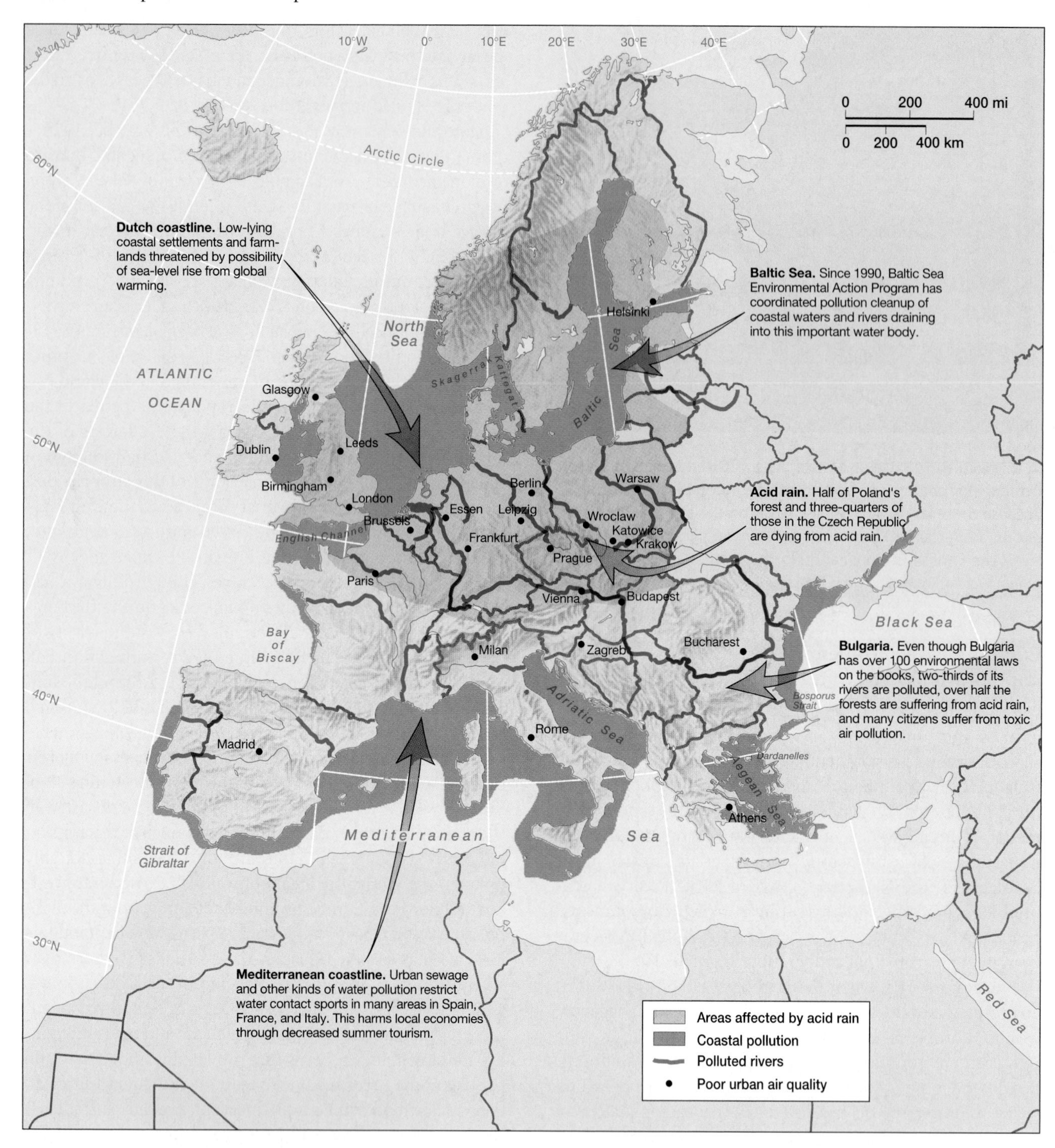

▲ **Figure 8.10 Environmental issues in Europe** In terms of environmental protection, there is a major disparity between western and eastern Europe. While the west has worked energetically over the last 30 years to solve problems such as air and water pollution, those same problems are still widespread in the east because of the long period of environmental neglect coincidental with the communist period. Because of the fabric of small nation-states, most environmental problems must be regional solutions, rather than one solved by one country alone.

Second, there is increasing evidence that capital is not flowing into certain areas or specific industrial sites because western businesses fear they will be charged for cleaning up toxic dumps or repairing environmental damage. Until eastern European governments show a willingness to share those cleanup costs, capital will stay away from these polluted areas and move instead into less-damaged regions (Figure 8.12).

Third, pollution control is intertwined with the privatization of former state-owned industries. Indeed, communist state-owned factories that had no wastewater treatment facilities or air emission controls caused many environmental

▲ **Figure 8.11 Acid rain and forest death** Acid precipitation has taken a devastating toll on eastern European forests, such as those shown here in Bohemia, Czech Republic. In this country, three-quarters of the forests are dead or injured from acid precipitation that is caused by industrial and auto emissions. *(Karol Kallay/Bilderberg Archiv der Fotografen)*

problems. To solve the problem, pollution controls must now be added. However, this will add to overhead costs as these industries are moved into the private sector.

Last, as is the case in western Europe, regional solutions that transcend political borders are necessary. Water pollution problems in the Danube River, for example, must be solved through cooperative agreements between the different states of the lower drainage basin, including Hungary, Yugoslavia, Bulgaria, and Romania. Similarly, regional air pollution problems must also be resolved through cooperation of different states.

Hope that regional cooperation can be extended into eastern Europe comes from recent action on the Baltic Sea. Before 1989, the **Iron Curtain** (the ideological border between democracy and communism) prevented cooperative solutions to pollution problems in this large brackish sea that is a major nursery for fish species. As a result, measures taken by Denmark, Sweden, Finland, and Western Germany to control water pollution were largely offset by unchecked water pollution from East Germany, Poland, and other Soviet states. A double standard in environmental pollution existed, to the detriment of Baltic ecosystems.

In 1990, however, immediately after the fall of communism in eastern Europe, and with the unification of Germany, a new Baltic Sea Joint Comprehensive Environmental Action Program was founded by all states sharing a shoreline with or drainage into the Baltic. Much progress has been made since then, as new standards and enforcement regulations have been put into place. Financial resources from the World Bank, along with the European Bank for Reconstruction and Development, have helped eastern European countries, such as Poland, Estonia, Belarus, and even Russia, to build sewage treatment plants, protect coastal wetlands, and clean up hazardous waste sites around the Baltic rim.

▲ **Figure 8.12 Toxic landscape in Romania** A troublesome legacy of Soviet communism in eastern Europe is the numerous toxic dump sites and polluted landscapes found in the region, such as this site in Romania. Not only do these environmental problems threaten public health, but the prospect of high cleanup costs also inhibits investment by western firms. *(Filip Horvat/SABA Press Photos, Inc.)*

Settlement and Population: Slow Growth and Rapid Migration

Europe contains more than half a billion people unevenly distributed across the landscape. From a densely populated and highly urbanized core in western Europe, settlement thins out to the east, north, and southwest, across its more rural, and often, poorer, periphery. Europe is unlike some world regions where high rates of natural growth are a problem. In fact, the opposite is true: Europe's population problem is one of slow or no growth, with rates that are, in some cases, far below replacement level.

Another population issue is migration into the region. Historically, the source area for out-migration to the Americas, Australia, and Africa, Europe has now become a target region for significant in-migration. Because of its low rates of natural increase, Europe's industrial heartland experienced a shortage of workers during the boom years of its postwar economic miracle. To fill this need, migrant workers were invited in from the rural and less well-developed periphery. As a result, thousands of workers from Italy, Turkey, and Yugoslavia labored in Germany's factories during the 1960s and 1970s.

Though the economy has cooled, this affluent European core is still a magnet for people from Europe's poorer periphery: eastern Europe, the former Soviet countries, as well as from former colonies in Africa. In addition, not all of these people are economic migrants. With its own long history of war, western Europe is sympathetic and reasonably welcoming to political refugees from troubled lands in the Balkans, eastern Europe, and Africa.

Population Density in Core and Periphery

The map of European population distribution (Figure 8.13) shows two densely settled axes of settlement that converge on London and southeastern England. One axis runs north from Italy's Po Valley through western Switzerland, the Paris Basin, and northern France to the Low Countries, and then across the Channel to London. The other axis starts to the east in Prague, but also includes the industrial area of southern Poland. From there, it runs west into Germany, through Frankfurt and the Ruhr industrial conurbation, and then meets the southern axis in the Netherlands before continuing to southeastern England.

These two axes are part of the densely settled industrial-urban heartland of Europe. While this generalization overlooks important population clusters in northern, eastern, and Mediterranean Europe, it does, nevertheless, convey an accurate sense of a densely settled European core set apart from a more rural, agricultural periphery. Many of Europe's economic and political tensions are played out against this back-

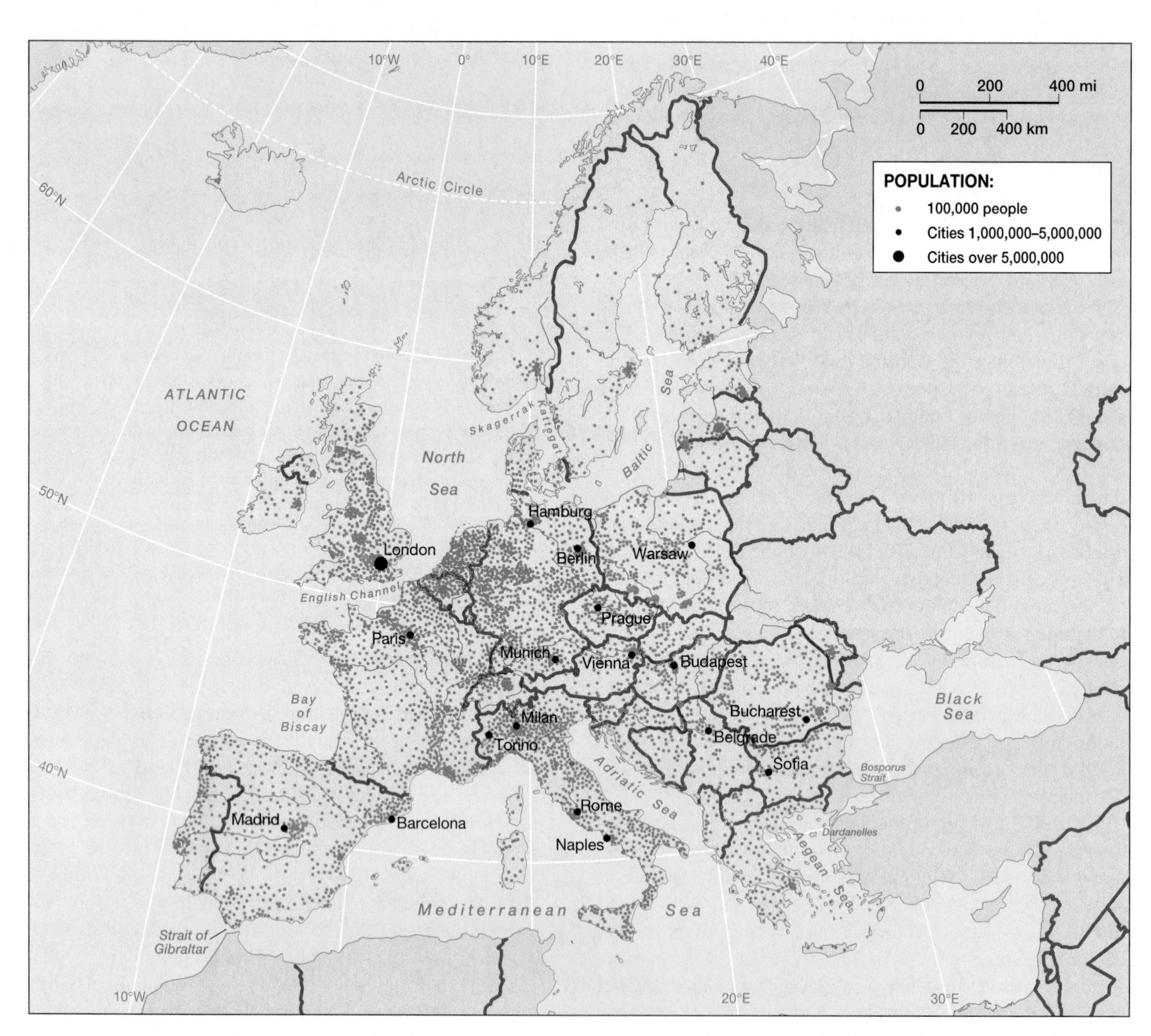

▲ **Figure 8.13 Population map of Europe** The European region includes more than 520 million people, many of them clustered together in large cities in both western and eastern Europe. As can be seen on this map, the most densely populated areas are in England, the Netherlands, Belgium, western Germany, northern France, and south across the Alps to northern Italy. There are also dense clusters of population scattered along the Mediterranean coast of Spain, where settlements are usually linked to intensive irrigated agriculture.

drop of a rich heartland that contrasts with a less well-to-do periphery. For example, economic subsidies from the core to the periphery have been central to the EU's development policies for several decades.

Natural Growth: Beyond the Demographic Transition

Probably the most striking characteristic of Europe's population is its continued slow natural growth (Table 8.1). More to the point, in many European countries, the death rate exceeds the birthrate, meaning that there is simply no growth at all. Instead, many countries are actually experiencing negative growth rates; were it not for in-migration from other countries and other world regions, these countries would record a decline in population over the next few decades. Italy, for example, currently has a population of 57.7 million. Yet if current natural growth holds true for 20 years, the population will decrease to 54.8 million in the year 2025.

There seem to be several causes for zero population growth in western Europe. First of all, recall that the concept of the demographic transition was formulated from the historical change in European growth rates as the population moved from rural settings to more urban and industrial locations. What we see today is an extension of that model, namely the continued expression of the low fertility–low mortality of the fourth stage of the demographic transformation. Some demographers suggest a fifth stage be added to the model, a "post-industrial" phase in which population falls below replacement levels. Evidence for this explanation comes from the highly urbanized and industrialized populations of Germany, France, and England, all of which are below zero population growth.

These low-birthrate countries have all attempted to increase their natural rate of growth. In Germany, for example, couples are given an outright gift of $650 if they produce a child. In Austria, the government gives a modest cash award to couples when they marry, and still more follows if they give birth. France provides monthly payments to families with small children that last until school age. Despite these incentives, most European countries continue with natural growth below replacement levels.

Additional explanations must be sought to explain the extraordinary negative growth in the eastern European countries. As seen in Table 8.1, the most striking examples of negative population growth are found in Bulgaria, Romania, Hungary, the Czech Republic, Latvia, Estonia, and Lithuania.

During the immediate postwar years, the growth of centralized planning and industrial development under the Soviet Union led to a labor shortage in eastern Europe, as well as in the Soviet Union itself. This was aggravated by the huge losses suffered during World War II. As a result, women were needed in the workforce. So that child-rearing would not conflict with jobs, Soviet policy was to make all forms of family planning and birth control available, including abortion. As a result, small families became the norm. Another factor was the widespread housing shortage that prevailed in most Soviet countries, a result of concentrating people in cities for industrial jobs, along with priorities placed on other forms of construction. Evidence from eastern Europe suggests that families experiencing a housing crunch tend to have few (if any) children.

When eastern European countries realized these policies were leading to negative population growth, several states tried to reverse the trend. Perhaps the most notorious illustration was in Romania, where the Ceausescu dictatorship outlawed all contraception and abortion. Reportedly, security police monitored pregnancies to make sure they reached full term. But this was to no avail. Instead, many women died from illegal abortions, and the infant mortality rate was the highest in Europe, probably because of parental neglect. When the communist government fell in 1989, some 14,000 unwanted children were living in orphanages under questionable conditions. Although one of the first acts of the new government was to legalize both contraception and abortion, this has simply institutionalized the existing low birthrate.

Finally, there is the intangible factor of outlook and mood in contemporary eastern Europe as it influences birthrates and family planning. Uncertainty about the future is commonplace, as one quickly learns by talking with eastern Europeans. In many of these countries, jobs, wages, food, and housing are scarcer now than before 1989. Given these factors, it is understandable that the birthrate remains very low.

Migration to and Within Europe

Migration is one of the most challenging population issues facing Europe today because it is caught between two conflicting policies. First, there is widespread social resistance to unlimited migration into Europe, ostensibly because of high unemployment in the western European industrial countries. Many Europeans argue that scarce jobs should go first to European citizens, not to "foreigners." Coming in conflict with these protective, nationalistic concerns is the EU's goal of reducing the barriers to the free movement of labor within Europe, a policy to make it possible for workers and professionals to move about freely, from job to job and country to country.

During the 1960s postwar recovery, when western Europe's economies were booming, many countries looked to migrant workers to alleviate labor shortages. Germany, for example, depended on workers from Europe's periphery, namely Italy, Yugoslavia, Greece, and Turkey, for industrial and service jobs. These *gastarbeiter*, or **guest workers,** arrived in the thousands. As a result, large ethnic enclaves of foreign workers became a common part of the German urban landscape (Figure 8.14). Today, for example, there are 1.5 million Turks in that country, most of whom are there because of the open-door foreign worker policies of past decades.

However, these foreign workers are now the target of considerable animosity and resentment. Economic recessions and stagnation have plagued western Europe for the last decade, with double-digit unemployment rates approaching 25 percent for young people. Many native Europeans protest that guest workers are taking jobs that could be filled by them. Complicating the issue is that many of first-generation guest

Table 8.1 Demographic Indicators

Country	Population[a]	Natural Increase	TFR[b]	%<15[c]	% Urban
Western Europe					
Austria	8.1	0.1	1.4	17	65
Belgium	10.2	0.1	1.6	18	97
France	58.8	0.3	1.7	19	74
Germany	82.3	–0.1	1.3	16	85
Liechtenstein	0.03	0.6	1.5	19	—
Luxembourg	0.4	0.4	1.8	19	86
Netherlands	15.7	0.3	1.5	18	61
Switzerland	7.1	0.3	1.5	18	68
United Kingdom	59.1	0.2	1.7	19	90
Eastern Europe					
Bulgaria	8.3	–0.5	1.2	17	68
Czech Republic	10.3	–0.2	1.2	18	77
Hungary	10.1	–0.4	1.4	18	63
Moldova	4.2	0.1	1.8	26	46
Poland	38.7	0.1	1.6	22	62
Romania	22.5	–0.2	1.3	19	55
Slovakia	5.4	0.2	1.5	22	57
Southern Europe					
Albania	3.3	1.2	2.0	34	37
Bosnia-Herzegovina	4.0	0.6	1.5	23	40
Croatia	4.2	0.1	1.6	19	54
Greece	10.5	0.0	1.3	16	59
Italy	57.7	0.0	1.2	15	67
Macedonia	2.0	0.8	2.1	26	60
Malta	0.4	0.6	2.1	22	89
Portugal	10.0	0.0	1.4	17	48
San Marino	0.03	0.4	1.3	15	89
Slovenia	2.0	0.0	1.3	18	50
Spain	39.4	0.0	1.2	16	64
Yugoslavia (Serbia)	10.6	0.2	1.8	22	51
Northern Europe					
Denmark	5.3	0.2	1.8	18	85
Estonia	1.4	–0.4	1.3	20	70
Finland	5.2	0.2	1.7	19	65
Iceland	0.3	0.9	2.0	24	92
Ireland	3.7	0.5	1.9	23	57
Latvia	2.4	–0.6	1.2	20	69
Lithuania	3.7	–0.1	1.4	21	68
Norway	4.4	0.3	1.8	20	74
Sweden	8.9	0.0	1.6	19	83

[a]Population in millions, 1996.
[b]Total fertility rate.
[c]Percentage of population younger than 15 years of age.

Source: *Population Reference Bureau World Population Data Sheet,* 1998.

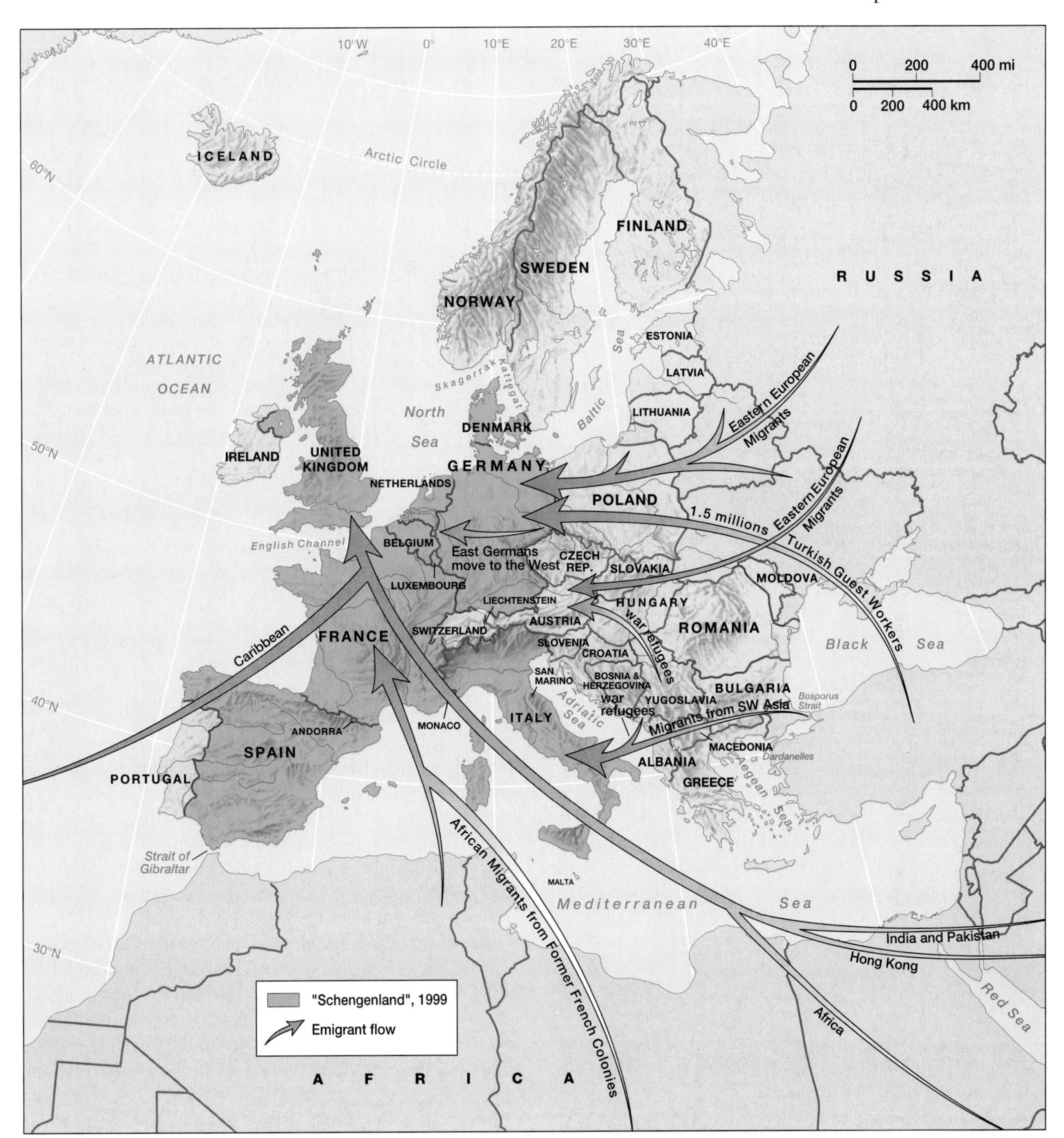

▲ **Figure 8.14 Migration into Europe** A labor shortage during the reconstruction of Europe following World War II opened the door for a large number of guest workers migrating into Germany. In the initial stage, these workers came from Italy and Yugoslavia; later, they came from Turkey and Greece. Most recently, economic stagnation in Europe's heartland has slowed the migration of guest workers. However, immigrants continue to migrate into Europe from former colonial countries in Asia and Africa, as well as from eastern Europe and the war-torn Balkans.

workers have now become citizens of their host country, and in many cases only ethnic distinctions separate different kinds of "native" workers. This subtlety opens the door for racially based discrimination (see "Immigrant Ghetto of the New Europe: Rinkeby, Sweden").

Additionally, the region has also witnessed a massive influx of migrants from former European colonies in Asia, Africa, and the Caribbean. The former colonial powers of England, France, and the Netherlands have been the major recipients of this immigration. England, for example, has inherited large

IMMIGRANT GHETTO OF THE NEW EUROPE Rinkeby, Sweden

Though Rinkeby could be somewhere near the Mediterranean given that most residents are from northern Africa, Turkey, and even Palestine, it is actually a suburb of Stockholm, only a short subway ride away from the central city. This utilitarian suburb of high-rise apartments was built in the early 1970s as a government project to provide one million new apartments for Swedes. However, today 80 percent of Rinkeby's 14,000 residents are immigrants, living in a troubled world in a country that only decades ago preached racial and ethnic tolerance to the rest of the world.

In the last two decades, Sweden—like so much of Europe—has seen its traditionally homogenous population changed by the influx of migrants, guest workers, and refugees from the world's many ethnic battles. Today, immigrants make up fully 10 percent of the Swedish population. Tragically, the assimilation and acceptance of ethnic diversity that Sweden loudly wished to see in other mixed societies such as the United States has not occurred. Instead, these immigrants live in a racially segregated, even discriminatory world.

Mazhar Goker came from Turkey in 1972 with his family. At that time they were the only immigrants on the block. Today, however, there are no native Swedes left. "When I first came here, we were exotic creatures, and people liked to look at us and feel our hair. Now they take detours not to see us." Because of the high number of foreigners, Rinkeby has been stigmatized as a haven for welfare cheats and a center of crime. Through the country, ill-spoken Swedish is known as "Rinkeby Swedish," used by urban toughs and middle-class youths eager for a little street credibility. If there is any kind of trouble in Stockholm, the newspapers usually point the finger at Rinkeby.

Recently, though, there were headlines of a different sort when one of Rinkeby's schools produced students with some of the best test scores in all Sweden. Headlines across the Sunday papers read, "It's a Sensation—Rinkeby Students Get Good Grades." This was no surprise to one of the teachers. Reciting the same formula heard over and over again in Europe's immigrant ghettos, the teacher said, "The reason they are so motivated is they feel they have to be twice as good as any other Swede." In the room, 13 teenagers, students from Bangladesh, Somalia, Iran, Turkey, Chile, and the Balkans, were taking a language class that would help them become fluent in three languages—Swedish, English, and French or Spanish. For them, this is a required passport into the New Europe.

Source: Adapted from "A Swedish Dilemma: The Immigrant Ghetto" by Warren Hoge, *New York Times*, October 6, 1998.

numbers of former colonials from India, Pakistan, Jamaica, and, most recently, Hong Kong. Indonesians (from the former Dutch East Indies) are common in the Netherlands, while emigrants in France are often from former colonies in both northern and Sub-Saharan Africa (Figure 8.15).

This ethnic presence has met with different degrees of acceptance throughout Europe. France, for example, seems uncomfortably challenged by the influx of Africans into Paris and other industrial cities. Currently, this finds expression in the growing numbers of National Front voters who advocate discriminatory policies against foreigners. In Germany, violence against foreigners by neo-Nazi and skinhead youth gangs has provoked outrage from citizens and the government alike. England, too, seems to be experiencing an increasing amount of ethnic tension and racism.

▲ **Figure 8.15 Immigrant ghettos in Europe** A large number of Africans from France's former colonies have migrated to large cities where they form distinct ethnic colonies—even ghettos—in the high-rise apartment buildings on the outskirts of Paris, Lyon, and Marseille. *(S. Elbaz/Sygma Photo News)*

Most recently, political and economic troubles in eastern Europe and the former Soviet Union have generated a new wave of migrants to Europe. Within Germany, for example, thousands of former East Germans have taken advantage of the Berlin Wall's demise by moving to the more prosperous and dynamic western parts of the unified Germany. With the total collapse of Soviet border controls in 1990, emigrants from Poland, Bulgaria, Romania, Ukraine, and other former Soviet satellite countries have poured into western Europe, looking for a better life. The major receiving countries have been Germany, Austria, Switzerland, Denmark, and Sweden.

This flight from the post-1989 turmoil also includes refugees from war-torn Yugoslavia, particularly from Bosnia and Kosovo. Europe's conscience was tested severely by the plight of these people who are so reminiscent of another generation's experience during and immediately after World War II.

The Geography of "Fortress Europe"

An important and perhaps even aggravating ingredient to European concerns with foreign migration is the agreement between the heartland countries of western Europe to facilitate free movement across borders without the traditional passport inspections and checks. This agreement is reshaping the political geography of Europe by creating divisions between insiders and outsiders. To those on the inside, it creates a long-dreamed-of "Europe without borders," while to those on the outside it presents a "Fortress Europe," a defensive perimeter hostile to migrants from Asia, Africa, and even eastern Europe (Figure 8.16).

Taking its name from the city of Schengen, Luxembourg, where the original declaration of intent was signed in 1985, the goals of the EU's **Schengen Agreement** are to gradually reduce border formalities for travelers moving between France, Germany, and the Benelux countries, Belgium, Netherlands, and Luxembourg. As other countries, such as England, Italy, Spain, and Denmark, have shown interest in joining this agreement, the emphasis on free travel between these states has been replaced with increasingly restrictive controls at the external borders of this new entity of "Schengenland." To illustrate, a French national (or "Schengen National" in the agreement's parlance) will be able to cross the border into Germany by auto without the usual traffic stop; instead, cars will simply slow down for a visual check by border police. However, on the Schengen perimeter—say, at the border between Germany and Poland—border checks will be much more rigorous and time consuming. Because of these procedures, critics of the Schengen Agreement refer to a "Fortress Europe," where outsiders will be subject to intimidating customs interrogation designed to weed out undesirable migrants (Figure 8.16).

Though reduced border formalities within western Europe seemed like reasonable and desirable goals in 1985, the situation is much more complicated today, both by migrants from the former Soviet lands and by emigrants from Africa and southwest Asia. Because of their long Mediterranean shorelines, Spain and Italy carry the burden for policing Schengen's outer limits, a task not always performed to the satisfaction of the other partners.

Italy, in particular, has had trouble stemming the flow of illegal immigrants into that country. Many emigrants travel from Istanbul by ship, paying thousands of dollars for the five- or six-day journey (often under steerage conditions) that simply provides them with an opportunity to be dumped on the beaches of Italy at night to take their chances with immigration police. However, since Italian law requires illegal immigrants to leave the country within 15 days, this provides illegal immigrants from Albania, Turkey, Iraq, Iran, and southeast Asia with a transit visa to more northern European countries, which are usually their destinations of choice anyway.

▲ **Figure 8.16 "Fortress Europe" border station** This border station between the Czech Republic and Germany has replaced the Iron Curtain. However, the border station now represents entry into "Schengenland," the group of western European countries that agreed to facilitate free movement by reducing border formalities. *(Linda Sykes)*

The Landscapes of Urban Europe

One of the major characteristics of Europe's population is its high level of urbanization. All countries except three (Albania, Portugal, and Yugoslavia) have more than half their population in cities, and several countries, such as the United Kingdom and Belgium, are more than 90 percent urbanized. Once again the gradient from the highly urbanized European heartland to the less-urbanized periphery reinforces the distinctions between the affluent, industrial core area and the more rural, less well-developed periphery. While historical factors explain much of this difference, given that the core countries have been industrialized for almost 200 years, it also speaks to the slower pace of contemporary economic development in the peripheral countries.

Historical Momentum While widespread urbanization is relatively recent in Europe, dating back only a century or two, the traditions of European urban life are much older. More specifically, the spread of cities into Europe can be associated with the classical empires of Greece and Rome, which makes many cities in Europe more than 2,000 years old. World cities such as London, Paris, Frankfurt, and Vienna, to name just a few, all sprouted from Roman roots. Historically, pre-Romans tended to settle on defensive sites, such as marshes, hilltops, or small islands, while the Romans located their early cities at river crossings (Frankfurt, Paris, London) to complement their expansive road network. These Roman locations were exceptionally well suited for the historical development of trade and the exercise of political power. For these reasons, the sites of these cities outlasted the Roman empire and provide continuity between today's urban geography and its classical antecedents.

However, this model explains the location only of those cities within the Roman settlement area, which generally included England and Europe south of the Rhine and Danube rivers. To the north and east of these Roman lands, the founding of true cities came later, during the Medieval period. Many of these northern cities, such as Berlin, Copenhagen, Hamburg, and Stockholm, arose with the expansion of regional trade that centered on the Baltic Sea. Even in these areas, most major cities were founded by 1400, thus providing Europe with a dynamic urban legacy that is still apparent in today's urban landscape, except where this older fabric was erased by the destruction of World War II.

The Past in the Present Three historical eras dominate most European city landscapes. The Medieval (roughly A.D. 900–1500), the Renaissance–Baroque (1500–1800), and the Industrial (1800–present) periods have each left their characteristic mark on the European urban scene. Learning to recognize these stages of historical growth provides visitors to Europe's cities with fascinating insight into both past and present landscapes (Figure 8.17).

The **medieval landscape** is one of narrow, winding streets, crowded by three- or four-story masonry buildings with little setback from the street. This is a dense landscape with few open spaces, except around churches or public buildings. Here and there, public squares or parks give clues to an earlier open-air marketplace where medieval commerce was transacted. As picturesque as we find medieval-era districts today, they nevertheless present challenges to modernization because of their narrow, congested streets and antiquated housing. Often, modern plumbing and heating are lacking, and rooms and hallways are small and cramped compared to our contemporary expectations. Because these medieval sectors usually lack modern facilities, they are often inhabited by low- and fixed-income residents. In many areas they are dominated by the elderly, university students, or ethnic migrants.

In contrast to the cramped and dense medieval landscape, those areas of the city built during the **Renaissance–Baroque** period produce a landscape that is much more open and spacious, with expansive ceremonial buildings, monuments, squares, ornamental gardens, and wide boulevards lined with elaborate residences. During this period (1500–1800), a new aesthetic and sense of urban planning arose in Europe that resulted in the restructuring of many European cities, particularly its large capitals. These changes were primarily for the benefit of the new urban elite: the royalty, aristocrats, and successful merchants. City dwellers of lesser means remained in the older medieval quarters, which became increasingly crowded and cramped, as city space was devoted to the space-extensive pleasures of its ruling classes.

Aggravating crowding during this time was the fact that defensive structures, such as city walls and other fortifications, constrained the outward spread of these growing cities. With the advent of high-powered assault artillery, European cities were forced to build an extensive system of defensive walls. Once girdled by these walls, the city could not expand laterally; instead, densities had to increase within the city confines. A common solution was to add several new stories to the medieval houses, which increased density and crowdedness.

These Renaissance–Baroque landscapes are still very much a part of the contemporary city and are, in fact, often the most prominent landmarks sought out by visitors. For example, after its devastating fire of 1666, London was completely re-

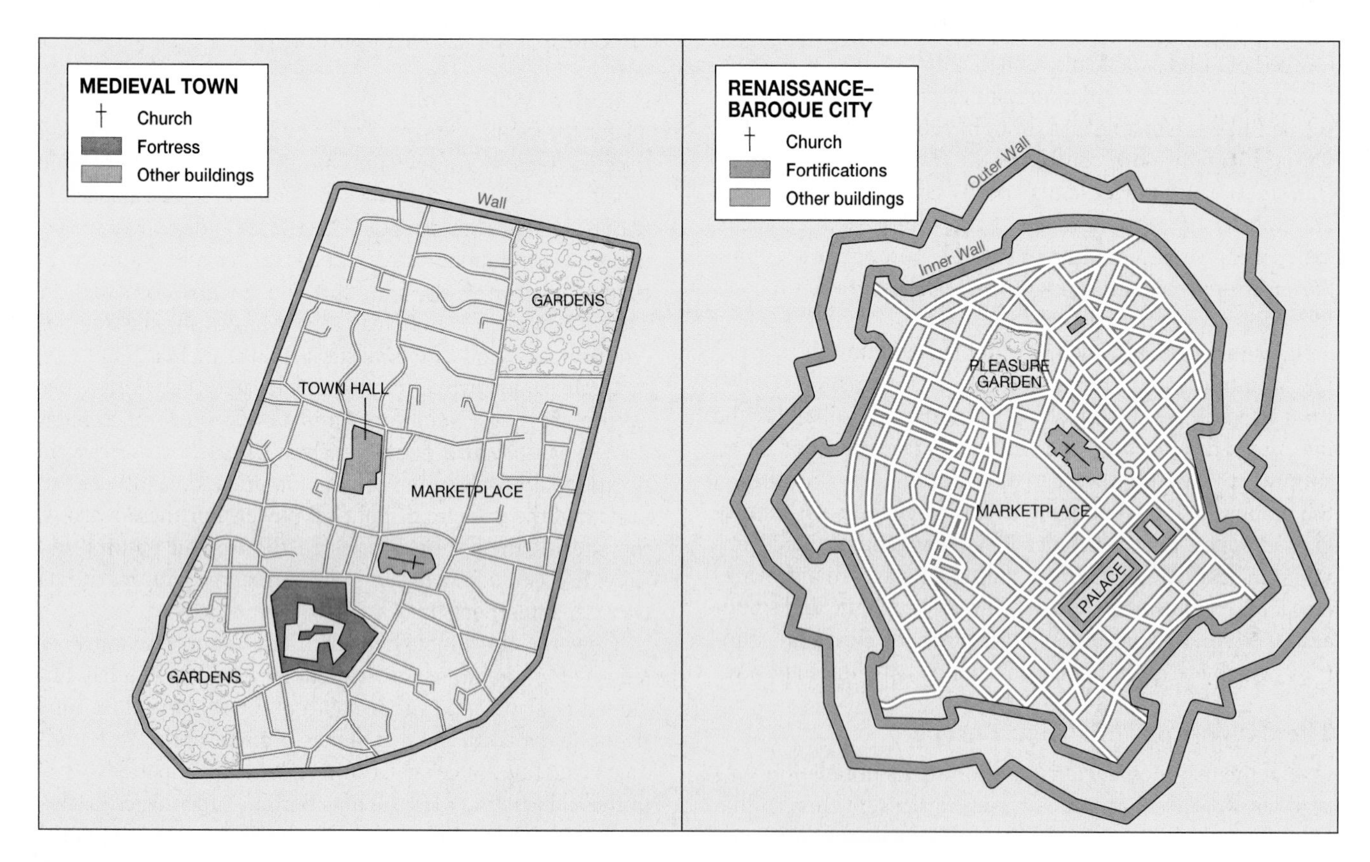

▲ **Figure 8.17 Historical urban landscapes** Contemporary European cities still exhibit landscapes shaped in the past. Most noticeable are the contrasts found between the narrow lanes, small market squares, and five-story houses of the medieval landscape, compared to the wide boulevards, open spaces, and elaborate palaces associated with the Renaissance–Baroque period. Today, medieval city quarters are often areas of historic preservation and renovation, and, as a result, some of these neighborhoods have become rather upscale.

built in the Baroque style with showpiece parks, boulevards, and monumental buildings, such as St. Paul's Cathedral. In Paris, although major boulevard-building took place in the mid-nineteenth century, it was still very much an expression of the Baroque mentality. In Vienna, the grand boulevard of the *Ringstrasse,* or ring-street, which encircles the inner city, was built in the late nineteenth century when city fortifications were declared obsolete and torn down.

Industrialization has dramatically altered the landscape of European cities. Historically, factories clustered together in cities beginning in the early nineteenth century, drawn to the city by their large markets and labor force and supplied by raw materials shipped by barge and railroad. Full-blown industrial districts of factories and worker tenements grew up around these transportation lines. In continental Europe, where many cities retained their defensive walls until the late nineteenth century, these new industrial districts were often located outside the fortifications, removed somewhat from the historic central city. In Paris, for example, when the railroad was constructed in the 1850s, it was not allowed to penetrate the city walls. As a result, terminals and train stations for the network of different tracks were located beyond the original fortifications. Although the walls of Paris are long gone, this pattern of outlying train stations persists today. As noted, the historical industrial areas were also along these tracks, outside the walls in what were essentially suburban locations. This contrasts somewhat with the geography of industrial areas in North American cities, where factories and working class housing often occupied central city locations.

In London, factories and workers located in the eastern part of the city, close to the busy docks of the Thames River. Early city planning (some would call it "social engineering") reinforced the distinction between the middle- and upper-class West End and London's impoverished blue-collar, industrial east side. This historical west–east difference has been lessened in postwar decades with massive urban renewal projects of office buildings, shopping malls, and middle-class housing. Much of the impetus for these recent projects came from the need to rebuild London's east side, which was heavily damaged by air raids in World War II. Postwar rebuilding has also been a mixed blessing for modernizing the landscape in the other European cities that suffered heavy war damage. In many cities, such as Berlin, Rotterdam, and Warsaw, the historic landscape was largely destroyed. In its place today one finds a landscape of modern office buildings, apartment houses, downtown highways, and shopping malls.

Protecting the Sense of Place Europeans take a strong sense of cultural identity from their cities and, as a result, have legislated a number of innovative programs designed to protect their urban landscapes. Many of these planning and preservation measures have now spread to other countries, including North America. Many cities have unique skylines that become signatures or symbols of that place, and some European cities have taken pains to protect that historic skyline. The best example is to contrast the skylines of Paris and London. In Paris, strong building limitations protect the inner-city skyline so that the historic bell towers of Notre Dame and the more modern Eiffel Tower remain visually prominent. As a result, the traditional skyline remains largely intact (Figure 8.18). New high-rise buildings are clustered together in outlying nodal points, such as in the high-density office and residential neighborhood of La Défense just to the west of the inner city. London, in contrast, has done little to protect its skyline, and it is now dominated by modern high-rise office buildings that obscure the landmark cathedral of St. Paul's and the spires of Westminster.

▲ **Figure 8.18 Landscape of central Paris** For several decades, Paris has attempted to preserve its historic skyline by banning high-rises in the central city, as shown in this photo. Instead, these modern structures are clustered in development zones just outside the inner city. *(Allard/REA/SABA Press Photos, Inc.)*

At the street level in European cities, it is common to find historic preservation measures that are designed to protect and renovate medieval neighborhoods while, at the same time, maintaining the traditional urban fabric and sense of place. These are often expensive projects, since restoring these old buildings often costs more than demolition and replacement with a modern structure. Still, most European cultures have decided that the benefits outweigh the costs. Once again, Paris was a leader in this historic preservation movement; in the 1960s, it allocated public funds to protect its medieval district. By the late 1970s, similar historic preservation measures were found throughout Europe; examples include Salzburg, Austria, and Regensburg, Germany.

Landscape protection is also common in the rural environments of Europe, reminding us of the link between a traditional sense of place and cultural identity. In England, Denmark, France, Austria, and Germany, for example, many rural areas and small settlements have enacted strong local ordinances to protect the visual qualities of their landscapes. In these formally designated landscape protection zones, new and renovated houses must use traditional building materials and design so that the structure fits into the existing environment.

▲ **Figure 8.19 Protecting traditional rural landscapes** Many countries attempt to protect their traditional rural landscapes by banning billboards, controlling architecture so that new buildings fit in with existing structures, and constraining the size and shape of new structures. This photo is of the Auvergne area of France. *(John Brooks/Liaison Agency, Inc.)*

Potential visual blemishes from billboards and convenience stores are tightly regulated and, in a few extreme cases, even the appearance of agricultural field patterns, fences, and land cover is regulated (Figure 8.19).

Though these regulations may seem odious and offensive to individuals and cultures emphasizing individual freedoms over communal values, the fact such landscape protection measures exist in Europe reminds us once again of the strong attraction people have to cities, villages, and countryside that exhibit a unique sense of place. For many, these preserved landscapes are welcome relief from the blandness of modern, globalized cities, shopping centers, and malls.

Cultural Coherence and Diversity: A Mosaic of Differences

The rich cultural geography of Europe demands our attention for several reasons. First, the highly varied and fascinating mosaic of different languages, customs, religions, ways of life, and landscapes that characterize Europe has also shaped strong local and regional identities that have all too often stoked the fires of separatism and nationalism. In many cases, this cultural fabric has been torn asunder by geopolitical conflicts that arose from competing cultural differences.

Second, European cultures are important to understand because they have played leading roles in processes of globalization. Through European colonialism and imperialism, regional languages, religion, economies, and values have transformed every corner of the globe. If you doubt this, think about a cricket game in Pakistan, high tea in India, Dutch architecture in South Africa, or the millions of French-speaking inhabitants of equatorial Africa. Without using the modern technology of satellite TV, the Internet, or Hollywood films and video, European culture spread across the world, changing the speech, religion, belief systems, dress, and habits of millions of people on every continent.

Today, though, many European countries now resist the varied expressions of global culture (Figure 8.20). France, for example, struggles against both U.S.-dominated popular culture and the multicultural influences of its large migrant population. In many ways the same is true of Germany and England. The cultural geography of contemporary Europe is also complicated by the fact that culture operates at many different scales and in many different ways simultaneously. While Europe fends off global culture with one hand, on the other it creates its own unique supranational culture through economic and political integration. As individual countries join together in political alliances and unions that transcend national sovereignty, local and regional cultures demand autonomy and independence. Indeed, the list of cultural contradictions is a long and fascinating one. Underlying the complexity of this dynamic cultural mosaic are the cultural fundamentals of language and religion, traits that form the basis of so many of Europe's cultural patterns.

Geographies of Language

Language has always been an important component of nationalism and group identity in Europe. This is true not only of the past, but also today as tensions between cultural convergence and divergence continue to play out. While some small ethnic groups, such as the Irish or the Bretons, work hard to preserve their local language in order to reinforce their cultural identity, millions of Europeans are busy learning multiple languages so they can communicate across cultural and national boundaries. In this age of globalization and world culture, rare is the European who does not speak at least two languages.

▲ **Figure 8.20 Global culture in Europe** U.S. popular culture is received with great ambivalence in Europe. While many people embrace everything from fast food to Hollywood movies, others protest the loss of Europe's traditional regional cultures. The French are particularly outspoken about resisting globalized culture. *(David R. Frazier Photolibrary, Inc.)*

As their first tongue, 90 percent of Europe's population speaks Germanic, Romance, or Slavic languages belonging to western linguistic groups of the Indo-European family. Germanic and Romance speakers each number almost 200 million in the European region. There are far fewer Slavic speakers (about 80 million) when Europe's boundaries are drawn to exclude Russia, Belarus, and Ukraine (Figure 8.21).

Germanic Languages Germanic languages dominate Europe north of the Alps. This linguistic region includes Germany, Scandinavia, two Baltic Republics (Latvia and Lithuania), and the English-speaking areas of the United Kingdom. Approximately 5,000 years ago, the earliest Germanic tribes were centered along the Baltic coast of what is today Germany and Denmark. From this prehistoric cultural hearth, Germanic peoples spread south, east, and west over the course of thousands of years, and with their diffusion, regional dialects evolved into distinct languages such as German, English, and Dutch.

Today, about 90 million people speak German as their first language. This is the dominant language of Germany, Austria, Liechtenstein, Luxembourg, eastern Switzerland, and several small areas in Alpine Italy. Until recently, there were also large German-speaking minorities in Romania, Hungary, and Poland, but many of these people left eastern Europe when the Iron Curtain was lifted in 1989. As "ethnic Germans," they were given automatic citizenship in Germany itself. As is true of most languages, there are very strong regional

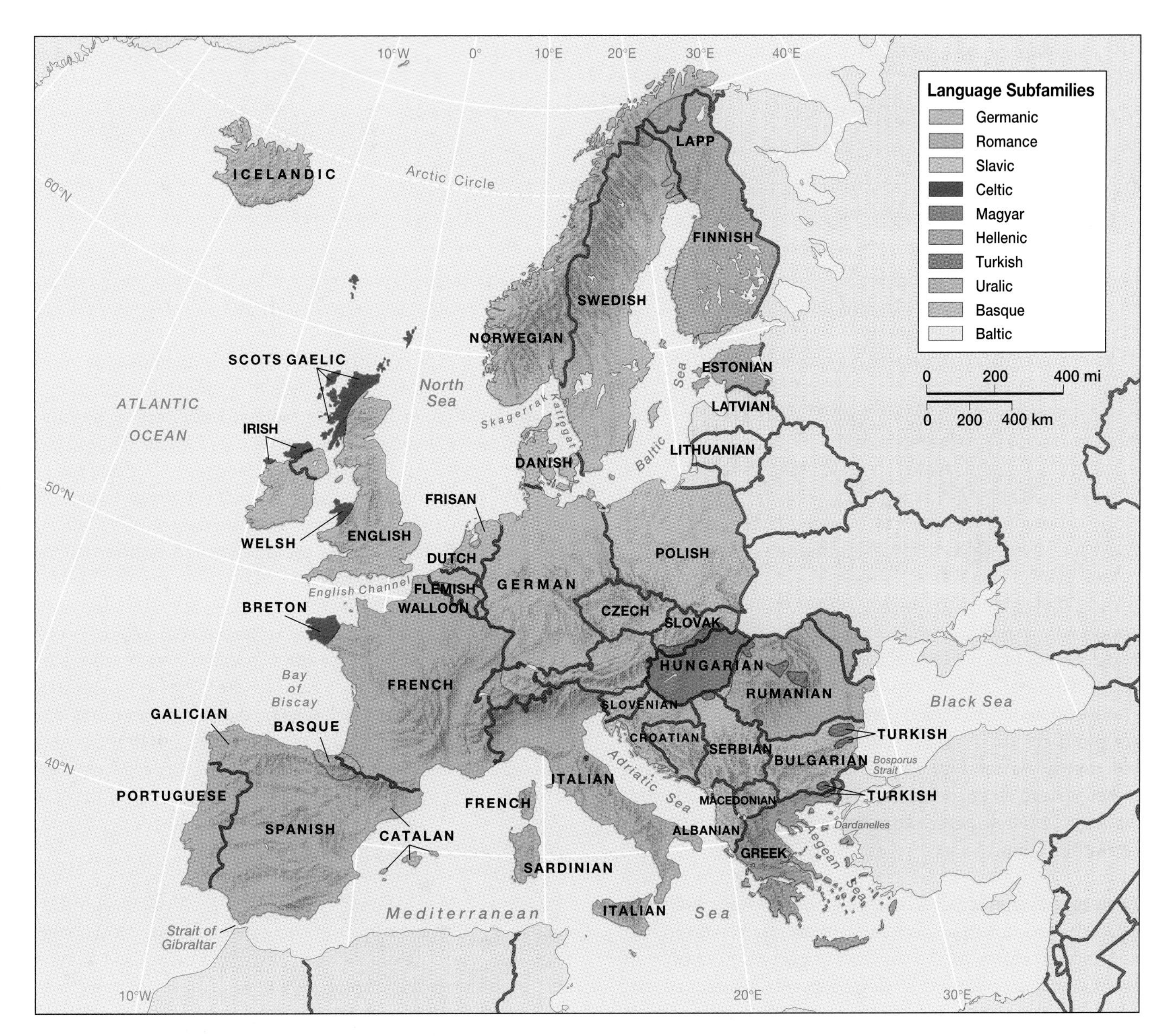

▲ **Figure 8.21 Language map of Europe** Ninety percent of Europeans speak an Indo-European language, grouped into the major categories of Germanic, Romance, and Slavic languages. As a first language, 90 million Europeans speak German, which places it ahead of the 60 million who list English as their native language. However, given the large number of Europeans who speak fluent English as a second language, one could make the case for English as the dominant language of modern Europe.

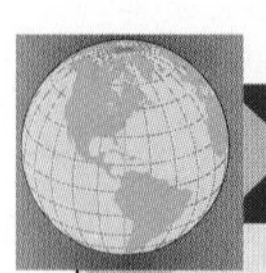

LANGUAGE AND CULTURAL IDENTITY "Denglish" in Germany

In Germany, a furor of sorts has broken out over the increasing use of English phrases instead of German. Never before has the country seen such a proliferation of foreign words.

In Berlin, the major morning newspaper came out with a new slogan in English, "Simply the best"; a major communication company bills its customers in Germany for three kinds of telephone calls: "Citycall," "Germancall," or "Globalcall." Volkswagen refers to its popular new car as the "New Beetle," while Audi talks about "die Power" of its new offering.

The target group for this new language, called "Denglish" by some, are the young: those with money, energy, and a future. As one commentator remarked, "Germany has a word for ambition, and it's English." Indeed, English is the language most studied by German students, and it has also dominated the Internet and American imported popular culture. But there might also be a larger issue at hand, one of cultural identity.

One critic charges that "Denglish" is a cowardly means for many Germans to disentangle themselves from German history, a way out of being German. "Shame explains why English does so much better here than in France, Italy, or Spain. But I refuse to let the history of our country be condensed into the 12 years of Hitler's terror." Others agree, saying that Germans seem just a bit too willing to disguise their nationality behind the mask of the new "Europeaness," or "citizens of the world," as some like to say.

Besides, if that Berlin newspaper had chosen the German phrase, "einfach besser" instead of "simply the best," it would have violated a national law that forbids sweeping assertions of superiority in advertising. "Better than what?" "How is it better?" Using a modern English phrase that is understood by most citizens is a clever way around this law.

Source: Adapted from "Berlin Has a Word for Its Ambitions: English," by Roger Cohen, *New York Times*, December 6, 1998.

dialects within German that set apart German-speaking Swiss, for example, from the "Letzeburgish" German spoken in Luxembourg. All dialects, though, are mutually intelligible (see "Language and Cultural Identity: 'Denglish' in Germany").

English is the second-largest Germanic language, with about 60 million speakers using it as their native tongue. Additionally, uncounted millions learn English as a second language, particularly in the Netherlands and Scandinavia, where many are as fluent as native speakers. Linguistically, English is closest to the Low German spoken along the coastline of the North Sea, which reinforces the notion that an early form of English evolved in the British Isles through contact with the coastal peoples of northern Europe. However, one of the distinctive traits of English that sets it apart from German is that almost a third of the English vocabulary is comprised of Romance words brought to England during the Norman–French conquest of the eleventh century.

Elsewhere in this region, Dutch (Netherlands) and Flemish (northern Belgium) account for another 20 million people, roughly the same number of Scandinavians who speak the closely related languages of Danish, Norwegian, and Swedish. Icelandic, though, is a distinct language because of its long separation from its Scandinavian roots.

Romance Languages Romance languages, such as French, Spanish, and Italian, evolved as regional dialects from their common ancestry in the vulgar (or everyday) Latin used within the Roman Empire. Today, about 60 million Europeans speak Italian as their first language. Italian is also an official language of Switzerland and is spoken on the French island of Corsica.

French is spoken in France, western Switzerland, and southern Belgium (where it is known as *Walloon*). Today, there are about 55 million native speakers in Europe. As with other languages, French also has very strong regional dialects, so strong that linguists differentiate between two forms of French in France itself, that spoken in the north (and the official form because of the dominance of Paris), and the language of the south, or *langue d'oc.* This linguistic divide expresses the long-standing tensions between Paris and southern France, where in the last decade the strong regional consciousness of the southwest (centered on Toulouse and the Pyrenees) has led to a resurrection of their own distinct language, Occitanian.

Spanish also has very strong regional variations. About 25 million people speak Castillian Spanish, the country's official language, which dominates the interior and northern areas of that large country. However, the Catalan form, which some argue is a completely separate language, is found along the eastern coastal fringe, centered on Barcelona, Spain's major city in terms of population and economy. This distinct language reinforces a strong sense of cultural separateness that has led to the state of Catalonia being given autonomous status within Spain. Portuguese is spoken by another 12 million speakers in that country and in the northwestern corner of Spain. Portugal's colonial legacy is suggested by the fact that far more people speak this language in its former colony of Brazil in Latin America than in Europe.

Finally, Romanian represents the most eastern extent of the Romance language family; it is spoken by 24 million in Romania and also in parts of the new state of Moldova. Though unquestionably a Romance language, Romanian also contains many Slavic words. Further, in Moldavia, the language is written in the **Cyrillic alphabet,** which is based on Greek letters and is used in Russian and other Slavic languages.

The Slavic Language Family Slavic is the largest western subfamily of the Indo-European language family if one includes European Russia and its neighbors.

Traditionally, Slavic speakers are separated into northern and southern groups, divided by the non-Slavic speakers of Hungary and Romania.

To the north, Polish has 35 million speakers and Czech and Slovakian about 14 million each. These numbers pale in comparison, however, with the number of northern Slav speakers in nearby Ukraine, Belarus, and Russia, where one can easily count more than 150 million. Southern Slav languages include three groups: 14 million Serbo-Croatian (now considered separate languages because of the political troubles between Serbs and Croats), 11 million Bulgarian-Macedonians, and two million Slovenian speakers.

The use of two different alphabets further complicates the geography of Slavic languages (Figure 8.22). In those countries with a strong Roman Catholic heritage, such as Poland or the Czech Republic, the Latin alphabet is used in writing. In contrast, countries with close ties to the Eastern or Orthodox church use the Greek-derived Cyrillic alphabet, as is the case in Bulgaria, Macedonia, Croatia, and Serbian areas of Yugoslavia.

Minor Indo-European Languages One of the most interesting minor language families is the *Celtic* group, which includes those languages native to Ireland, Scotland, and the Brittany region of France. Though Celtic languages were probably dominant in much of Europe several thousand years ago, today Celtic speakers number just over one million. The prehistoric cultural hearth for Celts, dating from 1500 B.C., is identified from archeological materials stretching along the northern foothills of the Alps, including portions of Austria, Switzerland, and France. One theory suggests Celts fled from this hearth as they were caught between the northward movement of Romans and the southward advance of Germanic peoples in A.D. 200 or 300. As a result, Celts moved to the western fringes of Europe, reestablishing their culture on the Brittany peninsula as well as in the highlands of what are now Wales, Scotland, and Ireland. These fringe areas then served as the new hearth area for the Celtic cultures that flourished for more than 1,000 years.

Modern Celtic languages, however, are struggling for survival. In western France, the Celtic Breton language counts fewer than half a million speakers, about one-third of its population at 1900. A major decline took place after World War II, when the Catholic clergy in Celtic Brittany switched to French. In Wales, about the same number of people speak Welsh, though several groups are trying to preserve the language and culture. In the highlands and islands of Scotland, the situation is even more desperate. Here, the number of Gaelic speakers is thought to be less than 100,000. In Ireland, the government has tried different strategies for promoting the use of Irish, but to little avail. Up until recently the teaching of Irish was mandatory in Irish public schools, yet that was dropped in the 1980s because it was thought to be highly impractical. Land grants and other forms of settlement subsidies were offered to Irish speakers, but that could not stem the tide against the increasing use of English. Today, only about 60,000 people use Irish in everyday conversation.

In southeastern Europe, modern residents trace their linguistic heritage to the *Hellenic* language of ancient Greece. About 10 million people in Greece and Cyprus speak this tongue. Much of the vitality of Hellenic languages comes from its use in the Greek Orthodox Church. It is said that modern Greek has the same relationship to classical Greek as contemporary Italian does to Latin. Finally, in the northeastern

◀ **Figure 8.22 The alphabet of ethnic tension** With the departure of many of Kosovo's Serbs, signs in the Cyrillic alphabet are being taken off many stores, shops, and restaurants as Kosovars of Albanian ethnicity restate their claims to the region. Here, the Albanian owner of a restaurant in the capital city of Pristina scrapes off Cyrillic letters following the recent war. *(David Brauchli/AP/Wide World Photos)*

corner of Europe, the Baltic languages include Lettish (spoken in Latvia) and Lithuanian. Together they number about four million speakers.

Non-Indo-European Languages Only 5 percent of Europe's population speak non-Indo-European languages, and with the exception of the Hungarians, they are found primarily on the outer reaches of the region. *Magyar* counts some 13 million, mainly in Hungary, but this linguistic region also spills over into nearby Romania, Slovakia, and northern Yugoslavia. Magyar tradition maintains their ancestors were a herding people who entered Europe from the Asian steppes around A.D. 900. Once settled in the central Danube basin, they created an important culture that, despite its linguistic differences, has played a major role in Europe's affairs. The Austro-Hungarian Empire, for example, dominated eastern Europe until World War I.

Far to the north, the *Uralic* linguistic family includes Estonian speakers, an even smaller number of Sami people (often called Lapps), and a larger number of Finns. In total, about five million people speak these related languages that possibly originated in the forest and tundra lands of northern Eurasia.

On the southeastern periphery of Europe, the *Altaic* language family includes Turkish. Because of the long rule of the Ottoman Turks in southeastern Europe, significant Turkish minorities remain in countries such as Bulgaria where, despite discrimination and attempts to expel them, they constitute almost 10 percent of the nation's population. Also important are the more than 1.5 million Turks currently in Germany, guest workers invited to that country to solve that country's labor problems, but now the focus of considerable discrimination and blatant racism.

Last, the Basque people on the Atlantic border of France and Spain speak *Euskara,* a language unlike any other in Europe. Some argue the Basques are ancestors of the original prehistoric Europeans. Perhaps more important than this uncertain past is the violent terrorism of today that pits many Spanish Basque people against the Madrid government as they seek political autonomy for their unique region at the western end of the Pyrenees mountain chain. Even though a large number of Basques have emigrated to North and South America, today there are still more than 600,000 speakers in this part of Europe.

Geographies of Religion, Past and Present

Religion is absolutely inseparable from the geography of cultural coherence and diversity in Europe because so many of today's cultural tensions are embedded in historical events. To illustrate, strong cultural borders in the Balkans and eastern Europe are drawn based upon the eleventh-century split into Christianity's eastern and western churches; in Northern Ireland, blood is still shed over the tensions between the sixteenth-century division into Catholicism and Protestantism; and much of the terrorism of ethnic cleansing in Yugoslavia comes from the historical struggle between Christianity and Islam in that part of Europe. Understanding these important contemporary issues thus involves a brief look back in time at the historical geography of Europe's religious complexity (Figure 8.23).

The spread of early Christianity into Roman Europe was slow until the Edict of Tolerance by the emperor Constantine in A.D. 313. After that, the new religion appears to have spread more rapidly, both within and outside of the Empire. The Celtic peoples of Britain adopted Christianity by A.D. 600, and their missionaries played a central role in spreading this religion to the Germanic tribes of central Europe by 800, then even farther east and north by the tenth century. This spread appears to have taken place by **hierarchical diffusion**, with local peoples converting because of the adoption of Christianity by a tribal leader or ruler. Centuries later, this same sort of hierarchical diffusion was important during the conflict between Catholics and Protestants.

The Schism Between Western and Eastern Christianity In southeastern Europe, early Greek missionaries spread Christianity through the Balkans and into the lower reaches of the Danube. Progress was slower than in western Europe, though, perhaps because of continued invasions by peoples from the Asian steppes. There were other problems as well, primarily the refusal of these Greek missionaries to accept the growing church hierarchy of Roman bishops.

This irreconcilable tension with western Christianity was formalized in 1054 with an official split of the eastern church from Rome. The eastern church increasingly splintered into sects closely linked to specific nations and states. Today, for example, we find Greek Orthodox, Bulgarian Orthodox, and Russian Orthodox churches, all of which have different rites and rituals, yet share cultural affinity. The current political ties between Russia and Serbian Yugoslavs, for example, grow from their shared language and religion.

Another distinction between east and west was the orthodox use of the Cyrillic alphabet instead of the Latin. Since Greek missionaries were primarily responsible for the spread of early Christianity in southeastern Europe, it is not surprising that they used an alphabet based upon Greek characters. More precisely, this alphabet is attributed to the missionary work of St. Cyril in the ninth century. As a result, the division between western and eastern churches, and between the two alphabets, remains one of the most problematic cultural boundaries in Europe.

Conflicts with Islam Both eastern and western Christian churches also struggled with incursions from Islamic empires to Europe's south and east. Even though historical Islam was reasonably tolerant of Christianity in its conquered lands, Christian Europe did not reciprocate by accepting Muslim imperialism. The first crusade to reclaim Jerusalem from the Turks took place in 1095. After the Ottoman Turks took Constantinople in 1453 and amassed control over the Straits of Bosporos and the Black Sea, they moved rapidly to spread their Muslim empire throughout the Balkans and arrived at the gates of Vienna in the middle of the sixteenth century. There, Christian Europe drew its line in the sand and turned back the Turks. However, Ottoman control of southeastern

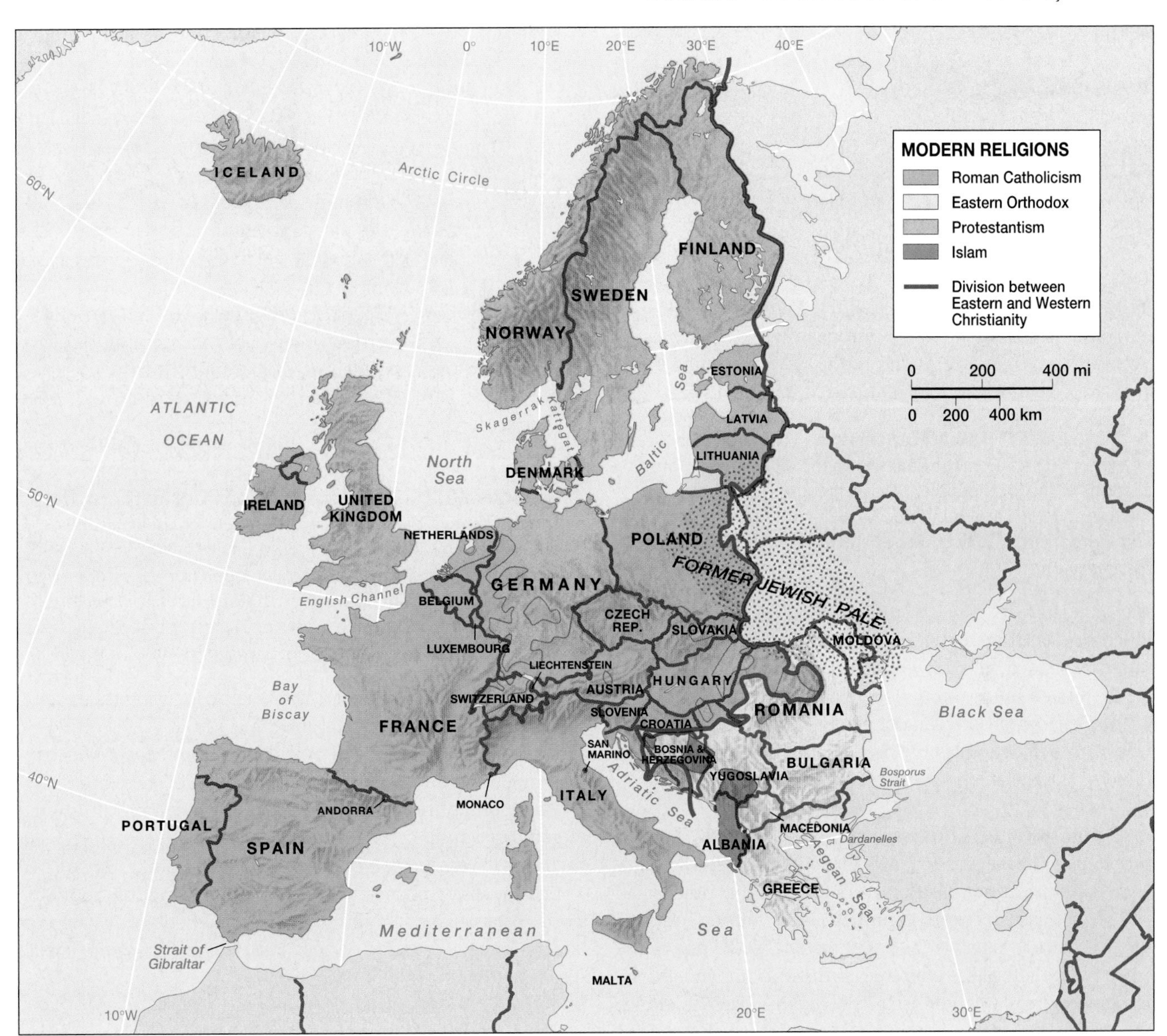

▲ **Figure 8.23 Religions of Europe** This map shows the divide in western Europe between the Protestant north and the Roman Catholic south. Historically, this distinction was much more important than it is today. Note the location of the former Jewish Pale that was devastated by the Nazis during World War II. Today, ethnic tensions with religious overtones are found primarily in the Balkans, where adherents to Roman Catholicism, Eastern Orthodoxy, and Islam are found in close proximity.

Europe lasted until the Empire's demise in the early twentieth century. This historical presence of Islam explains the current problematical mosaic of religions in the Balkans—intermixed areas of Muslims, Orthodoxy, and Roman Catholics. Commonly, ethnic boundaries in that part of Europe are drawn along religious lines (Figure 8.24).

Muslim incursions into Spain from North Africa left an enduring imprint of their Moorish heritage throughout the Iberian Peninsula in terms of architecture and technology. Irrigation and water-control techniques and strategies, for example, were brought by the Muslims, which provided great benefits not just to the Spanish, but were then diffused to other parts of the world colonized by Spain and Portugal.

The Protestant Revolt Besides the division between western and eastern churches, the other great split within Christianity occurred between Catholics and Protestants, a distinction that arose in Europe during the sixteenth century and has divided the region ever since. With the exception of the troubles in Northern Ireland, tensions today between these two major groups are far less damaging than in the past, when the European landscape was scarred by violent battles between these two groups.

Perhaps the low point came during the Thirty Years' War (1618–48), when central and western Europe were laid waste by war, famine, and the plague. Archival evidence suggests that more than 50 percent of the affected population was either

▲ **Figure 8.24 Balkan landscape** The urban landscape of Mostar, Bosnia-Herzegovina, shows the different religious affiliations of the local people. Muslim mosques, with their distinctive minarets, are found side by side with Roman Catholic and Eastern Orthodoxy churches. *(A. Ramey/Woodfin Camp & Associates)*

▲ **Figure 8.25 Religious tensions in Northern Ireland** One of the few places where tensions still exist between Roman Catholics and Protestants is in Northern Ireland. Here, members of the Protestant Orange Order are stopped by police from marching into a Roman Catholic area. *(John Giles/AP/Wide World Photos)*

killed or died from starvation or sickness. Fields lay fallow, forests expanded, villages were deserted as this religious struggle devastated Europe from Bohemia (now the Czech Republic) to Belgium on the Atlantic shore. Today, this contested landscape still delineates the north–south boundary between Protestant and Catholic Europe (Figure 8.23).

A Geography of Judaism Europe has long been a problematic homeland for Jews after their forced dispersal from Palestine during the Roman Empire. At that time, small Jewish settlements were found in cities throughout the Mediterranean. Later, by A.D. 900, about 20 percent of the Jewish population was clustered in the Muslim lands of the Iberian Peninsula, where Islam showed a greater tolerance than Christianity had for this religion. Furthermore, Jews played an important role in trade activities, both within and outside of the Islamic lands. After the Christian reconquest of Iberia, however, Jews once more faced severe persecution and fled from Spain to more tolerant countries in western and central Europe.

One focus for migration was the area in eastern Europe that became known as the Jewish Pale. In the late Middle Ages, at the invitation of the Kingdom of Poland, which offered an edict of tolerance, Jews settled in cities and small villages in what is now eastern Poland, Belarus, western Ukraine, and northern Romania (Figure 8.23). Jews collected in this region for several centuries with hopes for establishing a true European homeland, despite the poor natural resources of this marshy, marginal agricultural landscape.

Until emigration to North America began in the 1890s, 90 percent of the world's Jewish population lived in Europe. Most were clustered in the Pale. Even though many emigrants to the United States and Canada left this area, the Pale remained the largest aggregation of Jews in Europe until World War II. Tragically, Nazi Germany was able to use this ethnic clustering to their advantage by focusing their extermination activities on this area.

In 1939, on the eve of World War II, there were 9.5 million Jews in Europe, or about 60 percent of the world's Jewish population. During the war, German Nazis murdered some six million Jews during the horror of the Holocaust. Today, fewer than two million Jews live in Europe. Since 1990 and the lifting of constraints on Jewish emigration from Russia, Belarus, and Ukraine, more than 100,000 Jews have emigrated to Germany, giving it the fastest-growing Jewish population outside of Israel.

The Patterns of Contemporary Religion In Europe today, there are about 250 million Roman Catholics and fewer than 100 million Protestants. Generally, Catholics are found in the southern half of the region, except for significant numbers in Ireland and Poland, while Protestants dominate in the north (Figure 8.25). Additionally, since World War II, there has been a marked disinterest in organized religion, mainly in western Europe, which has led to plummeting church attendance in many areas. This trend is so marked that the term **secularization** is used, referring to the widespread movement away from the historically prominent organized religions of Europe.

Catholicism dominates the religious geography and cultural landscapes of Italy, Spain, France, Austria, Ireland, and southern Germany. In these areas large cathedrals, monasteries, iconic monuments to Christian saints, and religious placenames draw heavily upon pre-Reformation Christian culture. Since visible beauty is part of the Catholic tradition, ornate religious structures and monuments offer a more ornate landscape than in the northern Protestant lands.

Protestantism is most widespread in northern Germany, the Scandinavian countries, and England, and is intermixed with Catholicism in the Netherlands, Belgium, and Switzerland. Because of its reaction against the aesthetic excesses of the early Catholic church, the landscape of Protestantism is much more sedate and subdued. Large cathedrals and religious monuments in Protestant countries are associated primarily with the Church of England, which has strong historical ties to Catholicism; St Paul's Cathedral and Westminster Abbey in London are examples.

The emerging geography of religion in eastern Europe remains highly differentiated and complex. In Poland, where the Catholic Church was strong and remained an active opponent to communism during Cold War years, religion seems to be playing a prominent role in contemporary life. Three-quarters of the Polish population consider themselves practicing Catholics. If true, this reemergence of religious commitment far exceeds church attendance in some of the historically Catholic countries of western Europe. Historically, Catholicism also dominated in Hungary, and recent census figures suggest that more than two-thirds of the population considers itself Catholic in today's post-communism environment. This is also the case in nearby Slovakia. In contrast, religious adherence in the Czech Republic is evenly divided between Catholics (40 percent) and those who consider themselves atheist (also 40 percent).

In southern Europe, Catholicism has also been a defining trait of post-1989 culture and ethnicity in both Croatia and Slovenia. In fact, the differences between Serb and Croat in this troubled area are usually defined solely by religion. A Serb is usually Orthodox, while Croats are Catholic. In Yugoslavia overall, two-thirds of the population list themselves as Christian Orthodox, which is the official name given to Serbian Orthodoxy. About the same percentage of people list themselves as Orthodox in the neighboring countries of Romania and Bulgaria.

European Culture in Global Context

Europe, like all world regions, is caught up in a period of profound cultural change; in fact, many would argue that the pace of cultural change in Europe is accelerated because of the crosscurrents and complicated interactions between globalization and Europe's own internal agenda of political and economic integration. While newspaper pundits celebrate the "New Europe" of integration and unification, other critics refer to another, more tension-filled "New Europe," one of foreign migrants and guest workers, haunted by ethnic discrimination and racism.

Globalization and Cultural Nationalism Since World War II, Europe has been defensive, even outspoken, about cultural contamination from North America. Some countries are more reactionary than others. While a few large countries, such as England and Italy, seem to accept the onrush of American popular culture, other countries—most notably France—have expressed outright indignation over the corrupting impact of U.S. popular culture on speech, music, food, and fashion.

In response, France has taken cultural nationalism to new levels with protective legislation. Radio stations, for example, must devote at least 40 percent of their air play to French songs and musicians; French filmmakers are subsidized and supported by government subventions in hopes of staving off a complete takeover by Hollywood; the official body of the French Academy, whose job it is to legislate proper usage of the French language, has a long list of English and American words and phrases that are banned from use in official publications and from highly visible advertisements or billboards. Though the French commonly use "le weekend" or "le software" in their everyday speech, these words will never be found in official governmental speech or publications.

While France set the model for a continental brand of cultural nationalism, these ideas seem to be spreading to other European countries. Germany, for example, recently started public debate and discussion over banning certain English phrases and words in commercial advertisements and "proper" speech. There is a clear irony here, given that Europe's languages have borrowed heavily from each other. Examples include the high number of French and German cognates in the English language, the similarities between Romance languages, and the shared vocabulary of Scandinavian tongues.

A Common European Culture? Another key issue focuses on the emergence of a common, integrated European culture in place of traditional national identities. While some Europeans openly resist contamination of their national cultures by the tidal wave of U.S. popular culture, there appears to be little opposition to the melding together of different European cultures into a new homogenized culture of "Europeaness" (Figure 8.26).

While this new sense of cultural unity is a product of the integration of politics and economies through the EU, the

▲ **Figure 8.26 European student culture** European youth and students are increasingly blending their distinctive cultural styles into a common "Euro-style" that masks the formerly different national styles in dress, speech, music, and language. In this photo, students of four different nationalities are virtually indistinguishable. *(Bob Handelman/Tony Stone Images)*

specific roots are the increased mobility and travel enjoyed by contemporary Europeans. For citizens of EU member states, it is no longer difficult to visit, work, or study in another country. As a result, western Europeans are increasingly taking on cultural traits of other countries. Ten years ago only 20 percent of western Europeans vacationed in a country other than their own. Today, the figure has risen to almost 50 percent, and for the youth of Europe, it is more than 60 percent. Additionally, EU students can easily cross national borders to study in other countries without being hindered by visa requirements and red tape. The same is true of the EU workforce, which can move to higher-paying jobs where labor is in shorter supply. For example in the last several years, Ireland, the economic "tiger" of the EU, has been the destination of skilled workers from throughout the European Union.

This new mobility also requires multilinguality, so another trait of the new European culture is the ability to speak several languages. Of the student population of the EU, fully 70 percent speak a foreign language. Ironically, though, it is English, the language of globalization, that is the most popular second language. In all EU countries except for Spain, far more than half the student population has studied English. French lags behind in second place at about 40 percent; German, slightly less, depending on the country.

Other elements of culture, such as music, food, and clothing, are also being blended together into an Euro-synthesis. Ten years ago one could usually tell national origin based upon dress alone; no longer is this case as more and more people take on the same "Euro-look."

Migrants and Culture Migration patterns are also influencing the cultural mix. Historically, Europe spread its cultures worldwide through aggressive colonialization and imperialism. Today, however, the region is experiencing a reverse flow as millions of migrants move into Europe, bringing with them their own distinct cultures from the far-flung countries of Africa, Asia, and Latin America. Unfortunately, in some areas of Europe, the products of this cultural exchange are problematic, even destructive.

Ethnic clustering, even ghettoization, are now common in the cities and towns of western Europe. The high-density apartment buildings of suburban Paris, for example, are home to large numbers of French-speaking Africans and Arab Muslims caught in the cultural crossfire of high unemployment, poverty, and racial discrimination. The same is true in Germany, with more than 1.5 million Turks living in the country. Added to that are increasing numbers of emigrants from eastern Europe and the former Soviet Union. As a result, many European countries are experiencing significant culture wars that are fought on many fronts. French leaders, unsettled by the country's large Muslim migrant population, attempted to speed assimilation of female high school students into the mainstream culture by banning a key symbol of conservative Muslim life, the head scarf. This rule triggered riots, demonstrations, and counter-demonstrations.

The political landscape of many European countries now contains far-right, nationalistic parties with an exclusionary and discriminatory political agenda ("France is for the French," for example). Many of these political parties also celebrate the customs, symbols, and cultural icons associated with an idealized, nationalistic past, a construction that clearly does not include migrant foreigners. Most noxious, perhaps, is the cultural reaction of young German skinheads who purport to defend with terrorism and violence "true and pure" Aryan values historically associated with Nazi Germany (Figure 8.27).

▲ **Figure 8.27 Neo-Nazis in Germany** Purporting to embrace "pure and true" Aryan values, these neo-Nazi Germans provoke police at an anti-foreigner rally. This rise in extreme forms of nationalism has become increasingly problematic in Germany and is often directed against Turks and other foreign guest workers. *(Joanna B. Pinneo/Aurora & Quanta Productions)*

Geopolitical Framework: A Dynamic Map

One of Europe's unique characteristics is its dense fabric of 37 independent states found within a relatively small area. No other world region demonstrates the same kind of geopolitical fragmentation, although ethnic nationalists in parts of Asia and Africa strive to imitate such a geography by pointing to Europe as the model for small-scale autonomy. Historically, Europe invented the nation-state. More recently, it has fueled the flames of political independence and democracy worldwide, often promoting these engaging ideals as replacements for its own colonial rule in Asia, Africa, and the Americas.

But Europe's varied geopolitical landscape has been as much problem as promise. Twice in the last century Europe has shed blood to redraw its political geography. Within the last decade alone, seven new states have appeared in eastern Europe, more than half through war. The map of geopolitical troubles suggests still more political fragmentation may take place in the near future, much of it violently (Figure 8.28).

Most of Europe, however, sees a brighter geopolitical future. For many, this is based upon a widespread spirit of cooperation through unification rather than continued fragmentation. Now empowered by the European Union's recent successes at political and economic integration, many Europeans are still giddy over the sudden and unanticipated end of the Cold War in 1989. Many argue that the twentieth cen-

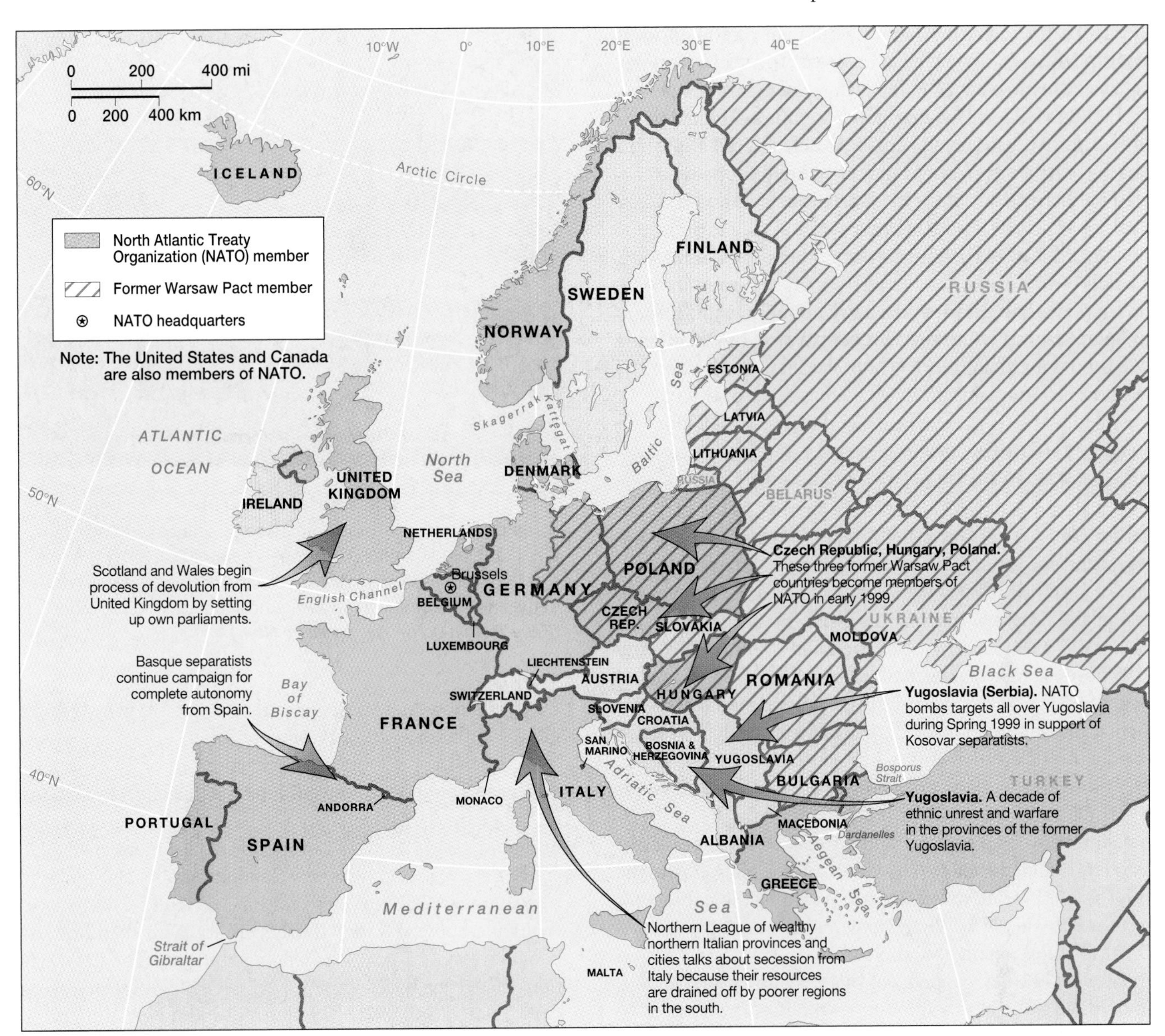

▲ **Figure 8.28 Geopolitical issues in Europe** While the major geopolitical issue of the early twenty-first century remains the integration of eastern and western Europe in the European Union, there are also numerous issues of micro- and ethnic nationalism that engender geopolitical fragmentation. In other parts of Europe, such as in Spain, France, and Great Britain, questions of autonomy within the nation-state structure challenge central governments.

tury was a disaster of Europe's own making. If true, this region seems determined to avoid those mistakes in the twenty-first century by giving the world a new geopolitical model for peace and prosperity. The first decade of the new century should give a clue as to whether this optimism is well founded.

Geopolitical Background: From Empire to the Nation-State

As might be expected, the political space of Europe has been organized in various ways through history. More importantly, these different forms have influenced the present geopolitical framework of the region. Broadly put, three different scales of political space have been apparent, including the empire, the feudal territory, and the nation-state. Though presented here chronologically, one can easily argue that all three of these geopolitical forms persisted in various expressions well into the twentieth century.

The Legacy of Rome The Roman Empire (in existence from approximately 300 B.C.–A.D. 400) had a profound and lasting influence on Europe's geography, so much so that its heritage remains apparent on the landscape today. Major western European cities such as London, Paris, and Frankfurt trace their origins and location back to the Romans. These Roman settlements were linked together with an impressive highway system that remained in use long after Rome had faded. Long, straight stretches of road cutting through a rolling landscape give clue to a highway's Roman origins. Bridges and aqueducts built by Roman engineers still span rivers and valleys in many parts of Europe, from Atlantic to Black Sea.

Geopolitically, the western Roman empire declined in the fifth century because it was unable to defend itself against the attacks of tribes from northern and central Europe who coveted the richer lands of the Mediterranean civilization. These invasions spelled the end of the spatial and civil organization that linked Rome's far-flung settlements and rural estates. As a result of this warfare in the west, political power and economic activity were reoriented away from Rome to Constantinople (Istanbul) in Turkey. This eastern or Byzantine Empire persisted until the fifteenth century, when it was replaced by the Islamic Ottoman Empire.

Feudal Territories After a period of decline in western Europe following the fall of Rome (the so-called "Dark Ages"), a revival and expansion of trade beginning in the eleventh century led to a resurgence of urban life, artisan guilds, and merchant associations. Trade based upon English wool moved across the Channel into the blossoming commercial cities of Ghent and Bruges in the Low Countries. In the Baltic Sea, the Hanseatic League of cities, anchored by Luebeck and Hamburg (Germany) linked together emerging economies in Scandinavia and northern Europe. In the Mediterranean, Genoa and Venice dominated trade as their merchants moved readily through settlements in Europe, Africa, and even Asia.

Contrasting with the urban power of a merchant class was the rural polity based upon **feudalism,** or the formal relationship between a superior and a vassal or person of lesser standing. Fundamental to this system was the obligation of the rural serf or peasant to farmland provided by the noble. Protection was offered by the regional noble in return for agricultural labor, obedience, and military service. As a result, there was complex geopolitical hierarchy of contracts with individuals linked to their masters, who in turn were tied to higher nobles. This network of privileges and rights, of obligations and duty, went right to the top: even emperors and the pope thoughts of themselves as "vassals of God." As a result of these feudal contracts, the geopolitical landscape of medieval Europe was dominated by regional aristocrats who expanded their holdings and power base through war, marriage, and even territorial exchange. Spatial and personal identities were oriented to religious and aristocratic authority, not to a shared sense of community, culture, or nation (Figure 8.29).

▲ **Figure 8.29 The feudal landscape** The imposing and picturesque fortress of Carcassonne, France, reminds visitors of the feudal period in Europe's geopolitical history. It was sited as a defensive structure on the major transportation routes between Mediterranean and Atlantic in southwestern France. *(Pierre Toutain Dorbec/Sygma Photo News)*

Rise of Nationalism and Nation-States Europe is considered the birthplace of the nation-state, a completely new geopolitical and spatial entity fostered by ethnic and cultural nationalism. While this new ideology began in Europe, it spread far beyond the region as it became the preferred modern model for independence from colonialism, authoritarianism, and other forms of unwanted political control. Today, as western Europe moves toward a supranational unity that plays down the power of individual nation-states, peoples in eastern and southern Europe continue to struggle, often violently, for nation-state status. Though some scholars argue that Europe is now in a "post-nation-state" phase, many ethnic groups in the Balkans beg to differ.

Between the fifteenth and eighteenth centuries, European society and politics underwent fundamental changes that undermined the feudal hierarchy. Overseas exploration and trade brought untold wealth to Europe. Urban life blossomed. Scientific discoveries brought new technologies, but also philosophies concerned with new notions of truth. Humanistic thought empowered individuals with tools for shedding superstition and, instead, challenging authority. The Reformation reacted against corrupt power in religious life, notions of democracy questioned absolute rule, and the printing press provided the means for spreading new ideas.

With this foundation, Europeans sought to construct political structures based upon much more than a single royal household or court. As European civilization became more stratified and complicated, a broad spectrum of institutions was needed to guide society, and as these evolved, new territorial identities also flourished. State political structures emerged that intermingled social control with unique cultural and territorial identities. The notion of congruence between polity and space became increasingly common by the eighteenth century.

Often, nation-state formation can be seen as a process of territorial accretion or growth out from a dominant core area. A congruence between a shared culture and political space results as core areas exert their own language, religion, social values, and ways of life over peripheral areas until there is an extensive common "national" culture occupying the territory within the state's political control. France is often used as an example of this process, expanding incrementally outward from the Paris Basin core area. The geopolitical expansion of our European countries, however, is not so tidy, though proponents of the core area model continue to make their case.

While the conceptual model of European nation-state formation probably is in need of revision, there is no question about the result. Over the course of several centuries, anchored by the aggregation of smaller regional polities into the "unified" nation-states of Italy and Germany in the second half of the nineteenth century, the political landscape of western Europe changed from a mosaic of small feudal fiefdoms into one of nation-states. Even though these different states were ruled in various ways, they shared a common trait: Their population found identity and common bonds in the fervor of nationalism, of a shared culture and of shared territory. This fiery spirit of nationalism, however, brought with it explosive fuel for the devastation of Europe through the twentieth century's two world wars.

Redrawing the Map of Europe Through War

Two world wars redrew the geopolitical maps of twentieth century Europe (Figure 8.30). Because of these conflicts, empires, nation-states, and even remnant feudal territories have appeared and disappeared all within the last 100 years. By the early twentieth century, Europe was divided into two opposing and highly armed camps with exaggerated national egos, testing each other for a decade before the tragic outbreak of World War I in 1914.

France, Britain, and Russia were allied against the new nation-states of Italy and Germany, along with the ineffectual Austria-Hungary or Hapsburg Empire that controlled a complex mosaic of ethnic groups in central Europe and the Balkans. Though at the time World War I was referred to as the "war to end all wars," it fell far short of solving Europe's geopolitical problems. Instead, according to many experts, it made a further European land war inevitable. Additionally, American entry into the war in 1917 ensured its continued involvement in European geopolitical affairs for the rest of the century.

When Germany, Italy, and Austria-Hungary surrendered in 1918, the Treaty of Versailles peace process set about redrawing the map of Europe with two goals in mind: First, to punish the losers through loss of territory and severe fiscal reparations, and, second, to recognize the nationalistic aspirations of certain peoples by creating new nation-states. As a result, the new states of Czechoslovakia and Yugoslavia were born. Additionally, Poland was reestablished, as were the Baltic states of Finland, Estonia, Latvia, and Lithuania.

Though the goals were lofty, the redrawn map created a fragmented Europe where few states were satisfied. New states were generally resentful that kinfolk were often left outside the new borders and became minorities in other new states. This created an epidemic of inter-war **irredentism**, or state policies for reclaiming lost territory (real or imagined) beyond their borders inhabited by people of the same ethnicity. An example would be the large German population located in the western portion of the new state of Czechoslovakia, or the Hungarians located in western Romania.

This imperfect geopolitical solution was aggravated greatly by the global economic depression of the 1930s that brought high unemployment, food shortages, and political unease to Europe. Three competing ideologies promoted their own solutions to Europe's pressing problems: Western democracy (and capitalism), communism from the Soviet revolution to the east, and a Fascist totalitarianism promoted by Mussolini in Italy and Hitler in Germany. With industrial unemployment commonly approaching 25 percent in western Europe, public opinion fluctuated wildly between extremist solutions. In 1936 Italy and Germany once again joined forces through the Rome-Berlin "axis" agreement. Once again this alignment was countered with mutual protection treaties between France, Britain, and the Soviet Union. When an imperialist Japan signed a pact with Germany, the scene was set for a second global war.

Nazi Germany tested Western resolve in 1938 by first annexing Austria, the country of Hitler's birth, and then Czechoslovakia, under the guise of providing protection for Germans located there. After signing a nonagression pact with the Soviet Union, Hitler invaded Poland on September 1, 1939. Two days later, France and Britain declared war on Germany. Within a month, the Soviet Union moved into eastern Poland, the Baltic states, and Finland to reclaim territories lost through the peace treaties of World War I. Nazi Germany then moved westward and occupied Denmark, the Netherlands, Belgium, and France. Preparations were made to invade England. In 1941 the war took some startling new turns. In June, Hitler broke the nonagression pact with the Soviet Union and, catching the Red Army by surprise, took the Baltic states and drove deep into Soviet territory. When Japan attacked Pearl Harbor (Hawaii) in December, the United States entered the war in both the Pacific and in Europe.

By early 1944, the Soviet army had recovered most of its territorial losses and moved against the Germans in eastern Europe, beginning a long communist hegemony in that region. By agreement with the Western powers, the Red Army stopped only when it reached Berlin in April 1945. At that time, Allied forces crossed the Rhine River and began their occupation of Germany. With Hitler's suicide, Germany signed an unconditional surrender on May 8, 1945, ending the war in Europe. But with Soviet forces firmly entrenched in the Baltics, Poland, Czechoslovakia, Bulgaria, Romania, Hungary, Austria, and eastern Germany, the military battles of World War II were quickly replaced by the geopolitical reality of an ideological Cold War that lasted more than 40 years until 1989.

A Divided Europe, East and West

From 1945 until 1989, Europe was divided into two geopolitical and economic blocs, east and west, separated by the infamous Iron Curtain that descended shortly after the armistice of World War II. In 1961 this division took concrete form when the Soviets built the Berlin Wall to stop Germans from fleeing to the west. East of this border, the Soviet Union imposed the heavy imprint of communism on all activities—political, economic, military, and cultural. To the west, as Europe rebuilt from the ravages of the war, new alliances and institutions were created to counter the Soviet presence in Europe.

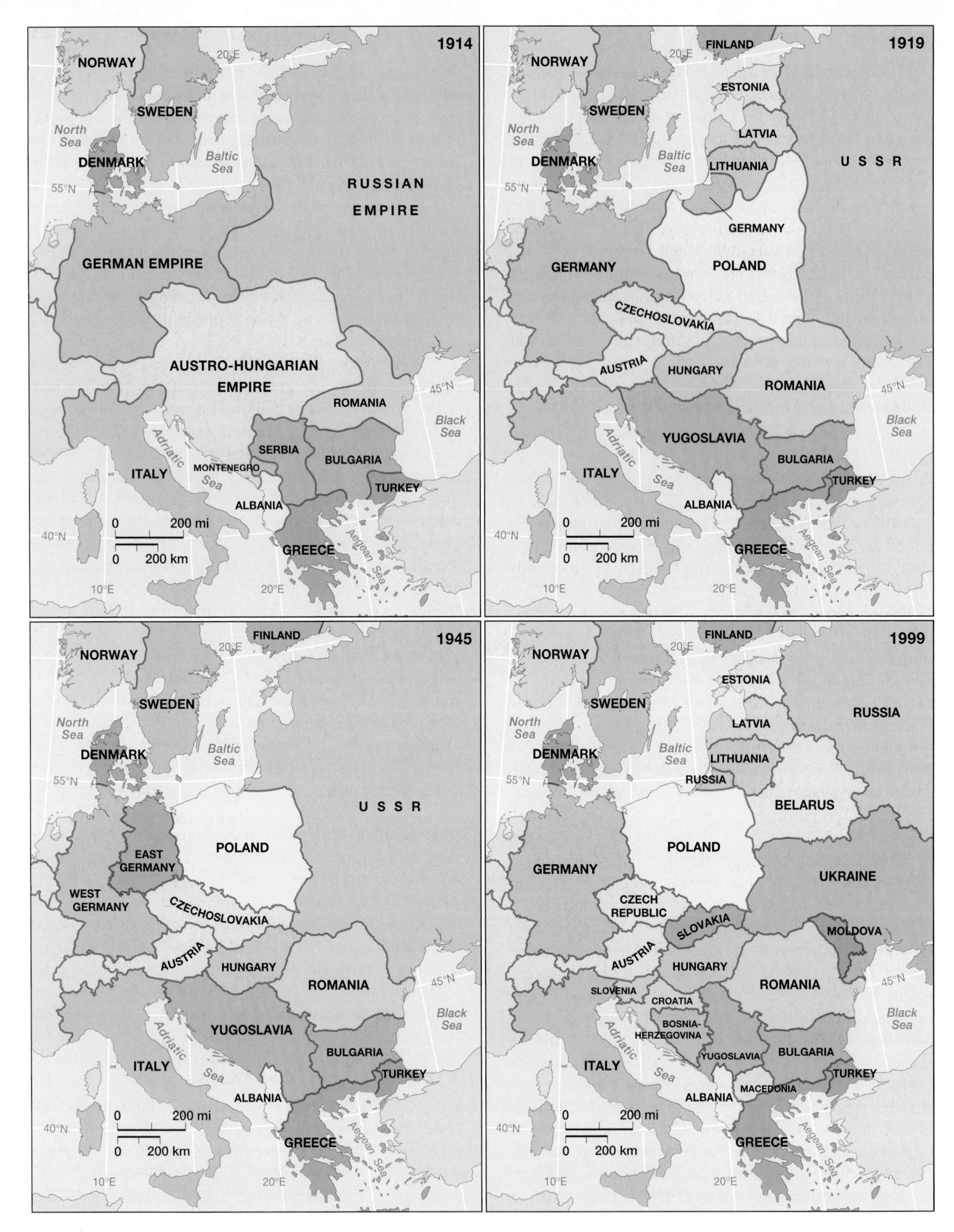

▲ **Figure 8.30 A century of geopolitical change** The twentieth century began with central Europe dominated by the German, Austro-Hungarians (or Hapsburg), and Russian empires. Following World War I, these empires were largely replaced by a mosaic of nation-states. More border changes followed World War II, largely as a result of the Soviet Union turning that area into a buffer zone between itself and western Europe. With the demise of Soviet hegemony in 1989, further political fragmentation took place.

Cold War Geography The seeds of the Cold War are commonly thought to have been planted at the Yalta Conference of February 1945, when Britain, the Soviet Union, and the United States met to plan the shape of postwar Europe. Since the Red Army was already in eastern Europe and moving quickly on Berlin, Britain and the United States accepted that the Soviet Union would occupy eastern Europe, just as the Western allies would occupy parts of Germany.

The larger geopolitical issue, though, was the Soviet desire for a **buffer zone** between its own territory and western Europe, an extensive bloc of satellite countries that would cushion the Soviet heartland against possible attack from western Europe. Some scholars argue this concern goes back to the Russian civil war (1918–1921), when western states attempted to intervene in hopes of influencing the outcome of the communist revolution. Regardless of its historical roots, the result was that former military allies ceased to be friends in 1945. Instead, Europe fell into two hostile blocs serving the interests of the two new global superpowers, the United States and the Soviet Union.

In the east the Soviet Union took control of the Baltic states, Poland, Czechoslovakia, Hungary, Bulgaria, Romania, Albania, and Yugoslavia. The latter two states later distanced themselves from the Soviet Union by taking a different pathway to communism. Austria and Germany were divided into occupied sectors by the four (former) allied powers. In both cases the Soviet Union dominated the eastern portion of each country, areas that contained the capital cities of Berlin and Vienna. Both capital cities, in turn, were divided into French, British, U.S., and Soviet sectors. In 1955, with the creation of an independent and neutral Austria, the Soviets withdrew from their sector, effectively moving the Iron Curtain eastward to the Hungary-Austria border. This was not the case with Germany, however, which quickly evolved into two separate states (West and East) that remained apart until their reunification in 1990.

Along the border between east and west, two hostile military forces faced each other for almost half a century. Both sides prepared for and expected an invasion by the enemy across the barbed wire of a divided Europe (Figure 8.31). In the west, NATO (North Atlantic Treaty Organization) forces, including the United States, were encamped from West Germany in the north to Turkey on the south. To the east, the Warsaw Pact forces were anchored by the Soviets but included token armies from most satellite countries. Both NATO and the Warsaw Pact countries were armed with nuclear weapons, making Europe a tinderbox for yet another world war.

Berlin was the flashpoint that brought these forces close to a fighting war on two occasions. In winter 1948, the Soviets imposed a blockade on the city by denying Western powers access to Berlin across its East Germany military sector. This attempt to starve the city into submission was thwarted by a nonstop airlift of food and coal by NATO. Then, in August 1961, the Soviets built the Berlin Wall to curb the flow of East Germans seeking political asylum in the west. The Wall became the concrete-and-mortar symbol of a firmly divided postwar Europe. For several days, while the Wall was built and the West agonized over destroying it, NATO and Warsaw Pact tanks and soldiers faced each other with loaded weapons at point-blank range. Though war was averted, the Wall remained for 28 years, until November 1989 (Figure 8.32).

▲ **Figure 8.31 The Iron Curtain** The former Czechoslovakia–Austrian border is marked by a barbed wire fence as part of an extensive border zone that included mine fields, tank barricades, watch towers, and a sanitized zone in which all trespassers were shot on sight. *(James Blair/NGS Image Collection)*

The Cold War Thaw The symbolic end of the Cold War in Europe came on November 9, 1989, as East and West Berliners joined forces to rip apart the Wall with jackhammers and hand tools. By October 1990, East and West Germany were officially reunified into a single nation-state. During this period, all other Soviet satellite states, from the Baltic to the Black Sea, also underwent major geopolitical changes that have resulted in a mixed bag of benefits and problems. While some, like the Czech Republic, appear to have made a successful transformation to democracy and capitalism, other east European states, such as Romania, appear stalled in their search for new directions.

The Cold War's end came as much from a combination of problems within the Soviet Union (discussed in Chapter 9) as from rebellion in eastern Europe. By the mid-1980s,

(a)

(b)

◄ **Figure 8.32 The Berlin Wall** In August 1961, the East German and Soviet armies built the concrete and barbed wire structure, known simply as "The Wall," to stem the flow of East Germans leaving the Soviet zone. It was the most visible symbol of the Cold War until November 1989 when Berliners physically dismantled the Wall after the Soviet Union renounced its control over eastern Europe. *[(a) Bettmann/Corbis; (b) Anthony Suau/Liaison Agency, Inc.]*

the Soviet leadership advocated an internal economic restructuring and also recognized the need for a more open dialogue with the West. Fiscal problems from supporting their huge military establishment, along with heavy losses from an unsuccessful war in Afghanistan, tempered the Soviet appetite for intervention.

In August 1989, Poland elected the first noncommunist government to lead an eastern European state since World War II. Following this, with just one exception, peaceful revolutions with free elections spread throughout eastern Europe, as communist governments renamed themselves and broke with doctrines of the past. In Romania, though, street fighting between citizens and military resulted in the violent overthrow and execution of their communist dictator, Nicolae Ceaucescu.

In October 1989, as Europeans nervously awaited a Soviet response, President Mikhail Gorbachev said the Soviet Union had no moral or political right to intervene in the domestic affairs of eastern Europe. Then Hungary opened its borders to the west, and eastern Europeans freely visited and migrated to western Europe for the first time in 50 years. Estimates are that more than 100,000 East Germans fled the country during this time.

As a result of the Cold War thaw, the map of Europe began changing once again. Germany reunified in 1990. Elsewhere, separatism and ethnic nationalism, long suppressed by the Soviets, once again was unleashed in central and southern Europe. Yugoslavia fragmented into the independent states of Slovenia, Croatia, and Bosnia. On January 1, 1993, Czechoslovakia was replaced by two separate states, the Czech and Slovak republics.

A New Geopolitical Stability? While the final shape of post-Cold War Europe is still far from clear, there is little question that geopolitical tensions have been reduced between former adversaries. Clearly, major steps have been taken to foster stability in the historically treacherous region of eastern Europe. In early 1999, three former Warsaw Pact enemies—Poland, the Czech Republic, and Hungary—became members of NATO. Most European countries wanted to extend the invitation further to include Romania and Slovenia, but the United States balked at the prospect of shouldering these additional costs and geopolitical obligations; consequently, the line was drawn with these three new members.

Russia resisted this encroachment into its former sphere of influence and buffer zone, so much so that NATO offered Moscow a special relationship of its own within the alliance. In a historic document, NATO affirmed in 1998 that it no longer viewed Russia as an enemy and, further, had no intention of deploying nuclear weapons on the territory of its new members, nor would it station permanent combat forces in these countries. Russia grudgingly accepted NATO expansion, since most European countries see this process as a way to finally establish geopolitical stability in eastern Europe. As it gets larger, NATO will probably become more European, with a reduced role for the United States. This may make its expansion into eastern Europe much more acceptable to Russia. A major stumbling block, however, is the tension over the former Yugoslavia.

Yugoslavia and the Balkans: Europe's Geopolitical Nightmare

Yugoslavia tests the mettle of the new, unified Europe. Because of the Balkan problem, both NATO and the EU are haunted by their inability to solve the complexities of ethnic conflict, micronationalism, and state disintegration. In 1918 the world responded to the problem of Balkan factionalism by creating the new state of Yugoslavia. However, by late in the twentieth century, it was painfully apparent that this was not a solution at all. Once again, ethnic fault lines in the Balkans threatened global stability.

That Yugoslavia is a global problem, and not merely a local or regional aberration, needs to be emphasized. Indeed, one of the complicating geopolitical dimensions to the Balkan issue is that these struggles and tensions play out at three different and sometimes contradictory scales. First, there are the local and regional tensions within Yugoslavia, its internal provinces (primarily Kosovo), and its neighboring states of Macedonia, Albania, Bosnia, Croatia, and Slovenia. Second, there are the involvements with the rest of Europe, with NATO and the EU, interactions that range from military actions to accepting refugees from the war-torn Balkans. Finally, and perhaps most vexatious, are the global-scale implications of the crisis. These now include the United Nations, which has committed peacekeeping monitors and observers; Russia, with its strong pro-Serb bias; and the Islamic countries (primarily Iraq, Iran, and Saudi Arabia), which support Muslim populations in the Balkans. Finally, the United States, which, as a NATO and UN member, has long committed military forces to enforce Balkan peace agreements (Figure 8.33; see also "Local Voices: The Internet and Ethnic Warfare in Kosovo").

Pundits commonly take refuge in Balkan history to explain away the current troubles, pointing out centuries of regional rivalries, of clan feuds, of religious intolerance, and of ethnic cleansing. Implied in this perspective is that the Balkans are a unique region absolutely unlike any other part of the world, and thus we should somehow accept as inevitable the violence and terrorism of ethnic separatism. Indeed, the term

▲ **Figure 8.33 Kosovar refugees** Thousands of Muslim Kosovars were forced to flee from the Yugoslavian province of Kosovo during a period of ethnic cleansing by Serbian forces in 1998–99. This violence led to aerial bombardment of Yugoslovia by NATO planes to force Serbian leaders to accept UN peacekeepers in the province. *(Buu/Liaison Agency, Inc.)*

LOCAL VOICES The Internet and Ethnic Warfare in Kosovo

The war in Kosovo is the first armed conflict in which a full range of local voices, from military to innocent victims, has an active presence on the Internet and can reach a worldwide audience within minutes. World Wide Web audiences who desire more information about the conflict can easily access information found only on the Net. This information includes daily human rights updates on the status of Kosovar refugees, NATO war briefings, eyewitness reports from Belgrade civilians being bombed by NATO forces, reports from Kosovar students under siege by Serb terrorists, or official proclamations from the Yugoslav government. If every major twentieth-century war had its own medium—film in World War II, TV footage in Vietnam, live TV from the Gulf War—then the Kosovo war has the laptop computer and the Internet.

One of the most poignant voices has come from a 16-year-old girl caught up in the violence of Kosovo who e-mails daily reports to her contact in Berkeley, California. He, in turn, has relayed her messages to National Public Radio and CNN. Her contact in California says, "She's like Anne Frank with a laptop. It's kind of like her diary, except in a real-time environment. Her messages get to me five minutes after she writes them."

Adona (not her real name), the Kosovar student, writes, "Just from my balcony, I can see people running with suitcases, and I can hear some gunshots. . . . A village just a few hundred meters from my home is all surrounded. As long as I have electricity, I will continue writing to you. Right now, I am trying to keep myself as calm as possible." Later, Adona writes about the irony of ethnic identity in the Balkans: "Did I tell you I am not a practicing Muslim and do you know why? . . . Because, if the Turks didn't force my grand-grandparents to change their religion, I might now be a Catholic or an orthodox" (thus, a Serb rather than a Kosovar).

Global Beat Syndicate, an online international news service based at New York University, also offers a wide range of viewpoints from local people who have found themselves involved in the Balkan war. One, an ethnic Albanian and resident of Pristina, the capital city of Kosovo, reported on her expulsion to Macedonia: "Many of us are still in shock. We're too proud to admit that we are refugees. People are using new expressions, like 'deportees.' . . . They want to pretend that the past few weeks didn't happen and that it can all be reversed. Even though many of us are dead. Even though we are, in fact, here in Macedonia."

A California company that normally offers privacy shields for a fee, launched a free service to help Web users in Yugoslavia conceal their identities from censors and police who track down dissenters by tracing e-mail or Web traffic. Several thousand sign up for the Kosovo shield each day, says the CEO of the company. As the war continues, the privacy of those local voices becomes increasingly vulnerable because they offer personal commentary on the horror of warfare and ethnic cleansing, nightmarish acts that governments prefer to keep secret.

balkanization has evolved to describe the fragmented geopolitical processes involved with small-scale independence movements and the phenomenon of mini-nationalism as it develops along ethnic fault lines.

While there can be some truth to this emphasis on Balkan history, it can also mislead and mask other relevant perspectives. As a result, some words of caution are needed. First, history is commonly reconstructed and reinvented to serve present-day actions and agendas. Thus, there is rarely agreement about the past by different Balkan groups. How history is used is usually more relevant than the content or facts. Second, historical currents run both deep and shallow. While some groups point to events and misdeeds that took place centuries ago (such as the Serb defeat at the hands of Muslims in 1389), other animosities are informed by twentieth-century events in the Balkans, as during World War II when an ethnic group's relationship to Nazi Germany remains a major issue.

That said, a general overview of recent Balkan history is helpful to understanding the current situation. As World War I ended in 1918, so did the Austro-Hungarian Empire's control of the Balkans. The independent kingdom of Serbia became the magnet for the creation of the new state of Yugoslavia. Attached to Serbia was a collection of other Hapsburg lands, including Slovenia, Croatia, Bosnia-Herzegovina, Montenegro, Macedonia, and Kosovo. The result was a new state of three mutually antagonistic religions (Islam, Catholicism, and Eastern Orthodox), a handful of different languages, and two distinct alphabets. From the outset, a Serbian king ruled over this new country and the Serbs dominated political and economic activity. Belgrade and the Serbian core amassed wealth at the expense of the ethnic periphery. As a result, Yugoslavia coexisted in a state of mutual hostility for two decades until invaded by Hitler in 1941.

German troops were welcomed as liberators in Croatia, a region that was particularly restless with Serbian rule. Major resistance to Nazi occupation, however, came from two fronts: first, Serbian royalists who supported the king and opposed Hitler; second, the Partisans, who were pan-Yugoslav communists with a distinct anti-Serb, anti-royalist cast who also resisted Nazi fascism. They were led by Marshall Tito, a Croat, and were supported by the Allies in their battle against Nazi Germany.

Following World War II, Tito was rewarded for his struggles by receiving backing from Britain, the United States, and the Soviet Union as the leader of a new socialist Yugoslavia. Conventional wisdom posits that the three major ethnic groups—the Serbs, Croats, and Muslims—coexisted peacefully under Tito's strong leadership by relegating their separatist agendas to the larger goal of a communism independent of the Soviet Union. While true to some degree, one must also consider that the economic challenges of rebuilding a country heavily damaged by war took precedent over historical ethnic animosities. This relatively peaceful coexistence continued for almost a decade after Tito's death in 1980. During that period, leaders of the different Yugoslavian states took turns serving as the country's president.

However, the region's geopolitical stability began to unravel in 1989. In concert with other changes taking place in eastern Europe, and a renewal of ethnic separatism within Yugoslavia, President Slobodan Milosevic, a Serbian nationalist, asked the people to vote on the future direction of Yugoslavia. At issue was whether the country should remain unified and whether it should continue with a communist government. Only Serbia and Montenegro voted for communist governments and unification, while the other four republics voted for independence (Figure 8.34).

Slovenia was the first to secede in 1990, followed by Croatia in 1991, Macedonia in January 1992, and, finally, Bosnia-Herzegovina in March 1992. A brutal civil war ensued in which Milosevic sought to keep all ethnically Serbian areas under Yugoslav control. The Slovenian army inflicted a humiliating defeat on the Yugoslavia Serbian army early on, and then moved ahead with its own independence. In Bosnia, the Serb-dominated Yugoslav army waged a ferocious war of ethnic cleansing to rid Bosnia of Muslims and Croats. A cease-fire came about in 1995. In Croatia, fighting continued between Serbs and Croats until September 1996. A fragile peace was achieved with the Dayton Accord of November 1995 that brought different ethnic factions to the negotiating table, with Serbia agreeing to respect a new set of boundaries between Yugoslavia and its new independent neighbors. Sixty thousand NATO troops, including 20,000 Americans, were brought in to enforce the peace treaty. Bosnia-Herzegovina was carved into two complex political entities, a Serb republic and a Muslim-Croat federation, both ruled by the same legislature and president. This legislature also appointed joint premiers in 1997, one a Serb, the other a Muslim.

Despite the Dayton Accord, the ethnic warfare plaguing the region is far from over. More specifically, the tense cease-fire in Bosnia could break into open warfare at any moment, inflamed by Serb resentment over loss of territory. Fuel for this fire is added by the outspoken Serb nationalism promoted by Yugoslav President Milosevic.

Kosovo, to the south, is the other trouble spot. This Muslim Albanian-dominated region of two million lost its autonomous status shortly after the 1989 elections. Kosovo rebels proclaimed an independent republic in July 1990. While some measures of autonomy were restored with the Dayton Accord, such as a return to school instruction in the Albanian language, this was not enough for the rebels. With the effective collapse of the Albanian government in 1997, looting of military bases provided the Kosovo rebels with a new and enlarged supply of arms. This resulted in escalating terrorism and warfare against the Serb-dominated Yugoslav army and police. Each act of violence set off a wave of retaliation between the two opposing groups. In an attempt to end this tragic situation, the UN sent 2,000 unarmed monitors into Kosovo in November 1998 to diffuse the situation, but this did not work; nor did a diplomatic solution that would have guaranteed autonomy for Kosovo within Yugoslavia because of outspoken rejection from President Milosevic. As a result of this diplomatic failure, NATO began an intensive air war against Yugoslavia in March 1999.

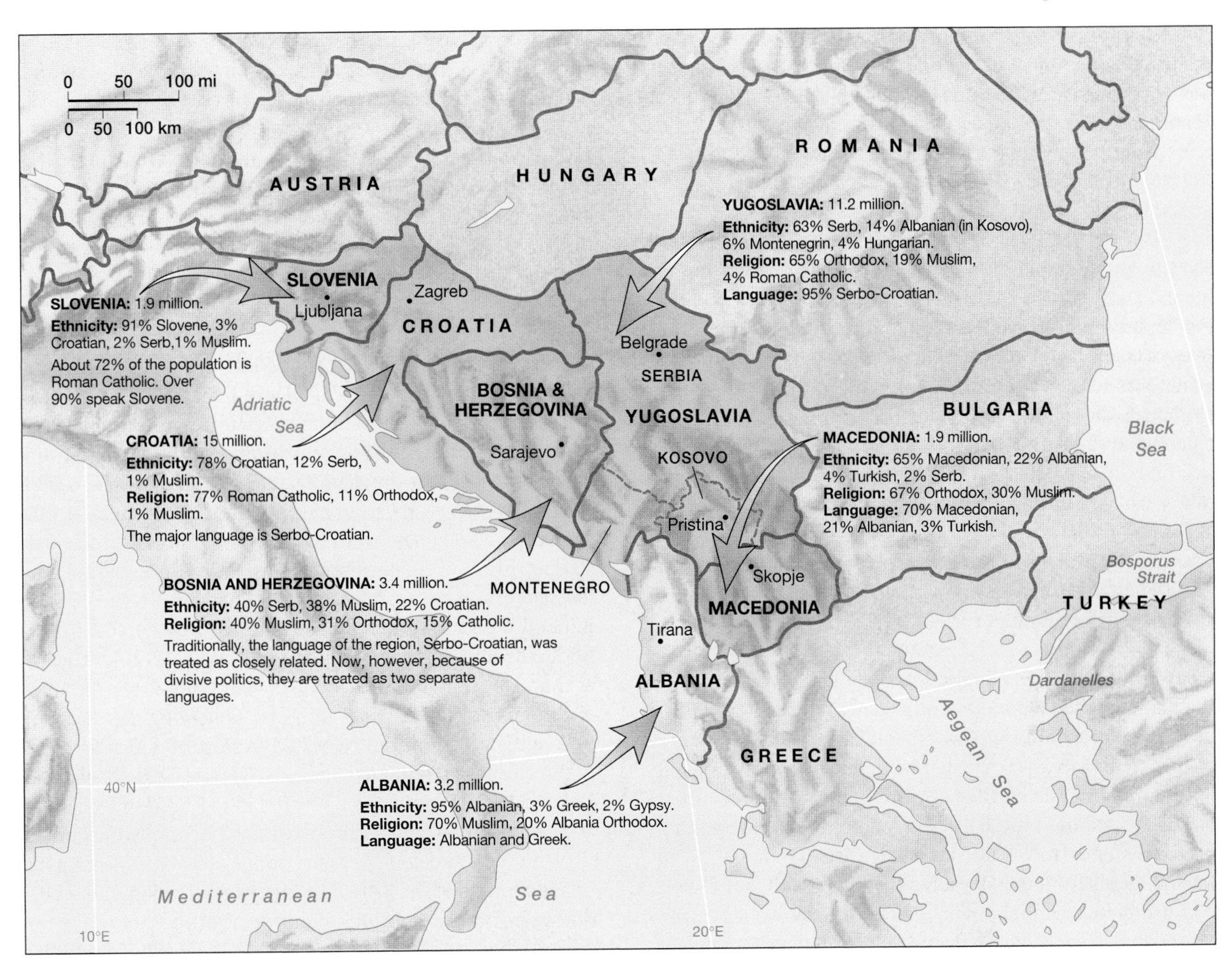

▲ **Figure 8.34 Ethnicity in the Balkans** The diverse and complicated mosaic of ethnic diversity in the Balkans has led to geopolitical fragmentation in the last decades. Not only is the area a meeting ground for the three major religions of Roman Catholicism, Eastern Orthodoxy, and Islam, but linguistic fault lines also complicate ethnic and national identity. Further, a long history of discrimination and retaliation between ethnic groups is embedded in ethnic consciousness.

In retaliation, the Serbian army and police have killed or driven out hundreds of thousands of ethnic Kosovars in a violent ethnic cleansing campaign designed to create a Serb-only province.

Economic and Social Development: Integration and Transition

As the acknowledged birthplace of the Industrial Revolution, Europe, in many ways, invented our modern economic system. Though Europe was the world's industrial leader in the early twentieth century, this position was eclipsed by Japan and the United States as Europe struggled to recover from two world wars, a decade of global depression, and the distracting schism of the Cold War. Additionally, Europe lost ground from internal competition between a handful of different national economies.

In the last 40 years, however, economic integration guided by the European Union has been increasingly successful. In fact, Europe's success at blending together national economies has given the world a new model for regional cooperation, an approach that could well be imitated in Latin America and Asia in the twenty-first century. Eastern Europe, however, has not fared so well. The results of four decades of Soviet economic planning were, at best, mixed. The total collapse of that system in 1990 cast eastern Europe into a period of chaotic economic, political, and social transition that may result in a stratified array of rich and poor regions, with some countries plugged into the affluent West, and others not.

Accompanying western Europe's economic boom has been an unprecedented level of social development, as measured by worker benefits, health services, education, literacy, and gender equality. Though these social services set a laudable standard for the world, they are currently under attack by cost-cutting politicians who argue these benefits add so much to the cost of business that European goods cannot compete in the globalized marketplace.

The early twenty-first century brings Europe renewed hope that it will once again be a world economic superpower. Yet this region also faces new and difficult challenges. In terms of economic and social development, the region is at a major crossroads. Though the different pathways are well marked, the benefits and costs of each are uncertain.

Europe's Industrial Revolution

Europe is the cradle of modern industrialism. Two fundamental traits are associated with this transformation. First, machines replaced human labor in many manufacturing processes and, second, inanimate energy sources such as water, steam, electricity, and petroleum powered these new machines. Though we commonly use the term *Industrial Revolution* for these transformations, a term implying rapid, overnight change, this was not the case. Instead, it took more than a century for the interdependent pieces of the industrial puzzle to come together. This happened first in England between 1730 and 1850. Once they were in place, however, rapid change ensued, as industrialization diffused to other parts of Europe and the world.

Centers of Change England's textile industry, located on the flanks of the Pennine Mountains, was the center of early industrial innovation. More specifically, innovation and change took place in the small towns and villages of Yorkshire and Lancashire, away from the rigid control of the urban guilds. Yorkshire, on the eastern side of the Pennines, had been a center of woolen textiles since medieval times, drawing raw materials from the extensive sheep herds of that region, and using the clean mountain waters to wash wool before it was spun in rural cottages (Figure 8.35). On the western flank of the mountains, Lancashire county workers used imported cotton from England's overseas colonies (including the United States) as the basis for its newer textile industry. By the 1730s, water wheels drove mechanized looms at the rapids and waterfalls of the Pennine streams.

▲ **Figure 8.35 Water power and textiles** Europe's industrial revolution began on the flanks of England's Pennine mountains where swift-running streams and rivers were used to power large cotton and wool looms in the textile industry. Later, many of these textile plants switched from water to coal power. *(James Marshall/Corbis)*

Locational Factors of Early Industrial Areas By the 1790s, the steam engine became the preferred source of energy because waterpower was hindered by low stream flow during stretches of dry weather. A problem, however, was that the steam engine required vast amounts of wood fuel, and England's forests had long been depleted for shipbuilding. Still, Great Britain did possess large amounts of coal, even though it was expensive to transport this bulky fuel long distances by horse-drawn cart. Only with the railroad (after 1820) could coal be moved long distances at reasonable cost.

Another fundamental innovation was the change from charcoal to coke in the manufacture of iron and steel. Until the early nineteenth century, iron foundries throughout Europe used wood charcoal in the smelting process, which created a pattern of small iron mills dispersed throughout rural forest areas, close to the source of charcoal. In the 1790s, mineral-based coke was first substituted for charcoal in the Midlands iron and steel industry. Not only was coke cheaper than charcoal in timber-poor England, but it also permitted the use of lower-grade iron ore. Following this change, iron and steel foundries relocated from rural forest areas to urban clusters adjacent to coalfields. By the 1820s, the world's first industrial landscapes appeared in the English Midlands and Scottish Lowlands. Europe's industrial era had begun.

While coal and coke were the primary locational factors for this first phase of the British Industrial Revolution, there was one important exception—London. Despite being far removed from the coalfields, this world city developed into the largest industrial center in the United Kingdom for other reasons. London had access to a large market, a large supply of skilled labor, a port for overseas trade; a merchant class with access to capital; and proximity to banks, insurance firms, investors, and other institutional support systems needed by industry.

Development of Industrial Regions in Continental Europe By the 1820s, the first industrial districts also began appearing across the channel in continental Europe. Most were, and still are, on coalfield locations (Figure 8.36). The first area outside of Britain was the *Sambre-Meuse* region, named for the two river valleys straddling the France-Belgian border. Like the English Midlands, it also had a long history of cottage-based wool textile manufacturing that quickly converted to the new technology of steam-powered mechanized looms. Additionally, a metalworking tradition drew upon charcoal-based iron forges in the nearby forests of the Ardenne mountains. Coke and coal were also found in these mountains; in 1823 the first coke blast furnace outside of Britain started operation in Liege, Belgium.

By the second half of the nineteenth century, the dominant industrial area in all Europe (including England) was the *Ruhr* district in northwestern Germany close to the Rhine River. Rich coal deposits close to the surface fueled the Ruhr's

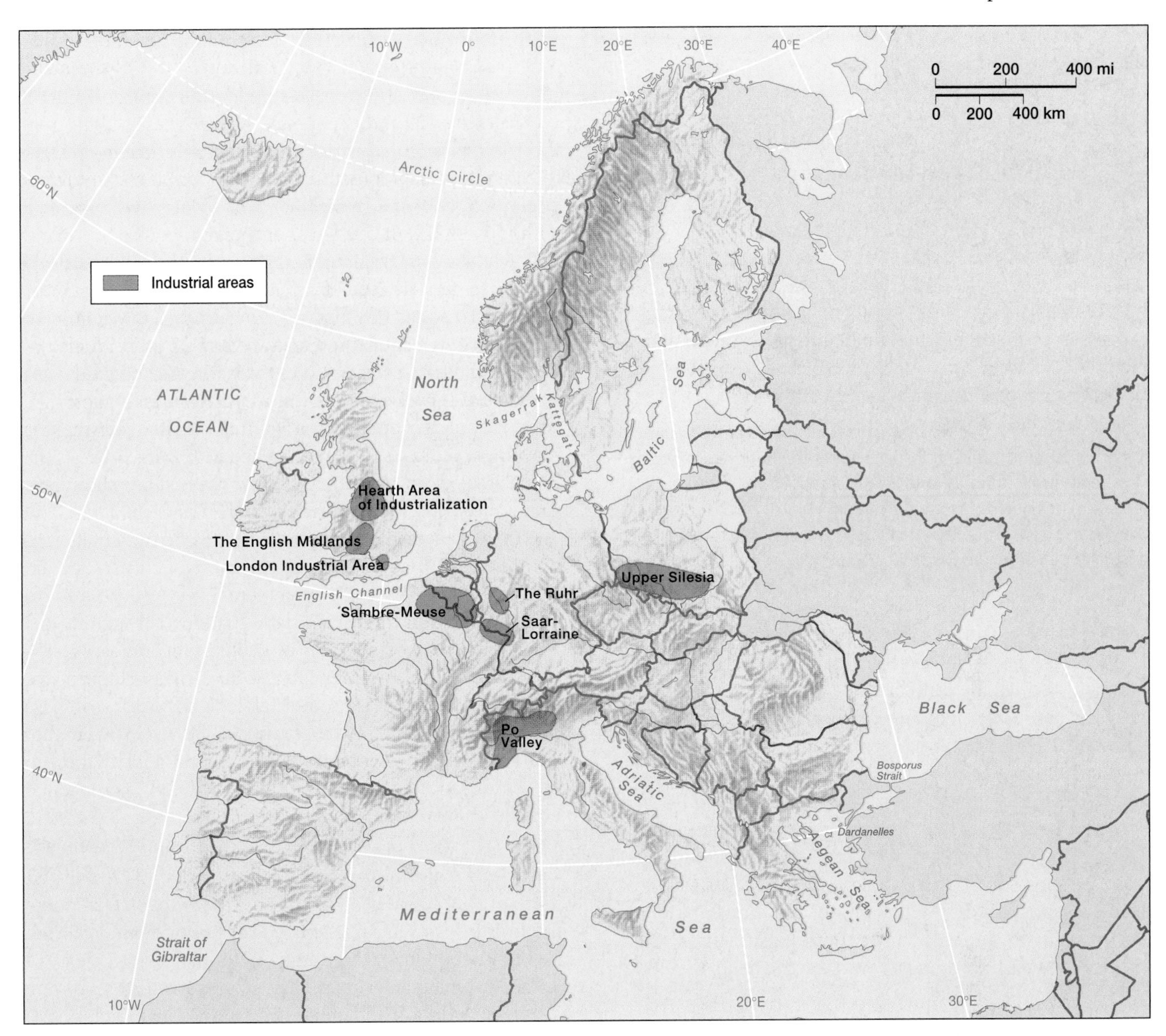

▲ **Figure 8.36 Industrial regions of Europe** From England, the Industrial Revolution spread to continental Europe, starting with the Sambre-Meuse region on the French-Belgium border, then diffusing to the Ruhr area in Germany. Readily accessible surface coal deposits powered these new industrial areas. Iron ore for steel manufacture was originally from local deposits, later imported from Sweden and other mines in the shield country of Scandinavia. Unlike most European manufacturing areas, the Po Valley of Italy was powered by electricity generated in the nearby Alps.

transformation from a small textile region to one oriented around heavy industry, particularly iron and steel manufacturing. By the early 1900s, the Ruhr had used up its modest iron ore deposits and was importing ore from Sweden, Spain, and France. Several decades later the Ruhr industrial region became synonymous with industrial strength behind the Nazi Germany war machine, and thus was bombed heavily in World War II (Figure 8.37).

To the southwest of the Ruhr, the industrial region of *Saar-Lorraine* straddles the border between Germany and France. Though this region contains rich deposits of both coal and iron ore, a much-contested political boundary separated these resources until after World War II. Most of the coal lies in the German Saarland, while iron ore is found primarily in the French Lorraine district. As the industrial strength of this region grew in the late nineteenth century, so did political strife over its control. Twice Germany annexed the iron deposits in France, and twice France attempted unsuccessfully to occupy the coal areas of the Saarland. It is only since 1953 that these resources have been integrated through trade agreements of the European Coal and Steel Community.

Historical boundary shifts and political tensions between states also plagued the *Upper Silesia* industrial district that now lies in southern Poland, west of Krakow. Historically a center of charcoal-iron smelting, this region was originally part of Prussia before German unification in the 1870s. German and Prussian financial interests developed the region in the late nineteenth century, combining the rich coalfields with imported

▲ **Figure 8.37 The Ruhr industrial landscape** Long the dominant industrial region in Europe, the Ruhr region was bombed heavily during World War II because of its central role in providing heavy arms to the German military. It has been rebuilt, has modernized, and remains competitive with other, newer industrial regions. *(Andrej Reiser/Bilderberg Archiv der Fotografen)*

coke and coal so that by 1900 Upper Silesia provided almost 10 percent of Germany's steel. After World War I, however, Germany lost half the region as the Polish border was adjusted westward. The remaining German half was awarded to Poland in 1945 with still another boundary shift.

While most early industrial development took place on or near coalfields, a noteworthy exception is the *Po Plain* in northern Italy. Here, hydroelectric power generated in the nearby Alps provided energy for industrialization of the region. Because Italy lacked extensive coal deposits, the government subsidized the generation and transmission of electrical energy. Indeed, the northern Italian city of Milan became the first European city with electric lights in 1883.

Rebuilding Postwar Europe: Economic Integration in the West

Because of its historical momentum, Europe was unquestionably the leader of the industrial world in the early twentieth century. Before World War I, European industry is estimated to have produced 90 percent of the world's manufactured output. However, four decades of political and economic chaos, punctuated by two devastating world wars, left Europe divided and in shambles. By mid-century, cities were in ruins; industrial areas were destroyed; vast populations were dispirited, hungry, and homeless; and millions of refugees moved about Europe looking for solace and stability.

In 1947 the European Recovery Program (ERP) was established to rebuild the region. Through the Marshall Plan, the United States was a major contributor to ERP. In eastern Europe, however, then controlled by the Soviet Union, this aid was rejected because of U.S. involvement. In its place the Soviets created in 1949 the **Council for Mutual Economic Assistance** (CMEA, or, alternatively, Comecon), which linked eastern European aid and recovery to the centralized, command economies of communism.

ECSC and EEC In 1950 western Europe began discussing a new form of economic integration that would avoid the historical pattern of nationalistic independence through tariff protection, as well as the economic inefficiencies resulting from the duplication of industrial effort. Robert Schuman, France's foreign minister, proposed that German and French coal and steel production be coordinated by a new supranational authority. In May 1952, the six nations of France, Germany, Italy, Netherlands, Belgium, and Luxembourg ratified a treaty that joined them into the European Coal and Steel Community (ECSC). This was the first step toward the now-dominant economic and political integration of the European Union. Because of the immediate success of the ECSC, these six states also agreed to work toward further integration by creating a larger European common market that would foster the free movement of goods, labor, and capital. In March 1957, the Treaty of Rome was signed, establishing the European Economic Community (EEC), popularly called the *Common Market*.

Britain remained outside the new EEC because of its economic obligations to former colonies (such as Canada) through Commonwealth agreements, but also because of French objection to English participation. Throughout the 1960s, EEC membership for the United Kingdom was resisted by France for fear the English economy was simply a Trojan horse for American interests. In response to these constraints, Britain joined together in 1958 with Switzerland, Austria, Portugal, Denmark, Sweden, and Norway to form the European Free Trade Association (EFTA), a group that became known as the "Outer Seven."

European Community and Union The EEC reinvented itself in 1965 with the Brussels Treaty that created the groundwork for adding a European political union to the successful economic community. With this "second Treaty of Rome," aspirations for more than economic integration were clearly stated with the creation of an EEC council, court, parliament, and commission. It also changed its name from the EEC to European Community (EC). To be fully successful, though, the EC had to expand its membership, and this could take place only if the French agreed to change their position on the United Kingdom. Eight years later it happened. In 1973 the EC added not just the United Kingdom, but also Denmark, Ireland, and Norway, though voters in that Scandinavian country subsequently rejected admission in a referendum. The EFTA, as a result, disintegrated. During the 1980s, Greece, Portugal, and Spain were added, with Austria, Finland, and Sweden joining in 1995 to round out the current membership of 15 member states (Figure 8.38).

In 1991 the EC changed its name once again to European Union (EU) and expanded its goals once again with the Treaty of Maastricht (named after the town in the Netherlands where delegates met). While economic integration remains an underlying theme in the new constitution, particularly with its commitment to a single currency through the European Monetary Union, the EU has also moved more explicitly into intergovernmental and supranational affairs with its stated goals of common foreign policies and mutual security agreements.

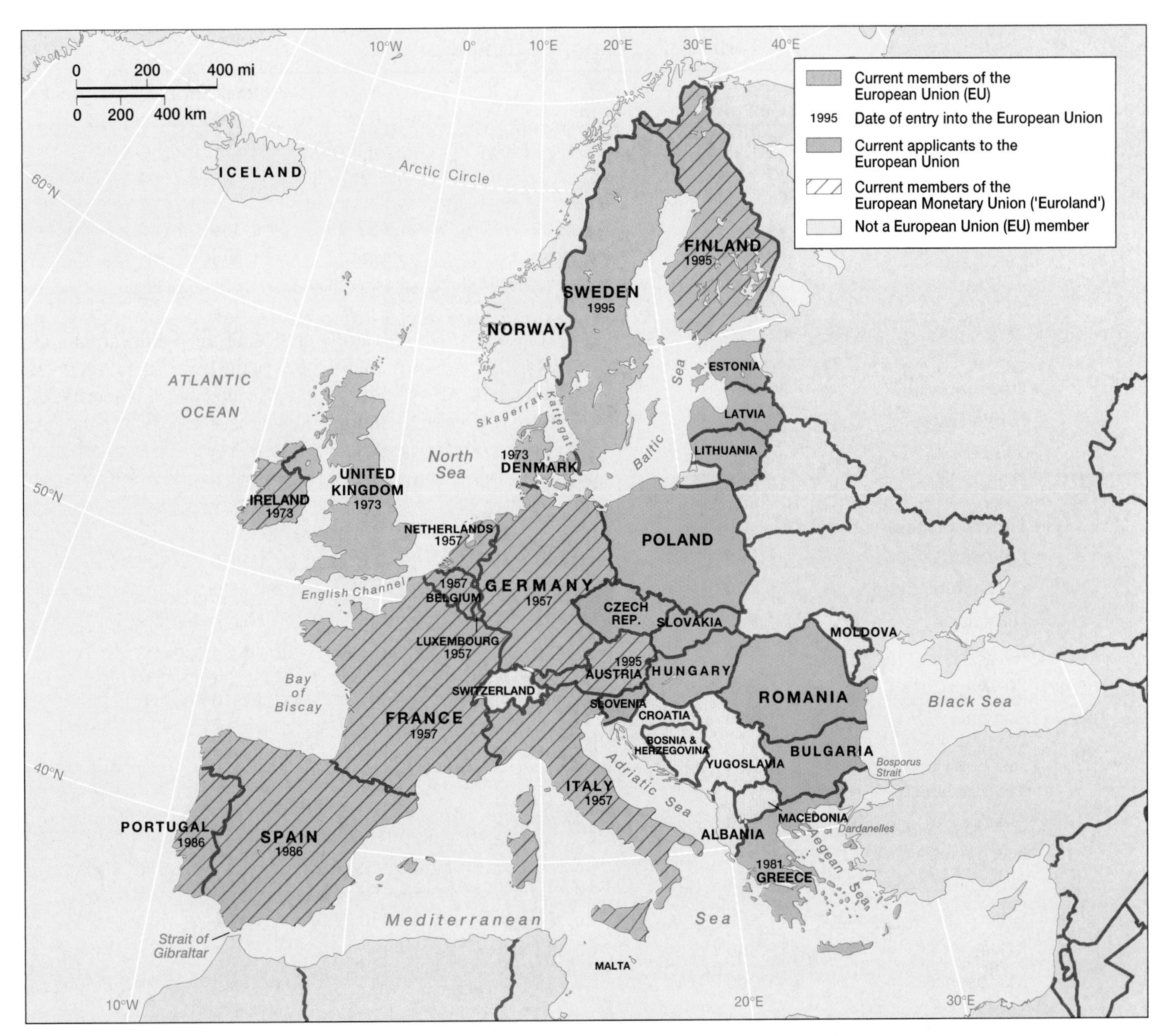

▲ **Figure 8.38 The European Union** The driving force behind Europe's economic and political integration has been the European Union (EU), which began in the 1950s as an organization focused solely on rebuilding the region's coal and steel industries. After that proved successful, the Common Market (as it was then called) expanded into a wider range of economic activities. Today, the EU has 15 members, with a handful of applicants, mostly from eastern Europe, anxiously waiting for membership.

The long list of EU applicants suggests that the advantages of EU membership outweigh any disadvantages from loss of national sovereignty. Currently the EU is entertaining applications from Bulgaria, Cyprus, Czech Republic, Estonia, Hungary, Malta, Poland, Romania, Slovakia, Slovenia, and Turkey. At least four of these countries may be accepted into the EU in the near future (Table 8.2).

Euroland: The European Monetary Union As any world traveler knows, each state usually has its own monetary system. Crossing a political border usually means changing money into different units and becoming familiar with new system of bills and coins. In Europe, this means buying pounds in England, marks in Germany, francs in France, and so on. Coining their own money has long been a fundamental component of state sovereignty.

However, this coinage system may become an artifact of the past as Europe moves away from individual state monetary systems and toward a common currency. On January 1, 1999, in what many saw as a historical advance for the notion of a united Europe, 11 of the 15 EU member states joined together in the European Monetary Union (EMU). As of that day, business and trade transactions began taking place in the new monetary unit of the euro. For three years the euro will remain a fiscal abstraction. Only in early 2002 will new euro coins and bills become available for consumer use. As a result, the different currencies of the "Euroland" member states are scheduled to disappear completely by July 2002.

Table 8.2 Economic Indicators

Country	GNP per Capita ($U.S., 1996)	Total GNP (Millions of $U.S., 1996)	PPP* ($Intl, 1996)	Real Annual Growth % per Capita, 1990–1996
Western Europe				
Austria	28,110	226,510	21,650	0.9
Belgium	26,440	268,633	22,390	1.2
France	26,270	1,553,619	21,510	0.7
Germany	28,870	2,364,632	21,110	0.7
Liechtenstein	—	—	—	—
Luxembourg	45,360	18,850	34,480	0.1
Netherlands	25,940	402,565	20,850	1.8
Switzerland	44,350	313,729	26,340	–1.0
United Kingdom	19,600	1,152,136	19,960	1.5
Eastern Europe				
Bulgaria	1,190	9,924	4,280	–1.8
Czech Republic	4,740	48,861	10,861	0.9
Hungary	4,340	44,274	6,730	–0.6
Moldova	590	2,542	1,440	–16.8
Poland	3,230	124,682	6,000	3.3
Romania	1,600	36,191	4,580	0.1
Slovakia	3,410	18,206	7,460	–1.2
Southern Europe				
Albania	820	2,705	—	2.2
Bosnia-Herzegovina	—	—	—	—
Croatia	3,800	18,130	4,290	2.2
Greece	11,460	120,021	12,730	1.3
Italy	19,880	1,140,484	19,890	0.9
Macedonia	990	1,956	—	–8.5
Malta	—	—	13,870	3.1
Portugal	10,160	100,934	13,450	1.5
San Marino	—	—	—	—
Slovenia	9,240	18,390	12,110	4.4
Spain	14,350	563,249	15,290	1.0
Yugoslavia (Serbia)	—	—	—	—
Northern Europe				
Denmark	32,100	168,917	22,120	2.1
Estonia	3,080	4,509	4,660	–4.9
Finland	23,240	119,086	18,260	–0.2
Iceland	26,580	7,175	21,710	0.5
Ireland	17,110	62,040	16,750	5.1
Latvia	2,300	5,730	3,650	–10.1
Lithuania	2,280	8,455	4,390	–6.0
Norway	34,510	151,198	23,220	3.7
Sweden	25,710	227,315	18,770	–0.2

*Purchasing power parity.

Source: *The World Bank Atlas,* 1998.

By joining together with a common currency, **Euroland** members expect to increase the efficiency and competitiveness of both domestic and international business. Formerly, when products were traded across borders, there were transaction costs associated with payments made in different currencies; these are now eliminated within Euroland. Germany, for example, exports two-thirds of its products to other EU members. With a common currency, this business now becomes essentially domestic trade, buffered from the fluctuations of many different currencies and without transaction costs. Euroland also creates benefits of scale from a mass market of 300 million people and a single economy of $6 trillion.

Besides a common currency, EMU member states are also committed to a common set of financial policies designed to control inflation, coordinate economic activity, and standardize wages and benefits. Though attractive in some ways, these common fiscal policies also reduce the traditional flexibility of member states to address the troublesome internal issues of inflation and unemployment. In the past, for example, if a state's economy stagnated, a country could always resort to the safety valve of devaluing its currency in order to make its products more competitive in international markets. But under the coordinated fiscal policies of EMU, this will not be an option. Some experts argue that Euroland will succeed only if the global economy picks up steam. If it cools to the point of a global recession, Euroland could be in big trouble. Indeed, uncertainty about the vitality of Euroland, along with concerns over loss of sovereignty, led the United Kingdom, Denmark, and Sweden to keep from participating in EMU.

Though primarily a tool to generate economic efficiency for Europe, there is little question that a common currency also has strong geopolitical overtones. If European states are bound together by a common economic policy and a common currency, fiscal issues should not get in the way of a coordinated and cooperative political interaction. Nationalism may actually diminish. However, should the Euroland economy falter and unemployment remain high, intra-EU competition may return through a rekindled nationalism that finds member states demanding a return to their historic rights of setting national economic policies. The next few years should provide a clearer sense of Euroland's future.

Economic Integration, Disintegration, and Transition in Eastern Europe

Eastern Europe has been historically less well developed economically than its western counterpart. Partial explanations come from the fact that eastern Europe is not particularly rich in natural resources. It has only modest amounts of coal, less iron ore and coke, and little to speak of in terms of oil and natural gas. Furthermore, what little resources were found in the region have been historically exploited by outside interests, including the Ottomans, Hapsburgs, Germans, and, more recently, the Soviet Russians. In many ways eastern Europe has been an economic colony of more-developed states to its west, south, and east.

The Soviet-dominated economic planning of the postwar period (1945–91) attempted to develop eastern Europe through a coordinated and planned regional economy that also served Soviet goals. However, the collapse of that centralized system in 1991 cast many eastern European countries into deep economic, political, and social chaos. Thus, we cannot understand the present-day problems of the region without some background about Soviet-bloc economic planning.

The Soviet Plan In 1949, with the creation of the Council for Mutual Economic Assistance (CMEA, or Comecon), the Soviet Union took control of the region's economies. Its goals were complete economic, political, and social integration through a **command economy,** one that was centrally planned and controlled. The methods were straightforward. In most settings the state owned all means of production, so the economy could be managed for the benefit of the people rather than private interests. As a result, most economic activity was nationalized. Supply and demand were coordinated at a regional level so that states would not compete with each other. Natural resources, such as coal, iron, coke, natural gas, and petroleum, were moved across state boundaries to serve the greater good. The Soviet Union also exported to eastern Europe those resources in short supply and would, in return, guarantee markets for the region's products. Additionally, each individual state subsidized industry to facilitate overall economic development and, as well, to provide manufacturing jobs for a reduced agricultural labor force. This industrial development was decentralized to upgrade the condition of backward regions (such as Slovakia and eastern Poland) as well as implanted in rural areas to provide employment to villagers and farmers. Finally, to increase food supplies, the agricultural sector was reshaped through **collectivization** (state ownership), mechanization, and centralized planning.

The Results After 40 years of communist economic planning, the results from CMEA were mixed and varied highly within the region. In Poland and Yugoslavia, for example, many farmers strongly resisted the nationalization of agriculture; thus, most arable and productive land remained in private hands. But in Romania, Bulgaria, Hungary, and Czechoslovakia, the story was different, with 80 to 90 percent of the agricultural sector converted to state and collective farms. Across eastern Europe, though, despite these radical changes, food production did not increase dramatically and food shortages, particularly for urban dwellers, were common. Agricultural workers fared somewhat better, being closer to the sources of food production, where most of the rural population tended subsistence gardens to meet their food needs (Figure 8.39).

Perhaps most notable during the Soviet period were dramatic changes to the industrial landscape, with many new factories built in both rural and urban areas and fueled by cheap energy and raw materials imported from the Soviet Union. For example, each eastern European country built its own steel mill even if it lacked the necessary mineral resources. The largest of these new mills was the Lenin Works in the new city of Nowa Huta just outside of Krakow, Poland.

▲ **Figure 8.39 Polish agriculture** One of the most problematic issues facing eastern European countries is the modernization of the agricultural sector. This is particularly pressing if eastern European countries are to gain membership in the European Union and compete with western European farmers. *(Hans Madej/Bilderberg Archiv der Fotografen)*

Though Poland had a modest supply of iron and coking coal, it grew increasingly dependent on the Soviet Union for raw materials.

In defining the pathway to economic development in eastern Europe, CMEA planners chose heavy industry over consumer goods. As in the Soviet Union, the bare shelves of retail outlets became both a signature of communist shortcomings and, perhaps more importantly, a source of considerable public tension. As western Europeans enjoyed an increasingly high standard of living with an abundance of consumer goods, eastern Europeans struggled along, trying to make ends meet as the utopian vision promised by Soviet communists became increasingly elusive. Though it is probably fair to say some good resulted during this period of CMEA economic planning, few envisaged the widespread problems that would result with the abrupt end to the economic subsidy provided by the Soviet Union.

Transition and Changes Since 1991 As Soviet political hegemony over eastern Europe disintegrated in the late 1980s, so did the integrated economies of CMEA. In its place has come a painful period of economic transition bridging on outright chaos in some places as eastern European countries rebuild and reorient their national economies. Most eastern Europe countries are redirecting their economic and political attention to western Europe with the explicit goal of joining the EU sometime in the twenty-first century. To link up with the west, many eastern European countries have moved already from a socialist-based economy of state ownership and control toward a capitalist economy predicated on private ownership and free markets. To achieve this, states such as the Czech Republic, Hungary, and Poland have gone through a period of **privatization**, which is the transferring to private ownership of those firms and industries previously owned and run by state governments.

Eastern Europeans have witnessed many changes as a result of this shift in economic policy. Price supports, tariff protection, and subsidies have been removed from consumer goods as countries transition to a free market. For the first time, goods from western Europe and other parts of the world are plentiful and common in eastern European retail stores. The irony, though, is that this prolonged period of economic transition has taken its toll on eastern European consumers; few can afford these long-dreamed-of products. Unemployment has often risen to more than 15 percent, and underemployment is common. Financial security is evasive, with irregular paychecks for those with jobs and uncertain welfare benefits for those without. Furthermore, for most people the basic costs of food, rent, and utilities are higher under a free market system than under the subsidized economies of communism. Table 8.2 shows the stark contrast in purchasing power parity between eastern and western Europe (see "From the Field: Traveling in Post-1989 Bulgaria").

The causes of this economic pain are complex. As the Soviet Union moved to address its own economic and political turmoil, it stopped exporting subsidized natural gas and petroleum to eastern Europe and, instead, sold it on the open global market to gain hard currency. Without cheap energy, many eastern European industries were unable to operate and shut down operations, laying off thousands of workers. In the first two years of the transition (1990–92), industrial production fell 35 percent in Poland and 45 percent in Bulgaria (Figure 8.40). In addition, the markets guaranteed under a command economy, many of them in the Soviet Union, dissolved overnight with the CMEA's collapse in 1991. Consequently many factories and services closed because they lacked a market or outlet for their goods. In some cases, products from the West, particularly beverages, clothes, and cigarettes, moved in quickly to fill this void.

▲ **Figure 8.40 Post-1989 hardship** With fall of communism in eastern Europe after 1989, economic subsidies and support from the Soviet Union ended. Accompanying this transition has been a high unemployment rate that results from the closure of many industries, such as this plant in Bulgaria. *(Rob Crandall/Rob Crandall Photographer)*

FROM THE FIELD Traveling in Post-1989 Bulgaria

Several of the authors visited Bulgaria recently at the invitation of the Institute of Geography in Sofia. During our visit, we met a wonderful cross section of people: professionals and workers, students and farmers, and traveled across this intriguing country, staying in large cities and small rural villages alike. It was a remarkable experience, seeing Bulgaria in the throes of change, yet one that was not always pleasant because of the many hardships caused by the profound and rapid changes that have affected every aspect of life since 1990.

Unemployment is high, underemployment even more common. Those with jobs do not get regular paychecks. Energetic optimism about the future alternates with a clouded and conditioned uncertainty that often degenerates into outright pessimism. In many ways what we saw in Bulgaria, a country of slightly more than eight million, is a microcosm for the rest of eastern Europe, though we appreciate the vast differences between places and people in this complex region. Here are a few of our impressions:

- For those of us who knew eastern Europe before 1989, the change in the cities is remarkable. In place of the monotonous, often drab, landscape of earlier years, Bulgaria's large cities are now dotted with the garish colors of western billboards and posters; everywhere are advertisements for soft drinks, cigarettes, and fashionable clothes. Further, the familiar names of western franchises dominate major shopping streets, often displacing traditional retail services, forcing booksellers and hardware stores to operate out of cartons and suitcases from open-air street stands because they cannot pay the high rents asked by the new free-market landlords.
- Traveling in the countryside, we see a large number of horse- or donkey-drawn carts, carrying not just farm products, but also serving as everyday transportation in many cases. This reminded us that even if people own cars, gasoline is scarce and expensive for the rural population. Additionally, little farm machinery is seen in the fields. This is because of uncertain access to formerly state-owned or communal farm equipment, along with the problem of expensive fuel. Instead, people have returned to hand tools and various forms of animate power. A rare mechanized wheat combine, for example, is surrounded not by trucks or tractors, but instead by pony-drawn wooden carts with people carrying heavy sacks of grain.
- Staying with a farm family in a small village of several hundred in northeastern Bulgaria, we see these changes close up. Farmers are probably the most disgruntled group; they resent the post-communist changes, though they are reluctant to openly advocate a return to communism before a group of American strangers. Most farmers are essentially unemployed because of the closure of state-owned farms, and few seem anxious to farm the small plots that would result from privatization. Furthermore, markets for their farm products are uncertain. Unlike their urban cousins, at least they have access to market gardens that provide food for the household. Walking about the village, we see small garden plots, intensively cultivated with vegetables of all sorts, yet usually protected by high fences and locked gates. Even the livestock are under lock and key, with guard dogs everywhere. We cannot determine where the threat lies, whether it might come from other village residents or from night-raiding urban-dwellers.
- Near this village lies one of the many closed factories we see throughout Bulgaria; shut for lack of raw materials or cheap fuel or because the organizational structure simply vanished. This factory formerly repaired large steel shipping containers used to transport goods across the Black Sea. Even though that large water body lies four hours east of this village, the factory was located here as part of the CMEA's decentralization philosophy of bringing work to the countryside. Now, however, there is no longer work in this factory or in this village.

Many of the shipping containers have been liberated from the factory and have been converted into toolsheds, livestock barns, and even small cottages. The most creative use, demonstrating, perhaps, the fledgling spirit of Bulgarian capitalism, was the conversion of several containers into shelters and food preparation structures for a village beer garden and coffeehouse. Behind the bar was a cheerful electrical engineer from the container factory who had created an attractive social focus for this dispirited village with his creative construction.

As with all of eastern Europe, change has not come easily to Bulgaria since the end of communism in 1990. While many people praise their new personal freedoms, harsh economic realities still cloud their skies and cast uncertainty on their future.

With the privatization of industry, services, and property, state subsidies ended, and thus the cost of living increased because of higher prices for goods, services, and rent. Free-market advocates point out that this should be a temporary stage and those prices should fall as stiffer competition drives prices down. Double-, even triple-digit, inflation has also vexed eastern European populations in recent years, aggravating the cost-of-living increases for the population. Food supplies remain problematic since agricultural production fell during the transition. A number of factors explain this decline, including fuel shortages to run farm equipment, uncertainty about land ownership and access to fields in former state or communal farms, and the inefficiency of operating units where privatization has taken place. In Bulgaria, for example, which previously had 90 percent of its productive land in state or communal farms, most of the new privatized farms are smaller than 2.5 acres (1 hectare) in size.

Regional Disparities in Eastern Europe Though certain generalizations hold true throughout eastern Europe, geographers are quick to appreciate significant differences between different countries and even subregions within that area. Perhaps more important, there is a troublesome pattern of a dual economy emerging in eastern Europe, characterized by a cluster of countries speeding ahead to reap the rewards of successful change, contrasted with those left in the dust, countries that appear to be falling ever-farther behind their neighbors (Table 8.3).

By most accounts, successful economic transitions are taking place in the Czech Republic, Slovenia, Hungary, and Poland. This transformation is supported by the purchasing power parity and per capita GNP data found in Table 8.2. Though far behind affluent western European states, these four countries have forged strong trade ties with western Europe, are receiving investment from the West, and are well on their way to privatization. Further, they have applied for membership in the EU and many experts agree they will be given serious consideration for the next round of EU expansion. What explains their success?

From a geographic perspective, it is important to note that all four countries are conterminous to western Europe, which gives them advantages of easier transportation links. Besides this spatial proximity to the west, these countries historically developed strong manufacturing sectors supported by a skilled workforce. The Czech Republic, for example, long led eastern Europe in industrial vitality, first as the workshop of the Hapsburg Empire during the late nineteenth century, and more recently as an industrial country in its own right. Its skilled labor enjoys an unemployment rate (currently 3.2 percent) well below that of most of western Europe. Each of these four countries also resisted centralized communism, either through outright revolt and rebellion (as in Czechoslovakia and Hungary), or by inventing its own unique brand of more flexible socialism, as was the case with Poland and Slovenia. As a result, privatization has proceeded well in all four countries. In Hungary, this began well before the collapse of Soviet control, and has continued with a number of western firms investing in Hungarian business. This country has become such a hotbed of capitalism that it recently opened its own national stock exchange, the first in eastern Europe.

At the other end of the spectrum, Macedonia, Moldova, and Albania rank lowest in most economic and social development measures. Common to these three countries is widespread political and ethnic strife that has inhibited development. Albania, for example, suffered near-anarchy for several years following the 1997 collapse of pyramid investment plans that destroyed the finances of tens of thousands of families. Violence and looting followed, and a multinational UN force of 7,000 was sent to restore order. Macedonia and Moldova both gained independence in the early 1990s by breaking away from larger states, the former from Yugoslavia, and the latter from Ukraine and Romania. As with Albania, however, political and ethnic strife inhibits economic and social development. Additionally, neighboring states and trading partners, such as Romania, Ukraine, Bulgaria, Yugoslavia, are also struggling with extremely weak economies.

In between these two opposing ends of the spectrum are other eastern European countries that make up the middle ground of economic and social change. Included are the three Baltic countries—Latvia, Estonia, and Lithuania—all formerly part of the Soviet Union. Also struggling are Bosnia-Herzegovina, Croatia, and Yugoslavia, weakened and distracted by their mutual animosities and strife; Slovakia, the rural and agricultural tail of the former Czechoslovakia; and, finally, the two Black Sea countries of Bulgaria and Romania. Though long independent, both countries were highly dependent on the Soviet Union for economic stability during the postwar years. Romania's problems, though, have been more internal than external. After a revolving-door series of governments with contrasting views about the country's future, stability appears to have set in finally, which has led to economic aid and loans from the World Bank and International Monetary Fund. This should end triple-digit inflation and jump-start the Romania economy (Figure 8.41).

Conclusion

The twentieth century brought unparalleled change to Europe. Disruption, even chaos, characterized many decades, with two world wars, a handful of minor ones, the political chaos accompanying a serious economic depression, and, then, almost half a century of political and military division during the Cold War. The wounds and scars of these hardships are still apparent, particularly in the east and southeast. Most of Europe, however, meets the twenty-first century with optimism and confidence. Even so, it is appropriate to add a few qualifications to our summary of geographic themes.

In terms of environmental issues, western Europe has made great progress in the last several decades. Not only have individual countries enacted strong environmental legislation, but also the EU has played an important role with its strong commitment to regional solutions that address trans-boundary concerns, such as air and water pollution, or ocean and coastal problems. Additionally, the EU has become an aggressive environmental advocate at the global level, as illustrated by its participation in global climate conferences and treaties where the EU has demonstrated a willingness to cut back its atmospheric emissions far more than any other major world region.

Serious environmental problems, however, plague eastern Europe, and it remains to be seen whether or not the EU can solve those daunting issues. Several variables will shape the East's environmental future. These new democracies need to enforce existing or new environmental legislation, pay the costs of toxic cleanup and new pollution controls, and cooperate regionally to solve air and water pollution problems. Lastly, eastern Europeans

Table 8.3 Social Indicators and Status of Women

Country	Life Expectancy at Birth		Under Age 5 Mortality, per 1,000 Live Births		Secondary School Enrollment %*		Female Labor Force Participation (% of total)
	Male	Female	1960	1993	Male	Female	
Western Europe							
Austria	74	80	43	8	109	104	41
Belgium	74	81	35	10	103	104	40
France	74	82	34	9	104	107	44
Germany	73	80	40	7	101	100	42
Liechtenstein	67	78	—	—	—	—	—
Luxembourg	73	79	—	—	72	73	37
Netherlands	75	80	22	8	126	120	40
Switzerland	76	82	27	8	93	89	40
United Kingdom	74	80	27	8	91	94	43
Eastern Europe							
Bulgaria	67	75	70	19	66	70	48
Czech Republic	71	77	—	10	85	88	47
Hungary	66	75	57	15	79	82	44
Moldova	62	70	—	36	67	72	49
Poland	68	77	70	15	82	87	46
Romania	65	73	82	29	83	82	44
Slovakia	69	77	—	18	87	90	48
Southern Europe							
Albania	70	76	151	41	84	72	41
Bosnia-Herzegovina	70	75	—	—	—	—	38
Croatia	69	76	—	—	80	86	44
Greece	75	80	64	10	100	98	37
Italy	75	81	50	9	81	82	38
Macedonia	69	73	—	—	53	55	41
Malta	75	80	—	—	91	84	37
Portugal	71	79	112	11	63	74	43
San Marino	73	79	—	—	—	—	—
Slovenia	71	78	—	—	88	90	46
Spain	73	81	57	9	107	120	36
Yugoslavia (Serbia)	70	75	—	—	64	65	42
Northern Europe							
Denmark	73	78	25	7	112	115	46
Estonia	62	74	—	23	87	96	49
Finland	73	81	28	5	110	130	48
Iceland	76	81	22	—	105	101	44
Ireland	72	78	36	7	101	110	33
Latvia	64	76	—	26	84	90	50
Lithuania	65	76	—	20	76	79	48
Norway	75	81	23	8	118	114	46
Sweden	77	82	20	6	99	100	48

*These data cover the percentage of people in a certain age group enrolled in secondary school. Older or younger people attending high school can boost the percentage above 100%.

Sources: *Population Reference Bureau Data Sheet,* 1998, Life Expectancy (M/F); World Resources Institute, *World Resources 1996–97,* Under-5 Mortality Rate; *Population Reference Bureau Data Sheet, 1996,* Secondary School Enrollment (M/F); *The World Bank Atlas, 1998,* Female Participation in Labor Force.

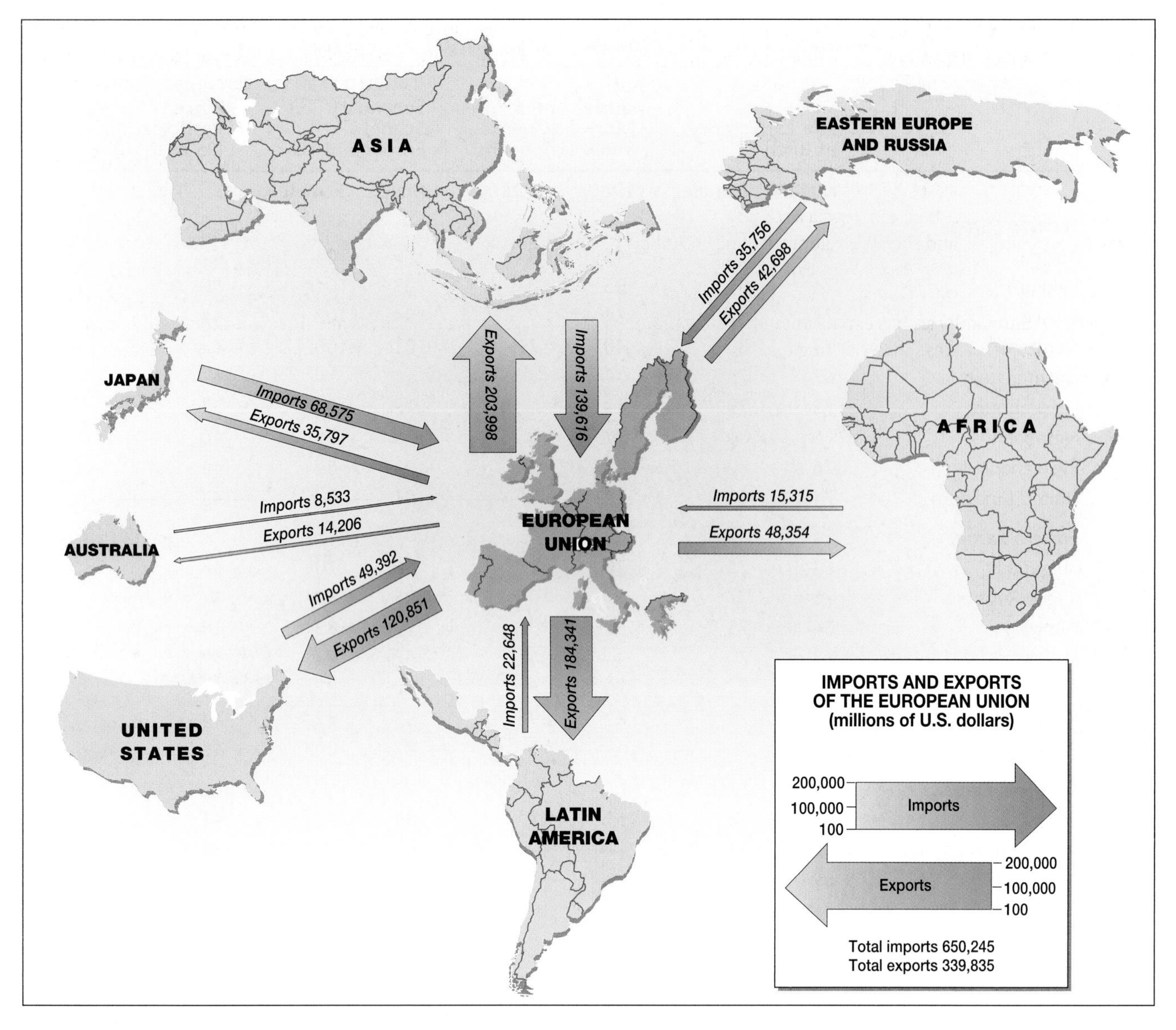

▲ **Figure 8.41 European Union trade map** Although trade wars between the United States and the European Union make daily headlines, this map demonstrates that the EU also has extensive trade ties with other world regions. For example, the EU imports more goods in dollar value from Asia than from the United States, and exports more to Latin America and Asia. EU trade with eastern Europe and Russia will surely increase once those economies stabilize. *(Data from* Euromonitor, International Marketing Data and Statistics, *1997)*

must control the air pollution and surface congestion resulting from an explosive increase in private automobile ownership.

The region also faces ongoing challenges related to population and migration. As for population and settlement issues, the phenomenon of slow and no growth is being appropriately confronted by individual countries as they determine what is the right amount of natural increase. The real problem, though, is how Europe deals with in-migration from Asia, Africa, Latin America, former Soviet lands, and from its own underdeveloped regions. If Europe's economy booms once again, this migration could be the solution to a new labor shortage, but if the economy stagnates, migration will continue to inflame a large number of issues throughout the region.

Continued cultural tension will be one of the problems caused by migration. As the bulk of Europeans meld into a common culture of sorts, the cultural differences between Europeans and outsiders will increase. While some European countries, such as France and even Britain, seem prepared to become multicultural societies, others appear less interested or willing, despite their taking explicit legal actions against ethnic and racial discrimination. While there is little doubt the cultural geography of Europe will change profoundly in the next few years, how Europeans may react to this social change is at the moment unclear.

With the end of the Cold War, Europe's geopolitical issues are continually redefined. The role and future direction of NATO is one of the important topics facing Europe. Many European states would like to replace NATO with a completely new military alliance, a European Security Force without a strong and dominating U.S. presence. Indeed, the United States may not resist this idea because NATO's future could be linked with costly Yugoslavia-like engagements, and the United States might prefer avoiding these actions. The Balkans, unfortunately, will remain problematic to Europe and the world beyond. Europe must address the internal geopolitics of this problem by deciding if full independence is appropriate for former Yugoslavia provinces such as Kosovo. In addition, Europe will have to continue its precarious balancing act between Russia and the Muslim world.

Issues of economic and social development must also be addressed by twenty-first century Europeans. Much uncertainty exists as the EU moves into new territory with its common monetary policy ("Euroland") and its expansion into eastern Europe. Prospective member states, namely Hungary, Poland, the Czech Republic, Slovenia, and Estonia, still have a difficult road to travel before they meet the fiscal and political requirements of the EU; should they fail and the invitations subsequently drop, a brighter economic future of eastern Europe could be in jeopardy. Even if these states are admitted, there remains the challenge of somehow alleviating what will surely be significant differences between rich and poor in eastern Europe. While it is easy to envision a positive future for an enlarged EU that includes new member states in eastern Europe, it is much more difficult to think of a positive economic, political, and social future for Moldova, Albania, Bulgaria, and Romania. Sharing the EU's twenty-first century dream with these countries will constitute the real challenge for this dynamic world region (Figure 8.41).

Key Terms

balkanization (page 346)
buffer zone (page 343)
Cold War (page 311)
collectivization (page 353)
command economy (page 353)
Continental climate (page 316)
Council of Mutual Economic Assistance (CMEA) (page 350)
Cyrillic alphabet (page 332)
Euroland (page 353)
European Union (EU) (page 311)
feudalism (page 340)
fjord (page 315)
guest worker (page 323)
hierarchical diffusion (page 334)
Iron Curtain (page 321)
irredentism (page 341)
Marine West Coast climate (page 316)
medieval landscape (page 328)
Mediterranean climate (page 316)
moraines (page 313)
polders (page 318)
privatization (page 354)
Renaissance–Baroque landscape (page 328)
Schengen Agreement (page 327)
secularization (page 336)
shield landscape (page 315)

Questions for Summary and Review

1. How does the European Lowland differ north and south of the Rhine River?
2. Describe the different upland and mountain regions of Europe.
3. What are the major climate controls influencing Europe's weather and climate?
4. Explain why eastern Europe has such serious environmental problems.
5. Make two lists: one of those European countries that are below replacement in natural population increase and, second, a list of those countries that will actually grow in the next 20 years. Explain the differences.
6. Explain "Schengenland" and "Fortress Europe." What are the geographic advantages and disadvantages of this new arrangement?
7. What are the major tongues of the Germanic language family? Which have the most speakers?
8. Map the north and south groups of the Slavic language family. Also, map those countries that do not speak Indo-European languages.
9. Map the interface boundaries, both historical and modern, between Christianity and Islam in different parts of Europe. Where is this still a problem? Why?
10. Make a list of the new nation-states that appeared on Europe's map at these dates: 1919, 1945, and 1992.
11. What were the major geopolitical institutions of the Cold War? Be able to map their areas of influence up to 1990.
12. Be able to describe the major ethnic groups as to language and religion in each of the Balkan countries.
13. What are the major components of the Industrial Revolution as seen in Europe? How did the location of industrial areas change through time?
14. Trace the evolution of the EU since 1952, noting how its goals and membership have changed. Be prepared to make a map of its membership at different points in time.
15. Describe the goals and agenda of CMEA, beginning in 1950.
16. Explain what has happened to the eastern European economy since 1990.

Thinking Geographically

1. To compare the scale of Europe to North America, draw a circle of 500 miles (800 kilometers), which is a day's journey, around Frankfurt, Germany. Then do the same for Chicago. Comment on the similarities and differences in mobility as a result of this exercise.
2. Discuss the strategic importance of Europe's seas and straits as it has changed during the twentieth century with changes in naval and shipping technology.
3. Investigate the implications of sea-level rise from global warming for different parts of Europe, including the Dutch coastline. How might this influence the European Union's policy on control of atmospheric emissions?
4. How and why has western Europe become more energy-efficient than North America?
5. Map the different rates of natural increase in Europe, noting which countries will grow and which will decline in the next 20 years. Then link this map to a discussion of migration in Europe. Given these two factors, how might the population map change in 20 years?
6. Find several good maps of mid-sized European cities. From these maps, infer the historical development of the city based upon different street patterns, arrangement of open spaces, boulevards, parks, and so on. Draw upon the differences between the medieval and Renaissance–Baroque for your interpretation.
7. Buy a French or German newsmagazine. Then go through the ads and make a list of "globalized" English words used. Discuss your findings in terms of what sectors of society seem most open to these foreign terms.
8. Choose an eastern European country and city, and then discuss how both the rural landscape and cityscape have changed in the last 10 years. One approach is to set a baseline of, say, 1985, and then compare it to the present.
9. Critically examine the geographic aspects of the Dayton Accord, and then offer an informed opinion as to whether you think the different boundaries were drawn correctly.
10. Inform yourself on what fiscal, political, and social changes must take place in Hungary and Poland before they are admitted to membership in the EU. Discuss the costs versus the benefits, the advantages compared to the disadvantages of membership. Who within each country will profit, and who might suffer?
11. Choose data from Table 8.2 (Economic Indicators) and Table 8.3 (Social Indicators) to map regional disparities and contrasts within eastern Europe and/or the Balkans. Discuss whether or not you think this pattern can be changed.

Regional Novels and Films

Novels

Jerzy Andrzejewski, *Ashes and Diamonds* (1962, Weidenfeld & Nicholson)

John Berger, *Pig Earth* (1979, Random House)

John Berger, *Once in Europa* (1983, Random House)

Emilie Carles, *A Life of Her Own: A Countrywoman in Twentieth-Century France* (1991, Rutgers University Press)

Natalie Zeamon Davis, *The Return of Martin Guerre* (1983, Harvard University Press)

James Joyce, *The Dubliners* (1967, Viking)

Milan Kundera, *The Unbearable Lightness of Being* (1984, Harper & Row)

W. S. Merwin, *The Lost Upland: Stories of Southwest France* (1992, Alfred Knopf)

Peter Nadas, *The Book of Memories* (1985, Farrar, Straus & Giroux)

V. S. Naipal, *The Enigma of Arrival* (1987, Alfred Knopf)

Charles Power, *In the Memory of the Forest* (1997, Charles Scribner's)

Peter Schneider, *The Wall Jumper: A Berlin Story* (1998, The University of Chicago Press)

Dimitur Talev, *The Iron Candlestick* (1964, Foreign Language Press)

Films

Angi Vera (1978, Hungary)

Antonia's Line (1995, The Netherlands)

Au Revoir Les Enfants (1987, France)

The Bicycle Thief (1948, Italy)

Christ Stopped at Eboli (1979, Italy)

Europa, Europa (1991, Germany)

The Firemen's Ball (1968, Czechoslovakia)

Hey, Babu Riba (1988, Yugoslavia)

Jean de Florette (1987, France)

Man of Iron (1981, Poland)

Manon of the Spring (1987, France)

My Life as a Dog (1985, Sweden)

Playtime (1967, France)

The Postman (1994, Italy)

The Remains of the Day (1994, United Kingdom)

The Stationmaster's Wife (1977, Germany)

Trainspotting (1995, United Kingdom)

Triumph of the Will (1934, Germany)

Underground (1995, Yugoslavia [Bosnia])

Weekend (1967, France)

Bibliography

Ardaghy, John. 1987. *Germany and the Germans: An Anatomy of Society Today.* New York: Harper & Row.

Barnes, Ian, and Robert Hudson. 1998. *The Historical Atlas of Europe. From Tribal Societies to a New European Unity.* New York: Macmillan.

Bowler, Ian. 1985. *Agriculture Under the Common Agricultural Policy.* Manchester: Manchester University Press.

Carter, Francis, and Turnock, David, eds. 1993. *Environmental Problems in Eastern Europe.* London: Routledge, Chapman & Hall.

Dawson, Andrew. 1993. *A Geography of European Integration: A Common European Home.* New York: John Wiley & Sons.

Denitch, Bogdan. 1996. *Ethnic Nationalism: The Tragic Death of Yugoslavia.* Minneapolis: University of Minnesota Press.

Doherty, Paul, and Poole, Michael. 1997. "Ethnic Residential Segregation in Belfast, Northern Ireland, 1971–1991." *The Geographical Review* 87(4), 520–536.

Engman, Max, ed. 1992. *Ethnic Identity in Urban Europe.* Aldershot, England: Dartmouth Publishing.

Fells, John, and Niznik, Jozef. 1992. "What Is Europe?" *International Journal of Sociology* 22, 201–207.

Glenny, Misha. 1993. *The Fall of Yugoslavia: The Third Balkan War.* New York: Penguin.

Goldfarb, Jeffrey. 1992. *After the Fall: The Pursuit of Democracy in Central Europe.* New York: Basic Books.

Hamilton, Kimberly. 1994. *Migration and the New Europe.* Boulder, CO: Westview Press.

Hoffman, Eva. 1993. *Exit into History: A Journey Through the New Eastern Europe.* New York: Viking.

Houston, James. 1953. *A Social Geography of Europe.* London: Ducksworth.

Jordan, Terry. 1996. *The European Culture Area,* 3rd edition. New York: HarperCollins.

Kaplan, Robert. 1993. *Balkan Ghosts: A Journey Through History.* New York: Vintage Books.

Lewis, Flora. 1992. *Europe: The Road to Unity.* New York: Touchstone.

Liefferink, J.; Lowe, P.; and Mol, A., eds. 1993. *European Integration and Environmental Policy.* New York: Belhaven Press.

McDonald, James. 1997. *The European Scene: A Geographical Perspective.* Upper Saddle River, NJ: Prentice Hall.

Netting, Robert. 1981. *Balancing on an Alp: Change and Continuity in a Swiss Mountain Community.* Cambridge: Cambridge University Press.

Oberhauser, Ann. 1991. "The International Mobility of Labor: North African Migrant Workers in France." *Professional Geographer* 43, 431–445.

Pells, Richard. 1997. *Not Like Us: How Europeans Have Loved, Hated, and Transformed American Culture Since World War II.* New York: Basic Books.

Renfrew, Colin. 1989. "The Origins of Indo-European Languages," *Scientific American* 261(4), 106–114.

Rosenberg, Tina. 1995. *The Haunted Land: Facing Europe's Ghosts After Communism.* New York: Random House.

Silber, Laura, and Little, Allan. 1995. *Yugoslavia: Death of a Nation.* New York: Penguin Books.

Slavenka, Drakulic. 1993. *The Balkan Express: Fragments from the Other Side of the War.* New York: W. W. Norton and Company.

Smith, A. D. 1986. *The Ethnic Origin of Nations.* Oxford: Blackwell.

Watts, Mary. 1971. *Reading the Landscape of Europe.* New York: Harper & Row.

Yarnal, Brent. 1995. "Bulgaria at the Crossroads." *Environment* 37(10), 7–32.

GREENLAND
(DENMARK)
North Pole
ARCTIC OCEAN
60°N
20°W
80°N
140°W
160°W
180°
160°E
140°E
120°E
80°E
60°E
40°E
0°
Arctic Circle
North Sea
Barents Sea
E U R O P E
Baltic Sea
Murmansk
Kaliningrad
RUSSIA
St. Petersburg
Archangel
BELARUS
Minsk
UKRAINE
Chernobyl
Kiev
Dnieper R.
Moscow
Yaroslavl'
Ivanovo
Nizhniy Novgorod
Kazan
Serov
Ob River
Norilsk
S I B E R I A
Verkhoyansk
Lena River
Yakutsk
R U S S I A
Yenisey River
Odessa
Kharkov
Dnepropetrovsk
Donetsk
Simferopol'
Sevastopol
Saratov
Samara
Yekaterinburg
Don R.
Volgograd
Magnitogorsk
Chelyabinsk
Black Sea
Volga R.
Omsk
Novosibirsk
Novokuznetsk
Krasnoyarsk
Baikal-Amur Mainline (BAM)
Trans-Siberian Railroad
Lake Baikal
Irkutsk
Amur
GEORGIA
Transcaucasia
Groznyy
Tbilisi
ARMENIA
Yerevan
Caspian Sea
A S I A
RUSSIAN REGION
Political Map
Over 1,000,000
500,000–1,000,000 (selected cities)
Selected smaller cities
Elevation in meters
4000+
2000–4000
500–2000
200–500
0–200
Below sea level
Sea Level
0 250 500 mi
0 250 500 km

CHAPTER 9

Russia and Its Neighbors

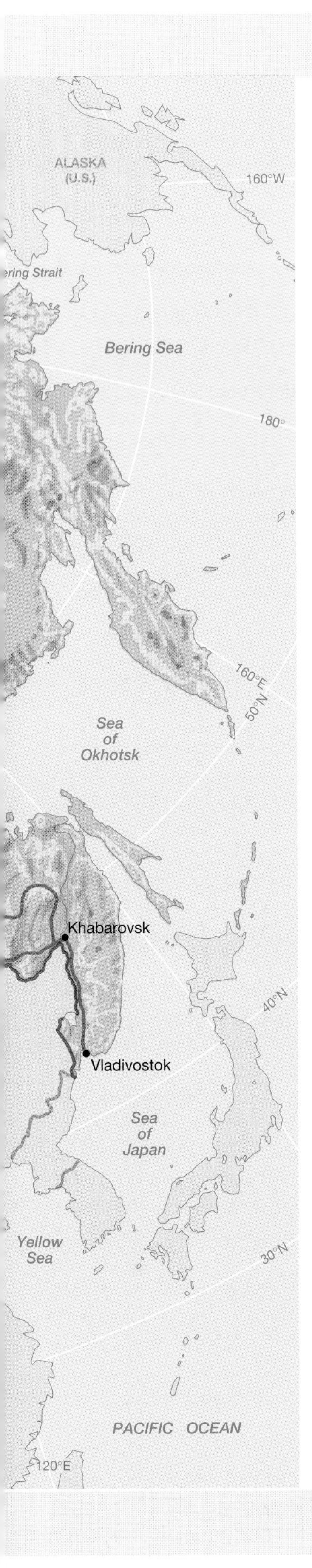

Russia and its neighbors dominate the northern half of Eurasia; the region includes not only Russia itself, but also the nations of Ukraine, Belarus, Georgia, and Armenia (see Figure 9.1; see also "Setting the Boundaries"). The land is rich with superlatives: vast Siberian spaces, unlimited natural resources, legends of ruthless Cossack warriors, and tales of epic wars and revolutions are all part of the region's geographical and historical mythology. Indeed, the rise of Russian civilization remarkably parallels the story of the United States. Both cultures grew from small beginnings to become imperial powers enriched by the fur trade, gold rushes, and transcontinental railroads during the nineteenth century and by dramatic industrialization in the twentieth century. Recently, however, the parallels have ended: incredible political and economic changes have rocked the Russian-dominated region and the near-term future remains uncertain. The key challenges are indeed daunting. Economic collapse across much of the region in the late 1990s has produced steep declines in living standards. Political instability includes both tensions between neighboring states as well as divisive pressures within particular countries. The region also faces some of the most-severe environmental challenges anywhere in the developed world.

The region's relationship with the rest of the world shifted dramatically within the last 15 years of the twentieth century. Until the end of 1991, all five countries belonged to the Soviet Union, the world's most powerful communist state. Under Soviet control, the region's economy saw large twentieth-century increases in industrial output that made the nation a major global producer of steel, weaponry, and petroleum products. Its communist system offered a powerful ideology that promised economic prosperity and hope to residents within the region and beyond. Indeed, the political and military reach of the Soviet Union spanned the globe, making it a superpower on a par with the United States. The Soviet presence dominated many eastern European countries, and nations from Cuba to Vietnam enjoyed close strategic ties with the country. Suddenly, as the old communist order evaporated early in the 1990s, the now independent republics of Russia, Ukraine, Belarus, Georgia, and Armenia had to carve out new regional and global relationships. With the breakdown of Soviet control, the region also felt the growing presence of western European and American influences. Westernized popular culture, as well as economic changes,

◀ **Figure 9.1 Russia and Its Neighbors** Russia and its neighboring states of Belarus, Georgia, Ukraine, and Armenia are a dynamic and unpredictable world region that has experienced tremendous change in the past decade. Sprawling from the Baltic Sea to the Pacific, the region includes huge industrial centers, vast farmlands, and almost empty stretches of tundra. Recent political and economic transformations place the region on a new, but highly uncertain course in the early years of the twenty-first century.

Setting the Boundaries

The boundaries of "Russia and Its Neighbors" or the "Russian region" have shifted over time. For decades, the task was relatively easy, as the highly centralized Soviet Union (or Union of Soviet Socialist Republics) dominated the region's political geography. The country, born in a communist revolution in 1917, dwarfed all other states in the world and was powerfully united by the Soviet government and largely controlled by the ethnic Russians. Most geographers agreed that the Soviet Union could thus be viewed as a single world region. Although the nation contained many cultural minorities, the Soviet Union wielded singular political and economic power from Leningrad on the Baltic Sea to Vladivostok on the Pacific Ocean. After World War II, some geographers even included much of Soviet-dominated eastern Europe within the region in response to the country's expanded military role in nations such as East Germany, Poland, and Hungary.

Suddenly, the maps were redrawn late in 1991. The once-powerful Soviet state was officially dissolved and in its place stood 15 former "republics" that had once been united under the Soviet Union. Now independent, each of these republics has tried to make its own way in a post-Soviet world. While some geographers initially treated the region as the "Former Soviet Union," it quickly became clear that diverse cultural forces, economic trends, and political orientations were taking the republics in different directions. Even so, the Russian Republic remained dominant in size and area and thus came to form a nucleus of a new "Russian region" that was considerably smaller than the Soviet Union and yet recognized some common ties shared by Russia and its neighbors.

The new regional definition reflects the changing political and cultural map since the breakup of the Soviet Union. Russia, Ukraine, and Belarus make up the core of the new region. Persisting cultural and economic ties closely connect these three countries. In addition, Georgia and Armenia, while more culturally distinctive, are still best classified within a zone of Russian influence. Two significant areas that were once a part of the Soviet Union have been eliminated. The mostly Muslim republics of Central Asia and the Caucasus (Kazakhstan, Uzbekistan, Kyrgyzstan, Turkmenistan, Tajikistan, and Azerbaijan) have become aligned with a "Central Asia" world region (Chapter 10), while the Baltic republics (Estonia, Latvia, and Lithuania) and Moldova are best grouped with Europe (Chapter 8).

both modest and radical, accompanied the political transformations within the realm. These new global relationships have not been easy. Social tensions have risen within the region as people grapple with fundamental economic changes. Political relationships with neighboring regions in Europe and Asia remain uncertain. The region's economic stability has also been threatened by its exposure to the competitive pressures of the global economy.

Slavic Russia (population 147 million) remains dominant within the region. Although only a fraction of the dimensions of the former Soviet Union, Russia's huge size still makes it the largest state on Earth. West of Moscow, the country's European front borders Finland and Poland, while far to the east the nations of Mongolia and China share a thinly peopled boundary with sprawling Russian Siberia. Its area of 6.6 million square miles (17 million square kilometers) dwarfs even Canada, and its 11 time zones are a reminder that dawn in Vladivostok on the Pacific Ocean is still only dinnertime in Moscow. In more ways than one, the Russians sit squarely between sunrise and sunset. With the recent end of the Soviet Union, the Russians ended almost 75 years of Marxist rule. What comes next is difficult to predict, and the country currently stands in an economic and political twilight between the old ways and the new. Will Russia, spurred by its numerous ethnic minorities, further fragment into smaller political units, or might it reassert its direct control over neighboring states such as Belarus and Ukraine? Will its economy achieve an appropriate balance between private control and state ownership?

Arguably, stability in the region may be a long way off, but two things are certain. First, by virtue of its size and location, change within Russia will inevitably shape political and economic geographies far beyond its borders. Second, even with its present problems, Russia possesses a long-established and coherent national identity as well as a rich assemblage of natural and human resources that can act to bind the nation together and provide the foundation for future political stability and economic development.

The bordering states of Ukraine, Belarus, Georgia, and Armenia will inevitably be linked to the evolution of their giant neighbor, but at present they are attempting to make their own way as newly independent nations. Emerging from the shadows of Soviet dominance has been difficult. Ukraine, in particular, has the size, population, and resource base to become a major European nation, but it has struggled to create real political and economic change since independence. With more than 50 million people and a rich storehouse of resources, Ukraine's size of 233,000 square miles (604,000 square kilometers) is similar to that of France. Nearby Belarus is smaller (80,000 square miles (208,000 square kilometers) and its population of 10 million is likely to remain more closely tied economically and politically to Russia. Presently, its strikingly authoritarian and anti-foreign leadership echoes remnants of the old Soviet Empire. South of Russia and beyond the bordering Caucasus Mountains (Figure 9.1), the Transcaucasian countries of Armenia and Georgia are smaller still. Their populations (3.8 million and 5.4 million, respectively) depart culturally from their Slavic neighbor to the north. In addition, these two nations each face significant political challenges: Armenia shares a hostile border with Azerbaijan (see Chapter 10) and Georgia's ethnic diversity threatens its political stability.

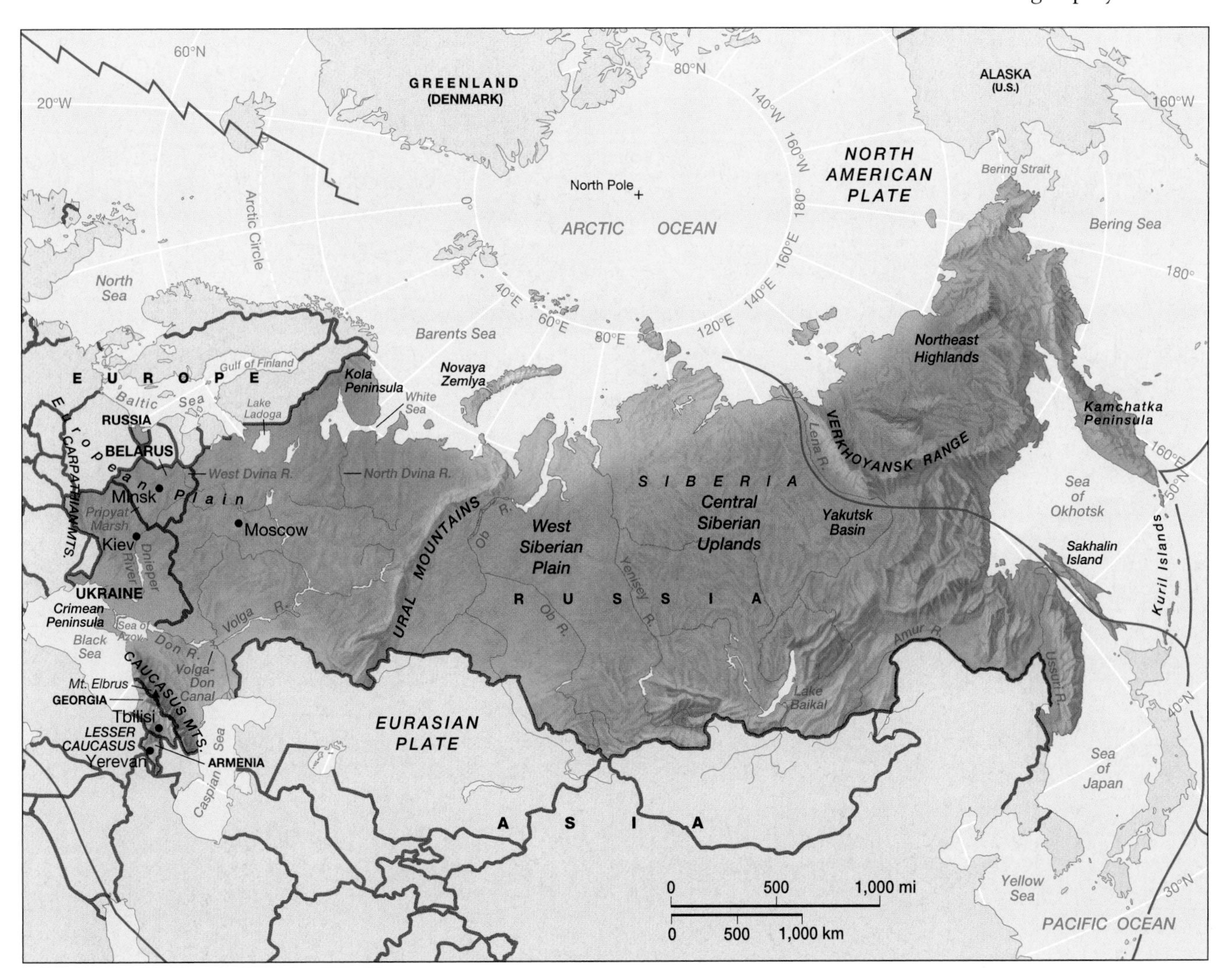

▲ **Figure 9.2 Physical geography of Russia and its neighbors** In the west, the European Plain stretches across Belarus and the Russian heartland. In the far south, the rugged Caucasus Mountains impose significant barriers to movement. East of the Urals, more large plains and plateaus alternate with mountains, particularly in eastern Siberia where higher peaks and volcanic uplifts further isolate the region from the world beyond.

Environmental Geography: A Vast and Varied Land

Size, latitude, and topography create a distinctive physical setting within the Russian region (Figure 9.2). The fact that the region occupies a major portion of the world's largest landmass means vast distances separate people and resources. Particularly within Russia, thousands of miles and many days of road or rail travel add immeasurable human and economic costs to the movement of people, goods, and services within the country. But size is also a blessing: varied natural resources are widely distributed across the vast Russian landscape and the region can boast some of Eurasia's best farmlands, metals resources, and petroleum reserves.

The region's northern latitudinal position is equally significant in affecting basic geographies of climate, vegetation, and agriculture. Indeed, the Russian region provides the world's largest example of a high-latitude continental climate where seasonal temperature extremes and short growing seasons profoundly limit opportunities for human settlement. Large, cold, dry arctic high-pressure systems become anchored over the Russian interior during the winter, providing some of the coldest temperature readings on the planet. Adding to the region's climatic extremes, the sprawling European Plain distances most of Russia from the modifying effects of the North Atlantic and Baltic Sea. Rugged mountains to the south and east further fragment the region and isolate it from other potential maritime influences. In terms of latitude, Moscow is positioned as far north as Ketchikan, Alaska, and even the Ukrainian capital of Kiev would sit north of the Great Lakes in Canada. Thus, apart from a subtropical zone near the Black Sea, much of the region experiences a classic continental climate with hard, cold winters and marginal agricultural potential (Figure 9.3).

The European West

European Russia, Belarus, and Ukraine cover the eastern portions of the vast European Plain that runs from southwest France to the Ural Mountains. The northern two-thirds of this area was covered by glaciers during the Pleistocene ice age

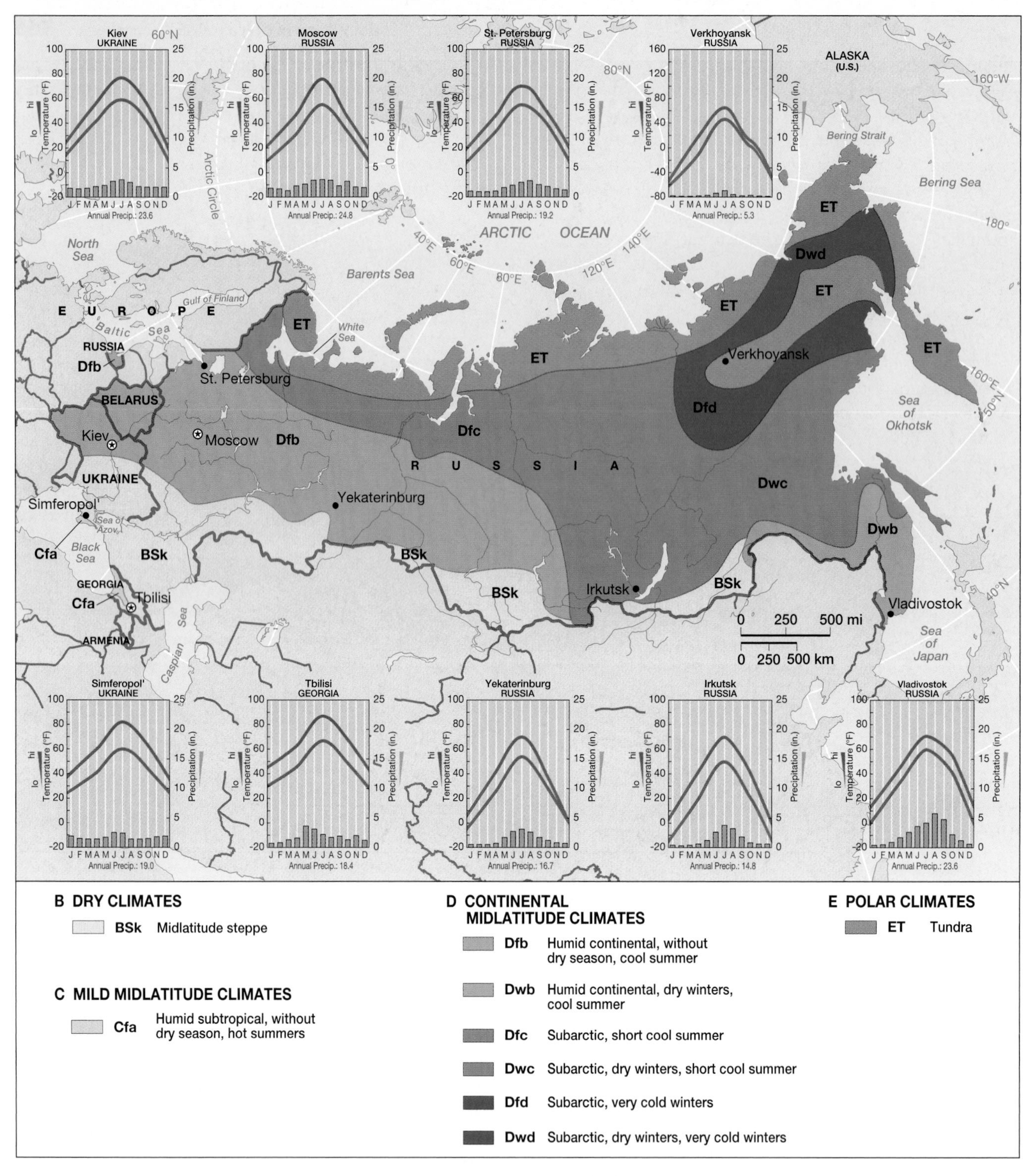

▲ **Figure 9.3 Climate map of Russia and its neighbors** The region's northern latitude and large land mass suggest that continental climates dominate. Indeed, farming is greatly limited by short growing seasons across much of the region. Aridity imposes limits elsewhere. Only a few small zones of mild midlatitude climates are found on the warming shores of the Black Sea in the far southwestern corner of the region, producing subtropical conditions in western Georgia.

and is characterized by low elevations, subdued topography, and extensive areas of poorly drained land. Europe's largest wetland, by a wide margin, is the Pripyat marsh of southern Belarus and northern Ukraine. The south was not glaciated and contains more hills, but it too is distinguished by low elevations. Mountains are located only in Russia's extreme south and in the far southwest of Ukraine (the Caucasus and Carpathian mountains, respectively). Ukraine also includes a small slice of territory, called Ruthenia, west of the Carpathian Mountains on the Hungarian plain.

The smoothly rolling terrain, low levels of evaporation, and moderately abundant precipitation of Ukraine, Belarus, and most of European Russia give rise to broad, gently flowing rivers. These waterways have long been important transportation arteries. One of the major geographical advantages of European Russia is the fact that different river systems, all now linked by canals, flow into four separate drainages. The Dnieper and Don rivers flow into the Black Sea; the West and North Dvina rivers drain into the Baltic and White seas, respectively; and the Volga River runs to the Caspian Sea (Figure 9.4). Russian civilization was historically integrated along these riverine passageways, and much of Russian history can be understood as the struggle to gain control not only of the rivers but also of the seas into which they flow.

Even though European Russia has a milder climate than Siberia, most of it still experiences cold winters and cool summers by North American standards. Moscow (Figure 9.3), for example, is about as cold as Minneapolis in January yet not nearly as warm in July. In Ukraine, Kiev is milder, however, and Simferopol' near the Black Sea offers wintertime temperatures that average more than 20° warmer than those of Moscow.

Three distinctive environments shape agricultural potential in the European West (Figure 9.5). North of Moscow and St. Petersburg, poor soils and cold temperatures severely limit farming, and much of the land remains in coniferous forest. Belarus and central portions of European Russia possess longer growing seasons, but acidic **podzol soils,** typical of northern forest environments, limit agricultural output and the ability of this region to support a highly productive farm economy. Still, diversified agriculture across the region includes grains (rye, oats, and wheat), potato cultivation, swine and meat production, and dairying. South of 50° latitude, agricultural conditions improve further across much of Ukraine and southern Russia. Forests gradually give way to steppe environments dominated by grasslands and by fertile "black earth" **chernozem soils** that have proven highly productive for commercial wheat, corn, and sugar beet cultivation and commercial meat production (Figure 9.6). As one approaches the Russian shoreline of the Caspian Sea, however, more desertlike conditions are encountered, limiting agriculture to irrigated tracts and to extensive livestock grazing.

▲ **Figure 9.4 Volga valley** The city of Nizhniy Novgorod lies along the shores of the Volga River east of Moscow. The huge Volga Basin remains a center of Russian settlement and economic development. It drains much of western Russia and empties into the Caspian Sea. *(Tass/Sovfoto/Eastfoto)*

The Ural Mountains and Siberia

The Ural Mountains mark European Russia's eastern edge, separating it from Siberia, or Asian Russia. Despite their geographical significance as the traditional division between continents, the Urals are not a particularly impressive range; several of their southern passes are less than 1,000 feet (305 meters) high, and railroad travelers sometimes fail to notice the passage. The setting is even less conducive to farming than farther west: the city of Yekaterinburg is distinctly colder and drier than Moscow, reflecting the increasingly continental climate of the Russian interior. Although not agriculturally productive, the range is still significant for two reasons: its ancient rocks are heavily mineralized and it once marked the eastern cultural boundary of the Russian region.

East of the Urals, the vastness of Russian Siberia unfolds across the landscape for thousands of miles. The great Arctic-bound Ob, Yenisey, and Lena rivers drain millions of square miles of northern country that includes the level and poorly drained West Siberian Plain, the hills and plateaus of the Central Siberian Uplands, and the rugged and isolated Northeast Highlands. Wintertime climatic conditions vary between legendary Verkhoyansk, with average January minimum temperatures of –58 °F (–50 °C) to the more moderate readings found in southern Siberian settlements, such as Irkutsk (–6°F/–21°C) near Lake Baikal. Most precipitation falls in the summer across Siberia, but much of the region receives less than 20 inches (51 centimeters) of moisture annually.

Siberian vegetation and agriculture reflect the climatic setting. The northern portion of the region is too cold for tree growth and instead supports tundra vegetation, which is characterized by mosses, lichens, and a few ground-hugging flowering plants. South of the tundra, the Siberian taiga, or coniferous forest zone, dominates a large portion of the Russian interior (Figure 9.7). Indeed, more than 20 percent of the world's forested area lies within the region. The trees of the taiga are generally small and slow-growing fir, spruce, and larch, as the climate is harsh and the soils generally acidic and poor in nutrients. Only localized agriculture (marginal wheat farming and potato cultivation) is possible even within favored portions of the zone. In the east, much of the tundra and taiga regions also are associated with **permafrost,** a cold-climate condition of unstable, seasonally frozen ground that limits the growth of vegetation and makes problematic the construction of even simple railroad tracks. Conditions moderate across southwestern Siberia: longer growing seasons and better soils offer more agricultural opportunities, although precipitation decreases along the border with Kazakhstan. Even with less dependable precipitation, southwest Siberia offers a narrowed eastward extension of the productive grain-growing belts of the western Russian steppe.

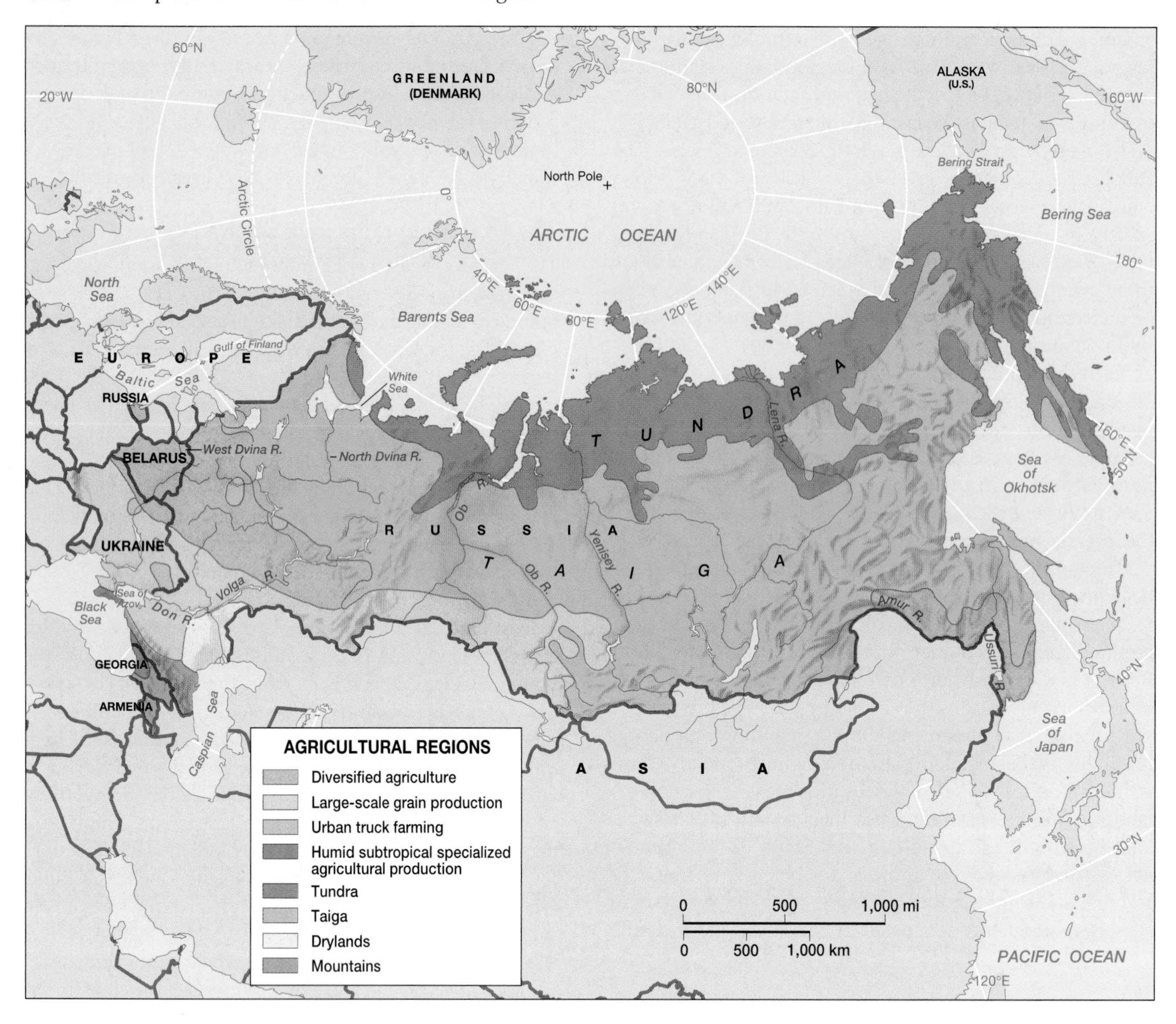

▲ **Figure 9.5 Agricultural regions** Climate and poor soils combine to limit agriculture across much of the Russian region. Better farmlands are found in Ukraine and in European Russia south of Moscow. Portions of southern Siberia support wheat production, but yield marginal results. In the Russian Far East, warmer climates and better soils translate into higher agricultural productivity. *(Modified from Clawson and Fisher, 1998,* World Regional Geography, *Upper Saddle River, NJ: Prentice Hall)*

The Russian Far East

Proximity to the Pacific Ocean, a more southerly latitude, and a pair of fertile river valleys create a distinctive subregion within the Russian Far East. About the same latitude as North America's New England, the region features longer growing seasons and milder climates than those found to the west or north. Here, the continental climates of the Siberian interior meet the seasonal monsoon rains of East Asia. Vladivostok is considerably warmer and wetter than Irkutsk, and the fertile Ussuri and Amur River valleys offer ample opportunities for mixed crop and livestock farming. The Amur forms a considerable portion of the Russian/Chinese border in the Far East and is the seventh longest river in the world. It is also a fascinating zone of ecological mixing, where conifers of the taiga mingle with Asian hardwoods and where reindeer, Siberian tigers, and leopards find common ground.

The Caucasus and Transcaucasia

In European Russia's extreme south, flat terrain gives way first to hills and then to the Caucasus Mountains, a large range stretching between the Black and Caspian seas. The highest point in the range, Mt. Elbrus, reaches higher than 18,000 feet (5,486 meters), and many peaks in the central and western Caucasus hold glaciers. The Caucasus marks Russia's southern boundary: farther south in Transcaucasia lie Georgia and Armenia as well as a distinctive natural setting. Extensive lowlands and low plateaus are found to the south of the Caucasus, and a second less-formidable mountain range, aptly named the Lesser Caucasus, runs through the southern part of Georgia and along the boundary between Armenia and Azerbaijan. South of the Lesser Caucasus, most of the rest of Armenia is a complex zone of diverse uplands.

▲ **Figure 9.6 Commercial wheat production** Ukraine possesses some of the region's best cropland, including a sizable zone of commercial wheat production. Highly mechanized operations improve farm productivity, both on state-controlled and privately managed acreage. *(Tass/Sovfoto/Eastfoto)*

▲ **Figure 9.7 Siberian taiga** Siberia's vast coniferous forests stretch from the Urals to the Pacific. Known as the taiga, these fir, spruce, and larch forests lie in a zone too cold for commercial agriculture. Lumbering and industrial pollution increasingly threaten this national resource. *(Andrey Zvoznikov/Hutchison Picture Library)*

Patterns of both climate and terrain in the Caucasus and Transcaucasia are tremendously complex. Rainfall is generally high in the western zone, with some slopes supporting dense forests. The area's eastern valleys, on the other hand, are semiarid or arid. In areas of adequate rainfall, or where irrigation is possible, agriculture can be quite productive. Georgia in particular has long been a noted producer of fruits, vegetables, flowers, and wines (Figure 9.8). Bordering the Black Sea, western Georgia is dominated by a fertile alluvial lowland, whereas central and eastern Georgia contain somewhat higher country wedged between the Caucasus and Lesser Caucasus ranges. The Georgian capital of Tbilisi enjoys a rare subtropical climate within the region that averages 30° warmer in the winter than bone-chilling Moscow. Armenia, located mostly to the south of the Lesser Caucasus, is somewhat drier and is dominated by grains, potatoes, and fruits.

A Devastated Environment

Another distinctive characteristic of the physical setting within the region is its extensive, often harmful modification by an increasingly urban and industrial population. The breakup of the Soviet Union and subsequent opening of the Russian region to international public scrutiny revealed some of the world's most severe environmental degradation (Figure 9.9). The frenetic pace of seven decades of Soviet industrialization took its toll across the region. Even in some of the most remote reaches of Russia, careless mining and oil drilling, the spread of nuclear contamination, and rampant forest cutting have resulted in frightening environmental damage. New Russian environmental and antinuclear movements have protested these ecological disasters, but to date these movements remain a minor political voice in a region dominated by concerns for economic growth. Indeed, the magnitude of many of these environmental challenges is so great that they have global implications and may affect world climate patterns, water quality, and nuclear safety. For example, since the 1980s, the global environmental costs of Siberian forests lost to lumbering and pollution may have exceeded the more widely publicized destruction of the Brazilian rain forest.

Air and Water Pollution Poor air quality plagues hundreds of cities and industrial complexes throughout the region. The Soviet penchant for building large clusters of industrial processing and manufacturing plants in concentrated areas, often with minimal environmental controls, has produced an ongoing legacy of fouled air that stretches from Belarus to

▲ **Figure 9.8 Subtropical Georgia** The modifying influences of the Black Sea and a more-southern latitude produce a small zone of humid subtropical agriculture in Georgia. The verdant landscapes of these tea plantations offer a sharp contrast to the colder country found north of the Caucasus. *(Sovfoto/Eastfoto)*

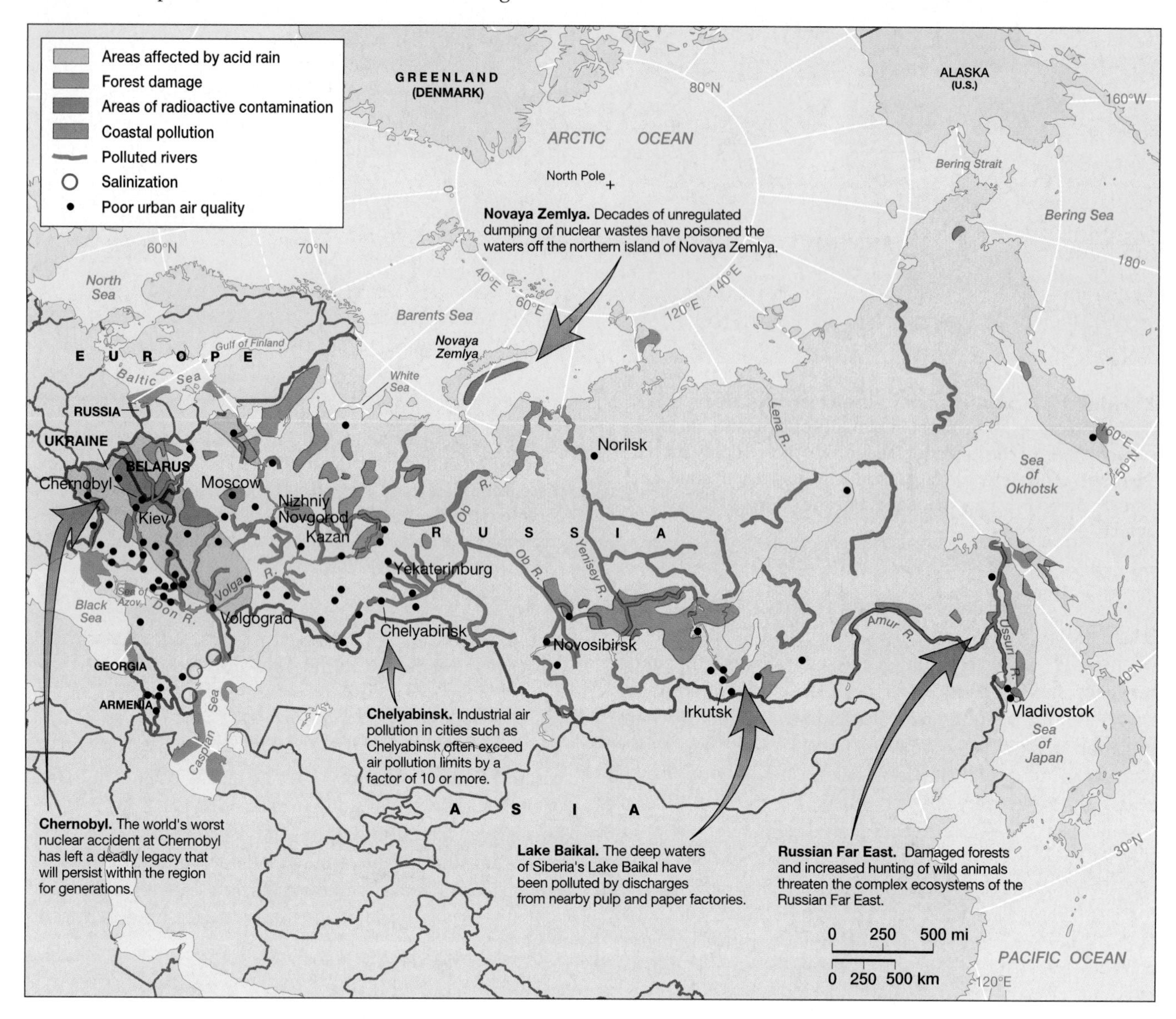

▲ **Figure 9.9 Environmental issues in Russia and its neighbors** Varied environmental hazards have left a legacy of a rapid, but often careless pattern of industrialization during the twentieth century. The landscape has been littered with nuclear waste, heavy metals, and air pollution. Fouled lakes and rivers pose additional problems in many localities. Present economic difficulties and political uncertainties only add to the costly challenge of improving the region's environmental quality in the twenty-first century.

Russian Siberia (Figure 9.10). A traditional reliance on abundant, but low-quality coal also contributes to pollution problems. The air in dozens of cities within the region typically fails to meet health standards for air quality, particularly in the winter when cold-air inversions trap the polluted atmosphere for days on end. Siberia's city of Norilsk is perhaps the largest single source of air pollution on the planet. Its smelters pump tons of sulfur dioxides and heavy metals into the atmosphere, creating a plume of poisonous air that can be traced to northern Canada.

Degraded water is another hazard that residents of the region must cope with daily. Municipal water supplies are constantly vulnerable to industrial pollution, flows of raw sewage, and demands that increasingly exceed capacity. Elsewhere, oil spills and seepage have harmed thousands of square miles in the tundra and taiga of the West Siberian Plain and along the Ob River estuary. One Russian scientist estimates that 10 percent of Russian oil production leaks everyday, the equivalent of an *Exxon Valdez* oil spill into Siberia every six hours. Water pollution is also significant along the northern coast of the Black Sea and along portions of the Caspian Sea shoreline.

The pollution of Siberia's Lake Baikal attracted international attention in the 1970s and 1980s. Lake Baikal, the world's largest reserve of fresh water, is a remarkable natural feature. Not only is the lake almost 400 miles (644 kilometers) long, but it is also 5,300 feet (1,615 meters) deep, occupying a structural rift in the continental crust (Figure 9.11). Lake Baikal is home to a large array of unique species, including the world's only freshwater seal. Until recently, Baikal's water was exceptionally pure. In the 1950s and 1960s, how-

▲ **Figure 9.10 Russian air pollution** The mineral-processing center of Norilsk achieved a dubious reputation during the Soviet period as one of the dirtiest cities in Siberia, if not the entire Northern Hemisphere. Toxic air and water pollution still plague the site today. *(Bilderberg Archiv der Fotografen)*

▲ **Figure 9.11 Lake Baikal** Southern Siberia's Lake Baikal is one of the world's largest deep-water lakes. Industrialization decimated water quality after 1950 as pulp and paper factories poured wastes into the lake. Recent cleanup efforts have helped, but many environmental threats remain. *(Digital Image © 1996 Corbis)*

ever, the Soviet government built pulp and paper factories along its shores, attracted by both the lake's clean water (useful for producing high-quality wood fibers) and the abundant forests in the surrounding uplands. With factory discharges, the lake's purity rapidly declined. By the 1970s, Russian ecologists warned that Baikal's entire ecosystem was threatened. Owing in part to the resulting international pressure, pollution from the paper mills was reduced. The lake, however, is by no means out of danger.

The Nuclear Threat The nuclear era brought its own particularly deadly dangers to the region. The Soviet Union's aggressive nuclear weapons and nuclear energy programs expanded greatly after 1950 and issues of environmental safety were often ignored. Northeast Siberia's Sakha (or Yakutia) region, for example, suffered regular nuclear fallout in the era of aboveground nuclear testing. In other areas, nuclear explosions were widely utilized for simple seismic experiments, oil exploration, and dam-building projects. The once-pristine Russian Arctic has also been poisoned. During the Soviet era, the area around the northern island of Novaya Zemlya served as a huge and unregulated dumping ground for nuclear wastes. Aging nuclear reactors also dot the landscape, often contaminating nearby rivers with plutonium leaks. Nuclear pollution is particularly pronounced in the northern Ukraine, where the Chernobyl nuclear power plant suffered a catastrophic meltdown in 1986 (see "Environmental History: The Legacy of Chernobyl"). Large areas of nearby Belarus were also devastated in the Chernobyl disaster, contaminating soils across much of the southern part of that country. Elsewhere, a nuclear accident within the Urals during the 1950s also devastated an area of several hundred square miles.

The Post-Soviet Paradox The end of Soviet control brings conflicting influences upon the region's environment. The demise of the Soviet Union brought about environmental improvement in some areas. Many factories shut down because they were no longer economically viable, which itself reduced pollution. The huge steel mills in the Russian city of Magnitogorsk, for example, produce less than half the raw steel they did a decade ago, and global competition threatens to reduce demand further. Ironically, cleaner air has been the result. Although costly, advanced pollution control equipment is also beginning to be imported from western Europe. Elsewhere in Russia, nuclear warhead storage facilities have been consolidated, and government authorities are responsible for maintaining control over the nation's 22,000 nuclear weapons. There is a growing environmental consciousness among young educated Russians and Ukrainians. Increased connections with environmental activists in North America and Europe have also raised public awareness of many environmental problems within the region.

In other areas, however, the breakdown of centralized authority has only contributed to more environmental degradation. Waste materials are often handled more casually than in the past, since the government no longer has much effective regulatory power. An especially vexing problem is the disposal of nuclear byproducts, now sometimes smuggled out of the country and possibly used in illegal weapons manufacturing. The current economic troubles of the former Soviet world also have brought about an accelerated exploitation

ENVIRONMENTAL HISTORY The Legacy of Chernobyl

Northern Ukraine's Chernobyl nuclear plant experienced a deadly meltdown on April 25, 1986 (Figure 1). The reactor burned for 16 days, pouring smoke two miles into the sky and spreading nuclear contaminants from southern Russia to northern Norway. It proved to be the world's worst nuclear accident and one of the greatest environmental disasters of the modern age, and it will continue to impact the ecological health of the region for decades to come. Vladimir Chernousenko, a Ukrainian nuclear physicist, led attempts to clean up the disaster. He describes the process:

As soon as the explosion happened, troops were placed around the area. The government put a lid on the event immediately, and millions were not evacuated in time. . . . I was called in by Mikhail Gorbachev to evaluate what had happened. When I concluded my investigation, I sent a three-volume report to Gorbachev. Immediately, it became a secret document. . . . 65 million people in Russia received a dose, 90 million people north of the Ukraine may have been contaminated, and as many as 7,000 died immediately. . . . A million and a half people in and around Chernobyl (including the people who cleaned up the site) received extremely high doses of radiation, and millions of others still receive internal radiation daily from food contamination. Prior to the Chernobyl disaster, Ukraine had been the breadbasket of Europe; now there is no way to clean up the soil.

Chernousenko reports on the discouraging aftermath of the disaster:

Since Chernobyl, childhood and animal diseases in my country have increased fourfold. . . . In the years to come, many people are going to die. . . . There have been an estimated 200 accidents in nuclear installations in the former USSR, with millions of curies released. In my country not one square inch is free of radioactive fall-out. People are losing their hair, and blood is coming out their mouths. Nuclear power stations are dangerous . . . even when they do not blow up. My assistants and I researched 10 plants, and we consistently found the water polluted and people around the plants sick.

▲ **Figure 1 Aerial view of Chernobyl** One of the world's greatest environmental nightmares unfolded in April 1986 when the Chernobyl nuclear reactor experienced a meltdown. Nearby portions of northern Ukraine, Belarus, and Russia remain a toxic testimony to the disaster. *(Tass/Sovfoto/Eastfoto)*

Chernousenko was dying of cancer as he recalled the event, one of thousands to perish, either directly or indirectly, from one of the greatest environmental disasters ever.

Source: Adapted from "Ten Years Later, Chernobyl Is as Deadly as Ever," *Utne Reader,* May–June, 1996.

of natural resources. Russia must now frantically export oil and other minerals, as well as timber, in order to obtain the cash necessary to operate in the global economy. In the resulting resource rush, few safeguards have been implemented or enforced. Russia also faces an impending crisis of wildlife extinction. The old Soviet regime had some success in protecting both endangered species and sizable areas of natural habitat noted for their biological diversity. Unregulated hunting and trapping are now on the rise, however, and in some areas are virtually uncontrolled. It is questionable whether such animals as the Siberian tiger and the Siberian leopard, both inhabiting the forests of Russia's Far East, will survive the next few decades.

Population and Settlement: Changing Geographical Patterns

The five states of the Russian region are home to more than 200 million residents, who are widely dispersed across a vast Eurasian landmass. The region's evolving political setting, distinctive distributions of natural resources, and changing migration patterns have shaped its population geography. The results produced a population strongly concentrated in the European West, periodic impulses to disperse beyond that traditional core, and an overall tendency toward rapid urbanization. Even as many cities have grown from migration, however, longer-term demographic trends reveal slowing population increases or even population declines in a variety of settings. Future settlement patterns will reflect ongoing processes of urbanization, but the precise regional settings for this urban growth and more detailed patterns of urban land-use and landscape changes will depend upon shifting political and economic conditions within the region.

Population Distribution

Striking differences in population densities exist between European and Asian portions of the Russian region. The more favorable agricultural setting of the European West historically encouraged higher densities of population versus the more inhospitable conditions found across central and northern Siberia. Although Russian efforts over the past century have encouraged a wider dispersal of their population, it remains heavily concentrated in the west (Figure 9.12). European Russia is home to 115 million persons, while Siberia, although far larger, holds only some 32 million. When one adds the 60 million inhabitants of Belarus and Ukraine, the imbalance between east and west becomes even more striking.

The European Core The region's largest cities, biggest industrial complexes, and most productive farms are located in the European Core, supporting population densities that greatly exceed those to the east, although they are modest by Northwest European or East Asian standards. Moscow's sprawling city and nearby satellite centers clearly dominate the settlement landscape with a metropolitan area of more than 8.5 million people (Figure 9.13). Within 250 miles, a series of other major urban centers are closely linked to Moscow. Largest of these is the industrial city of Nizhniy Novgorod (formerly named Gor'kiy in the Soviet period) (1.4 million), traditionally oriented around automobile and heavy equipment manufacturing.

Beyond Moscow's immediate orbit, three other areas of concentrated urban settlement dominate European Russia (Figure 9.12). On the shores of the Baltic Sea, St. Petersburg (formerly Leningrad in the Soviet period) (4.8 million), offers a major trading window to the West. Between 1712 and 1917, it served as the capital of the Russian Empire, and it acquired a rich skyline of baroque architecture and beautiful churches, which gave the city an urban landscape many have compared to the great cities of western Europe. Although industrialization and Soviet-style buildings took their toll on the city during the Communist era, St. Petersburg has recently seen an

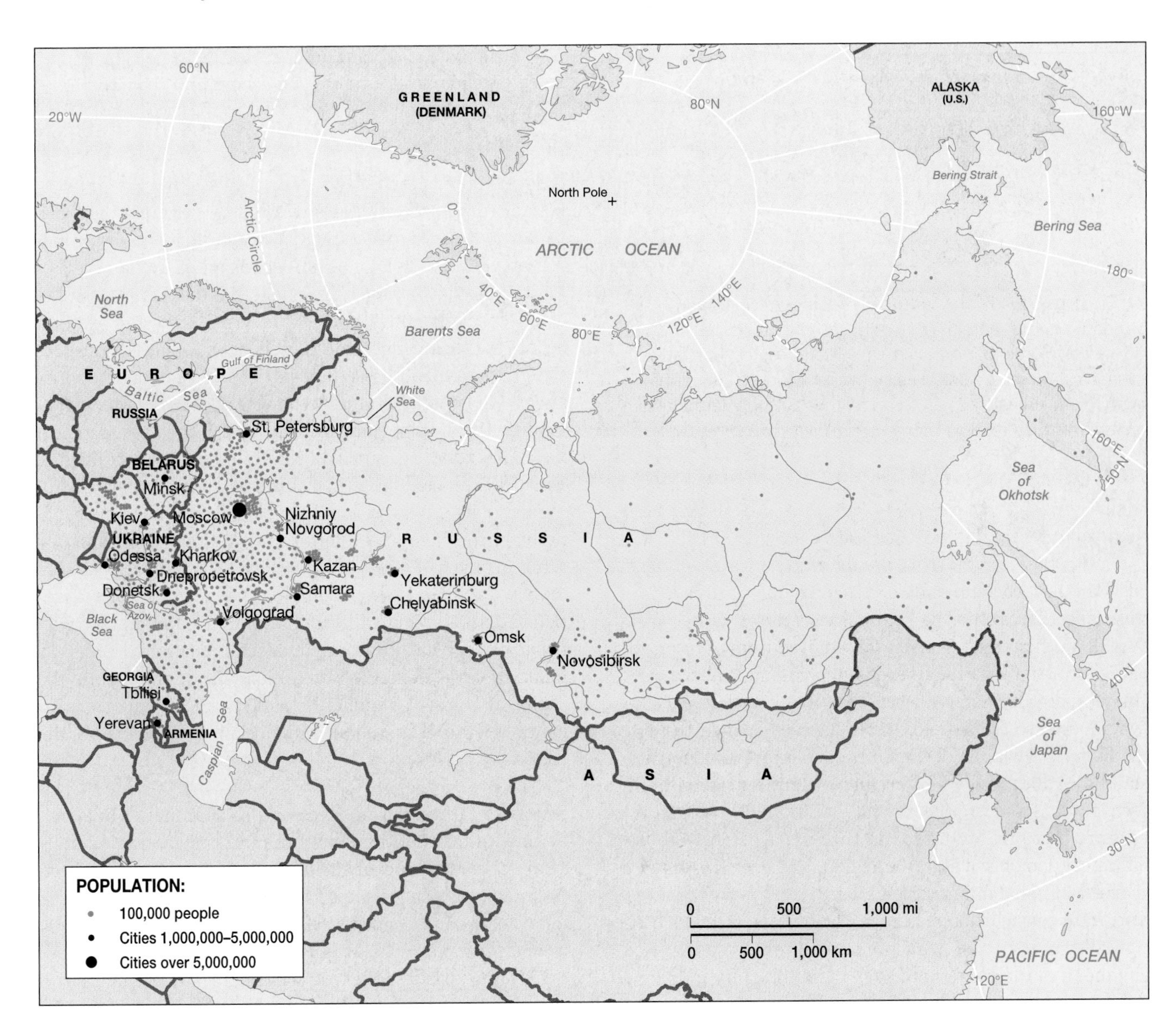

▲ **Figure 9.12 Population map of Russia and its neighbors** Population within the region is strongly clustered west of the Ural Mountains. Dense agricultural settlement, extensive industrialization, and large urban centers are found in Ukraine, much of Belarus, and across western Russia south of St. Petersburg and Moscow. A narrower chain of settlements follows the better lands and transportation corridors of southern Siberia, but most of Russia east of the Urals remains a sparsely settled land.

▲ **Figure 9.13 Metropolitan Moscow** Sprawling Moscow extends more than 50 miles (80 kilometers) beyond the city center. Home to more than 8.5 million people, the relative strength of Moscow's urban economy continues to attract migrants from elsewhere in the country, thus putting more pressure on its infrastructure. *(CNES/Spot Image/Photo Researchers, Inc.)*

▲ **Figure 9.14 St. Petersburg** Picturesque Canal Street captures some of the architectural character that makes St. Petersburg one of Russia's most beautiful cities. Despite the construction of many concrete buildings in the Soviet period, St. Petersburg retains a charm that echoes the urban landscapes of western Europe. *(Nick Nicholson/The Image Bank)*

architectural renaissance, interest in preserving its Russian Orthodox churches, and increased tourism (Figure 9.14).

Far southeast of Moscow, a second urban focus is oriented along the lower and middle stretches of the Volga River. Industrialization within the region accelerated greatly during World War II, as the region lay far removed from German advances in the West. Today, the highly commercialized river corridor, also blessed with nearby petroleum reserves, supports a diverse industrial base strategically located to serve the large populations of the European Core. From north to south, the four Volga Valley cities of Kazan (1.1 million), Samara (1.2 million), Saratov (900,000), and Volgograd (1 million; formerly Stalingrad in the Soviet period) are the largest settlements within the region, and each possesses a sizable industrial infrastructure.

A third constellation of key population centers on the eastern edge of the Core is anchored along the resource-rich alignment of the Ural Mountains. From the gritty industrial landscapes of Serov (1.0 million) and Yekaterinburg (1.3 million; formerly Sverdlovsk in the Soviet period) in the north to Chelyabinsk (1.1 million) and Magnitogorsk (425,000) in the south, the Urals region specializes in iron and steel manufacturing, metals smelting and refining, and heavy machinery construction. Outside of these industrial centers, however, the rural population density of the Urals is lower than in the better agricultural lands found farther west within the Core.

Beyond Russia, major population clusters within the European Core are also found in Belarus and Ukraine. The Belorussian capital of Minsk (1.7 million) is the dominant urban center in that country, and its drab Soviet-style appearance echoes the close economic and political ties that still exist between Belarus and Russia (Figure 9.15). Other industrial cities also dot Belarussia, another legacy of the considerable Soviet-era investment that transformed the regional landscape. In nearby Ukraine, the capital of Kiev (2.6 million) straddles the Dnieper River, and the city's rich architectural heritage is a reminder of its historic role in the political and economic geography of the European interior. In addition, large industrial centers, such as Kharkov (1.6 million), Dnepropetrovsk (1.2 million), and Donetsk (1.1 million), are located in resource-rich eastern Ukraine and benefit from their proximity to deposits of coal and iron ore. Smaller population clusters are found in Georgia's fertile Black Sea lowlands and around the capital of Tbilisi (1.3 million). To the east, the Armenian capital of Yerevan (1.3 million) contains almost one-third of the nation's population.

Siberian Hinterlands Leaving the southern Urals city of Chelyabinsk on the Siberia-bound train, one is acutely aware that the land ahead is ever more sparsely settled (Figure 9.12). The distance between cities grows, and the intervening countryside reveals a landscape shifting gradually from farms to forest. The Siberian hinterland is divided into two characteristic zones of settlement. To the south, an alignment of isolated, but sizable urban centers follows the **Trans-Siberian Railroad**, a key railroad corridor completed to the Pacific in 1904. The industrial settlements along the route benefit from good east–west connections as well as from proximity to natural resources such as oil, natural gas, coal, and iron ore. The eastbound traveler encounters Omsk

▲ **Figure 9.15 Minsk** Almost two million people reside in the Belorussian capital of Minsk. Its drab apartments and office buildings echo the strong shaping legacy of the Soviet period upon the urban scene. Indeed, Belarus retains many of its Soviet-style characteristics, trailing far behind Russia in the pace of economic reforms. *(Sovfoto/Eastfoto)*

(1.2 million) as the rail line crosses the Irtysh River, Novosibirsk (1.4 million) at its junction with the Ob River, and Irkutsk (600,000) near the southwest corner of Lake Baikal. More than 1,500 miles (2,415 kilometers) beyond, the better agricultural lands and industrial opportunities of the Russian Far East contribute to higher population densities. Khabarovsk (620,000) is the leading settlement of the Amur Valley, while the port city of Vladivostok (660,000) provides the leading Russian avenue to the Pacific. In addition, a thinner sprinkling of settlement appears along the more recently completed (1984) **Baikal-Amur Mainline (BAM) Railroad** that parallels the older line, but runs north of Lake Baikal to the Amur River.

North of the BAM line, however, the seemingly empty vastness of the Siberian hinterland dominates the scene. Settlements are few and far between, and larger urban areas are usually either regional administrative capitals such as Yakutsk (225,000) or resource extraction and processing centers such as Norilsk (270,000). Lying beyond the Arctic Circle, Norilsk is one of the world's least-livable mining centers, but its isolation and frigid winters are compensated by rich nearby deposits of copper, nickel, and platinum. Elsewhere, sprinklings of native population have long occupied the Siberian hinterland, but their numbers are miniscule when compared to the Russian-dominated population clusters farther to the south and west.

Regional Migration Patterns

Over the past 150 years, millions of people within the Russian region have been on the move. These major migration patterns, both forced and voluntary, reveal sweeping examples of human mobility that rival the great emigrations from Europe and Africa or the transcontinental spread of settlement across North America. Three migratory impulses have proven particularly formative in the region's modern human geography. First, the eastward expansion of Russian peoples has been an epic chapter within the region that spans the imperial, communist, and contemporary eras. Second, the Soviet period witnessed various politically motivated migration schemes designed to consolidate and enhance Russian communist control. Finally, as in many other global settings, residents of the region have gravitated toward cities, thus fundamentally reshaping geographies of settlement during the twentieth century.

Eastward Movement Just as European North Americans moved west across the continent, exploiting natural resources and displacing native peoples, European Russians moved east across the vast Siberian frontier to extend their influence and to rework basic population geographies within the Eurasian interior. Although the deeper historical roots of the movement extend back several centuries, the pace and volume of the eastward drift appreciably accelerated in the late nineteenth century once the Trans-Siberian Railroad was completed. Peasants were attracted to the region by its agricultural opportunities (in the south) and by greater political freedoms than they traditionally enjoyed under the **tsars** (or czars; Russian for *Caesar*), the authoritarian leaders who dominated politics during the pre-1917 Russian Empire. Almost one million Russian settlers moved into the Siberian hinterland between 1860 and 1914.

The eastward migration continued during the Soviet period, once communist leaders consolidated power during the late 1920s and saw the economic advantages of developing the region's rich resource base. The German invasion of European Russia during World War II demonstrated that there were also strategic reasons for settling the eastern frontier, and this propelled further migrations during and following the war. Indeed, by the end of the Communist era, 95 percent of Siberia's population was classified as Russian (including Ukrainians and other western immigrants). With the completion of the BAM Railroad in the 1980s, yet another focus of settlement opened in Siberia, prompting another new migratory stream into a region previously remote from the outside world.

Political Imperatives Political motives have also shaped migration patterns within the region. Particularly in the case of Russia, leaders from both the imperial and Soviet eras saw advantages in moving selective populations to new locations. Clearly, for example, the infilling of the southern Siberian hinterland had a political as well as an economic rationale. Both the tsars and the Soviet leaders saw their geopolitical fortunes rise as Russians moved into the resource-rich Eurasian interior. For some, however, the move to Siberia had a different connotation. The region became a repository for political dissidents and troublemakers. Especially in the Soviet period, uncounted millions were forcibly relocated to the region's infamous **Gulag Archipelago**, a vast collection of political prisons in which inmates often disappeared or spent years far removed from their families and home communities (Figure 9.16). The Communist regime of Joseph Stalin

▶ **Figure 9.16 Soviet prison camp** For decades under communism, political dissenters and disaffected intellectuals were sent to prison camps far from the mainstream of Soviet society. Dozens of camps were spread from European Russia to the far-flung corners of frozen Siberia. Thousands of prisoners were finally released as political conditions improved under Gorbachev during the 1980s. Recently uncovered security records reveal the vast extent of these prison camps during the Soviet period. *(Itar-Tass/Sovfoto/Eastfoto)*

(1928–53) is particularly noted for its forced migrations, including the removal of thousands of Jews to the Russian Far East between 1928 and 1958 and the involuntary relocation of numerous ethnic minorities during World War II. Toward the end of the Soviet era, less restrictive political policies brought an end to many of these abuses and allowed for the freer movement of people within and beyond the region. One recent result has been the greatly increased emigration of Jews from Russia to Israel.

Russification, the Soviet policy of resettling Russians into non-Russian portions of the Soviet Union, also had profound consequences for the region's human geography. Millions of Russians were given economic and political incentives to move elsewhere in the Soviet Union in order to increase Russian dominance in many of the outlying portions of the country. The migrations were geographically selective in that most of the Russians moved either to administrative centers or industrial complexes. As a result, by the end of the Soviet period, Russians made up significant minorities within former Soviet republics (now independent nations) such as Kazakhstan (38 percent Russian), Latvia (34 percent), and Estonia (30 percent). Among its Slavic neighbors, Belarus remains 13 percent Russian and Ukraine more than 22 percent Russian, with concentrations particularly high in the eastern portions of the country. Often, these Russian workers were given special employment and housing privileges in their new localities, thus provoking antipathy among the non-Russian population. Since the end of the Soviet Union, however, several of the newly independent non-Russian countries have imposed rigid language and citizenship requirements, which have reversed the historic Russian flow. By the late 1990s, more than five million Russians have returned to their homeland.

The Urban Attraction The Marxist philosophy embraced by Soviet planners encouraged urbanization. In 1917 the Russian Empire was still overwhelmingly rural and agrarian; fifty years later, the Soviet Union was primarily urban. Planners saw great economic and political advantages in efficiently clustering the population, and Soviet policies dedicated to large-scale industrialization obviously favored an urban orientation. Urbanization levels reached 66 percent in 1989, a lower figure than those of western Europe or North America, but one that included the more rural reaches of Central Asia. Today, Russian (73 percent), Ukrainian (68 percent), and Belarussian (69 percent) rates of urbanization are comparable to those of the industrialized capitalist countries (Table 9.1).

Soviet cities grew according to strict governmental plans. Planners selected different cities for different purposes. Some were designed for specific industries, while others were allocated primarily administrative roles. All cities were allotted set population levels. A system of internal passports prohibited people from moving freely from city to city; instead, people generally went where the government assigned them jobs. Moscow, the country's leading administrative city, thrived under the Soviet regime. It formed the undisputed core of Soviet bureaucratic power as well as the center of education, research, and the media. In fact, the Russian capital frequently exceeded its targeted population as both planned and unplanned migration gravitated to the nation's dominant metropolitan area. The impact of centralized administrative functions in the Soviet Union was also evident in the rise of the capital cities of its constituent republics. Yerevan, the capital of Armenia, was little more than a town in 1920, yet by the 1990s, it had grown to a major city of 1.3 million inhabitants. Minsk, the capital of Belarus, experienced a similar, although not quite as dramatic, expansion. Specialized industrial cities grew at an even faster pace. In the mining and metallurgical zone of the southern Urals, centers such as Yekaterinburg and Chelyabinsk mushroomed into major urban centers. Another cluster of specialized industrial cities, including Kharkov and Donetsk, emerged in the coal districts of eastern Ukraine.

With the end of the Soviet Union, people gained basic freedoms of mobility. This has led to some redistribution of pop-

Table 9.1 Demographic Indicators

Country	Population[a]	Natural Increase	TFR[b]	%<15[c]	% Urban
Armenia	3.8	0.6	1.6	28	67
Belarus	10.2	–0.4	1.3	21	69
Georgia	5.4	0.4	1.6	24	56
Russia	146.9	–0.5	1.2	20	73
Ukraine	50.3	–0.6	1.3	20	68

[a]Population in millions, 1998.
[b]Total fertility rate.
[c]Percentage of population younger than 15 years of age.
Source: *Population Reference Bureau. World Population Data Sheet,* 1998.

ulation from declining industrial areas to new centers of economic activity, but thus far, new mass migrations have not occurred. Many persons are reluctant to move because they often enjoy inexpensive, state-subsidized housing that they would risk losing (many forego paying rent and utilities altogether, yet remain secure in their lodgings). The police, moreover, severely punish vagrancy, and have discouraged the emergence of shantytowns or of a street-dwelling population in the more prosperous urban centers.

Inside the Russian City

Large Russian cities possess a core area, or center that features superior transportation connections; the best-stocked, upscale department stores and shops; the most desirable housing; and the most important offices (both governmental and private) (Figure 9.17). At the top of the urban hierarchy, large cities such as Moscow and St. Petersburg also feature extensive public spaces and examples of monumental architecture at the city center. Inner-city decay, so characteristic of the United States, is not a feature of the Russian central city. Russian cities also lack sprawling decentralized suburbs in the North American context, and thus tend to end abruptly. One often passes from a landscape of high-rise apartments into an essentially rural setting without traveling through an extensive zone of single-family dwellings. Indeed, recently this urban fringe zone has taken on a new look. One sobering reminder of the hard economic times are thousands of garden plots located just beyond the edge of most Russian cities. Doctors, teachers, and sales clerks have transformed the rural scene into intensely worked vegetable gardens that subsidize a dwindling ability to buy affordable food at the local grocery store.

Within the city, there is usually a distinctive pattern of concentric land-use zones, each of which was built at a later date as one moves outward from the center. Such a ringlike urban morphology is not a unique phenomenon, but it is probably more highly developed here than in any other part of the world, owing to the extensive power of government planners during the Soviet period. At their very center, the cores of many older cities predate the Soviet Union. Pre-1900 stone buildings often dominate older city centers. Some of these are former private mansions that were turned into government offices or subdivided into apartments during the socialist period but are now being reprivatized.

▲ **Figure 9.17 Downtown Moscow** The bustling traffic along central Moscow's Novy Arbat parallels urban scenes elsewhere in Europe and North America. The pace of new urban construction, however, slowed greatly in the late 1990s as the Russian economy struggled. *(Itar-Tass/Sovfoto/Eastfoto)*

Beyond this pre-socialist urban core, one can often find a ring of public housing projects and fully planned sotzgorods, or socialist neighborhoods. *Sotzgorods* are based on a close connection between workplace and home, and typically are supplied with spartan dormitory-style housing. These projects became most common in the heavy industrial cities that mushroomed during the early decades of the Soviet Union. Their age and their industrialized surroundings now make them among the least desirable places in which to live.

Another common urban zone, removed some distance from city cores, is that of the Chermoyuski. Chermoyuski are large, uniform apartment blocks built during the 1950s and 1960s. At this time a more prosperous Soviet Union could give each family its own small, private flat. Most of these nearly uniform apartment blocks are five stories or less in height, since they typically are not equipped with elevators. The main problem with the Chermoyuski is shoddy construction. Most were designed to last only a few decades, since Soviet planners expected that a booming economy would soon allow the construction of better, more permanent, dwellings. Today, many of these apartment complexes remain, but are now dilapidated with little money available for renovation.

Farther out from the city centers are the **mikrorayons,** the much-larger housing projects of the 1970s and 1980s (Figure 9.18). Mikrorayons are typically composed of massed blocks of standardized apartment buildings, ranging from 9 to 24 stories in height. Each mikrorayon was to form a self-contained community, with grocery stores and other basic services located within walking distance of each apartment building. The largest of these super-complexes contains up to 100,000 residents. While planners hoped that mikrorayons would foster a sense of community, most now serve largely as anonymous bedroom communities for urban cores. Although public transportation is well developed in most cities of the region, commuting time between city centers and these outer apartment complexes is usually 45 minutes or more.

▲ **Figure 9.18 Moscow housing** For many residents in larger Russian cities, home is a high-rise apartment house. Most of these satellite centers were built in the Soviet era. Poor construction and a lack of landscaping often yield a bleak suburban scene, but nearby stores, entertainment, and public transportation offer important amenities. *(Itar-Tass/Sovfoto/Eastfoto)*

The majority of the people of the Russian region live in an urban environment inherited from the Soviet period. Rents usually remain low, and a market in land and housing has been slow to develop. This does not mean, however, that no changes occurred in the urban fabric after the collapse of the Soviet Union. The most prominent transformations were symbolic: the renaming of streets (and even entire cities) and the destruction of public statuary. The old heroes of the Soviet Union, for the most part, were now to be reviled or forgotten. Other changes included the sudden proliferation of flea markets and other informal shopping areas, usually concentrated in public plazas or along major roads. For the political and economic elite, however, rural dachas, or country houses, continue to provide a quiet retreat from urban ills, much as they did during the Soviet period. For most urban Russians, however, a trip to the country is more likely to mean a visit to a small garden plot where a few potatoes or onions can be harvested for the next meal.

The Demographic Crisis

Declining populations, low birthrates, and rising mortality, particularly among middle-aged males, are all symptoms of a startling demographic crisis that has engulfed the region. While Georgia and Armenia are seeing very small increases in population, Russia, Belarus, and Ukraine are witnessing population declines (Table 9.1). Population changes were unpredictable in the region during the twentieth century. During World War II, large numbers of deaths combined with low birthrates to produce sizable population losses. While population increases accelerated in the 1950s, growth slowed by 1970, and death rates began exceeding birthrates in the early 1990s. If current trends continue, the population of Russia may decline by 30 million by the middle of the twenty-first century.

The region's demographic crisis appears related to the fraying social fabric and uncertain economic times. Very low birthrates within the region may reflect a lack of optimism about the future. In Russia, for example, total live births fell from a peak of 2.5 million in 1987 to 1.4 million in the middle 1990s, precisely during a period of tremendous economic and political disruption. Many Russian families report they simply do not have sufficient incomes to support children. The health of women of childbearing age has also declined, while problem pregnancies, maternal childbirth death rates, and birth defects are on the rise. Similarly, sharp increases in death rates, especially among Russian men, appear related to many stress-related diseases, such as alcoholism and heart disease. Murder and suicide rates have also climbed rapidly in the past 15 years. In addition, perhaps 20 to 30 percent of the rising death rates may be attributed to the region's increasingly toxic environment. Whatever the complex causes, the average life span for Russian women has fallen since the late 1980s from 75 to 73 years. For men, the drop has been truly unprecedented in the developed world, with life spans shrinking from 65 to 61 years. Until conditions in the region dramatically improve, it is doubtful that female fertility rates will rise from present levels.

Cultural Coherence and Diversity: The Legacy of Slavic Dominance

For hundreds of years, Slavic peoples speaking the Russian language expanded their influence from an early homeland in central European Russia. Eventually, this Slavic cultural imprint moved north to the Arctic Sea, south to the Black Sea and Caucasus, west to the shores of the Baltic, and east to the Pacific Ocean. In this process of diffusion, Russian cultural patterns and social institutions spread widely, and they also influenced scores of non-Russian ethnic groups that continued to live under the rule of the Russian Empire. The legacy of that Slavic expansion continues today. It offers Russians a rich historical identity, and it provides a meaningful context to understand how non-Russian cultures have evolved within the region.

The Heritage of the Russian Empire

The history of the Russian Empire is best understood within the context of early modern European imperial expansion. In the 1500s and 1600s, as Spain, Portugal, France, and Britain carved out empires in the Americas, the Russians expanded eastward across the great Eurasian landmass. In the 1800s, just as other European states were dividing up Africa and Southeast Asia among themselves, the Russians were advancing into Central Asia and along the Sea of Japan to the border of Korea. Unlike other European empires, however, that of the Russians formed one single territory, uninterrupted by oceans or seas. Partly because of its contiguous nature, the Russian Empire—after being transformed into the Soviet Union—was able to remain intact while the empires of the other European powers collapsed after World War II. Only with the fall of the Soviet Union in 1991 did this transformed empire finally begin to dissolve.

Origins of the Russian State The origin of the Russian state lies in the early history of the **Slavic peoples,** defined linguistically as a distinctive branch of the Indo-European language family. The Slavs originated in or near the Pripyat marshes of modern Belarus. Some 2,000 years ago they began to migrate to the east, extending as far as modern Moscow by A.D. 200. Somewhat later, westward migrations brought Slavs to central Europe, where they eventually gave rise to such nationalities as the Poles, Czechs, and Croatians. Slavic political power grew by A.D. 900 as they intermarried with southward-moving warriors from Sweden known as *Varangians,* or *Rus*. Within a century, their state of Rus extended from Kiev (the capital) in modern Ukraine, to Lake Ladoga near St. Petersburg. The new Russian state interacted with the rich and powerful Byzantine Empire of the Greeks. Byzantine influence brought Christianity, which the Russians began to accept shortly before the year A.D. 1000. Along with the new religion came many other aspects of Greek culture, including the Cyrillic alphabet. Even as the Russians converted to **Eastern Orthodox Christianity,** a form of Christianity historically linked to eastern Europe and church leaders in Constantinople (modern Istanbul), their Slavic neighbors to the west (the Poles, Czechs, Slovaks, Slovenians, and Croatians) accepted Catholicism. The resulting religious division split the Slavic-speaking world into two groups, one oriented to the west, the other to the east and south. Even today a certain cultural distance divides the Catholic Poles from the Eastern Orthodox Russians, just as the Catholic Croats differ from the Orthodox Serbs within eastern Europe.

The early Russian state soon faltered. It split into roughly a dozen independent principalities that then faced the common threat of an expanding Mongol Empire during the early thirteenth century. By the fourteenth century, a group of Turkish-speaking peoples (the Tatars) ruled the conquered Russians in a post-Mongol state centered in the lower Volga Valley.

Growth of the Russian Empire By the late 1400s, Slavic leaders in Moscow overthrew Tatar rule and established a new and unified Russian state. The new Russian state could no longer look to the Byzantine Empire for cultural and religious leadership, since that empire had itself fallen to Turkish forces a few decades earlier. Russia's rulers therefore claimed for themselves the imperial mantle. Henceforth Russia would be an empire ruled by a tsar seated in the capital city of Moscow (Figure 9.19). Under the influence of an evolving imperial ideology, nationalistic Russian thinkers concluded that they had a divine mandate to spread their power as widely as possible.

The core of the new Russian Empire lay near the eastern fringe of the old state of Rus. The former center around Kiev was now a war-torn borderland (or "Ukraine" in Russian) contested by the Orthodox Russians, the Catholic Poles, and the Muslim Turks. Gradually this area's language diverged from that spoken in the new Russian core, and Ukrainians and Russians therefore developed into two separate peoples. Eventually Ukraine was incorporated into the Russian Empire, but the identity of its inhabitants remained distinctive. A similar development took place among the northwestern Russians, who experienced several centuries of Polish rule and over time were transformed into a distinctive group known as the *Belorussians*. Thus the original Russians were divided into three distinctive but closely related peoples: the Russians proper, the Belorussians (of Belarus), and the Ukrainians.

The Russian Empire expanded remarkably in the sixteenth and seventeenth centuries (Figure 9.19). The arctic fringe of European Russia, inhabited primarily by non-Slavic peoples, provided furs as well as a sea route to western Europe (through the White Sea port of Archangel), and thus proved a valuable addition to the empire. Immediately east of Moscow in the Volga River Valley (near Kazan) lay Tatar territories incorporated into the Russian state in the mid-1500s. While the Volga Tatars themselves retained their Muslim religion and Turkish language, their lands were soon engulfed by an eastward-moving wave of Russian-speaking migrants.

The Russians also allied with the semi-nomadic **Cossacks,** Slavic-speaking Christians who had earlier migrated to the region to seek freedom in the ungoverned steppes (Figure 9.20). The Russian Empire granted them considerable privileges in exchange for their military service, an alliance that

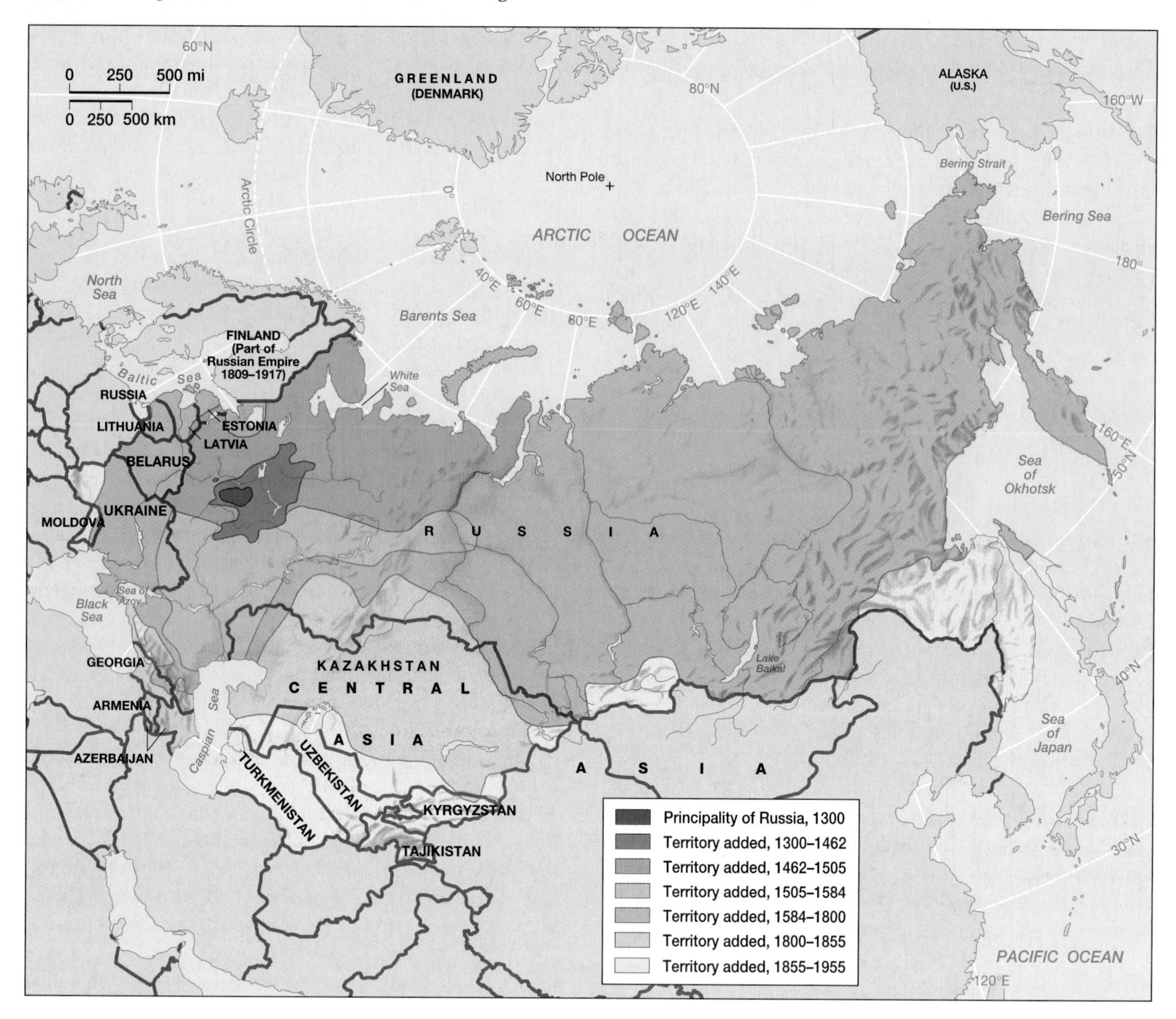

▲ **Figure 9.19 Growth of the Russian Empire** Beginning as a small principality in the vicinity of modern Moscow, the Russian Empire took shape between the fourteenth and sixteenth centuries. After 1600, Russian influence stretched from eastern Europe to the Pacific Ocean. Later, portions of the empire were added in the Far East, Central Asia, and near the Baltic and Black seas. *(Modified from Bergman and Renwick, 1999,* Introduction to Geography, *Upper Saddle River, NJ: Prentice Hall)*

greatly fortified the Russian Empire. In the late 1500s the Russian Cossacks commenced their conquest of Siberia. Equipped by wealthy Russian merchants, Cossack forces crossed the Urals and put their boats into Siberia's boundless rivers. Furs were the chief lure of this immense northern territory. Since cold climates produce high-quality furs, those of Siberia commanded premium prices in the developing global market for the resource. The initial Russian push into Siberia met the opposition of several Turkish states, including the Khanate of Sibir, which gave its name to the entire territory. Once these states were vanquished, however, the path to the Pacific lay open. By the 1630s, Russian power was entrenched in central Siberia, and by the end of the century it had reached the Pacific Ocean. Chinese resistance, however, delayed Russian occupation of the Far East region until 1858, and the imperial designs of the Japanese halted further expansion to the southeast when the Russians lost the Russo-Japanese War in 1905.

While the Russian Empire expanded to the east with great rapidity, its westward expansion was slow and halting. In the 1600s, Russia still faced formidable enemies in Sweden, Poland, and the Ottoman (Turkish) Empire. By the 1700s, however, all three of these states weakened, allowing the Empire to gain substantial territories. After defeating Sweden in the early 1700s, Tsar Peter the Great (1682–1725) obtained a foothold on the Baltic, where he built the new capital city of St. Petersburg, designed to be a window on the West. Later in the eighteenth century, Russia defeated both the Poles and the Turks and gained all of modern-day Belarus and Ukraine. Tsarina Catherine the Great (1762–96) was particularly piv-

▲ **Figure 9.20 Cossack** Natives of the Russian steppe, the highly mobile Cossacks played a pivotal role in aiding Russian expansion into Siberia during the sixteenth century. Many modern descendents retain their skills of horsemanship and are proud of their distinctive ethnic heritage. *(Julien Chatelin/Liaison Agency, Inc.)*

otal in colonizing Ukraine and in bringing the Russian Empire to the warm-water shores of the Black Sea.

The nineteenth century witnessed the Russian Empire's final major phase of expansion. The largest gains were made in Central Asia, where a group of once-powerful Muslim states was no longer able to resist the increasingly powerful and well-equipped Russian army. The Caucasus proved a greater challenge, as the peoples of this region had the advantage of rugged terrain in defending their lands. One sizable Caucasian ethnic group, the Circassians, were virtually exterminated in the process of resisting imperial rule. In Transcaucasia, on the other hand, the Christian Armenians and Georgians accepted Russian power with little struggle, since they found it preferable to rule by the Persian or Ottoman empires.

The Legacy of Empire The expansion of the Russian people was one of the greatest human movements Earth has ever witnessed. By 1900, one could travel from St. Petersburg on the Baltic to Vladivostok on the Sea of Japan, everywhere encountering peoples speaking the same language, following the same religion, and living under the rule of the same government. Nowhere else in the world did such a tightly integrated cultural region cover such a vast space. But despite this overriding unity, the Russian Empire as a whole, just like the Russian region of today, retained many pockets of linguistic, religious, and national diversity. Even in European Russia, non-Russian peoples retained their identity, while in much of the far north, indigenous peoples remained in the majority.

The history of the Russian Empire bequeathed several perennial sources of tension to the modern Russian world. One of these tensions centers on Russia's ambivalent relationship with western Europe. Russia shares with the West the historical legacy of Greek culture and Christianity and, since the time of Peter the Great, Russia has undergone several waves of intentional Westernization. At the same time, however, Russia has long been suspicious, even hostile, to European culture and social institutions. Highly nationalistic Russian thinkers have long viewed Western society as decadent and Western Christianity as heretical, and they have seen their country as the unique heir of the Greek and Roman traditions. Although the terms of this debate were transformed during the Soviet period, the central tension remained, and it continues to influence Russian political debates to this day.

Another significant source of tension in Russian culture stems from its long history of authoritarianism. From its earliest days, the Russian Empire was highly centralized, with the tsar and his advisors exercising tremendous power over the entire country. There has also been a countervailing history of resistance against overwhelming state power, as made evident in the dramatic breakup of the Soviet Union in 1991. Indeed, in early Russian history a republican state centered on the city of Novgorod offered a completely different Russian political model, one akin to those found in the most advanced sections of Europe at the time. In the late 1400s, however, Moscow vanquished Novgorod, establishing the foundations of a Russian authoritarianism that peaked during the Communist period in the 1930s under the cruel leadership of Joseph Stalin.

Geographies of Language

Slavic languages dominate the region (Figure 9.21). The distribution of Russian-speaking populations is complicated. Russian, Belorussian, and Ukrainian are closely related languages. Some linguists argue that they ought to be considered separate dialects of a single Russian language, as they are all mutually intelligible. Most Ukrainians, however, insist that Ukrainian is a distinct language in its own right, and there is a well-developed sense of national distinction between Russians and Ukrainians. Belorussians, on the other hand, are more inclined to stress their close kinship with the Russians.

The Belorussians and Ukrainians The geographic pattern of the Belorussian people is relatively simple. The vast majority of Belorussians reside in Belarus, and most people in Belarus are Belorussians (Figure 9.21). The country does, however, contain scattered Polish and Russian minorities. For the Russians this presents few problems, since Russians and Belorussians can relatively easily assume each other's ethnic identity.

The Ukrainian situation is more complex than that of the Belorussians. Parts of eastern Ukraine, as well as almost the entire Crimean Peninsula, are primarily Russian-speaking. Conversely, Ukrainians reside in scattered communities in southern Russia and southwestern Siberia. During the Soviet period, such geographical interspersion had little consequence, since the distinction between Russians and Ukrainians was not viewed as important in official circles. Now that Russia and Ukraine are separate countries, with a heightened sense of national distinction, this issue has emerged as a significant source of tension. Especially controversial is the Crimea region, which

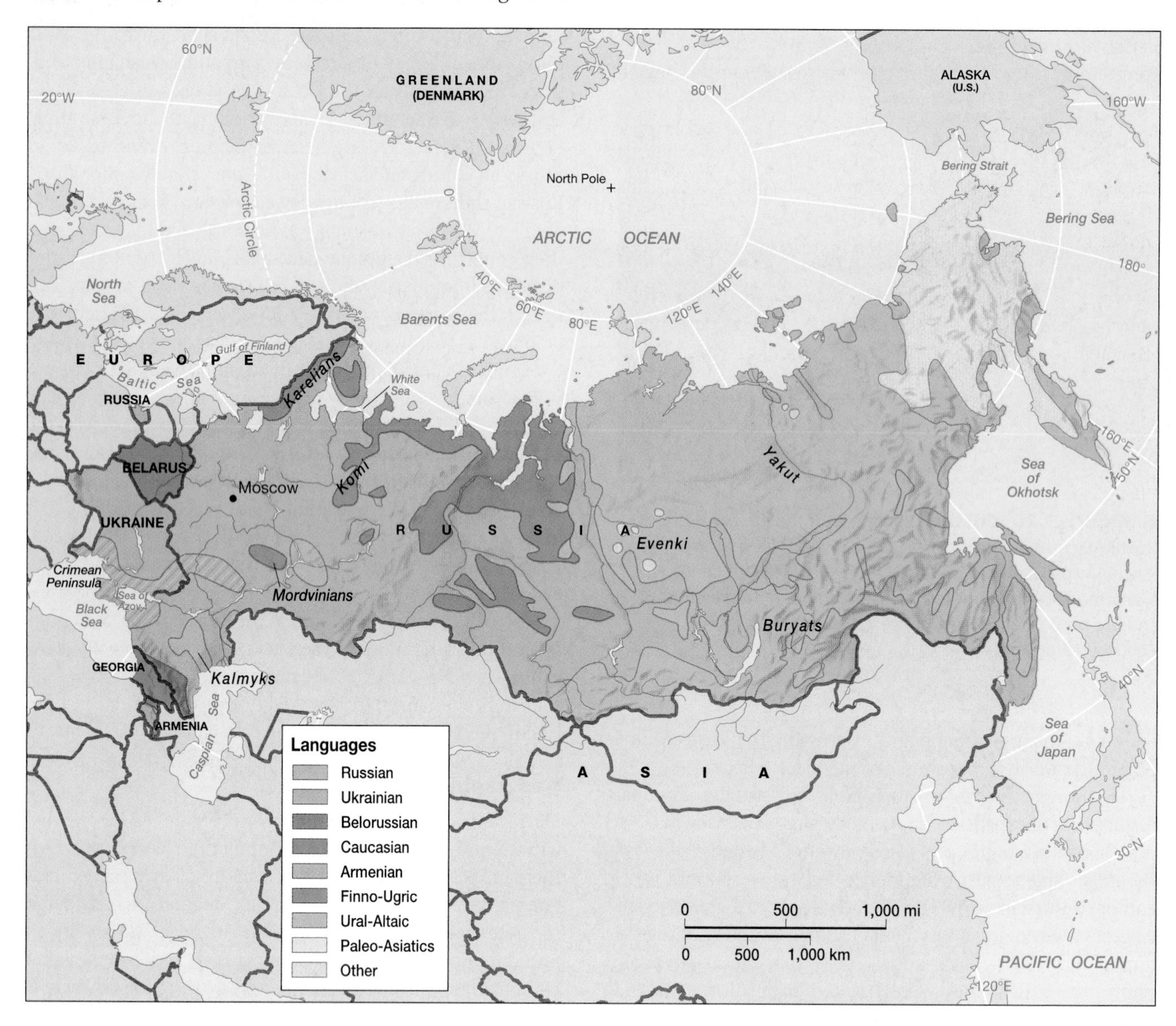

▲ **Figure 9.21 Languages of Russia and its neighbors** Slavic Russians dominate the realm, although many linguistic minorities are present. Siberia's diverse native peoples add cultural variety in that area. To the southwest, the Caucasus Mountains and the lands beyond contain the region's most complex linguistic geography. Ukrainians and Belorussians, while sharing a Slavic heritage with their Russian neighbors, add further variety in the west. *(Modified from Knox and Marston, 1998,* Human Geography, *Upper Saddle River, NJ: Prentice Hall)*

was once part of Russia before it was transferred to Ukraine in the 1950s. Many Russians want to reclaim the territory, while Ukrainians view it as an integral part of their country. The Russian language remains widely spoken in the Crimea and in other more urban portions of eastern Ukraine, creating tensions within a nation that appears determined to enforce the use of Ukrainian within its borders.

Patterns Within Russia Approximately 80 percent of Russia's population claims a Russian linguistic identity. Most of European Russia is inhabited primarily by Russians, but there are large enclaves of other peoples. The Russian zone extends beyond the Urals all the way across southern Siberia to the Sea of Japan. In the sparsely settled lands of central and northern Siberia, Russians are numerically dominant in many areas, but they share this territory with a large number of indigenous groups. While the map of languages suggests large areas remain beyond the Russian language realm, most of these regions are only sparsely populated. Still, their distinctive non-Russian cultural orientation has emboldened many of these groups to seek more political autonomy since the breakdown of the Soviet Union.

Finno-Ugric (Finnish-speaking peoples), though small in number, dominate sizable portions of the non-Russian north. The Finno-Ugric (or Uralic) speakers comprise an entirely different language family from the Indo-European Russians. Western Finns are found today within the separate nations of Estonia and Finland, but several eastern Finnish groups remain in Russia. While many have been culturally Russified in terms of intermarriage and religious adherence, distinct Finnish-speaking peoples such as the Karelians, Komi, and Mordvinians remain a part of Russia's modern cultural geography.

Ural-Altaic speakers also complicate the country's lingusitic geography. This language family includes the Volga Tatars, whose territory is centered around the city of Kazan in the middle Volga Valley. Numbering some five million, this distinctive group is a legacy of the Mongolian and Turkish pastoralists who overwhelmed the early Russian state and today the group, while retaining its ethnic identity, has extensively intermarried with and borrowed from its Russian neighbors. Elsewhere, the Yakut peoples of northeast Siberia represent another example of the Turkish group. The Yakuts, numbering around 300,000, are one of the most populous indigenous Siberian peoples, and their homeland is centered on the Lena River. Ancestors of the group migrated to the northern setting long ago, raising cattle and horses in the harsh Siberian setting. Other Ural-Altaic speakers in Russia belong to the Mongol group. In the west, examples include the Kalmyk-speaking peoples of the lower Volga Valley, who migrated to the region in the 1600s. Far to the east, the Buryats are also Mongolian peoples who live in the vicinity of Lake Baikal and represent an indigeneous Siberian group closely tied to the cultures and history of Central Asia (Figure 9.22).

Other non-Russian ethnic groups are sprinkled through the country, particularly in central and northern Siberia. These indigenous peoples are tremendously diverse and are divided into a number of unrelated linguistic groups. One entire language family, the Paleo-Asiatic, is limited to approximately 25,000 people widely dispersed through northeast Siberia. The most geographically widespread group within the region is the Evenki, whose traditional territory covers a large portion of central and eastern Siberia. A distant relation to the Altaic speakers, the Evenki bear close linguistic connections to the Manchu peoples of northeastern China. Only isolated islands of Evenki dominance remain, however. Many of these

▲ **Figure 9.22 Minority Buryats** Closely related to residents of Mongolia, Russia's Buryat population live in the vicinity of Lake Baikal. The Buryats enjoy some degree of political autonomy in recognition of their distinctive ethnic background. *(Hans-Jurgen Burkard/Bilderberg Archiv der Fotografen)*

Siberian peoples have seen their traditional ways challenged by the pressures of Russification, particularly during the socialist era. Today, many suffer from low levels of education and high rates of alcoholism, and their position parallels that of indigenous peoples of the United States and Canada.

Transcaucasian Languages Although small in size, Transcaucasia offers a bewildering variety of languages (Figure 9.23). From Russia, along the north slopes of the Caucasus, to Georgia and Armenia east of the Black Sea, a complex history and a fractured physical setting have combined to produce some of the most complex language patterns in the world. No fewer than three language families are spoken within a region smaller in size than Ohio, and many individual languages are represented by small, isolated cultural groups. In the Dagestan district of the northeast Caucasus, some 30 distinct languages are still spoken. The Caucasian language family, unrelated to any other on Earth, includes Chechnayan (Chechan) and Dagestani peoples north of the Caucasus and Georgians to the south. Altaic peoples appear on the scene, as well, complicating the cultural geographies of the region along the shores of the Caspian Sea and along the northwest border separating Russia and Georgia. Finally, Indo-Europeans dominate Armenia, and other Greek and Iranian peoples are sprinkled through the valleys of the Caucasus in central Georgia and southern Russia. Not surprisingly, language remains a pivotal cultural and political issue within the fragmented Transcaucasian subregion.

Geographies of Religion

Most Russians, Belorussians, and Ukrainians share a religious heritage of Eastern Orthodox Christianity. For hundreds of years the Orthodox Church formed a central cultural institution of the Russian Empire (Figure 9.24). Indeed, church and state were tightly fused up to the demise of the Empire in 1917. Under the Soviet Union, however, religion in all forms was severely discouraged and actively persecuted. Most monasteries and many churches were converted into museums or other kinds of public buildings, and schools disseminated the doctrine of atheism. Organized religion enjoyed something of a reprieve during World War II, when it was seen as helpful for the war effort, and religion was even more fully tolerated by the 1980s. Still, only a small minority of Soviet people remained actively religious. With the downfall of the Soviet Union, however, a religious revival swept much of the Russian region. Between 1988 and 1998, more than 11,000 Orthodox churches were returned to religious uses. Now, an estimated 75 million Russians are members of the Orthodox Church, including almost 400 monastic orders dispersed across the country.

Other forms of Western Christianity are also present in the region. For example, the people of western Ukraine, who experienced several hundred years of Polish rule, eventually joined the Roman Catholic Church. As members of the so-called *Uniate Church,* they were allowed to retain many of their Orthodox rituals and practices. Eastern Ukraine, on the other hand, remained fully within the Orthodox framework.

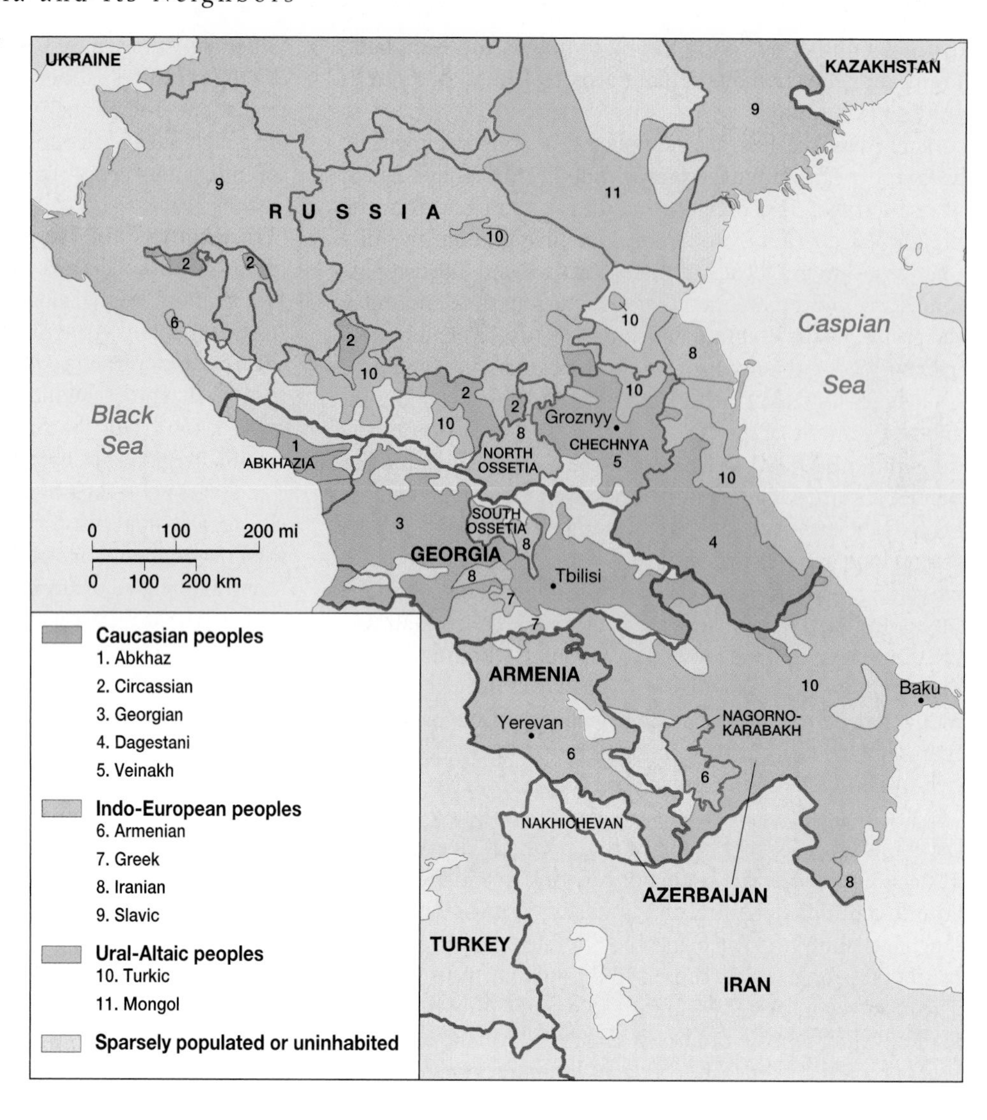

▶ **Figure 9.23 Languages of the Caucasus region** A bewildering mosaic of Caucasian, Indo-European, and Ural-Altaic languages characterize the Caucasus region of southern Russia and nearby Georgia and Armenia. Persisting political problems have erupted in the region as local populations struggle for more autonomy. Recent examples include independence movements in Chechnya and in nearby Dagestan. *(Modified from Martin Glassner,* Political Geography, *2nd Ed., 1996, p. 599. Reprinted by permission of John Wiley & Sons, Inc.)*

Because of this religious split, the Ukrainians themselves are a divided people. Western Ukrainians have generally been far more nationalistic, hence more firmly opposed to Russia, than eastern Ukrainians. Elsewhere, Christianity came early to the Caucasus, but modern Armenian forms, their roots dating to the fourth century A.D., differ on several doctrinal points from both the Eastern Orthodox and Catholic traditions. Georgian Christianity, however, is more closely tied to the Orthodox faith. Through the efforts of evangelical missionaries from the United States, evangelical Protestantism is also on the rise since the demise of the Soviet Union.

Non-Christian religions also appear and, along with language, shape ethnic identities and tensions within the region. The largest non-Christian group practices the religion of Islam, and probably numbers somewhere between 15 and 22 million adherents. Most are Sunni Moslems and they include peoples in the North Caucasus, the Volga Tatars, and Central Asian peoples near the Kazakhstan border. To date, Islamic fundamentalism has not become a major cultural or political issue in the region, although hard-core Russian Orthodox nationalists warn of its dangers. Elsewhere, Russia, Belarus, and Ukraine are home to more than one million Jews, and they are especially numerous in the larger cities of the European West. Jews suffered severe persecution under the tsars, although freer movements were allowed by the late nineteenth century, prompting migration to larger cities such as Moscow or Odessa. While a remote Jewish homeland was established in the Far East during the Soviet period, it never became a central focus of the nation's Jewish population. Recent out-migrations, prompted by new political freedoms, have further reduced numbers in Russia, Belarus, and Ukraine. Buddhists are also represented in the region, closely associated with the Kalmyk and Buryat peoples of the Russian interior. Indeed, the religion witnessed a renaissance during the greater religious freedom of the 1990s.

Russian Culture in Global Context

Russian culture has interacted in varied ways with the world beyond. Russian cultural norms have for centuries embodied both an inward orientation toward traditional forms of expression and an outward orientation directed primarily to western Europe. By the nineteenth century, even as Russian peasants interacted rarely with the outside world, Russian high culture had become thoroughly Westernized, and Russian composers,

▲ **Figure 9.24 Russian Orthodox Church, Siberia** A revival of interest in the Russian Orthodox Church followed the collapse of the Soviet Union in the early 1990s. The newly built Znamensky Cathedral in Kemerovo displays many of the faith's characteristic cultural landscape signatures. *(Novosti/ Sovfoto/Eastfoto)*

novelists, and dramatists gained considerable fame in Europe and the United States. By the turn of the twentieth century, Russian *avant guarde* artists were among the international leaders in devising the experimental styles of modernism.

During the Soviet period, a new mixture of cultural relationships unfolded within the socialist state. Initially, European-style modern art flourished in the Soviet Union, encouraged by the radical rhetoric of the new rulers. By the late 1920s, however, Soviet leaders turned against modernism, which they came to view as the decadent expression of a declining capitalist world. Many Soviet artists fled to the West and others were exiled to Siberian labor camps. Increasingly, state-sponsored Soviet artistic productions centered on **social realism**, a style devoted to the realistic depiction of workers harnessing the forces of nature or struggling against capitalism. Still, traditional high arts, such as classical music and ballet, continued to receive lavish state subsidies, and Russian artists, to this day, regularly achieve worldwide fame within their craft.

By the 1980s it was clear that the attempt had failed to fashion a new Soviet culture based on working-class solidarity and informed by the aesthetics of social realist art. The younger generation adopted instead a rebellious stance, turning for inspiration to fashion and rock music from the West. The mass-consumer culture of the United States proved immensely popular, symbolized above all by brand-name chewing gum, jeans, and cigarettes. The Soviet government attempted to ward off this perceived cultural onslaught, but with little success. Soviet officials could more easily censor books and other forms of written expression, but even here they were increasingly frustrated. By the end of the Soviet period, the persistent secrecy that separated the USSR from the West broke down, thus enabling more people and information to flow into the region.

After the fall of the Soviet Union in 1991, basic freedoms brought an inrush of global cultural influences, particularly to the region's larger urban areas such as Moscow. Shops were quickly flooded with Western books and magazines, people pondered the financial mysteries of home mortgages and condominium purchases, and they reveled in the newfound pleasures of fake Chanel handbags and McDonald's hamburgers (see "The Local and the Global: Globalization and the Russian Diet"). English language classes became even more popular in cities such as Moscow, where Russians hurried to embrace the world their former leaders had warned them about for generations. Cultural influences streaming into the country were not all Western in inspiration, either. Films from Hong Kong and Mumbai (Bombay), as well as the televised romance novels *(telenovelas)* of Latin America, for example, proved far more popular in the Russian region than in the United States.

Conversely, the opening of Russia's cultural doors also provoked problems within the region. The inflow of consumer products was unfortunately accompanied by a declining economy, putting many of the new luxuries beyond the reach of the average family, even as goods were stacked high on the shelves of the local department store. The integration of Russia and its neighbors into the global system has also provoked a strong cultural reaction among the more extreme nationalists. The rhetoric parallels Soviet-era concerns about the damaging effects of new cultural influences, particularly on the young. Organized resistance to foreign ideas and practices is also evident in the realm of religion. After an initial flurry of religious tolerance, many Russian Orthodox groups now work to limit the spread of such "non-Russian" faiths as Protestantism, Islam, and Tibetan Buddhism.

Geopolitical Framework: The Remnants of a Global Superpower

The geopolitical legacy of the former Soviet Union still weighs profoundly upon the modern Russian region. After all, the bold lettering of the "Union of Soviet Socialist Republics" dominated the Eurasian map for much of the twentieth century, and the country's global political reach left no corner of the world untouched. Many of the present political uncertainties that plague the region stem from the Soviet period. Former Soviet republics continue to struggle to define new geopolitical identities with one

THE LOCAL AND THE GLOBAL Globalization and the Russian Diet

Globalization has brought many changes to Russia, but perhaps none are more tangible to its residents than the shifts in daily diet experienced in the former Soviet Union. For those relatively few who can afford luxury items, grocery store shelves, particularly in Moscow, are now stacked high with Western-style convenience foods. Tyson chickens and Haagen-Dazs ice cream is there for the asking. New restaurants offer everything from fast-food pizza to premium steaks and haute cuisine.

Most Russians, however, lack the cash to indulge themselves. For many, in fact, the lean economic times of the post-Soviet era have literally been just that! Higher food prices and low wages have compelled people to slim down, skip meals, and increase their consumption of inexpensive bread, macaroni, and oatmeal. Garden plots at the city's edge offer opportunities for growing potatoes, beets, onions, and carrots. The current culinary challenge is quite different from the communist era when cheap food was available (under Soviet rule, office and factory workers got a free lunch), but variety was lacking. Even with these changes, half of all Russian adults are overweight, vitamin deficiencies are common, and heavy, high-fat foods dominate the daily routine.

For the average Moscow resident who cannot afford the finer luxuries of expensive Western dining, breakfast might include a glass of kefir (yogurt drink), a bowl of kasha (wheat porridge), a sausage sandwich (if there is cash for meat), and a cup of tea. An inexpensive lunch could feature a bowl of borscht (beet soup), bread, and boiled potatoes (Figure 1). Late evening supper would likely find more bread and potatoes on the table, as well as perhaps beef or vegetable cutlets, pickles, and more tea with veroniki (sweet berries). Future Russian diets will inevitably reflect the nation's place in the global economy: More hard times will yield an abundance of traditional bread and borscht, while affluence might bring with it more lowfat milk, Pringles potato chips, and Big Macs.

▲ **Figure 1 Russians eating traditional meal** Meager wages cannot stop these Russians from socializing over vodka and food in a concrete high rise in remote Murmansk. Russian diets reflect the economy: imported luxuries are fine when wages rise, but borscht and bread suffice in tougher times. *(Dominic Harcourt-Webster/Panos Pictures)*

Source: Adapted from "Lean Times at the Russian Dinner Table," *New York Times*, December 6, 1998.

another. Neighboring states persist in eyeing the region with trepidation, the legacy of former Soviet political and military power. Present demands for more local political control within countries such as Russia, Ukraine, and Georgia can still be understood in the context of the Soviet era when a highly centralized political apparatus gave little voice to regional dissent. For all of these reasons, future political geographies in the area also remain cartographic mysteries, more prone than most corners of the world to shift unpredictably as conditions rapidly evolve within the region.

Geopolitical Structure of the Former Soviet Union

The Soviet Union rose from the ashes of the Russian Empire, which collapsed abruptly in 1917. The ultraconservative policies of the Russian tsar generated opposition among businesspeople and workers, while the peasants, who formed the majority of the population, had always resisted the powerful land-owning aristocracy. After the fall of the tsar and the aristocracy, a broad-based coalition government assumed authority. Several months later, however, the **Bolsheviks**, a faction of Russian Communists representing the interests of the industrial workers, seized power within the country. The leader of these Russian communists was Vladimir Ilyich Ulyanov, usually known by his self-selected name of Lenin. Lenin was the main architect of the Soviet Union, the Russian Empire's successor state.

The new socialist state reconfigured Eurasian political geography. Although it resembled the territorial contours of the Russian Empire and centralized authority continued to be concentrated in European Russia, the spatial and economic structure of the country was radically transformed. When the Soviet Union emerged in 1917, Lenin and the other communist leaders were aware that they faced a major challenge in organizing the new state. The old empire contained many distinct nationalities, but Russians traditionally dominated these peoples. Lenin wanted to end such domination, which he considered contrary to the international and class-based spirit of socialism. At the same time, however, the new leaders wanted to retain as much of the territorial extent of the old empire as possible, and they certainly did not want unfriendly regimes to emerge in any of its lands. After winning a bloody civil war against anti-Communist Russian forces, the new authorities succeeded in reuniting almost all of the old imperial territory under the Soviet flag. Only a few marginal areas, such as Finland and Poland, gained and retained independence.

The Soviet Republics and Autonomous Areas Soviet leaders designed a geopolitical solution that maintained their country's territorial boundaries and that acknowledged, at least theoretically, the rights of its non-Russian citizens. Each major nationality was to receive its own "union republic," provided it was situated on one of the nation's external borders (Figure 9.25). Eventually 15 such republics were established, thus creating the Soviet Union. A number of these republics were quite small, while the massive Russian Republic sprawled over roughly three-quarters of the Soviet terrain. Each republic was to be administratively autonomous, vested with the right to secede from the union if it so desired. In practice, however, the Soviet Union remained a centralized state, with important decisions made in the capital of Moscow. Instead of a union of independent countries, the Soviet system evolved into a unitary state under partial Russian domination. It did, however, make a number of concessions to local administration and cultural identity.

The Soviets came up with different geopolitical solutions to acknowledge smaller ethnic groups and nationalities that were not situated on one of the country's external borders. Indeed, dozens of significant minority groups pressed for recognition. The supposed solution to this problem was the creation of **autonomous areas** at different spatial scales that recognized special ethnic homelands but did so within the structure of existing republics. Thus, within the Russian Republic the larger nationalities, such as the Yakut and the Volga Tatars, were granted their own "autonomous republics" (not to be confused with the 15 higher-level "union republics," such as Russia or Ukraine). Smaller nationalities, such as the Evenki, received

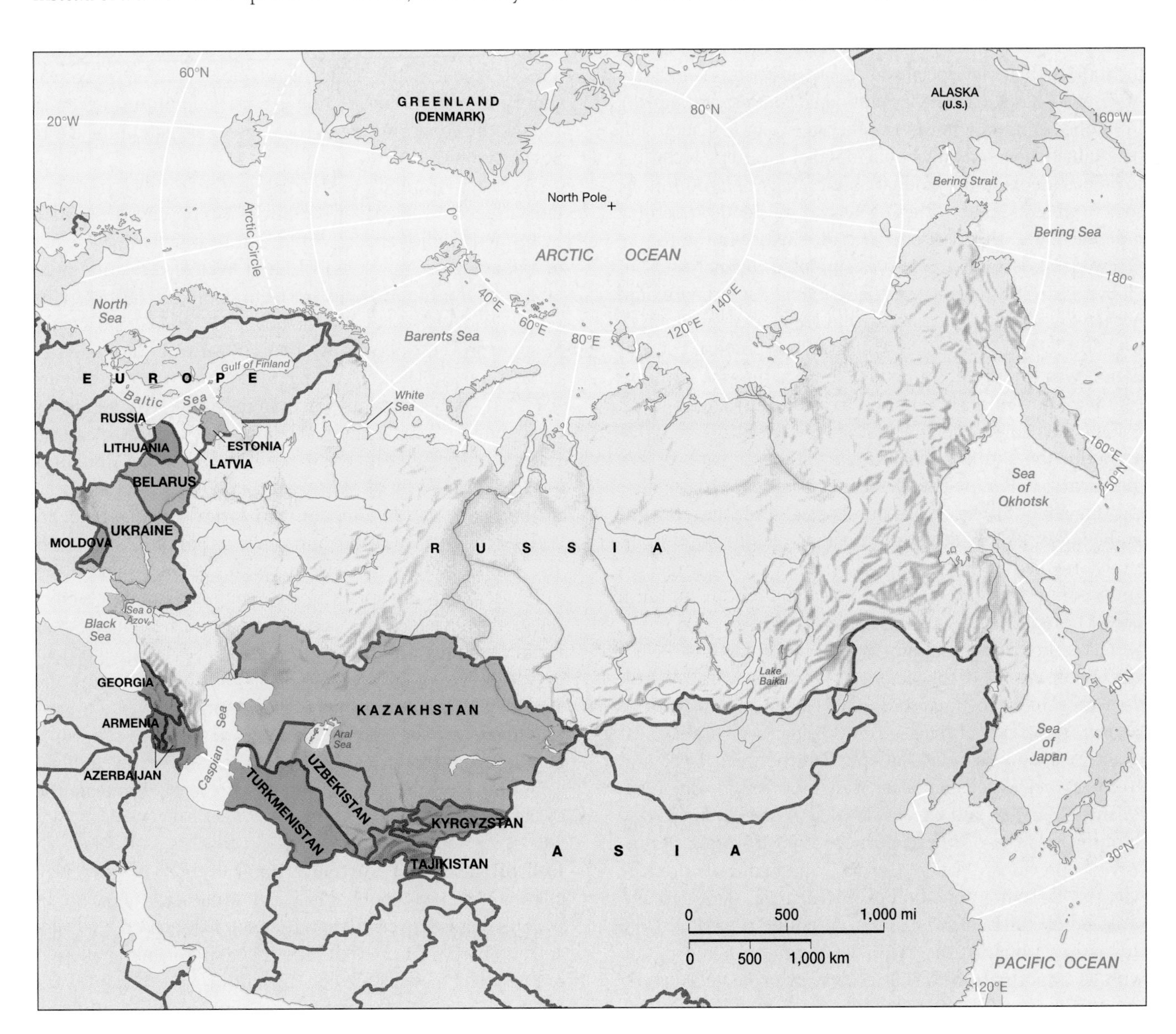

▲ **Figure 9.25 Soviet geopolitical system** During the Soviet period, the boundaries of the country's 15 internal republics often reflected major ethnic divisions. Ultimately, however, many of the ethnically non-Russian republics pressured the Soviet government for more political power. As the Soviet Empire disintegrated, the former republics became politically independent states and now form an uneasy ring of satellite nations around Russia. *(Modified from Rubenstein, 1999,* Introduction to Human Geography, *Upper Saddle River, NJ: Prentice Hall)*

less-important "autonomous regions," while still others were left with the lowest-order "national districts." The complex geography of indigenous Siberian peoples could not easily be contained within such a system, however, because the territories of different groups mingled with one another. While the Evenki received their own autonomous region, the group's dispersed geography meant that several lower-order Evenki national districts were also carved out of the Yakut autonomous republic. The Soviet system of national territoriality was perhaps the most complex geopolitical form ever devised. Its main deficiency was that the autonomy it was designed to provide proved to be more of a charade than a reality.

Centralization and Expansion of the Soviet State In the early years of the Soviet Union, it appeared that the framework of separate republics and autonomous areas might allow non-Russian peoples to protect their own cultures and to establish their own social and economic policies (provided, of course, that such policies embodied Marxist principles). From the beginning, however, it became clear that such self-determination would only be a temporary measure. According to official ideology, the gradual development of a communist society would see the withering away of all significant ethnic differences and the disappearance of religion. In the future, a new classless *Soviet society* was supposed to emerge.

By the 1930s, it was clear that national autonomy would not have any real significance within an increasingly centralized Soviet state. The chief architect of this political consolidation was Joseph Stalin, who did everything he could to centralize power in Moscow and to assert Russian authority. Stalin launched a ruthless plan of state-controlled agricultural production and industrialization. Although many people initially resisted his policies, Stalin never hesitated to use force to bring about his vision of a purer socialist revolution within the Soviet Union. Moreover, the Stalin administration directly attacked ethnic groups suspected of harboring anti-Soviet sentiments and further reduced any residual political power beyond the central authorities in Moscow.

The Stalin period also saw the final enlargement of the Soviet Union. As one of the victorious powers of World War II, the Soviet Union acquired southern Sakhalin and the Kuril Islands from Japan. It also regained the Baltic Republics (Lithuania, Latvia, and Estonia—independent between 1917 and 1940), as well as substantial territories formerly belonging to Poland, Romania, and Czechoslovakia. A small but strategic addition was the northern portion of East Prussia (Kaliningrad), previously part of Germany. Immediately after the war, the German population of this area was deported and replaced by Russian immigrants. Although separated from Russia by Lithuania, the Baltic Sea district of Kaliningrad, with its important port facilities, was incorporated directly into the Russian Republic. It still forms a small but strategic Russian **exclave,** which is defined as a portion of a country's territory that lies outside of its contiguous land area.

After World War II, the Soviet Union also gained significant authority, although not actual sovereignty, over a broad swath of eastern Europe (Figure 9.26). As they pushed the German army west, Soviet troops advanced across much of the region toward the end of the war, actively working toward the establishment of communist regimes thereafter. In the words of British leader Winston Churchill, the Soviets also extended an "**Iron Curtain**" between their eastern European allies and the more democratic nations of western Europe by restricting the free flow of people and information. Most eastern Europeans resented this Soviet presence, and many sought to regain genuine independence. Soviet leaders were determined to retain this informal empire, however, to the point of sending military force into Hungary in 1956 and Czechoslovakia in 1968. Such actions further discredited Soviet power in the eyes of most eastern Europeans. In addition, Yugoslavia carved its own pluralistic path, apart from the political dictates of the Soviets. By the 1980s, Romania also was able to forge its own foreign policy, while in Poland public demonstrations and labor unrest made the country almost ungovernable. The Soviet Union's gradual loss of authority in eastern Europe undermined communism even in Russia, thus contributing to the subsequent fall of the entire Soviet system.

As eastern Europe retreated behind the so-called Iron Curtain, it became clear that the Soviet Union and the United States, while allies during World War II, were to become potential antagonists in a global **Cold War** of escalating military competition that lasted from 1948 to 1991. In its post-World-War-II heyday, the Soviet Union became a global superpower, one of only two countries equipped with enough nuclear weapons to ensure global destruction (Figure 9.27). At the height of its power in the late 1970s, it enjoyed close military and economic alliances not only with eastern Europe but also with certain countries in Asia (Mongolia, North Korea, Vietnam, Laos, and Cambodia), the Caribbean region (Cuba and Nicaragua), and Africa (Angola, Somalia, and several others). India and other important Asian states also maintained cordial relations with the Soviet Union and purchased large quantities of Soviet military supplies. Across the globe, Soviet leaders vied with those of the United States for international influence and prestige. Soviet foreign policy was assertive and sometimes expansionist through the 1970s. In 1979, for example, the country launched a massive invasion of Afghanistan, a move designed to prop up a failing Afghani communist regime. The invasion failed, however, furthering the political and psychological disintegration of the Soviet system.

End of the Soviet System Ironically, Lenin's system of culturally defined republics helped sow the seeds of the Soviet Union's demise. Even though the nationally based republics and autonomous areas of the Soviet Union were never allowed real freedom, they did provide a persisting political framework for the perpetuation of distinct cultural identities. Indeed, contrary to the expectations of Soviet leaders, ethnic nationalism intensified in the post-World-War-II era as the Soviet system grew less repressive. As Soviet President Mikhail Gorbachev initiated his policy of **glasnost,** or greater openess, during the 1980s, several republics—most notably the Baltic

◀ **Figure 9.26 Soviet expansion in eastern Europe** The imperial leanings of the Soviet Empire are revealed in this map showing the expansion of Soviet influence across eastern Europe, particularly after victories achieved in World War II. Many eastern European countries became virtual satellite states of the USSR. With the collapse of Soviet power in the early 1990s, Russian influence has ebbed in eastern Europe. *(Modified from Knox and Marston, 1998,* Human Geography, *Upper Saddle River, NJ: Prentice Hall)*

▲ **Figure 9.27 Moscow's Red Square** Cold War politics often resulted in public displays of Soviet military might as shown in this view of Moscow's Red Square from the 1960s. While no longer a global superpower, modern Russia still maintains an extensive arsenal of nuclear weapons. *(Tass/Sovfoto/Eastfoto)*

states of Lithuania, Latvia, and Estonia—demanded outright independence. While it appeared for a time that the Soviet Army would crush these secession movements, the central government did little to halt the drift away from Soviet control.

Meanwhile, other forces were also working toward the political end of the Soviet regime. Worsening economic conditions, increasing food shortages, and the declining quality of life experienced by many Soviets after 1980 led to fundamental questions concerning the value of centralized planning within the country. In response, President Gorbachev's plan of **perestroika**, or planned economic restructuring, was aimed at making production more efficient and more responsive to the needs of Soviet citizens. In 1991, however, Gorbachev saw his authority slipping away amid rising pressures for political decentralization and more dramatic economic reforms. During the summer, Gorbachev's regime was further imperiled by the popular election of reform-minded Boris Yeltsin to the head of the Russian Republic and by a failed military coup by communist hardliners. By late December, all of the country's 15 constituent republics became independent states and the Soviet Union ceased to exist.

Modern Russian Geopolitics

Post-Soviet Russia and the nearby independent republics have radically rearranged their political geographical relationships since the collapse of the Soviet Union in 1991. All of the former republics still struggle to establish stable political relations with their neighbors, and increasing calls for further political decentralization in many settings threaten the internal integrity of Russia and nearby states (Figure 9.28).

Russia and the Former Soviet Republics For a time it seemed that a looser political union of most of the former republics, called the **Commonwealth of Independent States (CIS),** would emerge from the ruins of the Soviet Union. All the former republics, with the exception of the three Baltic states, joined the CIS soon after the dismemberment of the old union (Figure 9.28). By the late 1990s, however, the CIS had developed into little more than a forum for discussion, without real economic or political power. Newly independent countries, such as Uzbekistan, wanted to keep the CIS ineffectual for fear that Russia would otherwise use the organization to regain authority over the entire area. When it seemed that the CIS itself might dissolve, Russia made efforts to calm such fears, declaring 1998 to be "the year of the CIS." Even so, such proclamations failed to energize relations between the group of former republics.

Several other attempts have been made to reknit the bonds that were broken when the Soviet Union collapsed. Many Belorussians, for example, have pressed for closer ties with Russia, even though authoritarian leaders in the country have not paralleled the Russian moves toward democracy and increased political freedoms. Still, in 1996, Russia and Belarus declared their membership in a special two-country political and economic union, and the two countries moved even closer to reunification in 1999. Acting together, Russia and Belarus signed a separate customs union pact with Kazakhstan and Kyr-

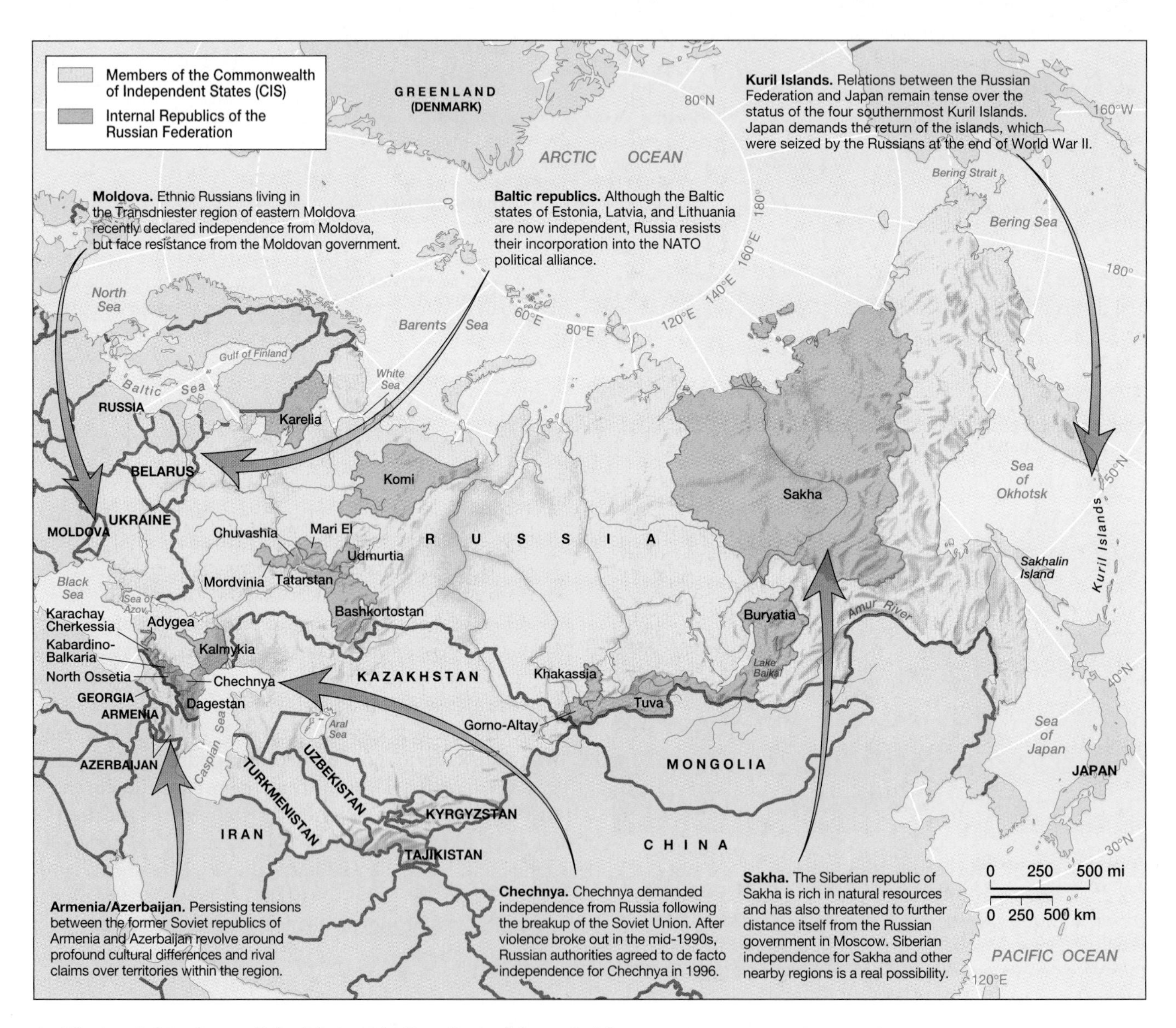

▲ **Figure 9.28 Geopolitical issues in Russia and its neighbors** The Russian Federation Treaty of 1992 created a new internal political framework that acknowledged many of the country's ethnic minorities. Political stability, however, continues to elude the country as many of these groups press for complete independence. Elsewhere, Armenia struggles with neighboring Azerbaijan, and Russia's relations with nearby states remain in flux. *(Modified from Bergman and Renwick, 1999,* Introduction to Geography, *Upper Saddle River, NJ: Prentice Hall)*

gyzstan, but the agreement has had little impact on trade between the countries. Russia and Ukraine also signed a 1998 agreement to develop trade relations and foster economic linkages. Elsewhere, Georgia, Ukraine, Azerbaijan, and Moldova formed the GUAM group, dedicated to facilitating trade, mainly of oil, through the Caspian–Black Sea corridor.

Certainly the most direct connection between Russia and its immediate neighbors are the millions of Russians living in the other former Soviet republics. In several countries Russians constitute almost half of the total population. Ethnic tensions intensified after several of these countries, most notably Latvia, adopted highly nationalistic social policies, such as denying citizenship to people of Russian background. This has created a large body of nationless persons, many of whom are unable to obtain passports. Ethnic tensions between Russians and non-Russians reached the point of war in the former Soviet state of Moldova. The Russian community in eastern Moldova's Transdniester region declared itself independent and exchanged gunfire with the Moldovan government. A stalemate resulted, and in the late 1990s tensions cooled somewhat. Elsewhere, Russians remaining in Ukraine and Kazakhstan have also elevated political and cultural tensions in those countries.

Another complicating factor in the post-Soviet period has been the ongoing military relationships between Russia and its former republics. **Denuclearization,** the return and partial dismantling of nuclear weapons from outlying republics to Russian control, was completed during the 1990s. The nations of Kazakhstan, Ukraine, and Belarus had their Soviet-era nuclear arsenals removed in the process. Elsewhere, Tajikistan invited the Russian army in during the early 1990s to quell its own internal ethnic struggles. While the action helped cement old ties between the Russian and the Tajik leadership, in other cases Russian involvement has not helped the cause of political cooperation. Georgians, for example, deeply resent Russian assistance to the Abkhazians, a breakaway Muslim group located in the northwestern part of their country. Naval bases of the former Soviet Union present another dilemma. Russia's Baltic fleet is headquartered in the Kaliningrad exclave, now separated from the rest of the country. The Soviet Black Sea fleet, on the other hand, is traditionally based in the Crimean port of Sevastopol, now in Ukraine. After lengthy negotiations, Russia and Ukraine agreed in 1997 to share the naval base.

The former Transcaucasian republics have witnessed other political tensions in the post-Soviet era. The area contains striking ethnic and religious diversity, and rivalries are accentuated by the geopolitical structure inherited from the Soviet period. The territories of the Christian Armenians and the Muslim Azeris, for example, interpenetrate one other in a complex fashion. The far southwestern portion of Azerbaijan (Nakhichevan) is separated from the rest of the country by Armenia, while the important Armenian-speaking district of Nagorno-Karabakh is officially an autonomous portion of Azerbaijan. Almost immediately after independence in 1991, Armenia and Azerbaijan once again went to war over these territories. After Armenia successfully joined Armenia proper with Nagorno-Karabakh, the fighting between the countries diminished. No peace treaty has been signed, however, and Azerbaijan demands the return of the territory. Meanwhile, 600,000 Azerbaijani refugees remain homeless. Armenia also has no international support for annexing Nagorno-Karabakh, which contributes to its own sense of isolation.

Devolution and the Russian Federation Within Russia, further pressures for devolution, or more localized political control, produced the March 1992 signing of a new Russian Federation Treaty. The treaty granted Russia's internal autonomous republics and its lesser administrative units greater political, economic, and cultural freedoms, including more control of their natural resources and foreign trade. Conversely, it weakened Moscow's centralized authority to collect taxes and to shape policies within its varied hinterlands. Defined essentially along ethnic lines, 21 regions possess status as republics within the Federation and now have constitutions that often run counter to national mandates. Some republics such as Tatarstan in the Volga Valley have even extracted agreements from Moscow to develop their own "foreign economic policy" (Figure 9.29). In addition, dozens of smaller regions and municipalities, including Moscow itself, have declared varying degrees of political autonomy from the federal reins of power.

Several clusters of autonomous republics are immediately apparent. One group, composed of a number of particularly small territories, is located along the northern slope of the Caucasus Mountains, an area noted for its extraordinary ethnic diversity. Another group of larger autonomous areas is found in the central Volga Valley, homeland of the Volga Tatars and of a number of other Turkish-speaking, as well as

▲ **Figure 9.29 Downtown Kazan** Capital of the internal Tatarstan Republic, Kazan is home to many Islamic residents who argue for even more freedom from distant Moscow. Despite these wishes, its urban landscape reflects long ties to Russian influences. *(Itar-Tass/Sovfoto/Eastfoto)*

GEOPOLITICAL SETTING A Separate Siberia?

The future map of the Russian Federation's internal republics is anyone's guess, but Siberia is one region that is particularly susceptible to redrawn political boundaries. Multiple forces are at work within that vast eastern territory that may encourage a break from Moscow's dwindling dominance. Distance itself makes governing the region difficult, particularly when central authorities have little spending power to devote to their Siberian hinterlands. Siberia's natural resources are another variable favoring potential separation: Why see revenues leave when more local control might bring benefits directly to those living in the region? Russia's new Regional Economic Associations (created in 1997) include Siberian and Far East representatives, and these groups are lobbying for more control of their own economic destinies. Ethnic minorities, already successful in achieving local autonomy, may press for complete independence.

Simply stated, political and economic self-interest may propel varied Siberian regions to go their own way. For example, within the present Russian Federation, northeast Siberia's sprawling Sakha Republic, based in Yakutsk, has already cut deliveries of tax revenues and natural resources to the central government in far-off Moscow, and other groups are doing the same (see Figure 9.28). Regional governors are increasingly allying with local generals, who pay little attention to the wishes of their superiors. They are making economic and political policies that directly challenge Moscow's sovereignty. Omsk, a region in southwestern Siberia, recently banned sales of food beyond its "borders," while nearby Krasnoyarsk is run like a virtual fiefdom by its strongman, Alexander Lebed. The future geopolitical map may well undergo what the Soviet Union witnessed in 1991: It might see the fragmentation of the existing Russian Federation and produce in the process an even smaller Russian state ringed by newly independent countries, including a constellation of Siberian and Far Eastern republics finally freed from their European shackles.

Finnish-speaking, peoples. A third cluster is located along the southern border of Siberia, homeland of the Mongolian Buryats and of several Turkish-speaking ethnic groups. Russia's largest autonomous areas are located in the far north, especially in Siberia. Sakha (or Yakutia), for example, sprawls over 1,200,000 square miles, an area larger than the territory of all but seven of the world's independent countries. But Sakha, like the other autonomous areas of the tundra and taiga zone, is a sparsely inhabited land.

The Russian trend toward more local political control has obvious advantages. Most clearly, after a long era of Soviet dominance, it allows people to feel as if they have a more direct role in the running of their own political affairs. They also can claim fuller control of their economic resources and more effectively manage public expenditures on infrastructure and economic development. For example, when the central government slowed progress on land reform, the regional governor in the city of Samara pressed ahead with his own program, privatizing 90 percent of the farmland. Other autonomous units within Russia have also gone ahead with their own development efforts, including offering special tax breaks to foreign companies trying to seek a foothold within the Russian economy.

Recent geopolitical tensions within Russia also reveal some of the hazards of these devolutionary impulses. After the Soviet Union disintegrated, several of Russia's internal autonomous areas threatened to secede. Initially, strong challenges came from Tatarstan on the Volga, an oil-rich area with a long history of national identity. For the time being, however, new concessions from Moscow have quelled demands for independence. Elsewhere, outlying regions in Siberia and in the Russian Far East have also voiced their sympathies for potential separation from Russia (see "Geopolitical Setting: A Separate Siberia?"). Most dramatically, the complex ethnic territories along the north slopes of the Caucasus made it clear that they resented too much centralized control from Moscow. By 1994, leaders of the Chechnyan Republic vowed to establish a genuinely independent state. Russia responded with a massive military invasion. After intensive fighting, the Russian Army took control of Chechnya's capital of Groznyy, located in the plains just to the north of the Caucasus (Figure 9.30). Chechnyan rebels held out in the rugged mountains to the south, eventually exhausting Russia's will to fight. In late 1996, the Russian government allowed Chechnya *de facto* independence in order to stop the fighting, but it refused to officially recognize the breakaway republic.

The Shifting Global Setting

Much of Russian geopolitical history has been focused on its creation of a vast land-based empire across much of Eurasia. As British geographer Halford Mackinder argued in 1904, Russia's control of the heartland, the huge interior landmass of Eurasia, gave it a global advantage over maritime powers such as Great Britain. Mackinder argued that Russia's continental interior essentially outflanked British sea power and would prove almost invulnerable to attack. Fifty years later, as the powerful Soviet Union gained control over eastern Europe and, for a while, influence over China, Mackinder's thesis seemed to be borne out.

The collapse of the Soviet Union, however, resulted in a simultaneous crash of Russian global power. Russia, with only a little more than half of the population of the former Soviet Union and with a teetering economy, could no longer afford to support and subsidize allies such as Cuba. Moreover, the communist ideology that had previously cemented the Soviet alliance rapidly declined in Russia and throughout most of the world. Russia did what it could to maintain the formidable military complex of the Soviet Union, but decay was in-

▲ **Figure 9.30 Devastated Groznyy** The recent war for independence in the Chechnyan Republic met stiff resistance from Russia's central government. Much of the capital city was caught in the crossfire and it is only now slowly rebuilding as Chechnya achieves de facto autonomy from Moscow. *(Tass/Sovfoto/Eastfoto)*

evitable. While the Russian army remained large and equipped with powerful weapons, its morale suffered as soldiers' paychecks failed to appear and as its equipment deteriorated amid dwindling defense budgets. In retrospect, Mackinder's thesis on the dominance of the heartland no longer seems supportable. The Russian example suggests that power in the modern world flows much more from economic productivity and technological prowess than from the mere control of continental territory.

Since the fall of the Soviet Union, Russian territorial conflicts beyond the former republics include tensions within East Asia. The boundary between Russia and China was imposed by the Russian Empire in 1858 and has never been fully accepted in Beijing. The two countries battled over the area in 1969, but tensions subsequently subsided. The leaders of both Russia and China now stress the need for mutual accommodation and cooperation, but the potential for renewed conflict remains. Territorial disagreements also complicate Russia's relationship with Japan. Japan has long demanded the return of the four southernmost islands of the Kuril Archipelago, seized by the Soviet Union in 1945. Although Russia could gain major financial benefits in any agreement, thus far it has refused to return any territory to Japan.

The relationship between Russia and the West also remains insecure. Russia worries, in particular, about the possible eastward expansion of the North Atlantic Treaty Organization. Although most Russian leaders accept the inclusion of Poland, Hungary, and the Czech Republic into NATO, virtually all strongly oppose the entry of any former Soviet territories. The Baltic nations of Estonia, Latvia, and Lithuania, however, have applied for admission into NATO, and they have found some support for this bid in the United States.

Still, Russia retains the remnants of its global reach. Its nuclear arsenal, while reduced in size, remains a powerful counterpoint to American, European, and Chinese interests. While Russia can no longer directly challenge the United States as it did in the days of the Soviet Union, it can act as a partial counterweight to the United States in international maneuverings. Russian leaders, for example, can offer themselves as intermediaries in diplomatic negotiations. Russia also retains a permanent seat on the United Nations Security Council, arguably the world's most important geopolitical body. Russia's recent inclusion in some of the G-7 economic meetings also signifies its remaining international clout.

Economic and Social Development: A Time of Turmoil

Recent economic declines have devastated the Russian region. Russia's gross national product declined by more than 40 percent between 1989 and 1998, and similar economic disasters have unfolded in nearby Belarus, Ukraine, Armenia, and Georgia (Table 9.2). They mark the most abrupt collapse within the industrialized world since the Great Depression of the early 1930s. The causes of the recent economic difficulties are complex, rooted both in the long-standing policies of the Soviet period as well as in the chaotic nature of economic reforms that have been unevenly carried out since 1991.

The true economic potential of the Russian region has always been difficult to gauge. Optimists point to the vast size, abundant natural resources, and well-educated, urbanized

Table 9.2 Economic Indicators

Country	GNP per Capita ($U.S., 1996)	Total GNP (Millions of $U.S., 1996)	PPP* ($Intl, 1996)	Real Annual Growth % per Capita, 1990–1996
Armenia	630	2,387	2,160	–15.0
Belarus	2,070	22,452	4,380	–8.6
Georgia	590	4,590	1,810	–19.6
Russia	2,410	356,030	4,190	–9.2
Ukraine	1,200	60,904	2,230	–13.4

*Purchasing power parity.

Source: *The World Bank Atlas,* 1998.

populations of the region as significant assets. Indeed, in its heyday, the Soviet Union rose to become one of the great industrial powers in the world, and did so in a remarkably short period of time. Skeptics note that size also brings its disadvantages, particularly by raising transportation costs within the region's economy. They also point out that Russia's northern location has always made food production problematic. Most notably, ever since the breakup of the Soviet Union, economies within the region have struggled to evolve in a stable fashion toward greater productivity and output. Instead, much of the 1990s witnessed rising prices and unemployment but declining investment and industrial output, hardly the recipe for economic health. In addition, economic troubles have had profound consequences for the quality of life within the region. The social fabric of families, the quality of health care, and even the psychological outlook of the population have all suffered amid the economic uncertainties of the period. The region's future path to prosperity remains clouded as it spends the early years of the twenty-first century reconfiguring an economy that was forged within a radically different political and ideological past.

The Legacy of the Soviet Economy

The birth of the Soviet Union in 1917 initiated a radical change within the region's economy. Under the Russian Empire, most people were peasants, farming the land much as they had done for centuries. Some changes were afoot by 1900, including a state-led industrialization program that saw the beginnings of the textile, steel, and chemical industries. Still, as Lenin assumed power, much of the Russian economy remained almost medieval, lagging far behind developments taking place in western Europe and the United States. Over the next 60 years, however, the Soviet Union quickly emerged to rival, and even surpass, the most powerful economies on Earth. During that era of unparalleled growth, much of the region's present economic infrastructure was established, including new urban centers and industrial developments, as well as a modern network of transportation and communication linkages. Thus, even though the Soviet Union has departed from the scene, much of its economic legacy remains to shape the progress and the challenges of its contemporary economy.

The Communists Come to Power Russia's Communist Revolution came quickly and unexpectedly in 1917. Mainstream Marxist thought argued that radical revolutions were most likely in the urban, industrialized societies of western Europe, not in a primarily agricultural country such as Russia. The Bolsheviks, however, were well organized and they received support from Russia's industrial workers and from a peasant population that bitterly resented the exploitative landlord class. Many members of the middle class also rejected the autocratic government of the tsar and thus initially supported the revolution. As Communist leaders consolidated power, they nationalized Russian industries, creating a system of **centralized economic planning** in which the state controlled production targets and industrial output. Limited markets persisted for a few years, but by the 1930s, virtually the entire economy lay under state control. Stalin's Soviet government stressed heavy, basic industries (steel, machinery, chemicals, and electrical generation), deferring demand for consumer goods to the future. Huge increases in production were realized, but at a significant human cost. Several mining and industrial areas, most of them in Siberia, were turned into virtual slave-labor camps.

Stalin also nationalized agriculture. By the late 1920s, individual holdings were combined into large-scale **collective farms** *(kolkhozes),* which organized agricultural production around state-mandated production goals. Elsewhere, larger state-farms *(sovkhozes)* became virtual rural factories in which large teams of laborers were put to work on increasingly mechanized operating units. Even the reindeer herds of Siberian tribal peoples were grouped into collective holdings. Most peasants resisted collectivization and many slaughtered their animals rather than turn them over to state ownership. Agricultural production declined precipitously, provoking governmental repression. More than 10 million peasants starved to death during the famines that resulted in the 1920s and 1930s, particularly in Ukraine. In addition, during the 1930s, Stalin killed many of the *kulaks,* or wealthier peasants, who were seen as leaders of the agrarian resistance.

A New Economic Geography By the 1930s, the Soviet Union was well on its way to fundamentally reorganizing the country's economic landscape in ways that still shape the modern scene. Economic decision-making became more centralized in Moscow, and huge investments were aimed at increasing agricultural output, expanding industrial production, and modernizing the nation's basic infrastructure.

Marxist ideology reconfigured the countryside. As collectivization spread, the agricultural landscape was reordered: field boundaries and cropping complexes reflected state-established production goals; rapid mechanization replaced human and animal labor with tractors; and entirely new areas were put into agricultural production to boost total food output. One of the most ambitious schemes to expand Soviet agricultural production was the sweeping Virgin and Idle Lands Project that put approximately 100 million acres (40.5 million hectares) of land east of the Volga River into new cultivation during the 1950s. Hopes ran high as pastures were converted to grain fields in the semiarid expanses of southern Siberia and northern Kazakhstan. Unfortunately, production increases were modest at best, as the region's marginal agricultural environment refused to live up to the expectations of Soviet planners. Ironically, the most significant increases in Soviet food production in the 1950s and 1960s did not come from grandiose production schemes such as the Virgin and Idle Lands Project. Rather, the liberalization of rules concerning private plots—small garden-sized fields of potatoes, vegetables, and pasturage—succeeded in significantly raising food output across the country. These private plots produced significant quantities of food and set the stage, during the post-Soviet period, for the selective privatization of the agricultural landscape.

Soviet-directed industrialization also redefined the country's economic geography. The nation's successful development of heavy industries during the 1930s proved vitally important in World War II. Although the German Army advanced to within miles of Moscow, new centers of heavy industry in the southern Urals and Siberia's Kuznetsk Basin remained untouched, supplying the Russian army with the materials necessary to conduct the war (Figure 9.31). Emerging victorious but exhausted in 1945, the Soviet Union also was able to ship large amounts of industrial equipment back home from occupied portions of Germany.

The industrial economy boomed in the 1950s and 1960s, establishing the industrial complexes that remain dominant today. Both the Urals and Kuznetsk industrial zones expanded further in the postwar period. While isolated from major population centers, these manufacturing zones enjoyed excellent access to raw energy (coal, oil, and natural gas) and metals (iron ore, bauxite, and copper) resources (Figure 9.32). As World War II demonstrated, they also provided the strategic advantage of dispersing industrial production to various parts of the country, thus lessening the vulnerability of the manufacturing sector in future conflicts. Cities such as the steel-production center of Magnitogorsk in the southern Urals appeared almost overnight and became showplaces of Soviet industrial dynamism. Farther east, Novosibirsk and Novokuznetsk boomed in southern Siberia, producing steel, farm machinery, transportation equipment, and chemicals.

Soviet industrial geography also expanded in European portions of the country after 1950. The energy-rich Volga Region dominated Soviet oil production until new Siberian fields were opened in the 1970s. Even today, petrochemical manufacturing remains important in the region, along with varied metallurgical and transportation-related industries. Moscow and its nearby cities (the Central Industrial Region) are less well-endowed with raw materials, but they benefit from some of the country's best infrastructure, a large skilled labor force, and domestic market demand for industrial goods. As a result, the country's oldest manufacturing region specializes in a variety of high value-added products that continue to be crucial to the nation's industrial output, particularly in consumer-oriented goods. Elsewhere, the Eastern Ukraine Industrial Region remains important to the south, benefiting from nearby access to raw materials. Major Soviet-era developments near the Krivoy Rog iron ore deposits and in the coal-rich Donetsk Basin continue to benefit the region's economy, although a lack of foreign investment and real economic reforms within the country have led to declining industrial production in the late 1990s.

Much of the Russian region's basic infrastructure—its roads, rail lines, canals, dams, and communications networks—also originated during the Soviet period and reflect the planning necessities of that era. Indeed, massive projects to improve Soviet infrastructure were pursued with patriotic fervor. While some proved to be a waste of resources, many have been excellent long-term investments that form the backbone of the post-Soviet Union's economic geography. Dam and canal construction, for example, turned the main rivers of European Russia into a virtual network of interconnected reservoirs. Invaluable links such as the Volga-Don Canal (completed in 1952), which connected those two key river systems, greatly facilitated the movement of industrial raw

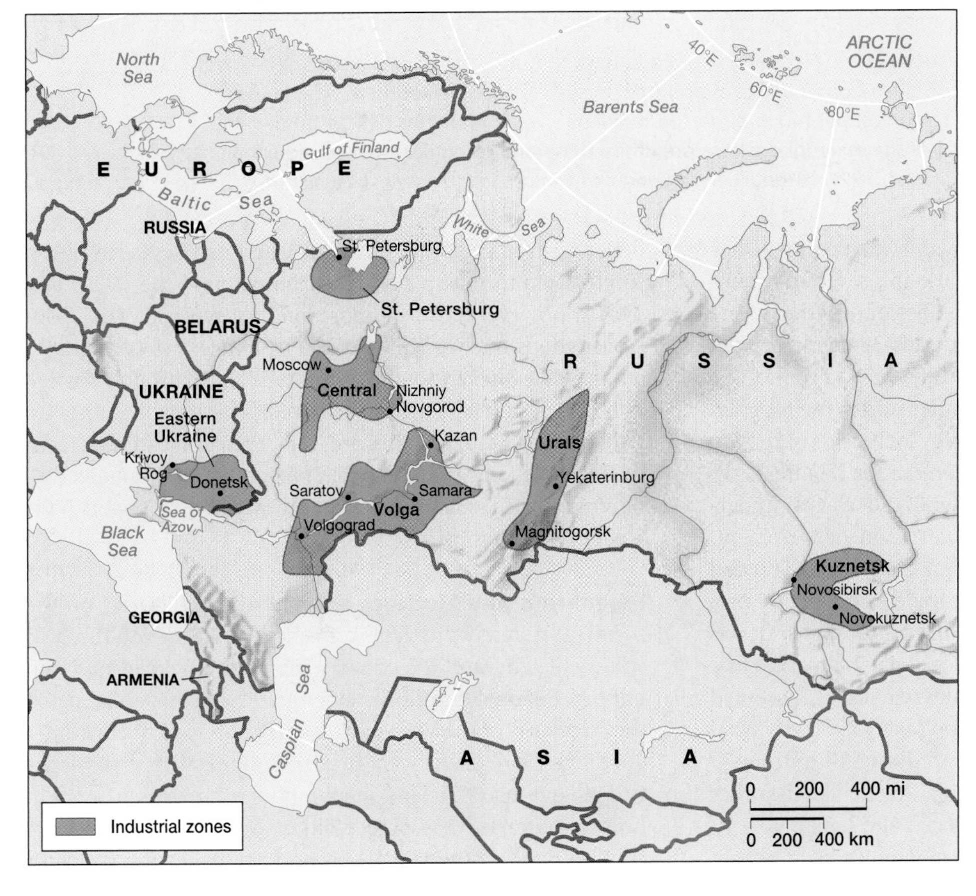

◀ **Figure 9.31 Major industrial zones** Most of the region's principal industrial complexes are located in the west, although the Ural Mountains and southern Siberia experienced great industrial expansion after World War II. Many of Ukraine's heavy industries are concentrated in its eastern mineral-rich zones. In Russia, the Volga and Central Industrial Regions remain dominant, producing most of the nation's manufacturing goods. *(Modified from Rubenstein, 1999,* Introduction to Human Geography, *Upper Saddle River, NJ: Prentice Hall)*

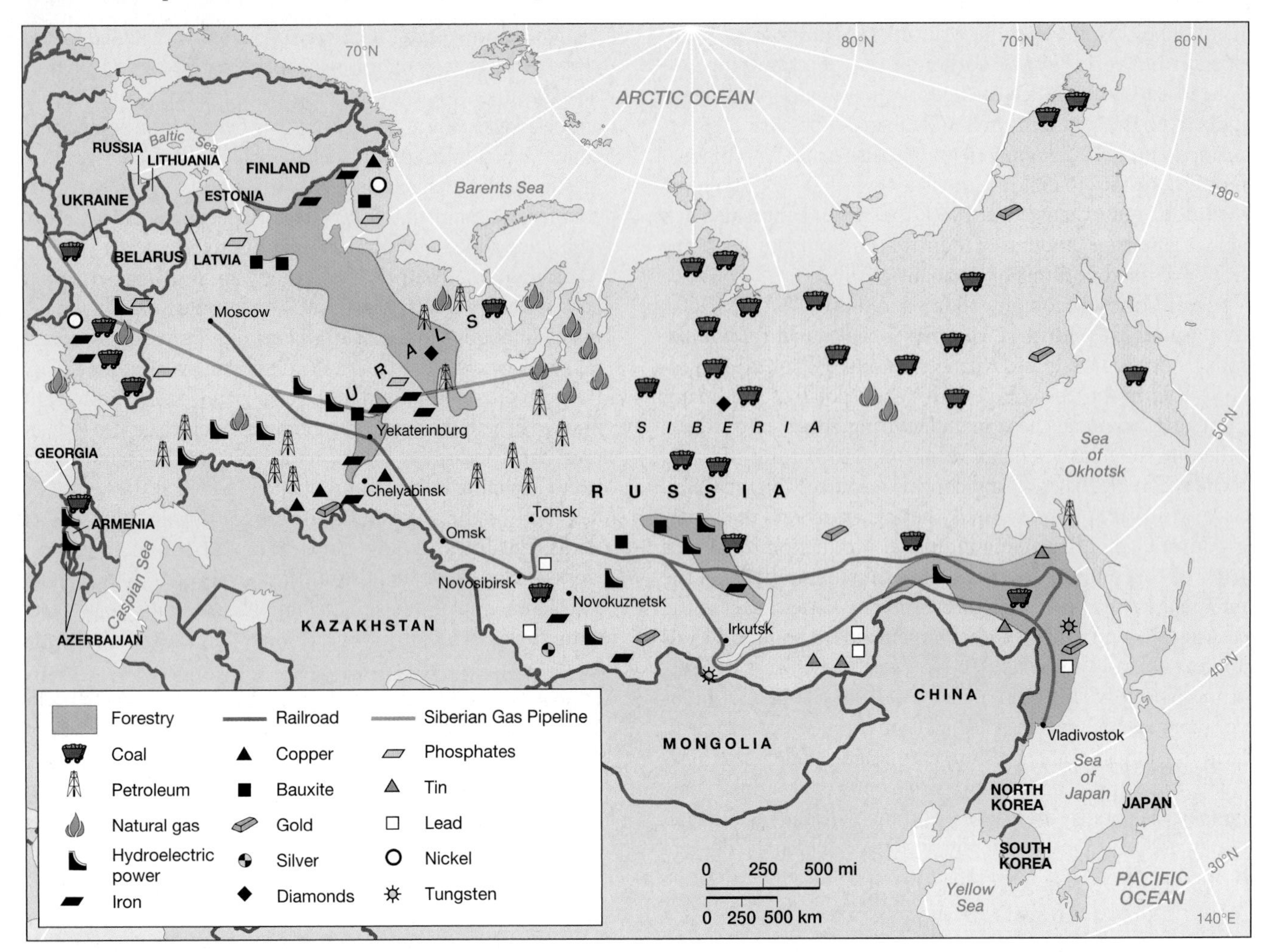

▲ **Figure 9.32 Major natural resources** The region's varied natural resources are widely distributed. Fossil fuels are in abundance, although their distance from markets often imposes special costs. In southern Siberia, rail corridors offer access to many mineral resources, but much of northern Siberia's wealth remains difficult to develop. The mineral-rich Urals and eastern Ukraine offer examples where proximity to natural resources sparked industrial expansion. *(Modified from Bergman and Renwick, 1999,* Introduction to Geography, *Upper Saddle River, NJ: Prentice Hall)*

materials and manufactured goods within the country. They continue to do so today. The Soviets also improved the country's railroad network. Thousands of miles of new track were added in the European West, and the Trans-Siberian Line was modernized and then complemented by the addition of the BAM link across central Siberia. Farther north, the Siberian Gas Pipeline was built to link the energy-rich fields of the Soviet Arctic with growing demand within Europe (Figure 9.33).

The Soviet economy boomed after World War II, underpinned by industrial growth and infrastructure development. During the 1950s, its rates of industrial expansion exceeded those of the United States, and Soviet leaders confidently predicted their country would be the world's most highly developed nation by 1980. The launch of *Sputnik*, the world's first artificial satellite, suggested that Soviet science even seemed to be outpacing American science during the 1950s. The postwar period also saw real economic improvements for the Soviet people. No longer a peasant nation, Russia, like the other Soviet republics, was now industrialized and largely urbanized. Before World War II, most urban families had to share housing; even the more fortunate often had access to only a single room. A massive housing campaign in the 1950s and 1960s, however, ensured most families a private flat. Consumer goods, such as televisions, telephones, and even private automobiles, began to spread more widely through society. The Soviet social system also achieved some notable successes. Literacy was virtually universal, quality health care was readily available, and every capable person was guaranteed employment. In terms of having eliminated dire poverty, the Soviet economy was a success.

Stagnation and Decline Despite these victories on the social front, many problems remained. Soviet agriculture was still inefficient, and the country increasingly relied on grain imports. Manufacturing efficiency and quality failed to match the standards of the West, particularly in regard to consumer goods. Soviet citizens were burdened by recurrent shortages and usually had to spend many hours each week waiting in line to purchase basic necessities. Such difficulties were compounded in the late 1970s and 1980s, as the country's

▲ **Figure 9.33 Siberian Gas Pipeline** Snaking its way across the remote Russian taiga, the Siberian Gas Pipeline connects arctic gas fields to needy consumers in far off Europe. The costly pipeline requires frequent maintenance in the harsh northern environment. *(Tass/Sovfoto/Eastfoto)*

overall economy slowed considerably. By that time, a number of Soviet leaders and scholars were beginning to conclude that their country was not only failing to catch up with the West, but that it was actually falling further behind. Rigid centralized planning simply did not have the flexibility necessary to provide adequate goods and services. Equally troubling was the fact that the Soviet Union was failing to participate fully in the technological revolutions that were transforming the United States, Europe, and Japan, a failure that had serious military consequences.

Disparities also visibly grew between the Soviet elite and its everyday citizens, despite the country's official egalitarianism. Political leaders shopped in private stores and avoided the long lines that most people faced to buy shoddier goods in the state-managed retail outlets. The same privileged individuals often had access to chauffeured limousines and vacation homes in the countryside. Workers who could exploit the shortages of the Soviet economy to their own advantage also gained privileges. Automobile mechanics, for example, could prosper by supplying spare parts only to those willing to pay a premium price. Such disparities of wealth, which sometimes seemed to have little to do with skill or education levels, further undermined faith in the socialist system.

There were also few gains in personal freedoms. Although not as repressive as it had been in the 1930s, the Soviet Union remained a totalitarian state in which writers and artists could be imprisoned or sent to Siberian exile merely for expressing views contrary to those held by the government. In the late 1970s and early 1980s, a protracted and seemingly pointless war in Afghanistan further undermined public support. By the late 1980s, the Soviet Union had reached both an economic and a political impasse.

The Post-Soviet Economy

Fundamental economic changes have shaped the Russian region since the demise of the Soviet Union. Particularly within Russia itself, much of the highly centralized state-controlled economy has been dismantled, replaced by a mixed economy of state-run operations and private enterprise. The changeover has been very difficult. Fundamental problems of unstable currencies, corruption, and changing government policies plague the system. The result has been an unprecedented economic decline for much of the Russian region during the 1990s. Russian GNP, for example, fell 9.2 percent annually between 1990 and 1996. Steel output declined from almost 70 million tons in 1992 to less than 50 million tons late in the decade. Indeed, the evolution of the Russian region's economy remains as clouded as the future of its political identity.

Persisting Economic Turmoil Even as hopes initially ran high for a smooth restructuring of the post-Soviet economy, recent years have been extraordinarily difficult for most residents of the region. In 1992 the Russian government suddenly freed prices from state control, encouraging more production and filling store shelves. As prices rose, however, inflation soared, wages failed to keep pace, and savings were wiped out as the value of the Russian currency (ruble) declined. Another dramatic move came in October 1993 when the government initiated a massive program to privatize the Russian economy. Millions of Russians were given the option to buy into newly privatized agricultural lands and industrial companies. These initiatives greatly opened the economy to more private initiative and investment. Unfortunately, however, a lack of legal and financial safeguards also invited many abuses and often resulted in mismanagement and corruption within the new system.

Russia's agricultural sector continues to struggle. No matter its economic system, Russia will always be challenged by the limitations of its physical setting. Short growing seasons, poor soils, and moisture deficiencies will always characterize much of the region. Russia's best farmlands remain limited to a small slice of southern territory wedged between the Black and Caspian seas, leaving most of the rest of the country on the agricultural margins. In addition, the recent drastic economic changes have not been easy for the nation's farmers. About two-thirds of the country's farmland was privatized by the late 1990s, with many farmers forming voluntary cooperatives or joint-stock associations to work the same acreage they did under the Soviet system. While crop prices have risen, costs have gone up even faster, and many farmers are not skilled in the uncertainties of a market-driven agricultural economy. Elsewhere, the government has held some agricultural land in the form of state farms, employing workers less willing to take risks in the fickle farm economy. Overall, agriculture continues to employ about 15 percent of Russia's workforce, and the basic geography of crops remains little changed from the Soviet era.

Within the industrial sector, the trend has also been toward rapid privatization of state-controlled assets. Initially, thousands of privatized retailing establishments appeared, and they now dominate that portion of the economy. Large portions of the natural resource and heavy industrial sectors of the economy were then privatized. By the end of 1994, the government reported more than 100,000 enterprises were in private hands, and three years later, about 70 percent of the nation's wealth was generated in the private sector, a radical departure from the Soviet period. Gazprom, the huge Russian natural gas company, exemplifies the process. Privatized in the spring of 1994, about one-third of the company was auctioned off to Russian citizens, 15 percent went to managers and workers, 10 percent remained in the company's treasury, and most of the rest was kept in government hands. Employing 385,000 people and controlling one-third of Earth's known natural gas reserves, Gazprom has been nicknamed "Russia, Inc." Indeed, the firm sponsors its own political party in the country and its international dealings often carry the same weight as Russian foreign policy.

Privatization, at scales both great and small, has also brought tremendous challenges. The privatization process has often been corrupt, liberally lining the pockets of company insiders and trade union officials who obtained controlling interest of many corporations. Special state decrees permitting privatization also funneled large sums of money to politicians and their Swiss bank accounts. The rise of the so-called *Russian mafia* has also complicated economic transactions. Some experts estimate that organized criminal elements, some with foreign connections and others based in the Caucasus, controlled more than 40 percent of the Russian economy in 1996 (see "Local Voices: The Russian Mafia"). Bribery and protection are part of the cost of doing business, and many entrepreneurs must turn to criminal elements to obtain loans in an economy where willing banks or foreign lending institutions are in short supply. Not surprisingly, one recent study of global corruption found that Russia ranked as one of the six most corrupt countries in the world.

Even though a small class of Russian elite has become wealthy with these economic changes, most workers have seen little real economic gain (Figure 9.34). In the current economic turmoil, corporate managers depend on barter transactions to obtain basic parts and materials for the products they manufacture. Cumbersome national and regional tax policies prevent efficient revenue collections from most private companies, and thus the government sees little benefit from such operations. For the average Russian worker, unemployment and homelessness are rising, wages often fail to be paid for months, and prices have skyrocketed for many imported consumer goods.

Beyond Russia itself, other nations of the region have also seen their economies struggle in the post-Soviet period. Although one of the wealthiest republics in the Soviet Union, Belarus saw its economy decline rapidly through the 1990s. The Belarussian government has resisted privatizing the country's agricultural lands. Collectives and state farms still account for 80 percent of the country's agricultural acreage. The country also has refused to embrace market reforms. Although economic ties with neighboring Russia remain strong, there has not been much outside investment from western Europe or the United States. Nearby Ukraine has taken a similar course. Less than 15 percent of the country's farmlands were in private hands in 1997, and most of its heavy industrial sector continued under inflexible state control. While some reforms have encouraged privatization of smaller businesses, the country's reluctance to proceed with similar measures for larger-scale industries has delayed its integration with the European economy. The nation's gross national product declined at an average rate of 13.5 percent annually between 1990 and 1997. Georgia and Armenia also suffered through tough economic times in the late 1990s. While both of these Transcaucasian countries have embarked on ambitious economic reform programs that may encourage more outside investment, persisting political problems continue to limit immediate hopes for a rapid economic turnaround.

▲ **Figure 9.34 Living in the lap of luxury** At one extreme, Russia's new entrepreneurial elite revel in the opportunities that have opened up amid recent economic reforms. Greater regulatory stability and a more stable currency, however, may be needed to offer more sustainable and widely shared paths to economic growth. *(B. Brecelj/Sygma Photo News)*

A Fraying Social Fabric Tough economic times and political uncertainties have contributed to a fraying social fabric within the Russian region (Table 9.3). Rates of violent crime increased late in the Soviet period, accelerating after the fall of communism. Organized criminal activities have profited from fewer state restrictions on economic activity, and street crime has escalated with growing urban poverty and insecurity. Rising unemployment, higher housing costs, and declining social welfare expenditures also have hit many families hard. Often, both husband and wife may have multiple, but low-paying jobs with few benefits and long hours.

Women are paying an especially heavy price amid the current economic and social problems. Increasing domestic violence is symptomatic of the recent turmoil that has particularly impacted women throughout the region. Divorce rates also continue to rise as families struggle to make ends meet. Economically, women often have suffered first as unemployment has risen. Forced out of traditional industrial and service sector jobs, many have been compelled to sell homemade products or services on city streets for extremely low wages (Figure

LOCAL VOICES The Russian Mafia

Victor A. Shabalin is a professor of political science at the Khabarvosk Graduate Militia College in Russia. A member of the Russian Criminological Association, he is widely considered one of the foremost experts on organized crime in Russia. The following passages, taken from a 1995 interview with Professor Shabalin, indicate his views on the role of the Russian Mafia:

In the summer of 1994, the Russian Ministry of Internal Affairs (MVD) estimated that 25% of the Russian gross national income was derived from organized criminal activities. The MVD also believed that 5,600 criminal groups were involved primarily in capital/money laundering, the drug business, and extortion. Consequently, President Yeltsin issued the decree "On The Urgent Measures To Defend The Population Against Gangsterism And Other Kinds of Organized Crime," on June 14. This decree suspended many of the existing laws that protected individual rights. The measure was controversial, but it was supported by the majority of the Russian people as a necessary evil to fight organized crime. The Russian criminal justice experts believe that these laws will help control the economic criminalization caused by organized criminal associations and will encourage the development of honest business. They also hope that this decree will mark the turning point in the fight against organized crime.

President Yeltsin's decree precipitated polemical debate in the Russian society. It caused heated discussions among lawyers, political scientists, and the political elite. Many of them think that the decree contradicts the Constitution and violates a number of rights and freedoms of the citizens. The majority of the commercial structures and a significant part of the press have a negative attitude toward the decree. The parliamentary faction "Choice of Russia, . . ." vehemently expressed the opinion to abolish it. Though there are many supporters of the decree, one of the extreme positions is taken by Zhirinovsky's parliamentary faction, the Liberal-Democratic Party [which is actually on the extreme right]. The L.D.P. considers the decree too mild and demands that the army participate in the fight against organized crime by using summary executions to shoot the mob leaders and the use of field-court marshal to purge the corrupt members of the MVD (national police).

In this case, the fight against organized crime is becoming part of the reform and recovery process of the Russian society and might be effective. If these laws are effective, organized crime little by little will gravitate to its traditional spheres (drugs, prostitution, pornography, gambling, etc.), where efficient criminal justice control can be taken. Time will show what scenario will develop from the fight against organized crime in Russia.

Several years later, Shabalin's hopes for better times have yet to be realized, and many elements of the Russian mafia still dominate major sectors of the struggling economy.

Table 9.3 Social Indicators and Status of Women

Country	Life Expectancy at Birth		Under Age 5 Mortality, per 1,000 Live Births		Secondary School Enrollment %		Female Labor Force Participation (% of total)
	Male	Female	1960	1993	Male	Female	
Armenia	69	76	—	33	80	90	48
Belarus	62	74	—	22	89	96	49
Georgia	69	76	—	28	—	—	46
Russia	61	73	—	31	84	91	49
Ukraine	62	73	—	25	65	95	49

Sources: *Population Reference Bureau Data Sheet, 1997,* Life Expectancy (M/F); *World Resources Institute, 1996–97,* Under-5 Mortality Rate; *Population Reference Bureau Data Sheet, 1997,* Secondary School Enrollment (M/F); *The World Bank Atlas, 1998,* Female Participation in Labor Force.

9.35). Many ordinary housewives and schoolteachers are forced into prostitution, simply as a way to survive. Politically, women in Russia, Belarus, and Ukraine serve in fewer governmental positions than their counterparts in western Europe or the United States, suggesting that women are not able to play a high-profile role in shaping major social and economic policies within the region. Still, reforms undertaken during the Communist era have produced high female literacy rates and school enrollments (Table 9.3) in all of the countries within the Russian region.

Overall, Russian students have seen a mix of costs and benefits associated with the end of the Soviet era. Although the Soviets invested heavily in schools and literacy rates remain high, declining government revenues are negatively impacting the quality of education within the region. State-supported institutions such as Moscow University are feeling the squeeze. Making matters worse, many top scholars and scientists are emigrating to other countries as they see salaries and academic opportunities disappear. Still, ordinary Russians are getting a fresh educational curriculum.

▲ **Figure 9.35 Life on the street** This young street vendor at the Moscow Food Market hopes to sell her pickled vegetables to passing consumers. Economic instability, higher unemployment, and falling currencies, however, make even modest purchases more difficult for many Russians. *(John Egan/ Hutchison Picture Library)*

Soviet history has been rewritten, once-banished works of literature are now available in libraries, and hundreds of private schools and universities have opened, offering new educational alternatives for thousands of Russian students.

Russia's health care crisis also illustrates the interplay between the region's troubled economy, dwindling government expenditures, and the declining quality of life. The crisis suggests that many of the roots of current social problems predate the fall of communism. For example, Soviet expenditures on health care averaged 6.6 percent of the country's gross national product in 1960 but declined to 2.3 percent by the end of the Soviet period. By the mid-1990s, Russia's annual per capita medical expenditures stood at $20.37, by far the lowest in the industrial world. Unfortunately, the nation's shrinking commitment to medical care comes at a time when several major health problems are on the rise. A shortage of vaccines and medical services has contributed to the return of diseases such as cholera, typhus, and the bubonic plague. Chronic illnesses, often related to lifestyle, produce the highest mortality rates in the developed world for cardiovascular disease, alcoholism, and smoking, particularly for Russian men. Indeed, one recent study suggests that half of all Russian men and one-third of Russian women are plagued by long-term drinking problems and that as many as 50 percent of Russian deaths are related to alcohol consumption. Most Russian men (66 percent) also smoke, creating a growth industry for multinational tobacco companies eager to increase sales. In addition, toxic environmental conditions have extracted a huge price from the Russian region, although precise estimates are difficult to make. Overall, the health-care crisis worsened in the late 1990s and will make economic recovery and improved social welfare that much more difficult to achieve during the early twenty-first century.

The Russian Region and the Global Economy The relationship between the Russian region and the world beyond has shifted greatly since the end of communism. During much of the Soviet era, the region was relatively isolated from the world economic system. By the 1970s, however, the Soviet Union began to export large quantities of fossil fuels to the West, while importing more food products. Connections with the global economy quickly intensified with the downfall of the Soviet Union. As Russia and its neighbors selectively embraced market economics, the region intersected with the global economy in several ways. Most visibly, a barrage of new consumer imports reached their markets. McDonald's hamburgers, Calvin Klein jeans, and other symbols of global capitalism were soon visible in the heart of Moscow. Luxury goods from the West found a small but enthusiastic market among the newly emergent Russian elite, a group noted for its devotion to BMW automobiles, Rolex watches, and other status emblems. Most Russians, however, found such goods far beyond their increasingly pressured budgets. Inexpensive consumer goods from East and Southeast Asia, and from industrializing neighbors such as Turkey, however, also flooded the region. Russia, moreover, remained a food-deficit country, importing grain, meat, and sugar.

While the Russian region has offered multinational companies new markets for their consumer goods, other global connections have been more tenuous. The widespread privatization of Russian natural resource and manufacturing operations has not prompted a flood of buyers to the Russian securities markets. One recent study suggests that Russian insiders control 80 percent of privatized Russian companies, offering a limited role for foreign investors. The instability of Russian currency, stock, and debt markets has also cooled interest in foreign investment. In addition, few foreign companies have come into the region to build their own manufacturing operations. Potential entrepreneurs have often been discouraged by the gyrating economy, government red tape, complex tax codes, and shrinking domestic demands for manufactured goods within the region. The political instability of the late 1990s was another disincentive to global investment. Although the countries of the region have grown closer to global institutions such as the World Bank and the International Monetary Fund, their recently unpredictable economies will probably slow future economic connections with western Europe or the United States. On the other hand, economic links between the former republics will continue to play a major role in patterns of trade within the region.

Trade patterns reflect the Russian region's economic relationships with the world beyond and suggest the challenges imposed by globalization (Figure 9.36). Russian exports are dominated by natural resources such as timber, petroleum products, and metals. Although these raw materials provide critical cash, fluctuating global prices for commodities make the Russian economy vulnerable to forces far beyond its control. Dramatic declines in the world price of oil, for example, greatly reduce the value of those exports for the Russian balance of trade. The largest recipients of Russian exports are Ukraine and Belarus in the former Soviet Union as well as

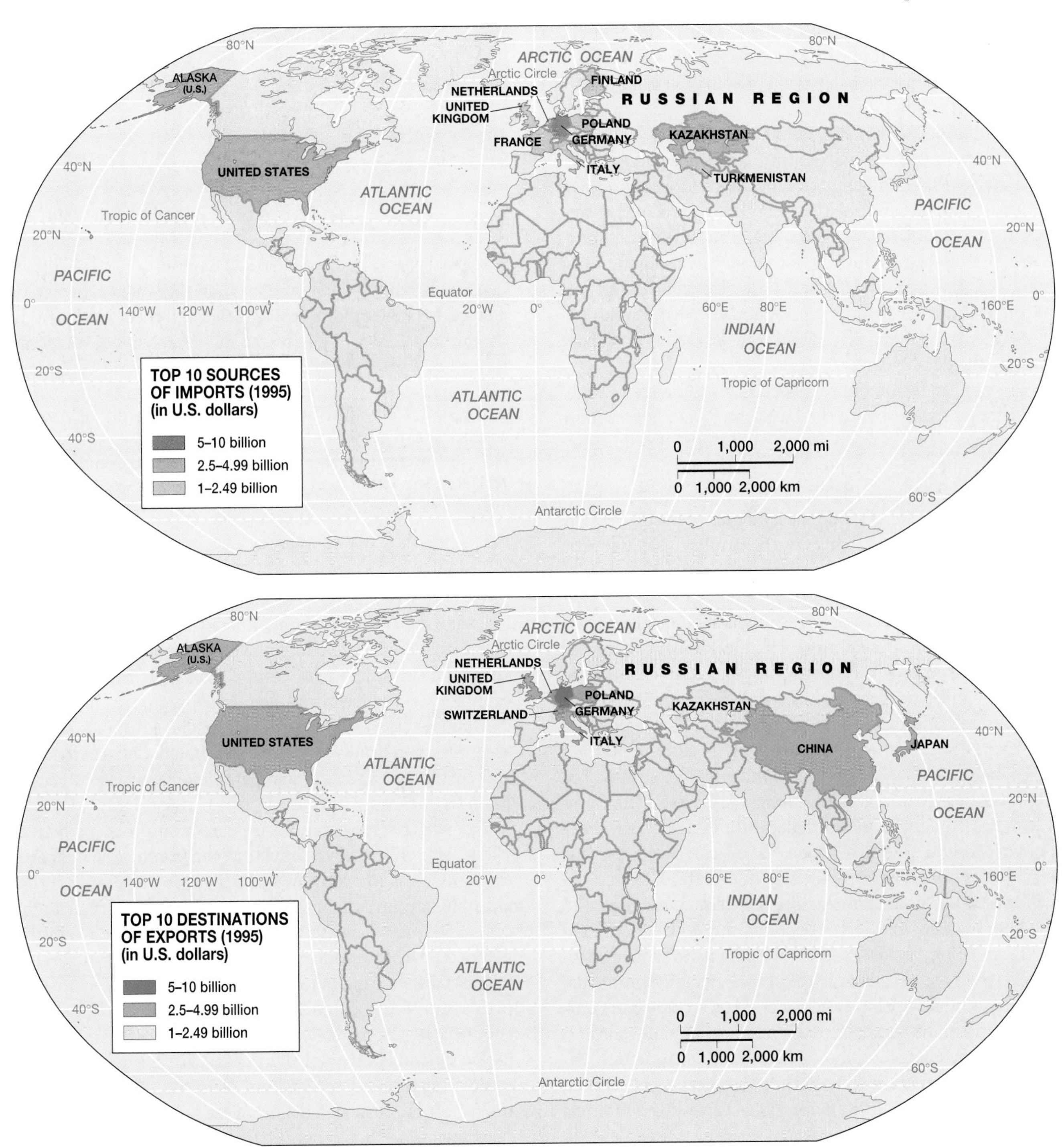

▲ **Figure 9.36 Global trade and the Russian region** Economic problems and political uncertainties have recently hurt the region's participation in international trade, but better economic times ahead may offer opportunities for expansion. Important trading partners include neighboring states, western Europe, Japan, China, and the United States. Natural resources dominate the flow of exports, while consumer goods are imported from the wealthy countries of the West. *(Data from* Europa Yearbook, *1998)*

western Europe, Japan, and the United States. The geography of Russian imports is oriented around these same countries, although manufactured goods and food products lead the incoming flow of trade. Imports are also a function of shifting global trends. For example, dramatic declines in the relative value of the Russian currency and in the ability of Russians to buy foreign goods led to a 55 percent decline in imports between the summer and fall of 1998.

Beyond Russia, the other four countries of the region maintain significant trading ties with former Soviet republics. Ukraine imports a great deal of raw materials (particularly oil) from Russia, and its exports (principally metals, food, and machinery) are led by return flows to the north. Still, the country is building stronger trading ties with western Europe and the United States. Belarus also retains close links with Russia, even though its trade with Europe tripled (both exports and imports) between 1993 and 1996. Given their location, the Transcaucasian countries of Georgia and Armenia generate somewhat different trading connections than those of Russia, Ukraine, and Belarus. Georgian exports (metals and food) and imports (energy and mineral products) are dominated by Russia and Turkey, but nearby Azerbaijan, Turkmenistan, Bulgaria, and Romania are also significant trade partners. Armenia is more closely linked with Russia, but trade ties to the United States, Iran, and western Europe have grown considerably since 1991.

The Russian region has witnessed many trade-related consequences associated with the breakup of the Soviet Union. The collapse of the communist state transformed planned internal exchanges between republics into less-predictable flows of foreign trade. Under the old regime, for example, government officials made certain that Ukraine received adequate supplies of oil from elsewhere in the country, just as Russia would obtain cotton for its textile industry from Uzbekistan. Under the new economy, Ukraine must buy oil, or barter other products for it, from Russia, just as Russia suddenly must purchase more cotton from abroad. The breakdown of the once tightly integrated Soviet economy has resulted in chaotic economic circumstances marked by numerous production bottlenecks and shortages, and has necessitated complex and sensitive negotiations among the former Soviet republics.

Globalization has also impacted different locations within the Russian region in very distinctive ways. In Russia, capitalism has brought its most dramatic, though selective benefits to Moscow. Luxury goods, fancy cars, and upscale restaurants cater to the city's newly rich. Indeed, more than half of new foreign investment into the country has gone into the Moscow area. Although it contains only 6 percent of Russia's population, almost 25 percent of the nation's tax collections (1997) are generated in the area. Aggressive municipal authorities in the city have also managed to become part owners in hundreds of local businesses and foreign joint ventures, thus guaranteeing a continuing flow of urban revenues. Beyond Moscow, cities such as Nizhniy Novgorod and Samara have also succeeded in attracting global capital by building a pro-business economic environment. Italy's Fiat Corporation, for example, recently committed $850 million for a new manufacturing plant in Nizhniy Novgorod, recognizing that the city and its leading private companies could offer a favorable economic setting for investment. In addition, port cities such as Vladivostok are well positioned to take advantage of their accessibility to nearby markets, even though political red tape and corruption have still hampered much growth in Pacific Basin trade (Figure 9.37).

▲ **Figure 9.37 Vladivostok** The busy harbor of Vladivostok remains Russia's leading trade center in the Far East. With easy access to markets in Japan, China, and even the United States, this Pacific port is poised to grow as Russia's economy recovers. *(Itar-Tass/Sovfoto/Eastfoto)*

Elsewhere, globalization has clearly imposed penalties. Older, less competitive industrial centers in the Urals and eastern Ukraine have been hit hard. Aging steel plants, for example, are no longer guaranteed markets for their high-cost, low-quality products as they were in the days of the planned Soviet economy. Instead, they must compete on the global market, a market that is increasingly prone to lower prices and weakening demand for many of the industrial goods the region produces. Russia's extractive centers are similarly vulnerable in a global economy that has recently driven down commodity prices. Indeed, the time has passed when the region can insulate itself from the larger world.

Conclusion

Many influences have shaped the human geographies of the Russian region. Much of its underlying cultural geography was forged centuries ago, the complex product of Slavic languages, Orthodox Christianity, and numerous ethnic minorities that continue to complicate the scene today. Further transforming the contemporary cultural milieu is an onslaught of global influences, a set of products, technologies, and attitudes that increasingly mingle with traditional cultural orientations. Much of the region's political legacy is rooted in the Russian Empire, a land-based system of colonial expansion that greatly enlarged Russian influence after 1600 and was then reasserted in somewhat different form during the Soviet period. Only large remnants of that Empire survive on the modern map, yet it still has stamped the geopolitical character of the region in lasting ways. Much of the region's economic geography remains defined by its abundant, yet fickle natural setting, as well as by the fundamental Marxist-guided changes wrought by twentieth-century industrialization and urbanization. The result is a human scene forged from diverse historical roots and a human geography crafted from forces that range from Tatar nationalism to twenty-first-century consumerism.

Looking toward the future, no other portion of the world faces uncertainties as deep as those confronting residents of the Russian region. Even its basic geopolitical structure remains in doubt. Will Belarus rejoin Russia? Even Ukraine and Russia could form a new federation, although if this were to happen, people in western Ukraine might decline to join. Northern Kazakhstan in Central Asia, home to some eight million Russians, is another target for possible Russian expansion. The most nationalistic Russian leaders would like to go several steps further and recreate the Soviet Union, if not the old tsarist Empire.

While partial reunification of the Russian region remains a possibility, there is also the potential for further fragmentation. Many non-Russian peoples within Russia might prefer independence. Secession also appeals to ethnic Russians who live in remote but resource-rich portions of the country. The Russian government appears determined to prevent political fragmentation, but its financial resources and even its military credibility were strained by the revolt of the small republic of Chechnya, which, in fact, seceded. One can only wonder what would happen if several new rebellions were to break out simultaneously.

Equally uncertain are the region's future political and economic systems, which will significantly influence its international relations. Russia's leaders in the immediate post-Soviet period promised a quick transition to European-style representative democracy and capitalism, as well as rapid integration into the global economic system. The subsequent economic decline and social chaos, however, have largely discredited this approach in the eyes of the Russian population. Meanwhile, both former communists and extreme nationalists have gained core political constituencies. The communists argue that more state control of the economy can return the region to greater stability. Right-wing populists stress Russian nationalism, social control, antipathy to the international capitalist system, and opposition to the United States. If either group gained power, the world could revert to Cold War conditions, although a much-weakened Russia would probably not pose the same threat to the West that the Soviet Union once did.

Even if Russia and its neighbors were to solve their political problems, serious economic and environmental difficulties would remain. In order to produce cash, Russia has moved backward from an economy based on industrial production to one oriented around the extraction and export of natural resources. The old industrial system was both inefficient and highly polluting, and its decline has had environmental benefits in some areas. Elsewhere, however, the intensified quest for raw materials has resulted in accelerating environmental degradation. The weakened condition of the government also has undermined efforts to conserve natural resources, while economic decline has made pollution control equipment difficult to obtain. In the absence of an economic miracle, it is difficult to see how the environmental ills of the Russian region can be cured in the near future. The vast size and generally low population density of the region does, however, offer some measure of environmental protection.

Despair remains the greatest challenge facing the people of the region. The disappointments they have faced have been monumental. Communism spectacularly failed in its promise to bring about a social and economic utopia. Unfortunately, with little experience in banking institutions and legal protections, Russia's experimentation with capitalism has brought not the freedom and prosperity its supporters promised but rather crime and poverty. Under such circumstances, it is not surprising that so many have sought solace in alcohol and other forms of escape. More than anything, the people of the Russian region need a renewal of hope, a prospect that tomorrow's world might be better than today's.

Key Terms

autonomous areas (page 387)
Baikal-Amur Mainline (BAM) Railroad (page 375)
Bolsheviks (page 386)
centralized economic planning (page 394)
chernozem soils (page 367)
Cold War (page 388)
collective farms (page 394)
Commonwealth of Independent States (CIS) (page 390)
Cossacks (page 379)
denuclearization (page 391)
Eastern Orthodox Christianity (page 379)
exclave (page 388)
glasnost (page 388)
Gulag Archipelago (page 375)
Iron Curtain (page 388)
mikrorayons (page 378)
perestroika (page 388)
permafrost (page 367)
podzol soils (page 367)
Russification (page 376)
Slavic peoples (page 379)
social realism (page 385)
Trans-Siberian Railroad (page 374)
tsars (page 375)

Questions for Summary and Review

1. Compare the climate, vegetation, and agricultural conditions of Russia's European West with those of Siberia and the Russian Far East.
2. Describe some of the high environmental costs of industrialization within the Russian region.
3. Discuss how major river and rail corridors have shaped the geography of population and economic development in the region. Provide specific examples.
4. How might the policy of Russification be seen as a Soviet-style expression of traditional Russian imperialism?
5. Describe some of the major land-use zones in the modern Russian city and suggest why it is important to understand the impact of Soviet-era planning within such settings.
6. What were the key phases of colonial expansion during the rise of the Russian Empire and how did each enlarge the reach of the Russian state?
7. What are some of the key ethnic minority groups within Russia and her neighboring states, and how have they been recognized in the region's geopolitical structure?
8. What was Halford Mackinder's heartland theory of Russian geopolitics? Has history borne out Mackinder's predictions?
9. Describe how centralized planning created a new economic geography across the former Soviet Union. How has it had a lasting impact?
10. Briefly summarize the key strengths and weaknesses of the post-Soviet Russian economy and suggest how globalization has shaped its evolution.

Thinking Geographically

1. How might it be argued that Russia's natural environment is one of its greatest assets as well as one of its greatest liabilities?
2. In the future, how might the forces of capitalism and free markets reshape the landscapes and land uses of cities within the Russian region?
3. What options do peoples such as the Volga Tatars or the Siberian Buryats have for preserving their cultural autonomy? What might be some of the advantages and disadvantages of such peoples pressing for greater political autonomy?
4. On a base map of the Russian region, suggest possible political boundaries 20 years from now. What forces will work for a larger Russian state? A smaller Russian state?
5. What were some of the greatest strengths and weaknesses of centralized Soviet-style planning between 1917 and 1991? How were Ukraine and Belarus impacted? Why did the system ultimately fail?
6. Does the threat of organized crime in the region justify curtailing basic human freedoms? Why is organized crime such a critical problem in this part of the world?
7. How have the growing forces of globalization impacted Russian culture?
8. From the perspective of a 22-year-old Russian college student and resident of Moscow, write a short essay that suggests how the economic and political changes over the past 10 to 15 years have changed your life.

Regional Novels and Films

Novels

Fyodor Dostoyevsky, *The Brothers Karamazov* (orig. pub., 1879; 1995, Bantam Books)

Fyodor Dostoyevsky, *Crime and Punishment* (orig. pub. 1866; 1984, Bantam Classics)

Eugenia Ginzburg, *Journey into the Whirlwind* (orig. pub. 1967; 1975, Harcourt Brace)

Nikolai Gogol, *Dead Souls* (orig. pub. 1842; 1961, Viking Press)

Mikhail Lermantov, *A Hero of Our Time* (orig. pub. 1840; 1966, Penguin)

Andrei Makine, *Dreams of My Russian Summers* (1998, Scribner's)

Aleksandr Solzhenitsyn, *One Day in the Life of Ivan Denisovich* (orig. pub. 1962; 1998, Signet Classics)

Leo Tolstoy, *War and Peace* (orig. pub. 1869; 1982, Viking)

Films

Burnt by the Sun (1994, France/Russia)

Dr. Zhivago (1965, U.S.)

Gorky Park (1983, U.S.)

Leo Tolstoy's Anna Karenina (1997, U.S.)

Little Vera (1988, USSR)

Rasputin (1985, USSR)

Reds (1981, U.S.)

Russia House (1990, U.S.)

War and Peace (1968, USSR)

Bibliography

Adams, A. E.; Matley, I. M.; and McCagg, W. O. 1966. *An Atlas of Russian and East European History*. New York: Praeger.

Bater, James H. 1996. *Russia and the Post-Soviet Scene: A Geographical Perspective.* New York: John Wiley & Sons.

Brawer, Moshe. 1994. *Atlas of Russia and the Independent Republics.* New York: Simon and Schuster.

Channon, John (with Robert Hudson). 1995. *The Penguin Historical Atlas of Russia*. New York: Penguin.

Cole, Jonathan P. 1984. *Geography of the Soviet Union*. London: Butterworths.

Demko, George J., et al., eds. 1996. *Population Under Duress: The Geo-demography of Post-Soviet Russia.* Boulder, CO: Westview.

Dewdney, John C. 1979. *A Geography of the Soviet Union*. Oxford: Pergamon.

Dukes, Paul. 1990. *A History of Russia: Medieval, Modern, Contemporary*. Durham, NC: Duke University Press.

"Facing Oblivion, Rust-Belt Giants Top Russian List of Vexing Crises." 1998. *New York Times,* November 8.

Forsyth, James. 1992. *A History of the Peoples of Siberia: Russia's North Asian Colony 1581–1990.* Cambridge: Cambridge University Press.

Gachechiladze, Revaz. 1995. *The New Georgia: Space, Society, Politics*. College Station: Texas A&M University Press.

Gooding, John. 1996. *Rulers and Subjects: Government and People in Russia, 1801–1991*. London: Arnold.

Halperin, Charles J. 1987. *Russia and the Golden Horde: The Mongol Impact on Medieval Russian History.* Bloomington: Indiana University Press.

Hauner, Milan. 1992. *What Is Asia to Us? Russia's Asian Heartland, Yesterday and Today.* London: Routledge.

Hutenbach, Henry. 1996. *The Caucasus: A Region in Crisis*. Boulder, CO: Westview.

"In Search of Spring: A Survey of Russia." 1997. *The Economist,* July 12.

Kaiser, Robert J. 1994. *The Geography of Nationalism in Russia and the USSR.* Princeton: Princeton University Press.

Khazanov, Anatoly M. 1996. *After the U.S.S.R.: Ethnicity, Nationalism, and Politics in the Commonwealth of Independent States.* Madison, WI: University of Wisconsin Press.

Lapidus, Gail W., ed. 1994. *The New Russia*. Boulder, CO: Westview.

Lincoln, W. Bruce. 1994. *The Conquest of a Continent: Siberia and the Russians.* New York: Random House.

Lydolph, Paul E. 1990. *Geography of the USSR.* New York: Wiley.

Medvedkov, Yuri; Medvedkov, Olga; Smith, W. Randy; and Krischynas, Raymond. 1993. "Cities of the Former Soviet Union." In Stanley D. Brunn and Jack F. Williams, eds., *Cities of the World: World Regional Urban Development*, pp.150–193. New York: HarperCollins.

Powell, David E. 1998. "The Dismal State of Health Care in Russia." *Current History* 97:335–341.

Pryde, Philip R., ed. 1995. *Environmental Resources and Constraints in the Former Soviet Republics.* Boulder, CO: Westview.

Rodgers, Allan, ed. 1990. *The Soviet Far East: Geographical Perspectives on Development.* London: Routledge.

Rorlich, Azade-Ayse. 1986. *The Volga Tatars: A Profile in National Resilience*. Stanford: Stanford University Press.

Shaw, Denis, ed. 1995. *The Post-Soviet Republics: A Systematic Geography*. New York: John Wiley & Sons.

Solzhenitsyn, Aleksandr I. 1997. *The Gulag Archipelago 1918–1956: An Experiment in Literary Investigation.* Boulder, CO: Westview.

Stewart, John Massey, ed. 1992. *The Soviet Environment: Problems, Policies and Politics*. Cambridge: Cambridge University Press.

Valencia, Mark J., ed. 1995. *The Russian Far East in Transition: Opportunities for Regional Economic Cooperation.* Boulder, CO: Westview.

"Will Russia Hold Together?" 1998. *The Economist,* September 12.

Wixman, Ronald. 1984. *The Peoples of the USSR: An Ethnographic Handbook*. Armonk, NY: M. E. Sharpe.

Wood, Alan, ed. 1991. *The History of Siberia: From Russian Conquest to Revolution.* London: Routledge.

CENTRAL ASIA
Political Map
Over 1,000,000
500,000–1,000,000
Selected smaller cities
RUSSIA
UKRAINE
GEORGIA
ARMENIA
AZERBAIJAN
Caspian Sea
Baku
Kara Bogaz Gol
Aral Sea
Lake Tengiz
Astana (Akmola)
KAZAKHSTAN
Lake Balkhash
Ob R.
Kyzyl Kum Desert
Syr Darya R.
UZBEKISTAN
TURKMENISTAN
Kara Kum Desert
Kara Kum Canal
Amu Darya R.
Ashgabat
Bukhara
Tashkent
Samarkand
Fergana Valley
KYRGYZSTAN
Bishkek
Almaty
TAJIKISTAN
Dushanbe
Pamir Mountains
Pamir Knot
Mazar-e Sharif
Hindu Kush
AFGHANISTAN
Kabul
Helmand R.
Indus R.
Karakoram Range
LADAKH
IRAN
PAKISTAN
Tien Shan
Dzungaria
Urumqi
Turfan Depression
XINJIANG
Tarim Basin
Taklamakan Desert
Kunlun Shan
Altai Mountains
MONGOLIA
Ulaanba
Go
Tsaidam Basin
Xining
QINGHAI
Tibetan Plateau
TIBET (XIZANG)
Yangtze R.
Mekong R.
Salween R.
Lhasa
Brahmaputra R.
HIMALAYAS
NEPAL
BHUTAN
Ganges R.
BANGLADESH
BURMA (MYANMAR)
INDIA
U.A.E.
SAUDI ARABIA
OMAN
Tropic of Cancer
Arabian Sea
Bay of Bengal
0 250 500 mi
0 250 500 km
60°E
70°E
90°E

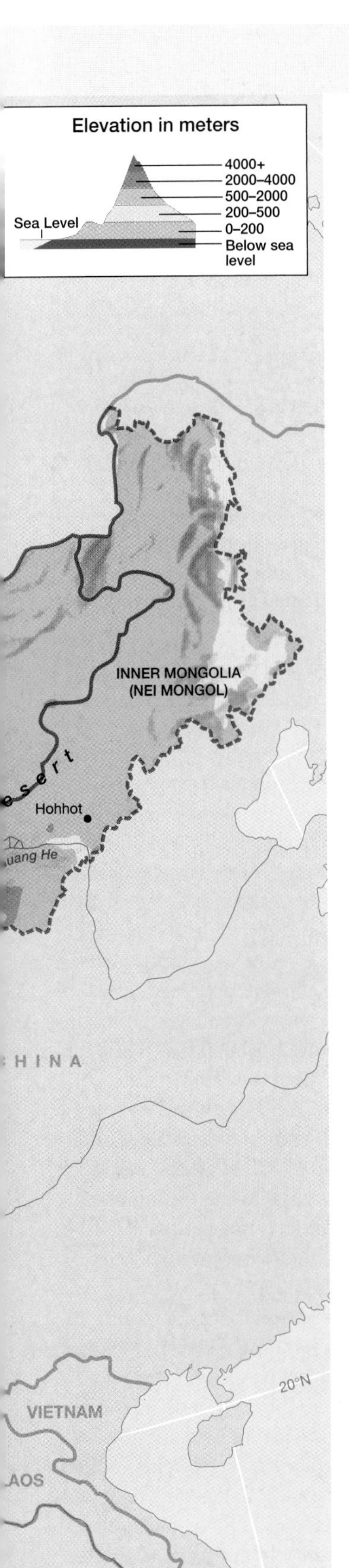

Central Asia

Central Asia does not appear in most books on world regional geography. Although the region covers a larger area than the United States, it is dominated by high mountains, barren deserts, and semiarid steppes (grasslands). It remains lightly populated, inhabited by only some 108 million persons. Central Asia has also been something of a geopolitical void, long dominated by external forces. Until 1991 it contained only two independent countries, Mongolia and Afghanistan. The rest of the region was at that time divided between the Soviet Union and China (Figure 10.1).

Although long absent from world maps, Central Asia began to reappear in discussions of global geography following the breakup of the Soviet Union. Suddenly, six new countries appeared on the international scene. The reemergence of independent states in this region has prompted scholars in many disciplines to reexamine the position of Central Asia in human affairs. Historians have convincingly shown that the region as a whole not only has a certain historical coherence, but also once played a pivotal role in the political drama of the entire Eurasian continent (see "Setting the Boundaries").

At present, Central Asia remains something of a backwater in the global system. It is poorly integrated into international trade networks, and it has often been overlooked by those concerned with global geopolitics. But there are signs of change. In the 1980s and 1990s, large oil and gas reserves were found, especially in Kazakhstan, Turkmenistan, and Azerbaijan; evidence is mounting that larger discoveries await. Estimates of the total oil reserves in the region run between 70 and 200 billion barrels, second only to the 600 billion barrels in the Persian Gulf area. As a result, Western oil companies are showing increasing interest in the area, although red tape and political complications have thus far prevented the realization of large profits. A number of important countries, moreover, are seeking to exert influence over Central Asia, including Iran, Pakistan, Turkey, the United States, and Russia. China's controversial control and periodic repression of its territories in Central Asia also highlights the significance of the region.

Central Asia forms a large, compact region in the center of the Eurasian landmass. Alone among all the world regions, it lacks ocean access. Owing to its continental position in the center of the world's largest landmass, Central Asia is noted for its rigorous climate. High mountains, deep basins, and extensive plateaus, as we shall see, accentuate its climatic extremes.

◀ **Figure 10.1 Central Asia** Central Asia, a vast, sprawling region in the center of the Eurasian continent, is dominated by arid plains and basins and lofty mountain ranges and plateaus. Eight independent countries—Kazakhstan, Turkmenistan, Uzbekistan, Kyrgyzstan, Tajikistan, Azerbaijan, Afghanistan, and Mongolia—form Central Asia's core. China's lightly populated far west and north are often placed within Central Asia as well, due to patterns of both cultural and physical geography.

Setting the Boundaries

The term "Central Asia" is defined differently by different writers. Most authorities agree that it includes five newly independent, former Soviet republics: Kazakhstan, Kyrgyzstan, Uzbekistan, Tajikistan, and Turkmenistan. This chapter, however, adds another post-Soviet state, Azerbaijan, in addition to Mongolia and Afghanistan, as well as the autonomous regions of western China (Tibet and Xinjiang). Several other provinces and regions of western China, such as Nei Mongol (Inner Mongolia) and Qinghai, are occasionally discussed.

The inclusion of these additional territories within Central Asia is controversial. Azerbaijan is often classified with its neighbors in the Caucasus region (Georgia and Armenia), western China is obviously part of East Asia by political criteria, and Mongolia is also often placed within East Asia because of both its location and its historical connections with China. Afghanistan, for its part, is just as often located within either South Asia or Southwest Asia and the Middle East. Indeed, some writers would include the entire former-Soviet zone within Southwest Asia—although others link it instead with a post-Soviet world region.

But considering Central Asia's historical unity, its common environmental circumstances, and its recent reentry onto the stage of global geopolitics, it deserves consideration in its own right. It also makes sense to define its limits rather broadly. Azerbaijan, for example, is linked by both cultural (language and religion) and economic (oil) factors more to the other countries of Central Asia than to Armenia and Georgia.

At the same time, however, any unity that Central Asia as a whole possesses is far from stable. Continuing Chinese political control over, and Han Chinese migration into, southeastern Central Asia threatens whatever claims may be made for regional coherence. Central Asia itself remains, moreover, deeply divided along cultural lines. Most of the region is Muslim in religious orientation and Turkish in language, but both the northeastern and southeastern section (Mongolia and Tibet) are firmly Buddhist. Only time will tell whether Central Asia will indeed merit recognition as a distinct world region in its own right.

Environmental Geography: Steppes and Deserts of the Eurasian Heartland

Roughly generalizing, the physical geography of Central Asia is dominated by grassland plains (or steppe) in the north, desert basins in the southwestern and central areas, and high plateaus and mountains in the south-center and southeast (Figure 10.2). Lofty mountains do extend, however, into the very heart of the region, dividing the desert zone into a series of separate basins.

The Central Asian Highlands

The highlands of Central Asia originated in one of the great tectonic events of Earth's history: the colliding of the Indian subcontinent into the Asian mainland. This ongoing collision has created the highest mountains in the world, the Himalayas, located where South Asia and Central Asia join together. The Himalayas, for all their grandeur and fame, are merely one portion of a much larger network of high mountains and plateaus. To the northwest they merge into the Karakoram Range and then the Pamir Mountains. From the so-called Pamir Knot, a complex tangle of mountains situated where Pakistan, Afghanistan, China, and Tajikistan converge, other towering ranges radiate outward in several directions. The Hindu Kush sweeps to the southwest through central Afghanistan, the Kunlun Shan extends to the east (along the northern border of the Tibetan Plateau), and the Tien Shan swings out to the northeast into China's Xinjiang province (Figure 10.3). All of these ranges have peaks higher than 20,000 feet (6,000 meters) in elevation. Much lower but still significant ranges are found in the southwestern reaches of the region, along Turkmenistan's boundary with Iran and along Azerbaijan's boundaries with Russia, Armenia, and Iran.

Much more extensive than these mountain ranges, however, is the Tibetan Plateau. This massive upland extends some 1,250 miles (2,000 kilometers) from east to west and 750 miles (1,200 kilometers) from north to south. More remarkable than its size is its elevation; virtually the entire area is higher than 12,000 feet (3,700 meters) above sea level, and its average height is about 15,000 feet (4,600 meters)—higher, in other words, that the highest mountains in the contiguous United States.

The Tibetan Plateau is ringed with extremely high mountains. The Himalayas to the south reach 29,000 feet (8,800 meters), while the Karakoram to the west reach 28,000 feet (8,500 meters). The eastern boundary of the plateau is marked not by a single mountain wall but rather a series of deep gorges alternating with high ranges (Figure 10.4). Many of these peaks are higher than 20,000 feet (6,000 meters), while the intervening valleys plunge many thousands of feet, making eastern Tibet one of the world's most topographically forbidding places.

Most of the large rivers of South, Southeast, and East Asia originate in the Tibetan Plateau and adjoining mountains, including the Indus, Ganges, Brahmaputra, Salween, Mekong, Yangtze, and Huang He. These rivers pass either through the canyon-lands of eastern Tibet or through gaps in the Himalayas to the south. Most of the Tibetan Plateau, however, drains internally. Runoff from mountain snowfields trickles down to evaporate from innumerable lakes and marshes. These lakes and wetlands are generally saline, since salt accumulates wherever bodies of water do not drain to the sea.

The greater part of the Tibetan Plateau, at about 15,000 feet (4,600 meters) of elevation, lies near the maximum elevation at which human life can exist. Rather than forming a flat, table-like surface, the plateau is punctuated with east-west running ranges alternating with undrained basins. The largest of these

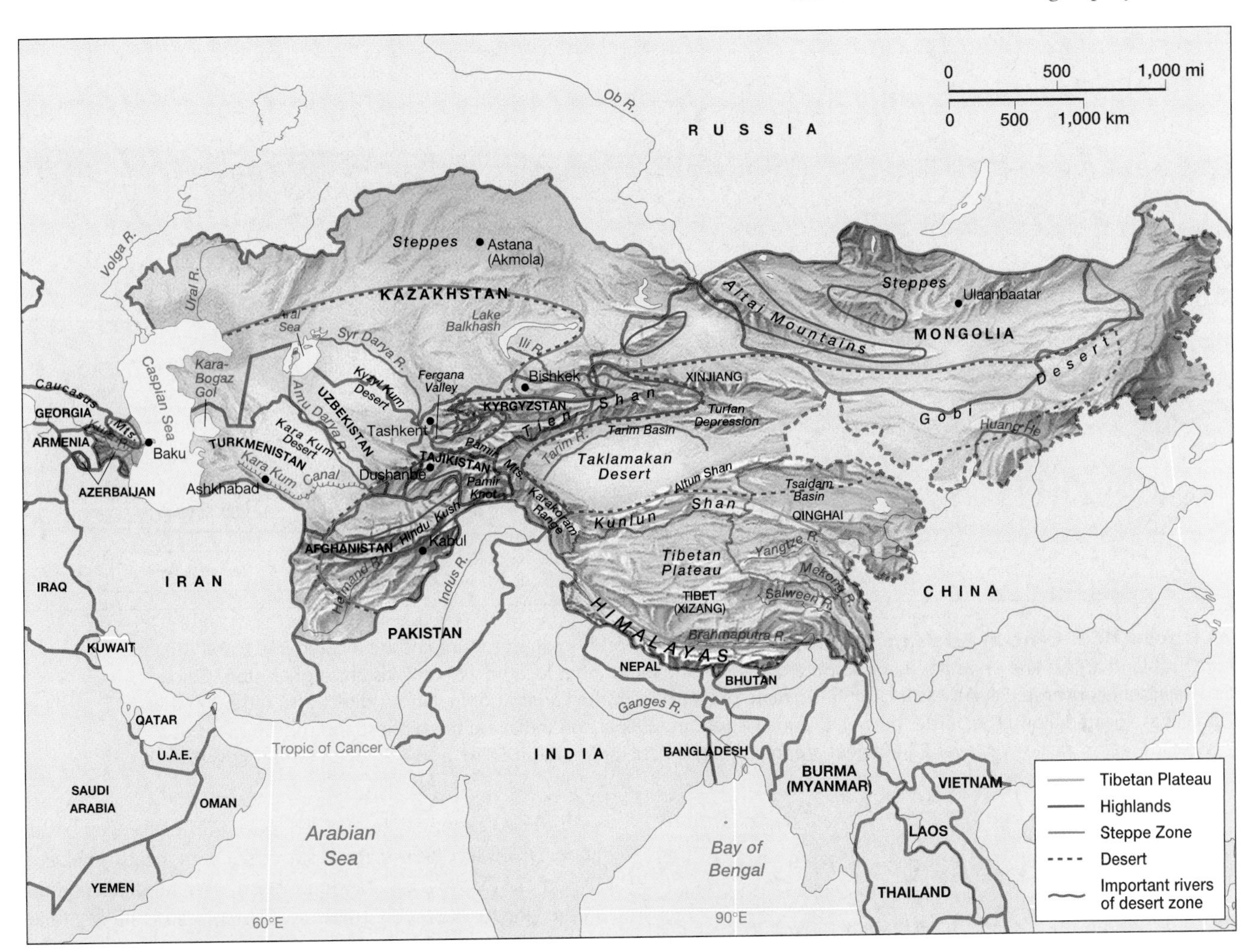

▲ **Figure 10.2 Physical regions of Central Asia** Central Asia is divided into three main regions based on physical geography. The north is dominated by relatively flat, grassy plains, known as the steppes. Most of the central portion of Central Asia is covered by desert plains and basins. Scattered throughout Central Asia, but particularly pronounced in the south, are the mountain ranges and plateaus of the highland zone. Since most of Central Asia is arid, rivers running out of the highlands have special significance for the region's human geography.

basins, the Tsaidam, is also the lowest, situated below 9,000 feet (2,700 meters). Many of Tibet's higher ranges contain glaciers, which owe their existence to cold temperatures more than to heavy snowfall. Although the southeastern sections of the plateau receive ample precipitation, most of Tibet is arid. Cut off by high ranges from any source of moisture, large areas of the plateau receive only a few inches of rain a year (Figure 10.5). Winters on the Tibetan Plateau are cold, and while summer afternoons can be warm, summer nights remain chilly.

The Plains and Basins

Although the mountains of Central Asia are higher and more extensive than those found anywhere else in the world, most of the region is actually characterized by plains and basins of low and intermediate elevation. The lower-lying portion of the region can be divided into two main areas: a central belt of deserts punctuated by verdant river valleys and a northern swath of semiarid steppe.

Central Asia's desert belt is itself divided into two discontinuous segments by the Tien Shan and Pamir Mountains. To the west lie the arid plains of the Caspian and Aral Sea basins, located primarily in Turkmenistan, Uzbekistan, and southern Kazakhstan. The most desolate areas, the Kara Kum and Kyzyl Kum deserts (in Turkish, the black and red sands, respectively), support meager vegetation. The Kara Kum contains the most extensive dune fields of Central Asia. Most of this area is relatively flat and very low. The Aral Sea lies only 135 feet (41 meters) above sea level, while the surface of the Caspian Sea is 92 feet (28 meters) lower than sea level. A significant proportion of the country of Azerbaijan, that along the lower Kura River, lies well below sea level. The climate of this region is strikingly continental; summers are dry and hot, while winter temperatures average well below freezing. Conditions along the Caspian Sea, however, are moderated by the massive lake.

Central Asia's eastern desert belt extends for almost 2,000 miles (3,200 kilometers), from the extreme west of China at the foot of the Pamirs to the southeastern edge of Inner Mongolia.

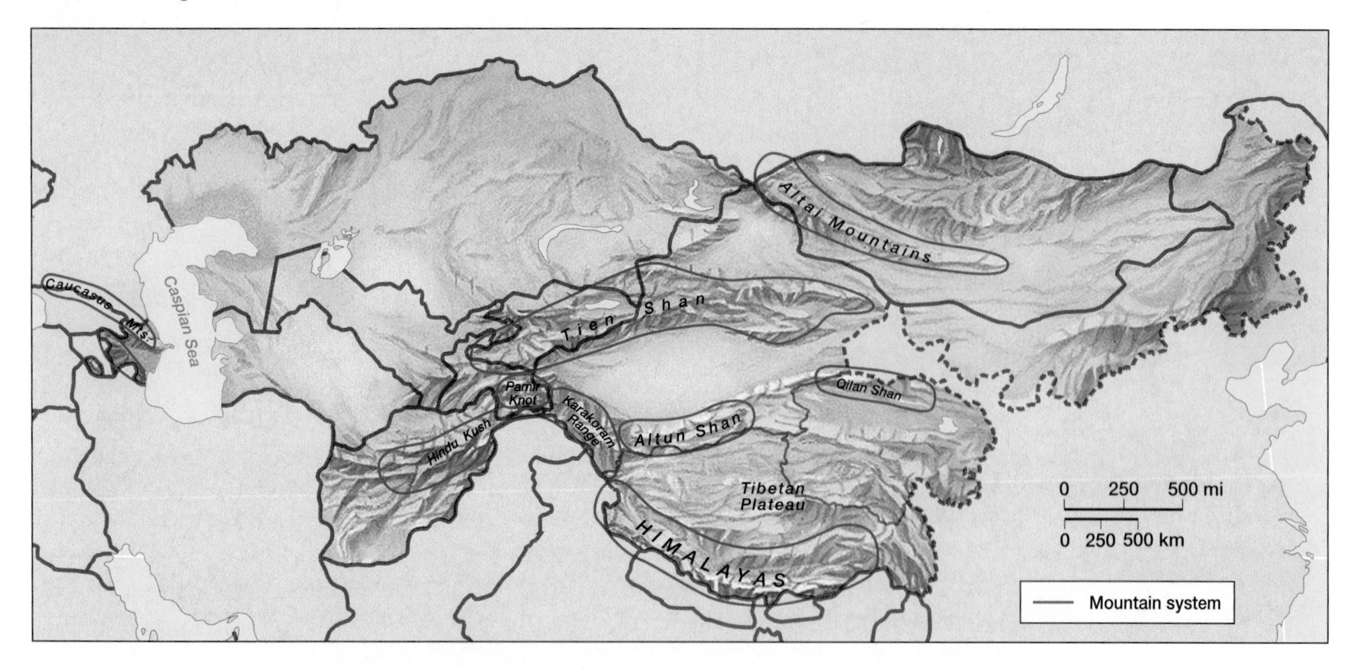

▲ **Figure 10.3 Central Asia's mountain systems** The world's highest mountains are located in Central Asia. Its core highland area is known as the Pamir Knot, a complex tangle of ranges located where Pakistan, Afghanistan, China, and Tajikistan converge. Several distinct ranges radiate outward from the Pamir Knot in various directions. Although the Himalayas, located along the border between China (Tibet) and Nepal and India, are the highest and most famous of these mountains, peaks well over 20,000 feet are found in the Karakoram, Altun Shan, and Tien Shan ranges as well.

▲ **Figure 10.4 Tibetan plateau** The Tibetan Plateau is dominated by alpine grasslands and tundra, interspersed with rugged mountains and saline lakes. In summer the sparse vegetation offers forage for the herds of nomadic Tibetan pastoralists. Much of the northern part of Tibet is too high to sustain such a low-intensity land use. *(Michel Peissel/SIPA Press)*

It is conventionally divided into several different deserts, even though the belt of aridity is continuous. The most important of these deserts are the Taklamakan, found in the Tarim Basin of Xinjiang, and the Gobi, which runs along the border between Mongolia proper and the Chinese region of Inner Mongolia. Much of the interior portion of the Taklamakan is covered by sand and rock and is virtually devoid of vegetation; the central Gobi is also extremely barren. The Tarim itself is a deep basin, about 3,000 feet (900 meters) above sea level, nearly enclosed by some of the world's highest mountains. Several smaller basins in the vicinity are much lower; the Turfan lies at 505 feet (154 meters) below sea level.

The environment of western **Turkestan** (in other words, the former Soviet zone) is distinguished from that of so-called eastern Turkestan (in other words, China's Xinjiang) in part by the much larger size of its rivers. Much more snow falls on the western than the eastern slopes of the Pamir Mountains, giving rise to more abundant runoff. The most important of these rivers, the Amu Darya and the Syr Darya, flow into the Aral Sea, a huge brackish lake. Others, such as the Helmand of Afghanistan, terminate in extensive marshes and salt flats.

Moving north from the deserts of Central Asia, rainfall gradually increases and desert eventually gives way to steppe. Near the region's northern boundary, trees begin to appear in favored locales, outliers of the great Siberian taiga (coniferous forest) of the north. A nearly continuous swath of grasslands extends some 4,000 miles (6,400 kilometers) east to west across the entire region, only partially broken by Mongolia's Altai Mountains. Particularly rich pastures are found in northern and eastern Kazakhstan, in the Dzungaria region of northern Xinjiang (north of the Tien Shan), and in northern and central Mongolia. Winters on the northern steppe are usually bitterly cold, but summer conditions are often quite pleasant.

Environmental Issues

Because of the low population density of Central Asia, much of the area has a relatively clean environment. Industrial pollution is a serious problem in a few of the cities of the former

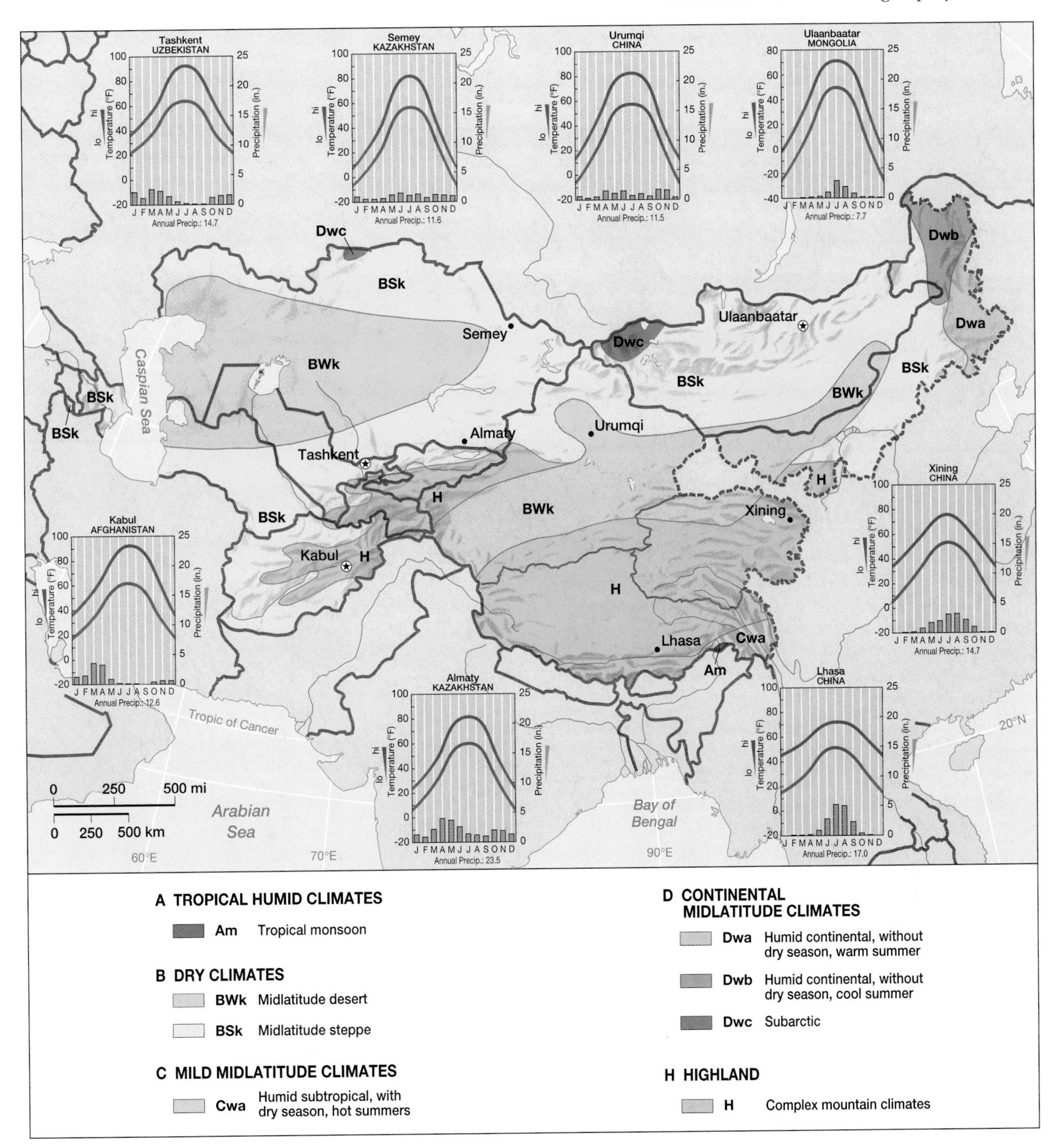

▲ **Figure 10.5 Climates of Central Asia** Central Asia is a dry region dominated by desert and steppe climates. Even in most of Central Asia's highlands, marked "H" on this map, arid conditions predominate. Truly humid areas in Central Asia are found only in limited areas of the far north and extreme southeast. As a midlatitude region located in the interior of a vast continent, Central Asia is marked by pronounced continentality, experiencing profound differences between winter and summer temperatures.

Soviet zone (such as Uzbekistan's Tashkent and Azerbaijan's Baku), but in most areas it is not significant. Some places, such as northwestern Tibet, remain practically pristine, with little human impact of any kind. Elsewhere, however, the typical environmental dilemmas of arid environments plague the region: desertification (the spread of deserts), salinization (the accumulation of salt in the soil), and desiccation (the drying up of lakes and wetlands).

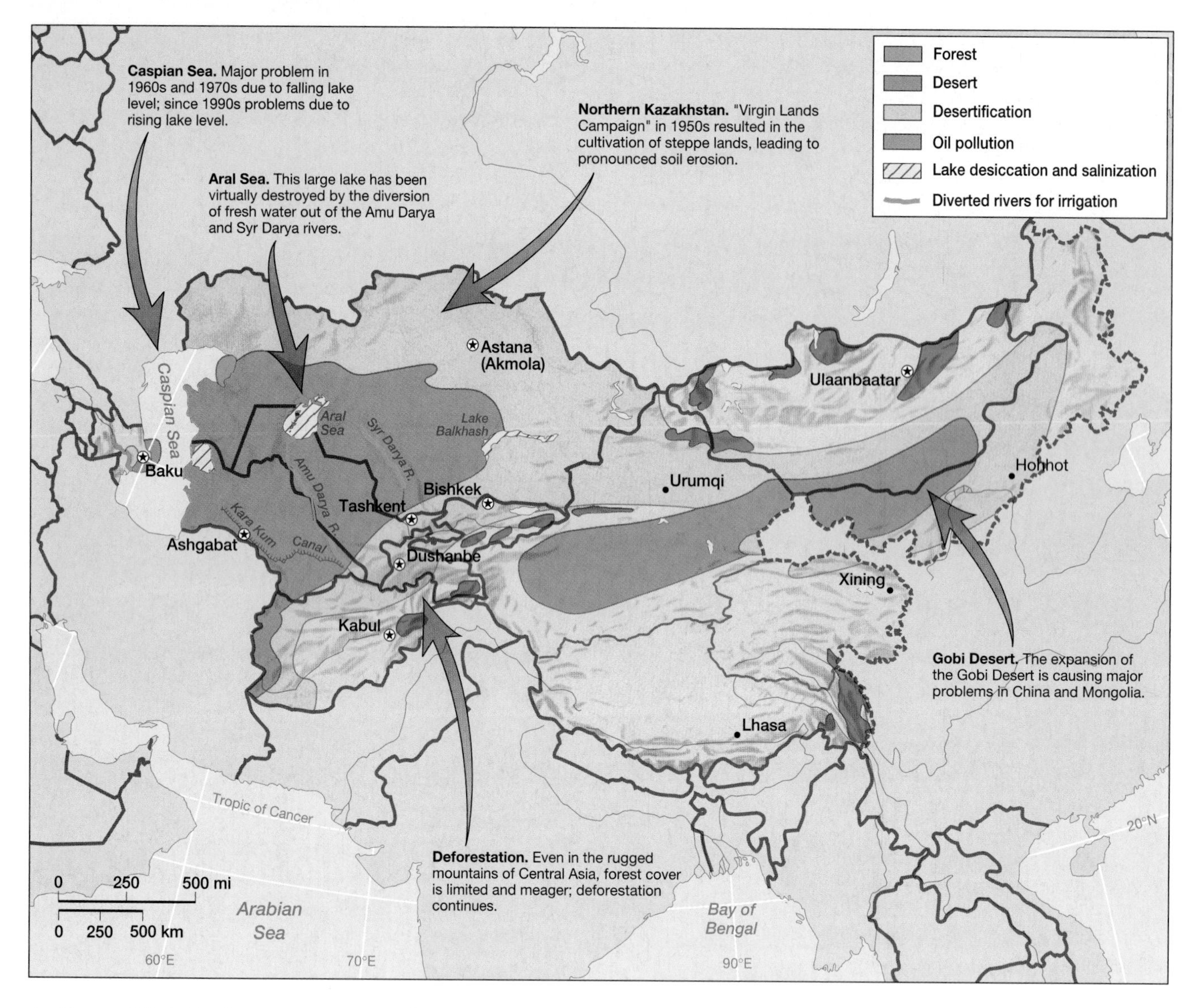

▲ **Figure 10.6 Environmental issues in Central Asia** Central Asia has experienced some of the most severe problems associated with desertification in the world. Soil erosion and overgrazing have led to the advance of desertlike conditions in much of western China and Kazakhstan. In western Central Asia, the most serious environmental problems are associated with the diversion of rivers for irrigation and the corresponding desiccation of lakes. Oil pollution is a particularly serious issue in the Caspian Sea area.

Desertification Desertification is a major concern in Central Asia (Figure 10.6). The Gobi Desert has gradually spread southward, encroaching on densely settled lands in northeastern China proper. The Chinese have tried to stabilize dune fields and to prevent the march of desert with massive tree and grass planting campaigns, but such efforts have only been partially successful. Northern Kazakhstan has also seen the deterioration of its soils. This area was one of the main sites of the ambitious Soviet "virgin lands campaign" of the 1950s, in which marginal areas were plowed and planted with wheat. Many of these lands have since returned to native grasses, but not before erosion stripped away much of their productivity.

Deforestation has also harmed much of the region. Although most of Central Asia is too dry to support forests, many of its mountains were once well wooded. Today, extensive forests can only be found in the wild gorge country of the eastern Tibetan Plateau, in some of the more remote slopes of the Tien Shan, Altai, and Pamir mountains, and in the highlands of northern Mongolia.

Shrinking and Expanding Seas Central Asia has the world's most serious problem of lake desiccation. The western part of the region, despite its overall aridity, contains three of the world's largest lakes and a number of smaller ones as well. As these lakes diminish in size, local fisheries are undermined and eventually destroyed. In the long run, the drying up of the major lakes may even result in a deterioration of climate. A certain (although debatable) proportion of the region's rainfall originates from evaporation from these lakes. Lake desiccation, therefore, could result in an overall decrease in precipitation.

Western Central Asia supports these large lakes because it forms a low-lying basin, without drainage to the ocean, that is virtually surrounded by mountains and other more humid areas. The world's largest lake, by a huge margin, is the Caspian Sea, located along the region's western boundary; the fourth largest is the Aral Sea, situated some 300 miles (480 kilometers) to the east, and the fifteenth largest is Lake Balkhash, found 600 miles (960 kilometers) farther east. The Caspian Sea is roughly the size of Montana, the Aral Sea is (or, more precisely, was) roughly the size of West Virginia, and Lake Balkhash is roughly the size of Connecticut and Rhode Island combined.

Despite their names, neither the Aral nor the Caspian are true seas, since they are not connected with the ocean. They are incorrectly called seas only because they are extremely large and somewhat salty. The Caspian is actually less salty than the ocean (particularly in the north), while until the 1970s the Aral was only slightly brackish (salty) Lake Balkhash is slightly brackish in the west but is quite salty in its long eastern extension.

The Caspian Sea receives most of its water from the large rivers of the north, the Ural and the Volga, which drain much of European Russia. Owing to the construction of large reservoirs and the development of extensive irrigation facilities in the lower Volga basin, the volume of fresh water reaching the Caspian began to decline in the second half of the twentieth century. With a reduced influx of water, the level of the great lake dropped, exposing as much as 15,000 square miles (39,000 square kilometers) of former lake bed. A reduced volume of water resulted in increased salinity levels, undermining fisheries and threatening the entire ecosystem. The Russian caviar industry, centered in the northern Caspian, suffered extensive damage.

The Caspian reached a low point in the late 1970s. At that point it began to rise, and by the late 1990s had risen some 8.2 feet (2.5 meters). This enlargement too has caused problems, inundating, for example, some of the new farmlands in the Volga Delta that had recently been reclaimed. The most serious current environmental threat to the Caspian, however, is probably pollution for the oil industry rather than fluctuation in size (see "Environmental Issues: Saving the Caspian Sea?").

The story of the Aral Sea is not nearly so encouraging as that of the Caspian. The only sources of water for this lake are the Amu Darya and Syr Darya rivers, both of which are intensively used for irrigation. Water has been taken out of the Amu Darya and Syr Darya rivers since antiquity, but the

ENVIRONMENTAL ISSUES Saving the Caspian Sea?

Like other lakes without an outlet to the ocean, the Caspian exhibits tremendous natural fluctuations in water level. During moist periods, the level of the lake rises; during drier times, it falls. In the early 1900s Russian geographer L. S. Berg investigated the history of the Caspian to see if there was any natural periodicity to this rise and fall. He discovered maximum levels in 1650, 1770, and 1900, and minimal levels in 1590, 1710, and 1840. Extrapolating into the future, he concluded that maximum levels would again be reached in 2020. Most climatologists today doubt that climatic fluctuations are as regular as Berg supposed. Still, the recent—and quite surprising—rise in the level of the Caspian does conform to his predictions.

Considering the economic importance of the Caspian, Soviet planners were eager to devise methods of stabilization during the earlier period of falling levels. There are two basic approaches that can be used to slow down or reverse the fall of nondrained lakes: increase the flow of fresh water or reduce evaporation.

The influx of fresh water into the Caspian could be significantly increased only by transferring water from another drainage system. This has already been done to a minor extent, as some water is diverted to the Volga from several rivers of northern European Russia. To make up for the irrigation loss in the lower Volga, however, much larger diversions would be required. In the 1960s a truly ambitious scheme was hatched, one that would have entailed tapping the flow of the huge Ob River, which crosses the sparsely populated wetlands of the West Siberian Plain.

Diverting water from the Ob to the Caspian would require building a canal some 1,500 miles (2,400 kilometers) long. Large amounts of water would be lost to evaporation and seepage along the way. Such a massive transfer project might also cause ecological changes in both the Ob Basin and in the Arctic Ocean, into which the river drains. Such fears, however, have been largely put to rest; by the 1980s, it was obvious that the Soviet Union could not afford the massive price tag that such a project would carry.

By the late 1970s, it had become clear that the second and much less expensive alternative, reducing evaporation, was more feasible. Evaporation can be reduced simply by decreasing the surface area of the lake, which in turn can be accomplished by diking off certain sections. Once this is accomplished, the elevation of the water surface could be stabilized even as the lake itself becomes smaller. The most easily diked area is the almost-enclosed gulf known as the Kara-Bogaz Gol in the Caspian's center-east. It loses large amounts of water to evaporation but receives little influx from either rainfall or streams.

In the early 1980s, construction commenced on a dike across the 1,800-foot-long (555-meter) mouth of the Kara-Bogaz-Gol. By the mid-1990s, the gulf was virtually dry—while the lake itself was beginning to cause problems by rising. Engineering projects obviously do not present a perfect solution for the problems of water control in basins characterized by internal drainage.

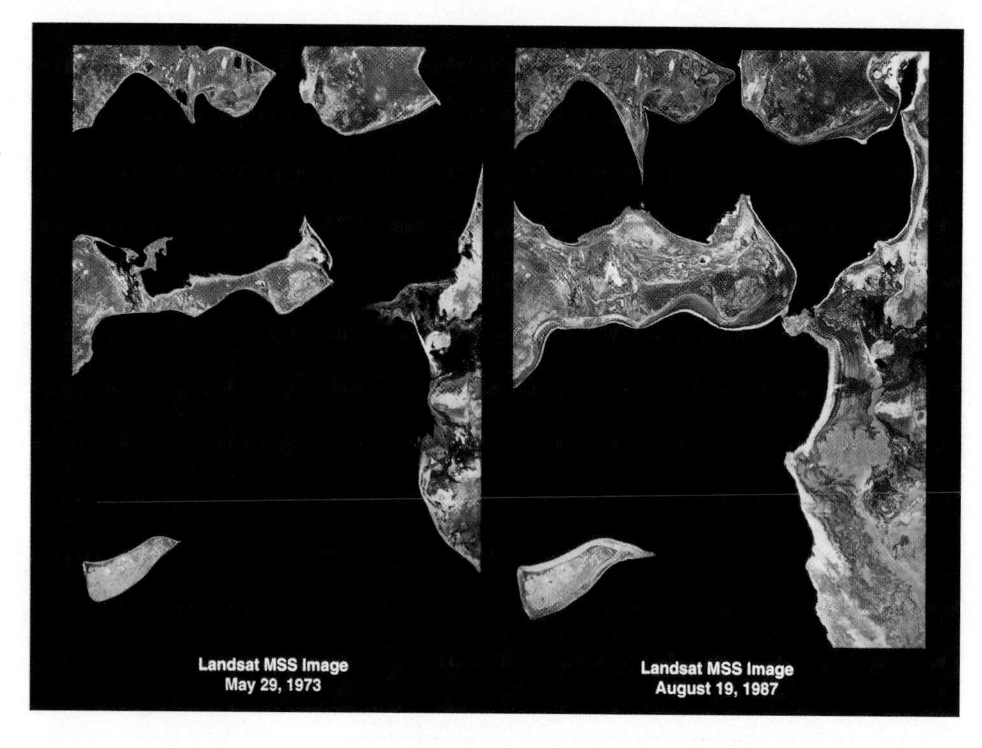

▶ **Figure 10.7 The shrinking of the Aral Sea** These two satellite images, the first taken on May 29, 1973, and the second on August 19, 1987, show the dramatic shrinkage of the Aral Sea—a process that that has continued in the years since the second image was taken. More than 60 percent of the lake's water has been lost since the 1960s, resulting in severe economic damage and environmental degradation. *(EROS Data Center, U.S. Geological Survey)*

scale of diversion vastly expanded after 1950. The valleys of these two rivers formed the southernmost farming districts of the Soviet Union, and thus became vital suppliers of warm-season crops. Cotton in particular emerged as the mainstay of Uzbekistan and Turkmenistan during the 1960s and the 1970s. The acreage devoted to rice—a very water-demanding crop—also increased. Soviet agricultural planners, much like those in the United States during the same period, favored huge engineering projects that could deliver water to arid lands and thus "make the deserts bloom." The biggest of these projects is the Kara Kum canal, which carries water from the Amu Darya across the deserts of southern Turkmenistan.

Unfortunately, the more crops the deserts produced, the less fresh water was available for the Aral Sea. The Aral proved to be much more vulnerable than the Caspian, since it is much smaller and shallower, and its tributaries flow across a larger, and drier, agricultural landscape. By the 1970s, the shoreline began to retreat at an unprecedented rate; eventually a number of "seaside" villages found themselves stranded up to 40 miles (64 kilometers) inland. An estimated 135—out of a total of 173—animal species in the lake have disappeared. New islands began to emerge, and by the 1990s the Aral Sea had been virtually divided into two separate lakes (Figure 10.7). Sixty percent of the lake's total volume of water is estimated to have disappeared.

The destruction of the Area Sea has resulted in economic as well as ecological damage. Fisheries, which were once large enough to support a canning industry, began to shut down as the lake grew increasingly salty. Even agriculture has suffered. The retreating lake (Figure 10.8) has left large salt flats on its exposed beds; windstorms pick up the salt, along with the agricultural chemicals that had accumulated in the lake's shallows, and deposit it in nearby fields. Yields have thus declined and desertification has accelerated.

Since the breakup of the Soviet Union, government planners have not pushed as hard to increase the cotton crop as they once did. But cotton is still a major foreign exchange earner for Uzbekistan and Turkmenistan, and these countries can ill afford to abandon all of the lands that were brought under cultivation during the Soviet period. Efforts to conserve water and to slow down the destruction of the lakes are being made, but conservation has not been granted a high priority.

▲ **Figure 10.8 A dying lake** Due to the shrinkage of the Aral Sea, former lakeside villages are now located far inland, as evidenced by these two beached ships. Not only have fishing economies been destroyed, but the desiccated lake bed itself is now a source of pollution, as desert winds deposit salt and agricultural chemicals on fields. *(David Turnley/Black Star)*

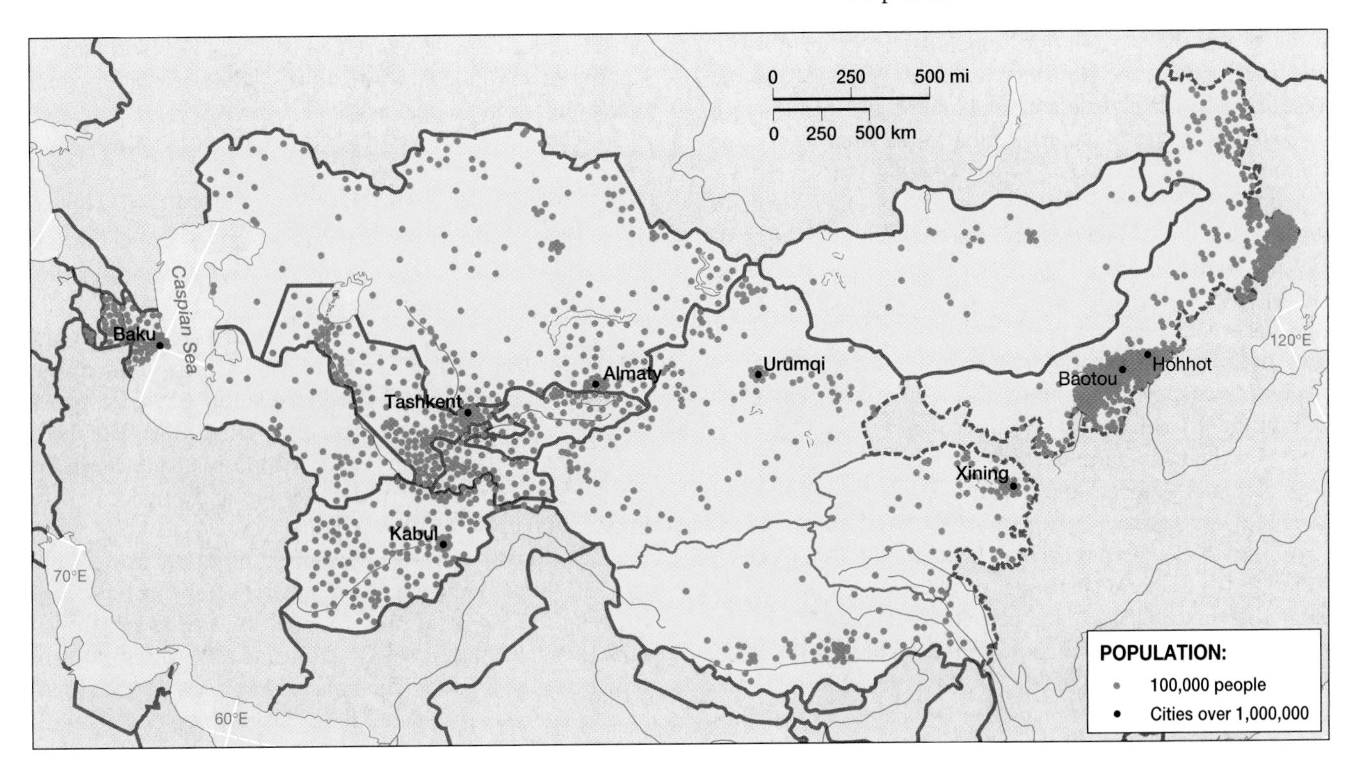

▲ **Figure 10.9 Population density in Central Asia** Central Asia as a whole remains one of the world's least densely populated regions. Large areas in northern Tibet, central Xinjiang, and southern Mongolia are virtually uninhabited. Concentrations of population are found in the river valleys of Uzbekistan and neighboring countries, in southern Inner Mongolia, and in Azerbaijan in the far west. Most of Central Asia's large cities are located near the region's periphery or in its major river valleys.

Population and Settlement: Densely Settled Oases Amid Vacant Lands

Most of Central Asia is sparsely populated (Figure 10.9). Large areas are essentially uninhabited, too arid, or too high in elevation to support human life. Even many of the more favorable areas are only populated by widely scattered groups of nomadic **pastoralists** (people who raise livestock for subsistence purposes). Mongolia, which is more than twice the size of Texas, has only 2.5 million inhabitants—fewer than live in the Dallas metropolitan area. But as is common in arid environments, those few lowland locales with good soil and dependable water supplies are thickly settled. Despite its overall aridity, Central Asia is actually well endowed with perennial rivers and fertile oases. Although the lowland basins receive little rain, abundant snows fall on many of the mountain slopes—and Central Asia has a wealth of mountains. While the nomadic pastoralists of the steppes and deserts have dominated the history of Central Asia, the sedentary peoples of the river valleys have always been more numerous.

Highland Population and Subsistence Patterns

The environment of the Tibetan Plateau is particularly harsh. Not only is the climate cold and water often scarce or brackish, but ultraviolet radiation, owing to the elevation, is always high. Only sparse grasses and herbaceous plants—so-called mountain tundra—can survive such rigors. Human subsistence is obviously difficult under such conditions. The only feasible way of life over most of the Tibetan Plateau is nomadic pastoralism based on the yak, an altitude-adapted relative of the cow. Several hundred thousand people manage to make a living in such a manner, roaming with their herds over vast distances. Much of northwestern Tibet, however, is too high even for yak pastoralism and is thus uninhabited.

Although most of the Tibetan Plateau can support only nomadic pastoralism, most Tibetans are sedentary farmers. Farming in Tibet is possible only in a few favorable locations, generally those that are *relatively* low in elevation and that have good soils and either adequate rain or a dependable irrigation system. The main zone of sedentary settlement lies in the far south, where protected valleys offer favorable conditions. Other agricultural areas include Ladakh in the far west (in India's Kashmir) and the far northeastern portion of Qinghai province, both of which only marginally belong within the Tibetan region.

The population of Tibet proper (the Chinese autonomous region of Xizang) is only 2.5 million, while that of China's Qinghai province (most of which lies on the plateau) is 4.2 million. A few hundred thousand persons in far northern

India and in the western third of China's Sichuan province also dwell on the greater Tibetan Plateau. Considering the vast size of this area, these are small numbers indeed. An area of comparable size in eastern China would hold nearly one billion human inhabitants.

Population densities are also low in the other highland areas of Central Asia, although settled agricultural communities can be found in the protected valleys of the southern ranges. Owing to their complex topography, the Pamirs in particular offer a large array of small and nearly isolated valleys that are suitable for agriculture and intensive human settlement. Not surprisingly, this area is marked by high levels of cultural and linguistic diversity. Many villages here are noted for their agricultural terraces and for their well-tended fruit orchards. Apricots are virtually a staple food in some of the more remote areas.

Central Asia's mountains are vitally important for people living in the adjacent lowlands, whether settled farmers or migratory pastoralists. Many herders use the highlands for summer pasture; when the lowlands are parched, the high meadows provide rich grazing. The Kyrgyz (of Kyrgyzstan) are noted for their traditional economy based on **transhumance,** moving their flocks from lowland pastures in the winter to highland meadows in the summer. The farmers of Central Asia rely on the highlands for their wood supplies (a few of these mountains still contain forests), and more importantly, for their water. Settled agricultural life in most of Central Asia is possible only because of the rivers and streams emanating from the region's mountains.

Lowland Population and Subsistence Patterns

Most of the inhabitants of Central Asia deserts live in the narrow belt where the mountains meet the basins and plains. Here water supplies are adequate and soils are neither salt- nor alkali-impregnated, as is often the case in the basin interiors. The population distribution pattern of the Tarim Basin forms an almost perfect ringlike structure (Figure 10.10). Streams flowing out of the mountains are diverted to irrigate fields and orchards in the narrow fertile band situated between the steep slopes of the mountains and the sandy, rocky, or salt-encrusted flats of the central basin.

The population west of the Pamirs, in former Soviet Central Asia, is also concentrated in the transitional zone nestled between the highlands and the plains. A series of **alluvial fans** (fan-shaped deposits of sediments dropped by streams flowing out of the mountains) have long been devoted to intensive cultivation (Figure 10.11). Fertile loess soil abounds (*loess* is a silty soil deposited by the wind), and in a few favored areas winter precipitation is high enough to allow rain-fed agriculture. Several large valleys in this area also offer fertile and easily irrigated farmland. The Fergana Valley of the

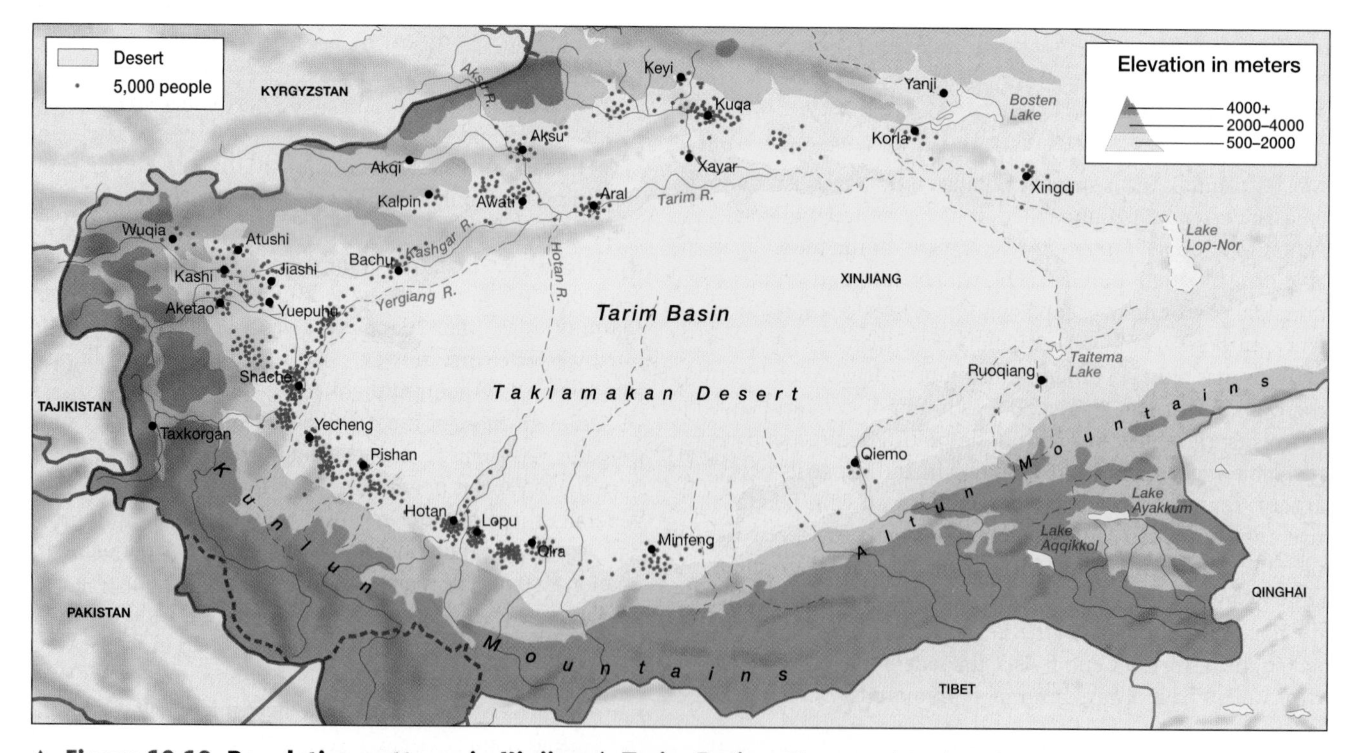

▲ **Figure 10.10 Population patterns in Xinjiang's Tarim Basin** The central portion of the Tarim Basin is a virtually uninhabited expanse of sand dunes and salt flats. Along the edge of the basin, however, dense agricultural and urban settlements are located where streams running out of the surrounding mountains allow for intensive irrigation. The largest of these oasis communities are found along the southwestern fringe of the basin.

▲ **Figure 10.11 Farmland in Uzbekistan** The fertile river valleys of Uzbekistan have been intensely cultivated for many centuries, producing large harvests of fruits and vegetables in addition to cotton and grain. Here tomatoes are harvested—a subtropical crop that has traditionally been exported to Russia and other areas. *(Jeremy Nicholl/Katz/SABA Press Photos, Inc.)*

▲ **Figure 10.12 Steppe pastoralism** The steppes of northern and central Mongolia offer lush pastures during the summer. Mongolians, some of the world's most skilled horseriders, have traditionally followed their herds of sheep and cattle, living in collapsible, felt-covered yurts. Many Mongolians still follow this way of life. *(Goussard/SIPA Press)*

upper Syr Darya River is particularly noted for its productivity. This narrow but important valley is partitioned between three different countries: Uzbekistan, Kyrgyzstan, and Tajikistan (see Figure 10.23). In the far west of the region, Azerbaijan's Kura river basin is another area of intensive agriculture (mostly cotton and rice) and concentrated settlement.

Unlike the other deserts of the region, the Gobi has few sources of permanent water. Rivers draining Mongolia's highlands flow to the north or terminate in interior basins, while only a few of the larger streams from the Tibetan Plateau reach the Gobi proper. (The Huang He, however, does swing north to reach the desert edge before turning south to flow through the Loess Plateau.) Owing to this paucity of **exotic rivers** (those originating in more humid areas), and to its own aridity, the Gobi remains one of Asia's least-populated areas.

The steppes of northern Central Asia are the classical land of nomadic pastoralism. Until the present century, virtually none of this area had ever been plowed and farmed. To this day, pastoralism remains a common way of life across the grasslands, particularly in Mongolia (Figure 10.12). In northwestern China and in the former Soviet republics, however, many pastoral peoples have been forced to adopt sedentary lifestyles. Central governments in the region, like those in most other parts of the world, find migratory people hard to control and difficult to provide with medical and other social services. In northern Kazakhstan the Soviet regime converted the most productive pastures into farmland in the mid-1900s in order to increase the country's supply of grain. Some of these lands have since reverted to steppe, but large areas remain under the plow, and Kazakhstan is a major producer of spring wheat. Consequently, northern Kazakhstan has the highest population density of the steppe belt. Mongolia has also developed a small agricultural sector since World War II, but only about 1 percent of the country is presently classified as arable land.

Population Issues

Although Central Asia remains a low-density environment, some portions of it are growing at a moderately rapid pace. In western China, much of the population growth over the past 30 years has stemmed from the migration of Han Chinese into the area—an influx much resented by many of the indigenous inhabitants. Population growth in the former Soviet zone, on the other hand, has stemmed not from immigration but rather from relatively high levels of fertility. This area, particularly Kazakhstan, has actually witnessed a substantial migration of people out of the region. These are largely ethnic Russians returning to the Russian homeland.

As Table 10.1 shows, population statistics are not readily obtainable for the Central Asian portions of China. The numbers available for the rest of the region show that its overall fertility rates are near the middle of those for the developing world as a whole. They are significantly higher, in other words, than the fertility rates of Sri Lanka or Kerala, but they are lower, on average, than those of Iran and most other Southwest Asian countries. During the final years of the Soviet Union, Central Asia's birthrates—which are substantially higher than those elsewhere in the country—were a major cause of concern for Russian nationalists and may actually have contributed to the breakup of the Soviet Union itself. Some observers attribute Central Asia's somewhat elevated birthrate to the fundamentalist beliefs of Islam, but others think that it rather reflects social and economic factors, including the region's relatively low levels of urbanization. Muslim Azerbaijan, they note, has a much lower birthrate than the less-urbanized but equally Muslim countries of Uzbekistan and Tajikistan.

Fertility patterns do vary substantially from one part of Central Asia to another. Afghanistan, the least-developed and most male-dominated country of the region, has the highest birthrate by a substantial margin. Although good data is difficult to find,

Table 10.1 Demographic Indicators

Country	Population[a]	Natural Increase	TFR[b]	%<15[c]	% Urban
Afghanistan	24.8	2.5	6.1	41	18
Azerbaijan	7.7	1.1	2.1	35	52
Kazakhstan	15.6	0.5	1.9	30	56
Kyrgyzstan	4.7	1.5	2.8	37	34
Mongolia	2.4	1.6	3.1	36	57
Tajikistan	6.1	1.7	2.9	40	28
Tibet	2.2	—	—	—	—
Turkmenistan	4.7	1.7	2.9	39	45
Uzbekistan	24.1	2.0	3.2	41	38
Xinjiang	16.0	—	—	—	—

[a]Population in millions, 1996.
[b]Total fertility rate.
[c]Percentage of population younger than 15 years of age.
Source: *Population Reference Bureau World Population Data Sheet,* 1998.

much evidence would suggest that Tibet's birthrate remains quite low (see "Demographic Issues: Polyandry in Tibet"). Kazakhstan's birthrate, just slightly over the natural replacement level, is also low for the region, reflecting the extremely low fertility level of the Russian-speakers in the north of the country.

Urbanization in Central Asia

Although the steppelands of northern Central Asia had no real cities before the modern age, the river valleys and oases of the desert have been partially urbanized for millennia. Such cities as Samarkand and Bukhara in Uzbekistan were famous even in medieval Europe for their riches and their lavish architecture (Figure 10.13). This early urban fluorescence was built upon the region's economic and political position. The Amu Darya and Syr Darya valleys lay near the midpoint of the trans-Eurasian silk route, and they formed the core of a number of empires based on the cavalry forces of the steppe.

The modern era of steamships and oceanic trade brought hardship to the cities of Central Asia. Isolated from maritime routes, these ancient mercantile centers could no longer compete well in world trade and began to diminish as a result.

The conquest by the Russian and Chinese empires resulted in further difficulties, but they also ushered in a new wave of urban formation. Cities slowly began to appear on the Kazakh

DEMOGRAPHIC ISSUES · Polyandry in Tibet

Earlier in this century some outsiders worried that the population of Tibet, never large to begin with, was in some danger of disappearing altogether owing to its extremely low birthrate. Today the main concern is over the influx of Han Chinese, but the Tibetan birthrate remains low. Historically speaking, this is partly the result of widespread monasticism. Another factor is the unusual institution of polyandry—in other words, of a single woman taking more than one husband. The opposite form of plural marriage, polygyny (whereby one man takes more than one wife) is historically common in many parts of the world. Polyandry, however, has almost always been very rare—except in Tibet. It has been especially widespread in western Tibet, that portion of the plateau farthest removed from Chinese cultural and political influences.

In traditional Tibetan society, only the eldest son inherits property, and only those inheriting property are allowed to marry. Younger brothers thus have had the option of joining a monastery or, in many cases, of sharing the eldest brother's wife. Such a custom not only avoided the endless subdivision of land, which could be a major problem in the harsh environment of Tibet, but also ensured that fertility rates remained low. Since large numbers of men have traditionally lived either in monasteries or in polyandrous households, many women have had no opportunity to marry, and thus have remained as spinsters in their parents' or brother's households. Since polyandrous women are not able to have any more children than monogamous ones (quite unlike polygynous men), a low birthrate is the unavoidable consequence of this custom.

Tibet's unique family structure has perhaps functioned in the past to maintain sustainable population levels in a harsh environment. Today, when Tibetans are competing against Han Chinese immigrants—especially in the growing urban sector—polyandry is perhaps no longer so adaptive.

▲ **Figure 10.13 Traditional architecture in Samarkand** Samarkand, Uzbekistan, is famous for its lavish Islamic architecture, some of it dating back to the 1400s. The city owes part of its rich architectural heritage to the fact that it was the capital of the great Medieval conqueror, Tamerlane. *(Haley/SIPA Press)*

steppes, where none had previously existed. The Manchu and Chinese conquerors of Inner Mongolia and Xinjiang similarly built new administrative and garrison cities, often placing them only a few miles away from indigenous urban sites. As is visible in Figure 10.14, this process created a dualistic urban framework that is still partially visible today. The old indigenous cities of the region are characterized by complex and almost maze-like networks of streets and alleyways, whereas the neighboring Manchu cities were constructed according to a strict geometrical order. The recent growth of urban populations in the area has obscured this old dualism, but several new "twin cities" have emerged in Xinjiang, where Han Chinese immigrants have built new settlements next to indigenous cities.

One may also distinguish Russian/Soviet cities from indigenous cities in the former Soviet zone, but this dichotomy is not so clear-cut. In Uzbekistan, for example, Tashkent is largely a Soviet creation, whereas many parts of Bukhara still reflect the older urban patterns. Several major cities, such as Kazakhstan's former capital of Almaty, did not exist before Russian colonization. In Azerbaijan, Baku emerged as a major city in the early twentieth century as the first Caspian oil fields began to be intensively exploited. Everywhere, moreover, one can see the ubiquitous effects of centralized Soviet urban planning and design.

Today, well-developed urban networks can be found in both the river and oasis zone of the south and in the agricultural areas of northern Kazakhstan. North central Kazakhstan

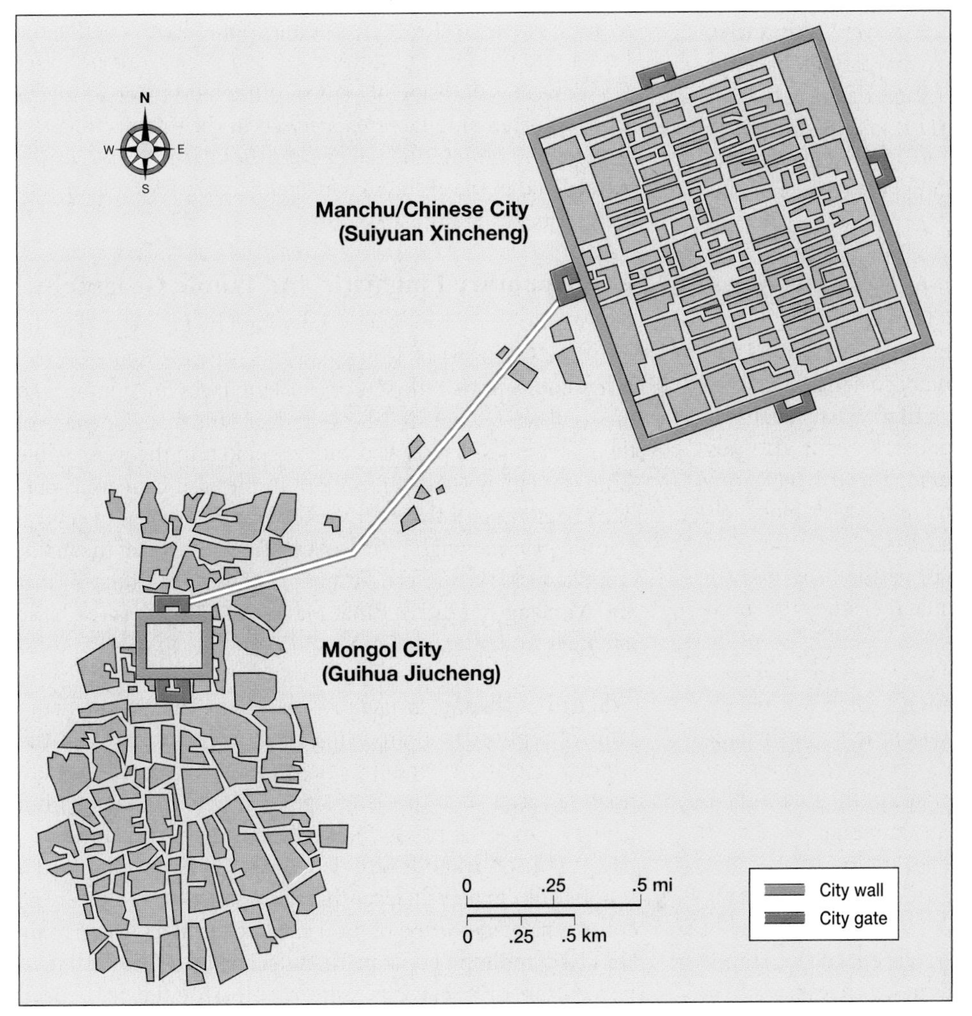

◀ **Figure 10.14 The dual city plan of traditional Hohot** Hohot, the main city of Inner Mongolia (the Chinese autonomous region of Nei Monggol), exemplifies the dual-city nature of many of the urban centers of western China. The Manchu/Chinese city shows the clear evidence of traditional Chinese geometrical city planning, while the Mongol city, at some distance, shows a more haphazard pattern that stemmed from spontaneous growth. Recent developments, however, have tended to fill in the intervening areas and thus obscure these traditional distinctions.

is also witnessing the rise of a new major city, Astana, which the government has designated as a new, centrally located, capital. Even Mongolia, long a land virtually without permanent settlements, now has more people living in cities than in the countryside. In some parts of the region, however, cities remain relatively few and far between. Only 28 percent of the people of Tajikistan, for example, are urban residents. Tibet similarly remains a predominantly rural society. But wherever war and ethic strife predominate, as they do in Afghanistan, cities may be expected to swell to uneconomic proportions as refugees seek safety from rural combat zones.

Cultural Coherence and Diversity: A Meeting Ground of Disparate Traditions

Although Central Asia has a certain environmental unity, its cultural coherence is more questionable. The western half of the region is largely Muslim and is often classified as part of Southwest Asia. Northeastern and southeastern Central Asia—Mongolia and Tibet—are rather characterized by a distinctive form of Buddhism sometimes called Lamaism. Tibet is culturally linked to both South and East Asia, and Mongolia is intimately associated with China, but neither fit easily within any world region.

Historical Overview: An Indo-European Hearth?

The river valleys and oases of Central Asia were early sites of sedentary, agricultural communities. Archaeologists have discovered abundant evidence of farming villages dating back to the Neolithic period (beginning circa 8000 B.C.) in the Amu Darya and Syr Darya valleys and along the rim of the Tarim Basin. After the domestication of the horse around 4000 B.C., nomadic pastoralism emerged in the steppe belt as a new human adaptation. Eventually, pastoral peoples gained power over the entire region, transforming not only the history of Central Asia but also that of virtually the entire Eurasian continent.

In the pre-modern period, pastoral nomads enjoyed profound military advantages over sedentary societies. They had access to large numbers of horses at a time when cavalry almost always held the edge over infantry. Nomads also possessed the benefits of mobility. If pursued by a larger army, they could merely withdraw to the more-inaccessible reaches of the steppe until the danger had passed. Not until the age of gunpowder were the benefits of pastoralism offset by the demographic and economic advantages held by the more-populous agriculturally based states.

The earliest recorded languages of Central Asia, spoken in both the oasis communities and among some of the pastoralists, were members of the Indo-European linguistic family. Indeed, Central Asia is often considered to be the birthplace of the Indo-European peoples. In the first millennium B.C., the inhabitants of the Amu Darya and Syr Darya valleys spoke languages closely related to ancient Persian. Many vestiges of this Persian heritage are still found in southwestern Central Asia.

Indo-European languages were replaced on the steppe roughly 2,000 years ago by languages in another major family: Altaic. Three great branches constitute the Altaic family: Tungusic (spoken by most of the indigenous peoples of Manchuria and Siberia), Mongolian, and Turkish. Altaic peoples spread as far west as southeastern Europe by the waning years of the Roman Empire. By the second century B.C., a powerful nomadic empire of Turkish-speaking peoples arose in what today is Mongolia—forcing the Chinese to begin building the Great Wall as a defensive measure. As Turkish power spread through most of Central Asia, Turkish languages gradually began to replace Indo-European tongues in the oasis communities. This process, however, has never been completed, and today southwestern Central Asia remains a meeting ground of Persian and Turkish languages.

The Turks were eventually replaced on the eastern steppes by another group of Altaic speakers, the Mongols. In the late 1100s the Mongols united all of the pastoral peoples of Central Asia and used the resulting force to conquer nearby sedentary societies. By the late-1200s, this Mongol Empire had grown into the largest contiguous political unit Earth had ever seen, stretching from Korea and southern China in the east to the Carpathian Mountains and the Euphrates River in the west (Figure 10.15).

Protected by mountain barriers and by the rigorous conditions of the plateau, Tibet has taken a different course from the rest of Central Asia. Tibet emerged a strong, unified kingdom around A.D. 700. Tibetan unity and power did not persist, however, and the region reverted to its former state of semi-isolation. Tibet was incorporated for a short period in the 1200s into the Mongol Empire, and in later centuries other Mongol states occasionally enjoyed limited powers over the Tibetans. These interactions resulted in the establishment of Mongolian communities in the northeastern portion of the plateau and in the eventual conversion of the Mongolian people to Tibetan Buddhism.

Contemporary Linguistic and Ethnic Geography

Today most of Central Asia is inhabited by peoples speaking the Altaic languages of Mongolian and Turkish (Figure 10.16). A few indigenous Indo-European languages are confined to the southwest, whereas Tibetan remains the main language of the plateau. Russian is also widely spoken in the west, while Chinese is increasingly important in the east. Chinese is perhaps beginning to threaten the long-term survival of several Central Asia languages, particularly Tibetan. One of the major complaints of the indigenous people of western China—Tibet and Xinjiang—is the fact that Mandarin Chinese is the basic language of higher education.

Tibetan Tibetan is usually placed in the Sino-Tibetan family, implying a shared linguistic ancestry between the Chinese and the Tibetan peoples. Many students of Tibetan, however, argue that this language is entirely distinctive from Chinese, and that no definite relationship between the two has ever been established. Tibetan itself is divided into a number of distinct dialects that are spoken over almost the entire inhabited portion of the Tibetan plateau. However, only about 1.5 million people in Tibet itself speak it, out of a total population of some 2.5 million (most of the rest speak

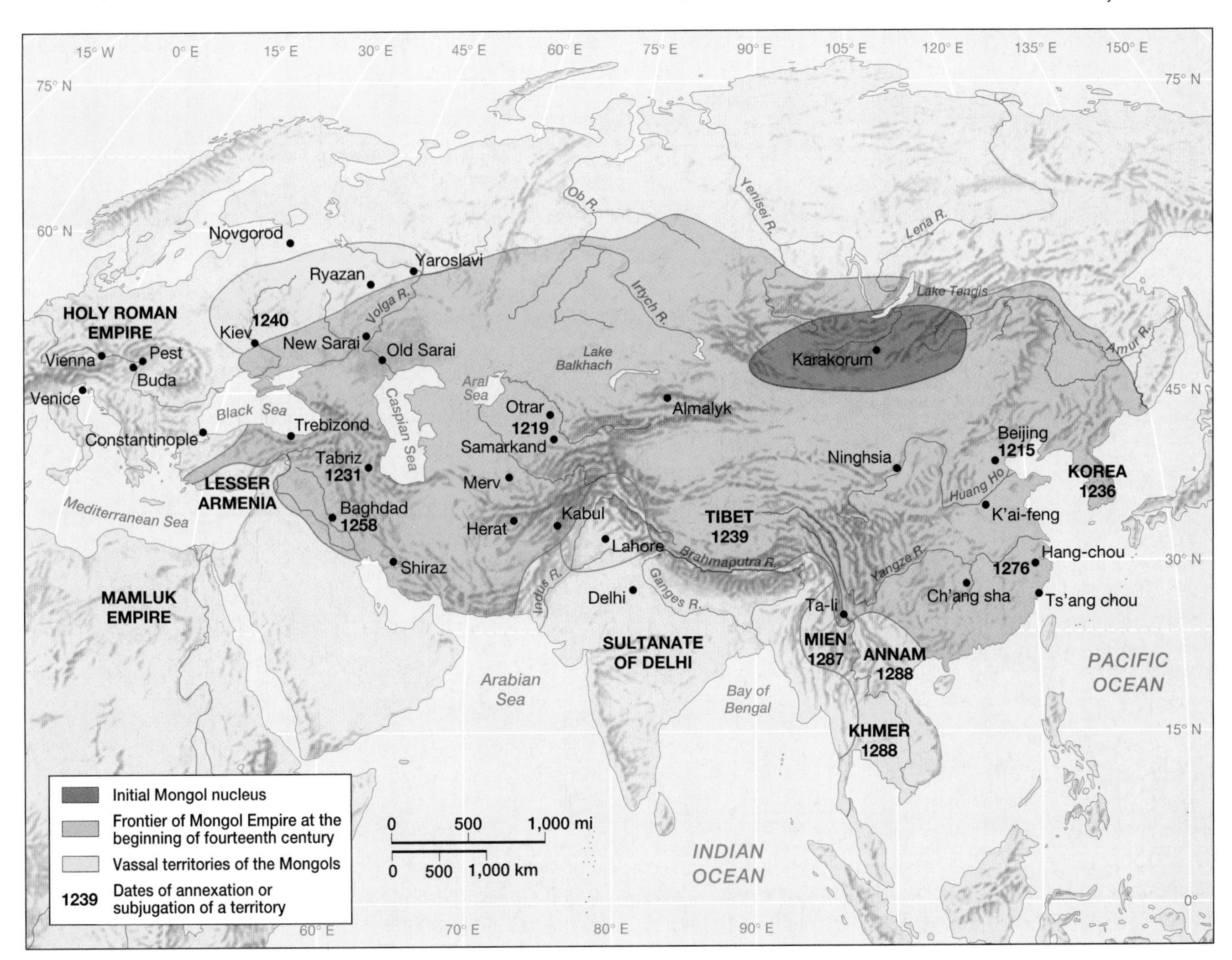

▲ **Figure 10.15 The Mongol Empire of the 1200s** In the 1200s the Mongols carved out the largest land-based empire Earth has ever seen. From a core area in modern-day Mongolia, the Mongol conquests extended as far as southern China in the southeast, the Ukraine in the west, and Iraq in the southwest. Although the empire did not long remain unified, it did have profound repercussions on the subsequent political and economic history of Eurasia.

Chinese). Perhaps another three million Tibetan speakers live in Qinghai and western Sichuan; smaller numbers may be found in the far northern Himalayan reaches of South Asia. Tibetan has an extensive literature written in its own script, most of which is devoted to religious topics.

Mongolian Mongolian forms a cluster of closely related dialects spoken by approximately five million persons. The standard Mongolian of both the independent country of Mongolia and of China's Inner Mongolia is called Khalkha; other Mongolian dialects include Buryat (found in southern Siberia) and Kalmyk (found in the extreme southeastern corner of Europe). Mongolian has its own distinctive script, which dates back some 800 years, but Mongolia itself adopted the Cyrillic alphabet of Russia in 1941. Efforts are now being made to revive the old script.

Mongolian speakers form about 90 percent of the population of Mongolia. In China's Inner Mongolian Autonomous Province, however, they have been almost submerged by a wave of Han Chinese migrants over the past 50 years. Today only about two million out of 22 million residents of Inner Mongolia speak Mongolian, making some observers wonder whether this area has been permanently lost to the Mongolian cultural sphere.

Turkish Languages Far more Central Asians speak Turkish languages than Mongolian and Tibetan combined. The Turkish linguistic sphere extends from Azerbaijan in the west through Xinjiang province in the east. The various Turkish languages are not as closely related to each other as are the dialects of Mongolian, but they are still obviously kindred tongues. Six main Turkish languages are found in Central Asia; five are associated with newly independent (former Soviet) republics of the west, while the sixth, Uygur, is the main indigenous language of China's Xinjiang province.

Uygur is an old language, dating back almost 2,000 years. The Uygur number about eight million, almost all of whom

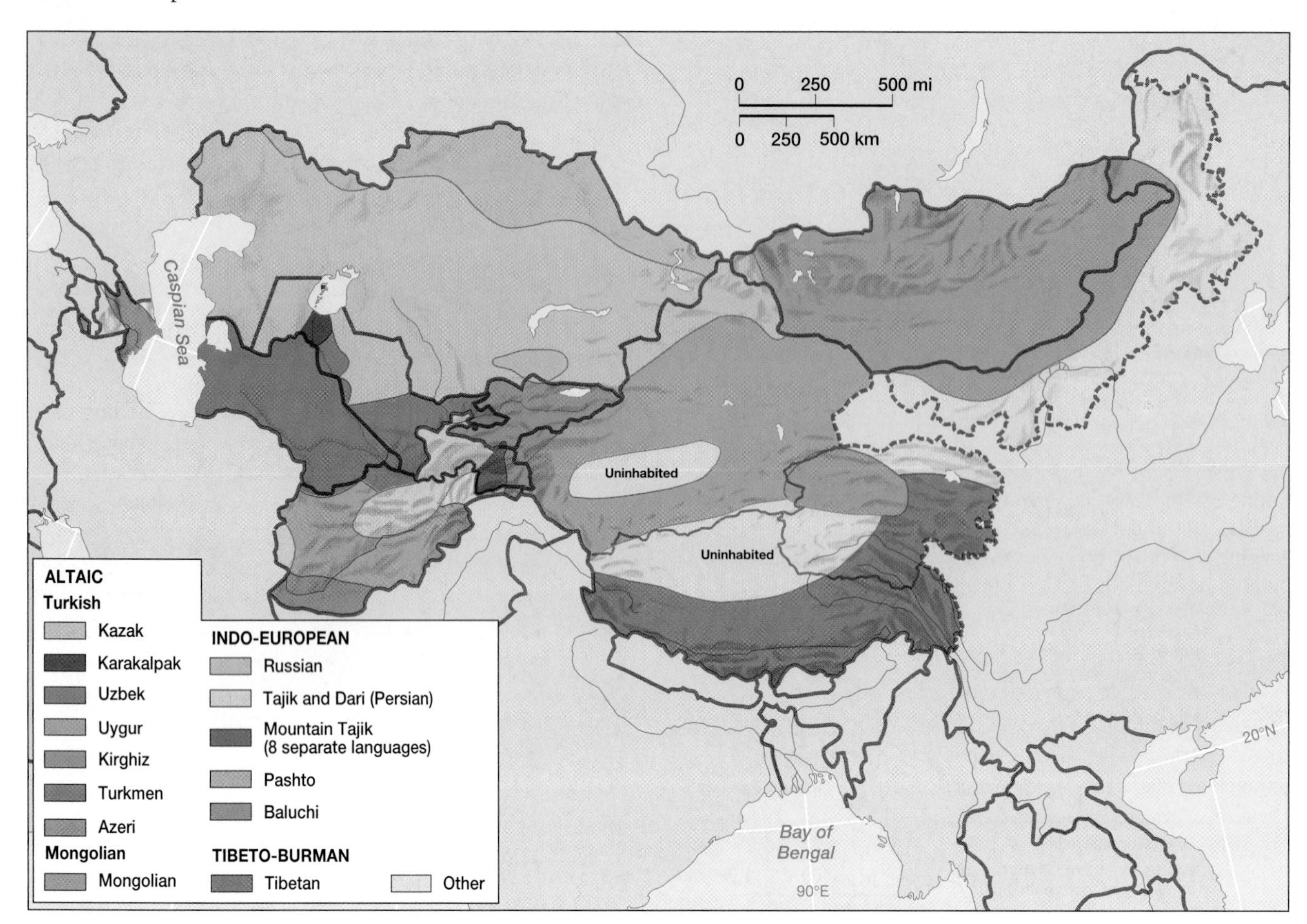

▲ **Figure 10.16 Linguistic geography of Central Asia** Most of Central Asia is dominated by languages in the Altaic family, which includes both the Turkish languages (found through most of the center and the west of the region) and Mongolian (found in Central Asia's northeast). Several Indo-European languages, however, are located in both the far northwest and the center-south, while the Tibeto-Burmese language of Tibetan covers most of the Tibetan Plateau in the southeast.

live in Xinjiang (Figure 10.17). As recently as 1953 the Uygur formed about 80 percent of the population of Xinjiang; now, because of Han Chinese immigration, they actually form a minority in their own homeland. There are also about one million Kazakh speakers in Xinjiang, as well as smaller numbers of people speaking Mongolian and other Central Asian languages.

Five of the six countries of the former Soviet Central Asia—Kazakhstan, Uzbekistan, Turkmenistan, Kyrgyzstan, and Azerbaijan—are named after the Turkish languages of their dominant native populations. In three of these countries, the indigenous people still form a clear majority. Some 82 percent of the people of Azerbaijan speak Azeri (there are more Azeris in northern Iran, however, than there are in Azerbaijan), some 70 percent of the people of Uzbekistan speak Uzbek as their native tongue, and some 73 percent of the people in Turkmenistan speak Turkmen. With more than 17 million speakers, Uzbek is the most widely spoken Central Asian language. In the Amu Darya delta in the far north of Uzbekistan, however, most people speak a different Turkish language called Karakalpak, and Kazakh speakers may be found in sparsely populated Uzbek deserts.

In the two other Turkish republics, the titular nationality forms only about half of the total population. Some 52 percent of the inhabitants of Kyrgyzstan speak Kyrgyz as their native language, while approximately 42 percent of the people of Kazakhstan speak Kazakh. The other residents of these countries speak Russian, Uzbek, Ukrainian, German, and a variety of other languages both indigenous and exogenous to Central Asia. In Kazakhstan the population is almost evenly split between speakers of Turkish languages, on the one hand, and European languages (Russian, Ukrainian, and German), on the other. In general, the Kazakhs and other Turks live in the center and south of the country, whereas the people of European descent live in the agricultural districts of the north and in the cities of the southeast.

Linguistic Complexity in the Former Soviet Zone The sixth republic of the former Soviet Central Asia, Tajikistan, is dominated by people who speak an Indo-European rather than a Turkish language. Tajik is so closely related to Persian that it is often considered to be a Persian dialect. Iran, the homeland of Persian, is, however, separated from Tajikistan by some

▲ **Figure 10.17 Uygur mosque** A small mosque, illustrating traditional Uygur architecture, survives amid blocks of modern apartments in Urumchi, Xinjiang. Traditional forms of housing and urban design can still be found in Uygur communities in northwestern China, but they are gradually disappearing. *(Chris Stowers/Panos Pictures)*

400 miles (640 kilometers) of Turkish-speaking territory. Roughly 3.5 million people in Tajikistan, about 65 percent of the total population, speak Tajik as their main language. The remote mountains of eastern Tajikistan are populated by peoples speaking a variety of distinctive Indo-European languages, sometimes collectively referred to as "Mountain Tajik."

Tajikistan, like much of the rest of former Soviet portion of Central Asia, is noted for its complex mixture of languages. About a quarter of its people, for example, are Uzbeks (whereas roughly 5 percent of the people of Uzbekistan are Tajiks). The peripheral portions of Azerbaijan—part of the fan. Caucasus "mountain of languages"—are also noted for thι ethnolinguistic complexity. Such ethnic mixing was actuall much greater in earlier decades, since Soviet policy resulted in gradual ethnic homogenization. The Soviet authorities also devised complex political boundaries among the different ethno-linguistic groups of Central Asia, partly in order to play one group off against another and thus bolster their own authority. Note in Figure 10.22, for example, the intricate political patchwork that they had established in the fertile Fergana Valley.

Language and Ethnicity in Afghanistan The linguistic geography of Afghanistan is even more complex than that of the former Soviet zone (Figure 10.18). Afghanistan was never colonized by outside powers, and it is one of the few countries of the world to have inherited the boundaries of a premodern, indigenous kingdom. This kingdom emerged in the 1700s on traditional dynastic lines that did not reflect ethnic or linguistic divisions ("dynastic" linkages are those based on the family of the monarch). The modern nation-state ideal—that each country should be identified with a single national group—never had much currency in Afghanistan.

The eighteenth-century creators of Afghanistan were mostly members of the Pathan ethnic group. They did not attempt, however, to build a nation-state around Pathan identity. Indeed, approximately half of the Pathan population (whose language is usually called Pashto or Pashtun), live not in Afghanistan but rather in Pakistan. In Afghanistan itself, estimates of the proportion of the populace speaking Pashto vary from 40 percent to 60 percent.

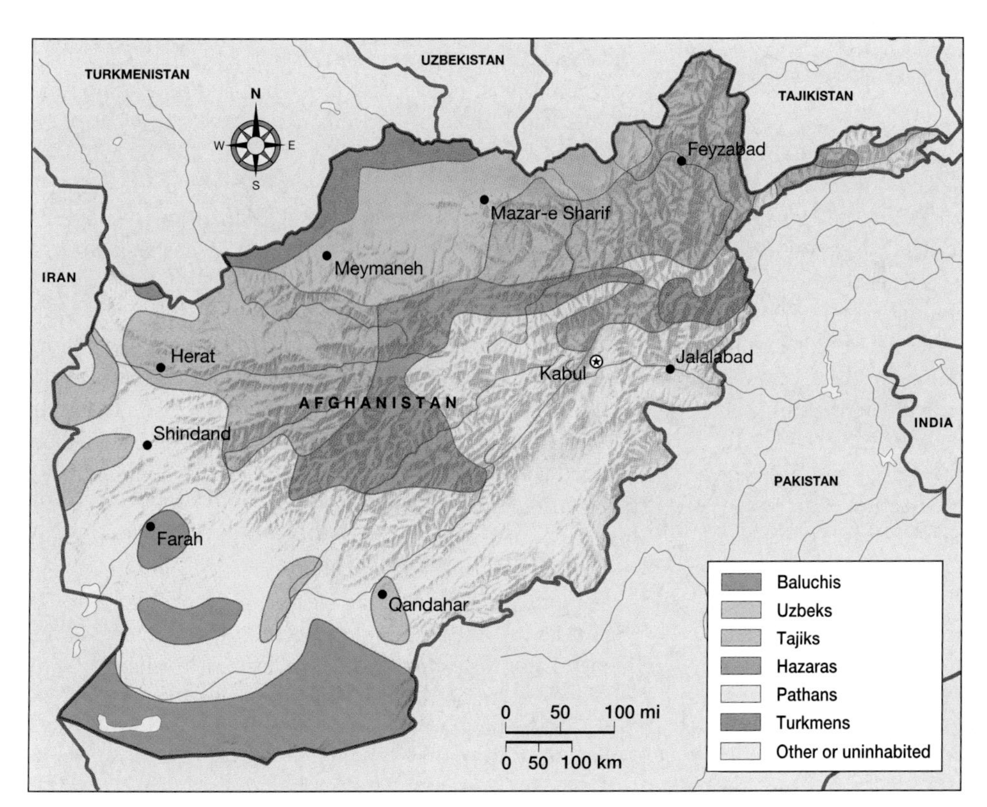

◀ **Figure 10.18 Afghanistan's ethnic patchwork** Afghanistan is one of the world's more ethnically complex countries. Its largest ethnic group is the Pathans, a people who inhabit most of the southern portion of the country as well as the adjoining borderlands of Pakistan. Northern Afghanistan, however, is mostly inhabited by Uzbeks, Tajiks, and Turkmens—whose main population centers are located in Uzbekistan, Tajikistan, and Turkmenistan, respectively. The Hazaras of Afghanistan's central mountains speak a form of Tajik (or Persian), but are considered to be a separate ethnic group in part because they, unlike other Afghans, follow Shiite rather than Sunni Islam.

Pashto speakers live primarily to the south of the Hindu Kush (along the Pakistani border) and in the far west. Almost as many people in Afghanistan speak Dari (although estimates vary widely), Afghanistan's variant of Persian. Dari speakers are concentrated in the cities of the far west, in the central mountains, and near the boundary with Tajikistan. Two separate ethnicities are ascribed to the Dari-speaking people; those in the west and north are considered to be Tajik, whereas those in the central mountains are called Hazaras, reputed to be descendants of Mongol conquerors who arrived in the twelfth century. Another 11 percent of the people of Afghanistan speak Turkish languages, mainly Uzbek. Uzbek speakers are concentrated near the Uzbekistan border.

The ethnic melange present in Afghanistan did not present a problem until the later part of the twentieth century. At present, however, the country is in serious danger of disintegrating along ethnolinguistic lines. As we shall see later in this chapter, the war currently being waged in Afghanistan has a complex ideological and geopolitical background, but it also has roots in the country's linguistic, as well as its religious, geography.

▲ **Figure 10.19 Afghan women in public** Especially in the Pathan areas of Afghanistan, women have traditionally been forced to cover their entire bodes when in public areas. In all areas now controlled by the Taliban, such dress codes are strictly enforced. Young girls have some freedom, but as they approach adolescence, extremely modest dress is demanded by the public authorities. *(Lemoyne/Liaison Agency, Inc.)*

Geography of Religion

Ancient and medieval Central Asia formed a religious hodgepodge. The major overland trading routes of pre-modern Eurasia crossed the region, giving easy access to both merchants and missionaries. Several varieties of Buddhism, Islam, Christianity, Judaism, Zoroastrianism, and several minor religions have all thrived, at various times and in various places, within the region. During the years of Mongol supremacy, adherents of many faiths mingled throughout the empire. Subsequently, however, religious lines hardened and the region was eventually divided into two opposed spiritual camps: Islam triumphed in the west and center, and Lamaist Buddhism prevailed in Tibet and Mongolia.

Islam in Central Asia As is true elsewhere in the Muslim world, different Central Asian peoples are known for their different interpretations of Islamic orthodoxy. The Pathans of Afghanistan are famed for their strict Islamic ideals—although critics contend that Pathan religious strictures, such as never allowing women's faces to be seen in public, are rather based on their own customs (Figure 10.19). The traditionally nomadic groups of the northern steppes, such as the Kazakhs, on the other hand, have often been considered lax in their religious observations. And whereas most of the region's Muslims are Sunnis, Shiism is widespread among both the Hazaras of central Afghanistan and the Azeris of Azerbaijan.

Under the communist rule of China, the Soviet Union, and Mongolia, all forms of religion were discouraged. Chinese authorities and student radicals attempted to suppress Islam in Xinjiang during the Cultural Revolution of the late 1960s and early 1970s. Mosques were destroyed or converted to museums, religious schools were closed, and people were sometimes forced to eat pork. Chinese Muslims now enjoy basic freedom of worship, but the state still closely monitors religious expression out of fear that it will lead to political separatism. Periodic persecution of Islam also occurred in Soviet Central Asia, and until the 1970s many observers thought that that religion was slowly disappearing from the region.

Religious expression was not, however, so easily repressed. Interest in Islam began to grow in former Soviet Central Asia in the 1970s and 1980s. In the post-Soviet period Islam continues to revive as people rediscover their indigenous heritage and identity. Thus far, however, there have been few signs of mass Islamic political fundamentalism. In Xinjiang, Islam does indeed seem to be emerging as a focal point of a nascent independence movement among the Uygur people. Most Uygur leaders, however, insist that their political beliefs are not fundamentalist. Only in Afghanistan is Islamic fundamentalism a powerful movement.

Islam is not the only religion represented in former Soviet Central Asia. Many Russian settlers belong to the Russian Orthodox Church, and Uzbekistan has a small Jewish population. There are even a few reports that Zoroastrianism has survived the centuries, hidden among small groups of people.

Lamaist Buddhism Mongolia and Tibet stand apart from the rest of Central Asia—and indeed, from the rest of the world—in the adherence of their people to Lamaist Buddhism. Buddhism entered Tibet from India many centuries ago, where it merged with the indigenous religion of the area, called Bon. The resulting hybrid, Lamaism, is sometimes said to be more oriented toward magic than are other forms of Buddhism, and it is certainly more hierarchically organized. Standing at the apex of Lamaist society is the Dalai Lama, considered to be the reincarnation of the Buddha. Ranking below him is the Panchen Lama, followed by other ecclesiastical officials. Until the Chinese conquest, Tibet was essentially a **theocracy** (or religious state), with the Dalai Lama enjoying political as well as religious authority.

Lamaism is noted for its dedication to monasticism (Figure 10.20). A substantial proportion of Tibet's male population has long been composed of monks, and monasteries once wielded both economic and political clout. Such widespread monasticism, which demanded that many people remain celibate, ensured that Tibet's population density remained low. Some scholars view Tibetan monasticism almost as an environmental adaptation, arguing that the plateau could not support dense human settlements and thus required some method of limiting the population.

Lamaism in Tibet suffered particularly brutal persecution after 1959. The Chinese hold on Tibet has never been as secure as that on Xinjiang, and Tibetan Buddhism is often viewed as a potent vehicle for political separatism. The Dalai Lama, who fled Tibet for India in 1959, has also long been a powerful advocate for the Tibetan cause in international circles. During the 1960s and 1970s, an estimated 6,000 Tibetan Buddhist monasteries were destroyed and thousands of monks were killed; the number of active monks today is only about 5 percent of what it had been before the Chinese occupation. Many monasteries have, however, been allowed to reopen, but they are kept under strict surveillance and their activities are severely limited. Still, the Lamaist faith continues to provide a bulwark of Tibetan identity, and in so doing helps keep alive the dream of independence.

In Mongolia the downfall of communism and hence of Russian influence has allowed the Lamaist Buddhist faith to experience a renaissance. Several monasteries have been refurbished and many people have returned to their national religion. The intensity of Buddhist belief, however, does not seem to be as strong in Mongolia as it is in Tibet.

▲ **Figure 10.20 Lamaist Buddhist monastery** Tibet is well known for its large Buddhist monasteries, buildings that in earlier years served as seats of political as well as religious authority. The Potala Palace in Lhasa, traditional seat of the Dalai Lama, is the largest, most important, and most famous of such monastic establishments. *(Alain le Garsmeur/Panos Pictures)*

Central Asian Culture in International and Global Context

Western Central Asia's closest external cultural relations are with Russia, while those of eastern Central Asia are with China. In the east, the main issue is the migration of Han Chinese, which has resulted in serious ethnic and political tensions. In the west, Russian influence is diminishing.

During the Soviet period, the Russian language spread widely through western Central Asia. Russian served both as a lingua franca (or common language) and as a means of instruction in higher education. One had to be fluent in Russian in order to reach any position of responsibility. The Cyrillic (or Russian) script, moreover, replaced the Arabic script that had previously been used for the indigenous languages. Russian speakers settled in all of the major cities and many became influential. Today, however, Russian speakers have begun migrating back to Russia, especially from the region's poorer countries, such as Tajikistan. The use of Russian in education, government, business, and the media is declining in favor of the local languages. Efforts are being made to abandon Cyrillic in favor of the Roman alphabet, the script of modern-day Turkey. Russian has long served an important role as the common international language of the area, however, and it can by no means be abandoned overnight.

Although Central Asia is remote and poorly integrated into global cultural circuits, it is hardly immune to the forces of globalization. The tensions existing throughout the region between religious and secular orientations, and between ethnic nationalism and multiethnic inclusion, are symptoms of the global condition at the turn of the third millennium. So, too, the increased usage of English and the influence of U.S. culture throughout Central Asia shows that this part of the world is not immune to the forces of cultural globalization. Such influences have been especially marked in the oil cities of the Caspian Basin, such as Baku (see "Global and Local: Louisiana Oil Culture in Central Asia"). Although the proportion of English speakers in Central Asia is low, most of the region's numerous Web pages are written in English. English-speaking Central Asians with computer skills are increasingly valued as the region strives to find a place and a voice in the global community (see "Local Voices: Virtual Tibet" later in this chapter).

Geopolitical Framework: Political Reawakening in a Power Void

Central Asia has played a marginal role in global political affairs for the past several hundred years. Before 1991, the entire region, except Mongolia and Afghanistan, lay under direct Soviet and Chinese control. Mongolia, moreover, had been little more than a Soviet satellite, and even Afghanistan came under Soviet domination in the late 1970s. The southeastern third of Central Asia, of course, is still an integral part of China. And although the breakup of the Soviet Union saw the emergence of six new Central Asian countries, all of them are economically troubled and geopolitically insecure (Figure 10.21).

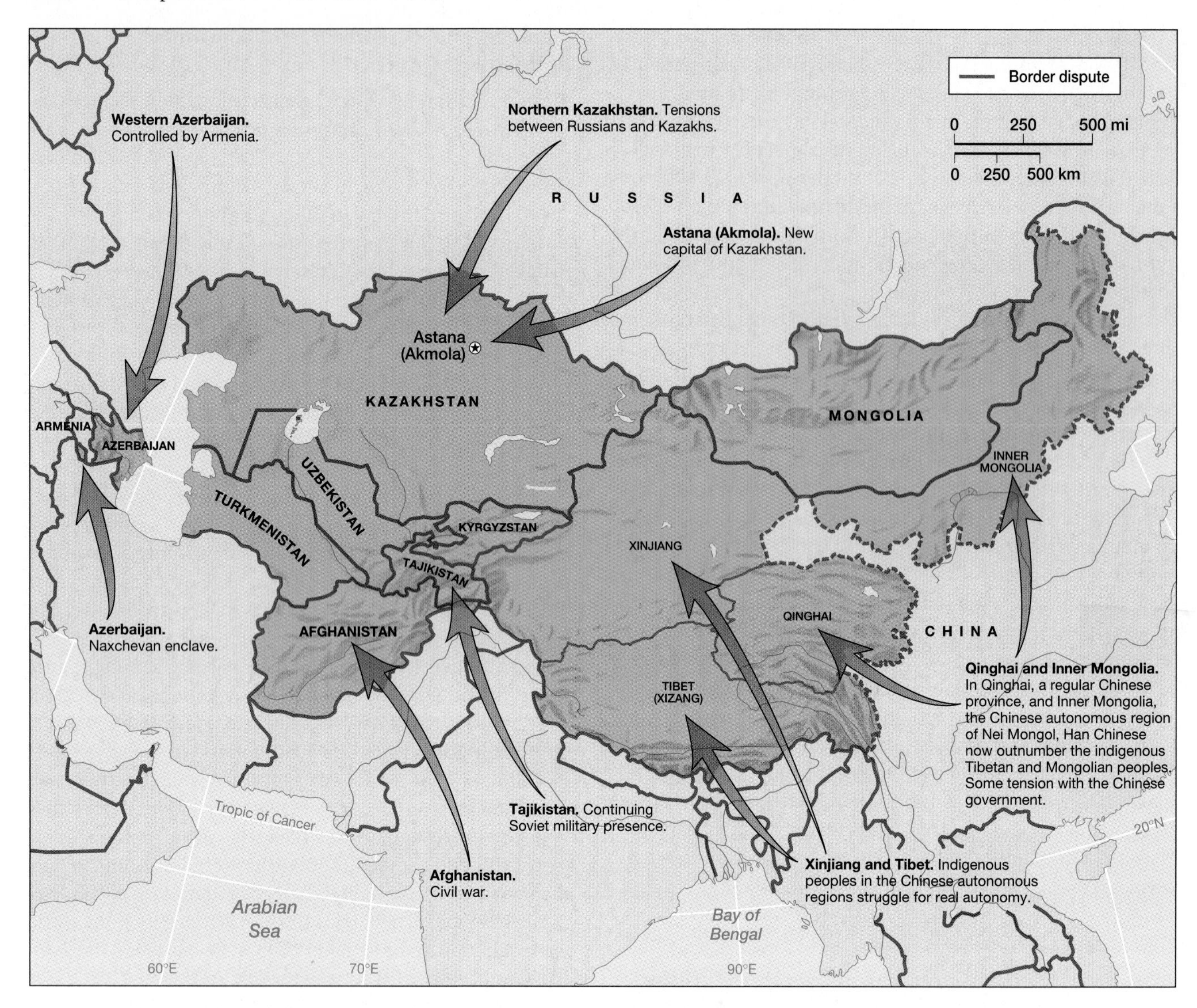

▲ **Figure 10.21 Central Asian geopolitics** Six of the eight independent states of Central Asia came into existence in 1991, with the dissolution of the Soviet Union. A number of border disputes and other geopolitical problems have been inherited from the period of Soviet rule. In western Central Asia, the most serious difficulties stem from China's maintenance of control over areas in which the indigenous peoples are not Chinese. Afghanistan, scene of a prolonged and brutal civil war, has experienced the most extreme forms of geopolitical tension in the region.

Partitioning of the Steppes

Although Central Asia in the twentieth century has been contested and controlled by outside powers, this had not always been the case. Before 1500 it was a power center, a region whose mobile armies periodically conquered and always threatened the far more populous sedentary states of the Eurasian rim. The development of gunpowder and of effective hand weapons changed the balance of power, however, allowing the wealthier and more tightly organized agricultural states to vanquish the nomadic pastoralists. By the 1700s, their armies had been defeated and their lands taken. The winners in this struggle were the two largest states bordering the steppes: Russia and China.

The Manchu conquest of China in 1644 undercut the autonomy of the steppe peoples. The Manchus came from the borderlands of Central Asia (in Manchuria) and were themselves adept at the arts of cavalry warfare. By the mid-1700s, the Chinese empire of the Manchus stood at its greatest territorial extent. Not only Mongolia and Xinjiang but also Tibet, and a slice of modern Kazakhstan lay within the empire.

From its height in the late 1700s, Manchu-ruled China declined. It was still able to retain most of its Central Asian dominions through the 1800s, largely because they were located far from the reaches of the imperialist European states. By the early 1900s, however, Chinese authority began to diminish in Central Asia as well. When the Manchu (or Qing) dynasty fell in 1911, Mongolia declared itself independent, although China did manage to keep the extensive borderlands of Inner Mongolia (Nei Mongol). Tibet had earlier gained *de facto* independence, and even Xinjiang lay beyond the reach of effective Chinese authority during the 1920s.

GLOBAL AND LOCAL Louisiana Oil Culture in Central Asia

Oil and gas production are highly specialized businesses, leading to a vigorous international trade in parts, services, and labor. Many U.S. oil workers, petroleum engineers, and geologists have long sought work overseas, a phenomenon that accelerated after falling prices in the late 1980s and 1990s brought recession to the oil fields of Texas, Louisiana, and Oklahoma. Most of the initial movement was to the Persian Gulf area, but after the breakup of the Soviet Union in 1991, the Caspian Sea emerged as a major new target.

Wherever large numbers of U.S. expatriates settle, U.S. culture soon follows. Central Asia, particularly the oil and gas center of Baku in Azerbaijan, is no exception. Most U.S. workers initially arriving in Baku found the lack of familiar foods and amenities discouraging. Soon, however, entrepreneurs began to fill the void. Particularly important have been Charlie and Marie Schroeder, Louisianans who had previously worked in the Middle East. While Mr. Schroeder's company, Caspian Sea Ventures, specializes in importing industrial parts, the couple has found particular success in the restaurant business. By early 1999 they were running four separate establishments, each with a different theme and flavor. Although distinctly Western, these establishments also reflect both the Gulf of Mexico flavor of much of the U.S. oil industry as well as the increasing cosmopolitanism of U.S. foodways. One restaurant, Ragin' Cajun, serves southern Louisiana cuisine (itself heavily influenced by French and African traditions); another, Margaritaville, specializes in Mexican-American food; and a third, Finnegan's, seeks to replicate the atmosphere of an Irish pub.

These "American" restaurants have not yet attracted many Azerbaijani customers, but they do provide employment opportunities. Mr. Schroeder, however, regrets that his employees do not have better options. "These kids are overqualified. Most of them speak anywhere between 4 to 13 languages, which makes me sick to my stomach because I can't even speak English. I speak New Orleans and a little bit of Texan."

Source: Adapted from Stephen Kinzer, "At a Crossroads of an Oil Boom, Everyone Comes to Charlie's." *The New York Times*, May 30, 1999.

Russia began to advance into Central Asia at roughly the same time as China. It already controlled Siberia, but its linkages with its far-flung Siberian outposts were threatened by the pastoral peoples of Kazakhstan. In the 1700s the Russian Empire therefore undertook the systematic conquest of the Kazakh steppes. When China weakened in the 1800s, Russia also advanced into former Chinese territory east of Lake Balqash. Russia's conquest of Kazakhstan proceeded fairly easily, since the Kazakh cavalry was no match for Russian guns. Expansion farther to the south, however, was blocked by the sedentary states of Uzbekistan. Only in the late 1800s, when European military techniques and materials raced ahead of those of Asia, was Russia able to conquer the Amu Darya and Syr Darya valleys. Its conquest of the area was not completed until the early 1900s, just before the Soviet Union replaced the Russian Empire.

One reason for the Russian advance into Central Asia was concern over possible British influence in the area. Britain did indeed attempt to conquer Afghanistan but was rebuffed by Afghan forces—and by the country's forbidding terrain. The British also sent a major military expedition to Tibet in the early twentieth century, and they almost created an autonomous Tibetan state under British "protection." This incident heightened China's determination to regain control over Tibet.

Central Asia Under Communist Rule

Western Central Asia came under communist rule after the foundation of the Soviet Union; Mongolia followed in 1924. Following the Chinese revolution of 1949, a communist system was also established in Xinjiang and Tibet. In all of these areas, major changes in the geopolitical order soon followed.

Soviet Central Asia During the chaos of the Russian revolutionary struggle from 1917 to the early 1920s, the Uzbeks and other Central Asian peoples attempted to regain their independence. They were soon crushed, however, by the Soviet Union's Red Army. The newly established Soviet Union thus inherited the Russian imperial dominion in Central Asia virtually intact. The policies that it directed toward this region, however, did not remain the same. The new regime sought not only to create a socialist economy, but also to build a new Soviet society that would eventually knit together all of the massive territories of the Soviet Union. Central Asia's leaders were replaced by Communist Party officials loyal to the new state, Russian immigration was encouraged, and local languages could no longer be written in Arabic script.

Although the early Soviet leaders foresaw the emergence of a single Soviet nationality, they realized that local ethnic diversity would not disappear overnight. Early Soviet leaders such as Vladimir Lenin also hoped to protect non-Russian peoples from Russian domination. They therefore divided the Soviet Union into a series of nationally defined union republics in which a certain degree of autonomy would be allowed. They were uncertain, however, about what the relevant units in Central Asia should be. Was there a single dominant Turkish-speaking nationality, or were the Turkish peoples themselves divided into a number of separate nationalities, with the Tajiks forming another? For several years boundaries shifted as new "republics" suddenly appeared on the map. Finally, in the 1920s, the modern republics of Kazakhstan, Kyrgyzstan, Tajikistan, Uzbekistan, Turkmenistan, and Azerbaijan assumed their present configurations. In certain areas, such as the fertile Fergana Valley, the political boundaries so drawn remained extremely complex (Figure 10.22).

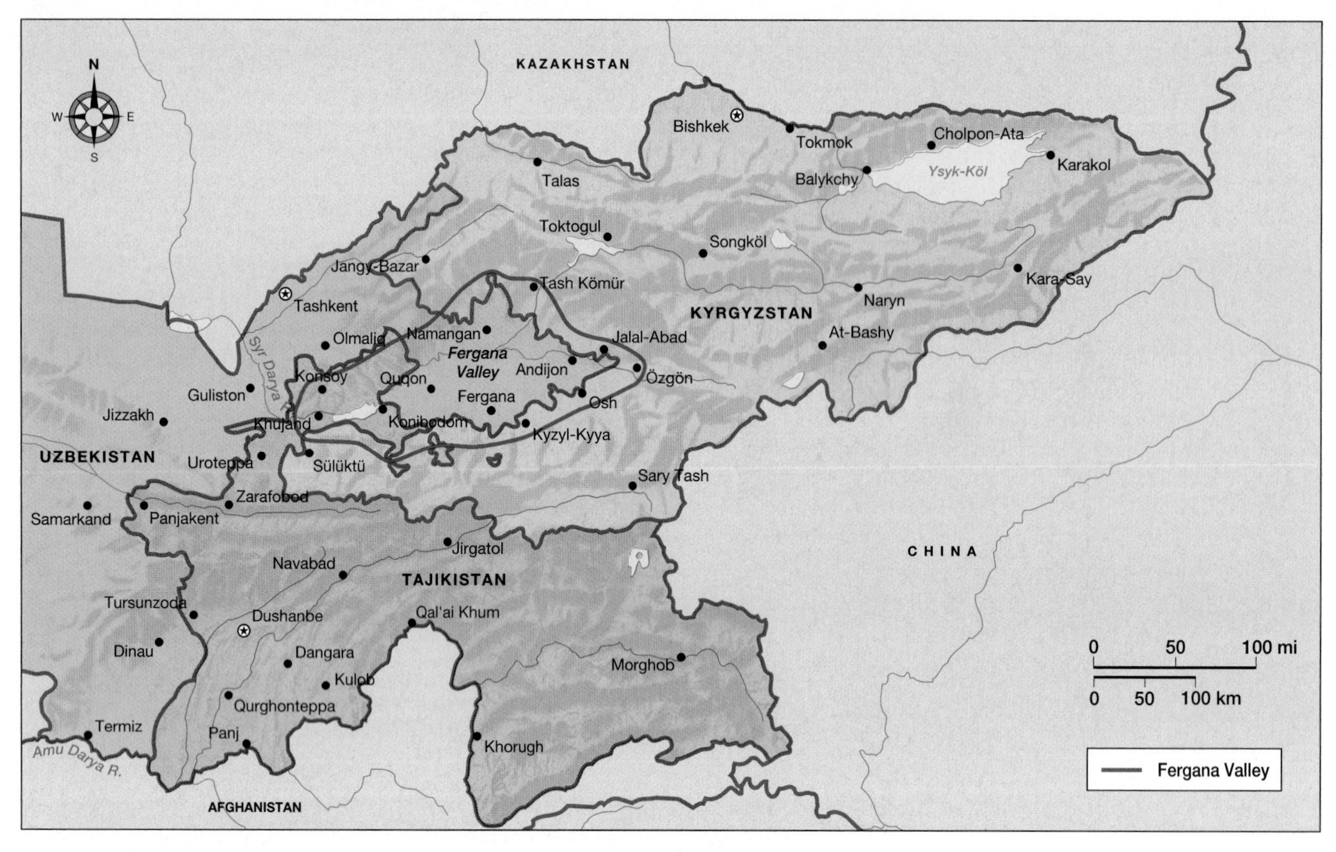

▲ **Figure 10.22 Political boundaries around the Fergana Valley** Some of the world's most convoluted political boundaries can be found in the vicinity of the Fergana Valley. The central portion of the valley belongs to Uzbekistan, which is otherwise separated from it by high mountains. The lower valley, on the other hand, is part of Tajikistan, the core area of which is likewise separated from the valley by highlands. The Fergana's upper periphery, finally, belongs to Kyrgyzstan.

Some scholars argue that the Soviet nationalities policy backfired rather severely. Rather than forming as a transitional step on the way to a Soviet identity, the constituent republics of the Soviet Union instead nurtured ideas of local nationalism that ultimately proved to be antithetical to the Soviet system. Identities such as Turkmen, Uzbek, and Tajik had been vague in the pre-Soviet period; now they were given real political significance.

Another problem undercutting Soviet unity was the fact the cultural and economic gaps separating Central Asians from Russians and other Europeans did not diminish as much as planned. Islamic beliefs remained entrenched in some quarters and began to revive elsewhere in the 1970s. Central Asia also remained poorer than most other parts of the Soviet Union, and by the 1980s it was becoming something of a burden on the national economy. Equally important were the higher birthrates found in Central Asia, leading many Russians to fear that the Soviet Union risked being overwhelmed by Turkish-speaking Muslims.

The Chinese Geopolitical Order After China reemerged as a united country in 1949, it too was able to reclaim most of its old Central Asia dominion. China's early communist leadership promised the non-Han peoples of the periphery a significant amount of political self-determination as well as cultural autonomy, and thus these leaders found much local support in Xinjiang. Tibet, isolated behind its mountain walls and virtually independent for the previous 150 years, presented a greater obstacle. China occupied Tibet in 1950, but the Tibetans launched a rebellion in 1959. When this was brutally crushed, the Dalai Lama, along with some 100,000 followers, found refuge in India (Figure 10.23).

Loosely following the Soviet nationalities model, China established autonomous regions in areas occupied primarily by non-Han peoples. Xinjiang, Tibet proper (called Xizang in Chinese), and Inner Mongolia were so classified, which implied that their indigenous peoples would retain certain cultural and administrative privileges. Such privileges, however, often turned out to be more theoretical than real, and they certainly did not prevent the massive immigration of Han Chinese into these areas. Nor were all parts of Chinese Central Asia granted autonomous status. The large and historically Tibetan and Mongolian province of Qinghai, for example, remained an ordinary Chinese province.

Current Geopolitical Tension

The former Soviet portion of Central Asia weathered the post-1991 transition to independence rather smoothly, but the region still suffers from a number of actual and potential ethnic

▲ **Figure 10.23 Chinese invasion of Tibet** In 1959 China launched a massive invasion of Tibet. Pursued by the Chinese army, the 23-year-old Dalai Lama (second from the lead) is shown here escaping over the Zsagola Pass, ultimately to find refuge in India. *(Hg/AP/Wide World Photos)*

conflicts. Much of China's Central Asian territory is seething, but China retains a firm grip on the region. Afghanistan, unfortunately, remains locked in what seems to be an interminable civil war.

Independence in Former Soviet Lands The breakup of the Soviet Union in 1991 generally proceeded peacefully in Central Asia as elsewhere. Suddenly the six republics found themselves able to pursue their own policies. All of them, however, had been dependent on the Soviet system, and it was no simple matter for these states to chart their own courses. They still had to cooperate with Russia over security issues, and all opted to remain part of the Commonwealth of Independent States, the rather hollow successors of the Soviet Union. In most cases, authoritarian rulers, rooted in the old order, retained power and sought to undermine oppositional groups. All told, democratic governance made less progress in Central Asia than in other parts of the former Soviet Union.

Two of these new republics, Kazakhstan and Tajikistan, have faced a particularly difficult transition. Kazakhstan is the largest and most resource-rich Central Asian state, and therefore seems to have the best chance for success. Potential ethnic strife, however, looms over the country. Many Kazakhs want to create a national state centered around Kazakh identity, and they resent the presence and power of the Russians, Ukrainians, Germans, and others of European background. These Europeans, for their part, fear the imposition of what they consider to be alien Central Asian cultural standards. The situation is complicated, moreover, by the fact that Kazakhstan inherited a sizable nuclear arsenal from the Soviet Union, although it has promised to dismantle all nuclear missiles within its territory.

Thus far, tensions in Kazakhstan have remained muted, but many fear an unstable future. The predominantly Russian population of northern Kazakhstan, the country's breadbasket, could well attempt to separate from the rest of the country to join their land with Russia. While extreme Russian nationalists would welcome such a move, moderates would reject it as destabilizing. In the late 1990s, Kazakhstan moved its capital from Almaty in the south to Astana in the Russian-dominated north, partly to forestall any plans for secession.

In Tajikistan war broke out almost immediately after independence in 1991. Many members of the smaller ethnic groups living in the mountainous east resented the imposition of authority by the lowland Tajiks and thus rebelled. They were joined by several Islamic groups seeking to overthrow the secular state. Although the civil war officially ended after several years, fighting continues to flare up periodically. New lines of tension are now emerging between the Tajiks and the country's large Uzbek minority, a conflict that threatens to involve Uzbekistan as well. The Tajikistan government has been able to retain control in most areas, thanks in part to its use of Russian troops. The presence of Russian military forces, however, open the question of Russia's long-term intentions in the region. Although the present Russian government probably does not plan to reabsorb any Central Asian territory, it does regard the entire area as lying within its zone of strategic interests.

Azerbaijan also experienced strife following the breakup of the Soviet Union. Armenia invaded, and still holds, the Armenia-speaking highlands in the western portion of the country. Hostile relations with Armenia make it difficult for Azerbaijan to retain control over its **exclave** of Naxcivan, a piece of Azerbaijani territory separated from the rest of the country by Armenia and Iran.

Strife in Western China Local opposition to Chinese rule in Central Asia increased during the 1990s, albeit without success. Any form of protest in Tibet is severely repressed, but the Tibetans have successfully brought the attention of the world to their struggle (see "Local Voices: Virtual Tibet"). China maintains several hundred thousand troops in a region that has only 2.4 million civilian inhabitants. Such an overwhelming military presence is considered necessary both because of Tibetan resistance and the geostrategic importance of the region. The border between China and India is still contested, with China controlling a small section of the Tibetan Plateau that India claims.

Until the late 1990s, China's control of Xinjiang seemed much more secure than its hold on Tibet. Although fragmented by deserts and mountains, Xinjiang is less forbidding than Tibet and is now the home of millions of Han Chinese immigrants. Xinjiang is also far more vital to China than is Tibet. It contains a variety of mineral deposits (including oil) essential for Chinese industry, and it has been the site of nuclear weapons tests. Many Uygurs, not surprisingly, chafe under such uses of their homeland, and they deeply resent the periodic suppression of their religion and culture by the communist regime. In 1990 and again in 1997, anti-government riots broke out in several major cities (Figure 10.24). During the latter event, Uygur separatists even planted bombs in Beijing. Immediate reprisals ended this overt display of secessionist

▲ **Figure 10.24 Ethnic tension in Xinjiang** Relations between Han Chinese Officials and Muslim Central Asians grew increasingly tense in the 1990s. In 1997, severe rioting broke out in several cities in Xinjiang. This photograph shows a Chinese official levying a trading tax on merchants in Kashgar's Sunday market. *(Chris Stowers/Panos Pictures)*

sentiment, but the underlying ideal of independence remains unbroken.

China's position is that all of its Central Asian lands are integral positions of its national territory. Those who advocate independence are viewed as traitors and are sometimes considered to be in league with Western political forces that have sought for the past 200 years to keep China weak and divided. Chinese officials have also linked separatist elements in Xinjiang to a global movement of radical Islamic fundamentalists. Leaders of Xinjiang's "Free Eastern Turkestan Movement," however, insist that their movement is founded on territory rather than religion or ethnicity, contending that it is based on an alliance of all non-Han peoples of Xinjiang, including Uygur Muslims, Chinese-speaking Muslims (Hui), Kazakh Muslims, Buddhist Mongols, Buddhist Uygurs (the small Yugur ethnic group), and even a minute group belonging to the Russian Orthodox Church.

War in Afghanistan None of the conflicts in western China or in the former Soviet republics compares in intensity to the struggle now being waged in Afghanistan. Afghanistan's troubles began in 1978, when a Soviet-supported military "revolutionary council" seized power. The new Marxist-oriented government soon began to suppress religion, which led almost immediately to widespread rebellion (Afghanistan is a very traditional and deeply Islamic country). When the government appeared to be on the verge of collapse, the Soviet Union responded with a massive invasion. Despite its power, the Soviet military was never able to gain control of the more rugged parts of the country. Both Pakistan and the United States, moreover, ensured that the anti-Soviet forces remained well armed.

The exhausted Soviets finally withdrew their troops in 1989. The puppet government that they installed remained to face the insurgents alone. It managed to hold the country's core around the city of Kabul for a few years, largely because the opposition forces were themselves divided. Local warlords increasingly grabbed power in the countryside, destroying any semblance of central authority or national unity.

In 1995–96, a new power called Taliban arrived on the Afghan scene. Taliban was founded by young Muslim religious students disgusted with the anarchy that was consuming their country. They were convinced that only the firm imposition of Islamic law could end corruption, quell the disputes among the country's different ethnic groups, and ulti-

LOCAL VOICES Virtual Tibet

The Tibetan movement for independence has inspired perhaps the largest number of sites on the World Wide Web of any political resistance movement, creating what is in effect a "virtual Tibet." Not only Tibetan activists and the Tibetan government in exile, but also a number of "Tibetan friendship committees" the world over run Web pages dedicated to gaining freedom for Tibet. A good source for locating many of these Web sites is "Tibet Online Resource Gathering" (www.tibet.org/). The political orientation of this site is evident on its home page, which states that "Tibet's ancient and fantastic civilization and ecosystem are faced with extinction due to 48 years of mismanagement and abuse under its colonial ruler, the People's Republic of China."

An equally important site is that of the official Tibetan Government in Exile, headed by the Dalai Lama and run out of Dharmasala, India (www.tibet.com/). One of the more interesting aspects of this site is its map of Tibet, which includes much more territory than is contained in Xizang, China's autonomous region of Tibet. As the Web page explains, "Tibet is comprised of the three provinces of Amdo (now split by China into the provinces of Qinghai and part of Gansu), Kham (largely incorporated into the Chinese provinces of Sichuan, Gansu, and Yunnan), and U-Tsang (which, together with western Kham, is today referred to by China as the Tibet Autonomous Region)."

Although the vast majority of Tibetan Web sites are devoted to Tibetan independence, the government of China does supply an opposing view. One important site here is run by the China News Organization (www.chinanews.org/Tibet/index.html). From this site one can open pages with such titles as "Exposing the Trickery of the Dalai Lama," "Dalai Sabotages Religious Order," and "Historical Records Prove Tibet Is Inseparable Part of China." The last-mentioned page contends that "History has proved time and again that only while in the embrace of the motherland [in other words, China] can Tibet achieve sound and rapid development, an improved living standard, and more freedom."

mately bring peace and unity to Afghanistan. Large numbers of soldiers began flocking to the Taliban standard almost immediately after it was raised. By 1997, Taliban forces had won control over all portions of Afghanistan except the most rugged mountains and the northernmost provinces.

Taliban acquired strength not only from its own religious nature but also from the ethnic divisions of Afghanistan. It is closely identified with the Pathan (or Pushto-speaking people). The Pathan are not only the most populous ethnic group of Afghanistan, but they also have a reputation for militarism—as well as a good connection for military supplies among their relatives across the Pakistan border. Pakistan is one of only a few countries to have recognized Taliban as forming the legitimate government of Afghanistan.

The main opposition to Taliban has come from the country's other ethnic groups, especially the Shiite Hazaras of the central mountains and the Uzbeks and Tajiks of the north. There Taliban met firm resistance and suffered several reversals, leading some to speculate that the country could irrevocable split into several ethnically defined territories (Figure 10.25). But it would be a mistake to regard this conflict as reducible to ethnic differences, as several noted Uzbeki commanders have joined forces with the Taliban. As we shall see later in this chapter, there is also a strong social dimension to this conflict, much of which revolves around the social position of women.

▲ **Figure 10.25 War in Afghanistan** Prolonged civil war in Afghanistan has been a social disaster. Many Afghans have sought refuge from the extreme fundamentalist forces of the Taliban, either by heading into other countries or by fleeing to areas of Afghanistan not under Taliban rule. These refugees have fled Kabul and are heading north. *(Martin Adler/Panos Pictures)*

Global Dimensions of Central Asian Tension

With the collapse of the Soviet Union in 1991, Central Asia emerged as a key arena of geopolitical tension. A number of important countries, including China, Russia, Pakistan, Iran, Turkey, and the United States, now vie for power and influence in the region. The revival of Islam has also generated geopolitical repercussions.

Islamic Fundamentalism? If Afghanistan were to dissolve altogether, it is conceivable that the northern parts of the country could unite with Uzbekistan and Tajikistan on the basis of ethnic solidarity. Such a reconfiguration of the geopolitical order, however, does not seem likely in the near term, especially considering the difficulties that Uzbekistan and Tajikistan face themselves. At present, these two countries are more concerned about the potential consequences of an outright Taliban victory, which could send millions of Uzbek and Tajik refugees fleeing northward across their boundaries.

All of the governments of Central Asia, except that of Mongolia, fear at some level that an Islamic-fundamentalist movement similar to that of Taliban might disrupt their own countries. But to a great extent, the problems of Afghanistan are uniquely rooted in its own history and cultural geography. The Afghan conflict, moreover, was ignited by the Cold War between the United States and the Soviet Union that is now over. A resurgence of radical Islamic fundamentalism in Central Asia thus seems unlikely.

Border Conflicts The border between China and the newly independent Central Asian republics presents another potential international issue. China's government does not formally accept the boundary now separating it from Tajikistan. It also objects to the fact that Kyrgyzstan supports radio broadcasts in Uygur, supposedly aimed at Kyrgyzstan's own small Uygur minority. One member of the Kyrgyz parliament responded by warning his country that China plans to overwhelm all of Central Asia with a massive Han migration scheme. Although his fears are no doubt overblown, it is true that some ardent Chinese nationalists have never accepted their country's loss of its former territories to the east of Lake Balqash. On the other side of the coin, some one million Kazaks live in northern Xinjiang, and many would like to be reunited with their co-nationals on the other side of the border. Leaders in Beijing and in the capitals of Kazakhstan, Kyrgyzstan, and Tajikistan, however, do not want to provoke overt conflicts over these unresolved issues.

China's boundary with Mongolia is also potentially troublesome. But while many Mongolian nationalists would like eventually to reclaim Inner Mongolia as part of their national patrimony, few are willing to provoke China. The Free Southern Mongolia pages on the World Wide Web (www.caccp.org/im/), however, are extremely forceful in their denunciations of China.

The Role of Russia, Iran, Pakistan, and Turkey Russia just barely touches northern Xinjiang and thus has little occasion to squabble with China over their common Central Asian border. Russia's immediate concerns are rather directed to its own former territories in the region. This is not just a matter of potential claims to Russian-speaking territories in northern Kazakhstan, or even, as extreme nationalists dream, of Russia's reclaiming a Central Asian empire; it is also a question of cultural and economic influence. Russia's economic ties to the region, as examined in the next section, were by no means erased when the Soviet Union collapsed. Russia's main transportation line linking Europe to central Siberia, moreover, cuts across northern Kazakhstan, while former Soviet Central Asia's rail links to the outside world are largely oriented toward Russia.

The Central Asian governments are themselves rather ambivalent about Russian influence. While all are eager to nurture their own sense of nationalism, the Russian connection brings undeniable benefits. Turkmenistan, for example, enacted measures in the late 1990s to discourage Russian speakers from moving back to Russia for fear that the resulting "brain drain" could undermine its economy.

Russia is not the only external power interested in influencing the new republics of Central Asia. Iran is now a major trading partner, and it offers the best route to the ocean. Since the completion of a rail link between Iran and Turkmenistan in 1996, part of Central Asia's global trade has been reoriented toward Iran's ports. Iran's cultural ties with the region are old and deep, particularly in Tajikistan and northern and central Afghanistan, where Persian is a major language. The Iranian desire for influence in Central Asia is partly religious in orientation, as Iran would like to export its own version of Islamic politics. Iran is seriously hampered in this quest, however, by the fact that the vast majority of Central Asian Muslims are Sunnis rather than Shiites.

Pakistan is also interested in gaining influence in Central Asia. In earlier centuries close religious and political bonds linked Central Asian and South Asian Muslim communities—bonds that many Pakistanis would like to revive. Extremely rugged topography, however, separates Pakistan from the Amu Darya and Syr Darya valleys; Afghanistan, moreover, occupies the intervening territory. While Pakistan enjoys close connections with the Taliban fighters who control most of Afghanistan, such influence does not win much support in Uzbekistan or Tajikistan.

Turkey's cultural connections with Central Asia are particularly close. Most Central Asians speak Turkish languages that are so closely related to that of Turkey that they can understand Turkish television and radio broadcasts. In the late 1800s and early 1900s, a political philosophy known as Pan-Turkism advocated that all Turkish-speaking peoples unite to form a single huge country. This school of thought no longer has many adherents, but a sense of kinship does indeed link Turkey to the Central Asian countries (including Xinjiang) known collectively as Turkestan.

Turkey offers itself as the model—contrary to that of Iran—of the modern state of Muslim heritage. In this secular model, religion continues to provide social and cultural glue, but politics and economics operate without religious content. Turkey itself, however, is having some difficulty maintaining this balancing act, and it is not obvious that the Turkish system can easily be exported, even to other Turkish-speaking states. Turkey is also separated from Central Asia by Armenia, Russia, and Iran—countries with which it does not enjoy good relations. But if Pan-Turkism were ever to catch fire, its potential for reconfiguring the map of the world is significant. The Turkish-speaking community extends from southeastern Europe almost to central China, a distance of more than 3,000 miles (4,800 kilometers).

Economic and Social Development: Abundant Resources, Devastated Economies

Central Asia is by most conventional measures one of the least prosperous regions of the world. One of its countries, Afghanistan, stands near the bottom of almost every list of economic and social indicators. Central Asia does, however, contain substantial natural resources, particularly of oil and natural gas. Much of the region also enjoys relatively high levels of health and education, a legacy of the social programs enacted by the communist regimes. These same regimes, however, built inefficient economic systems, and since the fall of the Soviet Union, the former Soviet zone has experienced a particularly severe economic decline.

The Post-Communist Economies

Soviet economic planners sought to spread the benefits of economic development widely across their country. This required building large factories even in such remote areas as Central Asia, regardless of the costs involved. Such Central Asian industries relied heavily on subsidies from the center. When those subsidies ended, the industrial base of the region began to collapse, leading to a massive economic decline and rapidly plummeting living standards. As is true elsewhere in the former Soviet Union, however, certain individuals living here have grown very wealthy since the fall of communism.

As Table 10.2 shows, no Central Asian country or region could be considered prosperous by any means. Kazakhstan stands as the most developed, and it may have the best prospects. Since Kazakhstan's agricultural base is potentially productive and its population density is low, it could emerge as a major food exporter. More important, it has one of the world's largest underutilized deposits of oil and natural gas, the vast Tengiz Field located to the east of the Caspian Sea, as well as sizable deposits of other minerals. Kazakhstan has signed agreements with Western oil companies to exploit its oil reserves (Figure 10.26), but work thus far has been slowed by bureaucratic entanglements.

Table 10.2 Economic Indicators

Country	GNP per Capita ($U.S., 1994)	Total GNP (Millions of $U.S., 1994)	PPP*	Real Annual Growth % per Capita, 1985–1994
Afghanistan	—	—	—	—
Azerbaijan	480	3,642	1,490	–18.7
Kazakhstan	1,350	22,213	3,230	–10.3
Kyrgyzstan	550	2,486	1,970	–12.7
Mongolia	360	902	1,820	–2.3
Tajikistan	340	1,964	900	–18.5
Turkmenistan	940	4,319	2,010	–13.1
Uzbekistan	1,010	23,490	2,450	–5.6

*Purchasing power parity.

Source: *The World Bank Atlas*, 1998.

Owing to its greater population, Uzbekistan has a larger total GNP than Kazakhstan, giving it the largest economy in the region. Uzbekistan, moreover, has not experienced the same degree of economic decline as have most of its neighbors. This is largely because it has retained many aspects of the old command economy (one run, in other words, by governmental planners rather than by private firms responding to the market), and has resisted economic liberalization. Critics contend, however, that Uzbekistan will grow increasingly uncompetitive unless industries are privatized and markets opened. At present, Uzbekistan remains a major exporter of cotton. It also has significant gold and natural gas deposits, and has inherited a number of chemical and machinery factories from the Soviet period. Cotton production, however, is threatened by environmental degradation, and agriculture in general remains under inefficient, Soviet-style state control.

▲ **Figure 10.26 Oil development in Azerbaijan** Although oil has brought a certain amount of wealth to Azerbaijan, it has also resulted in extensive pollution and visual blight. Since most of the petroleum is located either near or under the Caspian Sea, this sea—actually the world's largest lake—is now ecologically endangered. *(John Spaull/Panos Pictures)*

Kyrgyzstan, on the other hand, has moved aggressively to privatize former state-run industries. Its economy, however, is largely agricultural, and many of the industries that it does contain are not competitive. Since independence, Kyrgyzstan's economy has shrunk by almost 50 percent. It does, however, have the largest supply of fresh water in the region, a resource that will become increasingly valuable in years to come.

Turkmenistan also has a substantial agricultural base, due mainly to Soviet irrigation projects, and like Uzbekistan it remains a major cotton exporter. Turkmenistan, however, retains a state-run economy and has resisted pressure for economic liberalization. As late as 1999 it was still declining. Government planners, however, hope that the development of new oil and gas fields will bring prosperity, proclaiming that Turkmenistan will soon become the "Kuwait of natural gas."

The region's best-developed fossil fuel industry is located in Azerbaijan. Azerbaijan has attracted a great deal of international interest and investment, promising to revitalize its oil industry. Thus far, however, its economy has largely failed to respond, and Azerbaijan—despite its promises—remains a very poor country.

The most economically troubled of the former Soviet republics is Tajikistan. With a per capita GNP of only some $350, Tajikistan rates as one of the world's poorer countries. It has few natural resources. A prolonged civil war has undercut production, and Tajikistan is burdened by its remote location and rugged topography. Its government has done little to dismantle the old command economy, yet industrial output has still declined at an alarming rate. In a single year (1994), factory production is estimated to have declined some 31 percent.

Tajikistan's recovery probably depends not only on the resumption of peace and the enactment of reforms, but also on the stabilization of its neighbors. Tajikistan is one of the most remote countries of the world, with poor connections to the outside and high transportation costs (Figure 10.27). Most rail lines, inherited from the Soviet period, link the country ultimately to Russia. More efficient transportation routes would connect Tajikistan to the world economy through the

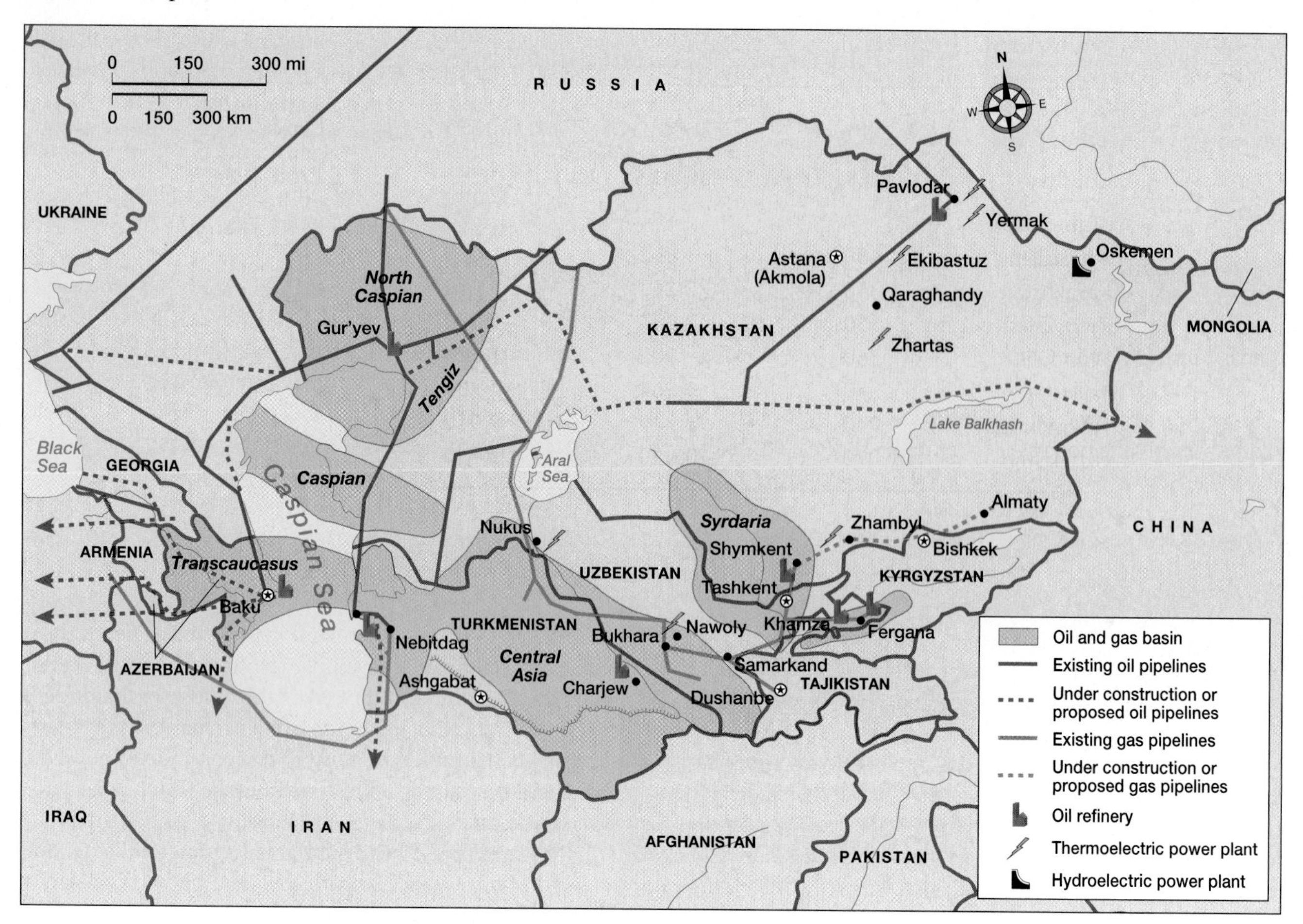

▲ **Figure 10.27 Oil and gas pipelines** Central Asia has some of the world's largest oil and gas deposits, and has recently emerged as a major center for drilling and exploration. Due to its landlocked location, Central Asia cannot easily export its petroleum products. Pipelines have been built to solve this problem, and a number of others are currently being planned. Pipeline construction is a contentious issue, however, since several of the potential pathways lie across Iran, a country that remains under U.S. sanctions, and Russia, a politically and economically unstable country.

Caspian and then the Black seas, or through Iran to the Arabian Sea. At present, the Iranian route is emerging as a favored alternative, thanks in part to the completion of a new rail line linking Iran to Turkmenistan.

Mongolia, although never part of the Soviet Union, was a close Soviet ally run by a communist party. It too suffered a partial economic collapse in the early 1990s. Mongolia no longer receives Soviet subsidies and its industries are not competitive in the global market. Its agricultural foundation is meager and its traditional trade in livestock products is not very profitable. Mongolia thus emerged in the post-communist period as a poor country indeed, and its economy continued to decline into the late 1990s. Isolation also plagues Mongolia. Mongolia probably has some substantial mineral reserves, however, and its low population density gives it a certain leeway. Subsistence production of meat and other animal products, moreover, helps its people maintain their livelihoods regardless of the country's official economic standing.

The Economy of Tibet and Xinjiang in Western China The Chinese portions of Central Asia have not suffered the same economic crash that has visited the other parts of the region. China as a whole has one of the world's fastest-growing economies, although its centers of dynamism are all located in the coastal zone—several thousand miles to the east of Xinjiang and Tibet. Still, the most remote parts of the country have at least maintained their positions. But China as a whole started out much less developed than the Soviet Union, and poverty remains widespread. Although reliable economic figures are difficult to obtain, it would seem that the overall levels of per capita economic production in western China are comparable to those of Tajikistan.

Tibet in particular remains one of the world's poorest places. Most of the plateau is relatively isolated from the Chinese economy, much less the global economy. Like the residents of other areas of subsistence production, however, the people of Tibet are at least able to provide most of their own basic needs. Tibet, and Xinjiang as well, do not suffer the overcrowding that strains the poorer parts of China proper. Then again, their harsh environments simply cannot support high population densities. This leads critics to contend that Han Chinese immigration threatens to upset the local balance between human numbers and the environment throughout western China.

Xinjiang does have tremendous mineral wealth, including a substantial portion of China's oil reserves. Its agricultural sector is also productive, although limited in extent. Many of the indigenous Muslim peoples of the area believe that the wealth of their land is being monopolized by the Chinese state and the Han immigrants. Some point to the China Xinjiang Construction Company, a firm that runs more than 340 industrial enterprises as well as 172 giant farms, 500 schools, and 200 hospitals. This huge company was initially formed by the army that rejoined Xinjiang to China in 1949 and is still run almost entirely by Han Chinese.

Economic Misery in Afghanistan Afghanistan is unquestionably the poorest country in the region, with one of the weakest economies in the world. Reliable statistics, however, are impossible to obtain, since the county has no real central government. Afghanistan has suffered nearly continuous war since the late 1970s, undercutting virtually all economic endeavors. Even before the war it was an impoverished country, with little industrial or commercial development. Its only significant legitimate exports are animal products, hand-woven carpets, and a few fruits, nuts, and semiprecious gemstones. Almost all fuel and consumer goods—and a good deal of basic food stocks—must be imported. Weapons imports have also burdened the Afghan economy in recent years.

One might well wonder how Afghanistan manages to pay for all of these imports and still keep afloat. Modern-day war, after all, is an expensive affair. The answer, it would appear, is simple: Afghanistan sells large quantities of illicit drugs on the global market. According to the CIA, it is now the world's largest producer of opium, and it is a significant exporter of *Cannabis* (marijuana and hashish) as well. Most of the opium produced in the country ends up being sold as heroin in the large cities of North America and Europe. As Figure 10.28 shows, opium

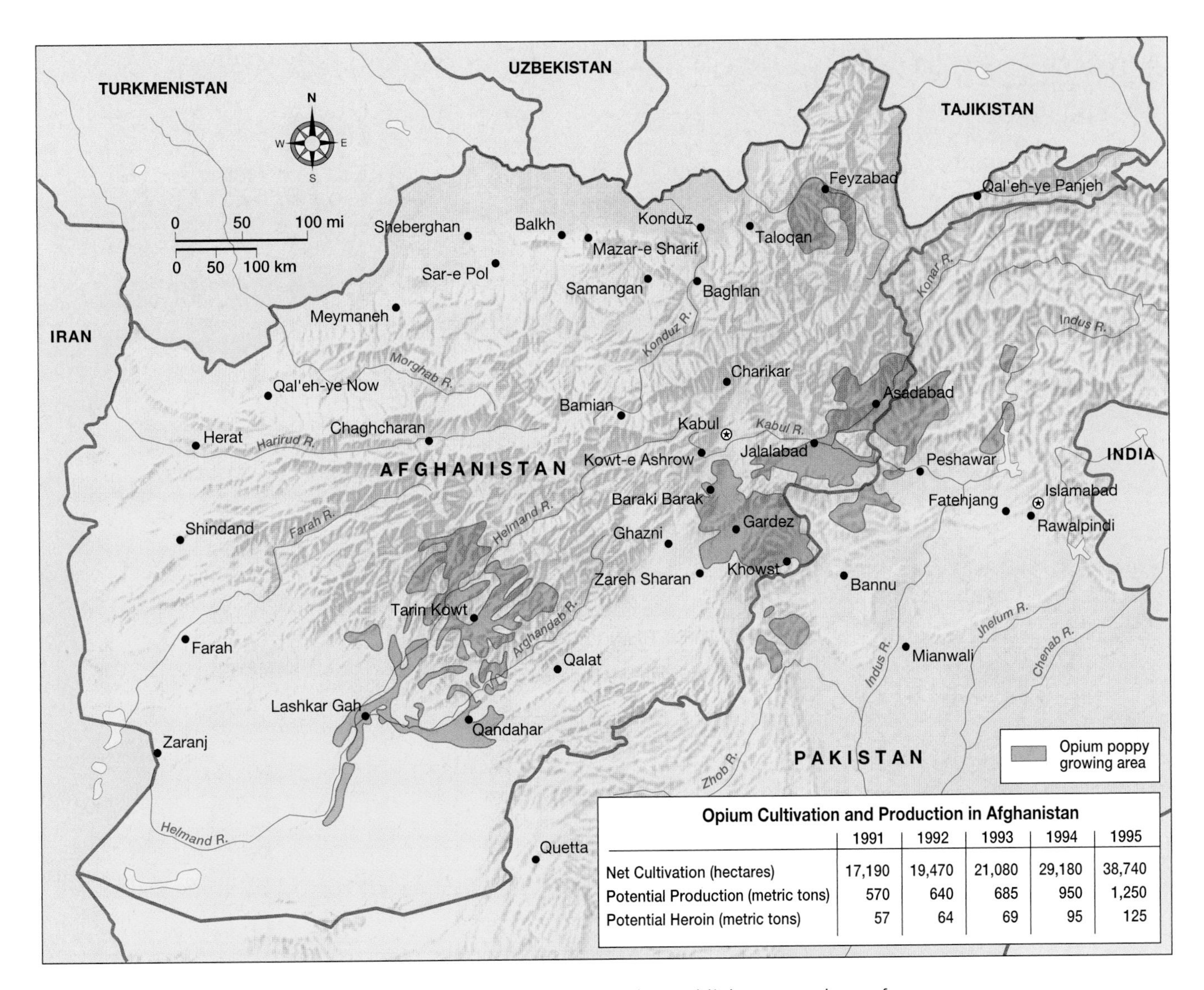

Opium Cultivation and Production in Afghanistan

	1991	1992	1993	1994	1995
Net Cultivation (hectares)	17,190	19,470	21,080	29,180	38,740
Potential Production (metric tons)	570	640	685	950	1,250
Potential Heroin (metric tons)	57	64	69	95	125

▲ **Figure 10.28 Opium growing areas** Afghanistan is now the world's largest producer of opium, from which heroin and other narcotics are manufactured. Most opium is produced in the southern portion of the country, an area controlled by the extreme Islamicist forces of the Taliban. Owing to Afghanistan's poverty, its lack of infrastructure, and its ongoing civil war, opium is one of the few products that it can successfully export to the world market.

growing is concentrated in the rugged central mountains. Although Taliban has denounced the opium trade, some evidence suggests that Taliban leaders have allowed production in the areas that they control. It is, after all, the only reliable source of hard currency at their disposal.

Central Asian Economies in Global Context As the discussion above indicates, despite its poverty and relative isolation, Afghanistan is thoroughly embedded in the global economy, albeit through illicit products. Illegal drugs are extremely lightweight and can thus be successfully exported from even the most isolated parts of the world. The non-opium producing parts of Central Asia, in contrast, cannot maintain such a close connection to such nerve centers of the world economy as New York and London.

In the former Soviet area, the most important international connections remain with Russia. Both Russia and the newly independent Central Asian countries are ambivalent about this relationship. Russia has sought to distance itself economically from its former Central Asian territories, most notably by forcing them to develop their own currencies and to stop using the Russian ruble. Turning away from Russia, the former Soviet zone is quickly developing economic as well as political ties to such countries as Iran, Pakistan, and Turkey.

The United States and other Western countries are increasingly drawn to the area because of its oil and natural gas deposits. Most large oil companies have established operations in Kazakhstan and Turkmenistan, which are perceived to contain the world's largest unexploited fossil fuel deposits open to Western firms (Figure 10.29). Western goods, in-

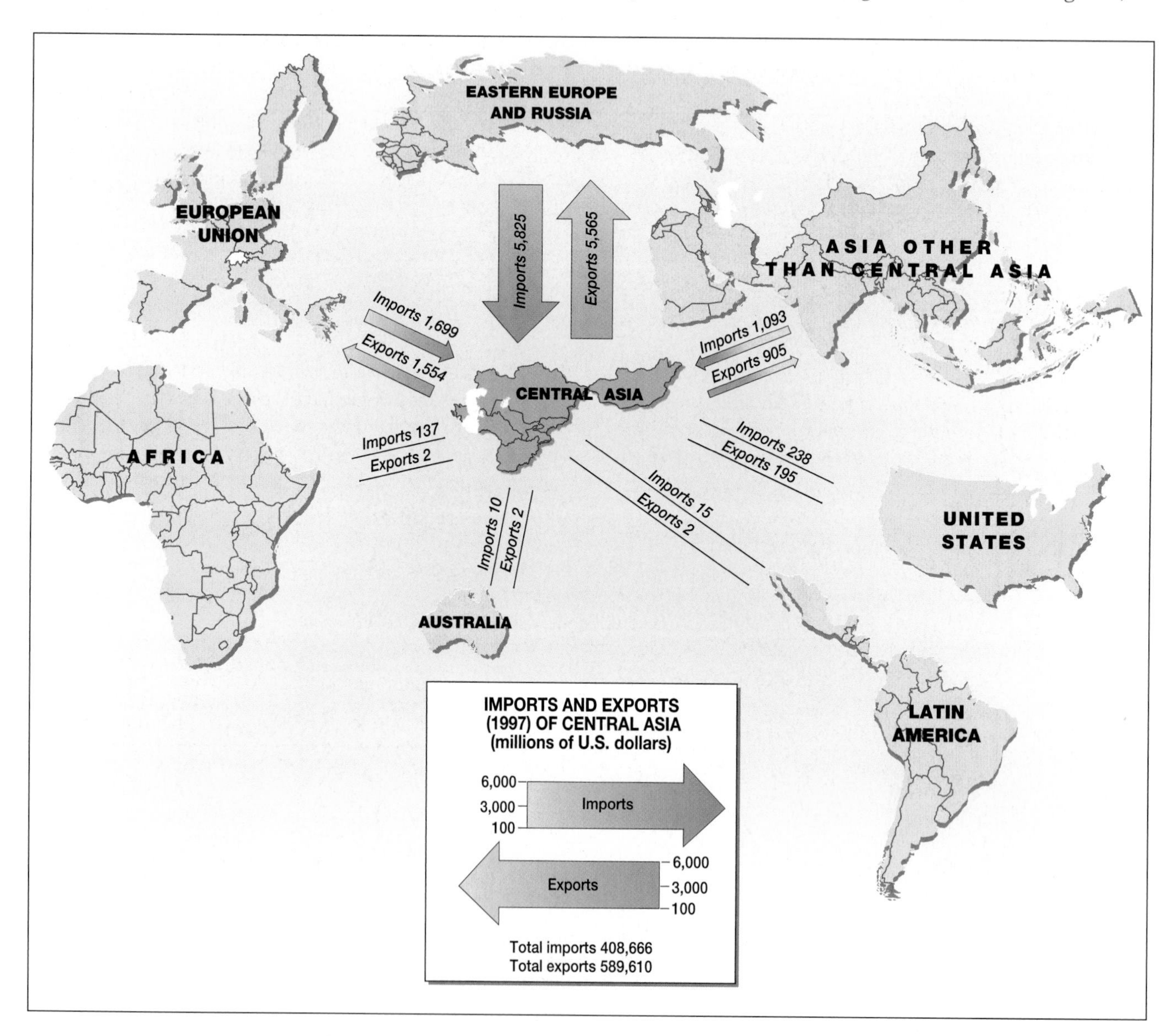

▲ **Figure 10.29 International trade flows** Central Asia is poorly integrated into global markets. Most of its trade remains oriented toward Russia and the other states of the former Soviet bloc. Trade with North America, East Asia, and especially the European Union has rapidly expanded in recent years. If the region's oil and gas deposits are as large as some geologists think, Central Asia will probably gain a much more prominent place in global commerce within the next several decades. *(Data from Euromonitor,* International Marketing Data and Statistics, 1998*)*

cluding luxury items such as sport utility vehicles, are now readily available for those profiting from oil wealth in cities such as Kazakhstan's Almaty, Turkmenistan's Ashgabat, and Azerbaijan's Baku.

These same fossil fuel reserves have created a very complex international economic environment. For the region's large oil fields to be economically viable, a vast pipeline system must be constructed. Existing pipelines, which are already inadequate, pass through Russia, which charges exorbitant prices. A large amount of oil is shipped out by rail through Russia and Georgia, an inefficient mode of export. International negotiations began in the late 1990s for the construction of new pipelines. The most efficient route for Turkmenistan would be through Iran, and in 1997 a small gas pipeline was opened between the two countries. The United States, however, steadfastly opposes any new pipelines passing through Iran. Another plan for a pipeline through Afghanistan to Pakistan has floundered, not surprisingly, due to unstable conditions in Afghanistan.

The United States has lobbied for a pipeline system under the Caspian Sea and then through Azerbaijan and Georgia to the Black Sea. The main problem with this scheme is that oil tankers would have to pass through the Bosporus in Turkey, a narrow passageway that is already overcrowded. Environmentalists fear that oil spills in the Bosporus could have devastating repercussions. A more expensive alternative plan would be to lay a pipeline across Turkey to the Mediterranean port of Ceyhan.

China is also interested in the fossil fuel reserves of the former Soviet zone. Kazakhstan and China have even discussed building a pipeline from eastern Kazakhstan all the way to eastern China. Such a project, however, would probably be prohibitively expensive, but China desires it for strategic reasons.

Social Development in Central Asia

Social conditions in Central Asia vary more than economic conditions. In the former Soviet territories, levels of health and education are relatively high. Afghanistan, not surprisingly, comes out at the bottom of the scale.

The Status of Women in Afghanistan Social conditions in Afghanistan are little if any better than its economic circumstances, although again reliable information is difficult to obtain. As can be seen in Table 10.3, the average life expectancy in the country is a mere 45 years, one of the lowest figures in the world. Infant and childhood mortality levels remain extremely high. Not only does Afghanistan suffer from constant warfare, but its rugged topography hinders the provision of basic social and medical services. Illiteracy is commonplace and is notably gender-biased. Afghanistan's 11 percent adult female literacy figure is one of the lowest in the world.

Women in traditional Afghan society—and especially in Pathan society—lead constrained lives. In many areas they must completely conceal their bodies, including their faces, when venturing into public space. Such restrictions have intensified in recent years. Dress controls are strictly enforced where Taliban has gained authority. In some cities, Taliban forces have actually prevented women from working and attending school. The prohibition on work has brought particular hardship to Afghanistan's war widows, leaving many in outright destitution. It has even apparently forced some younger women—against Taliban's intentions—into prostitution. Several international aid agencies have been forced to scale back their operations in Afghanistan after their female employees began to suffer from constant harassment (Figure 10.30).

The assault on women by Taliban has not gone uncontested in Afghanistan. Whereas Taliban leaders argue that they are upholding Islamic orthodoxy, their opponents contend

Table 10.3 Social Indicators and Status of Women

Country	Life Expectancy at Birth		Under Age 5 Mortality, per 1,000 Live Births		Secondary School Enrollment %*		Female Labor Force Participation (% of total)
	Male	Female	1960	1993	Male	Female	
Afghanistan	46	45	360	257	22	8	35
Azerbaijan	67	74	—	52	89	88	44
Kazakhstan	60	70	—	49	89	91	47
Kyrgyzstan	63	71	—	58	—	—	47
Mongolia	64	62	185	78	—	—	46
Tajikistan	65	71	—	83	98	101	44
Tibet	—	—	—	—	—	—	—
Turkmenistan	62	69	—	89	99	97	45
Uzbekistan	66	72	—	66	96	92	46
Xinjiang	—	—	—	—	—	—	—

Sources: *Population Reference Bureau Data Sheet, 1998,* except Mongolia (199), Life Expectancy (M/F); *World Resources Institute, 1996–97,* Under-5 Mortality Rate; *Population Reference Bureau Data Sheet, 1998,* except Mongolia (199), Secondary School Enrollment (M/F); *The World Bank Atlas, 1998,* Female Participation in Labor Force.

▲ **Figure 10.30 Taliban's urban militias** The Taliban's forces keep firm control over people living in the areas of Afghanistan that they rule, as illustrated by this security check. Women in particular have been subjected to severe harassment by Taliban forces if they are not properly dressed or if they attempt to work or attend school. *(Seamus Murphy/Panos Pictures)*

rather that they are actually enforcing Pathan customs. In the northern cities that remain beyond Taliban's grasp—which have swelled to huge proportions with the influx of refugees—women still enjoy certain freedoms. Many of the more prosperous Persian-speaking families in areas under Taliban control have sent their daughters to schools in Iran. Surprising as it may seem, the religious authorities of Iran are actually feminist in orientation when compared to Taliban's leaders, as they consistently support the right of women to study, work, and vote.

Female students seeking education in Iran actually account for a very small proportion of Afghanistan's total refugee flow. Conditions have been so miserable over the past decade that some six million persons—nearly one-fourth of the total population—have at various times sought refuge in other countries. Most have fled to Pakistan, which still holds an estimated one million Afghan refugees, and Iran, where up to 1.3 million Afghans remain. Another one million have moved within Afghanistan to urban areas, particularly those in the northern part of the country, such as Mazar-e Sharif.

Social Conditions in the Former Soviet Republics In marked contrast to the situation in Afghanistan, women elsewhere in Central Asia enjoy a relatively high social position. Traditionally, women had much more autonomy among the northern pastoral peoples (especially the Kazakh) than among the Uzbek and Tajik oasis dwellers of the south, but under Soviet rule the position of women everywhere improved. In the former Soviet republics today, women's educational rates are comparable to those of men, and women are well represented in the workplace. Note also that women can expect to live significantly longer than men in these countries. While women outlive men almost everywhere, the gap here is especially large. Some evidence would suggest, however, that the position of women has declined in many areas since the fall of the Soviet Union. In Kazakhstan, several leaders have blamed feminism for the country's low birthrate, and in much of the region newly rich men are increasingly practicing polygamy to ensure they will have more children.

The figures in Table 10.3 reveal generally favorable levels of social welfare overall. This is especially notable when one considers the dismal state of the region's economy—or contrasts conditions in the former Soviet zone with those of neighboring Afghanistan. Tajikistan, with a per capita GNP of only $350, has a female life expectancy of 71 years, as well as almost universal adult literacy. The countries of former Soviet Central Asia thus stand with Sri Lanka, Kerala in India, and Cuba as places that enjoy relatively high levels of social development despite economic impoverishment (Figure 10.31).

The social success of Tajikistan and its northern and western neighbors reflects investments made during the Soviet period. It is uncertain, however, whether local health and educational facilities can be maintained in the face of economic collapse. Certainly the region's relatively high levels of infant and childhood mortality do not bode well. While Tajikistan might hope that its educated workforce will attract foreign investment, its remote location and continued instability make such a scenario unlikely. Universal education may become a luxury that the country cannot afford.

Social Conditions in Western China It is particularly difficult to obtain reliable information about social conditions in the Chinese portion of Central Asia. Certainly China as a whole has made significant progress in health and education, but many reports suggest that the peoples of Tibet and Xinjiang have been left behind. According to the eastern Turkestan Web pages, some 60 percent of the non-Han people of Xinjiang are illiterate. China's minority peoples have been granted certain exemptions from the country's strict population control measures, and their populations have in general increased at a faster rate than the Han have over the past 15 years. There are, however, many reports from Xinjiang and elsewhere of forced sterilizations and abortions, adding significantly to the tensions present in the area.

▲ **Figure 10.31 Tarkent heliostation in Tashkent, Uzbekistan** Despite its generally poor economic statistics, Uzbekistan boasts a relatively well-developed medical, educational, and technological infrastructure. The Tarkent heliostation focuses the sun's rays to generate the extraordinarily high temperatures needed to produce super-pure materials such as aluminum titanate and magnesium oxide. *(Novosti/Liaison Agency, Inc.)*

Conclusion

Central Asia, long obscured by Russian and Chinese domination, has only recently reappeared on the map of the world. Many geographers remain unconvinced that it forms a world region in its own right. The six newly independent former Soviet republics, along with Afghanistan, could just as easily be classified within Southwest Asia. Tibet and Xinjiang, on the other hand, can be alternatively grouped with the rest of China as part of East Asia. Although the Mongol, Uygur, and Tibetan peoples of this area are struggling to maintain their lands and cultural autonomy, they may not be able to withstand the Han Chinese demographic tide. The Chinese government, moreover, shows no indication that it would even be willing to discuss the possibility of genuine autonomy for either Tibet or Xinjiang.

While China maintains a firm grip on Tibet and Xinjiang, the rest of Central Asia has emerged as a key area of geopolitical and economic competition. Russia, the United States, and China, along with Iran, Pakistan, and Turkey, all contend for influence. Kazakhstan, Uzbekistan, Tajikistan, Kyrgyzstan, Turkmenistan, and Azerbaijan have attempted, with some success, to play off power against each other in order to bolster their own positions. And while political structures throughout the area remain largely authoritarian, the economies of the region are gradually opening up to global connections.

Regardless of what political-economic model (or models) it follows, Central Asia is likely to face serious economic difficulties for some time. The region is not a significant participant in global trade, and it has attracted little foreign investments outside of the oil industry. Some economists argue, moreover, that a continental position by itself is a major economic liability. To thrive in the new economic global order, they contend, a country must have easy access to the sea lanes that form the main conduits of global exchange. Central Asia, unfortunately, is singularly lacking such oceanic connections. But whether a continental position will undercut economic prosperity in the twenty-first century remains to be seen. Certainly, railroads and even roadways have proved adequate to integrate certain interior portions of large countries, such as the U.S. Midwest, with the global economies.

Two countries in the region, Kazakhstan and Turkmenistan, are almost certain to emerge as major players in global trade, owing to their vast reserves of oil and natural gas. Whether such natural resources will allow these countries to become prosperous is another matter. Oil wealth—as Nigeria has proved—can be a curse as much as a blessing, since it often results in an inflated currency that undercuts both industry and agriculture. Relatively high levels of corruption in much of Central Asia do not bode well for the region's success. Relatively high levels of education, on the other hand, may help Kazakhstan and Turkmenistan use their oil riches to build a genuinely prosperous economic base.

Central Asia remains something of a question mark on the map of the world. A large part of the region has recently received its independence after more than 100 years of Russian control, and it remains to be seen what economic and political directions it will take. Much of the rest of the region—Tibet and Xinjiang—remains under the authority of what most indigenous residents consider to be a foreign power. China, however, regards these lands as an integral part of its own territory. Considering the demographic imbalance between the Han Chinese, on the one hand, and the Uygur and Tibetan peoples on the other, the future of indigenous Central Asian culture in Xinjiang and Tibet must be considered uncertain.

Key Terms

alluvial fan (p. 416)
exclave (p. 429)
exotic river (p. 417)
pastoralism (p. 415)
theocracy (p. 424)
transhumance (p. 416)
Turkestan (p. 410)

Questions for Summary and Review

1. Describe the three major regions of Central Asia as defined by physical geography.
2. Why is lake level a particularly serious concern in Central Asia?
3. Why is the distribution of population in Central Asia so uneven? Why are certain areas virtually uninhabited?
4. Describe the different agricultural patterns that are found on (a) the plains of northern Kazakhstan, (b) the major river valleys of Uzbekistan, and (c) the Tibetan Plateau.
5. Why is much of Central Asia often called "Turkestan"? Why does this label not apply to the entire region?
6. How does religion act as a political force in Central Asia? How does this vary in different parts of the region?
7. Why is the United States concerned about the fossil fuel resources of Central Asia? How do transportation routes play into this concern?
8. What role does Russia currently play in Central Asian geopolitics? How has Russian power historically influenced the region?
9. How do social conditions (health, longevity, literacy, etc.) vary across the border between Afghanistan and Uzbekistan? Why are the disparities so pronounced?
10. If western Central Asia is so rich in natural resources, why did the region experience such pronounced economic decline in the 1990s?

Thinking Geographically

1. How much of a disadvantage will Central Asia's "continental location" be in years to come? Is coastal access truly significant in determining a country's competitive position in the global economy?
2. Will the countries of former Soviet Central Asia be able to maintain their high levels of education now that they are independent? If so, will they be able use education to their own economic advantages?
3. Is Russia's influence in Central Asia bound to decline now that Russia has no direct political authority in the area?
4. What external connection will prove more important for Central Asia in years to come—those based on religion, language, economic ties, or geopolitical connections?
5. Is China's political control over eastern Central Asia justified? Should Han Chinese migration to the area be a concern?
6. Is Chinese control over Tibet a legitimate concern of U.S. foreign policy? Should the United States use trade sanctions to attempt to influence Chinese policy in the area?
7. How might the desiccation of Central Asia's great lakes best be addressed?

Regional Novels and Films

Novels

Rinjing Dorje, *Tales of Uncle Tompa, The Legendary Rascal of Tibet* (1997, Barrytown Ltd.)

Mark Frutkin, *Invading Tibet* (1993, Soho Press)

James Hilton, *Lost Horizon* (1996, William Morrow)

Amin Maalouf, *Samarkand: A Novel* (1998, Interlink Publishing)

Hilary Roe Metternich, ed., *Mongolian Folktales* (1996, Avery Press)

Films

Kundun (1997, U.S.)

Seven Years in Tibet (1997, U.S.)

Urga: Close to Eden (1991, Mongolia)

Bibliography

Adshead, S. A. M. 1993. *Central Asia in World History*. New York: St. Martin's Press.

Alworth, Edward, ed. 1989. *Central Asia: 120 Years of Russian Rule*. Durham: Duke University Press.

Bacon, Elizabeth E. 1980. *Central Asians Under Russian Rule: A Study in Culture Change*, 2nd ed. Ithaca: Cornell University Press.

Barfield, Thomas J. 1989. *The Perilous Frontier: Nomadic Empires and China from 221 B.C. to A.D. 1757*. Oxford: Basil Blackwell.

Batalden, Stephen K., and Batalden, Sandra L. 1993. *The Newly Independent States of Eurasia: A Handbook of Former Soviet Republics*. Phoenix: Oryx Press.

Beckwith, Christopher I. 1987. *The Tibetan Empire in Central Asia*. Princeton: Princeton University Press.

Bergholtz, Fred W. 1993. *The Partition of the Steppe: The Struggle of the Russians, Manchus, and the Zunghar Mongols for Empire in Central Asia*. New York: Peter Lang.

Brawer, Moshe. 1994. *Atlas of Russia and the Independent Republics*. New York: Simon and Schuster.

Chin, Jeff, and Kaiser, Robert. 1996. *Russians as the New Minority: Ethnicity and Nationalism in the Soviet Successor States*. Boulder, CO: Westview Press.

Christian, David. 1994. "Inner Eurasia as a Unit of World History." *Journal of World History* 5, 173–211.

Drompp, Michael. 1989. "Centrifugal Forces in the Inner Asian 'Heartland': History *Versus* Geography." *Journal of Asian History* 23, 135–155.

Frank, Andre Gunder. 1992. "The Centrality of Central Asia." *Bulletin of Concerned Asian Scholars* 24, 50–74.

Frye, Richard. 1996. *The Heritage of Central Asia: From Antiquity to the Turkish Expansion*. Princeton: Markus Wiener.

Fuller, Graham E. 1993. "Turkey's New Eastern Orientation." In Graham E. Fuller and Ian O. Lesser, eds., *Turkey's New Geopolitics: From the Balkans to Western China*, pp. 37–98. Boulder, CO: Westview Press.

Gaubatz, Piper Rae. 1996. *Beyond the Great Wall: Urban Form and Transformation on the Chinese Frontiers*. Stanford: Stanford University Press.

Gross, Jo-Ann, ed. 1992. *Muslims in Central Asia: Expressions of Identity and Change*. Durham: Duke University Press.

Grousset, Rene. 1970. *The Empire of the Steppes: A History of Central Asia*. New Brunswick, NJ: Rutgers University Press.

Hambly, Gavin. 1969. *Central Asia*. London: Weidenfeld and Nicolson.

Hauner, Milan. 1990. *What Is Asia to Us? Russia's Asian Heartland Yesterday and Today*. London: Routledge.

———. 1991. "Russia's Geopolitical and Ideological Dilemmas in Central Asia." In Robert Canfield, ed., *Turko-Persia in Historical Perspective*, pp. 189–216. Cambridge: Cambridge University Press.

Huntington, Ellsworth. 1907. *The Pulse of Asia*. New York: Houghton Mifflin.

Jagchid, Sechin, and Hyer, Paul. 1979. *Mongolia's Culture and Society*. Boulder, CO: Westview Press.

Kaiser, Robert J. 1994. *The Geography of Nationalism in Russia and the USSR*. Princeton: Princeton University Press.

Lattimore, Owen. 1988. *Inner Asian Frontiers of China*, 2nd ed. Oxford: Oxford University Press.

Lewis, Robert A., ed. 1992. *Geographical Perspectives on Soviet Central Asia*. London: Routledge.

Morgan, David. 1986. *The Mongols*. Oxford: Basil Blackwell.

Rossabi, Ralph A. 1990. "The 'Decline' of the Central Asian Caravan Trade." In James D. Tracy, ed., *The Rise of Merchant Empires: Long-Distance Trade in the Early Modern World, 1350–1750*, pp. 351–370. Cambridge: Cambridge University Press.

Shaw, Denis J. B., ed. 1995. *The Post-Soviet Republics: A Systematic Geography*. New York: Wiley/Longman.

Sinor, Denis, ed. 1990. *The Cambridge History of Early Inner Asia*. Cambridge: Cambridge University Press.

Tregear, Thomas R. 1980. *China: A Geographical Survey*. New York: John Wiley and Sons.

EAST ASIA
Political Map
Over 1,000,000
500,000–1,000,000 (selected cities)
Selected smaller cities
(National capitals shown in red)
Elevation in meters
Sea Level
RUSSIA
KAZAKHSTAN
KYRGYZSTAN
MONGOLIA
XINJIANG (SINKIANG)
Ürümqi
GANSU
QINGHAI
TIBET (XIZANG)
NEPAL
BHUTAN
INDIA
BANGLADESH
MYANMAR
Bay of Bengal
LAOS
THAILAND
VIETNAM
Gulf of Tonkin
HAINAN
Hainan (CHINA)
South China Sea
CHINA
Lhasa
Brahmaputra R.
Salween R.
Mekong R.
Yangtze R.
Hong (Red R.)
Xi R.
Huang He
Xining
Lanzhou
GANSU
NINGXIA
Yinchuan
INNER MONGOLIA
Hohhot
HEILONGJIANG
MANCHURIA
Harbin
Changchun
JILIN
Shenyang
LIAONING
HEBEI
BEIJING
Beijing
Tangshan
TIANJIN
Tianjin
HEBEI
Shijiazhuang
Taiyuan
SHANXI
SHANDONG
Jinan
Kaifeng
Zhengzhou
HENAN
Xi'an
SHAANXI
JIANGSU
ANHUI
Nanjing
Hefei
Suzhou
Shanghai
Hangzhou
HUBEI
Wuhan
ZHEJIANG
Chengdu
SICHUAN
CHONGQING
Chongqing
Nanchang
Changsha
JIANGXI
HUNAN
FUJIAN
Fuzhou
Guiyang
GUIZHOU
Kunming
YUNNAN
Guilin
GUANGXI
GUANGDONG
Guangzhou
Shenzhen
Nanning
MACAU
Hong Kong
TAIWAN
T'aipei
P'yongyang
Yellow Sea
90°E
110°E

C H A P T E R 11

East Asia

East Asia, composed of China, Japan, South Korea, North Korea, and Taiwan, is the most populous region of the world. China alone is inhabited by more than 1.2 billion persons, more than live in any other world region except South Asia. Although East Asia is historically one of the world's more culturally unified regions, in the latter half of the twentieth century it has been divided ideologically and politically, with the capitalist economies of Japan, South Korea, and Taiwan (as well as the former British colony of Hong Kong) separated from the communist bloc of China and North Korea. Disparities in levels of economic development also remained pronounced. As Japan has reached the pinnacle of the global economy, much of China has remained mired in extreme poverty (Figure 11.1).

During the 1990s, however, divisions with the East Asian region were muted to a certain extent. While China was still governed by the Communist Party, it embarked on a path of modified capitalist development. Ties between its booming coastal zone and Japan, South Korea, and Taiwan quickly strengthened. While mutual animosity still pervaded relations between North Korea and South Korea and between China and Taiwan, East Asia as a whole witnessed a gradual reduction in political tensions (see "Setting the Boundaries").

A thousand years ago, East Asia was not only the most populous but also the most economically advanced part of the world. In the 1800s, however, the fortunes of the region declined precipitously; China suffered repeated famine and revolution and was exploited by the European imperial powers. Japan managed to retain full independence and to build a powerful military, but it remained a poor country until the second half of the twentieth century. In 1945 at the end of World War II, Japan was utterly vanquished, while China remained gripped in a brutal civil war between nationalist and communist forces. The future of the region looked bleak.

Today, East Asia must be recognized as one of the core areas of the world economy, and it is quickly emerging as a center of political power as well. Japan, South Korea, and Taiwan are among the world's key trading states, although both Japan and South Korea experienced severe economic problems in the late 1990s. Tokyo, Japan's largest city, stands alongside New York and London as one of the main financial centers of the globe. Tokyo is also an increasingly cosmopolitan city, drawing immigrants from around the world

◀ **Figure 11.1 East Asia** This region includes China, Japan, North Korea, South Korea, and Taiwan. China, the world's largest country in terms of population, dominates East Asia with more than a billion people. The second largest country is Japan, with 126 million. Japan, South Korea, Taiwan, and Hong Kong (now once again part of China) have long dominated economically; however, China's recent development places that country solidly in the list of world players, both political and economically.

Setting the Boundaries

East Asia is easily marked off on the map of the world merely by the territorial extent of its constituent countries: China, North Korea, South Korea, Japan, and Taiwan (Figure 11.1). In political terms such a definition of the region is appropriate. If one turns to cultural considerations, however, the issue becomes more complicated.

The main cultural problem arises with regard to the western half of China. This is a huge but lightly populated space; some 95 percent of the residents of China live in the eastern half of the country (Figure 11.16). By certain criteria, only the populous eastern half of China (often referred to as "China proper") really fits into the East Asian world region. The indigenous inhabitants of western China are not Chinese by culture and language, and they have never accepted the religious and philosophical beliefs that have given historical unity to East Asian civilization. In the northwestern quadrant of China, called *Xinjiang* in Chinese, most indigenous inhabitants speak Turkish languages and are Muslim in religion. The indigenous culture of Tibet is highly distinctive, and the Tibetans in general resent Chinese authority. Xinjiang and Tibet are thus perhaps more appropriately classified within Central Asia, which is covered in Chapter 10. In this chapter we will examine western China only to the extent that it is politically part of China, the historical core of East Asia.

Vietnam presents another anomaly for the definition of East Asia. By most cultural and historical criteria, Vietnam clearly belongs within East Asia. Vietnam was actually part of China for almost a thousand years, and many of its cultural patterns are still similar to those of China. By modern political measures, however—as well as by its actual location—Vietnam fits much better within Southeast Asia. Vietnam will therefore be discussed occasionally in this chapter, but extended consideration will be reserved for Chapter 13.

(many, if not most of them, illegal), and participating as both a consumer and producer of global culture. China is still more economically and culturally self-contained, but it too is emerging as a global trading power. China's coastal zone, particularly the former British colony of Hong Kong, has been well integrated within global networks of information, commerce, and entertainment.

The global political position of East Asia is more ambiguous than its economic role. Since its defeat in World War II, Japan has maintained a low international profile, pledging itself to keep a relatively small armed force. China, on the other hand, is emerging out of poverty and relative isolation to become a major regional, if not global, military power. This development has caused much concern among China's neighbors—and even among distant countries such as the United States. Whether such concern is warranted is, however, debatable.

Environmental Geography: Rugged Uplands Interspersed with Alluvial Lowlands

East Asia is situated in the same general latitudinal range as the United States, although it extends considerably farther to the north and the south. The northernmost tip of China lies as far north as central Quebec, while China's southernmost point is at the same latitude as Mexico City (Figure 11.2). The climate of East Asia is thus similar to that of eastern North America; southern China is roughly comparable to southern Florida and the Caribbean, while northern China is not unlike Minnesota and neighboring areas of southern Canada (Figure 11.3). One difference, however, is that somewhat more extreme conditions—hotter summers and colder winters—are found in mainland East Asia. Another difference is that northern China and Korea tend to be dry in the winter. The islands of Japan, on the other hand, are characterized by a wetter and much more mild climate than mainland East Asia.

The insular belt of East Asia, extending from northern Japan through Taiwan, is situated at the intersection of three **tectonic plates** (the basic building blocks of Earth's crust): the Eurasian, the Pacific, and the Philippine. This area is therefore geologically active, experiencing numerous earthquakes and dotted with a number of volcanoes (Figure 11.4). The mainland is more geologically stable than the islands, although many parts of China are earthquake-prone. East Asia is not a particularly mineral-rich region. (Although the far western region of China has a number of important mineral deposits, this region will be discussed in Chapter 10 [Central Asia] rather than as part of East Asia proper [see "Setting the Boundaries"]). East Asia's petroleum reserves are moderate in size (concentrated in northern and northeastern China, and perhaps offshore in the South China Sea), and China does have vast deposits of coal.

Insular Environments

The islands of eastern Eurasia, associated with the convergence of tectonic plates, extend in a continuous line well to the north and the south of East Asia. Note in Figure 11.1 how the Kuril Islands of eastern Russia almost touch Japan's northernmost island of Hokkaido, and how the Ryukyu Islands form stepping-stones between Kyushu in southern Japan and Taiwan. South of Taiwan, the Bataan Islands form a link to the Philippines in Southeast Asia. Only the islands of Japan and Taiwan, however, are classified within East Asia.

Japan's Physical Environment Although slightly smaller than California, Japan is more elongated, extending farther to both the north and south. As a result, Japan's extreme south, in southern Kyushu and the Ryukyu Archipelago, is subtropical, while northern Hokkaido is almost subarctic. Most of the country, however, is distinctly temperate. The

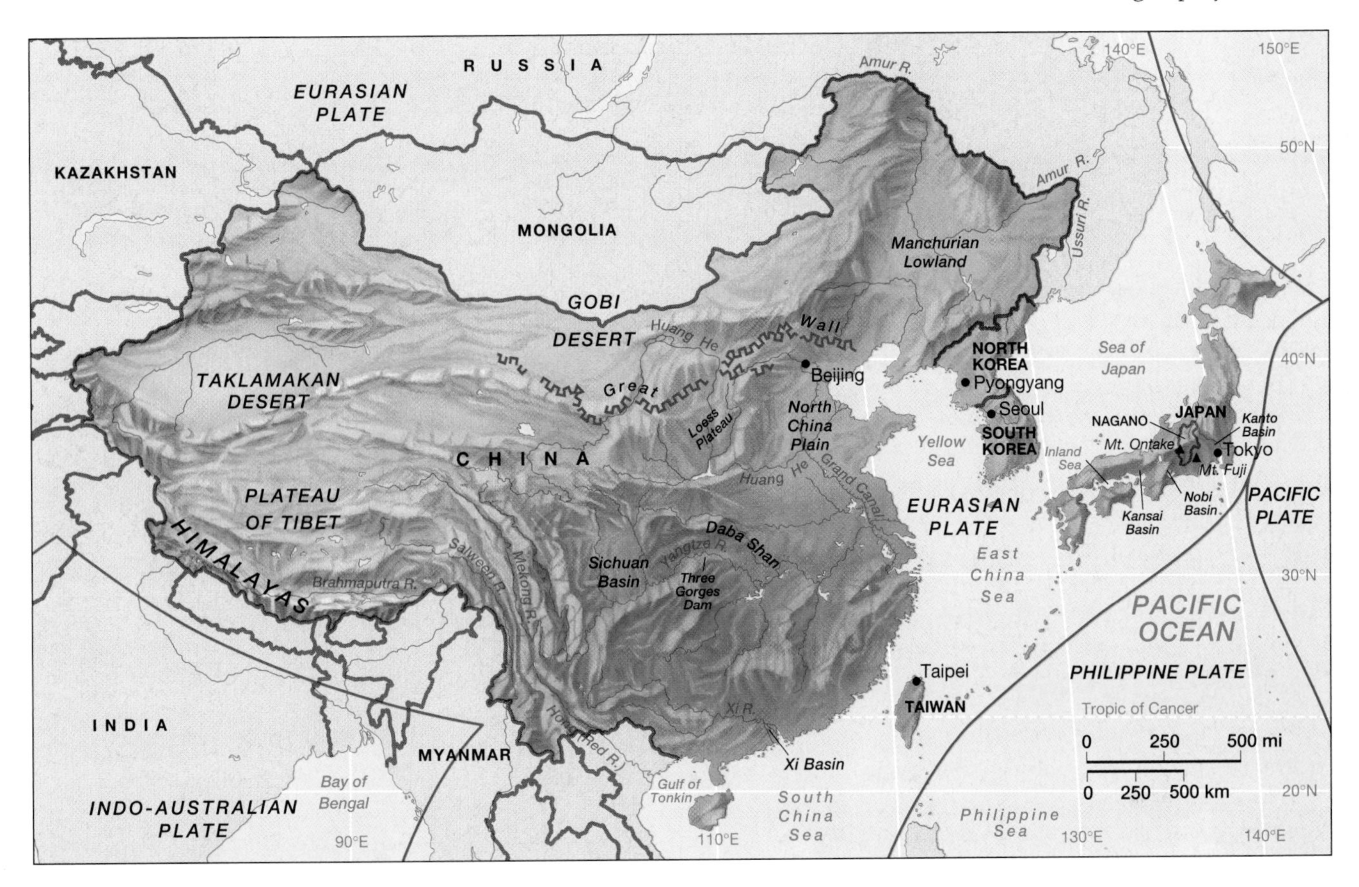

▲ **Figure 11.2 Physical geography of East Asia** On the mainland, China is the largest country, encompassing about the same east–west extent as the United States does in North America. The physical geography varies widely from high mountain plateaus in Tibet to the expansive deserts of western China. In the island region, landscapes are shaped by the convergence of three major tectonic plates: the Eurasian, Philippine, and Pacific plates.

climate of Tokyo is not unlike that of Washington, DC—although Tokyo is rainier.

Japan's climate varies not only from north to south but also from east to west (more precisely, southeast to northwest) across the main axis of the archipelago. In the winter the area facing the Sea of Japan is far cloudier and receives much more snow than the Pacific Ocean coastline. During this time of the year, cold winds from the Asian mainland blow across the relatively warm waters of the Sea of Japan. The air picks up moisture over the sea and deposits it, usually in the form of snow, when it hits land. This often cloud-shrouded northwestern coastline, sometimes called "the dark side of Japan," is well noted in Japanese literature. The Pacific coast of Japan, on the other hand, is far more vulnerable to the typhoons (hurricanes) that strike the country frequently.

The Pacific coast of Japan is separated from the Japan Sea coast by a series of mountain ranges. Japan is one of the world's most rugged countries, with mountainous terrain covering some 85 percent of its territory. As Figure 11.10 shows, most of these uplands are thickly wooded. By some measures, Japan is the world's second most heavily forested industrialized country (after Finland). Japan owes its lush forests both to its mild, rainy climate and to its long history of successful forest conservation. For hundreds of years, both the Japanese state and its village communities have enforced strict conservation rules, ensuring that timber and firewood extraction would be balanced in the long run by tree growth (Figure 11.6).

Along Japan's coastline and interspersed among its mountains are limited areas of low-lying alluvial plains (Figure 11.7). In prehistoric times these lowlands were covered by forests and wetlands. They have long since been cleared and drained and, for the past several thousand years, they have been used intensively for agriculture and human settlement. The largest Japanese lowland is the Kanto Plain to the north of Tokyo, but even it is only some 80 miles wide and 100 miles long (130 by 160 kilometers). The country's other main lowland basins are the Kansai, located around Osaka, and the Nobi, centered on Nagoya. In the mountainous Nagano prefecture, smaller basins are sandwiched between the imposing peaks of the Japanese Alps.

Taiwan's Environment Taiwan, an island about the size of Maryland, sits at the edge of the continental landmass. To the west, the Taiwan Strait is only about 200 feet (60 meters) deep; to the east, ocean depths of many thousands of feet are found only a few tens of miles offshore.

Taiwan itself forms a large tilted block. Its central and eastern regions are rugged and mountainous, while the west is dominated by a sizable alluvial plain. Bisected by the tropic of Cancer, Taiwan has a mild winter climate, but it is not

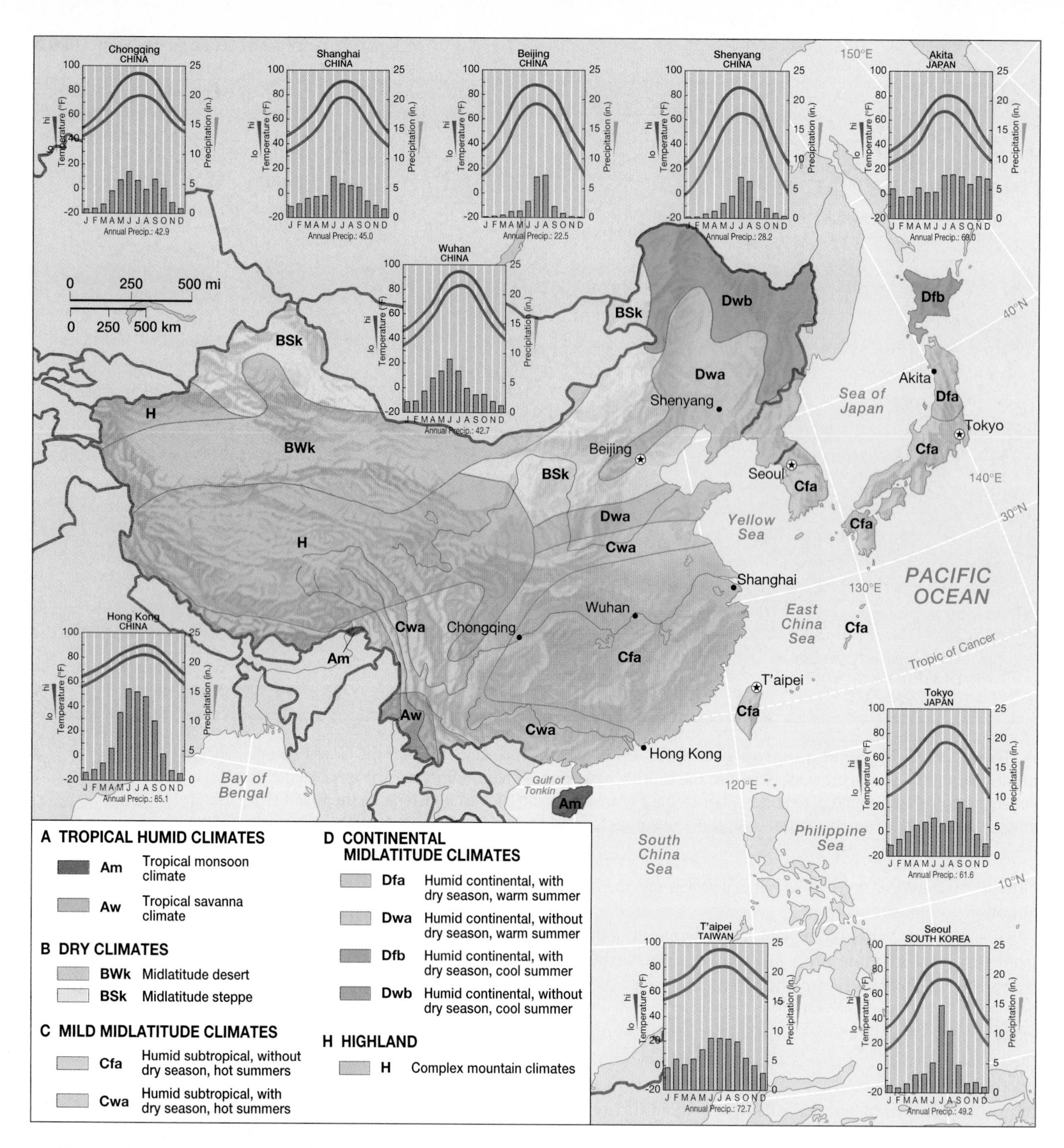

▲ **Figure 11.3 Climate map of East Asia** East Asia is located in roughly the same latitudinal zone as North America, so there are climatic parallels between the two world regions. The northernmost tip of China lies at about the same latitude as Quebec and shares a similar climate, whereas southern China approximates the climate of Florida. However, winters tend to be colder and summers hotter than in North America. In Japan, maritime influences moderate these extremes and produce a more mild climate.

infrequently battered by typhoons in the late summer and early autumn. Unlike nearby areas of China proper, Taiwan still has extensive forests in the more remote uplands of its central and eastern regions.

Mainland Environments

Mainland East Asia, constituted by Korea and the eastern half of China (China proper), is almost 10 times the size of the insular region, and its environmental circumstances are far more diverse. Several Chinese provinces are by themselves as large as Japan. For the sake of convenience, China proper can be divided into two main divisions, one lying to the north of the Yangtze River Valley, the other including the Yangtze and all areas to the south. As Figure 11.8 shows, each of these divisions, in turn, can be subdivided into a number of distinctive regions.

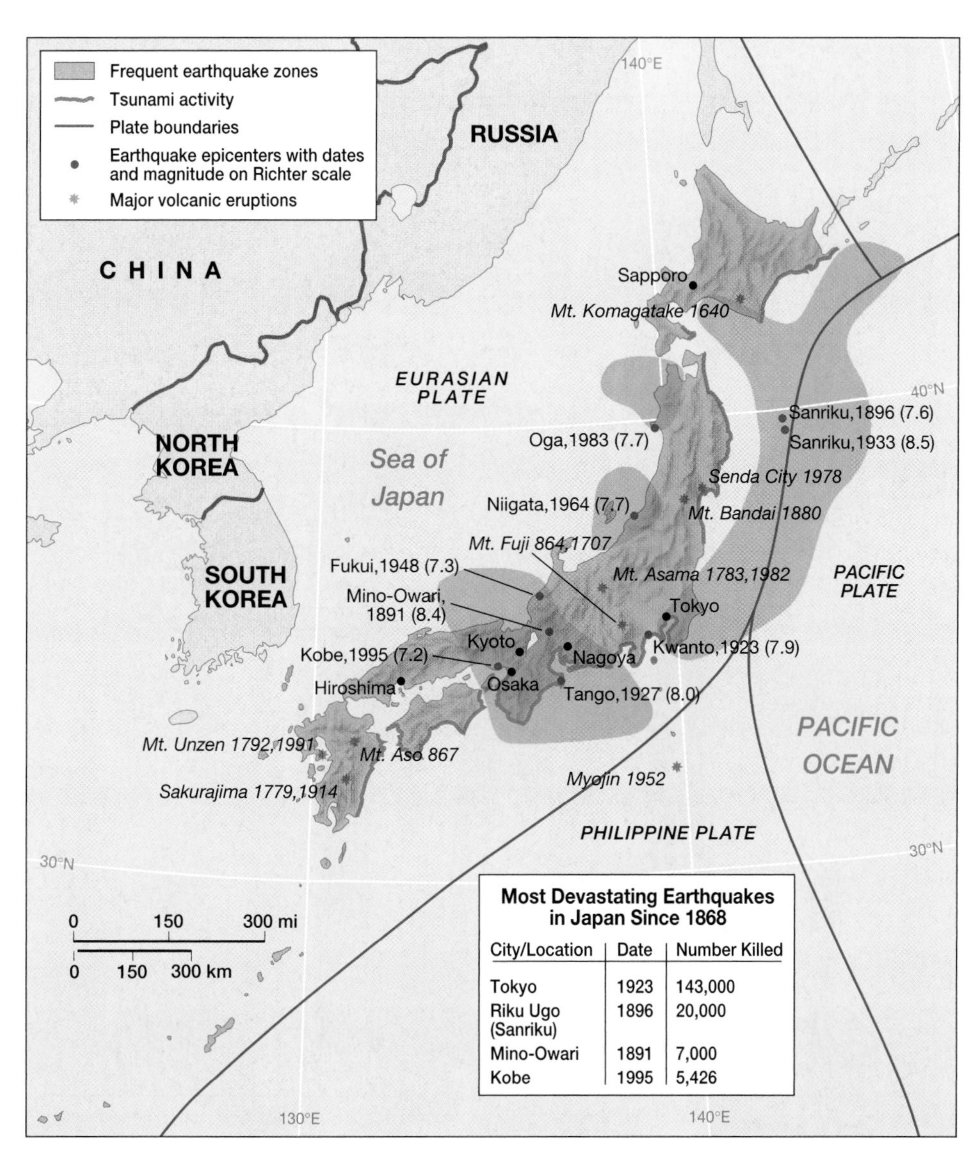

Most Devastating Earthquakes in Japan Since 1868

City/Location	Date	Number Killed
Tokyo	1923	143,000
Riku Ugo (Sanriku)	1896	20,000
Mino-Owari	1891	7,000
Kobe	1995	5,426

◀ **Figure 11.4 Natural hazards in Japan** Because of its location at the convergence of three major tectonic plates, earthquakes are a serious problem in Japan. Most recently, more than 5,000 people were killed in the 1995 earthquake that severely damaged Kobe. However, this disaster was no match for the killer quake of 1923 that destroyed Tokyo, killing 143,000 people. Volcanic eruptions can also be hazardous and are linked directly to Japan's location on tectonic plate boundaries. Additionally, much of Japan's coast is vulnerable to devastating tsunamis (tidal waves) caused by earthquakes in the Pacific basin.

▲ **Figure 11.5 Heavy snow in Japan's mountains** Moist air moving off the Sea of Japan, coupled with cold air from the Asian mainland, produces heavy snows in the mountainous spine of Japan. Numerous major ski areas dot the Japanese Alps, several of which have hosted world-class sports competitions. *(George Mobley/NGS Image Collection)*

▲ **Figure 11.6 Forested landscapes of Japan** Although much of Japan is forested and supports a viable wood products industry, the yield is not large enough to satisfy demand. As a result, Japan imports timber extensively from North America, Southeast Asia, and Latin America. *(Ric Ergenbright/Ric Ergenbright Photography)*

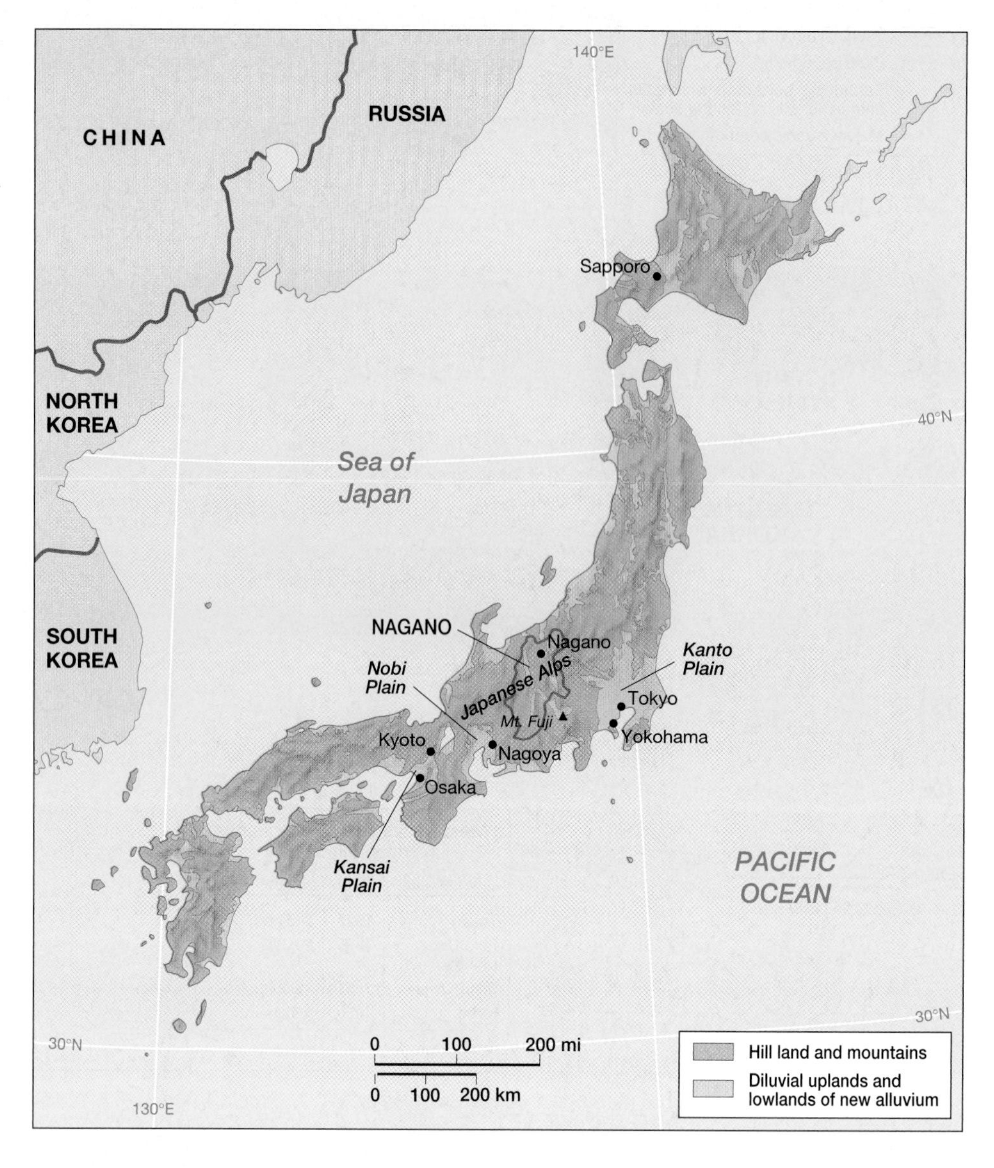

▶ Figure 11.7 Landscape regions of Japan There are several large lowland plains in Japan, primarily along the coastline but also interspersed among the mountains. These plains, originally covered by wetlands and forests, now are the centers of Japan's intensive agricultural system. Additionally, these lowlands are the sites of Japan's largest cities. The largest lowland is the Kanto Plain to the north of Tokyo, which is about 80 by 100 miles wide.

The Landscapes of Southern China Southern China is a land of rugged mountains and hills interspersed with river valleys and other lowland basins. The alluvial basins of southern China are far larger than those of Japan.

The largest of these basins lie along the Yangtze River (alternatively called the Chang Jiang). The Yangtze, the third largest (by volume) river in the world, emerges from the Tibetan highlands onto the rolling lands of the Sichuan (or Red) Basin. Surrounded by high mountains, the former lake bed of central Sichuan is climatically unique, virtually never experiencing frost. Passing south over the Daba Mountains in the winter a traveler might experience a mild shock; to the north, conditions are generally frigid and dry, whereas to the south, in the Red Basin, the land is green and wet. Eastward from Sichuan, the river passes through several other broad basins (in the "Middle Yangtze" region on Figure 11.8), partially separated from each other by hills and low mountains, before flowing into a large delta in the vicinity of Shanghai.

South of the Yangtze Valley, there are higher mountains (up to 7,000 feet, or 2,150 meters, in elevation), although they are intercut with alluvial lowlands. Larger valleys are found in the far south, the Xi Basin in Guangdong province being the most important. Here the climate is truly tropical, free of frost. West of Guangdong lies the moderate-elevation plateaus of Yunnan and Guizhou, the former noted for its perennial springlike weather. Finally, to the northeast of Guangdong lies the rugged coastal province of Fujian. Fujian's coastal plain is narrow and its coastline deeply indented, features which have encouraged the Fujianese people to seek a maritime orientation (Figure 11.9).

The Landscapes of Northern China North of the Yangtze Valley, the climate is both colder and drier than it is to the south. Summer rainfall is generally abundant except along the edge of the Gobi Desert, but the other seasons are often dry.

Immediately north of the lower course of the Yangtze lies the North China Plain, a large area of fertile soil crossed by the Huang He (or Yellow) River. Chinese civilization originated some 3,800 to 5,000 years ago along the western margin of this vast alluvial plain, at which time the eastern plain was

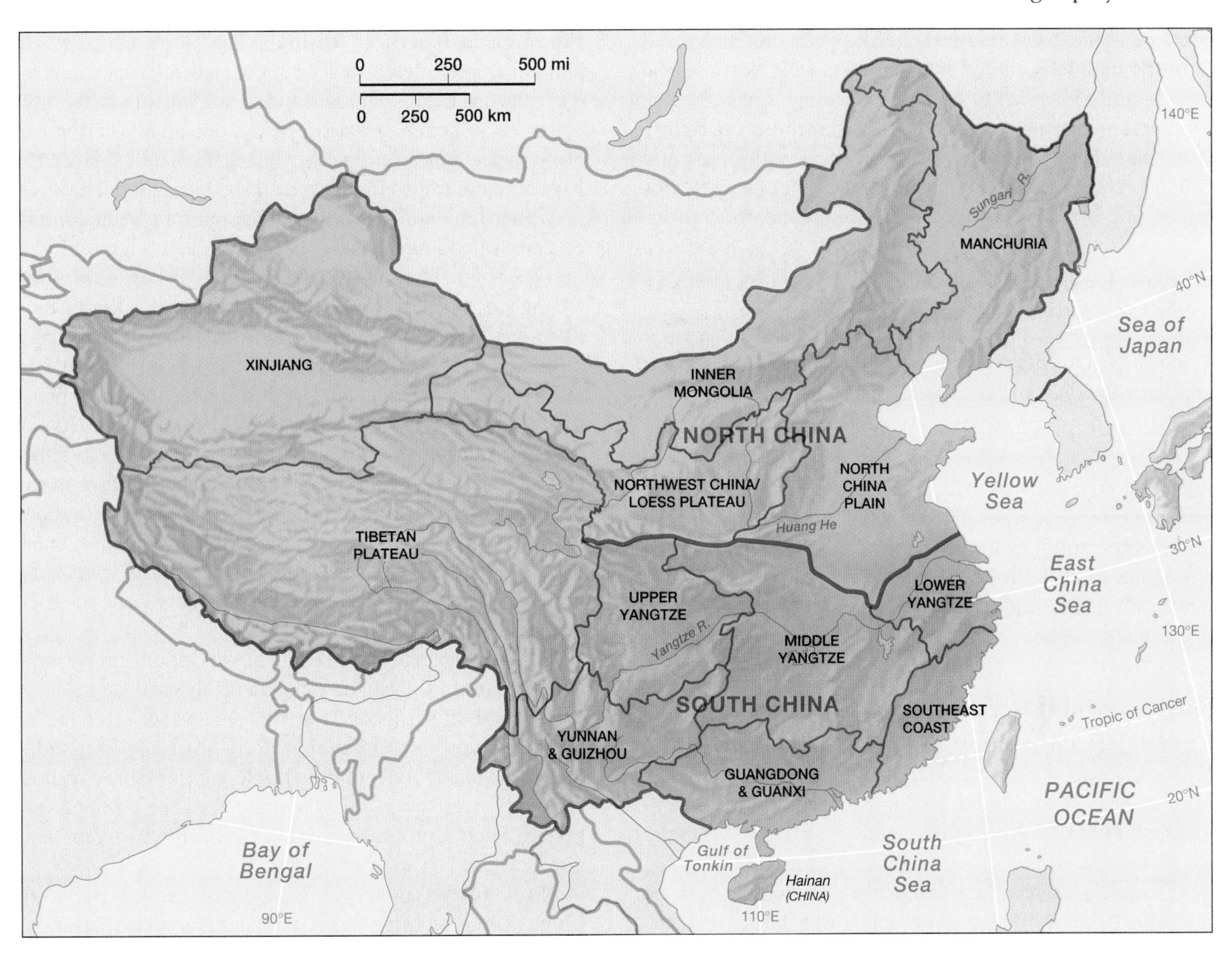

▲ **Figure 11.8 Landscape regions of China** The major regional dividing line in China is the Yangzte Valley, which divides the country into two general areas to its north and south. Immediately to the north is the large fertile area of the North China Plain, bisected by the Huang He (or Yellow) River. To the west is the Loess Plateau, an upland area of soil derived from wind-deposited silt blow after the prehistoric glacial period about 15,000 years ago. To the southwest lies the extensive Tibetan Plateau.

still heavily wooded. Before long, however, most of the North China Plain had been cleared and converted to agricultural fields. With the exception of a few uplands in Shandong province, the entire area is a virtually flat plain. The North China plain is cold and dry in winter and hot and humid in the summer. Overall precipitation is not high, however, and some areas are virtually semiarid and in some danger of **desertification** (or the spread of desert conditions)

West of the North China Plain sits the **Loess** Plateau (loess is a soil derived from wind-deposited silt). This is a fairly rough upland of moderate elevation and uncertain precipitation. It does, however, have fertile soil. As we shall see, some of the most severe environmental problems in East Asia have emerged through the overcultivation of the Loess region. West of the Loess region lie the semiarid plains and uplands of Gansu province, situated at the foot of the great Tibetan Plateau.

China's far northeastern region is called *Dongbei* in Chinese and *Manchuria* in English. Manchuria is dominated by

▲ **Figure 11.9 The Fujian coast of China** In southeast China lies the rugged coastal province of Fujian. Here, the coastal plain is narrow and the shoreline deeply indented, producing a picturesque landscape. Because of limited agricultural opportunities along this rugged coastline, many Fujian people work in maritime activities. *(The Image Bank)*

a broad, fertile lowland sandwiched between mountains and uplands stretching along China's borders with North Korea, Russia, and Mongolia. Although winters here can be brutal, summers are usually warm and moist. Manchuria's peripheral uplands have some of China's best-preserved forests and wildlife refuges. In the extreme northeast, extensive wetlands can still be found in the valleys of the Ussuri and Amur rivers.

Korean Landscapes Korea forms a well-demarcated peninsula, partially cut off from Manchuria by rugged mountains and sizable rivers. Like Japan, its latitudinal range is pronounced. The far north, which just touches Russia's Far East, has a climate not unlike that of Maine, whereas the southern tip of South Korea is more reminiscent of the Carolinas. Korea, again like Japan, is a mountainous country with scattered alluvial lowland basins. The lowlands of the southern portion of the peninsula are more extensive than those of the north, giving South Korea a distinct agricultural advantage over North Korea. The north, however, has much more abundant mineral deposits, as well as forest and hydroelectric resources, than the south.

Environmental Degradation in East Asia

Environmental problems in East Asia are particularly severe, owing to a combination of the region's heavy population burden, its extensive industrial development, and its unique conditions of physical geography (Figure 11.10). The wealthier countries of the region, particularly Japan, have, however, been able to invest heavily in environmental protection. China, on the other hand, not only has less money available for this purpose, but many of its basic forms of degradation, such as deforestation, flooding, and soil erosion, are more deeply rooted (see "Environment: China's Relationship with Nature").

Forests and Deforestation Most the uplands of China and South Korea support only grass, meager scrub, or stunted trees (Figure 11.11). China lacks the historical tradition of forest conservation that characterizes Japan. In earlier periods, hillsides were often cleared for fuelwood, and in some instances entire forests were burned merely for ash that could be used as fertilizer in rice fields. In much of southern China, sweet potatoes, maize, and other crops have been grown on steep and easily eroded hillsides for several hundred years. After centuries of exploitation, many upland districts have been so degraded that they cannot easily regenerate forests.

Although the Chinese government has initiated large-scale reforestation programs, few have been successful. As it now stands, substantial forests are found only in China's far north, where a cool climate prevents fast growth, and along the eastern slopes of the Tibetan Plateau, where rugged terrain restricts commercial forestry. As a result, China suffers a severe shortage of forest resources. If its economy continues to boom, China will likely become a major importer of lumber, pulp, and paper in the near future. One must wonder which part of the world could supply China's potentially vast demand for forest products.

Flooding in Northern China The North China Plain, which has been deforested for thousands of years, has long been plagued by both drought and flood. This area is dry most of the year, yet often experiences heavy downpours in summer. Since ancient times, large-scale works in hydraulic engineering have both controlled floods and allowed irrigation. However, no matter how much effort has been put into water control, disastrous flooding has never been prevented (Figure 11.12).

The worst floods in northern China are caused by the Huang He, or Yellow River, that cuts across the North China Plain. Owing to upstream erosion, the Huang He carries a huge **sediment load** (the amount of suspended clay, silt, and sand in the water), giving it the dubious distinction of being the muddiest river in the world. When the river exits the Loess Plateau and enters the low-lying plain, its velocity slows and its sediments begin to settle out and accumulate in the riverbed. As a result, the level of the river gradually rises above that of the surrounding lands. Eventually the river must break free of its course to find a new route to the sea over lower-lying ground. Twenty-six such course changes have been recorded in Chinese history.

Through the process of sediment deposition and periodic course changes, the Huang He has actually created the North China Plain. In prehistoric times the Yellow Sea extended far inland, with the hills of Shandong forming islands. Even today the sea is quickly retreating as the Huang He's delta expands; one study revealed a 6-mile advance of the land in a three-year period. Such a process occurs in other alluvial plains, but nowhere else is it so pronounced.

Nowhere else, moreover, is it so destructive. The North China Plain has been densely populated for millennia, and is now home to some 300 million persons. Since ancient times, the Chinese have attempted to keep the river within its banks by building progressively larger dikes. Eventually, however, the riverbed rises so high that the flow can no longer be contained. When this occurs, catastrophic flooding results. Such floods have been known to kill several million persons in a single episode. For this reason, the Huang He has been aptly called "the river of China's sorrow." While the river has not changed its course since the 1930s, most geographers agree that another correction is inevitable.

Erosion on the Loess Plateau The Huang He's heavy sediment burden is derived from the eroding soils of the Loess Plateau, located immediately west of the North China Plain. Loess is a fine, wind-blown material that was deposited on this upland area during the last Ice Age. The dust storms of the period must have been overwhelming, for in some places several hundred feet of loess has accumulated.

Loess forms fertile soil, but it washes away easily when exposed to running water. At the dawn of Chinese civilization, the semiarid Loess Plateau was covered with tough grasses and scrubby forests that helped retain the soil. Chinese farmers, however, began to clear the land for the abundant crops it yields when rainfall is adequate. Cultivation required plowing, which, by exposing the soil to runoff, exacerbated erosion. As the population of the region gradually increased, the remaining areas

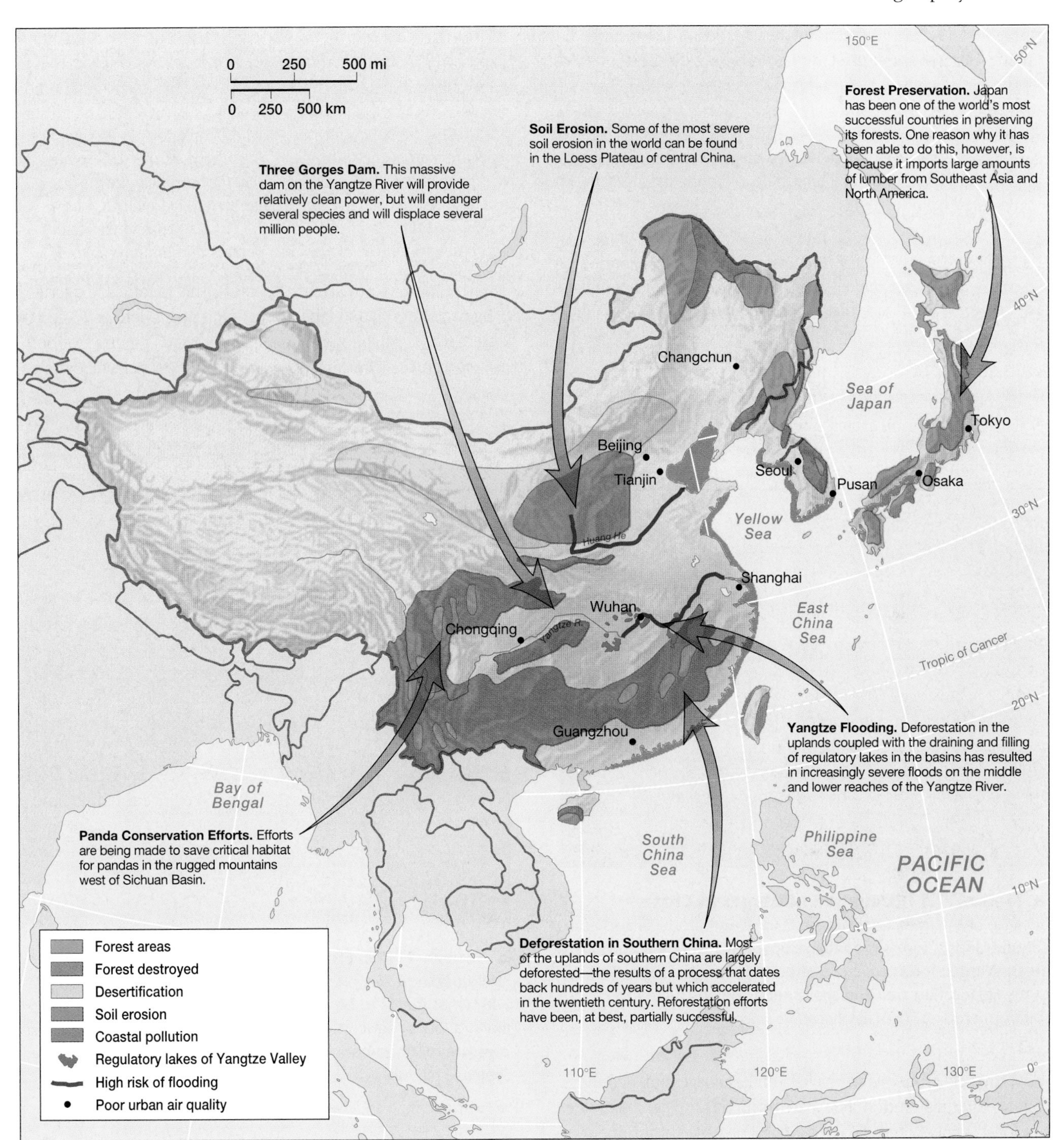

▲ **Figure 11.10 Environmental issues in East Asia** This vast world region has been transformed completely from its natural state and continues to have serious environmental problems. In China, some of the more pressing environmental issues deal with deforestation, flooding, water control, and soil erosion. The controversial Three Gorges Dam on the Yangzte is designed to alleviate flooding problems, yet in order to build the dams, several million people will have to be displaced.

of woodland diminished and then virtually disappeared, leading to ever-greater rates of soil loss. As the erosion process continued, great gullies—some of them hundreds of feet deep—cut across the plateau, steadily reducing the extent of arable land.

Today the Loess Plateau is one of the poorest parts of China, and thus one of the poorer parts of the world. Population is only moderately dense by Chinese standards, but good farmland is limited and drought is common. The population continues to expand, moreover, even while the extent of arable land declines. The Chinese government encourages the construction of terraces to conserve the soil, but such efforts have not been effective everywhere. Campaigns to plant woody vegetation on the

ENVIRONMENT China's Relationship with Nature

Vaclav Smil, a Czech-born geographer currently teaching in Canada, has written extensively on environmental degradation and energy use in China. The following passages are excerpted from his book, *The Bad Earth* (1984).

A reverence for nature runs unmistakably through the long span of Chinese history. The poet . . . found the mountains his most faithful companions; emperors . . . painted finches in bamboo groves and ascended sacred mountains; Buddhist monks sought their dhyana "midst fir and beech"; craftsmen located their buildings to "harmonize with the local currents of the cosmic breath"; painters were put through the rigors of mastering smooth, natural, tapering bamboo leaves and plum branches; and who couldn't admire the symphony of plants, rocks, and water in countless gardens.

*To stop here, however, . . . would be telling only the more appealing half of the story. There was also a clearly discernible current of destruction and subjugation: the burning of forests just to drive away dangerous animals; massive, total, and truly ruthless deforestation to create new fields, to get fuel and charcoal, and to obtain timber for fabulous palaces and ordinary houses, wood for cremation of the dead and (to no small effect) for making ink from the soot of burnt pines . . . the erection of sprawling rectilinear cities (fires would rage for days to consume the vast areas of wooden buildings) eliminating any trace of nature, save for some artificial gardens.**

*From *The Bad Earth* by Vaclav Smil. Armonk, NY: M. E. Sharp, 1984, pp. 6–7.

▲ **Figure 11.11 Denuded hillslopes in China** Because of the need to clear forests for wood products and agricultural lands, China's mountain slopes have long been problem areas. Without forest cover, soil erosion is a serious issue. This is in the Huangshan area and shows terraces constructed to keep soil from eroding. *(Ric Ergenbright/Ric Ergenbright Photography)*

▲ **Figure 11.12 Flooding on the Yangzte** Major river flooding takes place every year along the Yangzte River with what seems to be increasingly high tolls in both human life and loss of agricultural crops. As a result, there is increasing pressure for flood-control dams. *(Zhang Guoqiang/China Features/Sygma Photo News)*

most severely eroded lands have been less successful. Seedlings require careful attention if they are to survive the harsh climate, but local farmers often do not have the necessary equipment or knowledge to care for them, and they are seldom provided with adequate incentives.

China's government has also attempted to reduce the flow of silt out of the Loess Plateau by building small check-dams on the region's streams. Most of these dams were soon rendered ineffective by rapid siltation. A more ambitious flood-control project entailed the construction of a major dam on the Huang He itself. Within a few years, however, the reservoir behind this Sanmenxia Dam had simply filled in with silt, destroying its flood-control capability. Partly because of these environmental disasters, people have been migrating out of the Loess Plateau, seeking work in the burgeoning cities of coastal China.

River Control in Southern China The lower and middle stretches of the Yangtze also suffer periodically from devastating flooding. Flooding has been exacerbated over the past several decades by the progressive draining and filling of floodplain lakes and by deforestation in the mountains to the west of the Sichuan Basin. East of Sichuan, the Yangtze passes through several large mountain-rimmed basins (in Hunan, Hubei, and Jiangxi provinces), each containing a group of lakes. During flood periods, water flows from the river into these **regulatory lakes,** reducing the flood crest downstream (they are called *regulatory* because they reduce the flow of water during floods). As the lakes have been gradually converted to farmland, flooding downstream has grown more severe.

The government of China is thus attempting to control the Yangtze by building a series of large dams. The most prominent of these is a massive structure now being constructed in the deep canyon downstream from the Sichuan Basin. This $10–25 billion Three Gorges Dam will displace an estimated 1.2 to 1.8 million persons and will further jeopardize several endangered species, including the Yangtze River dolphin (Figure 11.13). The ecological consequences of the dam are so negative that the World Bank, which was funding the project, withdrew support. The Chinese government, however, decided that the dam was so important that it would finance it by itself. While the lake is scheduled to begin filling in 2003, few experts expect that the resettlement program will have been completed in time.

Pollution in China China insists on building the Three Gorges project and other major dams not only to control its rivers, but also to obtain the electricity that the dams will generate. As China industrializes, its demand for power is skyrocketing. At present, most Chinese electricity is generated by burning the country's abundant low-quality coal, which results in horrendous air pollution. Smog is particularly severe in the winter months when stagnant air masses often sit over northern and central China. Coal burning also forms acid rain, which is becoming a more severe problem throughout the region. While dam building may reduce air pollution somewhat, it will not be adequate to supply China's vast energy needs. Most environmentalists therefore argue that the damage caused by the large dams will be greater than the benefits that they confer. Not surprisingly, industrialists and government planners disagree.

As China's industrial base expands, other environmental problems, such as water pollution and toxic-waste dumping, are growing more acute, particularly in the booming coastal areas. Such a problem emerged earlier in other portions of East Asia. But Taiwan and South Korea, which have large chemical, steel, and other heavy industries, have responded by imposing more stringent environmental controls as they have grown wealthier. These two countries are also following in Japan's footsteps by setting up new factories in poorer countries—such as China—that have less exacting environmental standards. (Taiwan has even negotiated with North Korea, the poorest state in the region, for a nuclear-waste dumping site.) Such an option is not open to China. It thus remains to be seen whether China can successfully respond to the environmental degradation generated by a rapidly industrializing economy.

▲ **Figure 11.13 The Three Gorges of the Yangzte** The spectacular Three Gorges landscape of the Yangzte River is the site of the controversial flood control dam that will displace several million people. Not only are the human costs high to those displaced, but there may also be significant ecological costs to endangered aquatic species. *(Greg Baker/AP/Wide World Photos)*

Environmental Issues in Japan Considering its large population and intensive industrialization, Japan's environment is relatively clean. Actually, the very density of its population gives certain environmental advantages, allowing, for example, a very efficient public transportation system. In the 1950s and 1960s, Japan's most intensive period of industrial growth, the country did suffer from some of the world's worst water and air pollution, and several infamous toxic waste disasters killed and maimed thousands of persons. Soon afterward the Japanese government passed relatively stringent air and water pollution laws.

Japan's cleanup was aided by its insular location, since winds usually carry smog-forming chemicals out to sea. Equally important has been the phenomenon of **pollution exporting.** Because of both Japan's high cost of production and its relatively strict environmental laws, many Japanese companies have relocated their dirtier factories to other countries, particularly those of Southeast Asia. In effect, Japan's pollution has been partially displaced to poorer countries. Of course, the same argument can also be made about the United States and Europe, but to a somewhat lesser extent.

Endangered Species Another environmental question of truly global dimensions has been linked to the economic rise of East Asia. East Asia is unfortunately one of the main centers of trade in endangered species. Many forms of traditional Chinese and Korean medicine are based on products derived from rare and exotic animals. Deer antlers, bear gall bladders, snake blood, tiger penises, and rhinoceros horns are believed by many persons to have medical effectiveness, and certain individuals will pay fantastic sums of money to purchase them. As wealth has accumulated in the area, trade in such substances has expanded. China itself has relatively little remaining wildlife, but virtually all other areas of the world help supply its demand for wildlife products. China has, however, made good efforts to protect some of its remaining areas of wildlife habitat. Among the most important of these are the high-altitude forests and bamboo thickets of western Sichuan province, home of the panda bear (Figure 11.14). Efforts are also being made to preserve habitat in northern Manchuria, where evidence of a surviving population of Siberian tigers was discovered in 1997.

▲ **Figure 11.14 Endangered panda bear** One of the more successful Chinese environmental solutions has protected high-altitude forests and bamboo thickets in Sichuan province as panda bear habitat. Efforts are also being made in Manchuria to protect Siberian tigers, which were thought to be locally extinct until several were discovered in 1997. *(Keren Su/Pacific Stock)*

Population and Settlement: A Realm of Crowded Lowland Basins

East Asia, along with South Asia, is the most densely populated region of the world. The lowlands of Japan, Korea, and China are among the most intensely used portions of Earth, containing not only the major cities but also most of the agricultural lands of these countries.

Although the population density of East Asia is extremely high, the region's population growth rate has declined dramatically since the 1970s. In Japan, the current concern is one of impending population decline, and correspondingly of an aging populace that will need to be supported by a shrinking cohort of young and middle-aged workers. While China still has an expanding population, its rate of growth is, by global standards, relatively slow (see Table 11.1).

Japanese Settlement and Agricultural Patterns

Japan is a highly urbanized country, supporting two of the largest urban agglomerations in the world. Yet it is also one of the world's most mountainous countries, and in general its uplands are lightly inhabited. Agriculture must therefore share the limited lowlands with cities and suburbs, resulting in extremely intensive farming practices.

Japan's Agriculture Lands Japanese agriculture is largely limited to the country's restricted coastal plains and interior basins. Rice is Japan's major crop, and irrigated rice demands flat land. (Mountain slopes can be **terraced**, or sculpted into flat steps, for rice growing, but terracing is common only in limited areas of Japan.) Japanese rice farming has long been one of the most productive forms of agriculture in the world, helping support such a large population—some 126 million persons—on a relatively small and rugged land area. Although rice is grown in almost all Japanese lowlands, the country's premier rice-growing districts lie along the Japan Sea coast of central and northern Honshu.

Japanese agriculture is not wholly devoted to rice. Vegetables are grown intensively in all of the lowland basins, even on tiny patches within urban neighborhoods (Figure 11.15). The valleys of central and northern Honshu are famous for their temperate fruit, while citrus comes from the milder southwestern reaches of the country. Crops that thrive in a cooler climate, such as potatoes, are produced mainly in Hokkaido and northern Honshu, as are dairy products.

Settlement Patterns All Japanese cities—and the vast majority of the Japanese people—are located on these same basins and plains that support the country's agriculture. Not

Table 11.1 Demographic Indicators

Country	Population[a]	Natural Increase	TFR[b]	%<15[c]	% Urban
China	1,242.5	1.0	1.8	26	30
Hong Kong[d]	6.7	0.4	1.1	18	100
Japan	126.4	0.2	1.4	15	78
South Korea	46.4	1.0	1.7	22	79
North Korea	22.2	0.9	1.9	28	59
Taiwan	21.7	1.0	1.7	23	75

[a]Population in millions, 1996.
[b]Total fertility rate.
[c]Percentage of population younger than 15 years of age.
[d]Although still a part of China, Hong Kong's statistics are often compiled separately.
Source: *Population Reference Bureau Data Sheet,* 1998.

▲ **Figure 11.15 Japanese urban farm** Commonly, the landscape of Japan combines urban settlement with small patches of intensively farmed vegetable agriculture. Here, a cabbage farm coexists with an urban neighborhood. *(Kyodo News International, Inc.)*

surprisingly, the three largest metropolitan areas—Tokyo, Osaka, and Nagoya—sit near the centers of the three largest lowlands.

The overall population density of Japan is high (Figure 11.16). Japan as a whole contains some 870 persons per square mile (2,137 per square kilometer), a figure substantially higher than the 72 persons per square mile found in the United States. The fact that its settlements are largely restricted to roughly 15 percent of the country's land area means that Japan's effective population density—the actual crowding that the country experiences—is one of the highest in the world. This is especially true in the main industrial belt that extends in a west-southwest direction from Tokyo through Nagoya and Osaka and hence along the Inland Sea (the maritime region sandwiched between Shikoku, western Honshu, and Kyushu) to the northern coast of Kyushu. The lowland portion of this zone is perhaps the most intensively used part of the world in regard to human habitation, industry, and agriculture. It remains to be seen whether a balance between these different kinds of land uses can be maintained in such a limited land area.

Japan's Urban–Agricultural Dilemma Due to such space limitations, all Japanese cities are characterized by extremely dense settlement patterns. In the major urban areas, the

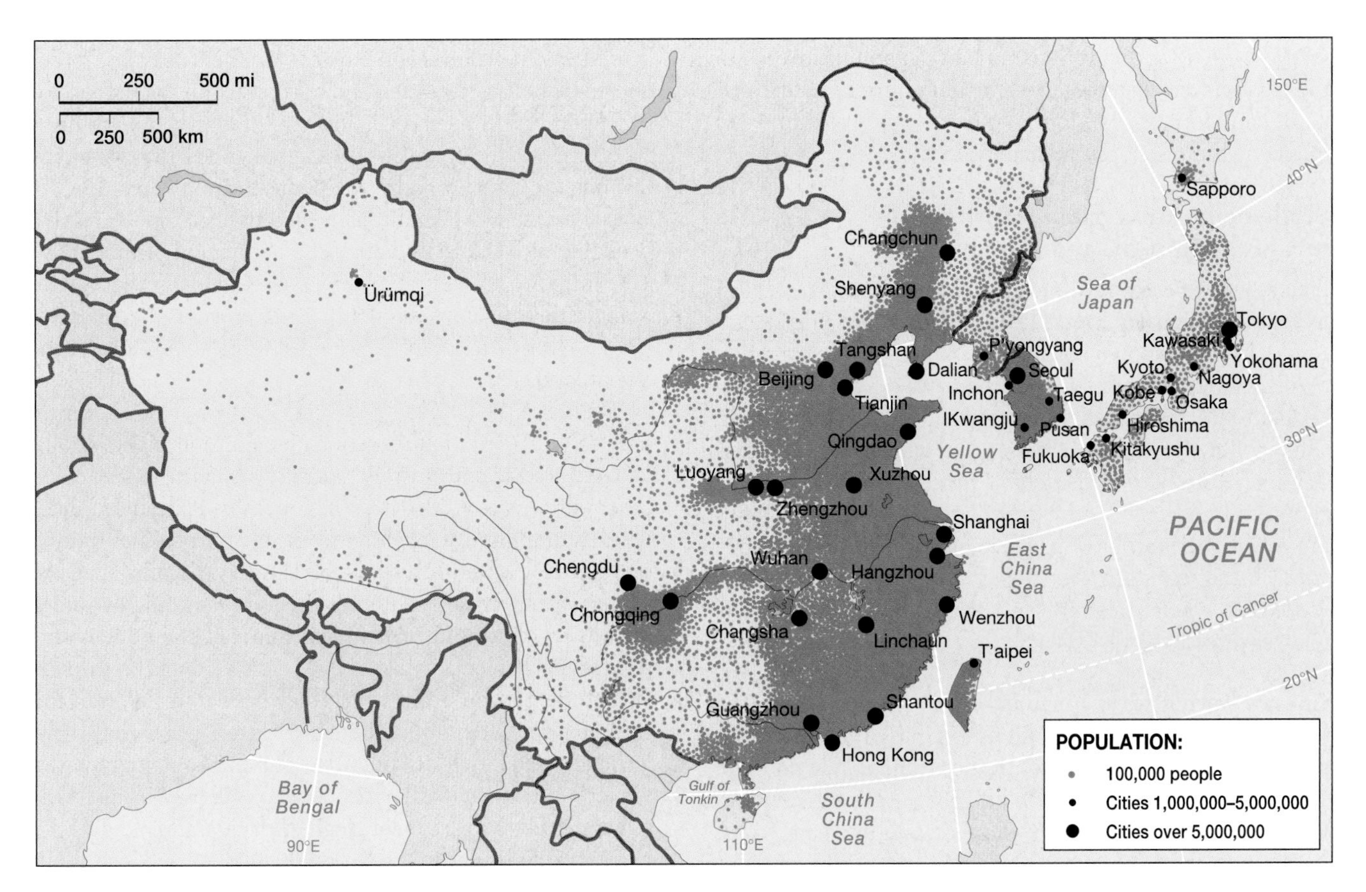

▲ **Figure 11.16 Population map of East Asia** Parts of East Asia are extraordinarily densely settled, particularly in the coastal lowlands of China and Japan. This contrasts with the sparsely settled lands of western China, North Korea, and northern Japan. Although the total population of this world region is high, as is the overall density, the rate of natural population increase has slowed rather dramatically in the last several decades because of several factors, primarily China's well-known "one child" policy.

amount of available living space is highly restricted for all but the most affluent families. In Tokyo an apartment the size of a large U.S. living room often shelters an entire middle-class family. Many observers—especially American ones—argue therefore that Japan should allow its cities and suburbs to expand into nearby rural areas. Since most upland areas are too steep for residential uses, such expansion would have to come at the expense of agricultural land.

As it now stands, the wholesale conversion of farmland to neighborhoods would be difficult. Although most farms are extremely small (the average is only several acres) and only marginally viable, farmers are tightly organized and politically powerful, and croplands are often protected by the tax code. Moreover, most Japanese citizens believe that it is vitally important for their country to remain self-sufficient at least in rice. Rice imports are therefore all but banned. Memories of wartime shortages linger, and many people consider rice cultivation to be an essential cultural or even spiritual aspect of the Japanese nation.

U.S. economic interests, along with certain Japanese consumer advocates, stress the fact that the present system forces Japanese consumers to pay up to six times the world market price for rice, their staple food. They also point out that Japan relies on imports for many of its other food needs, as well as for the energy that is required to grow rice in the first place. But whatever arguments are made, change will probably come gradually, if at all, and for the moment Japan must live with the tensions and contradictions resulting from the fact that its affluent population is forced to live in tight proximity and to pay high prices for basic staples.

Settlement and Agricultural Patterns in China, Taiwan, and Korea

Like Japan, Taiwan and Korea are essentially urban. Unlike Japan, China remains essentially rural, with only some 30 percent of its population living in cities. Chinese cities are rather evenly distributed across the plains and valleys of China proper. As a result, the overall pattern of population distribution in China follows closely with the geography of agricultural productivity. In other words, most Chinese people live in the countryside where they can be supported largely by local farming. China's cities are growing at a rapid pace, however, despite governmental efforts to keep them under control. China's population as a whole is also growing, again despite official efforts to achieve stability.

China's Agricultural Regions A line drawn just to the north of the Yangtze Valley divides China into two main agricultural regions. To the south, rice is the dominant crop. To the north, in the North China Plain, Loess Plateau, and Manchuria, wheat, millet, and sorghum are the main crops.

In southern and central China, population is highly concentrated in the broad valleys, which are famous for their fertile soil and intensive agriculture. More than 100 million persons live in the Sichuan Basin, while more than 70 million reside in deltaic Jiangsu, a province smaller than Ohio. (With 11 million people, Ohio is one of the most densely populated U.S. states.) Cropping occurs year-round in most of southern and central China; summer rice alternates with winter barley or vegetables in the north, while two rice crops can be harvested in the far south. In some of the most fertile and densely populated areas, as many as three crops can be produced in a single year. Southern China also produces a wide variety of tropical and subtropical crops, and moderate slopes throughout the area produce sweet potatoes, corn, and other upland crops.

The agricultural patterns found in the lowland basins of southern China exemplify the theory of agricultural intensification developed by economist Ester Boserup in the 1960s. Boserup argued that as population increases in a given area, agricultural lands will be cropped increasingly frequently. An area of low density is typically characterized by long periods of fallow, while more densely populated zones move to yearly cropping and eventually to producing, if possible, two or more crops in a single year. Boserup argued that this tendency to intensify agriculture under population pressure requires ever-increasing applications of labor, but that it can avoid the problems of overpopulation and help support economic development. The limits of Boserupian intensification, however, were reached in most Chinese lowlands long ago; further gains will have to come from larger yields per cropping session.

The North China Plain is one of the most thoroughly **anthropogenic landscapes** in the world (an anthropogenic landscape is one that has been heavily transformed by human activities). Virtually its entire extent is either cultivated or occupied by houses, factories, and other structures of human society. Too dry in most areas for rice, the North China Plain specializes in wheat, millet, and sorghum, the latter two grains forming the staple food for the poorest rural people.

Manchuria was a lightly populated frontier zone as recently as the mid-1800s. Today, with a population of more than 100 million, its central plain is thoroughly settled. Still, Manchuria remains less crowded than many other parts of China, and it is one of the few parts of China to produce a consistent food surplus.

The Loess Plateau is more thinly settled yet, supporting only some 70 million inhabitants. But considering its aridity and widespread soil erosion, this is a high figure indeed. Like other portions of northern China, the Loess Plateau produces wheat and millet, largely for subsistence purposes. A unique settlement feature in this part of China is the prevalence of subterranean housing. Although loess erodes quickly under the impact of running water, it coheres well in other circumstances. For millenia, villagers have excavated pits on the surface of the plateau, from which they have tunneled into the earth to form underground houses (Figure 11.17). These subterranean dwellings are cool in the summer and warm in the winter. Unfortunately, they also tend to collapse during earthquakes, leading to extremely high mortality rates. One quake in 1920 killed an estimated 100,000 persons, while another in 1932 killed some 70,000.

Settlement and Agricultural Patterns in Korea and Taiwan Like China and Japan, Korea is a densely populated country. It contains some 68 million persons (22 million in

▲ **Figure 11.17 Loess settlement** A typical subterranean dwelling carved out the soft loess sediment in central China. Approximately 70 million people live similarly in the Loess Plateau region. Unfortunately, this region is also prone to major earthquakes that take a high toll on the local population because of dwelling collapse. *(Christopher Liu/ChinaStock Photo Library)*

the north and 46 million in the south) in an area smaller than Minnesota. South Korea's population density is 1,150 per square mile (2,980 per square kilometer), a significantly higher figure than that of Japan. Most of these people are crowded into the alluvial plains and basins of the west and far south. The highland spine, extending from the far north to northeastern South Korea, remains relatively sparsely settled. South Korean agriculture, like that of Japan and southern China, is dominated by rice. North Korea, in contrast, also relies on corn and other upland crops that do not require irrigation.

Taiwan is actually the most densely populated state in East Asia. Roughly the size of the Netherlands, it contains more than 20 million inhabitants (5 million more than the Dutch homeland, one of the most crowded countries of Europe). Its overall population density is some 1,500 persons per square mile (3,885 per square kilometer), one of the highest figures in the world. Since central and eastern Taiwan are rugged and mountainous, virtually the entire population is concentrated in the narrow lowland belt in the north and along the western coast. Large cities and numerous factories are scattered here amid productive farmland.

East Asian Agriculture and Resource Procurement in Global Context

Although East Asian agriculture is highly productive, it is not productive enough to feed the huge number of people who live in the region. Japan, Taiwan, and South Korea are major food importers, and China is moving in the same direction. Other resources as well are being drawn in from all quarters of the world by the powerful economies of East Asia.

The Global Dimensions of Japanese Agriculture and Forestry Japan may be self-sufficient in its production of rice, but it is still one of the world's largest food importers. As the Japanese have grown more prosperous over the past 50 years, their diet has grown more diverse. That diversity is made possible largely by importing food from other parts of the world.

Japan imports food from a wide array of other countries. It procures both meat and the feed used in its domestic livestock industry from the United States, Canada, and Australia. These same countries supply wheat that is necessary to produce bread and noodles. Even soybeans, long a staple of the Japanese diet, must be purchased from such countries as Brazil and the United States. Japan has one of the highest rates of fish consumption in the world, and the Japanese fishing fleet scours the world's oceans to supply the demand. The people of Japan also purchase a large quantity of prawns and other seafood, much of it farm-raised in former mangrove swamps, from Southeast Asia.

Japan depends on imports to supply its demand for forest resources. While the forests of Japan produce high-quality cedar and cypress logs, the country obtains most of its construction lumber and pulp (for papermaking) from abroad. Western North America and Southeast Asia are its main traditional suppliers. Almost half of Southeast Asia's forest-product exports are shipped to Japan. As the rain forests of Malaysia, Indonesia, and the Philippines diminish, Japanese interests are beginning to turn to Latin America and Africa as sources of tropical hardwoods. Japanese and South Korean firms are also beginning to procure forest products from Siberia (eastern Russia), a nearby and previously little-exploited forest zone.

Japan is able to support its large and prosperous population on such a restricted land base only because it can purchase resources from abroad. Certainly all countries engage in trade, but Japan's basic resource dependence is particularly pronounced. Almost all of the oil, coal, and other minerals that it consumes are also imported. If forced to rely on its own resources, Japan would find itself in a economic and ecological bind. But because its industries are successful in the global market, Japan has avoided such a predicament.

Although geographers once looked at national resource endowments as a main support of every country's economic performance, such a view is no longer tenable in the increasingly interconnected global economy. As long as a country can export items of value, it can obtain whatever imports it requires. This also means, however, that the environmental degradation generated by a successful economy like that of Japan is no longer limited to its home territory. It too has been globalized.

The Global Dimensions of Chinese Agriculture Through the 1980s, China was essentially self-sufficient in its production of food, despite its huge population and crowded lands. Rapid economic growth, combined with changing dietary values, have, however, brought about an increased consumption of meat (which requires large amounts of feed grain) as well as the loss of agricultural lands to residential and industrial development. By the mid-1990s, China too had become a grain importer.

Optimists argue that China could produce much more food than it now does by increasing its use of fertilizers and by converting its "wastelands" of grass and scrub into agricultural fields. Chinese leaders are well aware of the current grain shortage, but they promise that agricultural production will soon increase. Even they concede, however, that by the year 2030 China will have to import 5–10 percent of its total grain demand. Noted U.S. environmentalist Lester Brown contends that China has already expanded its agricultural output almost to the maximum extent possible, and that in the entire world grain production may no longer keep pace with population growth. A more-affluent China, he argues, might drive up global grain prices, which could potentially put basic staples out of the reach of the world's poorest people. Few experts on Chinese agriculture, however, support Brown's position.

Korean Agriculture in a Global Context South Korea has already made the transition, as Japan and Taiwan have, to a global resource procurement pattern. In the mid-1990s, South Korea was the world's fifth leading importer of wheat (after China, Egypt, Japan, and Brazil) and the second leading importer of corn (after Japan). North Korea, on the other hand, has pursued a goal of strict self-sufficiency. While relatively successful for a number of years, in 1997 and 1998 this policy resulted in widespread famine after a series of floods followed by drought destroyed most of the country's rice and corn crops.

Urbanization in East Asia

China has one of the world's oldest urban foundations, dating back more than 3,500 years. In medieval and early modern times, East Asia as a whole possessed a well-developed system of cities that included some of the largest settlements on the planet. In the early 1700s, Tokyo, then called *Edo,* probably overshadowed all other cities, with a population of more than one million.

But despite this early start, East Asia was overwhelmingly rural at the end of World War II. Only some 10 percent of China's people lived in cities, and even Japan was only about 50 percent urbanized. As the region's economy began to grow, though, so did its cities. Japan, Taiwan, and South Korea are now between 70 and 80 percent urban, which is typical for advanced industrial countries. Some 70 percent of the Chinese people still live in rural areas, but this figure is decreasing steadily, and because China's population is immense, its urban foundation is still sizable. Twelve of the world's 100 largest cities are in China, and two Chinese cities, Beijing and Shanghai, rank among the top 15.

Chinese Cities Traditional Chinese cities were clearly demarcated from the countryside by defensive walls. Most were planned in accordance with strict geometrical principles that were believed to reflect the cosmic order (Figure 11.18). The old-style Chinese city was horizontal, dominated by low buildings and characterized by straight streets. Houses were typically built around courtyards, and narrow alleyways served both commercial and residential functions.

China's urban fabric began to change during the colonial period. A group of port cities was essentially taken over by European interests, which proceeded to build Western-style buildings situated in Western-style business districts. By far the most important of these semicolonial cities was Shanghai, built near the mouth of the Yangtze River—the main gateway to interior China.

When the Communists came to power in 1949, Shanghai, with a population of more than 10 million persons, was

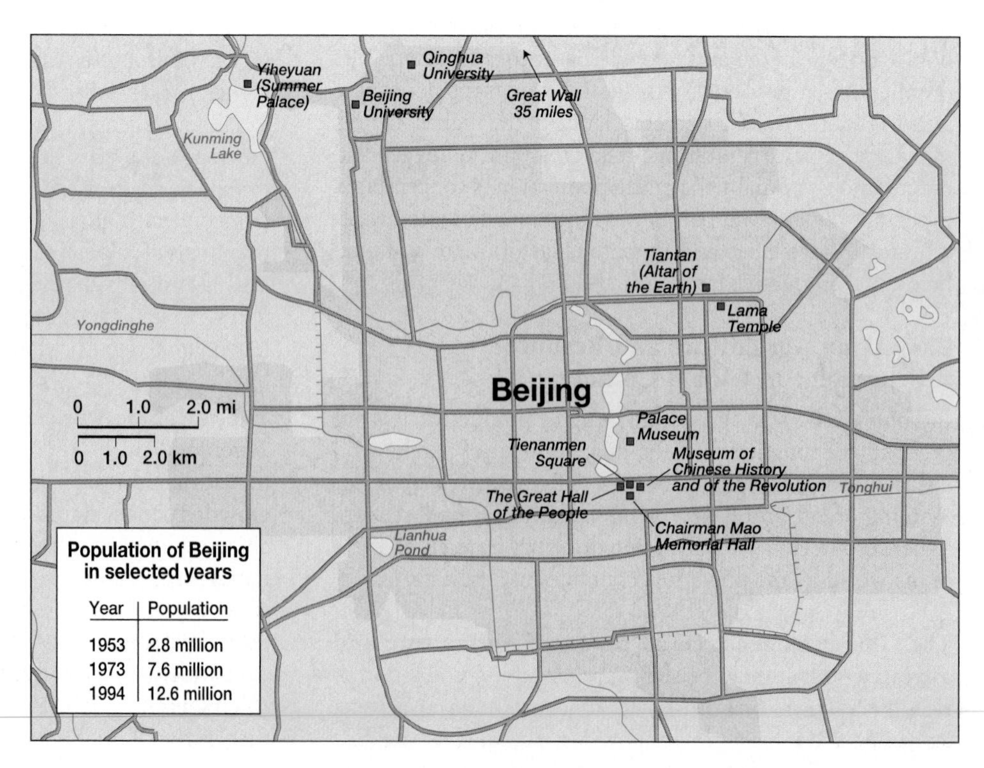

Population of Beijing in selected years

Year	Population
1953	2.8 million
1973	7.6 million
1994	12.6 million

▶ **Figure 11.18 Beijing, China** The historic capital during the Manchu period (1644–1912), Beijing regained its status as capital city in 1949. Under the communists, Beijing's historical landscape was largely razed and replaced with large blocks of government offices and massive apartment buildings. Only the historically significant Forbidden City, home to the Manchu rulers, was saved from this transformation of the urban landscape. Tienanmen Square, reputed to be the largest open square in any city, was created by clearing buildings away from the area in front of the Palace Museum.

the second-largest urban agglomeration in the world. The new authorities viewed the city as a foreign creation and as a rather decadent place. They therefore proceeded to milk it for taxes, which they largely invested elsewhere. Most of the old civic elite had fled to Hong Kong, and relatively few new migrants were allowed in. As a result, much of the city, and especially the old business district, began to decay.

Since the late 1980s, Shanghai has experienced a major revival and it is again in many respects the premier city of China. Migrants are now pouring into Shanghai, even through the state still tries to restrict the flow; building cranes crowd the skyline. Official statistics now put the population of the metropolitan area at some 14 million, but the actual number, if one were to include the large number of temporary migrants, may well be significantly larger. The new Shanghai, as is typical for the new China, is a city of massive high-rise apartments and concentrated industrial developments (Figure 11.19). Shanghai is also at the core of an old, and now revitalizing, network of urban places situated around the greater Yangtze Delta. But despite Shanghai's revived economic fortunes, the city remains politically secondary to Beijing, China's capital.

Beijing was China's capital during the Manchu period (1644–1912), a status it regained in 1949. Under communist rule, Beijing was radically transformed; old buildings were razed and broad avenues were plowed through old neighborhoods. Crowded residential districts gave way to large blocks of apartment buildings and massive government offices. Some historically significant buildings were saved; the buildings of the Forbidden City, for example, where the Manchu rulers once lived, survived as a complex of museums. The area immediately in front of the Palace Museum, however, was completely cleared of buildings. The resulting

▲ **Figure 11.19 Contemporary Shanghai** This vibrant city of more than 14 million embodies the new China with its massive high-rise apartments, industrial developments, and office towers. This photo shows the Nan Pu bridge and high-rise buildings in the newly developed Pu Dong area. *(Keren Su/Pacific Stock)*

plaza, Tienanmen Square, is reputed to be the largest open square in any city of the world. It makes a convenient display area for public spectacles and government-sponsored political rallies. Critics sometimes regard Beijing as a city designed primarily to express state power, and as such it is a rather inconvenient place in which to live.

China's urban system as a whole is fairly well balanced, with sizable cities relatively evenly spaced across the landscape and with no single city overshadowing all others. This balance stems from China's indigenous heritage of urbanism, its vast size and division into separate physiographical and economic regions, and its legacy of socialist planning. In the early 1960s noted China scholar G. William Skinner argued that the distribution of Chinese cities can best be explained through the use of **Central Place theory.** Central place theory, developed by German geographer Walter Christaller in the 1930s, holds that an evenly distributed rural population will give rise to a regular hierarchy of urban places, with uniformly spaced larger cities surrounded by constellations of smaller cities, each of which, in turn, is surrounded by smaller towns. Such a regular pattern, Central Place theorists contend, is generated largely by retail marketing; every family must have ready access to a nearby town to procure basic necessities, but larger urban areas, where more expensive but less-commonly purchased items are obtainable, can be situated at a more distant location. Critics of the theory, however, contend that the even distribution of cities and towns found in much of China and many other parts of the world stems more from political administration and from other kinds of economic activities than from retail marketing.

In the middle 1990s, Beijing and Shanghai vied for the first position among Chinese cities, with Tianjin, serving as Beijing's port, coming in at a close third. All three of these cities have long been removed from the regular provincial structure of the country and granted their own metropolitan territories and governments (in 1997 Chongqing in Sichuan was added to this list of province-level municipalities). In 1997 another major city, Hong Kong, was removed from British colonial control and granted a unique status as a largely autonomous "special administrative region" of China. While not as populous as Beijing or Shanghai, Hong Kong is far wealthier. The emerging greater metropolitan area of the Xi River Delta, composed of Hong Kong, Shenzhen, and Guangzhou (called *Canton* in the West), may now perhaps be regarded as China's premier urban area (see "Local Voices: China's Rural Poor and the Lure of the City").

Below these four primary cities are several dozen other large, and growing, urban centers. Some 32 of them had more than a million inhabitants each in 1993, and in the near future many more will pass that milestone. It is still questionable, however, whether China will become a predominantly urban society; at present, much industrial development is moving into rural areas, as factories sprout in a seemingly haphazard fashion amid rice fields and vegetable plots. The growth of rural industries, coupled with the expansion of urban areas, some scholars contend, might threaten China's livelihood by consuming ever-larger expanses of agricultural land.

LOCAL VOICES China's Rural Poor and the Lure of the City

One of the largest migrations in China's history is under way as at least 100 million migrant laborers roam the country looking for work; most are rural poor who have left their farms and villages to seek a better life in the city. The story of Sanzi, one family's 16-year-old third son who just graduated from junior high school, is typical. Here he speaks of his experiences:

My parents work a tiny plot of land that cannot support the costs of sending me to high school, so if I stay here in Qingdong I'm just another mouth to feed. There is no future other than working in the fields. Most of my friends have already left for the city; I feel like I'm choking to death in this village. I'm the only one in the family who has not tried the city.

My father went to Shanghai to work in a lumber business, but now he's back in the village working in the fields and the rest of us have to help pay off the debt from his failed attempt. My mother and two older brothers then went to Shanghai to work in a hotel kitchen but that lasted only a year because of mother's poor health.

One of the older brother's bosses has sent word to the village that there might be some sort of job for Sanzi in Guangzou, at the mouth of the Pearl River. The boss promised to pick up Sanzi at the train station, instructing him: "Wear your brother's white windbreaker and wait at the main exit."

The government has concentrated economic reform efforts in Guangzou, and every day thousands of peasants pour into the city hoping to cash in on the action. Greeting them at the train station, however, are an equal number of peasants on their way back to the countryside, having tired of—or failed at—their stint in the city.

Sanzi stumbles off of the train after a 26-hour ride in the crowded fourth-class car. He puts on the white windbreaker and waits. "I sat down in the shade to look at my new home. It was unbelievable. Blaring taxis, women in spiked heels and miniskirts, men with cell phones; all of it was so new."

After almost two hours of anxious waiting, the boss pulls up in an early-1990s Cadillac. He motions for Sanzi, who grabs his bag and jumps in the car—the first time he's been in an automobile. He asks about the job, but the boss, seeing Sanzi's youth and slight build, is noncommittal. All he says is that it will be with a construction crew working for a friend of his. "Can you handle hard labor?" asks the boss.

"Of course I can. I grew up in the village doing hard labor, working all day in the fields," Sanzi answers. The boss closes the door and the Caddy joins the stream of traffic heading into the city. Another of China's rural poor begins a new life in the city.*

*Adapted from "Desperate Journeys: China's Rural Poor Take Flight" by Tim Kao, *San Francisco Chronicle*, February 2, 1999.

City Systems of Japan, Taiwan, and South Korea The urban structures of East Asia's fully capitalist fringe are quite different from those of China. Both Taiwan and South Korea are noted for their pronounced **urban primacy** (the concentration of urban population in a single city), whereas Japan is the center of a new urban phenomenon, that of the **superconurbation** (a superconurbation, or megalopolis, is a huge zone of coalesced metropolitan areas).

Seoul, the historical power center of Korea and, since the postwar division, the capital of South Korea, overwhelms all other cities in the country. Seoul itself is home to more than 10 million persons, and its greater metropolitan area contains some 40 percent of South Korea's total population. All of South Korea's major governmental, economic, and cultural institutions are concentrated there. Seoul's explosive and generally unplanned growth has resulted in serious congestion. The South Korean government has attempted to promote industrial growth in other cities, but none of its actions has yet challenged the primacy of the capital.

Taiwan is similarly characterized by a high degree of urban primacy. The capital city of Taipei, located in the far north, mushroomed from some 300,000 people during the Japanese colonial period to some 6.2 million (in the metro area) in the late 1990s.

Japan has traditionally been characterized more by a kind of urban "bipolarity" than by urban primacy. Until the 1960s, Tokyo, the capital and main business and educational center, together with the neighboring port of Yokohama, was balanced by the mercantile center of Osaka and its neighboring port of Kobe. Kyoto, the former imperial capital and the traditional center of elite culture, is also situated in the greater Osaka region. A host of secondary and tertiary cities served to balance Japan's urban structure. Nagoya, with a metro area of 4.8 million persons, remains the center of the country's automobile industry—and is one of the few large Japanese cities in which travel by car is more efficient than travel by public transportation. Many other sizable and tightly packed cities dot the basins and coastal lowlands, most of which are little known outside of the country. For example, Kitakyushu, in far western Japan, is home to some 1.5 million persons and is a major center of the Japanese chemical industry.

As Japan's economy boomed in 1960s, 1970s, and 1980s, so did Tokyo. The capital city then outpaced all other urban areas in almost every urban function. The Greater Tokyo metropolitan area is now home to 25–28 million persons, depending on how one bounds the area. The Osaka–Kobe metro area stands at a distant second, with "only" some 10–14 million inhabitants. Concerned about the increasing primacy of Tokyo, the Japanese government has been attempting to steer new developments to other parts of the country. Such efforts, however, have been only moderately successful. Most firms and government agencies prefer to be near the center of power and wealth. The newly built "science city" of Tsukuba, for example, has failed to attract much private investment, and it sometimes seems almost deserted on weekends, when many of its residents seek the stimulation of Tokyo.

It should not be imagined, however, that Japan's other cities have begun to wither away to support Tokyo's growth. Most of them have expanded as well, albeit not at so rapid a pace. In general, urban growth has been supported by rural depopulation. Metropolitan expansion has been particularly pronounced in the cities linking Tokyo to Osaka, an area known as the Tokkaido corridor. Transportation connections are superb along this route, and proximity to Tokyo, Osaka, and Nagoya encourages development. As can be seen in Figure 11.20, the result has been the creation of a superconurbation (alternatively called a *megalopolis*), or connected series of metropolitan areas. One can travel from Tokyo to Osaka on the main rail line, a distance of almost 300 miles (480 kilometers), and virtually never leave the urbanized area. By some accounts, 65 percent of Japan's 126 million persons are crowded into this narrow supercity. Additional urban concentrations, moreover, extend west of Osaka along the Inland Sea to Fukuoka in northern Kyushu.

Japanese cities sometimes strike foreign visitors as rather gray and monotonous places, lacking historical interest. Little of the country's pre-modern architecture remains intact. Japanese buildings were traditionally made of wood, which survives earthquakes much better than stone or brick. Fires have thus been a long-standing hazard, and in World War II the U.S. Air Force fire-bombed most Japanese cities, virtually obliterating them (Hiroshima and Nagasaki were, on the other hand, completely destroyed by atomic bombs). The one exception was Kyoto, the old imperial capital, which was spared this devastation. As a result, Kyoto is famous for its beautiful (wooden) Buddhist monasteries and Shinto temples, which ring the basin in which central Kyoto lies. Other Japanese cities were largely reconstructed in the late 1940s and 1950s, a period when Japan was still poor and could afford only inexpensive concrete buildings. In the boom years of the 1980s, however, hyper-modern skyscrapers rose in many of the larger cities, and postmodernist architecture began to lend variety, especially in the wealthier urban districts.

Cultural Coherence and Diversity: A Confucian Realm?

East Asia is in some respects one of the world's more unified cultural regions. Although different East Asian countries, as well as the different regions within them, have their own unique cultural features, all parts of the region share certain historically rooted ways of life and systems of ideas (see "Food and Culture: East Asian Cuisine").

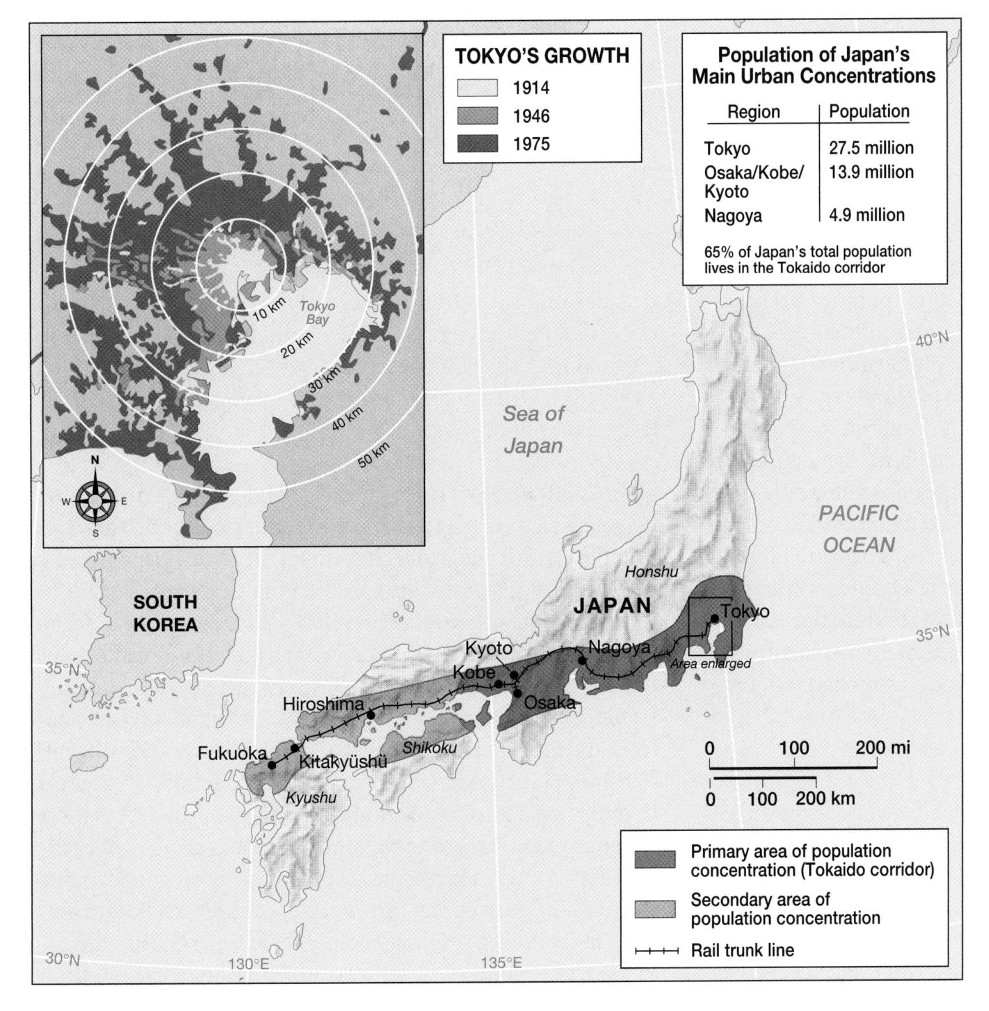

◀ **Figure 11.20 Urban concentration in Japan** The inset map shows the rapid expansion of Tokyo in the post-war decades. Today, the Greater Tokyo metropolitan area is home to almost 30 million people. The larger map shows the cluster of urban settlements along Japan's southeastern coast. The major area of urban concentration is between Tokyo and Osaka, a distance of 300 miles, known as the Tokkaido corridor. By some accounts, 65 percent of Japan's population lives in this area. The Osaka–Kobe metropolitan area ranks second to Tokyo, with approximately 14 million inhabitants.

FOOD AND CULTURE East Asian Cuisine

Americans tend to lump the various cuisines of East Asia into single category, based, perhaps, on their common focus on rice. Such a perception, however, is inaccurate, and East Asia as a whole cannot really be considered a coherent "food region." In most of northern China, for example, rice is a luxury food, and most people subsist largely on wheat as well as sorghum and other rough grains. Rice, moreover, is just as important in Southeast Asia and eastern India as it is in East Asia. Within East Asia there are also marked differences in seasoning and styles of preparation. Korean food, for example, is noted for its heavy use of hot (red) pepper and garlic, whereas Japanese food by comparison sometimes seems either bland or merely salty. In much of China, traditional cuisine revolves around vegetables and sometimes meat cooked in a sauce, which is then placed on rice; in Japan rice is usually served plain, offered alongside but not in combination with various side dishes.

Japanese cooking is, however, remarkably complex. The country has long borrowed liberally from other traditions (Chinese, European, and North American especially), while maintaining a distinct sense of what constitutes Japanese food. Its own cuisine is noted for the variety of its seafoods, pickles, and vegetables. Fruit is also popular in Japan, but is prohibitively expensive by American standards. Melons, for example, often cost more than $20 apiece and are considered appropriate gifts. One of the most popular Japanese foods is *sushi,* which literally means "vinegared rice." Sushi is often, but by no means always, eaten with *sashimi,* or raw fish. Before the late 1800s the Japanese ate virtually no meat, but that is no longer true. The consumption of beef especially has risen dramatically since the 1960s.

China has no single cuisine, but rather a variety of food styles associated with different provinces. The food of Guangdong, for example—the first Chinese food introduced to America—is seldom heavily spiced, while that of Sichuan, which has more recently gained favor in the United States, is noted for its spiciness. China has a long history of elite gastronomy, with many expensive delicacies based on rare and exotic plants or animals. A classic example is "bird's-nest" soup, the main constituent of which is the dried spittle that certain swallows use to glue their nests together. After the coming of communist rule, however, elite cooking traditions of all kinds nearly vanished from mainland China. During this period, little was available other than basic grains, vegetables, and small quantities of pork. The more elaborate forms of cuisine survived, however, in Taiwan, Hong Kong, and overseas Chinese communities, and they are now slowly returning to the Chinese mainland.

Despite the disparities of cooking traditions found across East Asia, the region does have a few dietary commonalties. For example, tofu, or soybean curd, is a traditional source of protein throughout the region—a region historically noted for its low levels of meat consumption. East Asia can also be considered the "chopstick region." In Japan, China, Korea, and Vietnam, food is generally eaten with chopsticks, and is cut and prepared to be easily handled with these implements. Chopsticks are not traditionally used in other parts of the world. Americans often ask for them in Thai restaurants, but in doing so they commit a minor error in cultural geographical categorization.

Most of these East Asian commonalties can be traced back to ancient Chinese civilization. China emerged roughly 4,000 years ago, largely in isolation from the Eastern Hemisphere's other early centers of civilization in the valleys of the Indus, Tigris-Euphrates, and Nile rivers. As a result, East Asian civilization developed along several unique lines. Before 1800, the entire region remained somewhat self-contained, with only secondary cultural and economic connections extending to the rest of Eurasia. The most prominent intellectual traditions of the region traditionally regarded China not only as the Middle Kingdom—or center of the world—but also as the world's only significant civilization. Such an inward focus was also characteristic of other civilizations, but probably not to the same extent as in East Asia.

Unifying Cultural Characteristics

The most important unifying cultural characteristics of East Asia are related to religious and philosophical beliefs. Throughout the region, Buddhism and especially Confucianism have traditionally shaped not only individual beliefs but also social and political structures. Although, in recent decades, the role of traditional belief systems has been seriously challenged, especially in China, historically grounded cultural patterns never disappear overnight. Even something as basic as written communication reveals a distinctly East Asian cultural background.

The Chinese Writing System The clearest distinction between East Asia and the world's other cultural regions is found in written language. All existing writing systems found elsewhere in the world are based on the alphabetic principle, in which each symbol represents a distinct sound. All of humankind's varied alphabetical scripts evidently can be traced back to a common point of origin in Syria and Lebanon. From there, alphabetical writing spread westward to Europe, eastward as far as the Philippines, and southward to Ethiopia. East Asia, on the other hand, evolved an entirely different system of **ideographic writing.** In an ideographic writing system, each symbol (or ideograph—more commonly called *character*) represents primarily an idea rather than a sound (although they can denote sounds in certain circumstances). As a result, ideographic writing requires the use of a large number of distinct symbols.

The East Asian writing system can be traced to the dawn of Chinese civilization. The first symbols were essentially pictures representing different words, but over time they grew more abstract. As the Chinese Empire expanded, and as the prestige of Chinese civilization carried its culture to other lands (such as Japan), the Chinese writing system spread. Japan, Korea, and Vietnam all came to use the same system, although in Japan it was substantially modified, while in Korea and Vietnam it was much later largely replaced by alphabetic systems.

The Chinese ideographic writing system has one major disadvantage and one major advantage when compared with alphabetic systems, both of which stem from the fact that it is largely divorced from spoken language. The disadvantage is that it is difficult to learn; to be literate, a person must memorize several thousand distinct characters. The main benefit is that two literate persons do not have to speak the same language to be able to communicate, since the written symbols that they use to express their ideas are the same.

This advantage was tremendously important in the creation of a unified Chinese culture. When the Chinese Empire expanded south of the Yangtze River beginning in about 200 B.C., ethnic groups speaking a variety of different languages were suddenly brought into the same political and cultural system. Because they could adopt the Chinese writing system, the peoples of southern China were able to integrate fully into the Middle Kingdom without adopting the spoken language of the north. To this day, the residents of Beijing in the north and Guangzhou in the south speak different languages, yet write them in identical form.

Korean Modifications In Korea, Chinese characters were adopted at an early date and were used exclusively for many hundreds of years. In the 1400s, however, Korean officials decided that Korea ought to have its own alphabet. They wanted to allow a larger number of people to be able to read and write, hoping also to more clearly differentiate Korean culture from that of China. The use of the new script spread quickly through the country, allowing Korea to enjoy higher literacy rates than were possible in pre-modern China. Korean scholars and officials, however, continued to use Chinese characters, regarding their own script as suitable only for popular purposes. Today the Korean script is used for almost all purposes, but scholarly works still contain occasional Chinese characters interspersed within Korean-alphabet texts.

Japanese Modifications The writing system of Japan is even more complex. Initially, the Japanese simply borrowed Chinese characters, referred to in Japanese as **kanji.** Owing to the profound grammatical differences between Japanese and the spoken languages of Chinese, the exclusive use of kanji resulted in awkward sentence construction. The Japanese solved this quandary by developing a unique quasi-alphabet, or more precisely a syllabary, known as **hiragana,** that allowed the expression of words and parts of speech not easily represented by Chinese characters. In hiragana, each symbol represents a distinct syllable, or combination of a consonant and a vowel sound. Because of the restricted sound system of Japanese, only 51 hiragana symbols are necessary. A different but essentially parallel syllabary, called *katakana,* is used in Japan for spelling words of foreign origin (Figure 11.21).

Use of the hiragana resulted in increased literacy in medieval Japan, especially for women. The greatest early works in Japanese literature were written by aristocratic women wholly in the hiragana script. Men with official duties and scholarly ambitions, however, continued to write in kanji. Eventually the two styles of writing merged, and written Japanese came to employ a complex mixture of symbols. In general, the more advanced a Japanese text is, the more kanji it contains. Japanese kanji today differ slightly from Chinese characters, as the latter have been simplified to some extent while the former have not. Some Japanese characters, moreover, have evolved with different meanings. Even so, it is still relatively easy for anyone literate in Japanese to learn how to read Chinese—but not to speak it!

Finally, it should be noted that Japanese, unlike Chinese, can be easily written in the roman alphabet. The resulting *romanji* style of writing is especially important in advertisements and computer use. Computer keyboards using roman letters are not only much less cumbersome than Japanese keyboards, but they are also a virtually universal component of modern

▲ **Figure 11.21 Japanese writing** Originally, the writing system of Japan was based on Chinese characters; however, because of grammatical differences, the Japanese developed two unique alphabets of syllables, known as *katakana* and *hiragana.* Here, two of these forms of writing are visible. *(Hiroshi Harada/DUNQ/Photo Researchers, Inc.)*

global culture. Computer technology, however, allows romanji entries to be easily converted into the traditional Japanese mixture of kanji, hiragana, and katakana symbols. Modern technology, in this case, helps spread global attributes while simultaneously allowing the continued florescence of local cultural forms.

The Confucian Legacy Just as the use of a common writing system helped forge strong cultural linkages throughout East Asia, so the idea-system of **Confucianism** (the philosophy developed by Confucius) came to occupy a significant position in all of the societies of the region. Indeed, so strong is the heritage of Confucius that some writers refer to East Asia as the "Confucian world." In Japan, however, Confucianism has never had the strength of influence that it has had in China and Korea.

The premier philosopher of Chinese history, Confucius (or Kung Fu Zi, in Mandarin Chinese) lived during the sixth century B.C., a period of marked political instability. The North China Plain and Loess Plateau, then the main areas of Chinese culture, were divided into a number of hostile states. Confucius's goal was to create a philosophy that could help generate social stability. While Confucianism is often considered to be a religion, Confucius himself was far more interested in the "here and now," focusing his attention on how to lead a proper life and organize a proper society. While Confucian thought does not deny the existence of deity or of an afterlife, neither does it give them much consideration. Ethics, however, does lie at the heart of the system.

Confucius stressed, on the one hand, deference to the properly constituted authority figures. On the other hand, he emphasized that authority has a responsibility to act in a benevolent manner. The lowest level of the traditional Confucian moral order is the family unit, considered to be the bedrock of society. The ideal family structure is somewhat patriarchal, and children are admonished to obey and respect their parents—especially their fathers—as well as their elder brothers. At the highest level of the traditional moral order sat the emperor of China, who was regarded as an almost godlike father-figure for the entire country.

Confucian philosophy also stresses the need for a well-rounded and broadly humanistic education. To a certain extent, Confucianism advocates a kind of meritocracy, holding that an individual should be judged on the basis of behavior and education, rather than merely on the basis of family background. The high officials of pre-modern China—the powerful **Mandarins**—were thus selected in accordance with their performance on competitive examinations. Only wealthy families, however, could afford to give their sons the education that was needed for success on those grueling tests.

Confucianism in Japan In Japan, Confucianism was never as important as it was on the mainland. Japanese officials were actually able to exclude certain Confucian beliefs that they considered dangerous. The most important of these was the revocable "mandate of heaven." According to this notion, the emperor derived his authority from the ordering principle of cosmic harmony, but such a mandate could be withdrawn if the emperor failed to fulfill his proper role. This idea was used both to explain and to legitimize the rebellions that occurred periodically in pre-modern China, occasionally resulting in the change of ruling dynasty. In Japan, on the other hand, there has been no need to claim that the mandate of heaven can be revoked since a single imperial dynasty has persisted throughout the entire period of written history. Although the emperor of Japan has had little if any real power for more than a thousand years, the absolute sanctity of his family lineage continues to be a basic principle of Japanese society.

The Modern Role of Confucian Ideology The significance of Confucianism in East Asian development has been hotly debated for most of this century. In the early 1900s, many observers believed that the inherent conservatism of the philosophy, derived from its respect for tradition and authority, was responsible for the economically backward position of China and Korea. Since East Asia has, during the latter decades of the 1900s, enjoyed the world's fastest rates of economic growth, such a position no longer seems supportable. New voices are now arguing that Confucianism's respect for education and the social stability that it supposedly generates give East Asia a tremendous advantage in international competition.

Other scholars remain skeptical of both views, preferring to cite material factors such as economic policy when explaining the region's rapid rates of economic growth. They also note that most interior portions of China have not participated fully in the current economic boom, even though these places share the Confucian legacy of the increasingly prosperous coastal districts. Confucianism has, moreover, lost much of the hold that it once had on public morality throughout the entire region.

Religious Unity and Diversity in East Asia

Certain explicitly religious beliefs have helped Confucianism cement together the East Asian region. The most important culturally unifying beliefs are associated with Mahayana Buddhism. Other religious practices, however, have had more of a disunifying role.

Mahayana Buddhism Buddhism, a religion that stresses the human soul's quest to escape an endless cycle of rebirths and reach a state of union with the divine cosmic principle (or nirvana), originated in India in the sixth century B.C. By the second century A.D., Buddhism had reached China, by way of Central Asia, and within a few hundred years it spread throughout East Asia. Today Buddhism remains widespread throughout the region, although it is far less significant there than it is in mainland Southeast Asia, Sri Lanka, and Tibet.

The variety of Buddhism practiced in East Asia—Mahayana, or Greater Vehicle—is distinct from the Therevada Buddhism of South and Southeast Asia. Most importantly, Mahayana Buddhism simplifies the quest for nirvana, in part by positing the existence of souls *(boddhisatvas)* who refuse divine union for themselves in order to help others experience spiritual progress. Mahayana Buddhism, unlike other forms of the religion, is nonexclusive; in other words, one may follow it while simultaneously professing the beliefs of other faiths. Thus, many Chinese consider themselves to be both Buddhists and Taoists (as well as Confucianists), while most Japanese are at some level both Buddhists and followers of Shinto (Figure 11.22).

As Mahayana Buddhism spread through East Asia in the medieval period, many different sects emerged and proliferated. Probably the best known of these is Japanese Zen, which demands that its followers engage in the rigorous practice of "mind emptying." At one time, Buddhist monasteries associated with Zen and other sects were rich and extremely powerful. In all East Asian countries, however, periodic reactions against Buddhism resulted in the persecution of monks and the suppression of monasteries. One reason is that Buddhism often was viewed by government officials as a foreign religion, one that places India—rather than China—at the center of the world. Despite such hardships, East Asian Buddhism has never been extinguished. But it also never became the focal point of society as it did in mainland Southeast Asia—or as Islam did in Southwest Asia or Christianity in Europe. In Japan, that position was partially captured by a different religion altogether, that of Shinto.

Shinto Shinto is so closely bound to the idea of Japanese nationality that it is questionable whether a non-Japanese person could even follow the religion. Shinto began as the animistic worship of nature spirits, but it was gradually refined into a subtle set of beliefs about the harmony of nature and its connections with human existence. Until the late 1800s, Buddhism and Shinto were complexly intertwined. Subsequently, the Japanese government began to disentangle them while elevating Shinto into a kind of nationalistic cult focused on the divinity of the Japanese imperial family. After World War II, the more excessive aspects of nationalism were removed from the Shinto faith, but a certain undercurrent remains.

Shinto is still a place- and nature-centered religion to a significant degree. Certain mountains, particularly two volcanoes—Fuji and Ontake—are considered sacred and are thus climbed by large numbers of people (Figure 11.23). Major Shinto shrines, often located in particularly picturesque places, still attract numerous pilgrims; the most notable of these is the Ise Shrine south of Nagoya, which is the central site of the cult of the emperor. Small local shrines, like Buddhist temples, offer verdant oases in otherwise largely treeless Japanese urban neighborhoods.

Taoism and Other Chinese Belief Systems The Chinese religion of Taoism (or Daosim) is similarly rooted in nature worship. Like Shinto, it stresses the acquisition of spiritual harmony and the pursuit of a balanced life. Taoism is indirectly associated with *feng shui,* commonly called **geomancy** in English, the Chinese and Korean practice of designing buildings in accordance with the spiritual powers that supposedly course through the local topography (Figure 11.24). Even in hyper-modern Hong Kong, skyscrapers worth millions of dollars have occasionally gone unoccupied because their construction failed to accord with geomantic principles. While Taoism is traditionally limited to areas of Chinese culture, many of its precepts have been embraced by a variety of "new age" faiths in North America and Europe.

Despite the commonalties imparted by Taoism and Buddhism, traditional religious practice in China has always

▲ **Figure 11.22 The Buddhist landscape** This Buddhist temple is located in Yunnan Province, China, and was used as the headquarters of the Yunnan provincial Buddhist association before the communist revolution of the middle twentieth century. Today it is preserved as a historical museum. *(Brian Vikander/Corbis)*

▲ **Figure 11.23 Mt Fuji, Japan** This picturesque volcanic mountain, sacred to Japan's Shinto religion, is climbed by a large number of people as part of a religious pilgrimage. In the foreground are tea fields. *(Pacific Stock)*

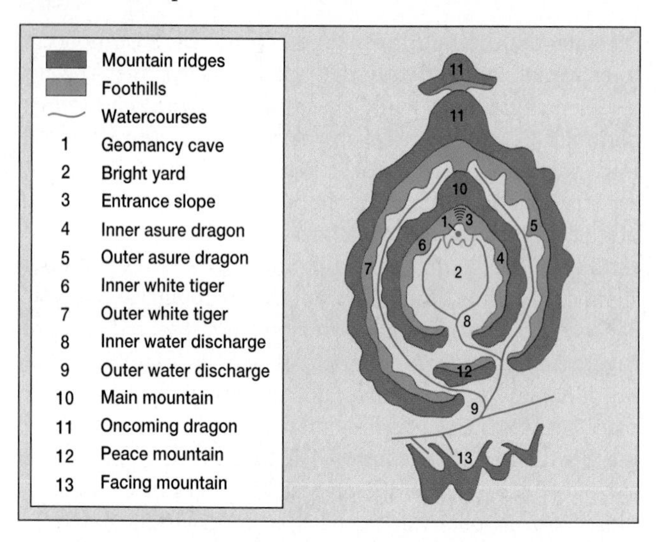

▲ **Figure 11.24 The landscape of geomancy** This diagram shows a particular favorable arrangement of natural features according to the doctrine of geomancy. Geomantic design is based on the idea that the formidable powers, both positive and negative, course through the topographical features of the physical landscape.

embraced local **particularism** (in other words, focusing on the unique attributes of particular places). Many minor gods were traditionally associated with single cities or other particular places. Such beliefs never had much significance for the country's educated elite, but in pre-modern China rituals associated with them were sometimes carried out by state officials. In modern-day rural China, village gods are often still honored.

Minority Religions Small numbers of adherents of virtually all world religions may be found in the increasingly cosmopolitan cities of East Asia. Millions of Chinese and Japanese belong to Christian churches—even though they constitute less than one percent of the population of either country. Far more South Koreans, some six million in total, are Christian, most of them Protestant. Some reports indicate that Christianity is growing rapidly in China—despite persecution—but reliable information is scarce.

Much larger than China's Christian population is its Muslim community. Several tens of millions of Chinese-speaking Muslims, called *Hui,* are concentrated in Gansu and Ningxia provinces in the northwest of China proper and in Yunnan province along the south-central border. Smaller clusters of Hui, often segregated in their own villages, live in virtually every province of China. The only Muslim congregations in Japan, on the other hand, are associated with recent and probably temporary immigrants from South, Southeast, and Southwest Asia. At one time there were also small Chinese Jewish communities in the cities of Kaifeng and Shanghai, but the former apparently disappeared in the 1800s, while the latter vanished in the 1900s.

Secularism in East Asia For all of these varied forms of religious expressions, East Asia must still be counted as one of the most secular regions of the world. In Japan, most people occasionally observe Shinto or Buddhist rituals and maintain a small shrine in their homes for their own ancestors; a small segment of the populace is devout. Japan also has a number of "new religions," sometimes called *cults,* a few of which are noted for the fanaticism of their adherents. For Japanese society as a whole, though, religion is simply not very important. Overall, Japan must be counted as one of the world's most secular societies.

Elite culture in China was formerly dominated by Confucianism, which is essentially a philosophy rather than a faith. After the communist regime took power in 1949, all forms of religion and traditional philosophy—including Confucianism—were discouraged and sometimes severely repressed. Under the new regime, atheistic **Marxist** philosophy (the communistic belief system developed by Karl Marx) became the official ideology. In the 1960s many observers believed that the traditional Chinese religious complex would survive only in overseas Chinese communities. With the easing off of Marxist orthodoxy in mainland China during 1980s and 1990s, however, many forms of religious expression have begun to return. This seems to be especially true in the more prosperous coastal areas of the country. In North Korea, Marxist orthodoxy is still rigidly enforced, and freedom to worship remains limited.

Linguistic and Ethnic Diversity in East Asia

While written languages may have helped unify East Asia, the same cannot be said for spoken languages. Japanese and Mandarin Chinese may partially share an ideographic system of writing, but the two languages bear no direct relationship. In their grammatical structures, Chinese and Japanese are as different from each other as are Chinese and English. Japanese, like Korean, has, however, adopted many words of Chinese origin, just as English has borrowed heavily from Greek and Latin.

Language and National Identity in Japan Japanese, by most accounts, is not related to any other language. Korean is also usually classified as the only member of its own language family (Figure 11.25). Many linguists, however, now believe that Japanese and Korean should be classified together in a new language family since they share many basic grammatical features. Only the mutual distrust of the Japanese and the Koreans, some suggest, has prevented this linguistic relationship from being acknowledged. A few linguists would go a step further to argue that Japanese and Korean are distantly related to the Altaic languages of Mongolia and Turkey, but this remains an uncommon view.

From many perspectives, the Japanese form one of the world's more homogeneous peoples, and they tend to conceptualize themselves in such a manner. To be sure, minor cultural distinctions are noted between the people of western Japan (centered on Osaka) and eastern Japan (centered on Tokyo), and many individual prefectures (particularly Nagano, in the mountainous heart of Honshu) are regarded as distinctive in their customs and lifeways. Overall, however, such differences are of little significance.

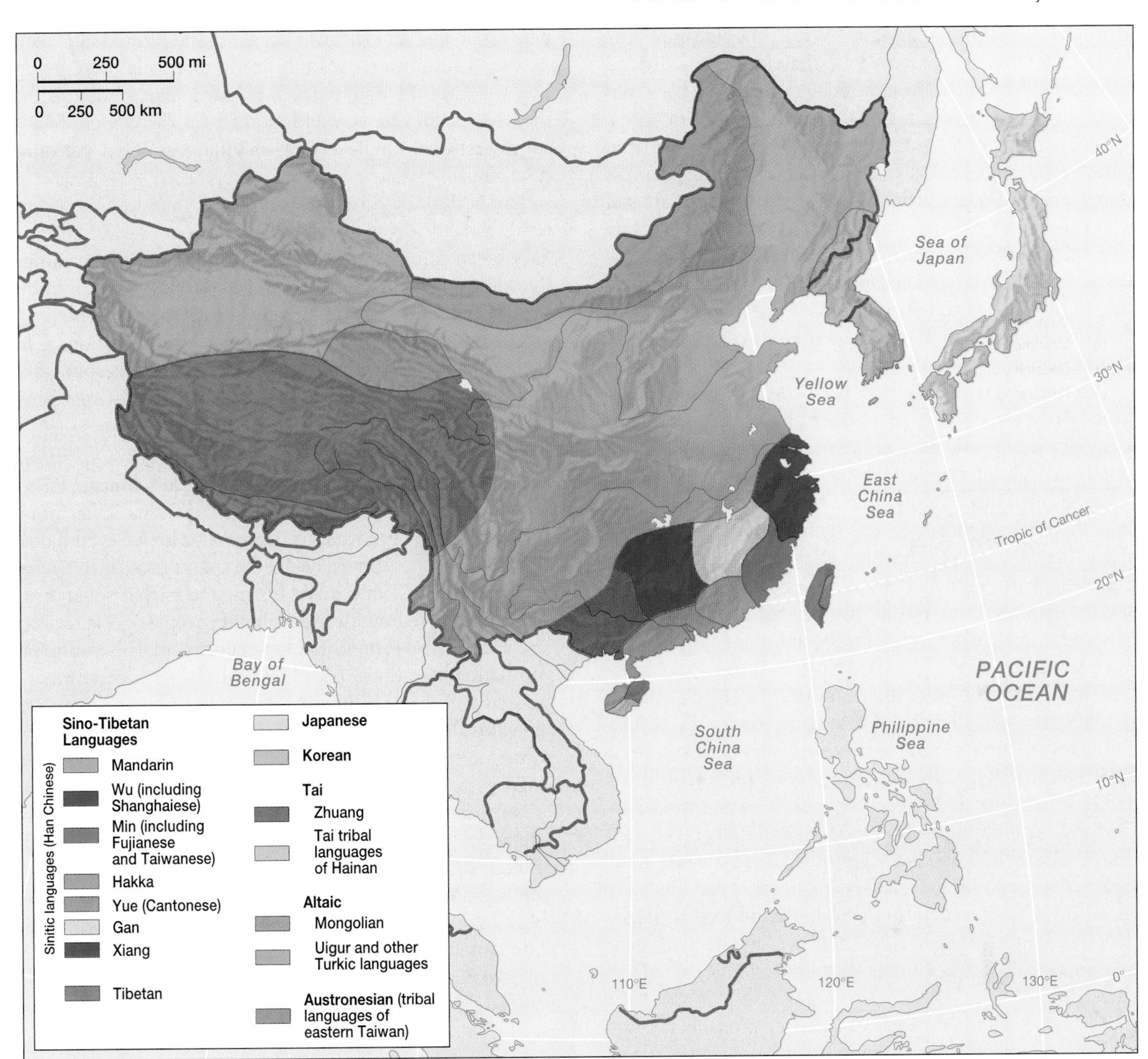

▲ **Figure 11.25 The language geography of East Asia** The linguistic geography of Korea and Japan is very straightforward, as the vast majority of people in those countries speak Korean and Japanese, respectively. In China, the dominant Han Chinese speak a variety of closely related *sinitic* languages, the most important of which is Mandarin Chinese. In the peripheral regions of China, a large number of languages—belonging to several different linguistic families—can be found.

In earlier centuries, the Japanese archipelago was divided between two very different groups: the Japanese, living to the south, and the Ainu, inhabiting the north. The Ainu are not only culturally distinct from the Japanese, possessing, for example, their own language, but they also have a markedly distinct physical appearance. Unlike Japanese men, Ainu men usually have heavy beards and abundant body hair, and were at one time commonly disparaged by their southern neighbors as "hairy barbarians." Because of this characteristic, as well as their general facial features, the Ainu were once categorized by anthropologists as members of the "Caucasian race." Few scholars, however, now believe that humankind is divided into discrete races, and Ainu do not in any event appear to be closely related to Europeans by genetic criteria.

For centuries the Japanese and the Ainu competed for land, and by the tenth century A.D. the Ainu had been largely driven off the main island of Honshu. Until the 1800s, however, Hokkaido, as well as Sakhalin (now part of Russia), remained Ainu territory. The Japanese people subsequently began to colonize Hokkaido, putting renewed pressure on the Ainu. Today only some 24,000 Ainu live on the northern Japanese island of Hokkaido, and most of them have mixed Japanese-Ainu ancestry. Ainu communities also survive on the Russian island of Sakhalin.

Minority Groups in Japan While the Japanese are themselves relatively homogeneous, their language is divided into several dialects. Tokyo and Osaka form the core territories of the two main dialect regions, with the speech of Tokyo considered standard. Osaka dialect is sometimes considered rather crude and is thus almost standard for Japanese comedians. Dialectal differences in the main islands are still relatively minor, and only in the Ryukyu Islands does one encounter a variant of Japanese so distinct that it might be said to constitute a separate language. Many Ryukyu people believe that they have not been considered full members of the Japanese nation, and they have suffered certain forms of discrimination.

Discrimination has also been keenly felt by approximately 700,000 persons of Korean descent living in Japan today. Many of these people were born in Japan (their parents and grandparents having left Korea early in the century) and speak Japanese rather than Korean as their primary language. But despite their deep bonds to Japan, such individuals are rarely able to obtain Japanese citizenship. Perhaps as a result of such treatment, many Japanese Koreans hold radical political views, and the community as a whole has sent substantial amounts of money to help support North Korea throughout the postwar period.

Starting in the 1980s, other immigrants began to arrive in Japan, most of them from the poorer countries of Asia and many of them without legal status. Men from southern Asian countries typically work in the construction industry and in other dirty and dangerous jobs; women from Thailand and the Philippines often work as entertainers or prostitutes. Almost 200,000 Brazilians of Japanese ancestry have returned to Japan for the relatively high wages they can earn there; however, immigration is less pronounced in Japan than it is in most other wealthy countries, and relatively few migrants are able to acquire permanent residency, let alone citizenship.

The most victimized people in Japan are probably not foreigners but rather the **Burakumin**, or *Eta*, an outcast group of Japanese whose ancestors worked in "polluting" industries such as leathercraft. While discrimination is now illegal, the Burakumin are still among the poorest and least-well-educated people in Japan. Private detective agencies do a brisk business checking prospective marriage partners and employees for possible Burakumin ancestries. The Burakumin, however, have banded together to demand their rights; the Buraku Liberation League is politically powerful and is reputed to have close connections with the Yakuza (the Japanese "mafia"). The Burakumin today usually live in their own neighborhoods, and are concentrated within the Osaka region of western Japan.

Language and Identity in Korea The Koreans are also a relatively homogeneous people. The vast majority of persons in both North and South Korea speak Korean and unquestioningly consider themselves to be members of the Korean nation. There is, however, a certain sense of regional identity and consciousness within the country, some of which may be traced back to the medieval period when the peninsula was divided into three separate kingdoms. The people of southwestern South Korea, especially those living in Kwangju and its environs, in particular tend to view themselves, and be viewed by others, as distinctive. Many southwesterners believe that they have suffered periodic discrimination.

Not all Koreans live in Korea. Several hundred thousand reside directly across the border in northern China. Substantial Korean communities can also be found in Kazakhstan in Central Asia, owing to the deportation strategy employed by the Soviet Union in the middle decades of the twentieth century. A more recent Korean **diaspora** has brought hundreds of thousands of people to the United States (a diaspora is a scattering of a particular group of people over a vast geographical area).

Language and Ethnicity Among the Han Chinese The geography of language and ethnicity in China is far more complex than that in Korea or Japan. This is true even if one considers only the eastern half of the country, so-called China proper. The most important distinction is that separating the Han Chinese from the non-Chinese peoples of the region. The Han, who form the vast majority of the country's population, are those people who have long been incorporated within the Chinese cultural and political systems, and whose languages are expressed in the Chinese ideographic writing system. They do not, however, all speak the same language.

Northern, central, and southwestern China—a vast area extending from Manchuria through the middle and upper Yangtze Valley to the valleys of Yunnan province in the far south—constitutes a single linguistic zone. The spoken language here is generally called *Mandarin Chinese,* with *mandarin* denoting the members of the elite bureaucracy in the imperial period before 1912. In China today, Mandarin is the national tongue, and is often called simply "the common languages." Mandarin itself is divided into a number of separate dialects, but these are closely related to each other and are mutually intelligible. The official Mandarin dialect of the Beijing area, moreover, is gradually spreading.

In southern China, from the Yangtze Delta to China's border with Vietnam, a variety of separate but closely related languages are spoken. Peoples speaking these languages are Han Chinese, but they are not native Mandarin speakers. The most important of these non-Mandarin languages are located along the coast. Traveling from south to north, one encounters Cantonese (or Yue) spoken in Guangdong, Fujianese (alternatively Hokkienese, or Min locally) spoken in Fujian, and Shanghaiese (or Wu), spoken in and around the city of Shanghai and in Zhejiang province. These are true languages, not dialects, since they are not mutually intelligible. They are usually called dialects, however, because they have no distinct written form.

One group of people speaking a southern Chinese language, the Hakka, is occasionally considered by others not to be true Han Chinese. The Hakka are sometimes disdained by other Chinese peoples as rootless wanderers. Evidently their ancestors fled northern China roughly a thousand years ago to settle in the rough upland area in the border zone where the Guangdong, Fujian, and Jiangxi provinces meet. As is evident in Figure 11.26, later migrations took them throughout southern China, where they typically settled in areas regarded by locals as hilly wastelands. The Hakka traditionally made their living by growing upland crops such as sweet potatoes and by working as lumberers, stonecutters, and metalworkers. The Hakka today still form one of the poorest communities of southern China.

Ironically, however, a large number of Hakkas have reached very high positions in the Chinese government (where they often conceal their Hakka origins). The Hakka have a tradition of rebellion against lowland landowners, and as such they quickly lent their support to the communist movement that eventually gained national power. Also paradoxically, the Hakka fiercely proclaim their own Han identity, and in fact often consider themselves to be exemplars of a universalistic Chinese culture. As such, they often disdain the sentiments of local particularism that are prevalent among their neighbors—just as they are disdained by others for their lack of local rootedness.

Despite their many differences, all of the languages of the Han Chinese (including Hakka) are closely related to each other, belonging to the same Sinitic language subfamily. Since their basic grammars and sound systems are similar, it is not particularly difficult for a person speaking one of these languages to learn another. It is usually difficult, however, for speakers of European languages—or of Japanese or Korean for that matter—to gain fluency in these tongues. All Sinitic languages are **tonal** and monosyllabic; their words are all composed of a single syllable (although compound words can be

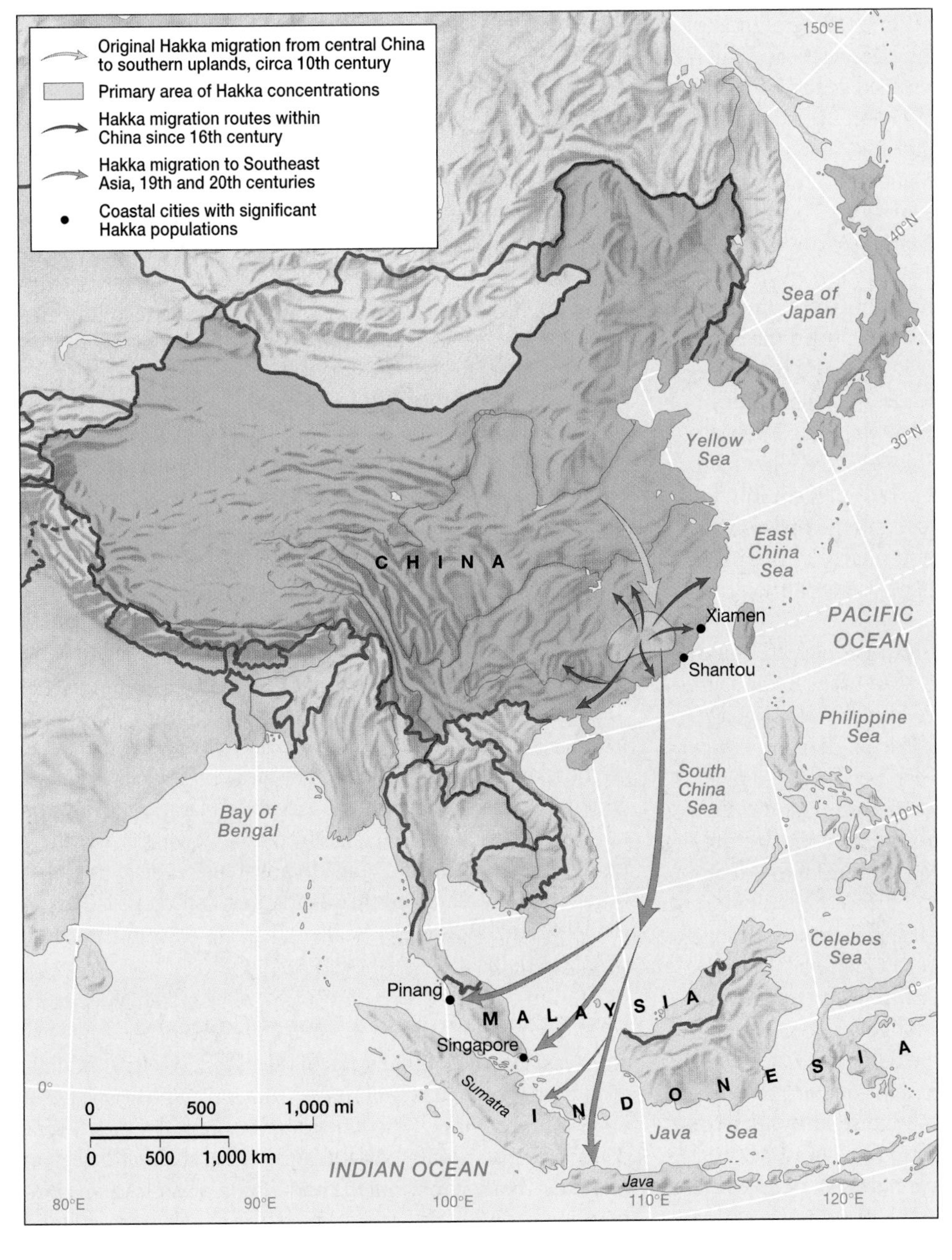

◀ **Figure 11.26 The Hakka diaspora** The Hakka form an important and very distinctive subgroup of the Han Chinese. The Hakka seem to have originated in north-central China but long ago migrated to the area where Fujian, Guangdong, and Jiangxi provinces converge. Here they developed distinctive agricultural patterns well suited to the rough uplands of the region. Later movements took Hakka communities to upland areas throughout much of southern China. In the 1800s and early 1900s, many Hakka people moved to Southeast Asia, where they still form important communities.

formed from several syllables), and the meaning of the each basic syllable changes completely in accordance with the pitch in which it is uttered.

The Non-Han Peoples Many of the more remote upland districts of China proper are inhabited by various groups of non-Han peoples speaking non-Sinitic languages. Such peoples are usually classified as **tribal,** implying that they have a traditional social order based on autonomous village communities. Such a view is not entirely accurate, however, since some of these groups once had their own kingdoms and are all now subject to the Chinese state. What they do have in common is a heritage of cultural and often political struggle against the Han Chinese, who have traditionally tended to regard non-Han peoples as backward, if not barbarian (Figure 11.27).

Over the course of many centuries, the territory occupied by these non-Han communities has been steadily reduced, both because of the continued expansion of the Han and because of their own outward emigration to the uplands of Southeast Asia. Acculturation into Chinese society and intermarriage with the Han have also reduced many non-Han groups. Their main concentrations today are in the rougher lands of the far north and the far south.

A few scattered communities of Manchus still live in the uplands of Manchuria. The Manchus speak several Tungusic languages related to those spoken by the tribal peoples of central and southeastern Siberia. They are few in number, and may be in danger of cultural and linguistic extinction. This is an ironic situation, since the Manchus ruled the entire Chinese Empire from 1644 to 1912. Until the end of this period, the Manchus prevented the Han from settling in central and northern Manchuria, which they hoped to keep as a kind of homeland preserve. Once Chinese were allowed to settle in Manchuria in the 1800s—in part to prevent Russian expansion into the area—the Manchus found themselves vastly outnumbered. As they began to intermarry with and to adopt the language and lifeways of the newcomers, their own culture began to disappear.

Much larger communities of non-Han peoples are found in the far south, especially in Guangxi. Most of the inhabitants of Guangxi's uplands and more remote valleys speak languages of the Thai family, closely related to those of Thailand. Since there are more than six million non-Han people in Guangxi, it has been designated an **autonomous region** (Figure 11.1). Such autonomy was designed to allow the non-Han peoples living here to experience socialist modernization at a different pace than that expected of the rest of the country. Critics contend that very little real autonomy has ever existed, despite the official designation. (In addition to Guangxi, there are four other autonomous regions in China. Three of these, Xizang [Tibet], Nei Monggol [Inner Mongolia], and Xinjiang are located in Central Asia and are thus discussed at greater length in Chapter 10. The final autonomous region, Ningxia, located in northwestern China, is distinguished by its large concentration of Hui [Mandarin-speaking Muslims].)

Other areas with sizable numbers of non-Han peoples are Yunnan and Guizhou, in southwestern China, and western

▲ **Figure 11.27 Tribal villages in south China** Non-Han people are usually classified as tribal in China; this assumes they have a traditional social order based upon autonomous village communities. Shown are Yi people at an open-air market in the village of Xhanghe in Yunnan province. *(Michael S. Yamashita/NGS Image Collection)*

Sichuan. Most tribal peoples here practice **swidden agriculture** (also called "slash and burn"; see Chapter 13) on rough slopes; valley bottoms and other flat areas are generally occupied by rice-growing Han Chinese. A wide variety of separate languages, falling into several linguistic families, are found among the scattered ethnic groups living in these uplands. Figure 11.28 shows that in Yunnan the resulting ethnic mosaic is staggeringly complex.

Language and Ethnicity in Taiwan Taiwan is also noted for its linguistic and ethnic complexity. In the island's mountainous eastern region, a few small groups of "tribal" peoples speak languages related to those of Indonesia and Malaysia (belonging to the Austronesian language family). These peoples resided throughout Taiwan before the sixteenth century. At that time, however, Han migrants began to arrive in large numbers. Most of these newcomers spoke Fujianese dialects, which eventually evolved into the distinctive language of Taiwanese.

Taiwan was transformed almost overnight in 1949, when nationalist forces, defeated by the communists, sought refuge on the island. Most of the nationalist leaders spoke Mandarin, which they immediately made the official language. Taiwan's new leadership discouraged Taiwanese, viewing it as a mere local dialect. As a result, considerable tension developed between the Taiwanese and the Mandarin communities. Only in the 1990s did Taiwanese speakers begin to reassert their own language rights.

East Asian Cultures in Global Context

East Asia, like most other parts of the world, has long exhibited tensions between an internal orientation and tendencies toward a more widely encompassing cosmopolitanism. This dichotomy has both a cultural and, as we shall later see, an economic dimension. Until the mid-1800s, all East Asian countries attempted to close their boundaries to Western cultural

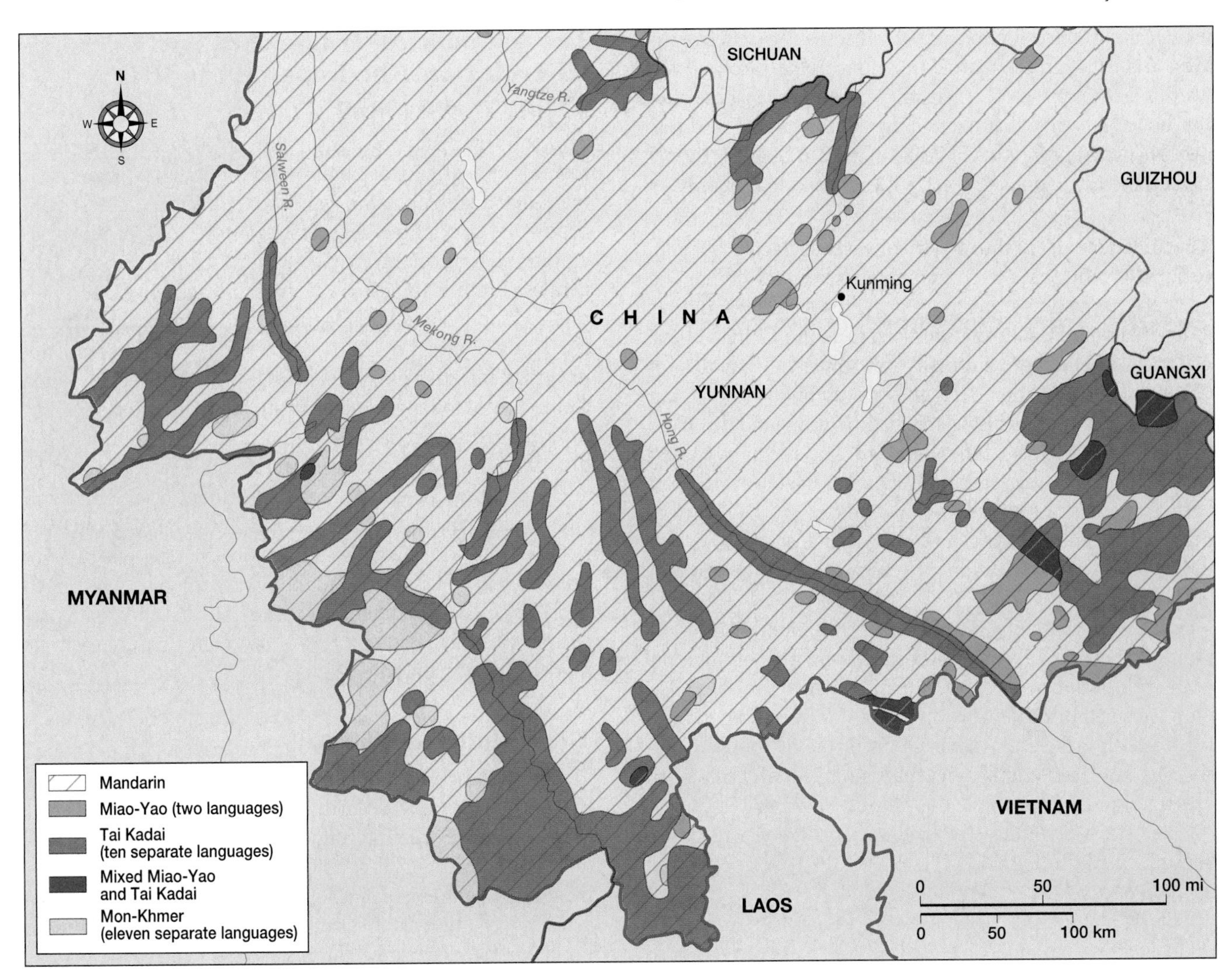

▲ **Figure 11.28 Language groups in Yunnan** China's Yunnan Province is the most linguistically complex area in East Asia. In Yunnan's valleys and relatively level plateau areas, and in its cities, most people speak Mandarin Chinese. In the hills and mountains, however, a wide variety of tribal languages, falling into several linguistic families, are spoken. In certain areas, several different languages can be found in very close proximity.

influences. Japan subsequently opened its doors, but the country as a whole remained highly ambivalent about foreign ideas. Only after its defeat in 1945 did Japan really opt for a globalist orientation. It was followed in this regard during the Cold War period by South Korea, Taiwan, and Hong Kong (then a British colony). The Chinese and North Korean governments at the same time decided to insulate themselves as much as possible from encounters with Western and global culture.

The Cosmopolitan Fringe The capitalist countries of East Asia are characterized by a vibrant cosmopolitan internationalism, especially in the large cities, which coexists with national and local cultural patterns. Virtually all Japanese, for example, study English for 6 to 10 years, and although relatively few learn to speak it fluently, most can read and understand a good deal. Business meetings among Japanese, Chinese, and Korean firms are more often than not conducted in English. Relatively large numbers of advanced students, especially from Taiwan, study in the United States and other English-speaking countries, and thus bring home a kind of cultural bilingualism. Internet usage, with its implicit globalism, is also widespread in East Asia's cosmopolitan fringe.

The current cultural flow is not merely from a globalist West to a previously isolated East Asia. Instead, the exchange is growing ever more reciprocal. Hong Kong's action films are immensely popular throughout most of the world, and with the success of director John Woo in the United States, they are beginning to influence filmmaking techniques in Hollywood. Significantly, Hong Kong's films are now some of the most multilingual in the world, often employing conversations in Cantonese, Mandarin, English, and even Japanese. Japan virtually dominates the world market in video games, and its ubiquitous comic-book culture and animation techniques are now following karaoke bars in their overseas march.

Cultural globalization is, of course, just as controversial in Japan as it is elsewhere. Japanese ultranationalists are few but vocal, calling their fellow citizens to resist the decadence of the West and to retain the martial traditions of the **samurai** (the warrior class of pre-modern Japan). Many other Japanese people, however, worry that their country is too insular, and that they still do not possess the English-language and global cultural skills necessary to operate effectively in the world economy.

The Chinese Heartland In one sense Japan is more culturally predisposed to cosmopolitanism than is China. The Japanese have always borrowed heavily from other cultures (particularly from China itself), whereas the Chinese have historically been more culturally self-sufficient. For most of Chinese history, cosmopolitanism has essentially implied an orientation to the norms established by the Mandarin class in the core of the vast Chinese Empire. The southern coastal Chinese have, however, sometimes embraced a different version of cosmopolitanism, one linked to the Chinese diaspora communities of Southeast Asia and the Pacific, and ultimately to maritime trading circuits extending over much of the globe.

In most periods of Chinese history, the interior orientation of the center has prevailed over the external orientation of the southern coast. After the communist victory of 1949, only the small British enclave of Hong Kong was able to pursue international cultural connections. In the rest of the country, a rather dour and somewhat puritanical cultural order was rigidly enforced. While this culture was largely founded on the norms of Chinese peasant society, it was also influenced by the Russian socialist system that had earlier emerged in the Soviet Union.

After China began to liberalize its economy and open its doors to foreign influences in the late 1970s and early 1980s, the southern coastal region suddenly assumed a new prominence. Through its doors global cultural patterns unevenly began to penetrate the rest of the country. The result has been the emergence of a vibrant but somewhat gaudy urban popular culture in China that is replete with such global features as nightclubs, karaoke bars, fast-food franchises, and theme parks.

The recent liberalization of the Chinese cultural order has also allowed the reemergence of regional and local identities. From the late 1960s to the late 1970s, all forms of localism were rigorously suppressed as China's leaders sought to build a nationally uniform working-class culture. Today, markers of local ethnicity are on the rise. This is again especially true in southern China, where most people do not speak Mandarin and have always remained culturally distinctive.

This resurgence of local identity, ironically, is also tied up with the process of globalization. Southern Chinese culture, particularly Cantonese culture, is now developing something of a national cachet precisely because it is identified as internationalist. Guangdong has been the main gateway for foreign culture to enter China, and is thus viewed as being at the forefront of the desirable process of globalization. Cantonese food, music, and films are now becoming popular in cities throughout China. With the return of Hong Kong to Chinese control, such a process may well accelerate in the near future.

The Geopolitical Framework and Its Evolution: The Imperial Legacies of China and Japan

The political history of East Asia revolves to a large extent around the centrality of China—and the ability of Japan to remain outside of China's grasp. The traditional Chinese conception of geopolitics was based on the idea of universal empire: all territories were either supposed to be a part of the Chinese Empire, to pay tribute to it and acknowledge its supremacy, or to stand outside the system altogether as barbarian lands. Until the 1800s, the Chinese state would not recognize any other government as its diplomatic equal. When China could no longer maintain its power in the face of European aggression, the East Asian political system fell into disarray. As European power declined in the 1900s, China and Japan became contenders for regional leadership, and after World War II, the region was split by larger **Cold War** rivalries (the "Cold War" was the global struggle carried out between the United States and the Soviet Union between 1946 and 1989). A legacy of tension between China and Japan persists to this day (Figure 11.29), although it is increasingly moderated by economic ties.

The Evolution of China

The original locus of Chinese civilization (dating back at least to 1800 B.C.) was the North China Plain and the adjacent Loess Plateau. For many centuries, periods of unification alternated with times of division into competing states. The most important episode of unification occurred in the third century B.C. Once political unity was achieved, the Chinese Empire began to expand vigorously to the south of the Yangtze Valley. Subsequently, the ideal of the imperial unity of a single Chinese state triumphed, and periods of political division into competing states were seen as indicating cosmic as well as earthly disorder. Such an ideology helped cement the Han Chinese into a single people. The potential for disunification, however, had always been present.

Several different Chinese dynasties rose and fell between 219 B.C. and 1912, most of them, as one can see in Figure 11.30, controlling roughly the same territory. The core of the Chinese Empire remained the area we have called China proper, excluding Manchuria. Other lands, however, were sometimes ruled as well. The most important of these was a western salient, or projecting territory, extending north of the Tibetan Plateau into the desert basins of Central Asia (modern Xinjiang). China valued this area because the vital trading route to western Eurasia (the "silk road") passed through it. But while China often ruled this territory (and indeed, does so today), it never was directly incorporated into the Chinese social and cultural systems.

Various Chinese dynasties attempted to conquer Korea and incorporate it into the empire. The Koreans, however, resisted Chinese expansion. Eventually China and Korea worked out a arrangement whereby Korea paid token tribute and acknowledged the supremacy of the Chinese Empire, and in return received trading privileges and retained independence. When

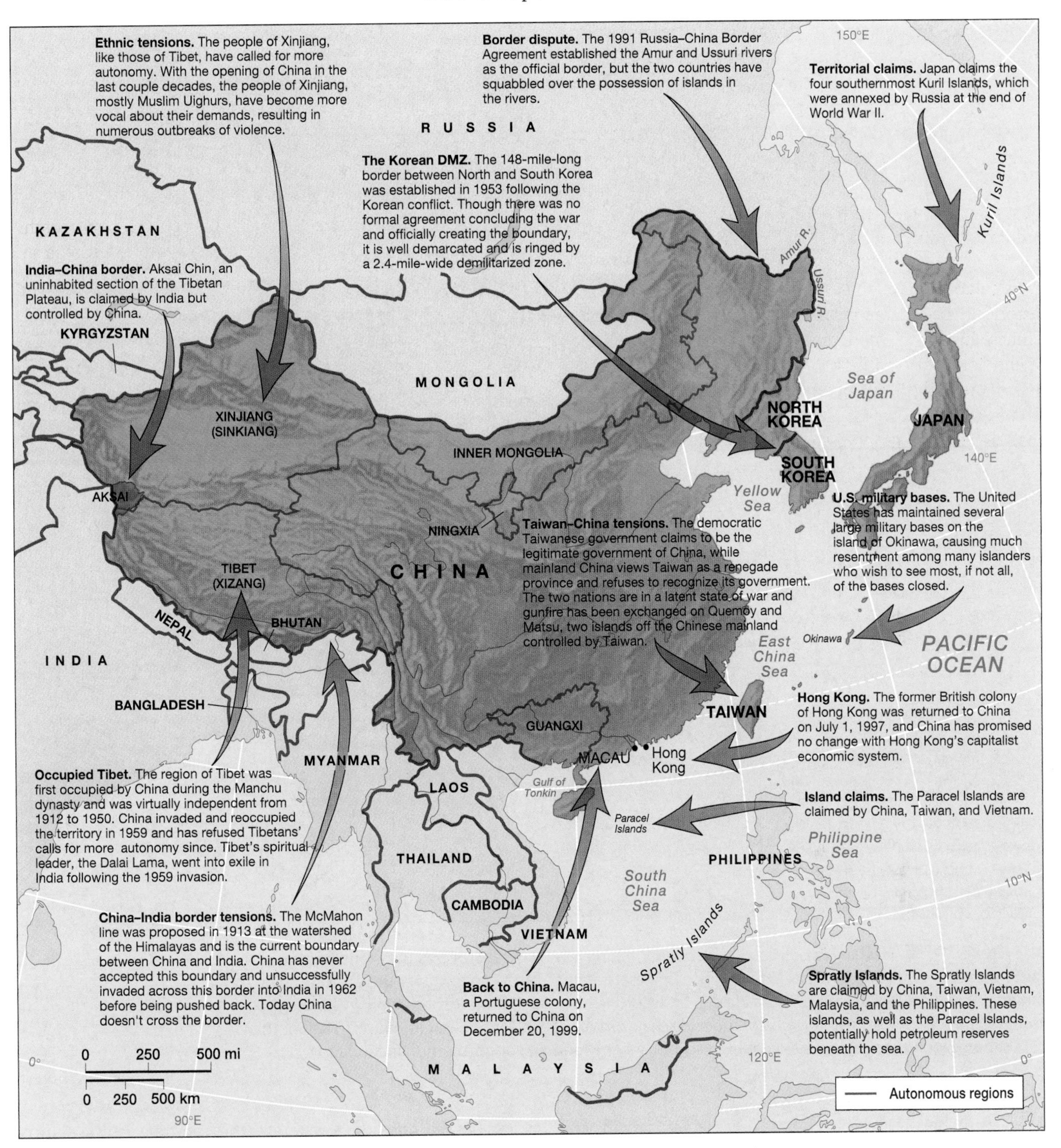

▲ **Figure 11.29 Geopolitical issues in East Asia** East Asia remains one of the world's geopolitical hot spots. Tensions are particularly severe between capitalist, democratic South Korea and the isolated communist regime of North Korea, and between China and Taiwan. China also has border disputes with a number of its neighbors that extend to a number of small islands in the South China Sea. Japan and Russia have not been able to resolve their quarrel over the southern Kuril Islands.

foreign armies invaded Korea—as did those of Japan in the late 1500s—China sent troops to support its "vassal kingdom."

For most of the past 2,000 years, the Chinese Empire was Earth's wealthiest and most powerful state. Its only real threat came from the pastoral peoples of Mongolia and Manchuria. Although vastly outnumbered by China, these societies were organized on a military basis and enjoyed the advantage of mobility conferred by their nomadic way of life. Usually the Chinese and the Mongols enjoyed a mutually beneficial trading relationship. Periodically, however, the two societies waged war, and on several occasions the northern pastoralists conquered the entire Chinese region (see Chapter 10). (The Great

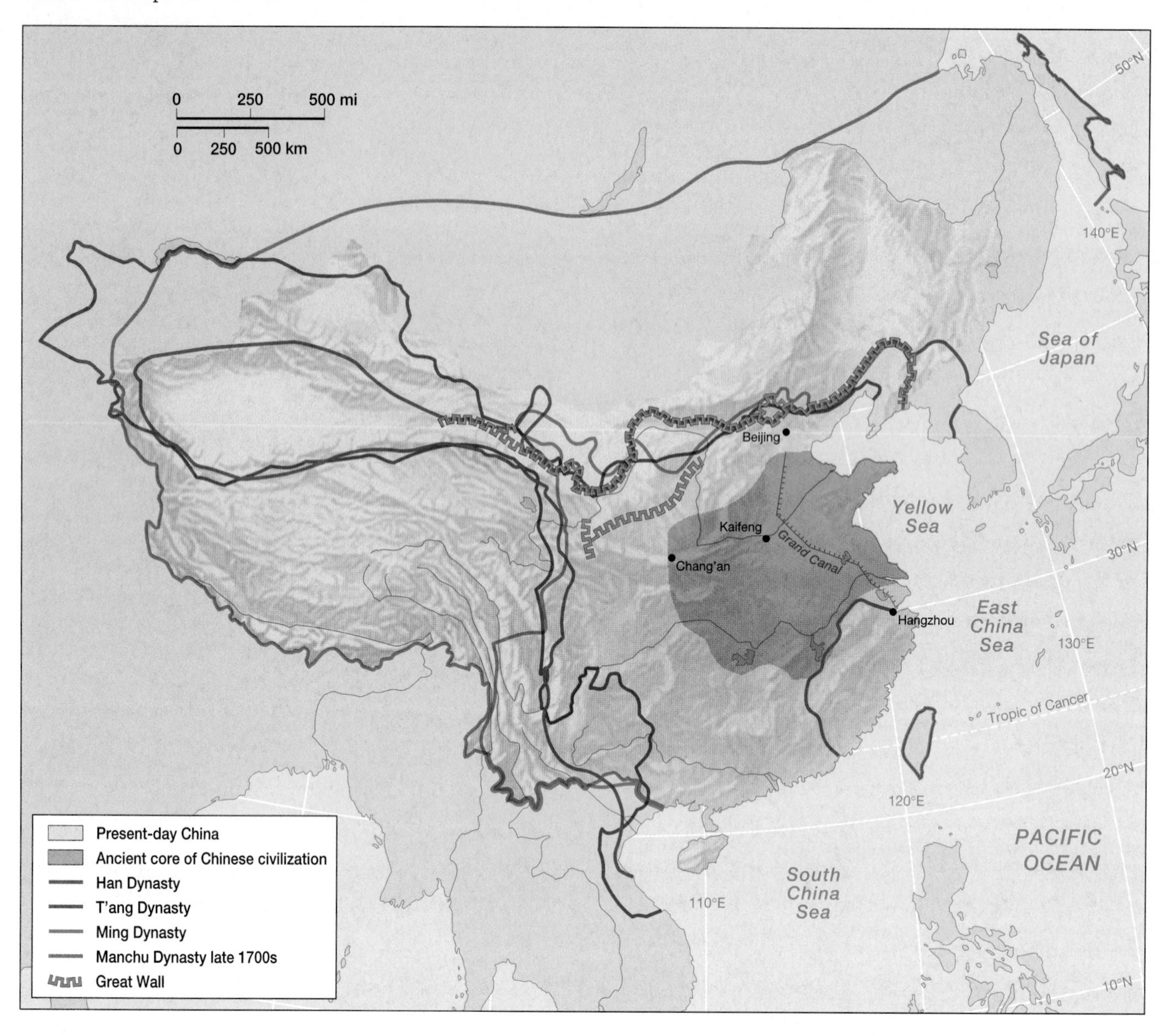

▲ **Figure 11.30 Historical extent of China** China is usually regarded as the world's oldest existing country, but the territorial extent of the Chinese state has varied greatly over the centuries. The earliest states were limited to the Loess Plateau and North China Plain, but most traditional Chinese dynasties controlled the entire core area of modern China as well as the Tarim Basin in Xinjiang. Before the 1600s, however, China seldom held control of Tibet, Inner Mongolia, or northern Manchuria.

Wall along China's border did not, in other words, provide adequate defense; Figure 11.31.) In time the conquering armies generally adopted Chinese customs in order to govern the far more numerous Han people, and in the long run they made a relatively small impact on Chinese society.

The Manchu Ch'ing Dynasty The final, and in many ways most significant, conquest of China occurred in 1644 when the Manchus toppled the indigenous Ming Dynasty and replaced it with the Ch'ing (or Manchu) Dynasty. As earlier conquerors did, the Manchus retained the Chinese bureaucracy and made few institutional changes. Their strategy was to adapt themselves to Chinese culture, yet at the same time to preserve their own identity as an elite military group. The system they established functioned well until the mid-nineteenth century, when the empire began to crumble before the onslaught of European and Japanese power.

China's most significant legacy from the Manchu Ch'ing Dynasty was an extension of its territory to include much of Central Asia. The Manchus subdued the Mongols, and by the mid-1700s established control over the entire eastern half of Central Asia, including Tibet. Even the states of mainland Southeast Asia sent tribute and acknowledged Chinese supremacy. Never before had the Chinese Empire been so extensive or so powerful.

The Modern Era From its height in the eighteenth century, the Chinese Empire descended rapidly in the nineteenth.

▲ **Figure 11.31 Great Wall of China** The Great Wall of China runs 1,500 miles (2,400 kilometers) east to west from the Yellow Sea to deep within Central Asia. The first parts of the Wall were built in the fourth century B.C.; however, most of the Wall was either rebuilt or finished in later times, mainly in the fifteenth and sixteenth centuries. *(Michael Howell/Pacific Stock)*

Unfortunately, it failed to keep pace with the technological and economic progress of Europe. Threats to the empire had always come from the north, and Chinese and Manchu officials saw little peril from petty European merchants operating along their coastline. The Europeans, for their part, were distressed by the amount of silver they had to pay to obtain Chinese silk, tea, and other products and by the fact that the Chinese disdained the manufactured goods that they offered. In response, the British began to sell opium, which Chinese authorities viewed as a moral and economic threat to their nation. When the imperial government tried to suppress the opium trade in the 1840s, the British attacked and quickly prevailed.

This first so-called "opium war" ushered in a century of political and economic chaos in China. The British demanded free trade in selected Chinese ports, and in the process overturned the traditional policy of managed exchange that had been a condition of the foreigners' acknowledgment of Chinese supremacy. As European economic enterprises increasingly penetrated China and undermined local economic interests, anti-Manchu rebellions began to break out. At first, all such uprisings were eventually crushed, but not before causing tremendous economic destruction. Meanwhile, European power continued to advance. In 1858 Russia annexed the northernmost reaches of Manchuria, and by 1900 China had been divided—as is shown in Figure 11.32—into separate "**spheres of influence**" where European economic power prevailed. (In a "sphere of influence" the colonial power had no formal political authority, but it did have a good deal of informal influence—as well as tremendous economic clout.)

A successful rebellion in 1911 finally toppled the Manchus, but subsequent efforts to establish a unified Chinese Republic were ineffectual. In many parts of the country, local military leaders ("warlords") usurped power for themselves. By the 1920s it appeared to many that the Chinese realm would be completely dismembered. The Tibetans had gained autonomy, Xinjiang had become a virtual Russian **protectorate,** and in China proper Europeans and local warlords vied with the weakly established Chinese Republic for power. (In a "protectorate" the indigenous rulers remain in power but are placed under the larger authority of the colonial power.) Japan was also increasing its demands and seeking to expand its territorial base. To understand this part of the story, it is necessary to turn away from the turmoil of China to examine the simultaneous rise of Japan.

The Rise of Japan

Japan did not emerge as a unified state, possessing a literate culture, until the seventh century, more than 2,000 years later than China. From its earliest days, Japan looked to China (and, at first, to Korea as well) for intellectual and political models. Its offshore location insulated Japan from the threat of actual rule by the Chinese Empire, but the Japanese state did periodically acknowledge Chinese supremacy. At the same time, the Japanese people conceptualized their own region as a separate empire, equivalent in certain respects to that of China. Between 1000 and 1580, however, Japan had no real unity, as the country was divided into a number of mutually antagonistic feudal realms.

The Closing and Opening of Japan In the early 1600s, Japan was reunited by the armies of the Tokugawa **Shogunate** (a shogun is a military leader who theoretically remains under the emperor but who actually holds political power). At this time, Japan asserted its absolute autonomy from Chinese civilization and attempted to isolate itself from the rest of the world. Until the 1850s, Japan traded with China only in an indirect manner through the Ryukyu islanders (who paid tribute to both China and Japan) and with Russia only through Ainu intermediaries. The only foreigners allowed to trade in Japan were the Dutch, and their activities were strictly limited.

Japan remained largely closed to foreign commerce and influence until U.S. gunboats sailed into Tokyo Bay to demand access in 1853. Aware that China was losing power to the Europeans and realizing that they could no longer keep the Westerners out, Japanese leaders quickly set about modernizing their economic, administrative, and military systems. This effort accelerated when the Tokugawa Shogunate was toppled in 1868, an event known as the Meiji Restoration. (It is called a *restoration* because it was carried out in the emperor's name, but it did not give the emperor any real power.) Unlike China, Japan successfully carried out most of these reform efforts.

The Japanese Empire Japan's new rulers realized that their country remained threatened by European imperial power. They nurtured the development of a silk export industry,

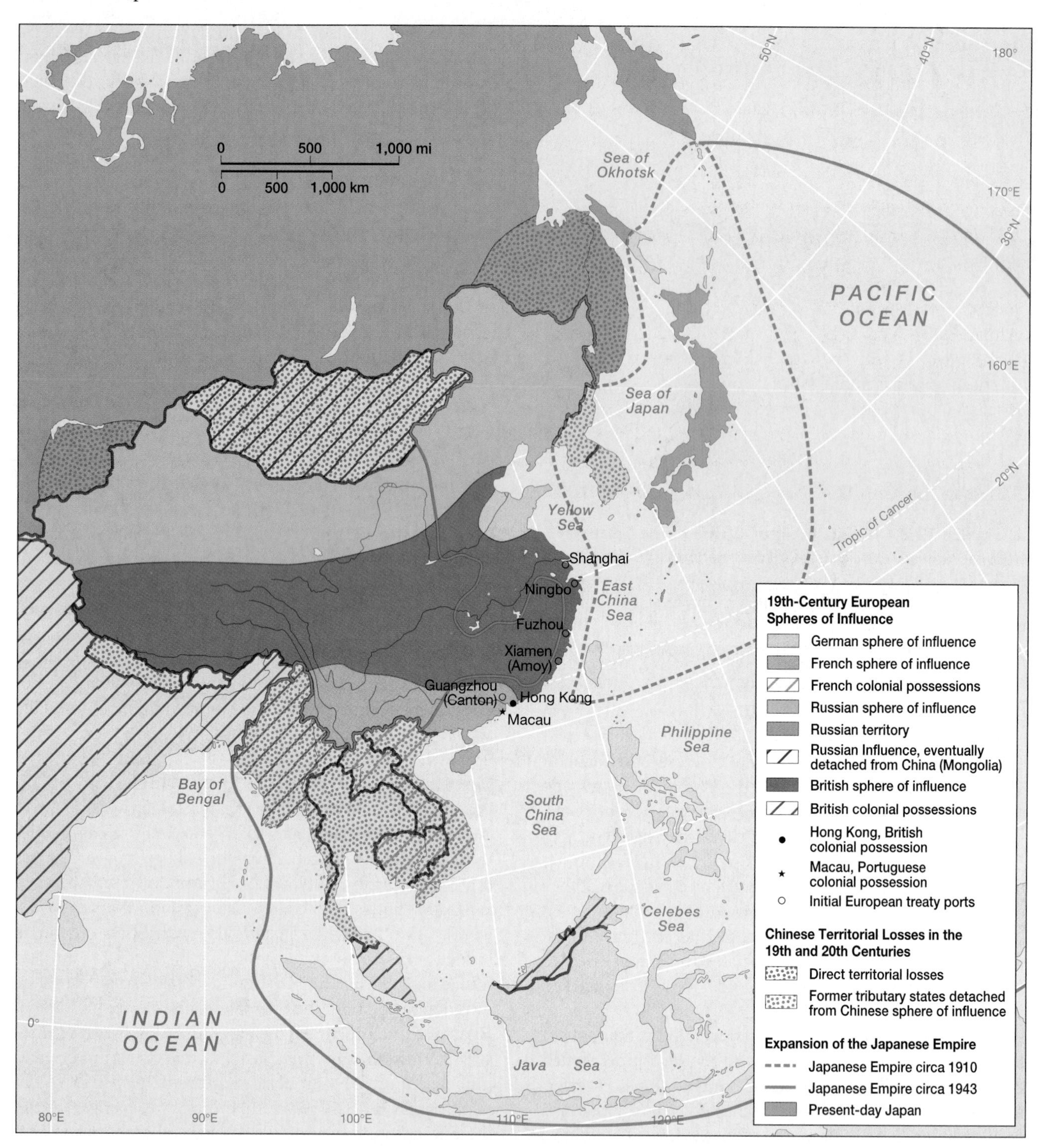

▲ **Figure 11.32 Nineteenth-century European spheres of influence** The Chinese lost influence and territory in the nineteenth century as European power expanded. Although China regained its autonomy and most of its territory in the 1900s, Russia retained large areas that were formerly under Chinese control. The first half of the twentieth century saw the rapid expansion of the Japanese Empire, which ended with the defeat of Japan in World War II.

which gave them the funds to buy modern equipment. Moreover, they decided that the only way to meet the European challenge was to become expansionistic themselves. Japan therefore quickly took control over the entire island of Hokkaido, and began to move farther north into the Kuril Islands and Sakhalin.

In 1895 the Japanese government tested its newly modernized army against China, winning a relatively quick (and profitable) victory that allowed Japan to take control of Taiwan. Tensions then mounted with Russia as the two countries vied for power in Manchuria and Korea. The Japanese defeated the Russians in 1905, giving them considerable influence in northern China. With no strong rival in the area, Japan annexed Korea in 1910. Alliance with Britain, France, and the United States during World War I brought further gains, as Japan was awarded Germany's former colonies in the north equatorial Pacific (Micronesia).

The 1930s brought a global depression, putting a resource-dependent Japan in a difficult situation. The country's leaders sought a military solution to their economic plight, and in 1931 Japan conquered resource-rich Manchuria. In 1937 Japanese armies moved south, occupying the North China Plain and the coastal cities of southern China. The Chinese government then withdrew to the relatively inaccessible Sichuan Basin to continue the struggle. During this period, Japan's relations with the United States deteriorated as the American government increasingly objected to Japanese imperialism. In response, the United States blocked the export of scrap iron and other resources vital to Japan.

In 1941 Japan's leaders decided to destroy the American Pacific fleet in order to clear the way for the conquest of Southeast Asia. Their grand strategy was to unite the East and Southeast Asian regions into an entity to be called "The Greater East Asia Co-Prosperity Sphere." This "sphere" was to be ruled by Japan, however, and was designed to keep the Americans and Europeans out of the area. Japanese forces sometimes engaged in brutal acts, earning their country the ill will of many other Asian peoples. In the infamous "Rape of Nanjing," Japanese troops systematically slaughtered up to several hundred thousand Chinese citizens. In Korea, which had been ruled by Japan for several decades, the colonial government actually planned to extinguish the Korean language in favor of Japanese.

Postwar Geopolitics

At the end of World War II, East Asia became a power vacuum, and hence an arena of rivalry between the United States and the Soviet Union. Initially, American interests prevailed in the maritime fringe while Soviet interests advanced on the mainland. Soon, however, East Asia began to experience its own revival.

Japan's Revival Japan lost its colonial empire when it lost the war. Its territory was reduced to the four main islands plus the Ryukyu Archipelago and a few minor outliers. In general, the Japanese government acquiesced to this loss of land. The only outstanding territorial conflict from the postwar settlement concerns the four southernmost islands of the Kuril chain, which were annexed by the Soviet Union in 1945. Although Japan still claims these islands, Russia refuses even to discuss relinquishing control, causing considerable strains in Russo-Japanese relations.

After losing its overseas possessions, Japan was forced to rely on trade to obtain the resources needed for its economy. Here it proved remarkably successful. Japan's military power, on the other hand, was limited by the constitution imposed on it by the United States. Because of this restriction of its own forces, Japan has relied to a large extent on the U.S. military for its defense needs. The U.S. Navy patrols many of its vital sea lanes, and U.S. armed forces maintain several sizable bases within Japan. This U.S. military presence in Japan, however, is becoming increasingly controversial. Many Japanese citizens believe that their country ought to provide its own defense, and they resent the presence of U.S. troops. Slowly but steadily, meanwhile, Japan's own military has emerged as a powerful regional force despite the constitutional limits imposed on it (see "Geopolitical Tensions: Military Bases in Okinawa").

The Division of Korea The end of World War II brought much greater changes to Korea than to Japan. As the end of the war approached, the Soviet Union and the United States agreed to divide the country; Soviet forces were to occupy the area north of the 38th parallel, whereas U.S. troops would occupy the area to the south. This soon resulted in the establishment of two separate regimes. In 1950 North Korea invaded the south, seeking to reunify the country. The United States, with support from the United Nations, supported the south, while China aided the north. The war ended in a stalemate, and Korea has remained a divided country—its two governments locked in a protracted cold war—ever since.

Large numbers of U.S. troops remained in the south after the war. South Korea in the 1960s was a poor agrarian country that was ill prepared to defend itself. Over the past 30 years, however, the south has emerged as a wealthy trading nation while the fortunes of the north have plummeted. Many South Korean students—a group famous for its radicalism—resent the presence of U.S. forces and seek rapprochement with the north. Their periodic and rather violent demonstrations have been a significant, and destabilizing, force in South Korean politics. The South Korean government, for its part, regards North Korea as a desperate rogue state that could attack at any time (Figure 11.33). Many U.S. policymakers concur, and some fear that North Korea has been on the verge of constructing nuclear arms. A near crisis was reached in 1994 when North Korea refused to allow international inspection of its nuclear power plants. Soon afterward, however, it relented in exchange for help in meeting its energy requirements.

The Division of China World War II brought tremendous economic destruction and loss of human life to China. Before the war began, China had already been engaged in a civil conflict between nationalists (who favored an authoritarian capitalist economic system) and communists, both of whom hoped to unify the country. The communists had originally been based in the middle Yangtze region, but in 1934 nationalists pressure forced them out. Under the leadership of Mao Zedong, they retreated in what came to be known as the "Long March," which took them to the Loess Plateau. This area is conveniently close to the traditional power center of northern China and to the more industrialized zones of Manchuria. After the Japanese invaded China proper in 1937, the two camps cooperated, but as soon as Japan was defeated, China again found itself embroiled in civil war. In 1949 the communists proved victorious, forcing the nationalists to retreat to Taiwan.

GEOPOLITICAL TENSIONS Military Bases in Okinawa

Okinawa, the largest and most important of Japan's Ryukyu Islands, was occupied by the United States from 1945 to 1972. Although Okinawa was returned to Japan in 1972, the United States retained control over several large military bases, concentrated in the northern half of the island. Some 20 percent of Okinawa's land area remains devoted to military purposes. The U.S. bases are also economically significant. In 1972, military spending accounted for almost half of the island's economic output. This figure had declined to less than 10 percent by 1997, but it still represented an important source of income for one of Japan's poorest areas.

Many Okinawans have long resented the presence of the U.S. military. Demonstrations and even riots occurred on numerous occasions before the island was returned to Japan. In 1962 Okinawa's legislature unanimously passed a resolution accusing the United States of colonialism; several months later President John F. Kennedy promised that the island would eventually revert to Japan.

Local opposition to the military bases began to mount again in the early 1990s. Several highly publicized crimes by U.S. servicemen, including one case of rape of a child, turned many Japanese against the U.S. military presence altogether. High levels of noise from military jets, as well as periodic aircraft accidents, further upset the Okinawans. In October 1995 mass demonstrations broke out in Okinawa, leading the island's governor to demand that all bases close by the year 2015. Okinawans later voted, by a substantial margin, to substantially reduce the area of land devoted to military operations.

While most Okinawans clearly want U.S. military activities—and personnel—to become less disruptive, they do not necessarily want the bases to be closed. Public polling, in fact, revealed that some three-quarters of Okinawa's residents support some form of U.S. military activity, largely for the economic advantages that it brings. The U.S. military presence in the far south of Japan thus remains secure for the moment, but its long-term future remains in doubt.

A latent state of war has persisted ever since between mainland China and Taiwan. Although no battles have been fought, gunfire has periodically been exchanged over Quemoy and Matsu, two small Taiwanese islands just offshore from the mainland. The Beijing government still claims Taiwan as an integral part of China and vows eventually to redeem it. The nationalists, for their part, maintain that they represent the true government of China, and they likewise vow (rather less realistically) to reclaim the mainland. It was actually made a crime in Taiwan to advocate Taiwanese independence, since the fiction had to be maintained that Taiwan itself is merely one province of a temporarily divided China. Taiwan is thus equipped with a provincial government that governs exactly the same territory that the national government does.

The idea of the intrinsic unity of China continues to be influential both in China and abroad. In the 1950s and 1960s, the United States recognized Taiwan as the only legitimate government of China, but its policy changed after U.S. leaders decided that it would be more useful—and more realistic—to recognize mainland China. Soon China entered the United Nations, and Taiwan found itself diplomatically isolated, virtually without international recognition. (A number of African and Caribbean countries, however, recognize Taiwan—unofficially—in exchange for Taiwanese aid.) In reality, however, Taiwan is very much a separate country, and increasing numbers of its citizens would like to proclaim it an independent republic. China, however, threatens to invade if Taiwan declares independence. Some observers think that China might indeed fight to reclaim Taiwan, pointing to its overwhelming military advantages; others view such a scenario as highly unlikely, pointing instead to the growing (un-

▲ **Figure 11.33 Demilitarized zone in Korea** North and South Korea were divided along the 38th parallel after World War II. Today, even after the conflict of the early 1950s, the DMZ separates these two different states. U.S. armed forces are active in patrolling the demilitarized zone. *(Yonhap/AP/Wide World Photos)*

official) economic links between the island and the mainland, and to the international complications that such an action would cause.

The Chinese Territorial Domain Despite the fact that it has been unable to regain Taiwan, China has been remarkably successful in retaining the Manchu territorial legacy. In the case of Tibet in particular, this has required considerable force; resistance by the Tibetans forced China to launch a full-scale invasion in 1959. The Tibetans, however, have continued to struggle for real autonomy if not actual independence, and they fear that the Han Chinese now moving to Tibet will eventually outnumber them and destroy their cultural foundations (Figure 11.34). Tibet proper (or *Xizang* in Chinese) is accorded the official status of an autonomous region by virtue of its non-Han indigenous society, but true autonomy has not been granted (see Chapter 10).

The postwar Chinese government also retained control over Xinjiang in the northwest, as well as Inner Mongolia (or Nei Monggol), a vast territory stretching along the Mongolian border. The indigenous inhabitants of both areas are not Han Chinese; they are, rather, Mongols in the latter case and Turkish-speaking Muslims in the former. The indigenous peoples of Xinjiang, as well as many Westerners, prefer to call the region "Eastern Turkestan" to emphasize its Turkish heritage; the Han Chinese, however, reject this term because it challenges the unity of China. Like Tibet, Nei Monggol and Xinjiang are classified as autonomous regions. The peoples of Xinjiang are increasingly asserting their religious and ethnic identities, and separatist sentiments are growing. Most Han Chinese, however, regard Nei Monggol and Xinjiang as integral parts of their country, and they regard any talk of succession as treasonous. In the case of Xinjiang, they cite the precedence of Chinese control dating back to the Han Dynasty some 2,000 years ago.

A few ardent Chinese nationalists dream of reclaiming Manchu territories that were taken by other powers. While any actual gains are unlikely, such claims complicate China's international relations. The most important potential conflict is with Russia, which controls a sizable territory north of Manchuria that was grabbed from China in the waning decades of the Manchu Dynasty. China also claims that some of its former territories in the Himalayas were illegally annexed by Britain when it controlled South Asia, resulting in several unresolved border disputes with India. The two countries went to war in 1962, when China successfully occupied a section of uninhabited highlands in far northeastern Kashmir. Tensions between India and China have eased in the 1990s, but their territorial disputes remain unresolved. Owing to this disagreement with India, China has maintained an informal military alliance with Pakistan.

China also claims a group of tiny islands—most of them submerged at high tide—in the South China Sea. The Paracel Islands, however, are also claimed by Taiwan and Vietnam, while the Spratly Islands are also claimed by Taiwan, Vietnam, Malaysia, and the Philippines. While the islands themselves have little importance, there is some evidence that they overlay considerable petroleum reserves. China has recently constructed buildings on some of these islands that it claims are fishing platforms but which the Philippines maintains are really military installations. The struggle over these small islands has made it difficult for China to enjoy cordial relations with the ASEAN countries of Southeast Asia (see Chapter 13).

▲ **Figure 11.34 Chinese soldiers in Tibet** Following a full-scale invasion of Tibet in 1959, China's presence continues to increase through its military forces, the relocation of migrants into the area from other parts of China, and rebuilding programs that mask the traditional Tibetan landscape. Here, Chinese soldiers observe a praying Tibetan while eating ice cream bars. *(Galen Rowell/Corbis)*

One additional territorial issue was finally resolved in 1997 when China reclaimed Hong Kong. In the isolationistic 1950s, 1960s, and 1970s, Hong Kong acted as China's window on the outside world, and it grew prosperous as a capitalist enclave and refuge for wealthy Chinese industrialists. As Chinese relations with the outer world opened in the 1980s, Britain decided to honor its earlier treaty provisions and return Hong Kong to China. China in turn promised that Hong Kong would retain its fully capitalist economic system for at least 50 years. The wealthy citizens of Hong Kong, however, are nervous about the transfer of power, and many have acquired residency rights in Canada and Australia as a fallback. The Hong Kong people also fear that their political freedoms—acquired only in the final years of British rule—will be curtailed. Some observers, however, predict that Hong Kong will act as a kind of "Trojan Horse," spreading liberal and capitalistic values throughout China. It is much too early to tell what the transfer of power will ultimately entail (Figure 11.35).

In 1999 Macao, the last colonial territory in East Asia, was returned to China. This small Portuguese enclave, located immediately across the estuary from Hong Kong, has functioned

▲ **Figure 11.35 China reclaims Hong Kong** Fireworks celebrate the return of Hong Kong to China in 1997 after a long period under British colonial rule. Britain agreed to honor its treaty provisions and return Hong Kong to China under the promise that Hong Kong's capitalistic system would remain for at least 50 years. *(D. Groshong/Sygma Photo News)*

partly as a gambling refuge. It remains to be seen whether the unusual function of this incongruous little piece of territory will survive under the new regime.

The Global Dimension of East Asian Geopolitics

In the early 1950s East Asia was divided into two hostile Cold War camps: China and North Korea were allied with the Soviet Union, while Japan, Taiwan, and South Korea were linked to the United States. The Chinese–Soviet alliance soon deteriorated into mutual hostility, however, and in the 1970s China and the United States found that they could accommodate each other, sharing as they did an enemy in the Soviet Union.

The end of the Cold War, coupled with the rapid economic growth of China, again reconfigured the balance of power in East Asia. The United States no longer needs China to offset the Soviet Union, and the U.S. military is now increasingly worried about the growing power of the Chinese army. China's neighbors are also concerned. China has the largest army in the world, nuclear capability, and sophisticated missile technology. It also has, by some measures, the world's third-largest economy.

China is coming of age as a major force in global politics. Whether it is a force to be feared by other countries is a matter of considerable debate. Chinese leaders insist that they have no expansionistic designs and no intention of interfering in the internal affairs of other countries. They do, however, regard concern by the United States and other countries with their own human rights record, as well as their activities in Tibet, as undue meddling in their own internal affairs. Many U.S. commentators agree, arguing that the best course would be for China and the United States to ignore their political differences and instead develop closer economic and cultural ties. In fact, in the mid- and late 1990s, the U.S. government transferred advanced missile technology to China, ostensibly to be used for peaceful purposes.

A number of influential international security experts see China in a different light. Noted political scientist Samuel Huntington, for example, views China as the core of the Confucian powerblock that will necessarily compete with the West for global influence. Huntington further fears the emergence of a potential Confucian–Islamic alliance, based largely on the Chinese–Pakistani military connection.

Huntington's critics see more paranoia than clairvoyance in his vision of Confucian unity. They point rather to the deep internal rifts existing within the Confucian world. Japan (which Huntington admittedly sees as more Western than Confucian) is nervous about China and has a long history of animosity toward Korea. Korea is itself bitterly divided, and South Korea is also wary of Chinese power. China itself remains split, and although Taiwan is small and diplomatically isolated, its economic and even its military power is not insignificant. Even mainland China itself is not as tightly unified as it might appear to be at first glance. As a final complicating factor, some observers claim that China itself welcomes the U.S. military presence in East Asia, seeing it as a counterweight to the potential remilitarization of Japan as well as to the possible resurgence of Russia.

Regardless of what one thinks about China's rise to political power, its economic ascent is undeniable. Yet for all of its remarkably rapid growth, China remains, as we shall see in the next section, a rather poor country.

Economic and Social Development: An Emerging Core of the Global Economy

East Asia exhibits extreme disparities of economic and social development. Japan's major urban belt contains one of the world's greatest concentrations of wealth, whereas many interior districts of China remain among the world's poorest places. Overall, however, East Asia experienced rapid rates of economic growth in the 1970s, 1980s, and 1990s, and two of its economies, those of Taiwan and South Korea, jumped from the ranks of underdeveloped to developed. Increasingly, East Asia functions as a global economic core and trade center (Figure 11.36). But again, growth has not been evenly distributed. North Korea, for example, has experienced rather a desperate decline during this same period. Moreover, in 1997 and 1998 the economies of Japan and especially South Korea entered a period of severe recession.

In regard to social development, the picture is more positive. Even in many of the poorer parts of China, most people are reasonably healthy and well educated. As China moves to a market economy, however, such "modern" problems as unemployment and homelessness are beginning to appear.

Japan's Economy and Society

Japan was the pacesetter of the world economy in the 1960s, 1970s, and 1980s. In the early 1990s, however, the Japanese economy experienced a major setback, and growth has been

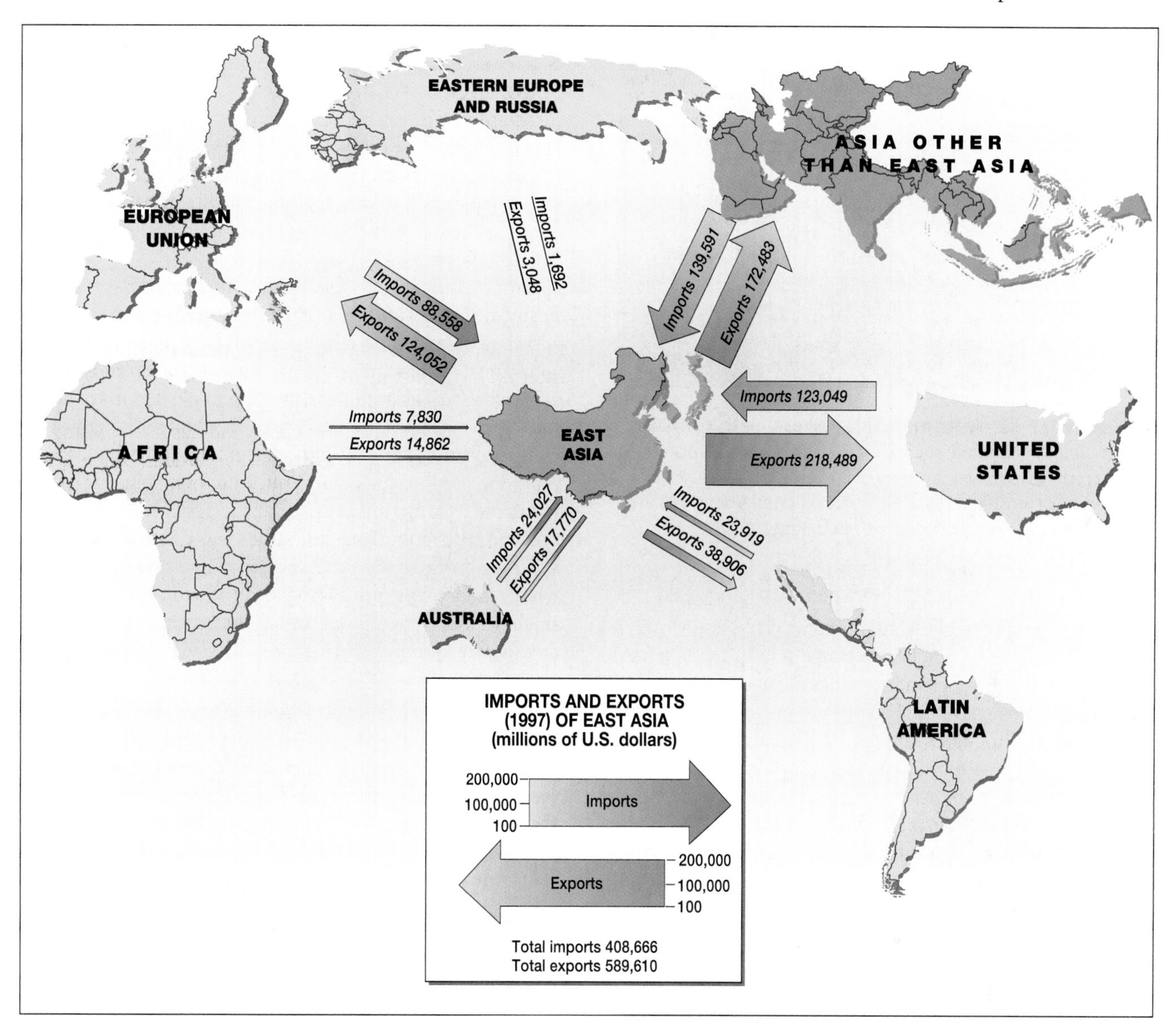

▲ **Figure 11.36 East Asian trade** Japan, South Korea, China, and Taiwan are among the world leaders in international trade. The United States is the most important export market for the region, with all four countries maintaining positive trade balances with the United States. Other important trading partners include the European Union and countries of Southeast Asia. East Asian trade is more balanced with other Asian countries than it is with either Europe or the United States. *(Data from Euromonitor,* International Marketing Data and Statistics, 1998*)*

slow ever since. Some analysts foresee an even more severe economic crisis in the years to come. But despite its current predicament, Japan is still the second largest economic power in the world.

Japan's Boom and Bust Although Japan's heavy industrialization began in the late 1800s, most of its people remained relatively poor. The 1950s, however, saw the beginnings of the Japanese "economic miracle." Shorn of its empire, Japan was forced to export manufactured materials. Beginning with inexpensive consumer goods such as clothing and toys, Japanese industry quickly moved to more sophisticated products such as automobiles, cameras, electronics, machine tools, and computer equipment. By the 1980s it was the unquestioned world leader in many segments of the global high-tech economy, and during this decade, its currency, the yen, was one of the world's strongest (Figure 11.37).

The early 1990s saw a collapse of Japan's hyper-inflated real estate market, bringing in turn a banking crisis. With high rates of savings and interest rates of virtually zero, Japan's financial institutions have remained inefficient and, according to critics, profligate in their dispensing of loans. At the same time, many Japanese companies discovered that producing labor-intensive goods in Japan had become too expensive. They therefore began to relocate factories to Southeast Asia and China. Because of these and related difficulties, Japan's economy slumped through the early and middle 1990s. In the mid- and late 1990s, the Japanese government made several attempts to revitalize the

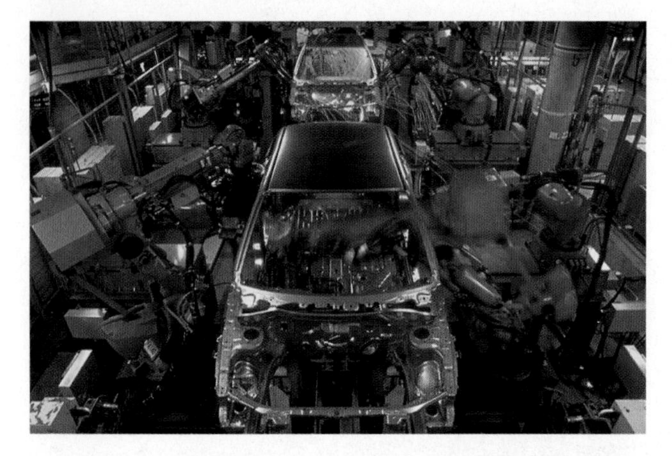

▲ **Figure 11.37 Automated Japanese auto factory** Part of Japan's economic success has resulted from automation of its factory assembly lines. Here, Mazda automobiles are assembled in a Hiroshima plant on Honshu island. These cars are destined for the east coast of the United States. *(Jodi Cobbings/ NGS Image Collection)*

economy through massive state spending. One result of this policy has been the accumulation of large deficits. In 1998 a serious recession began, coinciding with a sharp drop in value of the Japanese yen against the U.S. dollar.

Economic difficulties are particularly pronounced in Hokkaido, which is more dependent on resources than are the other main Japanese islands. The Osaka region, noted for its small and mid-sized manufacturing concerns, has also suffered. On the other hand, the Nagoya area—the center of Japan's internationally successful automobile manufacturing industry—has not experienced the same degree of economic stress as the rest of the country.

Despite its downturn in the 1990s, Japan remains a core country of the global economic system. Its economic system increasingly spans the globe, as Japanese multinational firms continue to invest heavily in production facilities in North America (both in the United States and in Mexico) and Europe, as well as in Third World countries. Japan remains a world leader in a large array of high-tech fields, including robotics, optics, and the manufacturing of machinery and tools for the semiconductor industry. It is also the world's largest creditor nation, owning a large percentage of U.S. government bonds.

The Japanese Economic System The business environments in Japan and the United States are in certain respects quite similar, but in others are very different. Underlying their discrepancies are two distinct versions of the capitalist economic system. In Japan the bureaucracy maintains far greater control over the economy than it does in the United States. In particular, the Ministry of International Trade and Industry (MITI) has been viewed by some scholars as the main engine behind Japan's industrial expansion in the 1950s, 1960s, 1970s, and 1980s; others, however, argue that MITI has sometimes been more of an obstacle to than a initiator of business success.

Japanese corporations are also structured differently from those of the United States. In Japan large groups of companies (called *keiretsu*) are complexly intertwined, owning each other's stock and preferentially buying products and services from each other. Because of these interconnections, Japanese firms are much less influenced by investors and stock brokers than are those of the United States. The connections between employers and employees are also much tighter in Japan. Japanese workers seldom switch companies, and the core workforce of each corporation is virtually never subjected to layoffs.

Proponents of the Japanese system argue that it creates business stability and encourages long-term planning. It also allows Japanese companies to retain their employees during recessions, since managers have little to fear from investors worried about declining profits. Opponents of the system, to the contrary, argue that it reduces flexibility, results in high consumer prices and low corporate profits, and will ultimately prove too costly to maintain. Critics also point out that despite its manufacturing prowess, Japanese agriculture, wholesaling, and distribution are still rather inefficient. Japan is only now beginning to see the rise of large-scale discount stores, which offer lower prices but also threaten the viability of small-scale merchants—as well as the continued existence of downtown shopping districts in towns and smaller cities.

Living Standards and Social Conditions in Japan Despite its difficulties in the early and mid-1990s, Japan still has a higher per capita **gross national product (GNP)** than the United States when calculated on the basis of currency equivalents (Table 11.2). Living standards are, however, somewhat lower in Japan, and America's per capita **gross domestic product (GDP)** remains larger when calculated on the basis of purchasing power parity. Housing, food, transportation, and services are particularly expensive. Certain amenities that are almost standard in the United States, such as central heating, remain relatively rare in Japan.

Although the Japanese may live in cramped quarters and pay high prices for basic products, they also enjoy many benefits unknown in the United States. Unemployment, depending on how it is measured, remains lower than in the United States; health care is universally provided by the government; and crime rates are extremely low. By such measures of social development as literacy, infant mortality, and average longevity, Japan surpasses the United States by a comfortable margin (Table 11.3). Probably the most important factor is that Japan lacks the extreme poverty found in certain pockets of American society. The disparities of wealth between the "haves" and the "have nots," while still substantial, are not nearly as great as in the United States or even in many European countries. Furthermore, while wealth is concentrated in Tokyo and a few other large cities, regional income disparities are rather minor.

Japan, of course, has its own share of social problems. Koreans and alien residents from other Asian countries suffer discrimination, as does the indigenous Japanese underclass, the Burakumin. Japan's more remote rural areas have few jobs, and many have seen prolonged population decreases. In small villages it often seems that most of the remaining people are

Table 11.2 Economic Indicators

Country	GNP per Capita ($U.S., 1996)	Total GNP (Millions of $U.S., 1996)	PPP* ($Intl, 1996)	Real Annual Growth % per Capita, 1990–1996
Japan	40,940	5,149,185	23,420	1.2
China	750	906,079	3,330	11.0
Taiwan	—	—	14,700**	5.0**
Hong Kong	24,290	153,288	24,260	3.7
South Korea	10,610	483,130	13,080	6.2
North Korea	—	—	900**	–4.5%**

*Purchasing power parity.

Source: *World Bank Atlas,* 1998, pp. 42–43, except those marked **, which notes data from the *CIA Factbook,* 1996.

elderly. Farming itself is an increasingly marginal occupation, and many farm families survive only because one family member works in a factory or office. Professional and managerial occupations in Japan's wealthy cities are noted for their long hours and high levels of stress. Overall, social regimentation is greater and civil liberties fewer in Japan than in the United States.

Women in Japanese Society Critics often contend, moreover, that Japanese women have not shared the benefits of their country's success. Advanced career opportunities remain limited for women, especially if they marry and have children. The expectation remains that mothers should devote themselves to their families and to their children's education. In fact, it is not uncommon for mothers to attend class when their children are ill in order to take notes. (The Japanese educational system is hierarchically organized, and poor performance in secondary school almost precludes career success.) Japanese businessmen often work, or socialize with their co-workers, until late every evening, and thus contribute little to child care. It has also been argued that one of the main outcomes of Japan's recession of the 1990s was a further reduction in career opportunities for women.

One response to the arduous conditions faced by Japanese women is a drop in the marriage rate. Many young Japanese women are delaying marriage, and a sizable number may be abandoning it altogether. Japan has seen an even more dramatic drop in its fertility rate. Whether this is due to the domestic difficulties faced by Japanese women or is merely the result of the pressures of a post-industrial society is an open question. Fertility rates have, after all, dropped even more dramatically in many parts of Europe.

But regardless of the cause, Japanese women are now bearing so few children that the country's population will soon begin to decline if the fertility rate remains constant. Economic planners are concerned about the increasing dependency burden that this will cause. A shrinking population means an aging population, and increasing numbers of retirees will have to be supported by shrinking numbers of workers. At present, most Japanese men retire at the relatively young age of 55 or 60, but many observers doubt that this practice can continue much longer.

Table 11.3 Social Indicators and Status of Women

Country	Life Expectancy at Birth		Under Age 5 Mortality, per 1,000 Live Births		Secondary School Enrollment %		Female Labor Force Participation (% of total)
	Male	Female	1960	1995	Male	Female	
Japan	77	84	40	6	95	97	41
China	69	73	209	47	60	51	45
Taiwan	72	78	—	—	94	98	—
Hong Kong	76	82	—	—	69	73	37
South Korea	70	77	124	9	97	96	41
North Korea	63	69	120	30	—	—	45

Sources: *Population Reference Bureau Data Sheet, 1998,* Life Expectancy (M/F); *World Resources: A Guide to the Global Environment, 1996–97,* pp. 194–195, Under-5 Mortality Rate; *Population Reference Bureau Data Sheet, 1996,* Secondary School Enrollment (M/F); *The World Bank Atlas, 1998,* pp. 8–9, Female Participation in Labor Force.

The Newly Industrialized Countries

The Japanese path to development has been generally, and successfully, followed by its former colonies: South Korea and Taiwan. Hong Kong also emerged as a newly industrialized economy in the same period, although its economic and political systems have remained quite distinctive.

The Rise of South Korea The recent success of South Korea has been even more remarkable than that of Japan. During the period of Japanese occupation, Korean industrial development was concentrated in the north, which has substantial deposits of coal and other natural resources. The south, in contrast, remained a densely populated, poor, agrarian region. South Korea emerged from the bloody Korean War as one of the world's least-developed countries.

In the 1960s the South Korean government initiated a program of export-led economic growth. In doing so, it guided the economy with a heavy hand and it denied basic political freedom to the Korean peoples. By the 1970s it was clear that such policies were highly successful, at least in the economic realm. Huge Korean industrial conglomerates, known as *chaebol,* moved from exporting inexpensive consumer goods to heavy industrial products to high-tech equipment.

At first, South Korean firms remained dependent on the United States and Japan for basic technology. By the 1990s, however, this was no longer the case. In fact, since the 1980s, South Korea has emerged as one of the world's main producers of semiconductors. South Korean wages have also risen at a rapid clip. The country has invested heavily in education (by some measures it has the world's most intensive educational system), which has served it well in the global high-tech economy. Increasingly, South Korean companies are themselves becoming **multinational**, building new factories in the low-wage countries of Southeast Asia and Latin America, as well as in the United States and Europe (a multinational firm operates and manufactures in more than one country).

▲ **Figure 11.38 Protests in South Korea** During the economic crisis of 1997, workers and students joined together to demonstrate against government financial and trade policies that had—in their minds, at least—led to the sudden downturn in the country's economic state. This protest rally is in the capital city, Seoul. *(Yun Jai-hyoung/AP/Wide World Photos)*

Contemporary South Korean Society The political and social development of South Korea has not been nearly as smooth as its economic progress. Throughout the 1960s and 1970s, student-led protests against the dictatorial government were brutally repressed (Figure 11.38). Dissension was particularly acute in the country's southwest region, an area that has suffered some discrimination. As the South Korean middle class expanded and prospered, pressure for democratization grew, and by the late 1980s it could no longer be denied. But even though free elections are now held and basic freedoms of expression allowed, political tension has not disappeared. In the mid-1990s several major scandals erupted, revealing substantial corruption at high levels of the government and business. In other words, the South Korean transition from an underdeveloped country ruled by a dictator to a prosperous democracy has been both rapid and troubled.

The South Korean political crisis of the late 1990s was accompanied by economic troubles. By 1997 the country's banking system was in chaos, and its economy entered a recession. As of September 1998, South Korea's economy was shrinking at an annual rate of 6.6 percent. Critics contend that the South Korean economy, like that of Japan, is too bureaucratic and needs substantial reforms, in particular the breaking up of the large industrial conglomerates *(chaebol)*. Others regard the South Korean economy as fundamentally sound, viewing its current slump merely as a temporary setback.

With recession looming, the South Korean people opted for a major change in 1998 when they elected Kim Dae Jung president. Kim not only had been a political prisoner during the period of authoritarian government, but he also represents the discriminated-against southwestern portion of the country. It is too early to tell what the Kim presidency will bring, but he has promised both economic and social reforms, as well as improved relations with the north.

Taiwan and Hong Kong Taiwan and Hong Kong have also experienced rapid economic growth since the 1960s. Both, in fact, have substantially higher per capita GDP levels than South Korea. The Taiwanese government, like that of South Korea and Japan, has guided the economic development of the country. Taiwan's economy, however, is organized not around large conglomerates and linked business firms, but is rather founded on small to mid-sized family firms. This characteristically Chinese form of business organization is sometimes said to give Taiwan greater economic agility and flexibility than its northern neighbors, but it has prevented it from entering certain industries that require huge concentrations of capital.

Hong Kong, unlike its neighbors, has been characterized by one of the most ***laissez-faire*** economic systems in the world (*laissez-faire* refers to complete market freedom, without governmental control). State involvement here has been minimal, which is one reason why the city's business elite are so nervous about the transition to Chinese rule. Hong Kong traditionally functioned mainly as a trading center, but in the 1960s and 1970s it emerged as a major producer of textiles, toys, and other low-tech consumer goods. By the 1980s, however, such cheap products could no longer be made in such an expensive city. Hong Kong industrialists subsequently began to move their plants to smaller towns and cities in southern China, while Hong Kong itself has increasingly specialized in business services, banking, telecommunications, and entertainment.

Both Taiwan and Hong Kong have close overseas economic connections. Linkages are particularly tight with Chinese-owned firms located in Southeast Asia and in North America. Taiwan's high technology businesses are also closely intertwined with those of the United States; there is a constant back-and-forth flow of talent, technology, and money between Taipei and Silicon Valley. Hong Kong's economy is also closely bound with that of the United States (as well as those of Canada and Britain), but its closest connections are, not surprisingly, with the rest of China.

Chinese Development

China dwarfs all of the rest of East Asia in both size and population. Its economic takeoff is thus reconfiguring the economy of the entire region. But despite its recent growth, China's economy has a number of serious weaknesses. The vast interior remains trapped in poverty, and many of its largest industries are not competitive. The future of the Chinese economy is thus one of the biggest uncertainties facing both East Asia and the world economy as a whole.

China Under Communism More than a century of war, invasion, and near-chaos in China finally ended in 1949 when the communist forces seized power. The new government, inheriting a weak economy, quickly set about nationalizing private firms and building heavy industries. Certain successes were realized, especially in Manchuria where large amounts of heavy industrial equipment were inherited from the Japanese.

In the late 1950s, however, China experienced an economic disaster ironically called the "Great Leap Forward." One of the main ideas behind this scheme was that small-scale village workshops could produce the large quantities of iron seen as necessary for sustained industrial growth. Communist party officials demanded that these inefficient workshops meet unreasonably high production quotas. In some cases, the only way they could reach their targets was to requisition peasants' agricultural tools and melt them down. Peasants were also forced to contribute such a large percentage of their crops to the state that they were often deprived of adequate food themselves. The result was a horrific famine that may have killed 20 million persons.

The early 1960s saw a return to more pragmatic policies, but toward the end of the decade a new wave of radicalism swept through China. This was the so-called "Cultural Revolution," which aimed at mobilizing young people to stamp out the country's remaining vestiges of capitalism. Experienced industrial managers, as well as scholars, accused of harboring capitalist views were expelled from their positions. Many were subsequently sent to villages to be "re-educated" through hard physical labor; others were simply killed. The economic consequences of such policies once again proved devastating.

Toward a Post-communist Economy When China's leader Mao Zedong, who had been revered as an almost superhuman being, died in 1976, the country faced a crucial turning point. Its economy was nearly stagnant and its people desperately poor, while the economies of most of its East Asian neighbors were booming. A political struggle ensued between pragmatists hoping for change and dedicated communists wishing to remain true to Mao's utopian vision. The former party emerged victorious, and by the late 1970s it was clear that China was going to embark on a different economic path. The new China would seek closer connections with the world economy and take an essentially capitalist road to development (Figure 11.39; see also "Economic Growth: Housing in China").

It would be wrong to suppose, however, that China transformed itself into a fully capitalist country. The Chinese state has continued to run most heavy industries, and the Communist Party retains a monopoly on political power. Instead

▲ **Figure 11.39 Industrial expansion in coastal China** One of the important economic reforms that has led to China's recent development was the creation of Special Economic Zones (SEZs) along its eastern coast. Here, workers in a coastal automobile plant assemble cars for the rapidly expanding domestic market. *(Serge Attal/Liaison Agency, Inc.)*

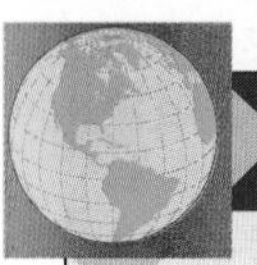

ECONOMIC GROWTH Housing in China

Since the imposition of communist rule in 1949, the residents of Chinese cities have been given a housing trade-off. The state offered them universal housing virtually free of charge, but it also used its housing monopoly to enforce social control. Those in charge of apartment buildings, for example, were expected to report all foreign visitors and suspicious activities to the local police. Standards, moreover, remained minimal. In 1990, the average person in Shanghai enjoyed only some 50 square feet (4.6 square meters) of living space, and as of 1998 approximately half of the households in the city—one of China's most prosperous—still used communal kitchens and bathrooms.

As China gradually moved from a socialistic to a capitalistic system, pressure began to mount on its housing system. In 1998 the government announced a major liberalization program; rents would rise, and housing stock would begin to revert to private control. Similar reform efforts had actually been implemented earlier in cities such as Shanghai, which are not only the pacesetters in the transition to market economics but which have also been burdened by high levels of immigration and crowding. By 1998 almost half of Shanghai's housing units were under private ownership, and a lively real estate market had emerged. Many if not most of Shanghai's residents now have access to improved housing stock.

The privatization of housing in China will present tremendous opportunities, but it will also come at a cost. The poorest people may be forced out of the market altogether, leading to the development of shantytown slums. More technical problems may emerge as well. China's banks, which are already burdened by a high level of bad loans, have no experience in mortgages. Some banks may well prosper in this new market, but others could easily fail.

of suddenly abandoning the communist model, as the former Soviet Union did, China allowed cracks to appear in the fabric of socialism in which capitalist ventures could take root and thrive.

One of China's first capitalist openings, in the late 1970s, was in agriculture, which had previously been dominated by large-scale communal farms. Individuals were suddenly allowed to act as agricultural entrepreneurs, selling foodstuffs for profit in the open market. Owing to this change, the incomes of many farmers rose dramatically. By the late 1980s, however, the focus of growth had shifted to the urban/industrial sector, and the government became increasingly concerned about inflation. After officials reinforced price lids on a variety of agricultural products, the growth of the rural economy began to slow down.

Industrial Reform One important early industrial reform involved opening **Special Economic Zones (SEZs)** where foreign investment was welcome and state interference minimal. The Shenzhen SEZ, adjacent to Hong Kong, proved particularly successful after Hong Kong manufacturers found it to be a convenient source of cheap land and labor. Additional SEZs were soon opened, most of them in the coastal region. The basic strategy was to attract foreign investment that could generate exports, the proceeds of which could supply China with the capital that it needed to build its basic infrastructure and thus achieve sustained economic growth.

Other capitalistic reforms followed. Former agricultural cooperatives and other rural concerns, for example, were increasingly allowed to transform themselves into quasi-capitalist entities. Many of these "township and village enterprises" also proved highly successful. By the early 1990s, the Chinese economy was growing at some 8 to 15 percent a year, perhaps the fastest rate of expansion Earth has ever seen. China emerged as one of the world's major trading nations, and by the mid-1990s it had amassed large trade surpluses, especially with the United States. Seeking to strengthen its connections with the global economic system, China made repeated bids to join the World Trade Organization (WTO), a body designed to facilitate free trade and provide ground rules for international economic exchange. Up to 1998, however, all such efforts had failed, largely because of opposition from the United States. China is, however, a WTO "observer country."

As of 1998, the economic crisis of the late 1990s that proved so debilitating in Japan and South Korea had had less effect on China. Its annual economic growth rate slowed to 6.8 percent, a relatively low figure for China, but high by global standards (Hong Kong, on the other hand, did enter a recession in 1998). Some evidence indicates, moreover, that China's banking system has equally serious problems, making the country vulnerable to recession. Critics contend that China must demolish its vestiges of centralized economic planning if its economic expansion is to continue. China's leadership, however, has made it clear that economic reform will be a gradual process.

Social and Regional Differentiation The Chinese economic surge unleashed by the reforms of the late 1970s and 1980s resulted in growing **social and regional differentiation.** In other words, certain groups of people—and certain portions of the country—prospered, while others faltered. Despite its official socialism, the Chinese state actually encouraged the formation of an economic elite, having concluded that only wealthy individuals can adequately transform the economy. The least-fortunate Chinese citizens were sometimes left without work, and many millions migrated from rural villages of the interior to seek employment in the booming coastal cities and towns. Through the middle 1990s, the government attempted to control the transfer of population, but with only partial success. Shantytowns, as well as homeless populations, began to emerge around some of China's major cities.

China, like all other countries, has always been divided into relatively rich and relatively poor areas. According to noted China scholar G. William Skinner, the country has long been arranged into a series of discrete economic regions, the relative fortunes of which have varied greatly over the centuries. Before the communist period, the Yangtze Delta was probably the most prosperous part of China, while the flood-prone area immediately to its north (in northern Jiangsu province) was one of the poorest. The communist government attempted to equalize the fortunes of its different regions, and it sometimes bestowed special privileges on individuals from poor places, such as northern Jiangsu. Such efforts were not wholly successful, and some provinces remained much poorer than others. Since the coming of market reforms, moreover, the process of regional economic differentiation has accelerated.

The Booming Coastal Region Most of the benefits from China's economic transformation have flowed to the coastal region and to the capital city of Beijing. At first the main beneficiaries were the southern provinces of Guangdong and Fujian. This region was to some extent predisposed to the new economy, since the southern Chinese have long been noted for their mercantile orientation. Guangdong and Fujian have also benefited from their close connections with the overseas Chinese communities of Southeast Asia and North America. (The vast majority of overseas Chinese emigrants originated from these provinces). Proximity to Taiwan and especially Hong Kong also proved beneficial. Huge amounts of capital have flowed to the south coastal region since the 1980s from foreign (and Hong Kong-based) Chinese business networks. American, Japanese, and European firms have also invested heavily in the region.

As can be seen in Figure 11.40, by the early 1990s the Yangtze Delta area, centered around the city of Shanghai, reemerged as the most dynamic region of China. The delta

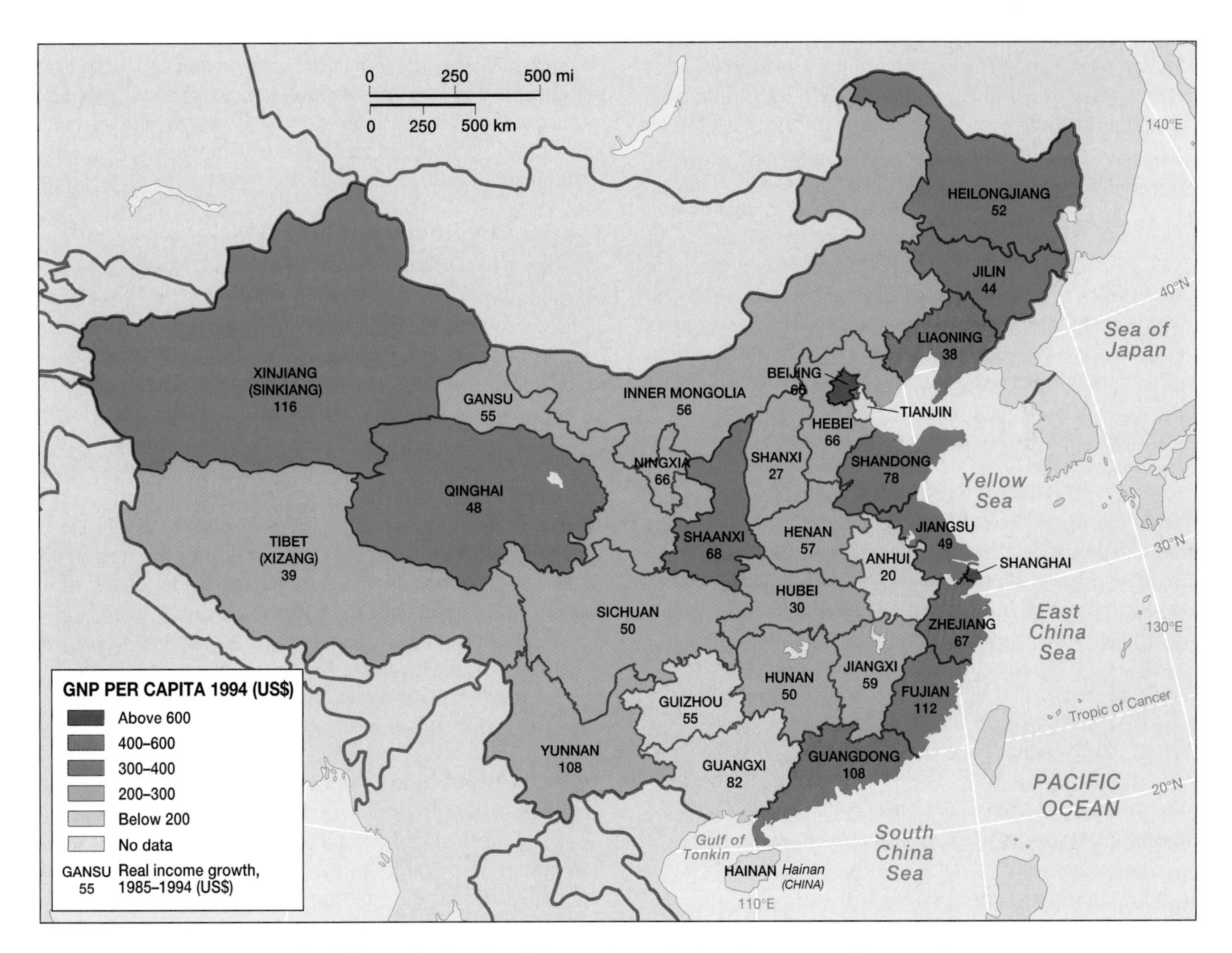

▲ **Figure 11.40 Economic differentiation in China** Although China has seen rapid economic expansion in recent years, the benefits of growth have not been evenly distributed throughout the country. Manchuria, China's industrial center, remains relatively prosperous, but recent development has been concentrated in the coastal zone and in Beijing. Much of the interior remains mired in poverty. One of the poorest parts of China is the upland region of Guizhou and Gaungxi in the south-central part of the country.

region was the traditional economic (and intellectual) core of China, and before the Communist takeover Shanghai had been the country's premier industrial and financial center. In this region much of the growth has been fueled by the township and village enterprises. The Chinese government, moreover, has encouraged the development of huge industrial, commercial, and residential complexes in the Shanghai area, hoping to take advantage of the region's dynamism. The Suzhou Industrial Park in nearby Jiangsu, if present plans are realized, will emerge as a hyper-modern city of 600,000 persons, thanks largely to a $20 billion investment, most of it Singaporean. Shanghai's own Pudong industrial development zone has attracted $10 billion, some of which will go the construction of a new airport and subway system.

The Beijing–Tianjin region has also taken a large part in China's economic boom. Its main advantage is its proximity to the seat of political power and its position as the "gateway" to northern China.

Interior and Northern China Most of the other parts of China, in contrast, have seen relatively little economic expansion. Manchuria remains more prosperous than most other parts of the country, owing to its fertile soils and early industrialization, but it has not participated much in the recent boom. The state-owned heavy industries of the Manchurian "**rust belt**," or zone of decaying factories, remain relatively inefficient, and little private enterprise has emerged.

Most of the interior provinces of China, which were relatively little-developed to begin with, have likewise missed out on the wave of growth that struck China in the 1980s and 1990s. In many areas, rural populations continue to grow while the natural environment continues to deteriorate. One consequence is high levels of underemployment and out-migration. Most interior provinces (especially Sichuan) experienced substantial out-migration from 1985 to 1990, whereas most coastal areas absorbed large numbers of migrants. By most measures, southwestern China (Guizhou, Guangxi, and Yunnan) is the poorest part of the country, followed by the Loess Plateau and other parts of interior northern China. In 1992 per capita GDP in Guizhou stood at 780 yuan (China's currency), whereas in Shanghai it had reached 5,570 yuan.

Scholars debate how bad conditions really are in the poorer parts of interior China. Some believe that China's official statistics are too positive, hiding a significant amount of hunger and destitution. Others think that the economic boom of the coastal zone is helping raise living standards even in the poorest districts.

Rising Tensions China's explosive but uneven economic growth has generated a number of problems. Inflation ran as high as 30 percent per year in the late 1980s and early 1990s, making planning difficult and creating hardship for those on fixed incomes. Corruption by state officials is by some accounts rampant; success often seems to depend on knowing the right people and having the proper connections. China's crime rate, moreover, which was extremely low in the 1950s and 1960s, has been rapidly mounting.

A more momentous issue, however, has been the struggle for free expression and democracy. The desire for democratic reform has been enhanced by rising incomes and the development of a sizable middle class. In 1989, however, the state crushed a movement for government accountability and democracy, and forced all opposition to go underground. Whether most Chinese people really want democracy is a controversial issue: Some scholars argue that a Confucian heritage predisposes China toward authoritarian government; others regard such a notion as little more than an apology for tyranny. If the latter camp is correct, tensions will probably mount as long as China's economy prospers while its ruling class retains a firm grip on power, denying basic freedoms.

China's political and human-rights policies have complicated its international relations. Sources of tension with the United States and other wealthy countries are also economic in nature. China's large and growing trade surplus and its supposed reluctance to enforce copyright and patent law irritate many of its trading partners, particularly the United States. Several U.S. firms have accused Chinese concerns of pirating music, software, and brand names. China has made efforts to stop such activities, but critics contend that its actions have been minimal. As the global market grows and as popular culture is itself increasingly globalized, copyright and trademark infringement can make a very lucrative, and thus difficult to control, illegal business.

Social Conditions in China

Despite its pockets of persistent poverty, China has achieved significant progress on a number of a social fronts. Since coming to power in 1949, the communist government has made large investments in medical care and education, and today China boasts impressive health and longevity figures. The illiteracy rate is still fairly high, but since 97 percent of children attend elementary school, it will drop substantially in the coming years (Table 11.3).

Human well-being in China is also geographically structured. The literacy rate, for example, remains relatively low in many of the poorer parts of China, including the uplands of Yunnan and Guizhou and some interior portions of the North China Plain, and relatively high in Manchuria, the Yangtze Delta, and most major urban areas. Such regional disparities may increase as China experiences heightened economic differentiation (in other words, as the gap between the wealthy and the poor grows larger).

China's Population Quandary Population policy also remains an unsettling issue for China. With more than 1.2 billion persons highly concentrated in less than half of its territory, China has one of the world's highest effective population densities. By the 1980s the government had become so concerned that it instituted the famous "one-child policy." Under this plan, couples in normal circumstances are expected to have only a single offspring and can suffer financial and other penalties if they do not comply (Figure 11.41). This strategy has been somewhat successful; the average fertility level is now only 1.8, and the population is

▲ **Figure 11.41 China's population policies** One aspect of China's population policy is the expansion of child-care facilities so that mothers can be near their child while at work. This way, females continue to participate in the working force soon after giving birth. This photo shows a typical day care center attached to an industrial plant in Guangdon province in coastal China. *(Xinhua/Liaison Agency, Inc.)*

growing at a relatively slow rate of 1.0 percent a year. Still, when one considers how large the population already is, even a 1.0 percent annual growth rate is worrisome. China will likely contain much more than 1.5 billion persons before stability is reached.

Fertility levels in China, as might be expected, vary from province to province. Birthrates are relatively low in most large cities (especially Shanghai), the Yangtze Delta, and in parts of the Sichuan Basin, and are relatively high in many upland areas of south China, in the Loess Plateau, and in the southern half of the North China Plain. Higher levels of fertility in these poorer and more rural areas may lead to increased migration, and perhaps also to heightened regional differentiation.

While China's population policy has reduced its growth rate, it has also generated social tensions and has apparently led to human rights abuse. Particularly troubling to many observers is the growing gender imbalance in the Chinese population. The number of baby boys in the country now far outweighs that of baby girls. This asymmetry reflects in part the practice of honoring one's ancestors; since family lines are traced through male offspring, it is necessary to produce a male heir if one wishes to maintain the practice. Many couples are therefore desperate to produce a son. Some opt to bear more than one child regardless of the penalties they may face. Another option is gender-selective abortion; if ultrasound tests reveal a female fetus, the pregnancy is sometimes terminated. Baby girls are also commonly born and abandoned, and well-substantiated rumors of female infanticide circulate. International women's organizations, as well as anti-abortion groups, are concerned over such ramifications of China's population policy. (Women's organizations are also concerned over the fact that China's population policies largely ignore men and their role in the reproductive process!) Many environmentalists, however, applaud China for its success in lowering its birthrate.

The Position of Women Women historically have had a low social position in Chinese society—as is true in most other world civilizations. One particularly blatant traditional manifestation of this was the practice of footbinding: the feet of elite and middle-class girls were usually deformed by breaking and binding them in order to produce a dainty appearance. This crippling and extremely painful practice was eliminated only in this century. In certain areas of southern China it was also common in traditional times for girls to be married, and hence leave their own families, when they were mere toddlers (such marriages, of course, would not be consummated for many years).

Not all women suffered such disabilities in pre-modern China. Among the Hakka, for example, women enjoyed a relatively high social position and were seldom subjected to footbinding. Both the nationalist and communist governments have, moreover, struggled to eliminate the worst practices and begin equalizing the relations between the sexes. Many of these measures have been successful, and women now have a relatively high level of participation in the Chinese workforce (Table 11.3). But it is still true that throughout East Asia—in Japan no less than in China—relatively few women have achieved positions of power in either business or government. As China modernizes, and as its urban economy grows, the position of its women will probably improve.

The Demographic Transition Reconsidered Population trends in China and elsewhere in East Asia conform relatively well to predictions generated by the **Theory of Demographic Transition.** This theory holds that a given country's population will begin to explode as it begins to modernize, since mortality rates will quickly drop while fertility rates remain high. Subsequently, as modernization proceeds, fertility rates will decline and population stability will be achieved. Japan thus experienced a rapid surge in population in the early 1900s, but by the 1960s its birthrate approached the replacement level. South Korea, which followed Japan in the development of its economy by several decades, similarly followed in its achievement of a stable birthrate. By the late 1990s it appeared that China was following the same course, both economically and demographically.

Critics of the theory argue that population expansion in the early stages of industrialization is more a function of increased fertility than of decreased mortality. In Japan in the late 1800s and early 1900s, for example, birthrates went up as opportunities expanded and as families sought to benefit from the labor of their children in unregulated factories. Another problem with the standard model of the demographic transition is that it does not adequately take into account social policy—the significance of which is especially evident in China. As the education level of women rises, birthrates typically decline even in the absence of other measures of economic development. Finally, a more complete model of the demographic transition would also have to consider the population decline that may well occur in the twenty-first century in such countries as Japan and South Korea.

Conclusion

East Asia is united by close cultural and historical bonds. Particularly important has been the role of the Chinese state, which at various times has encompassed virtually the entire region. Japan is exceptional in this regard, yet even it has profound historical connections to Chinese civilization. Because of these deep transregional ties, some contemporary observers predict of the formation of a "Confucian bloc" that will challenge the West for global primacy. Others, however, point to the tensions and mutual animosities that pervade the region, and to the economic linkages that increasingly connect all of East Asia, including China, to the world economy.

A few observers not only doubt the formation of a Confucian bloc, but even question whether China itself will persist as a unified political entity. The Tibetans and other non-Han peoples of western China long for independence, although this probably remains an unrealistic goal. Internal tensions also seem to be mounting within China proper. The relatively wealthy coastal provinces increasingly resent the control of Beijing and the flow of a large portion of their tax receipts to the national coffers. The old split between a more mercantile south China and a more bureaucratically focused north China may also be reemerging. Chinese provinces, moreover, are starting to set their own economic policies, rather than waiting for orders from the center. Contributing to this centrifugal process (in other words, one leading to a spreading out or a breaking apart) is the explosive growth of the private economy coupled with the near-stagnation of the state-owned sector. Some China-watchers believe that this growing imbalance may eventually sap the strength of the ruling Chinese Communist Party, undermining the country's central authority.

It is notoriously difficult, however, to predict China's political future, especially when present conditions themselves are not easily discerned. Just how strong is the popular sentiment for democratization in China? How intense are feelings of provincial, as opposed to national, loyalty? Such questions are difficult to address in a country in which freedoms of expression are limited. China does, however, have a several-thousand-year legacy of unity under a single government. It may therefore be reasonable to expect China to remain united. If it does, and if its economy continues to grow at current rates, China will clearly be one of the world's leading countries in the early twenty-first century.

Will China's booming economy lead to the emergence of a mass consumer culture like that of Japan? If so, would China's own territory be able to support the massive levels of consumption that this would generate? Since the answer to this query would likely be "no," one must then ask what the ramification on the global environment and the world economy would be if China were to follow Japan in provisioning its new needs largely from abroad. Perhaps more importantly, would global ecological systems be able to handle the increased output of carbon dioxide and other greenhouse gases that full-scale Chinese industrialization, especially if based on China's own abundant coal resources, would produce? Would a prosperous China necessarily be transformed into a democratic China? These are some of the most significant, but ultimately unanswerable, questions of global geography today.

The most important questions about East Asia's future are centered on China, but significant issues face the region's other countries. It remains to be seen, for example, whether South Korea will successfully manage the transition from being a relatively low-wage exporter to becoming a high-wage economic and technological powerhouse. South Korea also faces a daunting challenge to the north. As long as the present regime remains in power in Pyongyang, South Korea will feel threatened. But other dilemmas would arise if the North Korean government were to fail and the North Korean people sought unification with the south. The German experience shows how difficult and how expensive such reunification can be. In the case of Korea, where the two halves of the country are more evenly balanced in land area and population, reunification would be more expensive than it has been in Germany.

Japan, for its part, is now at something of a crossroads, brought on by the waning of its long-lasting economic miracle. For all of its economic power and prestige, Japan has yet to find its place in the world. Will Japan's political influence ever begin to match its economic clout? This question is hotly debated both in Japan and abroad. An equally important issue is whether Japan will be able to retain its unique economic and social institutions in the face of mounting global competition. There is a profound sense of uncertainty in the country today, as young people no longer feel assured about their futures and as leaders debate whether significant institutional changes are required.

One of the most intensive debates in Japan today centers around its educational system. Japanese education, beginning in junior high school, is noted for its severity and intense pressure. Standardized testing and rote memorization are emphasized, and if one does poorly on tests, one's future is virtually sealed. This system has served Japan well, resulting in high levels of basic knowledge and competence. Critics contend, however, that it stifles creativity—and that creativity will be the ultimate secret to success in the coming post-industrial global economy.

In an important sense, the education question cuts to the heart of Japan's current dilemma. One must wonder whether Japan's basic social and economic institutions will continue to be markedly different from those of the United States and other Western countries. Those who believe in enduring Confucian values usually answer in the affirmative. Others, however, believe that the process of globalization will force a certain degree of global economic and social convergence. Some thus predict that Japanese women will eventually enter the professions in large numbers, rigid hierarchies at work and school will begin to break down, and Japanese workers will find themselves increasingly threatened with layoffs and downsizing, just as U.S. workers are. Only the future, of course, can tell us who is right.

Key Terms

anthropogenic landscape (page 456)
autonomous region (page 470)
Burakumin (page 468)
Central Place theory (page 459)
Cold War (page 472)
Confucianism (page 464)
desertification (page 449)
diaspora (page 468)
geomancy (page 465)
gross domestic product (GDP) (page 482)
gross national product (GNP) (page 482)
hiragana (page 463)
ideographic writing (page 462)
kanji (page 463)
laissez-faire (page 465)
loess (page 449)
Mandarin (page 464)
Marxism (page 466)
multinational corporation (page 484)
particularism (page 466)
pollution exporting (page 453)
protectorate (page 475)
regulatory lakes (page 452)
rust belt (page 488)
samurai (page 472)
sediment load (page 450)
Shogun, Shogunate (page 475)
social and regional differentiation (page 486)
Special Economic Zones (page 486)
spheres of influence (page 475)
superconurbation (page 454)
swidden agriculture (page 470)
tectonic plates (page 444)
terraces (page 454)
Theory of Demographic Transition (page 489)
tonal language (page 469)
tribal peoples (page 470)
urban primacy (page 460)

Questions for Summary and Review

1. What is the major climatological difference between the Pacific coast of Japan and the coast facing the Sea of Japan?
2. How does the physiography of the Huang He River valley differ from that of the valley of the Yangtze?
3. Why is Japan so much more heavily forested than China?
4. How does the settlement pattern of North Korea differ from that of South Korea?
5. Why has Shanghai emerged as the most populous urban area in China?
6. What are the major ways in which Japanese cities differ from those of the United States?
7. What major cultural features are common to the entire East Asian world region?
8. Where are the non-Han peoples of China concentrated? Why are they concentrated in these areas?
9. In what ways have China and Japan reacted differently—or similarly—to the forces of global culture?
10. What have been the main consequences of the geographical division of Korea into two states?
11. Historically speaking, how did China and Japan act differently as imperial powers?
12. What role does the United States play in the contemporary geopolitics of East Asia?
13. How have the different countries of East Asia followed different paths to economic development?
14. Where in China would one find the most rapid economic development, and why would one find it there?
15. How does the position of women in Japan compare to the position of women in other wealthy, industrialized countries?

Thinking Geographically

1. Discuss the advantages and disadvantages of China's proceeding to build major dams.
2. Discuss the ramifications, both positive and negative, of Japan allowing the importation of rice and opening its agricultural lands to urban development.
3. What would be the consequences of China granting true autonomy to the Tibetans and other non-Han peoples? Independence?
4. What are the potential implications of Taiwan declaring itself an independent country?

5. Discuss the potential ramifications of the United States restricting the importation of Chinese goods in order to put pressure on the Chinese government for human rights reforms.
6. Discuss the advantages and disadvantages of China's current population policies.
7. Do you think that East Asia will emerge as the center of the world economy in the next century?

Regional Novels and Films

Novels

Sawako Ariyoshi, *The River Ki* (1982, Kodansha)

Kazuo Ishigura, *An Artist of the Floating World* (1989, Vintage Books)

Peter H. Lee, *Flowers of Fire* (1986, University of Hawaii Press)

Wang Shuo, *Playing for Thrills* (1998, Penguin)

Hsueh-Chin Tsao, *The Dream of the Red Chamber* (1958, Doubleday)

Films

Farewell My Concubine (1993, China)

The Gate of Heavenly Peace (1995, China)

Nomugi Pass (1979, Japan)

Seven Samurai (1954, Japan)

The Story of Qiu Ju (1992, China)

A Taxing Woman (1987, Japan)

Why Has Bodi-Darma Left for the East? (1989, Korea)

Bibliography

Adshead, Samuel A. M. 1988. *China in World History*. New York: St. Martin's Press.

Boserup, Ester. 1965. *The Conditions of Agricultural Growth: The Economics of Agrarian Change Under Population Pressure*. London: Allen & Unwin.

Cannon, Terry, and Jenkins, Alan, eds. 1990. *The Geography of Contemporary China: The Impact of Deng Xiaoping's Decade*. London: Routledge.

Chapman, Graham P., and Baker, Kathleen M., eds. *The Changing Geography of Asia*. London: Routledge.

Chiu, T. N. 1986. *A Geography of Hong Kong*. Oxford: Oxford University Press.

The Contemporary Atlas of China. 1988. Boston: Houghton Mifflin.

Cotterell, Arthur. 1993. *East Asia: From Chinese Predominance to the Rise of the Pacific Rim*. Oxford: Oxford University Press.

Cybriwsky, Roman A. 1991. *Tokyo: The Changing Profile of an Urban Giant*. Boston: G. K. Hall.

Gaubatz, Piper Rae. 1996. *Beyond the Great Wall: Urban Form and Transformations on the Chinese Frontiers*. Palo Alto, CA: Stanford University Press.

Gernet, Jacques. 1982. *A History of Chinese Civilization*. Cambridge, England: Cambridge University Press.

Hanley, Susan B., and Wolf, Arthur P. 1985. *Family and Population in East Asian History*. Palo Alto, CA: Stanford University Press.

Hoare, James, and Pares, Susan. 1988. *Korea: An Introduction*. London: Routledge.

Knapp, Ronald G., ed. 1992. *Chinese Landscapes: The Village as Place*. Honolulu: University of Hawaii Press.

Kolb, A. 1971. *East Asia, China, Japan, Korea, Vietnam: Geography of a Culture Region*. London: Methuen.

Kornhauser, David H. 1982. *Japan: Geographical Background to Urban-Industrial Development*. London: Longman.

Lee, Ki-baik. 1984. *A New History of Korea*. Cambridge, MA: Harvard University Press.

Leeming, Frank. 1993. *The Changing Geography of China*. Cambridge, MA: Blackwell.

Myers, Ramon H., and Peattie, Mark R. 1984. *The Japanese Colonial Empire, 1895–1945*. Princeton: Princeton University Press.

Perdue, Peter C. 1987. *Exhausting the Earth: State and Peasant in Hunan, 1500–1850*. Cambridge, MA: Harvard University Press.

Pomeranz, Kenneth. 1993. *The Making of a Hinterland: State, Society, and Economy in Inland North China, 1853–1937*. Berkeley, CA: University of California Press.

Rozman, Gilbert, ed. 1991. *The East Asian Region: Confucian Heritage and Its Modern Adaptation*. Princeton: Princeton University Press.

Rowe, William T. 1984. *Hankow: Commerce and Society in a Chinese City, 1796–1889*. Palo Alto, CA: Stanford University Press.

Skinner, G. William. 1964. "Marketing and Social Structure in Rural China." *Journal of Asian Studies* 24 (1), Part I: pp. 1–43; Part II, pp. 195–228.

Skinner, G. William, ed. 1977. *The City in Late Imperial China*. Palo Alto, CA: Stanford University Press.

Smil, Vaclav. 1984. *The Bad Earth: Environmental Degradation in China*. New York: M. E. Sharpe.

Smith, Christopher J. 1991. *China: People and Places in the Land of One Billion*. Boulder, CO: Westview.

Songqiao, Zhao. 1986. *Physical Geography of China*. New York: John Wiley & Sons.

Spence, Jonathan D. 1990. *The Search for Modern China*. New York: W. W. Norton.

Sun, Jingzhi, ed. 1988. *The Economic Geography of China*. Oxford: Oxford University Press.

Totman, Conrad. 1989. *The Green Archipelago: Forestry in Preindustrial Japan*. Berkeley, CA: University of California Press.

Tregear, T. R. 1965. *A Geography of China*. Chicago: Aldine.

Trewartha, Glenn T. 1965. *Japan: A Geography*. Madison: University of Wisconsin Press.

Veeck, Gregory, ed. 1991. *The Uneven Landscape: Geographical Studies in Post-Reform China*. Baton Rouge, LA: Geoscience Publications.

Wigen, Karen. 1995. *The Making of a Japanese Periphery, 1750–1920*. Berkeley, CA: University of California Press.

TURKMENISTAN
TAJIKISTAN
AFGHANISTAN
CHINA
Karakoram Range
Indus R.
Kashmir
Islamabad
Srinagar
JAMMU AND KASHMIR
HIMACHAL PRADESH
Lahore
Amritsar
PUNJAB
CHANDIGARH
HIMALAYAS
PAKISTAN
IRAN
HARYANA
Delhi
DELHI
New Delhi
BHUTAN
ARUNACHAL PRADESH
Thimphu
NEPAL
Kathmandu
RAJASTHAN
UTTAR PRADESH
SIKKIM
Brahmaputra R.
ASSAM
NAGALAND
Karachi
Sind
Ganges R.
MEGHALAYA
Tropic of Cancer
BIHAR
BANGLADESH
MANIPUR
Arabian Sea
MADHYA PRADESH
Dhaka
TRIPURA
MIZORAM
GUJARAT
Narmada R.
WEST BENGAL
Calcutta
Sundarbans
DAMAN AND DIU
20°N
INDIA
ORISSA
DADRA AND NAGAR HAVELI
Mumbai (Bombay)
MAHARASHTRA
Godavari R.
Bay of Bengal
Western Ghats
Deccan Plateau
Eastern Ghats
Krishna R.
Goa
GOA
KARNATAKA
ANDHRA PRADESH
PONDICHERRY
Malabar Coast
Andaman Islands (INDIA)
Madras
Bangalore
Lakshadweep (INDIA)
10°N
LAKSHADWEEP
TAMIL NADU
ANDAMAN AND NICOBAR ISLANDS
KERALA
Jaffna
Nicobar Islands (INDIA)
SRI LANKA
Colombo
Male
MALDIVES
INDIAN OCEAN
0°
Equator
70°E
80°E
90°E

C H A P T E R 1 2

South Asia

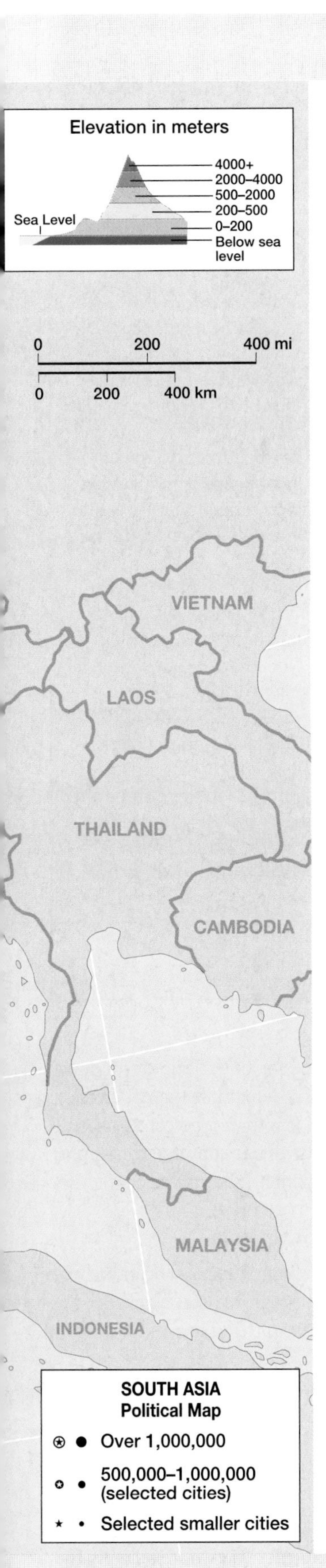

South Asia is a land of deep historical and cultural commonalities that has recently experienced intense political conflict. Dominated by British colonialism for several centuries, since independence in 1947 the two largest countries, India and Pakistan, have fought several wars and remain locked today in bitter animosity that contains a disturbing element of nuclear brinksmanship. This political tension reaches such high peaks that many arms control experts maintain South Asia is the leading candidate for a nuclear war. Religious divisions undercut this geopolitical turmoil, for India is primarily a Hindu country (with a large Muslim minority), while neighboring Pakistan and Bangladesh are both predominantly Muslim (Figure 12.1; see "Setting the Boundaries"). Even within India, cultural and political tensions, inflamed recently by the rise of Hindu nationalism, call into question the future of this huge federal state; some observers argue that India could possibly break apart as several of its states seek autonomy and independence.

Parallel to these geopolitical tensions in South Asia is a concern with its immense and rapidly growing population. Given its current rate of growth, South Asia could soon surpass East Asia as the world's most populous region. The underlying issue, though, is not simply the gigantic population of this region but whether it can support these people, given its economy and resource base. Although agricultural production has increased slightly faster than population growth in the last two decades, many experts feel these improvements have reached their limit; thus, population will once again outpace food resources. Compounding this serious situation is the widespread poverty of South Asia; it is, along with Sub-Saharan Africa, the poorest part of the world. More than half of India's population subsists on less than one dollar a day.

Compared to East and Southeast Asia, South Asia is far less connected to the contemporary globalized world. Few would use the term *economic tiger* to describe any South Asia country because of their slow rates of growth and inward orientation designed to meet internal needs rather than produce export goods. Many experts predict, however, that South Asia will soon make a significant global impact, given the impressive levels of scientific and technical skills found within the labor force, the international links the region already has because of South Asian migration, a fairly impressive resource base, and the enormous size of

◀ **Figure 12.1 South Asia** This region is the second most populous in the world, primarily because of India's population of almost one billion people. If population growth rates continue at current levels, India will overtake China as the world's most populous country sometime in the early twenty-first century. Bracketing India on the west and east are Pakistan and Bangladesh, two large countries with a predominantly Muslim population. The two Himalayan countries of Nepal and Bhutan, along with the island nations of Sri Lanka and the Maldives, round out the region.

Setting the Boundaries

Seven countries that vary tremendously in size and population constitute the South Asian world region (Figure 12.1). The unity of this region comes more from a shared history than from any contemporary political or cultural coherence. In the past, the region could be characterized as a meeting ground between the major religions of Hinduism, Islam, and, to a lesser degree, Buddhism. Additionally, the region also shared a colonial history dominated by British administration. Today, though, this region is now torn by ethnic and political tensions that have surfaced more intensely since postwar independence, and as often as not, have religious underpinnings.

India is by far the largest country, both in size and in population. Covering more than a million square miles, from the Himalayan mountains to the southern tip of the peninsula at Cape of Cochin, India is the world's seventh largest country in terms of area, and with almost a billion inhabitants, it is second only to China in population. Though ostensibly a Hindu country, India contains a large amount of religious, ethnic, linguistic, and political diversity within its borders, some of which is highly problematic.

Pakistan is the next largest country, in both size and population. Less than one-third the size of India, it also stretches from the high northern mountains to its arid coastline on the Arabian Sea; its population is 142 million, or about 15 percent the size of India's, yet these two countries—both with nuclear weapon systems—are locked in a tense power struggle over border issues that has the world on edge. Until independence in 1947, Pakistan was simply one portion of a larger undivided British colonial realm called *India,* but because of its strong ties to Islam, many Pakistanis argue their country is now more closely connected to its Muslim neighbors in Southwest Asia than to India and the other South Asian countries.

Bangladesh, to the east of India, is also a Muslim country and was originally created as "East Pakistan" in the hurried independence movement of 1947, then achieved independent status from greater Pakistan in a brief civil war in 1971. Though a small country in area (54,000 square miles), it is one of the most densely populated in the world—and also one of the poorest—with a population of 124 million in an area about the size of Wisconsin.

Nepal and Bhutan are both located in the Himalayan mountains, sandwiched between India and the Tibetan plateau of China. Nepal is the larger of the two in both area and population (23 .7 million) and, further, is more open to and engaged with the contemporary world—primarily though tourism—with the usual array of problems and benefits. Bhutan, on the other hand, has purposely remained disconnected to the modern world and remains a relatively pristine and isolated Buddhist kingdom of less than a million inhabitants.

Last, the two island countries of Sri Lanka (formerly Ceylon) and the Maldives round out the South Asia world region; each of these countries, though, has its own set of serious problems that clouds the future. Sri Lanka (population 18.9 million) has been mired in a civil war since 1983 that has taken a high toll on the once vibrant economy of the country. The problem facing the small island nation of the Maldives, in contrast, is not of their own making but instead speaks to the linkages of all countries in this contemporary, globalized world. If global warming continues because of atmospheric pollution, and if sea levels rise as predicted from this warming, this island nation—where the highest point is only 6 feet above sea level—will be flooded and its 300,000 inhabitants will have to seek higher ground somewhere else.

its local markets. Parts of India, for example, have recently emerged as major players in the global software industry. Other observers, though, are more skeptical and doubt that South Asia's global role will ever be very important because of its pervasive internal social conflicts and persistent poverty.

Environmental Geography: Diverse Landscapes, from Tropical Islands to Mountain Rim

As might be expected from a large world region, South Asia's environmental geography covers a wide spectrum that ranges from the highest mountains in the world to densely populated delta islands barely above sea level; from one of the wettest places on Earth to dry, scorching deserts; from tropical rain forests and dense forests to barren lands and coral reefs (Figure 12.2).

While South Asia offers a fascinating array of diverse physical environments as it stretches north from the equator to midlatitudes, an equally important theme is the interaction of people with these different environments, for this is one of the most densely settled regions of the world. Here, more than 20 percent of the world's population lives on less than 5 percent of Earth's land area. Whether the environment can continue to support such a huge population is a recurring theme of this chapter.

The Four Subregions of South Asia

To better understand this vast region, South Asia can be broken down into four physical subregions, starting with the high mountain ranges of its northern fringe and extending to the tropical islands of the south. Lying south of the mountains are the extensive river lowlands that form the heartland of both India and Pakistan; between river lowlands and the island countries, though, is the extensive area of peninsular India, an area more than 1,000 miles (1,600 kilometers) long from north to south.

Mountains of the North South Asia's northern rim of mountains is dominated by the great Himalayan Range, forming the northern borders of India, Nepal, and Bhutan.

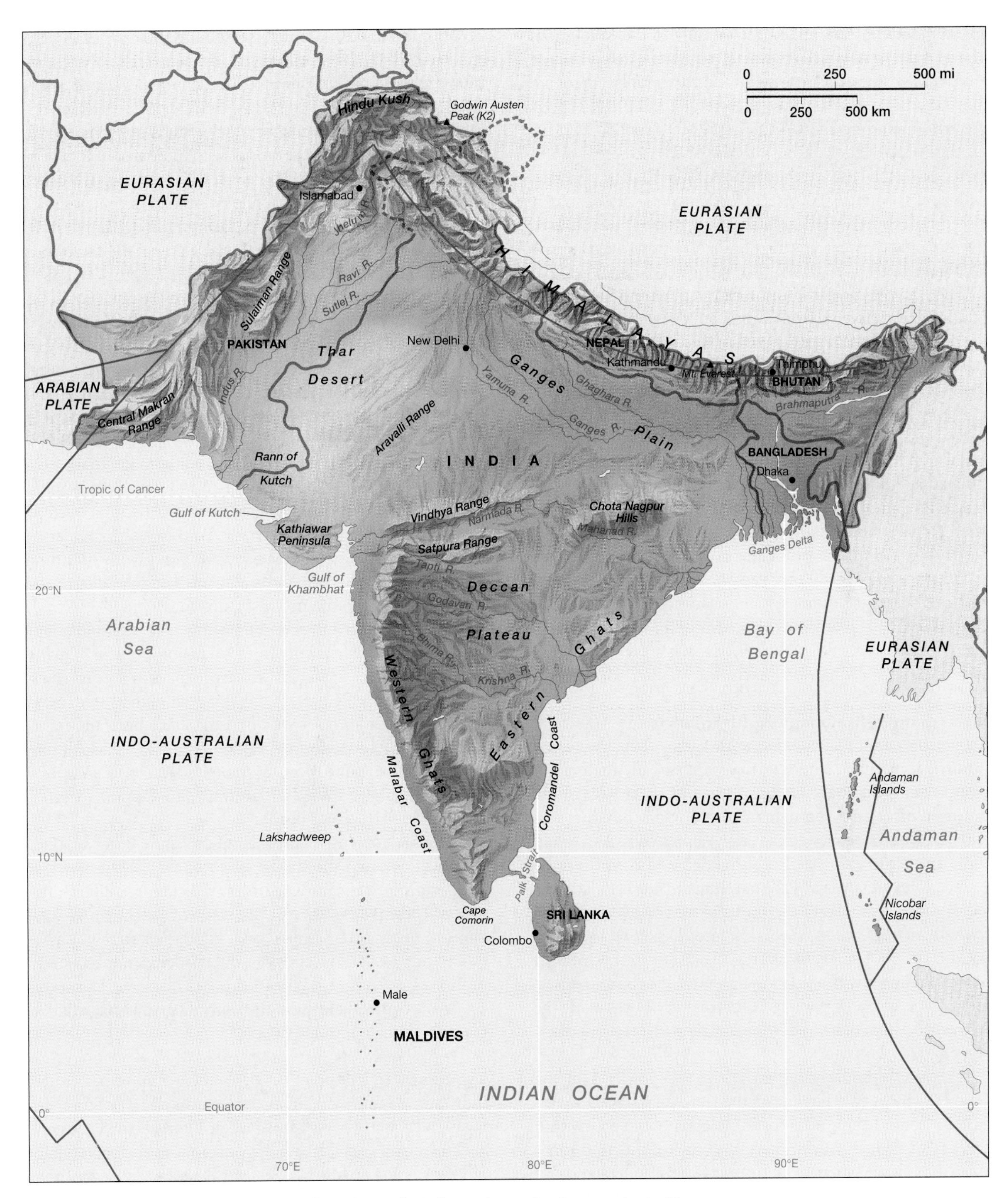

▲ **Figure 12.2 Physical geography of South Asia** This region is comprised of four extensive physical subregions: the high Himalayan mountains in the north; the expansive Indus-Ganges lowland that reaches from Pakistan in the west to the delta lands of Bangladesh; peninsular India, dominated by the volcanic Deccan Plateau; and, last, the island realm that includes Sri Lanka and the Maldives making up the southern rim of South Asia. Many of the region's landscapes are products of the slow northward movement of the Indo-Australian tectonic plate against the Eurasian plate. For example, this collision has created the Himalayan mountain chain.

These mountains are linked geologically to the Karakoram Range in the west, which are equally high and extend through Pakistan's northeast territories. In these two ranges there are more than two dozen peaks exceeding 25,000 feet (7,620 meters), including the world's highest mountain, Everest, on the Nepal–China (Tibet) border at 29,028 feet (8,848 meters). To the east are the lower Arakan Yoma mountains with peaks around 10,000 feet (3,300 meters), forming part of the border between India and Myammar (Burma) and, for our purposes, separating South Asia from neighboring Southeast Asia.

These impressive and formidable mountain ranges were produced by tectonic activity caused by peninsular India pushing northward into the larger Eurasia continental plate; as a result of the collision between these two tectonic plates, these east-west mountain ranges have been folded and upthrust.

While most of South Asia's northern mountains are too rugged and high to support dense human settlement, there are major population clusters in the Kathmandu Valley of Nepal situated at 4,400 feet (1,341 meters) and the Valley, or Vale, of Kashmir in northwestern India at 5,200 feet (1,585 meters). Nepal is home to the 85,000 mountain-dwelling Sherpa people, who have won fame as porters and guides for mountaineering expeditions. Historically, Sherpa porters carried rice across the Himalaya passes to exchange for salt in Tibet, but political tensions now restrict this movement, so Sherpa raise cattle and, additionally, make a livelihood from spinning and weaving wool products.

Indus-Ganges-Brahmaputra River Lowlands South of the northern mountains lie large lowlands, originally created through tectonic activity but now occupied by three major river systems that have carried sediments eroded off of the northern mountains through millions of years, building vast alluvial plains of rich fertile and easily farmed soils. As a result, these river lowlands are densely settled and constitute the population core areas of Pakistan, Indian, and Bangladesh.

Of these three rivers the Indus is the longest, covering more than 1,800 miles (2,880 kilometers) as it flows southward from the Himalayas through Pakistan to the Arabian Sea, providing much-need irrigation waters to the arid desert areas in the southern portions of that country. It was along the Indus Valley that one of the world's earliest civilizations arose some 5,000 years ago.

Even more densely settled is the vast lowland of the Ganges, which, after flowing off the Himalayas travels southeasterly some 1,500 miles (2,400 kilometers) to empty into the Bay of Bengal. Given the central role of this important river in South Asia's past and present, it is understandable why the river is considered sacred and holy. As with the Indus, the Ganges valley was also the hearth area of a formative prehistoric civilization, and, today, this river is the lifeblood of the most densely settled area of contemporary India.

Although this large South Asian lowland is often referred to as the Indus-Gangetic Plain, this term neglects the important role played by the Brahmaputra River. This river rises on the Tibetan Plateau and flows easterly, then southward over 1,700 miles (2,720 kilometers), joining the Ganges in central Bangladesh, then forming the vast delta region that forms more than half the land area for that densely populated country. Unlike the sparsely populated Indus delta in Pakistan, the low-lying Ganges-Brahmaputra delta lands are some of the most densely settled in the world, with more than 3,000 people per square mile (or 1,200 per square kilometer). Most of this delta is in Bangladesh. While river-borne sediments offer fertile soils to these farmers, devastating floods come from both river and ocean, making this also one of the hazardous regions of the world.

Peninsular India Jutting southward is the familiar shape of peninsular India, made up primarily of the elevated Deccan Plateau, which is bordered on each coast by narrow coastal plains backed by elongated north-south mountain ranges. On the west are the higher Western Ghats, which are generally about 5,000 feet (1,524 meters) but reach higher than 8,000 feet (2,438 meters) in the peninsula's southern tip; to the east, the Eastern Ghats are lower and discontinuous, thus forming far less of a transportation barrier to the broader eastern coastal plain of peninsular India. On both coastal plains, fertile soils and an adequate water supply, along with maritime activities, support population densities comparable to the Ganges lowland to the north (Figure 12.3).

But on the plateau itself, soil fertility is mixed, and a reliable water supply for agriculture is a major problem in many areas. Much of the western plateau region lies in the rain shadow of the Western Ghats, giving it a drier climate than areas farther east. Small reservoirs or tanks have been the traditional method for collecting monsoon rainfall for use during the dry season. More recently, though, deep wells and powerful pumps have mined groundwater to support irrigated crops and village water needs.

Because of overuse of these aquifers, future water supplies are uncertain, and the Indian government plans a series of large dams on plateau rivers to provide for agricultural irrigation. These plans, however, are controversial because—like dams in many parts of the world—the reservoirs will dislocate hundreds of thousands of rural residents in this densely settled region. A case in point is the Sardar Sarovar Dam project on the Narmada River in the state of Gujarat that will displace more than 100,000 people.

The Southern Islands At the southern tip of peninsular India, just 23 miles (37 kilometers) across the Palk Strait, lies the island country of Sri Lanka, where internal ethnic tensions and terrorism cast a troubling shadow over a potentially prosperous region. Much of Sri Lanka's earlier glory came from the tea plantations that covered the hillside flanks of this tropical island; in the interior, mountains of 8,000 feet (2,438 meters) provided cool, moist conditions perfect for tea plants, yet these plantations are mostly gone, casualties of the ethnic warfare that has plagued Sri Lanka for 25 years. Because the monsoons arrive from the southwest, that portion of the island is moister than the rain shadow region to the northeast and has been favored by settlement over the drier regions.

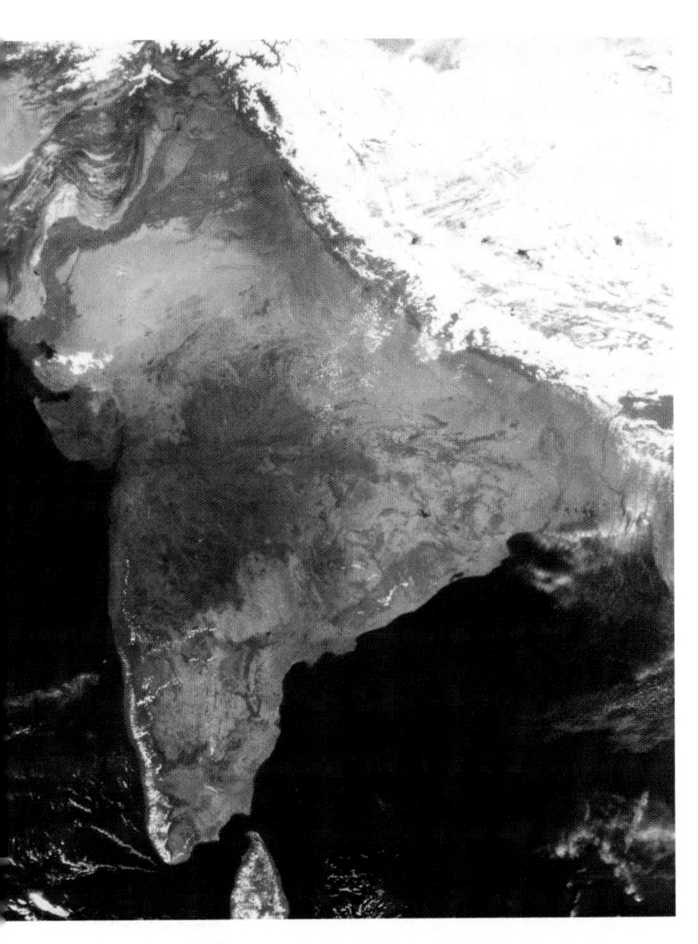

▲ **Figure 12.3 South Asia from space** The four physical subregions of South Asia are clearly seen in this satellite photograph, from the snow-clad Himalayan mountains in the north to the islands of the south. The Deccan Plateau is dark, fringed by white clouds as moist air is lifted over the uplands of the Western Ghats. *(Earth Satellite Corporation/Science Photo Library/Photo Researchers, Inc.)*

Forming a separate state within South Asia are the Maldives, a chain of more than 1,200 islands stretching south to the equator some 400 miles (640 kilometers) off the southwestern tip of India. The combined land area of these islands is about 116 square miles (290 square kilometers), though less than a quarter of these islands are actually inhabited.

In contrast to Sri Lanka and its picturesque mountain landscapes, the Maldives are flat and low coral atolls, with the highest elevation in the entire chain just over 6 feet (2 meters) above sea level, a fact that forces this island nation to play a prominent role in the international debate about global warming and the accompanying rise in worldwide sea levels (see Chapter 2). Should the worst-case global-warming-and-sea-level-rise scenario be true, the Maldives would be an endangered country.

South Asia's Monsoon Climates

The dominant climatic factor for most of South Asia is the **monsoon,** the distinct seasonal change between wet and dry periods. South Asians think of their region as having three very different seasons. First, and most important, is the warm and rainy season of the monsoon, which for most areas are the summer months, June through October; this is followed by a cool and dry season that extends from November until February. Only a few areas get rainfall during this otherwise dry time. The third season is the hot period from March to late May that builds up with heat and humidity until the monsoon's much-anticipated and rather sudden "burst" in early June (see "Environment: The Monsoon Arrives in Southern India").

This monsoon pattern is caused by large-scale meteorological processes that affect much of Southeast and East Asia as well (Figure 12.4). During the Northern Hemisphere winter, a large high-pressure system forms over the cold Asian landmass. From this high-pressure cell, cold, dry winds flow outward from Asia's interior, over the Himalaya Mountains and down across South Asia. As winter turns to spring, these winds diminish, which causes the hot, dry temperatures of March through May. Eventually this buildup of heat over South Asia, coupled with the warming of the interior landmass, produces a large thermal low-pressure cell. By early June this low-pressure cell is strong enough to draw in warm, moist air from the Indian Ocean. Usually the first monsoon rains arrive at the southern tip of peninsular India in late May; after this first "burst," it takes about 6 weeks for monsoon rains to travel northward to the Ganges lowland and into the Himalayas.

Along the Western Ghats, **orographic rainfall** results from the uplifting and cooling of moist monsoon clouds over these mountains; as a result, some hill stations receive more than 200 inches (508 centimeters) of rain during the 4-month wet season. On the climate map (Figure 12.5), these are the areas of Am climate, or tropical monsoon. Inland, though, a strong rain shadow effect dramatically reduces rainfall to the semiarid conditions of India's steppe climate.

Farther north, as the monsoon winds are forced up and over the Himalayan foothills, copious amounts of rainfall are characteristic. Simla, India, at 7,000 feet (2,134 meters), averages more than 16 inches (40 centimeters) in July alone. To the east, directly in line for moist air off the Bay of Bengal, Cherrapunji, India, at 4,000 feet (1,220 meters) is a strong contender for the title of world's wettest place (along with Mt. Waialeale on Kauai, Hawaii), with an average rainfall of 451 inches (1,128 centimeters).

However, not all of South Asia receives high rainfall totals from the monsoon. In Pakistan, particularly, precipitation is low enough to cause steppe, even desert climates (Figure 12.5). In Karachi, for example, the annual total is less than 10 inches (25 centimeters). Only in the northeastern regions of Pakistan, around the capital of Islamabad, does the monsoon bring adequate rainfall. Here, it is often July before the monsoon arrives.

Regardless of whether rainfall is heavy or light, the monsoon rhythm affects all of South Asia in many different ways, from the delivery of much-needed water for crops and villages, to the mood and disposition of millions of people as they eagerly await relief from the oppressively hot dry season

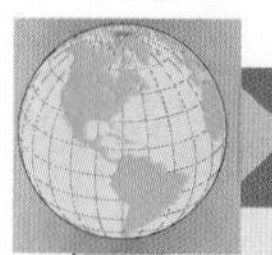

ENVIRONMENT The Monsoon Arrives in Southern India

In 1987 journalist Alexander Frater realized a lifelong dream of witnessing firsthand the most dramatic of meteorological events—the arrival and progress of the South Asia monsoon. He followed it from its first "burst" on the southern tip of India at Cape Comorin in June, northward to Mumbai (Bombay) and Delhi, then, to pick up the eastern arm of the monsoon, to Calcutta, and finally into the Himalaya foothills to experience the monsoon in Cherrapunji, one of the world's wettest places. Here he writes about the arrival of the monsoon in Cochin (or Kochi) on the west coast of southern India.

At l p.m. the serious cloud build-up started. Two hours fifty minutes later racing cumulus extinguished the sun and left everything washed in an inky violet light. At 4:40, announced by deafening ground-level thunderclaps, the monsoon finally rode into Cochin. The cloud-base blew through the trees like smoke; rain foamed on the hotel's harbourside lawn and produced a bank of hanging mist opaque as hill fog. In the coffee shop the waiters rushed to the windows, clapping and yelling, their customers forgotten.

Heaving a door open I stepped outside. Soaked to the skin within seconds I felt a wonderful sense of flooding warmth and invigoration; it was, indubitably, a little bit like being born again. Raindrops rang like coins on the flagstoned path and the air was filled with fusillades of crimson flowers from the flamboyant trees; they went arcing by like tracers and raked by an especially mean burst, I can testify that flamboyant blossoms hitting you at 60 k.p.h. cause pain and temporary loss of vision. At Fort Cochin they were ringing the bells in St. Francis Church.

Then, from the corner of an eye still watering from the flower strike, I witnessed an astonishing scene. Two straining waiters held the coffee-shop door open while a party of men and women filed into the storm. The men wore button-down shirts and smart business suits, the women best-quality silk saris and high-heeled shoes; as they emerged, they opened their arms and lifted their faces to the rain.

The Spices Board had come out to greet the monsoon. . . .

Buffeted by the gusts, unbalanced by the waves, the Spices Board executives clung to each other with water in their eyes and looks of sublime happiness on their faces. A young woman in a soaked and flapping gold-coloured sari laughed at me and clapped her hands. "Paradise will be like this!" she shouted.

Source: Adapted from Alexander Frater, 1990. *Chasing the Monsoon: A Modern Pilgrimage Through India.* New York: Henry Holt & Company.

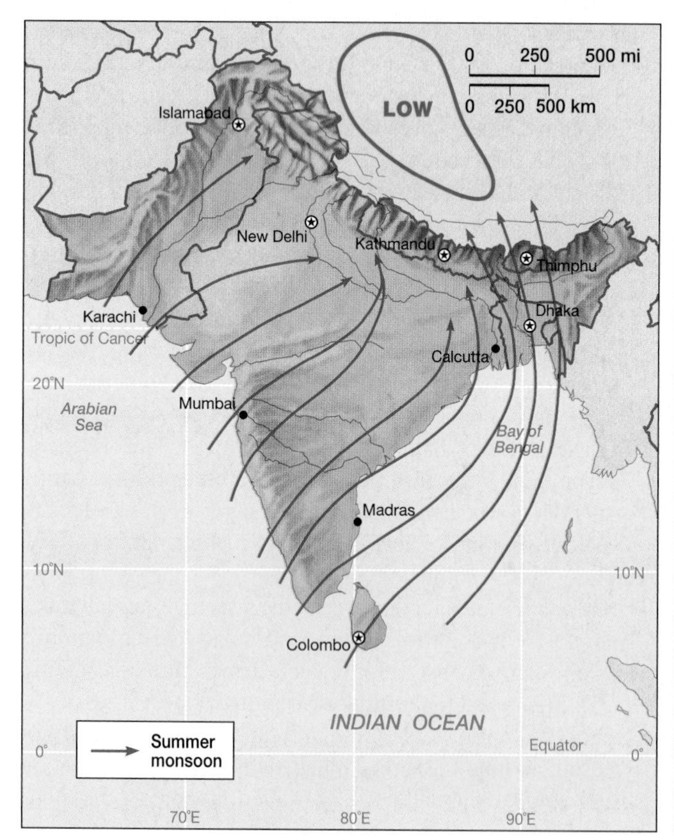

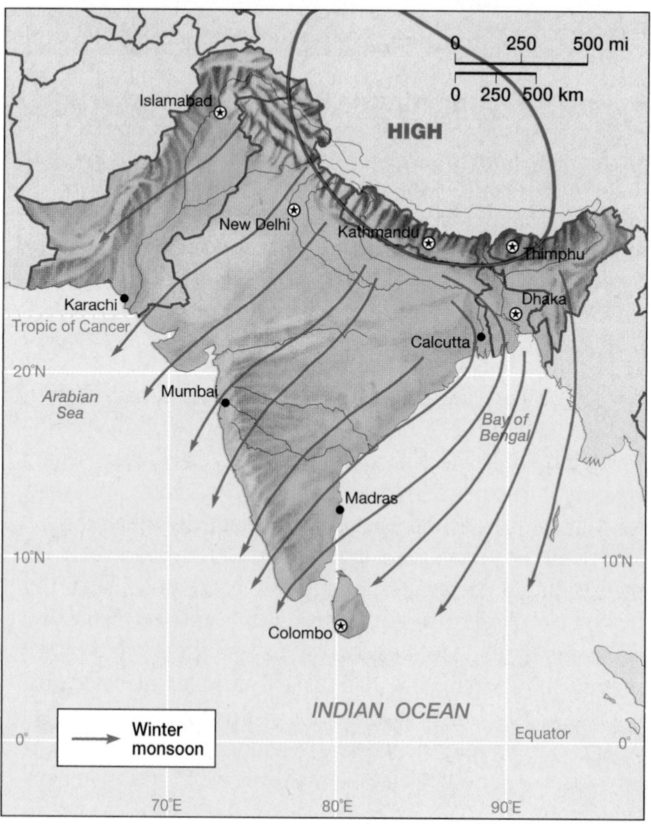

▲ **Figure 12.4 The summer and winter monsoons** Low pressure centered over the Tibetan Plateau and the Indo-Gangetic Plain draws in warm, moist air masses during the summer that bring heavy monsoon rains to most of South Asia. Usually these rains begin in June and last for several months. During the winter, high pressure forms over the Himalayan Mountains. As a result, winds are reversed from those of the summer. During this season only a few coastal locations along India's east coast and Sri Lanka receive rain.

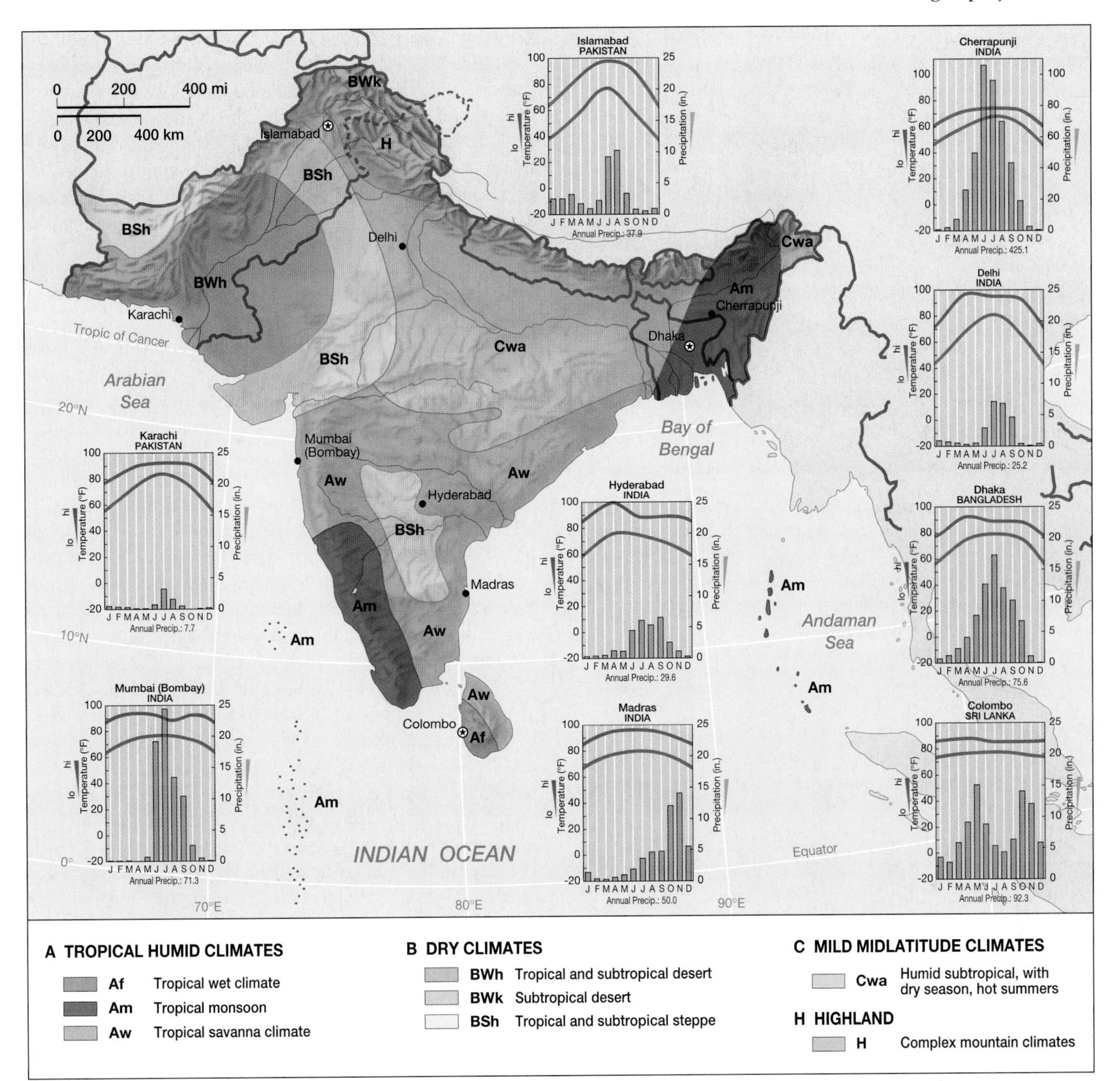

▲ **Figure 12.5 Climates of South Asia** Except for the extensive Himalayan mountain area, South Asia is dominated by tropical and subtropical climates. Many of these climates show a distinct summer rainfall season that is associated with the monsoon. The climographs for Mumbai (Bombay) and Delhi are excellent illustrations. However, the climographs for east coast locations such as Madras, India, and Colombo, Sri Lanka, show how some locations also receive winter monsoon rains.

(Figure 12.6). Some years the monsoon delivers its promise with abundant moisture; in other years, though, it brings only scant rainfall to South Asia, resulting in crop failure, famine, and hardship.

Environmental Issues in South Asia

As is true in most densely settled regions of the world where millions of rural people look to the land for their subsistence, there is a raft of serious environmental issues. Additionally, South Asia suffers from the usual environmental problems of water and air pollution that accompany early industrialization and manufacturing. Compounding all of these problems, though, are the immense numbers of new people added each year through natural population growth. Indeed, behind most environmental issues lie the dark shadow of South Asia's rapidly growing population (Figure 12.7).

Natural Hazards in Bangladesh Perhaps the link between population pressure and environment is nowhere clearer than in the delta area of Bangladesh, where the search for fertile

▲ **Figure 12.6 Monsoon rain** During the summer monsoon, some India cities such as Mumbai (Bombay) receive more than 70 inches of rain in just three months. These daily torrents cause floods, power outages, and daily inconvenience. However, these monsoon rains are crucial to India's agriculture. If the rains are late or abnormally weak, crop failure and famine often result. *(Sharad J. Devare/Dinodia Picture Agency)*

land has driven people into hazardous areas that now put millions at risk from seasonal flooding, as well as from powerful hurricanes (called *cyclones* in South Asia) from the Bay of Bengal.

For millennia, the drenching monsoon rains have eroded and transported huge quantities of sediment from the Himalayan slopes to the Bay of Bengal by the Ganges and Brahmaputra rivers, gradually building this low-lying and fertile deltaic environment. Until a few hundred years ago, most of this delta region was lightly inhabited. But with recent population growth, aggravated by the forced migration of Muslims into the area in 1947, people moved into the delta swamps to transform them into rice fields. While this agricultural activity has supported the large Bangladesh population, it exacerbated the area's natural hazards.

While periodic flooding is a natural, even beneficial, phenomenon that enlarges deltas by depositing fertile river-borne sediment over wide areas, this flooding has become a serious problem for those people who now inhabit these low-lying delta areas. In September 1998, for example, more than 22 million Bangladeshis were made homeless because of flooding that covered two-thirds of the country and killed more than 700 (Figure 12.8). Flood levels were the highest on record, even though the death toll was much reduced from the 1988 disaster that killed more than 3,000. (Bangladesh, it should be noted, was not the only country to suffer from the 1998 floods. In India, more than 1,000 died.)

With the population of both Bangladesh and northern India growing rapidly, there is a strong possibility that flooding will take even higher tolls in the next decade as desperate farmers locate in the hazardous floodplain and delta lands. Additionally, many environmental scientists see a close connection between deforestation of the Ganges and Brahmaputra watersheds and downstream flooding. Since forest cover and ground vegetation intercept rainfall and slow runoff, deforestation in the river headwaters could cause increased flooding. Unfortunately, the contentious political climate between India and Bangladesh has impeded the science necessary to better understand this relationship, as well as the policies necessary to forging solutions.

Forests and Deforestation Forests once covered most of South Asia, except for the desert areas in the northeast, but this forest cover has largely vanished as a result of human activities. The Ganges Valley and coastal plains of India, for example, were largely deforested thousands of years ago to make room for agriculture. Elsewhere, such as in the mountainous and hilly areas, trees have been cut for industrial needs; much of this deforestation resulted from railroad construction in the nineteenth century. More recently, hillslopes in the Himalayas have been logged for commercial purposes, serving both internal and export needs.

As a result of this deforestation, many villages of South Asia suffer from a fuelwood shortage for household cooking and heating, forcing people to use dung cakes from cows and oxen as fuel. While this low-grade fuel does provide some heat for fires, it also means that nutrients that could be used as natural fertilizers on farm fields are instead diverted to household fires. Where wood fuel is available, collecting it usually involves many hours of female labor because the remaining sources of wood are often far from the village, necessitating a long walk to and from the remnant forests.

Concern about India's fast-disappearing forests led women in northern India to engage in the historical Hindu practice of "tree-hugging" to protect ancient trees from logging. The modern **Chipko movement**, named after an Indian word for "hug" or "cling to," started in 1973 as a women's protest movement against deforestation and has spread throughout many Himalayan villages (Figure 12.9). Because of this movement, the Indian state of Uttar Pradesh recently banned commercial cutting of forests in its hills.

Wildlife: Extinction and Protection Although the environmental situation in South Asia might be grim at the regional level, there are a few issues that inspire guarded optimism. One of these is that the region has managed to retain a diverse assemblage of wildlife despite population pressure and intense poverty. For example, the only remaining Asiatic lions, which once roamed as far east as southern Europe, live in the Indian's Gujarat state. Additionally, Bangladesh has retained a viable population of tigers in the Sundarbans National Park in the Ganges delta. Wild elephants still live in several large reserves in India, Sri Lanka, and Nepal. This protection of wildlife in South Asia far exceeds that in other Asia regions; in China, for example, tigers, elephants, and many other large mammals have been extinct for centuries.

Some observers credit Hinduism's nonviolent credo and advocacy of vegetarianism for wildlife preservation. Further, there is less medicinal use of animal parts than in East Asia.

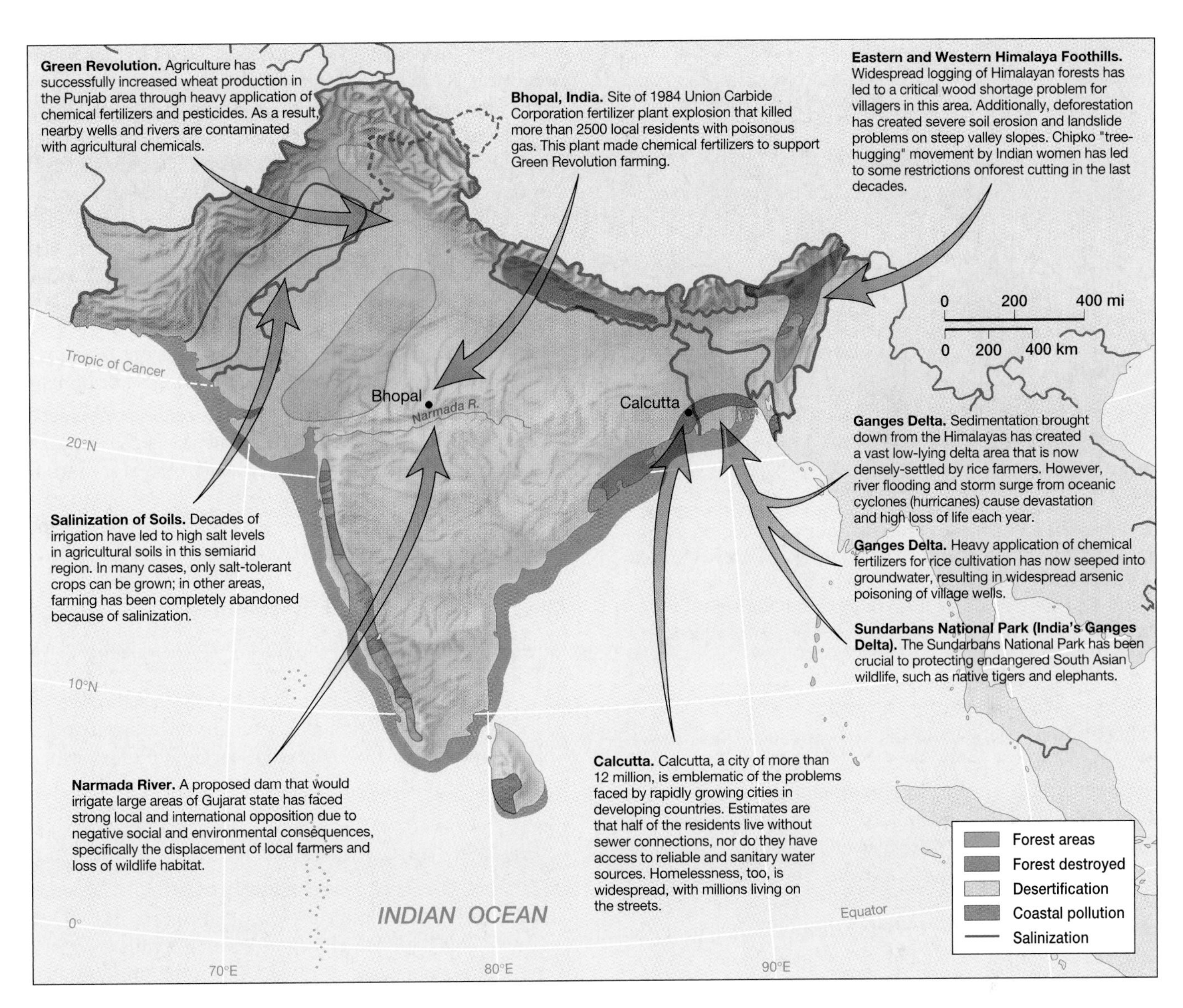

▲ **Figure 12.7 Environmental issues in South Asia** As might be expected in a highly diverse and densely populated region, there are a wide range of environmental issues and problems. These issues range from salinization of irrigated lands in the dry lands of Pakistan and western India, to groundwater pollution from Green Revolution fertilizers and pesticides in the verdant delta lands of Bangladesh. Additionally, issues of deforestation, erosion, and landslides are widespread in the mountainous lands of the north.

◀ **Figure 12.8 Flooding in Bangladesh** Devastating floods are common in the low-lying delta lands of Bangladesh. These problems result from two different sources. Not only is this area plagued by floods from rivers rising in the Himalaya mountains, but the delta is also vulnerable to flooding from strong tropical cyclones (hurricanes) from the Indian Ocean. *(Baldev/Sygma Photo News)*

▲ **Figure 12.9 Chipko tree-huggers** As a protest against deforestation in their countryside, women have resorted to the ancient Hindu practice of hugging trees to prevent them from being cut. Because of these protests, the state of Uttar Pradesh recently banned commercial logging. *(Rod Johnson/ Panos Pictures)*

Others, however, look to the traditional military elite's passion for hunting, which resulted in habitat protection and hunting reserves for large mammals. Hindu military castes, like the former Muslim and British rulers, were not vegetarians.

Though the current state of wildlife protection may be positive, the future is uncertain as population pressure mounts to convert wildlands to farmlands. The best remaining area of extensive wildlife habitat is in India's northeast, an area subjected to rapid immigration. Moreover, wild animals, particularly tigers and elephants, threaten crops, livestock, and even people living near the reserves. Adjacent to the Sundarbans reserve, almost 200 people were injured or killed by tigers in a recent five-year period. When a rogue elephant herd ruins a crop or a tiger kills livestock, government agents are usually forced to destroy the animal. As result, several key species now seem to be in decline, even though the government and conservation groups continue to struggle on their behalf.

Population and Settlement: The Demographic Dilemma

South Asia, given its current demographic situation, could soon surpass East Asia to become the world's most populous region. India alone is home to more than 988 million people, second only to China in size, while Pakistan and Bangladesh, at 142 and 124 million inhabitants, respectively, both rank among the world's 10 most populous countries (Table 12.1). Furthermore, the largest South Asia countries are growing almost twice as fast as those of East Asia. India, for example, has an annual rate of natural increase of 1.9 percent compared to China's 1.0 percent; at these rates India adds some 18 million people each year contrasted to China's 12 million.

This rapid population growth takes a high toll on South Asia, for it is the world region with the largest number of under- and malnourished people. About a third of India's population lives below the country's official poverty line; Bangladesh is poorer still, with two-thirds of its children classified by the World Bank as underweight, and thus, malnourished.

Though the region has made remarkable gains over the last several decades in its agricultural sector, there is still widespread concern over the region's ability to feed itself; in fact, the threat of famine is still very real because much of South Asia's agriculture remains highly vulnerable to the fickle monsoon rains or, often worse, disastrous flooding along rivers and in lowlands. Although recent gains in economic development have been heartening, they are largely neutralized by this region's rapid population growth; whether a vibrant industrial and export sector can provide the necessary bootstraps to pull up the whole region remains to be seen. Most experts are skeptical this can happen unless population growth is controlled.

The Geography of Family Planning

While all South Asian countries have some sort of family planning program in place, the commitment to these policies—along with the results—varies widely from place to place. A brief overview of population policies in the three largest South Asian countries provides important background for further discussion of the region.

India Widespread concern over India's population growth began in the 1960s and, to some extent, the measures taken over the last 40 years can be considered successful in that the TFR dropped from 6 in the 1950s to the current rate of 3.4 children. This national TFR figure, though, differs widely within India, from a low of 1.9 and 2.0 in the smaller states of Goa and Kerala, to a problematic high of 4.8 in Uttar Pradesh, which has more than 155 million people and would be the sixth largest country in the world were it independent. Because of the size of this Indian state, and because its TFR remains so high, much of the progress made in other areas of the country is effectively nullified. Besides the geographic differences between large and small states, national data on birthrates seems to vary by the important cultural variables of religion and education level. For example, within India, Muslim families appear to have larger families than do Hindu couples. Additionally, in many areas of South Asia there seems to be a strong relationship between women's education and family planning; in Sri Lanka, for example, as women's literacy increased, so did family planning.

Sterilization has been the predominant method of family planning, with 27 percent of India's married women now sterilized, which is the highest rate in the world. Male sterilization, as well, has been common. However, in both cases criticism has been directed at government agencies and family planners for setting up rigid quotas for sterilization, reneging on promises of agricultural loans and land grants in return for sterilization, and a high rate of infection, even death, resulting from hurried operations. Further criticism implies that the high rate

Table 12.1 Demographic Indicators

Country	Population[a]	Natural Increase	TFR[b]	%<15[c]	% Urban
India	988.7	1.9	3.4	36	26
Pakistan	141.9	2.8	5.6	41	28
Bangladesh	123.4	1.8	3.3	43	16
Nepal	23.7	2.2	4.6	43	10
Bhutan	0.8	3.1	5.6	43	15
Sri Lanka	18.9	1.3	2.2	35	22
Maldives	0.3	3.3	6.4	46	25

[a]Population in millions, 1996.
[b]Total fertility rate.
[c]Percentage of population younger than 15 years of age.
Source: *Population Reference Bureau, World Population Data Sheet,* 1998.

of sterilization was possible only because of the high rate of women's illiteracy, and that government family planners took advantage of illiterate rural women to achieve their quotas.

As is the case in China, there is a distinct cultural preference for male children in most of South Asia, a tradition that further complicates family planning in many areas. Where allowed (and they are now banned in some Indian states), sex determination clinics provide couples with information about the sex of the fetus through amniocentesis, and this seems to result in a higher rate of abortion for female fetuses. Consequently, in many areas a lower TFR is accompanied by a higher ratio of male infants compared to females.

Pakistan This large country of more than 141 million appears to have an ambivalent attitude toward family planning. While the government's official position is that the birthrate is too high, the country still lacks an effective, coordinated family planning program. As a result, the TFR remains very high at 5.6, and the current rate of natural increase of 2.8 percent adds about four million children each year. A partial explanation for the high birthrate may come from the fact that Pakistan still has a relatively high infant mortality rate: 91 per 1,000 children born, compared to 72 per 1,000 in neighboring India.

Another contributing factor to high population growth is that Pakistan has the lowest rates of female contraceptive usage for a major country in South Asia, with less than one woman in five using any sort of birth control method. While some attribute Pakistan's ambivalence toward family planning to its strong Muslim culture, the two are not necessarily mutually exclusive. The case of Islamic Bangladesh demonstrates this.

Bangladesh This country has one of the highest settlement densities in the world, with a population about half that of the United States packed into an area smaller than the state of Wisconsin. Although Bangladesh is predominantly Muslim, it has made significant strides in family planning; as recently as 1975 the TFR was 6.3, compared to the 3.3 found in the mid-1990s. Unlike India, most family planning comes from oral contraception, which is used by more than 50 percent of the women in Bangladesh (Figure 12.10).

The success of family planning can be attributed to strong support from the Bangladesh government that takes form in widespread media messages on radio and billboards and, as well, the presence of more than 35,000 women fieldworkers who take information about family planning into every village in the country.

The Settlement Landscape

South Asia is one of the least urbanized regions in the world, with only a quarter of its huge population living in settlements classified as cities; the majority of its immense population, then, lives in rural villages and small towns. Important to note, though, is that the Indian government classifies a settlement as a "city" only when it surpasses 5,000 inhabitants,

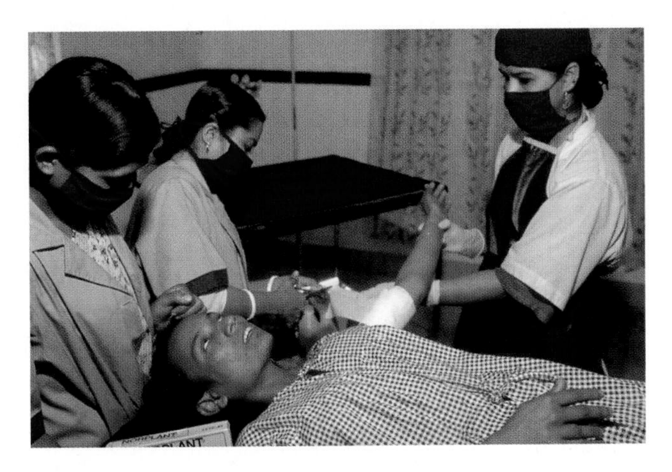

▲ **Figure 12.10 Family planning in Bangladesh** Bangladesh has been one of the most successful in South Asia in reducing its fertility rate through family planning. Many women in Bangladesh use oral contraceptives. This photo shows a woman receiving a contraceptive implant. *(Peter Barker/Panos Pictures)*

which is a relatively high threshold for an urban place; in Europe and North America, for example, settlements of more than 2,500 are commonly referred to as cities. Given India's definition of a city, millions of people living in small towns of 3,000 or 4,000 are classified as rural.

That said, one of the important themes in South Asia's settlement and population geography is the rapid migration from village to the region's large cities. This results more from the push forces of desperate conditions in the countryside than from the attraction of employment or security in the city. A primary cause of dislocation from the countryside are changes in agriculture: increased mechanization with resulting under- or unemployment; the higher costs of farming modern crops; expansion of large estate farms at the expense of subsistence farming; uncertain access to irrigation water; and the environmental deterioration of lands that have been cultivated for four and five centuries. Currently, fully two-thirds of India's population is employed in agriculture, which is a very high percentage for a developing country, and a figure that most economists feel will fall dramatically in the next decades with modernization of the agricultural sector.

The most densely settled areas of South Asia still coincide with agricultural activities associated with fertile soils and dependable water supplies (Figure 12.11). In India, the largest rural population is found in the core area of the Ganges River valley; following that, the coastal plains, both east and west, stand out with their dense village populations. Settlement is less dense on the Deccan Plateau, in rain shadow areas of the Western Ghats, and where water supplies are less dependable.

In Pakistan, the highest population densities are found in the more humid highland regions of the north, then along the Indus River where irrigation water is plentiful. Removed from this valley, though, the semiarid portions of the lower Indus Valley are sparsely populated, as is the delta itself. This situation is a striking contrast to the deltaic lands of the Ganges-Brahmaputra system in eastern India and Bangladesh, which has some of the highest population densities in the world.

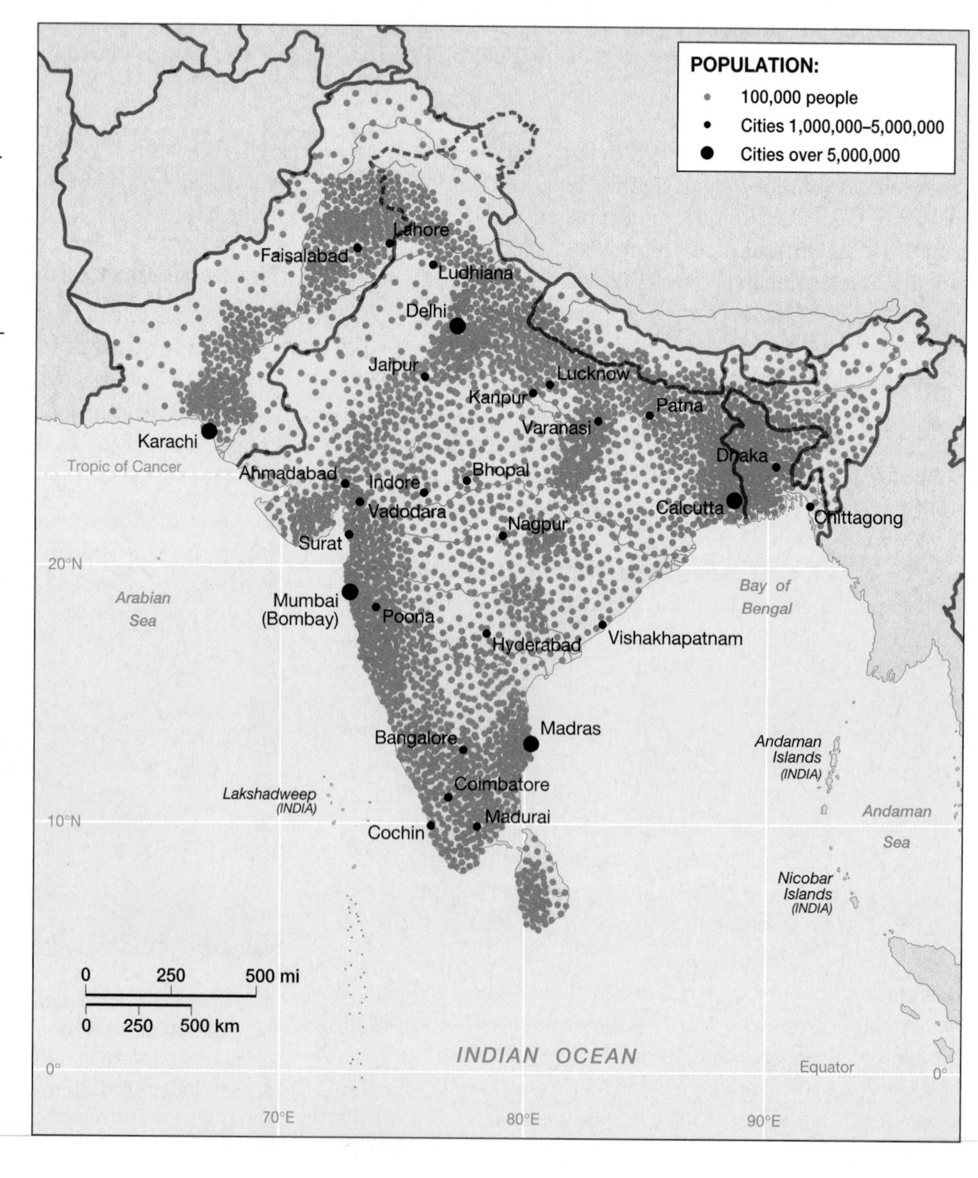

Figure 12.11 Population map of South Asia Except for the desert areas of the west and the high mountains of the north, South Asia is a densely populated region. Besides large urban areas, the highest densities are found associated with a landscape of agricultural villages that are often clustered around water sources, such as streams, wells, irrigation, or small tanks or reservoirs that store water between monsoon rains. This map also shows the dense populations found in lowland and coastal locations with reliable monsoon rainfall.

Agricultural Regions and Activities

South Asian agriculture has historically been relatively unproductive, especially when compared with that of East Asia. Although the reasons behind such poor production are complex, many experts cite the relatively low social status of cultivators and the fact that much farmland has long been controlled by elite social groups with little commitment to feeding the larger population. Others place the blame on the legacy of British colonialism, which emphasized exotic export crops for European markets or cotton and silk for England's textile industry.

Regardless of the causes, low agricultural yields in the context of a huge, rapidly growing population that is largely underfed constitute a pressing problem. Since the 1970s, however, agricultural production has grown faster than population growth primarily because of the **Green Revolution,** the term given to agricultural cultivation techniques using genetically altered seeds, irrigation, heavy use of industrial fertilizers, and chemical pesticides. Because the Green Revolution seems to also carry high social and environmental costs along with improved yields, this program has been rather controversial, as will be discussed shortly.

▲ **Figure 12.12 Rice cultivation** A great amount of irrigation water is needed to grow rice, as is apparent from this photo from Sri Lanka. Rice is also the main crop in the lower Ganges Valley, along the Indus River of Pakistan, and in the delta lands of Bangladesh. *(Mahaux Photography/The Image Bank)*

Crop Zones South Asia can be divided into a number of distinct agricultural regions, all with different problems and different potentials. In the most basic terms, the region can be divided into three broad agricultural zones based upon the three primary subsistence crops of rice, wheat, and millet.

Rice is the main crop and foodstuff in the lower Ganges Valley, along the lowlands of India's eastern and western coasts, in the delta lands of Bangladesh, along Pakistan's lower Indus Valley, and in Sri Lanka. This distribution expresses the high volume of irrigation water needed to grow rice. The sheer amount of rice grown in South Asia is impressive: India ranks only behind China in world rice production, and Bangladesh is the fourth largest producer after those two countries and Indonesia (Figure 12.12).

Wheat is the principal crop of the northern Indus Valley and in the western half of India's Ganges Valley. South Asia's "breadbasket" is the northwestern Indian state of Punjab, adjacent to Pakistan. Here the Green Revolution has been particularly successful in increasing grain yields. In the less fertile and drier areas of central India, millet and sorghum are the main crops, along with root crops such as manioc and sweet potatoes. In general, wheat and rice are the preferred staples throughout South Asia (and, indeed, most of the world) and it is generally the poorer people who consume the "rough" grains such as the various millets.

Many other crops are also widely cultivated in South Asia, some commercial, others for local subsistence. Oil seeds, such as sesame and peanuts, for example, are grown in semiarid districts, and the humid southwestern Malabar coast of Kerala state is noted for its coconut groves. In both Pakistan and west central India cotton is grown for export, while Bangladesh has long supplied most of the world's jute, a tough fiber used in the manufacture of rope. In general, though, the cultivation of stable foods for the local population has been of higher priority than export crops, particularly since the end of British colonialism.

Livestock Many if not most South Asians receive inadequate protein, and meat consumption is extremely low. Part of this is simply a reflection of poverty, since meat is expensive to produce and thus can be consumed only by wealthier people. In India the Hindu religion is an equally important factor because most Hindus (but not all) are vegetarians.

Despite this prohibition against meat-eating, animal husbandry is vitally important throughout South Asia. India, for example, has the world's largest cattle population, partially because cattle are sacred animals to Hindus, but also because milk is one of the South Asia's main sources of protein. While India's cattle have traditionally produced little milk, a so-called *white revolution* that has increased dairy efficiency has brought about larger milk supplies in recent decades. Cattle are also widely used as farm animals, though they are gradually being replaced by tractors. Additionally, cattle manure is a major source of household fuel for the poorer populations of South Asia, a practice that furthers the impoverishment of agriculture, since this prevents the use of animal manure for fertilizing soils.

The Green Revolution The only reason South Asian agriculture has been able to keep up with its rapid population growth is because of the Green Revolution, which originated during the 1960s in agricultural research stations established by international development agencies concerned with feeding the world's burgeoning population. One of the major problems researchers faced was the fact that higher yields could not be readily attained by simply fertilizing native seed

▲ **Figure 12.13 Green Revolution farming** Because of "miracle" wheat strains that have increased yields in the Punjab area, this region has become the breadbasket of South Asia. India more than doubled its wheat production in the last 25 years and has moved from a region of chronic food shortages to self-sufficiency. *(Earl Kowall/Corbis)*

strains, since the plants would simply grow taller, then fall to the ground before the grain could mature. The solution was to cross-breed new seeds that were essentially dwarf crop strains that would respond to heavy chemical fertilization by producing extra grain rather than longer stems and more abundant foliage (Figure 12.13).

By the 1970s it was clear that these efforts had succeeded in reaching their initial goals on increasing agricultural production. New "miracle wheat" varieties were quickly adopted by the more prosperous farmers of the Punjab area, which solidified its position as the breadbasket of South Asia. Green Revolution rice strains were also adopted in more humid areas with access to irrigation water. As a result, South Asia was transformed from a region of chronic food deficiency to one of self-sufficiency. As an example, India more than doubled its grain production between 1970 and the mid-1990s, from 80 to 191 million tons. Nonetheless, because of the region's poverty and the inability of poor people to participate in a market economy, the benefits of this Green Revolution were shared primarily by the more wealthy urbanites.

While the Green Revolution was clearly an agricultural success, many argue that it has been a ecological and social debacle. Serious environmental problems result from the chemical dependency of the new crop strains. Not only do they typically need large quantities of industrial fertilizer, which is expensive and also causes significant water pollution problems, but these "miracle" crops also require frequent pesticide applications since they lack natural resistance to local plant disease and pests.

Social problems have also followed the Green Revolution. In many areas, particularly where wheat is cultivated, only the more prosperous farmers are able to afford the new seed strains, irrigation equipment, farm machinery, fertilizers, and pesticides necessary to support this new high-technology agriculture. As a result, poorer farmers have been forced from their lands because the low-grade crops they produce cannot compete with the Green Revolution products. Many of these farmers either become wage laborers for their more successful neighbors or migrate to the region's already overcrowded cities.

Though the Green Revolution has been subjected to serious criticism, it also has been staunchly defended. Some advocates contend that its environmental dangers have been mitigated of late as farmers demonstrate they can grow new crop varieties using traditional methods of fertilizer and pest control. Others contend that overall poverty has actually decreased in the Punjab, overlooking for the moment that those most negatively affected by the Green Revolution have probably left the area. On a regional scale, the more important issue is whether South Asia could have fed its large and growing population during the last three decades without the Green Revolution. Most agree that traditional farming strategies could not have kept pace with population growth.

Future Food Supply While the Green Revolution has fed South Asia's expanding population over the past several decades, it remains unclear whether it will be able to continue doing so in the near future. Many of these crop improvements have seemingly exhausted their potential, though optimists believe that genetic engineering will provide another breakthrough that will see a second wave of increased crop yields.

Others place hope in expanded irrigation, since many fields remain unirrigated in South Asia's semiarid areas and even in the humid areas, dry-season fallow is often the norm. Irrigation, however, brings its own problems. In much of Pakistan and northwestern India, where irrigation has been practiced for generations, soil **salinization,** or the buildup of salt in agricultural fields, is already a major constraint that limits agricultural production. Additionally, water tables are falling in the Punjab, India's breadbasket, because double-cropping of Green Revolution wheat and rice has pushed water use beyond the sustainable yield of the underlying aquifers. Villagers and farmers without capital to deepen their wells are left high and dry, forcing them to abandon irrigated agriculture. Because of this, some experts see future food supplies actually decreasing.

Urban South Asia

Although South Asia is one of the least urbanized regions of the world, with about a quarter of its population living in cities (see Table 12.1), this does not suggest that cities are a minor part of the settlement landscape; on the contrary, South Asia has are some of the largest cities in the world—for example, Mumbai (Bombay) at more than 15 million—and India, alone, lists more than 30 cities with populations greater than a million. Perhaps more importantly, though, these South Asian cities are growing rapidly because of migration from the countryside.

Because of this rapid growth, most South Asian cities have staggering problems with homelessness, poverty, congestion, water shortages, air pollution, and sewage disposal. Calcutta's homeless are legend, with perhaps half a million sleeping on the streets each night. In that city and others, sprawling squatter settlements, or **bustees,** mushroom in and around South Asia cities, providing temporary shelter for the more fortunate urban migrants. A brief survey of the region's major cities gives clue to the problems and prospects of the region's urban areas.

By far the largest city in South Asia and, if current rates of growth continue, a city that will soon be one of the world's largest, Mumbai (Bombay) is India's financial, industrial, and commercial center; additionally, it is the major port on the Arabian Sea, the closest to Europe through the Suez Canal, and responsible for half the country's foreign trade (Figure 12.14). Long noted as the center of India's cotton industry, the city is also center to the country's large film industry.

▲ **Figure 12.14 Mumbai (Bombay) central city** The heart of this metropolitan area of 20 million is on the peninsula of Bombay Island, which is surrounded by the bay. Because space is limited, most recent growth has been to the east and to the north of this peninsula area. *(Steve McCurry/ Magnum Photos, Inc.)*

With about 10 million people within the city itself and another five or six million in the immediate urban area, Mumbai (Bombay) will soon challenge other megacities as one of the largest. This growth, however, brings a high cost, not just with congestion and other infrastructural problems, but, because of the city's economic vitality, ethnic groups from all over South Asia are drawn to this metropolitan area, resulting in ethnic tensions and strife in the sprawling suburbs.

Perhaps as reaction to this ethnic invasion, the nationalist party that governs the city wished to assert its local ethnic identity by officially changing the city's name from Bombay—a British colonial name—to Mumbai, the city's historical name in Marathi, which stems from the Hindu goddess Mumba. Mumbai was approved by the federal government in January 1996 as the city's official name. Though the political implications of the name change are highly charged, residents seem to take it in stride by using use both names interchangeably.

Mumbai (Bombay) occupies a peninsular site originally composed of seven small islands, but since the seventeenth century, drainage and reclamation projects, along with the construction of bridges and causeways, have joined the islands into a larger body known simply as Bombay Island; two parallel ridges of 100 feet (30 meters) form the spines of the inner city, one occupied by the historic fort, which is the commercial center, the other, Malabar Hill, an expensive residential area. Because of restricted space on the island, most growth has taken place to the north and east of the historic peninsula so that now the Mumbai (Bombay) metropolitan region comprises an area 10 times the size of the original island city.

Delhi, the sprawling capital of India, has more than 11 million people in its greater urban area and consists of two contrasting landscapes expressing its past: Delhi (or old Delhi) was the Muslim capital between the seventeenth and nineteenth centuries and contains the largest mosque in India, adjacent to the Red Fort on the hilly bank of the Yamuna River; New Delhi, in contrast, is a city of wide boulevards, monuments, parks, and expansive residential areas. It was born as a planned city when the British moved their colonial capital from Calcutta in 1911. Located here are the embassies, luxury hotels, government office buildings, and airline offices necessary to a vibrant political capital. South of the government area are more expensive residential areas with such names as Defence Colony, South Extension, Lodi Colony; often these neighborhoods are focused on expansive parks and ornamental gardens (see "The Urban Scene: Round-up Time in Delhi").

Rapid growth, along with the government's inability to control auto and industrial emissions, has caused Delhi to have an air pollution problem that ranks as one of the worst in the world. Simply breathing Delhi's air on a smoggy day is equivalent to smoking a pack of cigarettes. In a largely symbolic move, the city government attempted to address the problem in 1997 by banning smoking in public places.

To many, Calcutta is emblematic of the insurmountable problems faced by rapidly growing cities in developing countries. Not only is homelessness a major issue, but this city of more than 12 million also cannot supply its residents with water, power, or sewage treatment. Electrical power, for example, is so hopeless that every hotel, restaurant, shop, or small business has to have some sort of standby power generator or battery lighting system. No other Indian city has problems of this magnitude.

Aggravating the poverty, pollution, and congestion of Calcutta is the chronic labor unrest resulting from the decline of its suffering industry. As a British creation, Calcutta's primary role was as a trading center and port city on the Hooghly River. The river, though, is silting up, making navigation from Calcutta down to the sea increasingly difficult and limiting the size of ships that can use the port. Furthermore, the independence of colonial South Asia hurt Calcutta more than any other Indian city. Until partition, Calcutta was the jute-producing and export center of India, but with the drawing of the Bangladesh border just to the east, Calcutta became a city without a large trade area. This economic decline was also accompanied by a massive influx of Hindu refugees from

THE URBAN SCENE Round-up Time in Delhi

About 40,000 cows wander the streets of Delhi, taking full advantage of their sacred status by slowly crossing a busy highway or relaxing in a crowded intersection or snacking at an open-air vegetable stand. Revered as these cows may be, most live the life of the homeless, for they are usually unwanted animals, turned loose because they are old and milkless. Or, at best, milk producers who belong to city dwellers who do not feed them. With little grass to graze on in this paved-over landscape of 13 million people, these cows often scavenge through household trash looking for table scraps. Increasingly this causes problems for the cows because most garbage is packed into plastic bags that are not digestible in any of the cow's four stomachs. While animal rights groups have asked that these garbage bags be banned, that is not likely in the near future.

Dr. Vijay Chaudry, the veterinarian who runs a shelter that cares for Delhi's sacred animals, says that he has found glass, iron, wire, electrical cords, shoes, shirts, and razor blades inside cows. But the real killer is the plastic garbage bag. "We lose two or three cows a day," he says, "and when we cut them open, it is terrible what we find. For an animal so sacred, they die a bad death."

To try to keep these cows out of trouble, Delhi employs about 100 cow catchers, whose job it is to keep the cows out of trouble by herding them away from busy streets and, more troubling, away from the many trash piles found throughout the city. Usually it takes eight men to capture a street-smart cow; getting the cow into the small truck they use is a real challenge, aggravated by the fact that Delhi citizens who gather to watch the contest usually root for the cow. "It is necessary, if misunderstood work," says Raman Kumar Sharma, a crew chief for the cow catchers. "Sometimes people do not realize we have the cow's best interests at heart. We've had violence with the crowds."

Hindus venerate the cow as a symbol of motherhood and a source of life. Killing cows is banned in most of India's 27 states, though there seems to be no shortage of juicy steaks for those who can afford them. Beef is sold on the black market by butchers who deliver prime cuts door to door, yet never discuss how the meat was obtained.

Source: Adapted from Barry Bearak, 1998. "In New Delhi, Cows and Cow Catchers Run Risks," *International Herald Tribune,* October 22, 1998, page 12.

the eastern areas, first in 1947, and then again with Bangladeshi independence in 1971.

With continued rapid growth as migrants come from the countryside, a mixed Hindu-Muslim population that creates intense ethnic rivalry, a declining economic base, and an overloaded infrastructure, this city faces an uncertain and problematic future.

As the capital and major city of the Bangladesh, Dhaka (also spelled Dacca) has experienced rapid growth because of the migration from the surrounding countryside; in 1971, when the country gained independence from Pakistan, Dhaka had about a million inhabitants, yet today this urban area numbers close to 8 million. Like Calcutta, Dhaka shares a long history as a colonial trade center and river port city (Figure 12.15).

▲ **Figure 12.15 Dhaka street scene** This busy capital of Bangladesh was a former colonial river port but is now a global center for the manufacturing of clothing, shoes, and sports equipment. As seen in this photo, a common mode of transportation is the human-powered rickshaw. *(Dirk R. Frans/ Hutchison Picture Library)*

Unlike Calcutta, though, its economic vitality has ascended since independence as it combines the administrative functions of government with the largest industrial concentration in Bangladesh. Along with traditional products of muslin and embroidery, cheap and abundant labor in Dhaka has made the city a global center for clothing, shoe, and sports equipment manufacturing. North American shoppers are readily reminded of this new role by looking through clothing or sports equipment tags at a department store; these tags remind us Bangladeshi products are now everywhere.

Karachi, a sprawling port city of more than seven million people (and some estimates go to 11 million), is Pakistan's largest urban area and its commercial center, primarily because of the excellent harbor. It was also the country's capital until 1963 when the new city of Islamabad was created in Pakistan's northeast. The city, however, has suffered little from the exodus of government functions; it is, in many ways, the most cosmopolitan city in Pakistan, with its checkerboard street pattern lined with high-rise buildings, hotels, banks, and travel agencies. Indeed, in many ways the landscape conveys the sense that Karachi is a model metropolis for a developing country.

However, Karachi suffers from serious political and ethnic tensions that have turned parts of the city into armed camps as ethnic groups use the city as their battleground. During the worst of this violence in 1995, more than 200 people were killed on the city's streets each month, and army bunkers were in place at major urban crossroads. The major problems are between the Sindis, who are the region's indigenous inhabitants, and the Mohjirs, the Muslim refugees from India who settled in the region and city after independence and partition in 1947. There are also clashes between Sindis and Biharis, Urdu-speaking migrants from Bihar state. As a result, travel both within the city and beyond is usually a problem, particularly for international travelers, an unpleasant fact that tends to inhibit the city's commercial activity.

Though considerably smaller than Karachi with its five million inhabitants, Lahore is Pakistan's cultural, educational, and artistic center, and its landscape expresses its central role in the region's Muslim history with striking mosques, palaces, and monuments (Figure 12.16). After the British captured the city in 1846, they, too, added their indelible stamp to the cityscape with parks, monuments, and Victorian buildings. Migration to this city has been high in the last decades, with much of it linked to the ongoing geopolitical problems of the Punjab region, though the city has not suffered from the same level of violence as Karachi.

As noted earlier, upon independence Pakistan's largest city, Karachi, became the capital of the new country. That city, however, was thought to be too far from the cultural center of the country and shortly thereafter a new capital was planned from scratch, one that would make a statement through its name—Islamabad—about the religious foundation of Pakistan and, second, a capital city that would be located close to the contested region of Kashmir. In geographic terms, such a move is referred to as a **forward capital** since it makes a statement—both symbolically and geographically—about the intentions of the country. By building its new capital in the northeast, Pakistan sent a clear message to India that it was not giving up its claims to the contested region of Kashmir.

The new capital of Islamabad is closely linked to the historic military city of Rawalpindi, the major British encampment in the region until 1947, which is about 8 miles (13 kilometers) away. While these two cities form a metropolitan region of less than a million people, they are completely different in appearance and character, as one might expect with an explicitly planned, modern capital city. To avoid chaos and congestion, the Greek urban planner Doxiades designed Islamabad to grow along an axis in sectors toward Rawalpindi; each sector would be self-sufficient, with its own government buildings, residences, and shops. At this point only six sectors are complete, offering a rather homogenous (even bland) suburban landscape.

▲ **Figure 12.16 Lahore landscape** Lahore is Pakistan's cultural, educational, and artistic center, and plays a central role in the country's Islamic national identity. *(Nik Wheeler)*

Cultural Coherence and Diversity: A Common Heritage Divided by Religious Rivalries

Historically, South Asia is a relatively well-defined cultural region. A thousand years ago, virtually the entire area was united by the ideas and social institutions associated with Hinduism. The subsequent arrival of Islam added a new religious dimension without fundamentally undercutting the region's cultural unity (Figure 12.17). Since the mid-twentieth century, however, religious strife has intensified, leading some to question whether South Asia can still be conceptualized as a culturally coherent world region.

Though India has been a resolutely secular state since its inception, the Congress Party, its guiding political organization from the 1940s through the 1980s, has struggled to keep politics and religion separate, and it has relied heavily on Muslims as well as support from the lower "untouchable" castes. Since the 1980s, however, this secular political tradition has come under increasing pressure. India has also seen

▲ **Figure 12.17 Hindu-Muslim tensions** These two major religions give much of South Asia its cultural character, although tensions between the two groups remain high in some areas, such as within areas of India, and between India and neighboring Pakistan. *(John Moore/AP/Wide World Photos)*

the growth of radical Hindu "fundamentalism," which is probably better referred to as **Hindu nationalism;** these nationalists promote Hindu values as the essential and exclusive fabric of Indian society. Not only have Hindu nationalists gained considerable political power in many Indian states through the Bharatiya Janata Party (BJP), but open agitation against the country's Muslim minority has become rampant. In several high-profile instances, Hindu mobs have demolished Muslim mosques that had allegedly been built on the sites of ancient Hindu temples. To illustrate, the destruction of a Muslim mosque in Ayodhya located in the Ganges Valley galvanized the nationalistic BJP membership in 1992. This intensifying Hindu-Muslim strife is one of the great tragedies of modern South Asia.

Origins of South Asian Civilizations

Many scholars think that the roots of South Asian culture extend back to the Indus Valley civilization, which flourished in what is now Pakistan more than 5,000 years ago. This remarkable urban-oriented society vanished almost entirely in the second millennium B.C., after which the trace grows dim. By 800 B.C., however, a new urban focus had emerged in the middle Ganges Valley (Figure 12.18). The social, religious, and intellectual norms associated with this civilization eventually spread throughout the lowlands of South Asia. Although pronounced local diversity continued to exist, by the first millennium A.D., a traveler could have observed similar political institutions and religious practices throughout South Asia.

Hindu Civilization The religion, or more precisely religious complex, of this early South Asian civilization was Hinduism, which is tremendously complicated, incorporating diverse forms of worship and lacking any orthodox creed. Certain deities are recognized, however, by all believers, and the notion that these various gods are all manifestations of a single divine entity is widespread (Figure 12.19). All Hindus, moreover, share a common set of sacred epic stories, usually written in **Sanskrit,** which is the sacred language of this religion and culture. Additionally, the religion is noted for its mystical tendencies, which have long inspired many men (seldom women) to seek an ascetic lifestyle, renouncing property and sometimes all regular human relations. One of Hinduism's hallmarks is a belief in the transmigration of souls from being to being through reincarnation; the nature of one's acts in the physical world influences the course of these future lives.

Scholars once confidently argued that Hinduism originated from the fusion of two distinct religious traditions: the mystical beliefs of the subcontinent's indigenous inhabitants (including the people of the ill-fated Indus Valley civilization), and the sky-god religion of the Indo-European invaders who swept down onto the region sometime in the second millennium B.C. from central Asia. Such a scenario also proved convenient for explaining India's **caste system,** the strict division of society into different hierarchically ranked hereditary groups. The elite invaders, according to this theory, wished to remain separate from the people they had vanquished. Since the invading society itself was already divided into three main groups (warriors, priests, and traders), an elaborate system of social division was soon implemented. Recent research, however, has called this theory into question, and it now seems more likely that the caste system emerged through a more gradual process of social differentiation.

Buddhism While a caste system of some sort seems to have existed in the early Ganges Valley civilization emerging around 800 B.C., it was soon challenged from within by Buddhism. Prince Siddhartha Gautama, the Buddha, was born in 563 B.C. to an elite caste. He rejected the life of wealth and power that was laid out before him, however, and sought instead to attain enlightenment, or the mystical union with the cosmos. He preached that the path to such *nirvana* was open to all, regardless of social position. His followers eventually established Buddhism as a new religion based largely in monastic communities (Figure 12.20). Buddhism spread through most of South Asia, becoming something of an official faith under the Mauryan Empire, which ruled much of the subcontinent in the third century B.C. Later centuries saw Buddhism expand through most of East, Southeast, and Central Asia.

But for all of its successes abroad, Buddhism never replaced Hinduism in India. It remained focused on monasteries rather than permeating the wider society. Moreover, many Hindu priests struggled against the new faith. One of their techniques was to embrace many of Buddhism's philosophical ideas and enfold them within the malleable intellectual system of Hinduism. By A.D. 500, Buddhism was on the retreat throughout South Asia, and 500 years later, it had virtually disappeared from the South Asian region. The only exceptions were certain peripheral areas, notably the island of Sri Lanka and the high Himalayas, both of which remain mostly Buddhist to this day.

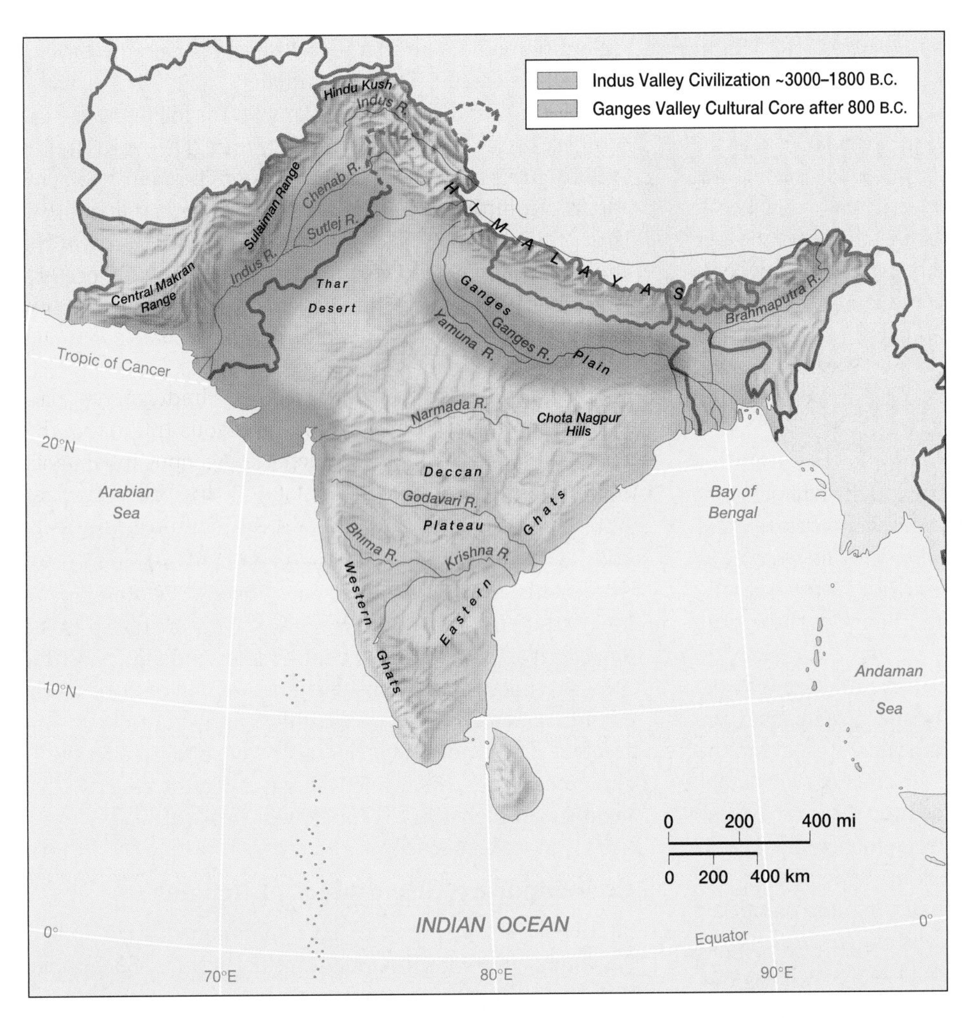

◀ **Figure 12.18 Early South Asian civilizations** The roots of South Asian culture may extend back 5,000 years to an Indus Valley civilization based on irrigated agriculture and vibrant urban centers. What happened to that civilization remains a topic of conjecture, because the archaeological record grows dim by 1800 B.C. However, later, a new urban focus emerged in the Ganges Valley, from which social, religious, and intellectual influences spread throughout lowland South Asia.

▲ **Figure 12.19 Hindu temple** Hindu temples dot India's landscape, taking many different forms, devoted to a wide range of deities. This is the Bhuthanatha temple at Lake Agastya, which is dedicated to the goddess Shiva. *(Ray Wood/ Panos Pictures)*

▲ **Figure 12.20 Buddhist monastic landscape** Buddhism arose from within Hinduism through the challenges of Prince Siddhartha Gautama, the Buddha, in 563 B.C. The role of monks and monasteries remains central still today, which may explain why its distribution is rather limited compared to religions such as Hindu and Islam. *(Linde Waidhofer/Liaison Agency, Inc.)*

Arrival of Islam The next major challenge to Hindu society—Islam—came from the outside. Arab armies conquered the lower Indus Valley (an area called *Sind*) around A.D. 700, but advanced no farther. Then around the year 1000 Turkish-speaking Muslims began to invade from Central Asia. At first they merely raided Hindu temples, but eventually they began to settle and rule on a permanent basis. By the 1300s, most of South Asia was under Muslim power, although Hindu kingdoms persisted in southern India and in the arid lands of Rajasthan. Later, during the sixteenth and seventeenth centuries, the **Mughal Empire,** the most powerful of the Muslim empires, dominated the region from its capital in Lahore, located in the upper Indus Valley.

At first, Muslims formed a small ruling elite, but over time increasing numbers of Hindus converted to the new religion, particularly those from lower castes who were then freed of the rigid social order of Hinduism. Conversions were most pronounced in the northwest and northeast, and eventually the areas now known as Pakistan and Bangladesh became predominantly Muslim.

At first glance, Islam and Hinduism are strikingly divergent faiths. Islam is resolutely monotheistic, austere in its ceremonies, and spiritually egalitarian (all believers stand in the same relationship to God). Hinduism, by contrast, is polytheistic, lavish in its rituals, and caste-structured. Because of these profound differences, there is a tendency to view the Hindu and Muslim communities in South Asia as utterly distinct, living in the same region, but not sharing the same culture or civilization. Increasingly, such a view is expressed in South Asia itself. Many residents of modern Pakistan, which has very few Hindu inhabitants—stress the Islamic nature of their country and its radical separation from India. In India, which has a Muslim minority some 80 million strong, religious animosity is threatening the political fabric of the country.

Yet by overemphasizing the separation of Hindu and Muslim communities, one risks missing much of what is historically distinctive about South Asia as a cultural region. Until the twentieth century, Hindus and Muslims usually coexisted on fairly amicable terms; the two faiths stood side by side for hundreds of years, during which time they came to influence each other in many ways. In earlier generations, especially in rural areas, many people actually participated in both Hindu and Muslim ceremonies. Moreover, certain aspects of caste organization have persisted in both religions.

The Caste System In the foregoing discussion, caste has been cited as one of the historically unifying features of the South Asian cultural region. Caste structure is not, however, and has never been, uniformly distributed across the subcontinent. It has never been significant in India's tribal areas, in modern Pakistan and Bangladesh its role is fading, and in the Buddhist Sinhalese society of Sri Lanka its influence has long been marginal. Even in India, caste is now being de-emphasized by many educated and professional segments of society. But for all of these disclaimers, it remains undeniably true that caste has been of tremendous social and historical significance over almost the entire region, and that it continues to structure day-to-day social existence for most Indians.

Caste is actually a rather clumsy term for denoting the complex social order of the Hindu world. The word itself is of Portuguese rather than indigenous origin; more problematic, it combines two distinct local concepts: *varna* and *jati*. *Varna* refers to the ancient fourfold social hierarchy of the Hindu world, whereas *jati* refers to the hundreds of local endogamous ("marrying within") groups that exist at each varna level (different jati groups are thus usually called *subcastes*). Jati, like varna, are hierarchically arranged, although the exact order of precedence is not clear-cut (see "Cultural Tensions: More About the Caste System in India").

It has often been argued that the essence of the entire caste system is the notion of social pollution. The lower one's position in the hierarchy, the more potentially polluting one's body is. Members of higher castes are not supposed to eat or drink with, or even use the same utensils, as members of lower castes. At the bottom of the hierarchy are the so-called *untouchables,* now sometimes called *dalits,* individuals whose mere presence is often considered defiling. In parts of southern India, some groups were until recently considered "unseeable." These unfortunate persons were obligated to loudly announce themselves while walking down the street so that members of higher castes could avert their gaze.

Contemporary Geographies of Religion

South Asia, as we have seen, has a predominantly Hindu heritage overlain by a substantial Muslim imprint. Such a picture fails, however, to capture the enormous complexity of modern religious expression in contemporary South Asia. The following discussion looks specifically at the geographical patterns of the region's main religions (Figure 12.21).

Hinduism Fewer than one percent of the people of Pakistan are Hindu, and in Bangladesh and Sri Lanka Hinduism is a distinctly minority religion. Almost everywhere in India, however—and in Nepal, as well—Hinduism is very much the religion of the majority. In east-central India, more than 95 percent of the population is Hindu. Hinduism is itself a geographically complicated religion, with different aspects of the faith, such as the worship of specific deities, varying highly between different parts of India. Even within a given state, the worship of deities differs greatly from region to region, even between neighborhoods in a large city.

Islam Islam may be considered a "minority" religion for the region as a whole, but such a designation obscures the tremendous importance of this religion in South Asia. With more than 300 million members, the South Asian Muslim community is one of the largest in the world. Bangladesh and especially Pakistan are overwhelmingly Muslim. India's Islamic community, although constituting only some 10 percent of the country's population, is still some 80 million strong—a figure larger than the total population of any country in the Muslim heartland of Southwest Asia and Northern Africa.

CULTURAL TENSIONS More About the Caste System in India

The caste system is a highly complicated social system that loses much through simplication and stereotyping. Here, we offer more detail on this important dimension of Indian society.

At the apex of the varna system sits the Brahmans, members of the traditional priestly caste. Brahmans perform the high rituals of Hinduism and they form the traditional intellectual elite of India. Almost all Brahman groups traditionally value education highly, and today they are disproportionally represented among India's professional classes. In many areas of India, Brahmans have historically controlled large expanses of land, providing them with substantial wealth.

Below the Brahmans in the traditional hierarchy lie the Kshatriyas, members of the warrior or princely caste. In premodern India this group actually had far more power and wealth than did the Brahmans; it was they who ruled the old Hindu kingdoms. Belonging to an expressly military group, Kshatriyas are exempted from many of the restrictions encountered by the other high-status groups. Whereas Brahmans are prohibited from handling weapons, for example, Kshatriyas traditionally considered this to be an occupational duty. Unlike members of other high castes, they are seldom vegetarians.

Next stand the Vaishyas, members of the traditional merchant caste. Although ranked below the Brahmans and the Kshatriyas, this is still an elite caste. In earlier centuries, a near monopolization of long-distance trade and money-lending gave many Vaishyas ample opportunities to accumulate capital. Like the Brahmans, Vaishyas are noted for their exacting standards of personal behavior. The Hindu precepts of vegetarianism and nonviolence, for example, are most highly developed among certain merchant subcastes of western India. One prominent representative of this tradition was Mohandas Gandhi, the founder of modern India and one of the twentieth century's greatest religious and political leaders.

The majority of India's population fits instead into the fourth varna category, that of the Shudras. The Shudra caste is composed of an especially large array of subcastes *(jati)*, most of which originally reflected occupational groupings. These varied endogamous groups are themselves ranked in a hierarchy of purity, although the distinctions between them are not always clear-cut. Most Shudra subcastes were traditionally associated with peasant farming, but others were based on craft occupations, including those of barbers, smiths, and potters.

While the Brahmans, Kshatriyas, Vaishyas, and Shudras form the basic fourfold scheme of caste society, another sizable group stands outside of the varna system altogether. These are the so-called *untouchables* (or *dalits*, as they are now sometimes called), individuals considered so polluting that they are not allowed to enter Hindu temples. Untouchables are also traditionally subdivided into more than 1,000 jati, some of which are considered more defiling than others. Such low status positions seem to have been derived in most instances from "unclean" occupations, such as those of leather workers (who dispose of dead animals), scavengers, latrine cleaners, and swine-herders. Their own traditional social codes often allowed behaviors considered repulsive by members of higher castes, such as eating pork or even beef.

Because "untouchables" are traditionally not even allowed to enter Hindu temples, it is sometimes questioned whether they can really be considered Hindus. Most untouchable groups, however, worship Hindu gods, and individuals among them even officiate at certain out-of-doors rituals (spiritually dangerous, or "unclean," rituals, that is) for the larger community. Not surprisingly, many dalits have converted to Islam, Christianity, and Buddhism in an attempt to escape from the caste system. Even so, they still suffer from discrimination.

The caste system is clearly in a state of flux in India today. Its original occupational structure has long been undermined by the necessities of a modern economy, and various social reforms have chipped away at the discrimination that it embodies. Mohandas Gandhi fought strenuously against caste prejudice, and consistently championed the cause of the untouchables. The dalit community itself has produced several notable national leaders who have waged partially successful political struggles. Owing to such efforts, the very concept of "untouchability" is now technically illegal in India, and a variety of affirmative-action plans have been enacted to give members of specific **scheduled castes** (like the "untouchables") special consideration in university admissions and government-sector employment. In several Indian states, moreover, successful alliances of low-caste groups have virtually excluded Brahmans from high-level government positions.

It would be a serious mistake, however, to conclude that caste discrimination has disappeared from modern India. A wide variety of caste restrictions also continues to structure the lives of the vast majority of Indian citizens—especially in matters of marriage. Only among a few professional groups in major Indian cities is it common to marry outside of one's caste or even subcaste.

As can be seen on Figure 12.21, Muslims live in almost every part of India. They are, however, concentrated in four main areas:

- In most of India's cities
- Kashmir, in the far north, particularly in the densely populated Valley of Kashmir around the capital city of Srinagar; 80 percent of the population here follows Islam
- The upper and central Ganges plain, the historical core of Islamic rule in South Asia; here, Muslims constitute 15 to 20 percent of the population
- The southwestern state of Kerala, which is approximately 25 percent Muslim

Interestingly, Kerala is one of the few parts of India that was never under Muslim rule. Islam in Kerala is historically

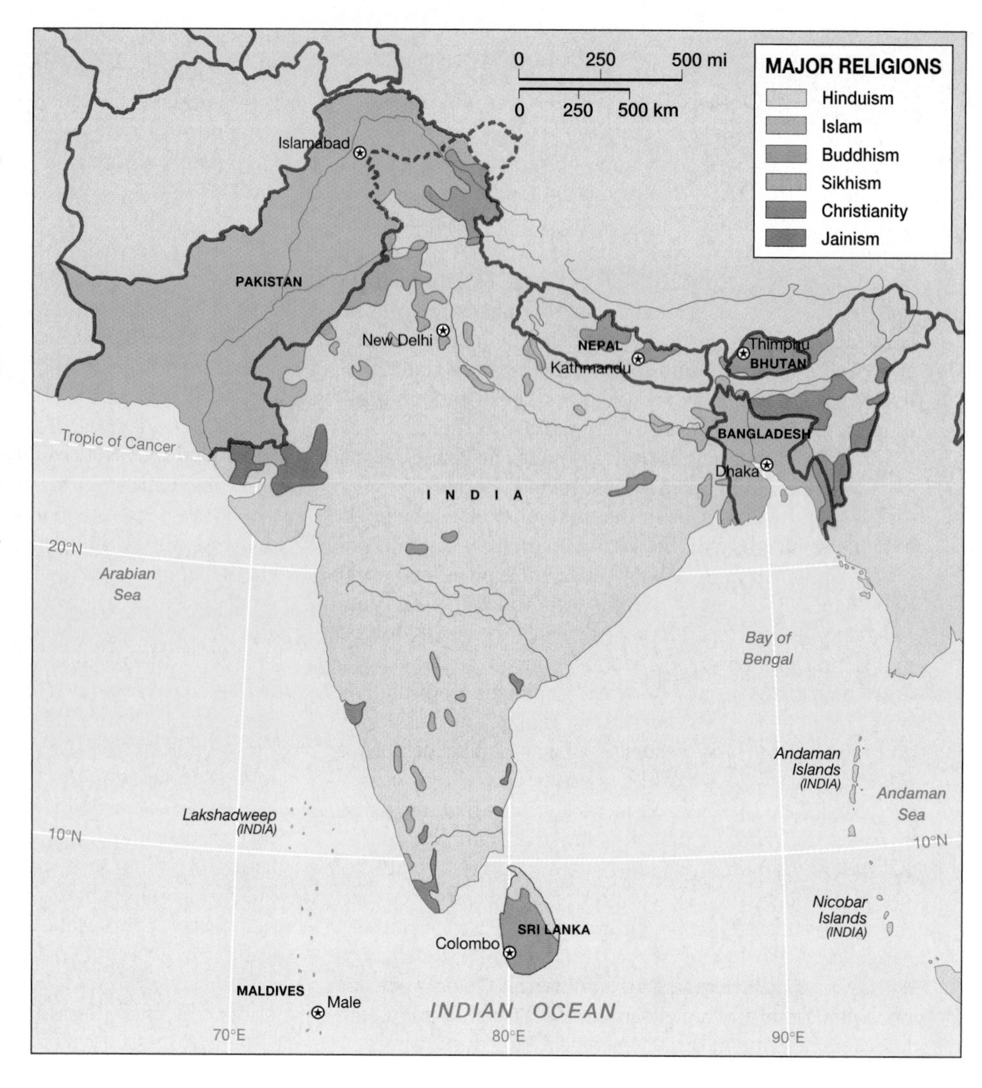

▶ Figure 12.21 Religious geography of South Asia This map shows several important patterns. First is the bracketing of Hindu-dominated India by the two Muslim states of Pakistan and Bangladesh. A second pattern, however, is of the 80 million Muslims found within India, a minority comprising some 10 percent of the total population. Of particular note are the Muslims in Northwest Kashmir and on the upper Ganges Plain near the Pakistan border. A third pattern is the Sikh presence in the Punjab area, where they constitute 60 percent of the population. Also note the Buddhist populations in Sri Lanka, Bhutan, and parts of Nepal.

connected not to Central Asia but rather to trade across the Arabian Sea. Kerala's Malabar Coast long supplied spices and other luxury products to Southwest Asia (and to Europe as well), enticing many Arabian traders to settle (Figure 12.22). Gradually many of Kerala's indigenous inhabitants converted to the new religion as well. Sri Lanka is approximately 9 percent Muslim and the Maldives is virtually entirely Muslim. Like Kerala in India, Islam in these two island countries is also rooted in the trade networks of the Arabian Sea.

Sikhism The tension between Hinduism and Islam in medieval northern South Asia gave rise to a new religion called **Sikhism.** Sikhism originated in the late 1400s in the Punjab, near the modern boundary between India and Pakistan. The Punjab was the site of religious fervor at the time; Islam was gaining converts and Hinduism was increasingly on the defensive. The new faith combined elements of both religions, and thus appealed to many who felt trapped between the competing claims of faith. Many orthodox Muslims, however, viewed Sikhism as a dangerous heresy, precisely because it incorporated elements of their own religion. Periodic bouts of persecution led the Sikhs to adopt a militantly defensive stance, and in the political chaos of the early 1800s they were able to carve out a large kingdom for themselves in northwestern South Asia. Even today, Sikhs are disproportionally represented in the Indian armed forces.

At present the Indian state of Punjab is approximately 60 percent Sikh. Small but often influential groups of Sikhs are scattered across the rest of India. Devout Sikh men are immediately visible, since they are not supposed to cut their hair or their beards. Instead, they wear their hair wrapped in a large turban and often tie their beards close to their faces with a small net.

Buddhism and Jainism Although Buddhism virtually disappeared from India in medieval times, it persisted in Sri Lanka. Among the island's dominant Sinhalese people, Theravada Buddhism developed into a virtual national religion, fostering close connections between Sri Lanka and mainland Southeast Asia. In the high valleys of the Himalayas, Buddhism also survived as the majority religion. Here one finds Lamaism, or Tibetan Buddhism. Tibetan Buddhism, with its esoteric beliefs and huge monastic complexes, has been better preserved in the small and isolationistic country of Bhutan and in the Ladakh region of northeastern Kashmir than in Chinese-controlled Tibet itself. The small town of

▲ **Figure 12.22 Mosque and minaret** The dominant features of the Islamic landscape are the centers of worship, the mosques, set off by the towers from which the faithful are called to prayer. Powerful loudspeakers that broadcast prayers throughout the city can be seen on the minarets. *(Dermot Tatlow/Panos Pictures)*

Dharmsala in the northwestern Indian state of Himachal Pradesh is the seat of Tibet's government-in-exile and the Dalai Lama, who fled Tibet in 1959 after an unsuccessful revolt against the communists.

At roughly the same time as the birth of Buddhism (circa 500 B.C.), another religion emerged out of the Hindu society in northern India as a protest against orthodox Hinduism: **Jainism.** This religion similarly stressed nonviolence, and indeed took this creed to its ultimate extreme. Jains are forbidden to kill any living creatures, and as a result the most devout adherents wear gauze masks to prevent them from inhaling small insects.

Agriculture is forbidden to Jains, since plowing destroys worms and other soil-inhabiting creatures. As a result, most members of the faith looked to trade for their livelihoods. Many of them prospered, aided, no doubt, by the frugal and abstemious lifestyles required by their religion. The Jain community is now one of the wealthiest in India, although it includes many poor individuals. Prosperous and devout Jains often contribute substantial sums of money to *pinjrapoles,* institutions devoted to caring for sick and injured animals, especially cattle. Today Jains are concentrated in northwestern India, particularly in Gujarat.

Other Religious Groups Even more prosperous than the Jains are the Parsis, or Zoroastrians, concentrated in the Mumbai (Bombay) area. The Parsis arrived as refugees, fleeing from Iran after it was conquered by Muslim armies in the seventh century. Zoroastrianism is an ancient religion that focuses on the cosmic struggle between good and evil. Although numbering only a few hundred thousand, the Parsi community has had a major impact on the Indian economy. Several of the country's largest industrial firms are still controlled by Parsi families.

Indian Christians are more numerous than either Parsis or Jains. Their religion arrived centuries ago; early contact between the Malabar Coast and Southwest Asia brought Christian as well as Muslim traders. A Jewish population also established itself but later languished, and today numbers only several thousand. Kerala's Christians, by contrast, are counted in the millions, constituting some 20 percent of the state's population. Several Christian sects are represented here, but the largest are affiliated with the Syrian Christian Church of Southwest Asia. As a result, Kerala is unique to India because the state is home for three major religious traditions—Hinduism, Islam, and Christianity—in almost equal numbers.

During the colonial period, British missionaries went to great efforts to convert South Asians to Christianity. They had very little success, however, in Hindu, Muslim, or Buddhist communities. Here people were largely content with their own traditions, and were wary of foreign missionaries associated with the ruling power. The only significant conversions came among a few untouchable groups in various parts of India. The remote tribal districts of British India, on the other hand, proved to be more receptive to missionary activity. In the uplands of India's extreme northeast, entire communities abandoned their traditional animist faith in favor of Protestant Christianity.

Geographies of Language

South Asia's linguistic diversity matches its religious diversity. In fact, one of the world's most important linguistic boundaries runs directly across India (Figure 12.23). North of the line, languages belong to the Indo-European family, the world's largest linguistic family. In southern India, on the other hand, most languages belong to the **Dravidian language** family, a linguistic group that is not Indo-European and is unique to South Asia. In early historical times it was spoken throughout the region, including northern India and what is today Pakistan. Along the mountainous northern rim of the region a third linguistic family, Tibeto-Burmese, prevails, but this area is marginal to the South Asian cultural sphere. Generally, South Asia as a whole can be divided into two major linguistic zones, the Indo-European-speaking north and a Dravidian-speaking south.

How or when Indo-European languages were introduced to South Asia is uncertain, but it has long been argued that they were brought in by pastoral peoples from Central Asia,

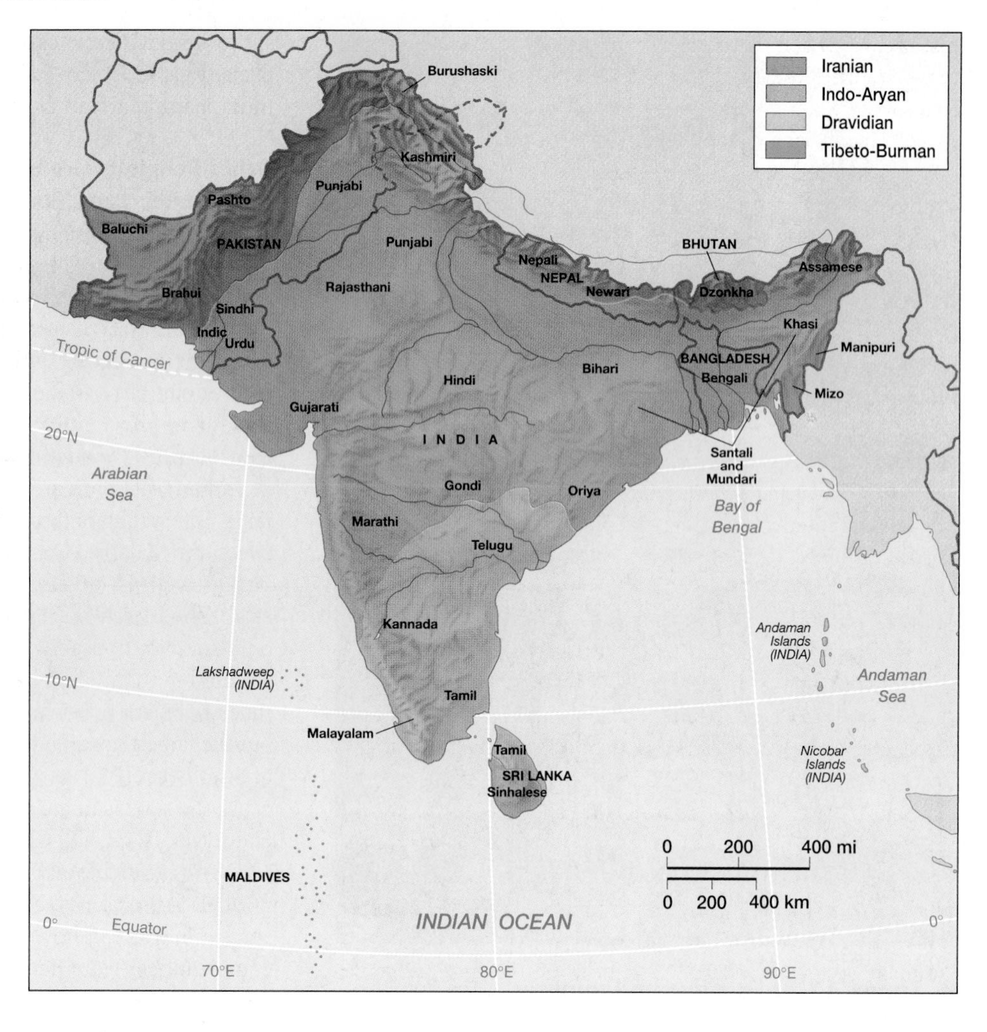

▶ **Figure 12.23 Language map of South Asia** A major linguistic divide is between the languages of the Indo-European family found in the north, and the Dravidian languages, which are not Indo-European, in the south. This remnant of Dravidian offers clue to its once-widespread distribution, before it was replaced by Indo-European languages. Of the Indo-European family, Hindi is the most widely spoken, with some 480 million speakers, which makes it the second-most widely spoken language in the world. Other regional languages are shown on the map and are closely associated with states in India.

who invaded the subcontinent in the second millennium B.C. According to this hypotheses, offshoots of the same original horse- and cattle-herding people also swept across both Iran and Europe, bringing their language, as well as their gods, to all three places. The ancestral Indo-European language introduced to India, according to this theory, was Sanskrit, or something very similar to it. This rather simplistic scenario is now regarded with some suspicion, and many scholars argue for a more gradual infiltration of Indo-European speakers from the northwest, perhaps from Asia Minor.

Any modern Indo-European language of India, such as Bengali, is more closely connected to English than it is to any Dravidian language of southern India, such as Tamil. That said, it should also be noted that South Asian languages on both sides of this linguistic divide have come to share a number of superficial features. Dravidian languages, for example, have borrowed many terms from Sanskrit, particularly as it is used in ancient Brahman scholarship.

The Indo-European North As can be seen in Figure 12.23, South Asia's Indo-European languages are themselves divided into two subfamilies: Iranian and Indo-Aryan. As might be expected, Iranian languages, such as Baluchi and Pushtu, are found only along the western margin of Pakistan, close to its border with Iran.

The much-larger group of Indo-Aryan languages are all closely related to each other, even though standardized written forms were not developed until the nineteenth century. Since educated South Asians formerly used "international" languages, such as Sanskrit or Persian (just as educated Europeans once used Latin), to communicate across linguistic boundaries, standardized regional languages were not necessary. Today all of South Asia's major languages are now standardized, though a great diversity of local dialects still exists.

Each of the major languages of India is associated with a particular Indian state, since the country deliberately structured its political subdivisions along linguistic lines after attaining independence in 1947. As a result, one finds Gujarati spoken in Gujarat state, Marathi in Maharashtra, and Oriya in Orissa. However, two of these languages, Punjabi and Bengali, span India's international boundaries to extend into Pakistan and Bangladesh, since these borders were established on religious rather than linguistic criteria. Nepali, the national language of Nepal, is largely limited to that country.

The most widely spoken language of South Asia is **Hindi**—not to be confused with the *Hindu* religion. In fact, with some 480 million speakers, Hindi is the second-most widely spoken language in the world. It occupies a prominent role in contemporary India, both because so many people speak it and because it is the main language of the Ganges Valley, the historical and demographic core of India. More specifically, it is the main tongue of Uttar Pradesh and Madhyra Pradesh states in central northern India. Many northerners would like to see Hindi become the country's common language, but many other Indians, particularly those living in the south, strenuously resist the idea.

Bengali is the second-most widely spoken language in South Asia. It is both the national language of Bangladesh and the main language of the Indian state of West Bengal. Today, Bengali is spoken by almost 200 million persons, making it the world's ninth-most widely spoken language. Bengali's significance extends beyond its official status in Bangladesh and its total numerical strength. Equally important is its extensive literature. The West Bengal area of India, particularly its capital of Calcutta, has long been one of South Asia's main literary and intellectual centers. Calcutta may be noted for its appalling poverty, but it still has one of the highest levels of cultural production in the world, as measured by the output of drama, poetry, fiction, and music.

The Punjabi-speaking zone in the west was similarly split at the time of independence, in this case between the Indian state of Punjab and Pakistan. While Punjabi is spoken by an estimated 100 million persons, it does not have the significance of Bengali. Although Punjabi has been the main vehicle of Sikh religious writings, it lacks an extensive literary tradition. More importantly, Punjabi did not become the national language of Pakistan upon independence, even though it is the day-to-day language of some two-thirds of Pakistan's inhabitants. Instead, that position was accorded to Urdu, a language so closely related to Hindi that the two are sometimes considered to form a single tongue.

Urdu, written in Arabic script because of its close ties to Persian, originated on the plains of northern India in the 1700s and 1800s. The difference between Urdu and Hindi was largely one of religion: Hindi was the language of the Hindu majority, Urdu that of the Muslim minority that included the ruling class. Owing to this distinction, Hindi and Urdu were from the start written differently—the former in the Devanagari script (derived from Sanskrit) and the latter in the Arabic script.

When independence was gained in 1947, millions of Urdu-speaking Muslims from the Ganges Valley fled to the new state of Pakistan. Since Urdu had a much higher status than Pakistan's indigenous languages, which were sometimes regarded as little more than local dialects, it was quickly established as the new country's official language, along with English.

Languages of the South Four thousand years ago, Dravidian languages were probably spoken across most of South Asia, even in the north. As Figure 12.23 indicates, a Dravidian tongue called Brahui is still found in the uplands of central-western Pakistan. It also seems likely that the language of the ancient Indus Valley civilization was Dravidian. The four main Dravidian languages are largely confined to India, although one of them, Tamil, is also found in Sri Lanka. As in the north, each language is closely associated with an Indian state: Kannadi in Karnataka, Malayalam in Kerala, Telugu in Andhra Pradesh, and Tamil in Tamil Nadu. Tamil is usually considered the most important member of the family because it has a longer history and a larger literature.

Although Tamil is spoken in northern and eastern Sri Lanka, the country's majority population, the Sinhalese, speak an Indo-European language. Apparently, the Sinhalese migrated from northern South Asia several thousand years ago. Although this movement itself is long lost to history, it is evident that the migrants settled on the island's fertile and moist southwestern coastal and central highland areas, which formed the core of a succession of Sinhalese kingdoms. These same Sinhalese speakers also migrated to the Maldives, where the national language, Divehi, is essentially a Sinhalese dialect. The drier north and east of Sri Lanka, on the other hand, were settled primarily by Tamils moving down from the southern tip of India. Later, British landowners imported Tamil peasants from the mainland to work on their tea plantations in the central highlands of Sri Lanka.

Linguistic Dilemmas Sri Lanka, Pakistan, and India are multilingual countries, and all have been troubled by linguistic conflicts. Such problems are most complex in India, simply because India is so large and has so many different languages. India's linguistic environment is changing in complicated ways, pushed along by modern economic and political imperatives.

Indian nationalists have long dreamed of a national language, one that could help forge the disparate communities of the country into a more unified nation. The main problem is that this **linguistic nationalism,** or the linking of a specific language with nationalistic goals, meets the stiff resistance of provincial loyalty, which itself is deeply intertwined with local languages. The obvious choice for a national language would be Hindi, and Hindi was indeed declared the new country's official language in 1947. Raising Hindi to such a position, however, quickly alienated speakers of important northern languages such as Bengali and Marathi, and even more so the speakers of the Dravidian tongues of the south. As a result of this cultural tension, in the 1950 Indian constitution Hindi was demoted somewhat to sharing the position of "official language" of India with 14 other indigenous tongues, which is where the situation lies today.

Regardless of such opposition, the role of Hindi is expanding—at least in the Indo-European-speaking portions of the country. Here, local languages are fairly closely related to Hindi, which can therefore be learned without too much difficulty. Hindi is spreading through educational channels, but even more significantly through popular media, especially television and motion pictures. Films and television programs are made in several Indo-Aryan languages, but Hindi remains the primary vehicle. In a poor but modernizing country such as India, where many people experience the wider world largely through moving images, the influence of a national film and television culture can be tremendous.

▲ **Figure 12.24 Persistence of English** In many ways, English, the colonial language of British rule, serves to bridge the gap between the many different languages of India. However, many Indian nationalists would like to see the English language discouraged. *(Jim Holmes/Panos Pictures)*

Even if Hindi is spreading, it still has a long way to go before it can be considered anything like a common national language, even in northern India (Figure 12.23). In the Dravidian south, more importantly, its role remains minimal. National-level political, journalistic, and academic communication thus cannot be conducted in Hindi, or in any other indigenous language. Only English, an "associate official language" of contemporary India and the language of administration across the subcontinent during colonial period, serves this function.

Before independence, many educated South Asians learned English for its political and economic benefits under colonialism. It therefore emerged as the *de facto* common tongue, albeit one largely limited to the upper and middle classes. Today, many nationalists wish to de-emphasize English, considering it to be the language of imperial oppression. Others, however, and particularly southerners, have advocated English as a neutral national language, since all parts of the country have an equal stake in it. Furthermore, English also confers substantial international benefits.

English is thus the main integrating language of India, and it remains widely used in Pakistan, Bangladesh, and Sri Lanka. Indeed, India is sometimes said to be the most populous English-speaking country in the world (Figure 12.24).

South Asians in a Global Cultural Context

The widespread use of English in South Asia has not only facilitated the spread of global culture into the region, but has also allowed South Asians' cultural production to reach a global audience. As early as the turn of the twentieth century, Rabindranath Tagore had gained international acclaim for his poetry and fiction, earning the Nobel Prize for Literature in 1913. In the 1980s and 1990s, such Indian novelists as Salman Rushdie and Vikram Seth have become major literary figures in Europe and North America.

The spread of South Asian culture abroad has been accompanied by the spread of South Asians themselves. There are now several million people of South Asian descent living in Britain (mostly Pakistani), and a similar number in North America (mostly Indian). Additionally, migration from South Asia has led to the establishment of large communities in such far-flung places as eastern Africa, Fiji (in the Pacific), and the southern Caribbean (Figure 12.25).

In South Asia itself, the globalization of culture has brought tensions as severe as those felt anywhere else in the world. Traditional Hindu and Muslim religious norms frown on any overt display of sexuality—a staple feature of global popular culture. While romance is a recurrent theme in the often melodramatic films of Mumbai (Bombay) (or "Bollywood," as it is commonly called), even kissing is considered risqué. Western films and videos are commonly denounced as immoral by religious leaders. Still, the pressures of internationalization are hard to resist. Many Indians were surprised in the mid-1990s when the Hindu-fundamentalist government in Mumbai (Bombay) relented and allowed Michael Jackson to perform in public.

▲ **Figure 12.25 South Asians in the world** A large number of South Asians live in other regions of the world, such as these inhabitants of Trinidad, in the Caribbean. About half of Trinidad's population are descendants of South Asian sugar workers who arrived in the nineteenth century. *(Pietro Cenini/Panos Pictures)*

Cultural globalization is beginning to allow South Asian novelists, particularly women, to explore themes that would otherwise be difficult for them to address. For example, Arundhati Roy, author of the 1997 international best-seller, *The God of Small Things,* examined the tragic romance between a high-caste woman (a Syrian Christian rather than a Hindu) and an untouchable man in her native state Kerala. Such a topic is virtually forbidden in the local cultural context, and Roy's novel was initially ignored in her homeland. Only its international acclaim allowed it widespread recognition in Kerala and elsewhere in India.

Geopolitical Framework: A Deeply Divided World Region

Before the coming of British imperial rule, South Asia was never politically united. While a few empires at various times ruled most of the subcontinent, none ever spanned its entire extent. Whatever unity the region had was cultural, not political. The British, however, did bring the entire region into a single political system by the middle of the nineteenth century. Then, 100 years later, independence in 1947 witnessed the traumatic separation of Pakistan from India; later, in 1971, Pakistan itself was divided with the independence of Bangladesh, which was formerly East Pakistan. While India is clearly the largest single country in South Asia, serious internal tensions began to rise in India itself in the 1980s, and today, many observers question whether this vast country can survive as a unified political state (Figure 12.26).

The major geopolitical issue with global implications, though, is the continuing tension between Pakistan and India, which is aggravated by the fact that both countries are now nuclear powers and have systems capable of delivering weapons of mass destruction to neighboring territory. Furthermore, both India and Pakistan seem infatuated with their recent nuclear status as it empowers and inflames their heightened sense of nationalism (see "Local Voices: South Asians Talk About Nuclear Weapons and War"). With tensions now reduced between the former Soviet Union, Europe, and the United States, many experts consider the highest possibility for a nuclear war now lies embedded in the geopolitical tensions of South Asia.

South Asia Before and After Independence in 1947

When Europeans began to arrive on the coasts of South Asia in the 1500s, most of the northern subcontinent was controlled by the Mughal Empire, a powerful Muslim state ruled by people of Central Asian descent. Southern India remained under the control of a Hindu kingdom called *Vijayanagara*. European merchants, keen to obtain spices, textiles, and other Indian products, soon established a series of coastal trading posts. The Mughals and other South Asian rulers were little concerned with growing European naval power, as their own focus was the control of land. The Portuguese established a sizable enclave in Goa, while the Dutch gained control over much of Sri Lanka in the 1600s, but neither was viewed as a significant threat.

The Mughal Empire grew stronger in the 1600s, while Hindu power declined until it was limited to the peninsula's far south. In the early 1700s, however, the Mughal Empire weakened rapidly. A number of contending states, some ruled by Muslims, others by Hindus, and a few by Sikhs, emerged in former Mughal territories. The largest of these was the Hindu state of the Marathas, centered in modern Maharashtra state. The Maratha rulers hoped to unify India, but failed after suffering several key defeats on the Ganges Plain. As a result, the 1700s was a century of political and military turmoil, often bordering on chaos.

The British Conquest These unsettled conditions of the eighteenth century provided an opening for European imperialism. The British and French, having largely displaced the Dutch and Portuguese, competed for trading posts (Figure 12.27). Before the Industrial Revolution, Indian textiles were considered the best in the world, and British and French merchants needed to obtain large quantities of these goods for their global trading networks. In competing against each other, the Europeans found themselves embroiled in the complex political struggles of the various South Asian states. With Britain's overwhelming victory over France in the Seven Years' War (1756–63), France was reduced to a few marginal coastal possessions. Britain, or more specifically the **British East India Company,** the private organization that acted as an arm of the British government, monopolized trade in the area and was now free to exploit the political chaos of the interior and stake out a real South Asian empire.

The Company's usual method was to make strategic alliances with Indian states in order to defeat the latter's enemies—most of whose territories it would then usurp for itself. As time passed, its army, largely composed of South Asian mercenaries, grew ever more powerful. Several Indian states put up heroic resistance, but none could ultimately resist the immense resources of the East India Company. With the defeat of the Sikh state in the Punjab in the 1840s, British control of South Asia was essentially completed. Valuable local allies, as well as a few former enemies, were allowed to remain in power, provided that they no longer threatened British interests. The territories of these indigenous states, however, were gradually whittled back, and their policies were increasingly dictated by British advisors.

From Company Control to British Colony The continuing reduction of the indigenous states, coupled with the growing arrogance of British officials, led to a rebellion in 1856 across much of South Asia. When this uprising (called the *Sepoy Mutiny* by the British) was finally crushed, a new political order was implemented. South Asia was now to be ruled directly by the British government, with the Queen of England as its head of state. The continuing usurpation of the indigenous states ended, and a stable political map of British India emerged.

As Figure 12.27 shows, the British enjoyed direct control over the region's most productive and most densely populated areas, including virtually the entire Indus-Ganges Valley and most of the coastal plains. The British also ruled Sri Lanka

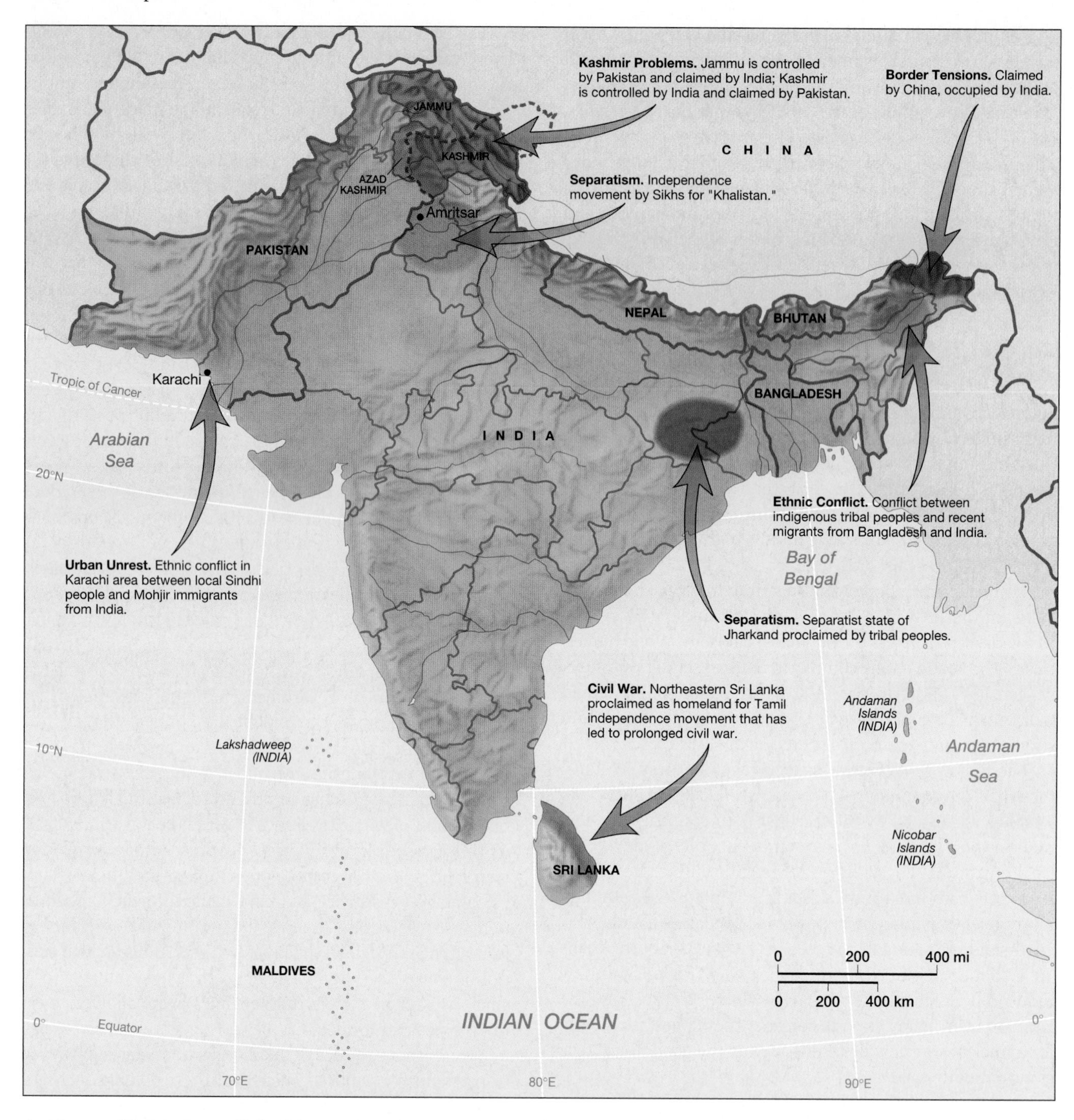

▲ **Figure 12.26 Geopolitical issues in South Asia** Given the cultural mosaic of South Asia, it is not surprising that ethnic tensions have created numerous geopolitical problems in the region. Particularly vexing are ethnic tensions in Sri Lanka, Kashmir, and the Punjab, where vying groups threaten the national fabric. Even within India, separatists struggle for independence and autonomy, such as in the tribal area of Jharkand.

directly, having supplanted the Dutch in the 1700s. The major areas of indirect rule, where Hindu, Muslim, and Sikh rulers retained their princely states under British advisors, were in Rajasthan, the uplands of central India, the Malabar coast of modern Kerala state, and along the frontiers. The British administered this vast empire through three coastal cities that were largely their own creation: Calcutta, Mumbai (Bombay), and Madras. In 1911 they established a new capital in New Delhi, near the divide between the Indus and Ganges drainage systems—a strategic site that had been used by the Mughals and many other earlier powers.

While the political geography of British India was stabilized after 1856, the empire's frontiers remained unsettled. British officials continually worried about threats to their immensely profitable colony, particularly from the Russians advancing across Central Asia. In response, they attempted to expand

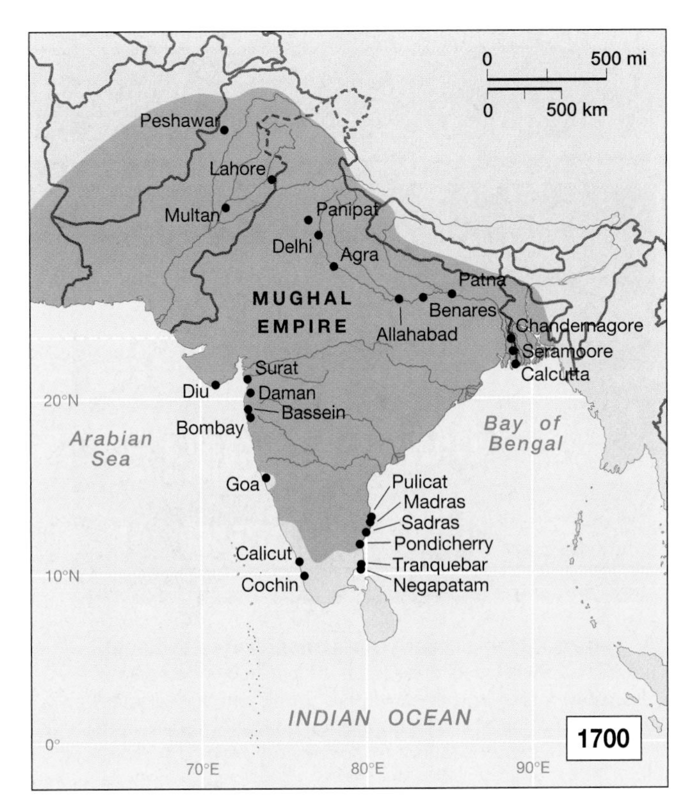

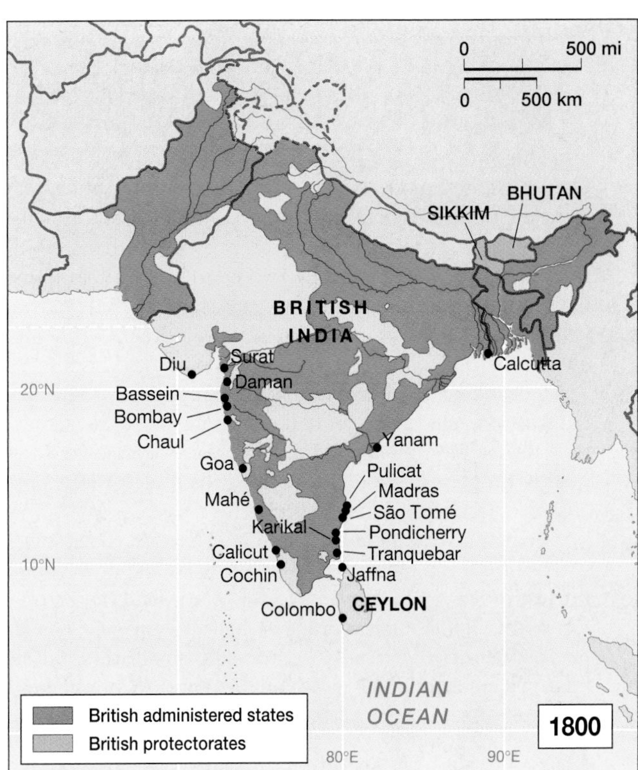

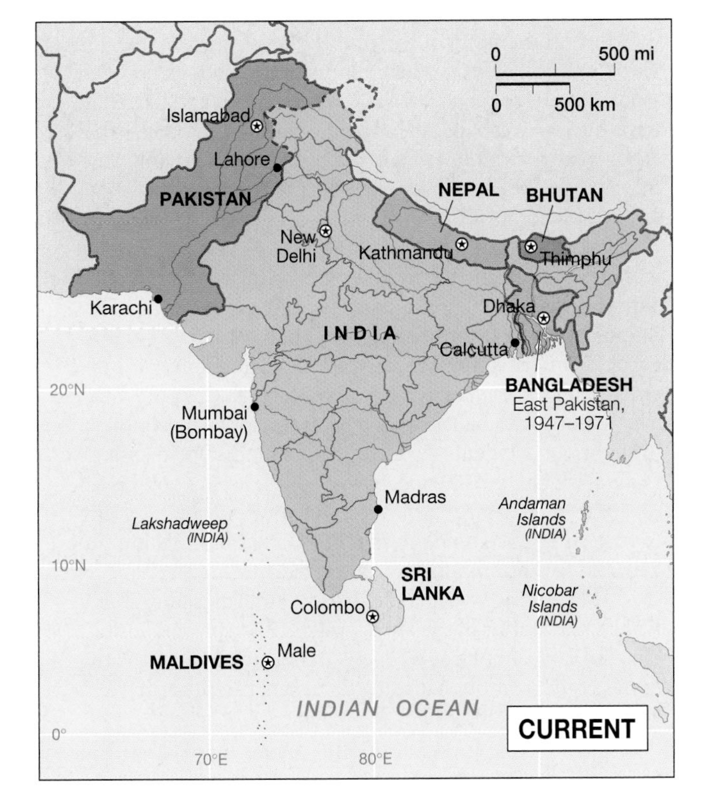

▲ **Figure 12.27 Geopolitical change** At the onset of European colonialism in 1700, much of South Asia was dominated by the Mughal Empire, a powerful Muslim state. However, this collapsed under British pressure and was replaced by a number of contending states, many of which were supported by the British colonial administration to further a balance of power. Independence for the region came after 1947 when the British abandoned their extensive colonial territory. Bangladesh, which was formerly East Pakistan, gained its independence in 1971 after a short struggle against centralized Pakistani rule from the west.

their authority as far to the north, and especially the northwest, as possible. In some cases this merely entailed the making of strategic alliances and agreements with local rulers. In such a manner Nepal and Bhutan retained their independence, although they would no longer be free of British interference. In the extreme northeast, a number of small states and tribal territories, most of which had never been part of the South Asian cultural sphere, were more directly brought into the British Empire. A similar policy was conducted on the vulnerable northwestern frontier. Here, however, local resistance was much more difficult to overcome, and the British-Indian army suffered defeat at the hands of the Afghans. Afghanistan thus retained its independence, forming an effective buffer between the British and the Russian empires.

LOCAL VOICES South Asians Talk About Nuclear Weapons and War

One of the most pressing geopolitical problems in South Asia is the continued tension between India and Pakistan; this already dangerous tension has recently escalated because of nuclear testing by each of the two countries. Indian journalist Amitav Ghosh talked with political and civil leaders in both countries about this new development, and the essence of some of his conversations is presented here.

A leading advocate of India's nuclear policies is K. Subrahmanyam, retired director of the Institute for Defence Studies and Analyses in New Delhi. He advocates an aggressive nuclear program based on the premise that nuclear weapons are the currency of global power.

"Nuclear weapons are not military weapons. Their logic is that of international politics and it is a logic of a global nuclear order. . . . India wants to be a player and not an object of this global nuclear order. A nuclear weapon . . . is of no apparent use. You can't use it to stop small wars. But it buys you credit, and that gives you the power to intimidate." He implies that if India plays its cards right, it can parlay its nuclear program into a seat on the United Nations Security Council and earn recognition as a "global player." Subrahmanyam, like many other supporters of the Indian nuclear program, sees little danger for the actual use of nuclear weapons. In New Delhi, it is widely believed that the very immensity of the destructive potential renders them useless as instruments of war, ensuring that their deployment can never be anything other than symbolic, which gives the weapons an aura of harmlessness.

"The bomb is a currency of self-esteem," argued Chandan Mitra, an influential newspaper editor. "Two hundred years of colonialism robbed us of our self-esteem . . . when you look at India today and ask how best you can overcome those feelings of inferiority, the bomb seems to be as good an answer as any."

Ram Vilas Paswan, a member of the lower house of India's parliament, is an outspoken critic of the nuclear program. "These nuclear tests were not in the Indian national interest. They were done in the interests of a party (the current ruling party of Hindu nationalists) to keep the present government from imploding. In the last election in Pakistan, Nawaz Sharif campaigned on a platform of better relations with India . . . the people of Pakistan want friendship with India. But how did our government respond? It burst a bomb in the face of a man who had reached out to us in friendship."

In Pakistan, Qazi Hussain Ahmed, the leader of the country's principal religious party, said "We are not for nuclear weapons. We are ourselves in favor of disarmament. But we don't accept that five nations (the United States. Russia, China, France, and Britain) should have nuclear weapons and others shouldn't. We say, 'Let the five also disarm.'"

▲ **Figure 1 Nuclear activists demonstrate** Although India's national leaders draw upon the prestige of being a nuclear power, there is, nevertheless, a strong anti-nuclear movement among the country's population, as is shown here in this recent Delhi demonstration. *(© Baldev/Corbis Sygma)*

But on the issue of the probability of nuclear war, his views were different: "When you have two nations between whom there is so much ill will, so much enmity, and they both have nuclear weapons, then there is always the danger that those weapons will be used if war breaks out. Certainly. And in war people become mad. And when a nation fears it is about to be defeated, it will do anything to spare itself the shame."

Pakistan's leading human-rights lawyer, Asma Jahangir, was also pessimistic when asked if nuclear war is possible. "Anything is possible," she said, "because our policies are irrational. Our decision-making is ad hoc. We are surrounded by disinformation. We have a historical enmity and the emotionalism of jihad against each other. And we are fatalistic nations who believe that whatever happens—famine, a drought, an accident—it is the will of God. Our decision-making is done by a few people on both sides. It's not the ordinary woman living in a village in Bihar whose voice is going to be heard, who's going to say, 'For God's sake—I want my cow and I want milk for my children.'"

After talking with people in India and Pakistan, Amitav Ghosh, the author of the article, offers this grim conclusion: "The pursuit of nuclear weapons in the subcontinent is the moral equivalent of civil war: the targets the rulers have in mind are, in the end, their own people."

Source: Adapted from Amitav Ghosh, 1988. "Countdown: Why Can't Every Country Have the Bomb?" *The New Yorker,* October 26 and November 2, 1998, p. 186–197.

Independence and Partition, 1947 The framework of British India began to unravel in the early twentieth century as the people of South Asia increasingly demanded independence. The British, however, were equally determined to stay, and by the 1920s South Asia was embroiled in massive political protests.

The rising nationalist movement's leaders faced a major dilemma in organizing a potentially independent regime. Many leaders, including Mohandas Gandhi—the father-figure of Indian independence—favored a unified state that would encompass all British territories in mainland South Asia. Most Muslim leaders, however, feared that a unified India would leave their people in a vulnerable position. They therefore argued instead for the division of British India into two new countries: a Hindu-majority India and a Muslim-majority Pakistan. In several parts of northern South Asia, however, Muslims and Hindus were settled in roughly equal proportions. A more significant obstacle was the fact that areas of clear Muslim majority were located on opposite sides of the subcontinent in modern-day Pakistan and Bangladesh.

No longer able to maintain their world empire after the high costs of World War II, the British withdrew from South Asia in 1947. As this occurred, the region was indeed quickly partitioned into two countries: India and Pakistan. Partition itself was a horribly disruptive event; not only were millions of persons displaced, but tens of thousands died in the process. Millions of Hindus fled from Pakistan, to be replaced by millions of Muslims fleeing India, especially from India's allotted portion of Punjab (Figure 12.28).

The Pakistan that emerged from partition was for several decades a difficult two-part country, its western section lying in the Indus Valley and its eastern section situated in the delta of the Ganges. This circumstance did not bode well for Pakistan. The Bengalis, occupying the poorer eastern half of the country, complained that they were treated as second-class citizens, with political and economic power remaining entrenched in the west. In 1971 they launched a rebellion, and, with the help of India, quickly prevailed. Bangladesh then emerged as a new country.

Geopolitical Structure of India The leaders of newly independent India faced a major challenge in organizing such a large and culturally diverse country. They decided to chart a middle ground between centralization and local autonomy. India itself was thus organized as a **Federal state**, with a significant amount of power being vested in its constitutive individual states. The national government, however, retained both control over foreign affairs, and also a significant degree of economic authority.

Following independence, India's constituent states were reorganized to accord with linguistic geography. The idea was that each major language group should have its own state and hence a certain degree of political and cultural autonomy; the large Hindi population, however, was to be divided between two states, Uttar Pradesh and Madya Pradesh. Yet only the largest ethnic groups received their own states, which has led to recurring demands from smaller groups that have felt politically excluded. The peoples of southern Bihar and neighboring areas, for example, are now pressing for the creation of a new state to be called *Jharkand*. In the mountains of northern Uttar Pradesh, pressure for the creation of another new state has also emerged.

▲ **Figure 12.28 Partition, 1947** Following Britain's decision to leave South Asia, violence and bloodshed broke out between Hindus and Muslims in the region. With the creation of the Islamic state of Pakistan (originally in two different sectors, west and east), millions of people relocated both to and from the new states. *(Bettmann/Corbis)*

Ethnic Conflicts in South Asia

The secession movement in Jharkand/southern Bihar is rooted in the ethnic tension existing between the majority Biharis of the Ganges Plain and the various tribal groups and other peoples of the southern uplands. Unfortunately, similar kinds of ethnic conflict have emerged in many different portions of South Asia. Some of these are far more extreme than that of Jharkand, threatening the territorial integrity of entire countries. Of these conflicts, the most complex is that of Kashmir.

Kashmir Relations between India and Pakistan were hostile from the start, and the situation in Kashmir has kept the conflict burning (Figure 12.29). During the British period, Kashmir was a large princely state with a primarily Muslim core joined to a Hindu district in the south (Jammu) and a Tibetan Buddhist district in the far northeast (Ladakh). Kashmir was then ruled by a Hindu **maharaja**, or a king subject to British advisors, whose ancestors had gained control during the chaotic waning years of the Mughal Empire. During partition, the maharaja managed to join Kashmir to India over the objection of most of its inhabitants. The leaders of Pakistan objected and ultimately decided to fight. In the brief war that ensued, Pakistan gained a slice of western and northern Kashmir, but the territory's core remained under Indian control.

Although the Indo-Pakistani boundary has subsequently remained fixed, fighting in Kashmir itself later intensified, reaching a peak in the late 1980s and early 1990s. Many

▶ **Figure 12.29 Conflict in Kashmir** Unrest in Kashmir inflames the continually hostile relationships between the two nuclear powers of India and Pakistan. Under the British, this region of predominantly Muslim population was ruled by a Hindu maharaja, who managed to join the province to India upon partition despite the objection of the Kashmir inhabitants. Today, many people still wish to join Pakistan while others argue for an independent state.

Muslim Kashmiris hope to join their homeland to Pakistan; others would rather see it emerge as an independent state. Hindu militants and other Indian nationalists, on the other hand, are adamant that Kashmir remain part of India, and the Indian government agrees. The result has been a low-level but periodically brutal war involving several different factions. Efforts have been made to reach a peaceful settlement, but they seem unlikely to succeed in the near future. Central Kashmir, with its lush valley nestled among some of the world's most spectacular mountains, was once one of South Asia's premier tourist destinations; now, however, it is a battle-scarred war zone.

The Punjab Religious conflict also lies at the root of political violence in India's Punjab. The original Punjab, an area of intermixed Hindu, Muslim, and Sikh communities, was itself divided between India and Pakistan in 1947. During partition, virtually all Hindu and Sikhs fled from Pakistan's allotted portion, just as Muslims left the zone awarded to India. Relations between Hindus and Sikhs had previously been relatively harmonious, but they began to deteriorate, and India's Punjab was itself later divided into two states: Hindu-majority Haryana and Sikh-majority Punjab.

But the division of the Indian state of Punjab did not solve the area's problems. Most Sikhs greatly resented the fact that the Indian government refused to recognize Sikhism as a separate religion, instead classifying it merely as a sect of Hinduism. In the 1970s and 1980s, as the area grew increasingly prosperous, Sikh leaders began to strive for autonomy, and Sikh radicals began to press for outright independence. In their vision, the state of Punjab should separate from India, renaming itself Khalistan, the "land of the pure."

Since many Sikh men had maintained the military traditions of their ancestors, this secession movement soon became a formidable fighting force. The Indian government reacted firmly, and tensions mounted. An open rupture occurred in 1984 when the Indian army raided the main Sikh temple at Amritsar, in which a group of militants had barricaded themselves (Figure 12.30). Shortly thereafter the president of India, Indira Gandhi, was assassinated by her own Sikh bodyguards. As hostility escalated, the Indian government placed the Punjab under martial law, instituting heavy repression. Such a policy has proved relatively successful in quelling political violence, at least in the short run. Renewed civil war in the area, however, remains a distinct possibility.

The Northeast Fringe A more complicated ethnic conflict emerged in the 1980s in the uplands of India's extreme northeast (particularly in the states of Arunachal Pradesh, Nagaland, Manipur, and portions of Assam). The underlying problem is rooted in demographic change and cultural collision. Much of this area is still relatively lightly populated, and as a result has attracted hundreds of thousands of migrants from Bangladesh and adjacent provinces of India. Many local inhabitants consider this movement a threat to

▲ **Figure 12.30 Sikh temple at Amritsar** Separatism in Indian's Punjab region is strong because of hostility between the Sikh majority and the Indian government. These tensions were magnified in 1984 when the Indian army raided the Amritsar temple to dislodge Sikh militants. *(Daniel O'Leary/Panos Pictures)*

both their lands and their cultural integrity. On several occasions indigenous war-bands have attacked newcomer villagers, and, in turn, have suffered reprisals from the Indian military. This is a remote area, however, and relatively little information from it reaches the outside world. The Indian government has placed travel restrictions over much of the northeastern frontier, making access by outsiders difficult.

Sri Lanka Inter-ethnic violence in Sri Lanka is, if anything, more severe than that faced by either India or Pakistan. Here the conflict stems from both religious and linguistic differences. Northern Sri Lanka and parts of its eastern coast are dominated by Hindu Tamils, while the island's majority group is Buddhist in religion and Sinhalese in language. Relations between the two communities have historically been fairly good, but tensions mounted soon after independence (Figure 12.31).

The basic problem is that Sinhalese nationalists have favored a unitary government, some of them going so far as to argue that Sri Lanka ought to be a Buddhist state. Most Tamils, on the other hand, favor local political and cultural autonomy, and they have accused the government of discriminating against them. In 1983 war erupted when the rebel force known as the *Tamil Tigers* attacked the Sri Lankan Army. Northern Sri Lanka has been embroiled in brutal fighting ever since. Both extreme Tamil and extreme Sinhalese nationalists remain unwilling to compromise, which makes a near-term solution seem unlikely.

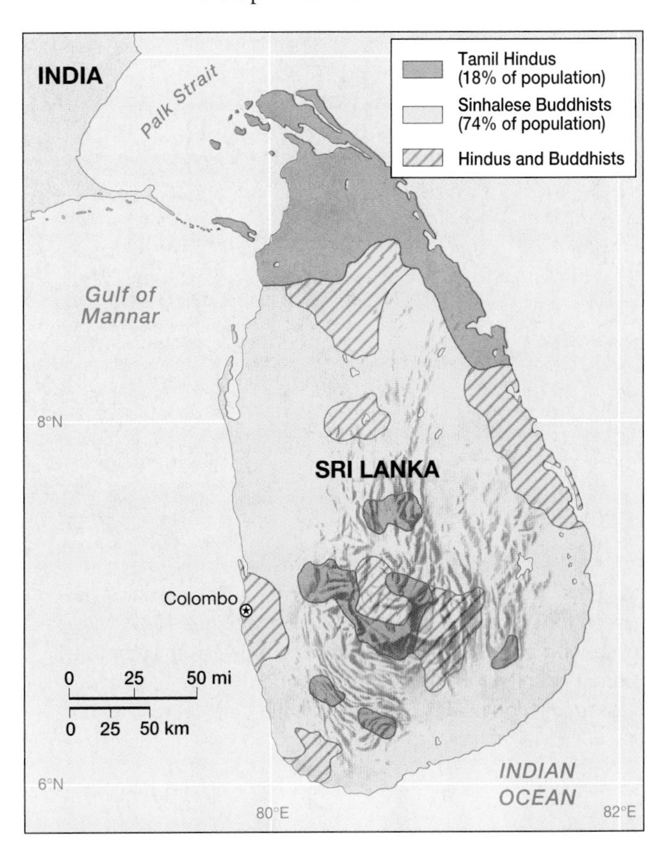

▲ **Figure 12.31 Civil war in Sri Lanka** The majority of Sri Lankans are Buddhist, and many maintain this country should be a Buddhist state. However, a Tamil-speaking Hindu minority in the northeast has waged a war of terrorism for several decades to prevent this from happening, since it would be a threat to their ethnic identity.

International and Global Geopolitics

South Asia's major international geopolitical problem is the continuing cold war between India and Pakistan. Since independence, these two countries have regarded each other as implacable enemies, and both maintain large military forces to neutralize the other. Today, however, the stakes are considerably higher because both India and Pakistan have nuclear capabilities and, more troubling, both countries seem to fuel nationalistic pride with nuclear testing and atomic saber-rattling.

During the global Cold War, Pakistan allied itself with the United States while India entered into a less-formal arrangement with the Soviet Union. However, both alliances fell apart with the end of the superpower conflict in the early 1990s and since then Pakistan has forged an informal alliance with China, from which it has obtained sophisticated military equipment. China's military connection to Pakistan is rooted in its own animosity toward India; it claims several territories along India's northern boundary, and in a brief war in 1962 China occupied a small uninhabited section of the Tibetan Plateau that formerly belonged to India. As a result, relations between China and India have remained icy ever since, although in the mid-1990s a slight thaw was detected.

The tension between India and Pakistan, and the triangular relationship that this entails with China, overshadows all other international geopolitical issues in South Asia. Elsewhere in the region, the power of India is simply overwhelming. Bangladesh, for example, owes its very existence to Indian

▲ **Figure 12.32 Border tensions** Relationships between India and Pakistan have been tense since 1947. Moreover, with both countries now nuclear powers, the fear that border hostilities will escalate into wider warfare has become a nightmarish possibility. *(Deepak Sharma/AP/Wide World Photos)*

support, and the two countries have enjoyed relatively cordial relations with each other despite their religious differences. India and Bangladesh have quarreled, however, over the Ganges River: India diverts large amounts of water out of the river during dry periods, yet lets virtually the full flow through during floods. Elsewhere in the region, India has used its power more forcefully. It simply annexed the formerly semi-independent country of Sikkim in the Himalayas in 1975 (albeit with substantial support from the Sikkimese), and later sent its army into northern Sri Lanka to try to quell the rebellion there. India's military leaders, though, quickly discovered how intractable Sri Lanka's problems are, and they withdrew their forces before their own losses mounted.

India's ambition has long been to assume the position of the dominant regional power in South Asia and, ultimately, in the Indian Ocean basin. Its army is large and capable, and its military technology is well advanced. It has been thwarted in these goals, however, not only by its struggle with Pakistan and its own internal conflicts, but also by its own severe economic problems (Figure 12.32).

Economic and Social Development: Burdened by Poverty

South Asia is a land of developmental paradoxes. It is, along with Sub-Saharan Africa, the world's poorest region, yet it is also the site of some immense fortunes. It has a sizable and growing middle class, yet large areas, and many social groups, remain virtually cut off from the developmental process. Many of South Asia's scientific and technological accomplishments are world-class; but this region also has some of the world's highest illiteracy rates. While South Asia's high-tech businesses are closely integrated with Silicon Valley in California and other centers of the global information economy, the South Asian economy as a whole remains one of the world's most self-contained and inward-looking. Predictions about South Asia's economic future vary accordingly. Some observers foresee intensified misery as the region's large and growing population undercuts natural systems and exhausts its resource base; others predict a regional renaissance as modern economic enterprises replace age-old subsistence systems.

South Asian Poverty

One of the clearest measures of human well-being is nutrition, and by this score South Asia ranks very low indeed. Nowhere else can one find so many undernourished and malnourished persons. More than 300 million Indian citizens live below their country's official poverty line, which is set at a very meager level. Bangladesh is poorer still. This country is so impoverished, and so beset by environmental and social problems, that some critics regard it as almost hopeless. According to World Bank statistics, some 67 percent of Bangladeshi children are underweight, and therefore malnourished. This is the highest figure in the world. By measures such as infant mortality and average longevity, Nepal and Bhutan are in an even worse condition, matched only by the poorer countries of Sub-Saharan Africa and, in other parts of Asia, by Laos, Cambodia, and Afghanistan (Figure 12.33).

In urban slums throughout South Asia, rapidly growing populations have little chance of finding decent housing or basic social services. In Calcutta, a survey found 500,000 people dwelling on the streets, but many experts argue that the actual figure is much higher. Moreover, it is estimated that roughly half a million South Asian children work as virtual slaves in carpet-weaving workshops and other small-scale factories.

Despite such deep and widespread poverty, South Asia should not be regarded as a zone of uniform misery. India especially has a large and growing middle class, as well as a small but wealthy upper class. Roughly 80 million Indians are able to purchase such modern consumer articles as televisions, motor scooters, and washing machines. This is a large market by any definition, and it has begun to excite the imaginations of corporate executives worldwide. India's economy has grown since the 1950s at a moderate but accelerating pace, and certain places and sectors are now virtually booming. But if several Indian states have shown marked economic vitality, others have seen only stagnation or even deterioration (Figure 12.34). Similarly, some parts of South Asia have made impressive gains in social well-being, far above what might be predicted on the basis of their economic production, but others have made only modest improvements, remaining burdened by high levels of illiteracy and poor health.

South Asia's developmental contradictions seem even more profound when considered in light of the region's economic history. From ancient times to the early modern period (1500–1700s), India was famed throughout the globe as a land of great riches. It was, of course, always afflicted with poverty—as were all civilizations—but it also probably held more raw accumulated wealth (especially in the form of gold and jewels) than any other world region.

▲ **Figure 12.33 Food aid** Because of the large population in South Asia, food problems have long plagued the region. In this photo, military personnel dispense wheat mix and dried skimmed milk to people in Nepal as aid against a famine. *(Jeremy Hartley/Panos Pictures)*

Not surprisingly, South Asia has long been the target of outsiders' greed. Central Asians began to pillage the region a thousand years ago, but in most respects the British did a much more thorough job in the last two centuries. Not only did they send much of India's hoarded wealth back to Europe, but they also systematically stifled Indian industrialization, since they wanted the colony to form a captive market for British exports.

Since independence, the governments of South Asia have attempted to create new economic systems that would benefit their own people rather than foreign countries or corporations. As in most other parts of the world, planners stressed heavy industry and economic autonomy. While some major gains have been realized, the overall pace of development remains slow in most parts of South Asia. As is true of most developing regions, there are cores of development and social progress, surrounded by large peripheral areas that lag behind, creating landscapes of striking economic disparity (Table 12.2).

Geographies of Economic Development

In this survey of South Asia's economic geography, we move up the ladder of economic development, moving from regions and countries of least development to those centers that are currently booming (Figure 12.34). By conventional economic and social measures, the Himalayan countries of Nepal and Bhutan form the bottom rungs.

The Himalayan Countries Both Nepal and Bhutan are disadvantaged by their rugged terrain and remote locations and by the fact that they have been isolated from modern technology and infrastructure. But such measurements are somewhat misleading for both Nepal and Bhutan, since they fail to convey the fact that these countries are still largely subsistence-oriented and are not participants in market economies.

Bhutan has purposely remained virtually disconnected from the modern world economy, and its small population lives in a relatively pristine natural environment. Indeed, Bhutan is so isolationistic that it has only recently allowed tourists to enter—provided that they pay some $1,000 to

Table 12.2 Economic Indicators

Country	GNP per Capita ($U.S., 1996)	Total GNP (Millions of $U.S., 1996)	PPP* ($Intl, 1996)	Real Annual Growth % per Capita, 1990–1996
India	380	357,759	1,580	3.8
Pakistan	480	63,567	1,600	1.1
Bangladesh	260	31,217	1,010	2.7
Nepal	210	4,710	1,090	2.3
Bhutan	390	282	—	2.0
Sri Lanka	740	13,475	2,290	3.4
Maldives	1,080	277	3,140	4.1

*Purchasing power parity.

Source: *The World Bank Atlas,* 1998.

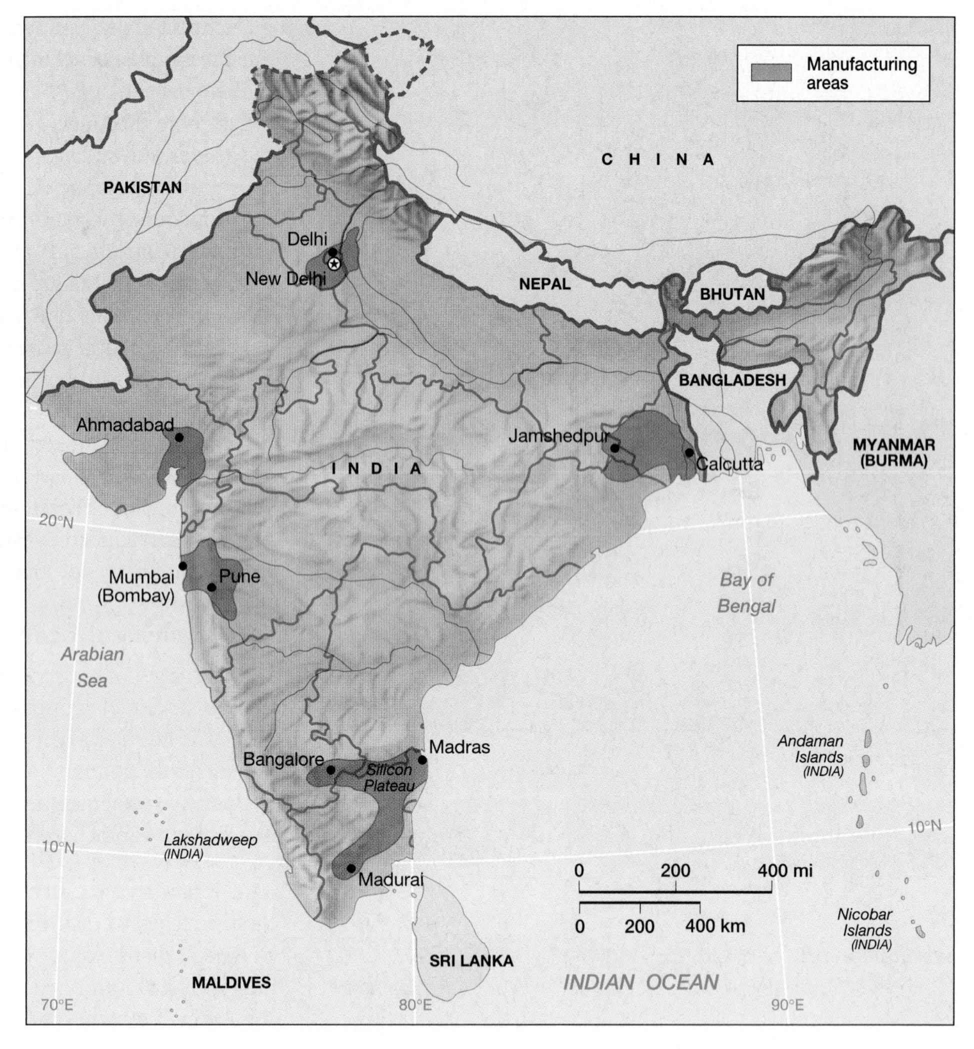

▶ **Figure 12.34 India's manufacturing areas** One of India's most pressing development problems is linked to its industrial sector. More specifically, many critics argue that the country's industrial output could be considerably more than is the case at present. This map shows the country's major industrial areas. Of these, perhaps the most promising are the Mumbai (Bombay) region, along with that of the so-called "Silicon Plateau" of the south, near Bangalore.

obtain a visa. Nepal, on the other hand, is more heavily populated and suffers much more severe environmental degradation. It is also more closely integrated with the Indian, and ultimately, the world, economy. Nepal relies heavily on international tourism. Tourism has brought some prosperity to a few favored locales, but often at the cost of heightened ecological damage.

Bangladesh The economic figures for Bangladesh are not quite as low as those of the Himalayan countries, but they are more indicative of widespread hardship, since most people there require cash to meet their basic needs. Unlike the Himalayan countries, most inhabitants of Bangladesh are involved with the market economies of commercial crops, such as rice and jute. Yet, largely because of the huge population, poverty is extreme and widespread.

Environmental degradation and the colonial legacy have contributed to the country's impoverishment, as did the partition of 1947. Most of pre-partition Bengal's businesses was located in the western area that became part of India. The division of Bengal tore apart an integrated economic region, much to the detriment of the poorer and mainly rural eastern section. Bangladesh has also suffered because of its agricultural emphasis on jute, a plant that yields tough fibers useful for making ropes and burlap bags. Synthetic materials have undercut the global jute market, and Bangladesh has not discovered any major alternative export crops.

Not all of the economic news coming from Bangladesh, however, is negative. The country is internationally competitive in textiles and clothing manufacture—in part because its wage rate is so abysmally low. Low-interest credit provided by the internationally acclaimed Grameen Bank has given hope to many poor women in Bangladesh and has allowed the emergence of a number of vibrant small-scale enterprises (Figure 12.35).

Pakistan Pakistan also suffered the effects of partition, as again a previously integrated economic system was destroyed by division. However, Pakistan, unlike Bangladesh, inherited a reasonably well-developed urban infrastructure. By measures of PPP and GNP, Pakistan today has a higher standard of living than either Bangladesh or India. The country has a productive agricultural sector (it shares the fertile Punjab with India); Pakistan also boasts a large and relatively prosperous textile industry, based in part on its large cotton crop. Many experts argue, however, that Pakistan's

▲ **Figure 12.35 Grameen Bank** This innovative institution loans money to rural women so they can buy land, purchase homes, or start cottage industries. In this photo, taken in Bangladesh, women proudly repay their loans to a bank official as testimony to their success. *(John Van Hasselt/Sygma Photo News)*

economy is far less dynamic than that of India and has less potential for growth. Part of the problem is that Pakistan is burdened by high levels of defense spending and, additionally, its best agricultural lands are controlled by a small but very powerful landlord class that pays virtually no taxes to the centralized government.

Sri Lanka and the Maldives As can be seen in Table 12.2, Sri Lanka's economy is the second most highly developed in South Asia by conventional criteria. Its exports are concentrated in textiles and agricultural products such as rubber and tea. By global standards, however, it is still a very poor country. Its progress, moreover, has been undercut in recent years by its seemingly perennial civil war. Were it not for this war between the Sinhalese and the Tamils, Sri Lanka would benefit more from the prime location of the port of Colombo.The Maldives is the most prosperous South Asian country based on per capita GNP, but its total economy, like its population, is very small. Most of its revenues are gained from fishing and from international tourism.

India's Lesser Developed Areas India's economy, like its population, dwarfs those of the other South Asian countries. While India's per capita GNP is lower than that of Pakistan, its total economy is more than five times larger. Being the largest country geographically, India also exhibits far more internal variation in economic development. South Asia's wealth was historically concentrated in the fertile lowlands of the Ganges Valley. Today, this pattern no longer holds true. The central and lower Ganges Valley is now actually one of the poorest parts of South Asia.

The most basic distinction in terms of economic well-being is that between India's more prosperous west and its poorer districts in the east. The so-called *tribal states* of the northeastern fringe generally form the bottom economic rungs as measured by per capita GNP, but the prevalence of subsistence economies make such statistics misleading. More extreme deprivation is found in the lower Ganges Valley, where cash economies prevail. To illustrate, Bihar is India's poorest state; indeed, it is often called the Mississippi of India because it ranks very low in most economic and social indices. Bihar does contain a huge iron and steel complex in its remote southern uplands, but this has not resulted in much genuine development.

Neighboring Uttar Pradesh, also in the Ganges Valley and India's most populous state, is also extremely poor. Like Bihar, it is densely populated and has experienced little industrial growth. While both Bihar and Uttar Pradesh have fertile soils, their agricultural systems have profited from the Green Revolution, as have those of the upper Gangetic Plain. Both states are also noted for their social conservatism; here the caste system is deeply entrenched, and opportunities for most farmers are extremely limited. Ironically, this unfortunate area was the original core area of Hindu civilization, and was once the richest and most powerful part of South Asia.

Other relatively poor states in eastern India include Orissa and West Bengal. The worst slums in India—and perhaps the world—are located in West Bengal's Calcutta. But Calcutta also supports a substantial and well-educated middle class, and it is the site of a large industrial complex. For most of the period of Indian independence, West Bengal has been governed by a socialist party that has fostered heavy, state-supported industry. In a dramatic turnaround in the 1990s, West Bengal's Marxist leaders began to advocate internationalization, encouraging large multinational firms to build new factories in the state.

Although western India is in general much more prosperous than eastern India, the large western state of Rajasthan still ranks among the country's poorest areas. Rajasthan suffers from an arid and drought-prone climate, and is also noted for its social conservatism. This large state was never completely brought under Muslim rule, and during the British period almost all of it remained outside of the sphere of direct British power. Here, in the courts of a number of maharajas, the military and political traditions of traditional Hindu India persisted up until recent times. Rajasthan's rulers not only maintained elaborate courts and fortifications, but also patronized many traditional Indian arts. Because of this political and cultural legacy, Rajasthan, despite its poverty, is one of India's most important tourist destinations.

India's Centers of Economic Growth North of Rajasthan lie the Indian states of Punjab and Haryana, showcases of South Asia's Green Revolution. Their economies rest largely on agriculture, but substantial investments have been recently made in food processing and other industries. On Haryana's eastern border lies the capital district of New Delhi. India's political power, along with much of its wealth, are concentrated here.

The west-central states of Gujarat and Maharashtra are noted for their industrial and financial clout, as well as for their agricultural productivity. Gujarat was the first place in

▲ **Figure 12.36 Mumbai (Bombay) stock exchange** Evidence of this city's central role in the globalizing South Asia economy is the new stock exchange, located on Bombay Island in the heart of the city. *(Sanjay M. Marathe/Dinodia Picture Agency)*

▲ **Figure 12.37 India's Silicon Plateau** With a highly educated population, India's workers are ideally suited for high-tech jobs, both in computer assembly and in programming and software development. Many of California's Silicon Valley firms draw upon facilities in India, which is 12 hours apart in time zones, to run nonstop operations. *(Chris Stowers/Panos Pictures)*

South Asia to experience substantial industrialization, and its textile mills are still among the most productive in the region. The fertile agricultural plain of central Gujarat has more recently been linked with India's "white revolution," which is increasing the efficiency of dairy production. Gujaratis have also long been famed as merchants and overseas traders, and they are disproportionally represented in the **Indian diaspora,** the migration of large numbers of Indians to foreign countries in Africa, North America, Europe, and the Pacific. As a result, cash remittances from these emigrants help to bolster the state's economy.

The state of Maharashtra is usually viewed as India's economic pacesetter. The huge city of Mumbai (Bombay) has long been the financial center and film-making capital of India; Mumbai (Bombay)'s port, moreover, carries some 25 percent of India's total foreign trade. Major industrial zones are located around Mumbai (Bombay) and in several other parts of Maharashtra. In recent years, Maharashtra's economy has grown more quickly than those of most other Indian states, reinforcing its primacy. Mumbai (Bombay) now has some of the highest office rents in the world, a distinction earned, admittedly, as much by its stringent building regulations as by its economic vitality (Figure 12.36). Some analysts now worry that the city's explosive growth is generating so much congestion and pollution that it is threatening to choke off any further development.

The center of India's fast-growing high-technology sector lies farther to the south, in Karnataka state's city of Bangalore. The Indian government selected the upland Bangalore area, which is noted for its pleasant climate and for its fledgling aviation industry in the 1950s. Other technologically sophisticated ventures soon followed. In the 1980s and 1990s, a quickly growing computer hardware and software industry emerged in Bangalore, earning it the label of "Silicon Plateau" (Figure 12.37). In the 1980s growth was spurred by the investments of U.S. and other foreign corporations eager to hire relatively inexpensive Indian scientific and technical talent. Since the 1990s, these multinational companies have been joined by an expanding group of locally owned firms, concentrated in software production.

India has proved especially competitive in software because it does not require a sophisticated infrastructure; computer code can be exported via wireless telecommunication systems without recourse to modern roads or port facilities. What is necessary, of course, is technical talent, and this India has in great abundance. Many Indian social groups have long been highly committed to education, and India has been a major scientific power for many decades. With the growth of the software industry, India's brain-power has finally begun to translate into economic gains. Whether such developments can spread benefits beyond the rather small high-tech enclaves they presently occupy remains to be seen.

Globalization and India's Economic Future

India's post-independence economic policy, and to a lesser extent those of the other South Asian nations as well, was based on widespread private ownership combined with governmental control of planning, resource allocation, and certain heavy industrial sectors. Independent India established high trade barriers to protect its economy from global competition, which was seen as biased in favor of the wealthy countries. This mixed socialist-capitalist system brought a fairly rapid growth of heavy industry and allowed India to become virtually self-sufficient even in technologically sophisticated goods. It also resulted in relatively steady economic growth, and, through the 1970s, a low level of foreign indebtedness.

By the late 1970s, however, problems with this model were becoming apparent, and frustration with India's economic progress was mounting among the business and political elite. While growth was persistent, it was never particularly fast, remaining in most years only a percentage point or two above the rate of population expansion. The percentage of Indians living below the poverty line, moreover, remained virtually constant. At the same time, it had become clear that countries such as China and Indonesia were experiencing far faster development after opening their economies to global forces. Many Indian businesspeople increasingly chafed at the governmental regulations that they believed thwarted their capacity for expansion. In the 1980s, foreign indebtedness began to mushroom, putting further pressure on the economy.

In response to these difficulties, the Indian government earnestly began to liberalize its economy in the early 1990s. Many regulations were modified and some were eliminated, and the economy was gradually opened to imports and multinational businesses. Other South Asian countries have followed a somewhat similar path. Pakistan, for example, began to privatize many of its state-owned industries in 1994 (Figure 12.38).

This gradual internationalization and deregulation of the Indian economy generated substantial opposition. Domestic firms are now being challenged by foreign competitors, threatening the livelihoods of their employees. India has a strong heritage of economic nationalism stemming from the colonial exploitation it long suffered. Several growing political parties have challenged the new internationalist policies, and by 1996 the nationalist right-wing BJP was on the verge of taking control of the government. It soon became apparent, however, that this group does not really want to disengage India from the global economy; rather, it wants to

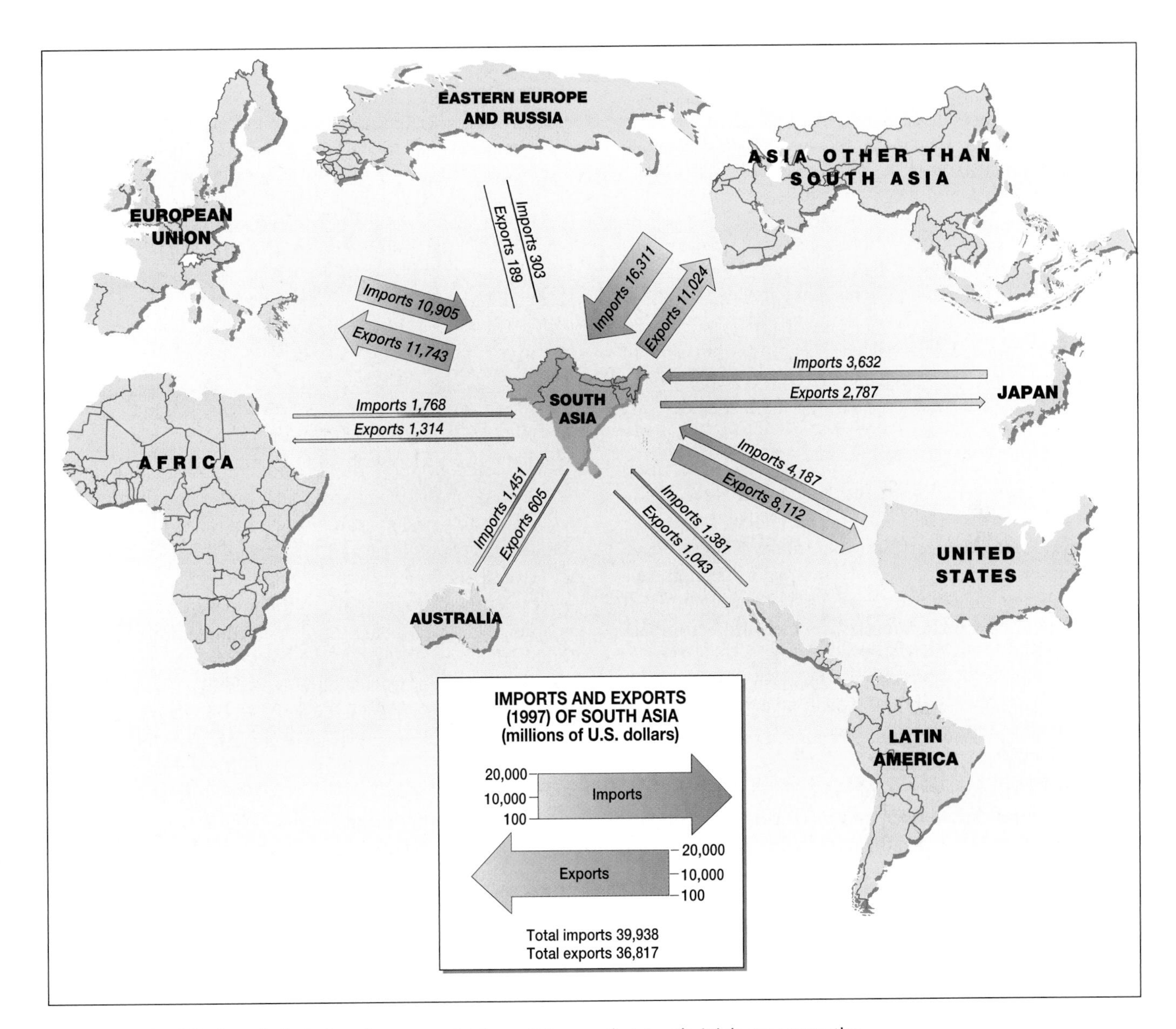

▲ **Figure 12.38 South Asia trade map** Note on this map that South Asia's exports to the European Union are slightly larger than the region's exports to other parts of Asia. Further, these exports are also greater than those to the United States. However, imports to South Asia from other world regions are dominated by those from other Asian countries. *(Data from Euromonitor,* International Marketing Data and Statistics, 1998*)*

THE LOCAL AND THE GLOBAL Supermarkets Versus Corner Grocery Stores in Indian Cities

One common aspect of economic globalization is the appearance of the Western-style supermarket in the cities of developing countries. Not only are these international food chains interested in selling food to the billions of people in these countries, but many international development workers maintain that supermarkets can actually play an important role in providing fresher, low-cost food to a population.

Not so in India, based upon the recent experience of several supermarket chains that have been attempting to change the shopping habits of people in Mumbai (Bombay) and Delhi. Instead, the local corner groceries—the "mom-and-pops," as they're called—are holding their own by providing shoppers with neighborhood convenience and good customer service. Home delivery for groceries is still common and free, as are zero-interest charge accounts. Additionally, these small markets cater to neighborhood shoppers by stocking whatever food preferences local people wish, unlike the supermarkets that offer only standard brands in international packaging. "I never go to the supermarket," says a Delhi shopper in a cramped corner grocery. "Here they have whatever I want. Further, if I ring him up, he'll send things over. He's very obliging."

While neighborhood stores don't provide the air-conditioned comfort or wide aisles of the supermarkets, they underscore once again the importance of local tastes, neighborhood traditions, and even a resistance to one-size-fits-all international marketing.

Supermarket chains in Indian cities also face problems of high land costs and an unreliable infrastructure. Because of old laws limiting real estate development in Delhi and Mumbai (Bombay), land prices for space-extensive shopping centers are sky high. In contrast, because corner grocery stores are family-owned and are passed on from one generation to another, land prices are not an issue. Additionally, since electrical service is unreliable in Indian cities, large supermarkets must have expensive backup diesel generators to keep their extensive food freezers going during frequent outages. With smaller stocks on hand, corner stores are not bothered by this problem.

Consequently, the familiar landscape of neighborhood grocery stores that dot Indian cities will persist, at least in the near term. International food chains, though, have made it quite clear they are not about to give up quite yet. Not when there is a potential market of more than a billion shoppers. This illustrates once again that tensions between the local and the global are everywhere, even at the neighborhood level.

Source: Adapted from Miriam Jordan, 1988. "India's 'Mom-and-Pops' Hold Off Supers," *International Herald-Tribune,* October 12, 1998, p. 14.

protect certain business sectors from foreign competition while negotiating better terms with foreign firms seeking to operate in India (see "The Local and the Global: Supermarkets Versus Corner Grocery Stores in Indian Cities"). Some of the most visible reactions against internationalization have been largely symbolic in nature, such as the closing down of a Kentucky Fried Chicken outlet in which inspectors had discovered several flies.

India's future economic policies remain uncertain. While globalization and liberalization can potentially generate faster growth, they also bring heightened insecurity and instability. Considering the country's history of fractious democratic politics, as well as its legacy of ethnic and religious discord, the current trend toward an open and market-driven economy could be quickly reversed in the near future.

Social Development

South Asia's social indices show relatively low levels of health and well-being, which is hardly surprising considering the region's poverty (Table 12.3). Levels of social well-being, not surprisingly, vary greatly across the region. As might be expected, people in the more prosperous areas of western India are healthier, live longer, and are better educated, on average, than the people of the poorer areas, such as the lower Ganges Valley. Bihar thus stands at the bottom of most social as well as economic indices, while Punjab stands near the top. Several measurements of social welfare are slightly higher in India than in Pakistan, despite Pakistan's higher level of per capita economic output.

Several discrepancies stand out, however, when one compares South Asia's map of economic development with its map of social well-being. Punjab's birthrate, for example, is higher than might be expected, considering its general prosperity. Portions of India's extreme northeast show relatively high literacy rates despite a general lack of development; in this case, the pattern may be explained by the educational efforts of Christian missionaries. Calcutta and its immediate environs clearly stand out as a relatively well-educated area, despite the general poverty of the lower Ganges Basin. Most pronounced discrepancies, however, strike the eye when one examines the far southern reaches of South Asia. In regard to health, longevity, and education, the south far outpaces the rest of the region.

The Educated South Southern South Asia's relatively high levels of social welfare are clearly visible when one examines Sri Lanka. Considering its meager economic resources and interminable civil war, Sri Lanka must be considered one of the world's great success stories of social development. It demonstrates that a country can achieve significant health and educational gains even in the context of an "undeveloped"

Table 12.3 Social Indicators and Status of Women

Country	Life Expectancy at Birth		Under Age 5 Mortality, per 1,000 Live Births		Secondary School Enrollment %		Female Labor Force Participation (% of total)
	Male	Female	1960	1995	Male	Female	
India	59	59	236	122	59	38	32
Pakistan	58	59	221	137	28	13	27
Bangladesh	59	58	247	122	25	13	42
Nepal	55	54	279	128	46	23	40
Bhutan	66 (no sex data)		324	197	8	2	40
Sri Lanka	70	74	130	19	71	78	36
Maldives	63	61	—	—	49	49	42

Sources: *Population Reference Bureau Data Sheet, 1998,* Life Expectancy (M/F); *World Resources Institute, 1996–97,* Under-5 Mortality Rate; *Population Reference Bureau Data Sheet, 1996,* Secondary School Enrollment (M/F); *The World Bank Atlas, 1998,* Female Participation in Labor Force.

economy. Sri Lanka's average longevity of 72 years stands in favorable comparison with many of the world's industrialized countries, as does its literacy rate. Equally impressive is the fact that Sri Lanka's fertility rate has been reduced almost to the replacement level. The Sri Lankan government has achieved these results by funding universal primary education and inexpensive medical clinics. With a well-educated, relatively healthy population growing at a slow rate, Sri Lanka could probably make rapid economic gains as well if only it could solve its political problems.

On the mainland, Kerala in southwestern India has achieved equally impressive results. Kerala is not a prosperous state; it is extremely crowded and has some difficulty feeding its population. Its overall economic figures are only average for India. However, Kerala's indices of social development, most notably in regard to longevity, literacy, and fertility, are the highest in India, comparable to those of Sri Lanka. Since Kerala is poorer than Sri Lanka, its social accomplishments must be considered all the more impressive.

Some observers attribute Kerala's social successes to its state policies. Kerala has long been led by a socialist party that has stressed mass education and community health care. While this has no doubt been an important factor, it still does not seem to offer a complete explanation. West Bengal, for example, also has a socialist political heritage, but it has not been nearly as successful in its social welfare programs. Kerala's neighboring state of Tamil Nadu has also made significant social progress despite having a different political environment. Some researchers suggest that one of the key variables for explaining the success of the south is the traditionally high social position of its women (Figure 12.39).

The Status of Women It is often said that South Asian women are accorded a very low social position in both the Hindu and Muslim traditions. In upper-class families of both religions throughout the Indus-Ganges basin, women were traditionally secluded to a large degree, their social relations with men outside of the family severely restricted.

To some extent, the Hindu tradition is traditionally more limiting to women than the Muslim tradition. Hindu women are expressly forbidden from engaging in certain economic activities (such as plowing), and in many areas they are excluded from inheriting land. Throughout northern India, women traditionally leave their own families shortly after puberty to join those of their husbands'; as outsiders, often in distant villages, young brides have few opportunities. Widows in the higher castes, moreover, are not supposed to remarry, and instead are encouraged to go into permanent mourning.

Several social indices show that women in the Indus-Ganges basin still suffer much discrimination (Table 12.3). In Pakistan, Bangladesh, and such Indian states as Rajasthan,

▲ **Figure 12.39 Women in Kerala** India's southwestern state of Kerala has the highest percentage of literate women and one of the lowest fertility rates. Because of the success of this region in lowering its birthrate, many argue that women's education and empowerment is the best and most enduring form of contraception. *(The Viesti Collection, Inc.)*

Bihar, and Uttar Pradesh, female levels of literacy are far lower than those of males. An even more telling statistic is that of gender ratios, the relative proportion of males and females in the population. All things being equal, there should be slightly more women than men in any population, since women generally have a longer life expectancy. Northern South Asia, however, contains many more men than women. Western Uttar Pradesh probably has the most male-biased population in the world.

An imbalance of males over females often results from what is known as "differential neglect." In poor families of the region, boys typically receive better nutrition and medical care than do girls, which results in higher rates of survival. Economics play a major role here. In rural households, boys are usually viewed as a blessing, since each typically remains with his family and works for its well-being. In the poorest groups, elderly people (especially widows) subsist only on what their sons can provide them. Girls, on the other hand, marry out of their families at an early age, at which time they must be provided with a dowry. They are thus seen as a net economic liability. Considering the economic desperation of the South Asian poor, it perhaps not surprising to see such unbalanced gender ratios.

Much evidence suggests that the social position of women is improving, especially in the more prosperous parts of northwestern India, where employment opportunities outside the family context are emerging. But even in many of the region's middle-class households, women still suffer major disabilities. Indeed, dowry demands seem to be increasing, and there have been a number of well-publicized murders of young brides whose own families failed to deliver an adequate dowry. In some areas, gender ratios may be growing even more male-biased, now that ultrasound technology allows the detection of sex prior to birth—hence the possibility of gender-selective abortion.

While the social bias against women across northern South Asia is striking, it is much less evident in Southern India and Sri Lanka. Geographic location, in other words, seems to play a much greater role than religion in determining the social position of women. In Kerala, for example, women have a relatively high status, regardless of whether they are Hindus, Muslims, or Christians. Here the gender ratio shows the normal pattern, with a slight predominance of females over males. Female literacy is very high in Kerala, which is one reason why the state's overall illiteracy rate is so low. Kerala's fertility rate is the lowest in India, which is another sign of women's social power.

Conclusion

South Asia, a large and complex area of more than a billion people, has in many ways been overshadowed by neighboring world regions; by the economic ascendance (and subsequent downturn) of Southeast Asia, for example, or by the giant size and political weight of East Asia; or to the west, by the geopolitical volatility and tensions of Southwestern Asia. In some ways, South Asia seemed quite willing to let other parts of the world dominate global headlines, following, perhaps, the nonalignment model set by India with its refusal to become closely involved with superpower squabbles during the Cold War.

Much of that has all changed, however, for South Asia now dominates discussions of world problems and issues. Not only is it quite possible that South Asia will surpass East Asia as the most populous world region, even if it does not and remains second in size to its eastern neighbor, South Asia demands attention over the daunting question of whether its huge population can be supported by the region's resource base. While most South Asian countries have shown remarkable progress in reducing population growth through state-sponsored family planning programs, even with these gains population growth will remain relatively high over the next several decades because of the large amount of young people in all countries. Except in India where that figure is slightly lower, fully 40 percent of the population in all other South Asia states is younger than age 15. With such a large cohort group entering their childbearing years, South Asian countries cannot afford to relax their family planning agendas.

Additionally, geopolitical tensions, both between countries and within them, remain highly problematic in South Asia. Because of the region's complex cultural geography, shaped as it was by people speaking two dozen different languages, complicated by the influences of four major religions and inflamed by several centuries of British colonial domination, cultural differences are often translated into geopolitical animosity. The longstanding malice between Pakistan and India, embedded deeply in the religious differences between Hindu and Islam, has escalated dangerously of late with reciprocal threats to use weapons of mass destruction. The outlook here is not good. International arms experts now rank South Asia is the world region most likely to initiate nuclear war.

Internal geopolitics, too, are a cause for concern. Sri Lanka's civil war continues, well into its second decade with no end in sight. Ethnic and clan warfare also rages in southern Pakistan, played out often on the streets of Karachi, making it one of the most dangerous cities in the world. But it is in giant India that internal politics demand world attention for, at times, the very future of that diverse federation seems in jeopardy. As with other parts of this complicated region, cultural differences often flare into violence. Religious strife between Muslims and Hindus continues; Hindu fundamentalists attack Christian missionaries and settlements; Sikhs clash with Hindus in the Punjab. Off the street and in the statehouse, Hindu fundamentalist politicians work to impose their agenda on education, the arts, even India's industry and economics. While successful to some degree in the past, at this writing their influence appears to be waning in the face of a more moderate multicultural backlash. India's geopolitical future, however, is notoriously difficult to predict.

Globalization affects South Asia in many different ways, reminding us of the many facets to this world phenomenon. Nepal, for example, has found a lucrative link to international tourism by putting its spectacular mountain scenery on sale to trekkers from all parts of the world. Though the cultural and environmental costs of this enterprise are often questionable, Nepal seems committed to this open-door policy for visitors. In contrast, Bhutan, Nepal's Buddhist neighbor in the Himalayas, is much more cautious about international visitors and uses quotas, strict guidelines, and high fees to limit the impact of outside cultural influences.

Bangladesh, a country that international aid workers often view with great pessimism because of its huge population, hazardous lowland environment, and limited agricultural resources, may actually be a country that has found temporary relief through economic globalization. As has been demonstrated countless times in the last decade, international industry will locate in those countries with a large low-paid labor force and a stable political environment. Currently, Bangladesh meets those needs. How long this will last remains to be seen.

Many argue that India is perfectly positioned to take advantage of economic globalization. Its labor force, for example, is better educated than most developing countries because of India's strong tradition of schooling; additionally, much of this labor force already speaks English, the major language of global commerce.

An illustration of how these skills connect with a globalized world comes from the increasing number of North American firms that have developed what are called "back-office" activities in India. These are labor-intensive record-keeping tasks that connected electronically to business headquarters. The credit records for millions of American consumers, for example, are kept in large computer files that demand constant updating and cross-referencing. Today, these electronic files are often kept in India, where they can be processed daily by skilled, low-wage workers, yet accessed instantaneously by electronic connections to North America. Often as not when you ask a customer service representative about a bank card transaction, the answer comes from files in India's "back offices."

But can these sorts of global connections make the difference in saving South Asia from what many experts predict is the inevitable disaster of famine and food shortages as population growth outraces the resource base? As with all complicated world problems, the answers vary from the resounding optimism of economic free trade advocates to the gloomy pessimism of ecologists and population experts, depending on one's point of view and bias. Perhaps it is fitting to conclude with the Hindu proverb: To make the gods laugh, attempt to predict the future.

Key Terms

British East India Company (page 521)
bustees (page 509)
caste system (page 512)
Chipko movement (page 502)
Dravidian language (page 517)
Federal state (page 525)
forward capital (page 511)
Green Revolution (page 507)
Hindi (page 518)
Hindu nationalism (page 512)
Indian diaspora (page 532)
Jainism (page 517)
linguistic nationalism (page 519)
maharaja (page 525)
monsoon (page 499)
Mughal Empire (also spelled *Mogul*) (page 514)
orographic rainfall (page 499)
salinization (page 508)
Sanskrit (page 512)
scheduled castes (page 515)
Sikhism (page 516)
Urdu (page 519)

Questions for Summary and Review

1. What are the four subregions of South Asia? Describe their similarities and differences.
2. What causes the South Asian monsoon? How does it affect different parts of South Asia?
3. What is orographic rainfall? Where is it important for South Asian agriculture?
4. How and why does the birthrate and TFR differ geographically within South Asia?
5. What are the similarities and differences between family planning programs in the major South Asian countries? With what results?
6. What are the considerations and variables when looking at the relationship between population growth and agricultural production? What is the outlook for the next several decades?
7. What are some of the "infrastructural" problems faced by South Asian cities? Give some specific examples.
8. Describe the geography of Islam within India. That is, where are the significant Muslim minorities located in India?
9. Describe the areas of origin in South Asia for Hinduism, Islam, Sikhism, and Buddhism.
10. What are the main features of the caste system? How has this system changed in India over the last several decades?
11. What are the major Indo-European languages in South Asia? Where are they located? Where are the non-Indo-European languages?
12. How has the political geography of South Asia changed in the post-colonial era?
13. Describe three different regions of geopolitical and ethnic tension within South Asia.
14. Where are the centers or core areas of economic development within the different South Asian countries? Conversely, what regions would be considered marginal or peripheral areas to these cores?
15. What kinds of relationships are seen between women's literacy and different aspects of economic and social development?
16. How does the geography of gender ratios (the number of males and females within a population) differ within South Asia? What does this tell us?

Thinking Geographically

1. As a geographer, suggest different strategies for solving (or at least lessening) the serious flooding problems of Bangladesh. Consider the fact that this crowded country must maximize most of its area for agricultural production.
2. What are the advantages and disadvantages of expanding irrigated agriculture in India and Pakistan? How can the disadvantages be reduced to acceptable levels?
3. What are the pros and cons of the Green Revolution as a means of increasing South Asia's food supplies? What is the outlook for the next decade?
4. Discuss the conflict between wildlife protection and rural villages in India by evaluating the tension between preserving habitat and the needs of the rural poor for agricultural and timberlands.
5. What are the drawbacks and benefits of using English as a national language in India? Might it help or hinder unity? Would this increase or decrease India's links to the contemporary world?
6. Choose one of the areas of geopolitical tension (Kashmir, Punjab, Sri Lanka, etc.) and, after becoming better acquainted with the complex issues that underlie this conflict, evaluate the different proposals currently offered for solving (or at least improving) the problem.
7. Conventional wisdom maintains that India will overtake China as the world's most populous country in the next several decades. First, using current rate of natural increase (RNI), calculate when this might be if rates of growth in each country do not change. Then explicate the different variables in both China and India that might change this outcome.
8. As a geographer, you work for an international arms reduction agency that is working to reduce tensions—which could lead to nuclear war—between Pakistan and India. What would you suggest?
9. Examine both the positive and negative aspects of the Indian diaspora. Begin with the less-than-obvious positive aspects of this migration by working through how this might benefit India in today's globalized context. In doing this exercise, pay some attention to what has been written about the personal characteristics and talents of those who migrate.
10. From a geographical point of view, what is the best course of action for near-term future economic development in Pakistan, India, and Bangladesh?
11. Acquaint yourself with the environmental, social, and economic implications of Nepal's open-door policy toward trekking and other forms of tourism. Do the benefits seem to outweigh the costs?
12. Is the state of Kerala a good model for economic and social development in other parts of India? Why or why not?

Regional Novels and Films

Novels

Herman Hesse, *Siddhartha* (1954, Vision Press)

Kamala Markandaya, *Nectar in a Sieve* (1954, J. Day)

Arundhati Roy, *The God of Small Things* (1997, Random House)

Salman Rushdie, *Midnight's Children* (1981, Alfred Knopf)

Paul Scott, *The Raj Quartet* (1980, Morrow)

Vikram Seth, *A Suitable Boy* (1993, HarperCollins)

Kushwant Singh, *Train to Pakistan* (1956, Chatto and Windus)

Films

Bandit Queen (1994, India)

City of Joy (1992, Great Britain)

Phantom India (1969, France)

Saalam Bombay (1988, India)

Surjo Dighal Bari (The House Along the Sun) (1979, Bangladesh)

Bibliography

Ahmad, Kazi. S. 1964. *A Geography of Pakistan*. Karachi: Oxford University Press.

Bayly, C.A. 1988. *Indian Society and the Making of the British Empire*. Cambridge: Cambridge University Press.

Bhardwaj, Surinder M. 1973. *Hindu Places of Pilgrimage in India: A Study in Cultural Geography*. Berkeley: University of California Press.

Carstairs, G. Morris. 1975. *The Twice-Born: A Study of a Community of High-Caste Hindus*. Bloomington: Indiana University Press.

Crossette, Barbara. 1995. *So Close to Heaven: The Vanishing Buddhist Kingdoms of the Himalayas*. New York: Knopf.

Dutt, Ashok K., and Geib, Margaret. 1987. *An Atlas of South Asia*. Boulder, CO: Westview.

Eaton, Richard M. 1993. *The Rise of Islam and the Bengal Frontier*, 1204–1760. Berkeley: University of California Press.

Er-Rashid, Haroun. 1977. *Geography of Bangladesh*. Boulder, CO: Westview.

Fox, Richard G., ed. 1977. *Realm and Region in Traditional India*. Durham, NC: Duke University, Program in Comparative Studies on Southern Asia.

Frater, Alexander. 1990. *Chasing the Monsoon*. New York: Henry Holt & Company.

Guha, Ramachandra. 1989. *The Unquiet Woods: Ecological Change and Peasant Resistance in the Himalaya*. Berkeley: University of California Press.

Inden, Ronald. 1990. *Imagining India*. Oxford: Blackwell.

Johnson, Basil L. 1983. *Development in South Asia*. New York: Penguin..

Kolanad, Gitanjali. 1994. *Culture Shock! A Guide to Customs and Etiquette in India*. Portland, OR: Graphic Arts Center Publishing Company.

Kothrai, Ashis, et al. 1995. "People and Protected Areas: Rethinking Conservation in India." *The Ecologist* 25(5): 188–194.

Lodrick, Deryck O. 1981. *Sacred Cows, Sacred Places: Origins and Survivals of Animal Homes in India*. Berkeley: University of California Press.

Malik, Yogendra K., and Singh, V.B. 1994. *Hindu Nationalists in India: The Rise of the Bharatiya Janata Party*. Boulder, CO: Westview.

McGowan, William. 1992. *Only Man Is Vile: The Tragedy of Sri Lanka*. New York: Farrar, Straus & Giroux.

Masica, Colin P. 1976. *Defining a Linguistic Area: South Asia*. Chicago: The University of Chicago Press.

Naipaul, V. S. 1991. *India. A Million Mutinies Now*. New York: Penguin Books.

Rothermund, Dietmar. 1993. *An Economic History of India*. London: Routledge.

Schwartzberg, Joseph E. 1992. *A Historical Atlas of South Asia* Oxford: Oxford University Press.

Singh, R. L., ed. 1968. *India: Regional Studies*. Calcutta: Indian National Committee for Geography.

Sopher, David E., ed. 1980. *An Exploration of India: Geographical Perspectives on Society and Culture*. Ithaca: Cornell University Press.

Spate, O. H. K., and Learmonth, A. T. A. 1967. *India and Pakistan: A General and Regional Geography*. London: Methuen.

Steven, Stanley F. 1993. *Claiming the High Ground: Sherpas, Subsistence, and Environmental Change in the Highest Himalaya*. Berkeley: University of California Press.

Tandon, Prakash. 1968. *Punjabi Century 1857–1947*. Berkeley: University of California Press.

von Furer-Haimendorf. 1982. *Tribes of India: The Struggle for Survival*. Berkeley: University of California Press.

Wolpert, Stanley. 1991. *India*. Berkeley: University of California Press.

Woodcock, George. 1967. *Kerala: A Portrait of the Malabar Coast*. London: Faber and Faber.

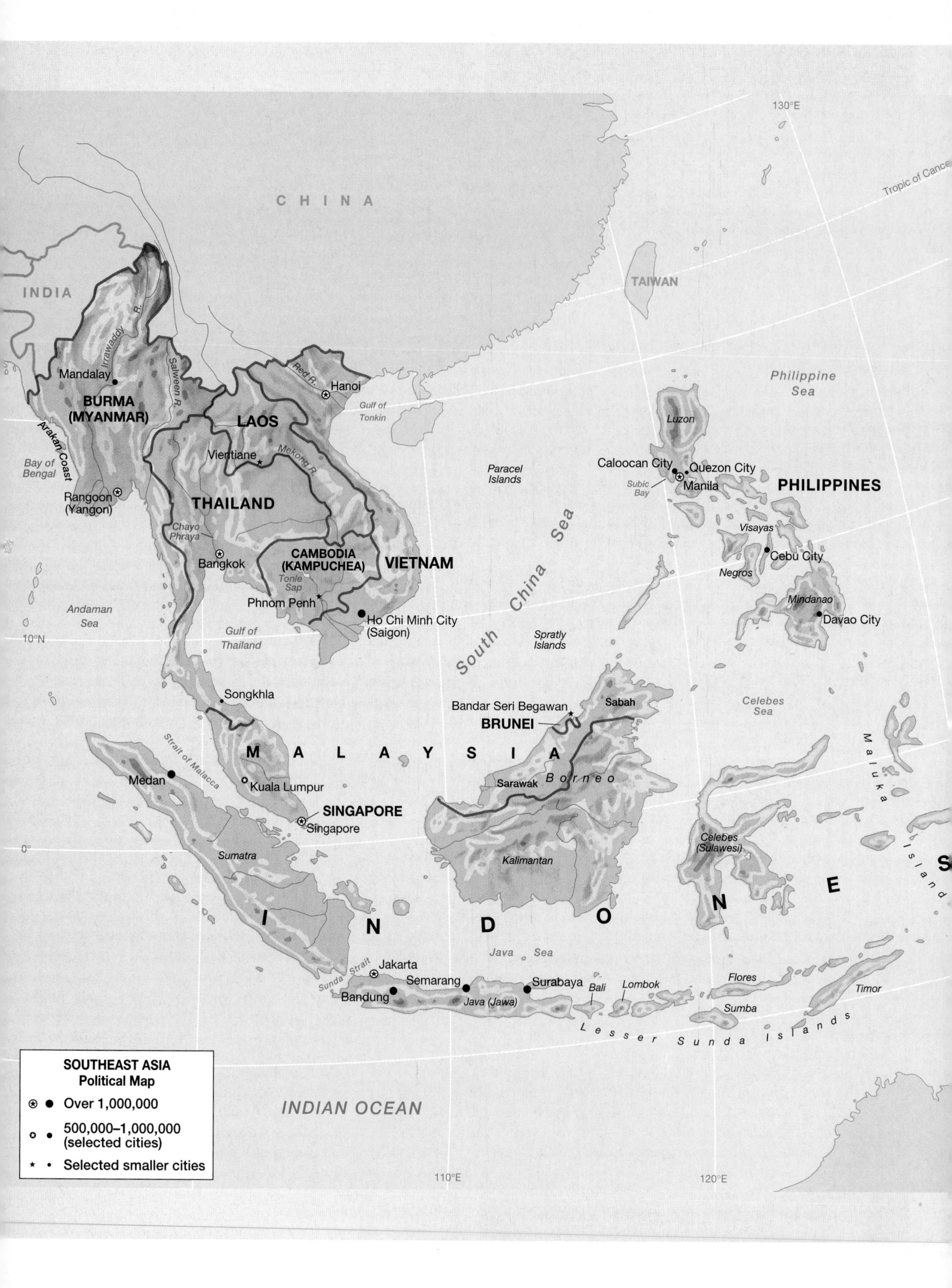

CHINA
INDIA
TAIWAN
BURMA (MYANMAR)
LAOS
THAILAND
CAMBODIA (KAMPUCHEA)
VIETNAM
PHILIPPINES
MALAYSIA
BRUNEI
SINGAPORE
INDONESIA
Mandalay
Rangoon (Yangon)
Hanoi
Vientiane
Bangkok
Phnom Penh
Ho Chi Minh City (Saigon)
Songkhla
Kuala Lumpur
Singapore
Medan
Jakarta
Semarang
Surabaya
Bandung
Bandar Seri Begawan
Caloocan City
Quezon City
Manila
Cebu City
Davao City
Irrawaddy R.
Salween R.
Red R.
Mekong R.
Chayo Phraya
Tonle Sap
Arakan Coast
Bay of Bengal
Andaman Sea
Gulf of Thailand
Gulf of Tonkin
Paracel Islands
Spratly Islands
South China Sea
Subic Bay
Philippine Sea
Celebes Sea
Luzon
Visayas
Negros
Mindanao
Sabah
Sarawak
Borneo
Kalimantan
Celebes (Sulawesi)
Maluka Island
Strait of Malacca
Sumatra
Sunda Strait
Java Sea
Java (Jawa)
Bali
Lombok
Flores
Sumba
Timor
Lesser Sunda Islands
Tropic of Cance
INDIAN OCEAN
10°N
0°
110°E
120°E
130°E
SOUTHEAST ASIA
Political Map
Over 1,000,000
500,000–1,000,000 (selected cities)
Selected smaller cities

C H A P T E R 13

Southeast Asia

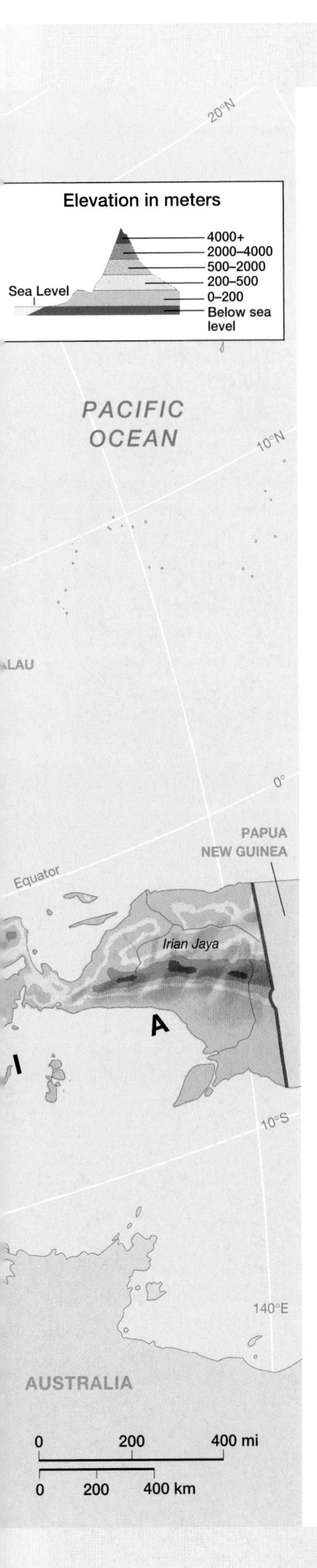

In many ways, Southeast Asia is a poster child for late-twentieth-century globalization. More specifically, within the last decade the region has experienced the roller-coaster ride of economic boom and burst, descending almost overnight from the giddy heights of fast-growing "tiger economies" to a serious economic downturn. Earlier, in the 1980s when Southeast Asia was the showpiece of global capitalism, it was fashionable to point to this region as a model for world economic development. By the middle 1990s, though, that economic model demanded revision as repercussions from Southeast Asia's economic woes affected countries throughout the globe. While some argue it is simply the Southeast Asia economies that need reorientation, others blame globalization itself, with its emphasis on free trade, global capital investment, and low-wage assembly lines. These critics use the Southeast Asian experience to remind us that globalization is a two-edged sword that cuts both ways.

Southeast Asia's involvement with the larger world is not new, for this region [which comprises the ten countries: Burma (Myanmar), Brunei, Cambodia, Indonesia, Laos, Malaysia, the Philippines, Singapore, Thailand, and Vietnam] has always been heavily influenced by external forces that have overwhelmed indigenous cultures. Centuries ago, the area was shaped by its giant neighbors to the north and west, China and India, with an influx of migrants, ideas, religion, and trade. Later still, commercial ties with the Middle East opened the doors to strong Muslim influences that are still found today in Indonesia and Malaysia. More recently, the heavy-handed imprint of Western colonialism has been felt, as Britain, France, the Netherlands, and the United States have administered large colonies within Southeast Asia. During this period, national territories were rearranged, populations relocated, and new cities and forts built to serve trade and military needs.

The region's resources and its strategic global location made Southeast Asia a major battlefield during World War II. Yet, long after world peace was restored in 1945, warfare of a different sort continued in Southeast Asia. As colonial powers withdrew and were replaced by newly independent countries, Southeast Asia once more became a battleground for world powers and their global ideologies. In Vietnam, Laos, and Cambodia, agents of communism, tacitly supported by the superpowers of China and the Soviet Union, waged

◀ **Figure 13.1 Southeast Asia** This region includes the large peninsula in the southeastern corner of Asia, as well as a vast number of islands scattered to the south and east. It is conventionally divided into two subregions: mainland Southeast Asia, which includes Burma (Myanmar), Thailand, Laos, Cambodia, and Vietnam, and insular (or island) Southeast Asia, which includes Indonesia, the Philippines, Malaysia, Brunei, and Singapore. Malaysia includes the tip of the mainland peninsula and most of the northern part of the island of Borneo.

Setting the Boundaries

The region of Southeast Asia consists of ten different countries that vary widely in spatial extent, population, cultural attributes, and levels of economic and social development. Geographically, these ten countries are commonly divided into those located on the Asian mainland and those located on islands, or the insular realm. The mainland includes Burma (called Myanmar by the current government), Thailand, Cambodia, Laos, and Vietnam. Of these countries, Burma is by the far the largest in area, about the same size as Texas. Thailand is next in size, covering an area somewhat larger than California. Though Burma is the largest in territory, Vietnam has the largest population of the mainland states, with almost 80 million people, about the same as Germany. Although usually classified as an insular state for convenience, Malaysia splits the difference between mainland and islands. Part of its national territory is on the mainland's Malay Peninsula, and part is on the large island of Borneo, which is some 300 miles removed from the mainland. This island of Borneo also includes the Muslim sultanate of Brunei, a small but oil-rich island state of 300,000 people, covering an area slightly larger than Rhode Island.

Indonesia is the quintessential island nation, stretching 3,000 miles (or about the same distance as from New York to San Francisco) from the large island of Sumatra in the west to New Guinea in the east, and comprising more than 13,000 different islands. Not only does Indonesia dwarf all other states in size, but it is by far the largest in population. With more than 200 million people, it is ranked as the world's fourth largest country. Additionally, because most of its inhabitants are Muslim, Indonesia is also considered the world's largest Islamic nation. Lying north of the equator is the Philippines, a state of 75 million people spread over a number of islands, both large and small.

Until the second half of the twentieth century, this world region was often referred to as "Indochina," a term that accurately reflects the strong historical influences of the large neighboring countries of India and China. Western colonial powers, including France, England, the Netherlands, and the United States, controlled the region until World War II, when the Japanese expanded their empire into the region. Because of the strategic importance of this area, many heated battles were fought on its territory, and it was during this period that the geographic term "Southeast Asia" replaced "Indochina." After World War II, with colonial powers gradually and, often reluctantly, withdrawing their hold on territory, new independent states appeared as the modern geopolitical map emerged. Today, because of its continued strategic value, coupled with close linkages to the dynamic world economy, Southeast Asia occupies a prominent place in the list of world regions.

a fierce and determined struggle for control of local territory and people. Resisting this were the United States and several of its allies, equally determined to defeat communism out of fear it would rapidly spread to the whole of Southeast Asia.

Today these competing ideologies have taken a back seat to other problems vexing Southeast Asia. Accompanying the recent decline of regional economies has been an increase in ethnic and social tensions within many countries, most notably Burma (called Myanmar by the current military government), Indonesia, and, to a lesser degree, the Philippines. Many argue that these countries must reinvent not only their economies but also their political systems before they can participate productively in the twenty-first century. Geopolitically, the **Association of Southeast Asian Nations (ASEAN)** has brought a new level of regional cooperation to the area, nurtured in part by the desire of these ten different countries to control—rather than to be controlled by—external global forces. In many ways, the ASEAN agenda captures the problematic geography of this diverse region as Southeast Asia forges its own identity within the context of world globalization (see "Setting the Boundaries").

Environmental Geography: Mainland and Islands

The most important physical characteristic of Southeast Asia is the separation of mainland or continental Southeast Asia from the insular, or island, portion of this world region. Figures 13.1 and 13.2 show clearly that this region is composed of a large peninsula in the northwest that is firmly anchored to the larger Asian landmass. In addition, there is an extensive series of islands, both large and small, extending to the south and east of the mainland. The former area constitutes *mainland Southeast Asia,* the latter *island or insular Southeast Asia.*

Island Southeast Asia is less geologically stable than the mainland, and thus it is a region of more diverse landforms. Four of Earth's tectonic plates converge here: the Pacific, the Philippine, the Indo-Australian, and the Eurasian (Figure 13.2). As a result of this tectonic structure, earthquakes occur frequently near the plate boundaries in island Southeast Asia. Large, often explosive volcanoes are another feature of these plate boundaries. A string of active volcanoes extends through the length of eastern Sumatra (in Indonesia) across Java and into the Lesser Sunda islands, which are the small islands lying east of Java. This was the site of the infamous Krakatau volcanic explosion of 1883 that killed more than 30,000 people and sent clouds of ash around the world. Numerous volcanoes also dot the Philippine archipelago, including Mt. Pinatubo, which erupted in 1991 and caused 800 deaths as well as temporary global cooling from its influence on Earth's atmosphere.

Mainland Environments

Mainland Southeast Asia is an area of rugged mountains interspersed with broad lowland associated with large rivers. The region's northern boundary lies in a cluster of mountains

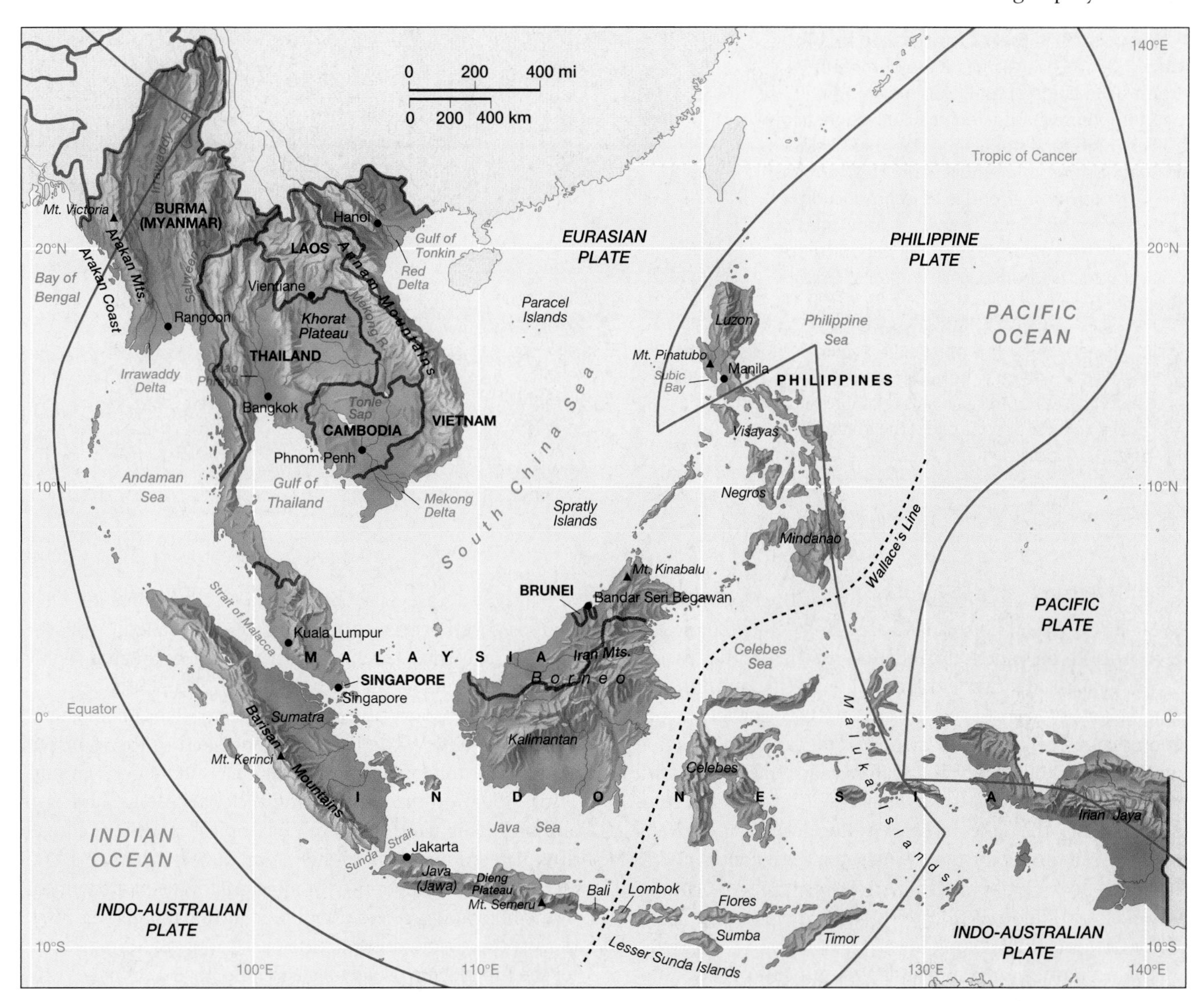

▲ **Figure 13.2 Physical geography of Southeast Asia** Southeast Asia is one of the world's most geologically active regions, owing to the intersection of several tectonic plates. As a result, earthquakes are a frequent occurrence in many areas, and volcanism and other forms of active mountain-building abound. Southeast Asia is divided into two distinct regions of faunal (or wildlife) geography by Wallace's Line. To the west of the line, wildlife is basically Asian in origin and affinity, whereas to the east it is essentially Australian.

connected to the highlands of western Tibet and south-central China. Here, in the far north of Burma (Myanmar), mountains reach 18,000 feet (5,500 meters). From this point, a series of distinct mountain ranges radiates out from this highland knot, extending through western Burma, along the Burma-Thailand border, and through Laos into southern Vietnam. Most of these rugged mountain ranges are less than 10,000 feet (3,500 meters) in height, and population is sparse throughout these mountains. The mountainous country of Laos, for example, has only 3.5 million people in an area as large as the United Kingdom in western Europe, which has almost 60 million people.

Highlands, River Valleys, and Deltas Several large rivers flow southward out of Tibet and its adjacent highlands into mainland Southeastern Asia. The longest is the Mekong, which is about as long as North America's Missouri River at 2,600 miles (4,190 kilometers). It flows through Laos and Thailand, then across Cambodia before entering the South China Sea with an extensive delta area in Vietnam. Second longest is the Irrawaddy at about 1,300 miles (2,100 kilometers), which flows through Burma's central plain before reaching the Bay of Bengal. This river, as well, has a large delta region. Two smaller rivers are equally significant: the Red River, which forms a sizable and heavily settled delta in northern Vietnam, and the Chao Phraya, a river that has created a fertile alluvial plain in central Thailand.

In many ways, the countries of mainland Southeast Asia are focused on the densely settled, agriculturally intensive valleys and deltas of these rivers (Figure 13.3). The population core of Thailand, for example, is formed by the valley and delta of the Chao Phraya River, just as Burma is focused on

▶ **Figure 13.3 Delta landscape in Vietnam** Southeast Asia has some of the world's largest delta landscapes. Deltaic environments are used for intensive, irrigated rice cultivation, allowing very high rural population densities. Delta wetlands are also used for aquaculture (fish farming) and other forms of intensive food production. Most of mainland Southeast Asia's large cities are located in delta areas, resulting in periodic flooding and other environmental problems. *(William Waterfall/Pacific Stock)*

the Irrawaddy. In Vietnam there are two distinct foci: the Red River delta in the far north and the Mekong delta in the south. In contrast to these densely settled areas, the middle reaches of the Mekong River provides only limited flatland in Laos, which is one of the reasons that country has a smaller population than its neighbors. In Cambodia the largest population historically has clustered around Tonle Sap, a large lake with a very unusual seasonal flow reversal. During the rainy summer months, the lake receives water from the Mekong drainage, but during the drier winter months it contributes to the river's flow. Cambodia's modern population core, and its capital city of Phnom Penh, are situated where the Tonle Sap drainage meets the Mekong River.

The centermost area of mainland Southeast Asia is Thailand's Khorat Plateau, which is neither a rugged upland or a fertile river valley. This low sandstone plateau averages about 500 feet (175 meters) and is noted for its thin and poor soils. Further, water shortages and droughts make this extensive area difficult for settlement; thus, population densities are generally low, with most people living in scattered villages.

Monsoon Climates Just about all of mainland Southeast Asia lies in the tropical monsoon zone, characterized by a distinct rainy season from May to October that is coincidental with the hot, sultry summer months. This is followed by dry conditions from November to April (Figure 13.4). Only the central highlands of Vietnam receive significant rainfall during this autumn and winter period, much of it coming from severe tropical storms that form in the western Pacific during these seasons.

As is the case in South Asia, the summer rainy season is caused by the moist southwest monsoon winds flowing off the Indian Ocean onto the extensive Asian landmass. In winter, as high pressure forms over Asia, these winds reverse, resulting in a dry season as the northeast monsoon brings cooler, dry air masses from the continental interior. Because the same monsoon climate affecting Southeast Asia is shared with most of South and East Asia, these three world regions are often lumped together as a superregion of sorts called "Monsoon Asia." (The climate mechanism that produces the monsoon is explained in more detail in Chapter 12, South Asia.)

Figure 13.4 shows two tropical climate regions in mainland Southeast Asia. While both are connected to the monsoon regime, the difference between the two is in the total amount of precipitation received during the year. Along the coasts and in the highlands, the Am, or tropical monsoon, climate dominates. Rainfall is higher immediately adjacent to the oceans, which is the source of moisture for the monsoon, and in the highlands, where orographic lifting of moist air masses over the mountains takes place. As a result, rainfall totals for the Am climate usually register more than 100 inches (254 centimeters) each year. In the uplands, this figure can easily reach more than 200 inches (508 centimeters).

Yangon (formerly Rangoon), the capital of Burma, located in the Irrawaddy River delta, is a good example of this wet monsoon climate. In April, just before the monsoon's arrival, daily temperatures reach their seasonal peak with one day after another approaching 100 °F (38 °C). On the average, only two days in April bring rain. May, however, is a different story. As the monsoon gathers force, two weeks of rain are common, bringing more than 12 inches (30.5 centimeters) in this short period. With the rain, temperatures drop by 10 °F (5.6 °C). During the height of the monsoon, in July, the city can count on rain almost every day. As a result, July averages more than 20 inches (51 centimeters) of precipitation. Then, in October, the monsoon wanes, with only 10 days of rain. By November, the dry season has returned with just three days of rain recorded. However, daytime high temperatures start to climb once again.

A second climate region, the Aw climate, covers much of mainland Southeast Asia. Here, annual rainfall totals are about half that of the wetter Am rainforest region. In most cases this can be explained by interior (rather than coastal) locations, removed from the oceanic source of moisture, along with the

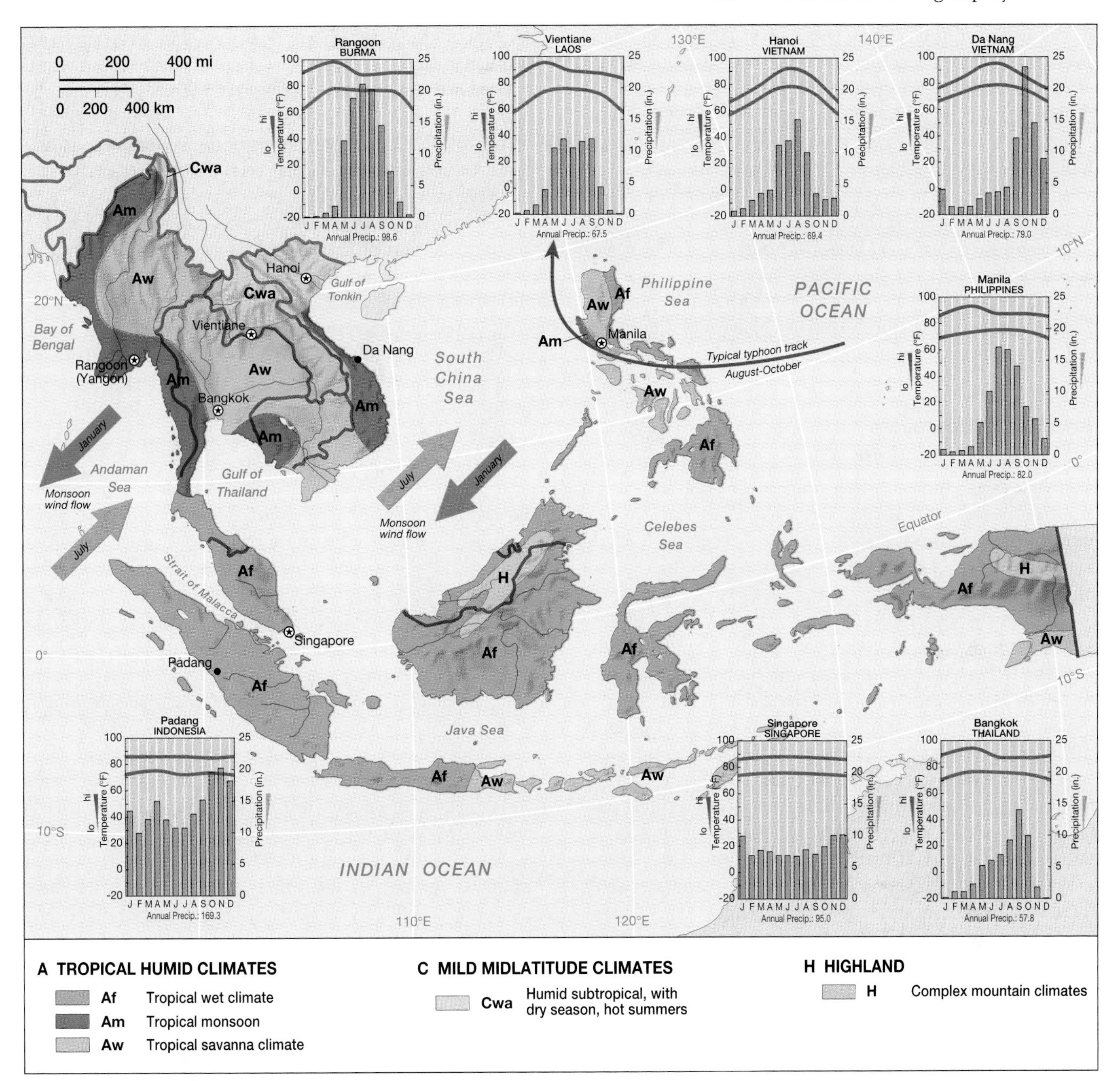

▲ **Figure 13.4 Climate map of Southeast Asia** Most of insular Southeast Asia is characterized by the constantly hot and humid climates of the equatorial zone. Mainland Southeast Asia, on the other hand, has the seasonally wet and dry climates of the tropical monsoon and tropical savanna types. Only in the far north are subtropical climates, with relatively cool winters, encountered. All of the northern half of the region is strongly influenced by the seasonally shifting monsoon winds, northeastern Southeast Asia—and especially the Philippines—often experiences typhoons from August to October.

lack of orographic lifting in these areas of more subdued topography. A good portion of Thailand, for example, receives only about 50 inches (127 centimeters) of rain during the year.

The Forest Landscape The original or native vegetation of the mainland was the tropical monsoon forest. Because of the dry winter season, this forest is not quite as dense or diverse as the equatorial rain forests of insular Southeast Asia that receive copious rainfall year-round. Until the late nineteenth century, this was one of the more heavily wooded portions of the world. Even the great deltas of the Irrawaddy and Mekong rivers remained largely uncultivated and covered with forests. Massive forest cutting, however, began in the late 1800s in mainland Southeast Asia. Lowland forests were gradually converted to farmland, both to feed the expanding local population and to supply rice for the growing world economy. Under British and French colonial influence, forests were cleared to create an export economy based upon rice cultivation. Within 30 years in the late nineteenth century, the Irrawaddy delta was deforested and converted into commercial rice paddies as the British relocated settlers from India and central Burma to work these plantations.

A second round of forest loss began in the mid-twentieth century, after World War II. This was focused on the forest of the upland areas. Monsoon forests in mainland Southeast Asia contain several valuable tree species, most notably teak. It is said that today one teak tree can be worth about $40,000. Further discussion on tropical forest logging is found later in this chapter under "Environmental Issues."

Insular Environments

The signal feature of insular Southeast Asia is its archipelagic environment. Indeed, this is a region of thousands of islands of all sizes and shapes. Borneo and Sumatra are the third and sixth largest islands in the world, respectively, while many thousands of other islands are little more than specks of land rising from a shallow sea.

Indonesia alone is said to comprise more than 13,000 islands, dominated by the four great islands of Sumatra, Borneo (the Indonesian portion is called Kalimantan), Java, and the oddly shaped Sulawesi. This island nation also includes the western half of New Guinea and the Lesser Sunda islands that extend to the east of Java, including Timor, which is roughly the same size as Taiwan off China's coast. A prominent mountain spine runs through these islands as a result of tectonic forces. In the western Indonesian islands volcanic peaks of 10,000 feet (3,500 meters) are common. The highest mountains, however, are on the island of New Guinea, where there are a handful of spectacular volcanic mountains around 15,000 feet (5,250 meters) in elevation.

The Philippines, which, like Indonesia, is an island nation, has more than 7,000 islands, centered around the two largest islands, Luzon (about the size of Ohio) in the north and Mindanao (the size of South Carolina) in the south. Sandwiched between them are the Visayan Islands, which number about a dozen. Again, because of tectonic forces that include active volcanoes, the topography of the Philippines includes mountainous landscapes that reach 10,000 feet (3,500 meters)(Figure 13.5).

Closely related to this impressive collection of islands is the fact that part of Southeast Asia also contains one of the world's largest expanses of shallow ocean. These waters cover the **Sunda Shelf**, which is an extension of the continental shelf, extending from the mainland through the Java Sea between Java and Kalimantan. Here, waters are generally less than 200 feet (70 meters) deep. Because of the rich marine life, many insular Southeast Asia peoples have adopted marine ways of life, essentially living on their boats and setting foot on land only on a temporary basis.

Equatorial Island Climates The climates of insular Southeast Asia are a bit more complex than the mainland because of three factors: a more complicated monsoon effect, the influence of Pacific typhoons, and the equatorial location of these islands. A note of caution is needed regarding the seasonal terms *summer* and *winter*. Since the Southeast Asia islands straddle the equator and are found in both the Northern and Southern hemispheres, seasons associated with specific months of the year can be confusing. Additionally, since there is little seasonal variation or temperature change, contrasting terms such as *summer* and *winter* are essentially meaningless. Nevertheless, those two terms are still commonly used—regardless of location—to explain the seasonal change in winds associated with the heating and cooling of the large Asian landmass. Because Asia lies entirely in the northern hemisphere, the period of southerly winds caused by the heating of this large landmass is referred to as the "summer monsoon." Conversely, the northerly winds that result from the cooling of the Asian continent are called the "winter monsoon" (Figure 13.4).

▶ **Figure 13.5 Volcanic eruption in the Philippines** The Philippines and Indonesia, sitting astride plate-tectonic boundaries, have a large number of volcanoes. Although several areas of insular Southeast Asia owe their fertile soils to past eruptions, volcanism remains a major threat to human life. This was demonstrated in the Philippines by the catastrophic eruption of Mt. Pinatubo in 1991. *(Durieux/SIPA Press)*

On the mainland, summer monsoon winds bring heavy rain because the air masses pick up moisture as they flow across the warm ocean waters into continental Asia. During the winter, these winds are generally dry because they flow off the Asian landmass. Most of insular Southeast Asia, on the other hand, does receive winter moisture because these same winter monsoon winds cross large areas of warm equatorial ocean waters, where they absorb moisture. As a result, these north winds can bring heavy rains as the saturated air masses are lifted over island uplands.

On Sumatra and Java, for example, where north winds blow between November and March, these air masses produce heavy rains on the northern side of these east–west running islands. Rainfall is particularly heavy on the northern slopes of the mountains. As an illustration, the climograph for Jakarta, in northern Java, shows the heaviest rainfall during January and February, with the lowest monthly rainfall coming during the May to September period (see Figure 13.4). But during the May–September period, the heaviest rains are found on the southern flanks of these same islands because of the south winds associated with the Asian summer monsoon.

A second factor in the climate pattern of insular Southeast Asia are the tropical hurricanes, or **typhoons** as they are called in the western Pacific, that bring heavy rainfall to the northeastern reaches of insular Asia during the period August to October. Similar to the hurricanes that strike the Caribbean and southeastern North America during the same period, these strong storms bring devastating winds and torrential rain. They develop east of the Philippines and then move westward into the South China Sea, where they intensify. Each year a number of typhoons affect parts of the Philippines with heavy damage and loss of life through flooding and landslides (Figure 13.6). These strong storms also commonly strike the eastern coast of Vietnam.

▲ **Figure 13.6 Typhoon damage** Typhoons are a major environmental threat in the Philippines and mainland Southeast Asia. Typhoons often result in wind damage, but flooding and mudslides cause the most destruction. Deforestation over the past 30 years has greatly increased the magnitude of such floods and mudslides. *(Vogel/Liaison Agency, Inc.)*

The third climate control, the equatorial influence, results from the low latitude location. More so than mainland Southeast Asia, the islands have very little seasonality, thus temperatures are warm throughout the year, with very little variation. In Jakarta, for example, the average daily high temperature varies only 4 °F (2.2 °C) during the year, from 84 °F (29 °C) in January and February (during the winter monsoon), to a high of 88 °F (31 °C) in September. The average low temperature varies even less during the night because high humidity retains heat. Similar temperature patterns with very little diurnal or seasonal variation are found throughout insular Southeast Asia.

Also associated with the equatorial influence is the fact that rainfall is both higher and more evenly distributed during the year than on the mainland. Although it is common to have a period of heavier rain linked to one of the two monsoon patterns, island climates close to the equator do not experience a distinct dry season as do those on the mainland. Instead, rain falls throughout the year from tropical thunderstorms and squalls common to equatorial climates. Parts of Indonesia have more thunderstorms than any other part of the world. As a result of this year-round precipitation, most of island Southeast Asia is placed into the Af, or tropical rain forest, climate category, as can be seen in Figure 13.4.

Wallace's Line and Island Biogeography Within insular Southeast Asia there is a dramatic difference in animal and plant life between western and eastern islands that has long fascinated scientists. On the western islands of Sumatra, Java, and Borneo, one finds large Asian mammals such as tigers, bears, elephants, and rhinoceros. Additionally, there are apes—orangutans and gibbons. Plants and birds, too, are familiar mainland species. But this wildlife is not found in the eastern islands. Large mammals and apes are missing completely. Instead, in the eastern islands one finds animal and bird species usually associated with Australia—marsupials such as opossums and wallabies, cockatoos, and arboreal kangaroos in the trees instead of apes.

This striking difference in flora and fauna was first documented by British naturalist Alfred Wallace, the cofounder with Charles Darwin of the theory of natural selection, who traveled extensively in this islands during the second half of the nineteenth century. He was one of the first to study **island biogeography,** which is the study of the distribution and ecology of plant and animal life unique to islands.

Wallace correctly surmised that the differences in plant and animal life can be explained by the last global ice age, which ended about 12,000 years ago. At that time, the world's oceans were about 300 feet (105 meters) lower because extensive glaciers and ice caps covered much of the Northern Hemisphere, and a significant amount of Earth's water was frozen in them. With a lower ocean level, the shallow Sunda Shelf was dry land, thus forming an extension of mainland Asia. The large islands of Sumatra, Java, and Borneo were connected to continental Asia by landbridges, thus providing a means for Asian mammals and primates to extend their range. In the east, this lower sea level connected New Guinea and the other islands to the Australian continent, thereby allowing birds, plants, and animals from that landmass to colonize new territory.

However, the advance of these species was apparently halted by open water between the islands of Borneo and Celebes, as it was in the narrow channel between Bali and Lombok in the Lesser Sunda Islands. As the world's ice sheets melted, sea level rose, reaching its current level about 5,000 years ago. With this higher ocean level, the islands of Southeast Asia were once again cut off from the mainland; thus, plants, animals, and birds were stranded in their modern locations. Because of Alfred Wallace's work on this fascinating ecological issue, the division between the two natural realms is called **Wallace's Line,** separating plants and animals from Asia from those originating in Australia.

Environmental Issues

Globalization has had a profound effect on the Southeast Asian environment. Multinational logging companies have reached deep into the region's forests, cutting trees, damaging watersheds, and adding tons of pollutants to the already overburdened atmosphere because of widespread burning of cut-over lands (Figure 13.7). This forest burning aggravates often intolerable air pollution within the rapidly growing major cities of Southeast Asia. Pollution levels in Jakarta, for

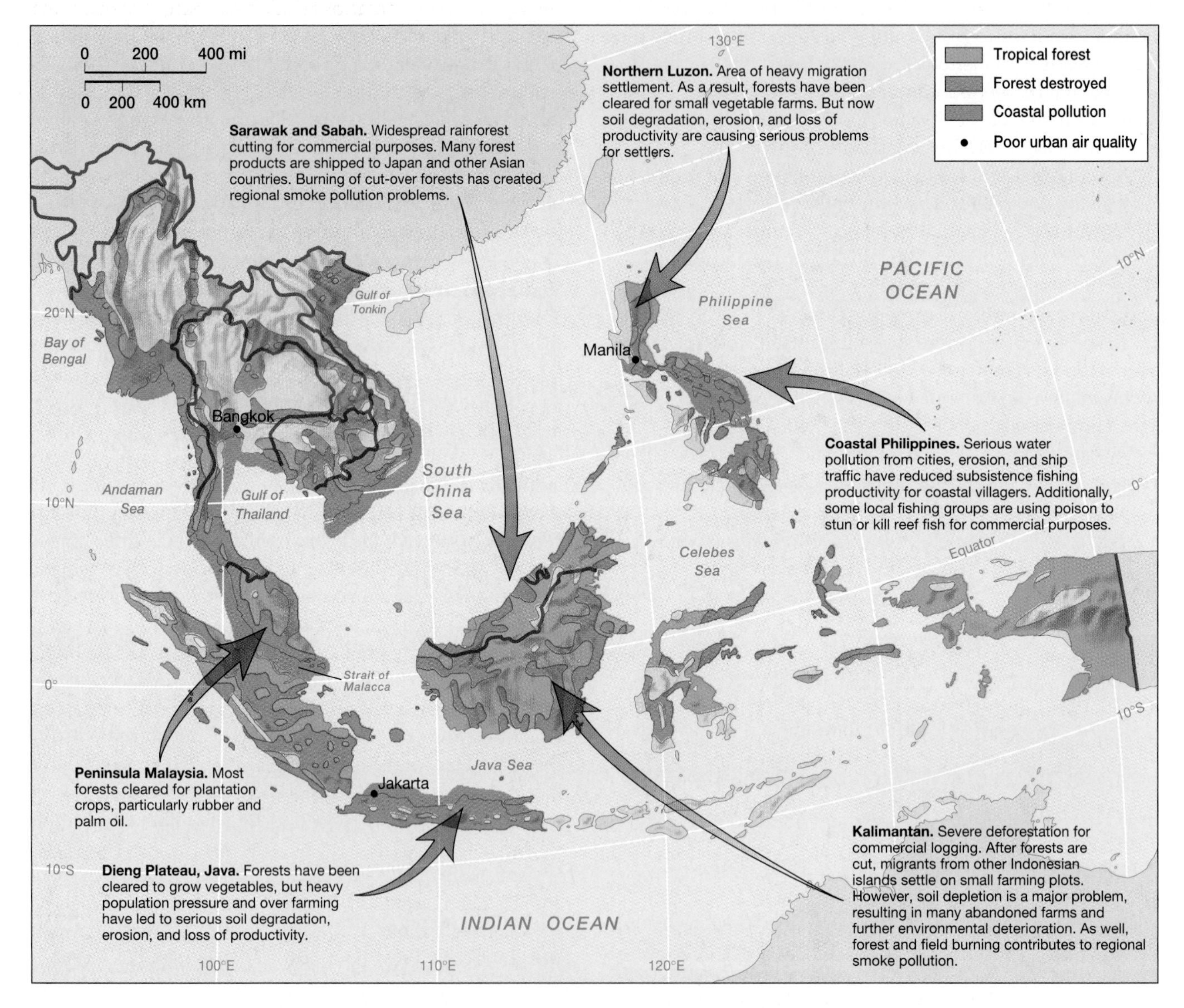

▲ **Figure 13.7 Environmental issues in Southeast Asia** Southeast Asia was once one of the most heavily forested regions of the world. Most of the tropical forests of Thailand, peninsular Malaysia, Sumatra, and Java, however, have been destroyed by a combination of commercial logging and agricultural settlement. The forests of Kalimantan (Borneo), Burma (Myanmar), Laos, and Vietnam, moreover, are now being rapidly cleared. Water and urban air pollution, as well as soil erosion, are other widespread environmental problems in Southeast Asia.

example, are almost three times higher than World Health Organization guidelines specify.

Tropical Forest Problems Although colonial powers cut Southeast Asian forests for tropical hardwoods and naval supplies, and indigenous peoples have long cleared small areas of forest for agricultural use, major environmental problems have come only in the last several decades with large-scale international commercial logging. This activity is driven by Asia's seemingly insatiable appetite for wood products such as plywood and paper pulp—the demand for paper pulp alone has increased at about 6 percent per year over the last decade throughout Asia.

While most forestry experts agree that Japan first globalized world forestry in the 1960s, other Asian countries such as Taiwan, South Korean, Malaysia, and China have followed suit with their own wood products firms. Although the headquarters, boardrooms, and profits of these companies may be global, the damaging effects of commercial logging are both local and global. Landscapes are denuded, watersheds destroyed, wildlife habitat devastated. Additionally, the costs are high on rural peoples who rely on forest resources for their traditional way of life. Beyond this local damage, the global effects of tropical forest clearing are also becoming increasingly problematic as more evidence links forest cutting to atmospheric warming and regional air pollution.

Two points are important for understanding forestry problems in Southeast Asia. First, while there is little question that countries such as Indonesia look to their rural forested areas as a way of increasing food supplies through expanded agriculture and relieving population pressure in their more densely settled areas, generally speaking, population growth is not the cause of deforestation. Most forests are cut for commercial uses so the wood products can be exported to other parts of the world. Only after forests are logged are cut-over lands areas then opened for settlement.

Second, most Southeast Asian countries now have widely publicized bans on the export of forest trees and logs. Some (Thailand is an example) have even gone so far as to ban forest cutting altogether. These restrictions, however, must be critically examined. Log export bans are common because countries now realize they can maximize their income from forest resources by milling and processing logs internally rather than simply cutting and exporting raw logs. In the past, it was common that logs were shipped to intermediary processing plants in Singapore or South Korea before entering the world market. Today, however, most Southeast Asian countries require international companies to co-invest in building internal processing plants. Often a quota is established. For example, in Indonesia one log can be exported for every four milled within the country. Though there can be considerable economic and social benefits from this arrangement, such a log ban does not necessarily mean either that the rate of forest cutting is less or that the environmental issues have been resolved. As for those countries with an outright ban on forest cutting, either it is simply not true, it is not enforced, or—as is the case in certain parts of Malaysia—there are no trees left to cut (Figure 13.8).

Malaysia has been the leading exporter of tropical hardwoods from Southeast Asia with about two-thirds of its total harvest leaving the country. In recent decades, 60 percent of these log exports went to Japan. Income from these exports puts Malaysian forest products on par with petroleum exports and larger-than-traditional plantation products such as rubber and agricultural crops. The cost of this cutting to the environment, however, has been high. Peninsular Malaysia was largely denuded by 1985 when a cutting ban was imposed. Since then, forest cutting has been concentrated in the states of Sarawak and Sabah on the island of Borneo, where the granting of logging concessions to foreign firms has caused considerable problems with local tribal people by disrupting their traditional resource base. International forestry experts estimate that at current rates of cutting, Malaysia will be almost completely deforested early in the twenty-first century.

▲ **Figure 13.8 Commercial logging** Southeast Asia has long been the world's most important supplier of tropical hardwoods. Unfortunately, most of the tropical forests of the Philippines and Thailand, as well as the Indonesian islands of Java and Sumatra, have been destroyed by the logging process. *(Jean-Leo Dugast/Panos Pictures)*

Thailand cut more than 50 percent of its forests in the 20-year period from 1960 to 1980. This loss was followed by a series of logging bans so that by 1995, no legal forest cutting took place in the country. Damage to the landscape, however, was severe, with heavy silting of irrigation works and hydroelectric facilities, increased flooding in lowlands areas, and heavy erosion on hillslopes. Increasingly, these cut-over lands are being reforested with fast-growing Australian eucalyptus trees that can be used for paper products and plywood. The ban on logging in Thailand was accompanied by agreements with neighboring Burma (Myanmar) to allow Thai companies to cut forests in that country, including the last of the teak trees in the northern mountains. Apparently this logging activity became a major source of revenue for Burma's military government, although it was reduced somewhat by a U.S. ban passed in 1990 on importing teak products from both Thailand and Burma. To get around this restriction, most recently Thai loggers have now moved into the forests of Laos in search of valuable teak trees.

Indonesia, the largest country in the Southeast Asia, has fully two-thirds of the region's forest area, including about 10 percent of the world's true tropical rain forests. In 1981 the country placed a ban on log exports for those companies that had not invested in Indonesia wood-products processing plants. The goal was to create jobs and infrastructural improvements within the country. This strategy seems to have worked, although unfortunately the recent downturn in Asia's economy along with political unrest in Indonesia make evaluation difficult because there is some evidence that wholesale forest cutting has resumed, to compensate for weakness in other sectors of the Indonesian economy.

Smoke and Air Pollution Until recently, Southeast Asia seemed oblivious to the widespread pall of air pollution covering most of the region that was created by a combination of urban smog and the smoke from tropical forest clearing. Then, late in the 1990s, the region suffered from two consecutive years of disastrous air pollution that served as a wake-up call. During that period, a commercial airliner crashed because of poor visibility, countless road accidents resulted, two ferries collided in smoke-laden conditions, and hundreds of thousands of people were admitted to hospitals with life-threatening respiratory problems. Additionally, because of the publicity about these pollution problems, billions of tourism dollars were lost when reservations were cancelled (Figure 13.9).

Several factors, both natural and economic, combined to produce the region's horrendous air pollution problems. First, large portions of insular Southeast Asia were dried out by drought caused by El Niño, turning the normally wet tropical forests into tinder boxes that were easily ignited. This drought also dried out the widespread peat bog landscape of Kalimantan that, once fired, continued to burn for months and months. Second, commercial forest cutting was—and is—responsible for most forest burning, even though the international logging firms commonly point the finger at small farmers as the culprits. In Sumatra's forest, 80 percent of the fires were from commercial forest plantations, with only 20 percent from small slash-and-burn agriculture. This ratio also seems to hold for other parts of the region. As commercial lumbering has increased over the region in the last decade, so has air pollution from post-logging fires.

▲ **Figure 13.9 Urban air pollution** Air pollution has reached a crisis stage in the rapidly industrializing cities of Southeast Asia, particularly in Bangkok, Manila, and Jakarta. People sometimes resort to using face masks to filter out soot and other forms of particulate matter, as is visible in this photograph. *(V. Miladinovic/Sygma Photo News)*

The third factor in this air pollution disaster comes from the region's large cities where cars, trucks, and factories emit huge amounts of pollutants. Even without the added pollution from forest burning, many cities, such as Bangkok, Jakarta, and Manila, already have an unhealthy level of pollution. The combination of these two sources, urban and rural, has been deadly. In Bangkok, rolling down a window during a taxi ride is considered a very serious breach of etiquette, since people protect themselves from street-level pollution by sequestering themselves in air-conditioned cars. Those who cannot afford taxis, however, are the ones who suffer most. As a general rule, lung cancer or pulmonary disease resulting from urban air pollution kills people at five times the rate as in the United States. Given the increase in auto traffic and other forms of urban pollution, coupled with the increasing presence of a pall of smoke from forest fires, this unfortunate statistic is liable to increase before any decisive action is taken.

Population and Settlement: Subsistence, Migration, Cities

The scale of Southeast Asia's population issues is quite different from those pressing problems faced by its giant neighbors China and India. With just more than 500 million people (East and South Asia each have more than twice that many people), Southeast Asia is relatively sparsely settled. Part of the reason can be attributed to extensive tracts of infertile soils and mountainous areas. Fertile alluvial soil derived from the region's large rivers has led to a contrast-

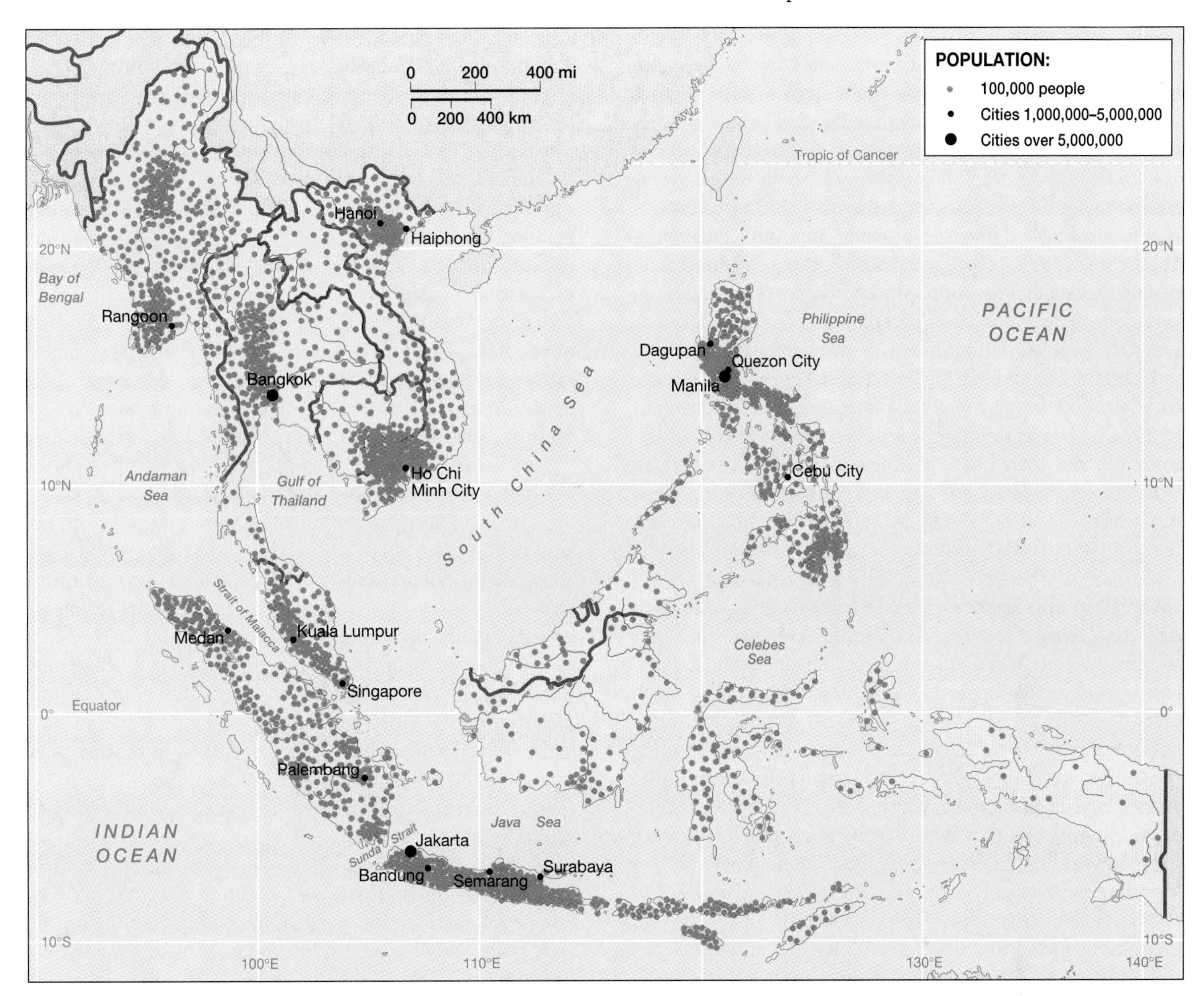

▲ **Figure 13.10 Population map of Southeast Asia** In mainland Southeast Asia, population is concentrated in the valleys and deltas of the region's large rivers. In the intervening uplands, population density remains relatively low. In Indonesia, extremely high density is found on Java, an island noted for its fertile soil and large cities. Some of Indonesia's outer islands, especially those of the far east, remain lightly settled. Overall, population density is high in the Philippines, especially in central Luzon.

ing pattern of dense pockets of population in deltas and coastal areas, with sparsely settled mountain areas (Figure 13.10).

In recent decades, however, population growth and family planning have become increasingly important, in different ways, to the Southeast Asian countries. While Indonesia promotes migration to its outer islands as a way to relieve population pressure in the core areas, other countries such as Thailand place considerable emphasis on family planning. Malaysia, on the other hand, concludes that its current population is too small, so the government advocates population growth by encouraging larger families and in-migration to the country.

As is the case throughout the world, one of the major changes in settlement and subsistence is the massive migration to cities. This has been particularly true in those Southeast Asian countries that experienced recent rapid economic growth and development during the boom years. Now, with these once-energetic tigers somewhat lethargic because of the downturn in economic globalization, Southeast Asian cities have a respite of sorts to address the environmental and infrastructural problems resulting from those boom years.

Settlement and Agriculture

Part of the reason for the historical paucity of population in Southeast Asia is the infertility of soils that are unable to support intensive agriculture and high population densities. The island rain forests, though lush and biologically rich, grow on poor soils. Plant nutrients are locked up in the vegetation itself rather than being stored in the soil where they would benefit agriculture. Further, the incessant rain of the equatorial zone tends to wash nutrients out of the soil. In these areas agriculture must be carefully adapted to this limited soil fertility by shifting crops and constant field rotation.

However, there are some notable exceptions to this generalization about soil fertility and settlement density. Unusually rich soils connected to volcanic activity are scattered throughout Southeast Asia. In Java, for example, which has more than 50 volcanoes, rich soils support both an array of tropical crops and a very high population density. More to the point, Java has well over 100 million people—more than half the total population of Indonesia—in an area smaller than the state of Iowa. Additionally, dense populations are found in pockets of fertile alluvial soils along the coasts, both on the mainland and in the islands, where people augment land-based farming with marine resources and trade activities. Within this complex environment of Southeast Asia, three general farming and settlement patterns are apparent.

Swidden in the Uplands Also known as shifting cultivation or even "slash-and-burn" agriculture, swidden is found throughout the rugged uplands of both mainland and island Southeast Asia (Figure 13.11). In the **swidden** system, small plots of several acres (5 hectares) of dense tropical brush are periodically cut or "slashed" by hand. Then, after this cutting, the fallen vegetation is burned to transfer nutrients to the soil before subsistence crops are planted. Yields remain high for several years, then drop off dramatically as the soil nutrients from burned vegetation are exhausted. These plots are then abandoned completely after about five years and allowed to revert to woody vegetation. The cycle of cutting, burning, and planting is moved to another small plot not far away—thus the term *shifting cultivation*. Families and small villages generally control a large amount of territory so that they can rotate their fields on a regular basis. After a period of between 10 and 50 years, the farmers will return to the original plot, which once again has nutrients in the dense vegetation. The cycle then continues with another round of slashing and burning.

Swidden is a sustainable form of agriculture when population densities remain relatively low and stable and when upland people control a large enough amount of territory to rotate fields regularly. Today, however, the swidden system is threatened in Southeast Asia for two reasons. First, it cannot support the increased population that usually results from the introduction of public health measures. With a higher population density, the swidden rotation period must be shortened, and this is generally unsuccessful because of finite soil resources. Second, the upland swidden system is often a casualty of commercial forest logging because of the devastation done to highland forest environments. Perhaps nothing illustrates the long arms of globalization better than the impact of international commercial forest cutting on the traditional agricultural system of upland tribal peoples.

When swidden can no longer support a population, upland people often adapt by switching to a cash crop that will allow them to participate in a rural trade economy. In the mountains of northern Southeast Asia, one of the main cash crops is opium, grown by swidden farmers for the global drug trade. This mountainous area is now referred to as the **"Golden Triangle."** Burma reportedly is now the world's second largest opium producer. Reportedly, 100 tons of opium leave the Golden Triangle each year for markets in Europe, Australia, and the United States. Many reports imply linkages between local military officials, teak logging, and opium growing in this area.

Plantation Agriculture With European colonization, Southeast Asia became a focus for commercial plantation agriculture, growing high-value specialty crops ranging from rice to rubber. Even in the nineteenth century, Southeast Asia was linked to a globalized economy through the plantation system. Forests were cleared and swamps drained to make room for these plantations; labor was supplied (often unwillingly)

▶ **Figure 13.11 Swidden agriculture** In the uplands of Southeast Asia, swidden (or "slash-and-burn") agriculture is widely practiced. When done by tribal peoples with low population densities, swidden is not environmentally harmful. When practiced by immigrants from the lowlands, however, swidden can result in deforestation and extensive soil erosion. *(Paula Bronstein/Liaison Agency, Inc.)*

either by indigenous people who were relocated from their traditional villages or by contract labor brought in to Southeast Asia from other parts of the world, often from India or China. These plantations were usually in the coastal lowlands, from which products could be easily shipped out by boat to Europe and North America. Over the years, with fluctuating world demand, competition from other world regions, and changing political conditions within the region, the array of plantation crops has changed (Figure 13.12).

Plantations are still an important part of Southeast Asia's geography and continue to play a major role in today's global economy. Most of the world's natural rubber, for example, is produced in the region, mainly in Malaysia, Indonesia, and Thailand. However, this natural rubber is a small fraction of the total world rubber consumption since today most rubber is made synthetically from petroleum. Cane sugar has long been a plantation crop of the Philippines. More recently, pineapple plantations have appeared in both the Philippines and Thailand. These two countries are now the world's leading exporters of this specialty crop. Indonesia is the region's leading producer of tea, and that country also dominates world production of coconut palm products such as oils and **copra,** which is dried coconut meat. Although plantations are still widespread in Southeast Asia, there is a general tendency for countries to depend less and less on agricultural exports and more on other forms of economic enterprise, such as mining, forestry, or industrial development that produce more favorable income. One exception appears to be Malaysia, a former British colony with long experience with plantation agriculture; it is country that can easily afford to buy food staples on the world market because of its trade surplus.

Rice in the Lowlands The lowland basins of mainland Southeast Asia are largely devoted to intensive rice cultivation. Throughout almost all of Southeast Asia, rice is the preferred staple food. In fact, in several local languages "to eat rice" means "to eat a meal." Traditionally, rice was cultivated on a subsistence basis by rural farmers, However, as the number of wage laborers in Southeast Asia has grown because of economic development, so has the demand for commercial rice cultivation. Rice harvests are no longer for local consumption only but, instead, are traded and exported to fulfill the food needs of the region's expanding urban laborers. Three riverine delta areas have been the focus for commercial rice cultivation, the Irrawaddy in Burma, the Chao Praya in Thailand, and the Mekong delta in Vietnam and Cambodia. The use of agricultural chemicals and high-yield crop varieties, along with improved water control and the expansion of irrigation facilities to facilitate dry-season cropping, have allowed production to keep pace with population growth. Thailand, in fact, has become a major rice exporter to other countries.

In those areas without irrigation and water control, yields remain relatively low. Rice growing on the Khorat Plateau, for example, depends largely on the uncertain rainfall, without the benefits of the more sophisticated water control methods available elsewhere in Thailand. In some lowland districts lacking irrigation, dry-field crops, especially sweet potatoes and manioc, form the staple foods of people too poor to buy market rice on a regular basis.

▲ **Figure 13.12 Tea harvesting in Indonesia** Plantation crops, such as tea, are major sources of exports for several Southeast Asian countries. Coconut, rubber, oil palms, and coffee are other major plantation crops. Many of these crops require large amounts of labor, particularly at harvest time. *(Dermot Tatlow/Panos Pictures)*

Recent Demographic Change

Because Southeast Asia is not facing the same kind of population pressure as East or South Asia, a wide range of different government population policies are found in this region. While several countries show concern about rapid population growth and thus have strong family planning programs, others believe their population is too small; hence, larger families are encouraged. Additionally, in those countries with rapid population growth, internal relocation away from densely populated areas to outlying districts is a common policy.

Population Contrasts The Philippines, the third largest country in Southeast Asia in terms of population, has a relatively high fertility and growth rate (Table 13.1). Further, complicated

Table 13.1 Demographic Indicators

Country	Population[a]	Natural Increase	TFR[b]	%<15[c]	% Urban
Burma (Myanmar)	47.1	2.0	3.8	36	25
Brunei	0.3	2.2	3.4	34	67
Cambodia	10.8	2.4	5.2	44	14
Indonesia	207.4	1.5	2.7	34	37
Laos	5.3	2.8	5.9	45	19
Malaysia	22.2	2.1	3.2	35	57
Philippines	75.3	2.3	3.7	38	47
Singapore	3.9	1.1	1.7	23	100
Thailand	61.1	1.1	2.0	27	31
Vietnam	78.5	1.2	2.3	40	20

[a]Population in millions.
[b]Total fertility rate.
[c]Percentage of population younger than 15 years of age.
Source: *Population Reference Bureau World Population Data Sheet,* 1998.

internal politics get in the way of an effective family planning effort. When a popular democratic government replaced the Marcos dictatorship in the 1980s, the Philippine Roman Catholic Church—which played an active role in the peaceful revolution—was able to pressure the new government to cut funding for family planning programs. As a result, the many clinics and centers that had earlier dispensed family planning information were closed. Though high birthrates are not always associated with Catholicism, particularly in more-developed countries such as Spain and Italy, the Church's outspoken stand on birth control and abortion clearly plays a role in the Philippines by inhibiting the dispersal of family planning information.

The highest total fertility rate (TFR) in Southeast Asia (5.9 children) is found in Laos, a country with a very different religious tradition (Buddhist) than Roman Catholicism. Here, the relationship between high birthrates and a low level of economic development is a better explanation than religion. Thailand, which shares cultural traditions with Laos, yet is considerably more developed because of recent economic growth, demonstrates the other end of that spectrum. Here, the TFR has dropped dramatically within the last 20 years from 5.4 (in 1970) to 2.0, a figure that will now lead to population stability. While economic growth may explain some of this decrease, the Thai government has also promoted a very visible family planning effort, both for population and health reasons—the latter because of the high incidence of AIDS in the country.

Indonesia, with the region's largest population at more than 200 million, also has seen a dramatic decline in fertility in recent decades, although its fertility rate still remains above the replacement level. However, if the present trend continues, Indonesia will reach population stability well before most other large developing countries. As with Thailand, this drop in fertility seems to have resulted from a strong government family planning effort, coupled with improvements in education.

Cambodia stands out (along with Laos, mentioned earlier) with its unusually high fertility and population growth rate. Connected with this, and surely part of the explanation, is that this country also has the lowest life expectancy and the highest infant mortality rate in Southeast Asia (see Table 13.3). As has been seen in other world regions, high birthrates are often found in association with high mortality. Much of this can be explained in the case of Cambodia by its recent civil strife and internal violence, factors that have inhibited both economic and social development. Now that some semblance of stability has returned to that country, it will be interesting to see whether the fertility rate declines in the next few years.

To a lesser degree, the relatively high fertility rates of both Burma and Vietnam can also be partially explained by decades of internal turmoil that has until recently restricted economic development. With recent economic development in Vietnam, coupled with the first signs of government concern about family planning, population growth in that country will probably slow in the near future. The situation in Burma, however, is far less certain.

Growth and Migration Indonesia is the country with the most explicit policy of **transmigration,** or relocation of its population from one region to another within its national territory. Primarily because of migration from densely populated Java, the population of the outer islands of Indonesia has grown rapidly since the 1970s. The province of East Kalimantan, for example, experienced an astronomical growth rate of 30 percent per year through the last two decades. As a result of this shift in population, many parts of Indonesia outside of Java now have moderately high population densities, although the more remote districts still remain lightly inhabited (Figure 13.13). If natural growth remains high on the main islands, these remote areas could well become targets for further transmigration.

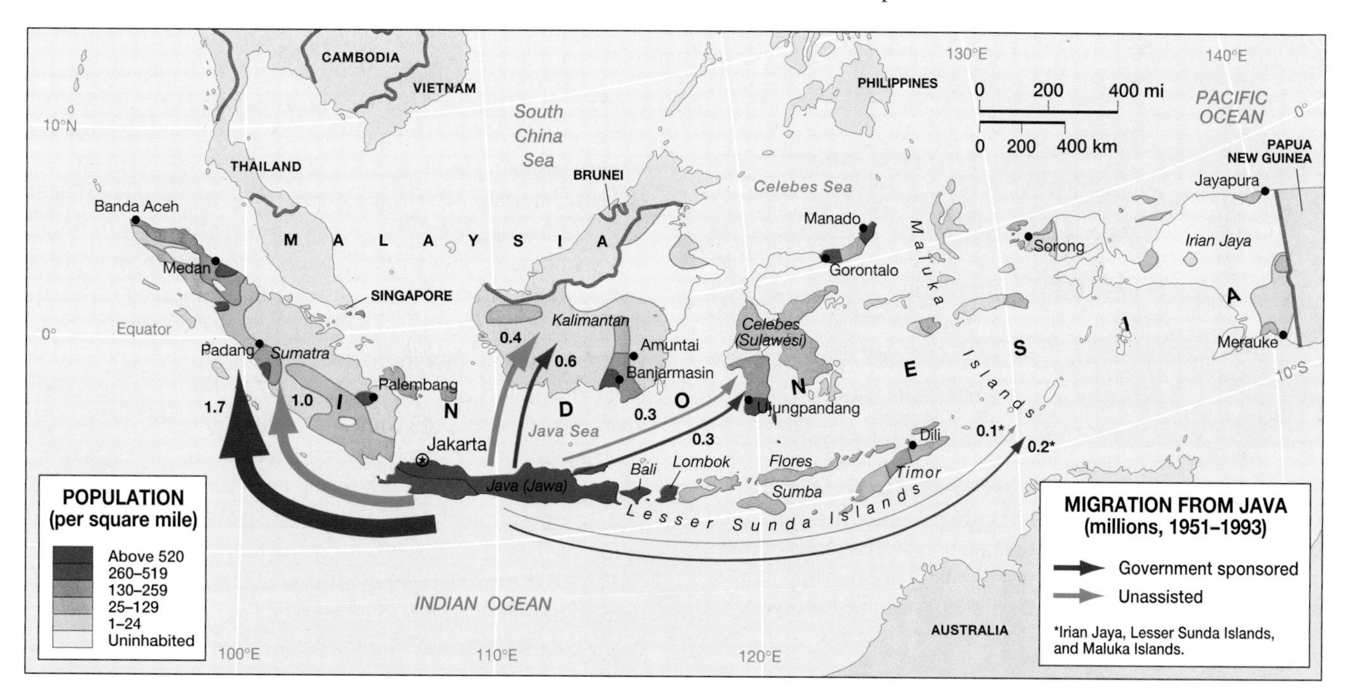

▲ **Figure 13.13 Indonesian transmigration** The distribution of population in Indonesia shows a marked imbalance; Java is one of the world's most densely settled places, whereas most of the country's other islands remain rather lightly populated. As a result, the Indonesian government has encouraged the resettlement of Javanese people to the outer islands, often paying the costs of relocation. This transmigration scheme has resulted in a somewhat more balanced population distribution pattern, but has also caused substantial environmental degradation and has intensified several ethnic conflicts.

As is the case worldwide, high social and environmental costs often accompany relocation schemes. Javanese peasants, accustomed to working the highly fertile soils of their home island, often fail in their attempts to grow rice in the former rain forest of Kalimantan. Field abandonment is high after repeated crop failures. In some areas farmers have little choice but to adopt a semi-swidden form of cultivation, moving to new sites once the old ones have been exhausted, a process that intensifies the processes of deforestation and soil degradation. The term **shifted cultivators** is often used for these rural migrants who are moved from one area to another through government relocation schemes (see "Local Voices: Land Rights and Indigenous Peoples").

The Philippines have also used subsidized migration as a means of alleviating population pressure on the main islands. Beginning in the late nineteenth century, Philippine society has responded to increasing population pressure in the core areas of central and northwestern Luzon by sending colonists to frontier areas. In the early twentieth century this settlement frontier still lay in Luzon; by the postwar years, it had moved south to the island of Mindanao. However, that frontier is currently saturated so the emphasis is now on international migration to other parts of Southeast Asia and the world. One unfortunate aspect of this new globalized migration is the thousands of young Philippine men who work as low-paid deckhands on ill-fitted merchant ships plying the oceans. These are migrants without a country to call home.

Slow Growth and Pro-Growth The city-state of Singapore stands out on the demographic charts with a fertility rate well below replacement levels. Unless this negative population rate is offset by immigration or a dramatic turnabout in the birthrate, this country's population will soon begin to decline. The government is concerned about this situation and is actively promoting marriage and childbearing, particularly among the most highly educated segment of its population. Ironically, it is this very segment that is the social group with the most pronounced fertility decline.

Malaysia, too, has a government that supports population growth even though its fertility rate is still well above replacement levels. Larger families are encouraged, as is in-migration to the country. Additionally, the government is particularly interested in increasing settlement in its Borneo island states of Sabah and Sarawak, which it considers vastly underpopulated (Figure 13.14). In fact, several government officials have recently suggested that Malaysia should double its population to around 50 million.

Urban Settlement

Despite the relatively high level of economic development found in Southeast Asia, the region is not heavily urbanized, with less than 30 percent of its population living in cities. Even Thailand retains its rural flavor, which is somewhat unusual for a country that has experienced so much recent industrialization. However, cities are growing rapidly throughout the region, so the rate of urbanization will undoubtedly increase over the next decade.

▲ **Figure 13.14 Migrant settlement in Indonesia** Migration from densely settled to sparsely settled areas of Southeast Asia has resulted in the creation of thousands of new communities. Many of these communities have minimal infrastructure and services, and some of them struggle to survive. *(Charly Flyn/Panos Pictures)*

Many Southeast Asian countries have **primate cities,** where a single, large urban settlement dominates all others. Thailand's urban system, for example, is dominated by Bangkok, just as Manila overshadows all other cities in the Philippines. Both cities have grown recently into megacities with more than 10 million residents. More than half of all city-dwellers in Thailand live in the Bangkok metropolitan area (Figure 13.15). The city's urban primacy is illustrated further by the fact that its population is 25 times larger than the country's second city, Nakhon Ratchasima. In both Manila and Bangkok, as is the case in most developing countries, explosive urban growth has led to housing problems, congestion, and pollution. With the recent arrival of mass automobile culture, Bangkok may now suffer from the worst traffic congestion in the world. In Manila, it is estimated that more than half of the city's population lives in squatter settlements, usually without the basic services of water and electricity.

Thailand, the Philippines, and Indonesia are all making efforts to encourage growth of secondary cities by decentralizing economic functions. The goal is to stabilize the population of these primate cities. In the case of the Philippines, the city of Cebu has emerged in recent years as a more dynamic economic center than Manila, leading to hopes that a more balanced urban system be soon emerge. Jakarta, Indonesia's primate city, is home to more than 13 million people. Its explosive growth started in the 1970s, at which time the government attempted to limit migration into the capital city. Most experts agree that this program had little success.

Primate city size is considerably less in other Southeast Asian countries. Vietnam, for example, has two primate cities,

LOCAL VOICES Land Rights and Indigenous Peoples

In the mountains of northern Luzon, as is true in many other parts of Southeast Asia, indigenous small-scale societies often have no legal rights to their own ancestral lands. This lack of basic land rights makes them vulnerable to outside interests eager to gain control of the their resources. The following dialogue, written by Filipino author Pedro Bundok, illustrates the tensions generated by the contradictions between two different legal systems of land right—one customary and indigenous, the other codified and identified with the modern state. Although the text was written more than 20 years ago, the issues that it brings up are still current. Here we find Pedro Bundok trying to explain the official land system to an elderly gentleman named Bugtong:

"Pedro, I have a problem. . . . They tell me that I should not farm my land anymore because I am a squatter. What do they mean by a squatter?"

"A squatter," I said, . . . "is someone who does not get permission to use land that belongs to someone else."

"How can the government say that I am a squatter? My father's bones and my grandfather's bones are buried on the land where my camote [sweet potatoes] are planted!"

"The government man says that you should have a piece of paper called a 'title,' Bugtong. If you have a paper it means that you own the land."

"Pedro, you act like you don't know very much. . . . We all know that the one from Heaven owns the land. We can only borrow the land. . . . How can that young boy from the government think that a paper is going to make me own my land."

"Bugtong, . . . the government wants every person to decide which land he is borrowing and then get a paper that they call a 'title' which stops the arguments about who is allowed to farm each part of the land."

"But Pedro, our ancestors have been here since Balitok and Bogan and they have not had any papers and they have not had any trouble. If anyone had an argument, they just called for a conference and the elders helped them to remember and they agreed what to do. Then they ate a pig and everyone went home happy. . . ."

"Bugtong, . . . the elders in the lowlands have forgotten how to do that, so now they have papers to help them remember. . . ."

"Pedro, I understand about the papers, but why did that man say that I am a squatter? This I cannot understand. . . ."

"The government says that any land which does not have papers now belongs to the government. You do not have papers for the land yet, so he thinks that you are farming the land which belongs to the government. . . ."

"Now I have two problems instead of one," he said. "First, why should I ask permission to farm the lands where the bones of my grandfather and brother and father are buried? Second, who is the government?"

Source: From *Democracy Among the Mountaineers* by Pedro Bundok. Quezon City, Philippines: New Day Publishers, 1973.

▲ **Figure 13.15 Bangkok** Bangkok saw the development of an impressive skyline during its boom years from the late 1970s through the late 1990s. Unfortunately, infrastructure did not keep pace with population and commercial growth, resulting in one of the most congested, and polluted, urban landscapes in the world. *(Robert Holmes/Robert Holmes Photography)*

▲ **Figure 13.16 Singapore** Singapore remains the economic and technological hub of Southeast Asia. It is famous for its clean, efficiently run, and hyper-modern urban environment. Some residents complain, however, that Singapore has lost much of its charm as it has developed. *(Adina Tovy/Photo 20-20, Inc.)*

Ho Chi Minh City (formerly Saigon) in the south with more than three million people, and the capital city of Hanoi in the north, which counts slightly more than a million residents in its population. Yangon (formerly Rangoon) is the capital and primate city of Burma. This city has doubled its population in the last two decades to more than four million residents. In Cambodia, the capital city of Phnom Penh has less than a million; the same is true of Vientiane, the capital and primate city of Laos. Kuala Lumpur is the largest city in Malaysia (and current capital, although a new one is under construction nearby) due to recent investment by both government and global business communities. This has produced a modern city of grandiose ambitions that is free of most infrastructural problems plaguing other Southeast Asian cities. As testimonial to its outlook, the Petronas Towers, owned by the country's national oil company, were the world's tallest buildings at almost 1,500 feet (450 meters) when completed in 1996. However, Shanghai, China, is currently building a taller structure, reminding us that skyscrapers are common symbols of national prestige and economic power.

The independent republic of Singapore is essentially a city-state of three million people on an island of 240 square miles (600 square kilometers), which is about three times the size of Washington, DC. While space is at a premium, Singapore has been very successful at developing high-tech industries that have brought prosperity to this sovereign city-state. Unlike other Southeast Asian cities, one finds no squatter settlements or slums in Singapore. Only in the fast-disappearing Chinatown and historic Colonial district does one find a landscape of older buildings. Otherwise Singapore is a city of modern high-rise skyscrapers and space-intensive industry (Figure 13.16).

Cultural Coherence and Diversity: A Meeting Ground of World Cultures

Unlike many other world regions, Southeast Asia lacks the historical dominance of a single civilization. Instead, the region has been a meeting ground for cultural diffusion from South Asia, China, the Middle East, Europe, and even North America. Abundant natural resources, along with the region's strategic location on oceanic highways connecting major continents has long made Southeast Asia attractive to outsiders. As a result, the contemporary cultural geography of this diverse region is best viewed as a product of borrowing and amalgamation from these external influences.

The Introduction and Spread of Major Cultural Traditions

In Southeast Asia contemporary cultural diversity is embedded in the historical influences connected to the major religions of Hinduism, Buddhism, Islam, and Christianity.

South Asian Influences The first major external influence arrived from South Asia some 2,000 years ago, when migrants from what is now India helped establish Hindu kingdoms in coastal locations in Burma, Thailand, Cambodia, central and southern Vietnam, Malaysia, and Indonesia. Although Hinduism faded away in most locations, this tradition is still found on the Indonesian islands of Bali and Lombok. Additionally, the ancient Indian script formed the basis for many Southeast Asian writing systems; in Muslim Java (Indonesia), the Hindu epic called the **Ramayana** remains a central cultural feature today.

A second wave of South Asian religious influence reached mainland Southeast Asia in the thirteenth century in the form

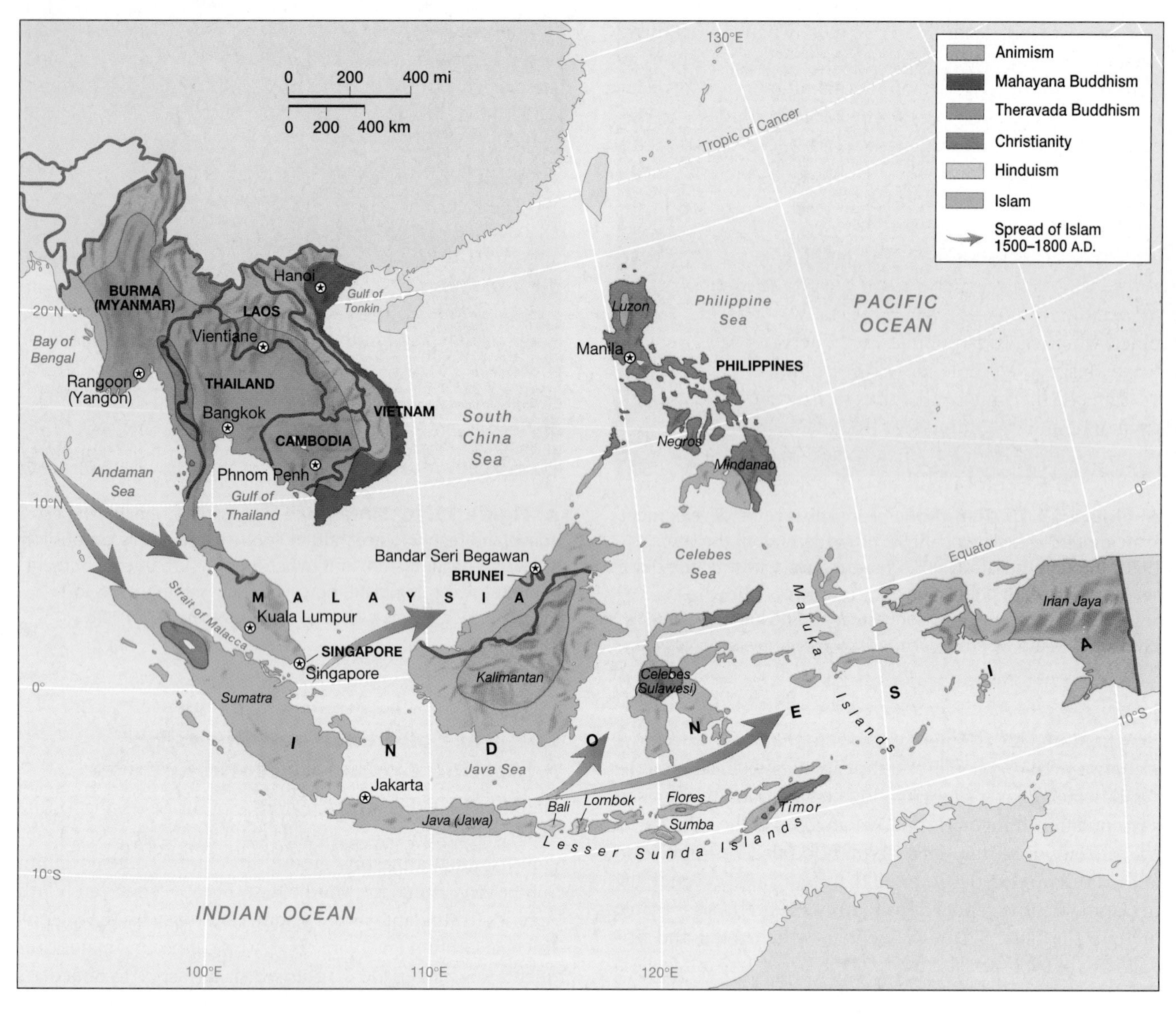

▲ **Figure 13.17 Religion in Southeast Asia** Southeast Asia is one of the world's most religiously diverse regions. Most of the mainland is predominantly Buddhist, with Theravada Buddhism prevailing in Burma (Myanmar), Thailand, Laos, and Cambodia, and Mahayana Buddhism prevailing in Vietnam. The Philippines is primarily Christian (Roman Catholic), but the rest of insular Southeast Asia is primarily Muslim. Substantial Muslim minorities are found in the Philippines, Thailand, and Burma. Animist minorities can be found in remote areas throughout Southeast Asia, especially in Indonesia's province of Irian Jaya.

of Theravada Buddhism, which is a religion more closely associated with Sri Lanka than China or Japan. Virtually all of the people in lowland Burma, Thailand, Laos, and Cambodia converted to this religion at that time (Figure 13.17), and today this religious tradition still forms the foundation for social institutions in those countries. Saffron-robed monks, for example, are a common sight in Thailand and Burma, where Buddhist temples abound in the landscape. In Thailand, that country's constitutional monarchy remains closely bound with Theravada Buddhism. While this form of Buddhism from South Asia shares many traits with the Mahayana Buddhism more closely associated with East Asia, nevertheless there are enough cultural and religious differences that the two are mapped separately in Figure 13.17.

Chinese Influences Unlike most other mainland people, the Vietnamese were not heavily influenced by South Asia civilization. Instead, their connections were to East Asia and, more specifically, to China. Vietnam was actually a province of China until about A.D. 1000, when the Vietnamese established a kingdom of their own. But while the Vietnamese rejected China's political rule, they nonetheless retained many attributes of Chinese culture. The traditional religious and philosophical beliefs of Vietnam are, for example, centered around Mahayana Buddhism and Confucianism. Further, until the French colonial period of the nineteenth century, the Vietnamese used Chinese ideographs or characters for their written communication. This contrasted with the other Southeast Asia countries that employ writing systems more closely associated with India.

East Asian cultural influences in many parts of Southeast Asia are directly linked to more recent historical immigration of southern Chinese. While this migration dates back hundreds of years, it reached a peak in the nineteenth and early twentieth centuries (Figure 13.18). China was then a poor and crowded country, which made sparsely populated Southeast Asia appear to many Chinese as a place of great opportunity. At first most migrants were single men. Many returned to China after accumulating money, but others married local women and established what are essentially mixed communities. This is especially true in the Philippines, where the elite population is often described as "Chinese Mestizo," people of mixed Chinese and Filipino descent. In the nineteenth century Chinese women begin to migrate in large numbers, which facilitated the creation of ethnically distinct Chinese settlements. Urban areas throughout Southeast Asia are still characterized today by large and cohesive Chinese communities. In Malaysia, the Chinese minority constitutes roughly one-third of the population, whereas in the city-state of Singapore, three-quarters of the people are of Chinese ancestry.

In many places within Southeast Asia, relationships between the Chinese minority and the indigenous majority are strained. Even though their ancestors arrived many generations ago, many Chinese are still considered resident aliens because they adamantly maintain their Chinese citizenship. Probably a more significant source of tension is the fact that most overseas Chinese communities are relatively prosperous. As a result, they usually exert tremendous economic influence on local affairs. Anti-Chinese riots have occurred throughout Southeast Asia, most recently in Indonesia. Chinese emigrants prospered because they were able to find a mercantile niche. Often this involved trading local products with China. To illustrate, in Thailand local people had historically concentrated on agriculture, crafts, and politics, leaving a somewhat underdeveloped commercial system that Chinese immigrants were able to dominate.

The Arrival of Islam Muslim merchants from South and Southwest Asia were also active in the region beginning in the thirteenth century. During this time, the cultural and religious practices of the Islamic world were introduced into Southeast Asia. Their focus was along the coasts and on the islands of the region's southern flank, from the Malay Peninsula, throughout the Indonesian islands, east to the Philippines. By 1650, Islam had largely replaced Hinduism and Buddhism throughout Malaysia and Indonesia. The only significant holdout was the small but fertile island of Bali, where thousands of Hindu musicians and artists fled from the courts of Java, giving the island a strong tradition of arts and crafts that has been maintained to the present day.

Today, the world's most populous Muslim country is Indonesia, where some 87 percent of the nation's 200 million inhabitants follow Islam. This figure, however, masks a significant amount of internal diversity. In some parts of Indonesia, such as in northern Sumatra, relatively orthodox forms of Islam took root (Figure 13.19). In others, such as central and eastern Java, a more lax form of worship emerged. As an illustration, while the vast majority of Javanese are Muslims, most of them also retain certain Hindu beliefs.

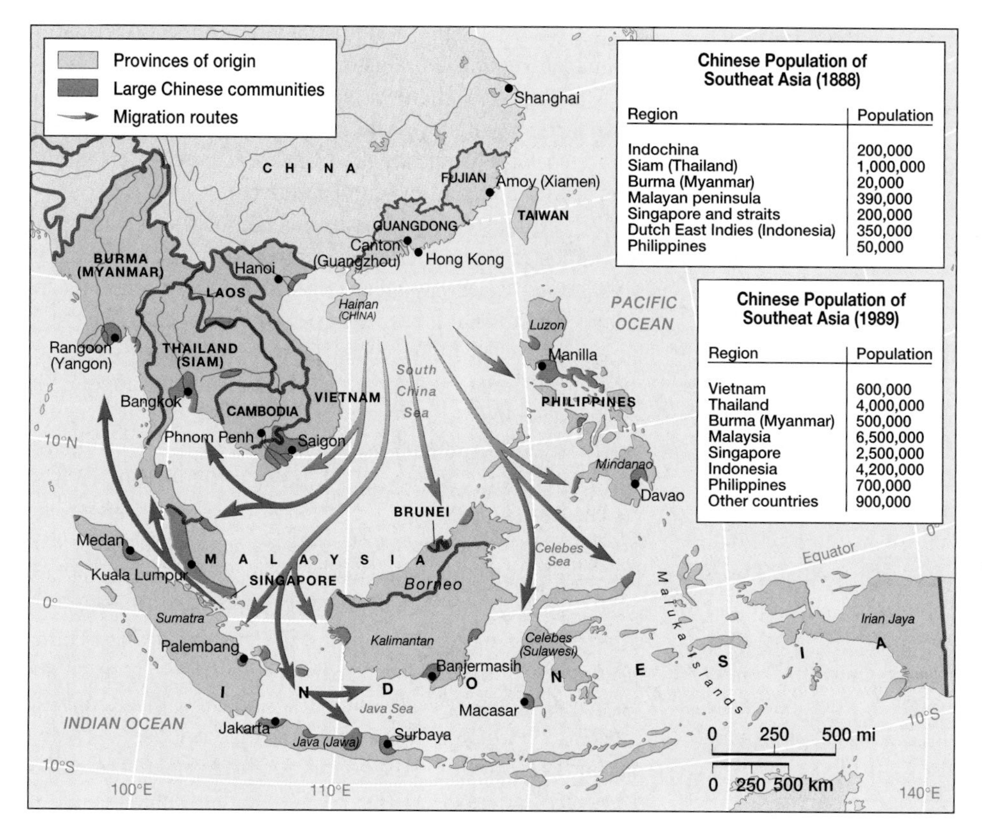

Chinese Population of Southeast Asia (1888)

Region	Population
Indochina	200,000
Siam (Thailand)	1,000,000
Burma (Myanmar)	20,000
Malayan peninsula	390,000
Singapore and straits	200,000
Dutch East Indies (Indonesia)	350,000
Philippines	50,000

Chinese Population of Southeast Asia (1989)

Region	Population
Vietnam	600,000
Thailand	4,000,000
Burma (Myanmar)	500,000
Malaysia	6,500,000
Singapore	2,500,000
Indonesia	4,200,000
Philippines	700,000
Other countries	900,000

◀ **Figure 13.18 Chinese in Southeast Asia** People from the southern, coastal region of China have been migrating to Southeast Asia for hundreds of years, a process that reached a peak in the late 1800s and early 1900s. Most Chinese migrants settled in the major urban areas, but, in peninsular Malaysia, sizable numbers were drawn to the countryside to work in the mining industry and in plantation agriculture. Today, Malaysia has the largest number of people of Chinese ancestry in the region. Singapore, however, proportionally has the largest Chinese population, and is the only Southeast Asian country with a Chinese majority.

▲ **Figure 13.19 Indonesia's largest mosque** Indonesia is often said to be the world's largest Muslim nation, since more Muslims reside here than in any other country. Islamic architecture in Indonesia is somewhat modernistic, especially when contrasted with the more traditional styles found in Southwest Asia and North Africa. *(Dana Downie/Photo 20-20, Inc.)*

Islamic fundamentalism also gained ground in Malaysia, where some 58 percent of the people consider themselves Muslim. While the current Malaysian government has generally supported the revitalization of Islam, it is wary of the growing power of the fundamentalist movement. Some Malaysian Muslims, moreover, view the fundamentalists as advocating social practices derived from the Arabian peninsula that are not necessarily religious in origin. These include women covering their heads or veiling themselves. As a result of these tensions, religious controversies abound in modern Malaysia. Recently, a national debate was waged after several contestants in a beauty pageant were arrested for having violated a state religious ban on such exhibitions. Subsequently, the national government ordered a review of all of Malaysia's Islamic laws, which had previously been enforced by local rather than the federal officials.

Historical Islam was still spreading eastward through insular Southeast Asia when the Europeans arrived in the sixteenth century. When the Spanish claimed the Philippine islands in 1570s, they found the southwestern portion of the archipelago thoroughly Islamic. Here, Muslims resisted the Christian religion introduced by the Spanish and, still today, derive their cultural identity from their Islamic ties. In the northern and central Philippines, Islam arrived only a few decades before the Europeans and was extinguished rather quickly by Spanish priests and soldiers. Today, the Philippines is about 85 percent Christian, making it the only predominantly Christian country in all Asia.

Christianity and Indigenous Cultures Christian missions spread throughout Southeast Asia in the late nineteenth and early twentieth century when European colonial powers controlled the region. While French priests converted many people in southern Vietnam to Catholicism, they had little influence elsewhere. Beyond Vietnam, missions failed to make headway in areas of Hindu, Buddhist, or Islamic heritage. However, these missionaries were far more successful in Southeast Asia's highland areas where they found a wide array of hill tribes who had never accepted the major lowland religions. Instead, these people retained their indigenous belief systems, which generally focus worship on nature's spirits and ancestors. The general name for such a religion is **animism.** While some modern hill tribes remain animist today, others were converted to Christianity. As a result, notable Christian concentrations are found in the Lake Batak area of north-central Sumatra, the mountainous borderlands between southern Burma and Thailand, and the northern peninsula of Sulawesi. Animism retains strongholds in the mountains of northern Southeast Asia, the highlands of central Vietnam, central Borneo, and the highlands of northern Luzon in the Philippines.

Geography of Language and Ethnicity

Along with religion, language is also an important part of Southeast Asia's diverse cultural geography since it is central to the construction of group identity and ethnicity. As with religion, language also expresses the long history of external cultural influences and migration. Though the language map of the region appears complicated (Figure 13.20), these different languages fit into just four major linguistic categories. These are *Austronesian,* which covers most of islands, from the Philippines to Indonesia, along with the Malay Peninsula; *Tibeto-Burmese,* which includes the languages of Burma; *Tai-Kadai,* centered on Thailand; and *Austro-Asiatic,* encompassing the languages of Vietnam and Cambodia.

The Austronesian Languages One of the world's most widespread language families is Austronesian, extending from Madagascar, off the coast of Africa, to Hawaii and Easter Island in the eastern Pacific. Linguistic geographers suggest this language originated prehistorically in Taiwan and adjacent areas of East Asia, then was spread widely across the Indian and Pacific oceans by seafaring people who migrated from island to island. Ironically, Austronesian languages are almost extinct in the area of origin near Taiwan.

Today, all insular Southeast Asian languages are within the Austronesian linguistic family. This means that elements of both grammar and vocabulary are widely shared across the insular realm; thus, it is easy for a person who speaks one to the languages to learn any other. Despite this linguistic commonality, there are more than 50 distinct languages spoken in Indonesia alone.

However, one language overshadows all others in the region, and that is Malay, which is native to the Malay Peninsula, eastern Sumatra, and coastal Borneo, yet was spread historically throughout the regions by merchants and seafarers. As a result, this language became a common trade language, or **lingua franca,** understood and used by people of different languages throughout the insular realm. Dutch colonists in Indonesia eventually employed Malay as an administrative language, although they wrote it with a Roman alphabet rather than in the Arabic-derived script used by native speakers. When Indonesia became an independent country in 1949, its leaders elected to use the lingua franca version of

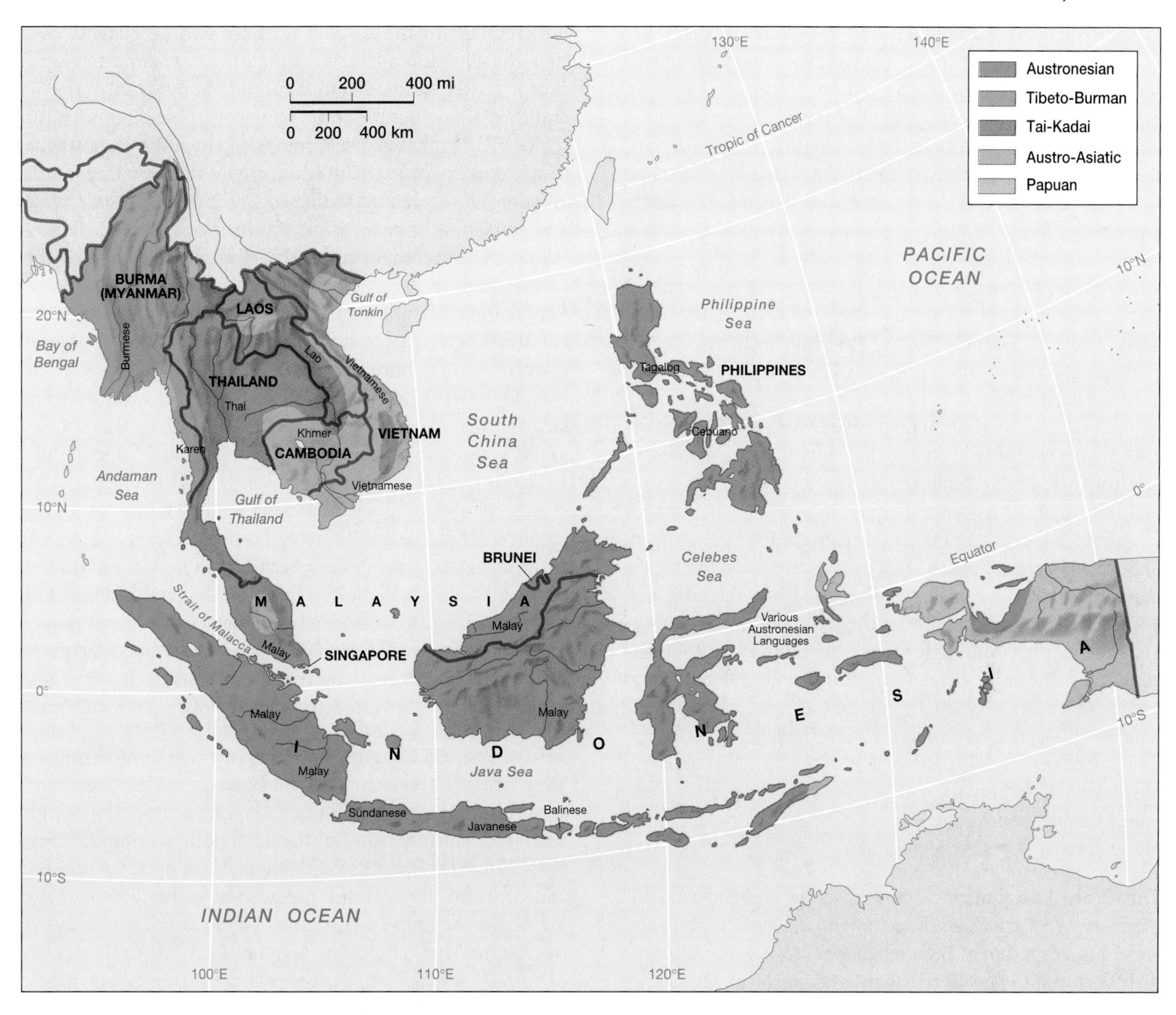

▲ **Figure 13.20 Language map of Southeast Asia** A vast number of languages are found in Southeast Asia, but most are tribal tongues spoken by only a few thousand persons. In mainland Southeast Asia—the site of three major language families—the central lowlands of each country are dominated by people speaking the national languages: Burmese in Burma, Thai in Thailand, Lao in Laos, and Vietnamese in Vietnam. Almost all languages in insular Southeast Asia belong to the Austronesian linguistic family. There were no dominant languages here before the creation of such national tongues as Filipino and Bahasa Indonesia in the mid-twentieth century.

Malay as the basis for a new national language called "Bahasa Indonesian." The goal of the new Indonesian government was to offer a common language that would overcome ethnic differences throughout the far-flung state. In general, this policy has been successful as a unifying force in that more than 80 percent of all Indonesians understand it as a second language and, further, it is widely used in government, education, and entertainment. Regionally based languages, such as Javanese, Balinese, or Sundanese, continue to be primary languages in most homes. For example, more than 75 million people speak Javanese, which makes it one of the world's major tongues.

The Philippines is not as linguistically cohesive as either Malaysia or Indonesia, even though the eight major languages spoken in the archipelago are all Austronesian. Despite more than 300 years of Spanish colonialism, the Spanish language never became a unifying force for the islands. During the American period (1900–1946), English served as the administrative language. After independence following World War II, Philippine nationalists decided to forge a national language that could replace English and, further, help unify the new country. They selected Tagalog, the tongue of Manila, and a language with a fairly well-developed literary tradition. The first task was to standardize and modernize Tagalog, which had many distinct dialects. After this was accomplished, it was renamed "Pilipino" (or "Filipino," although there is no "f" sound in Tagalog), and today, mainly because of its use in education, television, and movies, Pilipino is now on its way to becoming a unifying national language.

Tibeto-Burmese Languages Each country of mainland Southeast Asia is closely identified with the national language spoken in its core territory This does not mean, however, that all of the inhabitants of these countries speak these official languages on a daily basis. Often in the mountains and other remote districts, other non-national languages are common. This linguistic diversity reinforces ethnic diversity even in the face of national educational programs to force integration and unity.

A good example comes from Burma. There, the national language is Burmese, a language that is closely related to Tibetan, and perhaps more distantly to Chinese (hence the language family name, "Sino-Tibetan"). Once again this reminds us how Southeast Asia has been influenced by external forces and migration. Based upon linguistic evidence, the ancestors of Burmans probably lived in either Tibet or southern China centuries ago. Both Tibetan and Burmese are written in scripts of Indic origin. Today, some 22 million people speak Burmese (Figure 13.21). Though the nationalistic military government of Burma is determined to force their version of unity on the population, a major schism has developed with the three million Karen-speaking people of the uplands on the Burma-Thailand border. Although part of the Sino-Tibet linguistic family, the Karen language has a completely different word order than Burmese, which probably resulted from contact with neighboring Thai languages. Nevertheless, this language difference is enough to give the Karen people a separate ethnic identity apart from the core Burmese. Partly because of this ethnic distinctiveness, the Karen have been outspoken critics of the current Burmese government.

Tai-Kadai Languages The two major languages of this group are Thai and Lao, the national languages of Thailand and Laos, respectively. Both languages are written in scripts of Indic origin. The fact that this language group carries "Austro" as part of its title reminds us that linguists are uncertain as to the origins of these two languages. One hypothesis maintains that all Southeast Asian languages, except for the Sino-Tibetan family, are linguistically related, and thus offer clues to the ancient geographic origins for the region's early inhabitants. A competing hypothesis, however, drops "Austro" from the title and argues these two languages are more closely related to the languages of southwest China, and thus should be cousins within the Sino-Tibet family. While all agree there are some similarities, those of the "Austro" school say these similarities could easily have come from cultural exchange and borrowing. If true, these linguistic similarities give clues more to long-standing trade ties rather than prehistoric cultural origins.

Within Thailand, *Thai* language can refer to two different languages. Historically, it was restricted to a language spoken in the lower Chao Phraya valley that was called Siamese until the country changed its name in the 1930s to Thailand. This change was made to encompass all ethnic groups within the country, not just Siamese. Today, however, *Thai* usually refers to the amalgamation of closely related dialects spoken by most of the 60 million people of Thailand. More confusing still is that many linguists consider Lao as one of the contemporary Thai languages because even though it is the national language of Laos, more Lao speakers reside in Thailand than in Laos, where they form the majority population of the Khorat Plateau.

▲ **Figure 13.21 Burmese road signs** The Burmese language is written in a unique script, ultimately derived from South Asia. Roman letters, and the English language, are still used for some purposes, as is visible in this photo. Burma has, however, discouraged the use of English, viewing it as the language of colonial oppression. *(Alain Evrard/Liaison Agency, Inc.)*

Austro-Asiatic Languages This language family contains Vietnamese; Khmer, the national language of Cambodia; and a number of minor languages spoken by hill peoples scattered throughout Southeast Asia's mainland, most notably in the uplands of Vietnam, Thailand, Cambodia, and near the Burma border. As noted earlier, linguists are divided on whether the Austro-Asiatic languages (as with the Austro-Thai) constitute languages unique to the region or, instead, are distant relatives of various Chinese languages. Because of the historic Chinese influence in Vietnam, the Vietnamese language was written with Chinese characters until the French imposed a romanized alphabet during their colonial reign.

Southeast Asian Culture in Global Context

The imposition of European colonial rule ushered in a new era of globalization to Southeast Asia, bringing with it new governmental, economic, and educational systems; European languages; and western Christianity. This period also deprived the Southeast Asians of their own cultural autonomy. As a result, with decolonialization and political independence after World War II, many countries attempted to isolate themselves as much as possible from further cultural and economic influences of the emerging global system. Burma, for example, retreated into its own form of Buddhist socialism, placing strict limits on foreign tourism because the government viewed it as a source of cultural contamination. Though the door has opened a bit wider since the 1980s, Burma still remains wary of foreign practices and influences, a policy that has become a major source of tension within that country.

Other Southeast Asian countries, however, have been relatively receptive to foreign cultural influences. This is particularly true in the case of the Philippines, where American colonialism may have predisposed the country to many of the more popular forms of western culture. As a result, Filipino musicians and other entertainers are much in demand throughout East and Southeast Asia, in part because their performance styles echo those of the United States. Thailand, which was never subjected to colonial rule, appears to be the most receptive mainland southeastern Asian country to global culture, with its open policy toward tourism (including a notorious sex trade), mass media, and economic interdependence. But cultural globalization has been challenged in some Southeast Asian countries, most decisively in the Muslim areas of Malaysia and Singapore (Figure 13.22). There, one finds outspoken criticism of American films and satellite television, with government campaigns against the more explicit and crass Western expressions of Western entertainment. Censorship of global TV is common, as is the case with magazines, music videos, and movies.

English, as the global language, also causes ambivalent concern in many countries. On one hand, it is the language of questionable popular culture, yet on the other, countries recognize the necessity for maintaining high standards of English so that their citizens can participate in global business and politics. In Malaysia the widespread use of English grew increasingly controversial in the 1980s as nationalists stressed the importance of their native tongue. This distressed the business community, which considers English vital to Malaysia's competitive position, as well as the influential Chinese communities, for which Malay is not a native language.

In Singapore, the situation is more complex. Chinese, English, Malay, and Tamil (from southeastern India) are all official languages of this country. Furthermore, the languages and culture of southern China are common in home environments since 75 percent of Singapore's population is of Chinese ancestry. In recent years the Singapore government has actually encouraged the use of Mandarin Chinese, in large part because it wishes to inculcate the traditional Confucian cultural values supposedly associated with this language (Figure 13.23). Once again this seems to be a reaction to the pervasive spread of English and Western popular culture. Similarly, Philippine nationalists have decried the widespread use of English in that country, even though widespread fluency has proved beneficial to the millions of Filipinos who have migrated abroad for better economic conditions. Today, though, the government is trying to replace English with Pilipino and, as a result, competency in English is slowly declining. At the same time, as is the case with so many other languages, the official language of Tagalog is increasingly incorporating words and phrases from English that give rise to a hybrid dialect known as "Taglish."

▲ **Figure 13.22 Monks in a Burmese market** Buddhism remains important in mainland Southeast Asia, where many men become Buddhist monks. Religious traditionalism does not preclude engagement with modern, global culture, as is evident in this photograph of monks buying cassette tapes in a Burmese market. *(Andres Hernandez/Liaison Agency, Inc.)*

▲ **Figure 13.23 Chinese market in Singapore** All major Southeast Asian cities have "Chinatowns," areas in which Chinese people and language predominate. Such Chinatowns are often major retailing districts. In Singapore, where most citizens are of Chinese ancestry, Chinese is one of the country's official languages. *(Ted Streshinsky/Photo 20-20, Inc.)*

Geopolitical Framework: War, Ethnic Strife, and Regional Cooperation

Southeast Asia is readily defined as a geopolitical entity of ten different states that, after a long period of colonialism that shaped their current territories, have joined together since independence under the umbrella organization of the Association of Southeast Asian Nations, or ASEAN. Today, more than anything else, ASEAN gives Southeast Asia a geopolitical regional coherence. Within this framework, however, many states are still struggling with serious ethnic and regional tension. In some areas, these struggles could actually force a reshaping of the geopolitical landscape (Figure 13.24). This is particularly the case in Indonesia and Burma (Myanmar).

European Colonialism in Southeast Asia

The modern countries of mainland Southeast Asia all existed in one form or another as indigenous kingdoms before the onset of European colonialism. Cambodia emerged earliest, reaching its height in the twelfth century, when it controlled not only its current territory but also much of what is now Thailand and southern Vietnam. Later, in the 1300s, independent kingdoms were established by the Burmese, Siamese, and Vietnamese people that were centered on local river valleys and deltas. These kingdoms were reportedly in a constant state of war with each other, fighting more for labor than for territory. Victors would typically take home thousands of prisoners to settle their lands, which led to considerable ethnic mixing during this pre-colonial period.

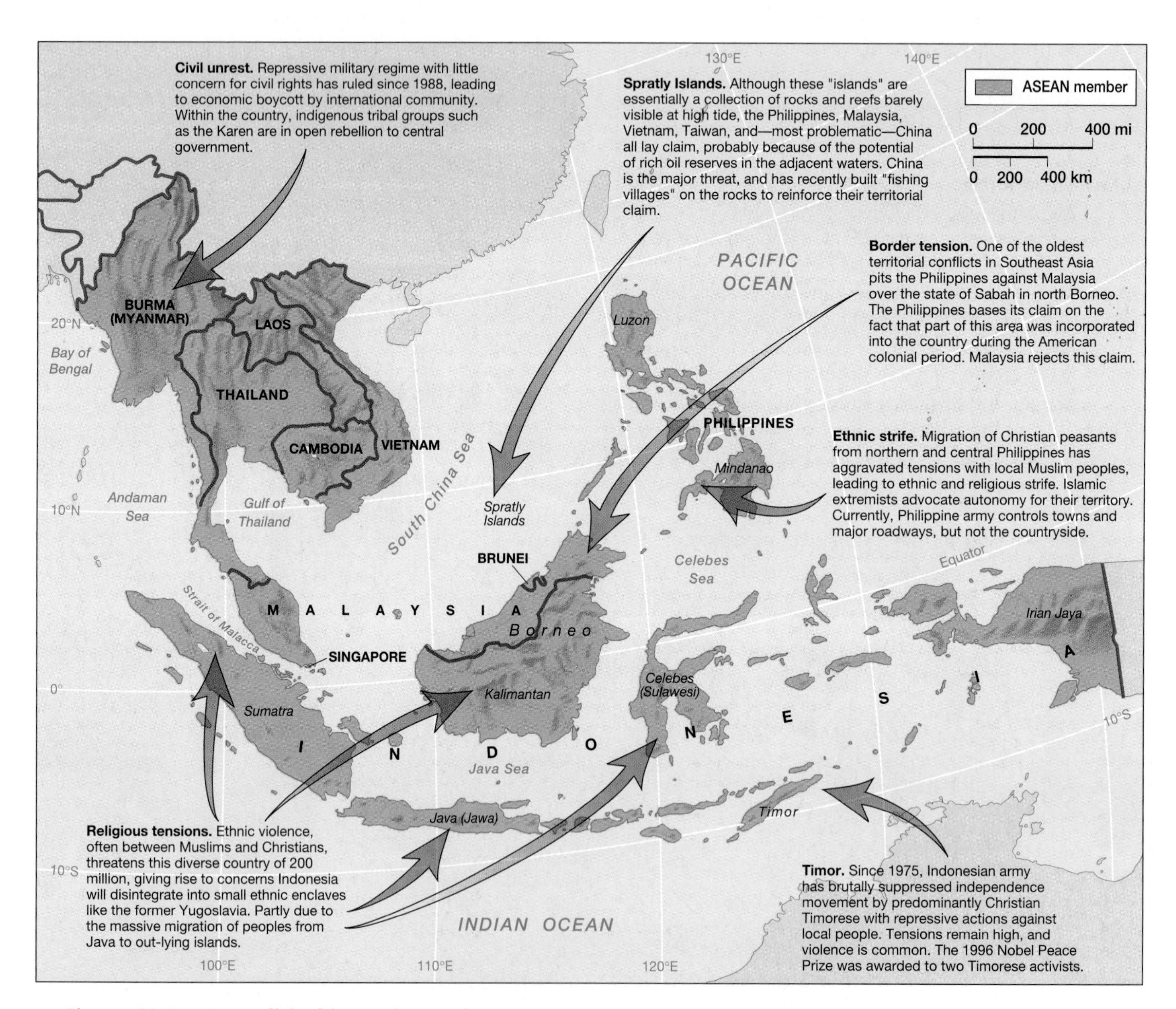

▲ **Figure 13.24 Geopolitical issues in Southeast Asia** The different countries of Southeast Asia have managed to solve most of their border disputes and other sources of potential conflicts through the ASEAN (Association of Southeast Asia Nations) forum. But internal disputes, mostly focused on issues of religious and ethnic diversity, continue to plague several of the region's states, particularly Indonesia and Burma (Myanmar). ASEAN continues to experience tension with China, an issue focused on the Spratly Islands of the South China Sea.

The situation in insular Southeast Asia was quite different from that of the mainland, with the map bearing no resemblance to modern nation-states. Instead of kingdoms, most island societies were organized at the village level. Only the Muslim areas of the Malay archipelago contained anything resembling a kingdom. Countries such as Indonesia, the Philippines, and Malaysia owe their territorial configuration almost wholly to European colonial powers (Figure 13.25).

The Portuguese were the first to arrive around 1500, lured by the cloves and nutmeg of the Maluku Islands (formerly Spice Islands) of what is now eastern Indonesia. About a century later, the Dutch started staking out territory, followed by the British. With superior weaponry, the Europeans were quickly able to conquer key ports and strategic trade locales. Yet for the first 200 years of colonialism, the Europeans made no major geopolitical changes. During this early period, trade in exotic items was more important than consolidated political power.

By the 1700s, the Dutch had become the most powerful force in the region. As a result, a Dutch Empire in the "East Indies" (or Indonesia) began appearing on world maps. This empire continued to grow right into the early twentieth century, when the Dutch vanquished their last main adversary, the powerful Islamic sultanate of northern Sumatra. Later, the Dutch invaded the western portion of New Guinea in response to German and British advances in the eastern half. In a subsequent treaty, these imperial powers sliced New Guinea in half, with the Dutch taking the western half, which was later joined to an independent Indonesia. There was, however, resistance to this expanding colonial rule. By the 1920s,

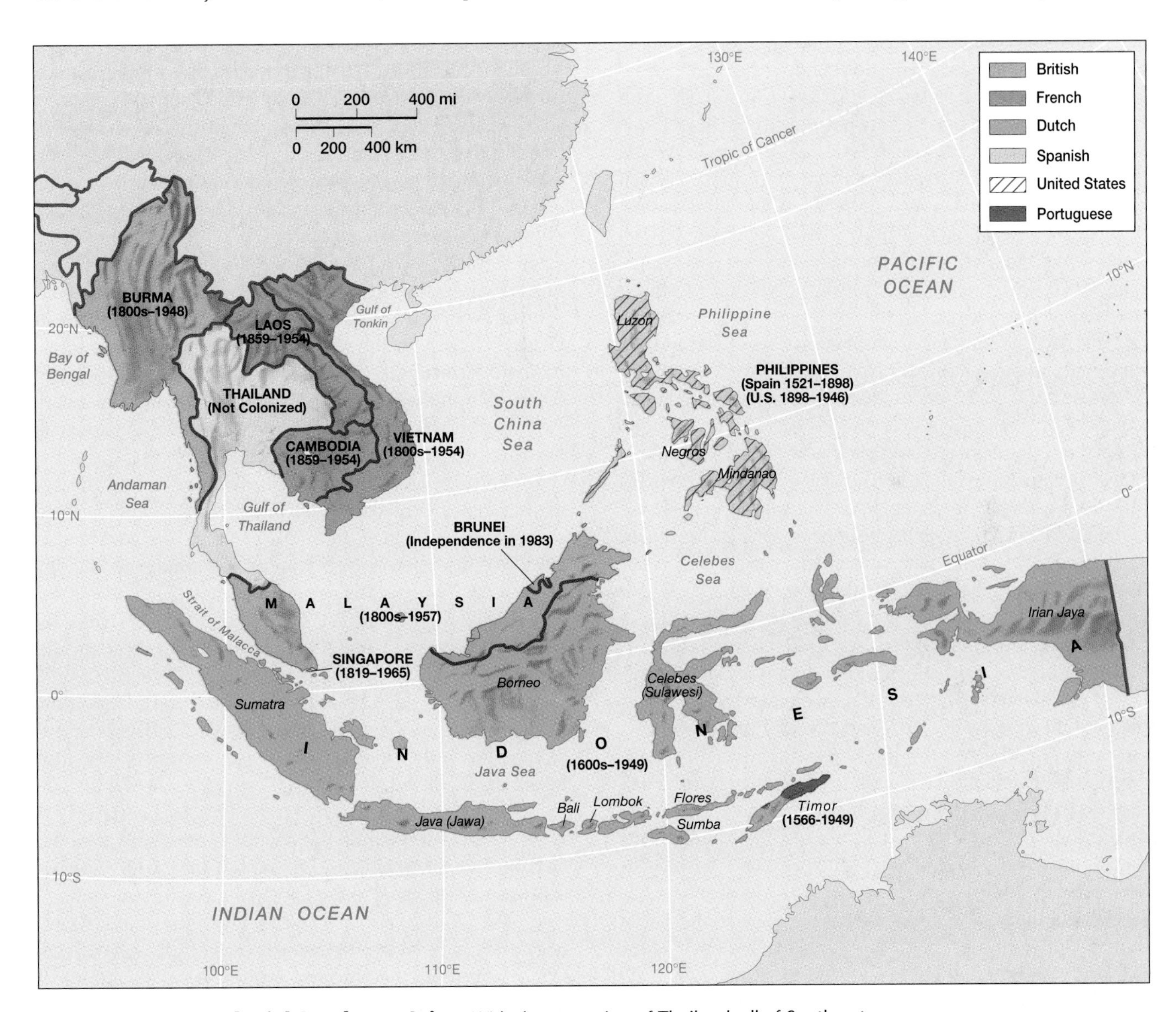

▲ **Figure 13.25 Colonial Southeast Asia** With the exception of Thailand, all of Southeast Asia was under Western colonial authority by the late 1800s. The Netherlands had the largest empire in the region, covering virtually all of the territory that later was to become Indonesia. France maintained a substantial imperial realm in Vietnam, Laos, and Cambodia, as did Britain in Burma and Malaysia (including Singapore and Brunei). The Philippines had been colonized by Spain but passed to the control of the United States in 1898.

several anticolonial groups advocated independence, primarily the Indonesia communist party, and a strong nationalist party led by Sukarno, the future ruler of newly independent Indonesia. During World War II, Indonesia was occupied by the Japanese, and upon their surrender, Indonesian independence was proclaimed. Though the Dutch attempted to reestablish their colonial rule, they finally acknowledged Indonesia's independence in 1949 and withdrew most territorial claims. This was finalized in 1963, when Indonesia gained control of the last Dutch outpost in Irian Jaya, or western New Guinea.

The British, preoccupied with their empire in India, concentrated their attention on the sea lanes linking South Asia to China. As a result, they established several fortified trading outposts along the vital Straits of Malacca, the most notable being on the island of Singapore. To avoid conflict, the British and Dutch agreed that the British would limit their attention to the Malay peninsula and the northern portion of the island of Borneo. Unlike the Dutch, the British allowed Muslim sultans to retain limited powers, much as they had done in India. When the British left this area in 1963, the country of Malaysia emerged in its wake. Two small portions of the former British sphere did not join the new country. In northern Borneo, the Sultanate of Brunei became an independent country in its own right, backed by riches from substantial oil reserves. Singapore, for its part, briefly joined Malaysia, but then withdrew and became fully independent in 1965. This divorce was carried out partly for ethnic reasons. Malaysia was to be a Malay state; with Singapore, this new state would have been almost 50 percent Chinese. To avoid this, the two states agreed to an amicable separation.

Farther into mainland Southeast Asia, European colonial power was unstoppable. The British, seeking to safeguard their South Asian empire, fought several wars against the indigenous Burma kingdom before annexing the entire area in 1885, including considerable upland territory that had never been under Burmese rule. During the same period, the French moved into Vietnam's Mekong Delta, gradually expanding their territorial control to the west into Cambodia and north to China's border. This colonial empire was formally consolidated as the Union of French Indo-China in 1887. Rubber plantations were established, and rice and timber exports were now under French control. Thailand was the only country to avoid colonial rule, though it did lose substantial territories to the British in Malaysia and the French in Laos. This independence was partly because it served both British and French interests to have a buffer state between their respective colonial empires.

Organized resistance to European rule began in the 1920s in mainland countries, but it took the Japanese occupation of World War II to shatter the myth of European colonial invincibility. After Japan's surrender in 1945, agitation for independence was renewed throughout Southeast Asia. As Britain realized it could no longer control its South Asia empire, it also withdrew from adjacent Burma so that it achieved independence in 1948. Far to the east, in the Philippines, the United States granted long-promised independence to that country on July 4, 1946, although it retained military bases, as well as considerable economic influence, for several decades.

The Vietnam War and Its Aftermath

Resistance to French colonial rule was organized primarily by communist groups that were deeply rooted in northern Vietnam. As the French reoccupied Indo-China in 1946 after Japan's defeat, the leader of this resistance movement, Ho Chi Minh, became president of a separatist government at Hanoi. France ceded local autonomy to this northern region but sought to retain the southern portion (Cochin China) as a colony. Open warfare between French soldiers and the Vietminh of the north went on for almost a decade, with the colonial French army suffering demoralizing losses against the indigenous guerilla troops. After the French suffered a decisive defeat at Dienbienphu in 1954, they agreed to withdraw. An international peace council in Geneva then determined that Vietnam would, like Korea, be divided into two countries. As a result, the leaders of the communist rebellion came to power in North Vietnam, and, not surprisingly, allied themselves with the Soviet Union and China. South Vietnam emerged as an independent capitalist-oriented state with close political ties to the United States.

However, this Geneva peace accord did not end the fighting. Communist guerrillas in South Vietnam fought to overthrow the new government and unite it with the north. For its part, North Vietnam sent troops and war materials across the border to aid the rebels. Most of these supplies reached the south over the Ho Chi Minh Trail, an ill-defined network of forest passages through Laos and Cambodia, thus steadily drawing these two countries into the conflict. In Laos, the communist Pathet Lao forces challenged the central government, while in Cambodia, the **Khmer Rouge** guerillas gained considerable influence. In South Vietnam, the government gradually lost control of key areas, including the Mekong Delta, a region perilously close to the capital city of Saigon.

By 1962, the United States was sending large numbers of military advisors to the region to train and organize forceful resistance to the communist guerillas. In Washington DC, the **Domino Theory** became accepted foreign policy. According to this notion, if Vietnam fell to the communists, then so would Laos and Cambodia; once those countries were lost, Burma and Thailand, and perhaps even Malaysia, would surely become members of the communist bloc. Indonesia and the Philippines, too, would fall. Soon all Southeast Asia would be part of the Soviet Union-Communist China world. Because this was unacceptable to the United States, the commitment to protect South Vietnam grew. In 1964, the United States began bombing North Vietnam. In 1965 thousands of U.S. troops began a ferocious land war against the communist guerrillas. Despite superiority in arms and troops, along with total control of the air by the United States, Vietcong guerrillas increased their control over the countryside. By 1969, there were more than half a million U.S. troops in the region (Figure 13.26). However, because U.S. casualties were unacceptably high, secret talks were initiated in search of a nego-

▲ **Figure 13.26 American soldier and Vietcong prisoners** The United States maintained a substantial military presence in Vietnam in the 1960s and early 1970s. Although U.S. forces claimed many victories, they were ultimately forced to withdraw, leading to the victory of North Vietnam and the reunification of the country. *(Hulton Getty/Liaison Agency, Inc.)*

▲ **Figure 13.27 Hmong refugee camp in Thailand** The tribal Hmong people of Laos sided with the United States during the Vietnam War. As a result, many Hmong were forced to flee the country after the victory of the Pathet Lao communist forces and languished for years in squalid refugee camps. *(Wendy Stone/Liaison Agency, Inc.)*

tiated settlement. U.S. troop withdrawals began in earnest by the early 1970s (see "Remembered Landscapes: War Tourism in Vietnam").

With the withdrawal of U.S. forces, the noncommunist governments began to collapse. In 1975, Saigon fell to the Viet Cong. The following year, Vietnam was officially reunited under the government of the north. Reunification was a traumatic event in southern Vietnam. Hundreds of thousands of people fled from the new regime to other countries, especially to the United States. The first wave of refugees consisted primarily of wealthy professionals and businesspeople, but later migrants included many relatively poor ethnic Chinese. Most of these refugees fled on small, rickety boats; large numbers suffered shipwrecks or pirate attack. Most found their way to squalid refugee camps elsewhere in Asia, where some still languish (Figure 13.27).

Vietnam proved fortunate compared to Cambodia. Here, the Khmer Rouge proved to be one of the most brutal regimes the world has ever seen. City-dwellers were forced into the countryside to become peasants, and most wealthy and highly educated persons were summarily executed. The Khmer Rouge's goal was to create a wholly new agrarian society by returning to what they called "year zero." After several years of this bloodshed, neighboring Vietnam invaded Cambodia and installed a far less brutal, but still repressive, communist regime. Fighting between different factions continued for more than a decade. Most Vietnamese troops were gone from Cambodia by 1989 as the United Nations worked to broker a settlement to this full-fledged civil war. Since that time, several coalition governments have brought a tenuous peace to the country.

Vietnam stationed significant numbers of troops in Laos after 1975. Large numbers of Hmong and other tribal peoples fled to Thailand. Elections were not held until 1989, when a people's assembly was elected to approve a new constitution. This was done in 1991. Single-party elections were held in 1992 but did little to alter the country's political landscape. Today, though, many observers predict changes are in the offering as militant communists begin passing from the scene. In 1997 Laos was admitted to the Association of Southeast Asian Nations (ASEAN), an act that has given this small country regional credibility.

Geopolitical Tensions in Contemporary Southeast Asia

A number of contemporary conflicts in Southeast Asia are rooted in the region's colonial past. In several instances, locally based ethnic groups are struggling against the centralized national governments that inherited territory from former colonial powers. Often in such a transfer, the rights of smaller groups are sacrificed for larger, national goals. Another source of tension is where small tribal or ethnic societies are attempting to preserve their homeland from the forces of globalization, be it international logging and mining, inter-regional migration, or some other form of spatial restructuring. Both forms of geopolitical tension are found in the large, multiethnic country of Indonesia.

Conflicts in Indonesia When Indonesia gained independence in 1949, it encompassed all of the former Dutch possessions in the region except Irian Jaya, or western New Guinea. The Netherlands retained this territory, arguing that its cultural background distinguished it from Indonesia. In 1962 the Dutch organized a plebiscite to see whether the people of Irian Jaya wished to join Indonesia or form an independent country. The vote went for union, but many observers believe that the election was rigged by the Indonesian government. Tensions increased in the following decades as Javanese immigrants, along with mining and lumber firms, arrived in western New Guinea. Faced with the

REMEMBERED LANDSCAPES War Tourism in Vietnam

War tourism is a booming business in Vietnam. As a result, the landscapes and experiences that were nightmares for hundreds of thousands during the 1960s and 70s are now attracting foreign tourists and U.S. veterans alike. For example, outside of Ho Chi Minh City (formerly Saigon), one can visit the Cu Chi tunnels, one of the most famous battlegrounds of the Vietnam War. There, visitors can tour the 75 miles of tunnels that once offered haven for Vietcong fighters. Underground, visitors crawl through narrow tunnels connecting living quarters, storerooms, kitchens, and hospitals. Above ground, young women dressed in the characteristic black pajamas of the Vietcong guerillas sell tourists T-shirts, pens made from bullets, and replicas of the coveted Zippo lighters carried by American GIs. Nearby, on a big screen TV, news footage from the war plays endlessly: B-52 bombers drop strings of bombs; villagers flee U.S. soldiers; and Vietcong guerillas fight heroically.

While foreign currency is the main goal of war tourism, few opportunities are missed for communist propaganda. Every now and then, however, sensibilities rule. In Ho Chi Minh City, the Museum of War Remnants attracts many visitors. Business seems to have improved since its name was changed from the "Museum of American War Crimes." Hue, the scene of heavy fighting during the Tet Offensive of 1968, has become a hub for war tours that list such attractions as "Khe Sanh, Dong Ha, Marble Mountain, China Beach, bombed-out church, and the DMZ." Even My Lai, the site of the infamous U.S. troop massacre of villagers, has been turned into something of a theme park. Guides escort tourists to a cemetery museum, and storytellers re-create the horrible events that destroyed the village and its people. Farther east, the government plans to open up for visitors parts of the Ho Chi Minh Trail.

In the last few years, an increasing number of U.S. veterans have come back to Vietnam in search of remembered landscapes and battlefield experiences. Why? One tour leader put it this way: "They get a sense of closure; that's the big benefit of going back as a veteran. We [the Americans] left suddenly. Now you know how the story has ended. All the Vietnamese are very friendly. It's a different country now." Often, however, the veterans are disappointed because the landscape has changed so dramatically. Forests have reclaimed the countryside, obscuring small hills that were previously major battlefield landmarks. Was the fire base here, or over there? Many veterans can't tell because the countryside looks so different. Instead of former military bases, one often finds coffee plantations and foreign trade warehouses. One enterprising tour operator uses GPS (global positioning system) coordinates and old military maps from the 1960s to guide U.S. veterans to specific spots as they revisit remembered landscapes from an earlier, more troubled era.

Source: Adapted from Seth Mydans, "Visit the Vietcong's World: Americans Welcome," *New York Times*, July 7, 1999, page A4.

loss of their land and the degradation of their environment, the indigenous inhabitants began to rebel. Rebel leaders demanded independence, or at least autonomy, but they faced a far stronger force in the Indonesian army. The war in Irian Jaya is still a smoldering, sporadic, and occasionally very bloody, guerilla affair.

An even more brutal war erupted in 1975 on the island of Timor, in southeastern Indonesia (Figure 13.28). The eastern half of this poor and rather dry island had been a Portuguese colony (the only survivor of Portugal's sixteenth-century foray into the region), and had therefore evolved into a largely Christian society. The Timorese expected independence when the Portuguese finally withdrew. Indonesia, however, viewed the area as rightfully its own, largely by virtue of its geographical position, and immediately invaded. A ferocious war ensued, which the Indonesian army won in part by starving the people of Timor into submission. Tensions remain high today. In August 1999, the people of Timor voted to sever their ties with Indonesia and become completely independent.

Secession struggles have occurred elsewhere in Indonesia as well. In the late 1950s central Sumatra and the northern portion of Sulawesi rebelled, but they were quickly defeated and eventually reconciled to Indonesian rule. In the 1980s fighting flared up in southern Maluku Islands. Yet another small-scale war was initiated in 1997, this one in western Kalimantan. Here the indigenous Dayaks, a tribal people now largely converted to Christianity, began to clash with Muslim migrants from Madura, a densely inhabited island north of Java. At first the Indonesian army stayed clear of the conflict, but it soon decided to pacify the aggrieved Dayaks.

Indonesia has obviously had difficulties creating a unified nation over its vast, sprawling extent and across its myriad distinct cultural and religious communities. As a creation of the colonial period, Indonesia has a weak historical foundation. Some critics have viewed it as something of a Javanese Empire, but in actuality many non-Javanese individuals have risen to high governmental positions. It is undeniable, however, that the Indonesian state has relied on repression. While it has some democratic aspects, the government remains quasi-dictatorial and the military's powers are vast. As the Indonesian economy grows, however, pressures for democratic reform are mounting. It is possible that a new political order will emerge in the near future. If this happens, one of its main challenges will be to address the underlying causes of ethnic conflict and regional secession movements.

Regional Tensions in the Philippines The Philippines has also suffered from regional secession movements, although not to the same extent as Indonesia. Its most persistent

▲ **Figure 13.28 East Timorese demonstration in Jakarta** Indonesia invaded and claimed East Timor in 1975 when the Portuguese left their last colonial outpost in Southeast Asia. The East Timorese resisted the Indonesian takeover; violence reached a climax in 1999 when the East Timorese voted for independence, after which pro-Indonesian militias attacked the Timorese leaders and large segments of Timorese society. *(V. Miladinovic/Corbis Sygma Photo News)*

problem area is the Islamic southwest. Recently, the government has made some headway in talks with the main rebel group by promising greater autonomy, but the more extreme Muslim factions continue to demand total independence. The migration of Christian peasants from northern and central Philippines to Mindanao has exacerbated local tensions. The Philippine army generally maintains control over the area's main cities and roadways, but it has little power in the more remote villages.

The Philippines also faces a communist-oriented nationwide rebellion. In the mid-1980s the NPA (New People's Army) controlled one quarter of the country's villages scattered across all the major islands. Since the mid-1990s, however, the strength of this insurgent group has declined considerably. The 1980s also saw the rebellion of tribal groups in northern Luzon, peoples whose lands and livelihoods have been threatened by dam building and forestry projects. This movement also diminished in the 1990s after the government offered significant concessions.

The Quagmire of Burma Burma (Myanmar) is perhaps the most war-ravaged country of Southeast Asia. Burma's simultaneous wars have pitted the central government, dominated by the Burmese-speaking ethnic group, against the country's varied non-Burmese-speaking societies. Fighting intensified gradually after independence in 1948, and by the 1980s almost half of the country's territory had become a combat zone. Burma's troubles, moreover, are by no means limited to the country's ethnic minorities (Figure 13.29). After 1988, the Burma has been ruled by a very repressive military regime, one that has little popular support even among the majority Burmese-speaking population.

The rebelling peoples of Burma want to maintain their cultural traditions, lands, and resources, and they generally see the national government as something of a Burman Empire seeking to impose its language and, in some cases, its religion, Theravada Buddhism, upon them. Most of the insurgent ethnic groups live in rugged and inaccessible terrain, and many of them are animists or Christians. But even some lowland groups have rebelled. The Muslim peoples of the Arakan coast in far western Burma have long faced discrimination because of their religious convictions. In the 1980s, many were forced to flee to neighboring Bangladesh.

The most serious challenge to the Burmese government has come from the Karen, a tribal people living along the border with Thailand. British missionaries converted many Karen to Protestant Christianity, and, as a result, a number of them obtained fairly high bureaucratic positions in Burma's colonial government. The Burmans resented this deeply, for they had long viewed the Karen as culturally inferior. After independence, the Karen lost their favored position and soon grew to resent what they saw as Burman domination.

By the 1970s, the Karen were in open rebellion and soon managed to establish an insurgent state of their own (Figure 13.30). They have supported their rebellion by smuggling goods between Thailand and Burma and by mining the gemstones of their territory. Since Burma has one of the world's most protected economies with a full range of governmental controls inhibiting commerce, smuggling is an especially lucrative occupation. Some estimates in the 1980s ranked the Karen economy as almost as large as the official Burmese economy. The Burmese army, however, began to make headway against the Karen in the early 1990s. Crucial to their success was an agreement made with Thailand, by which the Thai government agreed to prevent Karen soldiers from finding sanctuary on its side of the border. This agreement was in exchange for access by Thai timber interests to Burma's valuable teak forests.

Burma's other insurgencies have been financed largely by opium growing and heroin manufacture. This is especially true in regard to the Shan rebellion. The Shan are a Thai-speaking people inhabiting a rugged plateau area within the famous Golden Triangle of drug production. In the late 1980s and early 1990, a breakaway "Shan state" gained a firm financial basis through the narcotics trade and a strong political foundation through ethnic solidarity. For a number of years, this strategy proved to be successful, but by the mid-1990s it began to falter. Burmese agreements with Thailand

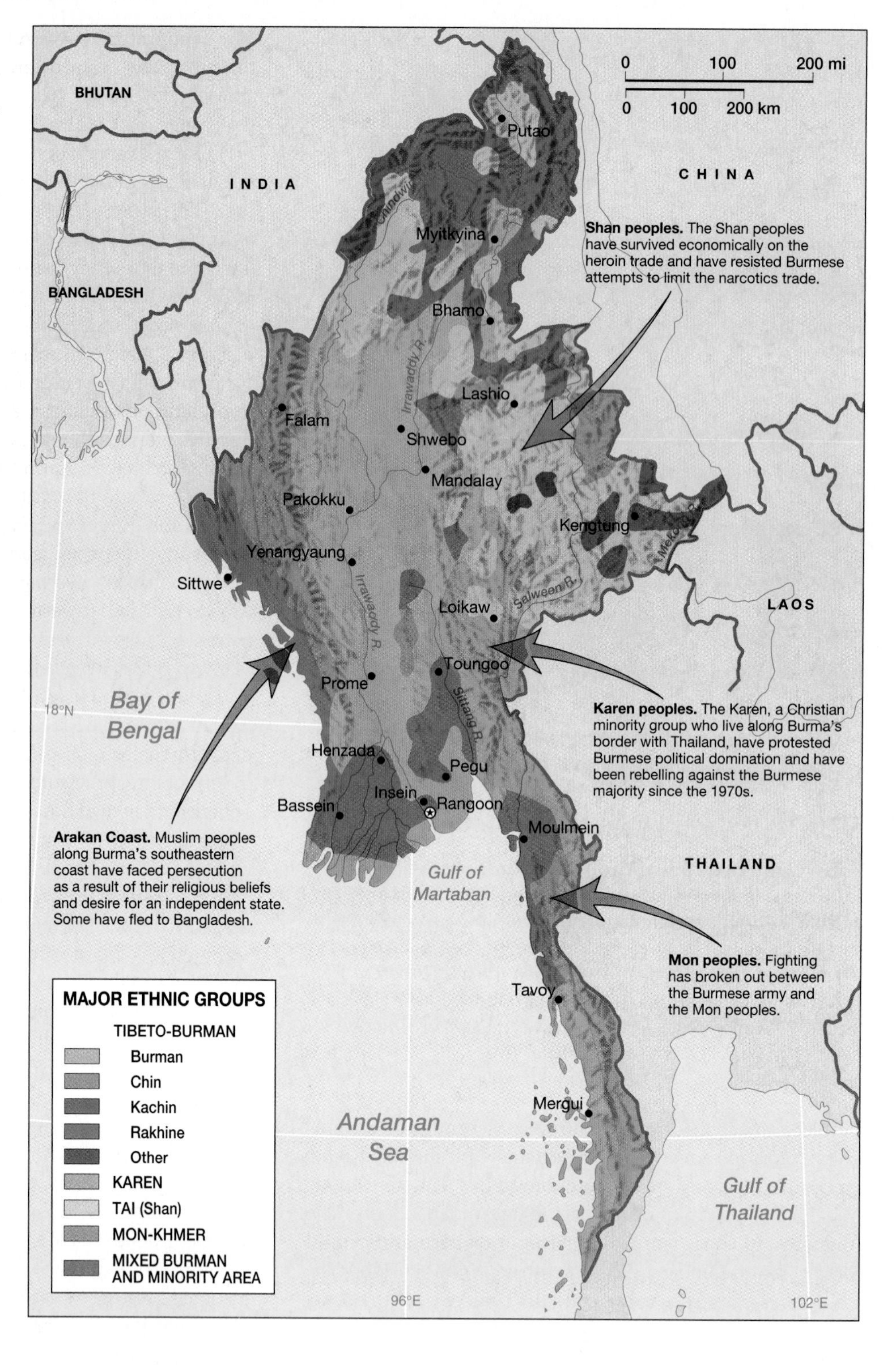

Figure 13.29 Ethnic conflict in Burma (Myanmar) Although the central lowlands of Burma are populated by people of the dominant Burman ethnic group, the peripheral highlands and the southeastern lowlands are the home of numerous non-Burman peoples. Most of these people have been in periodic rebellion against Burma since the 1970s, owing to their perception that the national government has been attempting to impose Burman cultural norms. Economic stagnation and political repression by the central government have intensified these conflicts.

reduced the Shan heroin trade, while military operations cut off the supply of raw opium reaching the Shan factories. But while Burma apparently triumphed over the Shan, other ethnic groups located in even more remote areas have been able to move into the void created in the drug trade. As a result, export of heroin from Burma has not abated.

As long as Burma retains its repressive, Burman-dominated state structure, social unrest and ethnic turmoil are not likely to diminish. Protests against the government are mounting even in the country's core. Additionally, the international community is putting some pressure for democratic change on Burma's government with trade boycotts and other sanctions. The combination of internal and external pressures make the future of this volatile state difficult to predict.

International Dimensions of Southeast Asian Geopolitics

Geopolitical conflicts in Southeast Asia have occurred not only within individual countries but also between different countries. As is typical, tensions have arisen where two coun-

▲ **Figure 13.30 Karen rebels** The Karen people have been in rebellion against Burma (Myanmar) since the 1970s. For several years they maintained a separate insurgent state, with its own capital city and regular army. The Burmese military advanced in the 1990s, however, forcing the Karen back to a guerilla-style war. *(Dean Chapman/Panos Pictures)*

tries claim the same piece of territory. Fortunately, these tensions have lessened as the Southeast Asian countries developed a framework for addressing their misunderstandings and highlighting their mutual interests through the Association of Southeast Asian Nations (ASEAN).

One of the oldest territorial debates in Southeast Asia pits the Philippines against Malaysia over the Malaysian state of Sabah in north Borneo. The Philippines base their claim on the fact that the Sulu Archipelago, which was incorporated into the Philippines during the American colonial period, at one time controlled part of Sabah. Malaysia, not surprisingly, has steadfastly rejected the Philippine claim. On another front, Malaysia has been interested in the Malay-speaking, Muslim region of Thailand located across its northern border. Malaysia has also quarreled with Indonesia over their boundary in Borneo.

In recent years, however, the various countries of Southeast Asia have largely agreed to drop these and other border disputes. With the rise of ASEAN and with the growth of trade relations, national leaders in the region have concluded that amicable relations with one's neighbors are more important in the long run than the gain of additional territory. This does not mean, however, that all territorial claims have been completely forgotten.

More intractable has been the dispute over the Spratly and Paracel islands in the South China Sea, two groups of insignificant rocks and reefs that are virtually submerged at high tide (Figure 13.31). The Philippines, Malaysia, and Vietnam have all advanced territorial claims here, as have China and Taiwan. In the mid-1990s China began to actively assert its claims by building ambiguous structures that it calls fishing shelters, but which its neighbors refer to as military posts. Increasingly, the dispute in the South China Sea pits Southeast Asia as a whole against China, clearly a far more powerful geopolitical and military entity. The reason for this keen display of interest in these two islands is that there is some evidence that substantial oil reserves lay under the sea in the vicinity of these two island groups.

It is partly because of mutual concern over growing Chinese power that the countries of Southeast Asia are banding together. This is especially visible in the development and enlargement of ASEAN. In its earlier days, ASEAN was an alliance of non-socialist countries fearful of the communist regimes that had come to power in Vietnam, Cambodia, and Laos. Through the 1980s, the United States maintained a strong military presence in the region through its control of naval and air bases in the Philippines, and U.S. military force helped bolster the anti-communist coalition. In the early 1990s, however, the United States, under pressure from Filipino nationalists, withdrew its forces from the Philippines.

By this time the struggle between communism and capitalism had become an increasingly moot issue, and it was apparent that Vietnam offered no threat to the ASEAN states.

▲ **Figure 13.31 The Spratly Islands** The Spratly Islands are small and barely above water at high tide, but they are geopolitically very important. Oil may exist in large quantities in the vicinity, heightening the competition over the islands. Southeast Asian countries are especially concerned about China's military activities in the Spratlys. *(Nouvelle Chine/Liaison Agency, Inc.)*

In 1995 Vietnam itself gained membership in the organization, followed by Laos. While ASEAN attempts to address economic issues, its main role has been to facilitate political cooperation and to present something of a united front against the rest of the world. Increasingly, the foreign power most at issue in Southeast Asia is China. Although Vietnam and China have similar political and economic systems—as well as cultural backgrounds—they also have a long history of antagonism. It is therefore not surprising that Vietnam would opt to join ASEAN.

While ASEAN is on friendly terms with the United States, it must also be recognized that one purpose of the organization is to prevent the United States, or any other country for that matter, from gaining undue influence in the region. For that reason, ASEAN leaders are keen to include all Southeast Asian countries within the association. In 1997 Burma gained membership, a move that disappointed the United States and other Western countries concerned about Burma's poor human rights record. The ASEAN states, however, were more concerned about Burma's incipient alliance with China, which they hoped might be forestalled if Burma were to gain admission into their organization. ASEAN's ultimate international policy, however, is also one of negotiation rather than direct confrontation. For this reason, the ASEAN Regional Forum (ARF) was established in 1994 as an annual conference in which ASEAN leaders could meet with the leaders of both the East Asian and the Western powers to attempt to ease tensions within the region.

ASEAN has not defused all of the tensions existing between the different countries of Southeast Asia. Minor disagreements within the ASEAN bloc have also surfaced over relations with the outside world. Malaysia in particular has pushed an anti-Western line on cultural and political (but not economic) issues, and it has resolutely opposed Australia's desire to gain closer association with its Asian neighbors. Thailand and the Philippines have been more willing to accommodate the West.

While ASEAN is concerned over Chinese power in the South China Sea, one must be careful not to overstate the potential for conflict between China and Southeast Asia. Economic ties between these two parts of the world are growing very quickly. Especially important are the networks of family business groups that increasingly link the various Chinese communities of Southeast Asia with Taiwan and China proper. Such networks, which ultimately extend throughout the so-called Pacific Rim (which includes all countries bordering on the Pacific, whether in the Americas, Asia, or Oceania), have been an important component of Southeast Asia's economic boom.

Economic and Social Development: The Roller-Coaster Ride of Tiger Economies

Until the recent downturn of commercial activity, economic development in Southeast Asia was held up to the world as a model for a new globalized capitalism. With investment capital flowing first from Japan, then later from international investment portfolios, countries such as Thailand, Malaysia, and Indonesia moved quickly into the ranks of booming "tiger" economies characterized by high annual growth rates. Because of the successful economic development of these countries, the notion of an "Asian model" evolved as a strategy that might also work in other world regions, such as Africa or South Asia.

More recently, however, Southeast Asian countries have suffered from the roller-coaster ride of globalized economic development. Within a decade most economies have experienced both unforeseen highs and lows (Figure 13.32). Since the summer of 1997, regional economies have been in a state of crisis. At that time, currencies in both Thailand and Indonesia were devalued almost 50 percent, and other countries have suffered similarly. Today, many regional economies stagnate, accompanied by high unemployment, social unrest, truncated real estate and industrial development, and lost fortunes. Many argue that the region's spectacular economic growth contained the seeds of its own destruction in high de-

▲ **Figure 13.32 Bank closings** The economic crisis of 1997–1998 hit Southeast Asia's banking sector particularly hard. Many banks have been closed, and many others may soon be closed, since their liabilities are much greater than their assets. The economic crisis has brought devastation to the poor and lower middle class. *(Marcus Rose/Panos Pictures)*

pendence on foreign investment. Because of this liability, Southeast Asia cannot simply wait for another round of global economic vitality but, instead, must restructure its economies in new ways. The exact shape of that future, though, is unclear.

Economic Leaders and Laggards

Southeast Asia today is a region of very uneven economic and social development. While some countries, such as the Philippines, have experienced both boom and bust, other countries, such as Burma and Cambodia, either hesitated or were not able to join the global economic bandwagon in the first wave. As a result, the regional economic geography remains uneven and fluid. As former boom economies struggle to regain their feet after the recent downturn, other countries simply struggle (Table 13.2).

The Philippine Decline Forty years ago, the Philippines was the most highly developed Southeast Asian country and was considered by many to have the brightest prospects in all of Asia. It boasted the best-educated populace in the region, and it seemed to be on the verge of sustained industrialization and economic development. Per capita GNP in 1960 was higher in the Philippines than in South Korea. By the late 1960s, however, Philippine development had been derailed. Through the 1980s and early 1990s, the country's economy grew more slowly than its population, resulting in declining living standards for both the poor and the middle class. The Philippine people are still well educated and reasonably healthy by world standards (see Table 13.3), but the country's educational and health systems declined during the dismal decade of the 1980s.

Why did the Philippines fail so spectacularly despite its initial promise? While there are no simple answers, it is clear that dictator Ferdinand Marcos (who ruled from 1968 to 1986) squandered—and perhaps even stole—hundreds of millions of dollars while failing to enact any programs conducive to genuine development. The Marcos regime instituted a kind of **crony capitalism** in which the president's many friends were granted huge sectors of the economy, while those perceived to be enemies had their properties expropriated. After Marcos declared martial law in 1972 and suspended Philippine democracy, revolutionary activity intensified, and the country began to fall into a downward spiral. Wealthy Filipinos increasingly lost faith in their country and began to send their money to investment havens in the United States and Europe.

While it is tempting to explain the failure of the Philippines on the policies of the Marcos regime, such an explanation must be regarded as only partially adequate. Indonesia has seen the development of a similar kind of crony capitalism based on the personal networks of President Suharto, yet in this case the companies in question generally proved competitive, at least until the financial collapse of 1997. In the case of the Philippines, however, it seems that far more money was actually stolen from the state by corrupt officials and well-connected businesspeople. Unfortunately, this kind of **kleptocracy** (or "government by thieves") is not uncommon in the poorer parts of the world.

In the Philippines, the Marcos dictatorship was finally replaced by an elected democratic government in 1986, but the Philippine economy continued to languish through the early 1990s. Many Filipinos have responded to the economic crisis by working overseas, either in the oil-rich countries of Southwest Asia, the wealthy cities of North America and Europe, or the newly industrialized nations of East and Southeast Asia. Men primarily work in the construction industry, women as nurses or domestic servants. Although foreign remittances have kept the Philippine economy afloat, this exodus of labor represents in many respects a tragedy for the

Table 13.2 Economic Indicators

Country	GNP per Capita ($U.S., 1996)	Total GNP (Millions of $U.S., 1996)	PPP* ($Intl, 1996)	Real Annual Growth % per Capita, 1990–1996
Burma (Myanmar)	—	—	—	3.9
Brunei	—	—	—	–1.5
Cambodia	300	3,088	—	2.9
Indonesia	1,080	213,384	3,310	5.9
Laos	400	1,895	1,250	3.9
Malaysia	4,370	89,800	10,390	6.1
Philippines	1,160	83,298	3,550	1.0
Singapore	30,550	92,987	26,910	6.6
Thailand	2,960	177,476	6,700	6.7
Vietnam	290	21,915	1,570	6.2

*Purchasing power parity.

Source: *The World Bank Atlas,* 1998.

country as a whole. Tens of thousands of Filipina teachers, no longer able to support themselves on their meager salaries, now work as maids and nannies in Singapore, Hong Kong, and the United Arab Emirates. With the loss of so many teachers, educational standards at home continue to decline (Figure 13.33).

In the mid-1990s, the Philippine economy finally began to revive. The government turned its attention to infrastructural problems, such as electricity provision, and foreign investments began to flow into the country. The Philippines is now seen as relatively stable, and multinational corporations are attracted by its low-paid but well-educated labor force. The most vibrant local economy in the country is now probably the former U.S. naval base of Subic Bay. Boasting both world-class facilities and a highly competent local government, Subic Bay has emerged as a major export-processing center. Cebu City, located in the central Visayan island of Cebu, has also expanded quickly, giving rise to its local nickname of "Ceboom." While the economic crisis of the late 1990s may have put an end to the Philippine recovery, it did not result in the massive declines seen in Indonesia, Malaysia, and Thailand.

Most Filipinos, however, have yet to realize any personal gains from the upturn of the mid-1990s. The Philippines still has one of the least equitable distributions of wealth in the region, if not the world. While many members of the elite are fantastically wealthy by any measurement, the poor are often simply destitute, subsisting for the most part on less than $1 a day. Although some efforts have been made toward land reform, vast estates of sugarcane and other plantation crops remain in the hands of a few super-elite families.

The Regional Hub: Singapore If the Philippines has been the biggest economic disappointment in Southeast Asia, Singapore and Malaysia have surely been the region's greatest developmental successes. Singapore has transformed itself from an **entrepôt** port city, a place where goods are imported, stored, and then transshipped, to one of the world's most prosperous and modern states. Besides continued port functions, Singapore is now the communications and financial hub of Southeast Asia, as well as a thriving high-tech manufacturing center. Although Singapore's economy did go into recession in the late 1990s, the economic crisis was relatively mild.

▲ **Figure 13.33 Filipina migrant workers in Kuwait** The long period of economic stagnation in the Philippines has resulted in an outflow of workers from the country. Women from the Philippines often work as domestic servants in the Persian Gulf region and in Singapore and Hong Kong. *(Penny Tweedie/Panos Pictures)*

The Singaporean government has played an active role in the development process, but has also allowed market forces freedom to operate. Singapore has encouraged investment by multinational companies (especially those involved in technology), and has itself invested heavily in housing, education, and social services. This government, however, is also repressive and generally undemocratic, allowing little freedom of the press or public expression. Chewing gum in public is illegal in Singapore, for example, and even minor acts of civil disobedience are sometimes severely punished. One is allowed to make money, but not to express one's political opinions. While many Singaporeans chafe under such repressive policies, others counter that they have brought fast growth as well as a clean, orderly, and safe society (Figure 13.34).

Much controversy surrounds this Singaporean model of development. Although the Singaporean government has thus far been able to repress dissent, its authoritarian form of capitalism confronts a new technological challenge in the form of the Internet. National leaders want the communication services that the Net provides, but they are worried about the free expression that it allows, fearing that it will lead to excessive individualism. The governments of China and Vietnam are impressed by, and may be seeking to emulate, the technocratic, authoritarian capitalism pioneered by Singapore.

The Malaysian Boom Although not nearly as prosperous as Singapore, Malaysia has recently experienced very rapid economic growth and has now broken into the ranks of the middle-income countries. Development was initially concentrated in the primary sector based on the extraction of natural resources, with Malaysia profiting from its exports of tropical hardwoods, plantation products (mainly palm oil and rubber), and tin. More recently, manufacturing, especially in labor-intensive high-tech sectors, has become the main engine of growth. Moreover, as Singapore prospers, many of its enterprises are spilling over into neighboring Malaysia. Increasingly, Malaysia's economy is multinational; many Western high-tech firms have established operations in the country, while several Malaysian companies are themselves beginning to operate in foreign lands.

Unfortunately, Malaysia was hard hit by the Asian economic crisis of the late 1990s. In 1998 alone its economy declined by some 8 percent. The Malaysian government's response to the crisis was different from that of its neighbors. Spurning the advice of the International Monetary Fund, Malaysia instituted currency controls, regulating the flow of international funds in and out of the country. Although it is too soon to tell, there are some indications that the

▲ **Figure 13.34 Social order in Singapore** Singapore has one of the world's most regulated societies. Freedom of expression is highly limited, and severe fines are imposed on activities that are considered to be disruptive or dirty, such as public eating, smoking, and gum chewing. *(Jean-Leo Dugast/ Panos Pictures)*

Malaysian response is working better than the more financially orthodox responses of other Southeast Asian countries, such as Thailand.

The modern economy of modern Malaysia is not uniformly distributed across the country. One disparity is geographical: most industrial development has occurred in a fairly narrow belt on the west side of peninsular Malaysia, with most of the rest of the country remaining largely dependent on agriculture and resource extraction. More important, however, have been disparities based on ethnicity. The industrial wealth generated in Malaysia has been concentrated in the Chinese community. Ethnic Malays remain less prosperous than Chinese-Malaysians, and the population of Tamil descent is poorer still. Many tribal peoples, moreover, have suffered as developmental proceeds and their lands are expropriated. Not surprisingly, such disparities of wealth have generated considerable ethnic tensions within the country.

To be sure, the disproportionate prosperity of the local Chinese community is a feature of most Southeast Asian countries. The problem is particularly acute in Malaysia, however, simply because its Chinese minority is so large (some 30 percent of the country's total population). The government's response has been one of aggressive "affirmative action," by which economic clout is transferred to the numerically dominant Malay, or **Bumiputra** ("sons of the soil"), community. This policy has been reasonably successful. Since the economy as a whole has grown during the 1980s and 1990s at some 8 percent a year, the Chinese community has been able to thrive even as its relative share of the country's wealth declines. Considerable resentment, however, is still felt by the Chinese. Especially galling is the fact that most seats in the country's universities are reserved for students of Malay background. Partly for this reason, Malaysia sends a larger proportion of student population abroad for education than any other Southeast Asian county.

Thailand: An Emerging Tiger Thailand, like Malaysia, has been rapidly climbing into the ranks of the world's newly industrialized countries. Yet it also experienced a major downturn in the late 1990s that negated much of this development through devalued currencies and the exodus of investment capital. Beginning from a low base, Thailand's economy virtually exploded in the late 1980s, reaching some of the fastest rates of growth ever experienced anywhere in the world. However, the Thai economy eventually overheated, suffering a dramatic drop in 1997 that triggered the more general Asian financial crisis. As of 1999, the Thai economy had still not experienced any real recovery.

Japanese companies were leading players in the earlier Thai boom. As Japan itself became too expensive for assembly and other manufacturing process, Japanese firms began to relocate factories to such places as Thailand. They have been particularly attracted by the country's low-wage, yet reasonably well-educated, workforce. Thailand is also seen as a politically stable country lacking the severe ethnic tensions found in many other parts of Asia. Although Thailand's Chinese population is large and economically powerful, relations between the Thai and the Chinese have generally been good, marked by much intermarriage and cultural synthesis. Thailand also has a democratic government and a thriving free press, although it does have a legacy of military coups followed by periods of authoritarian rule. In recent years, however, Thailand's widely revered constitutional monarchy has emerged as a democratic bulwark. Proponents of democracy can thus point to Thailand as a counterexample to the Singaporean notion that rapid growth requires authoritarian rule.

To repeat a familiar story, though, Thailand's economic boom has by no means benefited the entire country to an equal extent. Most industrial development has occurred in the country's historical core, especially in the city of Bangkok itself. Yet even in Bangkok the blessings of progress have been decidedly mixed. As the city begins to choke on its own growth, industrial growth has begun to spread outward. The entire Chao Phraya lowland area shares to some extent in the general prosperity, both because of its proximity to Bangkok and its own rich agricultural resources. In northern Thailand, the Chiang Mai area has also experienced some economic growth, due in part to the fact that it attracts many international tourists.

Thailand's Lao-speaking northeast (the Khorat Plateau), on the other hand, is the country's poorest region. Soils here are too thin to support intensive agriculture, yet the population is relatively large. Infrastructure remains rudimentary, discouraging industrial development. Due to the poverty of their homeland, northeasterners are often forced to seek employment in Bangkok. Men typically find work in the construction industry; northeastern women more often make their livings as prostitutes.

Thailand has actually earned the dubious distinction of being one of the world's prostitution centers. Thai prostitutes cater largely to a domestic clientele, but they also attract many foreign customers as well. Most of these "sex tourists" come from wealthy countries such as Japan and Germany, but a significant number are Malaysian. Prostitutes, especially those who are underage, are frequently coerced into their jobs. The "sex industry" of Thailand presents an ironic and tragic situation; in general, the social position of Thai women is relatively high, yet the exploitation of women in Thai brothels can be extraordinarily severe (see "Social Geography: Heroin, AIDS, and Prostitution in Thailand").

Recent Economic Expansion in Indonesia At the time of independence (1949), Indonesia was one of the poorest countries in the world. The Dutch had used their colony largely as an extraction zone for tropical crops and other resources and had invested little in infrastructure or education. The population of Java mushroomed in the nineteenth and early twentieth centuries, and serious land shortages burdened most peasant communities. Moreover, political instability inhibited economic development for the first decades of independence.

The Indonesian economy finally began to expand in the 1970s. Oil exports fueled the early growth, as did the logging of tropical forests. But unlike most other oil exporters, Indonesia continued to grow even after oil prices plummeted in the 1980s. Today, production is relatively low compared to overall economic growth, a scenario that may force Indonesia to import oil by 2005. Like Thailand and Malaysia, Indonesia proved attractive to multinational companies eager to export from a low-wage economy. Large Indonesian firms, some three-quarters of them owned by local Chinese families, have also capitalized on the country's low wages and abundant resources. The national government has nurtured an indigenous, technologically oriented business sector, and it is now moving into such fields as automobile and airplane manufacturing.

But despite relatively rapid growth, Indonesia remains a poor country. Its pace of economic expansion never matched that of Thailand, Singapore, and Malaysia, and it has remained much more dependent on the unsustainable exploitation of natural resources. The financial crisis of the late 1990s, moreover, hurt Indonesia more severely than any other country. In 1998 alone, industrial production declined by some 15 percent. Millions of Indonesians suddenly found themselves so destitute that they could no longer afford rice, the country's basic food staple. Political instability is also a continuing concern. While the local Chinese business leaders maintained good relations with the political and military elite, tensions between the Chinese community and the rest of the Indonesian populace continue to simmer.

As one would expect, social and economic development in Indonesia exhibits pronounced geographical disparities. Northwest Java, close to the capital city of Jakarta, has boomed, and much of the resource-rich and moderately populated island of Sumatra has long been relatively prosperous. In the overcrowded rural districts of central and east Java, moreover, many peasants have no land, or possess only tiny plots, and thus remain on the margins of subsistence. Far eastern Indonesia has experienced little economic or social development, and throughout the remote areas of the "outer islands," tribal peoples have suffered terribly as their lands have been lost to outsiders.

Persistent Poverty in Vietnam, Laos, and Cambodia The three countries of former French Indochina—Vietnam, Cambodia, and Laos—have experienced relatively little economic expansion. This area endured almost continual warfare between 1941 and 1975, and fighting persisted until

SOCIAL GEOGRAPHY Heroin, AIDS, and Prostitution in Thailand

Although Thailand has made considerable progress in providing health services to its citizens, the country faces one of the worst AIDS epidemics in the world. The prevalence of AIDS in Thailand stems from the combination of a high level of intravenous drug use and rampant prostitution (both female and male). Large quantities of opium are processed into heroin in the "Golden Triangle" area that includes parts of northern Thailand as well as Burma and Laos. Although opium smoking is an old cultural practice among the tribal peoples of the area, injecting heroin is new. Unfortunately, the use of heroin is now spreading rapidly through both the hills and adjacent lowlands. When narcotics are widely injected and needles shared, AIDS diffuses silently but rapidly.

As southern Thailand is beginning to prosper, its prostitution industry is being increasingly staffed by women from poorer areas. The Khorat Plateau supplies many prostitutes, but recruitment is also focused on the Golden Triangle drug zone, where the HIV virus is already widespread. Many of these women are forced into prostitution, and it is not unknown for addicted parents to sell their daughters, and sometimes their sons, into virtual sex slavery. Prostitutes from Burma, having been smuggled illegally into Thailand, are especially vulnerable to exploitation. Child prostitution is also widespread. Children and adolescents seem to be especially susceptible to infection, thereby hastening the spread of the disease.

Thailand's government has worked hard to encourage the use of condoms. It has done very little, however, to suppress—or even carefully regulate—prostitution. With its being such a huge international business in Thailand, prostitution ringleaders have found it worth their while—and easy—to circumvent the relatively weak Thai laws and regulation. Much evidence suggests that the AIDS crisis in Burma is almost as severe as that of Thailand. The government of Burma, however, has done very little to stop the epidemic, in part because it maintains—contrary to the facts—that a commercial sex industry cannot thrive in a Buddhist country such as Burma.

the late 1990s in Cambodia. Critics contend that the socialist economic system adopted by these countries forestalled economic growth. However, this debate is now moot, since a capitalist model of development is gradually supplanting socialism in all three countries.

Of these three countries, Vietnam is by most measures the most prosperous. Its per capita GNP of $290, however, still places it among the world's poorest economies. Post-war reunification in 1975 did not bring the anticipated growth, and economic stagnation ensued. Conditions grew worse in the early 1990s after the fall of the Soviet Union, Vietnam's main supporter and trading partner. Until the mid-1990s, Vietnam remained under embargo by the United States and thus somewhat isolated from the world economy. Frustrated with their country's economic performance, Vietnam's leaders began to embrace the market while retaining the political forms of a communist state. They have, in other words, followed the Chinese model. Vietnam now welcomes multinational corporations such as Nike, which are attracted by the extremely low wages received by its relatively well-educated workforce. In many cases these wages are lower than those of the Philippines and Indonesia. The government also hopes that a market economy will unleash the entrepreneurial potentials of the Vietnamese people, thus initiating a phase of rapid economic expansion (Figure 13.35). Planners recognize that such economic policies will result in increasing social differentiation, but this is seen as the price of economic growth.

These reforms did bring on an upsurge in economic activity in Vietnam, particularly in the formerly capitalist south. The world market, however, has been somewhat slow to invest, and exports remain dominated by primary products, especially rice and crude oil. U.S. firms were finally allowed into the country in 1994, after the reestablishment of diplomatic relations between the two countries, but many U.S. companies fear that the country's political climate remains too unstable. South Korean corporations, which are only now beginning to seek low-wage platforms outside of Korea, have been more eager to build factories. Labor tensions have arisen, however, owing to the rigid discipline standards imposed by Korean managers. Non-competitive, state-owned industries remain a burden on the Vietnamese economy.

▲ **Figure 13.35 Storefront capitalism in Vietnam** Although Vietnam is a communist state, it has—like China—embraced many forms of capitalism. Private shops are now allowed, and foreign investment is welcome. In general, the market economy is much more highly developed in the south than in the north. *(Liba Taylor/Panos Pictures)*

Vietnam is obviously undergoing a difficult social and economic transition. In the mid-1990s, some observers predicted a future of rapid growth leading to genuine development, seeing Vietnam as the next Thailand or even Malaysia. Such optimists highlight the supposed Confucian-derived work ethic of the Vietnamese people. Other writers counter that the Vietnamese government has not fully committed itself to a market economy, and that residual state control might thwart further expansion. By the late 1990s, however, the Asian crisis had partly undermined these arguments. Vietnam also saw its economy stumble, but not as severely as Thailand's and Indonesia's.

Laos and Cambodia face more serious problems than Vietnam. In Cambodia, the ravages of war have exacerbated an already unstable situation, while Laos faces special difficulties owing to its rough terrain and relative isolation. Both countries are also hampered by a lack of infrastructure; outside the few cities, paved roads and reliable electricity are rarities. As a result, the economies of both Cambodia and Laos remain largely agricultural in orientation. The Laotian government is pinning its economic hopes on hydropower development. The country is mountainous and has many large rivers and could therefore generate large quantities of electricity, which is in high demand in neighboring Thailand, if it can find funding to build the necessary dams.

Despite their basic lack of development, Cambodia and Laos are not as miserable as one might expect from the official economic statistics. Both countries experienced an upsurge of economic activity in the early and mid-1990s; both, moreover, have relatively low population densities and abundant resources. Their low per capita GNP figures, more importantly, partially reflect the fact that many of their people remain in a subsistence economy. While the highland peoples of Laos make few contributions to GNP, most of them do have adequate shelter and food. Although Laos, Cambodia, and Vietnam have lower per capita GNP figures than India, few of their people are seriously malnourished.

Burma's Troubled Economy Burma (Myanmar) stands with Cambodia and Laos at the bottom of the scale of Southeast Asian economic development. For all of its many problems, however, Burma has often been seen as a land of great potential. It has abundant natural resources (including oil and other minerals, water, and timber), as well as a large expanse of fertile farmland. Its population density is moderate, and its people are reasonably well educated. Burma's economy, however, has remained virtually stagnant since independence in 1948.

Burma's economic woes can be traced in part to the continual warfare the country has experienced. Most observers also blame economic policy. Beginning in earnest in 1962,

Burma attempted to isolate its economic system from global forces in order to achieve self-sufficiency under a system of Buddhist socialism. While its intentions may have been admirable, the experiment was not successful; instead of a creating a self-contained economy, Burma found itself burdened by endemic smuggling and black-market activities. Certainly its very lack of economic dynamism has made Burma's capital of Yangon (Rangoon) a much more pleasant place than its ancient rival of Bangkok, since it is not nearly so congested or polluted. Some critics of industrial development, moreover, argue that the Burmese people are wary of economic progress owing to their religious beliefs, and that they prefer the slow pace and reduced tensions of an agrarian economy. Other scholars counter that the Burmese people are dissatisfied with their limited economic opportunities and that the government has been able to maintain its power only through heavy repression.

Regardless of the debates over Buddhist socialism, the government of Burma has since the early 1990s been attempting to revitalize its economy by allowing more room for market forces and by gradually opening connections to the global system. By the mid-1990s Burma's economy was growing at an annual pace of some 6 to 9 percent, but per capita income was still lower than it had been in 1980. Burma remains a notoriously difficult place in which to conduct business, and its political future is cloudy. Several U.S. companies actually suspended their initial forays into Burma owing to protests over the country's abysmal human rights record. Trading connections with China, however, are growing.

The Southeast Asian Economy Reconsidered

As the discussion above shows, Southeast Asia as a whole has undergone rapid integration into the global economy (Figure 13.36). Singapore has thoroughly staked its future to the success of multinational capitalism, and several neighboring countries seem to be following suit. Even Marxist Vietnam and once-isolationist Burma are opening their doors to the global system, albeit with some hesitation.

Much debate has arisen among scholars over the roots of Southeast Asia's economic gains, as well as its more recent economic problems. Those who credit primarily the diligence, discipline, and entrepreneurial skills of the Southeast Asian peoples are optimistic that economies will rebound once again. Some skeptics argue that most of the region's growth has come merely from the application of large quantities of labor and capital, unsupported by real advances in productivity. Others contend that Southeast Asia moved too quickly from its traditional state-run economies to embrace free-market capitalism. The tiger boom, they argue, was an artificial blossoming resulting from external international capital that skimmed off the cream of regional economies. In hindsight, these skeptics claim to have predicted that the Southeast Asian boom was limited and would quickly run out of steam.

Regardless of what happens in coming years, global economic integration has already brought about a good deal of development in Singapore, Malaysia, Thailand, and even Indonesia. Over Southeast Asia as a whole, however, economic development has resulted in accelerating environmental degradation and growing social inequality. Outside of Singapore and Malaysia, moreover, successful development has generally been based on labor-intensive manufacturing in which workers are paid miserably low wages (often less than $1 a day) and subjected to harsh discipline. Consumers in the world's wealthy countries are increasingly aware that many of their basic purchases, such as sneakers and clothing, are produced under exploitative conditions. Movements have thus begun in Europe, the United States, and elsewhere to pressure both multinational corporations and Southeast Asian governments to improve the working conditions of laborers in the export industries. Some Southeast Asian leaders object. Malaysia's Prime Minister Mahathir, for example, accuses Western activists of wanting to preclude Southeast Asian development, and to protect their own home markets, under the misguided pretext of concern over worker rights.

Some critics of modern globalized development further worry that Southeast Asia will be even more thoroughly devastated if the world economy experiences a major depression such as the global economic collapse of the 1930s. The region's export-derived vitality, they argue, can also be viewed as vulnerability. Supporters of current policies, on the other hand, argue that export-oriented industries are currently the healthiest ones in Southeast Asia, owing in part to the currency devaluations that have accompanied the economic crisis. They also contend that economic development has gone hand in hand with social development, and that the increasingly healthy and well-educated population of Southeast Asia is well prepared to weather any global downturn.

Issues of Social Development

As might be expected, several key indicators of social development in Southeast Asia correlate relatively well with levels of economic development. Singapore thus ranks among the world leaders in regard to health and education, as does the small, yet oil-rich country of Brunei. Laos and Cambodia, not surprisingly, come out near the bottom of the chart (Table 13.3). Note, however, that the people of Vietnam are healthier and better educated than might be expected on the basis of their country's overall economic performance. Although Vietnam's Marxist government was not able to generate sustained economic growth, it did provide reasonably good social services.

With the exception of Laos, Cambodia, and Burma, Southeast Asia has achieved relatively high levels of social welfare. In Laos and Cambodia, however, life expectancy at birth hovers around a miserable 51 years (as compared to Thailand's 69 years), and literacy rates remain below 50 percent. Even the poorest countries of the region have made some improvements, as is evident in the figures in Table 13.3, for mortality under the age of 5. Note, however, that progress has been even more pronounced in prosperous countries such as

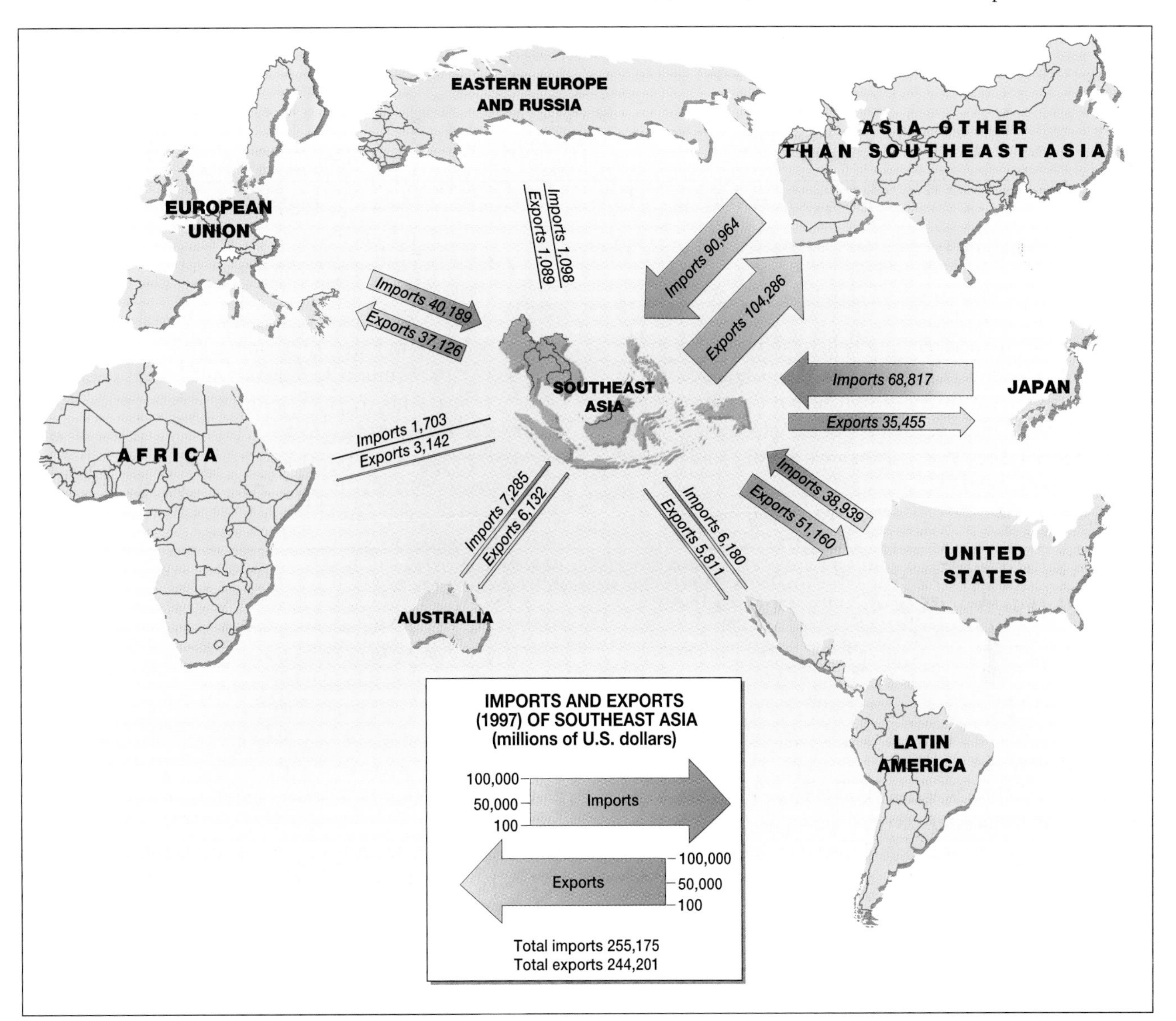

▲ **Figure 13.36 Southeast Asia trade** Southeast Asia has recently emerged as a major player in the global economy, although such international economic engagement is largely confined to Singapore, Malaysia, Thailand, Indonesia, and the Philippines. Most of the region's trade is focused on Japan and the rest of Asia, although the European Union and the United States are also important partners. Southeast Asia maintains a positive trade balance with the United States, but it imports far more from Japan than it exports to Japan.

Malaysia and Singapore, whereas war-torn Cambodia has made relatively small gains.

Most of the governments of Southeast Asia have placed a high priority on basic education. Literacy rates are relatively high in most countries of the region. Much less success, however, has been realized in university and technical education. As Southeast Asian economies continue to grow, this educational gap is beginning to have negative consequences, forcing many students to study abroad. In 1997, for example, more than 50,000 Malaysians were attending colleges in other countries. Economic and social development in Southeast Asia have also led to reduced birthrates. With population growing much more slowly now than before, economic gains are more easily translated into improved living standards.

Table 13.3 Social Indicators and Status of Women

Country	Life Expectancy at Birth		Under Age 5 Mortality, per 1,000 Live Births		Secondary School Enrollment %		Female Labor Force Participation (% of total)
	Male	Female	1960	1995	Male	Female	
Burma (Myanmar)	60	62	237	111	23	23	43
Brunei	70	73	—	—	67	74	34
Cambodia	50	53	217	181	—	—	53
Indonesia	60	64	216	111	48	39	40
Laos	52	55	233	141	31	19	47
Malaysia	70	75	105	17	56	61	37
Philippines	63	69	102	59	71	75	37
Singapore	74	80	40	6	69	71	38
Thailand	67	72	146	33	38	37	46
Vietnam	65	69	219	48	—	—	49

Sources: *Population Reference Bureau Data Sheet, 1998,* Life Expectancy (M/F); *World Resources Institute, 1998–99,* Under-5 Mortality Rate; *Population Reference Bureau Data Sheet, 1996,* Secondary School Enrollment (M/F); *The World Bank Atlas, 1998,* Female Participation in Labor Force.

Conclusion

In many ways, Southeast Asia is the world region most emblematic of this book's focus on diversity amid globalization. Unarguably, diversity within the region is pronounced, from the array of different cultural traditions to persistent and problematic disparities in economic and social development. Given this wide diversity, one might argue that modern globalization, for better or worse, actually helped Southeast Asia find a new sense of regional identity and cohesiveness through common struggles with economic development and revival, and, additionally, by the ten different countries coming together to address common geopolitical concerns. Most experts would agree that Southeast Asia has gained an unusual degree of solidarity in recent years through the regional forum of ASEAN; others would add that shared concerns about economic restructuring and integrated trading also nurtured regional identity. As a result, what Southeast Asia has lacked throughout history in cultural and ethnic unity may, in fact, be forced upon the region by twenty-first-century globalization.

That said, the role of ASEAN and the economic coherence of Southeast Asia should not be exaggerated. Most of the region's trade is still directed outward toward the traditional cores of the global economy—North America, Europe, and especially Japan. This sort of external orientation is not surprising, considering the export-focused economic policies embraced by most Southeast Asian countries. A significant question for Southeast Asia's future is whether the region will develop an integrated regional economy, or whether linkages with such external countries as Japan and the United States will remain more important. Certainly the phenomenal success of Singapore has led to some regional integration. Moreover, labor migration is also intensifying local connections as workers from low-wage countries such as the Philippines and Indonesia stream into higher-wage countries, such as Malaysia.

An equally important question concerns the future of the region's environment. Given the emphasis placed by global trade ties on exporting raw resources such as forest logs, it is perhaps understandable that Southeast Asia has sacrificed so much of its tropical forests for funds to support other kinds of economic development. Yet in most of the region's countries, the forests are now seriously depleted. One of the questions Southeast Asia must address is whether to cooperate with regional restraints on continued logging, or, instead, to cave in to those special interests promoting continued forest destruction. If the larger Asian economy rebounds as predicted, certainly market demand for Southeast Asian forest products will also increase. It is yet unclear whether the region can resist this temptation. Besides forest preservation, Southeast Asia must also meet the needs of its burgeoning cities by making significant expenditures in infrastructural improvement, specifically those that will improve traffic flow, end water pollution, and—somehow—reduce the air pollution problems that threaten public health in large and small cities.

Finally, some degree of geopolitical change is needed to bring stability to the region. More specifically, the national governments of Indonesia, Burma, and Cambodia must reinvent themselves so there is greater congruence and cooperation between the people of these states and their leaders. In June 1999, Indonesia held a multiparty national election that could recast that nation's government. Certainly one of the first tasks is to address ethnic and cultural tensions within the country, ranging from Timor in the eastern islands to the religious violence in Java and Sumatra. Cambodia must move beyond gratuitous recognition of Khmer

Rouge atrocities to genuine efforts at rebuilding and reconciliation. Last, Burma must achieve some sort of stability, coming to terms with both its separatist tribal peoples and with the urban population that is restless with the current military regime.

Southeast Asia will continue as a region heavily influenced by globalization, yet there will be more scrutiny and critique than in the past. This is because its recent economic downturn had widespread repercussions throughout the world. No one, be they in Southeast Asia, Latin America, Europe, or North America, wants to see a recurrence.

Key Terms

animism (page 560)
Association of Southeast Asia Nations (ASEAN) (page 542)
Bumiputra (page 575)
copra (page 553)
crony capitalism (page 573)
Domino Theory (page 566)
entrepôt (page 574)
Golden Triangle (page 552)
island biogeography (page 547)
Khmer Rouge (page 566)
kleptocracy (page 573)
lingua franca (page 560)
primate cities (page 556)
Ramayana (page 557)
shifted cultivators (page 555)
Sunda Shelf (page 546)
swidden (page 552)
typhoons (page 547)
transmigration (page 554)
Wallace's Line (page 548)

Questions for Summary and Review

1. Why are river deltas so important in the settlement pattern of Southeast Asia?
2. Explain how and why the monsoon climates of Southeast Asia differ between mainland and islands.
3. What is Wallace's Line? What is the significance, both prehistorically and today?
4. What do we mean when we refer to the "globalization of world forestry"?
5. Explain why smoke and air pollution are so bad in Southeast Asia.
6. Compare and contrast the three major types of agriculture in the region—swidden, plantation, and rice cultivation.
7. How might "transmigration" both solve, yet also aggravate, a country's population problems?
8. What major religions are found in Southeast Asia? Describe both the historical and contemporary patterns for each religion.
9. Why is there ambivalence about the English language in some Southeast Asia countries?
10. What are the ethnic tensions facing Indonesia? Locate these different problem areas on a map.
11. What are the goals of ASEAN and how have those goals changed in the last several decades?
12. What caused the recent financial downturn in Southeast Asian economies?
13. Why is Singapore an *entrepôt?*
14. Explain how Malaysia's financial reforms differ from those of, say, Thailand.
15. Why does the "Asian model" for economic development need rethinking?

Thinking Geographically

1. Discuss the ramifications of state-sponsored migration in Indonesia from areas of high population density to areas of low population density in both the "sending" and the "receiving" areas?
2. Why should—or should not—citizens of the United States be concerned about deforestation in Southeast Asia? If they should be, what would be the proper ways in which they might show such concern?
3. What might the fate of animism be in the new millennium? Consider whether it might be doomed to extinction before the forces of modern economics and national integration—or whether it may persist as tribal peoples struggle to retain their cultural identities.
4. What should be the position of the English language in the educational systems of Southeast Asia? What should be the position be of each country's national language? What about local languages?
5. How might ethnic tensions in countries such as Burma and Indonesia be reduced? Does Indonesia have a reasonable claim to such areas as Irian Jaya and eastern Timor?

6. What roles might ASEAN play in the coming years? Should it concentrate on economic or political issues?
7. How could Malaysia successfully integrate its economy into the global system and at the same time regulate the flow of global (or Western) culture? Evaluate whether Singapore can continue to experience economic growth while severely limiting basic freedoms.
8. Is the Southeast Asian economic path of integration into the global economy, marked by an openness to multinational corporations, going to prove wise in the long run, or do its potential hazards outweigh its benefits?

Regional Novels and Films

Novels

Anthony Burgess, *The Long Day Wanes: A Malaysian Trilogy* (1965, Norton)

Joseph Conrad, *Victory* (1936, Doubleday)

Graham Greene, *The Quiet American* (1956, Viking)

Mochtar Lubis, *A Road with No End* (1982, Graham Bush)

W. Somerset Maugham, *Borneo Stories* (1976, Heinemann)

Paul Theroux, *The Consul's Fire* (1977, Houghton Mifflin)

Films

The Deer Hunter (1978, U.S.)

Full Metal Jacket (1987, U.S.)

Heaven and Earth (1993, U.S.)

Indochine (1992, France)

Platoon (1987, U.S.)

The Scent of Green Papaya (1993, Vietnam)

The Ugly American (1962, U.S.)

Bibliography

Bello, Walden. 1998. "The Rise and Fall of South-east Asia's Economy." *The Ecologist* 28(1), 9–17.

Broad, Robin, and Cavanagh, John. 1993. *Plundering Paradise: The Struggle for the Environment in the Philippines*. Berkeley: University of California Press.

Broek, Jan O. 1944. "Diversity and Unity in Southeast Asia." *Geographical Review* 34, 175–195.

Dixon, Chris. 1991. *South East Asia in the World Economy: A Regional Geography*. Cambridge: Cambridge University Press.

Dobby, E.G.H. 1958. *Southeast Asia*. London: University of London Press.

Dutt, Ashok. K, ed. 1985. *Southeast Asia: Realm of Contrasts*. Boulder: Westview Press.

Freyer, Donald W. 1979. *Emerging Southeast Asia: A Study in Growth and Stagnation*. New York: John Wiley and Sons.

Hall, D.G.E. 1955. *A History of South-East Asia*. New York: St. Martin's Press.

Hanks, Lucien. 1972. *Rice and Man: Agricultural Ecology in Southeast Asia*. Chicago: Aldine Atherton.

Hardjono, Joan, ed. 1991. *Indonesia: Resources, Ecology, and Environment*. Singapore: Oxford University Press.

Hart, Gillian; Turton, Andrew; and White, Benjamin, eds. 1989. *Agrarian Transformations: Local Processes and the State in Southeast Asia*. Berkeley: University of California Press.

Hussey, Antonia. 1993. "Rapid Industrialization in Thailand." *Geographical Review* 83, 14–28.

Keyes, Charles F. 1977. *The Golden Peninsula: Culture and Adaptation in Mainland Southeast Asia*. New York: Macmillan.

Leach, E.R. 1954. *Political Systems of Highland Burma*. Boston: Beacon.

Leinbach, Thomas R., and Sien, Chia Lin. 1989. *South-East Asian Transport*. New York: Oxford University Press.

Leinbach, Thomas R., and Ulack, Richard. 1993. "Cities of Southeast Asia." In Brunn, Stanley D., and Williams, Jack F., eds., *Cities of the World: World Regional Urban Development*, pp. 389–429. New York: HarperCollins.

Lewis, Martin W. 1992. *Wagering the Land: Ritual, Capital, and Environmental: Degradation in the Cordillera of Northern Luzon, 1900–1986*. Berkeley: University of California Press.

McGee, Terence G. 1967. *The Southeast Asian City: A Social Geography*. New York: Praeger.

McLennan, Marshall S. 1980. *The Central Luzon Plain: Land and Society on the Inland Frontier*. Quezon City, Philippines: Alemar-Phoenix.

Peluso, Nancy Lee. 1992. *Rich Forests, Poor People: Resource Control and Resistance in Java*. Berkeley: University of California Press.

Pelzer, Karl J. 1945. *Pioneer Settlement in the Asiatic Tropics*. New York: American Geographical Society.

Pye, Lucian W. 1967. *Southeast Asia's Political Systems*. Englewood Cliffs: Prentice Hall.

Rambo, Terry, and Sajise, Percy, eds. 1984. *An Introduction to Human Ecological Research on Agricultural Systems in Southeast Asia*. Los Banos, Philippines: University of the Philippines.

Reid, Anthony. 1988. *Southeast Asia in the Age of Commerce. Volume I: The Lands Below the Winds*. New Haven: Yale University Press.

Rigg, Jonathan. 1991. *Southeast Asia—A Region in Transition: A Thematic Geography of the ASEAN Region*. New York: HarperCollins.

Steinberg, David Joel, ed. 1985. *In Search of Modern Southeast Asia: A Modern History*. Honolulu: University of Hawaii Press.

Taylor, Alice, ed. 1972. *Focus on Southeast Asia*. New York: Praeger.

Taylor, John, and Turton, Andrew, eds. 1988. *Sociology of "Developing Societies": Southeast Asia*. New York: Monthly Review Press.

Ulack, Richard, and Pauer, Gyula. 1989. *Atlas of Southeast Asia*. New York: Macmillan.

von der Mehden, Fred. 1986. *Religion and Modernization in Southeast Asia*. Syracuse: Syracuse University Press.

Wernstedt, Frederick, and Spencer, Joseph. 1967. *The Philippine Island World: A Physical, Cultural, and Regional Geography*. Berkeley: University of California Press.

Wyatt, David K. 1984. *Thailand: A Short History*. New Haven: Yale University Press.

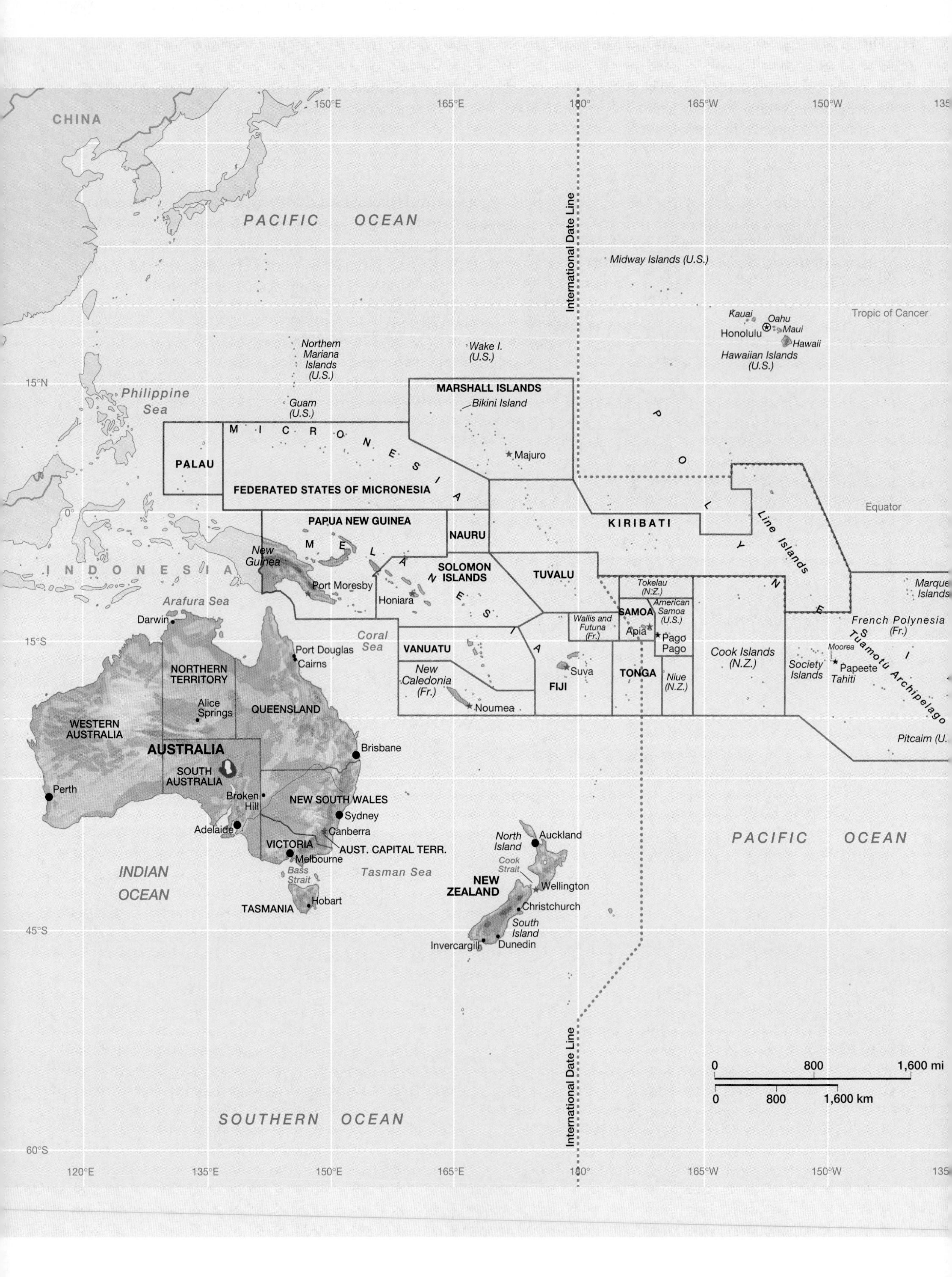

CHINA
PACIFIC OCEAN
Midway Islands (U.S.)
International Date Line
Tropic of Cancer
Kauai
Oahu
Honolulu
Maui
Hawaii
Hawaiian Islands (U.S.)
Northern Mariana Islands (U.S.)
Wake I. (U.S.)
Philippine Sea
Guam (U.S.)
MARSHALL ISLANDS
Bikini Island
Majuro
MICRONESIA
PALAU
FEDERATED STATES OF MICRONESIA
POLYNESIA
Equator
KIRIBATI
Line Islands
PAPUA NEW GUINEA
NAURU
MELANESIA
New Guinea
INDONESIA
SOLOMON ISLANDS
TUVALU
Port Moresby
Honiara
Tokelau (N.Z.)
American Samoa (U.S.)
SAMOA
Apia
Pago Pago
Marquesas Islands
Arafura Sea
Darwin
Wallis and Futuna (Fr.)
French Polynesia (Fr.)
Coral Sea
Port Douglas
Cairns
VANUATU
Cook Islands (N.Z.)
Moorea
Society Islands
Papeete
Tahiti
Tuamotu Archipelago
NORTHERN TERRITORY
New Caledonia (Fr.)
Suva
FIJI
TONGA
Niue (N.Z.)
Noumea
Alice Springs
QUEENSLAND
WESTERN AUSTRALIA
Pitcairn (U.
AUSTRALIA
Brisbane
SOUTH AUSTRALIA
Perth
Broken Hill
NEW SOUTH WALES
Sydney
Adelaide
Canberra
VICTORIA
AUST. CAPITAL TERR.
Melbourne
North Island
Auckland
PACIFIC OCEAN
INDIAN OCEAN
Bass Strait
Tasman Sea
Cook Strait
Wellington
NEW ZEALAND
TASMANIA
Hobart
Christchurch
South Island
Dunedin
Invercargill
SOUTHERN OCEAN
0
800
1,600 mi
1,600 km
15°N
0°
15°S
45°S
60°S
120°E
135°E
150°E
165°E
180°
165°W
150°W

CHAPTER 14

Australia and Oceania

Fashioned from the images of Crocodile Dundee and Paul Gauguin, Australia and Oceania hold a vivid, if peripheral, place in the European and North American imagination. This vast world region, dominated mostly by water, includes the island continent of Australia as well as **Oceania**, a sweeping collection of islands that reach from New Guinea and New Zealand to the far reaches of Polynesia in the mid-Pacific Ocean (Figure 14.1; see "Setting the Boundaries"). From the perspective of its European colonizers, the region was enlivened by its mythical character: exotic and otherworldly, its "lost islands" and "bush" were populated by strange plants and animals as well as by peoples free from the conventional cultural mores of European society (Figure 14.2). From the perspective of its indigenous inhabitants, the region was a world rich in natural and cultural complexity, a way of life rudely interrupted and changed forever by the arrival of European peoples. At the beginning of the twenty-first century, while the legacy of those earlier images still endures, the region is now caught up in powerful processes of globalization that are transforming the rural and urban landscape and producing new and sometimes unsettled cultural and political geographies (Figure 14.3).

Australia and New Zealand share many geographical characteristics and dominate the regional setting. Major population clusters in both countries are located in the middle latitudes (Figure 14.4). Australia's 18.7 million residents occupy a vast land area of 2.97 million square miles (7.69 million square kilometers), while New Zealand's combined North and South Islands (104,000 square miles, or 269,000 square kilometers) are home to 3.8 million people. Most residents of both countries live in urban settlements near the coast. Australia's huge and dry interior, often termed the **outback**, is as thinly settled as North Africa's Sahara Desert, and much of the New Zealand hinterland is a visually spectacular but sparsely occupied collection of volcanic peaks and rugged, glaciated mountain ranges. Taken together, the land areas of these two South Pacific nations almost equal that of the United States, but their populations total less than 10 percent of their distant North Pacific neighbor. All three countries, however, share a pervasive European cultural heritage, the product of common global-scale processes that sent Europeans far from their homelands over the past several centuries. The highly Europeanized populations of

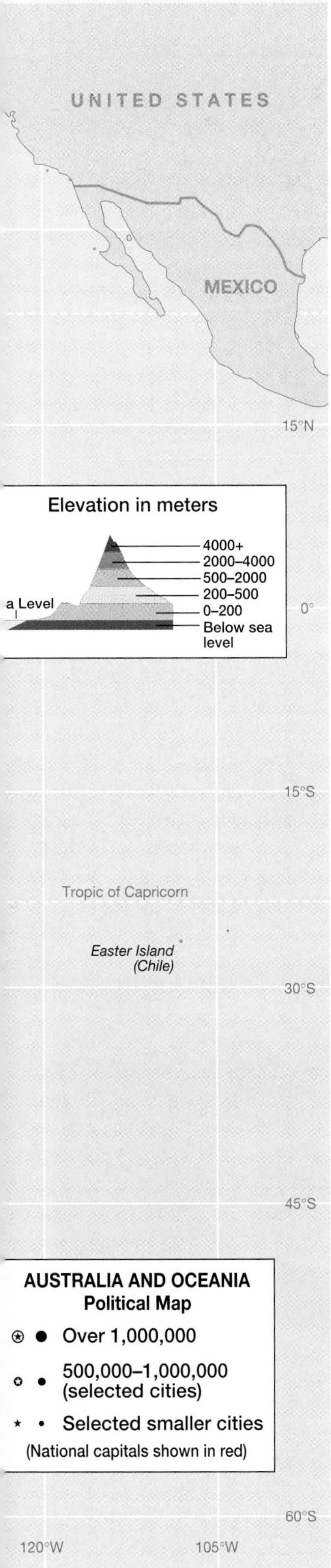

◀ **Figure 14.1 Australia and Oceania** More water than land, the Australia and Oceania region sprawls across the vast reaches of the western Pacific Ocean. Australia dominates the region, both in its physical size and in its economic and political clout. Along with New Zealand, Australia represents largely Europeanized settlement in the South Pacific. Elsewhere, however, the island subregions of Melanesia, Micronesia, and Polynesia contain important native populations that have mixed in varied ways with later European and American arrivals.

Setting the Boundaries

The vast, watery distances of the Pacific as well as shared elements of indigenous and colonial history help to define the boundaries of the Australia and Oceania region. Still, some of its regional boundaries are born from convenience while others remain ill-defined. Australia (or "southern land") forms a coherent political unit and subregion and clearly dominates the region in both area and population. Across the Tasman Sea, New Zealand, while usually considered part of "Oceania," is easily linked with Australia, particularly in their parallel histories of European—mostly British—colonization and development. Both countries also have significant native populations, although Australia's Aborigines are culturally distinct from New Zealand's Maori peoples.

More challenging questions arise when pondering how those larger landmasses (often termed "Meganesia") relate to the island worlds beyond as well as to nearby portions of Southeast Asia. Increasing economic ties and the common borders of the Pacific Ocean link Australia and New Zealand with the smaller islands of Oceania. Still, the area's environmental and cultural diversity does not offer any easy defining characteristics. More broadly, however, traditional island cultures often intermingled and they shared relationships with the sea and the varied fruits of the tropical world. Later, Europeans and Americans converged upon the scene. As they rushed to carve up the realm into colonial possessions, their strategic maneuverings often had lasting consequences. Polynesian Hawaii, once it was securely in the U.S. sphere, evolved to be a multicultural and political borderland in the central Pacific. To the south, the French produced their own distinctive island subregion, mostly scattered across the eastern reaches of Polynesia. On the island of New Guinea, however, complex colonial-era boundaries produced a persisting and problematic regional border. Today an arbitrary boundary line bisects the island and while Papua New Guinea (eastern half) is usually considered a part of Oceania, neighboring Irian Jaya (western half) is a part of Indonesia and thus typically grouped with Southeast Asia.

both Australia and New Zealand also retain particularly close cultural links with Britain, and they maintain relatively high levels of income and economic development.

Punctuated with isolated chains of sand-fringed and sometimes mountainous islands, the blue waters of the tropical Pacific dominate much of the rest of the region. Three major subregions of Oceania each contain a surprising variety of human settlements and political entities, although overall land areas and populations are small when compared with Australia. Farthest west, **Melanesia** (meaning "dark islands") contains the culturally complex, generally darker-skinned peoples of New Guinea, the Solomon Islands, Vanuatu, and Fiji. The largest of these countries, Papua New Guinea (179,000 square miles, or 463,000 square kilometers) includes the eastern half of the island of New Guinea (the western half is part of Indonesia) as well as nearby portions of the northern Solomon Islands. Its population of 4.3 million people is slightly higher than that of New Zealand. To the east,

▶ **Figure 14.2 Mythical Polynesia** The rich colors and idyllic setting of Paul Gauguin's painting *Nave Nave Moe* (Sacred Spring) suggest the romantic response of late nineteenth-century Europeans who were captivated by the societies and scenery of the South Seas. *[Paul Gauguin, French, (1848–1903),* Nave Nave Moe (Sacred Spring), *1894. Hermitage, St. Petersburg, Russia. Oil on canvas, 73 x 98 cm. © The Bridgeman Art Library International Ltd.]*

◀ **Figure 14.3 Sydney, Australia** Most Australians live in cities, and the country's urban landscapes often resemble their North American counterparts. This view of Sydney features its world-famed harbor and displays the site's dramatic interplay of land and water. *(Jean-Paul Ferrero/Auscape International Pty. Ltd.)*

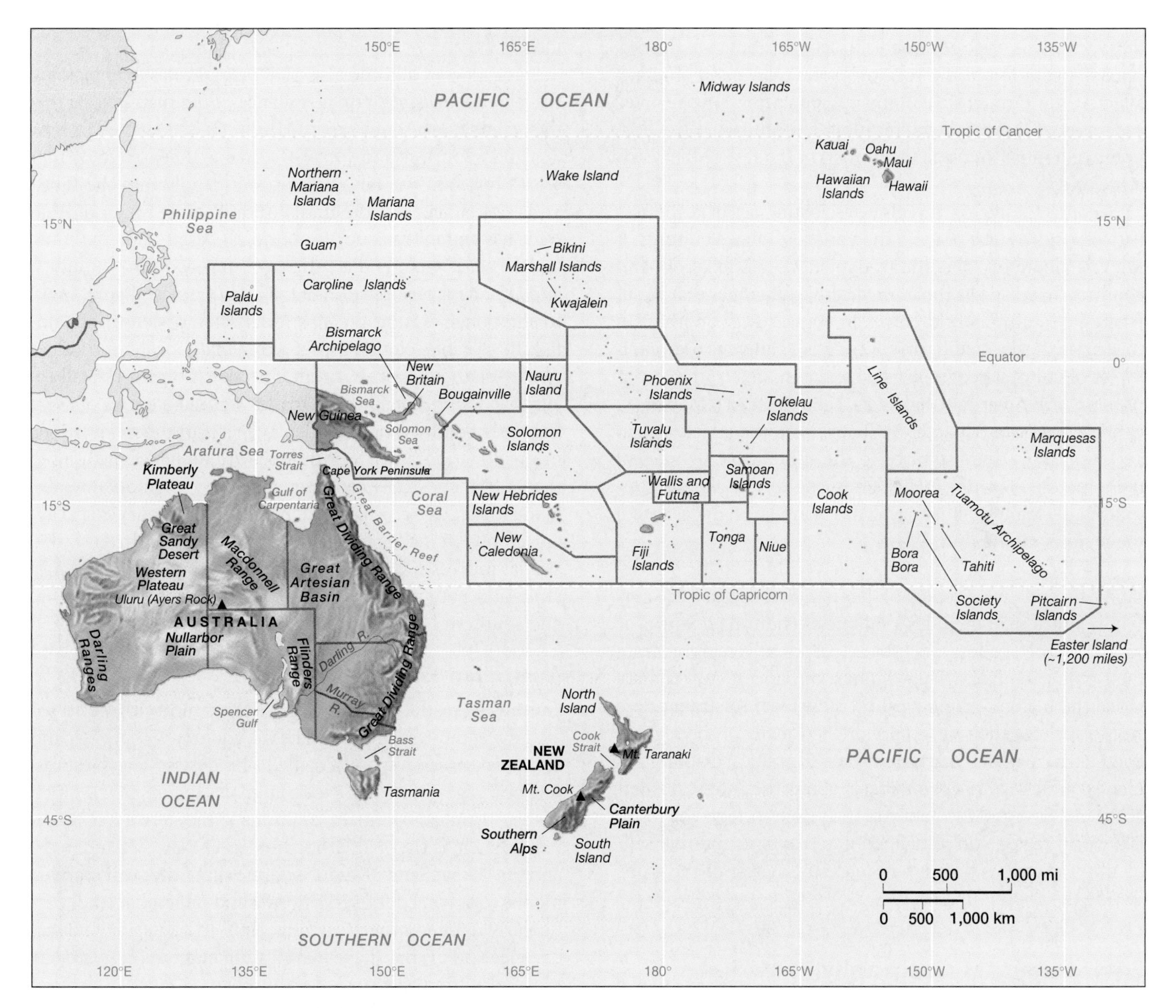

▲ **Figure 14.4 Physical geography of Australia and Oceania** Varied physical processes have shaped the Australia and Oceania region. Ancient shield rocks form the geological core of Australia, while active volcanoes from Hawaii to Papua New Guinea literally produce examples of Earth's newest landscapes. New Zealand is home to some of the region's most complex and varied physical settings. Dominant almost everywhere in the region, the omnipresent waters of the blue Pacific shape land and life in fundamental ways.

the small island groups, or **archipelagos,** of the central South Pacific are called **Polynesia** (meaning "many islands"), and this linguistically unified subregion includes French-controlled Tahiti in the Society Islands, the Hawaiian Islands, and smaller political states such as Tonga, Tuvalu, and Samoa. New Zealand is also often considered a part of Polynesia. Its native peoples, known collectively as the **Maori,** share many cultural and physical characteristics with the somewhat lighter-skinned peoples of the mid-Pacific region. Finally, the more culturally diverse region of **Micronesia** (meaning "small islands") is north of Melanesia and west of Polynesia and includes microstates such as Nauru and the Marshall Islands as well as the U.S. territory of Guam.

Three characteristics unite the Australia and Oceania region. First, the realm's unique physical setting has historically isolated it from much of the more densely settled parts of the world. This basic geographical fact contributed to the distinctive evolution of the region's exotic plants and animals, affected the timing of its human occupation, shaped later processes of European colonization, and impacted the economic potential for trade with much of the rest of the modern world.

Second, the region reveals a fascinating set of responses between indigenous peoples and varied European cultures. It is a vast laboratory of cultural adaptations, assimilation, and conflicts that displays the unpredictable outcomes that result when one culture world comes in contact with another. In this context, the Pacific world presents a different but parallel set of cultural geographies when compared to the examples of North America (Chapter 3), Latin America (Chapter 4), or the Caribbean (Chapter 5). Thus, from the largely Anglo and wealthy Sydney suburbs to the still-elusive peoples of the remote New Guinea Highlands, the region offers a continuum of examples illustrating the many-layered legacies of the region's multicultural roots.

Third, the region is united by its youthful political geography. Essentially a product of the twentieth century, the current political map includes a mosaic of enduring colonies as well as a varied collection of states that are still emerging from their political infancy. Thus, fluidity and uncertainty characterize the region's geopolitical identity. Even Australians are still torn about their ties to Britain, and many of the region's new states have been marked by internal political disputes, violent conflicts, and open debates about the future of their economic base and the nature of their civic society. The result is a world region still ill-defined in terms of its internal self-identity, as well as how it will relate to the rest of the globe during the twenty-first century.

Environmental Geography: A Varied Natural and Human Habitat

The region's physical setting is dramatic testimony to the power of space: the geology and climate of the seemingly limitless Pacific Ocean define much of the physical geography of Oceania, and the almost equally expansive interior of Australia shapes the basic physical setting of that island continent. Not coincidentally, the greatest human impacts upon the region and the densest clusters of population exist at the meeting place of land and sea: coastal strips and island shores share environmental settings where residents have opted to look inward as well as outward.

Globally, the region's peripheral position and its relative isolation have defined distinctive biogeographic realms that differ from those of any other on the planet (Figure 14.5). As Australia and the rest of the region became ever more separated from other plant and animal communities through geological time, unique species evolved that can be found nowhere else. Examples include Australia's diverse *Eucalyptus* species as well as its unique assemblage of marsupials. Recent exotic intrusions from the world beyond, carried mainly by Europeans, have radically altered these natural environments and highlighted their vulnerability to human modification.

Latitude and altitude also play pivotal roles in explaining the region's natural setting. A glance at the map reveals that the realm's north/south expanse (from Hawaii [20° N] to New Zealand's South Island [45° S]) exceeds that of many other world regions. The far south probes deeply into the moist and cool climates of the midlatitude Pacific, much of Australia lies within the zone of subtropical deserts, and the equatorial Pacific Islands offer the expected warmth, humidity, and abundant precipitation of the rainy tropics (Figure 14.6). In addition, elevation alters patterns of temperature and rainfall. In Oceania, clouds gather along many peaks and ridges of mountainous islands, producing much higher rainfall totals than found along nearby coasts. Mountain ranges on New Zealand's South Island also display incredible local variations in precipitation: west-facing slopes are drenched with more than 100 inches (254 centimeters) of precipitation annually, while lowlands to the east (such as the Canterbury Plain) average only 25 inches (64 centimeters) per year. In Australia, similar processes are at work to modify temperature and precipitation patterns in the hilly areas of Tasmania, Victoria, and southern New South Wales.

Australian Environments

Curiously, Australia is one of the world's most urbanized societies, yet most people associate the country with its vast and arid open spaces, a sparsely settled land of sweeping distances, scrubby vegetation, and exotic animals (Figure 14.7). Much of the Australian continent conforms to the stereotype: placenames such as the "Nullarbor [no trees] Plain," "Sandland," and the "Great Sandy Desert" suggest that Europeans exploring the region encountered a hostile and unfamiliar land, particularly as they probed its huge and arid interior. Indeed, distance and dryness have left a lasting mark upon Australia's physical setting and shaped the ways in which people have subsequently settled its more favored corners.

Regional Landforms Journey across Australia from west to east and three major landform regions dominate the continent's physical geography. The Western Plateau occupies more than half of the continent and geologically represents

(a)

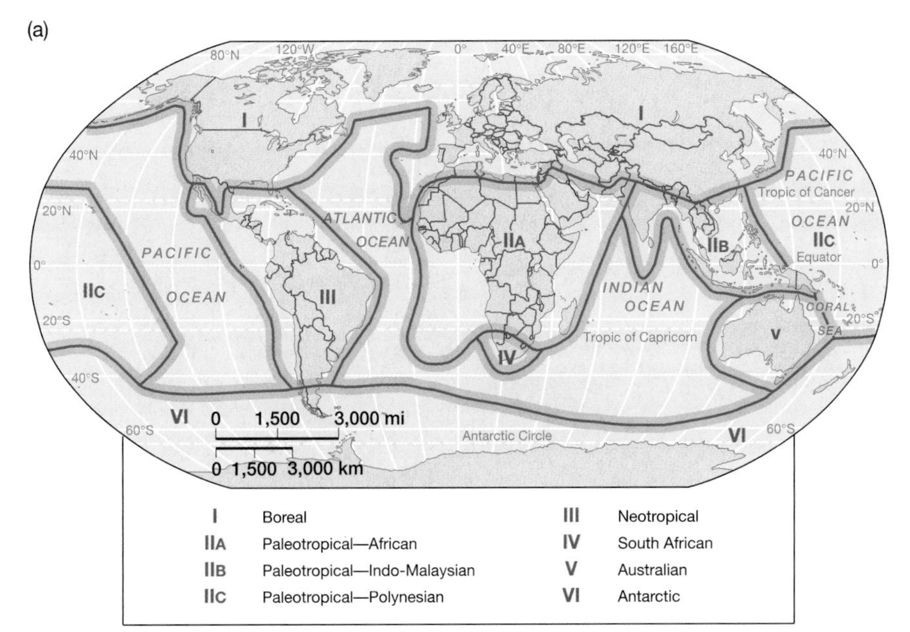

(b)

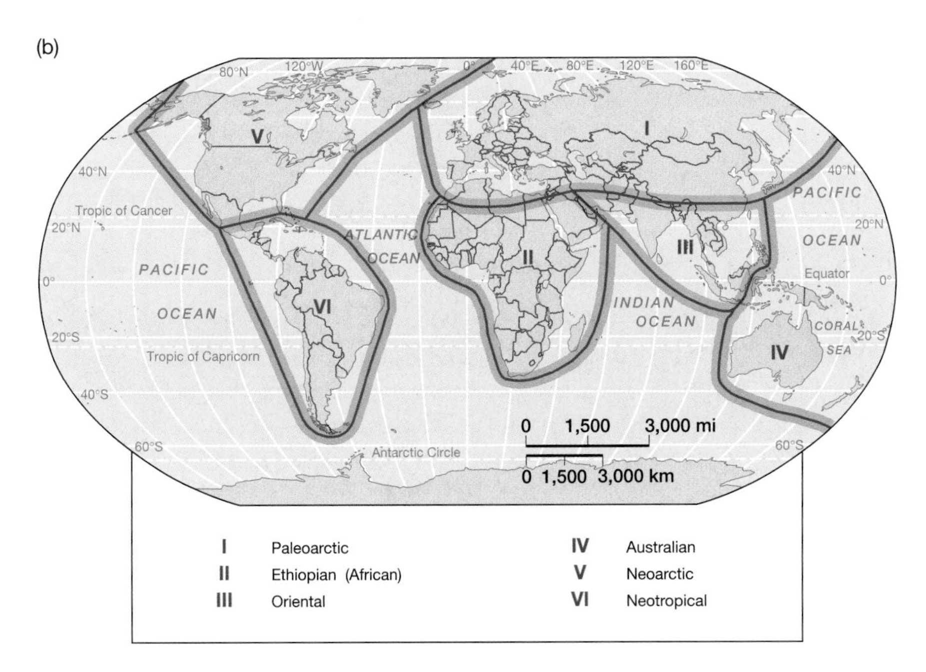

◄ **Figure 14.5 World biogeographic realms** The pattern of botanical (a) and zoological (b) realms highlights how the region's globally peripheral position contributes to its exotic plant and animal life that are unlike any other in the world. *(Source: Christopherson, 1999,* Geosystems, *Upper Saddle River, NJ: Prentice Hall)*

an ancient shield landmass that once joined Antarctica. Most of the region is a vast, irregular plateau that averages only 1,000 to 1,800 feet in height (305 to 550 meters). As one leaves the plateaus, the Interior Lowland Basins stretch north to south for more than 1,000 miles from the swampy coastlands of the Gulf of Carpentaria to the Murray and Darling valleys, Australia's largest river system. Most of the region is a flat, featureless plain, punctuated occasionally by dry lake beds and by stream valleys where water is only a rare visitor. The Great Artesian Basin of western Queensland is also included in this lowland region. Filled with sedimentary rocks, this vast basin, mostly below 500 feet (150 meters) in elevation, possesses a rich supply of underground water that has drained there from the higher country to the east. Finally, more forested and mountainous country defines the horizon as one arrives in the Eastern Highlands along Australia's Pacific Ocean rim. The Great Dividing Range extends from the Cape York Peninsula in northern Queensland to southern Victoria. Nearby Tasmania is also dominated by mountains. These ancient eroded highlands include rugged tablelands, but only a few peaks above 5,000 feet (1,525 meters). The east-facing edges of the mountains often feature steep escarpments and rugged, dissected canyons that slow easy interaction between the interior and the narrow, often densely settled coastal plains on the Pacific. Nearby, off the eastern coast of Queensland, the Great Barrier Reef offers a final dramatic subsurface feature: over the past 10,000 years, one of the world's most spectacular examples of coral reef-building has produced a living legacy now protected by the Great Barrier Reef Marine Park (Figure 14.8).

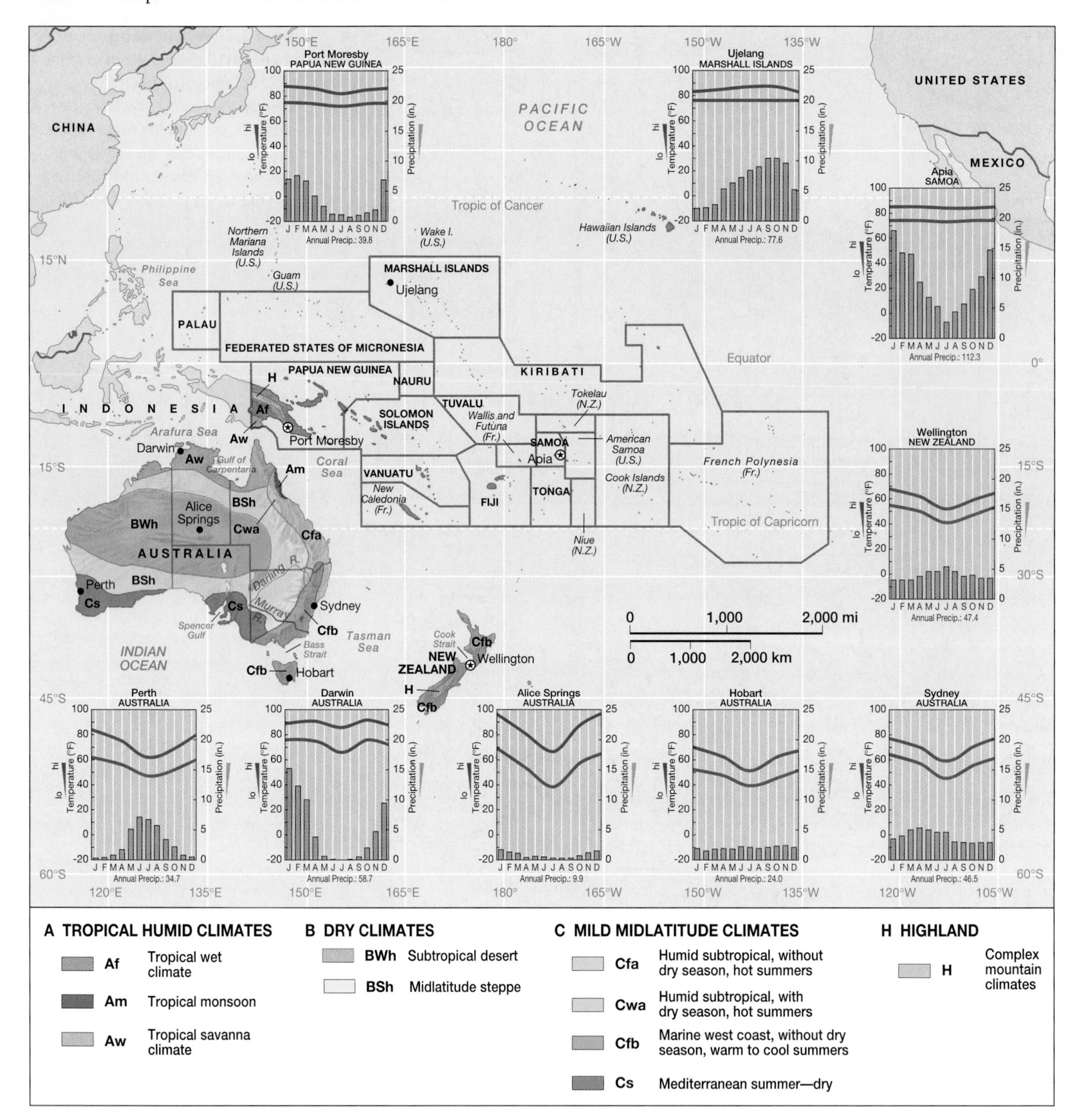

▲ **Figure 14.6 Climate map of Australia and Oceania** Latitude and altitude shape the climatic patterns of the region. Equatorial portions of the Pacific bask in all-year warmth and humidity, while the Australian interior is predictably dry and under the dominance of subtropical high pressure. Cool and moisture-bearing storms of the southern Pacific Ocean provide midlatitude conditions across New Zealand and portions of Australia. More locally, mountain ranges dramatically raise precipitation totals in many highland zones.

Climate and Vegetation Australia's varied climates strongly shape patterns of natural vegetation across the country. Generally, zones of somewhat higher precipitation encircle Australia's arid center. In the tropical low-latitude north, seasonal changes are dramatic and unpredictable. Localities such as Darwin can experience drenching monsoonal rains in the summer (December–March) followed by bone-dry winters (June–September). Indeed, life across the region is shaped by this annual rhythm of the "wet" and the "dry." Much of the north is clothed in a mix of tropical woodlands and thorn forests interspersed with open grasslands. As one ventures down the east coast of

◀ **Figure 14.7 The arid Australian Outback** The endless open spaces of Western Australia resemble some of the arid landscapes of the U.S. West. Even more thinly peopled than their North American counterparts, such settings are nevertheless often rich in mineral resources. *(Jean-Marc La Roque/Auscape International Pty. Ltd.)*

Queensland, precipitation remains high (60 to 100 inches, or 153 to 254 centimeters) but diminishes rapidly as one moves into the interior. Vegetation follows the pattern: coastal tropical rain forests give way to eucalyptus and acacia woodlands in the hill country and eventually the trees thin out west of the mountains, replaced by the short grass and desert scrub vegetation of southwest Queensland's arid outback. Indeed, some of the country's dry heartland is almost devoid of vegetation, and precipitation at interior locations such as the Northern Territory's Alice Springs averages less than 10 inches (25 centimeters) annually.

South of Brisbane, more midlatitude influences dominate eastern Australia's climate. Coastal New South Wales, southeastern Victoria, and Tasmania experience the country's most dependable year-round rainfall in a climatic regime that averages 40 to 60 inches (102 to 152 centimeters) of precipitation per year. Nearby mountains see frequent winter snows. Forests clothe most of the wetter southern coastlands and highlands, but tracts of more open heath dot the hills of southern Victoria, and drier conditions in the interior produce grasslands often suited to extensive grazing. Farther west, summers are hot and dry in much of South Australia and in the southwest corner of Western Australia, producing a distinctively Mediterranean climate. These zones of southern Mediterranean climate produce a mix of vegetation types. In moister hills near Perth, acacia, eucalyptus, and conifer forests cover the slopes, while portions of South Australia feature abundant **mallee** vegetation, a tough and scrubby eucalyptus woodland of limited economic value that has often been stripped away by determined farmers (Figure 14.9).

An Exotic Zoogeography Isolation and genetics have combined to produce Australia's animal kingdom, surely one of the most exotic on the planet. Long separated from Asia, animal species on the Australian landmass have evolved in unique ways. Marsupials, or animals that carry their young

▲ **Figure 14.8 Great Barrier Reef** Stretching along the eastern Queensland coast, the famed Great Barrier Reef is one of the world's most spectacular examples of coral reef-building. Threatened by varied forms of coastal pollution, much of the reef is now protected in a national marine park. *(Hilarie Kavanagh/Tony Stone Images)*

▲ **Figure 14.9 Mallee scrub vegetation** The interior of South Australia and New South Wales features thousands of square miles of a tough, scrubby eucalyptus vegetation known as mallee. Farmers often remove this native cover in order to plant exotic grass and grain crops. *(Jaime Plaza Van Roon/Auscape International Pty. Ltd.)*

in pouches, never faced much competition from other invading species and adapted well to the continent's varied natural settings. More than 120 species of marsupials inhabit the country. Europeans such as Captain Cook expressed amazement as they gazed upon leaping kangaroos, duck-billed platypus, hairy-nosed wombats, and snarling Tasmanian devils during their early exploratory visits. Many of these animals, such as the red and gray kangaroo, thrive on the drought-resistant vegetation of the Australian outback. Bird life is similarly varied. Flightless emus, shrieking kookaburras, rainbow lorikeets, and dozens of other parrot species celebrate the biological fruits of spatial isolation in an extensive and diverse natural environment.

New Zealand's Varied Landscape

Part of the Pacific Rim of Fire, New Zealand owes its geological origins to undersea mountain-building that produced two rugged and spectacular islands in the South Pacific more than 1,000 miles (1,600 kilometers) southeast of Australia (Figures 14.4 and 14.10). The active volcanic peaks and geothermal features of the North Island, in particular, reveal the country's fiery origins. Blessed with such indigenous names as Ruapehu, Ngauruhoe, and Tongariro, these volcanic peaks tower over nearby tablelands, reaching heights of more than 9,100 feet (2,775 meters). Even higher and more rugged mountains run down the western spine of the South Island. Mt. Cook (or Aoraki) is New Zealand's highest peak, cresting at more than 12,000 feet (3,660 meters). Often mantled by high mountain glaciers and surrounded by steeply sloping valleys, the Southern Alps are known to the Maori as *Te Tapu Nui* (the Peaks of Intense Sacredness). They are one of the world's most visually spectacular mountain ranges, complete with narrow, fjord-like valleys that intricately indent much of the South Island's isolated western coast. East of the mountains, small patches of coastal plain are found between Christchurch and Invercargill.

As in Australia, New Zealand's isolation offered opportunities for the development of unique plant and animal species. Eighty-five percent of the country's native trees and seed plants are found nowhere else on Earth. Bats are the region's only native mammals, while ancient tuatara reptiles, kiwi birds, and the now-extinct and flightless moas are telltale examples of the country's special biological legacy.

North Island Environments Most of New Zealand's North Island is distinctly subtropical. The coastal lowlands near Auckland are mild and wet year-round. Still, local variations can be striking, as the area's volcanic peaks create their own microclimates. For example, an ascent of 8,000-foot (2,440-meter) Mt. Taranaki, a volcano on the southwest side of the island, might include fern-enshrouded lowland forests, sagebrush-like subalpine slopes speckled with wildflowers, and the snow-clad peak itself, which is the haunt of high-country climbers and adventurous skiers. While much of the North Island's natural vegetation has been replaced with human-introduced species, portions of the region's ancient and towering kauri forests remain in government preserves (Figure 14.11). Elsewhere, European forest and grassland species have profoundly altered the look of the North Island landscape.

South Island Environments Across the narrow Cook Strait, New Zealand's South Island offers an equally complex natural setting. Although the encircling waters of the Pacific

▲ **Figure 14.10 Mt. Taranaki** New Zealand's North Island contains several volcanic peaks, including Mt. Taranaki. The 8,000-foot (2,440-meter) peak offers everything from subtropical forests to challenging ski slopes, and attracts both local and international tourists. *(Ken Graham/Ken Graham Agency)*

▲ **Figure 14.11 Kauri Forest** On New Zealand's North Island, the Waipoua Kauri Forest Reserve protects the nation's largest remaining example of this unique environment. Elsewhere, much of New Zealand's native vegetation has been replaced by exotic species. *(Paul A. Souders/Corbis)*

moderate the climate, conditions become distinctly cooler as one moves poleward south of the Cook Strait. Indeed, South Island's southern edge feels the seasonal breath of Antarctic chill as it lies more than 46° south of the equator (and more than 700 miles, or 1,120 kilometers, south of Auckland). It is no accident that the southern towns of Dunedin and Invercargill have a strong Scottish flavor, a function both of settlement history and climatic setting. To the north, milder conditions prevail on the Canterbury Plain in Christchurch, but nearby snow-covered mountains suggest the region's midlatitude character. Dramatic local differences in precipitation also have shaped patterns of vegetation on the South Island. The damp, west-facing or windward side of the Southern Alps, for instance, supports a dense, sometime impenetrable rain forest of misty, fern-covered canyons and forested slopes. In the challenging environments of the southern Fiordlands region and across the higher slopes of the Southern Alps, low-growing shrubs and grasslands replace the forests. East of the divide, however, much drier and less extreme conditions prevail. The Otago region, inland from Dunedin, sits partially in the rain shadow of the Southern Alps, and its rolling, open landscapes resemble the semiarid expanses of North America's Intermountain West (Figure 14.12).

▲ **Figure 14.12 Central Otago, South Island** On New Zealand's South Island, the Southern Alps capture rainfall on the west coast, but leave areas to the east in a drier rain shadow. As a result, the Central Otago region has a semiarid landscape resembling portions of the U.S. West. *(John Lamb/ Tony Stone Images)*

The Oceanic Realm

Leaving the region's large, midlatitude islands behind, the vast expanse of Pacific waters reveals another set of environmental settings that are as rich and complex as they are fragile and threatened with human alteration. Literally born from the sea, these thousands of mostly tiny, scattered landmasses are everywhere, shaped by their surrounding maritime environments. Oceanic currents and wind patterns have historically served as natural highways of movement between these island worlds. Those same forces define the rhythm of weather patterns across the region in a broad band that extends more than 20° north (Hawaiian Islands) and south (New Caledonia) of the equator.

Creating Island Landforms Much of Melanesia and Polynesia are part of the seismically active Pacific Rim. As a result, volcanic eruptions, major earthquakes, and **tsunamis**, or seismically induced sea waves, are not uncommon events across the region, and they impose major environmental hazards upon its inhabitants. For example, volcanic eruptions and earthquakes on the island of New Britain (Papua New Guinea) forced more than 100,000 people from their homes in 1994. Only four years later, a massive tsunami triggered by an offshore earthquake swept across the north coast of New Guinea, killing 3,000 residents and destroying numerous villages. Such events are unfortunately a part of life in this geologically active part of the world.

Island geology varies across the realm. The large islands of Melanesia, as well as New Zealand, are composed of fragments of continental rock and are thus geologically quite complex. New Guinea, for example, is dominated by multiple, generally east-west trending mountain ranges separated by rugged, elevated plateaus that remain difficult to traverse. Extensive coastal lowlands flank these intricate highland regions, particularly in the southern portion of the island. Indeed, the nearby islands of the Bismark Archipelago are an eastward extension of this continental mountain-building zone.

Most of the islands of Polynesia and Micronesia, however, are truly oceanic, having originated from volcanic activity on the ocean floor without any geological connection to larger landmasses. The larger active and recently active volcanoes form **high islands** that often rise to a considerable elevation and cover a substantial area. The island of Hawaii, the largest and youngest of the Pacific's high islands, is more than 80 miles (128 kilometers) across and rises to a height of more than 13,000 feet (3,980 meters). Indeed, the entire Hawaiian archipelago exemplifies a geological **hot spot** where moving oceanic crust passes over a supply of magma, thus creating a chain of volcanic uplifts. Many of the islands of French Polynesia, including Bora Bora, are smaller examples of high islands (Figure 14.13). Indeed, high islands are widely scattered throughout Micronesia and Polynesia. In tropical latitudes, most high islands are ringed by coral reefs, which quickly establish themselves in the shallow waters near the shore.

High islands have a limited lifespan in geological terms. Gradually the processes of volcanism that created them diminish, the islands subside, and erosion steadily wears them away. After a few hundred thousand years, only a few low peaks may rise out of a shallow lagoon surrounded by coral reefs. Indeed, nineteenth-century scientist Charles Darwin selected the island of Bora Bora to exemplify the geological evolution of these island settings (Figure 14.14). Eventually even remnant peaks erode away, leaving only a coral reef surrounding a shallow lagoon. Reefs tend to persist, however,

▶ **Figure 14.13 Bora Bora** Jewel of French Polynesia, Bora Bora displays many of the classic features of Pacific high islands. As the island's central volcanic core retreats, surrounding coral reefs produce a mix of wave-washed sandy shores and shallow lagoons. *(Paul Chesley/Tony Stone Images)*

because they are composed of living organisms that create new coral even as the (former) island base continues to subside. The top of the reef, therefore, tends to maintain its position at or just below sea level. **Low islands** are formed as large waves periodically break off and pulverize large pieces of coral, which are then deposited on adjacent sections of the reef to form narrow, wave-washed, and sandy islands.

The combination of narrow sandy islands, barrier coral reefs, and shallow central lagoons is also known as an **atoll** (Figure 14.14). The islands and reefs of the atoll characteristically form a circular or oval shape, although some are quite irregular. The world's largest atoll, Kwajalein in Micronesia's Marshall Islands, is 75 miles (120 kilometers) long and 15 miles (24 kilometers) wide. Polynesia and Micronesia are dotted with extensive atoll systems, and a number are found in Melanesia as well. Some extensive archipelagos, such as the Marshall Islands in Micronesia and the Tuamotus in Polynesia, are composed entirely of atolls.

Patterns of Climate Many Pacific islands receive abundant precipitation, and high islands in particular are often noted for their heavy rainfall and dense tropical forests. Much of the zone is located in the rainy tropics or in a tropical wet/dry climate region where abundant summer rains and even tropical cyclones can bring heavy seasonal precipitation. Apia, Samoa, for example, has a dry fall and winter (April–August), but the town receives almost 18 inches (46 centimeters) of precipitation in the rainy summer month of January. In Melanesia's Papua New Guinea, Port Moresby is distinctly drier but still averages almost 50 inches (127 centimeters) of rain per year, most of it coming in the Southern Hemisphere summer (December–March). In the nearby highlands, however, conditions are often wetter, and local microclimates provide varied precipitation patterns. Frequent snows even visit the higher 13,000-foot (3,980-meter) peaks. Some oceanic locales, however, experience significantly less precipitation. Low-lying atolls usually receive less precipitation than high islands and very often experience water shortages. This is partly because they have limited water-storage capacity, where a small "lens" of fresh water often "floats" above the salt water in the center of each sandy island. In dry periods, however, such stores are quickly depleted.

Environments at Risk

Even with relatively modest populations, many settings in Australia and Oceania face significant human-induced environmental problems. Some challenges are simply the outcome of natural events that increasingly impact larger and more widely distributed human populations. For instance, Pacific Rim-related seismic hazards, periodic Australian droughts, and unwelcome tropical cyclones now pose greater threats than they once did as new settlements, and developments have populated settings vulnerable to these problems. Other environmental issues, however, are even more directly related to human causes (Figure 14.15). Specifically, European colonization introduced many environmental threats, and recent economic globalization has further pressured the region's natural resource base.

Exotic Plants and Animals Exotic (non-native) species pose a major problem in many settings because they often threaten the viability of indigenous life forms. In Australia, some native marsupials could not compete with non-native animals. On the other hand, exotic rabbits successfully multiplied in an environment that lacked the diseases and predators that elsewhere kept their numbers in check. Before long, rabbit populations reached plague proportions, and

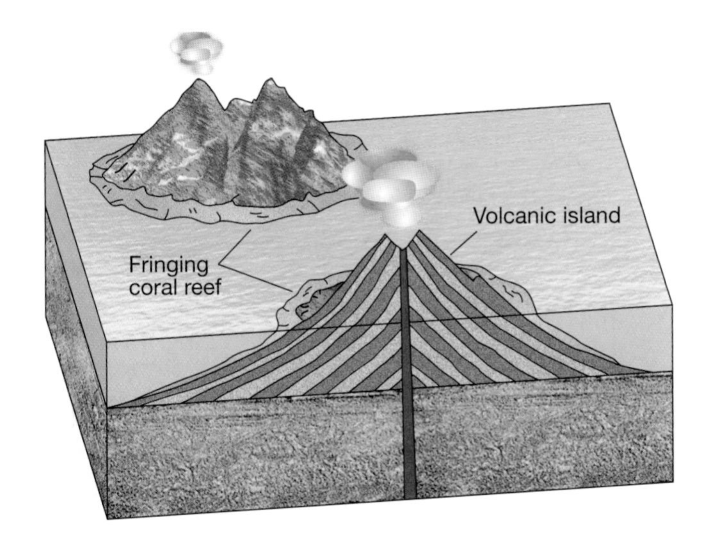

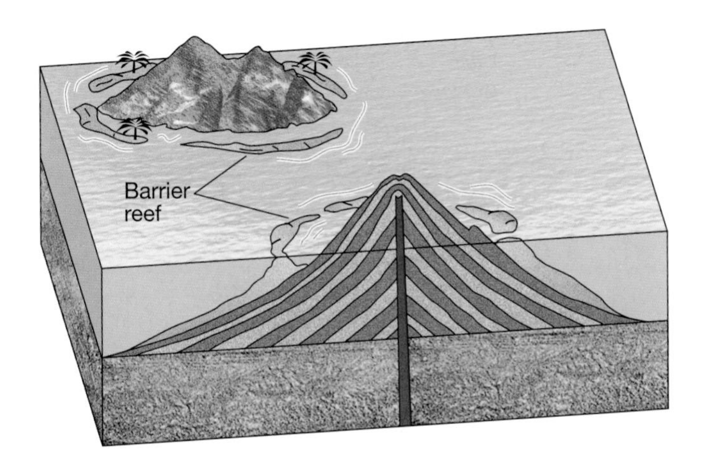

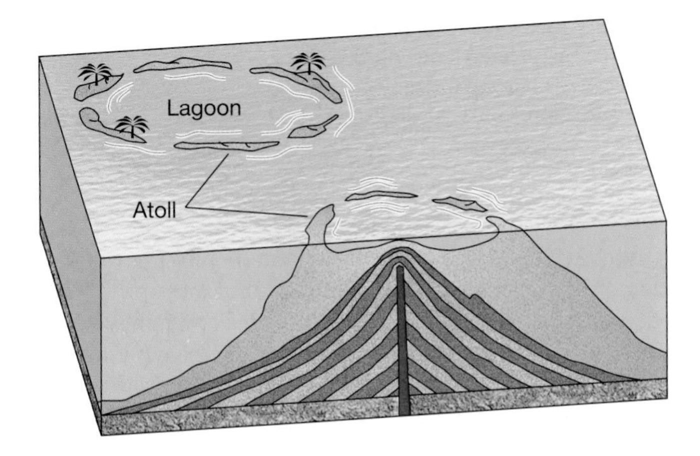

▲ **Figure 14.14 Evolution of an atoll** Many Pacific islands begin as rugged volcanoes (top). As the central core of the island subsides and erodes away, a fringing barrier reef of living coral grows along its edge (middle). Finally, all that remains is a coral atoll surrounding a shallow lagoon (bottom). *(Modified from Tarbuck and Lutgens, 1999,* Earth, *Upper Saddle River, NJ: Prentice Hall)*

large sections of land were virtually stripped of vegetation. The animal was brought under control only after the purposeful introduction of the rabbit disease myxomatosis. Introduced sheep and cattle populations have also stressed the region's environment by accelerating soil erosion and contributing to desertification. In addition, feral livestock have spread. In various parts of Australia, one can find large numbers of once-domestic but now wild goats, horses, pigs, cattle, sheep, water buffalo, and even camels. Exotic plants have caused similar problems. The North American prickly pear cactus, for example, has invaded large sections of the arid Australian range, reducing the stock-raising potential of many areas. Elsewhere, expanding agricultural operations and their associated irrigation projects have contributed to salinization and soil degradation.

Island settings have also witnessed the dramatic consequences of plant and animal invasion. For example, many small islands possessed no native land mammals, and both their native bird and plant species proved vulnerable to the ravages of introduced rats, pigs, and other animals. Polynesia has suffered a particularly large number of wildlife extinctions, especially of birds. The larger islands of the region, such as those of New Zealand, originally supported several species of large, flightless birds that filled some of the ecological niches held by mammals on the continents. The largest of these, the moas, were substantially larger than ostriches. In the first wave of human settlement in New Zealand some 1,500 years ago, moa numbers fell rapidly as they were hunted, their habitat burned, and their eggs consumed by invading rat populations. By 1800, the moas were completely exterminated.

A second wave of extinctions occurred in Polynesia and other portions of Oceania soon after the arrival of Europeans. As in Australia, introduced plants and animals have often competed successfully with native species and threatened to eliminate very valuable species that occur nowhere else on the planet. In particular, the Hawaiian Islands suffered devastating extinctions following the arrival of European and American influences. Today, many species of native Hawaiian plants and birds are severely threatened and survive only in diminishing refuges. Introduced plant and animal species, especially wild hogs, seem to be the main threat, along with the overwhelming pressures of economic development that have transformed much of the Hawaiian landscape. On New Zealand's South Island, imported red deer and elk have thrived and become a nuisance. Elsewhere, thanks to human introductions, pigs roam the Solomon Islands, wild horses graze in the Marquesas, and goats nibble away at Tahiti's vegetation.

Global Resource Pressures Processes of globalization have also exacted an environmental toll upon Australia and Oceania. Specifically, the region's considerable base of natural resources have been opened to development, much of it by outside interests. While gaining from the benefits of global investment, the region has also paid a considerable price for

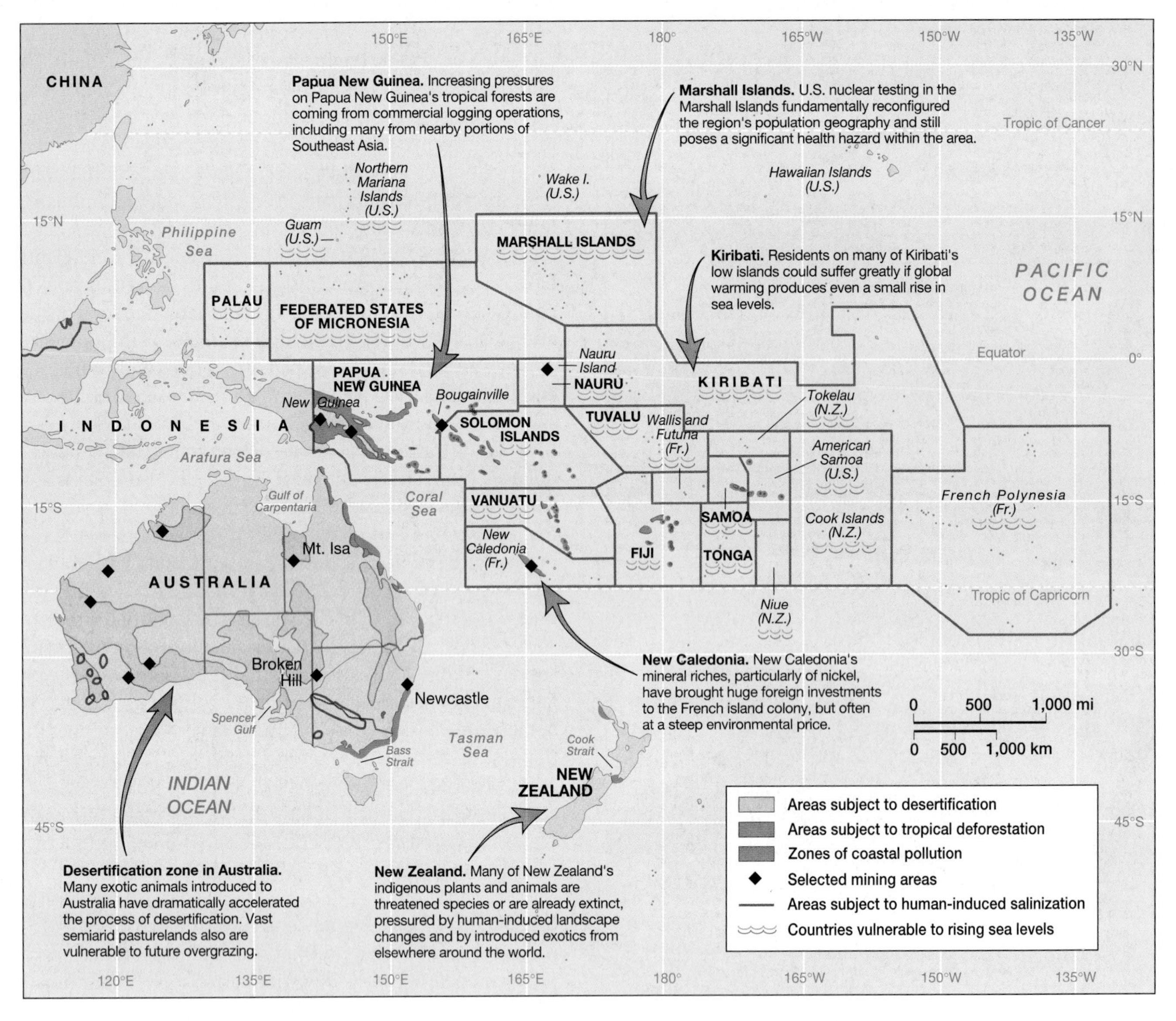

▲ **Figure 14.15 Environmental issues in Australia and Oceania** Modern environmental problems belie the region's myth as an earthly paradise. Tropical deforestation, extensive mining, and a long record of nuclear testing by colonial powers have brought varied challenges to the region. Human settlements have also extensively modified the pattern of natural vegetation. Future environmental threats loom on low-lying Pacific Islands if global sea levels rise.

encouraging development, and the result is an increasingly threatened environment. Major mining operations have profoundly impacted Australia, Papua New Guinea, New Caledonia, and Nauru. Some of Australia's largest gold, silver, copper, and lead mines are located in sparsely settled portions of Queensland and New South Wales, but watersheds in these semiarid regions are highly susceptible to metals pollution. In Western Australia, huge open-pit iron mines dot the landscape, unearthing ore that is often bound for global markets, particularly Japan. To the north, Papua New Guinea's Bougainville copper mine has transformed the Solomon Islands, while even larger gold mining ventures have raised increasing environmental concerns on the island of New Guinea. Elsewhere, Micronesia's tiny Nauru has been virtually turned inside out as much of the island's jungle cover was removed to get at some of the world's richest phosphate deposits. Former Australian and New Zealand mine owners have already paid millions of dollars to settle environmental damage claims.

Deforestation is another major environmental threat across the region. Vast stretches of Australia's eucalyptus woodlands, for example, have been burned or killed to produce better grazing pastures. In addition, coastal rain forests in Queensland are only a fraction of their original area, although a growing environmental movement in the region is fighting to save remaining forest tracts. Tasmania has also been an environmental battleground, particularly given the diversity of its midlatitude forest landscapes. While the island's earlier Eu-

ENVIRONMENT Saving the Samoan Rain Forest

Ethnobotanist Paul Alan Cox has spent years in the Samoan rain forest (Figure 1). His encounter with this remote Polynesian setting began in 1973 as a young Mormon missionary. He became fascinated with how native peoples utilized rainforest plants to treat disease. Eleven years later, he returned as a trained botanist, hoping to decipher the secrets held by native female healers as they worked among the local population. In his travels to remote villages far beyond the reach of Western-style cures, Cox discovered that these Samoan "medicine women" possessed an intimate knowledge of how dozens of plant species could be used to treat human ailments. Passed down from mother to daughter, this folk medicine included using the polo leaf for fighting infections, the ulu ma'afala root for diarrhea, and the bark of a local tree to combat hepatitis. Healers introduced Cox to 70 of these remedies, including one that may be effective against AIDS. Indeed, these are not magical native cures: 25 percent of prescription drugs made in the United States are extracted from flowering plants, many of which grow in tropical forest settings.

The Samoan treasure trove of botanical riches is currently under assault, however. Already more than 80 percent of Samoa's coastal rain forest has been logged, much of it just in the past 25 years. Cox sees this as a tragic waste, both environmentally and economically. He points out that research institutes and drug companies can pay handsome royalties to people who assist in the search for health remedies. In other words, the forests are worth much more standing than they are cut down. Awarded the Goldman Environmental Prize in 1997 for his forest-saving efforts, Cox fears that the island's many sawmills may complete their work before he and other researchers can even catalogue the area's vanishing riches. Such environmental races with time are commonplace elsewhere in Oceania and across the broader tropical world. No wonder the islanders have given Cox the name Nafanuna, one of their traditional protective Gods.

Source: Adapted from "Rainforest Pharmacist," *Audubon*, January 1999.

◀ **Figure 1 Samoan rain forest** Logging threatens Samoa's lush and diverse tropical vegetation. In addition to posing increased local environmental hazards, these operations may eliminate important medicinal plants useful for treating many diseases. *(Jean-Marc La Roque/Auscape International Pty. Ltd.)*

ropean and Australian development featured many lumbering and pulp mill operations, more than 20 percent of the island is now protected by national parks. The nation's Green Party is pushing hard for further preservation on the island. Deforestation is also a major problem in many parts of Oceania. With limited land areas, smaller islands are subject to rapid tree loss, which in turn often leads to soil erosion. Sizable expanses of rain forest are still to be found in most of the larger islands of Melanesia. Although rain forests still cover 70 percent of Papua New Guinea, more than 37 million acres (15 million hectares) have been identified as suitable for logging (Figure 14.16). Some of the world's most biologically diverse environments are being threatened in these operations, but landowners see the quick cash sales to loggers as attractive, even though the practice is a nonsustainable alternative to their traditional lifestyles. The global expansion of commercial lumbering also threatens nearby portions of the Solomon Islands as well as many unique environmental settings in Polynesia (see "Environment: Saving the Samoan Rain Forest"). Increasingly, residents question unbridled development: recently Solomon Islanders resisted a government resettlement plan that threw them off their land so that a Malaysian lumber company could clear the area of trees. Logging proceeded, despite continued protests.

▲ **Figure 14.16 Logging in New Guinea** Foreign logging companies have made large investments in tropical Pacific settings such as Papua New Guinea. While bringing new jobs, these ventures dramatically alter local environments, as precious hardwood forests are harvested for export. *(David Austen/ Woodfin Camp & Associates)*

Persisting Regional Hazards The region's peripheral economic and political status have often been costly. When nations such as the United States and France required atomic testing grounds for their nuclear weapons programs, the South Pacific was chosen as an ideal location. Local residents had little voice in such matters, although French tests in the 1990s provoked more widespread and sustained negative responses from the regional populations. The environmental consequences of such activities have been long lasting. In 1968 President Johnson confidently told former residents of Bikini and Rongelap in the Marshall Islands that they could return to their bomb-transfigured homelands. The area had been used for 67 nuclear weapons explosions between 1946 and 1958. Tragically, the natives returned, rebuilt their houses, and resettled their villages, only to find that radioactive cesium 137 was still concentrated in the soil and soon entered the food chain. Residents were later re-evacuated to other islands, but the health and environmental costs in such settings have been monumental.

Oceania's greatest future environmental threat may be global warming. If Earth grows significantly warmer in the near future, as many climate models predict, the impact on many low-lying atolls will be devastating. Higher global temperatures will partially melt polar ice caps, and resulting rises in sea levels could literally drown the region's low islands. Countries such as Kiribati and the Marshall Islands could simply disappear, forcing residents to flee elsewhere, probably to other overcrowded islands. Not surprisingly, the Pacific states have been major supporters of global conventions to limit the production of greenhouse gases. Nauru's leaders recently reported that their low island is seeing increased coastal erosion and that storm surges, elevated sea levels, and unusual drought conditions have damaged the island's supply of fresh water. Island residents, already displaced to the coast by inland mining operations, are being squeezed out of their homes from all directions. Unfortunately, these tiny island nations have little international clout.

Population and Settlement: A Diverse Cultural Landscape

Modern population patterns across the region reflect the combined influences of indigenous and European settlement. In settings such as Australia, New Zealand, and the Hawaiian Islands, European migrations have structured the distribu-

Table 14.1 Demographic Indicators

Country	Population[a]	Natural Increase	TFR[b]	%<15[c]	% Urban
Australia	18.7	0.7	1.8	21	85
Fed St. Micronesia	0.1	2.6	4.7	44	27
Fiji	0.8	1.8	2.8	38	46
French Polynesia	0.2	1.8	3.1	36	54
Guam	0.2	2.4	3.4	30	38
Marshall Islands	0.1	3.6	6.7	51	65
New Caledonia	0.2	1.7	2.8	31	71
New Zealand	3.8	0.8	2.0	23	85
Palau	0.02	1.6	2.5	30	69
Papua New Guinea	4.3	2.4	4.8	40	15
Samoa	0.2	2.4	4.2	41	21
Solomon Islands	0.4	3.2	5.4	47	13
Vanuatu	0.2	2.8	4.7	46	18

[a]Population in millions, 1998.
[b]Total fertility rate.
[c]Percentage of population younger than 15 years of age.
Source: *Population Reference Bureau World Population Data Sheet,* 1998.

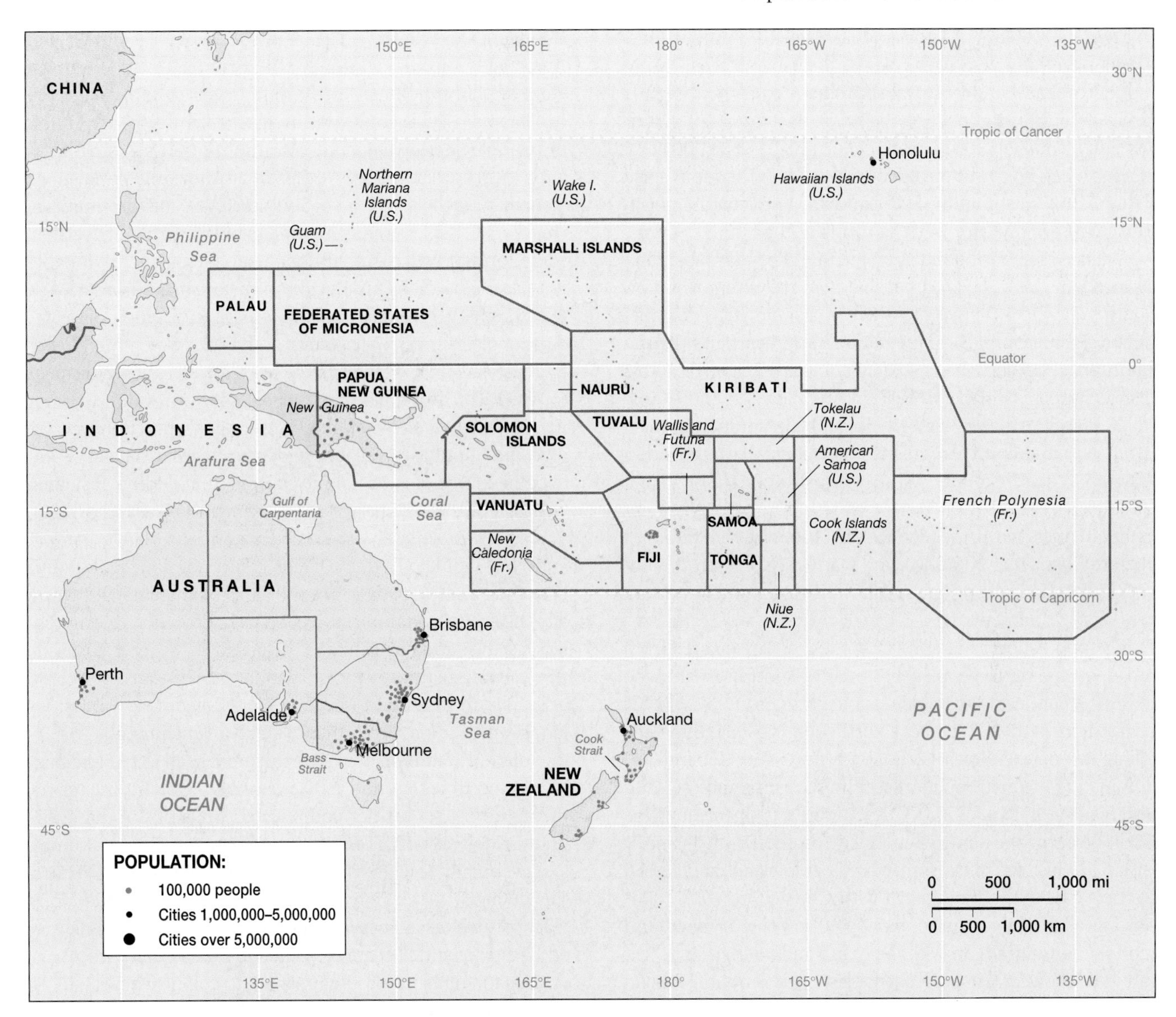

▲ **Figure 14.17 Population map of Australia and Oceania** Fewer than 30 million people occupy this world region. While Papua New Guinea and many Pacific islands feature mostly rural settlements, most regional residents live in the large urban areas of Australia and New Zealand. Sydney and Melbourne account for almost half of Australia's population, and most New Zealand residents live on the North Island, home to the cities of Auckland and Wellington.

tion and concentration of contemporary populations, while island settings elsewhere in Oceania display population geographies essentially determined by the movement and adaptations of native peoples. The cultural landscape also reflects these diverse influences: more urban and commercial agricultural settings are largely transplanted from Europe and North America, but traditional societies still leave their mark upon the scene in many localities, ranging from the New Guinea Highlands to Polynesia's Society Islands.

Contemporary Population Patterns

Despite the popular stereotypes of life in the outback, modern Australia is one of the most highly urbanized populations in the world (Table 14.1). Indeed, 40 percent of the country's residents live within either the Sydney or Melbourne metropolitan areas (Figure 14.17). Australia's eastern and southern rimland is home to the overwhelming majority of its almost 19 million people. Most residents of Queensland, for example, live along the well-watered, amenity-rich coast, culminating in the state's capital city of Brisbane (1.5 million). Inland, population densities decline as rapidly as the rainfall: semiarid hills west of the Great Dividing Range still contain significant rural settlement, but the state's southwestern periphery remains sparsely peopled. New South Wales is the country's most populous state, and its sprawling capital city of Sydney (3.9 million), focused around one of the world's most magnificent natural harbors, is the largest metropolitan area in the entire South Pacific.

In the nearby state of Victoria, Melbourne (3.3 million) residents have long vied with Sydney for urban status, claiming cultural and architectural supremacy over their slightly larger neighbor. In between these two metropolitan giants, the much smaller federal capital of Canberra (310,000) represents a classic geopolitical compromise in the same spirit that created Washington, D.C., midway between the populous southern and northern portions of the United States.

Smaller clusters of population are found beyond Australia's eastern rim. More favored farmlands in the interior of New South Wales and Victoria feature higher population densities than are found in most of the nation's arid heartland. Inland Aboriginal populations are widely but thinly scattered across districts such as northern Western Australia and South Australia, as well as in the Northern Territory. In addition to these rural settlements, urban clusters are focused around other Australian state capitals. Western Australia's growing metropolis of Perth (1.3 million) is now the largest of these peripheral cities, with Adelaide (South Australia, 1.1 million), Hobart (Tasmania, 200,000), and Darwin (Northern Territory, 85,000) accounting for smaller but regionally important centers of settlement.

The population geography of the rest of Oceania reflects a broad sprinkling of peoples, both native and European, who have clustered near favorable resource opportunities. In New Zealand, more than 70 percent of the country's 3.8 million residents live on the North Island, with Auckland (1.1 million) dominating the metropolitan scene in the north and the capital city of Wellington (350,000) anchoring settlement along the Cook Strait in the south. Settlement on the South Island is strikingly clustered in the somewhat drier lowlands and coastal districts east of the mountains, with Christchurch (337,000) serving as the largest urban center. Elsewhere, rugged and mountainous terrain on both the North and South islands feature much lower population densities. Such is not the case in Papua New Guinea: less than 20 percent of the country's population is urban and many people live in the isolated, interior highlands. The nation's largest city is the capital of Port Moresby (200,000), located along the narrow coastal lowland in the far southeastern corner of the country. The largest city on the northern periphery of Oceania is Honolulu (1 million) on the island of Oahu, where rapid metropolitan growth since World War II has occurred because of U.S. statehood and the scenic attractions of its mid-Pacific setting.

Legacies of Human Occupance

The historical settlement of the Pacific realm can never be precisely reconstructed, but several major human migrations succeeded in occupying the region over time. The region's peripheral position relative to many of the world's early population centers meant that it often lay beyond the dominant migratory paths of earlier peoples. Even so, settlers found their way to the isolated Australian interior and to the far reaches of the Pacific. The pace of new in-migrations accelerated greatly once Europeans identified the region and its resource potential.

Peopling the Pacific The large islands of New Guinea and Australia, given their closer proximity to the Asian landmass, were settled much earlier than the more distant islands of the Pacific, which could not be reached until the invention of sophisticated water craft. Around 40,000 years ago, the ancestors of today's native Australian or **Aborigine** populations were making their way out of Southeast Asia and into Australia (Figure 14.18). The first Australians most likely arrived via some kind of watercraft, although such boats were probably not very seaworthy, as the more distant islands remained inaccessible to humankind for tens of thousands of years. During glacial periods, however, sea levels were much lower than they are now, which would have allowed easier movement to Australia across relatively narrow spans of water. It is not known whether these original Australians came in one wave or in many, but the available evidence suggests that they soon occupied large portions of the continent, including Tasmania (which was then connected to the mainland by a land bridge).

Eastern Melanesia was settled much later than Australia and New Guinea. Distant islands could not be reached until better sail craft were developed. By approximately 3,500 years ago, however, certain Pacific peoples had mastered long-distance sailing and navigation, which eventually opened the entire oceanic realm to human habitation. In that era people gradually moved east to occupy New Caledonia, the Fiji Islands, and Samoa. From there, later movements took seafaring folk north into Micronesia, with areas such as the Marshall Islands occupied around 2,000 years ago. Continuing movements from Asia further complicated the story of these migrating Melanesians. Some of these migrants mixed culturally and eventually reached western Polynesia, where they formed the nucleus of the Polynesian people. By A.D. 1000, they reached such distant oceanic outposts as New Zealand, Hawaii, and Easter Island. Debate has centered on whether these Polynesians migrants purposefully set out to colonize new lands or whether they were blown off course during more routine voyages, only to end up later on new islands. Certainly, population pressures could quickly reach a crisis stage on relatively small islands, encouraging people to make spectacularly dangerous voyages. Equipped with sturdy outrigger sailing vessels and ample supplies of food, the Polynesians were quickly able to colonize most of the islands they discovered.

European Colonization Only about six centuries after the Maori brought New Zealand into the Polynesian realm, Dutch navigator Abel Tasman spotted the islands on his global reconnaissance of 1642. Tasman's initial sighting marked the beginning of a new chapter in the human occupation of the South Pacific. Late in the following century, more lasting European contacts cemented ties to the region. British sea captain James Cook surveyed the shorelines of both New Zealand and Australia between 1768 and 1780. Cook and others were convinced that these distant lands might be worthy of European development. In addition, other expeditions were probing the Pacific, and most of Oceania's major island groups were assuming their proper place on European maps by the end of the eighteenth century.

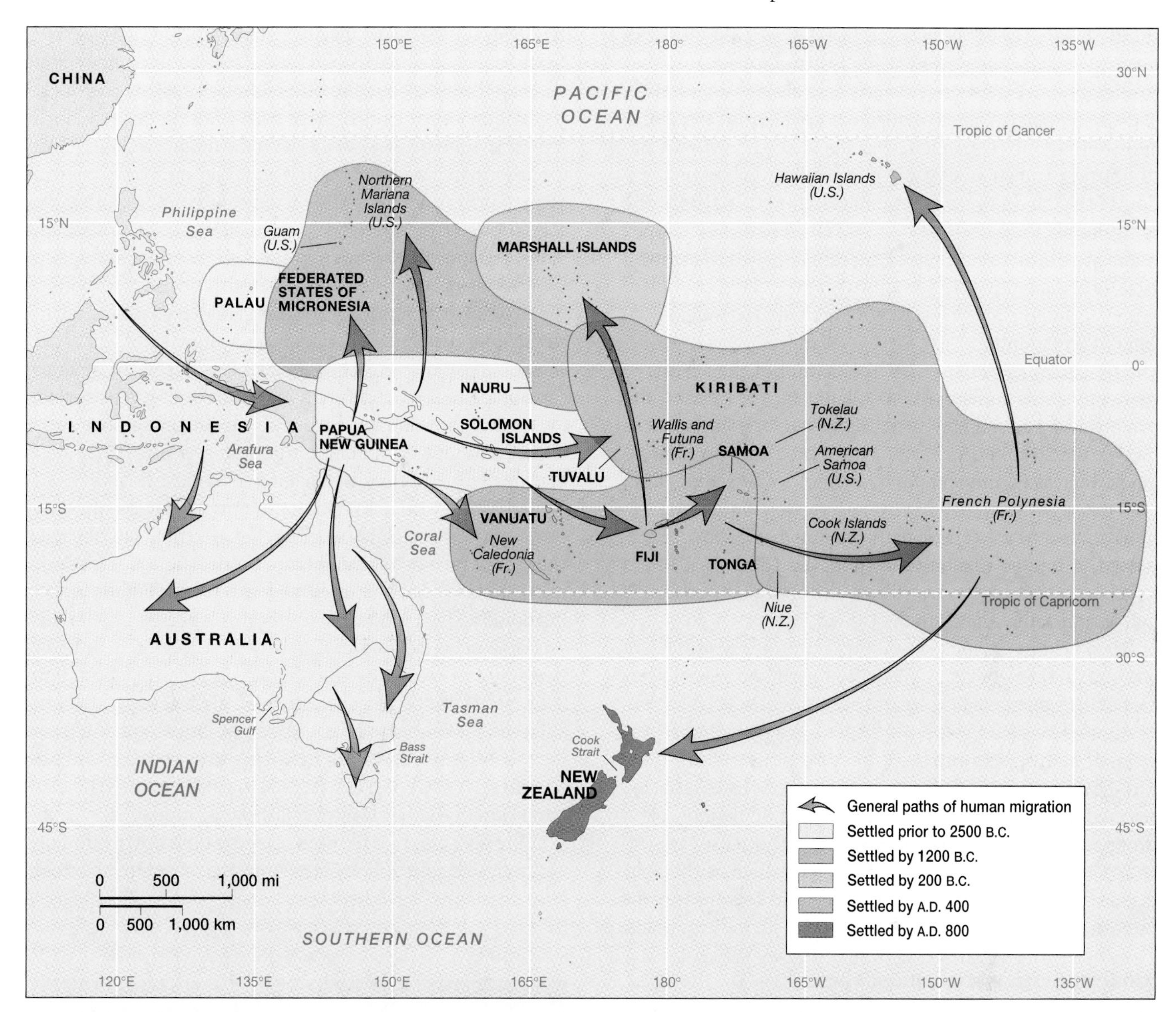

▲ **Figure 14.18 Peopling the Pacific** Ancestors of Australia's aboriginal population may have made their way into the island continent more than 40,000 years ago. Much more recent settlement of Pacific islands by Austronesian peoples from Southeast Asia shaped cultural patterns across the oceanic portions of the realm. Eastward migrations through the Solomon, Fiji, and Cook islands were followed by late movements to the north and south.

Actual European colonization of the region began in Australia. The British needed a new penal colony where convicts could be exiled. The southeastern coast of Australia was selected as an appropriate site, and in 1788 the First Fleet arrived in Botany Bay near modern Sydney with 750 prisoners. Other fleets and more convicts soon followed. Before long, however, free settlers outnumbered the convicts, who were themselves gradually gaining their freedom. The growing population of English-speaking people soon moved inland and settled other favorable coastal locales. British and Irish settlers were attracted by the agricultural and stock-raising potential of the distant colony and by the lure of gold and other minerals (a major gold rush occurred in the 1850s). The British government also encouraged the emigration of its own citizens, often paying the transportation fare of those too poor to afford it themselves.

The new settlers came into conflict with the Aborigines almost immediately after arriving. No treaties were signed, however, and, in most cases, Aborigines were simply expelled from their lands. In some places, most notably Tasmania, they were hunted down and killed. In mainland Australia, the Aborigines were greatly reduced in numbers by disease, dispossession, and exploitation, but they survived in substantial numbers. By the mid-nineteenth century, Australia was primarily an English-speaking land.

British settlers were also attracted to the lush and fertile lands of New Zealand. Europeans whalers and sealers arrived shortly before 1800, but more permanent agricultural settlement took shape after 1840 as the British formally declared sovereignty over the region. As new arrivals grew in numbers and the scope of planned settlement colonies on the North and South islands expanded, tensions increased with the native Maori population. Organized in small kingdoms or chiefdoms, the Maori were formidable fighters. In 1845 one indigenous group decided to resist further encroachments, leading to the more generalized "Maori Wars," which engulfed New Zealand until 1870. The British eventually prevailed, however, and the Maori lost most of their land as well as control of their country.

The native Hawaiians also lost control of their lands to more numerous immigrants. Hawaii emerged as a united and powerful kingdom in the early 1800s, and for many years its native rulers limited U.S. and European claims to their islands. Increasing numbers of missionaries and settlers from the United States were allowed in, however, and by the late nineteenth century, control of the Hawaiian economy largely passed to foreign plantation owners. By 1898, U.S. forces were strong enough to overthrow the Hawaiian monarchy and to annex the islands to the United States.

Elsewhere in Oceania, actual European or U.S. settlement was much less significant. While European and U.S. powers eventually gained political control over the entire area, few opportunities awaited prospective settlers. Whaling, timber harvesting, and opportunities for plantation agriculture drew Europeans to some islands, while mineral resources attracted selected colonization to others. Native populations often dropped as European diseases arrived, but in most cases recovery eventually occurred. Some islands, such as the Marquesas in Polynesia, however, have yet to rebound to the population levels they had prior to contact with the Europeans.

Modern Settlement Landscapes

The settlement geography of Australia and Oceania offers a fascinating juxtaposition of local and exotic influences. The contemporary cultural landscape still reflects the imprint of indigenous peoples in many settings where native populations have remained numerically dominant. Elsewhere, patterns of recent colonization have produced a modern scene profoundly shaped by processes of Europeanization. The result includes everything from German-owned vineyards in South Australia to houses on New Zealand's South Island that appear to be plucked directly from the British Isles. Further, processes of economic and cultural globalization have reshaped the settlement landscape, particularly in the twentieth century, resulting in urban forms that make cities such as Perth or Auckland look strikingly similar to places such as San Diego or Seattle.

The Urban Transformation Both Australia and New Zealand are highly urbanized, Westernized societies, and thus the vast majority of their populations live in urban and suburban environments. As in Europe and North America, much of this urban transformation came during the twentieth century as the rural economy became less labor-intensive and as opportunities for urban manufacturing and service employment grew. As urban landscapes evolved, they took on many of the characteristics of their largely European populations but blended these with a strong dose of North American influences as well as with the unique settings native to each urban place. The result is an urban landscape in which many North Americans, for example, are quite comfortable, even though the varied local accents heard on the street and many features of the metropolitan scene are reminders of the strong and enduring attachments to British traditions.

All of the major cities of Australia and New Zealand are focused around vibrant and dynamic downtown areas that often compare favorably with those of metropolitan North America. Coastal features, port settings, and waterfront districts shape the central city landscapes of every major urban area in the region and give each center a unique identity. Perth, Melbourne, Sydney, Brisbane, Wellington, and Auckland are anchored on the sea, a function of their important commercial connections to the global economy. Thus, central city skylines, complete with growing numbers of North American-style high-rises, are never far from their harbor settings (Figure 14.19). In addition, these urban places are often lauded for their modest crime rates, lack of slums, relatively clean streets, spacious parks and open spaces, and efficient public transportation. Sydney and Melbourne, in particular, are receiving millennial facelifts. Many new hotels, sports facilities, and entertainment complexes were built in Sydney in association with the Summer 2000 Olympics, and Melbourne's ambitious Federation Square Redevelopment Project is dramatically transforming 7.9 acres (3.2 hectares) of aging downtown buildings and railyards into a new entertainment, museum, and business district.

▲ **Figure 14.19 Downtown Melbourne** Metropolitan Melbourne lies along the Yarra River. Capital of the Australian state of Victoria, Melbourne resembles many growing North American cities with its high-rise office buildings, entertainment districts, and downtown urban redevelopment. *(Fritz Prenzel/ Peter Arnold, Inc.)*

▲ **Figure 14.20 Suburban Auckland** New Zealand's largest city increasingly sprawls miles from downtown. Excellent transportation links and preferences for suburban lifestyles have contributed to the city's increasingly North American appearance. *(Mike Yamashita/Woodfin Camp & Associates)*

▲ **Figure 14.21 Port Moresby, Papua New Guinea** Urban poverty and high crime haunt the city of Port Moresby, the capital of Papua New Guinea. The city's slums, many built out upon the water, reflect stresses of recent urban growth as rural residents emigrate from nearby highlands. *(Chris Rainier/Corbis)*

Suburbanization and peripheral commercial development are also changing these large cities of the Southern Hemisphere. For example, Sydney's metropolitan reach now extends west for more than 20 miles from the downtown area, and Melbourne's urban grasp is expanding inland toward the hills as well as further east along Port Phillip Bay. In addition, amenity-rich Perth (Western Australia) and Brisbane (Queensland) have seen some of the most rapid recent population gains, particularly in their expanding suburban fringe. Similarly, New Zealand's Auckland, given that city's water-bound limits between Manukau and Waitemata Harbors, is spreading well beyond the old boundaries of the traditional city (Figure 14.20). Land uses and settlement landscapes in these dynamic urban peripheries parallel patterns across North America: low-density residential areas are linked by modern commercial highways and punctuated by "edge city" collections of office centers, industrial parks, and entertainment complexes.

The affluent Western-style urban settings in Australia and New Zealand offer a stunning contrast with the urban landscapes found in less-developed settings elsewhere in the region. Walk the streets of Port Moresby in Papua New Guinea, and a very different urban landscape attests to the yawning gap between rich and poor within the South Pacific (Figure 14.21). Rapid growth in Port Moresby, the country's political capital and largest commercial center, has produced many of the classic challenges of urban underdevelopment: there is a shortage of adequate housing, the creation of basic infrastructure runs far behind the need, and street crime and alcoholism are on the rise. Elsewhere, urban centers such as Suva (Fiji), Noumea (New Caledonia), and Apia (Samoa) also reflect the economic and cultural tensions generated as Western influences are juxtaposed with indigenous populations. Rapid growth is a common problem in these smaller urban Pacific settings because many native peoples from neighboring rural areas and nearby islands gravitate toward the job opportunities available in these areas. In the past 50 years, the spectacular global growth of tourism in places such as Fiji and Samoa has also transformed the urban scene: nineteenth-century village life has often been replaced by a landscape dominated by souvenir shops, honking taxicabs, and crowded seaside resorts.

The Rural Scene Rural landscapes across Australia and the Pacific region also express a complex mosaic of cultural and economic influences. In some settings, Australian Aborigines or native Papua New Guinea Highlanders can still be found in their familiar homelands, their traditional life ways and settlements barely changed from pre-European times. Yet such settlement landscapes are becoming increasingly rare. Global influences penetrate the scene as the cash economy, foreign tourism and investment, and the accoutrements of popular culture work their way into the hinterlands. That still leaves relatively large areas of rural settlement across many areas of the Australian and New Zealand interiors, as well as smaller examples sprinkled among the islands of the Pacific. Even here, however, land uses are often shaped by the regional or global commercial agricultural economy or by the recreational values associated with such settings.

Much of rural Australia is too dry for farming or it serves as only marginally valuable agricultural land. Although modest in size, the area in crops actually has doubled since 1960 as increased use of fertilizers, more widespread irrigation, and more aggressive rabbit eradication efforts have opened up new areas for development. Much of the remainder of the interior, however, features range-fed livestock, areas beyond the pale of any agricultural potential, and isolated zones where remaining Aboriginal inhabitants still pursue their traditional forms of hunting and gathering (Figure 14.22).

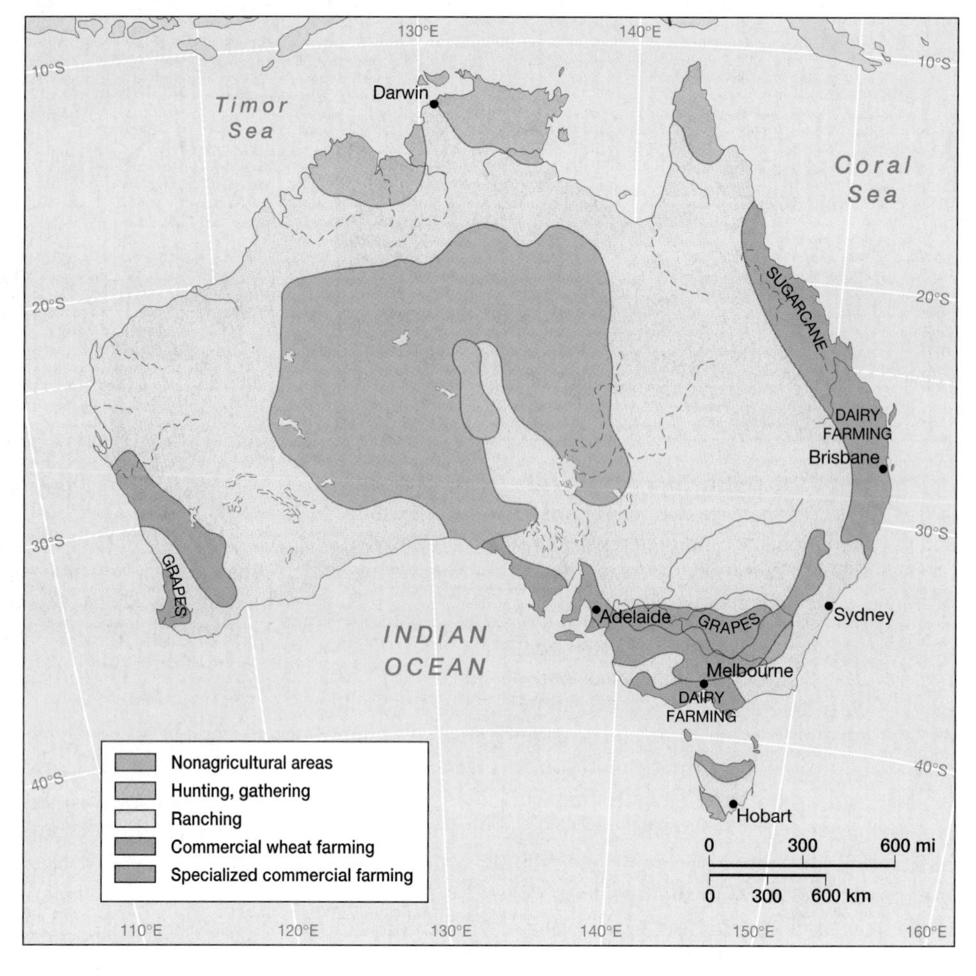

▶ **Figure 14.22 Australian agriculture** Australia's best farmlands are found along its eastern and southern rim where specialty crops such as sugarcane and grapes thrive in selected locations. Commercial wheat farming remains important in slightly drier districts. Much of the interior is too arid for agriculture or limited to extensive commercial livestock ranching. *(Modified from Clawson and Fisher, 1998,* World Regional Geography, *Upper Saddle River, NJ: Prentice Hall)*

Sheep and cattle dominate the livestock economy. Many rural landscapes in the interior of New South Wales, Western Australia, and Victoria, for example, are oriented around isolated sheep stations, or ranch operations that move the flocks from one vast pasture to the next. Cattle can sometimes be found in these same areas, although many of the more extensive, range-fed cattle operations are concentrated farther north in Queensland. Cattle are even grazed on the tough tropical grasslands and woodlands of northern Australia, although environmental conditions in these settings are often marginal for any agricultural activity. More specialized dairy cattle operations, often on much smaller family-owned farms, are found in moister settings from the Queensland coast to Victoria.

Croplands also vary across the region. Sometimes mingling with the sheep country, a band of commercial wheat farming includes southern Queensland; the moister interiors of New South Wales, Victoria, and South Australia; as well as a swath of more favored land east and north of Perth. Elsewhere, specialized sugarcane operations thrive along the narrow, warm, and humid coastal strip of Queensland. To the south and west, productive irrigated agriculture has developed in localities such as the Murray River Basin, allowing for the production of orchard crops and vegetables. **Viticulture,** or grape cultivation (and other specialized horticulture), increasingly shapes the rural scene in places such as South Australia's Barossa Valley, New South Wales' Riverina district, and Western Australia's Swan Valley. Indeed, the area under grape cultivation grew by 50 percent between 1991 and 1998 as the popular Chardonnay, Cabernet Sauvignon, and Shiraz varieties propelled wine production to revenues of more than $540 million per year.

Although much smaller in area, New Zealand's rural settlement landscape includes a variety of agricultural activities. Pastoral pursuits clearly dominate the New Zealand scene, with the vast majority of agricultural land devoted to livestock production, particularly sheep grazing and dairying. Commercial livestock outnumber people in New Zealand by a ratio of more than 20:1, and this is apparent everywhere on the rural scene. Sheep gnaw on largely imported European grasses from the isolated hinterlands of the South Island to the volcanic slopes and shorelines of the North Island. In addition, dairy operations dot the rural scene, mostly in the humid lowlands of the subtropical north, where they sometimes mingle with suburban landscapes in the vicinity of Auckland. One of the largest zones of more specialized cropping spreads across the fertile Canterbury Plain near Christchurch (Figure 14.23). This spectacular South Island setting, spotted early by British colonizer Edward Gibbon Wakefield, proved fertile ground for English settlement and continues to feature a varied landscape of pastures, grain fields, orchards, and vegetable gardens, all spread beneath the towering peaks of the Southern Alps.

▲ **Figure 14.23 Canterbury Plain** The varied agricultural landscape of South Island's Canterbury Plain offers a mix of grain fields, livestock, orchard crops, and vegetable gardens. The rugged Southern Alps are a dramatic backdrop to this productive region. *(Robert Frerck/Woodfin Camp & Associates)*

▲ **Figure 14.24 Yam harvest** These farmers on the Melanesian island of Vakuta (Papua New Guinea) are harvesting yams. Traditional tropical agriculture features a mix of crops, often grown in the same field. Where possible, fields are periodically rotated to maintain productivity. *(Peter Essick/Aurora & Quanta Productions)*

Elsewhere in Oceania, varied influences shape the rural scene. On better-watered high islands, denser populations take advantage of more diverse agricultural opportunities than are usually found on the more barren low islands, where fishing is often more important. Several types of rural settlement can be identified across the island realm. In rural New Guinea, village-centered shifting cultivation dominates: farmers clear a patch of forest and then, after a few years, shift to another patch, thus practicing a form of land rotation. Subsistence foods such as sweet potatoes, taro (another starchy root crop), coconut palms, bananas, and other garden crops often are intercropped in the same field, and growing numbers of planters also include commercial crops such as coffee (Figure 14.24). In other parts of Oceania, traditional agricultural patterns are similar: most rural settlements are organized around village-based societies surrounded by nearby fields in which varied crops are produced. Commercial plantation agriculture has also made its mark in many more accessible rural settings. In these localities, settlements consist of worker housing near crops that are typically controlled by absentee landowners. For example, copra (coconut), cocoa, and coffee operations have transformed many agricultural settings in places such as the Solomon Islands and Vanuatu. Sugarcane plantations have reshaped other island settings, particularly in Fiji and Hawaii.

Diverse Demographic Paths

Varied population-related issues face residents of the region today. In Australia and New Zealand, while populations grew rapidly (mostly from natural increases) early in the century, today's low birthrates parallel the pattern in North America. Just as in the United States and Canada, however, significant population shifts within these countries continue to impose challenges. For example, the exodus of farmers from Australia's wheat-growing and sheep-raising interior mirrors similar processes at work in the rural Midwest of the United States or in the Canadian prairies. Communities see many of their productive young people and professionals leave for the better employment opportunities of the city. Older urban and industrial areas, particularly near Sydney and Melbourne, have also lost population. Conversely, other portions of Australia and New Zealand face rapid growth. As in North America, amenity-rich settings, retirement communities, and new upscale suburbs draw ever more people, thus pressuring infrastructure and public services. Some people in these two countries also complain about liberal national immigration policies allowing in too many new residents and workers, although such frustrations are often more rooted in cultural fears than they are in concerns for overall population numbers.

Different demographic challenges grip many of the less-developed island nations of Oceania. Population growth rates remain above 2 percent per year and are substantially higher than that in localities such as Micronesia and the Solomon Islands. While the larger islands of Melanesia contain some room for settlement expansion, competitive pressures from commercial mining and logging operations limit the amount of new agricultural land that will probably be available in the future. On some of the smaller island groups in Micronesia and Polynesia, population growth is an even more pressing issue. Tuvalu (north of Fiji), for example, has fewer than 10,000 inhabitants, but they are crowded onto a land area of only 10 square miles (26 square kilometers), making it one of the world's more thickly populated countries. In atoll environments, a single island might be considered overpopulated even if it supports only 100 people. Thus, even relatively small population centers in the Pacific Islands have little internal flexibility in coping with higher birthrates. Making matters worse, many young people are also migrating to already crowded urban centers in these island nations. Reflecting the region's bizarre entanglements with nuclear testing, Marshall Islanders face the additional daunting task of crowding their country's 60,000 people onto atolls that were lucky enough to escape extensive U.S. bomb blasts. Today, more than half the nation's residents reside on Majuro Atoll, a narrow strip of sand crammed with tin shacks, heaps of nondisposable trash, and diaper-strewn beaches.

Cultural Coherence and Diversity: A Global Crossroads

Australia and Oceania offer superb laboratories to examine how cultural geographies are transformed as different groups migrate to a region, interact with one another, and evolve over time. Many general processes of culture change are exemplified. For example, some portions of the region saw multiple waves of initial *implantation* of indigenous cultural groups. Once in place, the relative isolation of cultural groups often made possible further cultural *differentiation* in which originally similar cultures evolved in different ways. In some cases, when different cultures came in fresh contact with one another, *assimilation* took place, and one culture absorbed important components of the other. As Europeans and other outsiders arrived in the region, colonization compelled cultural *accommodation* as native peoples adjusted to externally imposed forces of change. Finally, worldwide processes of *globalization* have also redefined the region's cultural geography, provoking fears of *homogenization* while at the same time promoting more coordinated cultural *preservation* efforts as threatened groups attempt to selectively protect their cultural heritage. A contemporary snapshot of the region reveals that all of these processes remain at work today and help account for the diverse cultural signatures found in Australia, New Zealand, and the other Pacific islands.

Multicultural Australia

Australia's cultural patterns illustrate many of these fundamental processes at work. Today, while still dominated by its colonial European roots, the country's multicultural character is becoming increasingly visible as native inhabitants assert their cultural identity and as varied immigrant populations play larger roles in society, particularly within major metropolitan areas.

Aboriginal Imprints For thousands of years, Australia's Aborigines dominated the cultural geography of the continent. These indigenous peoples never practiced agriculture, opting instead for a hunting-gathering way of life that persisted up to the time of the European conquest. As foragers and hunters living in a relatively dry land, settlement densities remained low, tribal groups were often isolated from one another, and overall populations probably never numbered more than 300,000 inhabitants. To survive, Australia's Aborigines developed great adaptive skills, often subsisting in harsh environments that Europeans avoided. Their low densities, however, meant linguistic fragmentation. Although precise counts vary, there were probably 250 languages spoken at the time of European contact, and recent totals suggest that almost 50 indigenous languages can still be found.

Radical cultural and geographical changes accompanied the arrival of Europeans, and Aboriginal populations were decimated in the process. The geographical results of colonization were striking as Aboriginal settlements were relegated to the sparsely settled interior, particularly in northern and central Australia, where fewer Europeans competed for land. In most cases, the European attitude toward the Aboriginal population was even more prejudicial than it was toward native peoples of the Americas. As hunter-gatherers lacking centralized political organization, Aborigines were usually considered less than human. No treaties were signed and no concessions were made. Typically, Aborigines were simply run off any lands desired by Europeans. Although decimated by these disruptions, as well as by European diseases, the Aborigines survived, particularly in the isolated outback, where they were far from entanglements with encroaching Europeans.

Today, Aboriginal cultures persevere in Australia, and a growing native peoples movement parallels similar activities in the Americas. Approximately 2 percent (or 400,000) of Australia's population comprises indigenous peoples, but their geographical distribution has changed dramatically in the past century. Aborigines account for almost 30 percent of the Northern Territory's population (many of these in Arnhemland near Darwin) and other sizable native reserves are located in northern Queensland and Western Australia (Figure 14.25). Most native peoples, however, live in the same large urban areas that dominate the country's overall population geography. Indeed, more than 70 percent of Aborigines live in cities, and very few of them still practice traditional hunting and gathering lifestyles. Processes of cultural assimilation are clearly at work: urban Aborigines are frequently employed in service occupations, Christianity has often replaced traditional animist religions, and only 13 percent of the native population still speaks an indigenous language. Recently, even native-language bilingual education programs have lost funding in the Northern Territory, mirroring the rejection of similar efforts in multicultural California.

Still, forces of diversity are at work, suggesting a growing Aboriginal interest in preserving traditional cultural values. Particularly in the outback, a handful of Aboriginal languages retain their vitality and have growing numbers of speakers. In addition, cultural leaders are selectively preserving Aboriginal spiritualism, and these religious practices often link local populations to surrounding places and natural features that are considered sacred. In fact, a growing number of these sacred locations are at the center of land-use and resource controversies between Aboriginal populations and Australia's European majority. The future for Aboriginal cultures remains unclear: pressures for cultural assimilation will be intense as many native peoples relocate to more Western-oriented urban

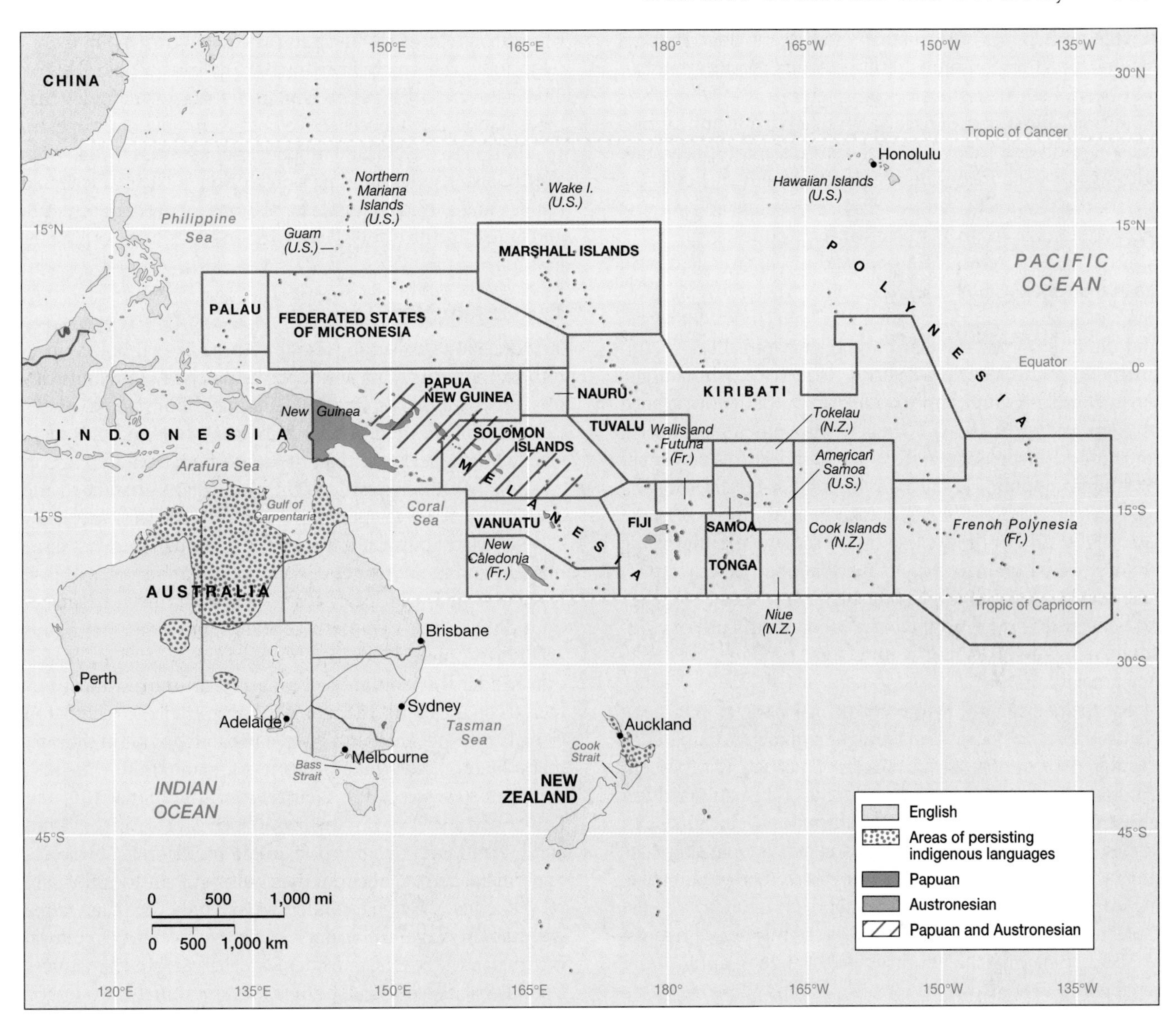

▲ **Figure 14.25 Language map of Australia and Oceania** While English is spoken by most residents, native peoples and their linguistic traditions remain an important cultural and political force in both Australia and New Zealand. Elsewhere, traditional Papuan and Austronesian languages dominate Oceania. The French colonial legacy also persists in select Pacific locations. Tremendous linguistic diversity has shaped the cultural geography of Melanesia, and more than 1,000 languages have been identified in Papua New Guinea.

settlements and lifestyles. At the same time, more rapid rates of natural increase (almost twice the national average) and a growing cultural awareness of Aboriginal traditions will work to preserve elements of the country's indigenous cultures.

A Land of Immigrants Most Australians reflect the continent's more recent European-dominated migration history, but even these patterns have become more complex as a rising tide of Asian cultures becomes important. Overall, more than 70 percent of Australia's population continues to reflect a British or Irish cultural heritage. These groups dominated many of the nineteenth- and early twentieth-century migrations into the country, and the close cultural ties to the British Isles remain fundamental. Sizable numbers of Italians, Greeks, and Germans have added diversity to the European mix. Some of these migrants became involved in specialized agricultural activities, particularly in southeastern Australia, while others moved to the cities, where they added ethnic variety to the mostly Anglo mix. A need for laborers along the fertile Queensland Coast also prompted European plantation owners to import inexpensive workers from the Solomons and New Hebrides. These Pacific Island laborers, known as **kanakas**, were spatially and socially segregated from their Anglo employers but further diversified the cultural mix of Queensland's "sugar coast." Historically, however, non-white migrations to the country were strictly limited by what is often termed a **White Australia Policy**, in which governmental guidelines promoted European and North American immigration at the expense of other groups.

Recent migration trends have reversed this historical bias and more diverse inflows of new workers and residents are adding to the country's multicultural character. Since the 1970s, the Migration Program has been dominated by a varied people chosen on the basis of their educational background and potential for succeeding economically in Australian society. For example, a growing number of families have come from places such as China, India, Malaysia, and the Philippines. Smaller numbers have qualified as migrants through their New Zealand citizenship, while others have arrived as refugees from troubled parts of the world such as Vietnam and Yugoslavia. The result is a much more varied foreign-born population in the country. Indeed, 22 percent of Australia's population are now immigrants, reflecting the country's global popularity as a migration destination. In the late 1990s, almost 40 percent of the settlers arriving in the country were from Asia. Major cities offer particularly attractive possibilities: Sydney's Asian population already exceeds 10 percent and is growing rapidly, while Perth's urban landscape increasingly displays cultural and economic links to its Asian neighbors.

Australian society, while enduringly Anglo, has been changed forever by its varied immigrant mix. For example, although the country is still largely Christian (27 percent Catholic and 22 percent Anglican), some of the nation's fastest growing religions are Islam, Buddhism, and Hinduism. In similar fashion, while English is Australia's national language, more than 2.5 million residents now speak another language, including more than 100,000 speakers each of Italian, Greek, Cantonese, Arabic, and Vietnamese. Economic ties with Japan are also reshaping the country's cultural geography: more Japanese travel to Australia for business and pleasure, and a growing number of Australians are learning Japanese. Indeed, a vocal minority of native-born whites resists the current cultural mosaic and advocates much more restrictive immigration policies, both on racial and economic grounds. The late 1990s popularity of Pauline Hanson's One Nation Party, though failing to dominate the national political agenda, demonstrates the tensions between some whites and the Aboriginal and immigrant populations they see as a threat to their cultural values and economic opportunities.

Australian society has also been shaped by global popular culture, and the relationship has often been a two-way street. The country's largely English-speaking population has embraced North American movies, music, and television. Many American television programs reappear down under, although the Australian media are mandated to show at least 50 percent local programming from 6 A.M. to midnight in order to "keep the Australia in Australian television." Australian influences have also shaped global media. Sydney's Summer Olympic Games in 2000 provided Australians with an ideal opportunity to share their culture and country with the world. The popular film *Crocodile Dundee* provided a humorous mix of fact and fancy in its portrayal of an Australian maverick in the outback. More significantly, Australian ownership (News Corporation) of many British newspapers as well as the Fox Television network in the United States is a reminder that global cultural influences move both in and out of the region. In addition to the cultural influences of the mass media, Australians also give leisure time a high priority. Their recreational habits parallel North American passions for outdoor activities and for competitive and spectator sports such as soccer, rugby, and horse racing.

Patterns in New Zealand

New Zealand's cultural geography broadly reflects the patterns seen in Australia, although the precise cultural mix differs slightly from its larger Pacific neighbor. Native Maori populations are more numerically important and culturally visible in New Zealand than their Aboriginal counterparts in Australia. While British colonization clearly mandated the dominance of Anglo cultural traditions by the late nineteenth century, Maori populations survived, although they lost most of their land in the process. Native populations, after declining with initial European contacts and conflicts, began rebounding in the twentieth century, and today the Maori account for more than 15 percent of the country's 3.8 million residents. Geographically, the Maori remain most numerous on the North Island, including a sizable concentration in metropolitan Auckland. While urban living is on the rise, many Maori, akin to their Aboriginal counterparts, are also committed to preserving their religion, traditional arts, and Polynesian lifeways (Figure 14.26; see "Local Voices: Land and Life in Maori Religion"). In addition, Maori is now a second official language within the country, along with English.

While many New Zealanders still identify with their largely British heritage, the country's twentieth century's cultural identity matured with an increasing sense of separateness from its exclusively British roots. Several things have forged New Zealand's special cultural character. As Britain tightened its own links with the European continent after World War II, New Zealanders increasingly forged a more independent and eclectic identity. In many ways, popular culture ties the country ever more closely to Australia, the United States, and continental Europe, a function of the increasingly global mass-media industry. For example, the most famous film ever made in New Zealand (*The Piano*) was produced by an Australian, financed with French capital, and reached its largest audiences in the United States. Indeed, by the late 1990s, less than 20 percent of New Zealand television offered locally produced programming.

Diversity also continues to shape the cultural setting. The nation's unique Polynesian roots impart a special regional character: in addition to its Maori population, more than 5 percent of its population are Pacific Islanders, and Auckland has the largest Polynesian population of any city in the world. Adding further complexity to the cultural mix are growing numbers of Asians who now make up another 5 percent of New Zealand's residents and who are making their own dynamic contribution to the country's culture and economy, particularly in its larger urban centers. The end result is a national

▲ **Figure 14.26 Maori artisans** New Zealand's native Maori population actively preserves its cultural traditions and has recently increased its political role in national affairs. These artisans are carving decorations for a traditional Maori canoe. *(Arno Gasteiger/Bilderberg/Aurora & Quanta Productions)*

character truly forged at a Pacific crossroads, an accumulation of varied and vastly different cultural influences that have been amalgamated into a uniquely New Zealand identity that still resists easy definition.

The Mosaic of Pacific Cultures

Native and exotic cultural influences produce a varied mosaic across the islands of the South Pacific. In more isolated locales, traditional cultures maintain their integrity largely insulated from outside influences. In most cases, however, modern life in the islands revolves around an intricate cultural and economic interplay of local and Western influences. One thing is certain: the relative cultural insularity of the past is gone forever and in its place is a Pacific realm rapidly adjusting to powerful forces of colonization, global capitalism, and popular culture.

Traditional Culture Worlds Defining the pre-European cultural setting is no simple task. Anthropologists were once confident that the division of Oceania into Melanesia, Micronesia, and Polynesian reflected clear cultural and racial distinctions. In the racist thinking of the early twentieth century, Polynesians were usually considered to be superior to the other peoples of the Pacific, and Melanesians inferior. According to the once-standard view, Polynesians were brown-skinned peoples (perhaps distantly related to Europeans) who possessed advanced political structures (chiefdoms and kingdoms) and complex systems of social stratification; Melanesians, on the other hand, were black-skinned peoples (probably related to Africans) living in simple, rather egalitarian village communities. Melanesians, moreover, were often regarded as savage cannibals, whereas Polynesians were sometimes considered to be noble exemplars of "natural" existence. Micronesians, by the same line of thinking, were usually placed in an intermediate position. European sailors in the 1800s were often known to jump ship in order to marry Polynesian women and remain in these idyllic islands. Indeed, many European artists and writers viewed Polynesia as a kind of natural and social paradise that had remained uncorrupted by the repressive social system of European society.

Today, such notions of cultural superiority and inferiority have been abandoned by serious scholars, and the division between Polynesian and Melanesian cultures is no longer as

LOCAL VOICES Land and Life in Maori Religion

Traditional Maori religion is closely bound to ideas about the natural world, and their environmental philosophy has attracted growing interest not only among Maori keen on preserving their culture, but also among Westerners sympathetic to its ideals. Maori envision each of the components of the natural world, including people, as possessing a harmonious and interrelated life force, or essence known as *mauri*. People must respect the integrity and vitality of this force, including its manifestation in mountains, rivers, trees, animals, and humans themselves. While understanding that one cannot live in the world without transforming it, the Maori believe the natural world must be damaged as little as possible and that interfering with rivers or other animals must be done respectfully and for justified reasons. As Maori writer Rangimarie Rose Pere suggests, the mauri of each creature interacts with Earth, and if people respect Earth, all of its inhabitants will prosper.

This Maori perspective on environment has many consequences. Recently, a local Maori leader protested the discharge of sewage into a river, claiming it upset the harmony, or mauri, of the stream. Indeed, such an attitude has often defined the rift separating Maori environmentalists from more development-minded New Zealanders. On the other hand, the Maori find it hard to conceive of the Western idea of wilderness, since it involves a fundamental split between people and nature. How can one segregate a wilderness space in the midst of one's forest home? Even Maori artwork is shaped by their environmental ethics. When artists such as wood carver Rangi Hetet work with their materials, they typically chose raw slabs rather than milled boards, suggesting that the natural wood better retains its original mauri, a quality that survives within the properly crafted carving. Europeans, conversely, are guilty of destroying the forest's mauri when they carelessly fell the timber and send it to the sawmill.

Source: Adapted from John Patterson, "Respecting Nature: The Maori Way," *The Ecologist* 29 (January–February 1999).

clear as it once was. Certainly most Polynesian societies were more politically centralized and class-based than those of Melanesia, but this was not always the case. Small-scale Melanesian societies, moreover, were sometimes linked to one another through very complex systems of trade and social exchange. Equally significant, upon close geographical inspection, the boundary of the two regions turns out to be a zone of pronounced cultural melding. The Melanesians of Fiji and the Polynesians of Tonga, for example, not only traded with each other, but often intermarried, and thus came to share a number of political and cultural institutions. Even in the heart of Melanesia, there are a number of atolls that are actually Polynesian in culture and language.

The modern language map reveals some significant cultural patterns that both unite and divide the region. Most of the indigenous languages of Oceania belong to the Austronesian language family that encompasses wide expanses of the Pacific, much of insular Southeast Asia, and Madagascar. Linguists hypothesize that the first great oceanic mariners spoke Austronesian languages, and thus disseminated them throughout this vast realm of islands and oceans. Within the broad Austronesian family, the Malayo-Polynesian subfamily includes most of the related languages of Micronesia and Polynesia, suggesting a common cultural and migratory history for these far-flung peoples.

Melanesia's language geography is more complex and still incompletely understood by outside experts: while coastal peoples often speak languages brought to the region by the seafaring Austronesians, more isolated highlander cultures, particularly on the island of New Guinea, speak varied Papuan languages. Indeed, the linguistic complexity of that island is so daunting—more than 1,000 languages have been identified—that many experts question whether or not they even comprise a unified "Papuan family" of related languages. Some scholars estimate that half of New Guinea's languages are spoken by fewer than 500 persons, suggesting the cultural role played by the region's rugged topography in isolating one cultural group from another. These New Guinea highlands may hold some of the world's few remaining **uncontacted peoples,** culture groups that have yet to be "discovered" by the Western world.

Traditional patterns of social life are as complex and varied as the language map. In many cases, however, life revolves around predictable settings. For example, across much of Melanesia, including Papua New Guinea, most people live in small villages often occupied by a single clan or family group. Many of these traditional villages contain fewer than 500 residents, although some larger communities may house more than 1,000 people. Life often revolves around the gathering and growing of food, an annual round of rituals and festivals, and complex networks of kin-based social interactions, often including the potential for conflict with other nearby groups.

Traditional Polynesian settings also feature village life, although there are often more class-based relationships between local elites (often religious leaders) and ordinary residents (Figure 14.27). Polynesian villages are also more likely linked to other islands by wider cultural and political ties. Despite the Western penchant for depicting Polynesian communities in idyllic terms, violent warfare was actually quite common across much of the region prior to European contact. Another cultural myth is the notion that Polynesians led a natural sex life uncomplicated by Western hang-ups. The truth of the matter seems to be that lower-class women were often ordered by Polynesian political elites to sleep with Western men so that they might obtain steel, knives, guns, and other goods unavailable in their own economy. Not surprisingly, such actions tended to be seriously misinterpreted by European men.

▲ **Figure 14.27 Tonga village** Although the economic and technological effects of globalization have arrived in Tonga, village life remains important in many Polynesian settings. Most village housing in these tropical environments reflects the use of locally available construction materials. *(Ted Streshinsky/Corbis)*

External Cultural Influences While traditional culture worlds persist in some settings, most Pacific islands have witnessed tremendous cultural transformations in the past 150 years. Outsiders from Europe, the United States, and Asia brought new settlers, values, and technological innovations that have forever changed Oceania's cultural geography and its place in the larger world. The result is a modern setting where Pidgin English has broadly supplanted native languages, Hinduism is practiced on remote Pacific Islands, and traditional fishing peoples now work at resort hotels and golf course complexes.

European colonialism transformed the cultural geography of the Pacific world by broadly if loosely incorporating it into new political and economic systems and by bringing in new peoples to the region who directly reconfigured its cultural makeup. Hawaii illustrates the pattern. By the mid-nineteenth century, Hawaii's King Kamehameha was already entertaining a varied assortment of whalers, Christian missionaries, traders, and navy officers from Europe and the United States. A small elite group of **haoles,** or light-skinned European and American foreigners, were successfully profiting from commercial sugarcane plantations and Pacific shipping contracts. Labor shortages on the islands, however, prompted the importation of Chinese, Portuguese, and Japanese workers who further

▲ **Figure 14.28 Multicultural Hawaiians** Many residents of the Hawaiian Islands represent a blend of Pacific Island, Asian, and European influences. These young women work to promote an appreciation of the island's rich ethnic legacy at the Polynesian Cultural Center on Oahu. *(Porterfield/Chickering/Photo Researchers, Inc.)*

complicated the region's cultural geography. By 1900, the Japanese had become a dominant part of the island workforce and the United States had formally annexed the islands (in 1898). The cultural mosaic revealed in the Hawaiian Census of 1910 suggests the magnitude of change: more than 55 percent of the population was Asian (mostly Japanese and Chinese), another 20 percent were native peoples, and about 15 percent (mostly imported European workers) were white. By the end of the twentieth century, the Asian population was less dominant but more ethnically varied, about 40 percent of Hawaii's residents were white, and the small number of remaining native Hawaiians were joined by an increasingly diverse group of other Pacific Islanders. In addition, ethnic mixing has produced a rich mosaic of Hawaiian creole cultures that offer a unique crossroads blend of North American, Asian, Pacific Island, and European influences (Figure 14.28).

Hawaii's story has been played out in many other Pacific island settings. In the Mariana Islands, Guam was absorbed into America's Pacific Empire as part of the Spanish American War in 1898. Thereafter, native peoples not only felt the effects of Americanization (the island remains a self-governing U.S. territory today), but thousands of Filipinos were moved there to supplement its modest labor force. To the southeast, the British-controlled Fiji Islands offered similar opportunities for fundamentally redefining Oceania's cultural mix. The same sugar plantation economy that spurred changes in Hawaii prompted the British to import thousands of South Asian laborers to Fiji. The descendents of these Indians (most practice Hinduism) now comprise almost half the island country's population and often come into sharp conflict with the native Fijians (Figure 14.29). In French-controlled portions of the realm, small groups of traders and plantation owners filtered into the Society Islands (Tahiti), but a larger contingent of French colonial settlers (many originally a part of a penal colony) had a major impact on the cultural makeup of New Caledonia. Still a French colony, New Caledonia's population is more than one-third French and its capital city of Noumea reveals a cultural setting curiously forged from French society, Melanesian traditions, and balmy South Pacific breezes.

Fundamental cultural changes have occurred in the Pacific even where fewer numbers of Europeans, Americans, or Asians have settled. Varied forms of **Pidgin English**, where English vocabulary and grammar is reworked and blended with native dialects, are spoken throughout the realm, particularly in island groups formerly colonized either by British or American interests. Many differences exist among Pidgin English dialects because the precise grammar, cadence, and sound systems originate locally. Even so, it remains relatively easy for English speakers to learn a language such as New Guinea's Pidgin. Much of its vocabulary is based on that of English, and it evolved in order to facilitate communication with a simple system of grammar and a limited number of words. The dialect is widely used across Papua New Guinea, particularly when different culture groups communicate with one another.

In many Pacific settings, Christianity has also been absorbed into the indigenous cultural mix. Roman Catholicism and various Protestant faiths are represented. Catholicism is more prevalent in areas of French colonization. Certain Protestant groups are heavily concentrated on islands where their missionaries were particularly active. Many Tongans, for example, were converted to Methodism in the nineteenth century. More recently, Mormon missionaries have been active through much of Oceania and have been particularly successful in parts of Polynesia. Several settings have also seen the emergence of indigenous Christian churches that often blend Western beliefs with traditional animism. One distinctive form of quasi-animist religion in lowland Melanesia was also stimulated by outside contacts and became known as the "**cargo cult**." Many areas of the western Pacific witnessed intense

▲ **Figure 14.29 South Asians in Fiji** British sugar plantation owners imported thousands of South Asian workers to Fiji during the colonial era. Today, almost half of Fiji's population is South Asian, including many urban residents. The group often comes into conflict with native Fijians. *(Frank Fournier/Woodfin Camp & Associates)*

fighting during World War II, and Allied forces received airdrops of war materials and other supplies. Many of these goods came into the hands of the local people, who were mystified by their sudden appearance. A complex set of ritual practices quickly emerged, designed to entice the cargo-dropping planes and ships to return. The main period of the cargo cult was in the immediate post-war period, but observations persist to this day in some areas.

A tidal wave of outside influences has engulfed the Pacific realm since World War II, producing further cultural changes as well as growing indigenous responses designed to preserve traditional values. Some groups, particularly in Melanesia, remain more isolated from the outside world, although even here growing demands for natural resources offer an avenue for increasing Western or Asian contacts. In many settings, however, the global growth of tourism has brought Oceania into the relatively easy reach of wealthier Europeans, North Americans, Asians, and Australians. The Hawaiian Islands, Fiji, French Polynesia, and American Samoa are being joined by an increasing number of other island tourist destinations. While offering tremendous economic benefits to certain localities, the onrush of tourists and their consumer-driven values has often come into sharp conflict with native cultures. Change is often seen in generational terms: grandparents may recall simpler times of more limited outside contact; parents might work long, hard hours at resort facilities or souvenir shops; and youngsters find themselves surrounded by fast food restaurants and the glittery attractions of television. Increasingly, however, selected native groups resist these pressures of cultural globalization. In Melanesia a growing number of countries have joined the "Spearhead Group" of nations dedicated to preserve cultural traditions. For example, to keep native dialects from dying, many primary schools in Papua New Guinea now teach early grades in local languages before switching to Pidgin English thereafter. Even more dramatically, some native Hawaiians, paralleling indigenous movements elsewhere in the world, are advocating the return of one or more of the major islands to Polynesian control, thus preserving their cultural heritage.

Geopolitical Framework: A Land of Fluid Boundaries

Pacific geopolitics reflect a complex interplay of local, colonial-era, and global-scale forces. Each political unit within the region reflects how these three forces have played out in particular places. Consider the story of Micronesia's Marshall Islands, and the complexities become apparent. This sprinkling of islands and atolls (covering 70 square miles, or 180 square kilometers, of land) historically held varied ethnic groups in essentially local political units before they were loosely incorporated into the Spanish and then German empires by the end of the nineteenth century. In 1914 the Japanese moved into the islands and the area remained under their control until 1944 when U.S. troops occupied the region. Following World War II, a United Nations Trust Territory (administered by the United States) was created across a wide swath of Micronesia, including the Marshall group. Demands for local autonomy grew during the 1960s and 1970s, resulting in a new constitution and independence for the Marshall Islanders by the early 1990s. Today, still benefiting from U.S. aid, government officials in the modest capital city on Majuro Atoll struggle to unite island populations, protect sprawling maritime sea claims, and adjudicate a generation of legal and medical problems that grew from U.S. nuclear bomb testing in the region (Figure 14.30). Such narratives are typical across the realm, suggesting a twenty-first century political geography that is still very much in the making.

Creating Geopolitical Space

Geopolitical space across Australia and Oceania has been reconfigured many times as different cultural groups and political powers have asserted themselves across the region. Present patterns of political organization are merely a snapshot in time and perhaps more prone than most areas of the world to be redefined on future maps.

▲ **Figure 14.30 Marshall Islands** The political control of Micronesia has shifted numerous times during the last two centuries. While now part of the independent Marshall Islands, these palm-fringed beaches were once claimed by Spanish, German, Japanese, and U.S. interests. *(Douglas Peebles Photography)*

Indigenous Patterns Prior to European contact, the region's political geography was an intricate and ever-changing mosaic of indigenous territories. The indigenous political forms of Australia and Oceania varied widely. Aboriginal Australian society was organized around fluid bands of 30 to 60 people, most of whom were related by blood or marriage. Thus, the periodic movements of hunters and gatherers defined the shifting political space of the Aboriginal world. Melanesia was oriented around more sizable and sedentary communities, also based largely on kinship ties. In Micronesia, Polynesia, and some coastal sections of Melanesia, however, larger-scale and more highly structured chiefdoms evolved. Centralized kingdoms, in which power was concentrated in the hands of a single monarch, existed on several of the larger volcanic islands of Polynesia. The most powerful of these kingdoms was based in Hawaii, which eventually encompassed all of the islands of the archipelago. While the days of ruling Hawaiian royalty have long since passed, these indigenous, frequently tribal affiliations still help native peoples define their political worlds, an orientation that sometimes runs counter to the boundaries of the region's modern states.

An Imposed Colonial Framework The European world turned the native political pattern on its head, replacing the nuanced, often fluid territorial boundaries of the indigenous world with more precise, yet unstable colonial borders. Great Britain claimed and colonized both Australia and New Zealand. By the 1860s, the various colonial footholds in Australia, focused on widely separated coastal settlements, were formally delineated by the simple geometrical boundaries still apparent today. These separate administrative units functioned as distinctive colonial entities until they joined in the formal Confederation of Australia in 1901 as an essentially independent commonwealth. New Zealand followed a similar path. Early on, isolated settlement clusters dotted the shoreline, marking the first tentative imprints of European political authority upon the Maori-controlled islands. After the Maori were conquered in the late nineteenth century, both islands were united in a formal and independent dominion under the British Crown in 1907.

The global reach of European colonialism also extended deep into the South Pacific during the nineteenth century. Interest in the area grew as early as 1842, when France claimed a "protectorate" over Tahiti. Most of Oceania was not formally colonized, however, until the end of the century, the same period that witnessed the scramble for African territory. By then, relatively few areas of the world were still open for annexation, and colonial powers were eager to acquire what remained before their rivals had a chance to do so. Certain oceanic islands were also seen as particularly desirable because they could serve as coaling supply points for steamships and act as relay stations for submarine telegraph cables.

Germany played a major role in the division of Oceania, just as it did in the partition of Africa. German imperialists viewed New Guinea and the larger islands of the nearby Bismarck Archipelago as the main prize. Other colonial powers opposed the German initiatives, however. By 1885 New Guinea was divided into three territories: the Dutch got the western half (near their empire in Indonesia), the British received the southeastern quarter (near Australia), and the Germans acquired the northeastern quarter, along with the Bismarck Archipelago and a portion of the nearby Solomon Islands chain. Portions of Micronesia and Polynesian Samoa rounded out Germany's colonial holdings.

France, Great Britain, and the United States also had major territorial ambitions in Oceania. By the end of the nineteenth century, virtually all of eastern Polynesia lay under French control. Their domain included the Society Islands, the Marquesas Islands, and the Tuamotu Archipelago. Farther west, France colonized New Caledonia in Melanesia and began to move into the nearby New Hebrides Islands (now Vanuatu). The British, however, having just acquired nearby territories to the north, were also interested in the archipelago. Rather than dividing it or fighting over it, the French and British simply agreed to an unprecedented "condominium" in the New Hebrides under which they shared power. Britain gained additional islands in the region, including the southern Solomon chain, the Fiji Islands, many of the smaller islands of western Polynesia, and a few outposts in eastern Polynesia (Pitcairn). Although most of these acquisitions became direct territories of the British Crown, Tonga (in western Polynesia) remained a protectorate with considerable local autonomy. Indeed, Tonga was the only indigenous state of Oceania to retain its own political structure, and it remains to this day a (constitutional) monarchy. For the United States, Hawaii was the main acquisition, but eastern Samoa also came under U.S. rule in 1900 as a result of an agreement signed with Britain and Germany. The United States also gained Guam, the largest island in Micronesia, as a result of its victory over Spain in the Spanish-American War.

Wars and treaty agreements among colonizing powers further changed Oceania's patterns of political geography. For example, Germany lost all of its foreign territories after World War I: those in Micronesia passed to Japan, those in Melanesia were ceded to Australia, and those in Polynesia went to New Zealand. Japan's gains proved crucial during World War II, as it provided the Japanese military with naval and air bases over a huge swath of the Pacific. After World War II, however, Japan's Micronesian empire evaporated, passing to the United States under UN auspices as the Trust Territory of the Pacific. Meanwhile, Australia inherited the British territory of southeastern New Guinea, to which it added the German territories in northeastern New Guinea and the Bismarck Archipelago. At this time the interior reaches of New Guinea were essentially unexplored, and the Australian government had little idea of what its large new colony actually contained. In the 1920s, however, gold was discovered and by the 1930s Australian prospectors penetrated the highlands. There they were shocked to discover broad highland plateaus inhabited by roughly a million persons. The New Guinea Highlanders, for their part, were probably even more shocked at the intrusive habits and new technologies of the light-skinned interlopers who set about establishing Australian rule over the entire area.

Roads to Independence The modern political states of the region have arrived at independence along many different paths. Other political units remain colonial entities to this day. The newness and fluidity of its political boundaries are remarkable: the region's oldest independent states are Australia and New Zealand, and both were twentieth-century creations that are only now pondering the desirability of completing their formal political separation from the British Crown. Elsewhere, political ties between colony and mother country are even more intimate and enduring. Even many of the newly independent Pacific **microstates,** known for their tiny overall land areas, retain special political and economic ties to countries such as the United States.

Independent Australia (1901) and New Zealand (1907) only gradually created their own political identities and still struggle with their ultimate political configuration. Although Australia became a commonwealth in 1901, it still acknowledges the British Crown as its sovereign (Figure 14.31). Although the Crown's role is strictly symbolic, a number of Australians prompted a national referendum (November 1999) that forced Australians to ponder whether or not they would like their country to drop this remaining tie to Britain and instead become a genuine republic. Australia, like the United States, is a federal country, with each of its six states retaining substantial powers. The Northern Territory, however, remains directly under the authority of the central government, although it does enjoy limited self-rule. In New Zealand, formal legislative links with Great Britain were only severed in 1947, and in 1994 some New Zealand officials began discussing the same formal break with the British Crown being debated by the Australians.

Elsewhere in the Pacific, colonial ties were severed even more slowly, and the process has not yet been completed. In the 1970s Britain and Australia began relinquishing their colonial empires in the Pacific. Fiji (Great Britain) gained independence in 1970, followed by Papua New Guinea (Australia) in 1975, and the Solomon Islands (Great Britain) in 1978. The small-island nations of Kiribati and Tuvalu (Great Britain) also became independent in the late 1970s, even though some observers argued that they did not have adequate populations or territorial bases to be viable states.

The United States has recently turned over most of its Micronesian territories to local governments, while retaining considerable influence in the area. After gaining these islands from Japan in the 1940s, the U.S. government supplied large subsidies to islanders but also utilized a number of islands for military purposes. Bikini Atoll was obliterated by nuclear tests, and the large lagoon of Kwajalein Atoll was used as a giant missile target. A major naval base, moreover, was established in Palau, the westernmost archipelago of Oceania. By the early 1990s, both the Marshall Islands and the Federated States of Micronesia (comprising the Caroline Islands) gained independence. Their ties to the United States, however, remain close. A number of other Pacific islands remain under loose U.S. sovereignty. Palau is an American "Trust Territory," although the Palauans have some local autonomy. The people of the Northern Marianas chose to become a "self-governing commonwealth in association with the United States," a rather ambiguous political position that allows them to become U.S. citizens. The residents of self-governing Guam and American Samoa are also U.S. citizens. Hawaii became a full-fledged U.S. state in 1959 and thus an integral part of the United States.

Other colonial powers were less inclined to relinquish their oceanic possessions. New Zealand still controls substantial territories in Polynesia, including the Cook Islands, Tokelau, and the island of Niue. France has even more extensive holdings in the region. Its largest maritime possession is French Polynesia, encompassing a vast expanse of mid-Pacific territory. To the west, France retains the much smaller territory of Wallis and Futuna in Polynesia and the larger island of New Caledonia in Melanesia.

▲ **Figure 14.31 Prince Charles in Australia** Australia became an independent Commonwealth in 1901, but it was slow to gain a distinctive political identity. This 1994 visit by Prince Charles was another reminder of the country's close and lingering historical ties with the British Crown. *(Tim Graham/ Sygma Photo News)*

Persisting Geopolitical Tensions

Cultural diversity, colonial legacy, youthful states, and a rapidly changing political map contribute to ongoing geopolitical tensions within the Pacific world (Figure 14.32). Indeed, some of these conflicts have consequences that extend far beyond the boundaries of the region. Others are more locally based but are still vivid reminders of the complexities involved as political space is radically redefined and refashioned across varied natural and cultural settings.

Native Rights in Australia and New Zealand Indigenous peoples in both Australia and New Zealand have used the political process to gain more control over land and resources in their two countries. Indeed, the strategies these native groups have used parallel similar efforts in the Americas and elsewhere. In Australia, Aboriginal groups are discovering newfound political power, both from more effective lobbying efforts by native groups, as well as from a more sympathetic federal government (see "Global and the Local: Aboriginal

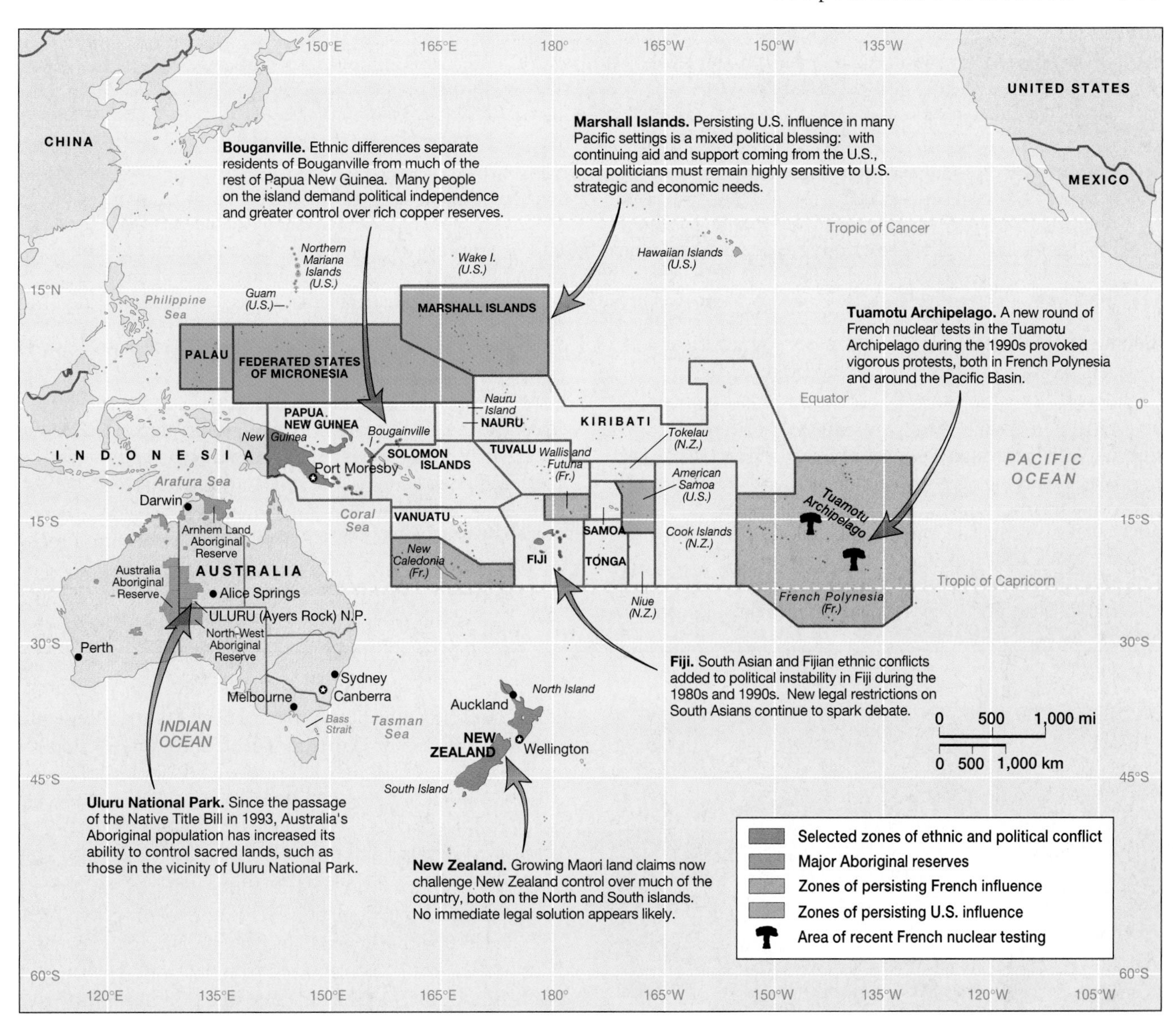

▲ **Figure 14.32 Geopolitical issues in Australia and Oceania** Native land claim issues increasingly shape domestic politics in Australia and New Zealand. Elsewhere, ethnic conflicts have raised political tensions in settings such as Fiji and Papua New Guinea. Colonialism's impact endures, as well: American and French interests remain particularly visible in the region, including a legacy of nuclear testing that continues to impact selected Pacific Island populations.

Videoconferencing in the Outback"). Since land treaties were generally not signed with Aborigines as whites conquered the continent, native peoples originally had no legal land rights whatsoever. More recently, the Australian government established a number of Aboriginal Reserves, particularly in the Northern Territory, and expanded Aboriginal control over sacred national parklands such as Uluru (Ayers Rock)(Figure 14.33). Further concessions to indigenous groups were made in 1993 as the government passed the **Native Title Bill**, which compensated Aborigines for already ceded lands, gave them the right to gain title to unclaimed lands they still occupied, and legally empowered them in dealings with mining companies in native-settled areas. Efforts to expand Aboriginal land rights have also met concerted opposition. In 1996 an Australian court ruled that pastoral leases (the form of land tenure held by the cattle and sheep "ranchers" who control most of the outback) do not necessarily extinguish Aboriginal land rights. Grazing interests were infuriated, which led the government to respond that Aboriginal claims might allow the visiting of sacred sites and some hunting and gathering, but not substantial economic control.

In New Zealand, Maori land claims have generated similar controversies in recent years. The Maori constitute a far larger proportion of the overall population and the lands that they claim tend to be much more valuable, further complicating the issue. Recent protests include periodic civic disobedience, growing Maori land claims over much of North and South islands, and a call to return the country's name to

GLOBAL AND THE LOCAL Aboriginal Videoconferencing in the Outback

Cultivating indigenous political consciousness, particularly in a sparsely settled land, is not an easy task. Australian Aborigines, increasingly interested in asserting their rights to tribal lands and local political power, have made ingenious use of twenty-first-century technologies in their quest for tribal unity and respect. Since 1993, the Northern Territory's Warlpiri Aborigines have made extensive use of a sophisticated videoconferencing system to more effectively unite the widely dispersed tribe and to link with Warlpiri expatriates who now live in cities such as Darwin, Alice Springs, and even far-off Sydney. Massachusetts-based PictureTel Corporation helped set up the system, networking together a maze of television monitors, video cameras, and satellite dish connections. The result has been the "Tanami Network," a series of videoconferencing sites that can link up to 16 participants in a single conversation.

The network has had a major personal as well as political impact upon many residents of the region. Since Aboriginal peoples depend heavily on extensive hand gestures in their communication, the interactive video component of the system has been essential. Family members can converse with distant relatives, increasing social cohesion among a group often fractured by distance and dislocation. Just as importantly, Aboriginal political activists have effectively used the system to foster a common political consciousness among their members and to provide more responsive government services to those in need. The Aborigines are also going global: they have had a series of videoconferences with other indigenous peoples from the Saami in northern Scandinavia to the Little Cree Nation in Alberta, Canada. Land rights and language preservation have been common topics of discussion, all made possible by a $1.5-million system largely financed by tribal mineral royalties and community funds.

Source: Adapted from Mark Hodges, "Online in the Outback," *Technology Review* 99 (April 1996).

▲ **Figure 14.33 Aborigines at Uluru National Park** Australian Aborigines gathered at Uluru National Park in 1985 to celebrate their increased political control over the region. Since then, the Native Title Bill has promoted numerous land cessions and further legal settlements. *(Michael Jensen/Auscape International Pty. Ltd.)*

the indigenous "**Aotearoa**" or "Land of the Long White Cloud." The government response, complete with a 1995 visit from Queen Elizabeth, has been to acknowledge increased Maori land and fishing rights as well as to propose a series of financial and land settlements that have yet to meet final agreement with the Maoris.

Conflicts in Oceania Other geopolitical issues simmer elsewhere in the Pacific, threatening periodically to further redefine the region's fluid territorial boundaries. For example, ethnic differences in Fiji have threatened to tear apart that small island nation. Indigenous Fijians and South Asian immigrants (from the British colonial period) are roughly equal in population. South Asians dominate most of the country's businesses and commercial settlements, but the Fijians maintain control of virtually all of the land (by law 83 percent of Fiji's territory is reserved for indigenous village communities) on both main islands of Viti Levu and Vanua Levu. Ethnic strife reached a peak in the late 1980s when Fijian military leaders staged a coup and took over control of the government after an election threatened to increase Indian control over the country. The coup was followed by a new and controversial constitution in 1990 that angered many Indians, furthered ethnic violence, and prompted a severe scolding from its former British rulers that included a 10-year exile from the British Commonwealth (it was readmitted in 1997).

Papua New Guinea (PNG) must also contend with ethnic tensions. The country is composed of different cultural groups, many of which have a long history of mutual animosity. Most of these peoples now get along with each other reasonably well, although tribal skirmishes occasionally break out in highland market towns. A much bigger problem for the national government is the rebellion on Bougainville. This sizable island, which has large reserves of copper and other minerals, is located in the Solomon archipelago, but it belongs to PNG because Germany colonized it in the late 1800s and it thus became politically attached to the eastern portion of New Guinea. Many of Bougainville's indigenous inhabitants believe that their resources are being exploited by foreign interests and by an unsympathetic national government, and they demand local control. Papua New Guinea has reacted with military force, about 5 percent of the island's entire population has already been killed in the conflict, and recent efforts have failed to create a more stable regional government.

The continued French colonial presence within the Pacific region has also created political uncertainties, both in relations with native peoples as well as between the French and other independent states within the region. Continued French rule in New Caledonia has provoked much local opposition. This large island has substantial mineral reserves (especially of nickel) and sizable numbers of French colonists. French settlement and exploitation of mineral resources angered many indigenous inhabitants. By the 1980s a local independence movement was gaining strength, but in 1987 and 1998 the island's inhabitants (indigenous and French immigrants alike) voted to remain under French rule, at least until 2018. Still, an underground independence movement continues to operate. Elsewhere, troubles have also reverberated through French Polynesia. Although the region receives substantial subsidies from the French and residents have voted to remain a colony, a large minority of the population opposes French control and demands independence. The independence movement was greatly strengthened in 1995 when France decided to resume nuclear testing in the Tuamotu Archipelago. Those recent activities angered not only the people of French Polynesia (including antigovernment rioting in Tahiti), but also those of other Pacific countries and territories. Australian leaders called the tests "an act of stupidity," New Zealand recalled its ambassador in protest, and other Pacific nations from Japan to Chile registered their disapproval. Future French nuclear policies in the Pacific will undoubtedly continue to impact geopolitical relations far beyond its Polynesian borders.

The Imprint of Maritime Claims Global policies on the legal status of maritime boundaries have fundamentally changed the political and economic importance of many small Pacific states. Under the International Law of the Sea, most nations now agree on an **Exclusive Economic Zone (EEZ),** which is an area that extends seaward for 200 nautical miles (371 kilometers) within which the coastal or island state has exclusive rights to develop or lease potential natural resources such as fishing and mining. The new global geography of EEZs has great importance even for Oceania's microstates, since clusters of small but far-flung atolls can effectively claim huge maritime domains for exclusive resource development. For example, tiny Tonga has a land area of only 270 square miles (700 square kilometers). The island group's wide areal coverage, however, secures it a large oceanic zone of exclusive economic development rights that it can use or sell to others (Figure 14.34). Nearby Fiji, Samoa, and other states that are within the 200-mile limit must divide the intervening waters equally, thus all but eliminating the extent of legally open high seas in large portions of the island-studded South Pacific. The result has frustrated many development interests globally, caused friction and confusion among the island nations themselves, but also stimulated interest in the enormous potential value of these newly defined maritime resources.

A Regional and Global Identity?

Australia and New Zealand have emerged to play key political roles in the South Pacific. Although these two countries sometimes disagree on strategic and military matters, their

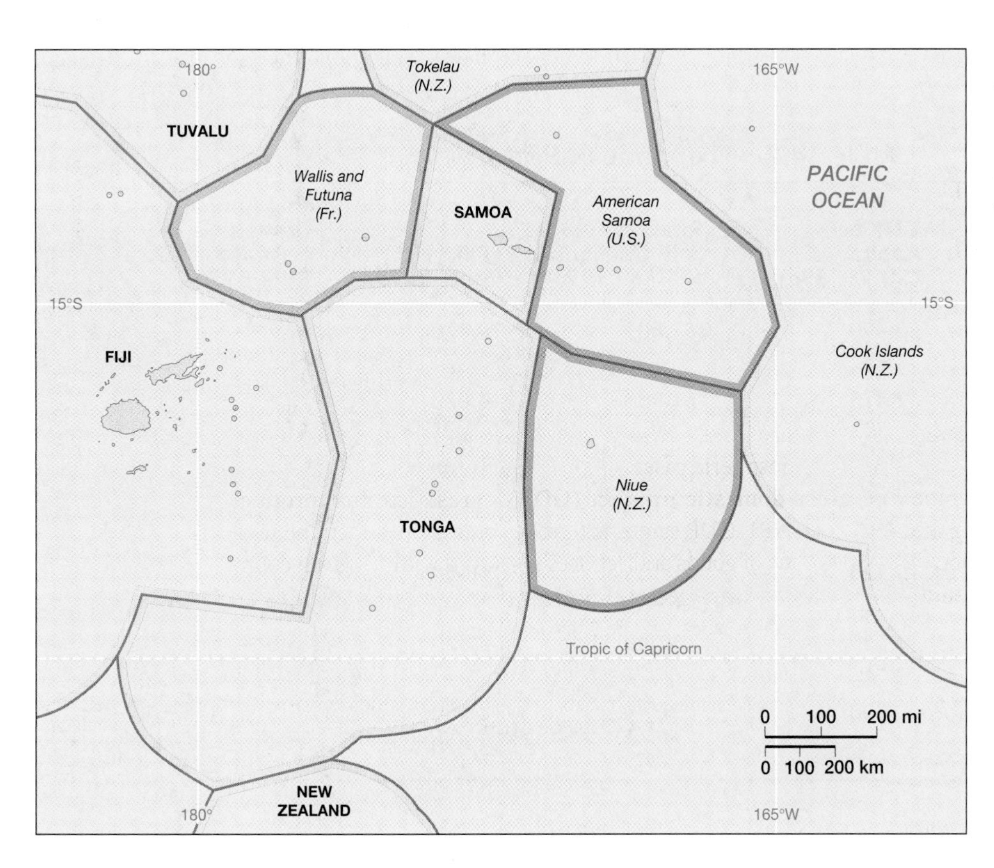

◀ **Figure 14.34 Tonga's EEZ** Tonga's Exclusive Economic Zone (EEZ) sprawls across vast stretches of the South Pacific. The 200-nautical mile (371-kilometer) limit defines an area where Tonga has rights to lease or develop potential natural resources such as fishing and mining. Where EEZs overlap, boundary lines are drawn to equally divide the intervening territory.

size, wealth, and collective political clout in the region make both countries important forces for political stability. Special colonial relationships still connect these nations with present and former Pacific holdings. Australia maintains close political ties with its former colony of Papua New Guinea, and New Zealand's continuing control over Niue, Tokelau, and the Cook Islands in Polynesia suggests its persisting political sphere of influence extends well outside its own borders. When political and ethnic conflicts arise elsewhere in Oceania, Australia and New Zealand are often involved in negotiating peace settlements. Recently, for example, both nations assisted in mediating ongoing disputes on Papua New Guinea's island of Bougainville. In general, the two countries enjoy close political and strategic relations and participate in joint military and reconnaissance efforts in the region. Given the other global interests in the region, however, it remains unclear whether or not these two nations can or wish to assert their own political dominance across the entire South Pacific.

Many other political connections tie the region to the world beyond. In 1951 Australia, New Zealand, and the United States forged the ANZUS strategic alliance, which provided a series of mutual security agreements between the three countries. New Zealand sorely tested those links in the 1980s with its outspoken opposition to nuclear weapons and the use of nuclear-powered warships and submarines. Since 1992, however, relations have warmed as the United States has modified its policy toward dispatching nuclear-armed warships to the region. Other strategic links are strengthening geopolitical ties with Southeast Asia. Australia, in particular, has played a growing role in the ARF (Association of South-East Asian Nations Regional Forum). This group is developing multilateral plans to deal with future defense issues, such as the possibility of nuclear conflict in nearby South Asia. In addition, colonial links to the region tie various settings to distant lands: New Caledonia and French Polynesia remain parts of France, while the Marshall Islands, Marianas, and Federated States of Micronesia are still connected politically and economically to Washington, D.C. Indeed, the persisting presence of French and U.S. interests in the region is not always welcomed by Australia and New Zealand, most notably with France's round of Polynesian nuclear tests in the 1990s. Whatever geopolitical balance of power exists in the future, one thing is certain: the region will endure on the global political periphery.

Economic and Social Development: A Hard Path to Paradise

Wealth and poverty coexist in the Pacific realm, but regional patterns are complex (Table 14.2). Affluent Australia and New Zealand, for example, also contain pockets of pronounced poverty. On the other hand, Oceania offers varied settings that include well-fed subsistence-based populations, relatively prosperous and more commercialized economies, and truly malnourished and impoverished peoples highly dependent on limited government assistance. The twenty-first century poses significant economic challenges for the entire region. Its nations have small populations and domestic markets, the realm retains an enduringly peripheral position in the global economy, and its diminishing resource base is set amid a vulnerable natural environment.

Uncertain Avenues to Affluence

The per capita gross national products in both Australia ($20,090) and New Zealand ($15,720) exemplify countries with living standards well above the global norm. Indeed,

Table 14.2 Economic Indicators

Country	GNP per Capita ($U.S., 1996)	Total GNP (Millions of $U.S., 1996)	PPP* ($Intl, 1996)	Real Annual Growth % per Capita, 1990–1996
Australia	20,090	367,802	19,870	2.7
Fed. St. Micronesia	—	—	—	—
Fiji	2,470	1,983	4,070	0.6
French Polynesia	—	—	—	—
Guam	—	—	—	—
Marshall Islands	1,890	108	—	–4.0
New Caledonia	—	—	—	—
New Zealand	15,720	57,135	16,500	1.7
Palau	—	—	—	—
Papua New Guinea	1,150	5,049	2,820	5.0
Samoa	—	—	—	—
Solomon Islands	900	349	2,250	1.3
Vanuatu	1,290	229	3,020	–1.1

*Purchasing power parity.

Source: *The World Bank Atlas*, 1998.

these two nations are generally grouped among the world's developed countries, along with much of Europe, North America, and Japan. Their many economic assets include highly educated populations, a diverse base of natural resources, and a modern, highly integrated urban and industrial infrastructure. Even so, the relative economic affluence of these two South Pacific nations declined late in the twentieth century due to slower economic growth that was still heavily dependent on the extraction and export of raw materials. Can the trend be reversed? Both nations are currently exploring ways to make the most of their natural resources while at the same time creating a more diversified economy integrated with the global economy.

The Australian Economy Much of Australia's past economic affluence has been built upon the cheap extraction and export of abundant raw materials. Export-oriented agriculture has long been one of the key supports of Australia's economy. Australian agriculture is highly productive in terms of labor input, and it produces a wide variety of both temperate and tropical crops, as well as huge quantities of beef and wool for world markets. While farm exports are still important to the economy, the mining sector has grown much more rapidly since 1970. Today, Australia is one of the world's mining superpowers (Figure 14.35). Beginning with the 1850s gold rush in Victoria, the past 150 years have seen a huge expansion in the nation's mineral output, a pattern that accelerated further in the 1960s. Among its many assets are coal; rich reserves of iron ore, particularly in Western Australia; and an assortment of other metals, such as bauxite (for aluminum), copper, gold, nickel, lead, and zinc. Indeed, the New South Wales-based Broken Hill Proprietary Company (BHP) is one of the world's largest mining corporations. Given the country's small domestic population, much of that mineral base is consumed elsewhere, making Australia a huge source for raw materials for developed nations, particularly Japan.

Although the nation's mineral wealth served it well in the commodity boom years of the 1970s, recent global economic trends have been less favorable for the country's economic fortunes. The nation's manufacturing sector continues to be a concern. Many of its industries are oriented around the simple processing of raw materials, but often these are not highly profitable, value-added operations. At the same time, small domestic markets discourage the blossoming of more specialized industries. What has been lacking is an entrepreneurial edge in fast-growing high-technology and information-based industries that have powered recent economic expansion in North America, Japan, and Europe. That may be changing: recent changes in government policy have encouraged more domestic investment, higher savings, and more rapid economic growth. Growing numbers of Asian immigrants and economic links with potential Asian markets also bode well for

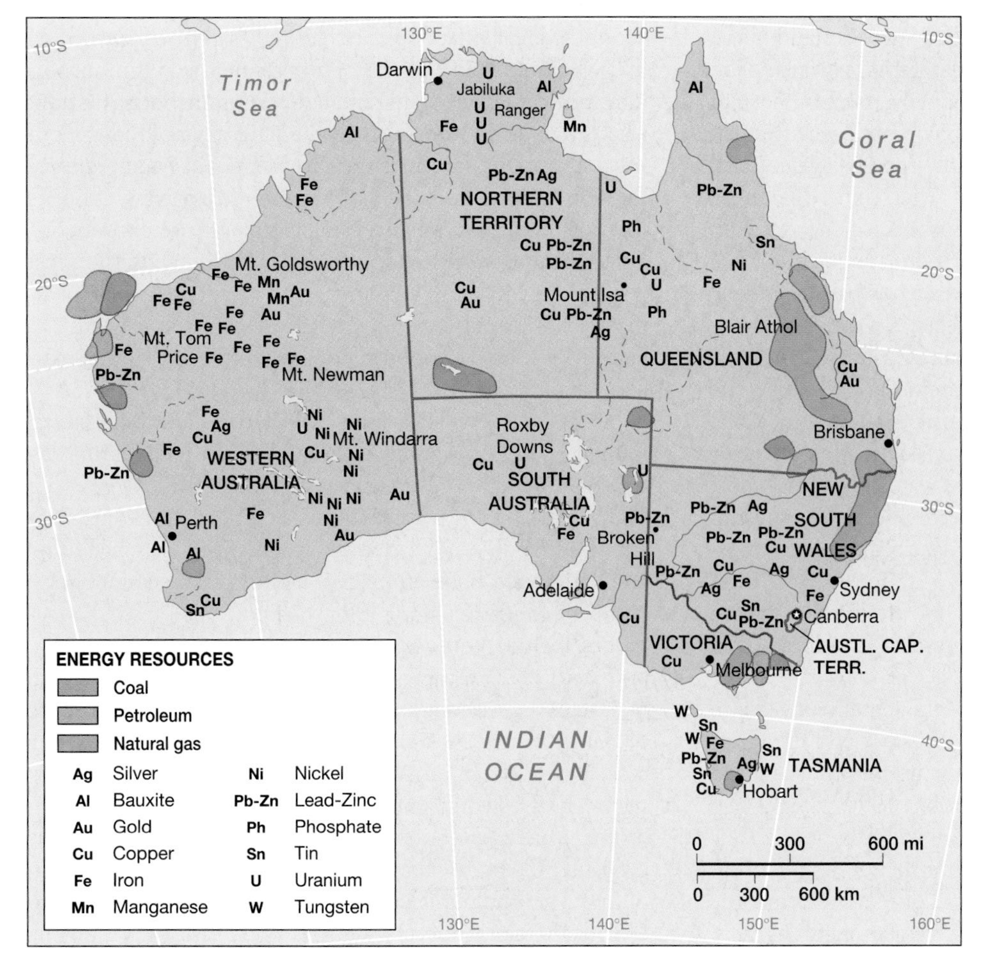

◄ **Figure 14.35 Australia's mineral resources** A treasure trove of natural resources, Australia has a sizable mining industry, much of it oriented to global markets. The country is a major producer of gold, lead-zinc, and iron ore. In addition, rich coal reserves as well as more modest supplies of oil and natural gas contribute to the national economy. *(Modified from Clawson and Fisher, 1998,* World Regional Geography, *Upper Saddle River, NJ: Prentice Hall)*

the future. In addition, an expanding tourism industry is helping to diversify the economy. More than 7 percent of the nation's workforce is now devoted to serving the needs of more than four million visitors annually. Popular destinations include Melbourne and Sydney, as well as recreational settings such as Queensland's resort-filled Gold Coast, the Great Barrier Reef, and the vastness of the arid outback. Asian visitation, particularly by the Japanese, rose sharply in the 1990s, and many tourist towns such as Queensland's Cairns and Port Douglas now feature bilingual resort settings tailored to their newly Nipponized clientele.

Australia's affluence is distributed widely but unevenly around the country. As is typical in the developed world, much of the wealth is concentrated in the major cities, especially those of the southeast. The high-income suburbs of Melbourne and Sydney parallel the North American patterns of Boston and Vancouver. Overall, however, no Australian state is significantly richer or poorer than any other, and the disparity of income between the most prosperous fifth of the population and the least prosperous fifth is relatively small in global terms. The Aborigine population, however, is often prone to poverty. Whether living in cities or in the outback, most Aborigines have low standards of living, and their incomes average only 65 percent of the national average. Although a small number of Aborigines still subsist off the land in the large reserves, most occupy a marginal place in the cash economy of the cities.

New Zealand's Economic Challenge New Zealand is also a wealthy country, but it is somewhat less prosperous than Australia. Its per capita GNP stands at only about 60 percent that of the United States. Before 1970, New Zealand relied heavily on exports to Great Britain, mostly agricultural products such as wool and butter. The strategy faltered, however, once Britain joined the European Union, which adopted stringent agricultural protection policies. Unlike Australia, however, New Zealand lacked a rich base of mineral resources to export to global markets. By the 1980s the country slipped into a serious recession. Eventually the New Zealand government enacted drastic reforms. Previously the country had been noted for its lofty taxes, high levels of social welfare, and state ownership of large economic concerns. Suddenly, "privatization" became the watchword, and most state industries were sold off to private parties. As a result, New Zealand has been transformed into one of the most market-oriented countries of the world. To diversify from its traditional export base, the nation has also promoted more aggressive development of its timber resources, fisheries, and tourist industry.

New Zealand's economic prospects remain uncertain. Whether its growth-oriented strategy will pay off in the long run remains to be seen, particularly as the country struggled again with economic stagnation in the late 1990s. Critics also worry that growing income disparities will undermine New Zealand's social foundation. While suburbanites in the north prosper from some of the country's economic reforms, the country's hinterland populations, still often bound to the land, continue to struggle as global prices for their products decline. In addition, the Maori population suffers some of the same economic disparities as Australia's Aborigines.

Oceania's Economic Diversity Varied economic activities shape the Pacific island nations. One way of life is still oriented around subsistence-based economies, such as shifting cultivation or fishing. In other settings, the commercial extractive economy dominates, with large-scale plantations, mines and timber activities often competing in land and labor with the traditional subsistence sector. Elsewhere, the tremendous growth in global tourism has fundamentally transformed the economic geographies of many island settings, forever changing the way that people make a living across much of the vast oceanic realm. Many island nations also benefit from direct subsidies and economic assistance that flow from present and former colonial powers, all designed to promote development and stimulate employment.

Melanesia is by most measures the least-developed and poorest part of Oceania. Melanesian countries have benefited less from tourism or from subsidies from wealthy colonial and ex-colonial powers. Most Melanesians live in remote villages that remain partially isolated from the modern economy. The Solomon Islands, for example, with few industries other than fish canning and coconut processing, has a per capita GNP of only $900 per year. Papua New Guinea's economy is somewhat more commercially oriented, supporting a per capita GNP of $1,150. Outbound shipments of coconut products and coffee have increasingly been supplemented with the rapid development of tropical hardwoods in the forest products sector. Gold- and copper-mining ventures have also dramatically transformed the landscape, although political instability has intermittently suspended mineral production in settings such as Bougainville. In New Guinea's interior highlands, however, many villages are entirely oriented around subsistence activities. Fiji remains the most prosperous Melanesian country with a per capita GNP of almost $2,500. Fiji is a major sugar producer and it has developed a tourist economy popular with North Americans and Japanese.

Among the smaller islands of Melanesia and Micronesia, mining economies dominate New Caledonia and Nauru. New Caledonia's nickel reserves, the world's second largest, are both a blessing and a curse: they sustain much of the island's export economy, yet they inevitably will dwindle in the future. Dramatic price fluctuations for the industrial economy also hamper economic planning for the French colony. Other activities include coffee growing, cattle grazing, and tourism. To the north the tiny, phosphate-rich island of Nauru also maintains its economic dependence on mining. The citizens of Nauru, for the most part, live directly off the royalties they receive from the mines. Much of the money gained in mining has been invested in a massive trust fund. The Nauruan government has assured its citizens that they will be prosperous even after the phosphate deposits have been exhausted. Much of the trust fund money, however, recently found its way into the Asian real estate market and these risky investments proved less than stellar. It now seems possible that the Nauruans will end up with little more than an environmentally devastated island.

Elsewhere in Micronesia and Polynesia, conditions depend upon the viability of local subsistence economies or upon economic linkages to the world beyond. Many archipelagos export a few food products, but native populations survive mainly on fish, coconuts, bananas, and yams. Some island groups, however, enjoy substantial subsidies from either France or the United States, although such support often comes with a political price. Change is also afoot in some of these island settings: In 1999 Japan agreed to build a spaceport for its future shuttlecraft on Micronesia's Christmas Island (Kiribati) and the Marshall Islands are the site of a planned industrial park financed by mainland Chinese.

Other island groups have been fundamentally transformed by tourism. In Hawaii, more than one-third of the state's economy flows directly from tourist dollars. With almost 5 million visitors annually (including more than 1.25 million from Asia), Hawaii represents all of the classic benefits and risks of the tourist economy. While job creation and economic growth have reshaped the island realm, congested highways, high prices, and the unpredictable spending habits of tourists have left the region vulnerable to future problems. Elsewhere, French Polynesia has long been a favored destination of the international jet set (Figure 14.36). More than 20 percent of French Polynesia's GNP is derived from tourism, making it one of the most prosperous parts of the Pacific. More recently, Guam has emerged as a favorite destination of Japanese and Korean tourists, especially those on honeymoons. Indeed, on a smaller scale, tourism is on the rise across much of the island realm, and many economic planners see it as the avenue to future prosperity. Critics, however, warn that tourist jobs tend to be low-paying (outsiders are often imported for managerial positions), the local quality of life may in fact decline with the presence of tourists, and the natural environments of small islands can quickly be overwhelmed by an influx of demanding visitors.

▲ **Figure 14.36 Tahitian resort** Luxury resort settings in Tahiti (near Papeete) exemplify a growth industry in the South Pacific. While bringing important investment capital, such ventures reorder the region's social and economic structure as well as refashion its cultural landscape. *(Bob & Suzanne Clemenz)*

The Global Economic Setting

Even as Australia and Oceania remain in the global economic hinterlands, their relationships with the world economy will increasingly shape the quality of life and the prospects for development within the region. Several intriguing questions remain. Will future trade patterns in the region shift away from Europe and North America in favor of closer links with Asia? Will there be a move away from the traditional extractive economies that have shaped economic development in the region everywhere from giant Australia to tiny Nauru? Can growing economic linkages within the region make a difference in a part of the world whose total population is less than that of California? None of these questions have ready answers, but they suggest how residents of the region will need to ponder their entry into the twenty-first century's global economy.

Many international trade flows link the area to the far reaches of the Pacific and beyond. Australia and New Zealand dominate global trade patterns in the region (Figure 14.37). In the past 30 years, ties to Great Britain, the British Commonwealth, and Europe have weakened in comparison with growing trade links with Japan, East Asia, the Middle East, and the United States. Australia, for example, now imports more manufactured goods from Japan and the United States than it does from Britain and Europe. Similarly, Australian exports, principally raw materials, flow increasingly toward destinations in Asia and North America, although Europe remains an important trading partner. Other global economic ties have come in the form of capital investment in the region. U.S. and Japanese banks and other financial institutions now dot the South Pacific landscape from Sydney to Suva. Both Australia and New Zealand also participate in the **Asia-Pacific Economic Co-operation Group (APEC),** an organization designed to foster economic development in Southeast Asia and in the Pacific Basin. The region's economic ties with Asia also carry risks, however; the recent Asian downturn in the late 1990s, for example, dampened the popular Korean tourist trade in New Zealand and slowed Asian demand for a variety of Australian raw material exports.

Economic integration within the region has also been promoted. In 1982 Australia and New Zealand signed the **Closer Economic Relationship (CER) Agreement,** which successfully slashed trade barriers between the two countries. New Zealand benefited because it opened larger Australian markets to New Zealand exports, and Australian corporate and financial interests gained new access to New Zealand business opportunities (see "Economic Change: The Awkward 'Australianization' of the New Zealand Economy"). Today, more than 20 percent of New Zealand's imports and

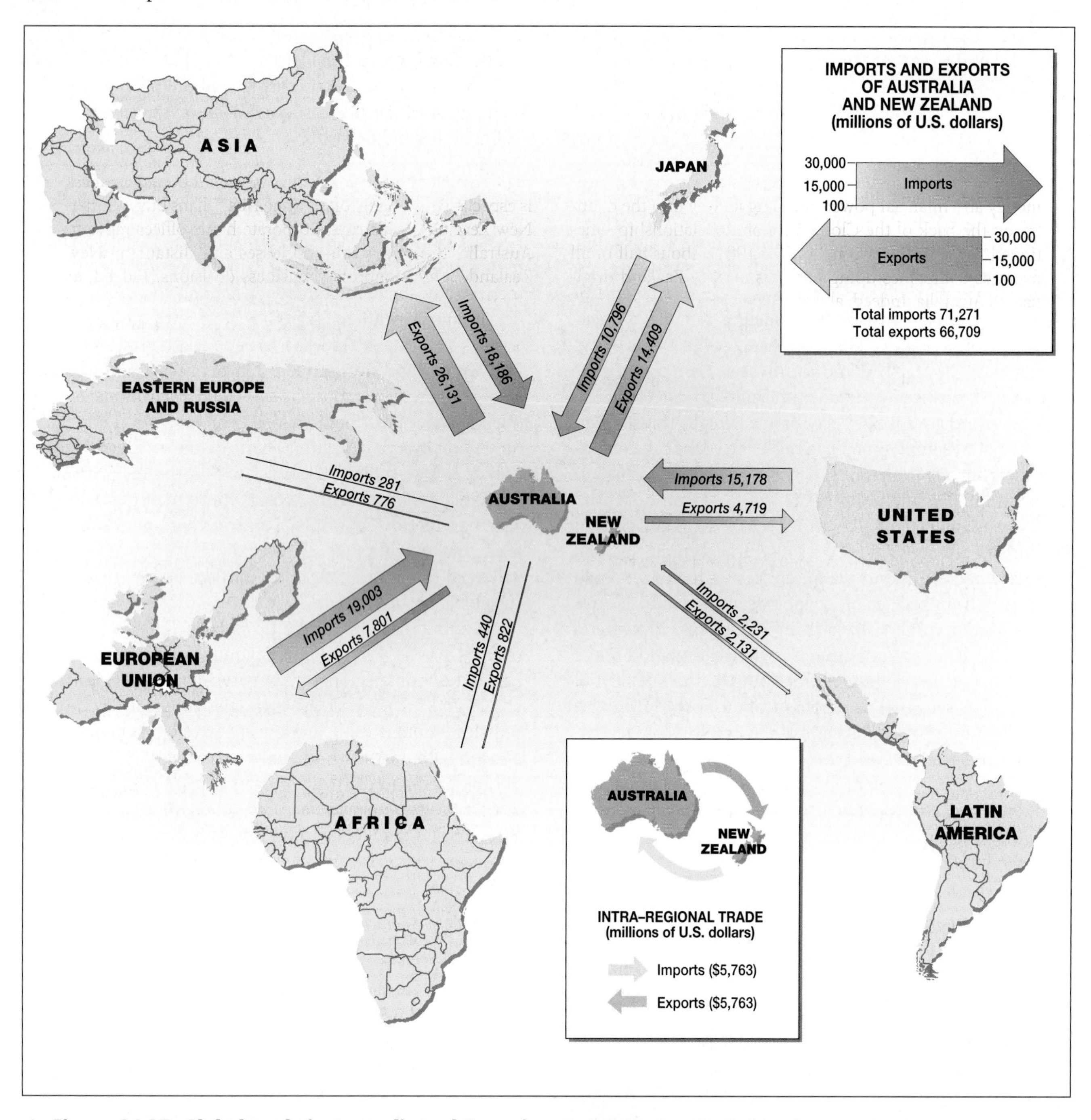

▲ **Figure 14.37 Global trade in Australia and Oceania** Trade flows from Australia and New Zealand demonstrate growing economic ties with nearby Asia. Persisting links with the United States and Europe (particularly Britain) also remain important, especially for importing needed manufactured goods. Important intraregional linkages also cement connections between Australia and New Zealand. *(Data from Euromonitor,* International Marketing Data and Statistics, 1997)

exports come from Australia and the pattern of regional free trade is likely to strengthen in the future. Smaller nations of Oceania, while often closely tied to countries such as Japan, the United States, and France, also benefit from their proximity to Australia and New Zealand. More than half of Fiji's imports come from those two nearby nations, and other countries such as Papua New Guinea, Vanuatu, and Solomon Islands enjoy a similarly close trading relationship with their more-developed Pacific neighbors.

Enduring Social Challenges

Australians and New Zealanders enjoy high levels of social welfare but face some of the same challenges evident elsewhere in the developed world (Table 14.3). Excellent schools

ECONOMIC CHANGE The Awkward "Australianization" of the New Zealand Economy

New Zealanders are nervous. Casting a glance across the Tasman Sea, they see an encroaching onslaught of Australian money and financial power muscling its way into the country on the back of the Closer Economic Relationship sanctioned between the two nations in 1982. About half of all overseas companies doing business in New Zealand originate in Australia. Indeed, almost 20 percent of New Zealand's largest 100 corporations are actually controlled by Australians. The trend is particularly pervasive in the all-important financial sector: 70 percent of New Zealand's banking assets are now managed by Australian institutions, and recent economic reforms have even allowed the Bank of New Zealand to be bought out by the National Australia Bank. As one business manager from Victoria put it, "I basically treat New Zealand as another state of Australia." Meanwhile, Australian mass-media magazines, movies, and television programs also flood the New Zealand market. Naturally, many New Zealanders are more than a little uneasy with this coercive brand of Australian "mateship"!

The current economic trend may have many long-lasting consequences. One outcome of the corporate coziness is especially unsettling: often when Australians buy out their New Zealand compatriots, corporate home offices move to Australia, thus triggering job losses and distancing New Zealanders from important business decisions. Indeed, as larger multinational corporations move into the region, they increasingly avoid setting up shop in New Zealand, preferring instead to establish a single South Pacific office in Sydney, Melbourne, or Brisbane. Some New Zealand investment bankers predict the New Zealand stock market will merge with its Australian counterpart and that a single Australian-dominated currency will follow. As in Europe, such moves make economic sense even as they ruffle this South Pacific nation's cultural and political feathers.

Source: Adapted from Simon Robinson, "Calling Australia Home . . . Control of the New Zealand Economy Is Shifting Across the Tasman," *Time International*, November 9, 1998.

Table 14.3 Social Indicators and Status of Women

Country	Life Expectancy at Birth		Under Age 5 Mortality, per 1,000 Live Births		Secondary School Enrollment %		Female Labor Force Participation (% of total)
	Male	Female	1960	1995	Male	Female	
Australia	75	81	24	8	83	86	43
Fed. St. Micronesia	65	67	—	—	—	—	—
Fiji	61	65	97	—	64	65	28
French Polynesia	68	72	—	—	68	87	—
Guam	72	76	—	—	—	—	—
Marshall Islands	60	63	—	—	—	—	—
New Caledonia	68	76	—	—	82	88	—
New Zealand	69	75	26	9	93	93	44
Palau	—	—	—	—	—	—	—
Papua New Guinea	56	57	248	95	15	10	42
Samoa	—	—	—	—	—	—	—
Solomon Islands	68	73	—	—	21	13	46
Vanuatu	—	—	—	—	23	18	—

Sources: *Population Reference Bureau Data Sheet, 1998,* Life Expectancy (M/F); *World Resources Institute, 1996–97,* Under-5 Mortality Rate; *Population Reference Bureau Data Sheet, 1996,* Secondary School Enrollment (M/F); *The World Bank Atlas, 1998,* Female Participation in Labor Force.

and universities offer comprehensive training opportunities in both nations and 80 to 93 percent of secondary-school-aged males and females remain enrolled. Lifespans average more than 70 years in both countries and rates of child mortality have fallen substantially since 1960. Paralleling patterns in North America and Europe, cancer and heart disease are leading causes of death, and alcoholism is a persisting social problem, particularly in Australia. Unfortunately, Australia's rate of skin cancer is among the world's highest, the result of having a largely fair-skinned, outdoors-oriented population from northwest Europe in a sunny, low-latitude setting. Overall, Australia's Medicare program (initiated in 1984) and New

Zealand's targeted system of social services provide high-quality health care to their populations. The position of women is also high in both countries, including participation in the workforce. In particular, women have recently played key political roles in New Zealand.

Not surprisingly, the social conditions of the Aborigines and Maoris are much less favorable than for the population overall. Schooling is irregular for many indigenous peoples, and levels of advanced post-secondary education for Aborigines (12 percent) and Maoris (14 percent) remain far below national averages (32 to 34 percent) in the two countries. Many other social measures reflect the pattern, as well. For example, less than one-third of Aboriginal households own their own home, while more than 70 percent of white Australian households are homeowners. Furthermore, considerable discrimination against native peoples persists in both countries, a situation that has been aggravated and more publicized with the recent assertion of indigenous political rights and land claims. As with North American African Americans, Hispanic, and Native American populations, no simple social policies offer solutions to these enduring disparities.

Levels of social welfare in Oceania are higher than one might expect based on the region's economic circumstances. Many of its countries and colonies have invested heavily in health and education services and have achieved considerable success. For example, the average life expectancy in the Solomon Islands, one of the world's poorer countries as measured by per capita GNP figures, is an impressive 71 years. By other social measures as well, the Solomon Islands and a number of other Oceania states have reached higher levels of human well-being than is the case in most Asian and African countries with similar levels of economic output. This is partly a result of successful policies, but it also reflects the relatively healthy natural environment of Oceania. Many of the tropical diseases that are so debilitating in Africa simply do not exist within the region.

Papua New Guinea (PNG) is the major exception to these relatively high levels of social welfare. Here the average life expectancy is only 56 years, and a recent study suggests that 34 percent of its young people suffer from malnutrition, particularly protein deficiencies. Adult illiteracy (28 percent) is also more prevalent in PNG than elsewhere in the region. Unfortunately, PNG has found it difficult to provide even basic educational and health services to its people. The country possesses the largest expanse of land in Melanesia, and much of the population lives in relatively isolated villages in the rugged central highlands.

Conclusion

Relative location and small populations contribute to the enduringly peripheral position of Australia and Oceania on the global stage. One of the last habitable portions of the planet to be occupied, the region was also a late chapter in the story of Europe's global colonial expansion. Similarly, the region's contemporary political geography reveals its still fluid character as countries struggle to disentangle themselves from colonial ties and assert their own political identities. Globalization complicates the process, both culturally and politically: new people, many from Asia, are adding ethnic variety as well as providing flows of investment capital that cut across old colonial relationships. In addition, the spatial isolation that water and distance once offered is fast disappearing, casting even the most insulated societies of the Australian outback and the Polynesian periphery rapidly into the postindustrial world of the twenty-first century. The transformation has not been easy or predictable: some native peoples have embraced the attractions of modern life, while others resist the cultural, economic, and political imperatives of the world beyond. Meanwhile, the natural environment, already broadly if unevenly transformed by earlier indigenous peoples and European colonists, has witnessed accelerating changes in the past 50 years as urbanization, extractive economic activities, and recent amenity-oriented investments reconfigure the landscape. Today, sprawling suburbs, open-pit copper and gold mines, and thatched-hut resorts signal the dynamic, sometimes intrusive signatures of a once-distant world now brought near by modern communications, improved air travel, and international trade flows.

Australia remains dominant in the greater Pacific realm, and should continue to play a pivotal role in the future as a principal economic entry point into the region. Its large land area, resource base, and population are complemented by its increasing emergence as a regional finance center for the South Pacific as well as nearby portions of Southeast and East Asia. Indeed, the Asian connection seems destined to play a key part in Australia's future economic and even cultural identity. Clearly, the country is no longer simply a distant outpost of Europe. Asian migrants, many of them skilled and affluent, add a dynamic and creative component to Australia's largely European population. In addition, intermingled flows of money, raw materials, and manufactured goods also bind Australia to its Asian neighbors. Will "Asianization" increasingly make Australia a part of that continent, or will political, economic, and cultural resistance to such linkages once again assert themselves? These important questions remain to be answered.

Even while Australia is the giant of the Pacific, it is a secondary player in the global arena. The entire population of the country is about half that of California, and its total economic output is only about as large as that of Texas. Australian optimists, however, have long contended that their country will achieve a more important position in the global economy in decades to come. While it remains to be seen whether Australia's current economic vigor will persist, or whether it can develop more high-growth industries, it seems likely that the nation can remain prosperous with enviable social and environmental conditions.

If Australia sits at the verge of Asia, New Zealand might be viewed as sitting on the threshold of Polynesia. New Zealand, of course, was once an integral part of Polynesia, but by the early 1900s it seemed to many to have been transformed into a little England exiled to the South Pacific. Today, however, its Maori population is growing quickly, joined by immigrants from other parts of Polynesia. New Zealand has taken an active role in the political affairs of the entire Pacific basin, and it is in many respects Oceania's leading state. Overall, its multicultural identity and its postindustrial economy continue to evolve. Along with the Polynesian links, closer ties with Australia and Asia will also shape its future cultural milieu, economic base, and political agenda.

Remaining islands of Oceania, while always wedded to the waters that surround and sustain them, have become ever more closely tied to the world beyond. Colonial links often persist. The French presence in the region is dominant, and sometimes unwelcome, particularly when it comes to the issue of nuclear testing. U.S. connections to present possessions and former territories attest to its sphere of influence across the region. In addition, globalization has brought fresh interconnections to these island worlds: Japanese-financed golf courses pop up along tropical shores, Canadian and Australian mining companies invest in the New Guinea Highlands, and Korean newlyweds honeymoon beneath the coconut palms. How are native cultures being transformed in this process, and what new cultural hybrids will emerge as Pacific, European, North American, and Asian peoples mingle among the atolls and archipelagos of the South Pacific? Will integration with the global economy irreparably harm the oceanic environment and diminish its appeal as a tourist destination? Can expanded Exclusive Economic Zones allow these island states to assert their broader economic and political role within the Pacific realm? These pivotal issues remain unresolved in the twenty-first century, but they share a common attribute. As quintessentially geographical concerns, they remind us that the South Pacific realm is a fascinating laboratory to study the interplay between land and life, and the often startling outcomes that result when one portion of the world encounters another in fresh and unanticipated ways.

Key Terms

Aborigine (page 600)
Aotearoa (page 616)
archipelagos (page 588)
Asia-Pacific Economic Co-operation Group (APEC) (page 621)
atoll (page 594)
cargo cult (page 611)
Closer Economic Relationship (CER) Agreement (page 621)
Exclusive Economic Zone (EEZ) (page 617)
haoles (page 610)
high islands (page 593)
hot spot (page 593)
kanakas (page 607)
low islands (page 594)
mallee (page 591)
Maori (page 588)
Melanesia (page 586)
Micronesia (page 588)
microstates (page 614)
Native Title Bill (page 615)
Oceania (page 585)
outback (page 585)
Pidgin English (page 611)
Polynesia (page 588)
tsunamis (page 593)
uncontacted peoples (page 610)
viticulture (page 604)
White Australia Policy (page 608)

Questions for Summary and Review

1. Why does it make sense to group together Australia, New Zealand, and the other Pacific islands as a region? What complexities are involved in doing so?
2. Briefly contrast the topographic settings and climates of Australia's Murray River Valley with that of New Zealand's North Island. What geological and climatic factors account for the similarities and differences?
3. For the following localities, identify a similar North American environment and defend your answer: Alice Springs, Adelaide, Nelson (New Zealand), and Brisbane.
4. Describe how a Pacific high island is created and then transformed into a low island. How do these environments produce distinctive settings for human settlement?
5. What might be some notable similarities between Australian cities and North American cities? Notable differences?
6. Describe the types of farming you might observe in (1) South Australia, (2) the New Guinea Highlands, and (3) Fiji. Explain the differences using both environmental and human variables in your discussion.
7. How do cargo cults and Pidgin English represent similar cultural processes at work in the Pacific world?
8. Discuss some of the key geopolitical issues faced by Pacific Islanders in the early twenty-first century.
9. Why is there such a wide economic gap separating affluent Australia and New Zealand from the rest of Oceania?
10. Why might it be argued that the social and environmental consequences of tourism are even more important than the economic changes it brings?

Thinking Geographically

1. Should New Zealand welcome or be wary of Australian and Japanese investments in the country? Defend your answer.
2. As a local economic development official in Papua New Guinea, make arguments both for and against a large new Malaysian-financed pulp mill being planned for your district.
3. Select a small Pacific island and briefly research its political history since 1800. What do your findings illustrate about more general processes of European colonization and political independence movements over the last 200 years?
4. Identify what you see as the three principal economic challenges facing Papua New Guinea. Discuss how the country's physical and human geography may affect its future prospects.
5. Does the continuing French colonial presence in the South Pacific help or hinder economic development within the region? Why?
6. As an Australian Aborigine in touch with your North American counterparts, what might you observe are some of the principal similarities and differences in the political and economic plight of your peoples?
7. It is 2042 and global warming has accelerated in the past 30 years. You are the ruler of a small Pacific island nation, now being deluged by rising sea levels and altered climatic patterns. What is your strategy to survive, considering these new geographical circumstances?
8. As a maker of educational films, you have been hired for a three-part series titled "The Essential South Pacific." The episodes are to be set in (1) Australia, (2) New Zealand, and (3) Polynesia. Budgets are limited, and only a single location in each of those settings can be used to communicate the personality of the region. Identify three specific spots for filming and explain your answers to the producer of the series.

Regional Novels and Films

Novels

Thea Astley, *The Beachmasters* (1986, Viking)

Eleanor Dark, *The Timeless Land* (1941, Collins)

Derek Hansen, *Sole Survivor* (1999, Simon and Schuster)

Xavier Herbert, *Capricornia* (1938, Publicist)

Keri Hulme, *The Bone People* (1985, Hodder and Stoughton)

Charlotte Jay, *Beat Not the Bones* (1952, Avon)

Herman Melville, *Typee* (1846, Wiley and Putnam)

James Michener, *Hawaii* (1959, Random House)

Arthur Upfield, *The Sands of Windee* (1958, Angus and Robertson)

Patrick White, *Riders in the Chariot* (1961, Viking) and *Voss* (1957, Viking)

B. Wongar, *Walg: A Novel of Australia* (1983, Dodd, Mead)

Films

The Adventures of Priscilla, Queen of the Desert (1994, Australia)

Crocodile Dundee (1986, U.S.)

A Cry in the Dark (1988, Australia)

Goodbye Pork Pie (1981, New Zealand)

The Man from Snowy River (1982, Australia)

Mauri (1988, New Zealand)

On the Beach (1959, U.S.)

Once Were Warriors (1994, New Zealand)

Oscar and Lucinda (1997, U.S.)

The Piano (1993, Australia, filmed in New Zealand)

South Pacific (1958, U.S.)

Thin Red Line (1998, U.S.)

Walkabout (1971, Australia)

Bibliography

Australia. Department of Immigration and Multicultural Affairs. 1998. "Birthplace and Related Data from the 1996 Census, Australia." Report C96.2.0.

Bambrick, Susan, ed. 1994. *The Cambridge Encyclopedia of Australia*. New York: Cambridge University Press.

Bellwood, Peter. 1979. *Man's Conquest of the Pacific: The Prehistory of Southeast Asia and Oceania*. New York: Oxford University Press.

Brookfield, Harold C., ed. 1973. *The Pacific in Transition: Geographical Perspectives on Adaptation and Change*. New York: St. Martin's Press.

Buchholz, Hanns J. 1987. *Law of the Sea Zones in the Pacific Ocean*. Singapore: Institute of Southeast Asian Studies.

Campbell, Ian C. 1990. *A History of the Pacific Islands*. Berkeley: University of California Press.

Cumberland, Kenneth B. 1968. *Southwest Pacific: A Geography of Australia, New Zealand and Their Pacific Island Neighbourhoods*. Christchurch: Whitcombe and Tombs.

Cumberland, Kenneth B., and Whitelaw, James S. 1970. *New Zealand*. Chicago: Aldine Publishing.

Forster, Clive A. 1995. *Australian Cities: Continuity and Change*. Melbourne: Oxford University Press.

Hall, Colin M. 1994. *Tourism in the Pacific Rim*. New York: John Wiley & Sons.

Heathcote, R. L. 1994. *Australia*. New York: John Wiley & Sons.

Hughes, Robert. 1986. *The Fatal Shore: The Epic of Australia's Founding*. New York: Alfred A. Knopf.

Kluge, P. F. 1991. *The Edge of Paradise: America in Micronesia*. New York: Random House.

Lines, William. 1992. *Taming the Great South Land: A History of the Conquest of Nature in Australia*. Berkeley: University of California Press.

McKnight, Tom L. 1995. *Oceania: The Geography of Australia, New Zealand, and the Pacific Islands*. Englewood Cliffs, NJ: Prentice Hall.

Marriott, Edward. 1998. "All Is (Most Definitely) Not Lost." *Geographical Magazine* 70(11), 10–16.

Meinig, D. W. 1962. *On the Margins of the Good Earth: The South Australian Wheat Frontier, 1869–1884*. Chicago: Rand McNally.

———. 1998. *The Shaping of America: Transcontinental America, 1850–1915, volume 3*. New Haven and London: Yale University Press.

Mitchell, Andrew. 1991. *The Fragile South Pacific: An Ecological Odyssey*. Austin: University of Texas Press.

Oliver, Douglas L. 1989. *The Pacific Islands*. Honolulu: University of Hawaii Press.

Pilger, John. 1991. *A Secret Country: The Hidden Australia*. New York: Alfred A. Knopf.

Powell, J. M. 1988. *An Historical Geography of Modern Australia: The Restive Fringe*. New York: Cambridge University Press.

"Reborn: Melbourne." 1997. *The Economist,* August 23.

Ross, Robert, ed. 1998. *Australia: A Traveler's Literary Companion*. San Francisco: Wherabouts Press.

Sekhran, Nik. 1997. "Green or Greed." *Geographical Magazine* 69(10), 75–81.

Shadbolt, Maurice. 1988. *Reader's Digest Guide to New Zealand*. Sydney: Reader's Digest.

Spate, O. H. K. 1968. *Australia*. New York: Praeger.

Terrill, Ross. 1987. *The Australians*. New York: Simon & Schuster.

Theroux, Paul. 1992. *The Happy Isles of Oceania: Paddling the Pacific*. New York: Ballantine Books.

Ward, Gerard, ed. 1972. *Man in the Pacific Islands: Essays on Geographical Change in the Pacific*. New York: Oxford University Press.

Woodard, Colin. 1998. "Marshall Islands: You Can't Go Home Again." *Bulletin of the Atomic Scientists* 54(5), 10–12.

GLOSSARY

Aborigine The indigenous inhabitants of Australia.

acid rain Harmful form of precipitation high in sulfur and nitrogen oxides. Caused by industrial and auto emissions, acid rain damages aquatic and forest ecosystems in regions such as eastern North America and Europe.

African diaspora The forced dispersion of African peoples from their native lands to various parts of the Americas and the Middle East through the slave trade.

agricultural density The number of farmers per unit of arable land is a country's agricultural density. This figure indicates the number of people who directly depend upon agriculture and it is an important indicator of population pressure in places where rural subsistence dominates.

alluvial fan A fan-shaped deposit of sediments dropped by a river or stream flowing out of a mountain range.

Altiplano The largest intermontane plateau in the Andes, it straddles Peru and Bolivia and ranges in elevation from 10,000 to 13,000 feet (3,000 to 4,000 meters).

altitudinal zonation The relationship between higher elevations, cooler temperatures, and changes in vegetation that result from the environmental lapse rate (averaging 3.5 °F for every 1,000 feet). In Latin America four general altitudinal zones exist: tierra caliente, tierra templada, tierra fria, and tierra helada.

animism A wide variety of tribal religions based upon the worship of nature's spirits and human ancestors.

anthropogenic An adjective for human-caused change to a natural system, such as the atmospheric emissions from cars, industry, and agriculture that are causing global warming.

anthropogenic landscape A landscape heavily transformed by human agency.

Aotearoa Maori name for New Zealand, meaning "Land of the Long White Cloud."

apartheid The policy of racial separateness that directed the separate residential and work spaces for whites, blacks, coloureds, and Indians in South Africa for nearly 50 years. In 1994 it was abolished when the African National Congress came to power.

archipelagos Island groups, often oriented in an elongated pattern.

areal differentiation The geographic description and explanation of spatial differences on Earth's surface; this includes both physical as well as human patterns.

areal integration The geographic description and explanation of how places, landscapes, and regions are connected, interactive, and integrated with each other.

Asia-Pacific Economic Co-operation Group (APEC) An international group of Asian and Pacific Basin nations that fosters coordinated economic development within the region.

Association of Southeast Asia Nations (ASEAN) A supranational geopolitical group linking together the ten different states of Southeast Asia.

atoll Low, sandy islands made from coral, often oriented around a central lagoon.

autonomous areas Minor political subunits created in the former Soviet Union and designed to recognize the special status of minority groups within existing republics.

autonomous region In the context of China, autonomous regions are provinces that have been granted a certain degree of political and cultural autonomy, or freedom from centralized authority, owing to the fact that they contain large numbers of non-Han Chinese people. Critics contend that they actually have little true autonomy.

Baikal-Amur Mainline (BAM) Railroad Key central Siberian railroad connection completed in Soviet era (1984), which links the Yenisey and Amur rivers and which parallels the Trans-Siberian Railroad.

Balfour Declaration Statement issued by Great Britain in 1917 pledging its support for establishing a home for the Jewish people in Palestine.

balkanization Geopolitical process of fragmentation of larger states into smaller ones through independence of smaller regions and ethnic groups. The term takes its name from the geopolitical fabric of the Balkan region.

barrios Urban Hispanic neighborhoods, often associated with low-income groups in North America.

Berlin Conference The 1884 conference that divided Africa into European colonial territories. The boundaries created in Berlin satisfied European ambition but ignored indigenous cultural affiliations. Many of Africa's civil conflicts can be traced to ill-conceived territorial divisions crafted in 1884.

biofuels Energy sources derived from plants or animals. Throughout the developing world, wood, charcoal, and dung are primary energy sources used for cooking and heating.

biome Ecologically interactive flora and fauna adapted to a specific environment. Examples would be a desert or a tropical rain forest.

bioregion A spatial unit or region of local plants and animals adapted to a specific environment, such as a tropical savanna.

Bolsheviks A faction within the Russian Communist movement led by Lenin that successfully took control of the country in 1917.

boreal forest Coniferous forest found in high-latitude or mountainous environments of the Northern Hemisphere.

brain drain Migration of the best-educated people from developing countries to developed nations where economic opportunities are greater.

British East India Company Private trade organization that acted as arm of colonial Britain—backed by the British army—in monopolizing trade in South Asia until 1857, when it was abolished and replaced by full governmental control.

buffer zone An array of non-aligned or friendly states that "buffer" a larger country from invasion. In Europe, keeping a buffer zone has been a long-term policy of Russia (and also of the former Soviet Union) to protect its western borders from European invasion.

Bumiputra The name given to native Malay (literally, "sons of the soil"), who are given preference for jobs and schooling by the Malaysian government.

Burakumin The indigenous outcast group of Japan, a people whose ancestors reputedly worked in leathercraft and other "polluting" industries.

bustees Settlements of temporary and often illegal housing in Indian cities; caused by rapid urban migration of poorer rural people and the inability of Indian cities to provide housing for this rapidly expanding population.

capital leakage The gap between the gross receipts an industry (such as tourism) brings into a developing area and the amount of capital retained.

Caribbean Community and Common Market (CARICOM) A regional trade organization established in 1972 that includes former English colonies as its members.

Caribbean diaspora The expulsion of Caribbean people to other parts of the Americas and Europe due to economic necessity.

cargo cult Quasi-animist Melanesian religious practice, originating with military cargo supply drops during World War II.

caste system Complex division of South Asian society into different hierarchically ranked hereditary groups. Most explicit in Hindu society, but also found in other cultures to a lesser degree.

Central Place theory A theory used to explain the distribution of cities, and the relationships between different cities, based on retail marketing.

centralized economic planning An economic system in which the state sets production targets and controls the means of production.

centrifugal forces Those cultural and political forces, such as linguistic minorities, separatists, and fringe groups, that pull away from and weaken an existing nation-state.

centripetal forces Those cultural and political forces, such as a shared sense of history, a centralized economic structure, or the need for military security, that promote political unity in a nation-state.

chain migration A pattern of migration in which a sending area becomes linked to a particular destination, such as Dominicans with Queens, New York.

chernozem soils A Russian term for a dark, fertile soil, often associated with grassland settings in southern Russia and Ukraine.

Chipko movement The "tree-hugging" movement of northern India in which women, drawing upon Hindu tradition, attempt to save forests from destruction by embracing the trees as loggers approach.

circular migration Temporary labor migration in which an individual seeks short-term employment overseas, saves money, and then returns home.

clan A social unit that is typically smaller than a tribe or ethnic group but larger than a family, based supposedly on descent from a common ancestor.

climate region A region of similar climatic conditions. An example would be the Marine West Coast climate regions, found on the west coasts of North America and Europe.

climograph Graphs of average annual temperature and precipitation data, by month and season.

Closer Economic Relationship (CER) Agreement An agreement signed in 1982 between Australia and New Zealand designed to eliminate all economic and trade barriers between the two countries.

Cold War The struggle between the United States and the Soviet Union that was conducted between 1946 and 1991.

collective farms Group-farmed agricultural units organized around state-mandated production goals.

collectivization The agglomeration of small, privately owned agricultural parcels into larger, state-owned farms. This was a central component of communism in eastern Europe and the Soviet Union.

colonialism The formal, established (mainly historical) rule over local peoples by a larger, imperialist government for the expansion of political and economic empire.

Coloureds A racial category used throughout South Africa to define people of mixed European and African ancestry.

Columbian Exchange An exchange of people, diseases, plants, and animals between the Americas (New World) and Europe/Africa (Old World) initiated by the arrival of Christopher Columbus in 1492.

comarca Provinces or homelands demarcated by indigenous groups in Panama that acknowledge their control over long-settled areas within the state.

command economy Centrally planned and controlled economies, generally associated with socialist or communist countries, in which all goods and services, along with agricultural and industrial products, are strictly regulated. This was done during the Soviet era in both the Soviet Union and its eastern European satellites.

Commonwealth of Independent States (CIS) A loose political union of former Soviet republics (without the Baltic states) established in 1992 after the dissolution of the Soviet Union.

concentric zone model A simplified description of urban land use: a well-defined central business district (CBD) is surrounded by concentric zones of residential activity, with higher-income groups living on the urban periphery.

Confucianism The philosophical system developed by Confucius in the sixth century B.C.

connectivity The degree to which different locations are linked with one another through transportation and communication infrastructure.

continental climate Climate regions in continental interiors, removed from moderating oceanic influences, that are characterized by hot summers and cold winters. At least one month must average below freezing.

convection cells Large areas of slow-moving molten rock in Earth's interior that are responsible for moving tectonic plates.

copra Dried coconut meat.

Cossacks Highly mobile, Slavic-speaking Christians of the southern Russian steppe that were pivotal in expanding Russian influence in sixteenth- and seventeenth-century Siberia.

Council of Mutual Economic Assistance (CMEA) The communist agency that coordinated economic planning and development between the Soviet Union and satellite countries in eastern Europe.

counterurbanization The movement of people out of metropolitan areas toward smaller towns and rural areas.

creolization The blending of African, European, and even some Amerindian cultural elements into the unique sociocultural systems found in the Caribbean.

crony capitalism A system in which close friends of a political leader are either legally or illegally given business advantages in return for their political support.

cultural assimilation The process in which immigrants are culturally absorbed into the larger host society.

cultural imperialism The active promotion of one cultural system over another, such as the implantation of a new language, school system, or bureaucracy. Historically, this has been associated primarily with European colonialism.

cultural landscape Primarily the visible and tangible expression of human settlement (house architecture, street patterns, field form, etc.) but also includes the intangible, value-laden aspects of a particular place and its association with a group of people.

cultural nationalism A process of protecting, either formally (with laws) or informally (with social values) the primacy of a certain cultural system against influences (real or imagined) from another culture.

cultural syncretism The blending of two or more cultures, which produces a synergistic third culture that exhibits traits from all cultural parents.

culture Learned and shared behavior by a group of people empowering them with a distinct "way of life"; culture includes both material (technology, tools, etc.) and immaterial (speech, religion, values, etc.) components.

culture hearth An area of historical cultural innovation.

Cyrillic alphabet Based on the Greek alphabet, this is used by Slavic languages heavily influenced by the Eastern Orthodox Church. Attributed to the missionary work of St. Cyril in the ninth century.

decolonialization The process of a former colony gaining (or regaining) independence over their territory and establishing (or reestablishing) an independent government.

demographic transition A four-stage model of population change derived from the historical decline of the natural rate of increase as a population becomes increasingly urbanized through industrialization and economic development.

denuclearization The process whereby nuclear weapons are removed from an area and dismantled or taken elsewhere.

dependency theory A popular theory to explain patterns of economic development in Latin America. Its central premise is that underdevelopment was created by the expansion of European capitalism into the region that served to develop "core" countries in Europe and to impoverish and make dependent peripheral areas such as Latin America.

desertification The spread of desert conditions into semiarid areas owing to improper management of the land.

diaspora The scattering of a particular group of people over a vast geographical area. Originally, the term *diaspora* referred to the migration of Jews out of their original homeland, but now it has been generalized to refer to any ethnic dispersion.

domestication The purposeful selection and breeding of wild plants and animals for cultural purposes.

Domino Theory A U.S. geopolitical policy of the 1970s that stemmed from the assumption that if Vietnam fell to the Communists, the rest of Southeast Asia would soon follow.

Dravidian language One of the earliest (from perhaps 4,000 years ago) language families and, unlike Hindi, not Indo-European. Once spoken throughout South Asia, Dravidian languages are now found only in southern India and part of Sri Lanka.

dual primacy A type of urban primacy in which a pair of cities dominates all others in a country in terms of size and economic importance: São Paulo and Rio de Janeiro are examples of dual primacy for Brazil.

Eastern Orthodox Christianity A loose confederation of self-governing churches in eastern Europe and Russia that are historically linked to Byzantine traditions and to the primacy of the patriarch of Constantinople (Istanbul).

ejido Communally held lands created after the Mexican Revolution in 1910.

El Niño An abnormally large warm current that appears off the coast of Ecuador and Peru in December. During an El Niño year, torrential rains can bring devastating floods along the Pacific coast and drought conditions in the interior continents of the Americas.

encomienda A social and economic system used in the early Spanish colonies where there were large native populations. Groups of Indians would be "commended" to a Spaniard, who would exact tribute from them in the form of labor or products. In return, the Spaniard was obligated to educate the Indians in the Spanish language and the Catholic faith.

entrepôt A city and port that specializes in transshipment of goods.

environmental lapse rate The decline in temperature as one ascends higher in the atmosphere. On average, the temperature declines 3.5 °F for every 1,000 feet ascended or 6.5 °C for every 1,000 meters.

ethnic religion A religion closely identified with a specific ethnic or tribal group, often to the point of assuming the role of the major defining characteristic of that group. Normally, ethnic religions do not actively seek new converts.

ethnicity A shared cultural identity held by a group of people with a common background or history, often as a minority group within a larger society.

ethnographic boundaries State and national boundaries that are drawn to follow distinct differences in cultural traits, such as religion, language, or ethnic identity.

Euroland The 11 EU states that form the European Monetary Union, with its common currency, the euro. This monetary unit will completely replace national currencies in July 2002.

European Union (EU) The current association of 15 European countries that are joined together in an agenda of economic, political, and cultural integration.

exclave A portion of a country's territory that lies outside of its contiguous land area.

Exclusive Economic Zone An agreement under the International Law of the Sea that provides for a 200-nautical-mile zone of exclusive legal offshore development rights that can be utilized or sold by coastal or island nations.

exotic river A river that issues from a humid area and then flows into a dry area otherwise lacking streams.

federal state Political system in which a significant amount of power is given to individual states; in India, these states were created upon independence in 1947 and were drawn primarily along linguistic lines so that today state power is often associated with specific ethnic groups within India.

Fertile Crescent An ecologically diverse zone of lands in Southwest Asia that extends from Lebanon eastward to Iraq and that is often associated with early forms of agricultural domestication.

feudalism The formal, power relationship that consists of well-defined responsibilities and obligations, between a superior (such as an aristocrat) and those lower in social status, such as serfs or vassals.

fjords Flooded, glacially carved valleys; in Europe, found primarily along Norway's western coast.

forward capital A capital city deliberately positioned near the international border of a contested territory, signifying the state's interest—and presence—in this zone of conflict.

fossil water Water supplies that were stored underground during wetter climatic periods.

free trade zones (FTZ) A duty-free and tax-exempt industrial park created to attract foreign corporations and create industrial jobs.

gentrification A process of urban revitalization in which higher-income residents displace lower-income residents in central city neighborhoods.

geomancy The traditional Chinese and Korean practice of designing buildings in accordance with the principles of cosmic harmony and discord that supposedly course through the local topography.

geometric boundaries Boundaries of convenience drawn along lines of latitude or longitude without consideration for any cultural or ethnic differences in an area.

ghettos Urban ethnic neighborhoods, often associated with low-income groups in North America.

glasnost A policy of greater political openness initiated during the 1980s by Soviet President Mikhail Gorbachev.

globalization The increasing interconnectedness of people and places throughout the world through converging processes of economic, political, and cultural change.

Golden Triangle An area of northern Thailand, Burma, and Laos that is known as a major source region for heroin and is plugged into the global drug trade.

Gondwanaland The ancient megacontinent that included Africa, South America, Antarctica, Australia, Madagascar, and Saudi Arabia. Some 250 million years ago it began to split apart due to plate tectonics.

Gran Colombia A short-lived Latin American state during the early years of independence (1822–30). A dream of Simon Bolívar's, it included Colombia, Venezuela, Ecuador, and Panama.

grassification The conversion of tropical forest into pasture for cattle ranching. Typically this processes involves introducing species of grasses and cattle, mostly from Africa.

Great Escarpment This landform rims southern Africa from Angola to South Africa. It forms where the narrow coastal plains meet the elevated plateaus in an abrupt break in elevation.

Greater Antilles The four large Caribbean islands of Cuba, Jamaica, Hispaniola, and Puerto Rico.

Green Line United Nations-patrolled dividing line separating Greek (southern) and Turkish (northern) portions of Cyprus.

Green Revolution Term applied to the development of agricultural techniques used in developing countries that usually combine new, genetically altered seeds that provide higher yields than native seeds when combined with high inputs of chemical fertilizer, irrigation, and pesticides.

greenhouse effect The natural process of lower atmosphere heating that results from the trapping of incoming and reradiated solar energy by water moisture, clouds, and other atmospheric gases.

gross domestic product (GDP)—gross national product (GNP) GDP stands for gross domestic product, the total value of goods and services produced within a given country (or other geographical unit) in a single year. Gross national product, or GNP, is similar to GDP but is a somewhat broader measure that includes the inflow of money from other countries in the form of the repatriation of profits and other returns on investments, as well as the outflow to other countries for the same purposes.

Group of 7 (G-7) A collection of powerful countries that confers regularly on key global economic and political issues.

It includes the United States, Canada, Japan, Great Britain, Germany, France, Italy, and sometimes Russia.

growth poles Planned industrial centers developed by states in order to increase manufacturing and stimulate economic growth in underdeveloped areas.

guest workers Workers from Europe's agricultural periphery—primarily Greece, Turkey, southern Italy, and the former Yugoslavia—solicited to work in Germany, France, Sweden, and Switzerland during chronic labor shortages of Europe's boom years (1950s to 1970s).

Gulag Archipelago A collection of Soviet-era labor camps for political prisoners, made famous by writer Aleksandr Solzhenitsyn.

Hajj An Islamic religious pilgrimmage to Makkah. One of the five essential pillars of the Muslim creed to be undertaken once in life, if physically and financially able to do so.

haoles Light-skinned Europeans or U.S. citizens in the Hawaiian Islands.

hierarchical diffusion The spread of an idea or cultural trait through adoption by leaders and other elite at the top of the social structure or hierarchy. In the case of religion, it was common for all members of a clan or tribe to convert if the leader adopted the new religion.

high islands Larger, more elevated islands, often focused around recent volcanic activity.

Hindi An Indo-European language with more than 480 million speakers, making it the second-largest language group in the world. In India it is the dominant language of the heavily populated north, specifically the core area of the Ganges Plain.

Hindu nationalism A contemporary "fundamental" religious and political movement that promotes Hindu values as the essential—and exclusive—fabric of Indian society. As a political movement, it appears to have less tolerance of India's large Muslim minority than other political movements.

hiragana The main Japanese syllabary, used for writing indigenous words. In hiragana, each symbol stands for a particular vowel consonant combination.

homelands Nominally independent ethnic territories created for blacks under the grand apartheid scheme. Homelands were on marginal land, overcrowded, and poorly serviced. In the post-apartheid era, they were eliminated.

Horn of Africa The northeastern corner of Sub-Saharan Africa that includes the states of Somalia, Ethiopia, Eritrea, and Djibouti. Drought, famine, and ethnic warfare in the 1980s and 1990s resulted in much political turmoil in this area.

hot spot A supply of magma that produces a chain of mid-ocean volcanoes atop a zone of moving oceanic crust.

houseyards A rural subsistence property that is matriarchal in organization.

hurricanes Storm systems with an abnormally low-pressure center sustaining winds of 75 mph or higher. Each year during hurricane season (July–October), a half dozen to a dozen hurricanes form in the warm waters of the Atlantic and Caribbean, bringing with them destructive winds and heavy rain.

ideographic writing A writing system in which each symbol represents not a sound but rather a concept.

import substitution A development policy that discourages imports (via high tariffs) and encourages the substitution of imports with domestically produced manufactured goods.

indentured labor Foreign workers (usually South Asians) contracted to labor on Caribbean agricultural estates for a set period of time, often several years. Usually the contract stipulated paying off the travel debt incurred by the laborers. Similar indentured labor arrangements have existed in most world regions.

Indian diaspora The historical and contemporary propensity of Indians to migrate to other countries in search of better opportunities. This has led to large Indian populations in South Africa, the Caribbean, and the Pacific islands, along with western Europe and North America.

informal sector A much-debated concept that presupposes a dual economic system consisting of formal and informal sectors. The informal sector includes self-employed, low-waged jobs that are usually unregulated and untaxed. Street vending, shoe shining, artisan manufacturing, and even self-built housing are considered part of the informal sector. Other scholars include illegal activities such as drug smuggling and prostitution in the informal economy.

insolation Incoming solar energy that enters the atmosphere adjacent to Earth.

internally displaced persons Groups and individuals that flee an area due to conflict or famine but still remain in their country of origin. These populations often live in refugee-like conditions but are harder to assist because they technically do not qualify as refugees.

Iron Curtain A term coined by British leader Winston Churchill during the Cold War that defined the western border of Soviet power in Europe. The notorious "Berlin Wall" was a concrete manifestation of the Iron Curtain.

irredentism A state or national policy of reclaiming lost lands or those inhabited by people of the same ethnicity in another nation-state.

Islamic fundamentalism A movement within both the Shiite and Sunni Muslim traditions to return to a more conservative, religious-based society and state. Often associated with a rejection of Western culture and with a political aim to merge civic and religious authority.

island biogeography The study of the ecology of plants and animals unique to island environments.

isolated proximity A concept that explores the contradictory position of the Caribbean states, which are physically close to North America and economically dependent upon that region. At the same time Caribbean isolation fosters strong loyalties to locality and limited economic opportunity.

Jainism A religious group in South Asia that emerged as a protest against orthodox Hinduism about the sixth century B.C. The ethical core of Janism is the doctrine of noninjury to all living creatures. Today, Jains are noted for their nonviolence, which prohibits them from taking the life of any animal.

Kanakas Melanesian workers imported to Australia, historically often concentrated along Queensland's "sugar coast."

kanji The Chinese characters, or ideographs, used in Japanese writing.

Khmer Rouge Literally, "Red (or communist) Cambodians." The left-wing insurgent group led by French-educated Marxists who rebelled against the royal Cambodian government, first in the early 1960s and then again in a peasants' revolt in 1967.

kibbutz A collective farm in Israel.

kleptocracy A state where corruption is so institutionalized that politicians and bureaucrats siphon off a huge percentage of a country's wealth.

laissez-faire An economic system in which the state has minimal involvement and in which market forces largely guide economic activity.

latifundia A large estate or landholding.

Lesser Antilles The arc of small Caribbean islands from St. Maarten to Trinidad.

Levant The eastern Mediterranean region.

lingua franca An agreed-upon common language to facilitate communication on specific topics such as international business, politics, sports, or entertainment.

linguistic nationalism The promotion of one language over others that is, in turn, linked to shared notions of nationalism. In India, some Hindu nationalists promote Hindi as the national language, yet this is resisted by many other groups where that language is either not spoken or does not have the same central role in their culture, as in the Ganges Valley. The lack of a national language in India remains problematic.

location factors The various influences that explain why an economic activity takes place where it does.

loess A fine, wind-deposited sediment that makes fertile soil but is very vulnerable to water erosion.

low islands Low, small, sandy islands formed from eroding coral reefs. Generally less fertile and populated than high islands.

machismo A stereotypical cultural trait of male dominance.

Maghreb A region in northwestern Africa, including portions of Morocco, Algeria, and Tunisia.

maharaja Regional Hindu royalty, usually a king or prince, who ruled specific areas of South Asia before independence, but who was usually subject to overrule by British colonial advisors.

mallee A tough and scrubby eucalyptus woodland of limited economic value that is common across portions of interior Australia.

Mandarin A member of the high-level bureaucracy of Imperial China (before 1911). Mandarin Chinese is the official spoken language of the country, and is the native tongue of the vast majority of people living in north, central, and southwestern China.

Maori Indigenous Polynesian people of New Zealand.

maquiladora Assembly plants on the Mexican border built by foreign capital. Most of their products are exported to the United States.

marianismo An idealized model for women that stresses the virtues of patience, deference, and working in the home.

Marine West Coast climate Moderate climates with cool summers and mild winters that are heavily influenced by maritime conditions. They are usually found on the west coasts of continents between latitudes of 45 to 50 degrees.

maritime climate Climates moderated by their proximity to oceans or large seas. They are usually cool, cloudy, wet, and lack the temperature extremes of continental climates.

maroons Runaway slaves who established communities rich in African traditions throughout the Caribbean and Brazil.

Marxism The philosophy developed by Karl Marx, the most important historical proponent of Communism. Marxism, which has many variants, presumes the desirability, and, indeed, the necessity, of a socialist economic system run through a central planning agency.

Medieval landscape Urban landscapes from A.D. 900 to 1500 characterized by narrow, winding streets, three- or four-story structures (usually in stone, but sometimes wooden), with little open space except for the market square. These landscapes are still found in the center of many European cities.

medina The original urban core of a traditional Islamic city.

Mediterranean climate A unique climate, found in only five locations in the world, that is characterized by hot, dry summers with very little rainfall. These climates are located on the west side of continents, between 30 to 40 degrees latitude.

megacity Urban conglomerations of more than 10 million people.

megalopolis A large urban region formed as multiple cities grow and merge with one another. The term is often applied to the string of cities in eastern North America that includes Washington, D.C.; Baltimore; Philadelphia; New York City; and Boston.

Melanesia Pacific Ocean region that includes the culturally complex, generally darker-skinned peoples of New Guinea, the Solomon Islands, Vanuatu, New Caledonia, and Fiji.

Mercosur The Southern Common Market established in 1991 that calls for free trade among member states and common external tariffs for nonmember states. Argentina, Paraguay, Brazil, and Uruguay are members; Chile is an associate member.

mestizo A person of mixed European and Indian ancestry.

Micronesia Pacific Ocean region that includes the culturally diverse, generally small islands north of Melanesia. Includes the Mariana Islands, Marshall Islands, and Federated States of Micronesia.

microstates Small, usually independent states in both area and population.

mikrorayons Large, state-constructed urban housing projects built during the Soviet period in the 1970s and 1980s.

minifundia A small landholding farmed by peasants or tenants who produce food for subsistence and the market.

mono-crop production Agriculture based upon a single crop.

monotheism A religious belief in a single God.

Monroe Doctrine A proclamation issued by U.S. President James Monroe in 1823 that the United States would not tolerate European military action in the Western Hemisphere. Focused on the Caribbean as a strategic area, the doctrine was repeatedly invoked to justify U.S. political and military intervention in the region.

monsoon The seasonal pattern of changes in winds, heat, and moisture in South Asia and other regions of the world that is a product of larger meteorological forces of land and water heating, the resultant pressure gradients, and jet stream dynamics. The monsoon produces a distinct seasonally of wet and dry seasons.

moraines Hilly topographic features that mark the path of Pleistocene glaciers. Moraines are composed of material eroded and carried by glaciers and ice sheets.

Mughal Empire (also spelled Mogul) The preeminent Islamic period of rule that covered most of South Asia during the early sixteenth to late-seventeenth centuries and attempted to unify both Muslim and Hindus into a large South Asia state. The capital of this empire was Lahore, in what is now Pakistan. The last vestiges of the Mughal dynasty were dissolved by the British following the uprisings of 1857.

multinational corporation A corporation that produces goods and services in a wide range of different countries. The classical multinational corporation, in contrast to the truly transnational corporation, does, however, remain solidly based in a single country.

nation-state A relatively homogeneous cultural group (a nation) with its own political territory (the state).

Native Title Bill Australian legislation signed in 1993 that provides Aborigines enhanced legal rights over land and resources within the country.

neocolonialism Economic and political strategies by which powerful states indirectly (and sometimes directly) extend their influence over other, weaker states.

neoliberal policies Economic policies widely adopted in the 1990s. They stress privatization, export production, and few restrictions on imports.

neotropics Tropical ecosystems of the Americas that evolved in relative isolation and support diverse and unique flora and fauna.

North America Free Trade Agreement (NAFTA) An agreement made in 1994 between Canada, the United States, and Mexico that established a 15-year plan for reducing all barriers to trade among the three countries.

Oceania A major world subregion that usually includes New Zealand and the major islands of Melanesia, Micronesia, and Polynesia.

offshore banking Islands or microstates that have become a specialized node in the geography of worldwide financial flows.

Organization of African Unity (OAU) Founded in 1963, the organization grew to include all the states of the continent except South Africa, which finally was asked to join in 1994. It is mostly a political body that has tried to resolve regional conflicts.

Organization of American States (OAS) Founded in 1948 and headquartered in Washington, DC, the OAS advocates hemispheric cooperation and dialogue. Most states in the Americas belong except Cuba.

Organization of Petroleum Exporting Countries (OPEC) An international organization of 12 oil-producing nations (formed in 1960) that attempts to influence global prices and supplies of oil. Algeria, Gabon, Indonesia, Iran, Iraq, Kuwait, Libya, Nigeria, Qatar, Saudi Arabia, UAE, and Venezuela are members.

orographic rainfall Enhanced precipitation over uplands that results from lifting (and cooling) of air masses as they are forced over mountains.

Ottoman Empire A large, Turkish-based empire (named for Osman, one of its founders) that dominated large portions of southeastern Europe, North Africa, and Southwest Asia between the sixteenth and nineteenth centuries.

outback Australia's large, generally dry and thinly settled interior.

overurbanization A process in which the rapid growth of a city, most often because of in-migration, exceeds the city's ability to provide jobs, housing, water, sewers, and transportation.

oxisols Reddish or yellowish soils found in the tropical shields of Latin America. They are formed over long periods of time and accumulate rusted or oxidized iron. When these fine soils are disturbed or compacted, a hard-pan surface results that is virtually impossible to farm.

Palestinian Authority (PA) A quasi-governmental body that represents Palestinian interests in the West Bank.

Pan-African Movement Founded in 1900 by U.S. intellectuals W. E. B. Du Bois and Marcus Garvey, this movement's slogan was "Africa for Africans" and its influence extended across the Atlantic.

particularism A mode of thought that emphasizes the uniqueness of different places and different phenomena. Particularism is opposed to universalism, which emphasizes locality-transcending commonalties.

pastoral nomadism A traditional subsistence agricultural system in which practitioners depend on the seasonal movements of livestock within marginal natural environments.

pastoralism A way of life and form of livelihood centered around raising large animals, usually cattle, sheep, goats, and horses, but also sometimes including camels, yaks, and other species.

pastoralists Nomadic and sedentary peoples who rely upon livestock (especially cattle, camels, sheep, and goats) for their sustenance and livelihood.

perestroika A program of partially implemented, planned economic reforms (or restructuring) undertaken during the Gorbachev years in the Soviet Union. Designed to make the Soviet economy more efficient and responsive to consumer needs.

permafrost A cold-climate condition in which the ground remains permanently frozen.

physiological densities A population statistic that relates the number of people in a country to the amount of arable land.

Pidgin English A version of English that also incorporates elements of other local languages, often utilized to foster trade and basic communication between different culture groups.

plantation America A cultural region that extends from midway up the coast of Brazil, through the Guianas and the Caribbean, and into the southeastern United States. In this coastal zone, European-owned plantations, worked by African laborers, produced agricultural products for export.

plate tectonics The theory that explains the gradual movement of large geological platforms (or plates) along Earth's surface.

podzol soils A Russian term for an acidic soil of limited fertility, typically found in northern forest environments.

polders Reclaimed agricultural areas along the Dutch coast that have been diked and drained. Many are at or below sea level.

pollution exporting The process of exporting industrial pollution and other waste material to other countries. Pollution exporting can be direct, as when waste is simply shipped abroad for disposal, or indirect, as when highly polluting factories are constructed abroad.

Polynesia Pacific Ocean region, broadly unified by language and cultural traditions, that includes the Hawaiian Islands, Marquesas Islands, the Society Islands, the Tuamotu Archipelago, Cook Islands, American Samoa, Samoa, Tonga, and Kiribati.

postindustrial economy An economy in which the tertiary and quaternary sectors dominate employment and expansion.

prairie An extensive area of grassland in North America. In the more humid eastern portions, grasses are usually longer than in the drier western areas, which are in the rain shadow of the Rocky Mountain range.

primate city The largest urban settlement in a country that dominates all other urban places, economically and politically. Often—yet not always—the primate city is also the country's capital.

privatization The process of moving formerly state-owned firms into the contemporary capitalist private sector.

protectorate During the period of global Western imperialism, a state or other political entity that remained autonomous but sacrificed its foreign affairs to an imperial power in exchange for "protection" from other imperial powers.

purchasing power parity (PPP) A method of reducing the influence of inflated currency rates by adjusting a local currency to a composite baseline of one U.S. dollar based upon its ability to purchase a standardized "market basket" of goods.

qanat system A traditional system of gravity-fed irrigation that uses gently sloping tunnels to capture groundwater and direct it to needed fields.

Quran (also spelled Koran) A book of divine revelations received by the prophet Muhammad that serves as a holy text in the religion of Islam.

rain shadow effect A weather phenomenon in which mountains block moisture, producing an area of lower precipitation on the leeward side of the uplift.

rate of natural increase (RNI) The standard statistic used to express natural population growth per year for a country, region, or the world based upon the difference between birth and death rates. RNI does not consider population change from migration. Though most often a positive figure (such as 1.7 percent), RNI can also be expressed as a negative (–.08) for no-growth countries.

refugee A person who flees his or her country because of a well-founded fear of persecution based on race, ethnicity, religion, ideology, or political affiliation.

remittances Money sent by immigrants to their country of origin to support family members left behind. For many countries, remittances are a principal source of foreign exchange.

Renaissance-Baroque landscape Urban landscapes generally constructed during the period from 1500 to 1800 that are characterized by wide, ceremonial boulevards, large monumental structures (palaces, public squares, churches), and ostentatious housing for the urban elite. A common landscape feature in European cities.

rift valley A surface landscape feature formed where two tectonic plates are diverging or moving apart. Usually this forms a depression or large valley.

rimland The mainland coastal zone of the Caribbean, beginning with Belize and extending along the coast of Central America to northern South America.

rural-to-urban migration The flow of internal migrants from rural areas to cities that began in the 1950s and intensified in the 1960s and 1970s.

Russification A policy of the Soviet Union designed to spread Russian settlers and influences to non-Russian areas of the country.

rust belt Regions of heavy industry that experience marked economic decline after their factories cease to be competitive.

Sahel The semi-desert region at the southern fringe of the Sahara, and the countries that fall within this region, which extends from Senegal to Sudan. Droughts in the 1970s and early 1980s caused widespread famine and dislocation of population.

salinization The accumulation of salts in the upper layers of soil, often causing a reduction in crop yields, resulting from irrigation with water of high natural salt content and/or irrigation of soils that contain a high level of mineral salts.

samurai The warrior class of traditional Japan. After 1600 the military role of the samurai declined as they assumed administrative positions, but their military ethos remained alive until the class was abolished in 1868.

Sanskrit The original Indo-European language of South Asia, introduced into northwestern India perhaps 4,000 years ago, from which modern Indo-Aryan languages evolved. Over the centuries it has become the classical literary language of the Hindus and is widely used as a scholarly second language, much like Latin in medieval Europe.

scheduled castes Part of the contemporary reform movement in India to lessen the social stigma of lower castes, particularly of the dalits (formerly untouchables), by reserving a certain number of places in universities, governmental offices, and seats in national and state legislatures for members of these "scheduled" castes.

Schengen Agreement The 1985 agreement between some—but not all—European Union member countries to reduce border formalities in order to facilitate free movement of citizens between member countries of this new "Schengenland." For example, today there are no border controls between France and Germany, or between France and Italy.

sectoral transformation The evolution of a labor force from one highly dependent on the primary sector to one oriented around more employment in the secondary, tertiary, and quaternary sectors.

secularization The widespread movement in western Europe away from regular participation and engagement with traditional organized religions such as Protestantism or Catholicism.

sediment load The amount of sand, silt, and clay carried by a given river.

shatterbelt A geopolitically unstable area where the superpowers vie for power. Eastern Europe was the classic shatterbelt until World War II. During the Cold War, Cuba's relationship with the Soviet Union helped to turn the Caribbean into a shatterbelt.

shield landscape Barren, mostly flat lands of southern Scandinavia that were heavily eroded by Pleistocene ice sheets. In many places, this landscape is characterized by large expanses of bedrock with little or no soil that resulted from glacial erosion.

shields Large upland areas of very old exposed rocks that range in elevation from 600–5,000 feet (200–1,500 meters). The three major shields in South America are the Guiana, Brazilian, and Patagonian.

shifted cultivators Migrants, with or without agricultural experience, who are transplanted by government relocation schemes.

Shiites Muslims who practice one of the two main branches of Islam; especially dominant in Iran and nearby southern Iraq.

Shogun, Shogunate The true ruler of Japan before 1868, as opposed to the emperor, whose power was merely symbolic.

Sikhism An Indian religion combining Islamic and Hindu elements, founded in the Punjab region in the late fifteenth century. A long tradition of militarism continues today with a large proportion of Sikh men in the Indian armed forces.

Slavic peoples A group of peoples in eastern Europe and Russia who speak Slavic languages, a distinctive branch of the Indo-European language family.

social and regional differentiation "Social differentiation" refers to a process by which certain classes of people grow richer when others grow poorer; "regional differentiation" refers to a process by which certain places grow more prosperous while others become less prosperous.

social realism An artistic style once popular in the Soviet Union that was associated with realistic depictions of workers in their patriotic struggles against capitalism.

Special Economic Zones (SEZ) Relatively small districts in China that have been fully opened to global capitalism.

spheres of influence In countries not formally colonized in the nineteenth and early twentieth centuries (particularly China and Iran), limited areas called "spheres of influence" were gained by particular European countries for trade purposes and more generally for economic exploitation and political manipulation.

squatter settlements Makeshift housing on land not legally owned or rented by urban migrants, usually in unoccupied open spaces within or on the outskirts of a rapidly growing city.

static frontiers Large, open areas believed to be unable to support permanent settlement because of poor soils, seasonal flooding, or inadequate rainfall. Areas such as the Llanos, the Chaco, and even the Amazon Basin itself are often described as static frontiers.

steppe Semiarid grasslands found in many parts of the world. Grasses are usually shorter and less dense than in prairies.

structural adjustment programs Controversial yet widely implemented programs used to reduce government spending, encourage the private sector, and refinance foreign debt. Typically, these IMF and World Bank policies trigger drastic cutbacks in government-supported services and food subsidies, which disproportionately affect the poor.

subduction zones Areas where two tectonic plates are converging or colliding. In these areas, one plate usually sinks below another. Subduction zones are characterized by earthquakes, volcanoes, and deep oceanic trenches.

subnational organizations Groups that form along ethnic, ideological, or territorial lines that can induce serious internal divisions within a state.

subsistence agriculture Farming that produces only enough crops or animal products to support a farm family's needs. Usually little is sold at local or regional markets.

Suez Canal Pivotal waterway connecting the Red Sea and the Mediterranean. Opened by the British in 1869.

Sunda Shelf An extension of the continental shelf from the Southeast Asia mainland to the outlying islands. Because of the shelf, the overlying sea is generally shallow (less than 200 feet deep).

Sunnis Muslims who practice the dominant branch of Islam.

superconurbation A massive urban agglomeration that results from the coalescing of two or more formerly separate metropolitan areas.

supranational organizations Governing bodies that include several states, such as trade organizations, and often involve a loss of some state powers to achieve the organization's goals.

sustainable development A vision of economic change and growth seeking a balance with environmental protection and social equity so that the short-term needs of contemporary society do not compromise needs of future generations. The operational scale of sustainable development is local rather than global.

swidden agriculture Also called "slash-and-burn agriculture." A form of cultivation in which forested or brushy plots are cleared of vegetation, burned, and then planted to crops, only to be abandoned a few years later as soil fertility declines.

syncretic religions The blending of different belief systems. In Latin America many animist practices were folded into Christian worship.

tectonic plates The basic building blocks of Earth's crust; large blocks of solid rock that very slowly move over the underlying semi-molten material.

terraces Flat areas carved across the face of slopes, usually in a steplike fashion.

theocratic state A political state led by religious authorities.

tierra caliente The hot zone from sea level to 3,000 feet (900 meters).

tierra fria The cold zone from 6,000–12,000 feet (1,800–3,600 meters).

tierra helada The freezing land above 12,000 feet (3,600 meters) where a few highland grains are produced and livestock graze.

tierra templada The temperate zone from 3,000–6,000 feet (900–1,800 meters).

tonal language Language in which the same set of phonemes (or basic sounds) may have very different meanings depending on the pitch in which a sound is uttered.

total fertility rate (TFR) The average number of children who will be borne by women of a hypothetical, yet statistically valid, population, such as that of a specific cultural group or with a particular country. Demographers consider TFR a more reliable indicator of population change than the crude birthrate.

township Racially segregated neighborhoods created for nonwhite groups under apartheid in South Africa. They are usually found on the outskirts of cities and classified as black, coloured, or South Asian.

transhumance A form of pastoralism in which animals are taken to high-altitude pastures during the summer months and then returned to low-altitude pastures during the winter.

transmigration The planned, government-sponsored relocation of people from one area to another within a state territory.

transnationalism Complex social and economic linkages that form between home and host countries through international migration. Unlike earlier generations of migrants, technological advances in the late twentieth century have allowed immigrants to maintain more enduring and complex ties to their home countries.

Trans-Siberian Railroad Key southern Siberian railroad connection completed during the Russian Empire (1904) that links European Russia with the Russian Far East terminus of Vladivostok.

Treaty of Tordesillas A treaty signed in 1494 between Spain and Portugal that drew a north-south line some 300 leagues west of the Azores and Cape Verde islands. Spain received the land to the west of the line and Portugal the land to the east.

tribal peoples Peoples who were traditionally organized at the village or clan level, without broader-scale political organization.

tribalism Allegiance to a particular tribe or ethnic group rather than to the nation-state. Tribalism is often blamed for internal conflict within Sub-Saharan states.

tribe A group of families or clans with a common kinship, language, and definable territory but not an organized state.

tsars A Russian term (same as *czar*) for *caesar,* or ruler; the authoritarian rulers of the Russian Empire before its collapse in the 1917 revolution.

tsetse fly A fly that is a vector for a parasite that causes sleeping sickness (typanosomiasis), a disease that especially affects humans and livestock. Livestock is rarely found in those areas of Sub-Saharan Africa where the tsetse fly is common.

tsunamis Very large sea waves induced by earthquakes.

tundra Arctic region with a short growing season in which vegetation is limited to low shrubs, grasses, and flowering herbs.

Turkestan That portion of Central Asia populated primarily by Turkish-speaking peoples; "eastern Turkestan" comprises those areas presently controlled by China and "western Turkestan" comprises areas formerly held by the Soviet Union.

typhoons Large tropical storms, similar to hurricanes, that form in the western Pacific Ocean in tropical latitudes and cause widespread damage to the Philippines and coastal Southeast and East Asia.

uncontacted peoples Cultures that have yet to be contacted and influenced by the Western world.

unitary state A political system in which power is centralized at the national level.

United Provinces of Central America Formed in 1823 to avoid annexation by Mexico, this union collapsed in the 1830s, yielding the independent states of Guatemala, Honduras, El Salvador, Nicaragua, and Costa Rica.

universalizing religion A religion, usually with an active missionary program, that appeals to a large group of people regardless of local culture and conditions. Christianity and Islam both have strong universalizing components. This contrasts with ethnic religions.

urban decentralization The process in which cities spread out over a larger geographical area.

urban form The physical arrangement or landscape of the city, made up of building architecture and style, street patterns, open spaces, housing types, etc.

urban primacy A state in which a disproportionately large city, such as London, Mexico City, or Bangkok, dominates a country's urban system and is the center of that country's economic, political, and cultural life.

urban realms model A simplified description of urban land use, especially descriptive of the modern North American city. It features a number of dispersed, peripheral centers of dynamic commercial and industrial activity linked by sophisticated urban transportation networks.

urban structure The distribution and pattern of land use, such as commercial, residential, or for manufacturing, within the city. Often commonalties are found that give rise to models of urban structure characteristic of the cities of a certain region or of a shared history, such as cities shaped by European colonialism.

urbanized population That percentage of a country's population living in settlements characterized as cities. Usually high rates of urbanization are associated with higher levels of industrialization and economic development since these activities are usually found in and around cities. Conversely, lower urbanized populations (less than 50 percent) are characteristic of developing countries.

Urdu Although Urdu originated in the region of northern India and arose from a similar colloquial base as Hindi, it is one of the official languages of Pakistan today because of its long association with Muslim culture. Urdu borrows heavily from Persian vocabulary and Arabic grammatical construction and, further, is written in a modified form of the Persian Arabic alphabet.

viticulture Grape cultivation.

Wallace's Line A line in Southeast Asia delineating the abrupt difference in flora and fauna from that found on the Asian mainland to plants and animals more common to Australia.

water stress An environmental planning tool used to predict areas that have—or will have—serious water problems based upon the per capital demand and supply of fresh water.

White Australia Policy Before 1975, a set of stringent Australian limitations on nonwhite immigration to the country. Largely replaced by a more flexible policy today.

World Trade Organization (WTO) Formed as an outgrowth of the General Agreement on Tariffs and Trade (GATT) in 1995, it is a large collection of member states dedicated to reducing global barriers to trade.

INDEX

Page numbers in *italic* indicate figures.

Stansfield/Westby
GeoTutor: Building Geographic Literacy (0-13-080644-7)
Prentice-Hall, Inc.

YOU SHOULD CAREFULLY READ THE TERMS AND CONDITIONS BEFORE USING THE CD-ROM PACKAGE. USING THIS CD-ROM PACKAGE INDICATES YOUR ACCEPTANCE OF THESE TERMS AND CONDITIONS.

Prentice-Hall, Inc. provides this program and licenses its use. You assume responsibility for the selection of the program to achieve your intended results, and for the installation, use, and results obtained from the program. This license extends only to use of the program in the United States or countries in which the program is marketed by authorized distributors.

LICENSE GRANT

You hereby accept a nonexclusive, nontransferable, permanent license to install and use the program ON A SINGLE COMPUTER at any given time. You may copy the program solely for backup or archival purposes in support of your use of the program on the single computer. You may not modify, translate, disassemble, decompile, or reverse engineer the program, in whole or in part.

TERM

The License is effective until terminated. Prentice-Hall, Inc. reserves the right to terminate this License automatically if any provision of the License is violated. You may terminate the License at any time. To terminate this License, you must return the program, including documentation, along with a written warranty stating that all copies in your possession have been returned or destroyed.

LIMITED WARRANTY

THE PROGRAM IS PROVIDED "AS IS" WITHOUT WARRANTY OF ANY KIND, EITHER EXPRESSED OR IMPLIED, INCLUDING, BUT NOT LIMITED TO, THE IMPLIED WARRANTIES OR MERCHANTABILITY AND FITNESS FOR A PARTICULAR PURPOSE. THE ENTIRE RISK AS TO THE QUALITY AND PERFORMANCE OF THE PROGRAM IS WITH YOU. SHOULD THE PROGRAM PROVE DEFECTIVE, YOU (AND NOT PRENTICE-HALL, INC. OR ANY AUTHORIZED DEALER) ASSUME THE ENTIRE COST OF ALL NECESSARY SERVICING, REPAIR, OR CORRECTION. NO ORAL OR WRITTEN INFORMATION OR ADVICE GIVEN BY PRENTICE-HALL, INC., ITS DEALERS, DISTRIBUTORS, OR AGENTS SHALL CREATE A WARRANTY OR INCREASE THE SCOPE OF THIS WARRANTY.

SOME STATES DO NOT ALLOW THE EXCLUSION OF IMPLIED WARRANTIES, SO THE ABOVE EXCLUSION MAY NOT APPLY TO YOU. THIS WARRANTY GIVES YOU SPECIFIC LEGAL RIGHTS AND YOU MAY ALSO HAVE OTHER LEGAL RIGHTS THAT VARY FROM STATE TO STATE.

Prentice-Hall, Inc. does not warrant that the functions contained in the program will meet your requirements or that the operation of the program will be uninterrupted or error-free.

However, Prentice-Hall, Inc. warrants the CD-ROM on which the program is furnished to be free from defects in material and workmanship under normal use for a period of ninety (90) days from the date of delivery to you as evidenced by a copy of your receipt.

The program should not be relied on as the sole basis to solve a problem whose incorrect solution could result in injury to person or property. If the program is employed in such a manner, it is at the user's own risk and Prentice-Hall, Inc. explicitly disclaims all liability for such misuse.

LIMITATION OF REMEDIES

Prentice-Hall, Inc.'s entire liability and your exclusive remedy shall be:
1. the replacement of any CD-ROM not meeting Prentice-Hall, Inc.'s "LIMITED WARRANTY" and that is returned to Prentice-Hall, or
2. if Prentice-Hall is unable to deliver a replacement CD-ROM that is free of defects in materials or workmanship, you may terminate this agreement by returning the program.

IN NO EVENT WILL PRENTICE-HALL, INC. BE LIABLE TO YOU FOR ANY DAMAGES, INCLUDING ANY LOST PROFITS, LOST SAVINGS, OR OTHER INCIDENTAL OR CONSEQUENTIAL DAMAGES ARISING OUT OF THE USE OR INABILITY TO USE SUCH PROGRAM EVEN IF PRENTICE-HALL, INC. OR AN AUTHORIZED DISTRIBUTOR HAS BEEN ADVISED OF THE POSSIBILITY OF SUCH DAMAGES, OR FOR ANY CLAIM BY ANY OTHER PARTY.

SOME STATES DO NOT ALLOW FOR THE LIMITATION OR EXCLUSION OF LIABILITY FOR INCIDENTAL OR CONSEQUENTIAL DAMAGES, SO THE ABOVE LIMITATION OR EXCLUSION MAY NOT APPLY TO YOU.

GENERAL

You may not sublicense, assign, or transfer the license of the program. Any attempt to sublicense, assign or transfer any of the rights, duties, or obligations hereunder is void.

This Agreement will be governed by the laws of the State of New York.

Should you have any questions concerning this Agreement, you may contact Prentice-Hall, Inc. by writing to:
Media Production
Engineering/Science/Math Division
Prentice-Hall, Inc.
1 Lake Street
Upper Saddle River, NJ 07458

Should you have any questions concerning technical support, you may write to:
Media Production
Engineering/Science/Math Division
Prentice-Hall, Inc.
1 Lake Street
Upper Saddle River, NJ 07458

YOU ACKNOWLEDGE THAT YOU HAVE READ THIS AGREEMENT, UNDERSTAND IT, AND AGREE TO BE BOUND BY ITS TERMS AND CONDITIONS. YOU FURTHER AGREE THAT IT IS THE COMPLETE AND EXCLUSIVE STATEMENT OF THE AGREEMENT BETWEEN US THAT SUPERSEDES ANY PROPOSAL OR PRIOR AGREEMENT, ORAL OR WRITTEN, AND ANY OTHER COMMUNICATIONS BETWEEN US RELATING TO THE SUBJECT MATTER OF THIS AGREEMENT.